T0189004

Lecture Notes in Computer Science   9032

Commenced Publication in 1973
Founding and Former Series Editors:
Gerhard Goos, Juris Hartmanis, and Jan van Leeuwen

Advanced Research in Computing and Software Science

Subline of Lecture Notes in Computer Science

More information about this series at http://www.springer.com/series/7407

Jan Vitek (Ed.)

Programming
Languages
and Systems

24th European Symposium on Programming, ESOP 2015
Held as Part of the European Joint Conferences
on Theory and Practice of Software, ETAPS 2015
London, UK, April 11–18, 2015
Proceedings

 Springer

Editor
Jan Vitek
Northeastern University
Boston
Massachusetts
USA

ISSN 0302-9743 ISSN 1611-3349 (electronic)
Lecture Notes in Computer Science
ISBN 978-3-662-46668-1 ISBN 978-3-662-46669-8 (eBook)
DOI 10.1007/978-3-662-46669-8

Library of Congress Control Number: 2015934000

LNCS Sublibrary: SL1 – Theoretical Computer Science and General Issues

Springer Heidelberg New York Dordrecht London

Printed on acid-free paper

Springer-Verlag GmbH Berlin Heidelberg is part of Springer Science+Business Media
(www.springer.com)

Foreword

ETAPS 2015 was the 18th instance of the European Joint Conferences on Theory and Practice of Software. ETAPS is an annual federated conference that was established in 1998, and this year consisted of six constituting conferences (CC, ESOP, FASE, FoSSaCS, TACAS, and POST) including five invited speakers and two tutorial speakers. Prior to and after the main conference, numerous satellite workshops took place and attracted many researchers from all over the world.

ETAPS is a confederation of several conferences, each with its own Program Committee and its own Steering Committee (if any). The conferences cover various aspects of software systems, ranging from theoretical foundations to programming language developments, compiler advancements, analysis tools, formal approaches to software engineering, and security. Organizing these conferences into a coherent, highly synchronized conference program enables the participation in an exciting event, having the possibility to meet many researchers working in different directions in the field, and to easily attend talks at different conferences.

The six main conferences together received 544 submissions this year, 152 of which were accepted (including 10 tool demonstration papers), yielding an overall acceptance rate of 27.9%. I thank all authors for their interest in ETAPS, all reviewers for the peer-reviewing process, the PC members for their involvement, and in particular the PC Co-chairs for running this entire intensive process. Last but not least, my congratulations to all authors of the accepted papers!

ETAPS 2015 was greatly enriched by the invited talks by Daniel Licata (Wesleyan University, USA) and Catuscia Palamidessi (Inria Saclay and LIX, France), both unifying speakers, and the conference-specific invited speakers [CC] Keshav Pingali (University of Texas, USA), [FoSSaCS] Frank Pfenning (Carnegie Mellon University, USA), and [TACAS] Wang Yi (Uppsala University, Sweden). Invited tutorials were provided by Daniel Bernstein (Eindhoven University of Technology, the Netherlands and the University of Illinois at Chicago, USA), and Florent Kirchner (CEA, the Alternative Energies and Atomic Energy Commission, France). My sincere thanks to all these speakers for their inspiring talks!

ETAPS 2015 took place in the capital of England, the largest metropolitan area in the UK and the largest urban zone in the European Union by most measures. ETAPS 2015 was organized by the Queen Mary University of London in cooperation with the following associations and societies: ETAPS e.V., EATCS (European Association for Theoretical Computer Science), EAPLS (European Association for Programming Languages and Systems), and EASST (European Association of Software Science and Technology). It was supported by the following sponsors: Semmle, Winton, Facebook, Microsoft Research, and Springer-Verlag.

The organization team comprised:

- General Chairs: Pasquale Malacaria and Nikos Tzevelekos
- Workshops Chair: Paulo Oliva
- Publicity chairs: Michael Tautschnig and Greta Yorsh
- Members: Dino Distefano, Edmund Robinson, and Mehrnoosh Sadrzadeh

The overall planning for ETAPS is the responsibility of the Steering Committee. The ETAPS Steering Committee consists of an Executive Board (EB) and representatives of the individual ETAPS conferences, as well as representatives of EATCS, EAPLS, and EASST. The Executive Board comprises Gilles Barthe (satellite events, Madrid), Holger Hermanns (Saarbrücken), Joost-Pieter Katoen (Chair, Aachen and Twente), Gerald Lüttgen (Treasurer, Bamberg), and Tarmo Uustalu (publicity, Tallinn). Other members of the Steering Committee are: Christel Baier (Dresden), David Basin (Zurich), Giuseppe Castagna (Paris), Marsha Chechik (Toronto), Alexander Egyed (Linz), Riccardo Focardi (Venice), Björn Franke (Edinburgh), Jan Friso Groote (Eindhoven), Reiko Heckel (Leicester), Bart Jacobs (Nijmegen), Paul Klint (Amsterdam), Jens Knoop (Vienna), Christof Löding (Aachen), Ina Schäfer (Braunschweig), Pasquale Malacaria (London), Tiziana Margaria (Limerick), Andrew Myers (Boston), Catuscia Palamidessi (Paris), Frank Piessens (Leuven), Andrew Pitts (Cambridge), Jean-Francois Raskin (Brussels), Don Sannella (Edinburgh), Vladimiro Sassone (Southampton), Perdita Stevens (Edinburgh), Gabriele Taentzer (Marburg), Peter Thiemann (Freiburg), Cesare Tinelli (Iowa City), Luca Vigano (London), Jan Vitek (Boston), Igor Walukiewicz (Bordeaux), Andrzej Wąsowski (Copenhagen), and Lenore Zuck (Chicago).

I sincerely thank all ETAPS SC members for all their hard work to make the 18th edition of ETAPS a success. Moreover, thanks to all speakers, attendants, organizers of the satellite workshops, and to Springer for their support. Finally, many thanks to Pasquale and Nikos and their local organization team for all their efforts enabling ETAPS to take place in London!

January 2015   Joost-Pieter Katoen

Preface

It is my distinct pleasure, and honor, to present you with the technical program of the 24th European Conference on Programming (ESOP) held during April 14–16 in London, UK as part of the ETAPS confederation. This year's program consisted of 33 papers selected from 113 submissions on topics ranging from program analysis of JavaScript to the semantics of concurrency in C11. The paper "A Theory of Name Resolution" by Neron, Tolmach, Visser, and Wachsmuth was nominated by the Program Committee for the ETAPS best paper award.

The process of selecting papers departed from previous years in several important respects. First, as papers were submitted, the deadline was October 17, 2014. Papers were checked for formatting, length, quality, and scope. Five papers were desk rejected at this point for reasons ranging from insufficient quality to double submission. The remaining papers were assigned a guardian, usually the Committee Member with most expertise. The guardian's role was to ensure each paper had at least one expert reviewer and to write the rejoinder. Each paper was then assigned three Program Committee reviewers (the guardian being one). External reviewers were invited when additional expertise was required. In some cases, several external reviewers were needed to give us confidence that all aspects of the work had been evaluated. Reviews and scores were forwarded to the authors on December 3. Authors were allowed to submit unlimited-length rebuttals. After reception of the rebuttals, the guardians wrote rejoinders that summarized the points for their papers, the main criticisms, and how the rebuttals addressed them. All rebuttals were thus carefully read and many were discussed among the reviewers. A live Program Committee meeting was held in London during December 11–12. In the meeting, every paper was presented by its guardian. Decisions were reached by consensus. Papers authored by committee members were held to a higher standard (namely, the absence of a detractor). Papers for which I had a conflict were handled by Peter Thiemann.

A conference such as this one is the product of the effort of many. Let me thank them, starting with the authors who entrusted us with their work, the external reviewers who provided much needed expertise, and the Program Committee members who produced timely reviews and managed to retain a positive attitude throughout. Guiseppe Castagna was instrumental in securing permission from the ETAPS Steering Committee to increase the page limit to 25 pages. Alastair Donaldson kindly hosted the committee meeting at Imperial College and provided tea and cookies. Eddie Kohler let us use the hosted version of the HotCRP software developed for SIGPLAN. Eelco Visser kindly donated a website built using Researchr and Elmer van Chastelet provided technical assistance. Lastly, Northeastern University provided financial support for the program Committee meeting.

January 2015   Jan Vitek

Organization

Program Committee

Umut Acar	Carnegie Mellon University, USA
Jade Alglave	University College London, UK
Gilles Barthe	IMDEA, Spain
Gavin Bierman	Oracle Labs, UK
Lars Birkedal	Aarhus University, Denmark
Luis Caires	FCT / Universidade Nova de Lisboa, Portugal
Adam Chlipala	MIT, USA
Charles Consel	University of Bordeaux / Inria / LaBRI, France
Delphine Demange	IRISA / University of Rennes 1, France
Isil Dillig	University of Texas, Austin
Alastair Donaldson	Imperial College London, UK
Derek Dreyer	MPI-SWS, Germany
Azadeh Farzan	University of Toronto, Canada
Cedric Fournet	Microsoft Research, UK
Giorgio Ghelli	Università di Pisa, Italy
Alexey Gotsman	IMDEA, Spain
Peter Müller	ETH Zurich, Switzerland
Luca Padovani	Università degli Studi di Torino, Italy
Keshav Pingali	University of Texas, Austin
Mooly Sagiv	Tel Aviv University, Israel
David Sands	Chalmers University of Technology, Sweden
Helmut Seidl	Technische Universität München, Germany
Armando Solar-Lezama	MIT, USA
Éric Tanter	University of Chile, Chile
Peter Thiemann	University of Freiburg, Germany
Hongseok Yang	University of Oxford, UK
Francesco Nardelli Zappa	Inria, France

External Reviewers

Vikram Adve	Ferruccio Damiani
Amal Ahmed	Ugo de'Liguoro
Jesper Bengtson	Pierre-Malo Deniélou
Johannes Borgström	R. Kent Dybvig
Ahmed Bouajjani	Jacques Garrigue
John Boyland	Adria Gascon
Arthur Chargueraud	Cinzia Di Giusto

Ganesh Gopalakrishnan

Arjun Guha

Reinhold Heckmann

Daniel Hedin

Fritz Henglein

Martin Hofmann

Sebastian Hunt

Atushi Igarashi

Danko Ilic

Thomas Jensen

Steffen Jost

Andres Löh

Ori Lahav

first last

Andrew Lenharth

Ben Lippmeier

Annie Liu

Etienne Lozes

Anders Möller

Geoffrey Mainland

Mark Marron

Ken McMillan

Donald Nguyen

Scott Owens

Matthew Parkinson

Madhusudan Parthasarathy

Kalyan Perumalla

Gustavo Petri

Michael Petter

Benjamin Pierce

Didier Rémy

Aseem Rastogi

Tiark Rompf

Claudio Russo

Andrey Rybalchenko

Sriram Sankaranarayanan

Susmit Sarkar

Christophe Scholliers

Ilya Sergey

Michael Sperber

Manu Sridharan

Martin Sulzmann

Eijiro Sumii

Josef Svenningsson

Jean-Pierre Talpin

Paul Tarau

Benoit Valiron

Dimitrios Vytiniotis

Meng Wang

Stephanie Weirich

Contents

Probabilistic Programs as Spreadsheet Queries[*]

Andrew D. Gordon[1,2], Claudio Russo[1], Marcin Szymczak[2], Johannes Borgström[3],
Nicolas Rolland[1], Thore Graepel[1], and Daniel Tarlow[1]

[1]Microsoft Research, Cambridge, United Kingdom
[2]University of Edinburgh, Edinburgh, United Kingdom
[3]Uppsala University, Uppsala, Sweden

Abstract. We describe the design, semantics, and implementation of a probabilistic programming language where programs are spreadsheet queries. Given
an input database consisting of tables held in a spreadsheet, a query constructs
a probabilistic model conditioned by the spreadsheet data, and returns an output
database determined by inference. This work extends probabilistic programming
systems in three novel aspects: (1) embedding in spreadsheets, (2) dependently
typed functions, and (3) typed distinction between random and query variables.
It empowers users with knowledge of statistical modelling to do inference simply
by editing textual annotations within their spreadsheets, with no other coding.

1   Spreadsheets and Typeful Probabilistic Programming

Probabilistic programming systems [11, 14] enable a developer to write a short piece of
code that models a dataset, and then to rely on a compiler to produce efficient inference
code to learn parameters of the model and to make predictions. Still, a great many of the
world's datasets are held in spreadsheets, and accessed by users who are not developers.
How can spreadsheet users reap the benefits of probabilistic programming systems?

Our first motivation here is to describe an answer, based on an overhaul of Tabular
[13], a probabilistic language based on annotating the schema of a relational database.
The original Tabular is a standalone application that runs fixed queries on a relational
database (Microsoft Access). We began the present work by re-implementing Tabular
within Microsoft Excel, with the data and program held in spreadsheets.

The conventional view is that the purpose of a probabilistic program is to define the
random variables whose marginals are to be determined (as in the query-by-missing-
value of original Tabular). In our experience with spreadsheets, we initially took this
view, and relied on Excel formulas, separate from the probabilistic program, for post-
processing tasks such as computing the mode (most likely value) of a distribution, or
deciding on an action (whether or not to place a bet, say). We found, to our surprise,
that combining Tabular models and Excel formulas is error-prone and cumbersome,
particularly when the sizes of tables change, the parameters of the model change, or we
simply need to update a formula for every row of a column.

In response, our new design contributes the principle that a probabilistic program defines a pseudo-deterministic query on data. The query is specified in terms of three sorts

---

[*] This work was supported by Microsoft Research through its PhD Scholarship Programme.

J. Vitek (Ed.): ESOP 2015, LNCS 9032, pp. 1–25, 2015.
DOI: 10.1007/978-3-662-46669-8_1

of variable: (1) deterministic variables holding concrete input data; (2) nondeterministic random variables constituting the probabilistic model conditioned on input data; and (3) pseudo-deterministic query variables defining the result of the program (instead of using Excel formulas). Random variables are defined by draws from a set of builtin distributions. Query variables are defined via an **infer** primitive that returns the marginal posterior distributions of random variables. For instance, given a random variable of Boolean type, **infer** returns the probability p that the variable is **true**. In theory, **infer** is deterministic—it has an exact semantics in terms of measure theory; in practice, **infer** (and hence the whole query) is only pseudo-deterministic, as implementations almost always perform approximate or nondeterministic inference. We have many queries as evidence that post-processing can be incorporated into the language.

Our second motivation is to make a case for *typeful probabilistic programming* in general, with evidence from our experience of overhauling Tabular for spreadsheets. Cardelli [5] identifies the programming style based on widespread use of mechanically-checked types as *typeful programming*. Probabilistic languages that are embedded DSLs, such as HANSEI [17], Fun [2], and Factorie [19], are already typeful in that they inherit types from their host languages, while standalone languages, such as BUGS [9] or Stan [28], have value-indexed data schemas (but no user-defined functions). Still, we find that more sophisticated forms of type are useful in probabilistic modelling.

We make two general contributions to typeful probabilistic programming.
(1) *Value-indexed function types usefully organise user-defined components, such as conjugate pairs, in probabilistic programming languages.*

We allow value indexes in types to indicate the sizes of integer ranges and of array dimensions. We add value-indexed function types for user-defined functions, with a grid-based syntax. The paper has examples of user-defined functions (such as Action in Section 6) showing their utility beyond the fixed repertoire of conjugate pairs in the original Tabular. An important difficulty is to find a syntax for functions and their types that fits with the grid-based paradigm of spreadsheets.
(2) *A type-based information-flow analysis usefully distinguishes the stochastic and deterministic parts of a probabilistic program.*

To track the three sorts of variable, each type belongs to a *space* indicating whether it is: (**det**) deterministic input data, (**rnd**) a non-deterministic random variable defining the probabilistic model of the data, or (**qry**) a pseudo-deterministic query-variable defining a program result. Spaces allow a single language to define both model and query, while the type system governs flows between the spaces: data flows from **rnd** to **qry** via **infer**, but to ensure that a query needs only a single run of probabilistic inference, there are no flows from **qry** to **rnd**. There is an analogy between our spaces and levels in information flow systems: **det**-space is like a level of trusted data; **rnd**-space is like a level of untrusted data that is tainted by randomness; and **qry** is like a level of trusted data that includes untrusted data explicitly endorsed by **infer**.

The benefits of spaces include: (1) to document the role of variables, (2) to slice a program into the probabilistic model versus the result query, and (3) to prevent accidental errors. For instance, only variables in **det**-space may appear as indexes in types to guarantee that our models can be compiled to the finite factor graphs supported by inference backends such as Infer.NET [20].

This paper defines the syntax, semantics, and implementation of a new, more typeful Tabular. Our implementation is a downloadable add-in for Excel. For execution on data in a spreadsheet, a Tabular program is sliced into (1) an Infer.NET model for inference, and (2) a C# program to compute the results to be returned to the spreadsheet.

The original semantics of Tabular uses the higher-order model-learner pattern [12], based on a separate metalanguage. Given a Tabular schema $\mathbb{S}$ and an input database DB that matches $\mathbb{S}$, our semantics consists of two algorithms.

(1) An algorithm CoreSchema($\mathbb{S}$) applies a set of source-to-source reductions on $\mathbb{S}$ to yield $\mathbb{S}'$, which is in a core form of Tabular without user-defined functions and some other features.
(2) An algorithm CoreQuery($\mathbb{S}', DB$) first constructs a probabilistic model based on the **rnd**-space variables in $\mathbb{S}'$ conditioned by DB, and then evaluates the **qry**-space variables in $\mathbb{S}'$ to assemble an output database DB'.

Our main technical results about the semantics are as follows.

(1) Theorem 1 establishes that CoreSchema($\mathbb{S}$) yields the unique core form $\mathbb{S}'$ of a well-typed schema $\mathbb{S}$, as a corollary of standard properties of our reduction relation with respect to the type system (Proposition 1, Proposition 2, and Proposition 3).
(2) Theorem 2 establishes pre- and post-conditions of the input and output databases when $DB' = $ CoreQuery($\mathbb{S}', DB$).

Beyond theory, the paper describes many examples of the new typeful features of Tabular, including a detailed account of Bayesian Decision Theory, an important application of probabilistic programming, not possible in the original form of Tabular. A language like IBAL or Figaro allows for rational decision-making, but via decision-specific language features, rather than in the core expression language. We present a numeric comparison of a decision theory problem expressed in Tabular versus the same problem expressed in C# with direct calls to Infer.NET, showing that we pay very little in performance in return for a much more succinct spreadsheet program.

2   Functions and Queries, by Example

Primer: Discrete and Dirichlet Distributions.   To begin to describe the new features of Tabular, we recall a couple of standard distributions. If array $V = [p_0; \ldots; p_{n-1}]$ is a *probability vector* (that is, each p_i is a probability and they sum to 1) then Discrete$[n](V)$ is the discrete distribution that yields a sample $i \in 0..n - 1$ with probability p_i. The distribution Discrete$[2]([\frac{1}{2}; \frac{1}{2}])$ models a coin that we know to be fair. If we are uncertain whether the coin is fair, we need a distribution on probability vectors to represent our uncertainty. The distribution Dirichlet$[n]([c_0; \ldots; c_{n-1}])$ on a probability vector V represents our uncertainty after observing a count $c_i - 1$ of samples of i from Discrete$[n](V)$ for $i \in 0..n - 1$. We omit the formal definition, but discuss the case $n = 2$.

A probability vector V drawn from Dirichlet$[2]([t + 1; h + 1])$ represents our uncertainty about the bias of a coin after observing t tails and h heads. It follows that $V = [1 - p; p]$ where p is the probability of heads. The expected value of p is $\frac{h+1}{t+h+2}$, and the variance of p diminishes as t and h increase. If $t = h = 0$, the expected value is $\frac{1}{2}$ and p is uniformly distributed on the unit interval. If $t = h = 10$ say, the expected value remains $\frac{1}{2}$ but p is much more likely near the middle than the ends of the interval.

Review: Probabilistic Schemas in Tabular. Suppose we have a table named Coins with a column Flip containing a series of coin flips and wish to infer the bias of the coin. (The syntax [**for** i < 2 → 1.0] is an array comprehension, in this case returning [1.0, 1.0].)

table Coins (original Tabular)		
V real[]	**static output**	Dirichlet[2]([**for** i < 2 → 1.0])
Flip **int**	**output**	Discrete[2](V)

The model above (in original Tabular up to keyword renaming) is read as a probabilistic recipe for generating the coin flips from the unknown parameter V, conditioned on the actual dataset. The first line creates a random variable $V = [1 - p; p]$ from Dirichlet[2]([1; 1]), which amounts to choosing the probability p of heads uniformly from the unit interval. The second line creates a random variable Flip from Discrete[n](V) for each row of the table and conditions the variable in each row to equal the actual observed coin flip, if it is present. Each Tabular variable is either at **static**- or **inst**-level. A **static**-variable occurs just once per table, whereas an **inst**-variable occurs for each row of the table. The default level is **inst**, so Flip is at **inst**-level.

Now, suppose the data for the column Flip is [1; 1; 0]; the prior distribution of V is updated by observing 2 heads and 1 tails, to yield the posterior Dirichlet[2]([2; 3]), which has mean $\frac{3}{5}$. Given our example model, the fixed queries of this initial form of Tabular compute the posterior distribution of V, and write the resulting distributions as strings into the spreadsheet, as shown below. The missing value in cell B6 of the Flip column is predicted by the distribution in cell M6: 60% chance of 1, 40% chance of 0. (Cells E2 and E3 show dependent types of our new design, not of the original Tabular.)

	A	B	C	D	E	F	G	H	I	J	K	L	M
1	Coins			Coins					posterior_Coins		Log Evidence	Coins	
2	ID	Flip		V	real[2]	static output	Dirichlet[2]([for i<2 -> 1.0])		V	Dirichlet(2 3)	-2.48490665	ID Flip	0 Discrete(1=1 0=0)
3	0	1		Flip	mod(2)	output	Discrete[2](V)						1 Discrete(1=1 0=0)
4	1	1											2 Discrete(0=1 1=0)
5	2	0											3 Discrete(1=0.6 0=0.4)
6	3												

New Features of Tabular. Our initial experience with the re-implementation shows that writing probabilistic programs in spreadsheets is viable but suggests three new language requirements, explained in the remainder of this section.

(1) User-defined functions for abstraction (to generalize the fixed repertoire of primitive models in the original design).

(2) User-defined queries to control how parameters and predictions are inferred from the model and returned as results (rather than simply dumping raw strings from fixed queries).

(3) Value-indexed dependent types (to catch errors with vectors, matrices, and integer ranges, and help with compilation).

(1) User-Defined Functions. The Coins example shows the common pattern of a discrete distribution with a Dirichlet prior. We propose to write a function for such a pattern as follows. It explicitly returns the ret output but also implicitly returns the V output.

fun CDiscrete		
N **int!det**	**static input**	
R **real!rnd**	**static input**	
V **real[N]!rnd**	**static output**	Dirichlet[N]([**for** i < N → R])
ret **mod(N)!rnd**	**output**	Discrete[N](V)

In a table description, **input**-attributes refer implicitly to fully observed columns in the input database. On the other hand, a function is explicitly invoked using syntax like CDiscrete(N = 2; R = 1), and the **input**-attributes N and R refer to the argument expressions, passed call-by-value.

(2) User-Defined Queries. To support both the construction of probabilistic models for inference, and the querying of results, we label each type with one of three *spaces*:

(1) **det**-space is for fully observed input data;
(2) **rnd**-space is for probabilistic models, conditioned by partially observed input data;
(3) **qry**-space is for deterministic results queried from the inferred marginal distributions of **rnd**-space variables.

We organise the three spaces via the least partial order given by **det** < **rnd** and **det** < **qry**, so as to induce a subtype relation on types. Moreover, to allow flows from **rnd**-space to **qry**-space, an operator **infer**.$D.y_i(E)$ computes the parameter y_i in **qry**-space of the marginal distribution $D(y_1, \ldots, y_n)$ of an input E in **rnd**-space.

For example, here is a new model of our Coins table, using a call to CDiscrete to model the coin flips in **rnd**-space, and to implicitly define a **rnd**-space variable V for the bias of the coin. Assuming our model is conditioned by data $[1; 1; 0]$, the marginal distribution of V is Dirichlet$[2]([2; 3])$ where $[2; 3]$ is the counts-parameter. Hence, the call **infer**.Dirichlet$[2]$.counts(V) yields $[2; 3]$, and the query returns the mean $\frac{3}{5}$.

table Coins

Flip	mod(2)!rnd	output	CDiscrete(N=2, R=1.0)(*returns Flip and Flip_V*)
counts	real[2]!qry	static local	infer.Dirichlet[2].counts(Flip_V)
Mean	real!qry	static output	counts[1]/(counts[1]+counts[0])

Our reduction relation rewrites this schema to the following core form.

table Coins

R	real!rnd	static local	1.0
Flip_V	real[2]!rnd	static output	Dirichlet[2]([for i < 2 → R])
Flip	mod(2)!rnd	output	Discrete[2](V)
counts	real[2]!qry	static local	infer.Dirichlet[2].counts(Flip_V)
Mean	real!qry	static output	counts[1]/(counts[1]+counts[0])

(3) Simple Dependent Types. Our code has illustrated dependent types of statically-sized arrays and integer ranges: values of $T[e]$ are arrays of T of size e, while values of **mod**(e) are integers in the set $0..(e-1)$. In both cases, the size e must be a **det**-space **int**. (Hence, the dependence of types on expressions is simple, and all sizes may be resolved statically, given the sizes of tables.) The use of dependent types for arrays is standard (as in Dependent ML [30]); the main subtlety in our probabilitic setting is the need for spaces to ensure that indexes are deterministic.

Primitive distributions have dependent types:

Distributions: $D_{spc} : [x_1 : T_1, \ldots, x_m : T_m](y_1 : U_1, \ldots, y_n : U_n) \rightarrow T$

Discrete$_{spc}$: $[N : \textbf{int}!\textbf{det}]$(probs : **real**!$spc[N]$) $\rightarrow$ **mod**(N)!**rnd**
Dirichlet$_{spc}$: $[N : \textbf{int}!\textbf{det}]$(counts : **real**!$spc[N]$) $\rightarrow$ (**real**!**rnd**)[N]

User-defined functions have dependent types written as grids, such as the following type $Q_{CDiscrete}$ for CDiscrete:

N	int!det	static input
R	real!rnd	static input
V	real[N]!rnd	static output
ret	mod(N)!rnd	output

Finally, the table type for our whole model of the Coins table is the following grid. It lists the **rnd**-space variables returned by CDiscrete as well as the explicitly defined Mean. Attributes marked as **local** are private to a model or function, are identified up to alpha-equivalence, and do not appear in types. Attributes marked as **input** or **output** are binders, but are not identified up to alpha-equivalence, and are exported from tables or functions. Their names must stay fixed because of references from other tables.

V	real[2]!rnd	static output
Flip	mod(2)!rnd	output
Mean	real!qry	static output

3   Syntax of Tabular Enhanced with Functions and Queries

We describe the formal details of our revision of Tabular in this section. In the next, Section 4, we show how features such as function applications may be eliminated by reducing schemas to a core form with a direct semantics.

Column-Oriented Databases. Let t range over table names and c range over attribute names. We consider a database to be a pair $DB = (\delta_{in}, \rho_{sz})$ consisting of a record of tables $\delta_{in} = [t_i \mapsto \tau_i{}^{i \in 1..n}]$, and a valuation $\rho_{sz} = [t_i \mapsto sz_i{}^{i \in 1..n}]$ holding the number of rows $sz_i \in \mathbb{N}$ in each column of table t_i. Each table $\tau_i = [c_i \mapsto a_i{}^{j \in 1..m_i}]$ is a record of attributes a_i. An *attribute* is a value V tagged with a *level* ℓ. An attribute is normally a whole *column* **inst**(V), where V is an array of length sz_i and the level **inst** is short for "instance". It may also be a single value, **static**(V), a *static attribute*. The main purpose of allowing static attributes is to return individual results (such as Mean in our Coins example) from queries.

Databases, Tables, Attributes, and Values:

$\delta_{in} ::= [t_i \mapsto \tau_i{}^{i \in 1..n}]$	whole database
$\tau ::= [c_i \mapsto a_i{}^{i \in 1..m}]$	table in database
$a ::= \ell(V)$	attribute value: V with level ℓ
$V ::= ? \mid s \mid [V_0, \dots, V_{n-1}]$	nullable value
$\ell, pc ::= \textbf{static} \mid \textbf{inst}$	level (**static** < **inst**)

For example, the data for our Coins example is $DB = (\delta_{in}, \rho_{sz})$ where $\delta_{in} = [\text{Coins} \mapsto [\text{Flip} \mapsto \textbf{inst}([1; 1; 0])]]$ and $\rho_{sz} = [\text{Coins} \mapsto 3]$.

In examples, we assume each table has an implicit primary key ID and that the keys are in the range $0..sz_i - 1$. A value V may contain occurrences of "?", signifying missing data; we write $\text{known}(V)$ if V contains no occurrence of ?. Otherwise, a value may be an array, or a constant s: either a Boolean, integer, or real.

Syntax of Tabular Expressions and Schemas. An *index expression* e may be a variable x or a constant, and may occur in types (as the size of an array, for instance). Given a database $DB = (\delta_{in}, \rho_{sz})$, **sizeof**$(t)$ denotes the constant $\rho_{sz}(t)$. Attribute names c (but not table names) may occur in index expressions as variables. A *attribute type T* can be a scalar, a bounded non-negative integer or an array. Each type has an associated *space* (which is akin to an information-flow level, but independent of the notion of level in Tabular, introduced later on). (The type system is discussed in detail in Section 5.)

Index Expressions, Spaces and Dependent Types of Tabular:

$e ::= x \mid s \mid \textbf{sizeof}(t)$	index expression
$S ::= \textbf{bool} \mid \textbf{int} \mid \textbf{real}$	scalar type
$spc ::= \textbf{det} \mid \textbf{rnd} \mid \textbf{qry}$	space
$T, U ::= (S\,!\,spc) \mid (\textbf{mod}(e)\,!\,spc) \mid T[e]$	(attribute) type

$$\text{space}(S\,!\,spc) \triangleq spc \quad \text{space}(T[e]) \triangleq \text{space}(T) \quad \text{space}(\textbf{mod}(e)\,!\,spc) \triangleq spc$$
$$spc(T) \triangleq \text{space}(T) = spc$$

We write $\textbf{link}(t)$ as a shorthand for $\textbf{mod}(\textbf{sizeof}(t))$, for foreign keys to table t.

The syntax of (full) *expressions* includes index expressions, plus deterministic and random operations. We assume sets of deterministic functions g, and primitive distributions D. These have type signatures, as illustrated for Discrete and Dirichlet in Section 2. In $D[e_1, \ldots, e_m](F_1, \ldots, F_n)$, the arguments $e_1, \ldots, e_m$ index the result type, while $F_1, \ldots, F_n$ are parameters to the distribution. The operator $\textbf{infer}.D[e_1, \ldots, e_m].y(E)$ is described intuitively in Section 2. We write $fv(T)$ and $fv(E)$ for the sets of variables occurring free in type T and expression E.

Expressions of Tabular:

$E, F ::=$	expression
$\quad e$	index expression
$\quad g(E_1, \ldots, E_n)$	deterministic primitive g
$\quad D[e_1, \ldots, e_m](F_1, \ldots, F_n)$	random draw from distribution D
$\quad \textbf{if } E \textbf{ then } F_1 \textbf{ else } F_2$	if-then-else
$\quad [E_1, \ldots, E_n] \mid E[F]$	array literal, lookup
$\quad [\textbf{for } x < e \to F]$	for loop (scope of index x is F)
$\quad \textbf{infer}.D[e_1, \ldots, e_m].y(E)$	parameter y of inferred marginal of E
$\quad E : t.c$	dereference link E to instance of c
$\quad t.c$	dereference static attribute c of t

A Tabular schema is a relational schema with each attribute annotated not just with a type T, but also with a *level ℓ*, a *visibility viz*, and a *model expression M*.

Tabular Schemas:

$\mathbb{S} ::= [(t_1 = \mathbb{T}_1); \ldots; (t_n = \mathbb{T}_n)]$	(database) schema
$\mathbb{T} ::= [\text{col}_1; \ldots; \text{col}_n]$	table (or function)
$\text{col} ::= (c : T \; \ell \; viz \; M)$	attribute c declaration
$viz ::= \textbf{input} \mid \textbf{local} \mid \textbf{output}$	visibility
$M, N ::= \varepsilon \mid E \mid M[e_{index} < e_{size}] \mid \mathbb{T} R$	model expression
$R ::= (c_1 = e_1, \ldots, c_n = e_n)$	function arguments

For $(c : T \; \ell \; viz \; M)$ to be well-formed, $viz = \mathbf{input}$ if and only if $M = \varepsilon$. We only consider well-formed declarations. The visibility viz indicates whether the attribute c is given as an input, defined locally by the model expression M, or defined as an output by the model expression M. When omitted, the level of an attribute defaults to $\mathbf{inst}$.

Functions, Models, and Model Expressions. A challenge for this paper was to find a syntax for functions that is compatible with the grid format of spreadsheets; we do so by re-interpreting the syntax $\mathbb{T}$ for tables as also the syntax of functions. A *function* is a table of the form $\mathbb{T} = [\mathsf{col}_1; \ldots; \mathsf{col}_n; (\mathsf{ret} : T \; \mathbf{output} \; E)]$. A *model* is a function where each col_i is a local or an output. A *model expression* M denotes a *model* as follows:

- An expression E denotes the model that simply returns E.
- A *function application* $\mathbb{T} \; (c_1 = e_1, \ldots, c_n = e_n)$ denotes the function $\mathbb{T}$, but with each of its inputs c_i replaced by e_i.
- An *indexed model* $M[e_{index} < e_{size}]$ denotes the model for M, but with any $\mathbf{rnd}$ $\mathbf{static}$ attribute c replicated e_{size} times, as an array, and with references to c replaced by the lookup $c[e_{index}]$.

Formally, functions are embedded within our syntax of function applications $\mathbb{T} \; R$. In practice, our implementation supports separate function definitions written as $\mathbf{fun} \; f \; \mathbb{T}$, such as CDiscrete in Section 1 and CG in Section 6. A function reference (within a model expression) is written $f \; R$ to stand for $\mathbb{T} \; R$.

Indexed models appear in the original Tabular, while function applications are new.

Binders and Alpha-Equivalence. All attribute names c are considered bound by their declarations. The names of **local** attributes are identified up to alpha-equivalence. The names of **input** and **output** attributes are considered as fixed identifiers (like the fields of records) that export values from a table, and are not identified up to alpha-equivalence, because changing their names would break references to them.

Let inputs($\mathbb{T}$) be the **input** attributes of table $\mathbb{T}$, that is, the names c in $(c : T \; \ell \; \mathbf{input} \; \varepsilon)$. Let locals($\mathbb{T}$) be all the **local** attributes of table $\mathbb{T}$, that is, the names c in $(c : T \; \ell \; \mathbf{local} \; M)$. Let outputs($\mathbb{T}$) be all the **output** attributes of table $\mathbb{T}$, that is, the names c in $(c : T \; \ell \; \mathbf{output} \; M)$ plus outputs(M), where the latter consists of the union of outputs($\mathbb{T}_i$) for any applications of $\mathbb{T}_i$ within M. Let dom($\mathbb{T}$) be the union inputs($\mathbb{T}$) $\cup$ locals($\mathbb{T}$) $\cup$ outputs($\mathbb{T}$). Hence, the free variables fv($\mathbb{T}$) are given by:

$$\mathrm{fv}((c : T \; \ell \; viz \; M) :: \mathbb{T}') \triangleq \mathrm{fv}(T) \cup \mathrm{fv}(M) \cup (\mathrm{fv}(\mathbb{T}') \setminus (\{c\} \cup \mathrm{outputs}(M))) \quad \mathrm{fv}([]) \triangleq \{\}$$

4   Reducing Schemas to Core Tabular

We define reduction relations that explain the meaning of function calls and indexed models by rewriting, and hence transforms any well-typed schema to a core form. The reduction semantics allows us to understand indexed models, and also function calls, within the Tabular syntax. Hence, this semantics is more direct and self-contained than the original semantics of Tabular [13], based on translating to a semantic metalanguage.

If all the attributes of a schema are simple expressions E instead of arbitrary model expressions, we say it is in *core* form:

Core Attributes, Tables, and Schemas:

$\text{Core}((c : T \; \ell \; \textbf{input} \; \varepsilon)) \quad \text{Core}((c : T \; \ell \; \textbf{local} \; E)) \quad \text{Core}((c : T \; \ell \; \textbf{output} \; E))$
$\text{Core}([\text{col}_1 ; \ldots ; \text{col}_n])$ if $\text{Core}(\text{col}_1), \ldots, \text{Core}(\text{col}_n)$
$\text{Core}([t_i = \mathbb{T}_i{}^{i \in 1..n}])$ if $\text{Core}(\mathbb{T}_i)$ for each $i \in 1..n$

To help explain our reduction rules, consider the following function definition.

fun CG

M	**real!det static input**		
P	**real!det static input**		
Mean	**real!rnd static output**	GaussianFromMeanAndPrecision(M,P)
Prec	**real!rnd static output**	Gamma(1.0,1.0)
ret	**real!rnd output**	GaussianFromMeanAndPrecision(Mean,Prec)

The following mixture model, for a dataset consisting of durations and waiting times for Old Faithful eruptions, uses three function applications and two indexed models. Each row of the model belongs to one of two clusters, indicated by the attribute cluster; the indexed models for duration and time give different means and precisions depending on the value of cluster. Since cluster is an **output**, Tabular allows missing values in that column (and indeed they are all missing), but the **qry**-space assignment returns the most likely cluster for each row as the result of the query.

table faithful

cluster	**mod(2)!rnd output**	CDiscrete(N=2, R=1.0)
duration	**real!rnd** output	CG(M=0.0, P=1.0)[cluster < 2]
time	**real!rnd** output	CG(M=60.0, P=1.0)[cluster < 2]
assignment	**mod(2)!qry output**	ArgMax(infer.Discrete[2].probs(cluster))

The relation $\mathbb{T} \vdash R \rightsquigarrow_o \mathbb{T}_1$ means that $\mathbb{T}_1$ is the outcome of substituting the arguments R for the **input** attributes of the function $\mathbb{T}$, within an attribute named o. For example, for the function application in the duration attribute, we have $\text{CG} \vdash [M = 0.0, p = 1.0] \rightsquigarrow_{\text{duration}} \text{CG}_1$, where CG_1 is as follows:

duration_Mean	**real!rnd static output**	GaussianFromMeanAndPrecision(0.0, 1.0)
duration_Prec	**real!rnd static output**	Gamma(1.0, 1.0)
duration	**real!rnd output**	GaussianFromMeanAndPrecision(duration_Mean, duration_Prec)

The inductive definition follows. Rule (APPLY INPUT) instantiates an **input** c with an argument e; (APPLY SKIP) prefixes **local** and **output** attributes with o; and (APPLY RET) turns the ret attribute of the function into name o of the call-site.

Inductive Definition of Function Application: $\mathbb{T} \vdash R \rightsquigarrow_o \mathbb{T}_1$

(APPLY RET)

$$\frac{}{[(\text{ret} : T \; \ell \; viz \; E)] \vdash [] \rightsquigarrow_o [(o : T \; \ell \; viz \; E)]}$$

(APPLY INPUT)

$$\frac{\mathbb{T}\{{}^e/_c\} \vdash R \rightsquigarrow_o \mathbb{T}_1 \quad \text{dom}(\mathbb{T}) \cap \text{fv}(e) = \varnothing}{(c : T \; \ell \; \textbf{input} \; \varepsilon) :: \mathbb{T} \vdash [c = e] :: R \rightsquigarrow_o \mathbb{T}_1}$$

(APPLY SKIP) $(viz \in \{\textbf{local}, \textbf{output}\})$

$$\frac{\mathbb{T}\{{}^{o\_c}/_c\} \vdash R \rightsquigarrow_o \mathbb{T}_1 \quad c \notin \text{fv}(R)}{(c : T \; \ell \; viz \; E) :: \mathbb{T} \vdash R \rightsquigarrow_o (o\_c : T \; \ell \; viz \; E) :: \mathbb{T}_1}$$

Next, we define $\text{index}_\sigma(\mathbb{T}, e_1, e_2)$ to be the outcome of indexing the **static rnd** or **qry** variables of a core table $\mathbb{T}$, that is, turning each declaration of such a variable into an array of size e_2, and each reference to such a variable into an array access indexed, at static level, by a local replication index i, or, at instance level, by the random indexing expression e_1 (or its mode at **qry** level). Both i and e_1 are integers bounded by e_2. Variables that require indexing are accumulated in the renaming substitution σ (which is initially empty). For instance, $CG_1[\text{cluster} < 2]$ expands to $\text{index}_\varnothing(CG_1, \text{cluster}, 2)$:

duration_Mean	real[2]!rnd	static output	[for $i < 2 \rightarrow$ GaussianFromMeanAndPrecision(0.0, 1.0)]
duration_Mean^	real!rnd	local	duration_Mean[cluster]
duration_Prec	real[2]!rnd	static output	[for $i < 2 \rightarrow$ Gamma(1.0, 1.0)]
duration_Prec^	real!rnd	local	duration_Prec[cluster]
duration	real!rnd	output	GaussianFromMeanAndPrecision(duration_Mean^, duration_Prec^)

Table Indexing: $\text{index}_\sigma(\mathbb{T}, e_1, e_2)$

$$\text{index}_\sigma([], e_1, e_2) \triangleq []$$
$$\text{index}_\sigma((c : T\ \ell\ \textbf{input}\ \varepsilon) :: \mathbb{T}, e_1, e_2) \triangleq (c : T\ \ell\ \textbf{input}\ \varepsilon) :: (\text{index}_\sigma(\mathbb{T}, e_1, e_2))$$
$$\text{index}_\sigma((c : T\ \ell\ viz\ E) :: \mathbb{T}), e_1, e_2) \triangleq$$
$\quad (c : T[e_2]\ \ell\ viz\ [\textbf{for}\ i < e_2 \rightarrow \rho(E)]) :: (\hat{c} : T\ \textbf{inst local}\ c[\hat{e}_1]) :: \text{index}_{\sigma[c \mapsto \hat{c}]}(\mathbb{T}, e_1, e_2)$
$\quad\quad \text{if } viz \neq \textbf{input},\ \ell = \textbf{static},\ \neg\textbf{det}(T)\ \text{where}$
$\quad\quad \rho = \{d \mapsto d[i] \mid d \in \text{dom}(\sigma)\},\ i \notin \text{fv}(E) \cup \text{fv}(\sigma),\ \hat{c} \notin \text{dom}(\mathbb{T}) \cup \text{fv}(\mathbb{T}, \sigma, c, e_1, e_2)$
$\quad\quad \text{and } \hat{e}_1 = e_1\ \text{if } \textbf{rnd}(T),\ \text{and } \hat{e}_1 = \text{ArgMax}(\textbf{infer}.\text{Discrete}[e_2].\text{probs}(e_1))\ \text{if } \textbf{qry}(T)$

$\quad (c : T\ \ell\ viz\ \sigma(E)) :: \text{index}_\sigma(\mathbb{T}, e_1, e_2)$
$\quad\quad \text{if } viz \neq \textbf{input}\ \text{and}\ (\ell = \textbf{inst}\ \text{or}\ \textbf{det}(T))$

Below, we give inductive definitions of reduction relations on schemas, tables, and model expressions. There are congruence rules, plus (RED INDEX) and (RED INDEX EXPR) for indexed models, and (RED APPL) for applications. The latter needs additional operations $\mathbb{T} \wedge \ell$ and $\mathbb{T} \wedge viz$, to adjust the model $\mathbb{T}$ of function body to the level ℓ and visibility viz of the call-site. These operators drop any **output** attributes to **local**, if the callsite is **local**, and drop any **inst**-level attributes to **static**, if the callsite is **static**.

- Consider the 2-point lattice **static** < **inst**. Let $\mathbb{T} \wedge \ell$ be the outcome of changing each $(c : T\ \ell_c\ viz\ M)$ in $\mathbb{T}$ to $(c : T\ (\ell_c \wedge \ell)\ viz\ M)$. Hence, $\mathbb{T} \wedge \textbf{inst}$ is the identity, while $\mathbb{T} \wedge \textbf{static}$ drops **inst** variables to **static** variables.
- Consider the 2-point lattice **local** < **output**. Let $\mathbb{T} \wedge viz$ be the outcome of changing each $(c : T\ \ell\ viz_c\ M)$ in $\mathbb{T}$ to $(c : T\ \ell\ (viz_c \wedge viz)\ M)$. Hence, $\mathbb{T} \wedge \textbf{output}$ is the identity, while $\mathbb{T} \wedge \textbf{local}$ drops **output** variables to **local** variables.

Reduction Relations: $\mathbb{S} \rightarrow \mathbb{S}',\ \mathbb{T} \rightarrow \mathbb{T}',\ M \rightarrow M'$

(RED SCHEMA LEFT)
$$\frac{\mathbb{T} \rightarrow \mathbb{T}'}{(t = \mathbb{T}) :: \mathbb{S} \rightarrow (t = \mathbb{T}') :: \mathbb{S}}$$

(RED SCHEMA RIGHT)
$$\frac{\mathbb{S} \rightarrow \mathbb{S}' \quad \text{Core}(\mathbb{T})}{(t = \mathbb{T}) :: \mathbb{S} \rightarrow (t = \mathbb{T}) :: \mathbb{S}'}$$

(RED MODEL)
$$\frac{M \rightarrow M'}{(c : T\ \ell\ viz\ M) :: \mathbb{T} \rightarrow (c : T\ \ell\ viz\ M') :: \mathbb{T}}$$

(RED TABLE RIGHT)
$$\frac{\mathbb{T} \rightarrow \mathbb{T}' \quad \text{Core}(col)}{col :: \mathbb{T} \rightarrow col :: \mathbb{T}'}$$

$$\frac{\text{(RED INDEX INNER)}}{M \to M'} \qquad M[e_{index} < e_{size}] \to M'[e_{index} < e_{size}]$$

$$\frac{\text{(RED INDEX)}}{\text{Core}(\mathbb{T}) \quad \text{fv}(e_{index}, e_{size}) \cap (\text{dom}(\mathbb{T})) = \varnothing}{(\mathbb{T}\, R)[e_{index} < e_{size}] \to (\text{index}_\varnothing(\mathbb{T}, e_{index}, e_{size}))\, R}$$

$$\frac{\text{(RED INDEX EXPR)}}{E[e_{index} < e_{size}] \to E}$$

$$\frac{\text{(RED APPL) (for Core}(\mathbb{T}))}{((\mathbb{T} \wedge \ell) \wedge viz) \vdash R \rightsquigarrow_o \mathbb{T}_1}{(\text{locals}(\mathbb{T}_1) \cup \text{inputs}(\mathbb{T}_1)) \cap (\text{fv}(\mathbb{T}') \cup \text{dom}(\mathbb{T}')) = \varnothing}{(o : T'\, \ell\, viz\, (\mathbb{T}\, R)) :: \mathbb{T}' \to \mathbb{T}_1 @ \mathbb{T}'}$$

Tables in core form have no reductions. Moreover, the reduction relation is deterministic (we include the Core($\mathbb{T}$) condition on the rules (RED SCHEMA RIGHT), (RED TABLE RIGHT), and (RED INDEX) to fix a particular reduction strategy).

By using the above rules to expand out the three function calls and the two model expressions in the Old Faithful example, we obtain the core model below:

table faithful			
cluster_V	real[2]!rnd	static output	Dirichlet[2]([for i < 2 →1.0])
cluster	mod(2)!rnd	output	Discrete[2](cluster_V)
duration_Mean	real[2]!rnd	static output	[for i < 2 →GaussianFromMeanAndPrecision(0.0, 1.0)]
duration_Mean^	real!rnd	local	duration_Mean[cluster]
duration_Prec	real[2]!rnd	static output	[for i < 2 →Gamma(1.0, 1.0)]
duration_Prec^	real!rnd	local	duration_Prec[cluster]
duration	real!rnd	output	GaussianFromMeanAndPrecision(duration_Mean^, duration_Prec^)
time_Mean	real[2]!rnd	static output	[for i < 2 →GaussianFromMeanAndPrecision(60.0, 1.0)]
time_Mean^	real!rnd	local	time_Mean[cluster]
time_Prec	real[2]!rnd	static output	[for i < 2 →Gamma(1.0, 1.0)]
time_Prec^	real!rnd	local	time_Prec[cluster]
time	real!rnd	output	GaussianFromMeanAndPrecision(time_Mean^, time_Prec^)
assignment	mod(2)!qry	output	ArgMax(infer.Discrete[2].probs(cluster))

Moreover, here are screen shots (best viewed in colour) of the data, model and inference results in Excel.

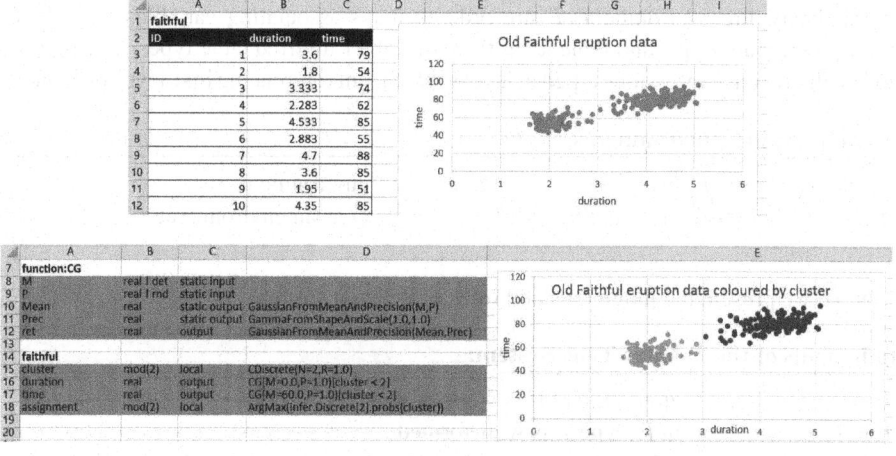

7	posterior_faithful				
8	cluster_V	Dirichlet(98.03 176)			
9	duration_Mean	[0] Gaussian(2.036, 0.0009324) [1] Gaussian(4.287, 0.001017)			
10	duration_Prec	[0] Gamma(49.51, 0.2231) [1] Gamma(88.49, 0.06344)			
11	time_Mean	[0] Gaussian(55.97, 0.2679) [1] Gaussian(74.65, 0.2673)			
12	time_Prec	[0] Gamma(49.51, 0.0005689) [1] Gamma(88.49, 0.000177)			
280					

faithful

ID	cluster	duration	time	assignment
	Discrete(1=0.999981673477424 0=1.83265225764259E-05)			
1	Discrete[0=1	Gaussian.PointMass(3.6)	Gaussian.PointMass(79)	1
2	1=5.99500933746004E-18) Discrete(1=0.999217126276896	Gaussian.PointMass(1.8)	Gaussian.PointMass(54)	0
3	0=0.000782873723104474) Discrete(0=0.999999999918014	Gaussian.PointMass(3.333)	Gaussian.PointMass(74)	1
4	1=8.19860393142277E-11) Discrete(1=0.999999994867373	Gaussian.PointMass(2.283)	Gaussian.PointMass(62)	0
272	0=5.13262696023921E-09)	Gaussian.PointMass(4.467)	Gaussian.PointMass(74)	1

5 Dependent Type System and Semantics

5.1 Dependent Type System

The expressions of Tabular are based on the probabilistic language Fun [2]. We significantly extend Fun by augmenting its types with the three spaces described in Section 1, adding value-indexed dependent types including statically bounded integers and sized arrays, and additional expressions including an operator for inference and operations for referencing attributes of tables and their instances.

We use *table types* Q both for functions and for concrete tables. When used to type a function Q must satisfy the predicate **fun**(Q), which requires it to use the distinguished name ret for the explicit result of the function (its final output). When used to type a concrete table Q must satisfy the predicate **table**(Q), which ensures that types do not depend on the contents of any input table t (except for the sizes of tables). We only need **table**(Q) to define a conformance relation on databases and schema types.

Table and Schema Types:

$$Q ::= [(c_i : T_i \ \ell_i \ viz_i)^{i \in 1..n}] \qquad \text{table type } (c_i \text{ distinct}, viz_i \neq \textbf{local})$$
$$Sty ::= [(t_i : Q_i)^{i \in 1..n}] \qquad \text{schema type } (t_i \text{ distinct})$$

fun(Q) iff $viz_n = $ **output** and $c_n = $ ret.
model(Q) iff **fun**(Q) and each $viz_i = $ **output**.
table(Q) iff for each $i \in 1..n$, $\ell_i = $ **static** $\Rightarrow$ **rnd**$(T_i) \vee$ **qry**(T_i).

Tabular typing environments Γ are ordered maps associating variables with their declared level and type, and table identifiers with their inferred table types. The typing rules will prevent expressions typed at level **static** from referencing **inst** level variables.

Tabular Typing Environments:

$$\Gamma ::= \varnothing \mid (\Gamma, x :^{\ell} T) \mid (\Gamma, t : Q) \qquad \text{environment}$$
$$\gamma([(c_i : T_i \ \ell_i \ viz_i)^{i \in 1..n}]) \triangleq c_i :^{\ell_i} T_i^{\ i \in 1..n} \qquad Q \text{ as an environment}$$

Next, we present the judgments and rules of the type system.

Judgments of the Tabular Type System:

$$\Gamma \vdash \diamond \qquad \text{environment } \Gamma \text{ is well-formed}$$
$$\Gamma \vdash T \qquad \text{in } \Gamma, \text{ type } T \text{ is well-formed}$$

$\Gamma \vdash^{pc} e : T$	in Γ at level pc, index expression e has type T
$\Gamma \vdash Q$	in Γ, table type Q is well-formed
$\Gamma \vdash Sty$	in Γ, schema type Sty is well-formed
$\Gamma \vdash T <: U$	in Γ, T is a subtype of U
$\Gamma \vdash^{pc} E : T$ path	in Γ at level pc, expression E is a path
$\Gamma \vdash^{pc} E : T$	in Γ at level pc, expression E has type T
$\Gamma \vdash^{pc}_o R : Q \to Q'$	R sends function type Q to model type Q' in column o
$\Gamma \vdash^{pc}_o M : Q$	model expression M has model type Q
$\Gamma \vdash^{pc} \mathbb{T} : Q$	table $\mathbb{T}$ has type Q
$\Gamma \vdash \mathbb{S} : Sty$	schema $\mathbb{S}$ has type Sty

The formation rules for types and environments depend mutually on the typing rules for index expressions. Only index expressions that are **det**-space and **static**-level may occur in types. We write $\mathrm{ty}(s)$ for the scalar type S of the scalar s.

Rules for Types, Environments, and Index Expressions: $\Gamma \vdash \diamond$ $\Gamma \vdash T$ $\Gamma \vdash^{pc} e : T$

(ENV EMPTY)

$$\varnothing \vdash \diamond$$

(ENV VAR)

$$\frac{\Gamma \vdash T \quad x \notin \mathrm{dom}(\Gamma)}{\Gamma, x :^{pc} T \vdash \diamond}$$

(ENV TABLE) (**table**(Q))

$$\frac{\Gamma \vdash Q \quad t \notin \mathrm{dom}(\Gamma)}{\Gamma, t : Q \vdash \diamond}$$

(TYPE SCALAR)

$$\frac{\Gamma \vdash \diamond}{\Gamma \vdash S \,!\, spc}$$

(TYPE RANGE)

$$\frac{\Gamma \vdash^{\textbf{static}} e : \textbf{int}\,!\,\textbf{det}}{\Gamma \vdash \textbf{mod}(e)\,!\, spc}$$

(TYPE ARRAY)

$$\frac{\Gamma \vdash T \quad \Gamma \vdash^{\textbf{static}} e : \textbf{int}\,!\,\textbf{det}}{\Gamma \vdash T[e]}$$

(INDEX VAR) (for $\ell \le pc$)

$$\frac{\Gamma \vdash \diamond \quad \Gamma = \Gamma_1, x :^{\ell} T, \Gamma_2}{\Gamma \vdash^{pc} x : T}$$

(INDEX SCALAR)

$$\frac{\Gamma \vdash \diamond \quad S = \mathrm{ty}(s)}{\Gamma \vdash^{pc} s : S\,!\,\textbf{det}}$$

(INDEX MOD)

$$\frac{\Gamma \vdash \diamond \quad 0 \le n < m}{\Gamma \vdash^{pc} n : \textbf{mod}(m)\,!\,\textbf{det}}$$

(INDEX SIZEOF)

$$\frac{\Gamma \vdash \diamond \quad \Gamma = \Gamma', t : Q, \Gamma''}{\Gamma \vdash^{pc} \textbf{sizeof}(t) : \textbf{int}\,!\,\textbf{det}}$$

Formation Rules for Table and Schema Types: $\Gamma \vdash Q$ $\Gamma \vdash Sty$

(TABLE TYPE [])

$$\frac{\Gamma \vdash \diamond}{\Gamma \vdash [] : []}$$

(TABLE TYPE INPUT)

$$\frac{\Gamma \vdash T \quad \Gamma, c :^{\ell} T \vdash Q}{\Gamma \vdash (c : T \,\ell\, \textbf{input}) :: Q}$$

(TABLE TYPE OUTPUT)

$$\frac{\Gamma \vdash T \quad \Gamma, c :^{\ell} T \vdash Q}{\Gamma \vdash (c : T \,\ell\, \textbf{output}) :: Q}$$

(SCHEMA TYPE [])

$$\frac{\Gamma \vdash \diamond}{\Gamma \vdash [] : []}$$

(SCHEMA TYPE TABLE)

$$\frac{\Gamma \vdash Q \quad \textbf{table}(Q) \quad \Gamma, t : Q \vdash Sty}{\Gamma \vdash (t : Q) :: Sty}$$

Subtyping allows **det**-space data to be used as **rnd**-space or **qry**-space data. The preorder $\le$ on spaces is the least reflexive relation to satisfy **det** $\le$ **rnd** and **det** $\le$ **qry**. The default space is **det**, so when we write S or $\textbf{mod}(e)$ as a type, we mean $S\,!\,\textbf{det}$ or $\textbf{mod}(e)\,!\,\textbf{det}$. We define a commutative partial operation $spc \vee spc'$, and lift this operation to types $T \vee spc$ to weaken the space of a type.

Least upper bound: $spc \vee spc'$ **(if** $spc \leq spc'$ **or** $spc' \leq spc$**)**

$spc \vee spc = spc$ $\mathbf{det} \vee \mathbf{rnd} = \mathbf{rnd}$ $\mathbf{det} \vee \mathbf{qry} = \mathbf{qry}$
(The combination $\mathbf{rnd} \vee \mathbf{qry}$ is intentionally not defined.)

Operations on Types and Spaces: $T \vee spc$

$(S\,!\,spc) \vee spc' \triangleq S\,!\,(spc \vee spc')$ $T[e] \vee spc \triangleq (T \vee spc)[e]$
$(\mathbf{mod}(e)\,!\,spc) \vee spc' \triangleq \mathbf{mod}(e)\,!\,(spc \vee spc')$

Given these definitions, we present the rules of subtyping and of typing expressions.

Rules of Subtyping: $\Gamma \vdash T <: U$

(SUB SCALAR)
$$\frac{\Gamma \vdash \diamond \quad spc_1 \leq spc_2}{\Gamma \vdash S\,!\,spc_1 <: S\,!\,spc_2}$$

(SUB MOD)
$$\frac{\Gamma \vdash^{\mathsf{static}} e : \mathbf{int}\,!\,\mathbf{det} \quad spc_1 \leq spc_2}{\Gamma \vdash \mathbf{mod}(e)\,!\,spc_1 <: \mathbf{mod}(e)\,!\,spc_2}$$

(SUB ARRAY)
$$\frac{\Gamma \vdash T <: U \quad \Gamma \vdash^{\mathsf{static}} e : \mathbf{int}\,!\,\mathbf{det}}{\Gamma \vdash T[e] <: U[e]}$$

The table below presents the typing rules for Tabular expressions, most of which are standard modulo the operations on spaces. For instance, in (DEREF INST), the type of the indexed column needs to be joined with the space of the index, because, for instance, an expression returning a deterministic value at a random index is random. Similarly, an expression returning an element of a deterministic array at a random index is random, hence the join in (INDEX).

Since deterministic parameters of random primitives can occur in the types of random arguments and the return type, they have to be substituted out in the (RANDOM) and (INFER) rules.

(Selected) Typing Rules for Expressions: $\Gamma \vdash^{pc} E : T$ path, $\Gamma \vdash^{pc} E : T$

(VARIABLE PATH)
$$\frac{\Gamma \vdash^{pc} x : T}{\Gamma \vdash^{pc} x : T \text{ path}}$$

(INDEXED PATH)
$$\frac{\Gamma \vdash^{pc} p_1 : T[e] \text{ path} \quad \Gamma \vdash^{pc} p_2 : \mathbf{mod}(e)\,!\,\mathbf{det} \text{ path}}{\Gamma \vdash^{pc} p_1[p_2] : T \text{ path}}$$

(SUBSUM)
$$\frac{\Gamma \vdash^{pc} E : T \quad \Gamma \vdash T <: U}{\Gamma \vdash^{pc} E : U}$$

(INDEX EXPRESSION)
$$\frac{\Gamma \vdash^{pc} e : T \quad (e \text{ is an index expression})}{\Gamma \vdash^{pc} e : T \quad (e \text{ seen as an expression})}$$

(DEREF STATIC)
$$\frac{\Gamma = \Gamma',t : Q,\Gamma'' \quad Q = Q'@[(c : T \text{ static } viz)]@Q''}{\Gamma \vdash^{pc} t.c : T}$$

(DEREF INST)
$$\frac{\Gamma \vdash^{pc} E : \mathbf{link}(t)\,!\,spc \quad \Gamma = \Gamma',t : Q,\Gamma'' \quad Q = Q'@[(c : T \text{ inst } viz)]@Q''}{\Gamma \vdash^{pc} E : t.c : T \vee spc}$$

(RANDOM) (where $\sigma(U) \triangleq U\{^{e_1}/_{x_1}\} \ldots \{^{e_m}/_{x_m}\}$)
$$\frac{\begin{array}{c} D_{\mathbf{rnd}} : [x_1 : T_1,\ldots,x_m : T_m](y_1 : U_1,\ldots,y_n : U_n) \to T \\ \Gamma \vdash^{\mathsf{static}} e_i : T_i \quad \forall i \in 1..m \quad \Gamma \vdash^{pc} F_j : \sigma(U_j) \quad \forall j \in 1..n \quad \Gamma \vdash \diamond \\ \{x_1,\ldots,x_m\} \cap (\bigcup_i \mathrm{fv}(e_i)) = \varnothing \quad x_i \neq x_j \text{ for } i \neq j \end{array}}{\Gamma \vdash^{pc} D[e_1,\ldots,e_m](F_1,\ldots,F_n) : \sigma(T)}$$

$$(\text{ITER}) \ (\text{where } x \notin \text{fv}(T))$$
$$\dfrac{\Gamma \vdash^{\text{static}} e : \textbf{int} \,!\, \textbf{det}}{\Gamma \vdash^{pc} [\textbf{for } x < e \to F] : T[e]} \quad \Gamma, x :^{pc} (\textbf{mod}(e) \,!\, \textbf{det}) \vdash^{pc} F : T$$

$$(\text{INDEX})$$
$$\dfrac{\text{space}(T) \leq spc \qquad \Gamma \vdash^{pc} E : T[e] \quad \Gamma \vdash^{pc} F : \textbf{mod}(e) \,!\, spc}{\Gamma \vdash^{pc} E[F] : T \vee spc}$$

$$(\text{INFER}) \ (\text{where } \sigma(U) \triangleq U\{^{e_1}/_{x_1}\} \dots \{^{e_m}/_{x_m}\})$$
$$D_{\textbf{qry}} : [x_1 : T_1, \dots, x_m : T_m](y_1 : U_1, \dots, y_n : U_n) \to T$$
$$\dfrac{\Gamma \vdash^{\text{static}} e_i : T_i \quad \forall i \in 1..m \quad \Gamma \vdash^{pc} E : \sigma(T) \text{ path} \quad j \in 1..n}{\{x_1, \dots, x_m\} \cap (\bigcup_i \text{fv}(e_i)) = \varnothing \quad x_i \neq x_j \text{ for } i \neq j}$$
$$\Gamma \vdash^{pc} \textbf{infer}.D[e_1, \dots, e_m].y_j(E) : \sigma(U_j)$$

For an example of (INFER), recall the expression $\textbf{infer}.\text{Dirichlet}[2].\text{counts}(V)$ from Section 1. Here $m = n = 1$, $y_1 = \text{counts}$ and $U_1 = \textbf{real}[N]\,!\,\textbf{qry}$ and $\sigma = \{^2/_N\}$ and the result type is $\sigma(U_1) = \textbf{real}[2]\,!\,\textbf{qry}$.

Below are the typing rules for function arguments. In (ARG INPUT), the level $\ell \wedge pc$ at which the argument needs to be checked is bounded both by the level pc of the function aplication and the level ℓ of the given column of the function. In (ARG OUTPUT), the level $\ell \wedge pc$ of the output column of the reduced application is bounded by the level pc at which the function was applied as well as the level ℓ of the given column.

Typing Rules for Arguments: $\Gamma \vdash_o^{pc} R : Q \to Q'$

$$(\text{ARG INPUT}) \dfrac{\Gamma \vdash^{\ell \wedge pc} e : T \quad \Gamma \vdash_o^{pc} R : Q\{^e/_c\} \to Q'}{\Gamma \vdash_o^{pc} ((c = e) :: R) : ((c : T \ \ell \ \textbf{input}) :: Q) \to Q'}$$

$$(\text{ARG OUTPUT}) \dfrac{\Gamma \vdash T \quad \Gamma \vdash_o^{pc} R : Q\{^{o\_c}/_c\} \to Q' \quad c \neq \text{ret}}{\Gamma \vdash_o^{pc} R : ((c : T \ \ell \ \textbf{output}) :: Q) \to ((o\_c : T \ (\ell \wedge pc) \ \textbf{output}) :: Q')}$$

$$(\text{ARG RET}) \dfrac{\Gamma \vdash T}{\Gamma \vdash_o^{pc} R : (\text{ret} : T \ \ell \ \textbf{output}) \to (\text{ret} : T \ (\ell \wedge pc) \ \textbf{output})}$$

For example, if $Q_{CDiscrete}$ is the function type of CDiscrete from Section 1 we can derive $b :^{\text{static}} \textbf{real}\,!\,\textbf{rnd} \vdash_{\text{Flip}}^{\text{inst}} (N = 2, \text{alpha} = b) : Q_{CDiscrete} \to Q'$ where Q', shown in the grid below, represents the outputs of the function call. Since the inputs N and alpha of CDiscrete are both **static**, arguments 2 and b are typed at level $\textbf{static} \wedge \textbf{inst} = \textbf{static}$.

Flip_V	real[2]!rnd	static output
ret	mod(2)!rnd	output

Next, we have rules for assigning a model type Q to a model expression M. (MODEL INDEXED) needs the following operation $Q[e]$ to capture the static effect of indexing:

Indexing a Table Type: $Q[e]$

$$\varnothing[e] \triangleq \varnothing$$

$$((c : T \ \textbf{inst } viz) :: Q)[e] \triangleq (c : T \ \textbf{inst } viz) :: (Q[e])$$

$$((c : T \ \textbf{static } viz) :: Q)[e] \triangleq (c : T \ \textbf{static } viz) :: (Q[e]) \quad \text{if } viz = \textbf{input} \text{ or } \textbf{det}(T)$$

$$((c : T \ \textbf{static } viz) :: Q)[e] \triangleq (c : T[e] \ \textbf{static } viz) :: (Q[e]) \quad \text{if } viz \neq \textbf{input} \text{ and } \neg\textbf{det}(T)$$

The vectorized c cannot appear in Q when $\mathbf{rnd}(T)$, so $Q[e]$ remains well-formed.

Typing Rules for Model Expressions: $\Gamma \vdash_o^{pc} M : Q$

(MODEL EXPRESSION)
$\Gamma \vdash^{pc} E : T$
$\overline{\Gamma \vdash_o^{pc} E : [(\text{ret} : T \; pc \; \mathbf{output})]}$

(MODEL APPL)
$\Gamma \vdash^{pc} \mathbb{T} : Q \quad \mathbf{fun}(Q) \quad \Gamma \vdash_o^{pc} R : Q \to Q'$
$\overline{\Gamma \vdash_o^{pc} \mathbb{T} R : Q'}$

(MODEL INDEXED)
$\Gamma \vdash_o^{pc} M : Q \quad \text{dom}(Q) \cap \text{fv}(e_{size}) = \varnothing \quad \Gamma \vdash^{pc} e_{index} : \mathbf{mod}(e_{size}) \;!\, \mathbf{rnd} \; \text{path}$
$\overline{\Gamma \vdash_o^{pc} M[e_{index} < e_{size}] : Q[e_{size}]}$

Finally, we complete the system with rules for tables and schemas.

Typing Rules for Tables: $\Gamma \vdash^{pc} \mathbb{T} : Q$

(TABLE [])
$\Gamma \vdash \diamond$
$\overline{\Gamma \vdash^{pc} [] : []}$

(TABLE INPUT)
$\Gamma, c :^{\ell \wedge pc} T \vdash^{pc} \mathbb{T} : Q$
$\overline{\Gamma \vdash^{pc} (c : T \; \ell \; \mathbf{input} \; \varepsilon) :: \mathbb{T} : (c : T \; (\ell \wedge pc) \; \mathbf{input}) :: Q}$

(TABLE OUTPUT)
$\Gamma \vdash_c^{\ell \wedge pc} M : Q_c @ [(\text{ret} : T \; (\ell \wedge pc) \; \mathbf{output})] \quad \Gamma, \gamma(Q_c), c :^{\ell \wedge pc} T \vdash^{pc} \mathbb{T} : Q$
$\overline{\Gamma \vdash^{pc} (c : T \; \ell \; \mathbf{output} \; M) :: \mathbb{T} : Q_c @ ((c : T \; (\ell \wedge pc) \; \mathbf{output}) :: Q)}$

(TABLE LOCAL) (where $(\text{dom}(Q_c) \cup \{c\}) \cap \text{fv}(Q) = \varnothing$)
$\Gamma \vdash_c^{\ell \wedge pc} M : Q_c @ [(\text{ret} : T \; (\ell \wedge pc) \; \mathbf{output})] \quad \Gamma, \gamma(Q_c), c :^{\ell \wedge pc} T \vdash^{pc} \mathbb{T} : Q$
$\overline{\Gamma \vdash^{pc} (c : T \; \ell \; \mathbf{local} \; M) :: \mathbb{T} : Q}$

Typing Rules for Schemas: $\Gamma \vdash \mathbb{S} : Sty$

(SCHEMA [])
$\Gamma \vdash \diamond$
$\overline{\Gamma \vdash [] : []}$

(SCHEMA TABLE)
$\Gamma \vdash^{\mathbf{inst}} \mathbb{T} : Q \quad \mathbf{table}(Q) \quad \Gamma, t : Q \vdash \mathbb{S} : Sty$
$\overline{\Gamma \vdash (t = \mathbb{T}) :: \mathbb{S} : (t : Q) :: Sty}$

5.2   Reduction to Core Tabular

Proposition 1 (Preservation).

(1) *If* $\Gamma \vdash \mathbb{S} : Sty$ *and* $\mathbb{S} \to \mathbb{S}'$ *then* $\Gamma \vdash \mathbb{S}' : Sty$.
(2) *If* $\Gamma \vdash^{pc} \mathbb{T} : Q$ *and* $\mathbb{T} \to \mathbb{T}'$ *then* $\Gamma \vdash^{pc} \mathbb{T}' : Q$.
(3) *If* $\Gamma \vdash^{pc} M : Q$ *and* $M \to M'$ *then* $\Gamma \vdash^{pc} M' : Q$.

Proposition 2 (Progress). *If* $\Gamma \vdash^{pc} \mathbb{S} : Q$ *either* $\text{Core}(\mathbb{S})$ *or there is* $\mathbb{S}'$ *such that* $\mathbb{S} \to \mathbb{S}'$.

Proposition 3 (Termination). *No infinite chain* $\mathbb{S}_0 \to \mathbb{S}_1 \to \dots$ *exists.*

Algorithm 1. Reducing to Core Schema: $\text{CoreSchema}(\mathbb{S})$

---

(1) Compute $\mathbb{S}'$ such that $\mathbb{S} \to^* \mathbb{S}'$ and $\text{Core}(\mathbb{S}')$.
(2) Output $\mathbb{S}'$.

As a corollary of our three propositions, we obtain:

Theorem 1. *If* $\varnothing \vdash \mathbb{S} : Sty$ *then* CoreSchema($\mathbb{S}$) *terminates with a unique schema* $\mathbb{S}'$ *such that* $\mathbb{S} \to^* \mathbb{S}'$ *and* Core($\mathbb{S}'$) *and* $\varnothing \vdash \mathbb{S}' : Sty$.

5.3 Semantics of Core Tabular (Sketch)

Following [2], we define a semantics based on measure theory for **det** and **rnd**-level attributes, plus a set of evaluation rules for **qry**-level variables. For the sake of brevity, we omit the precise definitions here, and instead sketch the semantics and state the key theoretical result, illustrating it by example. For full details, see [15].

The denotational semantics of a schema $\mathbb{S}$ with respect to an input database δ_{in} is a measure μ defined on the measurable space corresponding to this schema. In order to evaluate the queries in the schema, we need to compute marginal measures for all (non-**qry**) attributes of all tables.

More precisely, our semantics for Tabular factors into an idealised, probabilistic denotational semantics (abstracting the details of approximate inference algorithms such as Infer.NET and other potential implementations) and a mostly conventional operational semantics.

The denotational semantics interprets well-typed schema as inductively defined measurable spaces, $T[\![\mathbb{S}]\!]^{\rho_{sz}}$, and defines a (mathematical) function interpreting well typed schemas $P[\![\mathbb{S}]\!]^{\delta_{in}}_{(\tau,\delta)} \in T[\![\mathbb{S}]\!]^{\rho_{sz}}$ as sub-probability measures describing the joint distribution μ of random variables given the observed input database δ_{in}.

The relation $\delta \vdash_\sigma \mathbb{S} \Downarrow \delta_{out}$ of our operational semantics takes as input a nested map σ of marginal measures for each column in the database, and the current operational environment δ (a nested map from **qry** and **det** attributes to values), and a schema. It returns an (output) database value δ_{out}: a nested map that assigns values to each non-**rnd** attribute of the schema.

Algorithm 2. Query Semantics of Core Schema: CoreQuery($\mathbb{S}, DB$)

---
(1) Assume core($\mathbb{S}$) and $DB = (\delta_{in}, \rho_{sz})$.

(2) Let $\mu \triangleq P[\![\mathbb{S}]\!]^{\delta_{in}}_{[]}$ (that is, the joint distribution over all **rnd**-variables).

(3) Let $\sigma = $ marginalize($\mathbb{S}, \rho_{sz}, \mu$).

(4) Return $(\delta_{out}, \rho_{sz})$ such that $\varnothing \vdash_\sigma \mathbb{S} \Downarrow \delta_{out}$.

Theorem 2 below states that, given a well-typed schema and conforming database, the composition of the denotational semantics and the deterministic evaluation relation yields a well-typed output database (with the same dimensions). The notation $DB \models^{in} Sty$ means that the database DB is a well-formed input to Sty; dually, $DB \models^{out} Sty$ means that the database DB is a well-formed output of Sty.

Theorem 2. *Suppose* Core($\mathbb{S}$) *and* $\varnothing \vdash \mathbb{S} : Sty$ *and* $DB = (\delta_{in}, \rho_{sz})$ *and* $DB \models^{in} Sty$. *Then algorithm* CoreQuery($\mathbb{S}, DB$) *returns* $DB' = (\delta_{out}, \rho_{sz})$ *such that* $DB' \models^{out} Sty$.

To illustrate, consider the Old Faithful schema shown in Section 4, together with an input database $(\delta_{in}, [\text{faithful} \mapsto 272])$ with 272 rows (say) such as the following:

$$\delta_{in} = [\,\text{faithful} \mapsto [\,\text{duration} \mapsto \mathbf{inst}([1.9; 4.0; 4.9; \ldots)];\ \text{time} \mapsto \mathbf{inst}([50; 75; 80; \ldots])\,]\,]$$

The output of the marginalisation algorithm for the only table in this schema is:

$$[\text{cluster\_V} \mapsto \textbf{static}(\mu_{\text{Dirichlet}[2](1,1)});$$
$$\text{cluster} \mapsto \textbf{inst}([\mu_{20}; \mu_{21}; \mu_{22}; \ldots]);$$
$$\text{duration\_Mean} \mapsto \textbf{static}([\mu_{\text{Gaussian}(0,1)}, \mu_{\text{Gaussian}(0,1)}]);$$
$$\text{duration\_Prec} \mapsto \textbf{static}([\mu_{\text{Gamma}(1,1)}, \mu_{\text{Gamma}(1,1)}]);$$
$$\text{duration} \mapsto \textbf{inst}([\mu_{50}; \mu_{51}; \mu_{52}; \ldots]);$$
$$\text{time\_Mean} \mapsto \textbf{static}([\mu_{\text{Gaussian}(60,1)}, \mu_{\text{Gaussian}(60,1)}]);$$
$$\text{time\_Prec} \mapsto \textbf{static}([\mu_{\text{Gamma}(1,1)}, \mu_{\text{Gamma}(1,1)}]);$$
$$\text{time} \mapsto \textbf{inst}([\mu_{80}; \mu_{81}; \mu_{82}; \ldots])]$$

The output database is $(\delta_{out}, [\text{faithful} \mapsto 272])$ where δ_{out} contains those entries from such environments which correspond to non-random attributes (that is, are not measures). In our example, it is of the form:

$$\delta_{out} = [\text{faithful} \mapsto [\text{assignment} \mapsto \textbf{inst}([0; 1; 1; \ldots])]]$$

In this example, all of the **inst** arrays are of length 272.

6   Examples of Bayesian Decision Analysis in Tabular

To illustrate the value of query-space computations, we illustrate how they express a range of decision problems. Decisions such as these cannot be expressed in the original form of Tabular. Other probabilistic programming languages have built in constructs for decision-making, whereas Tabular does so using ideas of information flow.

We describe how three example decision problems are written as Tabular queries. The result of Bayesian inference is the posterior belief over quantities of interest, including model parameters such as the **rnd**-space variable V in our coins example. These inferences reflect a change of belief in light of data, but they are not sufficient for making decisions, which requires optimization under uncertainty.

In Bayesian Decision Analysis, the decision making process is based on statistical inference followed by maximization of expected utility of the outcome. Following Gelman et al. [8], Bayesian Decision Analysis can be described as follows:

(1) Enumerate sets D and X of all possible decision options $d \in D$ and outcomes $x \in X$.
(2) Determine the probability distribution over outcomes $x \in X$ conditional on each decision option $d \in D$.
(3) Define a utility function $U : X \to \mathbb{R}$ to value each outcome.
(4) Calculate the expected utility $E[U(x)|d]$ as a function of decision option d and make the decision with the highest expected utility.

(1) Optimal Betting Decisions. Consider a situation in which to decide whether or not to accept a given sports bet based on the TrueSkill model for skill estimation [16]. The following code shows the schema $\mathbb{S}_{TrueSkill}$. Following [13], the tables Players and Matches generate **rnd**-space variables for the results of matches between players, by comparison of their per-match performances, modelled as noisy per-player skills.

table Players			
Skill	real!rnd	output	Gaussian(25.0,100.0)
table Matches			
Player1	link(Players)!det	input	
Player2	link(Players)!det	input	
Perf1	real!rnd	output	Gaussian(Player1.Skill,100.0)
Perf2	real!rnd	output	Gaussian(Player2.Skill,100.0)
Win1	bool!rnd	output	Perf1 > Perf2
table Bets			
Match	link(Matches)!det	input	
Odds1	real!det	input	
Win1	bool!rnd	output	Match.Win1
p	real!qry	output	infer.Bernoulli[].Bias(Win1)
U	real[3]!det	output	[0.0;−1.0;Odds1 ∗ 1.0]
EU	real[2]!qry	output	[U[0];((1.0 − p)∗ U[1])+ (p ∗ U[2])]
PlaceBet1	mod(2)!qry	output	ArgMax(EU)

Table Bets represents the decision theoretic part of the code and refers to Matches together with the odds Odds1 offered for a bet on player 1 winning. Here, the two decision options in $D = \{0,1\}$ are to take the bet (PlaceBet1 = 1) or not (PlaceBet1 = 0), and the three possible outcomes in $X = \{0,1,2\}$ are abstain = 0, loss = 1 or win = 2. The optimal decision depends on the odds: a risky bet may be worth taking if the odds are good. The utility function U is given by money won for a fixed bet size of, say, \$1, so $U(\text{abstain}) = 0.0$, $U(\text{loss}) = -\$1.0$, and $U(\text{win}) = \text{Odds1} * \$1.0$. Variable p is obtained from **qry** expression **infer**.Bernoulli.Bias(Win1) and represents the inferred probability of a positive bet outcome. The **qry** variable PlaceBet1 is 1 if the expected utility $EU[1] = (1 - p) \cdot (-\$1.0) + p \cdot \text{Odds1} \cdot \$1.0$ is greater than $EU[0] = 0.0$, that is, betting is better than not betting. The ArgMax operator simply returns the first index (of type **mod**(n)) of the maximum value in its array argument (of type **real**[n]). It returns the decision delivering the maximum expected utility.

(2) Classes with Asymmetric Misclassification Costs. Consider the task of n-ary classi-fication with class-specific misclassification costs. We proceed by defining the schema for a Naive Bayes classifier (see, for instance, Duda and Hart [7]), in terms of the func-tion CG (from Section 4) which represents a Gaussian distribution, with **static** param-eters assuming natural conjugate prior distributions. (A prior is called conjugate with respect to a likelihood if it takes the same functional form.)

Hence, we can write down the Naive Bayes model (for a simple gender classification task) very succinctly as follows (using the indexed model notation from Section 3).

table People			
g	mod(2)!rnd	output	CDiscrete(N=2,R=1.0)
height	real!rnd	output	(CG(M=0.0,P=1.0))[g<2]
weight	real!rnd	output	(CG(M=0.0,P=1.0))[g<2]
footsize	real!rnd	output	(CG(M=0.0,P=1.0))[g<2]
Us	real[2][2]!qry	static output	[[0.0;−20.0];[−10.0;0.0]]
action	mod(2)!qry	output	Action(N=2,UPT=Us,**class**=g)

The first four lines define a Naive Bayes model with Gaussian features height, weight, and footsize, which are assumed to be distributed as independent Gaussians condi-tional on knowing gender g. At this point, we could simply return the probability vector **infer**.Discrete[2].probs(g): the probabilities that a person has either gender.

However, suppose we need to return a concrete gender decision and that for some reason the cost of false positives differs from the cost of false negatives. Below we encode how to decide whether to take the action of predicting the gender of 0 (female) or 1 (male), given that: A false positive (predict 1 but actually 0) costs 20. A false negative (predict 0 but actually 1) costs 10. A true positive or true negative costs 0. The costs, expressed as negative utilities, are in the matrix Us.

The query defined by the model computes an action column, classifying each row, taking into account the relative costs of false positives and false negatives. (It recommends an action for all rows, even those already labelled with a gender.)

fun Action			
N	int!det	static input	
UPT	real[N][N]!qry	static input	
class	mod(N)!rnd	input	
probs	real[N]!qry	output	infer.Discrete[N].probs(class)
EU	real[N]!qry	output	[for p < N →Sum([for t < N →(probs[t] * UPT[p][t])])]
ret	mod(N)!qry	output	ArgMax(EU)

We see that the function evaluates N different expected utilities, one for each decision option. ArgMax returns the option delivering the maximum expected utility.

In terms of Bayesian Decision Analysis, the outcome space X is all (predicted class (p), true class (t)) pairs, whose elements are given utilities by UPT. In the expected utility (EU) computations, the Action function only sums over the N outcomes that are consistent with the current p, that is, if the prediction is p, then the probability of any outcome (p',t) where $p' \neq p$ is 0 and can be dropped.

(3) F1 Score: Optimizing a more complex decision criterion. We introduce another model, the Bayes Point Machine, and use it to illustrate a more complicated utility function, namely the F1 score. The F1 score is a measure of accuracy for binary classification that takes into account both false positives and false negatives.

As can be seen from the Tabular code in Figure 1, in table Data (abbreviated here), the data schema consists of seven real-valued clinical measurements X0 to X6 and a Boolean outcome variable Y to be predicted. The model is an instance of the Bayes Point Machine [21], a Bayesian boolean classifier, in which the prior over the weight vector W is drawn from a VectorGaussian, and the label Y is generated by thresholding a noisy score Z which is the inner product between the input vector and the weight vector W. Attribute ProbTrue records the marginal predictive probability for the label, obtained by querying the bias of the Bernoulli random variable Y.

The set D of decision options is given in table Ts, which (we assume) enumerates a number of candidate thresholds Th used to decide the test results by thresholding the marginal predictive probability of each point against Th. For each threshold Th, attribute Decisions is an array, indexed by data point d, containing the candidate decision for d obtained by the thresholding expression d.ProbTrue > Th. The columns ETP, EFP, and EFN evaluate the expected number of true positives, false positives, and false negatives, respectively, by summing the relevant marginal probabilities over test data, which is valid due to linearity of the expectation operator. Finally, the approximate expected F1 score is calculated for each threshold using:

$$E[F_1] = E\left[\frac{2 \cdot \text{TP}}{2 \cdot \text{TP} + \text{FP} + \text{FN}}\right] \approx \frac{2 \cdot E[\text{TP}]}{2 \cdot E[\text{TP}] + E[\text{FP}] + E[\text{FN}]}.$$

table Data

X0	real!det	input	
X6	real!det	input	
Mean	vector!det	static output	VectorFromArray([for i < 7 →0.0])
CoVar	PositiveDefiniteMatrix!det	static output	IdentityScaledBy(7,1.0)
W	vector!rnd	static output	VectorGaussianFromMeanAndVariance(Mean,CoVar)
Z	real!rnd	output	InnerProduct(W,VectorFromArray([X0;X1;X2;X3;X4;X5;X6]))
Y	bool!rnd	output	Gaussian(Z,0.1)> 0.0
ProbTrue	real!qry	output	infer.Bernoulli[].Bias(Y)
Train	bool!det	input	

table Ts

Th	real!det	input	
Decisions	bool[SizeOf(Data)]	output	[for d < SizeOf(Data)→d.ProbTrue > Th]
ETP	real!qry	output	Sum([for d < SizeOf(Data)→ if (!d.Train)& Decisions[d] then d.ProbTrue else 0.0])
EFP	real!qry	output	Sum([for d < SizeOf(Data)→ if (!d.Train)& Decisions[d] then 1.0 − d.ProbTrue else 0.0])
EFN	real!qry	output	Sum([for d < SizeOf(Data)→ if (!d.Train)& (!Decisions[d])then d.ProbTrue else 0.0])
EF1	real!qry	output	(2.0 * ETP)/ ((2.0 * ETP)+ EFP + EFN)

table Decisions

ChosenThID	link(Ts)!qry	static output	ArgMax([for t < SizeOf(Ts)→t.EF1])
ChosenTh	real!qry	static output	ChosenThID.Th
DataID	link(Data)!det	input	
Decision	bool!qry	output	ChosenThID.Decisions[DataID]

Fig. 1. F1 computation in Tabular on mammography data

This is an approximation because the F1 score is a non-linear function in TP, FP, and FN, and is employed here because it allows us to express the expectation in terms of marginal probabilities which are available from our inference back end. Recent work by Nowozin [24] has shown that approximations of this form yield good results.

The final table Decisions determines the optimal threshold ChosenTh by finding the identity of the threshold t that maximises t.EF1. In addition, Decisions outputs, for each data point DataID, the labelling Decision obtained with the optimal threshold (assuming that column DataID enumerates the keys of table Data).

7 Tabular Excel: Implementing Tabular in a Spreadsheet

Public releases of the Tabular add-in for Excel are available from http://research. microsoft.com/tabular. The add-in extends Excel with a new task pane for authoring models, running inference and setting parameters of Infer.NET. A user authors the model within a rectangular area of a worksheet. Tabular parses and type-checks the model in the background, enabling the inference button when the model is well-typed. Tabular pulls the data schema and data itself from the relational Data Model of Excel 2013. The results of inference and queries are then reported back to the user as augmented Excel tables. Tabular Excel is able to concisely express a wide range of models beyond those illustrated here (see companion technical report [15]).

Type checking the Tabular schema results in a type-annotated schema. This is elaborated to core form, eliminating all function calls and indexed models. The core schema is then translated to an Infer.NET [20] factor graph, constructed dynamically with

Infer.NET's (imperative and weakly typed) modelling API. Our (type-directed) translation relies on and exploits the fact that all table sizes are known and that discrete random variables, which may be used to index into arrays, have known support. Moreover, the space of any (explicit or implicit) array indexing expression is used to insert the requisite Infer.NET *switch* construct when indexing through a **rnd**-space index (as demonstrated in an appendix to technical report [15]). The fruits of Infer.NET inference are approximate marginal distributions for the **rnd**-space bindings of the schema. Expressions in **det** and **qry**-space are evaluated by interpretation after inference, binding input to the concrete data and **rnd**-level variables to their inferred distributions. Thus **qry**-space expressions have access to the inputs, deterministic values and distributions on which they depend. For compilation, the type system ensures that the value of **qry**-space expression cannot depend on the particular value of a **rnd**-space variable (only its distribution) and that all **rnd**-space variables can be inferred prior to **qry** evaluation.

Users can also extract C# source code to compile and run their models outside Excel (see [15] for an example). This supports subsequent customization by Infer.NET experts as well as integration in standalone applications. One of our internal users has extracted code in this way to perform inference on a large dataset with approximately 42 million entities and 46 million relationships between them. Inference required 7.5 hours of processing time on a 2-core Intel Xenon L5640 server with 96 GB of RAM. The extracted code is also useful for debugging compilation and applications that need to separate learning (on training data) from prediction (on new data).

The following is direct comparison between the Tabular Excel form of the Mammography model (Figure 1) with code for the same problem written in C# using Infer.NET. We get the same statistical answers in both cases, though there are differences in code speed. Initially, Tabular queries were (naively) interpreted, not compiled; adopting simple runtime code generation techniques has allowed us to reduce the **qry** time from 1601ms (interpreted) to 29ms (compiled), a 55x fold increase. The handwritten C# model is slower on inference because it is effectively compiled and run twice, once for training and another time for prediction.

	data (LOC)	model (LOC)	decisions (LOC)	inference (ms)	query (ms)
Infer.NET	35	35	45	2968	6
Tabular	0	15	14	1529	1601/29

8 Related Work

Interest in probabilistic programming languages is rising as evinced by recent languages like Church [10], a Turing-complete probabilistic Scheme with inference based on sampling, and its relatives Anglican [29] (a typed re-imagination of Church) and Venture [18] (a variant of Church offering programmable inference). Other recent works include R2 [23], which uses program analysis to optimize MCMC sampling of probabilistic programming, Uncertain$<T>$ [3], a simple abstraction for embedding probabilistic reasoning into conventional programs that handle uncertain data, and Wolfe [27], where inference is expressed within a host language by providing a small set of primitives for writing distributons and operations for maximization and summation.

To the best of our knowledge, few systems offer explicit support for decision theory. IBAL's [25] impressive framework aims to combine Bayesian inference and decision theory "under a single coherent semantic framework". IBAL makes use of query information and only computes the quantities needed to answer specific queries. Other systems that extend probabilistic languages with dedicated decision theoretic constructs are described in [6, 4, 22]. The main difference in our approach is that while our post-processing can be used to implement decision theory strategies, decision theoretic constructs are not built into the language. This is a pragmatic choice. In general, decision theory involves two intractabilities: computing expected utilities, and optimizing over the decisions. IBAL and DT-ProbLog [4] have some general-purpose approximations, but often problem-specific approximations are needed as in our F1 optimization example or in [24]. It is not clear how these approximations fit into the above frameworks. Tabular's free-form post-processing design allows such bespoke approximations.

STAN [28] allows for post-processing of inference results, but only via separately declared code blocks, rather than being conveniently mingled with the model or abstracted in functions. Although STAN's facilities are expressive and can include arbitrary deterministic and stochastic computations, they are restricted to computing *per sample* quantities. In Tabular terms, this would correspond to computations restricted to **rnd**-space which prevents the computation of the aggregate **qry**-space quantities required for Bayesian decision theory.

Figaro [26] supports post-inference decision-making, but via separate, decision-specific language features, outside the core modelling language. Tabular, instead, uses types to distinguish between operations available in different spaces (or phases) (such as random draws in **rnd** space, optimization (ArgMax) and moments of distributions in **qry**-space). Embedded DSLs such as Infer.NET [20], HANSEI [17] and FACTORIE [19] enable arbitrary post-processing in the host language, but require knowledge of both the host and the embedded language, which is typically much simpler.

Tabular is, to our knowledge, the first probabilistic programming language with dependently typed abstractions. STAN and BUGS [9] do have value-indexed types, but cannot abstract over indexes appearing in types.

We advocate types to help catch errors in probabilistic queries on spreadsheets. There is a body of work on testing and discovering errors on spreadsheets. For example, Ahmad et al. [1] propose unit-based types as a means of catching errors. To the best of our knowledge, dependent types have not previously been applied to spreadsheets.

9   Conclusions

We recast Tabular as a query language on databases held in spreadsheets.

This paper presents a technical evaluation of the design consisting of theorems about its metatheory, demonstration of its expressiveness by example, and some numeric comparisons with the alternative of writing models directly in Infer.NET. Evaluating the usability by spreadsheet users is important, but we leave that task for future work.

We have in mind several lines of future development. One limitation of our current system is that data is modelled by map-style loops over data; to model time-series, it would be useful to add some form of iterative fold-style loops. Another limitation is that programs involve a single run of the underlying inference system: **rnd** space determines the model and its conditioning, and **qry** space determines how the results are processed. To support multiple runs of inference we might consider an indexed hierarchy of spaces where **infer** moves data from rnd_i space to qry_i space, and rnd_{i+1} space can depend on results computed in qry_i space.

Finally, our approach could be applied to add user-defined functions to languages such as BUGS or Stan, or to design typed forms of universal probabilistic languages such as those in the Church family.

Acknowledgement. Dylan Hutchison commented on a draft. We thank Natalia Larios Delgado and Matthew Smith for their feedback on our Excel addin.

References

[1] Ahmad, Y., Antoniu, T., Goldwater, S., Krishnamurthi, S.: A type system for statically detecting spreadsheet errors. In: 18th IEEE International Conference on Automated Software Engineering (ASE 2003), pp. 174–183 (2003)

[2] Borgström, J., Gordon, A.D., Greenberg, M., Margetson, J., Gael, J.V.: Measure transformer semantics for Bayesian machine learning. Logical Methods in Computer Science 9(3) (2013) preliminary version at ESOP 2011

[3] Bornholt, J., Mytkowicz, T., McKinley, K.S.: Uncertain<T>: A first-order type for uncertain data. In: Architectural Support for Programming Languages and Operating Systems (ASPLOS) (March 2014)

[4] Van den Broeck, G., Thon, I., van Otterlo, M., De Raedt, L.: DTProbLog: A decision-theoretic probabilistic Prolog. In: AAAI (2010)

[5] Cardelli, L.: Typeful programming. Tech. Rep. 52. Digital SRC (1989)

[6] Chen, J., Muggleton, S.: Decision-theoretic logic programs. In: Proceedings of ILP, p. 136 (2009)

[7] Duda, R.O., Hart, P.E.: Pattern Classification and Scene Analysis. John Wiley & Sons, New York (1973)

[8] Gelman, A., Carlin, J.B., Stern, H.S., Dunson, D.B., Vehtari, A., Rubin, D.B.: Bayesian Data Analysis, 3rd edn. Chapman & Hall (2014)

[9] Gilks, W.R., Thomas, A., Spiegelhalter, D.J.: A language and program for complex Bayesian modelling. The Statistician 43, 169–178 (1994)

[10] Goodman, N., Mansinghka, V.K., Roy, D.M., Bonawitz, K., Tenenbaum, J.B.: Church: a language for generative models. In: Uncertainty in Artificial Intelligence (UAI 2008), pp. 220–229. AUAI Press (2008)

[11] Goodman, N.D.: The principles and practice of probabilistic programming. In: Principles of Programming Languages (POPL 2013), pp. 399–402 (2013)

[12] Gordon, A.D., Aizatulin, M., Borgström, J., Claret, G., Graepel, T., Nori, A., Rajamani, S., Russo, C.: A model-learner pattern for Bayesian reasoning. In: POPL (2013)

[13] Gordon, A.D., Graepel, T., Rolland, N., Russo, C.V., Borgström, J., Guiver, J.: Tabular: a schema-driven probabilistic programming language. In: POPL (2014a)

[14] Gordon, A.D., Henzinger, T.A., Nori, A.V., Rajamani, S.K.: Probabilistic programming. In: Future of Software Engineering (FOSE 2014), pp. 167–181 (2014b)

[15] Gordon, A.D., Russo, C., Szymczak, M., Borgström, J., Rolland, N., Graepel, T., Tarlow, D.: Probabilistic programs as spreadsheet queries. Tech. Rep. MSR–TR–2014–135, Microsoft Research (2014c)

[16] Herbrich, R., Minka, T., Graepel, T.: TrueSkill^{tm}: A Bayesian skill rating system. In: Advances in Neural Information Processing Systems, NIPS 2006 (2006)

[17] Kiselyov, O., Shan, C.: Embedded probabilistic programming. In: Conference on Domain-Specific Languages, pp. 360–384 (2009)

[18] Mansinghka, V., Selsam, D., Perov, Y.: Venture: a higher-order probabilistic programming platform with programmable inference. arXiv preprint arXiv:1404.0099 (2014)

[19] McCallum, A., Schultz, K., Singh, S.: Factorie: Probabilistic programming via imperatively defined factor graphs. In: NIPS 2009, pp. 1249–1257 (2009)

[20] Minka, T., Winn, J., Guiver, J., Knowles, D.: Infer.NET 2.5 (2012), Microsoft Research Cambridge. http://research.microsoft.com/infernet

[21] Minka, T.P.: A family of algorithms for approximate Bayesian inference. Ph.D. thesis, Massachusetts Institute of Technology (2001)

[22] Nath, A., Domingos, P.: A language for relational decision theory. In: Proceedings of the International Workshop on Statistical Relational Learning (2009)

[23] Nori, A.V., Hur, C.K., Rajamani, S.K., Samuel, S.: R2: An efficient MCMC sampler for probabilistic programs. In: Conference on Artificial Intelligence, AAAI (July 2014)

[24] Nowozin, S.: Optimal decisions from probabilistic models: the intersection-over-union case. In: Proceedings of CVPR 2014 (2014)

[25] Pfeffer, A.: The design and implementation of IBAL: A general-purpose probabilistic language. In: Getoor, L., Taskar, B. (eds.) Introduction to Statistical Relational Learning. MIT Press (2007)

[26] Pfeffer, A.: Figaro: An object-oriented probabilistic programming language. Tech. rep., Charles River Analytics (2009)

[27] Riedel, S.R., Singh, S., Srikumar, V., Rocktäschel, T., Visengeriyeva, L., Noessner, J.: WOLFE: strength reduction and approximate programming for probabilistic programming. In: Statistical Relational Artificial Intelligence (2014)

[28] Stan Development Team: Stan: A C++ library for probability and sampling, version 2.2 (2014), http://mc-stan.org/

[29] Wood, F., van de Meent, J.W., Mansinghka, V.: A new approach to probabilistic programming inference. In: Proceedings of the 17th International conference on Artificial Intelligence and Statistics (2014)

[30] Xi, H., Pfenning, F.: Eliminating array bound checking through dependent types. In: Proceedings of the ACM SIGPLAN 1998 Conference on Programming Language Design and Implementation (PLDI), pp. 249–257 (1998)

Static Analysis of Spreadsheet Applications for Type-Unsafe Operations Detection*

Tie Cheng[1,2,3] and Xavier Rival[1,2,3]

[1] CNRS, Paris, France
[2] École Normale Supérieure, Paris, France
[3] INRIA Paris–Rocquencourt, France
{tie.cheng,xavier.rival}@ens.fr

Abstract. Spreadsheets are widely used, yet are error-prone. In particular, they use a weak type system, which allows certain operations that will silently return unexpected results, like comparisons of integer values with string values. However, discovering these issues is hard, since data and formulas can be dynamically set, read or modified. We propose a static analysis that detects all run-time type-unsafe operations in spreadsheets. It is based on an abstract interpretation of spreadsheet applications, including spreadsheet tables, global re-evaluation and associated programs. Our implementation supports the features commonly found in real-world spreadsheets. We ran our analyzer on the EUSES Spreadsheet Corpus. This evaluation shows that our tool is able to automatically verify a large number of real spreadsheets, runs in a reasonable time and discovers complex bugs that are difficult to detect by code review or by testing.

1 Introduction

Spreadsheet applications are ubiquitous in engineering, statistics, finance and management. They combine a flexible tabular representation of data in two-dimensional tables mixing formulas and values with associated programs (or macros), written in specific languages. For instance, Microsoft Excel includes a version of Visual Basic for Applications (VBA), whereas Google Spreadsheets have Google Apps Script and LibreOffice Calc has LibreOffice Basic.

Unfortunately, spreadsheet applications are subject to numerous defects, and often produce incorrect results that do not match user understanding as shown in [21,22]. In 2013, the Task Force Report [1] quoted losses of billions of dollars due to errors in spreadsheet applications used in JPMorgan's Chief Investment Office. More generally, spreadsheet defects may cause the release of wrong information, the loss of money or the taking of wrong decisions, therefore they now attract increasing attention from users, IT professionals, and from the research community. Approaches proposed so far include new languages [7] and

---

* The research leading to these results has received funding from the European Research Council under the European Union's seventh framework programme (FP7/2007-2013), grant agreement 278673, Project MemCAD.

© Springer-Verlag Berlin Heidelberg 2015
J. Vitek (Ed.): ESOP 2015, LNCS 9032, pp. 26–52, 2015.
DOI: 10.1007/978-3-662-46669-8_2

enhancements to functional features of spreadsheets with better language design [19,27] and implementation [24,25], model-driven engineering environments to allow only safe updates [14], and studies to detect code smells that indicate weak points in spreadsheets [13,17]. Additionally, type systems could be built [3,4,5,6,9] to capture value meanings such as physical units (e.g., apples, oranges) or dimensions (e.g., meters, kilometers) and to verify the correctness of formulas. Most existing works focus on the spreadsheet tables and ignore the associated programs despite them being a very important component of spreadsheet applications, which can have a significant impact on spreadsheet contents, either through function calls from a spreadsheet formula or through an execution of a subroutine launched by users. They are also massively present in industrial spreadsheet applications.

Verification Objective. Spreadsheet languages supply basic operators and functions to perform operations on values such as text, number, boolean, date and time to use in formulas and programs. The type system of spreadsheet languages is weak and rarely considers a type mismatch an error, even though that means unexpected or incorrect results may be produced instead. For instance, Microsoft Excel implicitly converts the empty value to **true** in expression AND(ϵ, **true**), whereas it converts it to **false** in expression IF(ϵ, 1, 0). It will also evaluate comparison "" $< n$ to **false**, yet the empty string does not have an obvious numeric value. More generally, type mismatches are common and rarely block the execution with an explicit error message such as #VALUE!. Thus, users develop and run spreadsheet applications in the environment where program defects can be hidden. Verifying a spreadsheet application is exempt of any such defect requires a *strong type discipline*, and *precise typing information* about formula operands should be inferred.

Static Analysis of Spreadsheets with Macros. Existing works focus on the spreadsheet contents, assume the data in the sheet are fully specified, and do not consider spreadsheet instances with different input data. Yet, industrial spreadsheet applications often handle non-deterministic or non-statically known input in the following cases: (1) input data may be left blank when the application is developed and entered at a later stage (Excel features "Data Validation" for such cases, which allows to specify restrictions on data before they are entered); (2) data may be defined dynamically, e.g., using functions generating random values, inserting values found on the Web or in external databases; (3) formulas and data edited in non-automatic calculation mode (i.e., when the spreadsheet environment does not always recalculate cell depending on the modified zone) may result in outdated values; (4) data and formulas may be set and manipulated by associated programs.

Therefore, in this paper, we propose a complete vision of spreadsheet applications, that includes spreadsheets storing *formulas* and *associated programs* (macros); they receive input data that is *unknown at verification time* (i.e., non-deterministic or read at run-time); their execution consists in *globally evaluating spreadsheet formulas* or *running an associated program*.

We propose a fully automatic and sound analysis, that discovers all type defects in spreadsheet applications. It features a strong type system, and an abstract domain that ties properties (like contents types) to *zones* in spreadsheet tables. It infers invariants by conservative abstract interpretation of spreadsheet applications. It either proves type correctness or displays potential issues to developers. Invariants also give a high-level view of program behaviors. The set of type-unsafe operations is a parameter of the analysis, so that users can select which behaviors are deemed unsafe and should be detected. Our analysis has the following benefits: (1) it unearths errors that dynamic tests may miss, as it computes an over-approximation of *all* the states executions can reach even in the presence of inputs at run-time; (2) it is efficient enough to be run during the development of a spreadsheet application. In this paper, we make the following contributions:

- We set up a concrete model for reasoning about spreadsheet applications, to be used as a basis for the definition and the proof of our analysis (Sect. 3);
- We propose an abstraction for spreadsheet applications, that takes the structure of formulas into account and is adapted to the type verification (Sect. 4);
- We define a static analysis, that takes into account both the contents of spreadsheet and the associated programs (Sect. 5), and is able to cope with global re-evaluation of spreadsheet contents (Sect. 5.2);
- We present our tool (Sect. 6) and report on results of verification of the EUSES Spreadsheet Corpus by our tool (Sect. 7).

2 Overview

In this section, we consider a realistic application, which silently produces wrong results that cannot be caught by the weak type system found in spreadsheet environments. This application is made up of a *spreadsheet table* shown in Fig. 1(a) and an *associated program* displayed in Fig. 1(b). The table contains several columns storing asset variations and values expressed in two currencies, and computes the number of weekdays where the total value was greater than a given amount. The area in the blue rectangle in Columns 1 and 2 is reserved for input data, which are the day name and the value variation for each weekday (no variation occurs on the weekend). The associated program shown in Fig. 1(b) and the spreadsheet formulas in the green rectangle are pre-coded. The associated program is run, upon user request, to eliminate meaningless empty weekend values of Column 2, and to populate the sequential list into Column 3. The formulas in Column 4 convert the variations stored in Column 3 into another currency. Last, the formulas in Column 5 compute the sequence of meaningful variations, and the number of weekdays where the total asset value was greater than 150, in the bottom right cell.

	1	2	3	4	5
1	Day	Delta	Delta	Delta	Total
2		(cur1)	(cur1)	(cur 2)	(cur 2)
3					100
4	Mon	-2	-2	$= \mathbf{C}[4,3]*1.3$	$= \mathtt{IF}(\mathtt{ISBLANK}(\mathbf{C}[4,3]),$ "", $\mathbf{C}[4,4]+\mathbf{C}[3,5])$
⋮					
9	Sat	0	-4	$= \mathbf{C}[9,3]*1.3$	$= \mathtt{IF}(\mathtt{ISBLANK}(\mathbf{C}[9,3]),$ "", $\mathbf{C}[9,4]+\mathbf{C}[8,5])$
10	Sun	0	5	$= \mathbf{C}[10,3]*1.3$	$= \mathtt{IF}(\mathtt{ISBLANK}(\mathbf{C}[10,3]),$ "", $\mathbf{C}[10,4]+\mathbf{C}[9,5])$
⋮					
33	Tue	8	20	$= \mathbf{C}[33,3]*1.3$	$= \mathtt{IF}(\mathtt{ISBLANK}(\mathbf{C}[33,3]),$ "", $\mathbf{C}[33,4]+\mathbf{C}[32,5])$
34	Wed	-3		$= \mathbf{C}[34,3]*1.3$	$= \mathtt{IF}(\mathtt{ISBLANK}(\mathbf{C}[34,3]),$ "", $\mathbf{C}[34,4]+\mathbf{C}[33,5])$
⋮					
43	Fri	20		$= \mathbf{C}[43,3]*1.3$	$= \mathtt{IF}(\mathtt{ISBLANK}(\mathbf{C}[43,3]),$ "", $\mathbf{C}[43,4]+\mathbf{C}[42,5])$
44					Number of days where asset > 150
45					$= \mathtt{SUM}(\mathtt{N}(\mathbf{C}[4,5] : \mathbf{C}[43,5] > 150))$

(a) Spreadsheet contents

```
1 Sub Macro()          7  j = 4;                    11        i = i + 1
2   INITIATE;          8  While (j < 44)            12    End;
3   Dim i As Int;      9    If C[j,1] <> "Sat"      13    j = j + 1
4   Dim j As Int;           And C[j,1] <> "Sun"     14  End;
5   CLEAR_ZONE(4,3,43,3);  Then                     15  Eval
6   i = 4;            10      C[i,3] = C[j,2];       16 End
```

(b) Associated program

Fig. 1. Erroneous behaviors in a spreadsheet application

In practice, data are filled either manually, or automatically (e.g., copying from somewhere else, or using another associated program INITIATE). Then, users launch the associated program to compute the values in Column 3, which, in turn, forces the re-evaluation of the formulas stored in Columns 4 and 5 using statement **Eval** in Line 15. The input data, their array size may be known only at run-time, whereas the spreadsheet formulas and the associated program are pre-coded.

The final result is computed in the bottom right cell. Its value is **incorrect**. We let $\mathbf{C}[i,j]$ denote the cell in row i and column j. In Fig. 1(a), the cells in region $\mathbf{C}[34,5] : \mathbf{C}[43,5]$ evaluate to the empty string, since the cells in $\mathbf{C}[34,3] : \mathbf{C}[43,3]$ are empty (Function ISBLANK checks whether a cell is empty). Comparison operator ">" always returns **true** when applied to a string and a numeric value, therefore, when cell $\mathbf{C}[i,5]$ is an empty string, the condition $\mathbf{C}[i,5] > 150$ evaluates to **true**. Then, built-in function N converts **true** into 1. **Therefore, the value produced when evaluating the formula in $\mathbf{C}[45,5]$ is off by 10.**

This incorrect result is produced without a warning, as it passes through the weak (and incorrect) spreadsheet type checking. Such issues are common in large

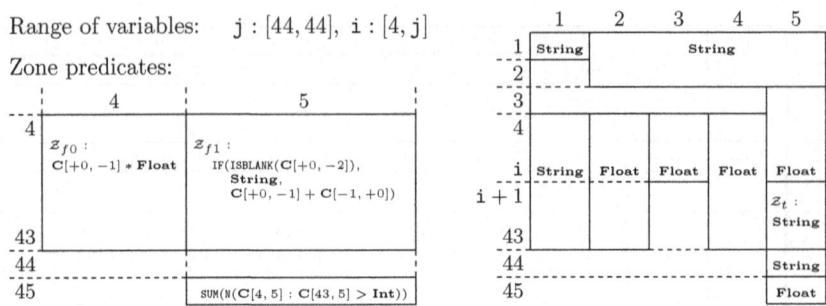

Fig. 2. Abstract state

applications, and hard to diagnose by non-expert users. In particular, testing is likely to miss such problems. In this case, any run with a data sample without empty cells in $\mathbf{C}[4,3] : \mathbf{C}[43,3]$ will produce no incorrect result. Therefore, checking the absence of defects by testing is not possible, especially when input data are made available at run-time, and detecting all such issues will require a *conservative* static analysis that raises a warning whenever an unsafe operation (such as the comparison of a string with a numeric value) might be executed. Different users may consider different sets of operations safe, thus the set of unsafe operations should be a *parameter* of the analysis.

Analysis of the Example. The properties of the application of Fig. 1 are shown in Fig. 2. After Line 14 of the associated program executes, j is always equal to 44, whereas i may take any value in $[4, 44]$ (as $4 \leq i \leq j$). The diagrams show properties that always hold for *zones* in the table: each rectangle accounts for a set of cells, and is labeled by a property of these cells. Cell properties consist either in *abstract formulas* or in *types*. Abstract formulas may use relative indexes (e.g., $\mathbf{C}[+0, -1]$) or absolute indexes (e.g., $\mathbf{C}[4, 5]$). The analysis of **Eval** at Line 15 will use abstract formulas to infer type **Float** for zone $\mathcal{Z}_{f0}$ (since the multiplication of *empty* value ϵ and a float value produces float value 0.0 as does the multiplication of two float values), and then split $\mathcal{Z}_{f1}$ into two sub-zones of type **Float** and **String**. We call the latter $\mathcal{Z}'_t$. Finally the type of cell $\mathbf{C}[45, 5]$ is inferred, which requires the types in Column 5 including zone $\mathcal{Z}'_t$, and thus involves the *unsafe comparison* (reported by the analysis) of a value of type **String** with a value of type **Float**.

Moreover, the analysis should not reject obviously correct applications. For instance, a corrected version of the example application would replace formulas in Column 5 with formulas of the form $\mathtt{IF(ISBLANK}(\mathbf{C}[4, 3]), 0.0, \mathbf{C}[4, 4] + \mathbf{C}[3, 5])$. Then, all results in Column 5 would have a floating point type, and no unsafe comparison of a string with a numeric value would occur. The same reasoning based on zones will allow to establish this.

$$\text{x}, \text{y}, \ldots \in \mathbb{X} \qquad v \in \mathbb{V}$$
$$t ::= \textbf{Bool} \mid \textbf{Float} \mid \textbf{Int} \mid \textbf{String} \mid \textbf{Empty} \mid \textbf{Currency} \mid \textbf{Date}$$
$$e ::= v \quad \mid \quad \text{x} \quad \mid \quad \mathbf{C}[e, e] \quad \mid \quad \mathbf{C}[\pm e, \pm e]$$
$$\mid \quad e \oplus e \quad \text{where } \oplus \in \{+, -, \star, \ldots\} \mid \mathbf{F}(e, \ldots, e) \quad \text{where } \mathbf{F} \text{ is a function symbol}$$
$$s ::= \text{x} = e \quad \mid \quad \mathbf{C}[e, e] = e \quad \mid \quad \mathbf{C}[e, e] = \text{`` }= e\text{''}$$
$$\mid \quad \textbf{Eval} \quad \mid \quad \textbf{If } e \textbf{ Then } s \textbf{ Else } s \textbf{ End} \quad \mid \quad \textbf{While}(e) \, s \textbf{ End} \quad \mid \quad s; \, s$$
$$a ::= \textbf{Dim x As } t; \ldots; \textbf{Dim x As } t; \, s$$

Fig. 3. Syntax: a core spreadsheet language

3   A Core Spreadsheet Language

In this section, we formalize a core language that incorporates both the spread-sheet table and the runnable code (the analyzer shown in Sections 6 and 7 supports a much wider feature set). This language has several distinctive features. First, a *spreadsheet application* comprises both the two-dimensional *spreadsheet table* itself (called for short *spreadsheet*) and *associated programs*, which may be run upon user request. Second, a spreadsheet cell contains both a *formula* and a *value*. The value is usually displayed. Cell formulas can be *re-evaluated* upon request. Automatic re-evaluation of the whole spreadsheet after cell modification is often deactivated in industrial applications; then, *re-evaluation* can be triggered by a specific command or instruction in the associated program (often used at the end of its execution).

Syntax. A basic value is either an integer $n \in \mathbb{V}_{\text{int}}$, a floating point $f \in \mathbb{V}_{\text{float}}$ or a string $s \in \mathbb{V}_{\text{string}}$. We write $\mathbb{V} = \mathbb{V}_{\text{int}} \uplus \mathbb{V}_{\text{float}} \uplus \mathbb{V}_{\text{string}} \uplus \{\epsilon, \Omega_{\text{e}}, \Omega_{\text{t}}\}$, where ϵ stands for value "undefined", and where Ω_{e} (resp., Ω_{t}) stands for an execution error (resp., a typing error). We let $\mathbb{X} = \{\text{x}, \text{y}, \ldots\}$ denote a finite set of variables. A variable or a cell content has a type. We assume a set of pre-defined data-types such as not only **Bool**, **Float**, **String**, but also **Date** or **Currency** (which exist in real spreadsheet languages). Moreover, ϵ is the only value of type **Empty**. The spreadsheet itself is a fixed size array of dimension two. Rows (resp., columns) are labeled in a range $\mathbb{R} = \{1, 2, \ldots, n_{\mathbb{R}}\}$ (resp., $\mathbb{C} = \{1, 2, \ldots, n_{\mathbb{C}}\}$). A cell address is referred to in absolute terms, by a pair (i, j) where $i \in \mathbb{R}$ and $j \in \mathbb{C}$.

An *expression* $e \in \mathbb{E}$ may be either a constant, the reading of a variable or of a cell, or the result of the application of a binary operator or of a built-in function (such as `ISBLANK`, `IF`, `SUM`, etc.). A statement s may be either a variable declaration (together with its type), or an assignment, or an evaluation statement or a control structure (sequence, condition test, loop). Assignments may modify either the contents of a variable or the contents of a cell. Assignment to a cell may store either an *evaluated expression value* as in $\mathbf{C}[e_0, e_1] = e_2$ or a *formula* and its currently evaluated value as in $\mathbf{C}[e_0, e_1] = \text{`` }= e_2\text{''}$: *unlike an expression, a formula may be re-evaluated in the future.* Cell reads in spreadsheet formulas should correspond to constant indexes, but may be relative to the position of the cell they appear in: for instance, formula $\mathbf{C}[-1, +0]$ in cell $\mathbf{C}[3, 4]$ corresponds to cell $\mathbf{C}[2, 4]$. Last, statement **Eval** causes a global re-evaluation of the *all* *formulas* in the spreadsheet (real spreadsheet software typically allows a finer-grained control of re-evaluation, which we do not model here, as its behavior is similar to our global **Eval**).

An Excel *spreadsheet application* comprises a spreadsheet and a set of associated programs, which may be run either immediately, or upon user request. In the following, and without a loss in generality, we assume that an application a is defined by a single program body s that includes the initialization of the spreadsheet by a series of assignments (and is preceded by the declaration of the variables used in the body of the program): the example of Sect. 2 would be represented by a single program filling in the spreadsheet with values and formulas prior to the body shown in Fig. 1(b). This allows us to describe many real spreadsheet applications with the core language shown in Fig. 3. Moreover, our implementation takes into account many additional features of spreadsheet environments such as data validation or circular references, which will be covered in Sect. 6.

Example 1 (Simple application). The application below declares one variable; it then fills in 4 cells, and modifies one (a global re-evaluation takes place in the middle of the process).

1 **Dim x As Int**;	4 **C**[2, 1] = " = **C**[1, 1]";	7 **C**[2, 2] = " = **C**[1, 1] + 8";
2 $x = -5$;	5 **Eval**;	8 **C**[3, 2] = " = **C**[2, 1] + **C**[2, 2]"
3 **C**[1, 1] = 6;	6 **C**[1, 1] = 24;	

States. At any time in the execution, the memory is defined by the values of variables, and the formulas and values stored in the spreadsheet. Thus, a *non-error state* consists of a 3-tuple $(\sigma^X, \sigma^{SE}, \sigma^{SV})$ where $\sigma^X \in X \to \mathbb{V}$ maps each variable to its value, $\sigma^{SE} \in \mathbb{S}_E = (\mathbb{R} \times \mathbb{C} \to \mathbb{E})$ maps each cell to the formula it contains and $\sigma^{SV} \in \mathbb{S}_V = (\mathbb{R} \times \mathbb{C} \to \mathbb{V})$ maps each cell to the value it contains. We write Σ for the set of such concrete states. For instance, the evaluation of the application of Example 1 produces the state shown below as a graphical view (we show only the results for cells in the first two columns and the first three rows as the others are empty):

$\sigma^X :$
$x \mapsto -5$

$\sigma^{SE} :$

	1	2
1	= 24	
2	= C[1, 1]	= C[1, 1] + 8
3		= C[2, 1] + C[2, 2]

$\sigma^{SV} :$

	1	2
1	24	ϵ
2	6	32
3	ϵ	38

Semantics of Expressions. The evaluation of an expression e is defined by its semantics $[\![e]\!]_E : \Sigma \to \mathcal{P}(\mathbb{V})$ (note that an expression may evaluate to several values in order to account for possible non-determinism and run-time inputs, which may arise due to calls to RAND, DATE, or other functions reading real-time data). We let $[\![\oplus]\!] : (\mathcal{P}(\mathbb{V}))^2 \to \mathcal{P}(\mathbb{V})$ denote the concrete mathematical function corresponding to operator $\oplus$, and $[\![F]\!] : (\mathcal{P}(\mathbb{V}))^n \to \mathcal{P}(\mathbb{V})$ denote the mathematical function associated with built-in (n-ary) function F (note that their arguments may also be non-deterministic). Then, $[\![e]\!]_{\mathbb{E}}$ can be defined as follows, by induction over the syntax:

$$[\![v]\!]_{\mathbb{E}}(\sigma) = \{v\}$$
$$[\![\mathbf{x}]\!]_{\mathbb{E}}(\sigma) = \{\sigma^{\mathrm{X}}(\mathbf{x})\}$$

$$[\![\mathbf{C}[e_0, e_1]]\!]_{\mathbb{E}}(\sigma) = \{\sigma^{\mathrm{SV}}(v_0, v_1) \mid \forall i, \ v_i \in [\![e_i]\!]_{\mathbb{E}}(\sigma)\}$$
$$[\![e_0 \oplus e_1]\!]_{\mathbb{E}}(\sigma) = [\![\oplus]\!]([\![e_0]\!]_{\mathbb{E}}(\sigma), [\![e_1]\!]_{\mathbb{E}}(\sigma))$$
$$[\![\mathbf{F}(e_1, \ldots, e_n)]\!]_{\mathbb{E}}(\sigma) = [\![\mathbf{F}]\!]([\![e_0]\!]_{\mathbb{E}}(\sigma), \ldots, [\![e_n]\!]_{\mathbb{E}}(\sigma))$$

We remark that this evaluation function uses the last evaluated value whenever it reads a cell. In particular, it does not evaluate the formulas of the cells it reads the value of, nor their ancestors. Therefore, (1) an update of an ancestor of a cell c will not cause the update of the value in c, which means the value in c may become "outdated"; (2) when a cell value is outdated, any evaluation function that uses its value returns a possibly outdated result. For instance, in Example 1, after the global re-evaluation in Line 5, $\sigma^{\mathrm{SV}}(2, 1) = 6$, since $\sigma^{\mathrm{SV}}(1, 1) = 6$. In Line 6, the value of $\mathbf{C}[1, 1]$ changes, then its descendant $\mathbf{C}[2, 1]$ becomes outdated. In Line 8, $[\![\mathbf{C}[2, 1] + \mathbf{C}[2, 2]]\!]_{\mathbb{E}}(\sigma) = \{38\}$ is calculated from the outdated value of $[\![\mathbf{C}[2, 1]]\!]_{\mathbb{E}}(\sigma) = \{6\}$; thus, $\mathbf{C}[3, 2]$ is outdated too.

Errors. The evaluation of some expressions may fail to produce a value. A common case is division by 0, or a cell read with invalid (e.g., negative) row and column indexes. These errors, represented by Ω_e, are treated by other techniques and are not studied in this paper. Instead, we are interested in typing errors that may arise when applying an operator or a function to arguments whose types do not match the convention or the expectation of that operator or function. We write Ω_t both for the value produced in case of a typing error and for the corresponding error state. For instance, as the comparison between a floating point value and a string is considered unsafe, we have $\forall v_f \in \mathbb{V}_{\mathrm{float}}, \ \forall v_s \in \mathbb{V}_{\mathrm{string}}, \ [\![>]\!](v_f, v_s) = \Omega_t$. Moreover, as a value, Ω_t has no type.

Semantics of Program Statements. The concrete semantics of a statement, program, or program fragment s is a function mapping an initial state to the set of final states that can be reached after executing it: $[\![s]\!]_{\mathbb{P}} : \Sigma \to \mathcal{P}(\Sigma)$. It can also be computed by induction over the syntax. For instance:
- $[\![s_0; \ s_1]\!]_{\mathbb{P}}(\sigma) = \bigcup\{[\![s_1]\!]_{\mathbb{P}}(\sigma_0) \mid \sigma_0 \in [\![s_0]\!]_{\mathbb{P}}(\sigma)\}$;
- $[\![\mathbf{If} \ e \ \mathbf{Then} \ s_0 \ \mathbf{Else} \ s_1 \ \mathbf{End}]\!]_{\mathbb{P}}(\sigma) = S_0 \cup S_1$ where $S_0 = [\![s_0]\!]_{\mathbb{P}}(\sigma)$ if $\mathbf{true} \in [\![e]\!]_{\mathbb{E}}(\sigma)$ and $S_0 = \emptyset$ otherwise (and the same for S_1, w.r.t. the second branch);
- as usual, the semantics of a loop involves a least-fixpoint computation.

Assignment statements (to a variable or to a cell) always trigger immediate evaluation. The semantics of assignment to a variable is straightforward: $[\![\mathbf{x} = e]\!]_{\mathbb{P}}(\sigma) = \{(\sigma^{\mathrm{X}}[\mathbf{x} \leftarrow v], \sigma^{\mathrm{SE}}, \sigma^{\mathrm{SV}}) \mid v \in [\![e]\!]_{\mathbb{E}}(\sigma)\}$. The two forms of assignments to a cell differ in the fact the formula is preserved *only in the formula assignment*:
- Assignment of a value to a cell: $[\![\mathbf{C}[e_0, e_1] = e_2]\!]_{\mathbb{P}}(\sigma) = \{(\sigma^{\mathrm{X}}, \sigma^{\mathrm{SE}}[(v_0, v_1) \leftarrow v_2], \sigma^{\mathrm{SV}}[(v_0, v_1) \leftarrow v_2]) \mid \forall i \in \{0, 1, 2\}, \ v_i \in [\![e_i]\!]_{\mathbb{E}}(\sigma)\}$
- Assignment of a formula: $[\![\mathbf{C}[e_0, e_1] = \text{``} = e_2\text{''}]\!]_{\mathbb{P}}(\sigma) = \{(\sigma^{\mathrm{X}}, \sigma^{\mathrm{SE}}[(v_0, v_1) \leftarrow e_2], \sigma^{\mathrm{SV}}[(v_0, v_1) \leftarrow v_2]) \mid \forall i \in \{0, 1, 2\}, \ v_i \in [\![e_i]\!]_{\mathbb{E}}(\sigma)\}$

Semantics of Global Spreadsheet Re-evaluation. The evaluation statement causes all formulas in all the cells of the spreadsheet to be re-evaluated. Therefore, the semantics of **Eval** involves a possibly large number of computation steps, and it boils down to a *fixpoint* computation over the whole spreadsheet that recalculates σ^{SV}.

In this section, we consider spreadsheet environments without circular references (which will be covered in Sect. 6). Any such spreadsheet has an acyclic cell dependency graph. By following a topological ordering of the cells, the formulas contained in cells are evaluated one by one. For instance, if we consider the state shown in Example 1, the dependencies are shown in the left figure below. Therefore, if we only take into account non-empty cells, total orderings $(1,1) \prec (2,1) \prec (2,2) \prec (3,2)$ and $(1,1) \prec (2,2) \prec (2,1) \prec (3,2)$ can be used for the computation (a non-total ordering could also be considered). Re-evaluation using any of these orders produces the state $(\sigma^{\mathrm{X}}, \sigma^{\mathrm{SE}}, \sigma_{\mathrm{res}}^{\mathrm{SV}})$ where $\sigma_{\mathrm{res}}^{\mathrm{SV}}$ is on the right:

	1	2
1		
2		
3		

$\sigma_{\mathrm{res}}^{\mathrm{SV}}:$

	1	2
1	24	ϵ
2	24	32
3	ϵ	56

As we intend to perform an abstract interpretation based static analysis of programs, and since abstract interpretation relies on *fixpoint transfer* theorems to derive sound analyses from the concrete semantics, we now formalize the definition of the semantics of **Eval** as a least-fixpoint. Following the intuitive calculation scheme defined above, we can define $\llbracket\mathbf{Eval}\rrbracket_{\mathbb{P}}(\sigma)$ as a fixpoint where each iterate computes exactly one cell.

In the following, we let $\prec$ denote a topological ordering over $\mathbb{R} \times \mathbb{C}$. A computation step calculates the lowest cell in ordering $\prec$ that has not been evaluated yet, and that can be evaluated. To distinguish cells whose value has been calculated from cells that remain to be re-evaluated, we introduce an additional $\bot$ value. We formalize this notion of computation step with a binary relation $\leadsto_{\prec}$, which is such that $\sigma_0^{\mathrm{SV}} \leadsto_{\prec} \sigma_1^{\mathrm{SV}}$ if and only if: $\sigma_1^{\mathrm{SV}}(i,j) \in \llbracket\sigma^{\mathrm{SE}}(i,j)\rrbracket_{\mathbb{E}}(\sigma^{\mathrm{X}}, \sigma^{\mathrm{SE}}, \sigma_0^{\mathrm{SV}})$ when $\sigma_0^{\mathrm{SV}}(i,j) = \bot$ and $\forall(i',j') \prec (i,j)$, $\sigma_0^{\mathrm{SV}}(i',j') \neq \bot$; otherwise $\sigma_1^{\mathrm{SV}}(i,j) = \sigma_0^{\mathrm{SV}}(i,j)$. We remark that $\sigma^{\mathrm{SV}} \leadsto_{\prec} \sigma^{\mathrm{SV}}$, when σ^{SV} is fully computed (i.e., when no unevaluated formula remains). Then, the iteration function $\mathcal{F}_{\prec} : \mathcal{P}(\mathbb{S}_{\mathrm{V}}) \to \mathcal{P}(\mathbb{S}_{\mathrm{V}})$ is defined as $\mathcal{F}_{\prec}(\mathcal{S}) = \{\sigma_1^{\mathrm{SV}} \in \mathbb{S}_{\mathrm{V}} \mid \exists\sigma_0^{\mathrm{SV}} \in \mathcal{S},\ \sigma_0^{\mathrm{SV}} \leadsto_{\prec} \sigma_1^{\mathrm{SV}}\}$.

We now need to set up a lattice structure where the computation of the least-fixpoint should take place. As the computation progresses by filling in more cells, we need an order relation over spreadsheets which captures property "σ_1^{SV} has more evaluated cells than σ_0^{SV} and they agree on common evaluated cells", allowing for the value of a cell to move from $\bot$ to any other value. First, we let $\sqsubseteq_{\mathrm{V}}$ be the relation over the set of values extended by a constant $\top$ (denoting the definition contradiction) defined by the lattice: $\forall v \in \{\ldots, -1, 0, 1, \ldots, \epsilon, \mathbf{true}, \mathbf{false}, \ldots\}$, $\bot \sqsubseteq_{\mathrm{V}} v \sqsubseteq_{\mathrm{V}} \top$. This relation extends to sets of spreadsheets: $\forall \mathcal{S}_0, \mathcal{S}_1 \in \mathcal{P}(\mathbb{S}_{\mathrm{V}})$, $\mathcal{S}_0 \sqsubseteq \mathcal{S}_1$ if and only if $\forall \sigma_0^{\mathrm{SV}} \in \mathcal{S}_0$, $\exists \sigma_1^{\mathrm{SV}} \in$

$\mathcal{S}_1$, $\forall(i,j) \in \mathbb{R} \times \mathbb{C}$, $\sigma_0^{\mathrm{SV}}(i,j) \sqsubseteq_{\mathrm{v}} \sigma_1^{\mathrm{SV}}(i,j)$. Moreover, we let $\sigma_\perp^{\mathrm{SV}} \in \mathbb{S}_{\mathrm{V}}$ be defined by $\forall(i,j)$, $\sigma_\perp^{\mathrm{SV}}(i,j) = \perp$. At this stage, we can define the semantics of the re-evaluation as the fixpoint of $\mathcal{F}_\prec$:

Theorem 1 (Definition of $[\![\mathbf{Eval}]\!]_{\mathbb{P}}$). *For all pairs of topological orders $\prec, \prec'$ compatible with the dependencies induced by formulas stored in cells (σ^{SE}), we have:* $\mathbf{lfp}_{\{\sigma_\perp^{\mathrm{SV}}\}}\mathcal{F}_\prec = \mathbf{lfp}_{\{\sigma_\perp^{\mathrm{SV}}\}}\mathcal{F}_{\prec'} = \bigsqcup\{(\mathcal{F}_\prec)^n(\{\sigma_\perp^{\mathrm{SV}}\}) \mid n \in \mathbb{N}\}$. *Thus, we define:* $[\![\mathbf{Eval}]\!]_{\mathbb{P}}(\sigma) = \{(\sigma^{\mathrm{X}}, \sigma^{\mathrm{SE}}, \sigma_{\mathrm{res}}^{\mathrm{SV}}) \mid \sigma_{\mathrm{res}}^{\mathrm{SV}} \in \mathbf{lfp}_{\{\sigma_\perp^{\mathrm{SV}}\}}\mathcal{F}_\prec\}$

Another property that follows from the absence of circular dependencies is the fact that value $\top$ never arises in the spreadsheets obtained in the set defined by this fixpoint. Moreover, all values are defined (i.e., not equal to $\perp$) and empty cells (with no formula) contain value ϵ. We can also remark that the least fixpoint is obtained after at most $n_{\mathbb{R}} \cdot n_{\mathbb{C}}$ iterations. Spreadsheet environments typically use a *total* topological order, in order to obtain a sequential computation of the fixpoint. This is not mandatory in Theorem 1, and this definition allows to perform "parallel computation" (i.e., in the same iterate) of cells that can be defined in the same time (but each cell is computed exactly once).

Semantics of a Spreadsheet Application. In order to reason about safety properties for a spreadsheet application a, we need to set up a semantics $[\![a]\!]_{\mathrm{A}} \subseteq \Sigma$ which collects *all* the states (not only final states, as $[\![s]\!]_{\mathbb{P}}$ does) that can be reached at any point in the execution of the application. The full definition of $[\![a]\!]_{\mathrm{A}}$ follows from that of $[\![s]\!]_{\mathbb{P}}$ and is based on a trivial fixpoint, starting from the initial state σ_{i} where all variables, formulas, and values are set to ϵ.

4   Abstraction

We now formalize the *abstraction* [10] used in our analysis (Sect. 5). It is based on *abstract formulas* (Sect. 4.1) that summarize the behavior of formulas depending on the type of their inputs and on *abstract zones* [8] that tie abstract predicates to sets of spreadsheet cells (Sect. 4.2).

4.1   Formula Abstraction

The computation of type information over zones requires the propagation of information not only through the associated program, but also through the formulas contained in the spreadsheet itself, to be able to analyze re-evaluation. Thus, the effect of formulas should be propagated through the analysis. However, dealing with all formulas stored in the spreadsheet would be too costly. Therefore, we propose an abstraction of the semantics of formulas, which expresses their effect on types, and replaces, e.g., constants with their type:

Definition 1 (Abstract formulas). *Abstract formulas are defined by:*

$$e^\sharp \ (\in \mathbb{E}^\sharp) ::= t \mid \mathbf{C}[n,n] \mid e^\sharp \oplus e^\sharp \mid \mathrm{F}(e^\sharp, \ldots, e^\sharp) \qquad where\ n \in \mathbb{V}_{\mathrm{int}},\ t \in \mathbb{T}$$

Example 2 (Abstract formulas). $\mathbf{Int} + \mathbf{Float}$, $\mathbf{Float} - \mathbf{C}[3,4]$, $\mathrm{ISBLANK}(\mathbf{C}[5,6])$ are all abstract formulas. Moreover, we also allow relative indexes in abstract formulas, as in $\mathbf{C}[+0,-1] \star \mathbf{Float}$ (Fig. 2).

Semantics of Abstract Formulas. We now give a semantics to abstract formulas, following a similar scheme as in Sect. 3, and where abstract formulas evaluate into *types.* We let a *type spreadsheet* be a function $\sigma^{\mathrm{T}} \in \mathbb{S}_{\mathrm{T}} = (\mathbb{R} \times \mathbb{C} \to \mathbb{T})$ mapping each cell to a type. Spreadsheet contents σ^{SV} has type σ^{T} (noted $\sigma^{\mathrm{SV}} : \sigma^{\mathrm{T}}$) if and only if $\forall (i,j) \in \mathbb{R} \times \mathbb{C}$, $\sigma^{\mathrm{SV}}(i,j) : \sigma^{\mathrm{T}}(i,j)$. To define the semantics of abstract formulas, we let each operator $\oplus$ (resp., built-in function $\mathbf{F}$) be abstracted by a partial function $[\![\oplus]\!]_{\mathrm{t}} : (\mathcal{P}(\mathbb{T}))^2 \to (\mathcal{P}(\mathbb{T}))$ (resp., $[\![\mathbf{F}]\!]_{\mathrm{t}} : (\mathcal{P}(\mathbb{T}))^n \to \mathcal{P}(\mathbb{T})$) that over-approximates its effect on types. For instance, $[\![+]\!]_{\mathrm{t}}(\{\mathbf{Int}\}, \{\mathbf{Int}\}) = \{\mathbf{Int}\}$ and $[\![*]\!]_{\mathrm{t}}(\{\mathbf{Int}\}, \{\mathbf{Float}\}) = \{\mathbf{Float}\}$. On the other hand, as noted in Sect. 2, comparing a string with an integer is unsafe, so $[\![<]\!]_{\mathrm{t}}(\{\mathbf{String}\}, \{\mathbf{Int}\})$ leads to Ω_{t}, as for all unsafe operations. The semantics of abstract formula $e^{\sharp}$ is a function $[\![e^{\sharp}]\!]_{\mathrm{T}} : \mathbb{S}_{\mathrm{T}} \to \mathcal{P}(\mathbb{T})$ mapping spreadsheets into sets of types.

Abstraction of Formulas. A spreadsheet formula can be translated into an abstract formula by replacing, e.g., all constants with types, this process is formalized in the definition of the translation function ϕ below:

$$\phi(\mathbf{C}[i,j]) = \mathbf{C}[i,j] \qquad\qquad \phi(v) = t \text{ where } t \text{ is the type of } v$$
$$\phi(e_0 \oplus e_1) = \phi(e_0) \oplus \phi(e_1) \qquad \phi(\mathbf{F}(e_1, \ldots, e_n)) = \mathbf{F}(\phi(e_1), \ldots, \phi(e_n))$$

Note that the translation applies only to formulas found in the spreadsheet (i.e. not to general expressions found in associated programs), thus ϕ is not defined for variables or cell accesses of the form $\mathbf{C}[e_0, e_1]$ where e_0 or e_1 is not a constant.

The intended effect on types is preserved by ϕ, and it satisfies the soundness condition: if $e \in \mathbb{E}$, $\sigma^{\mathrm{SV}} \in \mathbb{S}_{\mathrm{V}}$ and $\sigma^{\mathrm{T}} \in \mathbb{S}_{\mathrm{T}}$ are such that $\forall (i,j)$, $\sigma^{\mathrm{SV}}(i,j) : \sigma^{\mathrm{T}}(i,j)$, then $\forall v \in [\![e]\!]_{\mathbb{E}}(\sigma)$, $\exists t \in [\![\phi(e)]\!]_{\mathrm{T}}(\sigma^{\mathrm{T}})$, $v : t$.

Example 3 (Formulas abstraction). We have the abstractions $\phi(\mathbf{C}[4,4] * 1.3) = \mathbf{C}[4,3] * \mathbf{Float}$, and with relative indexes, $\phi(\mathbf{C}[+0, -1] * 1.3) = \mathbf{C}[+0, -1] * \mathbf{Float}$.

Simplification of Abstract Formulas. Some type formulas may be simplified, while still carrying the same information. For instance, the addition of two floating point values produces a new floating point value, thus type formula $\mathbf{Float} + \mathbf{Float}$ can be simplified into $\mathbf{Float}$. The concrete semantics of functions may allow for less trivial formula simplifications. For example, the function `ISERROR` checks if a value is of type **Error**; it always returns a boolean value whatever the argument is, then formula `ISERROR(`$\mathbf{C}[5,6]$`)` can be simplified into **Bool**.

Therefore, we use a *simplification function* $\mathbf{S} : \mathbb{E}^{\sharp} \to \mathbb{E}^{\sharp}$, defined by structural induction over formulas, that applies a set of local rules. It is sound with respect to the concrete semantics: $\forall e^{\sharp} \in \mathbb{E}^{\sharp}$, $[\![\mathbf{S}(e^{\sharp})]\!]_{\mathrm{T}} = [\![e^{\sharp}]\!]_{\mathrm{T}}$. Potentially unsafe operations should not be simplified (e.g., simplification rule $\mathbf{S}(\mathbf{Float} > \mathbf{String}) = \mathbf{Bool}$ is not admissible), as they are exactly what our analysis aims at discovering.

4.2 Spreadsheet Abstraction

Spreadsheet Zones Abstraction. To abstract spreadsheets, we need to tie abstract properties such as types or abstract formulas to table *zones*. In the following, we use the zone abstraction of [8], where a zone describes a set of cells in a compact manner. A zone abstraction is defined by a numeric abstract domain [10] $\mathbb{D}_{num}^\sharp$ over $\mathbb{X}$ (where $\mathbb{X}$ contains two special variable names $\bar{i}$ and $\bar{j}$ that cannot be used in the associated programs and that respectively denote the row and column of a cell), with a concretization function $\gamma_{num} : \mathbb{D}_{num}^\sharp \to \mathcal{P}(\mathbb{X} \to \mathbb{V})$. A set of cell coordinates S in concrete state $(\sigma^\mathbb{X}, \sigma^{SE}, \sigma^{SV})$ is abstracted by zone $\mathcal{Z} \in \mathbb{D}_{num}^\sharp$ if and only if $\forall (i,j) \in S$, $[\sigma^\mathbb{X}, \bar{i} = i, \bar{j} = j] \in \gamma_{num}(\mathcal{Z})$, i.e., the coordinates in S together with $\sigma^\mathbb{X}$ satisfy $\mathcal{Z}$. When not considering an associated program, no other variable than $\bar{i}, \bar{j}$ should appear in $\mathcal{Z}$. In this paper, we employ a variant of difference bound matrices (DBMs) which was used in [8], and inspired from the octagon abstract domain [20]. For clarity, we write bounds on $\bar{i}, \bar{j}$ using interval notation and let the zone defined by $\bar{i} \in [e_0, e_1] \wedge \bar{j} \in [e_2, e_3]$ be denoted by $[c_0, e_1] \times [e_2, e_3]$ (where $e_0, \ldots, e_3$ are linear expressions over the variables or constants).

Example 4 (Abstract zones). We define a few zones relevant to the example of Sect. 2. Zone $\mathcal{Z}_0 : \bar{i} \in [4, 43] \wedge \bar{j} = 2$ (or $[4, 43] \times [2, 2]$) describes a block in column 2, from row 4 till row 43. Similarly, zone $\mathcal{Z}_1 : [4, \mathtt{i}] \times [3, 3]$ describes a block in column 4, and spanning from row 4 till row $n_\mathtt{i}$, where $n_\mathtt{i}$ denotes the value of $\mathtt{i}$ in the current state. Last, zone $\mathcal{Z}_2 : [4, 43] \times [4, 4]$ describes a block in column 4.

State Abstraction. An abstract state encloses (i) numerical abstract properties of variables and (ii) a collection of abstract zone predicates, that is, abstract predicates that hold true over all cells that can be characterized by an abstract zone.

An *abstract predicate* is either a type or an abstract formula. This defines an abstract domain $\mathbb{D}_c^\sharp = \{\bot, \top\} \cup \mathbb{T} \cup \mathbb{E}^\sharp$. To distinguish an abstract type formula from a type, we insert "=" before the type (e.g., **String** $\in \mathbb{T}$, whereas "= **String**" $\in \mathbb{E}^\sharp$). Concretization function $\gamma_c : \mathbb{D}_c^\sharp \to \mathcal{P}(\mathbb{E} \times \mathbb{V})$ is defined by:
- $\forall t \in \mathbb{T}$, $\gamma_c(t) = \{(e, v) \in \mathbb{E} \times \mathbb{V} \mid v : t\}$;
- $\forall e^\sharp \in \mathbb{E}^\sharp$, $\gamma_c(e^\sharp) = \{(e, v) \in \mathbb{E} \times \mathbb{V} \mid \phi(e) = e^\sharp\}$.

We can now define abstract states as follows:

Definition 2 (Abstract zone predicate and abstract state). *An* abstract zone predicate *is a pair* $(\mathcal{Z}, \mathcal{P}) \in \mathbb{D}_z^\sharp$, *where* $\mathbb{D}_z^\sharp = \mathbb{D}_{num}^\sharp \times \mathbb{D}_c^\sharp$. *The concretization* $\gamma_z : \mathbb{D}_z^\sharp \longrightarrow \mathcal{P}(\Sigma)$ *is such that* $(\sigma^\mathbb{X}, \sigma^{SE}, \sigma^{SV}) \in \gamma_z(\mathcal{Z}, \mathcal{P})$ *if and only if:*

$$\forall (i,j) \in \mathbb{R} \times \mathbb{C}, \ [\sigma^\mathbb{X}, \bar{i} = i, \bar{j} = j] \in \gamma_{num}(\mathcal{Z}) \Longrightarrow (\sigma^{SE}(i,j), \sigma^{SV}(i,j)) \in \gamma_c(\mathcal{P})$$

An abstract state *is a pair* $(N^\sharp, P^\sharp) \in \mathbb{D}_\Sigma^\sharp = \mathbb{D}_{num}^\sharp \times \mathcal{P}_{fin}(\mathbb{D}_z^\sharp)$. *Moreover, concretization function* $\gamma_\Sigma : \mathbb{D}_\Sigma^\sharp \to \mathcal{P}(\Sigma)$ *is defined by:*

$$\gamma_\Sigma(N^\sharp, P^\sharp) = \{(\sigma^\mathbb{X}, \sigma^{SE}, \sigma^{SV}) \mid \sigma^\mathbb{X} \in \gamma_{num}(N^\sharp) \wedge (\sigma^\mathbb{X}, \sigma^{SE}, \sigma^{SV}) \in \bigcap_{p^\sharp \in P^\sharp} \gamma_z(p^\sharp)\}$$

We distinguish zone predicates attached to abstract formulas and zone predicates attached to types: we let $F^\sharp \in \mathcal{P}_{fin}(\mathbb{D}_{z,form}^\sharp)$ denote the abstract zone predicates for formulas and we let $T^\sharp \in \mathcal{P}_{fin}(\mathbb{D}_{z,type}^\sharp)$ denote those for types. Thus,

$P^\sharp = F^\sharp \uplus T^\sharp$, and $(N^\sharp, P^\sharp)$ is equivalent to $(N^\sharp, F^\sharp \uplus T^\sharp)$. The construction of Definition 2 utilizes the *reduced product* and *reduced cardinal power* of abstract domains [11]. It also extends the domain shown in [8] with abstract formulas.

Example 5 (Example 4 continued: abstract predicates over zones). The following abstract zone predicates are satisfied in the concrete state of Fig. 1(a):
- Zones $\mathcal{Z}_0$ and $\mathcal{Z}_1$ correspond to values of type **Float**, which are described by the predicates $(\mathcal{Z}_0, \textbf{Float})$ and $(\mathcal{Z}_1, \textbf{Float})$;
- All cells in zone $\mathcal{Z}_2$ contain a formula abstracted by $\mathbf{C}[+0, -1] * \textbf{Float}$, thus this zone can be described with abstract zone predicate $(\mathcal{Z}_2, \mathbf{C}[+0, -1] * \textbf{Float})$. Likewise, Fig. 2 displays an abstract state made of ten zones bound to types and three zones bound to abstract formulas.

5   Static Analysis Algorithms

We now set up a fully automatic static analysis, which computes an *over*-approximation of the set $[\![a]\!]_A$ or reachable states of an application a, expressed in the abstract domain defined in Sect. 4. It proceeds by *abstract interpretation* [10] of the body of a: the effect of each statement is over-approximated in a sound manner by some adequate transfer functions, and a widening operator enforces the convergence of abstract iterates whenever a concrete fixpoint needs to be approximated in the abstract level. We design two *sound* abstract semantics. The abstract semantics $[\![s]\!]_\mathbb{P}^\sharp : \mathbb{D}_\Sigma^\sharp \to \mathbb{D}_\Sigma^\sharp$ of statement s is a function which maps an abstract pre-condition into a conservative abstract post-condition (which is described by abstract states). The abstract semantics $[\![a]\!]_A^\sharp \subseteq \mathbb{D}_\Sigma^\sharp$ of application a is a finite set of abstract states. We defer the analysis of global re-evaluation to Section 5.2 and handle the others first in Section 5.1. Some abstract operations are common with [8] whereas others are deeply different, especially those related to *formulas*.

5.1   Abstract Interpretation of Basic Statements

Straight Line Code. The core language of Sect. 3 features several, rather similar forms of assignments (assignment to a variable, of an evaluated expression to a cell, or of a formula to a cell). Thus, the analysis defines three transfer functions $\textbf{assign}_\mathbb{X}^\sharp, \textbf{assign}_\mathbb{V}^\sharp, \textbf{assign}_\mathbb{E}^\sharp$ that share the same principles, thus we focus on formula assignment $\textbf{assign}_\mathbb{E}^\sharp : \mathbb{E} \times \mathbb{E} \times \mathbb{E} \times \mathbb{D}_\Sigma^\sharp \to \mathbb{D}_\Sigma^\sharp$. Given e_0, e_1, e_2, it should satisfy:

$$\forall \sigma^\sharp \in \mathbb{D}_\Sigma^\sharp, \ \forall (\sigma^\mathrm{X}, \sigma^\mathrm{SE}, \sigma^\mathrm{SV}) \in \gamma_\Sigma(\sigma^\sharp), \forall v_{i \in \{0,1,2\}} \in [\![e_i]\!]_\mathbb{E}(\sigma),$$
$$(\sigma^\mathrm{X}, \sigma^\mathrm{SE}[(v_0, v_1) \leftarrow e_2], \sigma^\mathrm{SV}[(v_0, v_1) \leftarrow v_2]) \in \gamma_\Sigma(\textbf{assign}_\mathbb{E}^\sharp(e_0, e_1, e_2, \sigma^\sharp))$$

To achieve this, both the type and the formula properties of zones may need to be updated. Information about the overwritten cell should be dropped from the abstract state, either by removing existing zone predicates, or by splitting zones into preserved / overwritten areas. Then, new type and formula information should be synthesized and attached to a zone corresponding only to the cell overwritten by the assignment. Type information is obtained by evaluating the semantics of abstract formulas; when this evaluation fails, a typing error should be reported.

Example 6 (Abstract assignment). Let us consider abstract state $\sigma^\sharp = ($i $<$ n$, \{([1, i-1]\times[2,2], e^\sharp), ([i,n]\times[2,2], $ " $=$ **String**"$)\})$, where $e^\sharp = $ **Int**$+$**C**$[+0, -1]$. Then, assignment **C**$[$i$, 2] = $ " $= 24 + $ **C**$[$i$, 1]$" replaces the constant formula (of type string) contained in cell **C**$[$i$, 2]$ with a formula that can be abstracted by **Int** $+$ **C**$[$i$, 1]$ (or equivalently **Int** $+$ **C**$[+0, -1]$), and it evaluates that formula, which returns a value of type **Int**. Thus, the string constant value that was previously stored in the cell is replaced, so the topmost cell of zone $[$i$,$n$] \times [2,2]$ should be removed from that zone. Therefore, we obtain abstract state $\sigma_0^\sharp = ($i $<$ n$, \{([1, i-1] \times [2,2], e^\sharp), (i\times[2,2], e^\sharp), ([i+1,n]\times[2,2], $ " $=$ **String**"$), ($i$\times [2,2], $**Int**$)\})$.

This update operation creates new zones, yet, when several adjacent zones have the same type and abstract formulas, they could be merged, with no loss of information. This operation is performed by an operator **reduce**$^\sharp : \mathbb{D}_\Sigma^\sharp \to \mathbb{D}_\Sigma^\sharp$ introduced in [8], and that satisfies soundness condition $\forall \sigma^\sharp \in \mathbb{D}_\Sigma^\sharp, \ \gamma_\Sigma(\sigma^\sharp) \subseteq \gamma_\Sigma($**reduce**$^\sharp(\sigma^\sharp))$.

Example 7 (Reduction). In abstract state $\sigma_0^\sharp$ of Example 6, $([1, i - 1] \times [2,2], e^\sharp)$ and $([$i$, i] \times [2,2], e^\sharp)$ can be merged into $([1, i] \times [2,2], e^\sharp)$. As $([4,4] \times [4,4], $**C**$[4,3] * $**Float**$)$ is equivalent to $([4,4] \times [4,4], $**C**$[+0, -1] * $**Float**$)$, and $([5,5] \times [4,4], $**C**$[5,3] * $**Float**$)$ is equivalent to $([5,5] \times [4,4], $**C**$[+0, -1] * $**Float**$)$, these two zones can be merged into $([4,5] \times [4,4], $**C**$[+0, -1] * $**Float**$)$.

We can now define the analysis of straight line code (sequences of assignments):

- $[\![s_0; \ s_1]\!]_\mathbb{P}^\sharp(\sigma^\sharp) = [\![s_1]\!]_\mathbb{P}^\sharp([\![s_0]\!]_\mathbb{P}^\sharp(\sigma^\sharp));$
- $[\![$x$ = e]\!]_\mathbb{P}^\sharp(\sigma^\sharp) = $ **reduce**$^\sharp($**assign**$_\mathbb{X}^\sharp($x$, e, \sigma^\sharp));$
- $[\![$**C**$[e_0, e_1] = e_2]\!]_\mathbb{P}^\sharp(\sigma^\sharp) = $ **reduce**$^\sharp($**assign**$_\mathbb{V}^\sharp(e_0, e_1, e_2, \sigma^\sharp));$
- $[\![$**C**$[e_0, e_1] = $ " $= e_2$"$]\!]_\mathbb{P}^\sharp(\sigma^\sharp) = $ **reduce**$^\sharp($**assign**$_\mathbb{E}^\sharp(e_0, e_1, e_2, \sigma^\sharp)).$

Control Structures. The analysis of control structures requires condition test, join and widening operators. Condition tests refine information on variable ranges (hence, refining zone bounds) and on cell types (due to operators testing the type of cell values, such as ISBLANK). They are analyzed by an operator **guard**$^\sharp :$ $\mathbb{E} \times \mathbb{D}_\Sigma^\sharp \to \mathbb{D}_\Sigma^\sharp$ that satisfies soundness condition $\forall \sigma \in \gamma_\Sigma(\sigma^\sharp), $ **true** $\in [\![e]\!]_\mathbb{E}(\sigma) \Rightarrow$ $\sigma \in \gamma_\Sigma($**guard**$^\sharp(e, \sigma^\sharp))$. Control flow joins are analyzed by a join operator $\sqcup^\sharp : \mathbb{D}_\Sigma^\sharp \times \mathbb{D}_\Sigma^\sharp \to \mathbb{D}_\Sigma^\sharp$ such that $\forall \sigma_0^\sharp, \sigma_1^\sharp \in \mathbb{D}_\Sigma^\sharp, \ \gamma_\Sigma(\sigma_0^\sharp) \cup \gamma_\Sigma(\sigma_1^\sharp) \subseteq \gamma_\Sigma(\sigma_0^\sharp \sqcup^\sharp \sigma_1^\sharp)$, whereas loops require a widening operator $\nabla^\sharp$, based on similar algorithms and that ensures the termination of abstract iterates. These operators *generalize* bounds on zones [8], hence play a critical role in the inference of non trivial zone invariants, such as those shown in Fig. 2:

Example 8 (Abstract join). Let us consider abstract states $\sigma_0^\sharp = (\mathbf{x} = 2, \{(\mathcal{Z}_0, e^\sharp)\})$ and $\sigma_1^\sharp = (\mathbf{x} = 3, \{(\mathcal{Z}_1, e^\sharp)\})$, where $\mathcal{Z}_0 = [1, 2] \times [2, 2]$ and $\mathcal{Z}_1 = [1, 3] \times [2, 2]$. In both zones, the upper bound on $\bar{\imath}$ is equal to $\mathbf{x}$. Thus, $\mathcal{Z}_0$ (resp., $\mathcal{Z}_1$) is semantically equivalent to $\mathcal{Z} = [1, \mathbf{x}] \times [2, 2]$. Therefore, $\sigma_1^\sharp \sqcup^\sharp \sigma_0^\sharp$ returns $(2 \le \mathbf{x} \le 3, \{(\mathcal{Z}, e^\sharp)\})$.

We can now define the analysis of condition statements and loops:

- $[\![\mathbf{If}\ e\ \mathbf{Then}\ s_0\ \mathbf{Else}\ s_1\ \mathbf{End}]\!]_\mathbb{P}^\sharp(\sigma^\sharp) = [\![s_0]\!]_\mathbb{P}^\sharp(\mathbf{reduce}^\sharp(\mathbf{guard}^\sharp(e, \sigma^\sharp))) \sqcup^\sharp$ $[\![s_1]\!]_\mathbb{P}^\sharp(\mathbf{reduce}^\sharp(\mathbf{guard}^\sharp(\neg e, \sigma^\sharp)))$;

- $[\![\mathbf{While}(e)\ s\ \mathbf{End}]\!]_\mathbb{P}^\sharp(\sigma^\sharp) = \mathbf{reduce}^\sharp(\mathbf{guard}^\sharp(\neg e, \mathbf{lfp}_\perp^\sharp\ F^\sharp))$ where $F^\sharp(\sigma_0^\sharp) = \sigma^\sharp \sqcup^\sharp [\![s]\!]_\mathbb{P}^\sharp(\mathbf{reduce}^\sharp(\mathbf{guard}^\sharp(e, \sigma_0^\sharp)))$ and $\mathbf{lfp}^\sharp$ computes abstract post-fixpoint, using classical abstract iteration techniques, using widening operator $\nabla^\sharp$.

We recall the abstract post-fixpoint operator is sound in the following sense: if $F : \mathcal{P}(\Sigma) \to \mathcal{P}(\Sigma)$ is continuous and $F^\sharp : \mathbb{D}_\Sigma^\sharp \to \mathbb{D}_\Sigma^\sharp$, $\mathcal{S} \subseteq \Sigma$ and $\sigma^\sharp \in \mathbb{D}_\Sigma^\sharp$ are such that, $F \circ \gamma_\Sigma \subseteq \gamma_\Sigma \circ F^\sharp$ and $\mathcal{S} \subseteq \gamma_\Sigma(\sigma^\sharp)$, then $\mathbf{lfp}_\mathcal{S} F \subseteq \gamma_\Sigma(\mathbf{lfp}_{\sigma^\sharp}^\sharp F^\sharp)$.

Applications. The analysis of an application a recursively computes the abstract semantics of all statements in the body of a from its initial state, and produces a finite set of abstract states $[\![a]\!]_\mathbb{A}^\sharp$. Our analysis is sound:

Theorem 2 (Soundness). *For all statements $s \in \mathbb{P}$, $[\![s]\!]_\mathbb{P}^\sharp$ is sound: $\forall \sigma^\sharp \in \mathbb{D}_\Sigma^\sharp$, $\forall \sigma \in \gamma_\Sigma(\sigma^\sharp)$, $[\![s]\!]_\mathbb{P}(\sigma) \subseteq \gamma_\Sigma([\![s]\!]_\mathbb{P}^\sharp(\sigma^\sharp))$. Thus, for all $a \in \mathbb{A}$, $[\![a]\!]_\mathbb{A}$ is sound, i.e., $[\![a]\!]_\mathbb{A} \subseteq \bigcup \gamma_\Sigma([\![a]\!]_\mathbb{A}^\sharp)$.*

Therefore, the whole analysis catches all typing errors following the definition given in Sect. 3, corresponding to the operations specified unsafe. In particular, it catches the error of the example shown in Sect. 2, and proves the fixed version safe.

5.2   Abstract Interpretation of Global Evaluation

The concrete semantics of **Eval** boils down to a fixpoint, that re-computes cell values in *the whole* spreadsheet while preserving formulas and variables values (Sect. 3). Thus, we assume $\sigma^\sharp = (N^\sharp, F^\sharp \uplus T^\sharp)$, and show the computation of $T_{\mathrm{res}}^\sharp$ (by fixpoint approximation) so as to let $[\![\mathbf{Eval}]\!]^\sharp(\sigma^\sharp) = (N^\sharp, F^\sharp \uplus T_{\mathrm{res}}^\sharp)$. We first show a very basic iteration strategy, and then discuss the analysis of **Eval**.

Cell-by-Cell Re-evaluation. The concrete semantics of **Eval** is based on function $\mathcal{F}_\prec$, defined by a cell ordering $\prec$ compatible with formula dependencies. In this paragraph, we show an abstract counterpart for $\mathcal{F}_\prec$ under the assumption that each abstract zone is reduced to a single concrete cell, thus elements of $T^\sharp$ (resp., $F^\sharp$) are equivalent to functions from $\mathbb{R} \times \mathbb{C}$ into $\mathbb{T}$ (resp., $\mathbb{E}^\sharp$). Abstract formulas follow the same dependencies as concrete formulas, thus the topological order $\prec$ can be retrieved from an abstract state, by topological sorting. Moreover, the analysis should support "not yet re-evaluated" cells, which are denoted by $\perp$:

- To extend type spreadsheets $\mathbb{S}_\mathbb{T}$, we let $\mathbb{S}_{\mathbb{T}\perp} = (\mathbb{R} \times \mathbb{C}) \to (\mathbb{T} \uplus \{\perp\})$, and let order relation $\sqsubseteq$ be defined by $\forall t \in \mathbb{T}$, $\perp \sqsubseteq t$. As each zone contains exactly one cell, an element $T^\sharp \in \mathcal{P}_{\mathrm{fin}}(\mathbb{D}^\sharp_{z,\mathrm{type}})$ is now equivalent to an element of $\mathbb{S}_{T\perp}$.

- We let $T^\sharp_\perp \in \mathbb{S}_{T\perp}$ be defined by $\forall(i,j),\ T^\sharp_\perp(i,j) = \perp$.
- Given an abstract formula $e^\sharp$, its abstract semantics $[\![e^\sharp]\!]_t$ can also be extended to compute a type (possibly $\top$) for an abstract element of $\mathcal{P}_{\mathrm{fin}}(\mathbb{D}^\sharp_{z,\mathrm{form}})$, using the type information available for each cell in the formula; we still use notation $[\![e^\sharp]\!]_t$ to denote that extended semantics.

We can now define the abstract counterpart $\mathcal{F}^\sharp_\prec : \mathcal{P}_{\mathrm{fin}}(\mathbb{D}^\sharp_{z,\mathrm{type}}) \to \mathcal{P}_{\mathrm{fin}}(\mathbb{D}^\sharp_{z,\mathrm{type}})$ of $\mathcal{F}_\prec$. It is such that for all $T^\sharp_0 \in \mathbb{S}_{T\perp}$, and for all i,j, $\mathcal{F}^\sharp_\prec(T^\sharp_0)(i,j) = [\![F^\sharp(i,j)]\!]_t(T^\sharp_0)$ when $T^\sharp_0(i,j) = \perp$ and $\forall(i',j') \prec (i,j),\ T^\sharp_0(i',j') \neq \perp$ (otherwise $\mathcal{F}^\sharp_\prec(T^\sharp_0)(i,j) = T^\sharp_0(i,j)$). It is sound: for all $\sigma^\sharp_0 = (N^\sharp, F^\sharp \uplus T^\sharp_0) \in \mathbb{D}^\sharp_\Sigma$, and for all $(\sigma^X, \sigma^{SE}, \sigma^{SV}_0) \in \gamma_\Sigma(\sigma^\sharp_0)$, we have $(\sigma^X, \sigma^{SE}, \mathcal{F}_\prec(\sigma^{SV}_0)) \subseteq \gamma_\Sigma(N^\sharp, F^\sharp \uplus \mathcal{F}^\sharp_\prec(T^\sharp_0))$. Therefore, the existence of the fixpoint follows from the continuity of $\mathcal{F}^\sharp_\prec$ (it is obtained after at most $n_\mathbb{R} \cdot n_\mathbb{C}$ iterations). Soundness is proved by fixpoint transfer:

Theorem 3 (Abstract interpretation of re-evaluation). $[\![\mathbf{Eval}]\!]^\sharp_\mathbb{P}(N^\sharp, F^\sharp \uplus T^\sharp) = (N^\sharp, F^\sharp \uplus T^\sharp_{\mathrm{res}})$ where $T^\sharp_{\mathrm{res}} = \mathrm{lfp}_{T^\sharp_\perp} \mathcal{F}^\sharp_\prec = \bigsqcup\{(\mathcal{F}^\sharp_\prec)^n(T^\sharp_\perp) \mid n \in \mathbb{N}\}$ defines a sound post-condition: $\forall \sigma^\sharp \in \mathbb{D}^\sharp_\Sigma,\ \forall \sigma \in \gamma_\Sigma(\sigma^\sharp),\ [\![\mathbf{Eval}]\!]_\mathbb{P}(\sigma) \subseteq \gamma_\Sigma([\![\mathbf{Eval}]\!]^\sharp_\mathbb{P}(\sigma^\sharp))$.

Moreover, this process will also allow us to prove no typing error (in the sense of Sect. 3) arises during re-evaluation.

Example 9 (Cell-by-cell re-evaluation). We illustrate this strategy with the abstraction of the spreadsheet studied in Sect. 3. The corresponding abstract formulas over zones are shown below, in the left hand side. Then, cells are treated following topological ordering $(1,1) \prec (2,1) \prec (2,2) \prec (3,2)$, and the type obtained for each cell is **Int**:

	1	2			1	2
1	= **Int**	= **Empty**		1	**Int**	(**Empty**)
2	= **C**[1,1]	= **C**[1,1] + **Int**		2	**Int**	**Int**
3	= **Empty**	= **C**[2,1] + **C**[2,2]		3	(**Empty**)	**Int**

Zone-by-Zone Strategy. The cell-by-cell strategy abstract interpretation of **Eval** would not be efficient in practice, as abstract states usually contain zones which are bounded, but possibly large and/or of variable size. However, since a whole zone is attached to a single abstract formula, type information for a whole zone can often be computed in a single step, which is much faster than cell-by-cell evaluation.

A zone can be re-evaluated as soon as the two following conditions are satisfied: (1) its abstract formulas induce no internal dependency, i.e., between its cells (in the example of Sect. 2, this holds for all zones, except the last column, which will be discussed in the next paragraph); (2) there exists a topological order $\prec$ compatible with formula dependencies, according to which all the cells lower than the cells in that zone have already been evaluated.

When a zone satisfies these two conditions, the analysis can re-evaluate its type by applying the abstract formula it corresponds to, since its arguments have already been re-evaluated. When an argument of the abstract formula may belong to several zones, it will be necessary to split the zone being re-evaluated. This will produce a set of zones with type information. The analysis will apply this efficient scheme whenever the topological ordering induced by abstract formulas zones allows it:

Example 10 (Zone-by-zone strategy). We assume the following abstract state:

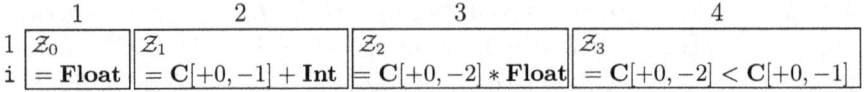

Then according to the formula dependencies, the abstract iteration can follow the order $\mathcal{Z}_0 \prec \mathcal{Z}_1 \prec \mathcal{Z}_2 \prec \mathcal{Z}_3$. It terminates after four iterations, and produces the type zones $\{(\mathcal{Z}_0, \mathbf{Float}), (\mathcal{Z}_1, \mathbf{Float}), (\mathcal{Z}_2, \mathbf{Float}), (\mathcal{Z}_3, \mathbf{Bool})\}$.

Abstract Iteration over Zones, Using Widening. When a zone contains internal dependencies (i.e., abstract formula using as arguments cell that belong to the zone itself), the zone-by-zone strategy does not apply. Such a *self-reference* occurs in the example of Sect. 2 since the evaluation of Column 5 requires types of Columns 3, 4 and Column 5 itself. *Inter-reference* among zones may also occur, e.g., when $\mathcal{Z}_0$ needs types of $\mathcal{Z}_1$, $\mathcal{Z}_1$ needs types of $\mathcal{Z}_2$, ..., and $\mathcal{Z}_n$ needs types of $\mathcal{Z}_0$.

Zones containing such patterns can be re-evaluated in the abstract level by simulating a cell-by-cell re-evaluation order, as part of an abstract fixpoint computation. To do this, under the assumption that there is no cycle in formulas (this case is discussed in Sect. 6), the analysis of **Eval** will consider a loop that computes the cells in the zone one-by-one, following the steps below:

1. introducing loop variable k, denoting the number of cells in the abstract formula zone that have been re-evaluated;
2. determining the first cell in the abstract formula zone dependency order;
3. splitting the abstract formula zones into two zones, respectively for cells that can be immediately re-evaluated, and for cells that cannot be re-evaluated yet;
4. iterating Steps 2 and 3 until the abstract formula zones are fully treated, and applying widening at each step to ensure termination, thanks to $\mathbf{lfp}^\sharp$;
5. synthesizing the final abstract state by restricting, when Step 4 produces stable type zones.

This strategy provides a way to compute the effect of **Eval** over large zones or zones of variable size. It does not need the full unrolling of the zone, thanks to the use of the widening operation over the cells that are generally well structured. Indeed, it effectively amounts to analyzing a loop with counter k that iterates over the zone in order to compute abstract types:

Example 11 (Abstract iteration over a zone). We consider the abstract formula zones below (which correspond to an excerpt of the example of Sect. 2), which define the dependencies shown in the right-hand side:

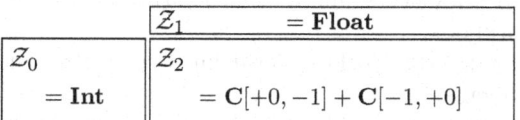
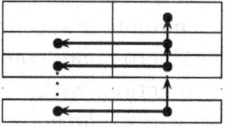

The first two iterations of the strategy described above produce the results below:

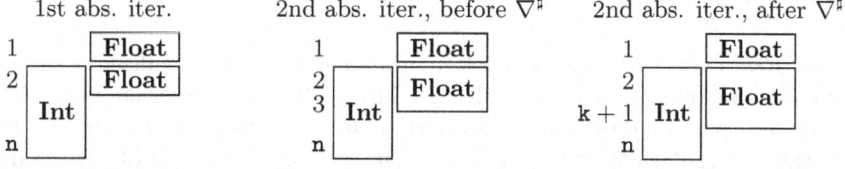

On the third iteration, abstract states are stable. Moreover, the analysis proves all the formulas evaluate without a typing error and produce a result of type **Float**.

Combined Strategies. In general, abstract states require the use of a combination of the strategies shown above. The zone-by-zone strategy is given priority in our analysis: it will always try to detect and to re-compute first the zones that can be evaluated as a whole. This strategy is the most efficient and turns out to be the most frequently used in practice. Remaining cases are dealt with by widening based and cell-by-cell strategies. After adding **Eval** statement to the set of the statements, the global soundness theorem (Theorem 2) still holds.

6   Implementation of an Excel VBA Analyzer

We have implemented our analysis. Our analyzer handles a large subset of Microsoft Excel functions and VBA, following the VBA specification [2]. Our tool consists of a frontend written in VBA, that parses Microsoft Excel spreadsheet tables (e.g., number formats, types, formulas, buttons) and VBA macros, and exports them to the static analyzer itself, which undertakes the verification (and includes 19000 lines of OCaml code). The verification of a spreadsheet application proceeds through two steps: (1) the verification of global re-evaluation; (2) the verification of the execution of any macro it contains, given the initial spreadsheet abstract state. The verification gradually infers invariants; finally, it either proves the correctness with regard to our typing system, or raises alarms by pointing out the location (e.g., the zone in spreadsheets and/or the line in macros) and the unsafe typing rule in question. The analyzer can also be launched over a set of Excel files and return a summary report for the whole set. We will present the analysis results for the EUSES Spreadsheet Corpus in Sect. 7.

Supported Features. In the previous sections, we formalized the analysis of the core spreadsheet language, but our analyzer supports many additional spreadsheet features, to be able to cope with real-world applications, including the following:

- A *workbook* may contain several worksheets, and formulas may refer to cells in another sheet or another workbook.
- Macros may contain *interprocedural calls*, other user-defined subroutine or function, with or without arguments.
- *Number formats* are options that Excel provides for displaying values such as percentages, currencies, dates, which impact value types in some cases. Therefore, we also abstract this information (using zones as well) and take it into account while typing.

Circular References. The spreadsheet environment we have formalized does not feature circular references among cells, yet Microsoft Excel allows circular references under certain circumstances. In particular, a number of iterations can be set so that a circular computation could terminate. In this case, both the starting cell and the ending cell of the evaluation can be identified. Following this order, the analyzer iterates the abstract evaluation until it reaches a fixpoint.

Data Validation. Excel users may define constraints on data to be entered in some areas, such as "empty or only date", "empty or only time", "only text of a certain length", etc. Such information constrains data to be written in some areas at run-time; thus, this information can be used in the analysis. Therefore, our analyzer parses areas with data validation constraints and uses the type information they provide in the initial abstract state. This allows a precise verification of spreadsheets that utilize data unknown at verification time / non-deterministic data.

Over-approximation of Empty Input Cells. Spreadsheet formulas may refer to empty cells where values will be entered by users later. If "Data Validation" is not available for these cells, we can still derive their "expected" type from the function that is applied to them. For instance, function SUM expects **Numeric** arguments, function AND expects **Bool** arguments, etc. To account for this, the analyzer will either treat these cells as empty or store a value of that type. This over-approximation helps better verify formulas / macros using those cells.

7   Experiments and Analysis Results

We evaluated the efficiency of our tool, and focused on the three following questions: (1) Does the analysis find real defects in spreadsheets & macros ? (2) How long does the analysis take ? (3) Is the analysis report precise enough ? Is it easy enough for users to diagnose analysis warnings, and adopt the analysis ?

Experimental Setup. We chose the EUSES Spreadsheet Corpus [15] as an experimental subject for two reasons. First, to the best of our knowledge, it is the largest publicly available sample of real-world spreadsheets. Secondly, it includes many macros that offer good candidates for evaluating our associated program analysis. The sizes of the files of the corpus range from several KB to dozens of MB. In general, the spreadsheets are no longer under development and are already operational.

A spreadsheet may contain zero, one or several macros. It may also not contain any formula. The following table presents the classification of the EUSES Spreadsheet Corpus. Category D corresponds to pure data-sheets without any formulas or macros: they are not meaningful for our analysis, as their analysis is trivial. Therefore, our sample was the 2120 spreadsheets of Categories A + B + C and the 1053 macros inside them.

A	# spreadsheets with ≥ 1 formulas & 0 macro	1959
B	# spreadsheets with ≥ 1 formulas & ≥ 1 macros	111
C	# spreadsheets with 0 formula & ≥ 1 macros	50
D	# spreadsheets with 0 formula & 0 macro	2532

We performed the experiments as follows. First, our tool parsed all the spreadsheets and the macros, and detected 27 macros and 59 spreadsheet tables that have syntactic bugs or are incomplete (e.g., users put evident annotations such as "not-available" in their spreadsheet where an analysis would not be relevant). Next, we launched the analyzer on the rest of the items, and it was able to analyze **Eval** for 1854 spreadsheets and 858 macros (the reason why 7.8% of the spreadsheet tables and 16.4% of the macros were not analyzed is due to the fact our tool currently does not handle all Excel & VBA features and built-in functions, which are quite complex and numerous). Last, we filtered out the items whose bugs are not type-related (e.g., calls to undefined macros). Our tool detected 15 such spreadsheets and 21 such macros, which is useful but orthogonal to our purpose. The rest of the analyzed items are either type-related safe or erroneous, we classify them by Category **TypeRSE** in the following table, which summaries the analyses of **Eval** of spreadsheets in Categories A + B and macros in spreadsheets of Categories B + C. From now on we shall focus on Category **TypeRSE** and discuss the core of the analysis.

	Syntactically Erroneous or Incomplete	Total		
		Syntactically Correct		
		Non-Analyzed	Analyzed	
			Type-Unrelated Erroneous	**TypeRSE**
Eval	59	157	15	1839
Macro	27	168	21	837

Real Defects. The analyzer was set up in such a way that, when an unsafe typing rule is applied, it raises an alarm and stops the analysis. Therefore, the number of alarms raised corresponds to the number of spreadsheets / macros in the USES Corpus that were considered potentially erroneous by our analysis. In total, the analyzer raised 69 type-related alarms for **Eval** and 73 for macros. For each alarm, the report specified its location (e.g., the zone in spreadsheets and/or the line in macros), the unsafe typing rule (bug pattern) it encountered, and an estimate rating of how severe the defect would be. We manually inspected the spreadsheet / macro for which an alarm was raised, to diagnose its cause and the consequence of the revealed problem.

The alarms of type-unsafe operations effectively led us to identify *real defects* in programs, part of which defects silently produce wrong results. We show some of them as examples:

Example 1. In "homework\processed\Finalgradebook.xls", an application of Function AVERAGE to an Empty zone was detected, whereas all of its other arguments were Double. We found formula "=AVERAGE(D4;F4;H4;J4;L40)", was referring to "L40" although it was an empty cell. This was probably due to a user's erroneous typing of "L40" instead of "L4" (which was a Double and should have been an argument of AVERAGE), whereas Excel considered the formula valid. This mistake will indeed result in the computation of an incorrect average grade.

Example 2. In "modeling\processed\2-26.xls", subroutine "do_assign" uses a two-level loop to copy a table of basic parameters into another sheet where biological simulations are performed. Our tool detected that the whole zone of the table was Double except the first line, which was Empty. However, this zone had been assigned to a Double zone in another sheet. Upon investigation, we noticed a one-line shift between the source table and the target table, because the range of the loop was wrong. This will result in the target table being incorrectly filled in (its first line filled in with 0s, the copy result of empty cells), and the simulations (run 100 times!) based on these parameters will generate incorrect results.

Example 3. In "homework\processed\pl_student2002.xls", the analyzer detected an application of Function SUM to a String value, whereas all of its other arguments were Double. Examining the spreadsheet, we observed that the String value was actually "I", whereas the other arguments were either 0 or 1. Clearly users had mistaken "I" and "1", which are visibly similar. As a result, Excel considers "I" as 0 by SUM, which leads to a different number from that originally intended.

As shown by these examples, our tool discovered defects that would be hard for users to spot. In total, among all the alarms raised by the analyzer, we identified 25 real defects for **Eval** and 20 for macros, corresponding to serious and harmful issues in spreadsheet applications.

Among patterns contributing to spreadsheet defects, we can cite: (1) binary operation on Numeric data and Non-Numeric data (e.g., String) (2) Non-Numeric data (e.g., String, Empty) among the arguments of SUM or AVERAGE.

Furthermore, the defects found in the programs can be classified into several major categories: (1) Formulas or statements are applied to a wrong sheet area, and consequently take unexpected arguments. In Example 1, it is the reference of an argument of the formula that is incorrect; in Example 2, the area of the copied table is wrongly set. This kind of defect typically occurs due to an inappropriate manipulation (e.g., mistyping, improper copy-paste). In total, we found 13 bugs of this category. (2) Formulas or statements have a certain assumption for the types or the values of input data, yet the assumption is not specified or will not always hold at run-time. For instance, in a macro of "homework\processed\RT_EvaluationWorkbook.xls", an addition of a String value and an Empty cell is involved, and the analyzer realized that the Empty cell (representing reference of products) could well be set to a number at run-time, which would block the execution of the macro. We detected 8 defects of this class.

Moreover, we observe that many real defects were found thanks to the abstraction of the initial state of the spreadsheets, since this abstraction takes into account data that will be entered at run-time (Data-Validation areas, functions reading external values, etc.). It is, for instance, the case of the error in "RT_EvaluationWorkbook.xls", where the over-approximation of an empty cell covers numeric data at run-time, which is not the current value of the given spreadsheet. This kind of error would not be discovered by verification techniques that rely on a single spreadsheet state, like testing.

Analysis Time. The analyzer succeeded in verifying 858 macros in 161 spreadsheets (Categories B + C in Table 7). The size of each macro ranges from a few LOCs to several hundred LOCs. As one LOC could well involve a complex abstract operation by executing a complex statement or calling another macro, the size of a macro is just one of the factors that have an impact on its analysis time. We can list other important factors such as the complexity of the abstract state (e.g., # formula zones, # type zones, # variables) and the number of complex abstract operations (e.g., join, widening, reduction, eval). By summarizing all of the 858 successfully analyzed macros, we observe that the analysis for macros is fast enough: only 2% of them lasted more than 3 seconds (the longest analysis takes 10.45 seconds), and 88% of them took less than 0.2 second. We note that all analyses with fewer than 100 abstract zones and no loop of nesting depth greater than 2 lasted less than 1.75 seconds.

Figure 4(a) indicates the analysis time for **Eval** against the number of cells for non-constant values in initial spreadsheets. We remark that the analyses were performed in a reasonable time frame: 99% of the analyses took less than 1 second. Thus, the analysis time is acceptable in practice, and the analysis would integrate in a seamless manner in development.

Additionally, Figure 4(b) shows the analysis time of **Eval**, against the number of abstract formula zones for non-constant values in initial spreadsheets. By comparing it with Fig. 4(a), we remark that the principal attribute for analysis time is the number of abstract zones, rather than the number of cells. This observation is consistent with our abstraction mechanism, which is based on a cardinal power of zone abstractions. Going further, we remark that, on average, the number of zones we have made is 0.1x as many as the number of cells for a spreadsheet. The larger a spreadsheet is, the lower this ratio is: for certain large spreadsheets, this ratio can be less than 0.01. This guarantees that our analysis based on zones is scalable and especially efficient for large spreadsheets.

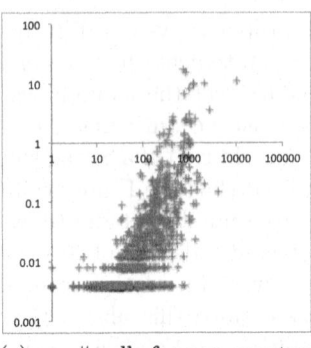

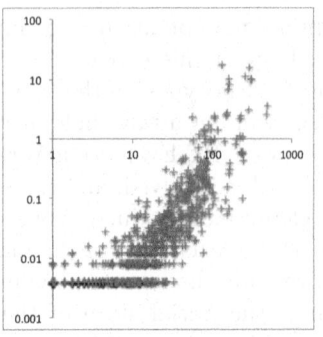

(a) vs. # cells for non-constant values

(b) vs. # abstract formula zones for non-constant values

Fig. 4. Analysis time for **Eval** in seconds

Precision and Diagnostics. With regard to our current typing system, the analyzer proved that global re-evaluations of 1770 spreadsheet tables in Categories A + B of Table 7 were correct and that 764 macros were correct.

When it raises an alarm, the analyzer issues a report including the context information (zone, macro line) and the category of the potential defect. In addition, Excel & Visual Studio provide an interactive debugging environment where the states of spreadsheets and program variables are highly visible. Thus, users can assess the alarm reports interactively with the help of this environment.

Besides the categories of real defects we presented previously, we can list several major categories of false alarms: (1) The first category is due to imprecisions in the analysis: the over-approximation causes the alarms corresponding to unsafe concrete states that will never be reached. We notice that the majority of the false alarms in this category come from the imprecisions in the analysis of certain VBA and Excel built-in functions. Few of the false alarms for **Eval** are due to the over-approximation of the initial state of the spreadsheet. This implies that a technique that relies on the given state of the spreadsheet would not reduce these false alarms. (2) The second category of false alarms is indeed related to type-unsafe operations, that are intended as such by the users. For example, sometimes users apply Function SUM to a column containing not only data, but also several titles. In this case, Function SUM will omit the titles and will thus still produce the correct result, summing the numeric data only. Yet, this pattern will result in false alarms due to Non-Numeric data among the arguments of SUM.

Therefore, diagnosing an alarm and triaging it as a false alarm or a real defect is fairly straightforward and typically takes a couple of minutes. We spent no more than 10 minutes on the most complex alarms. In total, we identified 44 false alarms for **Eval** and 53 for macros. The following table summarizes the core analyses.

	TypeRSE (Type-Related Safe or Erroneous)		
	alarm free	raise alarm	
		real defect	false alarm
Eval (1839)	1770	25	44
Macro (837)	764	20	53

Overall, the tool raised 142 alarms from 2676 analyses (**Eval** + macros), 45 of which alarms (i.e., approximately 30 %) were identified as real defects, which makes the false alarm number quite acceptable, considering that the defects found would be hard to spot by simple testing.

Summary. The experiments on the EUSES Corpus show that our analysis succeeds in detecting type-unsafe operations and can effectively be used to improve the quality of spreadsheets. It discovers defects that will cause unexpected results and that will not likely be found by testing. The diagnosis of alarms is not a tedious process with the guidance of the tool, and the false alarm number is reasonable. While the zone abstractions of a spreadsheet allows for the verification of type properties, it makes the analysis scalable for spreadsheets having a large number of cells. The analysis is efficient enough to be integrated within a development environment, as it could either be scheduled as a background task (e.g., scan systematically before saving), using reasonable resources, or launched upon user request in an interactive way.

8 Related Work

Unit Verification. The existing projects [3,9,4,5,6] resolve concrete units or dimensions with labels, headers and / or other annotations, build typing systems and reason about the correctness of formulas. We cannot find their experimental data or precision reports on comparable sets of benchmarks for a practical comparison with our results. Nevertheless, theoretically, our work is different from theirs in several ways: (1) we consider classical types in the programming language point of view, whereas their types refer to the concrete meaning of objects; thus, the built-in rules or bugs discovered by the two analyses are different; (2) we verify both the interface level and associated programs that the existing projects do not consider; (3) we evaluate formulas according to their order of precedence and thus support spreadsheets where data may be outdated; (4) our system covers a larger library of spreadsheet functions; (5) our classical types can always be retrieved from spreadsheets. By contrast, given a spreadsheet, the concrete meaning of objects are not always clear, and the retrieval of these meanings relies on annotation, though the analysis can be finer-grained if the retrieval is successful (e.g., they detect "adding apples and oranges", which our analysis does not regard as an error). Actually, combining our work with that of the existing projects would be a good direction for future work. By substituting other lattices with the type lattice and merging typing systems, we would be able to perform finer-grained analyses with various units and types.

Array Analysis and Zone Domain. Array analyses such as [23,12] also tie abstract properties to array regions; a notion of dependent types has been used to specify array properties such as array size [28]. One difference of our work is that we treat bi-dimensional arrays, whereas the existing works study uni-dimensional arrays.

Cheng and Rival [8] introduce an abstract domain to describe zones in two-dimensional arrays and apply it to analyze programs in a limited language, without formulas that can be re-evaluated after their inputs change. We aim at verifying real-world spreadsheets, which consist of associated programs *and* formulas. To this end, we formalize a larger spreadsheet language which includes formulas, and propose an abstraction that ties not only types but also abstract formulas to zones. Therefore, unlike [8], our analysis can cope with the re-evaluation of formulas (in automatic mode, upon user-request or from the associated programs), which is critical to handle real-world spreadsheets. Last, we evaluate the analysis and the implementation by analyzing a large set of real-world spreadsheets.

JavaScript, and Languages with Dynamic Evaluation. Thiemann [26] defines a type system that flags suspicious type conversions in JavaScript programs, which is a similar verification target to ours, albeit for a different language. Jensen *et al.* [18] address the `eval` function in JavaScript, which dynamically constructs code from text strings and executes it as if it were regular code in ways that obstruct existing static analyses. However, spreadsheet languages distinguish themselves from other scripting and dynamic languages by the way dynamicity is implemented: formulas are structured and organized in a two-dimensional array, whereas the `eval` function in JavaScript applies to strings and has a very different semantics. This led us to a very different abstraction based on zone and abstract formulas, than that of [18]. Moreover, Hammer *et al.* [16] propose a demand-driven incremental computation semantics of `eval` to provide speedups in spreadsheets, whereas our abstraction is based on the original concrete semantics of **Eval** in spreadsheets.

9   Conclusion

We have proposed a static analysis which is able to detect a significant class of subtle spreadsheet defects. It discovers inappropriate applications of operators and functions to arguments, which may produce unexpected results. To the best of our knowledge, our analysis is the first that can handle spreadsheet formulas, global re-evaluation and associated programs. Our evaluation on the EUSES Corpus has demonstrated that our analysis can effectively run on real-world spreadsheet applications and can verify a large number of them. It is able to discover defects that would be beyond the reach of both testing techniques and static analyses that would ignore the dynamic aspects of spreadsheets.

References

1. Report of JPMorgan Chase & Co. management task force regarding 2012 CIO losses (January 2013)

2. MS-VBAL: VBA language specification. Tech. rep., Microsoft Corporation (April 2014)
3. Abraham, R., Erwig, M.: Header and unit inference for spreadsheets through spatial analyses. In: Visual Languages and Human-Centric Computing. IEEE Computer Society (2004)
4. Abraham, R., Erwig, M.: UCheck: A spreadsheet type checker for end users. J. Vis. Lang. Comput. (2007)
5. Ahmad, Y., Antoniu, T., Goldwater, S., Krishnamurthi, S.: A type system for statically detecting spreadsheet errors. In: ASE (2003)
6. Antoniu, T., Steckler, P.A., Krishnamurthi, S., Neuwirth, E., Felleisen, M.: Validating the unit correctness of spreadsheet programs. In: International Conference on Software Engineering (2004)
7. Burnett, M., Atwood, J., Walpole Djang, R., Reichwein, J., Gottfried, H., Yang, S.: Forms/3: A first-order visual language to explore the boundaries of the spreadsheet paradigm (2001)
8. Cheng, T., Rival, X.: An abstract domain to infer types over zones in spreadsheets. In: Miné, A., Schmidt, D. (eds.) SAS 2012. LNCS, vol. 7460, pp. 94–110. Springer, Heidelberg (2012)
9. Coblenz, M.J., Ko, A.J., Myers, B.A.: Using objects of measurement to detect spreadsheet errors. In: Visual Languages and Human-Centric Computing (2005)
10. Cousot, P., Cousot, R.: Abstract interpretation: A unified lattice model for static analysis of programs by construction or approximation of fixpoints. In: Principles of Programming Languages. ACM (1977)
11. Cousot, P., Cousot, R.: Systematic design of program analysis frameworks. In: Principles of Programming Languages. ACM (1979)
12. Cousot, P., Cousot, R., Logozzo, F.: A parametric segmentation functor for fully automatic and scalable array content analysis. In: Principles of Programming Languages. ACM (2011)
13. Cunha, J., Fernandes, J.P., Ribeiro, H., Saraiva, J.: Towards a catalog of spreadsheet smells. In: Murgante, B., Gervasi, O., Misra, S., Nedjah, N., Rocha, A.M.A.C., Taniar, D., Apduhan, B.O. (eds.) ICCSA 2012, Part IV. LNCS, vol. 7336, pp. 202–216. Springer, Heidelberg (2012)
14. Cunha, J., Saraiva, J., Visser, J.: Model-based programming environments for spreadsheets. Science of Computer Programming (2014)
15. Fisher II, M., Rothermel, G.: The EUSES Spreadsheet Corpus: A shared resource for supporting experimentation with spreadsheet dependability mechanisms. In: Workshop on End-User Software Engineering (2005)
16. Hammer, M.A., Phang, K.Y., Hicks, M., Foster, J.S.: Adapton: Composable, demand-driven incremental computation. In: Programming Language Design and Implementation. ACM (2014)
17. Hermans, F., Pinzger, M., Deursen, A.V.: Detecting and visualizing inter-worksheet smells in spreadsheets. In: International Conference on Software Engineering (2012)
18. Jensen, S.H., Jonsson, P.A., Møller, A.: Remedying the eval that men do. In: International Symposium on Software Testing and Analysis. ACM (2012)
19. Jones, S.P., Blackwell, A., Burnett, M.: A user-centred approach to functions in Excel. In: International Conference on Functional Programming. ACM (2003)
20. Miné, A.: The octagon abstract domain. Higher-Order and Symbolic Computation (2006)
21. Panko, R.R.: What we know about spreadsheet errors. Journal of End User Computing (1998)

22. Rajalingham, K., Chadwick, D.R., Knight, B.: Classification of spreadsheet errors. In: EuSpRIG Symposium (2001)
23. Reps, T., Gopan, D., Sagiv, M.: A framework for numeric analysis of array operations. In: Principles of Programming Languages. ACM (2005)
24. Sestoft, P.: Online partial evaluation of sheet-defined functions. EPTCS (2013)
25. Sestoft, P.: Spreadsheet Implementation Technology. Basics and Extensions. MIT Press (2014)
26. Thiemann, P.: Towards a type system for analyzing javascript programs. In: Sagiv, M. (ed.) ESOP 2005. LNCS, vol. 3444, pp. 408–422. Springer, Heidelberg (2005)
27. Wakeling, D.: Spreadsheet functional programming. Journal of Functional Programming (2007)
28. Xi, H., Pfenning, F.: Eliminating array bound checking hrough dependent types. In: Programming Language Design and Implementation. ACM (1998)

Running Probabilistic Programs Backwards

Neil Toronto[1], Jay McCarthy[2], and David Van Horn[1]

[1] University of Maryland
{neil.toronto,jay.mccarthy}@gmail.com
[2] Vassar College
dvanhorn@cs.umd.edu

Abstract. Many probabilistic programming languages allow programs to be run under constraints in order to carry out Bayesian inference. Running programs under constraints *could* enable other uses such as rare event simulation and probabilistic verification—except that all such probabilistic languages are necessarily limited because they are defined or implemented in terms of an impoverished theory of probability. Measure-theoretic probability provides a more general foundation, but its generality makes finding computational content difficult.

We develop a measure-theoretic semantics for a first-order probabilistic language with recursion, which interprets programs as functions that compute preimages. Preimage functions are generally uncomputable, so we derive an abstract semantics. We implement the abstract semantics and use the implementation to carry out Bayesian inference, stochastic ray tracing (a rare event simulation), and probabilistic verification of floating-point error bounds.

Keywords: Probability, Semantics, Domain-Specific Languages.

1 Introduction

One key feature usually distinguishes a probabilistic programming language from general-purpose languages: finding the probabilistic conditions under which stated constraints are satisfied. Often, a probabilistic program simulates a real-world random process and the constraints represent observed, real-world outcomes. Running the program under the constraints *infers causes from effects*.

Inferring probabilistic causes from observed outcomes is called **Bayesian inference**, a technique used widely in artificial intelligence. It has been successful in analyzing phenomena at all scales, from genomes to celestial bodies. Automating it is one of the primary drivers of probabilistic language development.

One of the simplest probabilistic programs that allows us to demonstrate Bayesian inference simulates the following process of flipping two coins.

1. Flip a fair coin; call the outcome x.
2. If x is heads, flip another fair coin. If x is tails, flip an unfair coin with heads probability 0.3 (tails probability 0.7). In either case, call the outcome y.

© Springer-Verlag Berlin Heidelberg 2015
J. Vitek (Ed.): ESOP 2015, LNCS 9032, pp. 53–79, 2015.
DOI: 10.1007/978-3-662-46669-8_3

The following probabilistic program simulates this process.

$$
\begin{aligned}
\text{let } \ x &:= \text{flip } 0.5 \\
y &:= \text{flip (if } x = \text{heads then } 0.5 \text{ else } 0.3) \\
\text{in } \ &\langle x, y \rangle
\end{aligned}
\tag{1}
$$

Here, flip q returns heads with probability q and tails with probability $1 - q$.

The meaning of (1) is not the returned random value, but a **probability distribution** that describes the likelihoods of all possible returned random values. For discrete processes, this distribution can always be defined by a **probability mass function**: a mapping from possible values to their probabilities. These probabilities are computed by multiplying the probabilities of intermediate random values. For example, the probability of $\langle$heads, heads$\rangle$ is $0.5 \cdot 0.5 = 0.25$, and the probability of $\langle$tails, heads$\rangle$ (i.e. the second flip is unfair) is $0.5 \cdot 0.3 = 0.15$. The meaning of (1) is thus the probability mass function

$$
\begin{aligned}
\text{p} \ := \ \big[&\langle\text{heads, heads}\rangle \mapsto 0.25, \langle\text{heads, tails}\rangle \mapsto 0.25, \\
&\langle\text{tails, heads}\rangle \mapsto 0.15, \langle\text{tails, tails}\rangle \mapsto 0.35 \big]
\end{aligned}
\tag{2}
$$

Using p, we can answer any question about the process under constraints. For example, if we do not know x, but constrain y to be heads, what is the probability that x is also heads? We compute the answer by dividing the probability of the outcome we are interested in (i.e. $\langle x, y \rangle = \langle$heads, heads$\rangle$) by the total probability of outcomes in the constraint's corresponding subdomain $\{$heads, tails$\} \times \{$heads$\}$:

$$
\frac{\text{p} \ \langle\text{heads, heads}\rangle}{\sum_{z \in \{\text{heads,tails}\} \times \{\text{heads}\}} \text{p } z} = \frac{0.25}{0.25 + 0.15} = 0.625
\tag{3}
$$

Qualitatively, y being heads is a bit unusual if the second coin is unfair. Therefore, we infer that the second coin is most probably fair; i.e. x is most likely heads.

The time complexity of computing p is generally exponential in the number of random choices, which is intractable for all but the simplest processes. One popular way to avoid this exponential explosion is to use advanced Monte Carlo algorithms to sample according to p on the constraint's corresponding subdomain without explicitly enumerating that subdomain. The number of samples required is typically quadratic in the answer's desired accuracy [7, Sec. 12.2].

Probabilistic languages that are implemented using advanced Monte Carlo algorithms could be used not just for Bayesian inference, but for simulating **rare events** (i.e. very low-probability events) by encoding the events as constraints.

Stochastic ray tracing [30] is one such rare-event simulation task. As illustrated in Fig. 1, to carry out stochastic ray tracing, a probabilistic program simulates a light source emitting a single photon in a random direction, which is reflected or absorbed when it hits a wall. The program outputs the photon's path, which is constrained to pass through an aperture. Millions of paths that meet the constraint are sampled, then projected onto a simulated sensor array.

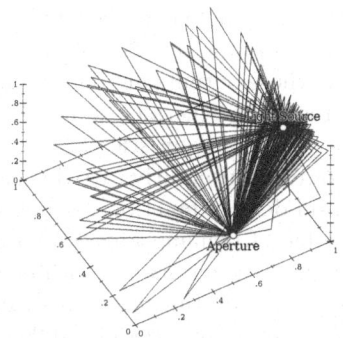

(a) Simulated photons from a single source, constrained to pass through an aperture.

(b) Simulated photons constrained to pass through the aperture, projected onto a plane and accumulated.

Fig. 1. Ray tracing by constraining the outputs of a probabilistic program

The program's main loop is a recursive function with two arguments: path, the photon's path so far as a list of points, and dir, the photon's current direction.

$$
\begin{aligned}
&\text{simulate-photon path dir} := &&(4)\\
&\quad \text{case (find-hit (fst path) dir) of}\\
&\quad\quad \text{absorb pt} &&\longrightarrow\ \langle \text{pt, path}\rangle\\
&\quad\quad \text{reflect pt norm} &&\longrightarrow\ \text{simulate-photon } \langle \text{pt, path}\rangle\ (\text{random-half-dir norm})
\end{aligned}
$$

Here, find-hit (fst path) dir finds the surface the photon hits. If the photon is absorbed, find-hit returns a data structure containing just the collision point pt. Otherwise, find-hit returns a data structure containing the collision point pt and surface normal norm, which random-half-dir uses to choose a new direction. Running simulate-photon ⟨pt, ⟨⟩⟩ dir, where pt is the light source's location and dir is a random emission direction, generates a photon path. The fst of the path (the last collision point) is constrained to be in the aperture. The remainder of the program is simple vector math that computes ray-plane intersections.

In contrast, hand-coded stochastic ray tracers, written in general-purpose languages, are much more complex and divorced from the physical processes they simulate, because they must interleave the advanced Monte Carlo algorithms that ensure the aperture constraint is met.

Unfortunately, while many probabilistic programming languages support random real numbers, none are capable of running a probabilistic program like (4) under constraints to carry out stochastic ray tracing. The reason is not lack of engineering or weak algorithms, but is theoretical at its core: they are all either defined or implemented using a naive theory of probability.

While probability mass functions cannot define distributions on $\mathbb{R}$ that give positive probability to uncountably many values, there is a near-universal substitute that can: probability *density* functions. Density functions map single values to probability-like quantities, which makes them intuitively appealing and apparently simple. Unfortunately, density functions are not general enough to be used as probabilistic program meanings without imposing severe limitations on probabilistic languages. In particular, programs whose outputs are deterministic functions of random values and programs with recursion generally cannot denote density functions. The program in (4) exhibits both characteristics.

Measure-theoretic probability is a more powerful alternative to this naive probability theory based on probability mass and density functions. It not only subsumes naive probability theory, but is capable of defining any computable probability distribution, and many uncomputable distributions. But while even the earliest work [15] on probabilistic languages is measure-theoretic, the theory's generality has historically made finding useful computational content difficult.

We show that measure-theoretic probability can be made computational by

1. Using measure-theoretic probability to define a compositional, denotational semantics that gives a valid denotation to every program.
2. Deriving an abstract semantics, which allows computing answers to questions about probabilistic programs to arbitrary accuracy.
3. Implementing the abstract semantics and efficiently solving problems.

In fact, our primary implementation, *Dr. Bayes*, produced Fig. 1b by running a probabilistic program like (4) under an aperture constraint.

The rest of this paper is organized as follows.

– Section 2 demonstrates why density functions are insufficient for interpreting probabilistic programs. It shows how measure-theoretic probability defines probability distributions using set-valued inverses, or *preimage functions*.
– Section 3 presents the categorical tools we use to derive many semantics from a single standard semantics in a way that makes them easy to prove correct.
– Section 4 defines the semantics of nonrecursive, nonprobabilistic programs, which interprets programs as preimage functions.
– Section 5 lifts this semantics to recursive, probabilistic programs.
– Section 6 derives a sound, implementable abstract semantics.
– Section 7 describes our implementations and gives examples, including probabilistic verification of floating-point error bounds.

In short, we show why and how to run probabilistic programs under constraints by computing preimage functions—that is, by running programs backwards.

2   Background

2.1   Probability Density Functions

Some distributions of real values can be defined by **probability density functions**: integrable functions $p : \mathbb{R}^n \to [0, \infty)$ that integrate to 1.

The simplest nontrivial probabilistic program is random, which returns a uniformly random value in the interval $[0, 1]$. The meaning of random is a probability distribution that can be defined by the density

$$p : \mathbb{R} \to [0, \infty) \qquad p\, x := \begin{cases} 1 & \text{if } x \in [0, 1] \\ 0 & \text{otherwise} \end{cases} \tag{5}$$

Though $p\, x$ for any x indicates x's relative frequency, $p\, x$ is not a probability. Probabilities are obtained by integration. For example, the probability that random returns a value in $[0, 0.5]$ is

$$\int_0^{0.5} (p\, x)\, dx \;=\; \int_0^{0.5} 1\; dx \;=\; \Big[x\Big]_0^{0.5} \;=\; 0.5 - 0 \;=\; 0.5 \tag{6}$$

Similarly, the probability of $[0.5, 0.5]$ or any other singleton set is zero. In fact, *every* probability density function integrates to zero on singleton sets.

This fact makes it trivial to write a probabilistic program whose distribution cannot be defined by a density. For example, consider max $\langle 0.5, \mathsf{random} \rangle$, where max $\langle a, b \rangle$ returns the greater of the pair $\langle a, b \rangle$. This program evaluates to 0.5 whenever random returns a number in $[0, 0.5]$. In other words, the value of max $\langle 0.5, \mathsf{random} \rangle$ is in $[0.5, 0.5]$ with probability 0.5. But if its distribution is defined by a density, then $[0.5, 0.5]$ must have probability zero—not 0.5.

A probabilistic language without the max function can still be useful. It is fairly easy to compute densities for the outputs of single-argument functions that happen to have differentiable inverses, such as exponentiation and square root. But two-argument functions such as addition and multiplication require evaluating integrals, which generally do not have closed-form solutions.

Perhaps the most constricting limitation of probability density functions is that the number of dimensions must be finite and fixed. This limitation rules out recursive data types, and makes recursion so difficult that few probabilistic languages attempt to allow it.

2.2 Measures, and Measures of Preimages

Measure-theoretic probability gains its expressive power by mapping sets directly to probabilities. Functions that do so are called **probability measures**. For example, the distribution of random is defined by the probability measure

$$P : \mathcal{P}\, [0, 1] \rightharpoonup [0, 1] \qquad P\, [a, b] \;=\; b - a \tag{7}$$

where $\mathcal{P}\, [0, 1]$ is the powerset of $[0, 1]$ and '$\rightharpoonup$' denotes a partial mapping. Though (7) apparently defines P only on intervals, it is regarded as defining P additionally on countable unions of intervals, their complements, countable unions of such, and so on. The resulting domain includes almost every subset of $[0, 1]$ that can be written down.

Probability measures can be defined on any domain, including domains with variable and infinite dimension. They can also map singleton sets to nonzero probabilities, which we will demonstrate shortly by deriving a probability measure for max $\langle 0.5, \mathsf{random} \rangle$.

Measure-theoretic probability takes great pains to separate random effects from the pure logic of mathematics. It does so in the same way Haskell and other purely functional programming languages allow random effects: by interpreting probabilistic processes as *deterministic functions* that operate on an assumed-random source. The probabilities of sets of outputs are uniquely determined by the probabilities of the corresponding sets of inputs.

Suppose we interpret max $\langle 0.5, \text{random} \rangle$ as the deterministic function

$$f := \lambda r \in [0,1].\, \text{max}\, \langle 0.5, r \rangle \tag{8}$$

and assume that r is its uniform random source; i.e. that its distribution is P as defined in (7). To compute the probability that max $\langle 0.5, \text{random} \rangle$ evaluates to 0.5, we apply P to the set of all r for which $f\, r \in [0.5, 0.5]$, and get, as expected,

$$P\,\{r \in [0,1] \mid f\, r \in [0.5, 0.5]\} \;=\; P\,[0, 0.5] \;=\; 0.5 - 0 \;=\; 0.5 \tag{9}$$

For any f and B, the set $\{a \in \text{domain } f \mid f\, a \in B\}$ is called the **preimage of B under f**. Functions that compute preimages are often denoted f^{-1} to emphasize that they are a sort of generalized inverse function. However, we find this notation confusing: inverse functions operate on *values* and may not be well-defined, whereas preimage functions operate on *sets* and are *always* well-defined.[1] Thus, we denote f's preimage function by preimage f. The probability that f outputs a value in B is therefore P ((preimage f) B), or P (preimage f B).

Though the distribution of max $\langle 0.5, \text{random} \rangle$, or the output of f, has no probability density function, its probability measure is defined by

$$P_f : \mathcal{P}\,[0.5, 1] \rightharpoonup [0, 1] \qquad P_f\,[a, b] \;=\; P\,(\text{preimage } f\,[a, b]) \tag{10}$$

An equivalent, more elegant definition is

$$P_f := P \circ (\text{preimage } f) \tag{11}$$

which clearly shows that P_f is factored into a part P that quantifies randomness, and a deterministic part preimage f that *runs f backwards on sets of outputs*.

This factorization confers the flexibility to interpret probabilistic programs by choosing any P and f for which P ∘ (preimage f) is the correct measure. For P, we choose uniform measures on cartesian products of $[0, 1]$ (e.g. $[0,1]^{\mathbb{N}}$) and interpret each random as a projection. Thus, for the remainder of this paper, we can concentrate solely on computing preimage f.

Because preimage f is deterministic, techniques to compute it have applications outside of probabilistic programming; for example, constraint-functional languages, type inference, and verification. More immediately, its determinism means that, for the bulk of this paper, *readers do not need to know anything about probability, let alone measure theory*—only basic set theory.

---

[1] If $f^{-1}\, b$ is undefined, then the preimage of $\{b\}$ under f is simply $\varnothing$.

2.3   Preimage Semantics

Several well-known identities suggest that preimages can be computed compositionally, which would make it possible to define a denotational semantics that interprets programs as preimage functions. For example, we have

$$
\begin{aligned}
\text{preimage id} &= \text{id} \\
\text{preimage } (f_2 \circ f_1) &= (\text{preimage } f_1) \circ (\text{preimage } f_2) \\
\text{preimage } \langle f_1, f_2 \rangle \, (B_1 \times B_2) &= (\text{preimage } f_1 \, B_1) \cap (\text{preimage } f_2 \, B_2)
\end{aligned}
\tag{12}
$$

where $\langle f_1, f_2 \rangle = \lambda a \in (\text{domain } f_1) \cap (\text{domain } f_2)$. $\langle f_1 \, a, f_2 \, a \rangle$ constructs pairing functions and id is the identity function.

It might seem we can easily use identities like those in (12) directly to define a semantic function $[\![\cdot]\!]_{\text{pre}}$ that interprets programs as preimage functions. Unfortunately, our task is not that simple, for the following reasons.

1. The preimage function requires its argument to have an observable domain. This includes **extensional** functions, which are sets of intput/output pairs (i.e. possibly infinite hash tables), but not **intensional** functions, which are syntactic rules for computing outputs from inputs (e.g. lambdas).[2]
2. We must ensure preimage f B is always in the domain of the chosen probability measure P. (Recall that probability measures are partial functions.) If this is true, we say f is **measurable**. Proving measurability is difficult, especially if f may not terminate.
3. The function app : $(X \to Y) \times X \to Y$, when restricted to measurable functions, is not generally measurable if we want good approximation properties [2]. This makes interpreting higher-order application difficult.

Implementing a language based on preimage semantics is complicated because

4. Ordinary set-based mathematics is unlike any implementation language.
5. It requires running programs written in a Turing-equivalent language backwards, efficiently, on possibly uncountable sets of outputs.

We address 1 and 4 by developing our semantics using λ_{ZFC} [29], an untyped, call-by-value λ-calculus with infinite sets, real numbers, extensional functions such as $\lambda r \in [0,1]. \max \langle 0.5, r \rangle$, intensional functions such as $\lambda r. \max \langle 0.5, r \rangle$, a computable sublanguage, and an operational semantics. It is essentially ordinary mathematics extended with lambdas and general recursion, or equivalently a lambda calculus extended with uncountably infinite sets and set operations.

We have addressed difficulty 2 by proving that all programs' interpretations as functions are measurable if language primitives are measurable, including uncomputable primitives such as limits and real equality, regardless of nontermination. The proof interprets programs as extensional functions and applies well-known theorems from measure theory such as the identities in (12). Unfortunately, the required machinery does not fit in this paper; see the first author's dissertation [28] for the entire development.

---

[2] The lambda $\lambda r. \max \langle 0.5, r \rangle$ is intensional, but $\lambda r \in [0,1]. \max \langle 0.5, r \rangle$ constructs an extensional function by pairing every $r \in [0,1]$ with its corresponding $\max \langle 0.5, r \rangle$.

We avoid difficulty 3 for now by interpreting a language with *first-order* functions and recursion. We address 5 by deriving and implementing a *conservative approximation* of the preimage semantics, and using its approximations to compute measures of preimages with arbitrary accuracy.

2.4   Abstract Interpretation, Categorically

We interpret nonrecursive, nonprobabilistic programs three different ways, using

1. A **standard semantics** $[\![\cdot]\!]_\perp$ that interprets programs that may raise errors (e.g. divide-by-zero) as intensional functions.
2. A **concrete semantics** $[\![\cdot]\!]_{\text{pre}}$ that interprets programs as preimage functions, which operate on uncountable sets, and are thus unimplementable.
3. An **abstract semantics** $[\![\cdot]\!]_{\widehat{\text{pre}}}$ that interprets programs as *abstract* preimage functions, which operate only on overapproximating, finite representations of uncountable sets, and thus *are* implementable.

Of course, we must prove for any program p, that $[\![p]\!]_{\text{pre}}$ correctly computes preimages under $[\![p]\!]_\perp$, and that $[\![p]\!]_{\widehat{\text{pre}}}$ is sound with respect to $[\![p]\!]_{\text{pre}}$.

For recursive, probabilistic programs, we define three more semantic functions analogous to $[\![\cdot]\!]_\perp$, $[\![\cdot]\!]_{\text{pre}}$ and $[\![\cdot]\!]_{\widehat{\text{pre}}}$, that have analogous proof obligations. We also prove that they correctly interpret nonrecursive, nonprobabilistic programs.

In the full development [28], two more semantic functions interpret programs as extensional functions, which are used to prove measurability. Another semantic function collects information needed for advanced Monte Carlo algorithms. In all, we have 9 related semantic functions, each defined by 11 or 12 rules, whose correctness and relationships must be proved by structural induction. Doing so is tedious and error-prone. We need a way to parameterize one semantic function on many meanings, where each "meaning" is simpler than a semantic function and ideally has exploitable properties.

Moggi [22] introduced monads as a categorical "metalanguage" for interpreting programs. Wadler [31] showed how to use monad categories in pure functional programming to encode and hide side effects such as mutation and randomness. Haskell programmers now primarily encode programs with side effects using **do-notation**, which is transformed into any monad. Essentially, Haskell has a built-in semantic function parameterized on a monad.

Other researchers have identified arrows [10] and idioms [19] as useful kinds of categories. Different kinds of categories are good for encoding different kinds of effects, and have different levels of expressiveness [16]. Arrows are good categories for interpreting first-order languages. We therefore interpret programs 9 different ways by parameterizing a semantic function on one of 9 arrow categories.

In our formulation, an arrow category consists of a type constructor and five combinators; each is thus half as complicated as the semantic function. Their categorical properties also allow two drastic simplifications. First, they allow proving the correctness of a semantic function $[\![\cdot]\!]_b$ with respect to $[\![\cdot]\!]_a$ by proving a simple theorem about arrows a and b. Second, they allow us to *derive* all the arrows for recursive, probabilistic programs at once, by lifting the arrows for nonrecursive, nonprobabilistic programs.

2.5   Types and Notation

Because some arrows carry out uncountably infinite computations, we must define their combinators in a sufficiently powerful λ-calculus. We use λ_{ZFC} [29].

Though λ_{ZFC} is untyped, it helps to use a manually checked, auxiliary type system. For example, the types of some of λ_{ZFC}'s primitives are those of membership $(\in) : x \rightarrow \mathsf{Set}\ x \rightarrow \mathsf{Bool}$, powerset $\mathcal{P} : \mathsf{Set}\ x \rightarrow \mathsf{Set}\ (\mathsf{Set}\ x)$, big union $\bigcup : \mathsf{Set}\ (\mathsf{Set}\ x) \rightarrow \mathsf{Set}\ x$, and the map-like image $: (x \rightarrow y) \rightarrow \mathsf{Set}\ x \rightarrow \mathsf{Set}\ y$. We allow sets to be used as types, as in $\max : \langle \mathbb{R}, \mathbb{R} \rangle \rightarrow \mathbb{R}$.

More precisely, types are characterized by these rules:

- $x \rightarrow y$ is the type of intensional, partial functions from type x to type y.
- $\langle x, y \rangle$ is the type of pairs of values with types x and y.
- $\mathsf{Set}\ x$ is the type of sets whose members have type x.
- An uppercase type variable such as X represents a set used as a type.

Because the inhabitants of the type $\mathsf{Set}\ X$ and $\mathcal{P}\ X$ (i.e. subsets of the set X) are the same, they are equivalent types. Similarly, $\langle X, Y \rangle$ is equivalent to $X \times Y$.

Type constructors are defined using ‘::=’; e.g. $X \leadsto_\perp Y ::= X \rightarrow (Y \cup \{\perp\})$.

The set X^J contains all extensional, total functions from set J to set X; i.e. vectors of X indexed by J. We use adjacency (i.e. f a) to apply both intensional and extensional functions. For example, the first element of $f : [0,1]^{\mathbb{N}}$ is f 0.

Proofs, which we elide to save space, are in the first author's dissertation [28].

3   Arrows and First-Order Semantics

This section presents the categorical tools we use to derive many semantics from a single standard semantics in a way that makes them easy to prove correct.

Arrows [10], like monads [31], thread effects through computations in a way that imposes structure. But arrow computations are always

- Function-like. The type constructor for arrow a is written $x \leadsto_a y$ to connote this. In fact, the *function arrow*'s type constructor is $x \leadsto y ::= x \rightarrow y$.
- First-order. There is no way to derive the higher-order application combinator $\mathsf{app} : \langle x \leadsto_a y, x \rangle \leadsto_a y$ from the combinators that define arrow a.

The first property makes arrows a good fit for a compositional translation from expressions to pure functions that operate on random sources. The second property makes arrows a good fit for the semantics of a first-order language.

3.1   Arrow Combinators and Laws

Arrows factor computation into the following tasks: (1) referring to pure, primitive functions, (2) applying primitive or first-order functions, (3) binding values to local variables and creating data structures, and (4) branching based on the results of prior computations. The first four arrow combinators correspond respectively with each of these tasks. A fifth combinator allows lazy branching in a call-by-value language such as λ_{ZFC}.

For laziness, we need a singleton type for thunks. We use the set $1 := \{0\}$.

Definition 1 (Arrow[3]). *A binary type constructor* $(\leadsto_a)$ *and the combinators*

$$\text{arr}_a : (x \to y) \to (x \leadsto_a y) \hspace{4cm} \text{lift}$$
$$(\ggg_a) : (x \leadsto_a y) \to (y \leadsto_a z) \to (x \leadsto_a z) \hspace{2cm} \text{compose}$$
$$(\&\&\&_a) : (x \leadsto_a y) \to (x \leadsto_a z) \to (x \leadsto_a \langle y, z \rangle) \hspace{2cm} \text{pair}$$
$$\text{ifte}_a : (x \leadsto_a \text{Bool}) \to (x \leadsto_a y) \to (x \leadsto_a y) \to (x \leadsto_a y) \hspace{1cm} \text{if-then-else}$$
$$\text{lazy}_a : (1 \to (x \leadsto_a y)) \to (x \leadsto_a y) \hspace{3cm} \text{laziness}$$

define an **arrow** *if certain monoid, homomorphism, and other laws hold [10].*

For example, the **function arrow** is defined by the type constructor $x \leadsto y ::= x \to y$ and the combinators

$$
\begin{aligned}
\text{arr } f &:= f \\
(f_1 \ggg f_2) \, a &:= f_2 \, (f_1 \, a) \\
(f_1 \,\&\&\&\, f_2) \, a &:= \langle f_1 \, a, f_2 \, a \rangle \hspace{3cm} (13) \\
\text{ifte } f_1 \, f_2 \, f_3 \, a &:= \text{if } f_1 \, a \text{ then } f_2 \, a \text{ else } f_3 \, a \\
\text{lazy } f \, a &:= f \, 0 \, a
\end{aligned}
$$

To demonstrate compositionally interpreting probabilistic programs as arrow computations, we interpret max $\langle 0.5, \text{random} \rangle$ as a function arrow computation $f : [0, 1] \leadsto \mathbb{R}$. For any random source $r \in [0, 1]$, the interpretation of 0.5 should return 0.5, so 0.5 means $\lambda r. \, 0.5$, or const 0.5 where const $v := \lambda\_. \, v$. Assuming $r \in [0, 1]$ is uniformly distributed, random means $\lambda r. \, r$, or id. We use $(\&\&\&)$ to apply each of these interpretations to the random source to create a pair, and $(\ggg)$ to send the pair to max. Thus, max $\langle 0.5, \text{random} \rangle$, interpreted as a function arrow computation, is $f := ((\text{const } 0.5) \,\&\&\&\, \text{id}) \ggg \text{max}$.

By substituting the definitions of const, id, $(\&\&\&)$ and $(\ggg)$, we would find that f is equivalent to $\lambda r. \, \text{max} \, \langle 0.5, r \rangle$, similar to the interpretation in (8).

Only the function arrow can so cavalierly use pure functions as arrow computations. In any other arrow a, pure functions must be *lifted* using arr_a, to allow the arrow to manage any state or effects. Therefore, the interpretation of max $\langle 0.5, \text{random} \rangle$ as an arrow a computation $f_a : [0, 1] \leadsto_a \mathbb{R}$ is

$$f_a := (\text{arr}_a \, (\text{const } 0.5) \,\&\&\&_a\, \text{arr}_a \, \text{id}) \ggg_a \text{arr}_a \, \text{max} \hspace{2cm} (14)$$

So far, we have ignored the many arrow laws, which ensure that arrows are well-behaved (e.g. effects are correctly ordered) and are useful in proofs of theorems that quantify over arrows (i.e. nothing else is known about them). Fortunately, we can prove all the laws for an arrow b by defining it in terms of an arrow a for which the laws hold, and proving two properties about the lift from a to b. The first property is that the lift from a to b is distributive.

---

[3] These are actually arrows *with choice*, which are typically defined using first$_a$ and left$_a$ instead of $(\&\&\&_a)$ and ifte$_a$. We find ifte$_a$ more natural for semantics than left$_a$, and $(\&\&\&_a)$ better matches the pairing preimage identity in (12).

$$p ::\equiv f := e; \dots ; e$$
$$e ::\equiv \text{let } e \; e \mid \text{env } n \mid \text{if } e \text{ then } e \text{ else } e \mid \langle e, e \rangle \mid f \; e \mid \delta \; e \mid v$$
$$f ::\equiv \text{(first-order function names)}$$
$$\delta ::\equiv \text{(primitive function names)}$$
$$v ::\equiv \langle v, v \rangle \mid \langle \rangle \mid \text{true} \mid \text{false} \mid \text{(other first-order constants)}$$

$$[\![f := e; \dots ; e_b]\!]_a \; :\equiv \; f := [\![e]\!]_a; \dots ; [\![e_b]\!]_a \qquad\qquad [\![\langle e_1, e_2 \rangle]\!]_a \; :\equiv \; [\![e_1]\!]_a \; \&\&\&_a \; [\![e_2]\!]_a$$

$$[\![\text{let } e \; e_b]\!]_a \; :\equiv \; ([\![e]\!]_a \; \&\&\&_a \; \text{arr}_a \; \text{id}) \ggg_a [\![e_b]\!]_a \qquad\qquad [\![f \; e]\!]_a \; :\equiv \; [\![\langle e, \langle \rangle \rangle]\!]_a \ggg_a f$$

$$[\![\text{env } 0]\!]_a \; :\equiv \; \text{arr}_a \; \text{fst} \qquad\qquad [\![\delta \; e]\!]_a \; :\equiv \; [\![e]\!]_a \ggg_a \text{arr}_a \; \delta$$

$$[\![\text{env } (n+1)]\!]_a \; :\equiv \; \text{arr}_a \; \text{snd} \ggg_a [\![\text{env } n]\!]_a \qquad\qquad [\![v]\!]_a \; :\equiv \; \text{arr}_a \; (\text{const } v)$$

$$[\![\text{if } e_c \text{ then } e_t \text{ else } e_f]\!]_a \; :\equiv \; \text{ifte}_a \; [\![e_c]\!]_a \; (\text{lazy}_a \; \lambda 0. [\![e_t]\!]_a) \; (\text{lazy}_a \; \lambda 0. [\![e_f]\!]_a)$$

$$\text{where} \quad \text{const } v := \lambda \_. v \qquad\qquad \text{subject to} \quad [\![p]\!]_a : \langle \rangle \leadsto_a y \text{ for some } y$$
$$\text{id} := \lambda v. v$$

Fig. 2. Interpretation of a let-calculus with first-order definitions and De-Bruijn-indexed bindings as arrow a computations. Here, '$::\equiv$' denotes definitional extension for grammars and '$:\equiv$' denotes definitional extension for syntax.

Definition 2 (Arrow Homomorphism). $\text{lift}_b : (x \leadsto_a y) \to (x \leadsto_b y)$ *is an* **arrow homomorphism** *from* a *to* b *if these distributive laws hold:*

$$\text{lift}_b \; (\text{arr}_a \; f) \equiv \text{arr}_b \; f \tag{15}$$

$$\text{lift}_b \; (f_1 \ggg_a f_2) \equiv (\text{lift}_b \; f_1) \ggg_b (\text{lift}_b \; f_2) \tag{16}$$

$$\text{lift}_b \; (f_1 \; \&\&\&_a \; f_2) \equiv (\text{lift}_b \; f_1) \; \&\&\&_b \; (\text{lift}_b \; f_2) \tag{17}$$

$$\text{lift}_b \; (\text{ifte}_a \; f_1 \; f_2 \; f_3) \equiv \text{ifte}_b \; (\text{lift}_b \; f_1) \; (\text{lift}_b \; f_2) \; (\text{lift}_b \; f_3) \tag{18}$$

$$\text{lift}_b \; (\text{lazy}_a \; f) \equiv \text{lazy}_b \; \lambda 0. \text{lift}_b \; (f \; 0) \tag{19}$$

where "$\equiv$" is an arrow-specific equivalence relation.

The second property is that the lift is right-invertible (i.e. surjective).

Theorem 1 (Right-invertible Homomorphism Implies Arrow Laws). *If* $\text{lift}_b : (x \leadsto_a y) \to (x \leadsto_b y)$ *is a right-invertible homomorphism from* a *to* b *and the arrow laws hold for* a, *then the arrow laws hold for* b.

3.2 First-Order Let-Calculus Semantics

Figure 2 defines a semantic function $[\![\cdot]\!]_a$ that interprets first-order programs as arrow computations for any arrow a. A program is a sequence of function definitions separated by semicolons (or line breaks), followed by a final expression. Function definitions may be mutually recursive because they are interpreted as definitions in a metalanguage in which mutual recursion is supported. (We thus do not need an explicit fixpoint operator.) Unlike functions, local variables are unnamed: we use De Bruijn indexes, with 0 referring to the innermost binding.

The result of applying $[\![\cdot]\!]_a$ is a λ_{ZFC} program in **environment-passing style** where the environment is a stack. The final expression has type $\langle \rangle \leadsto_a y$, where y is the type of the program's output and $\langle\rangle$ denotes the empty stack. A let expression uses pairing ($\&\&\&_a$) to push a value onto the stack and composition ($\ggg_a$) to pass the resulting stack to its body. First-order functions have type $\langle x, \langle\rangle\rangle \rightsquigarrow_a y$ where x is the argument type and y is the return type. Application passes a stack containing just an x using pairing and composition.

Using De Bruijn indexes, g x := g x is written g := g (env 0), which $\llbracket \cdot \rrbracket_a$ interprets as g := $\llbracket \langle$env 0, $\langle\rangle\rangle \rrbracket_a \ggg_a$ g. To disallow such circular definitions, and ill-typed expressions like max $\langle 0.5, \langle\rangle\rangle$, we require programs to be **well-defined**.

Definition 3 (Well-defined). *An expression (or program) e is **well-defined** under arrow a if $\llbracket e \rrbracket_a$ terminates and $\llbracket e \rrbracket_a : x \rightsquigarrow_a y$ for some x and y.*

Well-definedness guarantees that recursion is guarded by if expressions, as $\llbracket$if e_c then e_t else $e_f \rrbracket_a$ wraps $\llbracket e_t \rrbracket_a$ and $\llbracket e_f \rrbracket_a$ in thunks. It does *not* guarantee that *running* an interpretation always terminates. For example, the program g := if true then g (env 0) else 0; g 0 is well-defined under the function arrow, but applying its interpretation to $\langle\rangle$ does not terminate. Section 5 deals with such programs by defining arrows that take finitely many branches, or return $\bot$.

Most of our semantic correctness results rely on the following theorem.

Theorem 2 (Homomorphisms Distribute Over Expressions). *Let* lift_b : $(x \rightsquigarrow_a y) \rightarrow (x \rightsquigarrow_b y)$ *be an arrow homomorphism. For all e,* $\llbracket e \rrbracket_b \equiv \text{lift}_b \ \llbracket e \rrbracket_a$.

Much of our development proceeds in the following way.

1. Define an arrow a to interpret programs using $\llbracket \cdot \rrbracket_a$.
2. Define $\text{lift}_b : (x \rightsquigarrow_a y) \rightarrow (x \rightsquigarrow_b y)$ from arrow a to b with the property that if $f : x \rightsquigarrow_a y$, then lift_b f is correct.
3. Prove lift_b is a homomorphism; therefore $\llbracket e \rrbracket_b$ is correct (Theorem 2).
4. Prove lift_b is right-invertible; therefore b obeys the arrow laws (Theorem 1).

In shorter terms, *if b is defined in terms of a right-invertible homomorphism from arrow a to b, then* $\llbracket \cdot \rrbracket_b$ *is correct with respect to* $\llbracket \cdot \rrbracket_a$.

4   The Bottom and Preimage Arrows

The following commutative diagram shows the relationships between the arrows $X \rightsquigarrow_\bot Y$ and $X \rightsquigarrow_{pre} Y$ for interpreting nonrecursive, nonprobabilistic programs, and $X \rightsquigarrow_{\bot^*} Y$ and $X \rightsquigarrow_{pre^*} Y$ for interpreting recursive, probabilistic programs.

$$
\begin{array}{ccc}
X \rightsquigarrow_\bot Y & \xrightarrow{\ \text{lift}_{pre}\ } & X \rightsquigarrow_{pre} Y \\
{\scriptstyle \eta_{\bot^*}} \downarrow & & \downarrow {\scriptstyle \eta_{pre^*}} \\
X \rightsquigarrow_{\bot^*} Y & \xrightarrow[\ \text{lift}_{pre^*}\ ]{} & X \rightsquigarrow_{pre^*} Y
\end{array} \tag{20}
$$

In this section, we define the top row.

$$X \rightsquigarrow_\perp Y ::= X \rightarrow Y_\perp$$

$$arr_\perp\ f\ a := f\ a$$

$$(f_1 \ggg_\perp f_2)\ a := \text{case } f_1\ a \text{ of}$$
$$\perp \longrightarrow \perp$$
$$b \longrightarrow f_2\ b$$

$$lazy_\perp\ f\ a := f\ 0\ a$$

$$ifte_\perp\ f_1\ f_2\ f_3\ a := \text{case } f_1\ a \text{ of}$$
$$\text{true} \longrightarrow f_2\ a$$
$$\text{false} \longrightarrow f_3\ a$$
$$\perp \longrightarrow \perp$$

$$(f_1 \ \&\&\&_\perp\ f_2)\ a := \text{case } \langle f_1\ a, f_2\ a \rangle \text{ of}$$
$$\langle \perp, \_ \rangle \longrightarrow \perp$$
$$\langle \_, \perp \rangle \longrightarrow \perp$$
$$\langle b_1, b_2 \rangle \longrightarrow \langle b_1, b_2 \rangle$$

Fig. 3. Bottom arrow definitions

4.1 The Bottom Arrow

To use Theorem 2 to prove correct the interpretations of expressions as preimage arrow computations, we need to define the preimage arrow in terms of a simpler arrow with easily understood behavior. The function arrow (13) is an obvious candidate. However, we will need to represent possible nontermination as an error value, so we need a slightly more complicated arrow.

Fig. 3 defines the **bottom arrow**, which is similar to the function arrow but propagates the error value $\perp$. Its computations have type $X \rightsquigarrow_\perp Y ::= X \rightarrow Y_\perp$, where $Y_\perp ::= Y \cup \{\perp\}$.

To prove the arrow laws, we need coarse enough notion of equivalence.

Definition 4 (Bottom Arrow Equivalence). *Two computations* $f_1 : X \rightsquigarrow_\perp Y$ *and* $f_2 : X \rightsquigarrow_\perp Y$ *are equivalent, or* $f_1 \equiv f_2$, *when* $f_1\ a \equiv f_2\ a$ *for all* $a \in X$.

Using bottom arrow equivalence, it is easy to show that $(\rightsquigarrow_\perp)$ is isomorphic to the Maybe monad's Kleisli arrow. By Theorem 1, the arrow laws hold.

4.2 The Preimage Function Type and Operations

Before defining the preimage arrow, we need a type of preimage functions. Set $Y \rightarrow$ Set X would be a good candidate, except that the $(\ggg_{pre})$ combinator will require preimage functions to have observable domains, but instances of Set $Y \rightarrow$ Set X are intensional functions. We therefore define

$$X \rightarrow_{pre} Y ::= \langle \text{Set } Y, \text{Set } Y \rightarrow \text{Set } X \rangle \tag{21}$$

as the type of preimage functions. Fig. 4 defines the necessary operations on them. Operations $\langle \cdot, \cdot \rangle_{pre}$ and $(\circ_{pre})$ return preimage functions that compute preimages under pairing and composition, and are derived from the preimage identities in (12); $(\cup_{pre})$ computes unions and is used to define $ifte_{pre}$.

Fig. 4 also defines $image_\perp$ and $preimage_\perp$ to operate on bottom arrow computations: $image_\perp\ f\ A$ computes f's range (with domain A), and $preimage_\perp\ f\ A$ returns a function that computes preimages under f restricted to A. Together, they can be used to convert bottom arrow computations to preimage functions:

$$pre : (X \rightsquigarrow_\perp Y) \rightarrow \text{Set } X \rightarrow (X \rightarrow_{pre} Y)$$
$$pre\ f\ A := \langle image_\perp\ f\ A, preimage_\perp\ f\ A \rangle \tag{22}$$

$X \rightarrow_{pre} Y ::= \langle Set\ Y, Set\ Y \rightarrow Set\ X \rangle$

$pre : (X \rightsquigarrow_\perp Y) \rightarrow Set\ X \rightarrow (X \rightarrow_{pre} Y)$

$pre\ f\ A := \langle image_\perp\ f\ A, preimage_\perp\ f\ A \rangle$

$\varnothing_{pre} := \langle \varnothing, \lambda B. \varnothing \rangle$

$ap_{pre} : (X \rightarrow_{pre} Y) \rightarrow Set\ Y \rightarrow Set\ X$

$ap_{pre}\ \langle B', p \rangle\ B := p\ (B \cap B')$

$range_{pre} : (X \rightarrow_{pre} Y) \rightarrow Set\ Y$

$range_{pre}\ \langle B', p \rangle := B'$

$\langle \cdot, \cdot \rangle_{pre} : (X \rightarrow_{pre} Y_1) \rightarrow (X \rightarrow_{pre} Y_2) \rightarrow (X \rightarrow_{pre} \langle Y_1, Y_2 \rangle)$

$\langle \langle B_1', p_1 \rangle, \langle B_2', p_2 \rangle \rangle_{pre} :=$
$\quad let\ B' := B_1' \times B_2'$
$\quad\quad p := \lambda B. \bigcup_{\langle b_1, b_2 \rangle \in B} (p_1\ \{b_1\}) \cap (p_2\ \{b_2\})$
$\quad in\ \langle B', p \rangle$

$(\circ_{pre}) : (Y \rightarrow_{pre} Z) \rightarrow (X \rightarrow_{pre} Y) \rightarrow (X \rightarrow_{pre} Z)$

$\langle C', p_2 \rangle \circ_{pre} h_1 := \langle C', \lambda C. ap_{pre}\ h_1\ (p_2\ C) \rangle$

$(\cup_{pre}) : (X \rightarrow_{pre} Y) \rightarrow (X \rightarrow_{pre} Y) \rightarrow (X \rightarrow_{pre} Y)$

$\langle B_1', p_1 \rangle \cup_{pre} \langle B_2', p_2 \rangle :=$
$\quad \langle B_1' \cup B_2', \lambda B. ap_{pre}\ \langle B_1', p_1 \rangle\ B \cup ap_{pre}\ \langle B_2', p_2 \rangle\ B \rangle$

$image_\perp : (X \rightsquigarrow_\perp Y) \rightarrow Set\ X \rightarrow Set\ Y$

$image_\perp\ f\ A := (image\ f\ A) \backslash \{\perp\}$

$preimage_\perp : (X \rightsquigarrow_\perp Y) \rightarrow Set\ X \rightarrow Set\ Y \rightarrow Set\ X$

$preimage_\perp\ f\ A\ B := \{a \in A \mid f\ a \in B\}$

Fig. 4. Preimage functions and operations

$X \rightsquigarrow_{pre} Y ::= Set\ X \rightarrow (X \rightarrow_{pre} Y)$

$arr_{pre} := lift_{pre} \circ arr_\perp$

$(h_1 \ggg_{pre} h_2)\ A := let\ h_1' := h_1\ A$
$\quad\quad\quad\quad\quad h_2' := h_2\ (range_{pre}\ h_1')$
$\quad\quad\quad in\ h_2' \circ_{pre} h_1'$

$(h_1\ \&\&\&_{pre}\ h_2)\ A := \langle h_1\ A, h_2\ A \rangle_{pre}$

$ifte_{pre}\ h_1\ h_2\ h_3\ A :=$
$\quad let\ h_1' := h_1\ A$
$\quad\quad h_2' := h_2\ (ap_{pre}\ h_1'\ \{true\})$
$\quad\quad h_3' := h_3\ (ap_{pre}\ h_1'\ \{false\})$
$\quad in\ h_2' \cup_{pre} h_3'$

$lazy_{pre}\ h\ A := if\ A = \varnothing\ then\ \varnothing_{pre}\ else\ h\ 0\ A$

$lift_{pre} := pre$

Fig. 5. Preimage arrow definitions

Lastly, the ap_{pre} function in Fig. 4 applies a preimage function to a set.

Preimage arrow correctness depends on ap_{pre} and pre behaving like $preimage_\perp$.

Theorem 3 (ap_{pre} of pre Computes Preimages). *Let* $f : X \rightsquigarrow_\perp Y$. *For all* $A \subseteq X$ *and* $B \subseteq Y$, $ap_{pre}\ (pre\ f\ A)\ B \equiv preimage_\perp\ f\ A\ B$.

4.3 The Preimage Arrow

If we define the **preimage arrow** type constructor as

$$X \rightsquigarrow_{pre} Y ::= Set\ X \rightarrow (X \rightarrow_{pre} Y) \tag{23}$$

then we already have a lift $lift_{pre} : (X \rightsquigarrow_\perp Y) \rightarrow (X \rightsquigarrow_{pre} Y)$ from the bottom arrow to the preimage arrow: pre. If $lift_{pre}$ is pre, then by Theorem 3, lifted bottom arrow computations compute correct preimages, exactly as we should expect them to.

Fig. 5 defines the preimage arrow in terms of the preimage function operations in Fig. 4. For these definitions to make $lift_{pre}$ a homomorphism, preimage arrow equivalence must mean "computes the same preimages."

Definition 5 (Preimage Arrow Equivalence). *Two preimage arrow computations* $h_1 : X \rightsquigarrow_{pre} Y$ *and* $h_2 : X \rightsquigarrow_{pre} Y$ *are equivalent, or* $h_1 \equiv h_2$, *when* $ap_{pre}\ (h_1\ A)\ B \equiv ap_{pre}\ (h_2\ A)\ B$ *for all* $A \subseteq X$ *and* $B \subseteq Y$.

Theorem 4 (Preimage Arrow Correctness). $\mathsf{lift_{pre}}$ *is a homomorphism.*

Corollary 1 (Semantic Correctness). *For all* e, $[\![e]\!]_{\mathsf{pre}} \equiv \mathsf{lift_{pre}}\ [\![e]\!]_\perp$.

In other words, $[\![e]\!]_{\mathsf{pre}}$ always computes correct preimages under $[\![e]\!]_\perp$. Inhabitants of type $\mathsf{X} \leadsto_{\mathsf{pre}} \mathsf{Y}$ do not always behave intuitively; e.g.

$$\mathsf{unruly} : \mathsf{Bool} \leadsto_{\mathsf{pre}} \mathsf{Bool}$$
$$\mathsf{unruly}\ A\ :=\ \langle \mathsf{Bool} \backslash A, \lambda B.\ B \rangle \tag{24}$$

So $\mathsf{ap_{pre}}$ (unruly {true}) {false} = {false}$\cap$(Bool\{true}) = {false}—a "preimage" that does not even intersect the given domain {true}. Other examples show that preimage computations are not necessarily monotone, and lack other desirable properties. Those with desirable properties obey the following law.

Definition 6 (Preimage Arrow Law). *Let* $\mathsf{h} : \mathsf{X} \leadsto_{\mathsf{pre}} \mathsf{Y}$. *If there exists an* $\mathsf{f} : \mathsf{X} \leadsto_\perp \mathsf{Y}$ *such that* $\mathsf{h} \equiv \mathsf{lift_{pre}}\ \mathsf{f}$, *then* h *obeys the **preimage arrow law.***

By homomorphism of $\mathsf{lift_{pre}}$, preimage arrow combinators preserve the preimage arrow law. From here on, we assume all $\mathsf{h} : \mathsf{X} \leadsto_{\mathsf{pre}} \mathsf{Y}$ obey it. By Definition 6, $\mathsf{lift_{pre}}$ has a right inverse; by Theorem 1, the arrow laws hold.

5   The Bottom* and Preimage* Arrows

This section lifts the prior semantics to recursive, probabilistic programs.

We have defined the top of our roadmap:

$$\begin{array}{ccc} \mathsf{X} \leadsto_\perp \mathsf{Y} & \xrightarrow{\ \mathsf{lift_{pre}}\ } & \mathsf{X} \leadsto_{\mathsf{pre}} \mathsf{Y} \\ {\scriptstyle\eta_{\perp*}}\downarrow & & \downarrow{\scriptstyle\eta_{\mathsf{pre}*}} \\ \mathsf{X} \leadsto_{\perp*} \mathsf{Y} & \xrightarrow[\ \mathsf{lift_{pre*}}\ ]{} & \mathsf{X} \leadsto_{\mathsf{pre}*} \mathsf{Y} \end{array} \tag{25}$$

so that $\mathsf{lift_{pre}}$ is a homomorphism. Now we move down each side and connect the bottom, in a way that makes every morphism a homomorphism.

Probabilistic functions that may not terminate, but terminate with probability 1, are common. For example, suppose random retrieves numbers in $[0,1]$ from an implicit random source. The following probabilistic function defines the well-known geometric distribution by counting the number of times random $<$ p:

$$\mathsf{geometric}\ \mathsf{p}\ :=\ \mathsf{if}\ \mathsf{random} < \mathsf{p}\ \mathsf{then}\ 0\ \mathsf{else}\ 1 + \mathsf{geometric}\ \mathsf{p} \tag{26}$$

For any $\mathsf{p} > 0$, geometric p may not terminate, but the probability of not terminating (i.e. always taking the "else" branch) is $(1 - \mathsf{p}) \cdot (1 - \mathsf{p}) \cdot (1 - \mathsf{p}) \cdots = 0$.

Suppose we interpret geometric p as $\mathsf{h} : \mathsf{R} \leadsto_{\mathsf{pre}} \mathbb{N}$, a preimage arrow computation from random sources to $\mathbb{N}$, and we have a probability measure $\mathsf{P} : \mathsf{Set}\ \mathsf{R} \to [0,1]$. The probability of $N \subseteq \mathbb{N}$ is $\mathsf{P}\ (\mathsf{ap_{pre}}\ (\mathsf{h}\ \mathsf{R})\ N)$. To compute this, we must

- Ensure each $\mathsf{r} \in \mathsf{R}$ contains enough random numbers.
- Determine how random indexes numbers in r.
- Ensure $\mathsf{ap_{pre}}$ (h R) N terminates even though there are random sources in R for which geometric p does not terminate.

$$\mathsf{AStore}\ s\ (x \leadsto_a y) ::= J \to (\langle s, x\rangle \leadsto_a y)$$
$$x \leadsto_{a*} y ::= \mathsf{AStore}\ s\ (x \leadsto_a y)$$

$$\mathsf{arr}_{a*} := \eta_{a*} \circ \mathsf{arr}_a$$
$$(k_1 \ggg_{a*} k_2)\ j := (\mathsf{arr}_a\ \mathsf{fst}\ \&\&\&_a\ k_1\ (\mathsf{left}\ j)) \ggg_a k_2\ (\mathsf{right}\ j)$$
$$(k_1 \&\&\&_{a*} k_2)\ j := k_1\ (\mathsf{left}\ j)\ \&\&\&_a\ k_2\ (\mathsf{right}\ j)$$

$$\mathsf{ifte}_{a*}\ k_1\ k_2\ k_3\ j :=$$
$$\mathsf{ifte}_a\ (k_1\ (\mathsf{left}\ j))$$
$$(k_2\ (\mathsf{left}\ (\mathsf{right}\ j)))$$
$$(k_3\ (\mathsf{right}\ (\mathsf{right}\ j)))$$
$$\mathsf{lazy}_{a*}\ k\ j := \mathsf{lazy}_a\ \lambda 0.\, k\ 0\ j$$
$$\eta_{a*}\ f\ j := \mathsf{arr}_a\ \mathsf{snd} \ggg_a f$$

Fig. 6. AStore (associative store) arrow transformer definitions

The last task is the most difficult, but doing the first two will provide structure that makes it much easier.

5.1   Threading and Indexing

We need bottom and preimage arrows that thread a random source. To ensure random sources contain enough numbers, they should be infinite.

In a pure λ-calculus, random sources are typically infinite streams, threaded monadically: each computation receives and produces a random source. A little-used alternative is for the random source to be an infinite tree, threaded applicatively: each computation receives, but does not produce, a random source. Combinators split the tree and pass subtrees to subcomputations.

With either alternative, for arrows, the resulting definitions are large, conceptually difficult, and hard to manipulate. Fortunately, it is relatively easy to assign each subcomputation a unique index into a tree-shaped random source and pass the random source unchanged. For this, we need an indexing scheme.

Definition 7 (Binary Indexing Scheme). *Let* J *be the set of finite lists of* Bool. *Define* $j_0 := \langle\rangle$ *as the root node's index, and* left $: J \to J;$ left $j := \langle \mathsf{true}, j\rangle$ *and* right $: J \to J;$ right $j := \langle \mathsf{false}, j\rangle$ *to construct left and right child indexes.*

We define random-source-threading variants of both the bottom and preimage arrows at the same time by defining an **arrow transformer**: an arrow parameterized on another arrow. The AStore arrow transformer type constructor takes a store type s and an arrow $x \leadsto_a y$:

$$\mathsf{AStore}\ s\ (x \leadsto_a y)\ ::=\ J \to (\langle s, x\rangle \leadsto_a y) \tag{27}$$

Reading the type, we see that computations receive an index $j \in J$ and produce a computation that receives a store as well as an x. Lifting extracts the x from the input pair and sends it on to the original computation, ignoring j:

$$\eta_{a*} : (x \leadsto_a y) \to \mathsf{AStore}\ s\ (x \leadsto_a y)$$
$$\eta_{a*}\ f\ j := \mathsf{arr}_a\ \mathsf{snd} \ggg_a f \tag{28}$$

Fig. 6 defines the remaining combinators. Each subcomputation receives left j, right j, or some other unique binary index. We thus think of programs interpreted as AStore arrows as being completely unrolled into an infinite binary tree, with each expression labeled with its tree index.

5.2 Recursive, Probabilistic Programs

To interpret probabilistic programs, we put infinite random trees in the store.

Of all the ways to represent infinite binary trees whose nodes are labeled with values in $[0,1]$, the way most compatible with measure theory is to flatten them into vectors of $[0,1]$ indexed by J. The set of all such vectors is $[0,1]^J$.

Definition 8 (Random Source). *Define* $R := [0,1]^J$, *the set of infinite binary trees whose node labels are in* $[0,1]$. *A* ***random source*** *is any* $r \in R$.

To interpret recursive programs, we need to ensure termination. One ultimately implementable way is to have the store dictate which branch of each conditional, if any, is taken. If the store dictates that all but finitely many branches cannot be taken, well-defined programs must terminate (see Definition 3).

Definition 9 (Branch Trace). *A* ***branch trace*** *is any* $t \in (Bool_\perp)^J$ *such that* $t\ j = true$ *or* $t\ j = false$ *for no more than finitely many* $j \in J$.

Let $T \subset (Bool_\perp)^J$ *be the set of all branch traces.*

Let $X \leadsto_{a^*} Y ::= AStore \langle R, T \rangle (X \leadsto_a Y)$ denote the AStore arrow type that threads both random sources and branch traces through another arrow a. Thus, the type constructors for the **bottom*** and **preimage*** arrows are

$$X \leadsto_{\perp^*} Y ::= AStore \langle R, T \rangle (X \leadsto_\perp Y)$$
$$X \leadsto_{pre^*} Y ::= AStore \langle R, T \rangle (X \leadsto_{pre} Y) \tag{29}$$

For probabilistic programs, we define a combinator $random_{a^*}$ that returns the number at its tree index in the random source, and extend $[\![\cdot]\!]_{a^*}$ for arrows a^*:

$$random_{a^*} : X \leadsto_{a^*} [0,1] \qquad\qquad [\![random]\!]_{a^*} :\equiv random_{a^*}$$
$$random_{a^*}\ j := arr_a\ fst \ggg_a arr_a\ fst \ggg_a arr_a\ (\pi\ j) \tag{30}$$

where $\pi : J \to X^J \to X$, defined by $\pi\ j\ f := f\ j$, produces projection functions.

For recursive programs, we define a combinator that reads branch traces, and a new if-then-else combinator that yields $\perp$ when its test expression does not agree with the branch trace at its tree index:

$$branch_{a^*} : X \leadsto_{a^*} Bool$$
$$branch_{a^*}\ j := arr_a\ fst \ggg_a arr_a\ snd \ggg_a arr_a\ (\pi\ j)$$

$$ifte_{a^*}^{\Downarrow} : (x \leadsto_{a^*} Bool) \to (x \leadsto_{a^*} y) \to (x \leadsto_{a^*} y) \to (x \leadsto_{a^*} y) \tag{31}$$
$$ifte_{a^*}^{\Downarrow}\ k_1\ k_2\ k_3\ j := ifte_a\ ((k_1\ (left\ j)\ \&\&\&_a\ branch_{a^*}\ j) \ggg_a arr_a\ agrees)$$
$$\qquad\qquad\qquad (k_2\ (left\ (right\ j)))$$
$$\qquad\qquad\qquad (k_3\ (right\ (right\ j)))$$

where $agrees\ \langle b_1, b_2 \rangle := if\ b_1 = b_2\ then\ b_1\ else\ \perp$. We define a new semantic function $[\![\cdot]\!]_{a^*}^{\Downarrow}$ by replacing the if rule in $[\![\cdot]\!]_{a^*}$:

$$[\![if\ e_c\ then\ e_t\ else\ e_f]\!]_{a^*}^{\Downarrow} :\equiv ifte_{a^*}^{\Downarrow}\ [\![e_c]\!]_{a^*}^{\Downarrow}\ (lazy_{a^*}\ \lambda 0.\ [\![e_t]\!]_{a^*}^{\Downarrow})\ (lazy_{a^*}\ \lambda 0.\ [\![e_f]\!]_{a^*}^{\Downarrow}) \tag{32}$$

Suppose $f := (\llbracket p \rrbracket^{\Downarrow}_{\perp*}\ j_0) : X' \leadsto_{\perp} Y$ and $h := (\llbracket p \rrbracket^{\Downarrow}_{pre*}\ j_0) : X' \leadsto_{pre} Y$, where $X' = (R \times T) \times X$. For each $\langle \langle r, t \rangle, a \rangle \in X'$, we assume that only r is chosen randomly. Thus, the probability of $B \subseteq Y$ is

$$\begin{aligned} P\,(\text{image (fst} \circ \text{fst)}\,(\text{preimage}_{\perp}\ f\ X'\ B)) \\ = P\,(\text{image (fst} \circ \text{fst)}\,(\text{ap}_{pre}\,(h\ X')\ B)) \end{aligned} \tag{33}$$

if f and h always terminate and $\llbracket \cdot \rrbracket^{\Downarrow}_{pre*}$ is correct with respect to $\llbracket \cdot \rrbracket^{\Downarrow}_{\perp*}$.

5.3 Correctness and Termination

The proofs in this section require AStore arrow equivalence to be a little coarser.

Definition 10 (AStore Arrow Equivalence). *Two* AStore *arrow computations* k_1 *and* k_2 *are equivalent, or* $k_1 \equiv k_2$, *when* $k_1\ j \equiv k_2\ j$ *for all* $j \in J$.

Proving $\llbracket \cdot \rrbracket_{\perp*}$ and $\llbracket \cdot \rrbracket_{pre*}$ correct with respect to $\llbracket \cdot \rrbracket_{\perp}$ and $\llbracket \cdot \rrbracket_{pre}$, for programs without random, only requires proving η_{a*} homomorphic, using the arrow laws.

Theorem 5 (Pure AStore Arrow Correctness). η_{a*} *is a homomorphism.*

Corollary 2 (Pure Semantic Correctness). *For all pure* e, $\llbracket e \rrbracket_{a*} \equiv \eta_{a*}\ \llbracket e \rrbracket_{a}$.

We use a homomorphic lift to prove $\llbracket \cdot \rrbracket^{\Downarrow}_{pre*}$ correct with respect to $\llbracket \cdot \rrbracket^{\Downarrow}_{\perp*}$. If we define it in terms of $\text{lift}_b : (x \leadsto_a y) \to (x \leadsto_b y)$ as

$$\begin{aligned} \text{lift}_{b*} : (x \leadsto_{a*} y) \to (x \leadsto_{b*} y) \\ \text{lift}_{b*}\ f\ j := \text{lift}_b\,(f\ j) \end{aligned} \tag{34}$$

then we need only use the fact that a and b are arrows to prove the following.

Theorem 6 (Effectful AStore Arrow Correctness). *If* lift_b *is an arrow homomorphism from* a *to* b, *then* lift_{b*} *is an arrow homomorphism from* $a*$ *to* $b*$.

Corollary 3 (Effectful Semantic Correctness). *For all* e, $\llbracket e \rrbracket_{pre*} \equiv \text{lift}_{pre*}$ $\llbracket e \rrbracket_{\perp*}$ *and* $\llbracket e \rrbracket^{\Downarrow}_{pre*} \equiv \text{lift}_{pre*}\ \llbracket e \rrbracket^{\Downarrow}_{\perp*}$.

For termination, we need to define the largest domain on which $\llbracket e \rrbracket^{\Downarrow}_{a*}$ and $\llbracket e \rrbracket_{a*}$ computations should agree.

Definition 11 (Maximal Domain). *Let* $f : X \leadsto_{\perp*} Y$. *Its **maximal domain** is the largest* $A^* \subseteq (R \times T) \times X$ *for which* $A^* = \{a \in A^* \mid f\ j_0\ a \neq \perp\}$.

Because $f\ j_0\ a \neq \perp$ implies termination, all inputs in A^* are terminating.

Theorem 7 (Correct Termination Everywhere). *Let* $\llbracket e \rrbracket^{\Downarrow}_{\perp*} : X \leadsto_{\perp*} Y$ *have maximal domain* A^*, *and* $X' := (R \times T) \times X$. *For all* $a \in X'$, $A \subseteq X'$ *and* $B \subseteq Y$,

$$\begin{aligned} \llbracket e \rrbracket^{\Downarrow}_{\perp*}\ j_0\ a\ &= \text{if } a \in A^* \text{ then } \llbracket e \rrbracket_{\perp*}\ j_0\ a \text{ else } \perp \\ \text{ap}_{pre}\,(\llbracket e \rrbracket^{\Downarrow}_{pre*}\ j_0\ A)\ B\ &= \text{ap}_{pre}\,(\llbracket e \rrbracket_{pre*}\ j_0\,(A \cap A^*))\ B \end{aligned} \tag{35}$$

In other words, $\llbracket \cdot \rrbracket^{\Downarrow}_{pre*}$ computations always terminate, and the sets they yield are correct preimages.

6   Abstract Semantics

This section derives a sound, implementable abstract semantics. Most preimages of uncountable sets are uncomputable. We therefore define a semantics for approximate preimage computation by

1. Choosing abstract set types that can be finitely represented, and operations that overapproximate concrete set operations.
2. Replacing concrete set types and operations with abstract set types and operations in the definitions of the preimage and preimage* arrows.
3. Proving termination, soundness, and other desirable properties.

In a sense, this is typical abstract interpretation. However, not having a fixpoint operator in the language means there is no abstract fixpoint to compute, and abstract preimage arrow computations actually apply functions.

6.1   Abstract Sets

We use the abstract domain of rectangles with an atypical extension to represent rectangles of X^J (i.e. infinite binary trees of X).

Definition 12 (Rectangular Sets). *For a type* X *of language values,* Rect X *denotes the type of **rectangular sets** of* X: *a bounded lattice of sets in* Set X *ordered by* $(\subseteq)$; *i.e. it contains* $\varnothing$ *and* X, *and is closed under meet* $(\cap)$ *and join* $(\sqcup)$. *Rectangles of cartesian products are defined by*

$$\text{Rect } \langle X_1, X_2 \rangle \ ::= \ \{A_1 \times A_2 \mid A_1 : \text{Rect } X_1, A_2 : \text{Rect } X_2\} \tag{36}$$

Rectangles of infinite binary trees (i.e. products indexed by J) *are defined by*

$$\text{Rect } X^J \ ::= \bigcup_{J' \subset J \ finite} \left\{ \prod_{j \in J} A_j \ \middle| \ A_j : \text{Rect } X, \ j \notin J' \iff A_j = X \right\} \tag{37}$$

i.e. for $A : \text{Rect } X^J$, *only finitely many axes of* A *are proper subsets of* X. *Joins of products are defined by*

$$(A_1 \times A_2) \sqcup (B_1 \times B_2) \ = \ (A_1 \sqcup B_1) \times (A_2 \sqcup B_2) \tag{38}$$

$$\left(\prod_{j \in J} A_j\right) \sqcup \left(\prod_{j \in J} B_j\right) \ = \ \prod_{j \in J}(A_j \sqcup B_j) \tag{39}$$

The lattice properties imply that $(\sqcup)$ overapproximates $(\cup)$; i.e. $A \cup B \subseteq A \sqcup B$. For non-product types X, Rect X may be any bounded sublattice of Set X. Interpreting conditionals requires $\{\text{true}\}$ and $\{\text{false}\}$; thus Rect Bool $::=$ Set Bool.

Intervals in ordered spaces can be implemented as pairs of endpoints. Products in Rect $\langle X_1, X_2 \rangle$ can be implemented as pairs of type $\langle \text{Rect } X_1, \text{Rect } X_2 \rangle$. By (37), products in Rect X^J have only finitely many axes that are proper subsets of X, so they can be implemented as *finite* binary trees. All operations on products proceed by simple structural recursion.

$X \to_{\widehat{\mathsf{pre}}} Y ::= \langle \mathsf{Rect}\ Y, \mathsf{Rect}\ Y \to \mathsf{Rect}\ X \rangle$

$\langle \cdot, \cdot \rangle_{\widehat{\mathsf{pre}}} : (X \to_{\widehat{\mathsf{pre}}} Y_1) \to (X \to_{\widehat{\mathsf{pre}}} Y_2)$
$\qquad\qquad \to (X \to_{\widehat{\mathsf{pre}}} \langle Y_1, Y_2 \rangle)$

$\varnothing_{\widehat{\mathsf{pre}}} := \langle \varnothing, \lambda B.\ \varnothing \rangle$

$\langle \langle Y_1', p_1 \rangle, \langle Y_2', p_2 \rangle \rangle_{\widehat{\mathsf{pre}}} :=$
$\quad \langle Y_1' \times Y_2', \lambda B.\ p_1\ (\mathsf{proj}_1\ B) \cap p_2\ (\mathsf{proj}_2\ B) \rangle$

$\mathsf{ap}_{\widehat{\mathsf{pre}}} : (X \to_{\widehat{\mathsf{pre}}} Y) \to \mathsf{Rect}\ Y \to \mathsf{Rect}\ X$
$\mathsf{ap}_{\widehat{\mathsf{pre}}}\ \langle Y', p \rangle\ B := p\ (B \cap Y')$

$(\circ_{\widehat{\mathsf{pre}}}) : (Y \to_{\widehat{\mathsf{pre}}} Z) \to (X \to_{\widehat{\mathsf{pre}}} Y) \to (X \to_{\widehat{\mathsf{pre}}} Z)$
$\langle Z', p_2 \rangle \circ_{\widehat{\mathsf{pre}}} h_1 := \langle Z', \lambda C.\ \mathsf{ap}_{\widehat{\mathsf{pre}}}\ h_1\ (p_2\ C) \rangle$

$\mathsf{range}_{\widehat{\mathsf{pre}}} : (X \to_{\widehat{\mathsf{pre}}} Y) \to \mathsf{Rect}\ Y$
$\mathsf{range}_{\widehat{\mathsf{pre}}}\ \langle Y', p \rangle := Y'$

$(\sqcup_{\widehat{\mathsf{pre}}}) : (X \to_{\widehat{\mathsf{pre}}} Y) \to (X \to_{\widehat{\mathsf{pre}}} Y) \to (X \to_{\widehat{\mathsf{pre}}} Y)$
$\langle Y_1', p_1 \rangle \sqcup_{\widehat{\mathsf{pre}}} \langle Y_2', p_2 \rangle :=$
$\quad \langle Y_1' \sqcup Y_2', \lambda B.\ \mathsf{ap}_{\widehat{\mathsf{pre}}}\ \langle Y_1', p_1 \rangle\ B \sqcup \mathsf{ap}_{\widehat{\mathsf{pre}}}\ \langle Y_2', p_2 \rangle\ B \rangle$

(a) Definitions for abstract preimage functions, which compute rectangular covers.

$X \rightsquigarrow_{\widehat{\mathsf{pre}}} Y ::= \mathsf{Rect}\ X \to (X \to_{\widehat{\mathsf{pre}}} Y)$

$\mathsf{ifte}_{\widehat{\mathsf{pre}}}\ h_1\ h_2\ h_3\ A :=$
$\quad \mathsf{let}\ h_1' := h_1\ A$
$\qquad h_2' := h_2\ (\mathsf{ap}_{\widehat{\mathsf{pre}}}\ h_1'\ \{\mathsf{true}\})$
$\qquad h_3' := h_3\ (\mathsf{ap}_{\widehat{\mathsf{pre}}}\ h_1'\ \{\mathsf{false}\})$
$\quad \mathsf{in}\ h_2' \sqcup_{\widehat{\mathsf{pre}}} h_3'$

$(h_1 \ggg_{\widehat{\mathsf{pre}}} h_2)\ A := \mathsf{let}\ h_1' := h_1\ A$
$\qquad\qquad\qquad\quad h_2' := h_2\ (\mathsf{range}_{\widehat{\mathsf{pre}}}\ h_1')$
$\qquad\qquad\qquad \mathsf{in}\ h_2' \circ_{\widehat{\mathsf{pre}}} h_1'$

$(h_1 \&\!\&\!\&_{\widehat{\mathsf{pre}}} h_2)\ A := \langle h_1\ A, h_2\ A \rangle_{\widehat{\mathsf{pre}}}$

$\mathsf{lazy}_{\widehat{\mathsf{pre}}}\ h\ A := \mathsf{if}\ A = \varnothing\ \mathsf{then}\ \varnothing_{\widehat{\mathsf{pre}}}\ \mathsf{else}\ h\ 0\ A$

(b) Abstract preimage arrow, defined using abstract preimage functions.

$\mathsf{id}_{\widehat{\mathsf{pre}}}\ A := \langle A, \lambda B.\ B \rangle$

$\mathsf{const}_{\widehat{\mathsf{pre}}}\ b\ A := \langle \{b\}, \lambda B.\ \mathsf{if}\ B = \varnothing\ \mathsf{then}\ \varnothing\ \mathsf{else}\ A \rangle$

$\mathsf{fst}_{\widehat{\mathsf{pre}}}\ A := \langle \mathsf{proj}_1\ A, \mathsf{unproj}_1\ A \rangle$

$\pi_{\widehat{\mathsf{pre}}}\ j\ A := \langle \mathsf{proj}\ j\ A, \mathsf{unproj}\ j\ A \rangle$

$\mathsf{snd}_{\widehat{\mathsf{pre}}}\ A := \langle \mathsf{proj}_2\ A, \mathsf{unproj}_2\ A \rangle$

$\mathsf{proj}_1 := \mathsf{image}\ \mathsf{fst}$

$\mathsf{proj} : J \to \mathsf{Set}\ X^J \to \mathsf{Set}\ X$

$\mathsf{proj}_2 := \mathsf{image}\ \mathsf{snd}$

$\mathsf{proj}\ j\ A := \mathsf{image}\ (\pi\ j)\ A$

$\mathsf{unproj}_1\ A\ B := A \cap (B \times \mathsf{proj}_2\ A)$

$\mathsf{unproj} : J \to \mathsf{Set}\ X^J \to \mathsf{Set}\ X \to \mathsf{Set}\ X^J$

$\mathsf{unproj}_2\ A\ B := A \cap (\mathsf{proj}_1\ A \times B)$

$\mathsf{unproj}\ j\ A\ B := A \cap \prod_{i \in J}\ \mathsf{if}\ j = i\ \mathsf{then}\ B\ \mathsf{else}\ \mathsf{proj}\ j\ A$

(c) Explicit instances of $\mathsf{arr}_{\widehat{\mathsf{pre}}}\ f$ (e.g. $\mathsf{arr}_{\widehat{\mathsf{pre}}}\ \mathsf{id}$) needed to interpret probabilistic programs.

$X \rightsquigarrow_{\widehat{\mathsf{pre}^*}} Y ::= \mathsf{AStore}\ \langle R, T \rangle\ (X \rightsquigarrow_{\widehat{\mathsf{pre}}} Y)$

$\mathsf{ifte}_{\widehat{\mathsf{pre}^*}}^{\Downarrow} : (X \rightsquigarrow_{\widehat{\mathsf{pre}^*}} \mathsf{Bool}) \to (X \rightsquigarrow_{\widehat{\mathsf{pre}^*}} Y) \to (X \rightsquigarrow_{\widehat{\mathsf{pre}^*}} Y)$
$\qquad\qquad \to (X \rightsquigarrow_{\widehat{\mathsf{pre}^*}} Y)$

$\mathsf{random}_{\widehat{\mathsf{pre}^*}} : X \rightsquigarrow_{\widehat{\mathsf{pre}^*}} [0, 1]$

$\mathsf{ifte}_{\widehat{\mathsf{pre}^*}}^{\Downarrow}\ k_1\ k_2\ k_3\ j :=$
$\quad \mathsf{let}\ \langle C_k, p_k \rangle := k_1\ (\mathsf{left}\ j)\ A$

$\mathsf{random}_{\widehat{\mathsf{pre}^*}}\ j :=$
$\quad \mathsf{fst}_{\widehat{\mathsf{pre}}} \ggg_{\widehat{\mathsf{pre}}} \mathsf{fst}_{\widehat{\mathsf{pre}}} \ggg_{\widehat{\mathsf{pre}}} \pi_{\widehat{\mathsf{pre}}}\ j$

$\qquad \langle C_b, p_b \rangle := \mathsf{branch}_{\widehat{\mathsf{pre}^*}}\ j\ A$
$\qquad C_2 := C_k \cap C_b \cap \{\mathsf{true}\}$
$\qquad C_3 := C_k \cap C_b \cap \{\mathsf{false}\}$

$\mathsf{branch}_{\widehat{\mathsf{pre}^*}} : X \rightsquigarrow_{\widehat{\mathsf{pre}^*}} \mathsf{Bool}$

$\qquad A_2 := p_k\ C_2 \cap p_b\ C_2$

$\mathsf{branch}_{\widehat{\mathsf{pre}^*}}\ j :=$
$\quad \mathsf{fst}_{\widehat{\mathsf{pre}}} \ggg_{\widehat{\mathsf{pre}}} \mathsf{snd}_{\widehat{\mathsf{pre}}} \ggg_{\widehat{\mathsf{pre}}} \pi_{\widehat{\mathsf{pre}}}\ j$

$\qquad A_3 := p_k\ C_3 \cap p_b\ C_3$
$\quad \mathsf{in}\ \mathsf{if}\ C_b = \{\mathsf{true}, \mathsf{false}\}$

$\mathsf{fst}_{\widehat{\mathsf{pre}^*}} := \eta_{\widehat{\mathsf{pre}^*}}\ \mathsf{fst}_{\widehat{\mathsf{pre}}}; \ \cdots$

$\qquad\quad \mathsf{then}\ \langle Y, \lambda B.\ A_2 \sqcup A_3 \rangle$
$\qquad\quad \mathsf{else}\ k_2\ (\mathsf{left}\ (\mathsf{right}\ j))\ A_2 \cup_{\widehat{\mathsf{pre}}} k_3\ (\mathsf{right}\ (\mathsf{right}\ j))\ A_3$

(d) Abstract preimage* arrow combinators for probabilistic choice and guaranteed termination. Fig. 6 defines $\eta_{\widehat{\mathsf{pre}^*}}$, $(\ggg_{\widehat{\mathsf{pre}^*}})$, $(\&\!\&\!\&_{\widehat{\mathsf{pre}^*}})$, $\mathsf{ifte}_{\widehat{\mathsf{pre}^*}}$ and $\mathsf{lazy}_{\widehat{\mathsf{pre}^*}}$.

Fig. 7. Implementable arrows that approximate preimage arrows

6.2 Abstract Arrows

To define the abstract preimage arrow, we start by defining abstract preimage functions, by replacing set types in ($\rightharpoonup_{\mathsf{pre}}$) with abstract set types:

$$X \rightharpoonup_{\widehat{\mathsf{pre}}} Y ::= \langle \mathsf{Rect}\ Y, \mathsf{Rect}\ Y \rightarrow \mathsf{Rect}\ X \rangle \tag{40}$$

Fig. 7a defines the necessary operations on abstract preimage functions by replacing set operations with *abstract* set operations—except for $\langle \cdot, \cdot \rangle_{\widehat{\mathsf{pre}}}$, which is greatly simplified by the fact that preimage distributes over pairing and products (12). (Compare Fig. 4.) Similarly, Fig. 7b defines the abstract preimage arrow by replacing preimage function types and operations in the preimage arrow's definition with *abstract* preimage function types and operations. (Compare Fig. 5.) The lift $\mathsf{arr}_{\widehat{\mathsf{pre}}} : (X \rightarrow Y) \rightarrow (X \rightharpoonup_{\widehat{\mathsf{pre}}} Y)$ exists, but $\mathsf{arr}_{\widehat{\mathsf{pre}}}$ f is not always unique (because by definition, Rect X^J is an incomplete lattice) nor computable.

Fortunately, implementing $[\![\cdot]\!]_{\widehat{\mathsf{pre}}}$ as defined in Fig. 2 requires lifting only a few pure functions: id, fst, snd, const v for any literal constant v, and primitives δ. According to (30) and (31), implementing the extended semantics $[\![\cdot]\!]_{\widehat{\mathsf{pre}}^*}^{\Downarrow}$, which supports random choice and guarantees termination, requires lifting only π j for any $j \in J$. Fig. 7c gives explicit definitions for $\mathsf{id}_{\widehat{\mathsf{pre}}}$, $\mathsf{fst}_{\widehat{\mathsf{pre}}}$, $\mathsf{snd}_{\widehat{\mathsf{pre}}}$, $\mathsf{const}_{\widehat{\mathsf{pre}}}$ and $\pi_{\widehat{\mathsf{pre}}}$.

Fig. 7d defines the abstract preimage* arrow using the AStore arrow transformer (see Fig. 6), in terms of the abstract preimage arrow, and defines $\mathsf{random}_{\widehat{\mathsf{pre}}^*}$ and $\mathsf{branch}_{\widehat{\mathsf{pre}}^*}$ using the manual lifts in Fig. 7c.

Guaranteeing termination requires some care. The definition of $\mathsf{ifte}_{\widehat{\mathsf{pre}}^*}^{\Downarrow}$ in Fig. 7d is obtained by expanding the definition of $\mathsf{ifte}_{\mathsf{pre}^*}^{\Downarrow}$, and changing the case in which the set of branch traces allows both branches. Instead of taking both branches, it takes neither, and returns a loose but sound approximation.

6.3 Correctness and Termination

Let $h := [\![e]\!]_{\mathsf{pre}^*}^{\Downarrow} : X \rightsquigarrow_{\mathsf{pre}^*} Y$ and $\widehat{h} := [\![e]\!]_{\widehat{\mathsf{pre}}^*}^{\Downarrow} : X \rightsquigarrow_{\widehat{\mathsf{pre}}^*} Y$ for some expression e.

Theorem 8 (Terminating, Monotone, Sound and Decreasing). *For all* A : Rect $\langle \langle R, T \rangle, X \rangle$ *and* B : Rect Y,

- $\mathsf{ap}_{\widehat{\mathsf{pre}}}\ (\widehat{h}\ \mathsf{j}_0\ A)\ B$ *terminates.*
- $\lambda A'. \mathsf{ap}_{\widehat{\mathsf{pre}}}\ (\widehat{h}\ \mathsf{j}_0\ A')\ B$ *and* $\lambda B'. \mathsf{ap}_{\widehat{\mathsf{pre}}}\ (\widehat{h}\ \mathsf{j}_0\ A)\ B'$ *are monotone.*
- $\mathsf{ap}_{\mathsf{pre}}\ (h\ \mathsf{j}_0\ A)\ B \subseteq \mathsf{ap}_{\widehat{\mathsf{pre}}}\ (\widehat{h}\ \mathsf{j}_0\ A)\ B \subseteq A$ *(i.e. sound and decreasing).*

Given these properties, we might try to compute preimages of B by computing preimages restricted to the parts of increasingly fine discretizations of A.

Definition 13 (Preimage Refinement Algorithm). *Let* B : Rect Y. *Define*

$$\mathsf{refine} : \mathsf{Rect}\ \langle \langle R, T \rangle, X \rangle \rightarrow \mathsf{Rect}\ \langle \langle R, T \rangle, X \rangle$$
$$\mathsf{refine}\ A := \mathsf{ap}_{\widehat{\mathsf{pre}}}\ (\widehat{h}\ \mathsf{j}_0\ A)\ B \tag{41}$$

Define partition : Rect $\langle\langle R, T\rangle, X\rangle \to$ Set (Rect $\langle\langle R, T\rangle, X\rangle$) *to produce positive-measure, disjoint rectangles, and define*

$$\begin{aligned} &\text{refine}^* : \text{Set (Rect } \langle\langle R, T\rangle, X\rangle) \to \text{Set (Rect } \langle\langle R, T\rangle, X\rangle) \\ &\text{refine}^* \; \mathcal{A} := \text{ image refine } (\textstyle\bigcup_{A\in\mathcal{A}} \text{partition A}) \end{aligned} \tag{42}$$

For any A : Rect $\langle\langle R, T\rangle, X\rangle$, *iterate* refine* *on* {A}.

Monotonicity ensures refining a partition of A never does worse than refining A itself, decreasingness ensures refine A $\subseteq$ A, and soundness ensures the preimage of B is covered by the partition refine* returns. Ideally, the algorithm would be complete, in that covering partitions converge to a set that overapproximates by a measure-zero subset. Unfortunately, convergence fails on some examples that terminate with probability less than one. We leave completeness conditions for future work, and for now, use algorithms that depend only on soundness.

7   Implementations and Examples

This section describes our implementations and gives examples, including probabilistic verification of floating-point error bounds.

We have three implementations: two direct implementations of the abstract semantics, and a less direct but more efficient one called **Dr. Bayes**. All of them can be found at https://github.com/ntoronto/drbayes.

Given a library for operating on rectangular sets, the abstract preimage arrows defined in Figs. 6 and 7 can be implemented with few changes in any practical λ-calculus. We have done so in Typed Racket [27] and Haskell [1]. Both implementations are almost line-for-line transliterations from the figures.

Dr. Bayes is written in Typed Racket. It includes $[\![\cdot]\!]_{a^*}$ (Fig. 2), its extension $[\![\cdot]\!]_{a^*}^{\Downarrow}$, the bottom* arrow (Figs. 3 and 6), the abstract preimage and preimage* arrows (Figs. 7 and 6), and other manual lifts to compute abstract preimages under real functions such as arithmetic, sqrt and log. The abstract preimage arrows operate on a monomorphic rectangular set data type, which includes tagged rectangles and disjoint unions for ad-hoc polymorphism, and floating-point intervals to overapproximate real intervals.

Definition 13 outlines preimage refinement, a discretization algorithm that repeatedly shrinks and repartitions a program's domain. *Dr. Bayes does not use this algorithm directly* because it is inefficient: good accuracy requires fine discretization, which is exponential in the number of discretized axes. Instead of *enumerating* covering partitions of the random source, Dr. Bayes *samples parts* from the covering partitions and then *samples a point* from each sampled part, with time complexity linear in the number of samples and discretized axes. It applies bottom* arrow computations to the random source samples to get output samples, rejecting those outside the requested output set.

In short, Dr. Bayes uses preimage refinement only to reduce the rate of rejection when sampling under constraints, and thus relies only on its soundness.

We have tested Dr. Bayes on a variety of Bayesian inference tasks, including Bayesian regression and model selection [28]. Some of our Bayesian inference tests use recursion and constrain the outputs of deterministic functions, suggesting that Dr. Bayes and future probabilistic languages like it will allow practitioners to model real-world processes more expressively and precisely.

Recent work in probabilistic verification recasts it as a probabilistic inference task [9]. Given that Dr. Bayes's runtime is designed to sample efficiently under low-probability constraints, using it to probabilistically verify that a program does not exhibit certain errors is fairly natural. To do so, we

1. Encode the program in a way that propagates and returns errors.
2. Run the program with the constraint that the output is an error.

Sometimes, Dr. Bayes can determine that the preimage of the constrained output set is ∅, which is a proof that the program never exhibits an error. Otherwise, the longer the program runs without returning samples, the likelier it is that the preimage has zero probability or is empty; i.e. that an error does not occur.

As an extended example, we consider verifying floating-point error bounds.

While Dr. Bayes's numbers are implemented by floating-point intervals, semantically, they are real numbers. We therefore cannot easily represent floating-point numbers in Dr. Bayes—but we do not want to. We want *abstract* floating-point numbers, each consisting of an exact, real number and a bound on the relative error with which it is approximated. We define the following two structures to represent abstract floats.

```
(struct/drbayes float-any ())
(struct/drbayes float (value error))
```

An abstract value (float v e) represents every float between (* v (- 1 e)) and (* v (+ 1 e)) inclusive, while (float-any) represents NaN and other catastrophic error conditions. Abstract floating-point functions such as flsqrt compute exact results and use input error to compute bounds on output error:

```
(define/drbayes (flsqrt x)
  (if (float-any? x)
      x
      (let ([v  (float-value x)]
            [e  (float-error x)])
        (cond [(negative? v)  (float-any)]    ; NaN
              [(zero? v)       (float 0 0)]    ; exact case
              [else  ; v is positive
               (float (sqrt v)                          ; exact square root
                      (+ (- 1 (sqrt (- 1 e)))    ; relative error
                         (* 1/2 epsilon)))])))))  ; rounding error
```

We have similarly implemented abstract floating-point arithmetic, comparison, exponentials, and logarithms in Dr. Bayes.

Suppose we define an abstract floating-point implementation of the geometric distribution's inverse CDF using the formula $(\log u)/(\log (1 - p))$:

```
(define/drbayes (flgeometric-inv-cdf u p)
  (fl/ (fllog u) (fllog (fl- (float 1 0) p))))
```

We want the distribution of $\langle u, p \rangle$ in $(0,1) \times (0,1)$ with the value of

```
(float-error (flgeometric-inv-cdf (float u 0) (float p 0)))
```

constrained to $(3 \cdot \varepsilon, \infty)$, where $\varepsilon \approx 2.22 \cdot 10^{-16}$ is floating-point epsilon for 64-bit floats. That is, we want the distribution of inputs for which the floating-point output may be more than 3 epsilons away from the exact output.

Dr. Bayes returns samples of $\langle u, p \rangle$ within about $(0,1) \times (\varepsilon, 0.284)$, a fairly large domain on which error is greater than 3 epsilons. Realizing that the rounding error in $1 - p$ is magnified by log's relative error when p is small, we define

```
(define/drbayes (flgeometric-inv-cdf u p)
  (fl/ (fllog u) (fllog1p (flneg p))))
```

where `fllog1p` (abstractly) computes $\log(1 + x)$ with high accuracy. Dr. Bayes reports that the preimage of $(3 \cdot \varepsilon, \infty)$ is $\varnothing$. In fact, the preimage of $(1.51 \cdot \varepsilon, \infty)$ is $\varnothing$, so `flgeometric-inv-cdf` introduces error of no more than 1.51 epsilons.

We have used this technique to verify error bounds on the implementations of hypot, sqrt1pm1 and sinh in Racket's `math` library.

8   Related Work

Probabilistic languages can be approximately placed into two groups: those defined by a semantics, and those defined by an implementation.

Kozen's seminal work [15] on probabilistic semantics defines two measure-theoretic, denotational semantics, in two different styles: a **random-world semantics** [18] that interprets programs as deterministic functions that operate on a random source, and a **distributional semantics** that interprets programs as probability measures. It seems that all semantics work thereafter is in one of these styles. For example, Hurd [11] develops a random-world semantics in HOL and uses it to formally verify randomized algorithms such as the Miller-Rabin primality test. Ours is also a random-world semantics.

Jones [12] defines the probability monad as a categorical metatheory for interpreting probabilistic programs as distributions. Ramsey and Pfeffer [25] reformulate it in terms of Haskell's `return` and '$\gg=$', and use it to define a distributional semantics for a probabilistic lambda calculus. They implement the probability monad using probability mass functions, show that computing certain queries is inefficient, and devise an equivalent semantics that is more amenable to efficient implementation, for programs with finite probabilistic choice.

To put Infer.NET [21] on solid footing, Borgström et al. [4] define a distributional semantics for a first-order probabilistic language with bounded loops and constraints, by transforming terms into arrow-like combinators that produce measures. But Infer.NET interprets programs as probability density functions,[4] so they develop a semantics that does the same and prove equivalence.

---

[4] More precisely, as factor graphs, which represent probability density functions.

The work of Borgström et al. and Ramsey and Pfeffer exemplify a larger trend: while *defining* probabilistic languages can be done using measure theory, *implementing* them to support more than just evaluation (such as allowing constraints) has seemed hopeless enough to necessitate using a less explanatory theory of probability that has more obvious computational content. Indeed, the distributional semantics of Pfeffer's IBAL [24] and Nori et al.'s R2 [23] are defined in terms of probability mass and density functions in the first place. R2 lifts some of the resulting restrictions and speeds up sampling by propagating constraints toward the random values they refer to.

Some languages defined by an implementation are probabilistic Scheme [14], BUGS [17], BLOG [20], BLAISE [3], Church [8], and Kiselyov's embedded language for OCaml [13]. Recently, Wingate et al. [32] define nonstandard semantics that enable efficient inference, but do not define the languages. All of these languages are implemented in terms of probability mass or density functions.

Our work is similar in structure to monadic abstract interpretation [26,6], which also parameterizes a semantics on categorical meanings.

Cousot's probabilistic abstract interpretation [5] is a general framework for static analyses of probabilistic languages. It considers only random-world semantics, which is quite practical: because programs are interpreted as deterministic functions, many existing analyses easily apply. Our random-world semantics fits in this framework, but the concrete domain of preimage functions does not appear among Cousot's many examples, and we do not compute fixed points.

9   Conclusions and Future Work

To allow arbitrary constraints and recursion in probabilistic programs, we combined the power of measure theory with the unifying elegance of arrows. We (a) defined a transformation from first-order programs to arbitrary arrows, (b) defined the bottom arrow as a standard translation target, (c) derived the uncomputable preimage arrow as an alternative target, and (d) derived a sound, computable approximation of the preimage arrow, and enough computable lifts to transform programs. We implemented the abstract semantics and carried out Bayesian inference, stochastic ray tracing, and probabilistic verification.

In the future, we intend to add expressiveness by adding lambdas (possibly via closure conversion), explore ways to use static or dynamic analyses to speed up Monte Carlo algorithms, and explore preimage computation's connections to type checking and type inference. More broadly, we hope to advance probabilistic inference by providing a rich modeling language with an efficient, correct implementation, which allows general recursion and arbitrary constraints.

Acknowledgments. Special thanks to Mitchell Wand for careful review and helpful feedback. This material is based on research sponsored by DARPA under the Automated Program Analysis for Cybersecurity (FA8750-12-2-0106) project. The U.S. Government is authorized to reproduce and distribute reprints for Governmental purposes notwithstanding any copyright notation thereon.

References

1. Haskell 98 language and libraries, the revised report (December 2002), http://www.haskell.org/onlinereport/
2. Aumann, R.J.: Borel structures for function spaces. Illinois Journal of Mathematics 5, 614–630 (1961)
3. Bonawitz, K.A.: Composable Probabilistic Inference with Blaise. Ph.D. thesis, Massachusetts Institute of Technology (2008)
4. Borgström, J., Gordon, A.D., Greenberg, M., Margetson, J., Van Gael, J.: Measure transformer semantics for Bayesian machine learning. In: Barthe, G. (ed.) ESOP 2011. LNCS, vol. 6602, pp. 77–96. Springer, Heidelberg (2011)
5. Cousot, P., Monerau, M.: Probabilistic abstract interpretation. In: Seidl, H. (ed.) ESOP 2012. LNCS, vol. 7211, pp. 169–193. Springer, Heidelberg (2012)
6. Darais, D., Might, M., Van Horn, D.: Galois transformers and modular abstract interpreters. CoRR abs/1411.3962 (2014), http://arxiv.org/abs/1411.3962
7. DeGroot, M., Schervish, M.: Probability and Statistics. Addison Wesley Publishing Company, Inc. (2012)
8. Goodman, N., Mansinghka, V., Roy, D., Bonawitz, K., Tenenbaum, J.: Church: A language for generative models. In: Uncertainty in Artificial Intelligence (2008)
9. Gulwani, S., Jojic, N.: Program verification as probabilistic inference. In: Principles of Programming Languages, pp. 277–289 (2007)
10. Hughes, J.: Generalizing monads to arrows. Science of Computer Programming 37, 67–111 (2000)
11. Hurd, J.: Formal Verification of Probabilistic Algorithms. Ph.D. thesis. University of Cambridge (2002)
12. Jones, C.: Probabilistic Non-Determinism. Ph.D. thesis, Univ. of Edinburgh (1990)
13. Kiselyov, O., Shan, C.: Monolingual probabilistic programming using generalized coroutines. In: Uncertainty in Artificial Intelligence (2008)
14. Koller, D., McAllester, D., Pfeffer, A.: Effective Bayesian inference for stochastic programs. In: 14th National Conference on Artificial Intelligence (August 1997)
15. Kozen, D.: Semantics of probabilistic programs. In: Foundations of Computer Science (1979)
16. Lindley, S., Wadler, P., Yallop, J.: Idioms are oblivious, arrows are meticulous, monads are promiscuous. In: Workshop on Mathematically Structured Functional Programming (2008)
17. Lunn, D.J., Thomas, A., Best, N., Spiegelhalter, D.: WinBUGS – a Bayesian modelling framework. Statistics and Computing 10(4) (2000)
18. McAllester, D., Milch, B., Goodman, N.D.: Random-world semantics and syntactic independence for expressive languages. Tech. rep. MIT (2008)
19. McBride, C., Paterson, R.: Applicative programming with effects. Journal of Functional Programming 18(1) (2008)
20. Milch, B., Marthi, B., Russell, S., Sontag, D., Ong, D., Kolobov, A.: BLOG: Probabilistic models with unknown objects. In: International Joint Conference on Artificial Intelligence (2005)
21. Minka, T., Winn, J., Guiver, J., Webster, S., Zaykov, Y., Yangel, B., Spengler, A., Bronskill, J.: Infer.NET 2.6 (2014), http://research.microsoft.com/infernet
22. Moggi, E.: Computational lambda-calculus and monads. In: IEEE Symposium on Logic in Computer Science, pp. 14–23 (1989)
23. Nori, A.V., Hur, C.K., Rajamani, S.K., Samuel, S.: R2: An efficient MCMC sampler for probabilistic programs. In: AAAI Conference on Artificial Intelligence (2014)

24. Pfeffer, A.: The design and implementation of IBAL: A general-purpose probabilistic language. In: Statistical Relational Learning. MIT Press (2007)
25. Ramsey, N., Pfeffer, A.: Stochastic lambda calculus and monads of probability distributions. In: Principles of Programming Languages (2002)
26. Sergey, I., Devriese, D., Might, M., Midtgaard, J., Darais, D., Clarke, D., Piessens, F.: Monadic abstract interpreters. In: Programming Language Design and Implementation, pp. 399–410 (2013)
27. Tobin-Hochstadt, S., Felleisen, M.: The design and implementation of typed Scheme. In: Principles of Programming Languages, pp. 395–406 (2008)
28. Toronto, N.: Trustworthy, Useful Languages for Probabilistic Modeling and Inference. Ph.D. thesis, Brigham Young University (2014), http://students.cs.byu.edu/~ntoronto/dissertation.pdf
29. Toronto, N., McCarthy, J.: Computing in Cantor's paradise with λ_{ZFC}. In: Functional and Logic Programming Symposium, pp. 290–306 (2012)
30. Veach, E., Guibas, L.J.: Metropolis light transport. In: ACM SIGGRAPH, pp. 65–76 (1997)
31. Wadler, P.: Monads for functional programming. In: Jeuring, J., Meijer, E. (eds.) Advanced Functional Programming (2001)
32. Wingate, D., Goodman, N.D., Stuhlmüller, A., Siskind, J.M.: Nonstandard interpretations of probabilistic programs for efficient inference. In: Neural Information Processing Systems, pp. 1152–1160 (2011)

A Verified Compiler
for Probability Density Functions

Manuel Eberl, Johannes Hölzl, and Tobias Nipkow

Fakultät für Informatik, Technische Universität München, Germany

Abstract. Bhat *et al.* developed an inductive compiler that computes density functions for probability spaces described by programs in a probabilistic functional language. We implement such a compiler for a modified version of this language within the theorem prover Isabelle and give a formal proof of its soundness w. r. t. the semantics of the source and target language. Together with Isabelle's code generation for inductive predicates, this yields a fully verified, executable density compiler. The proof is done in two steps: First, an abstract compiler working with abstract functions modelled directly in the theorem prover's logic is defined and proved sound. Then, this compiler is refined to a concrete version that returns a target-language expression.

1 Introduction

Random distributions of practical significance can often be expressed as probabilistic functional programs. When studying a random distribution, it is often desirable to determine its *probability density function* (PDF). This can be used to e. g. determine the expectation or sample the distribution with a sampling method such as *Markov-chain Monte Carlo* (MCMC).

Bhat *et al.* [5] presented a compiler that computes the probability density function of a program in the probabilistic functional language Fun. Fun is a small functional language with basic arithmetic, Boolean logic, product and sum types, conditionals, and a number of built-in discrete and continuous distributions. It does not support lists or recursion. They evaluated the compiler on a number of practical problems and concluded that it reduces the amount of time and effort required to model them in an MCMC system significantly compared to hand-written models. A correctness proof for the compiler is sketched.

Bhat *et al.* [4] stated that their eventual goal is the formal verification of this compiler in a theorem prover. We have verified such a compiler for a similar probabilistic functional language in the interactive theorem prover Isabelle/HOL [18, 19]. Our contributions are the following:

- a formalisation of the source language, target language (whose semantics had previously not been given precisely), and the compiler on top of a foundational theory of measure spaces
- a formal verification of the correctness of the compiler
- executable code for the compiler using Isabelle's code generator

© Springer-Verlag Berlin Heidelberg 2015
J. Vitek (Ed.): ESOP 2015, LNCS 9032, pp. 80–104, 2015.
DOI: 10.1007/978-3-662-46669-8_4

In the process, we uncovered an incorrect generalisation of one of the compiler rules in the draft of an extended version of the paper by Bhat *et al.* [6].

The complete formalisation is available online [13].

In this paper, we focus entirely on the correctness proof; for more motivation and applications, the reader should consult Bhat *et al.* [5].

1.1 Related Work

Park *et al.* [20] developed a probabilistic extension of Objective CAML called $\lambda_\bigcirc$. While Bhat *et al.* generate density functions of functional programs, Park *et al.* generate *sampling functions*. This approach allows them to handle much more general distributions, even recursively-defined ones and distributions that do not have a density function, but it does not allow precise reasoning about these distributions (such as determining the exact expectation). No attempt at formal verification is made.

Several formalisations of probabilistic programs already exist. Hurd [16] formalises programs as random variables on infinite streams of random bits. Hurd *et al.* [17] and Cock [8, 9] both formalise pGCL, an imperative programming language with probabilistic and non-deterministic choice. Audebaud and Paulin-Mohring [1] verify probabilistic functional programs in Coq [3] using a shallow embedding based on the Giry monad on discrete probability distributions. All these program semantics support only discrete distributions – even the framework by Hurd [16], although it is based on measure theory.

Our work relies heavily on a formalisation of measure theory by Hölzl [15] and some material developed for the proof of the Central Limit Theorem by Avigad *et al.* [2].

1.2 Outline

Section 2 explains the notation and gives a brief overview of the mathematical basis. Section 3 defines the source and target language. Section 4 defines the abstract compiler and gives a high-level outline of the soundness proof. Section 5 explains the refinement of the abstract compiler to the concrete compiler and the final correctness result and evaluates the compiler on a simple example.

2 Preliminaries

2.1 Typographical Notes

We will use the following typographical conventions in mathematical formulæ:

- Constants, functions, datatype constructors, and types will be set in slanted font.
- Free and bound variables (including type variables) are set in italics.
- Isabelle keywords are set in bold font: **lemma**, **datatype**, etc.
- σ-algebras are set in calligraphic font: $\mathcal{A}$, $\mathcal{B}$, $\mathcal{M}$, etc.
- File names of Isabelle theories are set in a monospaced font: `PDF_Compiler.thy`.

2.2 Isabelle/HOL Basics

The formalisations presented in this paper employ the Isabelle/HOL theorem prover. We will aim to stay as close to the Isabelle formalisation syntactically as possible. In this section, we give an overview of the syntactic conventions we use to achieve this.

The term syntax follows the λ-calculus, i.e. function application is juxtaposition as in $f\ t$. The notation $t :: \tau$ means that the term t has type τ. Types are built from the base types *bool*, *nat* (natural numbers), *real* (reals), *ereal* (extended reals, i.e. $real \cup \{+\infty, -\infty\}$), and type variables ($\alpha$, β, etc) via the function type constructor $\alpha \to \beta$ or the set type constructor α *set*. The constant *undefined* :: α describes an arbitrary element for each type α. There are no further axioms about it, expecially no defining equation. $f\ `X$ is the image set of X under f: $\{f\ x \mid x \in X\}$. We write $\langle P \rangle$ for the indicator of P: 1 if P is true, 0 otherwise.

Because we represent variables by de Bruijn indices [7], variable names are natural numbers and program states are functions of type *nat* $\to \alpha$. As HOL functions are total, we use *undefined* to fill in the unused places, e.g. ($\lambda x.$ *undefined*) describes the empty state. Prepending an element x to a state $\omega :: nat \to \alpha$ is written as $x \bullet \omega$, i.e. $(x \bullet \omega)\ 0 = x$ and $(x \bullet \omega)\ (n+1) = \omega\ n$. The function *merge* merges two states with given domains:

$$\text{merge } V\ V'\ (\rho, \sigma) = \begin{cases} \rho\ x & \text{if } x \in V \\ \sigma\ y & \text{if } x \in V' \setminus V \\ undefined & \text{otherwise} \end{cases}$$

Notation. We use Γ to denote a type environment, i. e. a function from variable names to types, and σ to denote a state.

The notation $t \bullet \Gamma$ (resp. $v \bullet \sigma$) denotes the insertion of a new variable with the type t (resp. value v) into a typing environment (resp. state). We use the same notation for inserting a new variable into a set of variables, shifting all other variables, i. e.: $0 \bullet V = \{0\} \cup \{y + 1 \mid y \in V\}$

2.3 Measure Theory in Isabelle/HOL

We use Isabelle's measure theory, as described in [15]. This section gives an introduction of the measure-theoretical concepts used in this paper. The type α *measure* describes a measure over the type α. Each measure μ is described by the following three projections: *space* $\mu :: \alpha$ *set* is the space, *sets* $\mu :: \alpha$ *set set* are the measurable sets, and *measure* $\mu :: \alpha$ *set* $\to$ *ereal* is the measure function valid on the measurable sets. The type α *measure* guarantees that the measurable sets are a σ-algebra and that *measure* μ is a non-negative and σ-additive function. In the following we will always assume that the occuring sets and functions are measurable. We also provide integration over measures:

$\int$ :: $(\alpha \rightarrow ereal) \rightarrow \alpha \ measure \rightarrow ereal$, we write $\int x.\ f\ x\ \partial\mu$. This is the *non-negative Lebesgue integral*; for this integral, most rules do not require integrable functions – measurability is enough.

We write $(A, \mathcal{A}, \mu)$ for a measure with space A, measurable sets $\mathcal{A}$ and the measure function μ :: $\alpha\ set \rightarrow ereal$. If we are only intersted in the measurable space we write $(A, \mathcal{A})$. When constructing a measure in this way, the measure function is the constant zero function.

Sub-probability Spaces. A sub-probability space is a measure $(A, \mathcal{A}, \mu)$ with $\mu\ A \leq 1$. For technical reasons, we also assume $A \neq \emptyset$. This is required later in order to define the *bind* operation in the Giry monad in a convenient way within Isabelle. This non-emptiness condition will always be trivially satisfied by all the measure spaces used in this work.

Constructing Measures. The semantics of our language will be given in terms of measures. We have the following functions to construct measures:

Counting: $measure\ (count\ A)\ X = |X|$
Lebesgue-Borel: $measure\ borel\ [a; b] = b - a$
With density: $measure\ (density\ \mu\ f)\ X = \int x.\ f\ x \cdot \langle x \in X \rangle \partial\mu$
Push-forward: $measure\ (distr\ \mu\ \nu\ f)\ X = measure\ \mu\ \{x \mid f\ x \in X\}$
Product: $measure\ (\mu \otimes \nu)\ (A \times B) = measure\ \mu\ A \cdot measure\ \nu\ B$
Indexed product: $measure\ (\bigotimes_{i \in I} \mu_i)\ (\times_{i \in I} A_i) = \prod_{i \in I} measure\ \mu_i\ A_i$
Embedding: $measure\ (embed\ \mu\ f)\ X = measure\ \mu\ \{x \mid f\ x \in X\}$

The push-forward measure and the embedding of a measure have different measurable sets. The σ-algebra of the push-forward measure *distr* $\mu\ \nu\ f$ is given by ν. The measure is only well-defined when f is μ-ν-measurable. The σ-algebra of *embed* $\mu\ f$ is generated by the sets $f[A]$ for A μ-measurable. The embedding measure is well-defined when f is injective.

2.4 Giry Monad

The category theory part of this section is based mainly on a presentation by Ernst-Erich Doberkat [11]. For a more detailed introduction, see his textbook [10] or the original paper by Michèle Giry [14]. Essentially the Giry monad is a monad on measures.

The category $\mathfrak{Meas}$ has measurable spaces as objects and measurable maps as morphisms. This category forms the basis on which we will define the Giry monad. In Isabelle/HOL, the objects are represented by the type *measure*, and the morphism are represented as regular functions. When we mention a measurable function, we explicitly need to mention the measurable spaces representing the domain and the range.

The *sub-probability* functor $\mathbb{S}$ is an endofunctor on $\mathfrak{Meas}$. It maps a measurable space $\mathcal{A}$ to the measurable space of all sub-probabilities on $\mathcal{A}$. Given a measurable space $(A, \mathcal{A})$, we consider the set of all sub-probability measures on $(A, \mathcal{A})$:

$$M = \{\mu \mid \mu \text{ is a sub-probability measure on } (A, \mathcal{A})\}$$

The measurable space $\mathbb{S}(A, \mathcal{A})$ is the smallest measurable space on M that fulfils the following property:

For all $X \in \mathcal{A}$, $(\lambda\mu. \; measure \; \mu \; X)$ is $\mathbb{S}(A, \mathcal{A})$-Borel-measurable

A $\mathcal{M}$-$\mathcal{N}$-measurable function f is mapped with $\mathbb{S}(f) = \lambda\mu. \; distr \; \mu \; \mathcal{N} \; f$, where all μ are sub-probability measures on $\mathcal{M}$.

The Giry monad naturally captures the notion of choosing a value according to a (sub-)probability distribution, using it as a parameter for another distribution, and observing the result.

Consequently, *return* yields a Dirac measure, i. e. a probability measure in which all the "probability" lies in a single element, and *bind* (or $\ggg$) integrates over all the input values to compute one single output measure. Formally, for measurable spaces $(A, \mathcal{A})$ and $(B, \mathcal{B})$, a measure μ on $(A, \mathcal{A})$, a value $x \in A$, and a $\mathcal{A}$-$\mathbb{S}(B, \mathcal{B})$-measurable function f:

$$return :: \alpha \to \alpha \; measure$$
$$return \; x := \lambda X. \; \begin{cases} 1 & \text{if } x \in X \\ 0 & \text{otherwise} \end{cases}$$

$$bind :: \alpha \; measure \to (\alpha \to \beta \; measure) \to \beta \; measure$$
$$\mu \ggg f := \lambda X. \int x. \; f(x)(X) \, \partial\mu$$

The actual definitions of *return* and *bind* in Isabelle are slightly more complicated due to Isabelle's simple type system. In informal mathematics, a function typically has attached to it the information of what its domain and codomain are and what the corresponding measurable spaces are; with simple types, this is not directly possible and requires some tricks in order to infer the carrier set and the σ-algebra of the result.

The "do" Syntax. For better readability, we employ a Haskell-style "**do** notation" for operations in the Giry monad. The syntax of this notation is defined recursively, where M stands for a monadic expression and $\langle text \rangle$ stands for arbitrary "raw" text:

$$\textbf{do} \; \{M\} \; \hat{=} \; M \qquad \textbf{do} \; \{x \leftarrow M; \; \langle text \rangle\} \; \hat{=} \; M \ggg (\lambda x. \; \textbf{do} \; \{\langle text \rangle\})$$

3   Source and Target Language

The source language used in the formalisation was modelled after the language Fun described by Bhat *et al.* [5]; similarly, the target language is almost identical to the target language used by Bhat *et al.* However, we have made the following changes in our languages:

- Variables are represented by de Bruijn indices.
- No sum types are supported. Consequently, the **match-with** construct is replaced with an *IF-THEN-ELSE*. Furthermore, booleans are a primitive type rather than represented as *unit + unit*.
- The type *double* is called *real* and it represents a real number with absolute precision as opposed to an IEEE 754 floating point number.

In the following subsections, we give the precise syntax, typing rules, and semantics of both our source language and our target language.

3.1 Types, Values, and Operators

The source language and the target language share the same type system and the same operators. Figure 1 shows the types and values that exist in our languages.[1] Additionally, standard arithmetical and logical operators exist.

All operators are *total*, meaning that for every input value of their parameter type, they return a single value of their result type. This requires some non-standard definitions for non-total operations such as division, the logarithm, and the square root. Non-totality could also be handled by implementing operators in the Giry monad by letting them return either a Dirac distribution with a single result or, when evaluated for a parameter on which they are not defined, the null measure. This, however, would probably complicate many proofs significantly.

To increase readability, we will use the following abbreviations:

- *TRUE* and *FALSE* stand for *BoolVal True* and *BoolVal False*, respectively.
- *RealVal, IntVal*, etc. will be omitted in expressions when their presence is implicitly clear from the context.

---

datatype *pdf_type* =
    UNIT | $\mathbb{B}$ | $\mathbb{Z}$ | $\mathbb{R}$ | *pdf_type* × *pdf_type*
datatype *val* =
    UnitVal | *BoolVal bool* | *IntVal int* | *RealVal real* | <|*val, val*|>
datatype *pdf_operator* =
    Fst | *Snd* | *Add* | *Mult* | *Minus* | *Less* | *Equals* | *And* | *Or* | *Not* | *Pow* |
    Fact | *Sqrt* | *Exp* | *Ln* | *Inverse* | *Pi* | *Cast pdf_type*

Fig. 1. Types and values in source and target language

---

[1] Note that *bool*, *int*, and *real* stand for the respective Isabelle types, whereas $\mathbb{B}$, $\mathbb{Z}$, and $\mathbb{R}$ stand for the source-/target-language types.

Table 1. Auxiliary functions

FUNCTION	DESCRIPTION
op_sem op v	semantics of operator *op* applied to *v*
op_type op t	result type of operator *op* for input type *t*
dist_param_type dst	parameter type of the built-in distribution *dst*
dist_result_type dst	result type of the built-in distribution *dst*
dist_measure dst x	built-in distribution *dst* with parameter *x*
dist_dens dst x y	density of the built-in distribution *dst* w. parameter *x* at value *y*
type_of Γ e	the unique *t* such that $\Gamma \vdash e : t$
val_type v	the type of value *v*, e. g. *val_type (IntVal 42) = INTEG*
type_universe t	the set of values of type *t*
countable_type t	*True* iff *type_universe t* is a countable set
free_vars e	the free variables in the expression *e*
e det	*True* iff *e* does not contain *Random* or *Fail*
extract_real x	returns *y* for *x = RealVal y* (analogous for int, pair, etc.)
return_val v	*return (stock_measure (val_type v)) v*
null_measure M	measure with same measurable space as *M*, but 0 for all sets

3.2 Auxiliary Definitions

A number of auxiliary definitions are used in the definition of the semantics; Table 1 lists some simple auxiliary functions. Additionally, the following two notions require a detailed explanation:

Stock Measures. The *stock measure* for a type *t* is the "natural" measure on values of that type. This is defined as follows:

- For the countable types *UNIT*, $\mathbb{B}$, $\mathbb{Z}$: the count measure over the corresponding type universes
- For type $\mathbb{R}$: the embedding of the Lebesgue-Borel measure on $\mathbb{R}$ with *RealVal*
- For $t_1 \times t_2$: the embedding of the product measure

$$\text{stock\_measure } t_1 \otimes \text{stock\_measure } t_2$$

with $\lambda(v_1, v_2). <|v_1, v_2|>$

Note that in order to save space and increase readability, we will often write $\int x. \, f \, x \, \partial t$ instead of $\int x. \, f \, x \, \partial \text{stock\_measure } t$ in integrals.

State Measure. Using the stock measure, we construct a measure on states in the context of a typing environment Γ. A state σ is *well-formed* w. r. t. to *V* and

Γ if it maps every variable $x \in V$ to a value of type $\Gamma\ x$ and every variable $\notin V$ to *undefined*. We fix Γ and a finite V and consider the set of well-formed states w.r.t. V and Γ. Another representation of these states are tuples in which the i-th component is the value of the i-th variable in V. The natural measure that can be given to such tuples is the finite product measure of the stock measures of the types of the variables:

$$state\_measure\ V\ \Gamma := \bigotimes_{x \in V} stock\_measure\ (\Gamma\ x)$$

3.3   Source Language

---

datatype *expr* =

    Var nat | *Val val* | *LET expr IN expr* | *pdf_operator* $ *expr* | *<expr, expr>* |

    Random pdf_dist | *IF expr THEN expr ELSE expr* | *Fail pdf_type*

Fig. 2. Source language expressions

Figure 2 shows the syntax of the source language. It contains variables (de Bruijn), values, *LET*-expressions (again de Bruijn), operator application, pairs, sampling a parametrised built-in random distribution, *IF-THEN-ELSE* and failure. We omit the constructor *Val* when its presence is obvious from the context.

Figures 3 and 4 show the typing rules and the monadic semantics of the source language.

$$\frac{}{\Gamma \vdash Val\ v\ :\ val\_type\ v} \qquad \frac{}{\Gamma \vdash Var\ x\ :\ \Gamma\ x} \qquad \frac{}{\Gamma \vdash Fail\ t\ :\ t}$$

$$\frac{\Gamma \vdash e\ :\ t \qquad op\_type\ op\ t = Some\ t'}{\Gamma \vdash op\ \$\ e\ :\ t'} \qquad \frac{\Gamma \vdash e_1\ :\ t_1 \qquad \Gamma \vdash e_2\ :\ t_2}{\Gamma \vdash <e_1, e_2>\ :\ t_1 \times t_2}$$

$$\frac{\Gamma \vdash b\ :\ \mathbb{B} \qquad \Gamma \vdash e_1\ :\ t \qquad \Gamma \vdash e_2\ :\ t}{\Gamma \vdash IF\ b\ THEN\ e_1\ ELSE\ e_2\ :\ t} \qquad \frac{\Gamma \vdash e_1\ :\ t_1 \qquad t_1 \bullet \Gamma \vdash e_2\ :\ t_2}{\Gamma \vdash LET\ e_1\ IN\ e_2\ :\ t_2}$$

$$\frac{\Gamma \vdash e\ :\ dist\_param\_type\ dst}{\Gamma \vdash Random\ dst\ e\ :\ dist\_result\_type\ dst}$$

Fig. 3. Typing rules of the source language

$$expr\_sem \; :: \; state \to expr \to val \; measure$$

$$expr\_sem \; \sigma \; (Val \; v) \; = \; return\_val \; v$$

$$expr\_sem \; \sigma \; (Var \; x) \; = \; return\_val \; (\sigma \; x)$$

$$expr\_sem \; \sigma \; (LET \; e_1 \; IN \; e_2) \; =$$

$$\textbf{do} \; \{v \leftarrow expr\_sem \; \sigma \; e_1; \; expr\_sem \; (v \bullet \sigma) \; e_2\}$$

$$expr\_sem \; \sigma \; (op \; \$ \; e) \; =$$

$$\textbf{do} \; \{v \leftarrow expr\_sem \; \sigma \; e; \; return\_val \; (op\_sem \; op \; v)\}$$

$$expr\_sem \; \sigma \; <e_1, e_2> \; =$$

$$\textbf{do} \; \{v_1 \leftarrow expr\_sem \; \sigma \; e_1; \; v_2 \leftarrow expr\_sem \; \sigma \; e_2;$$

$$return\_val \; <|v_1, v_2|>\}$$

$$expr\_sem \; \sigma \; (IF \; b \; THEN \; e_1 \; ELSE \; e_2) \; =$$

$$\textbf{do} \; \{b' \leftarrow expr\_sem \; \sigma \; b;$$

$$expr\_sem \; \sigma \; (\textbf{if} \; b' = TRUE \; \textbf{then} \; e_1 \; \textbf{else} \; e_2)\}$$

$$expr\_sem \; \sigma \; (Random \; dst \; e) \; =$$

$$\textbf{do} \; \{p \leftarrow expr\_sem \; \sigma \; e; \; dist\_measure \; dst \; p\}$$

$$expr\_sem \; \sigma \; (Fail \; t) \; = \; null\_measure \; (stock\_measure \; t)$$

Fig. 4. Semantics of the source language

Figure 5 shows the built-in distributions of the source language, their parameter types and domains, the types of the random variables they describe, and their density functions in terms of their parameter. When given a parameter *outside* their domain, they return the null measure. We support the same distributions as Bhat *et al.*, except for the Beta and Gamma distributions (merely because we have not formalised them yet).

3.4 Deterministic Expressions

We call an expression e *deterministic* (written as "e det") if it contains no occurrence of *Random* or *Fail*. Such expressions are of particular interest: if all their free variables have a fixed value, they return precisely one value, so we can define a function $expr\_sem\_rf$[2] that, when given a state σ and a deterministic expression e, returns this single value. This function can be seen as a non-monadic analogue to *expr_sem* and its definition is therefore analogous and is not shown here. The

---

[2] In Isabelle, the expression *randomfree* is used instead of *deterministic*, hence the "rf" suffix. This is in order to emphasise the syntactical nature of the property. Note that a syntactically deterministic expression is not truly deterministic if the variables it contains are randomised over, which can be the case.

DISTRIBUTION	PARAM.	DOMAIN	TYPE	DENSITY FUNCTION
Bernoulli	$\mathbb{R}$	$p \in [0;1]$	$\mathbb{B}$	$\lambda x. \begin{cases} p & \text{for } x = TRUE \\ 1-p & \text{for } x = FALSE \end{cases}$
UniformInt	$\mathbb{Z} \times \mathbb{Z}$	$p_1 \le p_2$	$\mathbb{Z}$	$\lambda x. \dfrac{\langle x \in [p_1;p_2]\rangle}{p_2 - p_1 + 1}$
UniformReal	$\mathbb{R} \times \mathbb{R}$	$p_1 < p_2$	$\mathbb{R}$	$\lambda x. \dfrac{\langle x \in [p_1;p_2]\rangle}{p_2 - p_1}$
Gaussian	$\mathbb{R} \times \mathbb{R}$	$p_2 > 0$	$\mathbb{R}$	$\lambda x. \dfrac{1}{\sqrt{2\pi p_2^2}} \cdot \exp\left(-\dfrac{(x-p_1)^2}{2p_2^2}\right)$
Poisson	$\mathbb{R}$	$p \ge 0$	$\mathbb{Z}$	$\lambda x. \begin{cases} \exp(-p) \cdot p^x / x! & \text{for } x \ge 0 \\ 0 & \text{otherwise} \end{cases}$

Fig. 5. Built-in distributions of the source language.
The density functions are given in terms of the parameter p, which is of the type given in the column "parameter type". If p is of a product type, p_1 and p_2 stand for the two components of p.

function $expr\_sem$ has the following property (assuming that e is deterministic and well-typed and σ is a valid state):

$$expr\_sem\ \sigma\ e\ =\ return\ (expr\_sem\_rf\ \sigma\ e)$$

This property will enable us to convert deterministic source-language expressions into "equivalent" target-language expressions.

3.5   Target Language

The target language is again modelled very closely after the one by Bhat *et al.* [5]. The type system and the operators are the same as in the source language. The key difference is that the *Random* construct has been replaced by an integral. As a result, while expressions in the source language return a measure space, expressions in the target language always return a single value.

Since our source language lacks sum types, so does our target language. Additionally, our target language differs from the one by Bhat *et al.* in the following respects:

– Our language has no function types; since functions only occur as integrands and as final results (as the compilation result is a density function), we can simply define integration to introduce the integration variable as a bound variable and let the final result contain a single free variable with de Bruijn index 0, i. e. there is an implicit λ abstraction around the compilation result.

- Evaluation of expressions in our target language can never fail. In the language by Bhat *et al.*, failure is used to handle undefined integrals; we, on the other hand, use the convention of Isabelle's measure theory library, which returns 0 for integrals of non-integrable functions. This has the advantage of keeping the semantics simple, which makes proofs considerably easier.

- Our target language does not have *LET*-bindings, since, in contrast to the source language, they would be semantically superfluous here. However, they are still useful in practice since they yield shorter expressions and can avoid multiple evaluation of the same term; they could be added with little effort.

Figures 6, 7, and 8 show the syntax, typing rules, and semantics of the target language.

The matter of target-language semantics in the papers by Bhat *et al.* is somewhat unclear. In the 2012 POPL paper [4], the only semantics given for the target language is a vague denotational rule for the integral. In the 2013 TACAS paper [5], no target-language semantics is given at all; it is only said that "standard CBV small-step evaluation" is used. The extended version of this paper currently submitted for publication [6] indeed gives some small-step evaluation rules, but only for simple cases. In particular, none of these publications give the precise rules for evaluating integral expressions. It is quite unclear to us how small-step evaluation of integral expressions is possible in the first place. Another issue is how to handle evaluation of integral expressions where the integrand evaluates to $\perp$ for some values.[3]

Converting deterministic expressions. The auxiliary function *expr_rf_to_cexpr*, which will be used in some rules of the compiler that handle deterministic expressions, is of particular interest. We mentioned earlier that deterministic source-language expressions can be converted to equivalent target-language expressions.[4] This function does precisely that. Its definition is mostly obvious, apart from the *LET* case. Since our target language does not have a *LET* construct, the function must resolve *LET*-bindings in the source-language expression by substituting the bound expression.

expr_rf_to_cexpr satisfies the following equality for any deterministic source-language expression e:

$$cexpr\_sem \ \sigma \ (expr\_rf\_to\_cexpr \ e) \ = \ expr\_sem\_rf \ \sigma \ e$$

---

[3] We do not have this problem since in our target language, as mentioned before, evaluation cannot fail.

[4] Bhat *et al.* say that a deterministic expression "is also an expression in the target language syntax, and we silently treat it as such" [5]

---

datatype *cexpr* =
  CVar nat | *CVal val* | *pdf_operator* $\$_c$ *cexpr* | $<cexpr, cexpr>_c$ |
  IF$_c$ cexpr THEN cexpr ELSE cexpr | $\int_c cexpr\, \partial pdf\_type$

Fig. 6. Target language expressions

$$\overline{\Gamma \vdash_c CVal\ v\ :\ val\_type\ v} \qquad\qquad \overline{\Gamma \vdash_c CVar\ x\ :\ \Gamma\ x}$$

$$\frac{\Gamma \vdash_c e\ :\ t \qquad op\_type\ op\ t = Some\ t'}{\Gamma \vdash_c op\ \$_c\ e\ :\ t'} \qquad \frac{\Gamma \vdash_c e_1\ :\ t_1 \qquad \Gamma \vdash_c e_2\ :\ t_2}{\Gamma \vdash_c <e_1, e_2>_c\ :\ t_1 \times t_2}$$

$$\frac{\Gamma \vdash_c b\ :\ \mathbb{B} \qquad \Gamma \vdash_c e_1\ :\ t \qquad \Gamma \vdash_c e_2\ :\ t}{\Gamma \vdash_c IF_c\ b\ THEN\ e_1\ ELSE\ e_2\ :\ t} \qquad \frac{t \bullet \Gamma \vdash_c e\ :\ \mathbb{R}}{\Gamma \vdash_c \int_c e\ \partial t\ :\ \mathbb{R}}$$

Fig. 7. Typing rules for target language

$cexpr\_sem\ ::\ state \rightarrow cexpr \rightarrow val$
$cexpr\_sem\ \sigma\ (CVal\ v)\ =\ v$
$cexpr\_sem\ \sigma\ (CVar\ x)\ =\ \sigma\ x$
$cexpr\_sem\ \sigma\ <e_1, e_2>_c\ =\ <| cexpr\_sem\ \sigma\ e_1, cexpr\_sem\ \sigma\ e_2 |>$
$cexpr\_sem\ \sigma\ op\ \$_c\ e\ =\ op\_sem\ op\ (cexpr\_sem\ \sigma\ e)$
$cexpr\_sem\ \sigma\ (IF_c\ b\ THEN\ e_1\ ELSE\ e_2)\ =$
    (**if** $cexpr\_sem\ \sigma\ b = TRUE$ **then** $cexpr\_sem\ \sigma\ e_1$ **else** $cexpr\_sem\ \sigma\ e_2$)
$cexpr\_sem\ \sigma\ (\int_c e\ \partial t)\ =$
    $RealVal\ (\int x.\ extract\_real\ (cexpr\_sem\ (x \bullet \sigma)\ e)\ \partial stock\_measure\ t)$

Fig. 8. Semantics of target language

4 Abstract Compiler

The correctness proof is done in two steps using a refinement approach: first, we define and prove correct an *abstract compiler* that returns the density function as an abstract mathematical function. We then define an analogous *concrete compiler* that returns a target-language expression and show that it is a *refinement* of the abstract compiler, which will allow us to lift the correctness result from the latter to the former.

4.1 Density Contexts

First, we define the notion of a *density context*, which holds the context data the compiler will require to compute the density of an expression. A density context is a tuple $\Upsilon = (V, V', \Gamma, \delta)$ that contains the following information:

- The set V of random variables in the current context. These are the variables that are randomised over.
- The set V' of parameter variables in the current context. These are variables that may occur in the expression, but are not randomised over but treated as constants.
- The type environment Γ
- A density function δ that returns the common density of the variables V under the parameters V'. Here, δ is a function from *space* (*state_measure* ($V \cup V'$) Γ) to the extended real numbers.

A density context (V, V', Γ, δ) describes a parametrised measure on the states on $V \cup V'$. Let $\rho \in$ *space* (*state_measure* V' Γ) be a parameter state. We write

$$\text{dens\_ctxt\_measure } (V, V', \Gamma, \delta) \ \rho$$

for the measure obtained by taking *state_measure* V Γ, transforming it by merging a given state σ with the parameter state ρ and finally applying the density δ on the resulting image measure. The Isabelle definition of this is:

$$\text{dens\_ctxt\_measure} \ :: \ \text{dens\_ctxt} \to \text{state} \to \text{state measure}$$
$$\text{dens\_ctxt\_measure} \ (V, V', \Gamma, \delta) \ \rho \ = \ \text{density } (\text{distr } (\text{state\_measure } V \ \Gamma)$$
$$(\text{state\_measure } (V \cup V') \ \Gamma) \ (\lambda\sigma. \ \text{merge } V \ V' \ (\sigma, \rho))) \ \delta$$

Informally, *dens_ctxt_measure* describes the measure obtained by integrating over the variables $\{v_1, \ldots, v_m\} = V$ while treating the variables $\{v'_1, \ldots, v'_n\} = V'$ as parameters. The evaluation of an expression e with variables from $V \cup V'$ in this context is effectively a function

$$\lambda v'_1 \ \ldots \ v'_n. \int v_1. \ \ldots \int v_m. \ \text{expr\_sem } (v_1, \ldots, v_m, v'_1, \ldots, v'_n) \ e \ \cdot$$
$$\delta \ (v_1, \ldots, v_m, v'_1, \ldots, v'_n) \ \partial\Gamma \ v_1 \ldots \partial\Gamma \ v_m \ .$$

A density context is *well-formed* (predicate *density_context* in Isabelle) if:

- V and V' are finite and disjoint
- $\delta \, \sigma \geq 0$ for any $\sigma \in space \; (state\_measure \; (V \cup V') \; \varGamma)$
- δ is Borel-measurable w. r. t. $state\_measure \; (V \cup V') \; \varGamma$
- the measure $dens\_ctxt\_measure \; (V, V', \varGamma, \delta) \; \rho$ is a sub-probability measure for any $\rho \in space \; (state\_measure \; V' \; \varGamma)$

4.2 Definition

As a first step, we have implemented an abstract density compiler as an inductive predicate $\varGamma \vdash_d e \Rightarrow f$, where $\varGamma$ is a density context, e is a source-language expression and f is a function of type $val \; state \to val \to ereal$. Its first parameter is a state that assigns values to the free variables in e and its second parameter is the value for which the density is to be computed. The compiler therefore computes a density function that is parametrised with the values of the non-random free variables in the source expression.

The compilation rules are very similar to those by Bhat *et al.* [5], except for the following adaptations:

- Bhat *et al.* handle *IF-THEN-ELSE* with the "match" rule for sum types. As we do not support sum types, we have a dedicated rule for *IF-THEN-ELSE*.
- The use of de Bruijn indices requires shifting of variable sets and states whenever the scope of a new bound variable is entered; unfortunately, this makes some rules somewhat technical.
- We do not provide any compiler support for *deterministic LET*-bindings. They are semantically redundant, as they can always be expanded without changing the semantics of the expression. In fact, they *have* to be unfolded for compilation, so they can be regarded as a feature that adds convenience, but no expressivity.

The following list shows the standard compilation rules adapted from Bhat *et al.*, plus a rule for multiplication with a constant.[5] The functions $marg\_dens$ and $marg\_dens2$ compute the marginal density of one and two variables by "integrating away" all the other variables from the common density δ. The function $branch\_prob$ computes the probability of being in the current branch of execution by integrating over *all* the variables in the common density δ.

$$\text{HD\_VAL} \quad \frac{countable\_type \; (val\_type \; v)}{(V, V', \varGamma, \delta) \vdash_d Val \; v \Rightarrow \lambda \rho x. \; branch\_prob \; (V, V', \varGamma, \delta) \; \rho \cdot \langle x = v \rangle}$$

$$\text{HD\_VAR} \quad \frac{x \in V}{(V, V', \varGamma, \delta) \vdash_d Var \; x \Rightarrow marg\_dens \; (V, V', \varGamma, \delta) \; x}$$

---

[5] Additionally, three congruence rules are required for technical reasons. These rules are required because the abstract and the concrete result may differ on a null set and outside their domain.

$$\text{HD\_PAIR} \quad \frac{x \in V \qquad y \in V \qquad x \neq y}{(V, V', \Gamma, \delta) \vdash_{\mathrm{d}} <\!\mathit{Var}\ x, \mathit{Var}\ y\!> \Rightarrow \mathit{marg\_dens2}\ (V, V', \Gamma, \delta)\ x\ y}$$

$$\text{HD\_FAIL} \quad \frac{}{(V, V', \Gamma, \delta) \vdash_{\mathrm{d}} \mathit{Fail}\ t \Rightarrow \lambda \rho\, x.\ 0}$$

$$\text{HD\_LET}$$
$$\frac{\begin{array}{c}(\emptyset, V \cup V', \Gamma, \lambda x.\ 1) \vdash_{\mathrm{d}} e_1 \Rightarrow f \\ (0 \bullet V, \{x + 1 \mid x \in V'\}, \mathit{type\_of}\ \Gamma\ e_1 \bullet \Gamma, \\ \lambda \rho.\ f\ (\lambda x.\ \rho\ (x + 1))\ (\rho\ 0) \cdot \delta\ (\lambda x.\ \rho\ (x + 1))) \vdash_{\mathrm{d}} e_2 \Rightarrow g\end{array}}{(V, V', \Gamma, \delta) \vdash_{\mathrm{d}} \mathit{LET}\ e_1\ \mathit{IN}\ e_2 \Rightarrow \lambda \rho.\ g\ (\mathit{undefined} \bullet \rho)}$$

$$\text{HD\_RAND} \quad \frac{(V, V', \Gamma, \delta) \vdash_{\mathrm{d}} e \Rightarrow f}{\begin{array}{c}(V, V', \Gamma, \delta) \vdash_{\mathrm{d}} \mathit{Random}\ dst\ e \Rightarrow \\ \lambda \rho\, y.\ \int x.\ f\ \rho\ x \cdot \mathit{dist\_dens}\ dst\ x\ y\ \partial\, \mathit{dist\_param\_type}\ dst\end{array}}$$

$$\text{HD\_RAND\_DET}$$
$$\frac{e\ \mathit{det} \qquad \mathit{free\_vars}\ e \subseteq V'}{\begin{array}{c}(V, V', \Gamma, \delta) \vdash_{\mathrm{d}} \mathit{Random}\ dst\ e \Rightarrow \\ \lambda \rho\, x.\ \mathit{branch\_prob}\ (V, V', \Gamma, \delta)\ \rho \cdot \mathit{dist\_dens}\ dst\ (\mathit{expr\_sem\_rf}\ \rho\ e)\ x\end{array}}$$

$$\text{HD\_IF} \quad \frac{\begin{array}{c}(\emptyset, V \cup V', \Gamma, \lambda \rho.\ 1) \vdash_{\mathrm{d}} b \Rightarrow f \qquad (V, V', \Gamma, \lambda \rho.\ \delta\ \rho \cdot f\ \mathit{TRUE}) \vdash_{\mathrm{d}} e_1 \Rightarrow g_1 \\ (V, V', \Gamma, \lambda \rho.\ \delta\ \rho \cdot f\ \mathit{FALSE}) \vdash_{\mathrm{d}} e_2 \Rightarrow g_2\end{array}}{(V, V', \Gamma, \delta) \vdash_{\mathrm{d}} \mathit{IF}\ b\ \mathit{THEN}\ e_1\ \mathit{ELSE}\ e_2 \Rightarrow \lambda \rho\, x.\ g_1\ \rho\ x + g_2\ \rho\ x}$$

$$\text{HD\_IF\_DET} \quad \frac{\begin{array}{c}b\ \mathit{det} \\ (V, V', \Gamma, \lambda \rho.\ \delta\ \rho \cdot \langle \mathit{expr\_sem\_rf}\ \rho\ b = \mathit{TRUE}\rangle) \vdash_{\mathrm{d}} e_1 \Rightarrow g_1 \\ (V, V', \Gamma, \lambda \rho.\ \delta\ \rho \cdot \langle \mathit{expr\_sem\_rf}\ \rho\ b = \mathit{FALSE}\rangle) \vdash_{\mathrm{d}} e_2 \Rightarrow g_2\end{array}}{(V, V', \Gamma, \delta) \vdash_{\mathrm{d}} \mathit{IF}\ b\ \mathit{THEN}\ e_1\ \mathit{ELSE}\ e_2 \Rightarrow \lambda \rho\, x.\ g_1\ \rho\ x + g_2\ \rho\ x}$$

$$\text{HD\_FST} \quad \frac{(V, V', \Gamma, \delta) \vdash_{\mathrm{d}} e \Rightarrow f}{(V, V', \Gamma, \delta) \vdash_{\mathrm{d}} \mathit{fst}\ e \Rightarrow \lambda \rho\, x.\ \int y.\ f\ \rho\ <\!|x, y|\!>\ \partial\, \mathit{type\_of}\ \Gamma\ (\mathit{snd}\ e)}$$

$$\text{HD\_SND} \quad \frac{(V, V', \Gamma, \delta) \vdash_{\mathrm{d}} e \Rightarrow f}{(V, V', \Gamma, \delta) \vdash_{\mathrm{d}} \mathit{snd}\ e \Rightarrow \lambda \rho\, y.\ \int x.\ f\ \rho\ <\!|x, y|\!>\ \partial\, \mathit{type\_of}\ \Gamma\ (\mathit{fst}\ e)}$$

$$\text{HD\_OP\_DISCR}$$
$$\frac{\mathit{countable\_type}\ (\mathit{type\_of}\ (op\ \$\ e)) \qquad (V, V', \Gamma, \delta) \vdash_{\mathrm{d}} e \Rightarrow f}{(V, V', \Gamma, \delta) \vdash_{\mathrm{d}} op\ \$\ e \Rightarrow \lambda \rho\, y.\ \int x.\ \langle \mathit{op\_sem}\ op\ x = y\rangle \cdot f\ \rho\ x\ \partial\, \mathit{type\_of}\ \Gamma\ e}$$

$$\text{HD\_NEG} \quad \frac{(V, V', \Gamma, \delta) \vdash_{\mathrm{d}} e \Rightarrow f}{(V, V', \Gamma, \delta) \vdash_{\mathrm{d}} -e \Rightarrow \lambda \rho\, x.\ f\ \rho\ (-x)}$$

$$\text{HD\_ADDC} \quad \frac{e' \; \text{det} \qquad \text{free\_vars } e' \subseteq V' \qquad (V, V', \Gamma, \delta) \vdash_{\mathrm{d}} e \Rightarrow f}{(V, V', \Gamma, \delta) \vdash_{\mathrm{d}} e + e' \Rightarrow \lambda \rho \, x. \; f \; \rho \; (x - \text{expr\_sem\_rf } \rho \; e')}$$

$$\text{HD\_MULTC} \quad \frac{c \neq 0 \qquad (V, V', \Gamma, \delta) \vdash_{\mathrm{d}} e \Rightarrow f}{(V, V', \Gamma, \delta) \vdash_{\mathrm{d}} e \cdot \text{Val } (\text{RealVal } c) \Rightarrow \lambda \rho \, x. \; f \; \rho \; (x/c) \, / \, |c|}$$

$$\text{HD\_ADD} \quad \frac{(V, V', \Gamma, \delta) \vdash_{\mathrm{d}} {<}e_1, e_2{>} \Rightarrow f}{(V, V', \Gamma, \delta) \vdash_{\mathrm{d}} e_1 + e_2 \Rightarrow \lambda \rho \, z. \; \int x. \; f \; \rho \; {<}|x, z - x|{>} \; \partial \, \text{type\_of } \Gamma \; e_1}$$

$$\text{HD\_INV} \quad \frac{(V, V', \Gamma, \delta) \vdash_{\mathrm{d}} e \Rightarrow f}{(V, V', \Gamma, \delta) \vdash_{\mathrm{d}} e^{-1} \Rightarrow \lambda \rho \, x. \; f \; \rho \; (x^{-1}) \, / \, x^2}$$

$$\text{HD\_EXP} \quad \frac{(V, V', \Gamma, \delta) \vdash_{\mathrm{d}} e \Rightarrow f}{(V, V', \Gamma, \delta) \vdash_{\mathrm{d}} \exp \; e \Rightarrow \lambda \rho \, x. \; \textbf{if } x > 0 \textbf{ then } f \; \rho \; (\ln x) \, / \, x \textbf{ else } 0}$$

Consider the following simple example program:

IF Random *Bernoulli* 0.25 THEN 0 ELSE 1

Applying the abstract compiler yields the following HOL function:

$$\text{branch\_prob } (\emptyset, \emptyset, \Gamma, \lambda \rho. \; \text{branch\_prob } (\emptyset, \emptyset, \Gamma, \lambda \rho. \; 1) \; \rho \; *$$
$$\text{dist\_dens } Bernoulli \; 0.25 \; True) \; \rho \cdot \langle x = 0 \rangle +$$
$$\text{branch\_prob } (\emptyset, \emptyset, \Gamma, \lambda \rho. \; \text{branch\_prob } (\emptyset, \emptyset, \Gamma, \lambda \rho. \; 1) \; \rho \; *$$
$$\text{dist\_dens } Bernoulli \; 0.25 \; False) \; \rho \cdot \langle x = 1 \rangle$$

Since the *branch_prob* in this example is merely the integral over the empty set of variables, this simplifies to:

$$\lambda \rho \; x. \; \text{dist\_dens } Bernoulli \; 0.25 \; \rho \; True \cdot \langle x = 0 \rangle +$$
$$\text{dist\_dens } Bernoulli \; 0.25 \; \rho \; False \cdot \langle x = 1 \rangle$$

4.3   Soundness Proof

We proved the following soundness result for the abstract compiler:[6]

---

[6] Note that since the abstract compiler returns parametrised density functions, we need to parametrise the result with the state $\lambda x.$ *undefined*, even if the expression contains no free variables.

---

lemma expr_has_density_sound:

assumes $(\emptyset, \emptyset, \Gamma, \lambda\rho.\ 1) \vdash_d e \Rightarrow f$ **and** $\Gamma \vdash e : t$ **and** free_vars $e = \emptyset$

shows has_subprob_density (expr_sem σ e) (stock_measure t) (f ($\lambda x.$ undefined))

---

where has_subprob_density M N f is an abbreviation for the following four facts: applying the density f to N yields M, M is a sub-probability measure, f is N-Borel-measurable, and f is non-negative on its domain.

The lemma above follows easily from the following generalised lemma:

---

lemma expr_has_density_sound_aux:

assumes $(V, V', \Gamma, \delta) \vdash_d e \Rightarrow f$ **and** $\Gamma \vdash e : t$ **and**

   density_context V V' Γ δ **and** free_vars $e \subseteq V \cup V'$

shows has_parametrized_subprob_density (state_measure V' Γ)

   ($\lambda\rho.$ **do** $\{\sigma \leftarrow$ dens_ctxt_measure (V, V', Γ, δ) ρ; expr_sem σ $e\}$)

   (stock_measure t) f

---

where has_parametrized_subprob_density R M N f means that f is Borel-measurable w. r. t. $R \otimes N$ and that for any parameter state ρ from R, the predicate has_subprob_density $(M\ \rho)$ N $(f\ \rho)$ holds.

The proof is by induction on the definition of the abstract compiler. In many cases, the monad laws for the Giry monad allow restructuring the induction goal in such a way that the induction hypothesis can be applied directly; in the other cases, the definitions of the monadic operations need to be unfolded and the goal is essentially to show that two integrals are equal and that the output produced is well-formed.

The proof given by Bhat *et al.* [6] (which we were unaware of while working on our own proof) is analogous to ours, but much more concise due to the fact that side conditions such as measurability, integrability, non-negativity, and so on are not proven explicitly and many important (but uninteresting) steps are skipped or only hinted at.

It should be noted that in the draft of an updated version of their 2013 paper [6], Bhat *et al.* added a scaling rule for real distributions similar to our HD_MULTC rule. However, in the process of our formal proof, we found that their rule was too liberal: while our rule only allows multiplication with a fixed constant, their rule allowed multiplication with any deterministic expression, even expressions that may evaluate to 0, but multiplication with 0 always yields the Dirac distribution, which does not have a density function. In this case, the compiler returns a PDF for a distribution that has none, leading to unsoundness. This shows the importance of formal proofs.

5   Concrete Compiler

5.1   Approach

The concrete compiler is another inductive predicate, modelled directly after the abstract compiler, but returning a target-language expression instead of a HOL function. We use a standard refinement approach to relate the concrete compiler to the abstract one. We thus lift the soundness result on the abstract compiler to an analogous result on the concrete compiler. This shows that the concrete compiler always returns a well-formed target-language expression that represents a density for the sub-probability space described by the source language.

The concrete compilation predicate is written as

$$(vs, vs', \Gamma, \delta) \vdash_c e \Rightarrow f$$

Here, vs and vs' are lists of variables, Γ is a typing environment, and δ is a target-language expression describing the common density of the random variables vs in the context. It may be parametrised with the variables from vs'.

5.2   Definition

The concrete compilation rules are, of course, a direct copy of the abstract ones, but with all the abstract HOL operations replaced with operations on target-language expressions. Due to the de Bruijn indices and the lack of functions as explicit objects in the target language, some of the rules are somewhat complicated – inserting an expression into the scope of one or more bound variables (such as under an integral) requires shifting the variable indices of the inserted expression correctly. For this reason, we do not show the rules here; they can be found in the Isabelle theory file PDF_Compiler.thy [13].

5.3   Refinement

The refinement relates the concrete compilation

$$(vs, vs', \Gamma, \delta) \vdash_c e \Rightarrow f$$

to the abstract compilation

$$(\text{set } vs, \text{set } vs', \Gamma, \lambda\sigma.\ \text{expr\_sem } \sigma\ \delta) \vdash_c e \Rightarrow \lambda\rho x.\ \text{cexpr\_sem } (x \bullet \rho)\ f$$

In words: we take the abstract compilation predicate and

- the variable sets are refined to variable lists
- the typing context and the source-language expression remain unchanged
- the common density in the context and the compilation result are refined from HOL functions to target-language expressions (by applying the target language semantics)

The main refinement lemma states that the concrete compiler yields a result that is equivalent to that of the abstract compiler, modulo refinement. Informally, the statement is the following: if e is ground and well-typed under some well-formed concrete density context Υ and $\Upsilon \vdash_c e \Rightarrow f$, then $\Upsilon' \vdash_d e \Rightarrow f'$, where Υ' and f' are the abstract versions of Υ and f.

The proof for this is conceptually simple – induction over the definition of the concrete compiler; in practice, however, it is quite involved. In every single induction step, the well-formedness of the intermediary expressions needs to be shown, the previously-mentioned congruence lemmas for the abstract compiler need to be applied, and, when integration is involved, non-negativity and integrability have to be shown in order to convert non-negative Lebesgue integrals to general Lebesgue integrals and integrals on product spaces to iterated integrals.

Combining this main refinement lemma and the abstract soundness lemma, we can now easily show the concrete soundness lemma:

---

lemma expr_has_density_cexpr_sound:

assumes $([], [], \Gamma, 1) \vdash_c e \Rightarrow f$ **and** $\Gamma \vdash e : t$ **and** free_vars $e = \emptyset$

shows has_subprob_density (expr_sem σ e) (stock_measure t)

$\qquad\qquad\qquad (\lambda x.\ cexpr\_sem\ (x \bullet \sigma)\ f)$

$\qquad \Gamma'\ 0 = t \Longrightarrow \Gamma' \vdash_c f : \text{REAL}$

$\qquad$ free_vars $f \subseteq \{0\}$

Informally, the lemma states that if e is a well-typed, ground source-language expression, compiling it with an empty context will yield a well-typed, well-formed target-language expression representing a density function on the measure space described by e.

5.4   Final Result

We will now summarise the soundness lemma we have just proven in a more concise manner. For convenience, we define the symbol $e : t \Rightarrow_c f$ (read "e with type t compiles to f"), which includes the well-typedness and groundness requirements on e as well as the compilation result:[7]

$e : t \Rightarrow_c f \longleftrightarrow$

$\qquad (\lambda x.\ \textit{UNIT}) \vdash e : t\ \wedge\ \text{free\_vars } e = \emptyset\ \wedge\ ([], [], \lambda x.\ \textit{UNIT}, 1) \vdash_c e \Rightarrow f$

The final soundness theorem for the compiler, stated in Isabelle syntax:[8]

---

[7] In this definition, the choice of the typing environment is completely arbitrary since the expression contains no free variables.

[8] The lemma statement in Isabelle is slightly different: for better readability, we unfolded one auxiliary definition here and omitted the type cast from *real* to *ereal*.

lemma expr_compiles_to_sound:

assumes $e : t \Rightarrow_c f$

shows $expr\_sem\ \sigma\ e = density\ (stock\_measure\ t)\ (\lambda x.\ cexpr\_sem\ (x \bullet \sigma')\ f)$

$\quad\quad \forall x \in type\_universe\ t.\ cexpr\_sem\ (x \bullet \sigma')\ f \geq 0$

$\quad\quad \Gamma \vdash e : t$

$\quad\quad t \bullet \Gamma' \vdash_c f : REAL$

$\quad\quad free\_vars\ f \subseteq \{0\}$

In words, this result means the following:

Theorem

Let e be a source-language expression. If the compiler determines that e is well-formed and well-typed with type t and returns the target-language expression f, then:

- the measure obtained by taking the stock measure of t and using the evaluation of f as a density is the measure obtained by evaluating e
- f is non-negative on all input values of type t
- e has no free variables and indeed has type t (in any type context Γ)
- f has no free variable except the parameter (i.e. the variable 0) and is a function from t to $REAL$[9]

5.5   Evaluation

Isabelle's code generator allows us to execute our inductively-defined verified compiler using the **values** command[10] or generate code in one of the target languages such as Standard ML or Haskell. As an example on which to test the compiler, we choose the same expression that was chosen by Bhat *et al.* [5]:[11]

$$LET\ x = Random\ UniformReal\ <0,1>\ IN$$
$$LET\ y = Random\ Bernoulli\ x\ IN$$
$$IF\ y\ THEN\ x\ +\ 1\ ELSE\ x$$

---

[9] Meaning if its parameter variable has type t, it is of type REAL.

[10] Our compiler is inherently non-deterministic since it may return zero, one, or many density functions, seeing as an expression may have no matching compilation rules or more than one. Therefore, we must use the **values** command instead of the **value** command and receive a set of compilation results.

[11] *Val* and *RealVal* were omitted for better readability and symbolic variable names were used instead of de Bruijn indices.

We abbreviate this expression with e. We can then display the result of the compilation using the following Isabelle command:

$$\textbf{values } "\{(t, f) \mid t \; f. \; e : t \Rightarrow_c f\}"$$

The result is a singleton set which contains the pair $(REAL, f)$, where f is a very long and complicated expression. Simplifying constant subexpressions and expressions of the form $fst <e_1, e_2>$ and again replacing de Bruijn indices with symbolic identifiers, we obtain:

$$\int b. \; (IF \; 0 \leq x - 1 \wedge x - 1 \leq 1 \; THEN \; 1 \; ELSE \; 0) \cdot (IF \; 0 \leq x - 1 \wedge x - 1 \leq 1 \; THEN$$
$$IF \; b \; THEN \; x - 1 \; ELSE \; 1 - (x - 1) \; ELSE \; 0) \cdot \langle b \rangle \; +$$
$$\int b. \; (IF \; 0 \leq x \wedge x \leq 1 \; THEN \; 1 \; ELSE \; 0) \cdot (IF \; 0 \leq x \wedge x \leq 1 \; THEN$$
$$IF \; b \; THEN \; x \; ELSE \; 1 - x \; ELSE \; 0) \cdot \langle \neg b \rangle$$

Further simplification yields the following result:

$$\langle 1 \leq x \leq 2 \rangle \cdot (x - 1) + \langle 0 \leq x \leq 1 \rangle \cdot (1 - x)$$

While this result is the same as that which Bhat *et al.* have reached, our compiler generates a much larger expression than the one they printed. The reason for this is that they β-reduced the compiler output and evaluated constant subexpressions. While such simplification is very useful in practice, we have not automated it yet since it is orthogonal to the focus of our work, the compiler.

6 Conclusion

6.1 Effort

The formalisation of the compiler took about three months and roughly 10000 lines of Isabelle code (definitions, lemma statements, proofs, and examples) distributed as follows:

Type system and semantics	2900 lines
Abstract compiler	2600 lines
Concrete compiler	1400 lines
General measure theory	3400 lines

As can be seen, a sizeable portion of the work was the formalisation of results from general measure theory, such as integration by substitution and measure embeddings.

6.2 Difficulties

The main problems we encountered during the formalisation were:

Missing background theory. As mentioned in the previous section, a sizeable amount of measure theory and auxiliary notions had to be formalised. Most notably, the existing measure theory library did not contain integration by substitution. We proved this, using material from a formalisation of the Central Limit Theorem by Avigad *et al.* [2].

Proving side conditions. Many lemmas from the measure theory library require measurability, integrability, non-negativity, etc. In hand-written proofs, this is often "hand-waved" or implicitly dismissed as trivial; in a formal proof, proving these can blow up proofs and render them very complicated and technical. The measurability proofs in particular are ubiquitous in our formalisation. The measure theory library provides some tools for proving measurability automatically, but while they were quite helpful in many cases, they are still work in progress and require more tuning.

Lambda calculus. Bhat *et al.* use a simply-typed λ-calculus-like language with symbolic identifiers as a target language. For a paper proof, this is the obvious choice. We chose de Bruijn indices instead; however, this makes handling target language terms less intuitive and requires additional lemmas. Urban's nominal datatypes [21] would have allowed us to work with a more intuitive model, but we would have lost executability of the compiler, which was one of our aims.

6.3 Summary

We formalised the semantics of a probabilistic functional programming language with predefined probability distributions and a compiler that returns the probability distribution that a program in this language describes. These are modelled very closely after those given by Bhat *et al.* [5]. Then we formally verified the correctness of this compiler w.r.t. the semantics of the source and target languages.

This shows not only that the compiler given by Bhat *et al.* is correct (apart from the problem with the scaling rule we discussed earlier), but also that a formal correctness proof for such a compiler can be done with reasonable effort and that Isabelle/HOL in general and its measure theory library in particular are suitable for it. A useful side effect of our work was the formalisation of the Giry Monad, which is useful for formalisations of probabilistic computations in general.

Possible future work includes support for sum types, which should be possible with little effort, and a verified postprocessing stage to automatically simplify the density expression would be desirable.

Acknowledgements. We thank the referees for their helpful comments. We also thank Kristina Magnussen for proofreading the Master's Thesis upon which this work is based [12] and Andrei Popescu for categorical advice. David Cock and Jeremy Avigad helped us with their quick and concise answers to our questions about their work.

References

[1] Audebaud, P., Paulin-Mohring, C.: Proofs of randomized algorithms in Coq. In: Uustalu, T. (ed.) MPC 2006. LNCS, vol. 4014, pp. 49–68. Springer, Heidelberg (2006), http://dx.doi.org/10.1007/11783596_6

[2] Avigad, J., Hölzl, J., Serafin, L.: A formally verified proof of the Central Limit Theorem. CoRR abs/1405.7012 (2014)

[3] Bertot, Y., Castéran, P.: Interactive Theorem Proving and Program Development. Coq'Art: The Calculus of Inductive Constructions. Springer (2004)

[4] Bhat, S., Agarwal, A., Vuduc, R., Gray, A.: A type theory for probability density functions. In: Proceedings of the 39th Annual ACM SIGPLAN-SIGACT Symposium on Principles of Programming Languages, POPL 2012, pp. 545–556. ACM, New York (2012), http://doi.acm.org/10.1145/2103656.2103721

[5] Bhat, S., Borgström, J., Gordon, A.D., Russo, C.: Deriving probability density functions from probabilistic functional programs. In: Piterman, N., Smolka, S.A. (eds.) TACAS 2013. LNCS, vol. 7795, pp. 508–522. Springer, Heidelberg (2013), http://dx.doi.org/10.1007/978-3-642-36742-7_35

[6] Bhat, S., Borgström, J., Gordon, A.D., Russo, C.: Deriving probability density functions from probabilistic functional programs (full version, submitted for publication)

[7] de Bruijn, N.G.: Lambda calculus notation with nameless dummies, a tool for automatic formula manipulation, with application to the Church-Rosser theorem. Indagationes Mathematicae 34, 381–392 (1972)

[8] Cock, D.: Verifying probabilistic correctness in Isabelle with pGCL. In: Proceedings of the 7th Systems Software Verification, pp. 1–10 (November 2012)

[9] Cock, D.: pGCL for Isabelle. Archive of Formal Proofs Formal proof development (July 2014), http://afp.sf.net/entries/pGCL.shtml

[10] Doberkat, E.E.: Stochastic relations: foundations for Markov transition systems. Studies in Informatics. Chapman & Hall/CRC (2007)

[11] Doberkat, E.E.: Basing Markov transition systems on the Giry monad (2008), http://www.informatics.sussex.ac.uk/events/domains9/Slides/Doberkat_GiryMonad.pdf

[12] Eberl, M.: A Verified Compiler for Probability Density Functions. Master's thesis, Technische Universität München (2014), https://in.tum.de/~eberlm/pdfcompiler.pdf

[13] Eberl, M., Hölzl, J., Nipkow, T.: A verified compiler for probability density functions. Archive of Formal Proofs, Formal proof development (October 2014), http://afp.sf.net/entries/Density_Compiler.shtml,

[14] Giry, M.: A categorical approach to probability theory. In: Mosses, P.D., Nielsen, M. (eds.) CAAP 1995, FASE 1995, and TAPSOFT 1995. LNCS, vol. 915, pp. 68–85. Springer, Heidelberg (1995), http://dx.doi.org/10.1007/BFb0092872, doi:10.1007/BFb0092872

[15] Hölzl, J.: Construction and stochastic applications of measure spaces in Higher-Order Logic. PhD thesis, Technische Universität München, Institut für Informatik (2012)

[16] Hurd, J.: Formal Verification of Probabilistic Algorithms. Ph.D. thesis, University of Cambridge (2002)

[17] Hurd, J., McIver, A., Morgan, C.: Probabilistic guarded commands mechanized in HOL. Electron. Notes Theor. Comput. Sci. 112, 95–111 (2005), http://dx.doi.org/10.1016/j.entcs.2004.01.021

[18] Nipkow, T., Klein, G.: Concrete Semantics with Isabelle/HOL. Springer (2014), http://www.concrete-semantics.org

[19] Nipkow, T., Paulson, L.C., Wenzel, M.: Isabelle/HOL. LNCS, vol. 2283. Springer, Heidelberg (2002)

[20] Park, S., Pfenning, F., Thrun, S.: A probabilistic language based upon sampling functions. In: Proceedings of the 32Nd ACM SIGPLAN-SIGACT Symposium on Principles of Programming Languages, POPL 2005, pp. 171–182. ACM, New York (2005), http://doi.acm.org/10.1145/1040305.1040320

[21] Urban, C.: Nominal techniques in Isabelle/HOL. Journal of Automated Reasoning 40, 327–356 (2008)

Appendix

Notation	Name / Description	Definition
$f\ x$	function application	$f(x)$
$f\ {}^{\backprime}X$	image set	$f(X)$ or $\{f(x) \mid x \in X\}$
$\lambda x.\ e$	lambda abstraction	$x \mapsto e$
undefined	arbitrary value	
Suc	successor of a natural number	$+1$
case_nat $x\ f\ y$	case distinction on natural number	$\begin{cases} x & \text{if } y = 0 \\ f(y-1) & \text{otherwise} \end{cases}$
[]	Nil	empty list
$x \mathbin{\#} xs$	Cons	prepend element to list
$xs \mathbin{@} ys$	list concatenation	
map $f\ xs$	applies f to all list elements	$[f(x) \mid x \leftarrow xs]$
merge $V\ V'\ (\rho, \sigma)$	merging disjoint states	$\begin{cases} \rho\ x & \text{if } x \in V \\ \sigma\ y & \text{if } x \in V' \\ undefined & \text{otherwise} \end{cases}$
$y \bullet f$	add de Bruijn variable to scope	see Sect. 2.2
$\langle P \rangle$	indicator function	1 if P is true, 0 otherwise
$\int x.\ f\ x\,\partial\mu$	Lebesgue integral on non-neg. functions	
$\mathfrak{Meas}$	category of measurable spaces	see Sect. 2.4
$\mathbb{S}$	sub-probability functor	see Sect. 2.4
return	monadic return (η) in the Giry monad	see Sect. 2.4
$\ggeq$	monadic bind in the Giry monad	see Sect. 2.4
do $\{\dots\}$	monadic "do" syntax	see Sect. 2.4
density $M\ f$	measure with density	result of applying density f to M
distr $M\ N\ f$	push-forward/image measure	$(B,\ \mathcal{B},\ \lambda X.\ \mu(f^{-1}(X)))$ for $M = (A, \mathcal{A}, \mu),\ N = (B, \mathcal{B}, \mu')$

Segment Abstraction for Worst-Case Execution Time Analysis*

Pavol Černý[1], Thomas A. Henzinger[2], Laura Kovács[3],
Arjun Radhakrishna[4], and Jakob Zwirchmayr[5]

[1] University of Colorado Boulder
[2] IST Austria
[3] Chalmers University of Technology
[4] University of Pennsylvania
[5] Institut de Recherche en Informatique de Toulouse

Abstract. In the standard framework for worst-case execution time
(WCET) analysis of programs, the main data structure is a single instance of integer linear programming (ILP) that represents the whole
program. The instance of this NP-hard problem must be solved to find an
estimate for WCET, and it must be refined if the estimate is not tight. We
propose a new framework for WCET analysis, based on abstract segment
trees (ASTs) as the main data structure. The ASTs have two advantages.
First, they allow computing WCET by solving a number of independent
small ILP instances. Second, ASTs store more expressive constraints,
thus enabling a more efficient and precise refinement procedure. In order to realize our framework algorithmically, we develop an algorithm
for WCET estimation on ASTs, and we develop an interpolation-based
counterexample-guided refinement scheme for ASTs. Furthermore, we
extend our framework to obtain parametric estimates of WCET. We experimentally evaluate our approach on a set of examples from WCET
benchmark suites and linear-algebra packages. We show that our analysis, with comparable effort, provides WCET estimates that in many
cases significantly improve those computed by existing tools.

1 Introduction

Worst-case execution time (WCET) analysis [18] is important in many classes
of applications. For instance, real-time embedded systems have to react within a
fixed amount of time. For another example, consider computer algebra libraries
that provide different implementations for the most heavily-used methods. Users
have to choose the most suitable method for their particular system architecture.

---

* This research was supported in part by the European Research Council (ERC) un-
der grant agreement 267989 (QUAREM), the Austrian National Research Network
RiSE (FWF grants S11402-N23), the Swedish VR grant D0497701, the WWTF
PROSEED grant ICT C-050, the French National Research Agency grant W-SEPT
(ANR-12-INSE-0001), NSF Expeditions award CCF 1138996, DARPA under agree-
ment FA8750-14-2-0263, and a gift from the Intel Corporation.

© Springer-Verlag Berlin Heidelberg 2015
J. Vitek (Ed.): ESOP 2015, LNCS 9032, pp. 105–131, 2015.
DOI: 10.1007/978-3-662-46669-8_5

In both cases, a tool that soundly and tightly approximates the WCET of a program on a given architecture would thus be very helpful.

State of the Art. Most state of the art WCET estimation tools proceed in three phases (see for instance the survey [18]):

- *First phase: Architecture-independent flow analysis,* which computes invariants, loop bounds, and finds correlations between the number of times different basic blocks in the program are executed. Such facts are called flow facts in the WCET literature, and the analysis is called flow analysis.
- *Second phase: Architecture-dependent low-level analysis,* which finds WCET for each basic block, using a model of a particular architecture, and abstract interpretation over domains that model for instance caches and pipelines.
- *Third phase: Path analysis,* which combines the results of the previous two phases. The commonly-used algorithm is called Implicit Path Enumeration Technique (IPET) [15]. It constructs an Integer Linear Programming (ILP) problem using WCET estimates for each basic block and constraints arising from flow facts to rule out some infeasible paths.

Recently, several approaches to refining WCET estimates were proposed. These works [14,3] use an approach named "WCET squeezing" in [14], which adds constraints to the ILP problem arising from IPET.

The main data structure used is a single ILP problem for the whole program. We make two observations about the standard approach: first, flow facts (gathered in the first phase, or obtained by refinement) lead to *global* constraints in the ILP constructed in the third phase. Hence, the ILP problem (an instance of an NP-hard problem) cannot be decomposed into smaller problems. Second, the current approaches to refinement add constraints to the ILP; and in this way eliminate only one path at a time.

Our Thesis. The main thesis of this paper is that hierarchical segment abstraction [8] is the right framework for WCET analysis. Segments are sequences of program instructions. Segment abstractions are those where an abstract state represents a set of segments, rather than set of concrete states. We represent hierarchical segment abstraction in a data structure called abstract segment trees (ASTs). The concept of segment abstraction [8] and its quantitative version [6] were introduced only very recently. We believe that WCET analysis is a prime application for segment abstraction.

We give two main arguments in support of the thesis. First, hierarchical segment abstraction allows us to compute the WCET by solving a number of independent ILP problems, instead of one large global ILP problem. This is because ASTs allow storing constraints *locally*. Second, hierarchical segment abstraction enables us to develop a more precise and efficient refinement procedure. This is because ASTs store more *expressive* constraints than ILP.

Algorithm. In order to substantiate our thesis, we develop an algorithm for producing increasingly tight WCET estimates. There are three key ingredients to the algorithm. First, we define an abstraction of programs that contains quantitative information necessary to estimate WCET. Second, we develop an algorithm for WCET estimation on ASTs. Third, we develop counterexample-guided refinement for segment abstraction based on interpolation.

Abstractions for WCET: Abstract Segment Trees. The main idea of hierarchical segment abstraction is that an abstract state corresponds to a set of concrete segments, rather than to a set of concrete states, as in state-based abstractions. Reasoning about segments is suitable for WCET estimation, where the analysis needs, for example, to distinguish execution times for different paths through a loop body, and store the relative number of times these paths are taken. Consider a simple example of a loop through which there are two paths, P_1 and P_2. Let us assume that the path P_1 takes a long time to execute, but is taken only once every 4 iterations of the loop; otherwise, a cheaper path P_2 is taken. To obtain a precise estimate of WCET, we thus need to store two types of numerical facts: the first one is the current estimates of execution time of P_1 and P_2, and the second fact is that an iteration taking path P_1 is followed by 3 iterations taking path P_2. We show that both these facts about paths can be stored *locally* if the basic object in the representation is a set of segments. To contrast with the standard approach to WCET analysis, note that the second quantitative fact is stored as a global constraint in the ILP.

The abstraction is hierarchical in order to capture the hierarchical nature of traces of structured programs with loops and procedures. For example, we split the set of traces through a nested loop into repeated iterations of the outer loop, and each outer loop iteration is split into repeated inner loop iterations. The hierarchical abstraction is represented by an *abstract segment tree.*

Each node of the abstract segment tree represents a set of segments. The nodes in our abstract segment trees contain quantitative information so that the WCET of the program can be estimated using only the abstraction.

Evaluation of WCET on Abstractions. After constructing the abstraction of the program (an AST), the next step is to evaluate the WCET on the AST. The AST is a hierarchical structure. Each node of the AST gives rise to a problem that can be solved using an ILP encoding. The key difference to the standard IPET approach is that IPET constructs one global ILP problem for the whole program, and the ILP cannot be decomposed due to global constraints. segment abstractions and the hierarchical nature of ASTs enable us to decompose the solving into smaller problems, with one such problem for each node.

We propose a new encoding of the constraint-solving problem that arises at each node into ILP. The problem at each node could be reduced to ILP by a technique presented in [16], but this would lead to a possibly exponential number of constraints, whereas our encoding produces linear number of constraints.

Abstraction-Refinement for WCET. After evaluating the WCET on ASTs, we obtain a *witness* trace, that is, a trace through the AST that achieves the WCET. It might be that the trace is feasible in the current abstraction but is infeasible in the program. Hence, we need a refinement as the next phase, in order to obtain a more precise WCET estimate. The algorithm refines the abstraction based on the current WCET estimate and the corresponding worst-case path.

We use the classical abstraction-refinement loop approach, adapted to ASTs. If the witness trace is feasible in the original program, the current WCET estimate is tight and we report it. Otherwise, we refine the abstraction using a novel interpolation-based approach for refinement of segment predicates. The refinements are monotonic w.r.t. WCET estimates, i.e., they are monotonically decreasing. Having expressive constraints (relational predicates) stored in ASTs allows us to perform more efficient and more precise refinement than state-of-the-art WCET refinement techniques that add constraints to the ILP problem. Our refinement is more efficient, as we can potentially eliminate many traces at a time, and it is more precise, as the constraints that determine how the set represented by abstract nodes can be combined are more expressive than the constraints in ILP.

Parameters. Furthermore, we extend our framework to provide parametric estimates of WCET, following [4,1]. In many cases, a single number as a WCET estimate is a pessimistic over-estimate. For instance, for a program that transposes a $n \times n$ matrix, a single numeric WCET is the WCET for the largest possible value of n. We adapt our evaluation algorithm to compute parametric WCET estimates as disjunctive linear-arithmetic expressions.

Experimental Evaluation. Our goal was to evaluate the idea of using ASTs as the basic data structure for WCET estimation. We built a tool IBART for computing (parametric) WCET estimates for C programs. For obtaining WCET estimates for basic blocks of programs, we used two low-level analyzers from existing frameworks, r-TuBound [13] and OTAWA [2]. The low-level analyzers CalcWCET167 and owcet included in r-TuBound and OTAWA respectively provide basic block WCET estimates for the Infineon C167 processor and the LPC2138 ARM7 processor. We re-used the low-level analyzers of these frameworks, and show that our high-level analyzer provides more precise constraints that our solver uses to compute tighter WCET estimates than these two frameworks.

We evaluated IBART on challenging examples from WCET benchmark suites and open-source linear algebra packages. These examples were parametric (with parameters such as array sizes and loop bounds), and our tool provided parametric estimates. All the examples we considered were solved under 20 seconds. To compare our estimates with the non-parametric results provided by r-TuBound and OTAWA, we instantiated the parameters to a number of sample values.

The results show that IBART provides better WCET estimates, though based on the same low-level analyzers. This demonstrates that our segment algorithm improves WCET estimates independently of the low-level analyzer. We thus expect that ASTs and our framework could be used by other WCET tools.

2   Illustrative Examples

This section illustrates our approach to WCET computation. We use Example 1 to demonstrate the main differences between our approach and the standard approach

```
for (i=0;i<1000;i++)
    if ((i mod 4) == 0)
        logValues()  cost=50
        work1()  cost=3
    else
        work2()  cost=3
```

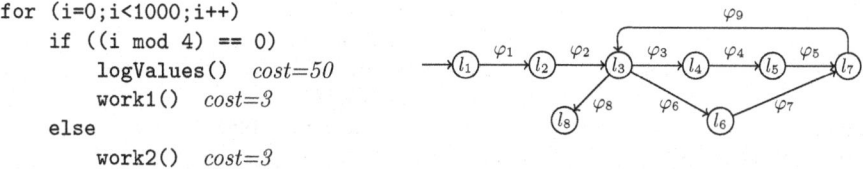

Fig. 1. Example 1   **Fig. 2.** CFG of Example 1

to WCET analysis. We then use Example 2 to present the main steps of our method: segment abstraction, estimation of WCET on ASTs, and counterexample-guided abstraction refinement with interpolation.

Example 1. The program in Figure 1 performs operation `work()` (of execution cost 3 time units) within a loop. Every 3 loop iterations, it logs some values into a file, by using operation `logValue` whose execution takes 50 time units.

Consider the CFG of Example 1 in Figure 2 where edges have been marked by instructions. For instance, the edge labels φ_4, φ_5, and φ_7 correspond to instructions `logValues()`, `work1()`, and `work2()`, respectively. The standard IPET algorithm would construct an ILP as follows. For each instruction φ_i, the variable X_i represents the number of times the instruction is executed in the worst-case path. The objective function for the ILP is to maximize $\sum_i X_i \cdot cost(\varphi_i)$, where $cost(\varphi_i)$ is the time taken to execute φ_i. The ILP constraints correspond to either conservation of flow, for instance, $X_7 + X_5 = X_9$ and $X_2 + X_9 = X_3 + X_6 + X_8$, or loop bounds, for instance $1000X_2 = X_3 + X_6$. Solving the ILP gives a way to estimate the WCET. However, the estimate would be imprecise, as it would find a solution that takes the expensive branch every time.

We could add a constraint specifying that the expensive branch is taken once, every 4 iterations: $3X_3 = X_6$. However, two important points are to be noted:

- First, this type of constraint that relates edges in different branches of the program is non-local. In general, the two branches that need to be related, can be far apart in the CFG. This makes decomposing the single large ILP (representing the whole program) into smaller problems hard, and to the best of our knowledge, no existing tool attempts this.
- Second, consider a version where the if-condition is replaced by `((i mod 30) == 0)`. We cannot use the constraint $29X_3 = X_6$, as this would not have a solution: the number of iterations (1000) is not divisible by 30. Alternatively, we could use less precise constraints like $29X_3 \geq X_6 \geq 29(X_3 - 1)$. Most current WCET tools would not handle the example precisely.

Consider in contrast how we obtain *local* bounds by reasoning about hierarchies of sets of segments. Let B_2 be the set of segments representing a single iteration of the loop. For example, B_2 contains segments that start at l_3, and go through l_4, l_5, l_7 (see Figure 2). This represents an iteration that goes through the expensive branch. The set of segments B_2 is represented by the regular expression $l_3(l_4l_5 \vee l_6)l_7$. Let B_1 denote a set of segments through the loop - the set of segments that start at l_3 and exit the loop. The set B_1 can be over-approximated

```
                              11: if (*)
                                    assume a<b;  (φ₁)
                              12:     i:=0;  (φ₂)
                              13:     while (*)
 if (a<b)                               assume (i<n);  (φ₃)
    for (i=0;i<n;i++)        14:        if (*)
       if (i<⌊n/2⌋)                         assume i<⌊n/2⌋;  (φ₄)
           op1();  cost=10   15:            op1();  (φ₅), cost=10
       else                              else
           op2();  cost=1                   assume i≥⌊n/2⌋;  (φ₆)
 else                        16:            op2();  (φ₇), cost=1
    op3();  cost=50          17:         i:=i+1;  (φ₈)
                                        assume (i≥n);  (φ₉)
                                     else
                                        assume a≥ b;  (φ₁₀)
                              18:        op3();  (φ₁₁), cost=50
```

Fig. 3. Example 2 (above); written in a while-language (right)

by B_2^*. We also store the loop bound with B_1, and thus the over-approximation effectively becomes B_2^{1000}.

The counterexample-guided refinement will then refine the segment set of B_2, by splitting B_2 into two sets: the set (B_2^t) where the formula $((i \bmod 4) = 0)$ holds (the expensive iteration), and the set (B_2^f) where the formula $((i \bmod 4) = 0)$ does not hold (the cheap iteration). The set of segments B_2^t is represented by the regular expression $l_3 l_4 l_5 l_7$, and the set of segments B_2^t is represented by the regular expression $l_3 l_6 l_7$. The over-approximation B_1 will therefore become $((B_2^t)^1 (B_2^f)^3)^*$. Note that this keeps the information locally: it says that one expensive iteration is followed by 3 cheap iterations. The node B_2 is hence refined into 1 iteration of B_2^t, followed by 3 iterations of B_2^f. The loop bound of 1000 would still be stored locally with B_1, requiring the total number of calls to B_2^t and B_2^f be 1000. This information is enough to obtain a precise WCET estimate. The same approach would work for the variant considered above.

Example 2 (Running example). We now explain our approach in detail using the program in Figure 3, which will be our running example through the paper. Program blocks op1(), op2(), and op3() are operations whose executions take 10, 1, and 50 time units, respectively (these costs are derived from a low-level timing analysis tool). In this example, we assume that program conditionals and simple assignments take 1 time unit.

It is not hard to see that for small values of the *loop bound* n, the WCET path of this program visits the outermost else branch containing op3() – when n is small, the execution cost of op3() dominates the cost of the loop. However, for larger values of n, the WCET path visits the then branch of the outermost if and the for-loop. The WCET of this example thus depends on n. Our approach discovers this fact, and infers the WCET of the program as a function of n as follows: *if $n \leq 5$; 51 else $3 + 4n + 9\lfloor n/2 \rfloor$*. The computation proceeds as follows.

Control-flow graph. We construct the control-flow graph (CFG) of the program in Figure 3. First, for clarity of presentation, we transform the program in a `while`-loop language with `assume` statements — see Figure 3 (right column). We have labeled the assumptions and the transition relations (i.e. transition predicates) of instructions. For example, (φ_1) denotes the assumption $a < b$; and (φ_8) represents the transition predicate $i' = i + 1$ of the assignment $i := i + 1$.

Hierarchical segment-base abstraction. We next apply segment abstraction on the CFG of Figure 4. The initial abstraction is given by the abstract segment tree (AST) (Figure 5). The tree structure arises from the hierarchical nature of the CFG. Nodes of the tree (denoted by A_k) represent a set of execution segments, i.e., parts of program executions. Each node stores a *shape predicate* (denoted Shape) describing the paths of the segments through the CFG, and a *transition predicate* (denoted Trans) characterizing the transition relation of the segments. A shape predicate is an extended regular expression over either the children of the node, or over the CFG nodes. It is an extended regular expression, as it may contain symbolic exponents obtained, for example, from loop bounds. The transition predicate is a formula over the values of program variables at the beginning and end of segments. Note that in the formal definitions, the shape is a transition system rather than a regular expression and the nodes store more detailed information. Here, we use a regular expression for better readability.

We describe node A_2 in more detail—the other nodes are constructed similarly. The construction of A_2 in the initial abstraction is done syntactically. Node A_2 represents all segments corresponding to the then-branch of the outermost `if`. It is split into three sets of segments: (a) node A_3 denoting the set of segments before the loop, i.e., segments through the CFG nodes $l_1 l_2 l_3$; (b) node A_4 denoting the set of segments given by the loop of the CFG; and (c) node A_5 representing the set of segments after the loop of the CFG. Take n as the bound on the number of loop iterations in the CFG. For building A_4 we use node A_6 describing all segments in one iteration of the loop in the CFG. The segments in A_6 can be concatenated to cover all segments in A_4. For computing loop bounds, we use [13]. The loop bound n is noted in the shape predicate of A_4.

WCET estimation on ASTs. For each node in Figure 5, we next calculate the cost of the segments represented by it, i.e., its WCET. As each node is defined in terms of its children, we traverse the tree bottom-up. The root contains then a WCET estimate of the complete set of segments, and hence of the program.

In order to evaluate WCET on an AST, we need to supply an ILP at each node of the AST. Here, instead of presenting each ILP and solving it, we give only a simple explanation tailored to the example under consideration. To estimate the WCET of a node, we consider the graph represented by the node. The vertices in this graph correspond to children of the node. We use the shape predicate of the node to construct the graph, and use the WCET of the children nodes to estimate the cost of the node. For example, for node A_2, we construct a graph with three nodes, with directed edges from A_3 to A_4 and from A_4 to A_5. For

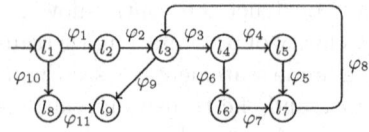

Fig. 4. CFG of Example 2

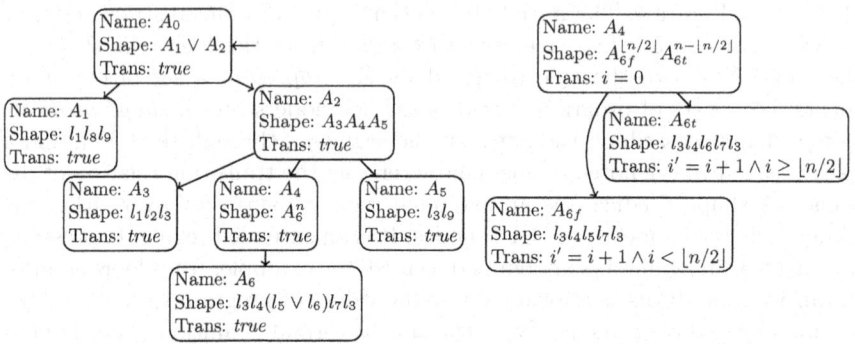

Fig. 5. Initial abstraction for Example 2   **Fig. 6.** Partial structure of the
refined tree of Figure 5

node A_4, we obtain a graph with one node (A_6) that can repeat at most n times.
The costs of the AST nodes are calculated as:

- $cost(A_6) = cost(\varphi_3) + \max(cost(\varphi_4) + cost(\varphi_5), cost(\varphi_6 + cost(\varphi_7)) + cost(\varphi_8) = 13$
- $cost(A_4) = n \cdot A_6 = 13n$
- $cost(A_3) = cost(\varphi_1) + cost(\varphi_2) = 2$
- $cost(A_5) = cost(\varphi_9) = 1$
- $cost(A_2) = cost(A_3) + cost(A_4) + cost(A_5) = 3 + 13n$
- $cost(A_1) = cost(\varphi_{10}) + cost(\varphi_{11}) = 51$
- $cost(A_0) = \max(cost(A_1), cost(A_2)) = \max(51, 3 + 13n) = if\ n \leq 3\ ;\ 51\ else\ 3 + 13n$

The WCET estimate of our running example is given by $cost(A_0)$, and depends
on the value of n, i.e., when $0 \leq n \leq 3$ the WCET is different than in the case
when $n > 3$. To ensure that the computed WCET estimate is precise, we need to
ensure that our abstraction did not use an infeasible program path to derive the
current WCET estimate. We therefore pick a concrete value of n for each part
of the WCET estimate, and check whether the corresponding witness worst-case
path is feasible. If it is, the derived WCET estimate is actually reached by the
program and we are done. Otherwise, we need to refine our AST. In our example,
we thus have the following two cases:

- *Case 1:* $n \leq 3$. We pick $n = 1$. The WCET estimate of A_0 is then 51. Here,
  the witness trace is $l_1 l_8 l_9$. This trace is a feasible trace of Figure 3.
- *Case 2:* $n > 3$. We pick $n = 4$ and the witness trace is $l_1 l_2\ (l_3 l_4 l_5 l_7)^4 l_3 l_9$, which
  is infeasible, and we proceed to the refinement step.

Counterexample-guided refinement using interpolation. We refine the AST of Figure 5 using the infeasible trace. We traverse the tree top-down to refine each node of the counterexample. We refine the children of the node corresponding to a node in the counterexample with new context information obtained from the counterexample, via interpolation. We detail our refinement approach only for A_4, the rest of the nodes are refined in a similar way. By analyzing the predecessor segments of A_4 in the counterexample, we derive $i = 0$ as a useful property for our refinement. This property is obtained using the same refinement process that we now describe for A_4.

To refine A_4, we analyze its children, that is n repetitions (i.e., iterations) of A_6. In what follows, we denote by i_k the value of the variable i after the k-th iteration of A_6, for $0 \leq k \leq n$. Let i_0 denote the value of i before A_4. We compute the property $i_1 = i_0 + 1$ summarizing the first iteration of A_6, where the summarization process includes interpolation-based refinement. Similarly, from the second iteration of A_6 we compute $i_2 = i_1 + 1$. Hence, at the second iteration of A_6 the formula $i_0 = 0 \wedge i_1 = i_0 + 1 \wedge i_2 = i_1 + 1 \wedge n = 4$ is a valid property of the witness trace; let us denote this formula by A (recall that we fixed $n = 4$ above). However, after the second iteration of A_6 we have $(i_2 < n) \wedge (i_2 < \lfloor n/2 \rfloor)$ as a valid property of the witness trace; we denote this formula by B. Observe that $A \wedge B$ is unsatisfiable, providing hence a counterexample to the feasibility of the current witness trace. From the proof of unsatisfiability of $A \wedge B$, we then compute an interpolant I such that $A \implies I$, $I \wedge B$ is unsatisfiable, and I uses only symbols common to both A and B. We derive $i_2 \geq \lfloor n/2 \rfloor$ as the interpolant of A and B.

We now use the interpolant $i_2 \geq \lfloor n/2 \rfloor$ to refine the segment abstraction of A_6, as follows. The interpolant $i_2 \geq \lfloor n/2 \rfloor$ is mapped to the predicate $i \geq \lfloor n/2 \rfloor$ over the program variables. We then split A_6 into two nodes: node A_{6f} denoting segments where $i \geq \lfloor n/2 \rfloor$ does not hold, and node A_{6t} describing segments where $i \geq \lfloor n/2 \rfloor$ holds. The interpolants $i_1 = i_0 + 1$ and $i_2 = i_1 + 1$ computed from the first and second iteration of A_6 yield the transition predicate $i' = i + 1$; this formula holds for every segment in A_6, and hence also in A_{6f} and A_{6t}. The transition predicates of A_{6t} and A_{6f} are then used to compute the new shape predicate $A_{6f}^{\lfloor n/2 \rfloor} A_{6t}^{n - \lfloor n/2 \rfloor}$ for A_4. The resulting (partial) refined AST is given in Figure 6. This refined AST yields the WCET estimate *if* $n \leq 5$; 51 *else* $3 + 4n + 9\lfloor n/2 \rfloor$, which is a precise WCET estimate for the program in Figure 3.

3   Problem Statement

Instruction and Predicate Language. We express program instructions, predicates, and assertions using standard first-order logic. Let $\mathcal{F}(X)$ represent the set of linear integer arithmetic *formulae* over integer variables X. We represent an instruction of a program as a formula from $\mathcal{F}(V \cup V')$. Intuitively, a variable $v \in V$ and its primed version $v' \in V'$ represent the values of the program variable v before and after the execution of the instruction, respectively. For example, an instruction i := i + j in a C-like language would be represented as $i' = i + j$.

Program Model. We model programs with assignments, conditionals, and loops, over a finite set of scalar integer variables V. While we do not handle procedure calls, our techniques can be generalized to non-recursive procedure calls. We represent programs by their control-flow graphs. A *control-flow graph* (CFG) is a graph $G = \langle \mathcal{C}, E, V, \Delta, \iota_0, init, F \rangle$, where (a) $\mathcal{C}$ is a set of nodes (representing control-flow locations); (b) $E \subseteq \mathcal{C} \times \mathcal{C}$ is a set of edges; (c) V is the set of program variables; (d) $\iota_0 \in \mathcal{C}$ is an initial control-flow location; (e) $init \in \mathcal{F}(V)$ is an initial condition on variables; (f) $F \subseteq \mathcal{C}$ is a set of final program locations; and (g) $\Delta : E \rightarrow \mathcal{F}(V \cup V')$ maps edges to the instruction that is executed when the edge is taken. We denote *program states* by pairs of the form (l, σ) where $l \in \mathcal{C}$ and σ is a valuation of program variables V.

Semantics. The semantics $[\![G]\!]$ of a CFG G is the set of *finite* sequences of program states (called *traces*) $(l_0, \sigma_0) \ldots (l_k, \sigma_k)$ such that: (a) $l_0 = \iota_0$ and $\sigma_0 \models init$, (b) $l_k \in F$, and (c) $\forall 0 \leq i < k. (l_i, l_{i+1}) \in E \wedge (\sigma_i, \sigma_{i+1}) \models \Delta((l_i, l_{i+1}))$. Note that we assume that the program represented by G is terminating.

Cost Model. We assume a simple cost model for instructions given by a function $cost : E \rightarrow \mathbb{N}$ where $cost((l_1, l_2))$ is the maximum execution time of the instruction from l_1 to l_2. We also refer to costs as *weights*. The weight $cost(\pi)$ of a trace $\pi = (l_0, \sigma_0) \ldots (l_k, \sigma_k)$ is $\Sigma_{i=0}^{k-1} cost((l_i, l_{i+1}))$. In practice, costs of edges are obtained from a low-level architecture dependent analyzer. Note that even this simple cost model can already capture some information about the context of an instruction's execution such as some cache hit/miss information. For example, if the low-level analysis determines that an instruction will always be a cache hit, it can provide a lower cost accordingly.

Problem Statement. The *worst-case execution time* $WCET(G)$ of a CFG G is defined by $WCET(G) = \max_{\pi \in [\![G]\!]} cost(\pi)$. The task of the *WCET estimation problem* is: Given a CFG G, compute a number e such that $e \geq WCET(G)$. The additional aim is to compute an estimate e that is tight, i.e., close to $WCET(G)$.

The rest of this paper describes the main steps of our approach to solving this problem: segment abstraction (Section 4), WCET estimation for segment abstractions (Section 5), and counterexample-guided refinement (Section 6). We summarize our algorithm in Section 7, describe the parametric extension in Section 8, and present our tool and experimental results in Section 9.

4   Segment Abstraction for Flow Analysis

Our abstraction technique for flow analysis is based on the hierarchical segment abstraction of [8,6]. We adapt the definitions of hierarchical segment abstraction from [6] to the setting of worst-case execution time analysis.

Let us fix a CFG $G = \langle \mathcal{C}, E, V, \Delta, \iota_0, init, F \rangle$. A *segment* is a finite sequence of program states (i.e., pairs of control-flow locations and variable valuations).

Abstract Segment Trees (ASTs). An *abstract segment tree* T is a rooted tree, where each node represents a set of segments. Intuitively, the segments of each node are composed from the segments of its children. Each node is a tuple $(segPred, children, shape, Init, Exit, slMin, slMax, gMax)$ where:

- $segPred \in \mathcal{F}(V \cup V')$ is a relational predicate satisfied by the initial and final variable valuations of all the segments represented by the current node;
- For internal nodes, the set *children* is the set of its children in T, and for leaf nodes, *children* is a subset of the control-flow edges E of the CFG G;
- *shape* $\subseteq$ *children* $\times$ *children* is a transition relation on *children* — for leaf nodes, where *children* $\subseteq E$, we have that $((l_0, l_1), (l_2, l_3)) \in$ *shape* if $l_1 = l_2$;
- $Init, Exit \subseteq children$ are a set of initial child nodes and exit child nodes;
- $gMax \in \mathbb{N} \cup \{\infty\}$ is a bound on the maximal number of segments of child nodes in a segment of the current node; and
- $slMin, slMax : children \to \mathbb{N} \cup \{\infty\}$ are functions that map each child to the minimum and maximum possible consecutive repetitions of segments represented by the child in a segment represented by the current node.

Table 1. Definition of ASTs

Component	Value in A_1 (Fig. 5)	Value in A_4 (Fig. 6)
segPred	*true*	$i = 0$
children	$\{A_1, A_2\}$	$\{A_{6t}, A_{6f}\}$
shape	$\emptyset$	$\{(A_{6f}, A_{6f}), (A_{6f}, A_{6t}), (A_{6t}, A_{6t})\}$
Init	$\{A_1, A_2\}$	$\{A_{6f}\}$
Exit	$\{A_1, A_2\}$	$\{A_{6t}\}$
gMax	1	4
slMax	$slMax(A_1) = 1$ $slMax(A_2) = 1$	$slMax(A_{6f}) = 2$ $slMax(A_{6t}) = 2$
slMin	$slMin(A_1) = 1$ $slMin(A_2) = 1$	$slMin(A_{6f}) = 2$ $slMin(A_{6t}) = 2$

We use the functions $gMax$, $slMin$, and $slMax$ to store information about bounds on the number of times certain iterations of a loop can be repeated. In practice, these are computed using standard loop bound computation techniques.

Remark 1. Note that the quantitative information stored in the AST, i.e., $slMin$, $slMax$, and $gMax$, is different from the quantitative information in [6]. In [6], the interest was in limit-average estimation where storing bounds on segment length is useful; while here, bounds on the number of segments is more useful.

Example 3. We clarify the definition of ASTs using Figures 5 and 6. In node A_0 from Figure 5 and node A_4 from Figure 6 (with parameter $n = 4$), the AST representation is in Table 1. In Figure 5 and Figure 6, the components *shape*, *Init*, *Exit*, *slMin*, and *slMax* have been combined into one regular expression.

AST Semantics. We define $[\![A]\!]$ for a node A in terms of its children. The semantics $[\![T]\!]$ of the AST T is then the semantics of the root node. AST T is a *sound abstraction* for a CFG G if $[\![G]\!] \subseteq [\![T]\!]$. We need two notions to aid the definition. Let $s_1 = (l_0^1, \sigma_0^1) \ldots (l_n^1, \sigma_n^1)$ and $s_2 = (l_0^2, \sigma_0^2) \ldots (l_m^2, \sigma_m^2)$ be two segments.

- the function $form(s_1)$ represents the serial composition of formulas $\Delta((l_i, l_{i+1}))$ for $0 \leq i < k$, i.e., it is the relation on the initial and final program states of s_1 implied by the instructions of s_1; and
- the segment $s_1 \oplus s_2$ is the concatenation of s_1 and s_2 where the last state of s_1 is substituted for the first state of s_2, i.e., $s_1 \oplus s_2 = (l_0^1, \sigma_0^1) \ldots (l_n^1, \sigma_n^1)(l_1^2, \sigma_1^2) \ldots (l_m^2, \sigma_m^2)$.

Let $A = (children, shape, segPred, Init, Exit, slMin, slMax, gMax)$ be a node in T. Segment s is in $[\![A]\!]$ iff $s = s_0 \oplus \ldots \oplus s_n$ and there exist $c_0, \ldots c_n$ such that:

(a) for each i, $s_i \in [\![c_i]\!]$ where $c_i \in children$—for leaf nodes, c_i is a control-flow edge (say (l_i, l_j)) and we let $((l_i, \sigma_i), (l_j, \sigma_j)) \in [\![c_i]\!]$ if $(\sigma_i, \sigma_j) \models \Delta((l_i, l_j))$;
(b) for all $0 \leq i < n$, we have that $(c_i, c_{i+1}) \in shape$;
(c) initial and final variable valuations of s satisfy $segPred$: $form(s) \Rightarrow segPred$;
(d) $c_0 \in Init$ and $c_n \in Exit$, and for each maximal contiguous sequence $c_p c_{p+1} \ldots c_q$ of the same child, $slMin(c_p) \leq q - p + 1 \leq slMax(c_p)$, and
(e) $n \leq gMax$.

Example 4. Consider the CFG from Figure 4, and its AST in Figure 5. A segment π passing through locations $l_1 l_2 l_3 l_4 l_5 l_7 l_3 l_9$ is in the semantics of the node A_2, as it can be split into three segments: (a) the prefix π_1 through $l_1 l_2 l_3$; (b) the middle π_2 through $l_3 l_4 l_5 l_7 l_3$; and (c) the suffix π_3 through $l_3 l_9$. As π_1 is in $[\![A_3]\!]$, π_2 is in $[\![A_4]\!]$, and π_3 is in $[\![A_5]\!]$, we have that $\pi \in [\![A_2]\!]$.

Reducibility of CFGs and the Initial Abstraction. We assume that CFGs are reducible, i.e., that every loop has a unique entry. This assumption holds for programs in high-level programming languages. The function $InitAbs(G)$ takes a CFG an input, and constructs an AST T such that $[\![G]\!] \subseteq [\![T]\!]$. The construction is simple (see, for example, Figure 5). The main point to note is that each maximal strongly connected component (i.e., a loop) corresponds to a node with just one child. The child represents segments corresponding to individual iterations. The $segPred$ predicate for each node is initially set to true.

Proposition 1. *Let G be a CFG. If $T = InitAbs(G)$, we have that $[\![G]\!] \subseteq [\![T]\!]$.*

5   Evaluating WCET on ASTs

In the previous section, we discussed segment abstractions (ASTs). Here, we present a method to compute WCET estimates from ASTs. Given an AST T, let $WCET(T) = \sup_{\pi \in [\![T]\!]} cost(\pi)$. If T is a sound abstraction of G, $[\![T]\!] \supseteq [\![G]\!]$ and hence, $WCET(T) \geq WCET(G)$. Therefore, if an AST T is a sound abstraction of a CFG G, $WCET(T)$ is an over-approximation of $WCET(G)$.

5.1   Maximum-Weight Length-Constrained Paths

We take a recursive approach to computing $WCET(T)$ for an AST T. The WCET of each node is computed using the WCET values of its children by reducing the problem to the the *length-constrained maximum-weight path* problem.

Let $\langle V, E \rangle$ be a graph with vertices V and edges E. Given initial and final vertices v_{in} and v_{out}, cost function $cost : V \rightarrow \mathbb{N}$, global length bound $g_{max} \in (\mathbb{N} \cup \{\infty\})$, and local bounds $l_{min}, l_{max} : V \rightarrow \mathbb{N} \cup \{\infty\}$, the length-constrained maximum-weight path problem asks for the maximum weight path: (a) starting at v_{in} and ending at v_{out}; (b) of length at most g_{max}; and (c) with every maximal contiguous repetition of a vertex v in the path having length at least $l_{min}(v)$ and at most $l_{max}(v)$. Without loss of generality, we assume that $v_{in} \neq v_{out}$ and that v_{in} and v_{out} have only outgoing and incoming edges, respectively.

Reduction. Given a node $A = (children, shape, segPred, Init, Exit, slMin, slMax, gMax)$, we define a graph with vertices being $children \cup \{v_{in}, v_{out}\}$, edges being $shape \cup \{(v_{in}, v) \mid v \in Init\} \cup \{(v, v_{out}) \mid v \in Exit\}$, starting and ending vertices being v_{in} and v_{out}, and the global bound being $gMax$, respectively. For a child c, we have $cost(c) = WCET(c)$, $l_{min}(c) = slMin(c)$, and $l_{max}(c) = slMax(c)$. We call this graph with the corresponding functions the *semantic structure graph* for A and denote it by $Gr(A)$.

Theorem 1. *For each node A in an AST, $WCET(A)$ is equal to the weight of the length-constrained maximum-weight path in $Gr(A)$.*

Hardness. The length-constrained maximum-weight path problem is at least as hard as UNAMBIGUOUS-SAT. Hence, a PTIME algorithm implies that NP = RP, i.e., non-deterministic and randomized polynomial time are the same. However, considering g_{max} as a parameter, the problem is fixed parameter tractable FPT.

5.2   Encoding Optimal Paths

We first discuss the standard technique used in WCET tools to find optimal paths in graphs—the implicit path enumeration technique (IPET) [15,16]. We emphasize that the graph in standard techniques for WCET estimation is the control-flow graph (i.e., a syntactic object), while in our technique it is the semantic structure graph. However, similar principles apply for the graph problem in both cases and we briefly recall the IPET approach as a starting point.

Implicit Path-Enumeration Technique. IPET encodes paths in a graph as an integer linear program (ILP). The encoding uses variables X_v and $X_{(u,v)}$ to represent the number of times vertex v and edge (u, v) occur in the path.

Objective function. The weight of a path is given by $\sum_{v \in V} cost(v) \cdot X_v$. Hence, the objective of the ILP is to maximize $\sum_{v \in V} cost(v) \cdot X_v$.

Kirchhoff's law. To ensure that X_v and $X_{(u,v)}$ values correspond to a real path, we have: for each $v \in V$, we have $X_v = \sum_{(u,v)} X_{u,v} + start_v = \sum_{(v,w)} X_{v,w} + end_v$ where $start_v = 1$ (resp. $end_v = 1$) for $v = v_{in}$ (resp. $v = v_{out}$); otherwise, $start_v = 0$ (resp. $end_v = 0$). Intuitively, for each vertex, the number of incoming edges is equal to the number of outgoing edges, except for v_{in} and v_{out}.

Connectivity. However, Kirchhoff's laws are not sufficient to ensure that the values for X_v and $X_{(u,v)}$ form a feasible path. This is because the disconnected components problem, i.e., the values may correspond to a feasible path along with additional cycles that are disconnected from the path.

Example 5. Consider a graph with vertices $\{v_{in}, v_{out}, v_1, v_2\}$ and edges $\{(v_{in}, v_{out}), (v_{in}, v_1), (v_1, v_2), (v_2, v_1)\}$. The values $X_{v_{in}} = X_{v_{out}} = 1$, $X_{v_1} = X_{v_2} = 10$, $X_{(v_{in}, v_{out})} = 1$, $X_{(v_{in}, v_1)} = 0$, and $X_{(v_1, v_2)} = X_{(v_2, v_1)} = 10$ satisfy the Kirchhoff's law. However, these values do not correspond to a path as the cycle $(v_1 \rightarrow v_2 \rightarrow v_1)$ is disconnected from the rest of the path.

Standard IPET formulations overcome this problem through loop bounds—constraints are added to ensure that a loop is executed at most a constant multiple of times an edge to enter the loop is taken. In Example 5, we would add $X_{(v_1, v_2)} \leq c \cdot X_{(v_{in}, v_1)}$ where c is the loop bound for the cycle $v_1 \rightarrow v_2 \rightarrow v_1$. For structured (reducible) graphs, this approach works very well as each cycle has a unique entry. However, for irreducible graphs, a cycle may not have a unique entry—instead, we need to write such constraints for each subset of vertices which may form a cycle, and each entry to such a cycle, adding an exponential number of constraints just to ensure connectivity (see, for example, [16]).

Semantic Structure of Loops. While the IPET approach works well in the standard WCET analysis framework even for irreducible graphs, the simple loop bound approach to handling connectivity does not work directly as the vertices in the semantic structure graph may represent not only instructions, but also more complex segments (such as different iterations of a loop).

- While CFGs and graphs arising from real programs may be irreducible in the IPET approach, the "degree of irreducibility" is usually low, i.e., only a few additional constraints are necessary to ensure connectivity. On the other hand, since the graphs arising from AST correspond to the semantic structure of loops, they may be highly irreducible (for example, a clique) and an exponential number of additional constraints may be necessary.
- A further reason why an IPET-like approach is not possible for semantic graphs is that there may not exist bounds on cycles in the semantic graphs.

Example 6. Consider the logging example from Section 2 with the modification of the if condition from i mod 4 == 0 to i mod 4 == 0 ∧ started_logging. The boolean variable started_logging is set to false initially, and is non-deterministically set to true at some point during the execution of the loop. Now, consider the three segments sets corresponding to the iterations where the following hold: (a) ¬*started_logging* (say I_1), (b) *started_logging* ∧ $i\%4 == 0$ (say I_2), and (c) *started_logging* ∧ $i\%4 \neq 0$ (say I_3). In the semantic structure graph, there is a cycle containing vertices corresponding to I_2 and I_3; and an entry to this cycle from the vertex corresponding to I_1. However, there is no bound on the number of times this cycle can be executed in terms of the number of times the entry is taken.

The LC-IPET Encoding. We now present our ILP encoding for the length-constrained maximum-weight path problem. This encoding works for (a) firstly, irreducible graphs with only linearly many constraints; and (b) secondly, no bounds on the execution of cycles are required. Hence, this encoding is of interest for WCET analysis independent of the rest of our framework. Given a graph G, we denote the encoding into ILP by LC-IPET(G).

Objective and Kirchhoff's laws. The objective function and Kirchhoff's law constraints are as in the classical IPET approach.

Global and local bounds. The global bound and the local bounds can be ensured using $\sum_{v \in V} X_v \leq g_{max}$ and $l_{min}(v) \cdot \sum_{u|(u,v) \wedge v \neq u} X_{(u,v)} \leq X_v \leq l_{max}(v) \cdot \sum_{u|(u,v) \wedge v \neq u} X_{(u,v)}$ for each vertex v in the graph.

Connectivity flow. We ensure connectivity of the path generated by ILP using an auxiliary flow that goes through only the edges in the path. Intuitively, we ensure that some flow is lost at each visited vertex (i.e., it is a partial sink) except the start vertex which may generate flow (i.e., only the start vertex can be a source). Hence, flow in a component of the path is feasible if and only if it is connected to the start vertex. We use the variables $F_{(u,v)}$ to represent the auxiliary flow through an edge. We have the following:

- Flow is non-negative only in visited edges: for all edges, $F_{(u,v)} \geq 0$ and for all edges, $|V| X_{(u,v)} \geq F_{(u,v)}$. Hence, if $X_{(u,v)}$ is zero, we have $F_{(u,v)} = 0$.
- Every visited vertex other than the start vertex loses some flow: for all vertices $v \neq v_{in}$, $\sum_{(u,v)} F_{(u,v)} - \sum_{(v,w)} F_{(v,w)} \geq X_v$. If X_v is positive (i.e., v is visited), $(\sum_{(u,v)} F_{(u,v)} - \sum_{(v,w)} F_{(v,w)})$ is positive, i.e., the incoming flow to v is greater than the outgoing flow from v.

Theorem 2. *Given graph G, initial and final vertices, and local and global length bounds, the optimal value of LC-IPET(G) is the cost of the length-constrained maximum weight path in G.*

Proof. Clearly, the objective function of LC-IPET(G) corresponds exactly to the weight of a set of nodes in the graph. Hence, it is sufficient to show that every feasible solution of LC-IPET(G) corresponds to a feasible path in G, and vice versa. That every feasible solution of LC-IPET(G) encodes at least one path that follows the local and global bounds is easy to check.

Given a solution to LC-IPET(G), we show that the X_v and $X_{(u,v)}$ values form a path. By Kirchhoff's laws, the value of X_v and $X_{(u,v)}$ consist of a path along with some possibly disconnected components of visited vertices. We show that there cannot be any disconnected components using the auxiliary flow. Suppose $\overline{V}$ and $\overline{E}$ are the subset of vertices and subset of edges which form a disconnected component. Now, we have that for each vertex $v \in \overline{V}$, $\sum_{(u,v)} F_{(u,v)} - \sum_{(v,w)} F_{(v,w)} > 0$. Hence, we have $\sum_{v \in \overline{V}} (\sum_{(u,v)} F_{(u,v)} - \sum_{(v,w)} F_{(v,w)}) > 0$ or equivalently, $\sum_{v \in \overline{V}} \sum_{(u,v)} F_{(u,v)} - \sum_{v \in \overline{V}} \sum_{(v,w)} F_{(v,w)} > 0$ However, note that for edges (u,v) or (v,w) that enter or leave the component we have $F_{(u,v)} = 0$ as the flow is positive if and only if (u,v) is visited. Therefore, we have that both $\sum_{v \in \overline{V}} \sum_{(u,v)} F_{(u,v)}$ and $\sum_{v \in \overline{V}} \sum_{(v,w)} F_{(v,w)}$ are equal to the sum of flow through all edges in $\overline{E}$. This leads to a contradiction as we need $\sum_{v \in \overline{V}} \sum_{(u,v)} F_{(u,v)} - \sum_{v \in \overline{V}} \sum_{(v,w)} F_{(v,w)} > 0$.

Now, given a length-constrained path π in G, we provide a satisfying assignment to the variables in LC-IPET(G). Clearly, the X_v and $X_{(u,v)}$ variables are assigned to the number of times v and (u,v) are visited in π, respectively. We need to find satisfying assignments for $F_{(u,v)}$. For this, we construct separate simple paths π_v from v_{in} to v for each visited vertex v through edges in π. We let $F_{(x,y)} = \sum_v X_v \cdot \pi_v[(x,y)]$ where $\pi_v[(x,y)]$ is equal to 1 if (x,y) occurs is π_v and 0 otherwise. It is easy to see that these $F_{(u,v)}$ values satisfy the auxiliary flow constraints. Intuitively, the auxiliary flow is composed of separate flows of magnitude X_v going from v_{in} to v; call each such flow the flow to v. Now, for every $v' \neq v$, the flow to v' enters and leaves v in the same magnitude. However, the flow to v stops at v, ensuring that the incoming flow to each visited vertex is greater than the outgoing flow by 1.

Given an optimal solution to LC-IPET($Gr(A)$) the worst-case path can be computed using an algorithm for finding Eulerian paths in multi-graphs. Summarizing the approach, given an AST T, we compute the WCET of each node (using its children's WCET values) by reducing to the length-constrained maximum-weight problem. This problem is then solved through the LC-IPET encoding, and the worst-case path can be computed from the solution to the ILP.

Optimizations and Practicality. The semantic structure graphs that arise in practice often allow us to avoid the flow variables in the LC-IPET encoding.

The first case where we can avoid the flow variables is the case where we have no global bound. In our methodology, global bounds arise due to loop bounds in the input program and hence, in each AST node that does not deal with loops, there is no global bound ($g_{max} = \infty$). In this case, either no cycle is reachable in the graph $Gr(A)$, or the worst-case path has weight ∞. Hence, in this case, one can avoid solving the ILP and instead use simpler polynomial time algorithms.

If we have a global bound, we are analyzing different kinds of iterations of a loop, i.e., branches through a loop. While the semantic structure of the loops may be arbitrarily complicated, most of the programs generate semantic structures that fall into several common easily analyzable patterns:

- Progressive phases: These are cases where the execution of the loop is divided into phases, i.e., in each phase only one particular branch through the loop is taken, and this branch is never taken after the completion of the phase. For example, the loop from Example 2 instantiated with $n = 10$ is divided into two phases, one for $i < 5$ and one for $i \geq 5$. The evaluation is easy in such cases as the semantic structure graph is a directed acyclic graph and there cannot be any disconnected cycles. A large number of loops in practice fall into this class (see [17] for an empirical study).
- Cyclic phases: These are cases where the execution of the loop is divided into phases, which repeat in a cycle. For example, the loop in the program from Example 1 is divided into two phases, one for (i % 4 == 0) and one for (i % 4 ≠ 0). Again, in such cases, auxiliary flow variables are unnecessary as there is exactly one cycle in the semantic structure graph.

6   Interpolation for AST Refinement

Once the WCET path is computed from an AST, we check if the computed worst-case path is feasible in the CFG. If such a path is infeasible, we call it an infeasible witness trace. Formally, a *witness trace* (*wit*) is a sequence of CFG nodes that witnesses the current WCET estimate. It is obtained from the techniques presented above. We now describe our interpolation-based AST abstraction refinement algorithm for an infeasible witness trace.

AST Refinement Algorithm. The main idea of Algorithm 1 is to trace the *infeasible* witness trace (*wit*) through the abstract segment tree (AST), and refine the AST nodes touched by *wit*. For each node N, we discover the segment predicates that are important at the interface of the subtree rooted at N and the rest of the AST. When processing an AST node, we split each child (visited by the *wit*) with some new "context" information, obtained via interpolation. Algorithm 1 takes four inputs: (a) an AST T, (b) a node N in T, (c) an infeasible witness trace *wit* that is a segment in N, and (d) a formula SumAbove that summarizes the part of the original witness trace *wit* outside of the subtree rooted at N. Initially, the algorithm is called with N being the root of the AST, and the formula SumAbove is set to *true*.

Algorithm 1. Procedure Refine

Input: AST T, node N in T, witness trace *wit* and formula SumAbove
Output: Refined AST T
1: $s \leftarrow \text{TraceWit}(N, wit)$
2: **for all** $i \in \{0, \ldots, |s|\}$ **do**
3:      context $\leftarrow$ SumAbove $\wedge$ SumLR(s, wit, N) $\wedge$ $segPred(N)$
4:      child $\leftarrow form(\text{projection}(wit, s_i))$
5:      $I \leftarrow \text{Interpolate}(\text{child}, \text{context})$ ▷ context $\wedge$ child unsat
6:      $r_t \leftarrow \text{addToTree}(s_i, I)$; $r_f \leftarrow \text{addToTree}(s_i, \neg I)$
7:      REFINE($T, r_t, I\wedge segPred(s_i)$, projection($wit, s_i$))   ▷ Recursively refine.
8:      StrengthenDown(r_f)
9:      RemoveFromTree(T, s_i)
10: StrengthenUp(N)

Refinement Procedure. We now detail the REFINE procedure of Algorithm 1 and illustrate it on our running example. For a node N, the procedure REFINE obtains a sequence $s = s_0 s_1 \ldots s_k$ of children of N that the witness trace *wit* passes through (line 1 of Alg. 1). Note that a child can be repeated in s. The *wit* can be split into segments, where the i-th segment wit_i of *wit* belongs to the i-th child s_i. Recall that the infeasible *wit* of Example 2 was $wit = l_1 l_2 (l_3 l_4 l_5 l_7)^4 l_3 l_9$ for $n = 4$. The CFG of Example 2 is Figure 4, and its AST is in Figure 5. Consider the node A_4 in Figure 5 as the node N. The node A_4 represents a loop and the node A_6 a single iteration. The sequence s is then A_6^4.

Algorithm 2. Precision Refinement for WCET

1: **Input:** Program $\mathcal{P}$;   **Output:** WCET of $\mathcal{P}$
2: Build the CFG G of $\mathcal{P}$;
3: Construct the AST T corresponding to G;   // *Abstraction*
4: **for** each node A in T (post-order traversal) **do**   // *Evaluation*
5:     construct the LC-IPET($Gr(A)$);
6:     $WCET(A) \leftarrow$ optimum of LC-IPET($Gr(A)$);
7: $w_i \leftarrow$ witness trace corresponding to $WCET(T)$
8: **if** w_i is infeasible **then**
9:     Refine(T,w_i,root(T),true) (Algorithm 1)   // *Refinement*
10:     **go to** line 4;
11: **return** $WCET(T)$.

Next, each child s_i is refined using wit_i (loop at line 2). The variable *context* stores a formula that summarizes what we know about *wit* outside of s_i (line 3). It is obtained as a conjunction of the formula SumAbove, the segment predicate of N, and the information computed by the function SumLR(). The function SumLR() computes information about the trace *wit* as it passes through the children of N other than s_i. When refining s_i, SumLR() returns a formula $\wedge_{k<|s|\wedge(k\neq i)}J_k$, where J_k is $form(wit_k)$ and wit_k is the part of the *wit* going through the node s_k. (*form* was defined in Section 4.) The variable *child* stores a formula that summarizes what we know about the *wit* inside of s_i (line 4). It is computed as $form(wit_i)$, where wit_i was obtained by the projection of *wit* to the node s_i. For our running example, consider the third iteration of A_6. In this case, the value of *context* is $i_0 = 0 \wedge \bigwedge_{k=0}^{1} i_{k+1} = i_k + 1$ (we show only the relevant part of the formula) and the value of *child* is $i_2 < 4 \wedge i_2 < 2$ (4 and 2 are n and $n/2$, respectively).

Note that *child* $\wedge$ *context* is unsatisfiable, as (a) the original *wit* is infeasible, and (b) *context* and *child* summarize the *wit*. We hence can use interpolation to infer a predicate explaining the infeasibility at the boundary of the subtree of child s_i and the rest of the AST. We compute an interpolant I from the proof of unsatisfiability *context* $\wedge$ *child* (line 5) such that *context* $\implies I$ and *child* $\wedge I \implies \bot$, where I is over only those variables that are common to both *context* and *child*. In our running example, we obtain the interpolant $i_2 \geq 2$.

Using the computed interpolant I, we next replace the node s_i by two nodes r_t and r_f (line 6). The node r_t is like s_i (in terms of its children in the AST), but has a transition predicate equal to $segPred(s_i) \wedge I$. Similarly, for r_f we take its transition predicate $segPred(r_f)$ as $segPred(s_i) \wedge \neg I$. In this way, each of these nodes has more information about its context than s_i had. We can further refine these two nodes and use them in the AST T instead of s_i.

Observe that for r_t we added the predicate I to its transition predicate. As *child* $\wedge I$ is unsatisfiable, the trace *wit* is not represented in r_t. The node r_t can thus be refined by a recursive call to the REFINE procedure (line 7). As *child* $\wedge \neg I$ is satisfiable, for r_f there is nothing more to learn from the *wit*. We simply strengthen the node, that is, propagate the new predicate, $\neg I$, to the children of r_f. This is done by calling the *StrengthenDown*() function (line 8), which propagates the new information $\neg I$ to the children t of the node r_f. To this end, it checks whether it finds a segment in the node t which is excluded from the node by $\neg I$, and then calls the REFINE procedure to perform refinement with the discovered segment used as a witness trace. In our running example r_f corresponds to A_{6f} and r_t is A_{6t}. Finally, the function $StrengthenUp()$ uses the information discovered during the refinement process for the children of a node N, and strengthens the $segPred$ predicate of N (line 10).

7 WCET Computation Algorithm

Algorithm 2 describes our approach to computing precise WCET estimates. Given a program $\mathcal{P}$, we first construct its CFG (line 2), and build the corresponding initial AST T (line 3). For the AST T, we compute the $WCET(T)$ (lines 4-6), using the LC-IPET approach detailed in Section 5. The WCET is precise if a feasible program path exhibits the WCET. We therefore check if the witness trace exhibiting the WCET is indeed feasible (line 8). If not, we refine our current AST using Algorithm 1 (line 9).

8 Parametric WCET Computation

We now extend our techniques to handle parametric programs and return a parametric WCET estimate, i.e., one that may depend on the parameter values.

Parameters. *Program parameters* $P \subseteq V$ are program variables whose values do not change in any execution. Given a valuation $val(P) : P \to \mathbb{N}$ of P, the CFG $G_{val(P)}$ is obtained by replacing variables in P by their values given by $val(P)$.

Solution Language. Let $\mathcal{A}(P)$ be the set of arithmetic expressions over P. The language of *disjunctive expressions* $\mathcal{E}(P)$ consists of sets of pairs $W = \{D_0 \mapsto N_0, \dots, D_k \mapsto N_k\}_i$ where D_i and N_i are boolean and arithmetic expressions over P, respectively; and we have $\bigvee_i D_i = true$ and $\forall i \neq j.D_i \wedge D_j \implies \bot$. Intuitively, the value of W is N_i when D_i holds. Given a valuation $val(P)$, we write $W[val(P)]$ for the explicit integer value of $N_i[P]$ where $D_i(val(P))$ holds.

It is easy to define standard arithmetic, comparison, and max operators over $\mathcal{E}(V)$. For example, if $W^1 = \{D_i \mapsto N_i\}_i$ and $W^2 = \{D_j \mapsto N_j\}_j$, then $W^1 + W^2 = \bigcup_{i,j}\{(D_i \wedge D_j \mapsto N_i + N_j)\}$, and $\max(W^1, W^2) = \{D_i \wedge D_j \wedge N_i > N_j \mapsto N_i\} \cup \{D_i \wedge D_j \wedge N_i \leq N_j \mapsto N_j\}$.

Problem Statement. A *parametric WCET estimate* $WCET_p(G, P)$ of a CFG G is an expression in $\mathcal{E}(P)$, such that for all valuations $val(P)$ of parameters, $WCET_p(G, P)[val(P)] \geq WCET(G_{val(P)})$. The parametric WCET estimate, $WCET_p(G, P)$, is an over-approximation $WCET(G_{val(P)})$ for each valuation $val(P)$. The task of our *parametric WCET estimation problem* is: Given a CFG G and a set of parameters P, compute $WCET_p(G, P)$, the parametric WCET.

The Parametric Framework. We describe the changes necessary to adapt our WCET estimation framework to the parametric case.

Abstraction. A *parametric AST* is similar to an AST, except that that for each node, *gMax* has type $\mathcal{A}(P)$, and *slMax* and *slMin* have type *children* $\rightarrow \mathcal{A}(P)$.

Evaluation. The evaluation of $WCET(T)$ for a parametric AST is more involved than for a standard AST. This procedure is detailed in Section 8.1.

Refinement. The refinement procedure from Section 6 can be used directly in the parametric framework. However, an important aspect is that the procedure works best if the interpolants generated are independent of the parameter valuations. In our implementation, the theorem prover was tuned to produce such interpolants.

The Parametric WCET Estimation Algorithm. The parametric WCET estimation algorithm follows Algorithm 2 with the major difference being the feasibility checking of worst-case paths. As parametric WCET estimates are disjunctive, we generate worst-case paths for each disjunct by choosing appropriate parameter valuations and use them for feasibility analysis and refinement as in Algorithm 2.

8.1   Parametric Maximum-Weight Length-Constrained Paths

For evaluating the WCET of parametric ASTs, we proceed recursively as in the non-parametric case. At each level, we reduce the problem to the parametric version of the length-constrained maximum-weight paths in a graph. Let $\langle V, E \rangle$, v_{in}, and v_{out} be as in the non-parametric case. Given a cost function $cost : V \rightarrow \mathcal{E}(P)$, a global bound expression $g_{max} \in \mathcal{A}(P)$, and local bound functions $l_{min}, l_{max} : V \rightarrow \mathcal{A}(P)$, the *parametric length-constrained maximum-weight path problem* asks for an expression $W \in \mathcal{E}(P)$ such that for every valuation of parameters $val(P)$, we have that $W[val(P)]$ is equal to the cost of the length-constrained maximum-weight path in the graph where $cost$, g_{max}, l_{min}, and l_{max} have been instantiated with $val(P)$.

Restrictions. The problem is hard even in the case where l_{min}, l_{max} and g_{max} range over polynomial expressions. Hence, we place restrictions on the expressions and assume that g_{max}, l_{min}, and l_{max} are all linear expressions over a single parameter. Further, we present our techniques for the case where $cost(v)$ is a numeric value instead of a disjunctive expression. The algorithm where $cost(\cdot)$ yields disjunctive expressions is similar with all max and $+$ operations over integers being replaced by max and $+$ operations over $\mathcal{E}(P)$. Note that restricting l_{min}, l_{max}, and g_{max} to expressions in one parameter does not restrict the CFG and the AST to one parameter—multiple parameters may appear in different nodes of the AST. Before we present our algorithm for the parametric length-constrained maximum-weight path problem, we need the following lemmata.

Lemma 1 (One non-extremal node). *For every parametric length-constrained maximum-weight problem and $val(P)$, there is an optimal path* $\pi = v_0^{k_0} v_1^{k_1} \ldots v_n^{k_n}$ *such that $l_{min}(v_i) < k_i < l_{max}(v_i)$ for at most one i.*

The lemma holds as for any path having two non-extremal nodes (say v_{i_1} and v_{i_2} with $cost(v_{i_1}) \geq cost(v_{i_2})$), we can build another path of equal or greater weight where v_{i_1} is taken more number of times and v_{i_2} fewer times.

While the previous lemma bounds the number of repetitions of self-loops we need to consider, the next one does the same for other cycles. The cycle decomposition of a path π is given by $\langle \sigma, (L_0, n_0), (L_1, n_1) \ldots, (L_k, n_k) \rangle$ where: (a) σ is a simple path from v_{in} to v_{out}; (b) each L_i is a simple cycle; (c) together, the multi-set of visited nodes in σ and the cycles L_i's each taken n_i times is the same as the visited nodes in π. Note that L_i's are not self-loops and that the classification "simple" does not take into account self-loops. Every path has a cycle decomposition and further, for every cycle decomposition where the simple path and simple cycles are connected, there is a path for which it is a cycle decomposition. In any worst-case path, the *heaviest cycle* is taken most often.

Lemma 2 (One heavy loop). *For every parametric length-constrained maximum-weight problem and* $val(P)$, *there exists an optimal path* π *with cycle decomposition* $\langle \sigma, (L_0, n_0), (L_1, n_1), \ldots, (L_k, n_k) \rangle$ *such that for all* $i > 0$: *(a)* $cost(L_0)/|L_0| \geq cost(L_i)/|L_i|$; *and (b)* $n_i|L_i| < lcm(|L_0|, |L_i|)$.

The Algorithm. We describe the algorithm for the restricted version of the parametric length-constrained maximum weight problem. Intuitively, the algorithm considers cycle decompositions $\langle \sigma, (L_0, n_0), (L_1, n_1), \ldots (L_k, n_k) \rangle$ where n_0 is a linear expression in $\mathcal{A}(P)$, and each $n_i < lcm(|L_0|, |L_i|)/|L_i|$ is an integer for $i > 0$, and a non-extremal node v and builds the disjunctive expression $\{cond \mapsto wt, \neg cond \mapsto 0\}$ where *cond* and *wt* are explained below. The solution is the maximum of such disjunctive expressions. Note that σ and L_i's can be restricted to sequences where vertices only occur either l_{max} or l_{min} times; and further, it can be assumed that $|L_0|$ is not a parametric expression, but an integer. The expression *wt* is the expression over the parameters P obtained as the sum of weights in the guessed path, i.e., $cost(\sigma) + \sum_i n_i \cdot cost(L_i)$. The condition *cond* expresses that $\langle \sigma, (L_0, n_0), (L_1, n_1), \ldots (L_k, n_k) \rangle$ is a valid cycle decomposition that respects Lemma 2, and that the total length is less than g_{max}. The correctness of the algorithm depends on the above lemmata and the fact that there are only a finite number of such parametric cycle decompositions.

Theorem 3. *The restricted parametric length-constrained maximum-weight problem can be solved in* EXPSPACE *in the size of the inputs on a computing model where operations on disjunctive expressions have constant cost.*

Practical Cases. As in the non-parametric case, we provide efficient algorithms for the most commonly occurring practical cases.

- *No global bounds.* If the graph has $g_{max} = \infty$, we can use the standard dynamic programming longest-path algorithm for DAGs with the integer max and + operations being replaced by max and + operations over $\mathcal{E}(P)$.
- *Progressive phases.* In the progressive phases case, the same maximum-weight longest-path algorithm can be used with the modification of accumulating the length of the path along with the weight, and then constraining the final result with the condition that the length is at most the global bound.

9   Experimental Evaluation

We implemented our approach in a tool called IBART. It takes C programs (with no procedure calls) as input, and returns a parametric WCET estimate.

```
n:=0;
while(n < iters)
 if(health==round0)
  HighVoltageCurrent(health)
  UpdatePeriod(temp, 5)
  if(hit_trigger_flag==0)
   ResetPeakDetector()
 if(health==round1)
   ...
 if(health==round4)
   LowVoltageCurrent()
 ...
 if(health!=0)
   health--
 else
   health=9
 n++
```

Fig. 7. from ex2 from Debie suite

Low-Level Analysis. IBART analyzes WCET for Infineon C167 and LPC2138 ARM7 processors, using CalcWcet167 [12] and owcet [2] respectively, as low-level analyzers to compute basic block execution costs. These costs are then mapped from the binary to the source level and used in our analysis. Thus, IBART is platform-aware; it can be easily extended to other architectures by supplying the architecture-dependent basic block execution times on source level. However, we note that due to this approach, we cannot refine the WCET estimates in the case the infeasibility is due to caching or pipeline effects between basic blocks, and are beyond what is analyzable with the low-level tools. This is an orthogonal issue, as we focus on better estimates of WCET by having a better approximation of feasible paths. However, in the future, we plan to alleviate this issue by using our framework to automatically discovering predicates about caches and pipelines, and by performing the loop unrolling on-demand to aid the low-level analyzers.

Dependent Loops. We implement a slight extension of the parametric algorithm presented in Section 8 to handle dependent loops. Consider two loops `for(i=0;i<n;i++) for(j=0;j<i;j++){...}`. The worst-case cost of an outer loop iteration is $n \cdot k$ (where k is the cost of an iteration of the inner loop). Using this worst-case cost, we get that the worst-case cost of the outer loop is $n^2 \cdot k$. However, the inner-loop costs only $k \cdot i$ in the ith iteration. In this case, we incorporate the precise cost of the child node while computing the cost of the parent node, i.e., the more precise estimate for the outer loop is $\sum_{i=0}^{n-1} i \cdot k$. Intuitively, when the child node costs a polynomial (say $p(i)$) in the ith repetition, we can compute the more precise estimate as $\sum_{i=1}^{n} p(i)$. Note that this extension is equivalent to considering the loop counter of the outer loop a parameter while evaluating the inner loop.

Benchmarks. We evaluated IBART on 10 examples (examples 2 to 11 in Table 2) taken from WCET benchmark suites and open-source linear algebra packages. Of the 10 examples, 3 are small functions with less than 30 lines of code; the remaining 7 have between 34 and 109 lines of code. While small, the examples were chosen to be challenging for WCET analysis, due to two features: (a) branching statements within loops, leading to iterations with different costs, and (b) nested loops, whose inner loops linearly depend on the outer loops.

Table 2. Parametric WCET computation for the C167 architecture

Ex	Source/File	Parametric WCET (C167)
ex1	Section 2, Figure 3	$n \leq 5 \mapsto 24940,$ $n \geq 6 \mapsto 5040 + 2800\lfloor n/2 \rfloor + 1900n$
ex2	Debie/ health	$n \leq 0 \mapsto 2620,$ $n > 0 \mapsto 2620 + \lfloor n/10 \rfloor 59100 + (n\%10) * 6800$
ex3	Mälardalen/ adpcm	$dlt \neq 0 \mapsto 4180 + 5060n,$ $dlt = 0 \mapsto 4260 + 2500n$
ex4	Mälardalen/ crc	$jrev > 0 \mapsto 5560 + 3860len,$ $jrev \leq 0 \mapsto 4320 + 3380len$
ex5	Mälardalen/ crc	$len \geq -1 \wedge init = 0 \mapsto 7800 + 3840len,$ $init \neq 0 \mapsto 3060$
ex6	Mälardalen/ lcdnum	$n \geq 0 \mapsto 1740 + 2460n,$ $n < 0 \mapsto 1740$
ex7	Jampack/ Inv	$2 > n \wedge n \geq 0 \mapsto 13540 + 6420n,$ $0 > n \mapsto 13380,$ $n > 2 \mapsto 13380 - 3100n + 9480n^2$
ex8	Jampack/ Zsvd	$nc \leq nr \wedge r \geq c \mapsto 3840,$ $nc > nr \wedge c > r > b \mapsto 18260 + 18820(r - b),$ $nc \leq nr \wedge c < r \mapsto 3920$ ...
ex9	JAMA/ Cholesky- Decomposition	$1 = n \mapsto 44880,$ $1 > n \mapsto 14260,$ $n > 1 \mapsto 14260 + 15447n + 13419n^2 + 1754n^3$
ex10	JAMA/ Eigenvalue-Decomposition	$1 > n \mapsto 11780,$ $n \geq 1 \mapsto -11784 + 17602n - 5146n^2 + 11108n^3$
ex11	Jampack / Eigenvalue-Decomposition	$n < 0 \mapsto 25460,$ $n \geq 0 \mapsto 25460 + 28400n + 9500n^2 + 11220n^3$

WCET Benchmark Suites. We used the Debie and the Mälardalen benchmark suite from the WCET community [18], which are commonly used for evaluating WCET tools. We analyzed one larger example (109 lines) from the Debie examples (ex2 in Table 2) and 4 programs from the Mälardalen suite. The parametric timing behavior of these examples comes from the presence of symbolic loop bounds. An excerpt from the Debie example is shown in Figure 7.

Note that in Figure 7, different paths in the loop have different execution times. Moreover, every conditional branch is revisited at every tenth iteration of the loop. Computing the WCET of the program by taking the most expensive conditional branch at every loop iteration would thus yield a pessimistic overestimate of the actual WCET. Our approach derives a tight parametric WCET by identifying the set of feasible program paths at each loop iteration.

Linear Algebra Packages. We used 5 examples from the open-source Java linear algebra libraries JAMA and Jampack. These packages provide user-level classes for matrix operations including inverse calculation (ex7), SVD (ex8), triangularization (ex9), and eigenvalue decomposition (ex10, ex11) of matrices. We manually translated them to C. These benchmarks contain nested loops, often with conditionals, and with inner loops linearly depending on outer loops.

Results. We evaluated IBART for parametric WCET computation, and compared IBART with state-of-the-art WCET analyzers. All experiments were run on a 2.2 GHz Intel Core i7 CPU with 8 GB RAM and took less than 20 seconds.

IBART Results. Our results are summarized in Table 2. Column 3 shows the parametric WCET (in the solution language of Section 8) calculated by IBART with basic block execution times provided by CalcWCET167. In all cases, the number of refinements needed was between 2 and 6.

Comparison with WCET Tools. We compared the precision of IBART to r-TuBound [13] supporting the Infineon C167 processor and OTAWA, supporting the LPC2138 processor. Note that r-TuBound and OTAWA can only report a single numeric value as a WCET estimate. Therefore, to allow a fair comparison of the WCET results we use the basic block execution times of the respective low-level analyzer in IBART, and instantiate the symbolic parameters in the flow facts with concrete values when analyzing the WCET with r-TuBound, respectively OTAWA. To this end, parameters were supplied to OTAWA by means of (high-level) input annotations. r-TuBound does not support input annotations, therefore parameters were encoded directly in the ILP, if possible.

Our results, summarized in Table 3, show that IBART provides significantly better WCET estimates than the respective framework. For larger values of parameters, the difference increases rapidly. This is because r-Tubound and OTAWA over-approximate each iteration much more than IBART; so if the number of iterations increases, the difference grows. Column 2 lists the values of parameters. Columns 3 and 4 show the WCET computed by IBART and r-TuBound for the C167 architecture, while columns 5 and 6 show the WCET computed by IBART and OTAWA for the LPC2138 architecture. Note that Columns 3 and 4 are in nanoseconds, while Columns 5 and 6 are in cycles.

IBART reports a parametric formula instead of a single number. Instantiating with concrete parameter values (see Table 3), often gives a tighter WCET estimate. In cases when the WCET estimate of IBART overlaps with the estimate of r-TuBound or OTAWA, IBART usually allows to infer tighter estimates for specific parameter configurations. For example, for ex4, in both architectures the estimates are identical when $jrev < 0$. IBART automatically discovers the predicate $jrev \geq 0$ to specialize cases where a tighter estimate is possible. On the other hand, this information cannot be used in r-TuBound, while OTAWA fails to exploit the supplied input-annotations leading to over-estimation.

10   Related Work

We briefly summarize the large body of related work here.

Segment abstraction. Segment abstraction was introduced in [8] and was shown to subsume a large class of program analysis techniques. In [6], it was extended to quantitative properties. This paper brings a key contribution: a systematic way for computing WCET using computation of local ILPs at each node of AST, instead of one large global ILP. Furthermore, (a) we adapt segment abstraction for the timing analysis by using global and local bound functions; (b) we refine using a novel interpolation based technique (c) we propose the LC-IPET encoding, and (d) the parametric bounds are novel.

Asymptotic analysis. Computing bounds automatically was explored (e.g., [9]). Our work differs both conceptually and methodologically from these as we compute

Table 3. WCET comparisons for the C167 and the LPC2138

Ex	Parameter assignments	IBART	r-TuBound	IBART	OTAWA
		C167 (ns)		LPC2138 (cycles)	
ex1	$n = 5$	**22,300**	26,060	**393**	393
	$n = 100$	**388,040**	48,0160	**6,888**	8,338
ex2	$n = 10$	**62,020**	124,920	**2,115**	2,258
	$n = 50$	**298,420**	612,920	**10,435**	11,098
	$n = 200$	**1184,920**	2442,920	**41,635**	44,248
ex3	$n = 6, dlt = 0$	**19,260**	345,040	**246**	295
	$n = 0, dlt \neq 0$	**4,180**	4,260	56	56
	$n = 0, dlt = 0$	4,260	4,260	**55**	56
ex4	$len = 5, jrev < 0$	24,860	24,860	520	520
	$len = 5, jrev \geq 0$	**21,220**	24,860	**447**	520
	$len = 0, jrev \geq 0$	**4,320**	5,560	**112**	140
ex5	$init = 0, len = 255$	987,000	987,000	**18,534**	18,534
	$init = 1, len = 255$	**2,920**	987,000	**102**	18,534
ex6	$n = 10$	**21,660**	26,340	**349**	404
	$n = 5$	13,960	13,960	214	214
ex7	$n = 5$	**243,880**	519,280	**2,123**	2,538
	$n = 1$	**19,760**	19,960	**263**	346
ex8	$r = 4, c = 5,$	**93,540**	100,480	**837**	880
	$nr < nc, b = 0$				
ex9	$n = 5$	**646,220**	1545,760	**2,902**	6,993
	$n = 1$	44,880	44,880	**317**	357
ex10	$n = 5$	**1335,920**	2606,180	**6,059**	23,089
	$n = 0$	11,620	11,620	**209**	445
ex11	$n = 5$	1799,620	1799,620	**6,371**	22,946
	$n = 1$	74,580	74,580	**582**	582

the worst-case execution time rather than asymptotic complexity, and we infer predicates using interpolation rather than using template-based methods.

Static WCET analysis. Most state-of-the-art static WCET tools, see e.g. [13,2], compute a constant WCET, requiring numeric upper bounds for all loops. Our parametric WCET is computed only once and replacing parameters with their values yields the precise WCET for each set of concrete values, without rerunning the WCET analysis as in [13,2]. These and other WCET tools use ILP as the basic data structure. Our basic data structure is ASTs, which leads to more efficient, and more precise, algorithms. All these approaches to WCET estimation (including ours) are dependent on low-level analysis developed for modelling timing related features of architectures. For a survey of techniques in this area, see [18].

Parametric WCET estimates. Parametric WCET calculation is also described in [5,1,10], where polyhedra-based abstract interpretation is used to derive integer constraints on program executions. These constraints are solved as parametric integer linear programming problem, and a parametric WCET is obtained. In [10], various heuristics are applied in order to approximate program paths by small linear expressions over execution frequencies of program blocks. In [4], the authors describe an efficient, but approximate method of solving parametric constraints—in contrast, our approach solves parametric constraints exactly. Compared to [5,10], our segment abstraction lets us to reason about the WCET as a property of a sequence of instructions rather than a state property.

Refinement for WCET. We are aware of two recent works for refining WCET estimates [14,3]. WCET squeezing [14] is based on learning ILP constraints from infeasible paths—the constraints learned are based purely on syntactic methods. The recent work [3] also automatically discovers additional ILP constraints using minimal unsatisfiable cores of infeasible paths. These approaches have the disadvantage of being susceptible to requiring many refinements to eliminate related infeasible paths that can be eliminated with one segment predicate.

Symbolic Execution based WCET analysis. The most relevant work to our approach is [7]. In [7], the authors use symbolic execution to explore the program as a tree, and find and merge segments that have similar timing properties (for example, different loop iterations). More precisely, by merging segments generated by symbolic execution, one can obtain trees equivalent to ASTs (generated in our approach by splitting nodes in the initial tree). However, the cost of generating these objects can be significantly different. The key difference is that parts of the tree that do not occur in the worst-case path can be analyzed quickly in our case (we do not need to split nodes when even the over-approximation shows that it is not part of WCET path). On the other hand, in the symbolic execution case, even the parts of the tree not occurring in the worst-case path have to be explicitly explored and merged. However, if all paths have similar execution times, symbolic execution has an advantage as in our case the tree will eventually be split up completely. However, many of the advantages such as having locality of constraints also apply in the case of [7].

11   Conclusion

Our approach to WCET analysis is based on the hypothesis that segment abstraction is the ideal framework for WCET estimation. This is intuitively clear, as WCET is a property that accumulates over segments, i.e., sequences of instructions, and is not a state property, and therefor not being fully amenable to standard state-based abstractions. Our approach based on abstract segment trees provides two clear advantages. First, ASTs allow us to decompose the problem into multiple smaller ones. In particular, it allows us to decompose the integer linear program (ILP) for path analysis to multiple smaller integer linear programs. Second, it allows us to compute more precise refinements compared to existing techniques. This is because ASTs can encode more expressive constraints than ILPs.

A possible direction for future work is to explore is to extend our techniques to additional cost models—including aspects such as cache-persistence [18] (where only the first access to a location is a cache-miss). In fact, the segment abstraction is rich enough to incorporate hard constraints such as scoped persistence [11]. Further, segment abstraction and refinement can be used to refine the low-level timing analysis—segment abstractions can be used to drive CFG transformations to enhance the precision of low-level timing analysis. Extending our method with interpolation in first-order theories, e.g., the theory of arrays, could yield transition predicates over data structures. Finally, estimating WCET could be used in synthesis of optimal programs.

References

1. Altmeyer, S., Humbert, C., Lisper, B., Wilhelm, R.: Parametric timing analysis for complex architectures. In: RTCSA 2008, pp. 367–376 (2008)
2. Ballabriga, C., Cassé, H., Rochange, C., Sainrat, P.: OTAWA: An Open Toolbox for Adaptive WCET Analysis. In: Min, S.L., Pettit, R., Puschner, P., Ungerer, T. (eds.) SEUS 2010. LNCS, vol. 6399, pp. 35–46. Springer, Heidelberg (2010)
3. Blackham, B., Liffiton, M., Heiser, G.: Trickle: automated infeasible path detection using all minimal unsatisfiable subsets. In: RTAS, pp. 169–178 (2014)
4. Bygde, S., Ermedahl, A., Lisper, B.: An efficient algorithm for parametric WCET calculation. Journal of Systems Architecture 57(6), 614–624 (2011)
5. Bygde, S., Lisper, B.: Towards an Automatic Parametric WCET Analysis. In: Proc. of WCET (2008)
6. Černý, P., Henzinger, T., Radhakrishna, A.: Quantitative Abstraction Refinement. In: Proc. of POPL, pp. 115–128 (2013)
7. Chu, D., Jaffar, J.: Symbolic simulation on complicated loops for WCET path analysis. In: EMSOFT, pp. 319–328 (2011)
8. Cousot, P., Cousot, R.: An abstract interpretation framework for termination. In: Proc. of POPL, pp. 245–258 (2012)
9. Gulwani, S., Zuleger, F.: The reachability-bound problem. In: Proc. of PLDI, pp. 292–304 (2010)
10. Huber, B., Prokesch, D., Puschner, P.: A Formal Framework for Precise Parametric WCET Formulas. In: Proc. of WCET, pp. 91–102 (2012)
11. Huynh, B.K., Ju, L., Roychoudhury, A.: Scope-aware data cache analysis for WCET estimation. In: RTAS 2011, pp. 203–212. IEEE (2011)
12. Kirner, R.: The WCET Analysis Tool CalcWcet167. In: Margaria, T., Steffen, B. (eds.) ISoLA 2012, Part II. LNCS, vol. 7610, pp. 158–172. Springer, Heidelberg (2012)
13. Knoop, J., Kovács, L., Zwirchmayr, J.: r-TuBound: Loop Bounds for WCET Analysis (Tool Paper). In: Bjørner, N., Voronkov, A. (eds.) LPAR-18 2012. LNCS, vol. 7180, pp. 435–444. Springer, Heidelberg (2012)
14. Knoop, J., Kovács, L., Zwirchmayr, J.: WCET squeezing: on-demand feasibility refinement for proven precise WCET-bounds. In: RTNS, pp. 161–170 (2013)
15. Li, Y., Malik, S.: Performance analysis of embedded software using implicit path enumeration. In: DAC, pp. 456–461 (1995)
16. Puschner, P., Schedl, A.: Computing Maximum Task Execution Times – A Graph-Based Approach. Real-Time Systems 13(1), 67–91 (1997)
17. Sharma, R., Dillig, I., Dillig, T., Aiken, A.: Simplifying loop invariant generation using splitter predicates. In: Gopalakrishnan, G., Qadeer, S. (eds.) CAV 2011. LNCS, vol. 6806, pp. 703–719. Springer, Heidelberg (2011)
18. Wilhelm, R., Engblom, J., Ermedahl, A., Holsti, N., Thesing, S., Whalley, D., Bernat, G., Ferdinand, C., Heckmann, R., Mitra, T., Mueller, F., Puaut, I., Puschner, P., Staschulat, J., Stenström, P.: The Worst-Case Execution-Time Problem - Overview of Methods and Survey of Tools. ACM Trans. Embedded Comput. Syst. 7(3) (2008)

Automatic Static Cost Analysis
for Parallel Programs

Jan Hoffmann and Zhong Shao

Yale University

Abstract. Static analysis of the evaluation cost of programs is an extensively studied problem that has many important applications. However,
most automatic methods for static cost analysis are limited to sequential
evaluation while programs are increasingly evaluated on modern multi-
core and multiprocessor hardware. This article introduces the first automatic analysis for deriving bounds on the worst-case evaluation cost of
parallel first-order functional programs. The analysis is performed by a
novel type system for amortized resource analysis. The main innovation
is a technique that separates the reasoning about sizes of data structures
and evaluation cost within the same framework. The cost semantics of
parallel programs is based on call-by-value evaluation and the standard
cost measures *work* and *depth*. A soundness proof of the type system
establishes the correctness of the derived cost bounds with respect to
the cost semantics. The derived bounds are multivariate resource polynomials which depend on the sizes of the arguments of a function. Type
inference can be reduced to linear programming and is fully automatic.
A prototype implementation of the analysis system has been developed
to experimentally evaluate the effectiveness of the approach. The experiments show that the analysis infers bounds for realistic example programs
such as quick sort for lists of lists, matrix multiplication, and an implementation of sets with lists. The derived bounds are often asymptotically
tight and the constant factors are close to the optimal ones.

Keywords: Functional Programming, Static Analysis, Resource Consumption, Amortized Analysis.

1 Introduction

Static analysis of the resource cost of programs is a classical subject of computer
science. Recently, there has been an increased interest in formally proving cost
bounds since they are essential in the verification of safety-critical real-time and
embedded systems.

For sequential functional programs there exist many automatic and semi-
automatic analysis systems that can statically infer cost bounds. Most of them
are based on sized types [1], recurrence relations [2], and amortized resource analysis [3, 4]. The goal of these systems is to automatically compute easily-understood
arithmetic expressions in the sizes of the inputs of a program that bound resource
cost such as time or space usage. Even though an automatic computation of cost

© Springer-Verlag Berlin Heidelberg 2015
J. Vitek (Ed.): ESOP 2015, LNCS 9032, pp. 132–157, 2015.
DOI: 10.1007/978-3-662-46669-8_6

bounds is undecidable in general, novel analysis techniques are able to efficiently compute tight time bounds for many non-trivial programs [5–9].

For functional programs that are evaluated in parallel, on the other hand, no such analysis system exists to support programmers with computer-aided derivation of cost bounds. In particular, there are no type systems that derive cost bounds for parallel programs. This is unsatisfying because parallel evaluation is becoming increasingly important on modern hardware and referential transparency makes functional programs ideal for parallel evaluation.

This article introduces an automatic type-based resource analysis for deriving cost bounds for parallel first-order functional programs. Automatic cost analysis for sequential programs is already challenging and it might seem to be a long shot to develop an analysis for parallel evaluation that takes into account low-level features of the underlying hardware such as the number of processors. Fortunately, it has been shown [10, 11] that the cost of parallel functional programs can be analyzed in two steps. First, we derive cost bounds at a high abstraction level where we assume to have an unlimited number of processors at our disposal. Second, we prove once and for all how the cost on the high abstraction level relates to the actual cost on a specific system with limited resources.

In this work, we derive bounds on an abstract cost model that consists of the *work* and the *depth* of an evaluation of a program [10]. Work measures the evaluation time of sequential evaluation and depth measures the evaluation time of parallel evaluation assuming an unlimited number of processors. It is well-known [12] that a program that evaluates to a value using work w and depth d can be evaluated on a shared-memory multiprocessor (SMP) system with p processors in time $O(\max(w/p, d))$ (see Section 2.3). The mechanism that is used to prove this result is comparable to a scheduler in an operating system.

A novelty in the cost semantics in this paper is the definition of work and depth for terminating and non-terminating evaluations. Intuitively, the non-deterministic big-step evaluation judgement that is defined in Section 2 expresses that *there is a (possibly partial) evaluation with work n and depth m*. This statement is used to prove that a typing derivation for bounds on the depth or for bounds on the work ensures termination.

Technically, the analysis computes two separate typing derivations, one for the work and one for the depth. To derive a bound on the work, we use multivariate amortized resource analysis for sequential programs [13]. To derive a bound on the depth, we develop a novel multivariate amortized resource analysis for programs that are evaluated in parallel. The main challenge in the design of this novel parallel analysis is to ensure the same high compositionality as in the sequential analysis. The design and implementation of this novel analysis for bounds on the depth of evaluations is the main contribution of our work. The technical innovation that enables compositionality is an analysis method that separates the static tracking of size changes of data structures from the cost analysis while using the same framework. We envision that this technique will find further applications in the analysis of other non-additive cost such as stack-space usage and recursion depth.

We describe the new type analysis for parallel evaluation for a simple first-order language with lists, pairs, pattern matching, and sequential and parallel composition. This is already sufficient to study the cost analysis of parallel programs. However, we implemented the analysis system in Resource Aware ML (RAML), which also includes other inductive data types and conditionals [14]. To demonstrate the universality of the approach, we also implemented NESL's [15] parallel list comprehensions as a primitive in RAML (see Section 6). Similarly, we can define other parallel sequence operations of NESL as primitives and correctly specify their work and depth. RAML is currently extended to include higher-order functions, arrays, and user-defined inductive types. This work is orthogonal to the treatment of parallel evaluation.

To evaluate the practicability of the proposed technique, we performed an experimental evaluation of the analysis using the prototype implementation in RAML. Note that the analysis computes worst-case bounds instead of average-case bounds and that the asymptotic behavior of many of the classic examples of Blelloch et al. [10] does not differ in parallel and sequential evaluations. For instance, the depth and work of quick sort are both quadratic in the worst-case. Therefore, we focus on examples that actually have asymptotically different bounds for the work and depth. This includes quick sort for lists of lists in which the comparisons of the inner lists can be performed in parallel, matrix multiplication where matrices are lists of lists, a function that computes the maximal weight of a (continuous) sublist of an integer list, and the standard operations for sets that are implemented as lists. The experimental evaluation can be easily reproduced and extended: RAML and the example programs are publicly available for download and through an user-friendly online interface [16].

In summary we make the following contributions.

1. We introduce the first automatic static analysis for deriving bounds on the depth of parallel functional programs. Being based on multivariate resource polynomials and type-based amortized analysis, the analysis is compositional. The computed type derivations are easily-checkable bound certificates.
2. We prove the soundness of the type-based amortized analysis with respect to an operational big-step semantics that models the work and depth of terminating and non-terminating programs. This allows us to prove that work and depth bounds ensure termination. Our inductively defined big-step semantics is an interesting alternative to coinductive big-step semantics.
3. We implemented the proposed analysis in RAML, a first-order functional language. In addition to the language constructs like lists and pairs that are formally described in this article, the implementation includes binary trees, natural numbers, tuples, Booleans, and NESL's parallel list comprehensions.
4. We evaluated the practicability of the implemented analysis by performing reproducible experiments with typical example programs. Our results show that the analysis is efficient and works for a wide range of examples. The derived bounds are usually asymptotically tight if the tight bound is expressible as a resource polynomial.

The full version of this article [17] contains additional explanations, lemmas, and details of the technical development.

2   Cost Semantics for Parallel Programs

In this section, we introduce a first-order functional language with parallel and sequential composition. We then define a big-step operational semantics that formalizes the cost measures *work* and *depth* for terminating and non-terminating evaluations. Finally, we prove properties of the cost semantics and discuss the relation of work and depth to the run time on hardware with finite resources.

2.1   Expressions and Programs

Expressions are given in let-normal form. This means that term formers are applied to variables only when this does not restrict the expressivity of the language. Expressions are formed by integers, variables, function applications, lists, pairs, pattern matching, and sequential and parallel composition.

$$e, e_1, e_2 ::= n \mid x \mid f(x) \mid (x_1, x_2) \mid\mid \mathsf{match}\, x \,\mathsf{with}\, (x_1, x_2) \Rightarrow e$$
$$\mid \mathsf{nil} \mid \mathsf{cons}(x_1, x_2) \mid \mathsf{match}\, x \,\mathsf{with}\, \langle \mathsf{nil} \Rightarrow e_1 \mid \mathsf{cons}(x_1, x_2) \Rightarrow e_2 \rangle$$
$$\mid \mathsf{let}\, x = e_1 \,\mathsf{in}\, e_2 \mid \mathsf{par}\, x_1 = e_1 \,\mathsf{and}\, x_2 = e_2 \,\mathsf{in}\, e$$

The parallel composition $\mathsf{par}\, x_1 = e_1 \,\mathsf{and}\, x_2 = e_2 \,\mathsf{in}\, e$ is used to evaluate e_1 and e_2 in parallel and bind the resulting values to the names x_1 and x_2 for use in e.

In the prototype, we have implemented other inductive types such as trees, natural numbers, and tuples. Additionally, there are operations for primitive types such as Booleans and integers, and NESL's parallel list comprehensions [15]. Expressions are also transformed automatically into let normal form before the analysis. In the examples in this paper, we use the syntax of our prototype implementation to improve readability.

In the following, we define a standard type system for expressions and programs. Data types A, B and function types F are defined as follows.

$$A, B ::= \mathsf{int} \mid L(A) \mid A * B \qquad\qquad F ::= A \rightarrow B$$

Let $\mathcal{A}$ be the set of data types and let $\mathcal{F}$ be the set of function types. A signature $\Sigma : \mathrm{FID} \rightharpoonup \mathcal{F}$ is a partial finite mapping from function identifiers to function types. A context is a partial finite mapping $\Gamma : \mathit{Var} \rightharpoonup \mathcal{A}$ from variable identifiers to data types. A simple type judgement $\Sigma; \Gamma \vdash e : A$ states that the expression e has type A in the context Γ under the signature Σ. The definition of typing rules for this judgement is standard and we omit the rules.

A *(well-typed) program* consists of a signature Σ and a family $(e_f, y_f)_{f \in \mathrm{dom}(\Sigma)}$ of expressions e_f with a distinguished variable identifier y_f such that $\Sigma; y_f{:}A \vdash e_f{:}B$ if $\Sigma(f) = A \rightarrow B$.

2.2   Big-Step Operational Semantics

We now formalize the resource cost of evaluating programs with a big-step operational semantics. The focus of this paper is on time complexity and we only define the cost measures *work* and *depth*. Intuitively, the work measures the time that is needed in a sequential evaluation. The depth measures the time that is needed in a parallel evaluation. In the semantics, time is parameterized by a metric that assigns a non-negative cost to each evaluation step.

$$\frac{V,H \;\vdash^{M}\; e_1 \Downarrow \circ \mid (w,d)}{V,H \;\vdash^{M}\; \mathsf{let}\, x = e_1 \,\mathsf{in}\, e_2 \Downarrow \circ \mid (M^{\mathsf{let}}{+}w, M^{\mathsf{let}}{+}d)}\;(\text{E:Let1}) \qquad \frac{(\text{E:Abort})}{V,H \;\vdash^{M}\; e \Downarrow \circ \mid (0,0)}$$

$$\frac{V,H \;\vdash^{M}\; e_1 \Downarrow (\ell, H') \mid (w_1, d_1) \qquad V[x \mapsto \ell], H' \;\vdash^{M}\; e_2 \Downarrow \rho \mid (w_2, d_2)}{V,H \;\vdash^{M}\; \mathsf{let}\, x = e_1 \,\mathsf{in}\, e_2 \Downarrow \rho \mid (M^{\mathsf{let}}{+}w_1{+}w_2, M^{\mathsf{let}}{+}d_1{+}d_2)}\;(\text{E:Let2})$$

$$\frac{V,H \;\vdash^{M}\; e_1 \Downarrow \rho_1 \mid (w_1, d_1) \qquad V,H \;\vdash^{M}\; e_2 \Downarrow \rho_2 \mid (w_2, d_2) \qquad \rho_1 {=} \circ \vee \rho_2 {=} \circ}{V,H \;\vdash^{M}\; \mathsf{par}\, x_1 = e_1 \,\mathsf{and}\, x_2 = e_2 \,\mathsf{in}\, e \Downarrow \circ \mid (M^{\mathsf{Par}}{+}w_1{+}w_2, M^{\mathsf{Par}}{+}\max(d_1, d_2))}\;(\text{E:Par1})$$

(E:Par2)

$$\frac{\begin{array}{c}V,H \;\vdash^{M}\; e_1 \Downarrow (\ell_1, H_1) \mid (w_1, d_1) \qquad (w', d') {=} (M^{\mathsf{Par}}{+}w_1{+}w_2{+}w, M^{\mathsf{Par}}{+}\max(d_1, d_2){+}d) \\ V,H \;\vdash^{M}\; e_2 \Downarrow (\ell_2, H_2) \mid (w_2, d_2) \qquad V[x_1 {\mapsto} \ell_1, x_2 {\mapsto} \ell_2], H_1 \uplus H_2 \;\vdash^{M}\; e \Downarrow (\ell, H') \mid (w, d)\end{array}}{V,H' \;\vdash^{M}\; \mathsf{par}\, x_1 = e_1 \,\mathsf{and}\, x_2 = e_2 \,\mathsf{in}\, e \Downarrow (\ell, H') \mid (w', d')}$$

Fig. 1. Interesting rules of the operational big-step semantics

Motivation. A distinctive feature of our big-step semantics is that it models terminating, failing, and diverging evaluations by inductively describing finite subtrees of (possibly infinite) evaluation trees. By using an inductive judgement for diverging and terminating computations while avoiding intermediate states, it combines the advantages of big-step and small-step semantics. This has two benefits compared to standard big-step semantics. First, we can model the resource consumption of diverging programs and prove that bounds hold for terminating and diverging programs. (In some cost metrics, diverging computations can have finite cost.) Second, for a cost metric in which all diverging computations have infinite cost we are able to show that bounds imply termination.

Note that we cannot achieve this by step-indexing a standard big-step semantics. The available alternatives to our approach are small-step semantics and coinductive big-step semantics. However, it is unclear how to prove the soundness of our type system with respect to these semantics. Small-step semantics is difficult to use because our type-system models an intentional property that goes beyond the classic type preservation: After performing a step, we have to obtain a refined typing that corresponds to a (possibly) smaller bound. Coinductive derivations are hard to relate to type derivations because type derivations are defined inductively.

Our inductive big-step semantics can not only be used to formalize resource cost of diverging computations but also for other effects such as event traces. It is therefore an interesting alternative to recently proposed coinductive operational big-step semantics [18].

Semantic Judgements. We formulate the big-step semantics with respect to a stack and a heap. Let Loc be an infinite set of $locations$ modeling memory addresses on a heap. A value $v ::= n \mid (\ell_1, \ell_2) \mid (\mathsf{cons}, \ell_1, \ell_2) \mid \mathsf{nil} \in Val$ is either an integer $n \in \mathbb{Z}$, a pair of locations (ℓ_1, ℓ_2), a node $(\mathsf{cons}, \ell_1, \ell_2)$ of a list, or nil.

A $heap$ is a finite partial mapping $H : Loc \rightharpoonup Val$ that maps locations to values. A $stack$ is a finite partial mapping $V : Var \rightharpoonup Loc$ from variable identifiers to locations. Thus we have boxed values. It is not important for the analysis whether values are boxed.

Figure 1 contains a compilation of the big-step evaluation rules (the full version contains all rules). They are formulated with respect to a resource metric M. They define the evaluation judgment

$$V, H \vdash^{\underline{M}} e \Downarrow \rho \mid (w, d) \qquad \text{where} \qquad \rho ::= (\ell, H) \mid \circ .$$

It expresses the following. In a fixed program $(e_f, y_f)_{f \in \text{dom}(\Sigma)}$, if the stack V and the initial heap H are given then the expression e evaluates to ρ. Under the metric M, the work of the evaluation of e is w and the depth of the evaluation is d. Unlike standard big-step operational semantics, ρ can be either a pair of a location and a new heap, or $\circ$ (pronounced *busy*) indicating that the evaluation is not finished yet.

A resource metric $M : K \to \mathbb{Q}_0^+$ defines the resource consumption in each evaluation step of the big-step semantics with a non-negative rational number. We write M^k for $M(k)$.

An intuition for the judgement $V, H \vdash^{\underline{M}} e \Downarrow \circ \mid (w, d)$ is that there is a partial evaluation of e that runs without failure, has work w and depth d, and has not yet reached a value. This is similar to a small-step judgement.

Rules. For a heap H, we write $H, \ell \mapsto v$ to express that $\ell \notin \text{dom}(H)$ and to denote the heap H' such that $H'(x) = H(x)$ if $x \in \text{dom}(H)$ and $H'(\ell) = v$. In the rule E:PAR2, we write $H_1 \uplus H_2$ to indicate that H_1 and H_2 agree on the values of locations in $\text{dom}(H_1) \cap \text{dom}(H_2)$ and to a combined heap H with $\text{dom}(H) = \text{dom}(H_1) \cup \text{dom}(H_2)$. We assume that the locations that are allocated in parallel evaluations are disjoint. That is easily achievable in an implementation.

The most interesting rules of the semantics are E:ABORT, and the rules for sequential and parallel composition. They allow us to approximate infinite evaluation trees for non-terminating evaluations with finite subtrees. The rule E:ABORT states that we can partially evaluate every expression by doing zero steps. The work w and depth d are then both zero (i.e., $w = d = 0$).

To obtain an evaluation judgement for a sequential composition $\text{let } x = e_1 \text{ in } e_2$ we have two options. We can use the rule E:LET1 to partially evaluate e_1 using work w and depth d. Alternatively, we can use the rule E:LET2 to evaluate e_1 until we obtain a location and a heap (ℓ, H') using work w_1 and depth d_1. Then we evaluate e_2 using work w_2 and depth d_2. The total work and depth is then given by $M^{\text{let}} + w_1 + w_2$ and $M^{\text{let}} + d_1 + d_2$, respectively.

Similarly, we can derive evaluation judgements for a parallel composition $\text{par } x_1 = e_1 \text{ and } x_2 = e_2 \text{ in } e$ using the rules E:PAR1 and E:PAR2. In the rule E:PAR1, we partially evaluate e_1 or e_2 with evaluation cost (w_1, d_1) and (w_2, d_2). The total work is then $M^{\text{Par}} + w_1 + w_2$ (the cost for the evaluation of the parallel binding plus the cost for the sequential evaluation of e_1 and e_2). The total depth is $M^{\text{Par}} + \max(d_1, d_2)$ (the cost for the evaluation of the binding plus the maximum of the cost of the depths of e_1 and e_2). The rule E:PAR2 handles the case in which e_1 and e_2 are fully evaluated. It is similar to E:LET2 and the cost of the evaluation of the expression e is added to both the cost and the depth since e is evaluated after e_1 and e_2.

2.3   Properties of the Cost-Semantics

The main theorem of this section states that the resource cost of a partial evaluation is less than or equal to the cost of an evaluation of the same expression that terminates.

Theorem 1. *If* $V, H \vdash^{\underline{M}} e \Downarrow (\ell, H') \mid (w, d)$ *and* $V, H \vdash^{\underline{M}} e \Downarrow \circ \mid (w', d')$ *then* $w' \leqslant w$ *and* $d' \leqslant d$.

Theorem 1 can be proved by a straightforward induction on the derivation of the judgement $V, H \vdash^{\underline{M}} e \Downarrow (\ell, H') \mid (w, d)$.

Provably Efficient Implementations. While work is a realistic cost-model for the sequential execution of programs, depth is not a realistic cost-model for parallel execution. The main reason is that it assumes that an infinite number of processors can be used for parallel evaluation. However, it has been shown [10] that work and depth are closely related to the evaluation time on more realistic abstract machines.

For example, *Brent's Theorem* [12] provides an asymptotic bound on the number of execution steps on the shared-memory multiprocessor (SMP) machine. It states that if $V, H \vdash^{\underline{M}} e \Downarrow (\ell, H') \mid (w, d)$ then e can be evaluated on a p-processor SMP machine in time $O(\max(w/p, d))$. An SMP machine has a fixed number p of processes and provides constant-time access to a shared memory. The proof of Brent's Theorem can be seen as the description of a so-called *provably efficient implementation*, that is, an implementation for which we can establish an asymptotic bound that depends on the number of processors.

Classically, we are especially interested in non-asymptotic bounds in resource analysis. It would thus be interesting to develop a non-asymptotic version of Brent's Theorem for a specific architecture using more refined models of concurrency [11]. However, such a development is not in the scope of this article.

Well-Formed Environments and Type Soundness. For each data type A we inductively define a set $[\![A]\!]$ of values of type A. Lists are interpreted as lists and pairs are interpreted as pairs.

$$[\![\text{int}]\!] = \mathbb{Z} \qquad [\![A * B]\!] = [\![A]\!] \times [\![B]\!]$$
$$[\![L(A)]\!] = \{[a_1, \ldots, a_n] \mid n \in \mathbb{N}, a_i \in [\![A]\!]\}$$

If H is a heap, ℓ is a location, A is a data type, and $a \in [\![A]\!]$ then we write $H \vDash \ell \mapsto a : A$ to mean that ℓ defines the semantic value $a \in [\![A]\!]$ when pointers are followed in H in the obvious way. The judgment is formally defined in the full version of the article.

We write $H \vDash \ell : A$ to indicate that there exists a, necessarily unique, semantic value $a \in [\![A]\!]$ so that $H \vDash \ell \mapsto a : A$. A stack V and a heap H are *well-formed* with respect to a context Γ if $H \vDash V(x) : \Gamma(x)$ holds for every $x \in \text{dom}(\Gamma)$. We then write $H \vDash V : \Gamma$.

Simple Metrics and Progress. In the reminder of this section, we prove a property of the evaluation judgement under a simple metric. A *simple metric* M assigns the value 1 to every resource constant, that is, $M(x) = 1$ for every $x \in K$. With a simple metric, work counts the number of evaluation steps.

Theorem 2 states that, in a well-formed environment, well-typed expressions either evaluate to a value or the evaluation uses unbounded work and depth.

Theorem 2 (Progress). *Let M be a simple metric, $\Sigma; \Gamma \vdash e : B$, and $H \vDash V : \Gamma$. Then $V, H \vdash^{M} e \Downarrow (\ell, H') \mid (w, d)$ for some $w, d \in \mathbb{N}$ or for every $n \in \mathbb{N}$ there exist $x, y \in \mathbb{N}$ such that $V, H \vdash^{M} e \Downarrow \circ \mid (x, n)$ and $V, H \vdash^{M} e \Downarrow \circ \mid (n, y)$.*

A direct consequence of Theorem 2 is that bounds on the depth of programs under a simple metric ensure termination.

3   Amortized Analysis and Parallel Programs

In this section, we give a short introduction into amortized resource analysis for sequential programs (for bounding the work) and then informally describe the main contribution of the article: a multivariate amortized resource analysis for parallel programs (for bounding the depth).

Amortized Resource Analysis. Amortized resource analysis is a type-based technique for deriving upper bounds on the resource cost of programs [3]. The advantages of amortized resource analysis are compositionality and efficient type inference that is based on linear programming. The idea is that types are decorated with resource annotations that describe a potential function. Such a potential function maps the sizes of typed data structures to a non-negative rational number. The typing rules ensure that the potential defined by a typing context is sufficient to pay for the evaluation cost of the expression that is typed under this context and for the potential of the result of the evaluation.

The basic idea of amortized analysis is best explained by example. Consider the function mult : int $*$ L(int) $\rightarrow$ L(int) that takes an integer and an integer list and multiplies each element of the list with the integer.

```
mult(x,ys) = match ys with | nil  → nil
                           | (y::ys')  → x*y::mult(x,ys')
```

For simplicity, we assume a metric M^* that only counts the number of multiplications performed in an evaluation in this section. Then $V, H \vdash^{M^*}$ mult(x, ys) $\Downarrow (\ell, H') \mid (n, n)$ for a well-formed stack V and heap H in which ys points to a list of length n. In short, the work and depth of the evaluation of mult(x, ys) is $|ys|$.

To obtain a bound on the work in type-based amortized resource analysis, we derive a type of the following form.

$$x{:}\text{int}, ys{:}L(\text{int}); Q \vdash^{M^*} \text{mult}(x, ys) : (L(\text{int}), Q')$$

Here Q and Q' are *coefficients* of multivariate resource polynomials $p_Q : [\![\text{int} * L(\text{int})]\!] \rightarrow \mathbb{Q}_0^+$ and $p_{Q'} : [\![L(\text{int})]\!] \rightarrow \mathbb{Q}_0^+$ that map semantic values to non-negative rational numbers. The rules of the type system ensure that for every evaluation context (V, H) that maps x to a number m and ys to a list a, the potential $p_Q(m, a)$ is sufficient to cover the evaluation cost of mult(x, ys) and the potential $p_{Q'}(a')$ of the returned list a'. More formally, we have $p_Q(m, a) \geqslant w + p_{Q'}(a')$ if $V, H \vdash^{M^*}$ mult(x, ys) $\Downarrow (\ell, H') \mid (w, d)$ and ℓ points to the list a' in H'.

In our type system we can for instance derive coefficients Q and Q' that represent the potential functions

$$p_Q(n, a) = |a| \qquad \text{and} \qquad p_{Q'}(a) = 0 \ .$$

The intuitive meaning is that we must have the potential $|ys|$ available when evaluating $\mathsf{mult}(x, ys)$. During the evaluation, the potential is used to pay for the evaluation cost and we have no potential left after the evaluation.

To enable compositionality, we also have to be able to pass potential to the result of an evaluation. Another possible instantiation of Q and Q' would for example result in the following potential.

$$p_Q(n, a) = 2 \cdot |a| \qquad \text{and} \qquad p_{Q'}(a) = |a|$$

The resulting typing can be read as follows. To evaluate $\mathsf{mult}(x, ys)$ we need the potential $2|ys|$ to pay for the cost of the evaluation. After the evaluation there is the potential $|\mathsf{mult}(x, ys)|$ left to pay for future cost in a surrounding program. Such an instantiation would be needed to type the inner function application in the expression $\mathsf{mult}(x, \mathsf{mult}(z, ys))$.

Technically, the coefficients Q and Q' are families that are indexed by sets of base polynomials. The set of base polynomials is determined by the type of the corresponding data. For the type $\mathsf{int} * L(\mathsf{int})$, we have for example $Q = \{q_{(*,[])}, q_{(*,[*])}, q_{(*,[*,*])}, \ldots\}$ and $p_Q(n, a) = q_{(*,[])} + q_{(*,[*])} \cdot |a| + q_{(*,[*,*])} \cdot \binom{|a|}{2} + \ldots$. This allows us to express multivariate functions such as $m \cdot n$.

The rules of our type system show how to describe the valid instantiations of the coefficients Q and Q' with a set of linear inequalities. As a result, we can use linear programming to infer resource bounds efficiently.

A more in-depth discussion can be found in the literature [3, 19, 7].

Sequential Composition. In a sequential composition $\mathsf{let}\, x = e_1 \,\mathsf{in}\, e_2$, the initial potential, defined by a context and a corresponding annotation (Γ, Q), has to be used to pay for the work of the evaluation of e_1 and the work of the evaluation of e_2. Let us consider a concrete example again.

```
mult2(ys) = let xs = mult(496,ys) in
            let zs = mult(8128,ys) in (xs,zs)
```

The work (and depth) of the evaluation of the expression $\mathsf{mult2(ys)}$ is $2|ys|$ in the metric M^*. In the type judgement, we express this bound as follows. First, we type the two function applications of mult as before using

$$x{:}\mathsf{int}, ys{:}L(\mathsf{int}); Q \vdash^{M^*} \mathsf{mult}(x, ys) : (L(\mathsf{int}), Q')$$

where $p_Q(n, a) = |a|$ and $p_{Q'}(a) = 0$. In the type judgement

$$ys{:}L(\mathsf{int}); R \vdash^{M^*} \mathsf{mult2(ys)} : (L(\mathsf{int}) * L(\mathsf{int}), R')$$

we require that $p_R(a) \geqslant p_Q(a) + p_Q(a)$, that is, the initial potential (defined by the coefficients R) has to be shared in the two sequential branches. Such a sharing can still be expressed with linear constraints. such as $r_{[*]} \geqslant q_{(*,[*])} + q_{(*,[*])}$. A valid instantiation of R would thus correspond to the potential function $p_R(a) = 2|a|$. With this instantiation, the previous typing reflects the bound $2|ys|$ for the evaluation of $\mathsf{mult2(ys)}$.

A slightly more involved example is the function dyad : $L(\text{int}) * L(\text{int}) \rightarrow L(L(\text{int}))$ which computes the dyadic product of two integer lists.

```
dyad (u,v) = match u with | nil → nil
           | (x::xs) → let x' = mult(x,v) in
                       let xs' = dyad(xs,v) in x'::xs';
```

Using the metric M^* that counts multiplications, multivariate resource analysis for sequential programs derives the bound $|u|\cdot|v|$. In the cons branch of the pattern match, we have the potential $|xs|\cdot|v| + |v|$ which is shared to pay for the cost $|v|$ of mult(x, v) and the cost $|xs|\cdot|v|$ of dyad(xs, v).

Moving multivariate potential through a program is not trivial; especially in the presence of nested data structures like trees of lists. To give an idea of the challenges, consider the expression e that is defined as follows.

```
let xs = mult(496,ys) in
let zs = append(ys,ys) in dyad(xs,zs)
```

The depth of evaluating e in the metric M^* is bounded by $|ys| + 2|ys|^2$. Like in the previous example, we express this in amortized resource analysis with the initial potential $|ys| + 2|ys|^2$. This potential has to be shared to pay for the cost of the evaluations of mult(496, ys) (namely $|ys|$) and dyad(xs, zs) (namely $2|ys|^2$). However, the type of dyad requires the quadratic potential $|xs|\cdot|zs|$. In this simple example, it is easy to see that $|xs|\cdot|zs| = 2|ys|^2$. But in general, it is not straightforward to compute such a conversion of potential in an automatic analysis system, especially for nested data structures and super-linear size changes. The type inference for multivariate amortized resource analysis for sequential programs can analyze such programs efficiently [7].

Parallel Composition. The insight of this paper is that the potential method works also well to derive bounds on parallel evaluations. The main challenge in the development of an amortized resource analysis for parallel evaluations is to ensure the same compositionality as in sequential amortized resource analysis.

The basic idea of our new analysis system is to allow each branch in a parallel evaluation to use all the available potential without sharing. Consider for example the previously defined function mult2 in which we evaluate the two applications of mult in parallel.

```
mult2par(ys) = par xs = mult(496,ys)
               and zs = mult(8128,ys) in (xs,zs)
```

Since the depth of mult(n, ys) is $|ys|$ for every n and the two applications of mult are evaluated in parallel, the depth of the evaluation of mult2par(ys) is $|ys|$ in the metric M^*.

In the type judgement, we type the two function applications of mult as in the sequential case in which

$$x{:}\text{int}, ys{:}L(\text{int}); Q \vdash^{M^*} \text{mult}(x, ys) : (L(\text{int}), Q')$$

such that $p_Q(n, a) = |a|$ and $p_{Q'}(a) = 0$. In the type judgement

$$ys{:}L(\text{int}); R \vdash^{M^*} \text{mult2par}(ys) : (L(\text{int}) * L(\text{int}), R')$$

for mult2par we require however only that $p_R(a) \geqslant p_Q(a)$. In this way, we express that the initial potential defined by the coefficients R has to be sufficient to cover the cost of each parallel branch. Consequently, a possible instantiation of R corresponds to the potential function $p_R(a) = |a|$.

In the function dyad, we can replace the sequential computation of the inner lists of the result by a parallel computation in which we perform all calls to the function mult in parallel. The resulting function is dyad_par.

```
dyad_par (u,v) = match u with | nil  → nil
                | (x::xs)  → par x' = mult(x,v)
                             and xs' = dyad_par(xs,v) in x'::xs';
```

The depth of dyad_par is $|v|$. In the type-based amortized analysis, we hence start with the initial potential $|v|$. In the cons branch of the pattern match, we can use the initial potential to pay for both, the cost $|v|$ of mult(x, v) and the cost $|v|$ of the recursive call dyad(xs, v) without sharing the initial potential.

Unfortunately, the compositionality of the sequential system is not preserved by this simple idea. The problem is that the naive reuse of potential that is passed through parallel branches would break the soundness of the system. To see why, consider the following function.

```
mult4(ys) = par xs = mult(496,ys)
   and zs = mult(8128,ys) in (mult(5,xs), mult(10,zs))
```

Recall, that a valid typing for xs = mult(496, ys) could take the initial potential $2|ys|$ and assign the potential $|xs|$ to the result. If we would simply reuse the potential $2|ys|$ to type the second application of mult in the same way then we would have the potential $|xs| + |zs|$ after the parallel branches. This potential could then be used to pay for the cost of the remaining two applications of mult. We have now verified the unsound bound $2|ys|$ on the depth of the evaluation of the expression mult4(ys) but the depth of the evaluation is $3|ys|$.

The problem in the previous reasoning is that we doubled the part of the initial potential that we passed on for later use in the two parallel branches of the parallel composition. To fix this problem, we need a separate analysis of the sizes of data structures and the cost of parallel evaluations.

In this paper, we propose to use cost-free type judgements to reason about the size changes in parallel branches. Instead of simply using the initial potential in both parallel branches, we share the potential between the two branches but analyze the two branches twice. In the first analysis, we only pay for the resource consumption of the first branch. In the second, analysis we only pay for resource consumption of the second branch.

A cost-free type judgement is like any other type judgement in amortized resource analysis but uses the cost-free metric cf that assigns zero cost to every evaluation step. For example, a cost-free typing of the function mult(ys) would express that the initial potential can be passed to the result of the function. In the cost-free typing judgement

$$x{:}\mathrm{int}, ys{:}L(\mathrm{int}); Q \vdash^{\mathrm{cf}} \mathrm{mult}(x, ys) : (L(\mathrm{int}), Q')$$

a valid instantiation of Q and Q' would correspond to the potential

$$p_Q(n, a) = |a| \qquad \text{and} \qquad p_{Q'}(a) = |a| .$$

The intuitive meaning is that in a call zs = mult(x, ys), the initial potential $|ys|$ can be transformed to the potential $|zs|$ of the result.

Using cost-free typings, we can now correctly reason about the depth of the evaluation of mult4. We start with the initial potential $3|\text{ys}|$ and have to consider two cases in the parallel binding. In the first case, we have to pay only for resource cost of $\text{mult}(496, \text{ys})$. So we share the initial potential and use $2|\text{ys}|$: $|\text{ys}|$ to pay the cost of $\text{mult}(496, \text{ys})$ and $|\text{ys}|$ to assign the potential $|\text{xs}|$ to the result of the application. The reminder $|\text{ys}|$ of the initial potential is used in a cost-free typing of $\text{mult}(8128, \text{ys})$ where we assign the potential $|\text{zs}|$ to the result of the function without paying any evaluation cost. In the second case, we derive a similar typing in which the roles of the two function calls are switched. In both cases, we start with the potential $3|\text{ys}|$ and end with the potential $|\text{xs}| + |\text{zs}|$. We use it to pay for the two remaining calls of mult and have verified the correct bound.

In the univariate case, using the notation from [3, 19], we could formulate the type rule for parallel composition as follows. Here, the coefficients Q are not globally attached to a type or context but appear locally at list types such as $L^q(\text{int})$. The sharing operator $\Gamma \curlyvee (\Gamma_1, \Gamma_2, \Gamma_3)$ requires the sharing of the potential in the context Γ in the contexts Γ_1, Γ_2 and Γ_3. For instance, we have $x{:}L^6(\text{int}) \curlyvee (x{:}L^2(\text{int}), x{:}L^3(\text{int}), x{:}L^1(\text{int}))$.

$$\frac{\Gamma \curlyvee (\Delta_1, \Gamma_2, \Gamma') \quad \Gamma \curlyvee (\Gamma_1, \Delta_2, \Gamma') \quad \Gamma_1 \vdash^{M} e_1 : A_1 \quad \Delta_2 \vdash^{cf} e_2 : A_2}{\Delta_1 \vdash^{cf} e_1 : A_1 \quad \Gamma_2 \vdash^{M} e_2 : A_2 \quad \Gamma', x_1{:}A_1, x_2{:}A_2 \vdash^{M} e : B}{\Gamma \vdash^{M} \text{par } x_1 = e_1 \text{ and } x_2 = e_2 \text{ in } e : B}$$

In the rule, the initial potential Γ is shared twice using the sharing operator $\curlyvee$. First, to pay the cost of evaluating e_2 and e, and to pass potential to x_1 using the cost-free type judgement $\Delta_1 \vdash^{cf} e_1 : A_1$. Second, to pay the cost of evaluation e_1 and e, and to pass potential to x_2 via the judgement $\Delta_2 \vdash^{cf} e_2 : A_2$.

This work generalizes the idea to multivariate resource polynomials for which we also have to deal with mixed potential such as $|x_1| \cdot |x_2|$. The approach features the same compositionality as the sequential version of the analysis. As the experiments in Section 7 show, the analysis works well for many typical examples.

The use of cost-free typings to separate the reasoning about size changes of data structures and resource cost in amortized analysis has applications that go beyond parallel evaluations. Similar problems arise in sequential (and parallel) programs when deriving bounds for non-additive cost such as stack-space usage or recursion depth. We envision that the developed technique can be used to derive bounds for these cost measures too.

Other Forms of Parallelism. The binary parallel binding is a simple yet powerful form of parallelism. However, it is (for example) not possible to directly implement NESL's model of sequences that allows to perform an operation for every element in the sequence in constant depth. The reason is that the parallel binding would introduce a linear overhead.

Nevertheless it is possible to introduce another binary parallel binding that is semantically equivalent except that it has zero depth cost. We can then analyze more powerful parallelism primitives by translating them into code that uses this cost-free parallel binding. To demonstrate such a translation, we implemented NESL's [15] parallel sequence comprehensions in RAML (see Section 6).

4   Resource Polynomials and Annotated Types

In this section, we introduce multivariate resource polynomials and annotated types. Our goal is to systematically describe the potential functions that map data structures to non-negative rational numbers. Multivariate resource polynomials are a generalization of non-negative linear combinations of binomial coefficients. They have properties that make them ideal for the generation of succinct linear constraint systems in an automatic amortized analysis. The presentation might appear quite low level but this level of detail is necessary to describe the linear constraints in the type rules.

Two main advantages of resource polynomials are that they can express more precise bounds than non-negative linear-combinations of standard polynomials and that they can succinctly describe common size changes of data that appear in construction and destruction of data. More explanations can be found in the previous literature on multivariate amortized resource analysis [13, 7].

4.1   Resource Polynomials

A resource polynomial maps a value of some data type to a nonnegative rational number. Potential functions and thus resource bounds are always resource polynomials.

Base Polynomials. For each data type A we first define a set $P(A)$ of functions $p : [\![A]\!] \to \mathbb{N}$ that map values of type A to natural numbers. These *base polynomials* form a basis (in the sense of linear algebra) of the resource polynomials for type A. The resource polynomials for type A are then given as nonnegative rational linear combinations of the base polynomials. We define $P(A)$ as follows.

$$P(\text{int}) = \{a \mapsto 1\} \quad P(A_1 * A_2) = \{(a_1, a_2) \mapsto p_1(a_1) \cdot p_2(a_2) \mid p_i \in P(A_i)\}$$

$$P(L(A)) = \{\Sigma\Pi[p_1, \ldots, p_k] \mid k \in \mathbb{N}, p_i \in P(A)\}$$

We have $\Sigma\Pi[p_1, \ldots, p_k]([a_1, \ldots, a_n]) = \sum_{1 \leqslant j_1 < \cdots < j_k \leqslant n} \prod_{1 \leqslant i \leqslant k} p_i(a_{j_i})$. Every set $P(A)$ contains the constant function $v \mapsto 1$. For lists $L(A)$ this arises for $k = 0$ (one element sum, empty product).

For example, the function $\ell \mapsto \binom{|\ell|}{k}$ is in $P(L(A))$ for every $k \in \mathbb{N}$; simply take $p_1 = \ldots = p_k = 1$ in the definition of $P(L(A))$. The function $(\ell_1, \ell_2) \mapsto \binom{|\ell_1|}{k_1} \cdot \binom{|\ell_2|}{k_2}$ is in $P(L(A) * L(B))$ for every $k_1, k_2 \in \mathbb{N}$ and $[\ell_1, \ldots, \ell_n] \mapsto \sum_{1 \leqslant i < j \leqslant n} \binom{|\ell_i|}{k_1} \cdot \binom{|\ell_j|}{k_2} \in P(L(L(A)))$ for every $k_1, k_2 \in \mathbb{N}$.

Resource Polynomials. A *resource polynomial* $p : [\![A]\!] \to \mathbb{Q}_0^+$ for a data type A is a non-negative linear combination of base polynomials, i.e., $p = \sum_{i=1,\ldots,m} q_i \cdot p_i$ for $q_i \in \mathbb{Q}_0^+$ and $p_i \in P(A)$. $R(A)$ is the set of resource polynomials for A.

An instructive, but not exhaustive, example is given by $R_n = R(L(\text{int}) * \cdots * L(\text{int}))$. The set R_n is the set of linear combinations of products of binomial coefficients over variables $x_1, \ldots, x_n$, that is, $R_n = \{\sum_{i=1}^m q_i \prod_{j=1}^n \binom{x_j}{k_{ij}} \mid q_i \in \mathbb{Q}_0^+, m \in \mathbb{N}, k_{ij} \in \mathbb{N}\}$. Concrete examples that illustrate the definitions follow in the next subsection.

4.2 Annotated Types

To relate type annotations in the type system to resource polynomials, we introduce names (or indices) for base polynomials. These names are also helpful to intuitively explain the base polynomials of a given type.

Names For Base Polynomials. To assign a unique name to each base polynomial we define the *index set* $\mathcal{I}(A)$ to denote resource polynomials for a given data type A. Essentially, $\mathcal{I}(A)$ is the meaning of A with every atomic type replaced by the *unit index* $\circ$.

$$\mathcal{I}(\text{int}) = \{\circ\} \qquad \mathcal{I}(A_1 * A_2) = \{(i_1, i_2) \mid i_1 \in \mathcal{I}(A_1) \text{ and } i_2 \in \mathcal{I}(A_2)\}$$
$$\mathcal{I}(L(A)) = \{[i_1, \ldots, i_k] \mid k \geq 0, i_j \in \mathcal{I}(A)\}$$

The *degree* $\deg(i)$ of an index $i \in \mathcal{I}(A)$ is defined as follows.

$$\deg(\circ) = 0 \qquad \deg(i_1, i_2) = \deg(i_1) + \deg(i_2)$$
$$\deg([i_1, \ldots, i_k]) = k + \deg(i_1) + \cdots + \deg(i_k)$$

Let $\mathcal{I}_k(A) = \{i \in \mathcal{I}(A) \mid \deg(i) \leq k\}$. The indices $i \in \mathcal{I}_k(A)$ are an enumeration of the base polonomials $p_i \in P(A)$ of degree at most k. For each $i \in \mathcal{I}(A)$, we define a base polynomial $p_i \in P(A)$ as follows: If $A = \text{int}$ then $p_\circ(v) = 1$. If $A = (A_1 * A_2)$ is a pair type and $v = (v_1, v_2)$ then $p_{(i_1, i_2)}(v) = p_{i_1}(v_1) \cdot p_{i_2}(v_2)$. If $A = L(B)$ is a list type and $v \in [\![L(B)]\!]$ then $p_{[i_1, \ldots, i_m]}(v) = \Sigma \Pi[p_{i_1}, \ldots, p_{i_m}](v)$. We use the notation 0_A (or just 0) for the index in $\mathcal{I}(A)$ such that $p_{0_A}(a) = 1$ for all a. We have $0_{\text{int}} = \circ$ and $0_{(A_1 * A_2)} = (0_{A_1}, 0_{A_2})$ and $0_{L(B)} = []$. If $A = L(B)$ for a data type B then the index $[0, \ldots, 0] \in \mathcal{I}(A)$ of length n is denoted by just n. We identify the index (i_1, i_2, i_3, i_4) with the index $(i_1, (i_2, (i_3, i_4)))$.

Examples. First consider the type int. The index set $\mathcal{I}(\text{int}) = \{\circ\}$ only contains the unit element because the only base polynomial for the type int is the constant polynomial $p_\circ : \mathbb{Z} \to \mathbb{N}$ that maps every integer to 1, that is, $p_\circ(n) = 1$ for all $n \in \mathbb{Z}$. In terms of resource-cost analysis this implies that the resource polynomials can not represent cost that depends on the value of an integer.

Now consider the type $L(\text{int})$. The index set for lists of integers is $\mathcal{I}(L(\text{int})) = \{[], [\circ], [\circ, \circ], \ldots\}$, the set of lists of unit indices $\circ$. The base polynomial $p_{[]} : [\![L(\text{int})]\!] \to \mathbb{N}$ is defined as $p_{[]}([a_1, \ldots, a_n]) = 1$ (one element sum and empty product). More interestingly, we have $p_{[\circ]}([a_1, \ldots, a_n]) = \sum_{1 \leq j \leq n} 1 = n$ and $p_{[\circ, \circ]}([a_1, \ldots, a_n]) = \sum_{1 \leq j_1 < j_2 \leq n} 1 = \binom{n}{2}$. In general, if $i_k = [\circ, \ldots, \circ]$ is as list with k unit indices then $p_{i_k}([a_1, \ldots, a_n]) = \sum_{1 \leq j_1 < \cdots < j_k \leq n} 1 = \binom{n}{k}$. The intuition is that the base polynomial $p_{i_k}([a_1, \ldots, a_n])$ describes a constant resource cost that arises for every ordered k-tuple $(a_{j_1}, \ldots, a_{j_n})$.

Finally, consider the type $L(L(\text{int}))$ of lists of lists of integers. The corresponding index set is $\mathcal{I}(L(L(\text{int}))) = \{[]\} \cup \{[i] \mid i \in \mathcal{I}(L(\text{int}))\} \cup \{[i_1, i_2] \mid i_1, i_2 \in \mathcal{I}(L(\text{int}))\} \cup \cdots$. Again we have $p_{[]} : [\![L(L(\text{int}))]\!] \to \mathbb{N}$ and $p_{[]}([a_1, \ldots, a_n]) = 1$. Moreover we also get the binomial coefficients again: If the index $i_k = [[], \ldots, []]$ is as list of k empty lists then $p_{i_k}([a_1, \ldots, a_n]) = \sum_{1 \leq j_1 < \cdots < j_k \leq n} 1 = \binom{n}{k}$. This describes a cost that would arise in a program that computes something of constant cost for tuples of inner lists (e.g., sorting with respect to the smallest head elements). However, the base polynomials can also refer to the lengths of the inner lists. For instance, we have $p[[\circ, \circ]]([a_1, \ldots, a_n]) = \sum_{1 \leq i \leq n} \binom{|a_i|}{2}$, which represents a quadratic cost for every inner list (e.g, sorting the inner lists). This is not to be confused with the base polynomial $p_{[\circ, \circ]}([a_1, \ldots, a_n]) = \sum_{1 \leq i < j \leq n} |a_i||a_j|$, which can be used to account for the cost of the comparisons in a lexicographic sorting of the outer list.

Annotated Types and Potential Functions. We use the indices and base polynomials to define type annotations and resource polynomials. We then give examples to illustrate the definitions.

A *type annotation* for a data type A is defined to be a family

$$Q_A = (q_i)_{i \in \mathcal{I}(A)} \text{ with } q_i \in \mathbb{Q}_0^+$$

We say Q_A is of *degree (at most) k* if $q_i = 0$ for every $i \in \mathcal{I}(A)$ with $\deg(i) > k$. An *annotated data type* is a pair (A, Q_A) of a data type A and a type annotation Q_A of some degree k.

Let H be a heap and let ℓ be a location with $H \models \ell \mapsto a : A$ for a data type A. Then the type annotation Q_A defines the *potential* $\Phi_H(\ell : (A, Q_A)) = \sum_{i \in \mathcal{I}(A)} q_i \cdot p_i(a)$. If $a \in \llbracket A \rrbracket$ and Q is a type annotation for A then we also write $\Phi(a : (A, Q))$ for $\sum_i q_i p_i(a)$.

Let for example, $Q = (q_i)_{i \in L(\text{int})}$ be an annotation for the type $L(\text{int})$ and let $q_{[]} = 2$, $q_{[\circ]} = 2.5$, $q_{[\circ, \circ, \circ]} = 8$, and $q_i = 0$ for all other $i \in \mathcal{I}(L(\text{int}))$. The we have $\Phi([a_1, \ldots, a_n] : (L(\text{int}), Q)) = 2 + 2.5n + 8\binom{n}{3}$.

The Potential of a Context. For use in the type system we need to extend the definition of resource polynomials to typing contexts. We treat a context like a tuple type. Let $\Gamma = x_1 : A_1, \ldots, x_n : A_n$ be a typing context and let $k \in \mathbb{N}$. The index set $\mathcal{I}(\Gamma)$ is defined through $\mathcal{I}(\Gamma) = \{(i_1, \ldots, i_n) \mid i_j \in \mathcal{I}(A_j)\}$.

The degree of $i = (i_1, \ldots, i_n) \in \mathcal{I}(\Gamma)$ is defined through $\deg(i) = \deg(i_1) + \cdots + \deg(i_n)$. As for data types, we define $I_k(\Gamma) = \{i \in I(\Gamma) \mid \deg(i) \leq k\}$. A *type annotation* Q for Γ is a family $Q = (q_i)_{i \in \mathcal{I}_k(\Gamma)}$ with $q_i \in \mathbb{Q}_0^+$. We denote a *resource-annotated context* with $\Gamma; Q$. Let H be a heap and V be a stack with $H \models V : \Gamma$ where $H \models V(x_j) \mapsto a_{x_j} : \Gamma(x_j)$.

The potential of an annotated context $\Gamma; Q$ with respect to then environment H and V is $\Phi_{V,H}(\Gamma; Q) = \sum_{(i_1, \ldots, i_n) \in \mathcal{I}_k(\Gamma)} q_{\vec{i}} \prod_{j=1}^n p_{i_j}(a_{x_j})$. In particular, if $\Gamma = \varnothing$ then $\mathcal{I}_k(\Gamma) = \{()\}$ and $\Phi_{V,H}(\Gamma; q_{()}) = q_{()}$. We sometimes also write q_0 for $q_{()}$.

5   Type System for Bounds on the Depth

In this section, we formally describe the novel resource-aware type system. We focus on the type judgement and explain the rules that are most important for handling parallel evaluation. The full type system is given in the extended version of this article [17].

The main theorem of this section proves the soundness of the type system with respect to the depths of evaluations as defined by the operational big-step semantics. The soundness holds for terminating and non-terminating evaluations.

Type Judgments. The typing rules in Figure 2 define a *resource-annotated typing judgment* of the form

$$\Sigma; \Gamma; \{Q_1, \ldots, Q_n\} \vdash^{\underline{M}} e : (A, Q')$$

where M is a metric, $n \in \{1, 2\}$, e is an expression, Σ is a resource-annotated signature (see below), $(\Gamma; Q_i)$ is a resource-annotated context for every $i \in \{1, \ldots, n\}$, and (A, Q') is a resource-annotated data type. The intended meaning of this judgment is the following. If there are more than $\Phi(\Gamma; Q_i)$ resource units available for every $i \in \{1, \ldots, n\}$ then this is sufficient to pay for the depth of the evaluation of e under the metric M. In addition, there are more than $\Phi(v : (A, Q'))$ resource units left if e evaluates to a value v.

In outermost judgements, we are only interested in the case where $n = 1$ and the judgement is equivalent to the similar judgement for sequential programs [7]. The form in which $n = 2$ is introduced in the type rule E:PAR for parallel bindings and eliminated by multiple applications of the sharing rule E:SHARE (more explanations follow).

The type judgement is affine in the sense that every variable in a context Γ can be used at most once in the expression e. Of course, we have to also deal with expressions in which a variable occurs more than once. To account for multiple variable uses we use the sharing rule T:SHARE that doubles a variable in a context without increasing the potential of the context.

As usual Γ_1, Γ_2 denotes the union of the contexts Γ_1 and Γ_2 provided that $\mathrm{dom}(\Gamma_1) \cap \mathrm{dom}(\Gamma_2) = \varnothing$. We thus have the implicit side condition $\mathrm{dom}(\Gamma_1) \cap \mathrm{dom}(\Gamma_2) = \varnothing$ whenever Γ_1, Γ_2 occurs in a typing rule. Especially, writing $\Gamma = x_1 : A_1, \ldots, x_k : A_k$ means that the variables x_i are pairwise distinct.

Programs with Annotated Types. *Resource-annotated first-order types* have the form $(A, Q) \to (B, Q')$ for annotated data types (A, Q) and (B, Q'). A *resource-annotated signature* Σ is a finite, partial mapping of function identifiers to *sets of* resource-annotated first-order types. A program with resource-annotated types for the metric M consists of a resource-annotated signature Σ and a family of expressions with variables identifiers $(e_f, y_f)_{f \in \mathrm{dom}(\Sigma)}$ such that $\Sigma; y_f : A; Q \vdash^{\underline{M}} e_f : (B, Q')$ for every function type $(A, Q) \to (B, Q') \in \Sigma(f)$.

Sharing. Let $\Gamma, x_1 : A, x_2 : A; Q$ be an annotated context. The *sharing operation* $\curlyvee Q$ defines an annotation for a context of the form $\Gamma, x : A$. It is used when the potential is split between multiple occurrences of a variable. Details can be found in the full version of the article.

Typing Rules. Figure 2 shows the annotated typing rules that are most relevant for parallel evaluation. Most of the other rules are similar to the rules for multivariate amortized analysis for sequential programs [13, 20]. The main difference it that the rules here operate on annotations that are singleton sets $\{Q\}$ instead of the usual context annotations Q.

In the rules T:LET and T:PAR, the result of the evaluation of an expression e is bound to a variable x. The problem that arises is that the resulting annotated context $\Delta, x : A, Q'$ features potential functions whose domain consists of data

$$\frac{\Sigma; \Gamma_1, \Gamma_2; R \vdash^{\underline{M}} e_1 \rightsquigarrow \Gamma_2, x{:}A; R' \qquad \Sigma;, \Gamma_2, x{:}A; \{R'\} \vdash^{\underline{M}} e_2 : (B, Q') \qquad Q = R + M^{\mathsf{let}}}{\Sigma; \Gamma_1, \Gamma_2; \{Q\} \vdash^{\underline{M}} \mathsf{let}\, x = e_1 \, \mathsf{in}\, e_2 : (B, Q')} \text{(T:Let)}$$

$$\frac{\begin{array}{c} \Sigma; \Gamma_1, \Gamma_2, \Delta; P \vdash^{\underline{\mathsf{cf}}} e_1 \rightsquigarrow \Gamma_2, \Delta, x_1{:}A_1; P' \\ \Sigma; \Gamma_2, \Delta, x_1{:}A_1; P' \vdash^{\underline{M}} e_2 \rightsquigarrow \Delta, x_1{:}A_1, x_2{:}A_2; R \\ \Sigma; \Gamma_2, \Delta, x_1{:}A_1; Q' \vdash^{\underline{\mathsf{cf}}} e_2 \rightsquigarrow \Delta, x_1{:}A_1, x_2{:}A_2; R \\ \Sigma; \Gamma_1, \Gamma_2, \Delta; Q \vdash^{\underline{M}} e_1 \rightsquigarrow \Gamma_2, \Delta, x_1{:}A_1; Q' \qquad \Sigma; \Delta, x_1{:}A_1, x_2{:}A_2; R \vdash^{\underline{M}} e : (B, R') \end{array}}{\Sigma; \Gamma_1, \Gamma_2, \Delta; \{Q + M^{\mathsf{Par}}, P + M^{\mathsf{Par}}\} \vdash^{\underline{M}} \mathsf{par}\, x_1 = e_1 \, \mathsf{and}\, x_2 = e_2 \, \mathsf{in}\, e : (B, R')} \text{(T:Par)}$$

$$\frac{\Sigma; \Gamma, x_1{:}A, x_2{:}A; \{P_1, \ldots, P_m\} \vdash^{\underline{M}} e : (B, Q') \qquad \forall i\, \exists j : Q_j = \curlyvee P_i}{\Sigma; \Gamma, x{:}A; \{Q_1, \ldots, Q_n\} \vdash^{\underline{M}} e[x/x_1, x/x_2] : (B, Q')} \text{(T:Share)}$$

$$\bullet \quad \bullet \quad \bullet$$

$$\frac{\forall j \in \mathcal{I}(\Delta): \quad j{=}\vec{0} \implies \Sigma; \Gamma; \pi_j^{\Gamma}(Q) \vdash^{\underline{M}} e : (A, \pi_j^{x{:}A}(Q')) \\ \qquad\qquad\quad j{\neq}\vec{0} \implies \Sigma_j; \Gamma; \pi_j^{\Gamma}(Q) \vdash^{\underline{\mathsf{cf}}} e : (A, \pi_j^{x{:}A}(Q'))}{\Sigma; \Gamma, \Delta; Q \vdash^{\underline{M}} e \rightsquigarrow \Delta, x{:}A; Q'} \text{(B:Bind)}$$

Fig. 2. Selected novel typing rules for annotated types and the binding rule for multivariate variable binding

that is referenced by x as well as data that is referenced by Δ. This potential has to be related to data that is referenced by Δ and the free variables in e.

To express the relations between mixed potentials before and after the evaluation of e, we introduce a new auxiliary binding judgement of the from

$$\Sigma; \Gamma, \Delta; Q \vdash^{\underline{M}} e \rightsquigarrow \Delta, x{:}A; Q'$$

in the rule B:Bind. The intuitive meaning of the judgement is the following. Assume that e is evaluated in the context Γ, Δ, that $\mathrm{FV}(e) \in \mathrm{dom}(\Gamma)$, and that e evaluates to a value that is bound to the variable x. Then the initial potential $\Phi(\Gamma, \Delta; Q)$ is larger than the cost of evaluating e in the metric M plus the potential of the resulting context $\Phi(\Delta, x{:}A; Q')$.

The rule T:Par for parallel bindings $\mathsf{par}\, x_1 = e_1 \, \mathsf{and}\, x_2 = e_2 \, \mathsf{in}\, e$ is the main novelty in the type system. The idea is that we type the expressions e_1 and e_2 twice using the new binding judgement. In the first group of bindings, we account for the cost of e_1 and derive a context $\Gamma_2, \Delta, x_1{:}A_1; P_1'$ in which the result of the evaluation of e_1 is bound to x_1. This context is then used to bind the result of evaluating e_2 in the context $\Delta, x_1{:}A_1, x_2{:}A_2; R$ without paying for the resource consumption. In the second group of bindings, we also derive the context $\Delta, x_1{:}A_1, x_2{:}A_2; R$ but pay for the cost of evaluating e_2 instead of e_1. The type annotations Q_1 and Q_2 for the initial context $\Gamma = \Gamma_1, \Gamma_2, \Delta$ establish a bound on the depth d of evaluating the whole parallel binding: If the depth of evaluating e_1 is larger than the depth of evaluating e_2 then $\Phi(\Gamma; Q_1) \geq d$. Otherwise we have $\Phi(\Gamma; Q_2) \geq d$. If the parallel binding evaluates to a value v then we have additionally that $\max(\Phi(\Gamma; Q_1), \Phi(\Gamma; Q_2)) \geq d + \Phi(v{:}(B, Q'))$.

It is important that the annotations Q_1 and Q_2 of the initial context $\Gamma_1, \Gamma_2, \Delta$ can defer. The reason is that we have to allow a different sharing of potential in the two groups of bindings. If we would require $Q_1 = Q_2$ then the system would be too restrictive. However, each type derivation has to establish the equality of the two annotations directly after the use of T:PAR by multiple uses of the sharing rule T:SHARE. Note that T:PAR is the only rule that can introduce a non-singleton set $\{Q_1, Q_n\}$ of context annotations.

T:SHARE has to be applied to expressions that contain a variable twice (x in the rule). The sharing operation $\curlyvee P$ transfers the annotation P for the context $\Gamma, x_1{:}A, x_2{:}A$ into an annotation Q for the context $\Gamma, x{:}A$ without loss of potential . This is crucial for the accuracy of the analysis since instances of T:SHARE are quite frequent in typical examples. The remaining rules are affine in the sense that they assume that every variable occurs at most once in the typed expression.

T:SHARE is the only rule whose premiss allows judgements that contain a non-singleton set $\{P_1, \ldots, P_m\}$ of context annotations. It has to be applied to produce a judgement with singleton set $\{Q\}$ before any of the other rules can be applied. The idea is that we always have $n \leqslant m$ for the set $\{Q_1, \ldots, Q_n\}$ and the sharing operation $\curlyvee_i$ is used to unify the different P_i.

Soundness. The operational big-step semantics with partial evaluations makes it possible to state and prove a strong soundness result. An annotated type judgment for an expression e establishes a bound on the depth of all evaluations of e in a well-formed environment; regardless of whether these evaluations diverge or fail. Moreover, the soundness theorem states also a stronger property for terminating evaluations. If an expression e evaluates to a value v in a well-formed environment then the difference between initial and final potential is an upper bound on the depth of the evaluation.

Theorem 3 (Soundness). *If* $H \models V{:}\Gamma$ *and* $\Sigma; \Gamma; \mathcal{Q} \vdash e{:}(B, Q')$ *then there exists a* $Q \in \mathcal{Q}$ *such that the following holds.*
1. *If* $V, H \vdash^{\underline{M}} e \Downarrow (\ell, H') \mid (w, d)$ *then* $d \leqslant \Phi_{V,H}(\Gamma; Q) - \Phi_{H'}(\ell{:}(B, Q'))$.
2. *If* $V, H \vdash^{\underline{M}} e \Downarrow \rho \mid (w, d)$ *then* $d \leqslant \Phi_{V,H}(\Gamma; Q)$.

Theorem 3 is proved by a nested induction on the derivation of the evaluation judgment and the type judgment $\Gamma; Q \vdash e{:}(B, Q')$. The inner induction on the type judgment is needed because of the structural rules. There is one proof for all possible instantiations of the resource constants.

The proof of most rules is very similar to the proof of the rules for multivariate resource analysis for sequential programs [7]. The main novelty is the treatment of parallel evaluation in the rule T:PAR which we described previously.

If the metric M is simple (all constants are 1) then it follows from Theorem 3 that the bounds on the depth also prove the termination of programs.

Corollary 1. *Let* M *be a simple metric. If* $H \models V{:}\Gamma$ *and* $\Sigma; \Gamma; Q \vdash e{:}(A, Q')$ *then there are* $w \in \mathbb{N}$ *and* $d \leqslant \Phi_{V,H}(\Gamma; Q)$ *such that* $V, H \vdash^{\underline{M}} e \Downarrow (\ell, H') \mid (w, d)$ *for some* ℓ *and* H'.

Type Inference. In principle, type inference consists of four steps. First, we perform a classic type inference for the simple types such as nat array. Second, we fix a maximal degree of the bounds and annotate all types in the derivation of the simple types with variables that correspond to type annotations for resource polynomials of that degree. Third, we generate a set of linear inequalities, which express the relationships between the added annotation variables as specified by the type rules. Forth, we solve the inequalities with an LP solver such as CLP. A solution of the linear program corresponds to a type derivation in which the variables in the type annotations are instantiated according to the solution.

In practice, the type inference is slightly more complex. Most importantly, we have to deal with resource-polymorphic recursion in many examples. This means that we need a type annotation in the recursive call that differs from the annotation in the argument and result types of the function. To infer such types we successively infer type annotations of higher and higher degree. Details can be found in previous work [21]. Moreover, we have to use algorithmic versions of the type rules in the inference in which the non-syntax-directed rules are integrated into the syntax-directed ones [7]. Finally, we use several optimizations to reduce the number of generated constraints. See [7] for an example type derivation.

6   Nested Data Parallelism

The techniques that we describe in this work for a minimal function language scale to more advanced parallel languages such as Blelloch's NESL [15].

To describe the novel type analysis in this paper, we use a binary binding construct to introduce parallelism. In NESL, parallelism is introduced via built-in functions on sequences as well as parallel sequence comprehension that is similar to Haskell's list comprehension. The depth of all built-in sequence functions such as *append* and *sum* is constant in NESL. Similarly, the depth overhead of the parallel sequence comprehension is constant too. Of course, it is possible to define equivalent functions in RAML. However, the depth would often be linear since we, for instance, have to sequentially form the resulting list.

Nevertheless, the user definable resource metrics in RAML make it easy to introduce built-in functions and language constructs with customized work and depth. For instance we could implement NESL's *append* like the recursive append in RAML but use a metric inside the function body in which all evaluation steps have depth zero. Then the depth of the evaluation of append(x, y) is constant and the work is linear in $|x|$.

To demonstrate this ability of our approach, we implemented parallel list comprehensions, NESL's most powerful construct for parallel computations. A list comprehension has the form $\{\ e\ :\ x_1 \text{ in } e_1\ ;\dots;\ x_n \text{ in } e_n\ |\ e_b\ \}$. where e is an expression, $e_1,\dots,e_n$ are expressions of some list type, and e_b is a boolean expression. The semantics is that we bind $x_1,\dots,x_n$ successively to the elements of the lists $e_1,\dots,e_n$ and evaluate e_b and e under these bindings. If e_b evaluates to true under a binding then we include the result of e under that binding in the resulting list. In other words, the above list comprehension is equivalent to the Haskell expression $[\ e\ |\ (x_1,\dots,x_n) \leftarrow \text{zip}_n\ e_1\ \dots\ e_n\ ,\ e_b\ ]$.

The *work* of evaluating $\{\ e\ :\ x_1 \text{ in } e_1\ ;\dots;\ x_n \text{ in } e_n\ |\ e_b\ \}$ is sum of the cost of evaluating $e_1,\dots,e_{n-1}$ and e_n plus the sum of the cost of evaluating e_b

Table 1. Compilation of Computed Depth and Work Bounds

Function Name / Function Type	Computed Depth Bound / Computed Work Bound	Run Time	Asym. Behav.
dyad	$10m + 10n + 3$	0.19 s	$O(n+m)$
$L(\text{int})*L(\text{int}) \rightarrow L(L(\text{int}))$	$10mn + 17n + 3$	0.20 s	$O(nm)$
dyad_all	$1.\bar{6}n^3 - 4n^2 + 10nm + 14.\bar{6}n + 5$	1.66 s	$O(n^2+m)$
$L(L(\text{int})) \rightarrow L(L(L(\text{int})))$	$1.\bar{3}n^3 + 5n^2m^2 + 8.5n^2m + \ldots$	0.96 s	$O(n^3+n^2m^2)$
m_mult1	$15xy + 16x + 10n + 6$	0.37 s	$O(xy)$
$L(L(\text{int}))*L(L(\text{int})) \rightarrow L(L(\text{int}))$	$15xyn + 16nm + 18n + 3$	0.36 s	$O(xyn)$
m_mult_pairs $[M := L(L(\text{int}))]$	$4n^2 + 15nmx + 10nm + 10n + 3$	3.90 s	O(nm + mx)
$L(M)*L(M) \rightarrow L(M)$	$7.5n^2m^2x + 7n^2m^2 + n^2mx \ldots$	6.35 s	$O(n^2m^2x)$
m_mult2 $[M := L(L(\text{int}))]$	$35u + 10y + 15x + 11n + 40$	2.75 s	$O(z+x+n)$
$(M*\text{nat})*(M*\text{nat}) \rightarrow M$	$3.5u^2y + uyz + 14.5uy + \ldots$	2.99 s	$O(nx(z+y))$
quicksort_list	$12n^2 + 16nm + 12n + 3$	0.67 s	$O(n^2+m)$
$L(L(\text{int})) \rightarrow L(L(\text{int}))$	$8n^2m + 15.5n^2 - 8nm + 13.5n + 3$	0.51 s	$O(n^2m)$
intersection	$10m + 12n + 3$	0.49 s	$O(n+m)$
$L(\text{int})*L(\text{int}) \rightarrow L(\text{int})$	$10mn + 19n + 3$	0.28 s	$O(nm)$
product	$8mn + 10m + 14n + 3$	1.05 s	$O(nm)$
$L(\text{int})*L(\text{int}) \rightarrow L(\text{int}*\text{int})$	$18mn + 21n + 3$	0.71 s	$O(nm)$
max_weight	$46n + 44$	0.39 s	$O(n)$
$L(\text{int}) \rightarrow \text{int}*L(\text{int})$	$13.5n^2 + 65.5n + 19$	0.30 s	$O(n^2)$
fib	$13n + 4$	0.09 s	$O(n)$
nat $*$ nat $\rightarrow$ nat	$-\,-\,-$	0.12 s	$O(2^n)$
dyad_comp	13	0.28 s	$O(1)$
$L(\text{int})*L(\text{int}) \rightarrow L(L(\text{int}))$	$6mn + 5n + 2$	0.13 s	$O(nm)$
find	$12m + 29n + 22$	0.38 s	$O(m+n)$
$L(\text{int})*L(\text{int}) \rightarrow L(L(\text{int}))$	$20mn + 18m + 9n + 16$	0.41 s	$O(nm)$

and e with the successive bindings to the elements of the results of the evaluation of $e_1, \ldots, e_n$. The *depth* of the evaluation is sum of the cost of evaluating $e_1, \ldots, e_{n-1}$ and e_n plus the maximum of the cost of evaluating e_b and e with the successive bindings to the elements of the results of the e_i.

7   Experimental Evaluation

We implemented the developed automatic depth analysis in Resource Aware ML (RAML). The implementation consists mainly of adding the syntactic form for the parallel binding and the parallel list comprehensions together with the treatment in the parser, the interpreter, and the resource-aware type system. RAML is publically available for download and through a user-friendly online interface [16]. On the project web page you also find the source code of all example programs and of RAML itself.

We used the implementation to perform an experimental evaluation of the analysis on typical examples from functional programming. In the compilation of our results we focus on examples that have a different asymptotic worst-case behavior in parallel and sequential evaluation. In many other cases, the worst-case behavior only differs in the constant factors. Also note that many of the classic examples of Blelloch [10]—like quick sort—have a better asymptotic average behavior in parallel evaluation but the same asymptotic worst-case behavior in parallel and sequential cost.

Table 1 contains a representative compilation of our experimental results. For each analyzed function, it shows the function type, the computed bounds on the work and the depth, the run time of the analysis in seconds and the actual asymptotic behavior of the function. The experiments were performed on an iMac with a 3.4 GHz Intel Core i7 and 8 GB memory. As LP solver we used IBM's CPLEX and the constraint solving takes about 60% of the overall run time of the prototype on average. The computed bounds are simplified multivariate resource polynomials that are presented to the user by RAML. Note that RAML also outputs the (unsimplified) multivariate resource polynomials. The variables in the computed bounds correspond to the sizes of different parts of the input. As naming convention we use the order n, m, x, y, z, u of variables to name the sizes in a depth-first way: n is the size of the first argument, m is the maximal size of the elements of the first argument, x is the size of the second argument, etc.

All bounds are asymptotically tight if the tight bound is representable by a multivariate resource polynomial. For example, the exponential work bound for fib and the logarithmic bounds for bitonic_sort are not representable as a resource polynomial. Another example is the loose depth bound for dyad_all where we would need the base function $\max_{1 \leqslant i \leqslant n} m_i$ but only have $\sum_{1 \leqslant i \leqslant n} m_i$.

Matrix Operations. To study programs that use nested data structures we implemented several matrix operations for matrices that are represented by lists of lists of integers. The implemented operations include, the dyadic product from Section 3 (dyad), transposition of matrices (transpose, see [16]), addition of matrices (m_add, see [16]), and multiplication of matrices (m_mult1 and m_mult2).

To demonstrate the compositionality of the analysis, we have implemented two more involved functions for matrices. The function dyad_all computes the dyadic product (using dyad) of all ordered pairs of the inner lists in the argument. The function m_mult_pairs computes the products $M_1 \cdot M_2$ (using m_mult1) of all pairs of matrices such that M_1 is in the first list of the argument and M_2 is in the second list of the argument.

Sorting Algorithms. The sorting algorithms that we implemented include quick sort and bitonic sort for lists of integers (quicksort and bitonic_sort, see [16]).

The analysis computes asymptotically tight quadratic bounds for the work and depth of quick sort. The asymptotically tight bounds for the work and depth of bitonic sort are $O(n \log n)$ and $O(n \log^2 n)$, respectively, and can thus not be expressed by polynomials. However, the analysis computes quadratic and cubic bounds that are asymptotically optimal if we only consider polynomial bounds.

More interesting are sorting algorithms for lists of lists, where the comparisons need linear instead of constant time. In these algorithms we can often perform the comparisons in parallel. For instance, the analysis computes asymptotically tight bounds for quick sort for lists of lists of integers (quicksort_list, see Table 1).

Set Operations. We implemented sets as unsorted lists without duplicates. Most list operations such as intersection (Table 1), difference (see [16]), and union (see [16]) have linear depth and quadratic work. The analysis finds these asymptotically tight bounds.

The function product computes the Cartesian product of two sets. Work and depth of product are both linear and the analysis finds asymptotically tight bounds. However, the constant factors in the parallel evaluation are much smaller.

Miscellaneous. The function max_weight (Table 1) computes the maximal weight of a (connected) sublist of an integer list. The weight of a list is simply the sum of its elements. The work of the algorithm is quadratic but the depth is linear.

Finally, there is a large class of programs that have non-polynomial work but polynomial depth. Since the analysis can only compute polynomial bounds we can only derive bounds on the depth for such programs. A simple example in Table 1 is the function fib that computes the Fibonacci numbers without memoization.

Parallel List Comprehensions. The aforementioned examples are all implemented without using parallel list comprehensions. Parallel list comprehensions have a better asymptotic behavior than semantically-equivalent recursive functions in RAML's current resource metric for evaluation steps.

A simple example is the function dyad_comp which is equivalent to dyad and which is implemented with the expression $\{\{x * y : y \ in \ ys\} : x \ in \ xs\}$. As listed in Table 1, the depth of dyad_comp is constant while the depth of dyad is linear. RAML computes tight bounds.

A more involved example is the function find that finds a given integer list (needle) in another list (haystack). It returns the starting indices of each occurrence of the needle in the haystack. The algorithm is described by Blelloch [15] and cleverly uses parallel list comprehensions to perform the search in parallel. RAML computes asymptotically tight bounds on the work and depth.

Discussion. Our experiments show that the range of the analysis is not reduced when deriving bounds on the depth: The prototype implementation can always infer bounds on the depth of a program if it can infer bounds on the sequential version of the program. The derivation of bounds for parallel programs is also almost as efficient as the derivation of bounds for sequential programs.

We experimentally compared the derived worst-case bounds with the measured work and depth of evaluations with different inputs. In most cases, the derived bounds on the depth are asymptotically tight and the constant factors are close or equal to the optimal ones. As a representative example, the full version of the article contains plots of our experiments for quick sort for lists of lists.

8   Related Work

Automatic amortized resource analysis was introduced by Hofmann and Jost for a strict first-order functional language [3]. The technique has been applied to higher-order functional programs [22], to derive stack-space bounds for functional programs [23], to functional programs with lazy evaluation [4], to object-oriented programs [24, 25], and to low-level code by integrating it with separation logic [26]. All the aforementioned amortized-analysis–based systems are limited to linear bounds. The polynomial potential functions that we use in this paper were introduced by Hoffmann et al. [19, 13, 7]. In contrast to this work, none of the previous works on amortized analysis considered parallel evaluation. The main technical innovation of this work is the new rule for parallel composition that is not straightforward. The smooth integration of this rule in the existing framework of multivariate amortized resource analysis is a main advantages of our work.

Type systems for inferring and verifying cost bounds for sequential programs have been extensively studied. Vasconcelos et al. [27, 1] described an automatic analysis system that is based on sized-types [28] and derives linear bounds for higher-order sequential functional programs. Dal Lago et al. [29, 30] introduced linear dependent types to obtain a complete analysis system for the time complexity of the call-by-name and call-by-value lambda calculus. Crary and Weirich [31] presented a type system for specifying and certifying resource consumption. Danielsson [32] developed a library, based on dependent types and manual cost annotations, that can be used for complexity analyses of functional programs. We are not aware of any type-based analysis systems for parallel evaluation.

Classically, cost analyses are often based on deriving and solving recurrence relations. This approach was pioneered by Wegbreit [33] and has been extensively studied for sequential programs written in imperative languages [6, 34] and functional languages [35, 2].

In comparison, there has been little work done on the analysis of parallel programs. Albert et al. [36] use recurrence relations to derive cost bounds for concurrent object-oriented programs. Their model of concurrent imperative programs that communicate over a shared memory and the used cost measure is however quite different from the depth of functional programs that we study.

The only article on using recurrence relations for deriving bounds on parallel functional programs that we are aware of is a technical report by Zimmermann [37]. The programs that were analyzed in this work are fairly simple and more involved programs such as sorting algorithms seem to be beyond its scope. Additionally, the technique does not provide the compositionality of amortized resource analysis.

Trinder et al. [38] give a survey of resource analysis techniques for parallel and distributed systems. However, they focus on the usage of analyses for sequential programs to improve the coordination in parallel systems. Abstract interpretation based approaches to resource analysis [5, 39] are limited to sequential programs.

Finally, there exists research that studies cost models to formally analyze parallel programs. Blelloch and Greiner [10] pioneered the cost measures work and depth that we use in this work. There are more advanced cost models that take into account caches and IO (see, e.g., Blelloch and Harper [11]), However, these works do not provide machine support for deriving static cost bounds.

9   Conclusion

We have introduced the first type-based cost analysis for deriving bounds on the depth of evaluations of parallel function programs. The derived bounds are multivariate resource polynomials that can express a wide range of relations between different parts of the input. As any type system, the analysis is naturally compositional. The new analysis system has been implemented in Resource Aware ML (RAML) [14]. We have performed a thorough and reproducible experimental evaluation with typical examples from functional programming that shows the practicability of the approach.

An extension of amortized resource analysis to handle non-polynomial bounds such as max and log in a compositional way is an orthogonal research question that we plan to address in the future. A promising direction that we are currently studying is the use of numerical logical variables to guide the analysis to derive non-polynomial bounds. The logical variables would be treated like regular variables in the analysis. However, the user would be responsible for maintaining and proving relations such as $a = \log n$ where a is a logical variable an n is the size of a regular data structure. In this way, we would gain flexibility while maintaining the compositionality of the analysis.

Another orthogonal question is the extension of the analysis to additional language features such as higher-order functions, references, and user-defined data structures. These extensions have already been implemented in a prototype and pose interesting research challenges in there own right. We plan to report on them in a forthcoming article.

Acknowledgments. This research is based on work supported in part by NSF grants 1319671 and 1065451, and DARPA grants FA8750-10-2-0254 and FA8750-12-2-0293. Any opinions, findings, and conclusions contained in this document are those of the authors and do not reflect the views of these agencies.

References

1. Vasconcelos, P.: Space Cost Analysis Using Sized Types. PhD thesis, School of Computer Science, University of St Andrews (2008)
2. Danner, N., Paykin, J., Royer, J.S.: A Static Cost Analysis for a Higher-Order Language. In: 7th Workshop on Prog. Languages Meets Prog. Verification (PLPV 2013), pp. 25–34 (2013)
3. Hofmann, M., Jost, S.: Static Prediction of Heap Space Usage for First-Order Functional Programs. In: 30th ACM Symp. on Principles of Prog. Langs. (POPL 2003), pp. 185–197 (2003)
4. Simões, H.R., Vasconcelos, P.B., Florido, M., Jost, S., Hammond, K.: Automatic Amortised Analysis of Dynamic Memory Allocation for Lazy Functional Programs. In: 17th Int. Conf. on Funct. Prog. (ICFP 2012), pp. 165–176 (2012)
5. Gulwani, S., Mehra, K.K., Chilimbi, T.M.: SPEED: Precise and Efficient Static Estimation of Program Computational Complexity. In: 36th ACM Symp. on Principles of Prog. Langs. (POPL 2009), pp. 127–139 (2009)
6. Albert, E., Arenas, P., Genaim, S., Puebla, G., Zanardini, D.: Cost Analysis of Java Bytecode. In: De Nicola, R. (ed.) ESOP 2007. LNCS, vol. 4421, pp. 157–172. Springer, Heidelberg (2007)
7. Hoffmann, J., Aehlig, K., Hofmann, M.: Multivariate Amortized Resource Analysis. ACM Trans. Program. Lang. Syst. (2012)
8. Brockschmidt, M., Emmes, F., Falke, S., Fuhs, C., Giesl, J.: Alternating Runtime and Size Complexity Analysis of Integer Programs. In: Ábrahám, E., Havelund, K. (eds.) TACAS 2014. LNCS, vol. 8413, pp. 140–155. Springer, Heidelberg (2014)

9. Sinn, M., Zuleger, F., Veith, H.: A Simple and Scalable Approach to Bound Analysis and Amortized Complexity Analysis. In: Biere, A., Bloem, R. (eds.) CAV 2014. LNCS, vol. 8559, pp. 745–761. Springer, Heidelberg (2014)

10. Blelloch, G.E., Greiner, J.: A Provable Time and Space Efficient Implementation of NESL. In: 1st Int. Conf. on Funct. Prog. (ICFP 1996), pp. 213–225 (1996)

11. Blelloch, G.E., Harper, R.: Cache and I/O Efficent Functional Algorithms. In: 40th ACM Symp. on Principles Prog. Langs. (POPL 2013), pp. 39–50 (2013)

12. Harper, R.: Practical Foundations for Programming Languages. Cambridge University Press (2012)

13. Hoffmann, J., Aehlig, K., Hofmann, M.: Multivariate Amortized Resource Analysis. In: 38th ACM Symp. on Principles of Prog. Langs. (POPL 2011) (2011)

14. Hoffmann, J., Aehlig, K., Hofmann, M.: Resource Aware ML. In: Madhusudan, P., Seshia, S.A. (eds.) CAV 2012. LNCS, vol. 7358, pp. 781–786. Springer, Heidelberg (2012)

15. Blelloch, G.E.: Nesl: A nested data-parallel language (version 3.1). Technical Report CMU-CS-95-170, CMU (1995)

16. Aehlig, K., Hofmann, M., Hoffmann, J.: RAML Web Site (2010-2014), http://raml.tcs.ifi.lmu.de

17. Hoffmann, J., Shao, Z.: Automatic Static Cost Analysis for Parallel Programs. Full Version (2014), http://cs.yale.edu/~hoffmann/papers/parallelcost2014.pdf

18. Charguéraud, A.: Pretty-Big-Step Semantics. In: Felleisen, M., Gardner, P. (eds.) ESOP 2013. LNCS, vol. 7792, pp. 41–60. Springer, Heidelberg (2013)

19. Hoffmann, J., Hofmann, M.: Amortized Resource Analysis with Polynomial Potential. In: Gordon, A.D. (ed.) ESOP 2010. LNCS, vol. 6012, pp. 287–306. Springer, Heidelberg (2010)

20. Hoffmann, J., Shao, Z.: Type-Based Amortized Resource Analysis with Integers and Arrays. In: Codish, M., Sumii, E. (eds.) FLOPS 2014. LNCS, vol. 8475, pp. 152–168. Springer, Heidelberg (2014)

21. Hoffmann, J., Hofmann, M.: Amortized Resource Analysis with Polymorphic Recursion and Partial Big-Step Operational Semantics. In: Ueda, K. (ed.) APLAS 2010. LNCS, vol. 6461, pp. 172–187. Springer, Heidelberg (2010)

22. Jost, S., Hammond, K., Loidl, H.W., Hofmann, M.: Static Determination of Quantitative Resource Usage for Higher-Order Programs. In: 37th ACM Symp. on Principles of Prog. Langs. (POPL 2010), pp. 223–236 (2010)

23. Campbell, B.: Amortised Memory Analysis using the Depth of Data Structures. In: Castagna, G. (ed.) ESOP 2009. LNCS, vol. 5502, pp. 190–204. Springer, Heidelberg (2009)

24. Hofmann, M.O., Jost, S.: Type-Based Amortised Heap-Space Analysis. In: Sestoft, P. (ed.) ESOP 2006. LNCS, vol. 3924, pp. 22–37. Springer, Heidelberg (2006)

25. Hofmann, M., Rodriguez, D.: Automatic Type Inference for Amortised Heap-Space Analysis. In: Felleisen, M., Gardner, P. (eds.) ESOP 2013. LNCS, vol. 7792, pp. 593–613. Springer, Heidelberg (2013)

26. Atkey, R.: Amortised Resource Analysis with Separation Logic. In: Gordon, A.D. (ed.) ESOP 2010. LNCS, vol. 6012, pp. 85–103. Springer, Heidelberg (2010)

27. Vasconcelos, P.B., Hammond, K.: Inferring Costs for Recursive, Polymorphic and Higher-Order Functional Programs. In: Trinder, P., Michaelson, G.J., Peña, R. (eds.) IFL 2003. LNCS, vol. 3145, pp. 86–101. Springer, Heidelberg (2004)

28. Hughes, J., Pareto, L., Sabry, A.: Proving the Correctness of Reactive Systems Using Sized Types. In: 23th ACM Symp. on Principles of Prog. Langs. (POPL 1996), pp. 410–423 (1996)

29. Lago, U.D., Gaboardi, M.: Linear Dependent Types and Relative Completeness. In: 26th IEEE Symp. on Logic in Computer Science (LICS 2011), pp. 133–142 (2011)
30. Lago, U.D., Petit, B.: The Geometry of Types. In: 40th ACM Symp. on Principles Prog. Langs. (POPL 2013), pp. 167–178 (2013)
31. Crary, K., Weirich, S.: Resource Bound Certification. In: 27th ACM Symp. on Principles of Prog. Langs. (POPL 2000), pp. 184–198 (2000)
32. Danielsson, N.A.: Lightweight Semiformal Time Complexity Analysis for Purely Functional Data Structures. In: 35th ACM Symp. on Principles Prog. Langs. (POPL 2008), pp. 133–144 (2008)
33. Wegbreit, B.: Mechanical Program Analysis. Commun. ACM 18(9), 528–539 (1975)
34. Alonso-Blas, D.E., Genaim, S.: On the limits of the classical approach to cost analysis. In: Miné, A., Schmidt, D. (eds.) SAS 2012. LNCS, vol. 7460, pp. 405–421. Springer, Heidelberg (2012)
35. Grobauer, B.: Cost Recurrences for DML Programs. In: 6th Int. Conf. on Funct. Prog. (ICFP 2001), pp. 253–264 (2001)
36. Albert, E., Arenas, P., Genaim, S., Gómez-Zamalloa, M., Puebla, G.: Cost Analysis of Concurrent OO Programs. In: Yang, H. (ed.) APLAS 2011. LNCS, vol. 7078, pp. 238–254. Springer, Heidelberg (2011)
37. Zimmermann, W.: Automatic Worst Case Complexity Analysis of Parallel Programs. Technical Report TR-90-066, University of California, Berkeley (1990)
38. Trinder, P.W., Cole, M.I., Hammond, K., Loidl, H.W., Michaelson, G.: Resource Analyses for Parallel and Distributed Coordination. Concurrency and Computation: Practice and Experience 25(3), 309–348 (2013)
39. Zuleger, F., Gulwani, S., Sinn, M., Veith, H.: Bound Analysis of Imperative Programs with the Size-change Abstraction. In: Yahav, E. (ed.) Static Analysis. LNCS, vol. 6887, pp. 280–297. Springer, Heidelberg (2011)

Sound, Modular and Compositional Verification of the Input/Output Behavior of Programs

Willem Penninckx, Bart Jacobs, and Frank Piessens

iMinds-DistriNet, KU Leuven, 3001 Leuven, Belgium

Abstract. We present a sound verification approach for verifying input/output properties of programs. Our approach supports defining high-level I/O actions on top of low-level ones (compositionality), defining input/output actions without taking into account which other actions exist (modularity), and other features. As the key ingredient, we developed a separation logic over Petri nets. We also show how with the same specification style we can elegantly modularly verify "I/O-like" code that uses the Template Pattern. We have implemented our approach in the VeriFast verifier and applied it to a number of challenging examples.

1   Introduction

Many software verification approaches are based on Hoare logic. A Hoare triple [6] consists of a precondition, a program, and a postcondition. If a Hoare triple is true, then every execution of the program starting from any state satisfying the precondition results (if it terminates) in a state satisfying the postcondition. Hoare logic has been extended to support various features, e.g. aliasing and concurrency. But a certain limitation is often left untackled. Indeed, the pre- and postcondition of a Hoare triple typically constrain the behavior of a program by only looking at the initial and final state of memory. This makes it possible to prove e.g. that a quicksort implementation sorts properly, but it does not prevent that e.g. an incorrect result is printed on the screen. For the user of a program, the proofs about the state of memory of a program are useless if the result visible on the screen is incorrect. In the end, the performed input/output must be correct, a problem typically left untouched.

There are some conceptual differences between verifying memory state and verifying I/O behavior. One difference is that, when verifying memory state, we only care about the final state. Indeed, if the function that sorts has an intermediate state that looks like garbage, but then cleans up and still gives a correctly sorted output, we are happy. In contrast, when verifying input/output we do care about the intermediary state. If e.g. the calculator displays the wrong output on the screen and then the right output, this is a bad calculator, even though the final image displayed on the screen is correct.

Another difference is that termination is usually a desired property of programs not performing I/O, but often undesired for programs performing I/O. For example, a quicksort implementation should always terminate, but a text editor should not.

J. Vitek (Ed.): ESOP 2015, LNCS 9032, pp. 158–182, 2015.
DOI: 10.1007/978-3-662-46669-8_7

However, just verifying I/O properties is not the interesting challenge itself. The interesting challenges are the side constraints such as compositionality and modularity:

Modularity. A programmer of a library typically does not consider all possible other libraries that might exist. Still, a programmer of an application can use multiple libraries in his program, even though these libraries do not know of each other's existence. Similarly, we want to write specifications of a library without keeping in mind existence of other libraries.

Compositionality. In regular software development, a programmer typically does not call the low-level system calls. Instead, he calls high-level libraries, which might be implemented in terms of other libraries, themselves implemented on top of yet other libraries, and so on. This is the concept of compositionality. The verification approach for I/O should support programs written in a compositional manner. Furthermore, it should be possible to write the formal I/O specifications themselves in a compositional manner, i.e. in terms of other libraries' I/O actions instead of in terms of the low-level system calls.

Other. Besides compositionality and modularity, the I/O verification should also

– be static, i.e. detecting errors at compile time, not at run time.
– be sound, i.e. not searching for most bugs, but proving absence of bugs.
– blend in well with existing verification techniques that solve other problems like aliasing
– support non-deterministic behavior (e.g. operations can fail, or return unspecified values, like reading user input)
– support impricise specifications (e.g. the specifications describe two possibilities and the implementor can choose freely). The number of possibilities can be large, e.g. "print a number less than 0".
– support arguments for operations (e.g. when writing to a file, the content and the filename are arguments that should be part of the specifications)
– support unspecified ordering of operations. If the order is unimportant, the specification should not fix them such that the implementor can choose freely.
– support specifying ordering of operations, also if the operations are specified and implemented by independent teams. For example, it might be necessary that the put-shield-on operation happens before the start-explosion operation.
– support both non-terminating and terminating programs: a non-terminating program can still only do the allowed I/O operations in the allowed order, and a terminating program is only allowed to terminate after it has performed all desired I/O operations.
– support operations that depend on the outcome of the previous operation, e.g. a specification like "read a number, and then print a number that is one higher than the read number".

This paper proposes an elegant way to perform input/output verification based on separation logic. It supports all the requirements explained above. We consider soundness, compositionality, and modularity the more interesting properties.

This paper does not include an approach to prove liveness properties. For both non-terminating and terminating programs, the approach presented in this paper allows to prove that all performed I/O is correct and in the correct order, but for non-terminating programs it does not provide a way to prove that any I/O happens. For terminating programs, the approach allows to prove that the intended I/O has happened upon termination.

The remainder of this paper is organized as follows. Section 2 defines a basic programming language supporting I/O. Section 3 uses this language to explain the verification approach. Section 4 proves soundness of this approach. Section 5 gives some examples. Section 6 provides a quick look at what is different and common in verifying I/O behavior and memory state. Section 8 concludes and points out future work.

2   The Programming Language

We define a simple programming language that supports performing I/O.

$v \in \text{VarNames}$, $n, r \in \mathbb{Z}$, $bio \in \text{BioNames}$, $f \in \text{FuncNames}$

$$e ::= n \mid v \mid e + e \mid e - e \mid \text{head}(e) \mid \text{nil} \mid e :: e \mid e {+}{+} e \mid \text{tail}(e) \mid \text{true} \mid \neg e \mid e = e$$
$$\quad \mid e \wedge e \mid e < e$$
$$c ::= \textbf{skip} \mid v := e \mid c; c \mid \textbf{if } e \textbf{ then } c \textbf{ else } c \mid \textbf{while } e \textbf{ do } c \mid v := bio(\bar{e}) \mid$$
$$\quad v := f(\bar{e})$$

For lists, we use the infix functions $+{+}$ for concatenation and $::$ for cons. We write the empty list as nil. We frequently notate lists with overline, e.g. $\bar{e}$ denotes a list of expressions. We leave the technical parts implicit, e.g. when two lists are expected to have the same length. Sometimes we use such a list as a set. We use simple mathematical functions for lists with their expected meaning, e.g. tail and distinct.

The language is standard except that it supports doing Basic Input Output actions (BIOs). A BIO can be thought of as a system call, but for readability we use names ($bio \in$ BioNames) as their identifiers instead of numbers. The arguments of a BIO can be considered as data that is output to the outside world, while the return-value can be considered as data that is input from the outside world. This way, a BIO allows doing both input and output.

We define Commands as the set of commands creatable by the grammar symbol "c" and quantify over it with c. Stores $=$ VarNames $\rightharpoonup (\mathbb{Z} \cup \mathbb{Z}^*)$, quantified over by s. Here, $\mathbb{Z}^*$ denotes the set of lists of integers. The partial function Stores maps the program variables to their current value.

We assume a set FuncDefs $\subset \{(f, \bar{v}, c) \mid f \in \text{FuncNames} \wedge \bar{v} \in \text{VarNames}^* \wedge c \in \text{Commands} \wedge \text{mod}(c) \cap \bar{v} = \emptyset \wedge \text{distinct}(\bar{v})\}$. Here, $\text{mod}(c)$ returns the set

$$\text{ASSIGN} \over s, v := e \Downarrow s[v := [\![e]\!]_s], \text{nil}, \textbf{done}$$

$$\text{IFTHEN} \quad [\![e]\!]_s = \text{true} \quad s, c_{then} \Downarrow s', \tau, \kappa \over s, \textbf{if } e \textbf{ then } c_{then} \textbf{ else } c_{else} \Downarrow s', \tau, \kappa$$

$$\text{IFELSE} \quad [\![e]\!]_s \neq \text{true} \quad s, c_{else} \Downarrow s', \tau, \kappa \over s, \textbf{if } e \textbf{ then } c_{then} \textbf{ else } c_{else} \Downarrow s', \tau, \kappa$$

$$\text{WHILEIN} \quad [\![e]\!]_s = \text{true} \quad s, c \,; \textbf{while } e \textbf{ do } c \Downarrow s', \tau, \kappa \over s, \textbf{while } e \textbf{ do } c \Downarrow s', \tau, \kappa$$

$$\text{WHILEOUT} \quad [\![e]\!]_s \neq \text{true} \over s, \textbf{while } e \textbf{ do } c \Downarrow s, \text{nil}, \textbf{done}$$

$$\text{SKIP} \over s, \textbf{skip} \Downarrow s, \text{nil}, \textbf{done}$$

$$\text{SEQ} \quad s_1, c_1 \Downarrow s_2, \tau_2, \textbf{partial} \over s_1, c_1 ; c_2 \Downarrow s_2, \tau_2, \textbf{partial}$$

$$\text{SEQ2} \quad s_1, c_1 \Downarrow s_2, \tau_1, \textbf{done} \quad s_2, c_2 \Downarrow s_3, \tau_2, \kappa \over s_1, c_1 ; c_2 \Downarrow s_3, \tau_1 \mathbin{++} \tau_2, \kappa$$

$$\text{EMPTY} \over s, c \Downarrow s, \text{nil}, \textbf{partial}$$

$$\text{BIO} \quad i \in \mathbb{Z} \over s, v := bio(\bar{e}) \Downarrow s[v := i], bio([\![\bar{e}]\!]_s, i) :: \text{nil}, \textbf{done}$$

$$\text{FUNCCALL} \quad \emptyset[\bar{v} := [\![\bar{e}]\!]_s], c \Downarrow s_f, \tau, \kappa \quad (f, \bar{v}, c) \in \text{FuncDefs} \over s, v := f(\bar{e}) \Downarrow s[v := [\![\text{result}]\!]_{s_f}], \tau, \kappa$$

Fig. 1. Step semantics

of variables that command c writes to. FuncDefs represents the functions of the program under consideration. Note that we disallow functions for which the body assigns to a parameter of the function. We also disallow overlap in parameter names. For simplicity, we only consider functions and programs without (mutual) recursion.

For better readability, we use abbreviations with the expected meaning, e.g. $e_1 \neq e_2$ means $\neg(e_1 = e_2)$ and $f(\bar{e})$ means $v := f(\bar{e})$ for a fresh v.

Evaluation of the expression e using store s is written as $[\![e]\!]_s$. We write $\overline{[\![e]\!]}$ as $[\![\bar{e}]\!]$. Evaluation of the expressions consisting of a variable is defined as $[\![v]\!]_s = s(v)$ (if v defined in s, otherwise unspecified). Evaluation of the other expressions is defined as $[\![op(\bar{e})]\!]_s = op([\![\bar{e}]\!]_s)$ where op is an operator with zero arguments (for constants, e.g. true, 2, or nil) or more. For example, $[\![\text{tail}(e)]\!]_s = \text{tail}([\![e]\!]_s)$, $[\![e_1 + e_2]\!]_s = [\![e_1]\!]_s + [\![e_2]\!]_s$ and $[\![2]\!]_s = 2$. Expressions that are not well-typed evaluate to an unspecified value, e.g. head(0) can evaluate to nil and to true.

Step Semantics. We define Traces as the set of lists over the set $\{bio(\bar{n}, r) \mid bio \in \text{BioNames} \land \bar{n} \in \mathbb{Z}^* \land r \in \mathbb{Z}\}$. An element of the list, $bio(\bar{n}, r)$, expresses the BIO bio has happened with arguments $\bar{n}$ and return value r. The order of the items in the list expresses the order in time in which they happened. We quantify over Traces with τ.

Continuations $= \{\mathbf{partial}, \mathbf{done}\}$, quantified over by κ.

Figure 1 shows the step semantics, relating a store and a command to a new store, a trace and a continuation. In Figure 1, $f[a := b]$ denotes the (partial) function obtained by updating the (partial) function f:

$$f[a := b] = \{ (x, y) \mid (x \neq a \wedge x \in dom(f) \wedge y = f(x)) \\ \vee (x = a \wedge y = b) \}$$

Note that the step semantics do not only support terminating runs, but also partial runs. This allows us to verify the input/output behavior of programs that do not always terminate, e.g. a server.

3   Verification Approach

We present an approach to verify input/output-related properties of programs. The first subsection gives an informal intuition of how the approach works. Subsection 3.2 defines the approach formally.

3.1   Informal Introduction

This subsection describes an intuitive understanding of the I/O verification approach using simple examples.

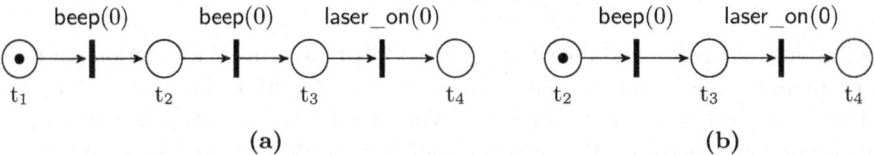

Fig. 2. Visual representation of a heap and one execution step thereof

Figure 2a shows a Petri net. The circles are called places, the black bars actions[1], and the dots in a circle are called tokens. This figure expresses that the program can beep twice, and afterwards (not earlier) turn on a laser beam. Instead of a graphical notation, we can describe the same as a multiset: $\{\mathbf{token}(t_1),$ $\mathsf{beep}(t_1, 0, t_2), \mathsf{beep}(t_2, 0, t_3), \mathsf{laser\_on}(t_3, 0, t_4)\}$. We call such a multiset a heap. The Petri nets in this subsection are graphical representation of heaps. You can ignore the zeroes for now.

In general, an assertion describes constraints. The assertion $x > 10$ constrains the program variable x to be greater than 10. Therefore, it constrains the store which maps program variables to values. Besides the store, an assertion also constrains the heap.

---

[1] In Petri net terminology actions are called transitions.

token(T_1) $*$ beep$(T_1, \_, T_2)$ $*$ beep$(T_2, \_, T_3)$ $*$ laser_on$(T_3, \_, T_4)$ is an assertion satisfied by the heap given earlier (together with a store and an interpretation mapping the logical variable T_1 to the place t_1, T_2 to t_2 and so on). You can ignore the underscore arguments for now.

Let's look at how we use an assertion. We write it in a Hoare triple as follows:

$\{$ **token**(T_1) $*$ beep$(T_1, \_, T_2)$ $*$ beep$(T_2, \_, T_3)$ $*$ laser_on$(T_3, \_, T_4)$ $\}$
 beep();
 beep();
 laser_on()
$\{$ **token**(T_4) $\}$

One might be surprised that the intended I/O behavior is written in the precondition, while in the mindset of verifying memory state we have the habit of writing what the program does in the postcondition. Programs performing I/O are relatively often not intended to always terminate, and thus do not always reach the postcondition. You can consider the subassertions such as beep$(T_2, \_, T_3)$ as permissions: a permission to execute an action under certain constraints.

You could associate with each point during execution of a program a heap, a store and an interpretation that describe the state of the program: e.g. if before the instruction $x := x + 1$, the store is $\{(x, 0)\}$, then after that instruction the store is $\{(x, 1)\}$. We do not use a heap during concrete execution of a program, but you can apply the same reasoning for the heap: after executing an I/O action, the permission to do so disappears from the heap[2], the token disappears, and a new token appears; see Figure 2b. After completely executing the example program starting from the example heap, the heap is thus $\{$**token**$(t_4)\}$. Now look at the postcondition: it exactly constrains this to be the heap. In other words, the program is only allowed to terminate after having performed the actions.

Consider a program that reads one byte, and then outputs the same byte. Two of the many possible heaps that describe this behavior are $\{$**token**(t_1), read$(t_1, 'x', t_2)$, write$(t_2, 'x', 0, t_3)\}$ and $\{$**token**(t_1), read$(t_1, 'y', t_2)$, write$(t_2, 'y', 0, t_3)\}$. Since the world or environment can "return" different bytes when reading a byte, multiple heaps are required to describe all possible behaviors of the program.

The precondition **token**(T_1) $*$ read$(T_1, 'x', T_2)$ $*$ write$(T_2, 'x', \_, T_3)$ constrains the byte read to be 'x', and the byte written to be 'x'. Of the heaps given in the previous paragraph, only the first one models this precondition. Assertions can constrain the environment or the program, or both. read$(T_1, 'x', T_2)$ constrains the value read from the world will be 'x'. This constrains the world or the environment. write$(T_2, 'x', \_, T_3)$ states that the program must write value 'x'. This constrains the program. In both heaps and assertions, the last argument (that is not a place) of an action constrains the environment (input argument)

---

[2] This differs from the standard way of executing Petri nets where one usually does not remove parts of the Petri net during execution.

and the others constrain the program (output arguments). In the previous examples, the arguments with value zero in heaps and the arguments written as underscores in assertions were input arguments we were not interested in.

While constraining the environment can be useful, it is undesired in this example. A precondition that describes the intended behavior of the program that reads one byte and then prints that byte is: $\textbf{token}(T_1) * \textsf{read}(T_1, X, T_2) * \textsf{write}(T_2, X, \_, T_3)$. In assertions, arguments of I/O actions can be any expression, and expressions in assertions can refer to program variables and logical variables such as X.

Quite often, the behavior of programs depends on the behavior of the environment. It is possible to write preconditions that take this into account, for example: $\textbf{token}(T_1) * \textsf{read}(T_1, X, T_2) * \textbf{if } X > 10 \textbf{ then } \textsf{write}(T_2, X+1, \_, T_3) \textbf{ else } T_3 = T_2$. This assertion specifies the behavior of a program that reads a number and outputs one number higher if the number read is greater than ten. Note that writing this assertion is more convenient than writing or drawing all intended heaps.

For most nontrivial preconditions, a large (possibly infinite) number of heaps satisfy the precondition. This is necessary, because when thinking of the heaps as something executable, we will not know which of these heaps we will execute: it depends on the behavior of the environment, which we do not know in advance.

Let us try to write a contract for a program that reads a whole file. It needs read-permissions until end of file (EOF) has been read. Reading a negative number indicates reading EOF. We do not want the program to read past EOF, so we should not give more read-permissions than necessary. How many read-permissions we should write in the precondition thus depends on the behavior of the environment. We can write such a contract by defining a high-level I/O action as follows:

> $\textbf{predicate } \textsf{reads}(T_1, \textit{Text}, T_3) = \exists R. \exists T_2.$
> $\quad \textsf{read}(T_1, R, T_2) *$
> $\quad \textbf{if } R < 0 \textbf{ then } \textit{Text} = \text{nil} * T_3 = T_2$
> $\quad \textbf{else } (\textsf{reads}(T_2, \textit{Sub}, T_3) * \textit{Text} = R :: \textit{Sub})$

Note that this definition uses recursion. As we will see in the formal explanation, we allow infinite recursion (contrary to [10]), which is useful for e.g. a program that reads temperature sensor values forever.

Low-level actions that are not defined on top of other actions are called BIO actions. When verifying a program that uses an unverified library, the BIO actions can be actions provided by that library. If all libraries are verified but the kernel is not, BIO actions are system calls to the kernel.

In a contract both can be used, so

$$\{\, \textbf{token}(T_1) * \textsf{reads}(T_1, \textit{Text}, T_2) \,\} \; c \; \{\, \textbf{token}(T_2) \,\}$$

is a valid contract. In heaps only BIO actions are present. So $\{\textbf{token}(t_1), \textsf{read}(t_1, \text{'x'}, t_2), \textsf{read}(t_2, -1, t_3)\}$ is a valid heap for the precondition of this contract (together with an interpretation mapping T_1 to t_1, T_2 to t_3 and $\textit{Text}$ to the list 'x' :: -1 :: nil).

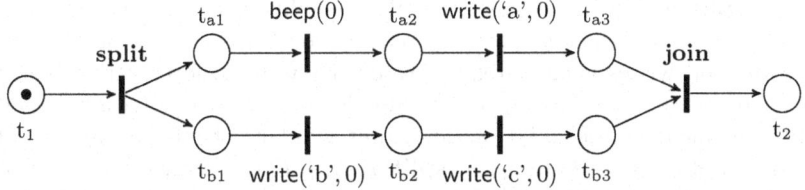

Fig. 3. Interleaving of actions

Figure 4 allows underspecification of actions: it allows to write 'a' and 'b' (but not both). A precondition of which this is a heap is $\mathbf{token}(T_1)$ * $\mathsf{write}(T_1, \text{'a'}, \_, T_2)$ * $\mathsf{write}(T_1, \text{'b'}, \_, T_2)$. It simply contains both permissions.

Instead of two write permissions, we can also write a contract with a write permission and a "dummy" I/O permission $\mathbf{no\_op}$, like this: $\mathbf{token}(T_1)$ * $\mathbf{no\_op}(T_1, T_2)$ * $\mathsf{write}(T_1, \text{'a'}, \_, T_2)$. This precondition expresses the program is allowed to write 'x' or terminate without performing any I/O.

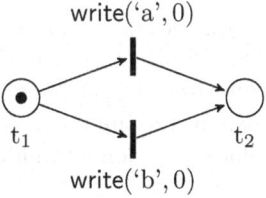

Fig. 4. Underspecification for actions

Figure 3 allows arbitrary interleavings of beeping and writing 'a', and, writing 'b' and 'c'. The transition labeled **split** does not perform I/O. When executing this heap, the **split** transition splits the token of place t_1 into two tokens: one for t_{a1} and one for t_{b1}. Both tokens can then be used to execute a transition. This allows interleaving of actions[3]. The transition labeled **join** does the inverse of **split**: it merges two tokens into one. It only does this when both t_{a3} and t_{b3} have a token.

A Hoare triple with a precondition of which Figure 3 is a model is:

$$
\left\{
\begin{array}{l}
\mathbf{token}(T_1) * \mathbf{split}(T_1, T_{a1}, T_{b1}) \\
* \; \mathsf{beep}(T_{a1}, \_, T_{a2}) * \mathsf{write}(T_{a2}, \text{'a'}, \_, T_{a3}) \\
* \; \mathsf{write}(T_{b1}, \text{'b'}, \_, T_{b2}) * \mathsf{write}(T_{b2}, \text{'c'}, \_, T_{b3}) \\
* \; \mathbf{join}(T_{a3}, T_{b3}, T_2)
\end{array}
\right\}
$$
  write('b');
  beep();
  write('c');
  write('a')
$\{\, \mathbf{token}(T_2) \,\}$

The following is an incorrect program for the contract of the Hoare triple:
write('b'); write('a'); write('c'); beep().

You might have noticed the precondition of the above contract can also be written without **split** and **join** by writing all possible interleavings by hand.

---

[3] It is also worth mentioning that **split** and **join** allow one to write contracts for multi-threaded programs.

However, **split** and **join** are very useful in combination with high-level I/O actions. Suppose we want to write a contract for Unix's cat, a program that reads a file and writes what it reads. We do not want to force cat to read the whole file before it starts writing, and we also do not want to enforce a fixed buffer size. If we define a high-level I/O action writes, similar to reads, we can write the precondition of cat as $\textbf{token}(T_1) * \textbf{split}(T_1, T_{r1}, T_{w1}) * \text{reads}(T_{r1}, Xs, T_{r2}) * \text{writes}(T_{w1}, Xs, T_{r2}) * \textbf{join}(T_{r1}, T_{w1}, T_2)$.

3.2   The Verification Approach from a Formal Point of View

Heaps We use countably infinite multisets where an element can have a countably infinite number of occurrences. We define $\text{NatInf} = \mathbb{N} \cup \{\infty\}$. We represent a multiset of a set X as $X \to \text{NatInf}$. This allows infinitely many occurrences of items in the multiset. We define addition of any elements a and b in NatInf as follows ($+_{\mathbb{N}}$ denotes addition of natural numbers):

$$a + b = \begin{cases} a +_{\mathbb{N}} b & \text{if } a \neq \infty \wedge b \neq \infty \\ \infty & \text{otherwise} \end{cases}$$

For multisets A and B, $A + B$ yields the multiset such that $(A + B)(x) = A(x) + B(x)$. Example: $\{1, 1, 2\} + \{1, 2, 3\} = \{1, 1, 1, 2, 2, 3\}$

Let X be a multiset of multisets. We define the union of X, written ΣX, as follows.

$$(\Sigma X)(y) = \begin{cases} \sum_{u \in X} u(y) & \text{if } |\{u \in X | u(y) > 0\}| \in \mathbb{N} \wedge \forall u \in X.u(y) \in \mathbb{N} \\ \infty & \text{otherwise} \end{cases}$$

We associate with each point during the execution of the program a multiset of permissions which we call the program's *heap* at that point. Such a permission is e.g. a permission to write to a file. For simplicity, the programming language used in this paper does not support dynamic memory allocation, so we do not use the heap for memory footprint or for the state of memory and the concrete execution does not use a heap. If desired, support for dynamic memory allocation with classic separation logic can easily be integrated.

We consider a set Places containing an infinite number of places.

We define Chunks as $\{\textbf{token}(t) \mid t \in \text{Places}\} \cup \{\textbf{join}(t_1, t_2, t_3) \mid t_1, t_2, t_3 \in \text{Places}\} \cup \{\textbf{split}(t_1, t_2, t_3) \mid t_1, t_2, t_3 \in \text{Places}\} \cup \{bio(t_1, \overline{n}, r, t_2) \mid bio \in \text{BioNames} \wedge \overline{n} \in \mathbb{Z}^* \wedge r \in \mathbb{Z} \wedge t_1, t_2 \in \text{Places}\}$ and Heaps as $\text{Chunks} \to \text{NatInf}$.

The intuitive meaning of a heap is given in Sec. 3.1.

Assertions. Assertions, P, and assertion expressions, E, are written in the following grammar:

$V \in \text{LogicalVarNames}, p \in \text{PredNames}.$

$E ::= n \mid v \mid V \mid E + E \mid E - E \mid \text{head}(E) \mid E ++ E \mid E :: E \mid \text{tail}(E) \mid \text{true} \mid$
$\quad \text{nil} \mid E = E \mid \neg E \mid E < E$
$P, Q, R ::= E \mid \textbf{emp} \mid P * P \mid bio(E, \overline{E}, E, E) \mid \textbf{split}(E, E, E) \mid \textbf{join}(E, E, E) \mid$
$\quad \textbf{no\_op}(E, E) \mid \textbf{token}(E) \mid p(\overline{E}) \mid P \vee P \mid \exists V.\ P \mid \circledast_{V \in \mathbb{Z}} P \mid \circledast_{V \in \mathbb{Z}^*} P$

Assertions can refer to both program variables and logical variables. Logical variables do not appear in programs and remain constant. An interpretation maps logical variables to their value, similar to a store for program variables, but a value can also be a place in Places.

Interpretations = LogicalVarNames $\rightharpoonup$ ($\mathbb{Z} \cup$ Places $\cup$ $\mathbb{Z}^*$). We quantify over interpretations with i. We start logical variable names with a capital.

$[\![v]\!]_{s,i} = s(v)$ and $[\![V]\!]_{s,i} = i(V)$ (unspecified if s and i does not contain v and V respectively). Evaluation of the other assertion expressions is defined as $[\![op(\overline{e})]\!]_{s,i} = op([\![\overline{e}]\!]_{s,i})$ where op is an operator with zero or more arguments.

For a formula P, $P[e/v]$ is the formula obtained by replacing all free occurrences of the variable v with the expression e. We write $P[\overline{e}/\overline{v}]$ for multiple simultaneous replacements. We also use this notation for replacing logical variables with logical expressions.

We use underscore, _, for assertion expressions we are not interested in; the meaning of $P[\_/V]$ is $\exists V.P$.

We write assertions of the form **if** E **then** P **else** $Q * R$; this is shorthand notation for $((E * P) \vee (\neg E * Q)) * R$. We abbreviate multiple existential quantifications, e.g. $\exists A, B$ is shorthand for $\exists A.\exists B$. We use standard abbreviations for boolean expressions with their expected meaning, e.g. $E_1 > E_2 \vee E_1 = E_2$ is abbreviated as $E_1 \geq E_2$, and $\neg(E_1 = E_2)$ as $E_1 \neq E_2$.

Assertions constrain the store, the heap and the interpretation. Figure 5 formally defines the semantics of assertions. It uses addition and union of multisets as defined in Sec. 3.2 (p. 166).

emp asserts that the heap is empty and $*$ is the separating conjunction [12].

The meaning of **token**, BIO, **no_op**, **split** and **join** assertions is explained in Sec. 3.1.

Predicates. We use predicates based on and similar to [10]. A predicate can be considered as a named parameterized assertion, but the assertion can contain predicate names, including the name of the predicate itself. They are used for defining new input/output actions on top of BIO actions, other predicates, and the predicate itself. As example of a predicate definition that defines the action of reading repeatedly until an error or end of file is encountered is given in Section 3.1 (p. 164). Our definition of predicates is nonstandard, as we will see later, and allows infinite recursion.

We write PredDefs for the set of predicate definitions for the program under consideration. This is the set of definitions for (the contracts of) a particular program, not the set of all possible definitions. PredDefs $\subset$ PredNames $\times$ LogicalVarNames* $\times$ P.

$$I, s, h, i \models E \iff [\![E]\!]_{s,i} = \text{true} \land h = \{\}$$

$$I, s, h, i \models bio(E_1, \overline{E}, E_r, E_2) \iff h = \{bio([\![E_1, \overline{E}, E_r, E_2]\!]_{s,i})\}$$

$$I, s, h, i \models \mathbf{join}(E_1, E_2, E_3) \iff h = \{\mathbf{join}([\![E_1, E_2, E_3]\!]_{s,i})\}$$

$$I, s, h, i \models \mathbf{split}(E_1, E_2, E_3) \iff h = \{\mathbf{split}([\![E_1, E_2, E_3]\!]_{s,i})\}$$

$$I, s, h, i \models \mathbf{no\_op}(E_1, E_2) \iff h = \{\mathbf{no\_op}([\![E_1, E_2]\!]_{s,i})\}$$

$$I, s, h, i \models \mathbf{token}(E) \iff h = \{\mathbf{token}([\![E]\!]_{s,i})\}$$

$$I, s, h, i \models \mathbf{emp} \iff h = \{\}$$

$$I, s, h, i \models P * Q \iff \exists h_1, h_2 . \quad h_1 + h_2 = h \ \land \ I, s, h_1, i \models P \ \land \ I, s, h_2, i \models Q$$

$$I, s, h, i \models p(\overline{E}) \iff (p, ([\![\overline{E}]\!]_{s,i}), h) \in I$$

$$I, s, h, i \models P \lor Q \iff (I, s, h, i \models P) \lor (I, s, h, i \models Q)$$

$$I, s, h, i \models \exists V. \ P \iff \exists x \in \mathbb{Z} \cup \mathbb{Z}^* \cup \text{Places}. \ I, s, h, i \models P[x/V]$$

$$I, s, h, i \models \circledast_{V \in \mathbb{Z}} P \iff \exists f : \mathbb{Z} \to \text{Heaps} . \ h = \Sigma_{n \in \mathbb{Z}} f(n) \land \forall n \in \mathbb{Z} . \ I, s, f(n), i \models P[n/V]$$

$$I, s, h, i \models \circledast_{V \in \mathbb{Z}^*} P \iff \exists f : \mathbb{Z}^* \to \text{Heaps} . \ h = \Sigma_{l \in \mathbb{Z}^*} f(l) \land \forall l \in \mathbb{Z}^* . \ I, s, f(l), i \models P[l/V]$$

Fig. 5. Satisfaction relation of assertions

We define J as PredNames $\times$ $(\mathbb{Z} \cup \mathbb{Z}^* \cup \text{Places})^* \times \text{Heaps}$. For predicates, we define a context $I_{\text{fix}} \subseteq J$ which expresses, for a given predicate name and argument values, the heap chunks a predicate assertion covers.

Let us define I_{fix}. Consider the function f:

$$f : \mathcal{P}(J) \to \mathcal{P}(J) : j \mapsto \{ \ (p, (\overline{x}), h) \in J \mid \exists \overline{V}, P. \ (p, (\overline{V}), P) \in \text{PredDefs} \land j, \emptyset, h, \emptyset \models P[\overline{x}/\overline{V}] \ \}$$

We will prove that f has a greatest fixpoint and define I_{fix} as the greatest fixpoint of f. The reason we take the greatest fixpoint instead of the least, is such that we can specify I/O behavior of programs that do not have a condition (such as: user clicks exit button) to terminate. Consider for example the following predicate definition: **predicate** $\text{inf\_print}(T_1, X) = \exists T_2 . \ \text{print}(T_1, X, \_, T_2) * \text{inf\_print}(T_2, X + 1)$. In case we use the greatest fixpoint, this expresses the action of printing a sequence of numbers (e.g. 1, 2, 3, ...). If we would have chosen to take the least fixpoint, the predicate inf_print would be equivalent to false, and we would not be able to specify the I/O behavior of this never ending program.

To prove that the greatest fixpoint of f exist, we will apply Knaster-Tarski's theorem which states that any monotone function on a complete lattice has a greatest and a least fixpoint.

We consider the partial order relation $\subseteq$. Note that $J, \subseteq$ is a complete lattice.

We have to show that f is monotone, i.e. that for any $j_1, j_2 \in J$ such that $j_1 \subseteq j_2$, then $f(j_1) \subseteq f(j_2)$. Take such a j_1 and j_2. Let $y \in f(j_1)$ (in case this is impossible, i.e. $f(j_1) = \emptyset$, it immediately follows that $f(j_1) \subseteq f(j_2)$). Because $y \in f(j_1)$, $y \in \{(p, (\overline{x}), h) \mid \exists \overline{V}, P.\ (p, (\overline{V}), P) \in \text{PredDefs} \wedge j_1, \emptyset, h, \emptyset \models P[\overline{x}/\overline{V}]\}$. It suffices to show that $\forall j_1, j_2, s, h, i, P.\ j_1 \subseteq j_2 \Rightarrow j_1, s, h, i \models P \Rightarrow j_2, s, h, i \models P$. This can easily be proven by induction on P. Note that negation of assertions is syntactically disallowed.

Big Star. The big star operator, $\circledast$, allows to easily express an infinite number of permissions. If one would accept '...' in formulae, we could write $\circledast_{V \in \mathbb{Z}} P$ as $P[0/V] * P[1/V] * P[-1/V] * P[2/V] * P[-2/V] * \ldots$. The following example expresses the permission to print any number greater than 20.

$$\textbf{token}(T_1) *\ \circledast_{V \in \mathbb{Z}}\ (\textbf{if } V > 20 \textbf{ then } \text{print}(T_1, V, \_, T_2) \textbf{ else emp})$$

Besides the big star operator, predicates also allow us to express an infinite number of permissions. The big star operator therefore does not increase expressiveness, but it increases convenience.

Validity of Hoare Triples. Intuitively, the Hoare triple $\{P\}\, c\, \{Q\}$ expresses that the program c satisfies the contract with precondition P and postcondition Q. We give a simple example of a Hoare triple:

$\{\, \textbf{token}(T_1) * \text{print}(T_1, 1, \_, T_2) * \text{print}(T_2, 2, \_, T_3)\,\}$
    $\text{print}(1); \text{print}(2)$
$\{\, \textbf{token}(T_3)\,\}$

The contract of this program states that the program can write the numbers 1 and 2 in this order. If the program terminates, it has performed these actions.

Note that a program that satisfies the contract cannot perform any other I/O operations, cannot do them in another order, cannot do them more than once, etc. For more interesting examples, see Section 5.

We define a relation traces $\subseteq$ (Heaps $\times$ Traces $\times$ (Heaps $\cup \{\bot\}$)) in Figure 6. $\{h\}\, \tau\, \{g\}$ denotes $(h, \tau, g) \in$ traces.

A heap expresses a permission to execute a (potentially infinite) sequence of BIO actions (with certain arguments). Multiple sequences of actions can be allowed by a heap. (h, τ, g), where $g \neq \bot$, expresses that h allows to perform the sequence of actions τ followed by a sequence of actions allowed by the heap g.

An element in the heap can make a prediction about the environment, e.g. $bio(t_1, \overline{n}, r, t_2)$ predicts performing the BIO bio with arguments $\overline{n}$ (starting at place t_1) will have return-value r. If a programs performs a BIO where the environment violates a prediction, the program can then perform any sequence of BIOs. In that case, we write $(h, \tau, \bot)$. This expresses the sequence of actions τ is allowed by h: τ will consists of a (potentially empty) list of allowed actions where the environment did not break a prediction, followed by an action where the environment broke a prediction, followed by any (finite, infinite or empty) sequence of actions.

TRACEBIO

$$\frac{}{\{\{\textbf{token}(t_1), bio(t_1, \overline{n}, r, t_2)\}\}\ bio(\overline{n}, r) :: \text{nil}\ \{\{\textbf{token}(t_2)\}\}}$$

TRACENIL

$$\frac{}{\{\{\}\}\ \text{nil}\ \{\{\}\}}$$

TRACEFRAME

$$\frac{\{h\}\ \tau_1\ \{h'\}}{\{h + h_r\}\ \tau_1\ \{h' + h_r\}}$$

TRACECOMPOSITION

$$\frac{\{h\}\ \tau_1\ \{h_0\}\qquad \{h_0\}\ \tau_2\ \{h'\}}{\{h\}\ \tau_1 ++ \tau_2\ \{h'\}}$$

TRACELEAK

$$\frac{\{h\}\ \tau\ \{h' + h_r\}}{\{h\}\ \tau\ \{h'\}}$$

TRACESPLIT

$$\frac{}{\{\{\textbf{token}(t_1), \textbf{split}(t_1, t_2, t_3)\}\}\ \text{nil}\ \{\{\textbf{token}(t_2), \textbf{token}(t_3)\}\}}$$

TRACEJOIN

$$\frac{}{\{\{\textbf{token}(t_1), \textbf{token}(t_2), \textbf{join}(t_1, t_2, t_3)\}\}\ \text{nil}\ \{\{\textbf{token}(t_3)\}\}}$$

TRACECONTRADICT

$$\frac{r \neq r' \qquad \{h_1\}\ \tau_1 ++ bio(\overline{n}, r) :: \tau_2\ \{h_2\}}{\{h_1\}\ \tau_1 ++ bio(\overline{n}, r') :: \tau_3\ \{\bot\}}$$

Fig. 6. Definition of the traces relation. h quantifies over Heaps, not Heaps $\cup \bot$.

Also note that a heap can contradict itself, e.g. $\{\textbf{token}(t_1), \textsf{somebio}(t_1, 2, t_2), \textsf{somebio}(t_1, 3, t_2)\}$. It contradicts itself because it says performing the BIO somebio (starting at place t_1) will return 2 and will return 3.

We define validity of a Hoare triple. Intuitively, it expresses that any execution starting from a state (a store s, a heap h and an interpretation i) that satisfies the precondition, results in a trace that is allowed by the heap h. In case the execution is a finished one (i.e. the program terminated) and the environment did not violate a prediction expressed in h, the state at termination must satisfy the postcondition.

$$\forall P, c, Q.\ \models \{P\}\, c\, \{Q\} \iff$$
$$\forall s, h, i.\ I_{\text{fix}}, s, h, i \models P \Rightarrow$$
$$\forall s', \tau, \kappa.\ s, c \Downarrow s', \tau, \kappa \Rightarrow$$
$$\exists g \in \text{Heaps} \cup \{\bot\}.\ \{h\}\, \tau\, \{g\} \wedge$$
$$(\kappa = \textbf{done} \wedge g \neq \bot \Rightarrow I_{\text{fix}}, s', g, i \models Q)$$

In case you expected a universal quantifier for g, note that the concrete execution does not use a heap. If it would use a heap, g would be introduced in the universal quantification together with s', τ and κ.

Proof Rules. The proof rules are listed in Figure 7. Here, fpv(E) and fpv(P) returns the set of free program variables of the expression E and formula P respectively. The Frame rule is copied from separation logic [12].

Note that the programming language does not support recursive function calls. The structure of the proof tree is similar to the structure of the call graph.

ASSIGNMENT

$$\frac{}{\{P[e/v]\}\, v := e\, \{P\}}$$

WHILE

$$\frac{\{e * P\}\, c\, \{P\}}{\{P\}\, \textbf{while}\ e\ \textbf{do}\ c\, \{\neg e * P\}}$$

COMPOSITION

$$\frac{\{P_1\}\, c_1\, \{P_2\} \qquad \{P_2\}\, c_2\, \{P_3\}}{\{P_1\}\, c_1; c_2\, \{P_3\}}$$

CONSEQUENCE

$$\frac{P_1 \Rightarrow P_2 \qquad \{P_2\}\, c\, \{P_3\} \qquad P_3 \Rightarrow P_4}{\{P_1\}\, c\, \{P_4\}}$$

IF

$$\frac{\{P * e\}\, c_{\text{then}}\, \{Q\} \qquad \{P * \neg e\}\, c_{\text{else}}\, \{Q\}}{\{P\}\, \textbf{if}\ e\ \textbf{then}\ c_{\text{then}}\ \textbf{else}\ c_{\text{else}}\, \{Q\}}$$

SKIP

$$\frac{}{\{P\}\, \textbf{skip}\, \{P\}}$$

DISJUNCTION

$$\frac{\{P_1\}\, c\, \{Q\} \qquad \{P_2\}\, c\, \{Q\}}{\{P_1 \vee P_2\}\, c\, \{Q\}}$$

SUBSTITUTION

$$\frac{\{P\}\, c\, \{Q\} \qquad \text{fpv}(\overline{E}) \cap \text{mod}(c) = \emptyset}{\{P[\overline{E}/\overline{V}]\}\, c\, \{Q[\overline{E}/\overline{V}]\}}$$

NOOP

$$\frac{\{P * \textbf{token}(V_{t1}) * \textbf{no\_op}(V_{t1}, V_{t2})\}\, c\, \{Q\}}{\{P * \textbf{token}(V_{t2})\}\, c\, \{Q\}}$$

LEAK

$$\frac{\{P\}\, c\, \{Q * R\}}{\{P\}\, c\, \{Q\}}$$

SPLIT

$$\frac{\{P * \textbf{token}(V_{t2}) * \textbf{token}(V_{t3})\}\, c\, \{Q\}}{\{P * \textbf{token}(V_{t1}) * \textbf{split}(V_{t1}, V_{t2}, V_{t3})\}\, c\, \{Q\}}$$

JOIN

$$\frac{\{P * \textbf{token}(V_{t3})\}\, c\, \{Q\}}{\{P * \textbf{token}(V_{t1}) * \textbf{token}(V_{t2}) * \textbf{join}(V_{t1}, V_{t2}, V_{t3})\}\, c\, \{Q\}}$$

BIO

$$\frac{v \notin \text{fpv}(E_r)}{\{bio(V_{t1}, \overline{e}, E_r, V_{t2}) * \textbf{token}(V_{t1})\}\, v := bio(\overline{e})\, \{v = E_r * \textbf{token}(V_{t2})\}}$$

FRAME

$$\frac{\{P\}\, c\, \{Q\} \qquad \text{fpv}(R) \cap \text{mod}(c) = \emptyset}{\{P * R\}\, c\, \{Q * R\}}$$

EXISTS

$$\frac{\{P\}\, c\, \{Q\}}{\{\exists V.\ P\}\, c\, \{\exists V.\ Q\}}$$

FUNCCALL

$$\frac{\{P\}\, c\, \{Q\} \quad \text{fpv}(P) \subseteq \overline{v} \quad \text{fpv}(Q) \subseteq \overline{v} \cup \{\text{result}\}}{v \notin \text{fpv}(\overline{e}) \qquad (f, (\overline{v}), c) \in \text{FuncDefs}} \\ \frac{}{\{P[\overline{e}/\overline{v}]\}\, v := f(\overline{e})\, \{Q[\overline{e}, v/\overline{v}, \text{result}]\}}$$

Fig. 7. Proof rules

We say a Hoare triple $\{P\}\, c\, \{Q\}$ is derivable, written $\vdash \{P\}\, c\, \{Q\}$, if it can be derived using these proof rules.

For the Consequence rule, we define implication of assertions as

$$\forall P, Q\, .\, P \Rightarrow Q \iff \forall s, h, i\, .\, I_{\text{fix}}, s, h, i \models P \Rightarrow I_{\text{fix}}, s, h, i \models Q$$

Note that $\circledast_{V \in \mathbb{Z}} P \Rightarrow P[E/V] * \circledast_{V \in \mathbb{Z}} (\textbf{if}\ V = E\ \textbf{then}\ \textbf{emp}\ \textbf{else}\ P)$ if $V \notin \text{fv}(E)$, where $\text{fv}(E)$ returns the set of free variables of E.

4   Soundness

Theorem 1 (Soundness). $\forall P, c, Q. \vdash \{P\} c \{Q\} \Rightarrow \models \{P\} c \{Q\}.$

Proof. Due to space limits, we only outline some cases of the induction on $\vdash \{P\} c \{Q\}$ which is nested in an induction on $s, c \Downarrow s', \tau, \kappa$. In these cases we know $I_{\mathrm{fix}}, s, h, i \models P$.

- **Bio:** Because of the Bio proof rule, we know there is some E_r, v, bio, V_{t1}, $\bar{e}$, E_r, V_{t2} such that $P = bio(V_{t1}, \bar{e}, E_r, V_{t2}) * \mathbf{token}(V_{t1})$, $c = v := bio(\bar{e})$, and $Q = (v = E_r * \mathbf{token}(V_{t2}))$.
  We consider the case where the step rule that applies is Bio (the other case, where the step rule is Empty, is trivial). We know $s, c \Downarrow s', \tau, \kappa$. Thus there is some n such that $\tau = bio([\![\bar{e}]\!]_s, n) :: \mathrm{nil}$.
  If $n \neq [\![E_r]\!]_{s,i}$ we need to prove that there is some g such that $\{h\} \tau \{g\}$. Note that $\kappa = \mathbf{partial}$. Let $g = \bot$. By applying the TraceBio and the TraceContradict rule, we obtain $\{h\} \tau \{g\}$.
  If $n = [\![E_r]\!]_{s,i}$, we know $h = \{bio([\![V_{t1}]\!]_{s,i}, [\![\bar{e}]\!]_s, n, [\![V_{t2}]\!]_{s,i}), \mathbf{token}([\![V_{t1}]\!]_{s,i})\}$. Let $g = \{\mathbf{token}([\![V_{t2}]\!]_{s,i})\}$. Because of the TraceBio rule, we obtain $\{h\} \tau \{g\}$, which we wanted to prove. Because of the Bio step rule we know $\kappa = \mathbf{done}$. $I_{\mathrm{fix}}, s', g, i \models Q$ follows from the equalities of Q and g given above, and from $s' = s[v := n]$, which we know because of the Bio step rule.
- **Frame:** Because of the Frame proof rule, there is some P_0, Q_0, R such that $P = P_0 * R$, $Q = Q_0 * R$ and $\vdash \{P_0\} c \{Q_0\}$. Because $P = P_0 * R$ and $I_{\mathrm{fix}}, s, h, i \models P$, there is some h_0, h_r such that $h_0 + h_r = h \wedge I_{\mathrm{fix}}, s, h_0, i \models P_0 \wedge I_{\mathrm{fix}}, s, h_r, i \models R$.
  Because of the induction hypothesis, we know $\models \{P_0\} c \{Q_0\}$. Thus for some h'_0, it holds that $\{h_0\} \tau \{h'_0\}$ and $\kappa = \mathbf{done} \Rightarrow I_{\mathrm{fix}}, s', h'_0, i \models Q_0$.
  Let $g = h'_0 + h_r$. By applying the TraceFrame rule, we obtain $\{h\} \tau \{g\}$, which is what we wanted to prove.
  If $\kappa = \mathbf{done}$, we need to prove that $I_{\mathrm{fix}}, s', g, i \models Q$.
  Because $\mathrm{fpv}(R) \cap \mathrm{mod}(c) = \emptyset$ and $s, c \Downarrow s', \tau, \kappa$ and $I_{\mathrm{fix}}, s, h_r, i \models R$, we obtain $I_{\mathrm{fix}}, s', h_r, i \models R$.
  Combined with $I_{\mathrm{fix}}, s', h'_0, i \models Q_0$ we know $I_{\mathrm{fix}}, s', g, i \models Q$ by definition of $*$.
- **Split:** Because of the Split proof rule, there is some $P_1, V_{t1}, V_{t2}, V_{t3}$ such that $P = P_1 * \mathbf{token}(V_{t1}) * \mathbf{split}(V_{t1}, V_{t2}, V_{t3})$. Let $P_0 = P_1 * \mathbf{token}(V_{t2}) * \mathbf{token}(V_{t3})$. We know $h = h_{p1} + \{\mathbf{token}([\![V_{t1}]\!]_{s,i}), \mathbf{split}([\![V_{t1}, V_{t2}, V_{t3}]\!]_{s,i})\}$ for some h_{p1}. Let $h_0 = h_{p1} + \{\mathbf{token}([\![V_{t2}]\!]_{s,i}), \mathbf{token}([\![V_{t3}]\!]_{s,i})\}$.
  Because of the induction hypothesis, $\models \{P_0\} c \{Q\}$. Combined with $I_{\mathrm{fix}}, s, h_0, i \models P_0$, we obtain there is some g such that $\{h_0\} \tau \{g\}$.
  By using the TraceSplit and TraceFrame rule, we know $\{h\} \mathrm{nil} \{h_0\}$. Now we can apply the TraceComposition rule to obtain $\{h\} \tau \{g\}$, which we wanted to prove.
  Because $\models \{P_0\} c \{Q\}$ and $I_{\mathrm{fix}}, s, h_0, i \models P_0$, we know $\kappa = \mathbf{done} \Rightarrow I_{\mathrm{fix}}, s', g, i \models Q$.

5   Examples

We give some examples of contracts.

5.1   Tee

Figure 8 lists the implementation and specification of a program that reads from standard input (until end-of-file), and writes what it reads to both standard output and standard error. The contract is written in a compositional manner: an action tee_out represents the action of writing to both standard output (stdout) and standard error (stderr). The action that represents the whole program, tee, is built upon the tee_out action. The specifications allow a read-buffer of any size. The implementation chooses a read-buffer of size 2. The reads predicate defined in Section 3.1 (p. 164) is used.

Figure 9 gives a proof outline for the tee_out function and the main function.

5.2   Read Files Mentioned in a File

The specification of Figure 10 allows the program to read a file, "f", which contains filenames (of length one character). The files mentioned in this file are read. The program prints to standard output the contents of these files in order. To make the example more interesting, the specifications allow the program to choose whether to output a read character as soon as possible or postpone it, and whether to read a file as soon as possible or postpone it after or while reading "f". Filenames that consist of the zero character or non 7bit-ASCII are ignored.

---

predicate tee_out$(T_1, C, T_2) = \exists T_{p1}, T_{p2}, T_{r1}, T_{r2}.$
   split(T_1, T_{p1}, T_{r1})
   $*$ stdout$(T_{p1}, C, \_, T_{p2})$
   $*$ stderr$(T_{r1}, C, \_, T_{r2})$
   $*$ **join**(T_{p2}, T_{r2}, T_2)

function tee_out$(c) =$
   $\{\, \textbf{token}(T_1) * \textbf{tee\_out}(T_1, c, T_2) \,\}$
   stdout(c);
   stderr(c)
   $\{\, \textbf{token}(T_2) \,\}$

predicate tee_outs$(T_1, Text, T_2) = \exists T_{out}.$
   if $Text = \textbf{nil}$ **then** $T_2 = T_1$ **else** (
   tee_out$(T_1, \text{head}(Text), T_{out})$
   $*$ tee_outs$(T_{out}, \text{tail}(Text), T_2)$)

predicate tee$(T_1, Text, T_2) = \exists T_{r1}, T_{w1}, T_{r2}, T_{w2}.$
   split(T_1, T_{r1}, T_{w1})

   $*$ reads$(T_{r1}, Text, T_{r2})$
   $*$ tee_outs$(T_{w1}, Text, T_{w2})$
   $*$ **join**(T_{r2}, T_{w2}, T_2)

function main$() =$
   $\{\, \textbf{token}(T_1) * \textbf{tee}(T_1, Text, T_2) \,\}$
   $c_2 := 0$;
   while $c_2 \geq 0$ **do** (
   $c_1 := \text{read}()$;
   if $c_1 \geq 0$ **then**
   $c_2 := \text{read}()$;
   tee_out(c_1);
   if $c_2 \geq 0$ **then**
   tee_out(c_2)
   else skip
   else
   $c_2 := -1$)
   $\{\, \textbf{token}(T_2) \,\}$

Fig. 8. Specification and implementation of the Tee program

```
function tee_out(c) =
{ token(T₁) * tee_out(T₁, c, T₂) }
⎧ token(T_p1 * stdout(T_p1, c, _, T_p2)    ⎫
⎨ * token(T_r1) * stderr(T_r1, c, _, T_r2) ⎬
⎩ * join(T_p2, T_r2, T₂)                   ⎭
stdout(c);
⎧ token(T_p2) * token(T_r1)  ⎫
⎨ * stderr(T_r1, c, _, T_r2) ⎬
⎩ * join(T_p2, T_r2, T₂)     ⎭
stderr(c)
⎧ token(T_p2) * token(T_r2) ⎫
⎩ * join(T_p2, T_r2, T₂)    ⎭
{ token(T₂) }

predicate invariant(C₂, T₂) =
∃T_r1, T_r2, T_w1, T_w2.
  if C₂ ≥ 0 then
    token(T_r1)
    * reads(T_r1, Text, T_r2)
    * token(T_w1)
    * tee_outs(T_w1, Text, T_w2)
    * join(T_r2, T_w2, T₂)
  else token(T₂)

function main() =
{ token(T₁) * tee(T₁, FullText, T₂) }
c₂ := 0;
{ invariant(c₂, T₂) }
while c₂ ≥ 0 do (
  { invariant(c₂, T₂) }
  c₁ := read();
  if c₁ ≥ 0 then (
    ⎧ token(T_rb1)             ⎫
    ⎪ * reads(T_rb1, Sub, T_r2) ⎪
    ⎨ * token(T_w1)            ⎬
    ⎪ * tee_outs(T_w1, c₁ :: Sub, T_w2) ⎪
    ⎩ * join(T_r2, T_w2, T₂)   ⎭
```

```
    c₂ := read();
    ⎧ token(T_rb2) * token(T_w1)         ⎫
    ⎪ * if c₂ ≥ 0 then                    ⎪
    ⎪   reads(T_rb2, SubSub, T_r2)        ⎪
    ⎪   * tee_outs(T_w1,                  ⎪
    ⎨       c₁ :: c₂ :: SubSub, T_w2)     ⎬
    ⎪ else (                              ⎪
    ⎪   T_r2 = T_rb2                       ⎪
    ⎪   * tee_outs(T_w1, c₁ :: nil, T_w2) ) ⎪
    ⎩ * join(T_r2, T_w2, T₂)              ⎭
    tee_out(c₁);
    if c₂ ≥ 0 then (
      ⎧ token(T_rb2)                       ⎫
      ⎪ * reads(T_rb2, SubSub, T_r2)       ⎪
      ⎨ * token(T_wb1)                     ⎬
      ⎪ * tee_outs(T_wb1, c₂ :: SubSub, T_w2) ⎪
      ⎩ * join(T_r2, T_w2, T₂)             ⎭
      tee_out(c₂)
      { invariant(c₂, T₂) }
    ) else (
      { c₂ < 0 * token(T₂) }
    )
    { invariant(c₂, T₂) }
  ) else (
    { token(T₂) }
    c₂ := -1;
    { invariant(c₂, T₂) }
  )
  { invariant(c₂, T₂) }
)
{ token(T₂) }
```

Fig. 9. Proof outline of the tee_out and main function of the tee program

predicate freads($T_1, F, Text, T_{end}$) =
$\exists C, Sub, T_2.$
  fread(T_1, F, C, T_2)
  $*$ **if** $C \geq 0$ **then**
    freads(T_2, F, Sub, T_{end})
    $*$ $Text = C :: Sub$
  else (
    $T_{end} = T_2$
    $*$ $Text = $ nil)

predicate get_file($T_1, Name, Text,$
T_{end}) =
$\exists Handle, T_2, T_3.$
  fopen($T_1, Name, Handle, T_2$)
  $*$ freads($T_2, Handle, Text, T_3$)
  $*$ fclose($T_3, Handle, \_, T_{end}$)

predicate prints($T_1, Text, T_{end}$) =
$\exists T_2.$
  if $Text = $ nil **then**
    $T_{end} = T_1$
  else (
    print($T_1,$ head($Text$), $\_, T_2$)
    $*$ prints($T_2,$ tail($Text$), T_{end}))

predicate get_files($T_1, FNames,$
$Text, T_{end}$) =
$\exists Text1, Text2, T_2, Fname, SubNames.$
  if $FNames = $ nil **then**
    $T_{end} = T_1$
    $*$ $Text = $ nil
  else (
    get_file($T_1, Fname :: $ nil$, Text1, T_2$)
    $*$ get_files($T_2, SubNames, Text2, T_{end}$)
    $*$ $Fnames = Fname :: SubNames$
    $*$ $Text = Text1 ++ Text2$)

predicate read_fname_list($T_1, Handle,$
$FNames, T_{end}$) =
$\exists C, T_2, Sub.$
  fread($T_1, Handle, C, T_2$)
  $*$ **if** $C \geq 0$ **then**
    read_fname_list($T_2, Handle, Sub, T_{end}$)
    $*$ **if** $C > 0 \wedge C \leq 127$ **then**
      $FNames = C :: Sub$
    else
      $FNames = Sub$
  else (
    $T_{end} = T_2$
    $*$ $FNames = $ nil)

predicate main($T_1, Fname, T_{end}$) =
$\exists T_2, T_{meta}, T_{rw}, FNames, T_{meta2}, T_r, T_w,$
$T_{meta3}, T_{r2}, T_{w2}, T_{rw2}, Handle .$
  fopen($T_1, Fname, Handle, T_2$)
  $*$ **split**(T_2, T_{meta}, T_{rw})
  $*$ read_fname_list($T_{meta}, Handle,$
    $FNames, T_{meta2}$)
  $*$ fclose($T_{meta2}, Handle, T_{meta3}$)
  $*$ **split**(T_{rw}, T_r, T_w)
  $*$ get_files($T_r, FNames, Text, T_{r2}$)
  $*$ prints($T_w, Text, T_{w2}$)
  $*$ **join**(T_{r2}, T_{w2}, T_{rw2})
  $*$ **join**($T_{meta3}, T_{rw2}, T_{end}$)

function main() =
  $\{$ **token**(T_1) $*$ main($T_1,$ 'f' $::$ nil$, T_2$) $\}$
  $\dots$
  $\{$ **token**(T_2) $\}$

Fig. 10. Specification of a program that prints the contents of all files whose filenames are in a given list. This list is not static, it is read from a file "f".

The predicates representing actions are written on top of standard library actions like fopen for opening a file. These could be BIOs, but they can also be predicates built on top of lower-level actions, e.g. system calls.

The implementation is left out. As mentioned earlier, a verified implementation is shipped as an example with VeriFast.

5.3 Print Any String of the Grammar of Matching Brackets

predicate brackets(T_1, T_2) =
$\exists T_{open}, T_{center}, T_{close}.$
  no_op(T_1, T_2)
  * print($T_1,$ '(', _, T_{open})
  * brackets(T_{open}, T_{center})
  * print($T_{center},$ ')', _, T_{close})
  * brackets(T_{close}, T_2)

function main() =
  { **token**(T_1) * brackets(T_1, T_2) }
    print('(');
    print(')')
  { **token**(T_2) }

Fig. 11. Specification of a program that is allowed to output any string of the matching brackets grammar

The specification of Figure 11 states that the program is only allowed to output a string of the grammar of matching brackets. Note that the specifications do not specify which string: any string of the grammar is allowed. The grammar under consideration is clearly visible in the specification, and is as follows:

$B ::= (B)B \mid \epsilon$

Here, ϵ denotes the empty string. In the specification, '(' and ')' are shorthand for 40 and 41 respectively (the ASCII number of these characters).

5.4 Turing Machine

This example's only purpose is to illustrate the expressiveness of the contracts, i.e. it is not an example of a typical contract. It is possible to define a predicate

predicate tm(T_1, *EncodingOfTM, InitialState, TapeLeft, TapeRight*, T_2)

that expresses that the allowed I/O actions are the actions a Turing machine, given as second argument, performs. In other words, the program under consideration is allowed to perform the I/O actions that the Turing machine (TM) performs. The TM is encoded as a list of integers by serializing the table representation of the TM's transition function. The states and symbols are represented using integers.

Normally, the transition function of a TM maps the state and the symbol read from the tape, to the symbol to write on the tape, a new state, and an action:

whether the TM's tape should move left, right, or not move. Typically, a TM accepts an input if the machine ends in an accepting state when launched from the full input on its tape. To make it more interactive, we add one input and one output action. The output action prints (i.e. writes to the world) the symbol on the tape without moving the tape. The tape does not move. The input action reads a symbol from the world and puts it on the tape at the current position.

A benefit of this approach is that it supports nonterminating programs naturally. The TM does not have to terminate. If the program terminates, it must have performed all the I/O actions the TM does (otherwise it will not obtain its postcondition **token**(T_2)), and the TM must have terminated as well.

Besides interactive and nonterminating I/O we also want underspecified input/output. The specifications should thus allow multiple behaviors. This is easy to deal with by making the TM non deterministic.

The code of this example is left out to save space, but an annotated C version is shipped with VeriFast (see next section).

5.5 Mechanical Verification of the Examples

C versions of all examples in this paper have been mechanically verified using the VeriFast [7] tool. This increases confidence that the approach is usable in practice and not only on paper. The input of VeriFast is the C source code of the program, the contracts of the functions, and extra annotations. With this input, VeriFast outputs whether it is "convinced" the C implementation conforms to the contracts. In case VeriFast says yes, we are sure the implementation is free of bugs violating the contract (and basic properties like memory safety). Note that this is not just detecting bugs: it proves absence of bugs.

The examples in this paper are included in the directory `examples/io` in the VeriFast distribution. VeriFast is available from `http://distrinet.cs.kuleuven.be/software/VeriFast/`.

VeriFast only performs very limited automated proof; non-trivial proof steps must be explicitly specified in annotations. In particular, VeriFast does not encode predicates as SMT solver axioms; the user must explicitly fold and unfold predicates through 'open' and 'close' annotations. Therefore, VeriFast typically has a high annotation overhead and a short execution time. This is also the case with the presented examples:

Example	LoC	Lines of annotations	Time (ms)
Tee	24	54	192
Read files mentioned in a file	20	90	101
Matching brackets	4	20	96
Turing machine	7	58	103
Template method (Sec. 6)	39	77	60

The reported timings are using the Redux SMT solver on a Intel Core i5 CPU (max value of 10 runs).

6 Verifying Memory State Using I/O Style Verification

One might wonder whether there really is a difference between verifying I/O behavior and verifying memory-state. For example, we could consider writing to a file both as I/O and as performing memory operations. If the filesystem is in memory, and we consider not only the process that writes the file but also the kernel, then we are only manipulating memory. Therefore, verifying I/O and verifying memory state can be considered as another point of view, and not necessarily as technically different.

The regular approach for verification using the memory-state point of view can be insufficient for verifying applications for which the I/O behavior point of view is "natural". For example, the memory-state point of view usually only cares about the state when the program has reached the postcondition. This is insufficient for verifying I/O behavior. First, for I/O applications, nontermination is common and often not undesired, hence it is normal that the postcondition is never reached. Second, by looking only at (memory) state when the program has reached the postcondition, we ignore intermediate state. For I/O applications intermediate state is important: for a movie player, not only the last frame but all frames of the movie should be displayed correctly.

```
abstract class A {
  void template() {
    m1();
    m2();
  }
  abstract void m1();
  abstract void m2();
}
```

Fig. 12. Minimalistic example of the Template Method design pattern (in Java)

While the memory-state point of view can be insufficient from an I/O style point of view, one might wonder what happens when we try the other way around: what if we want to verify an application for which the memory-based point of view is expected at first sight, from an I/O point of view.

This section will take a quick look at this question, by looking at one example or use case: the Template Method design pattern.

Template Method [5] is an object-oriented design pattern in which an abstract class has a method implementing an algorithm of which a number of steps are delegated to subclasses. This delegation happens by calling abstract methods, which subclasses can implement.

How can we write the contract of this template method? The method must perform what the subclass's hook methods (m1 and m2 in Figure 12) will do, but we do not know what that will be. Furthermore, we do not want to change the contract or perform verification again of the template method when new subclasses are added. In this section, we will write an easy contract for the template method using the approach of this paper.

predicate token$(A; T_1) = $ **emp**
predicate m1_io$(A; T_1, T_2) = $ true
predicate m2_io$(A; T_1, T_2) = $ true
class A $=$

   method A.template() $=$

$$\left\{ \begin{array}{l} \text{token}(\textbf{this}; T_1) \\ * \; \text{m1\_io}(\textbf{this}; T_1, T_2) \\ * \; \text{m2\_io}(\textbf{this}; T_2, T_3) \end{array} \right\}$$

      m1();
      m2()
  $\{\,\text{token}(\textbf{this}; T_3)\,\}$

method A.m1() $=$

$$\left\{ \begin{array}{l} \text{token}(\textbf{this}; T_1) \\ * \; \text{m1\_io}(\textbf{this}; T_1, T_2) \end{array} \right\}$$

$\{\,\text{token}(\textbf{this}; T_2)\,\}$

method A.m2() $=$

$$\left\{ \begin{array}{l} \text{token}(\textbf{this}; T_1) \\ * \; \text{m2\_io}(\textbf{this}; T_1, T_2) \end{array} \right\}$$

$\{\,\text{token}(\textbf{this}; T_2)\,\}$

Fig. 13. A template method and its hook methods, with I/O style contracts. m1_io, m2_io and token are predicate families. For a subclass with implementation of m1 and m2, see Figure 14.

The language described so far does not support object-oriented programming, so we extend the language in a standard way to support object field access, method calls, casts, writing to object fields, object allocation and object deallocation. We assume a set Classes quantified over with C, and set MethodNames quantified over with m.

$$e ::= \ldots \mid \textbf{this}$$
$$c ::= \ldots \mid v := e.v \mid v := v.m(\overline{e}) \mid v := (C)v \mid e.v := e \mid v := \textbf{new}(C) \mid$$
$$\textbf{dispose}(v)$$

We assume a set MethodDefs $\subset C \times m \times \overline{v} \times c$ and AbstractMethodDecls $\subset C \times m \times \overline{v}$ that describe the methods and abstract methods of the program under consideration. We only consider valid sets: there is no overlap in arguments $(\overline{v})$, an (abstract) method cannot appear twice in the same class, and a method cannot write to its arguments. We assume a partial function from Classes to Classes expressing inheritance. We only support non-circular single inheritance.

The step semantics is extended as expected. Note that it will need a concrete heap, to keep track of (1) the values of object fields to support memory (de)allocation and field access, and (2) the dynamic type of objects to support dynamic binding of method calls.

The assertion language needs to be extended to support assertions describing the fields of objects. We also need predicate families, i.e. predicates indexed by class. This allows multiple versions of a predicate. For the semantics of predicate families and proof rules, we refer to [10]. Furthermore, the assertions can describe that an object is an instance of a class C (not a subclass of C). We drop the keyword "**token**". We drop support for **split** and **join**.

$$E ::= \ldots \mid \textbf{this}$$
$$P ::= \ldots \mid E.v \mapsto E \mid p(E; \overline{E}) \mid E : C$$

predicate $\text{token}(\text{B}; T_1) = \textbf{this}.x \mapsto T_1$
predicate $\text{m1\_io}(\text{B}; T_1, T_2) = T_2 = T_1 + 1$
predicate $\text{m2\_io}(\text{B}; T_1, T_2) = T_2 = T_1 + 10$
class B **extends** A =
    field $B.x$

    method B.m1() =
    $\left\{ \begin{array}{l} \text{token}(\textbf{this}, T_1) \\ * \textbf{this} : \text{B} \\ * \text{m1\_io}(\textbf{this}; T_1, T_2) \end{array} \right\}$
    $y := \textbf{this}.x;$
    $\textbf{this}.x := y + 1$
    $\{\, \text{token}(\textbf{this}; T_2) \,\}$

method B.m2() =
$\left\{ \begin{array}{l} \text{token}(\textbf{this}; T_1) \\ * \textbf{this} : B \\ * \text{m2\_io}(\textbf{this}; T_1, T_2) \end{array} \right\}$
$y := \textbf{this}.x;$
$\textbf{this}.x := y + 10$
$\{\, \text{token}(\textbf{this}; T_2) \,\}$

method B.getValue() =
$\{\, \text{token}(\textbf{this}; T_1) \,\}$
$result := \textbf{this}.x$
$\{\, \text{token}(\textbf{this}; T_1) * result = T_1 \,\}$

Fig. 14. A subclass implementing hook methods m1 and m2

Figure 13 shows how the contract of the template method can then be written. Figure 14 shows an example of a subclass.

This section presented an approach for verifying memory, while Sec. 3 presented an approach for verifying I/O properties. Both are instantiations of the same specification style. Note that the approach of Sec. 3 supports some more features: nondeterminism/underspecification and interleaving.

An annotated Java version of the example of this section is shipped with VeriFast.

7   Related Work

Approaches for verifying input/output behavior and case studies doing so have been developed and performed before.

The Verisoft email client [1] has a verified fullscreen text-based user interface. The approach identifies points in the main loop of the program, and restricts I/O to the screen to only these points. This approach is elegant but does not scale. A program of reasonable size typically uses libraries that also perform I/O. A contribution of our approach is compositional I/O verification. This allows verified libraries to perform I/O.

The heaps (not the assertions) of the approach described by this paper can be represented using Petri nets. The assertions are more expressive by using features such as using actions composed out of other actions. These features are also present in coloured Petri nets [8], which is a generalization of Petri nets where tokens are represented as data values. By modeling the contracts as a coloured Petri net, one could analyse the contracts using techniques to analyze coloured Petri nets. Note that the goal of our contribution is not just to model input/output behavior, but to verify input/output behavior of programs.

Nakata and Uustalu [9] define a Hoare logic for a programming language with tracing semantics. The assertion language is inspired by interval temporal logic.

The programming language is defined using big step semantics, but relates a program and a state to a trace, instead of to a state as is usually done. An assertion can express properties of this (potentially infinite) trace. The paper rather provides a framework to build upon than proposing a final assertion language. In every example of the paper, the assertion language is extended to support the example. Using such an extension it is possible to prove liveness properties, which our approach does not support. The paper uses Coq and not a verification tool specialized for software verification. It is unclear how well the approach blends in with solutions for other problems, e.g. aliasing.

Linear Time Calculus (LTC) provides a methodology for modeling dynamic systems in general in an extension of first order logic. In LTC, an action has an argument which represents a point in (linear) time when the action happens. Such a point in time is a natural number and is clearly before or after another point in time. This differs from our approach where an action has two arguments, each representing a place. A place is not always clearly before or after another place. LTC is a generic approach, while our approach focuses on software verification. For an explanation of LTC we refer to [2], which also shows many tool-supported practical applications.

Model checking [3,11,4] allows checking whether properties written in a temporal language such as LTL and CTL hold for a model. Such models can be created automatically from the software subject to verification. Expressing temporal properties in temporal languages is natural, making it a good candidate for expressing input/output-related properties. Furthermore, liveness properties can be expressed. The approach suffers from the state explosion problem, a problem that remains after major improvements made in the last three decades [4].

Wisnesky, Malecha, and Morrisett [13] verify I/O properties by constraining the list of performed actions in the postcondition, but this approach does not seem to prevent nonterminating programs to perform undesired I/O.

8   Conclusions and Future Work

We identified several requirements for approaches that verify the input/output behavior of computer programs, including modularity, compositionality, soundness and non-determinism of the environment. We created a verification approach that meets these requirements.

Because the approach is designed to work compositionally and modularly, we hope the approach works well for bigger applications, but to confirm this, a real-world case study of considerable size should be carried out. Such a case study is future work.

Acknowledgements. We would like to thank Amin Timany for many useful discussions. This work was funded by Research Fund KU Leuven and by EU FP7 FET-Open project ADVENT under grant number 308830.

References

1. Beuster, G., Henrich, N., Wagner, M.: Real world verification – Experiences from the Verisoft email client. In: Sutcliffe, G., Schmidt, R., Schulz, S. (eds.) Proceedings of the FLoC 2006 Workshop on Empirically Successful Computerized Reasoning (ESCoR 2006). CEUR Workshop Proceedings, vol. 192, pp. 112–125. CEUR-WS.org (August 2006)
2. Bogaerts, B., Jansen, J., Bruynooghe, M., De Cat, B., Vennekens, J., Denecker, M.: Simulating dynamic systems using linear time calculus theories. Theory and Practice of Logic Programming 14, 477–492 (2014)
3. Clarke, E.M., Emerson, E.A.: Design and synthesis of synchronization skeletons using branching-time temporal logic. In: Logic of Programs, Workshop, London, UK, pp. 52–71. Springer (1981)
4. Clarke, E.M., Klieber, W., Nováček, M., Zuliani, P.: Model checking and the state explosion problem. In: Meyer, B., Nordio, M. (eds.) LASER 2011. LNCS, vol. 7682, pp. 1–30. Springer, Heidelberg (2012)
5. Gamma, E., Helm, R., Johnson, R., Vlissides, J.: Design Patterns: Elements of Reusable Object-Oriented Software. Addison-Wesley (1995)
6. Hoare, C.A.R.: An axiomatic basis for computer programming. Communications of the ACM 12(10), 576–580 (1969)
7. Jacobs, B., Smans, J., Philippaerts, P., Vogels, F., Penninckx, W., Piessens, F.: VeriFast: A powerful, sound, predictable, fast verifier for C and Java. In: Bobaru, M., Havelund, K., Holzmann, G.J., Joshi, R. (eds.) NFM 2011. LNCS, vol. 6617, pp. 41–55. Springer, Heidelberg (2011)
8. Kristensen, L.M., Christensen, S., Jensen, K.: The practitioner's guide to coloured Petri nets. International Journal on Software Tools for Technology Transfer 2, 98–132 (1998)
9. Nakata, K., Uustalu, T.: A Hoare logic for the coinductive trace-based big-step semantics of While. In: Gordon, A.D. (ed.) ESOP 2010. LNCS, vol. 6012, pp. 488–506. Springer, Heidelberg (2010)
10. Parkinson, M., Bierman, G.: Separation logic and abstraction. In: Proceedings of the 32nd Symposium on Principles of Programming Languages, pp. 247–258. ACM (2005)
11. Queille, J.-P., Sifakis, J.: Specification and verification of concurrent systems in CESAR. In: Dezani-Ciancaglini, M., Montanari, U. (eds.) Programming 1982. LNCS, vol. 137, pp. 337–351. Springer, Heidelberg (1982)
12. John, C.: Reynolds. Separation logic: A logic for shared mutable data structures. In: Proceedings of the 17th Symposium on Logic in Computer Science, pp. 55–74. IEEE, Washington (2002)
13. Wisnesky, R., Malecha, G., Morrisett, G.: Certified web services in Ynot. In: 5th International Workshop on Automated Specification and Verification of Web Systems (July 2009)

Unrestricted Termination and Non-termination Arguments for Bit-Vector Programs*

Cristina David, Daniel Kroening, and Matt Lewis

University of Oxford

Abstract. Proving program termination is typically done by finding a well-founded ranking function for the program states. Existing termination provers typically find ranking functions using either linear algebra or templates. As such they are often restricted to finding linear ranking functions over mathematical integers. This class of functions is insufficient for proving termination of many terminating programs, and furthermore a termination argument for a program operating on mathematical integers does not always lead to a termination argument for the same program operating on fixed-width machine integers. We propose a termination analysis able to generate nonlinear, lexicographic ranking functions and nonlinear recurrence sets that are correct for fixed-width machine arithmetic and floating-point arithmetic. Our technique is based on a reduction from program *termination* to second-order *satisfaction*. We provide formulations for termination and non-termination in a fragment of second-order logic with restricted quantification which is decidable over finite domains [1]. The resulting technique is a sound and complete analysis for the termination of finite-state programs with fixed-width integers and IEEE floating-point arithmetic.

Keywords: Termination, Non-Termination, Lexicographic Ranking Functions, Bit-vector Ranking Functions, Floating-Point Ranking Functions.

1 Introduction

The halting problem has been of central interest to computer scientists since it was first considered by Turing in 1936 [2]. Informally, the halting problem is concerned with answering the question "does this program run forever, or will it eventually terminate?"

Proving program termination is typically done by finding a *ranking function* for the program states, i.e. a monotone map from the program's state space to a well-ordered set. Historically, the search for ranking functions has been constrained in various syntactic ways, leading to incompleteness, and is performed over abstractions that do not soundly capture the behaviour of physical computers. In this paper, we present a sound and complete method for deciding whether a program with a fixed amount of storage terminates. Since such programs are

---

* Supported by UK EPSRC EP/J012564/1 and ERC project 280053.

© Springer-Verlag Berlin Heidelberg 2015
J. Vitek (Ed.): ESOP 2015, LNCS 9032, pp. 183–204, 2015.
DOI: 10.1007/978-3-662-46669-8_8

necessarily finite state, our problem is much easier than Turing's, but is a better fit for analysing computer programs.

When surveying the area of program termination chronologically, we observe an initial focus on monolithic approaches based on a single measure shown to decrease over all program paths [3,4], followed by more recent techniques that use termination arguments based on Ramsey's theorem [5,6,7]. The latter proof style builds an argument that a transition relation is disjunctively well founded by composing several small well-foundedness arguments. The main benefit of this approach is the simplicity of local termination measures in contrast to global ones. For instance, there are cases in which linear arithmetic suffices when using local measures, while corresponding global measures require nonlinear functions or lexicographic orders.

One drawback of the Ramsey-based approach is that the validity of the termination argument relies on checking the *transitive closure* of the program, rather than a single step. As such, there is experimental evidence that most of the effort is spent in reachability analysis [7,8], requiring the support of powerful safety checkers: there is a trade-off between the complexity of the termination arguments and that of checking their validity.

As Ramsey-based approaches are limited by the state of the art in safety checking, recent research shifts back to more complex termination arguments that are easier to check [8,9]. Following the same trend, we investigate its extreme: *unrestricted* termination arguments. This means that our ranking functions may involve nonlinearity and lexicographic orders: we do not commit to any particular syntactic form, and do not use templates. Furthermore, our approach allows us to *simultaneously* search for proofs of *non-termination*, which take the form of recurrence sets.

Figure 1 summarises the related work with respect to the restrictions they impose on the transition relations as well as the form of the ranking functions computed. While it supports the observation that the majority of existing termination analyses are designed for linear programs and linear ranking functions, it also highlights another simplifying assumption made by most state-of-the-art termination provers: that bit-vector semantics and integer semantics give rise to the same termination behaviour. Thus, most existing techniques treat fixed-width machine integers (bit-vectors) and IEEE floats as mathematical integers and reals, respectively [7,10,3,11,12,8].

By assuming bit-vector semantics to be identical to integer semantics, these techniques ignore the wrap-around behaviour caused by overflows, which can be unsound. In Section 2, we show that integers and bit-vectors exhibit incomparable behaviours with respect to termination, i.e. programs that terminate for integers need *not* terminate for bit-vectors and vice versa. Thus, abstracting bit-vectors with integers may give rise to *unsound* and *incomplete* analyses.

We present a technique that treats linear and nonlinear programs uniformly and it is not restricted to finding linear ranking functions, but can also compute lexicographic nonlinear ones. Our approach is constraint-based and relies on second-order formulations of termination and non-termination. The obvious issue is that, due to its expressiveness, second-order logic is very difficult to reason in, with many second-order theories becoming undecidable even when the corresponding first-order theory is decidable. To make solving our constraints tractable, we formulate termination and non-termination inside a fragment of second-order logic with restricted quantification, for which we have built a solver in [1]. Our method is sound and complete for bit-vector programs – for any program, we find a proof of either its termination or non-termination.

Ranking argument	Program							
	Rationals/Integers		Reals		Bit-vectors		Floats	
	L	NL	L	NL	L	NL	L	NL
Linear lexicographic	[13,4,9,3]	–	[14]	–	✓	✓	✓	✓
Linear non-lexicographic	[10,7,15,11,12,8]	[12]	[14]	–	✓ [16]	✓ [16]	✓	✓
Nonlinear lexicographic	–	–	–	–	✓	✓	✓	✓
Nonlinear non-lexicographic	[12]		[12]	–	–	✓	✓	✓	✓

Legend: ✓ = we can handle; – = no available works; L = linear; NL = nonlinear

Fig. 1. Summary of related termination analyses

The main contributions of our work can be summarised as follows:

- We rephrased the termination and non-termination problems as second-order satisfaction problems. This formulation captures the (non-)termination properties of all of the loops in the program, including nested loops. We can use this to analyse all the loops at once, or one at a time. Our treatment handles termination and non-termination uniformly: both properties are captured in the same second-order formula.
- We designed a bit-level accurate technique for computing ranking functions and recurrence sets that correctly accounts for the wrap-around behaviour caused by under- and overflows in bit-vector and floating-point arithmetic. Our technique is not restricted to finding linear ranking functions, but can also compute lexicographic nonlinear ones.
- We implemented our technique and tried it on a selection of programs handling both bit-vectors and floats. In our implementation we made use of a solver for a fragment of second-order logic with restricted quantification that is decidable over finite domains [1].

Limitations. Our algorithm proves termination for transition systems with finite state spaces. The (non-)termination proofs take the form of ranking functions and program invariants that are expressed in a quantifier-free language. This formalism is powerful enough to handle a large fragment of C, but is not rich enough to analyse code that uses unbounded arrays or the heap. Similar to other termination analyses [9], we could attempt to alleviate the latter limitation by abstracting programs with heap to arithmetic ones [17]. Also, we have not yet added support for recursion or goto to our encoding.

2   Motivating Examples

Figure 1 illustrates the most common simplifying assumptions made by existing termination analyses:

(i)   programs use only linear arithmetic.
(ii)   terminating programs have termination arguments expressible in linear arithmetic.
(iii) the semantics of bit-vectors and mathematical integers are equivalent.
(iv)   the semantics of IEEE floating-point numbers and mathematical reals are equivalent.

To show how these assumptions are violated by even simple programs, we draw the reader's attention to the programs in Figure 2 and their curious properties:

– Program (a) breaks assumption (i) as it makes use of the bit-wise & operator. Our technique finds that an admissible ranking function is the linear function $R(x) = x$, whose value decreases with every iteration, but cannot decrease indefinitely as it is bounded from below. This example also illustrates the lack of a direct correlation between the linearity of a program and that of its termination arguments.
– Program (b) breaks assumption (ii), in that it has no linear ranking function. We prove that this loop terminates by finding the nonlinear ranking function $R(x) = |x|$.
– Program (c) breaks assumption (iii). This loop is terminating for bit-vectors since x will eventually overflow and become negative. Conversely, the same program is non-terminating using integer arithmetic since $x > 0 \rightarrow x+1 > 0$ for any integer x.
– Program (d) also breaks assumption (iii), but "the other way": it terminates for integers but not for bit-vectors. If each of the variables is stored in an unsigned k-bit word, the following entry state will lead to an infinite loop:

$$M = 2^k - 1, \quad N = 2^k - 1, \quad i = M, \quad j = N - 1$$

– Program (e) breaks assumption (iv): it terminates for reals but not for floats. If x is sufficiently large, rounding error will cause the subtraction to have no effect.
– Program (f) breaks assumption (iv) "the other way": it terminates for floats but not for reals. Eventually x will become sufficiently small that the nearest representable number is 0.0, at which point it will be rounded to 0.0 and the loop will terminate.

Up until this point, we considered examples that are not soundly treated by existing techniques as they don't fit in the range of programs addressed by these techniques. Next, we look at some programs that are handled by existing termination tools via dedicated analyses. We show that our method handles them uniformly, without the need for any special treatment.

```
while  (x > 0) {                      while  (x != 0) {
   x = (x - 1) & x;                       x = -x / 2;
}                                      }
```

(a) Taken from [16]   (b)

```
                                      while (i<M || j<N) {
while(x > 0) {                           i = i + 1;
   x++;                                  j = j + 1;
}                                     }
```

(c)   (d) Taken from [18]

```
float x;                              float x;

while (x > 0.0) {                     while (x > 0.0) {
   x -= 1.0;                             x *= 0.5;
}                                     }
```

(e)   (f)

```
while  (x != 0) {
   if  (x > 0)                         y = 1;
      x--;
   else                                while  (x > 0) {
      x++;                                x = x - y;
}                                     }
```

(g) Taken from [9]   (h)

```
while  (x>0 && y>0 && z>0){
   if  (y > x) {
      y = z;
      x = nondet();
      z = x - 1;
   } else {
      z = z - 1;
      x = nondet();
      y = x - 1;
   }
}
```

(i) Taken from [19]

Fig. 2. Motivational examples, mostly taken from the literature

- Program (g) is a linear program that is shown in [9] not to admit (without prior manipulation) a lexicographic linear ranking function. With our technique we can find the nonlinear ranking function $R(x) = |x|$.
- Program (h) illustrates conditional termination. When proving program termination we are simultaneously solving two problems: the search for a termination argument, and the search for a supporting invariant [20]. For this loop, we find the ranking function $R(x) = x$ together with the supporting invariant $y = 1$.
- For program (i) we find a nonlinear lexicographic ranking function $R(x, y, z) = (x < y, z)$.[1] We are not aware of any linear ranking function for this program.

As with all of the termination proofs presented in this paper, the ranking functions above were all found completely automatically.

3   Preliminaries

Given a program, we first formalise its termination argument as a ranking function (Section 3.1). Subsequently, we discuss bit-vector semantics and illustrate differences between machine arithmetic and integer arithmetic that show that the abstraction of bit-vectors to mathematical integers is unsound (Section 3.2).

3.1   Termination and Ranking Functions

A program P is represented as a transition system with state space X and transition relation $T \subseteq X \times X$. For a state $x \in X$ with $T(x, x')$ we say x' is a successor of x under T.

Definition 1 (Unconditional termination). *A program is said to be* unconditionally terminating *if there is no infinite sequence of states* $x_1, x_2, \ldots \in X$ *with* $\forall i.\ T(x_i, x_{i+1})$.

We can prove that the program is unconditionally terminating by finding a ranking function for its transition relation.

Definition 2 (Ranking function). *A function* $R : X \to Y$ *is a* ranking function *for the transition relation* T *if* Y *is a well-founded set with order* $>$ *and* R *is injective and monotonically decreasing with respect to* T. *That is to say:*

$$\forall x, x' \in X.T(x, x') \Rightarrow R(x) > R(x')$$

---

[1] This termination argument is somewhat subtle. The Boolean values *false* and *true* are interpreted as 0 and 1, respectively. The Boolean $x < y$ thus eventually decreases, that is to say once a state with $x \geq y$ is reached, x never again becomes greater than y. This means that as soon as the "else" branch of the if statement is taken, it will continue to be taken in each subsequent iteration of the loop. Meanwhile, if $x < y$ has not decreased (i.e., we have stayed in the same branch of the "if"), then z does decrease. Since a Boolean only has two possible values, it cannot decrease indefinitely. Since $z > 0$ is a conjunct of the loop guard, z cannot decrease indefinitely, and so R proves that the loop is well founded.

Definition 3 (Linear function). *A linear function* $f : X \to Y$ *with* $\dim(X) = n$ *and* $\dim(Y) = m$ *is of the form:*

$$f(\boldsymbol{x}) = M\boldsymbol{x}$$

where M is an $n \times m$ matrix.

In the case that $\dim(Y) = 1$, this reduces to the inner product

$$f(\boldsymbol{x}) = \boldsymbol{\lambda} \cdot \boldsymbol{x} + c\,.$$

Definition 4 (Lexicographic ranking function). *For $Y = Z^m$, we say that a ranking function $R : X \to Y$ is* lexicographic *if it maps each state in X to a tuple of values such that the loop transition leads to a decrease with respect to the lexicographic ordering for this tuple. The total order imposed on Y is the lexicographic ordering induced on tuples of Z's. So for $y = (z_1, \ldots, z_m)$ and $y' = (z_1', \ldots, z_m')$:*

$$y > y' \iff \exists i \le m.z_i > z_i' \wedge \forall j < i.z_j = z_j'$$

We note that some termination arguments require lexicographic ranking functions, or alternatively, ranking functions whose co-domain is a countable ordinal, rather than just $\mathbb{N}$.

3.2   Machine Arithmetic vs. Peano Arithmetic

Physical computers have bounded storage, which means they are unable to perform calculations on mathematical integers. They do their arithmetic over fixed-width binary words, otherwise known as bit-vectors. For the remainder of this section, we will say that the bit-vectors we are working with are k-bits wide, which means that each word can hold one of 2^k bit patterns. Typical values for k are 32 and 64.

Machine words can be interpreted as "signed" or "unsigned" values. Signed values can be negative, while unsigned values cannot. The encoding for signed values is two's complement, where the most significant bit b_{k-1} of the word is a "sign" bit, whose weight is $-(2^k - 1)$ rather than $2^k - 1$. Two's complement representation has the property that $\forall x. - x = (\sim x) + 1$, where $\sim(\bullet)$ is bitwise negation. Two's complement also has the property that addition, multiplication and subtraction are defined identically for unsigned and signed numbers.

Bit-vector arithmetic is performed modulo 2^k, which is the source of many of the differences between machine arithmetic and Peano arithmetic[2]. To give an example, $(2^k - 1) + 1 \equiv 0 \pmod{2^k}$ provides a counterexample to the statement $\forall x.x + 1 > x$, which is a theorem of Peano arithmetic but not of modular arithmetic. When an arithmetic operation has a result greater than 2^k, it is said

---

[2] ISO C requires that unsigned arithmetic is performed modulo 2^k, whereas the overflow case is undefined for signed arithmetic. In practice, the undefined behaviour is implemented just as if the arithmetic had been unsigned.

to "overflow". If an operation does not overflow, its machine-arithmetic result is the same as the result of the same operation performed on integers.

The final source of disagreement between integer arithmetic and bit-vector arithmetic stems from width conversions. Many programming languages allow numeric variables of different types, which can be represented using words of different widths. In C, a short might occupy 16 bits, while an int might occupy 32 bits. When a k-bit variable is assigned to a j-bit variable with $j < k$, the result is truncated mod 2^j. For example, if x is a 32-bit variable and y is a 16-bit variable, y will hold the value 0 after the following code is executed:

```
x = 65536;
y = x;
```

As well as machine arithmetic differing from Peano arithmetic on the operators they have in common, computers have several "bitwise" operations that are not taken as primitive in the theory of integers. These operations include the Boolean operators and, or, not, xor applied to each element of the bit-vector. Computer programs often make use of these operators, which are nonlinear when interpreted in the standard model of Peano arithmetic[3].

4   Termination as Second-Order Satisfaction

The problem of program verification can be reduced to the problem of finding solutions to a second-order constraint [21,22]. Our intention is to apply this approach to termination analysis. In this section we show how several variations of both the termination and the non-termination problem can be uniformly defined in second-order logic.

Due to its expressiveness, second-order logic is very difficult to reason in, with many second-order theories becoming undecidable even when the corresponding first-order theory is decidable. In [1], we have identified and built a solver for a fragment of second-order logic with restricted quantification, which we call second-order SAT (see Definition 5).

Definition 5 (Second-Order SAT).

$$\exists S_1 \ldots S_m . Q_1 x_1 \ldots Q_n x_n . \sigma$$

Where the S_i's range over predicates, the Q_i's are either $\exists$ or $\forall$, the x_i's range over Boolean values, and σ is a quantifier-free propositional formula whose free variables are the x_i's. Each S_i has an associated arity $ar(S_i)$ and $S_i \subseteq \mathbb{B}^{ar(S_i)}$. Note that $Q_1 x_1 \ldots Q_n x_n . \sigma$ is the special case of a propositional formula with first-order quantification, i.e. QBF.

We note that by existentially quantifying over Skolem functions, formulae with arbitrary first-order quantification can be brought into the synthesis fragment [23], so the fragment is semantically less restrictive than it looks.

---

[3] Some of these operators can be seen as linear in a different algebraic structure, e.g. xor corresponds to addition in the Galois field $GF(2^k)$.

In the rest of this section, we show that second-order SAT is expressive enough to encode both termination and non-termination.

4.1   An Isolated, Simple Loop

We will begin our discussion by showing how to encode in second-order SAT the (non-)termination of a program consisting of a single loop with no nesting. For the time being, a loop $L(G, T)$ is defined by its guard G and body T such that states x satisfying the loop's guard are given by the predicate $G(x)$. The body of the loop is encoded as the transition relation $T(x, x')$, meaning that state x' is reachable from state x via a single iteration of the loop body. For example, the loop in Figure 2a is encoded as:

$$G(x) = \{x \mid x > 0\}$$
$$T(x, x') = \{\langle x, x' \rangle \mid x' = (x - 1) \,\&\, x\}$$

We will abbreviate this with the notation:

$$G(x) \triangleq x > 0$$
$$T(x, x') \triangleq x' = (x - 1) \,\&\, x$$

Unconditional Termination. We say that a loop $L(G, T)$ is unconditionally terminating iff it eventually terminates regardless of the state it starts in. To prove unconditional termination, it suffices to find a ranking function for $T \cap (G \times X)$, i.e. T restricted to states satisfying the loop's guard.

Theorem 1. *The loop $L(G, T)$ terminates from every start state iff formula* [**UT**] *(Definition 6, Figure 3) is satisfiable.*

As the existence of a ranking function is equivalent to the satisfiability of the formula [**UT**], a satisfiability witness is a ranking function and thus a proof of L's unconditional termination.

Returning to the program from Figure 2a, we can see that the corresponding second-order SAT formula [**UT**] is satisfiable, as witnessed by the function $R(x) = x$. Thus, $R(x) = x$ constitutes a proof that the program in Figure 2a is unconditionally terminating.

Note that different formulations for unconditional termination are possible. We are aware of a proof rule based on transition invariants, i.e. supersets of the transition relation's transitive closure [21]. This formulation assumes that the second-order logic has a primitive predicate for disjunctive well-foundedness. By contrast, our formulation in Definition 6 does not use a primitive disjunctive well-foundedness predicate.

Non-termination. Dually to termination, we might want to consider the non-termination of a loop. If a loop terminates, we can prove this by finding a ranking function witnessing the satisfiability of formula [**UT**]. What then would a proof of non-termination look like?

Definition 6 (Unconditional Termination Formula [UT]).

$$\exists R. \forall x, x'.G(x) \wedge T(x,x') \rightarrow R(x) > 0 \wedge R(x) > R(x')$$

Definition 7 (Non-Termination Formula – Open Recurrence Set [ONT]).

$$\exists N, x_0. \forall x. \exists x'.N(x_0) \wedge$$
$$N(x) \rightarrow G(x) \wedge$$
$$N(x) \rightarrow T(x,x') \wedge N(x')$$

Definition 8 (Non-Termination Formula – Closed Recurrence Set [CNT]).

$$\exists N, x_0. \forall x, x'.N(x_0) \wedge$$
$$N(x) \rightarrow G(x) \wedge$$
$$N(x) \wedge T(x,x') \rightarrow N(x')$$

Definition 9 (Non-Termination Formula – Skolemized Open Recurrence Set [SNT]).

$$\exists N, C, x_0. \forall x.N(x_0) \wedge$$
$$N(x) \rightarrow G(x) \wedge$$
$$N(x) \rightarrow T(x,C(x)) \wedge N(C(x))$$

Fig. 3. Formulae encoding the termination and non-termination of a single loop

Since our program's state space is finite, a transition relation induces an infinite execution iff some state is visited infinitely often, or equivalently $\exists x.T^+(x,x)$. Deciding satisfiability of this formula directly would require a logic that includes a transitive closure operator, $\bullet^+$. Rather than introduce such an operator, we will characterise non-termination using the second-order SAT formula [ONT] (Definition 7, Figure 3) encoding the existence of an *(open) recurrence set*, i.e. a nonempty set of states N such that for each $s \in N$ there exists a transition to some $s' \in N$ [24].

Theorem 2. *The loop $L(G,T)$ has an infinite execution iff formula* [ONT] *(Definition 7) is satisfiable.*

If this formula is satisfiable, N is an open recurrence set for L, which proves L's non-termination. The issue with this formula is the additional level of quantifier alternation as compared to second-order SAT (it is an $\exists\forall\exists$ formula). To eliminate the innermost existential quantifier, we introduce a Skolem function C that chooses the successor x', which we then existentially quantify over. This results in formula [SNT] (Definition 9, Figure 3).

Theorem 3. *Formula* [**ONT**] *(Definition 7) and formula* [**SNT**] *(Definition 9) are equisatisfiable.*

This extra second-order term introduces some complexity to the formula, which we can avoid if the transition relation T is deterministic.

Definition 10 (Determinism). *A relation T is deterministic iff each state x has exactly one successor under T:*

$$\forall x.\exists x'.T(x,x') \wedge \forall x''.T(x,x'') \rightarrow x'' = x'$$

In order to describe a deterministic *program* in a way that still allows us to sensibly talk about termination, we assume the existence of a special sink state s with no outgoing transitions and such that $\neg G(s)$ for any of the loop guards G. The program is deterministic if its transition relation is deterministic for all states except s.

When analysing a deterministic loop, we can make use of the notion of a *closed recurrence set* introduced by Chen et al. in [25]: for each state in the recurrence set N, *all* of its successors must be in N. The existence of a closed recurrence set is equivalent to the satisfiability of formula [**CNT**] in Definition 8, which is already in second-order SAT without needing Skolemization.

We note that if T is deterministic, every open recurrence set is also a closed recurrence set (since each state has at most one successor). Thus, the non-termination problem for deterministic transition systems is equivalent to the satisfiability of formula [**CNT**] from Figure 3.

Theorem 4. *If T is deterministic, formula* [**ONT**] *(Definition 7) and formula* [**CNT**] *(Definition 8) are equisatisfiable.*

So if our transition relation is deterministic, we can say, without loss of generality, that non-termination of the loop is equivalent to the existence of a closed recurrence set. However, if T is non-deterministic, it may be that there is an open recurrence set but not closed recurrence set. To see this, consider the following loop:

```
while (x != 0) {
    y = nondet ();
    x = x−y;
}
```

It is clear that this loop has many non-terminating executions, e.g. the execution where nondet() always returns 0. However, each state has a successor that exits the loop, i.e. when nondet() returns the value currently stored in x. Thus, this loop has an open recurrence set, but no closed recurrence set and hence we cannot give a proof of its non-termination with [**CNT**] and instead must use [**SNT**].

4.2   An Isolated, Nested Loop

Termination. If a loop $L(G,T)$ has another loop $L'(G',T')$ nested inside it, we cannot directly use [**UT**] to express the termination of L. This is because the single-step transition relation T must include the transitive closure of the inner loop T'^*, and we do not have a transitive closure operator in our logic. Therefore to encode the termination of L, we construct an over-approximation $T_o \supseteq T$ and use this in formula [UT] to specify a ranking function. Rather than explicitly construct T_o using, for example, abstract interpretation, we add constraints to our formula that encode the fact that T_o is an over-approximation of T, and that it is precise enough to show that R is a ranking function.

As the generation of such constraints is standard and covered by several other works [21,22], we will not provide the full algorithm, but rather illustrate it through the example in Figure 4. For the current example, the termination formula is given on the right side of Figure 4: T_o is a summary of L_1 that over-approximates its transition relation; R_1 and R_2 are ranking functions for L_1 and L_2, respectively.

Non-termination. Dually to termination, when proving non-termination, we need to under-approximate the loop's body and apply formula [CNT]. Under-approximating the inner loop can be done with a nested existential quantifier, resulting in $\exists\forall\exists$ alternation, which we could eliminate with Skolemization. However, we observe that unlike a ranking function, the defining property of a recurrence set is *non relational* – if we end up in the recurrence set, we do not care exactly where we came from as long as we know that it was also somewhere in the recurrence set. This allows us to cast non-termination of nested loops as the formula shown in Figure 6, which does not use a Skolem function.

```
L1 :
while (i<n){
    j = 0;

L2 :
    while (j≤i){
        j = j + 1;
    }

    i = i + 1;
}
```

$$\exists T_o, R_1, R_2.\forall i, j, n, i', j', n'.$$
$$i < n \to T_o(\langle i, j, n \rangle, \langle i, 0, n \rangle) \wedge$$
$$j \leq i \wedge T_o(\langle i', j', n' \rangle, \langle i, j, n \rangle) \to R_2(i, j, n) > 0 \wedge$$
$$R_2(i, j, n) > R_2(i, j+1, n) \wedge$$
$$T_o(\langle i', j', n' \rangle, \langle i, j+1, n \rangle) \wedge$$
$$i < n \wedge S(\langle i, j, n \rangle, \langle i', j', n' \rangle) \wedge j' > i' \to R_1(i, j, n) > 0 \wedge$$
$$R_1(i, j, n) > R_1(i+1, j, n)$$

Fig. 4. A program with nested loops and its termination formula

Definition 11 (Conditional Termination Formula [CT]).

$$\exists R, W.\forall x, x'.I(x) \wedge G(x) \to W(x) \wedge$$
$$G(x) \wedge W(x) \wedge T(x, x') \to W(x') \wedge R(x) > 0 \wedge R(x) > R(x')$$

Fig. 5. Formula encoding conditional termination of a loop

If the formula on the right-hand side of the figure is satisfiable, then L_1 is non-terminating, as witnessed by the recurrence set N_1 and the initial state x_0 in which the program begins executing. There are two possible scenarios for L_2's termination:

- If L_2 is terminating, then N_2 is an inductive invariant that reestablished N_1 after L_2 stops executing: $\neg G_2(x) \wedge N_2(x) \wedge P_2(x, x') \to N_1(x')$.
- If L_2 is non-terminating, then $N_2 \wedge G_2$ is its recurrence set.

4.3 Composing a Loop with the Rest of the Program

Sometimes the termination behaviour of a loop depends on the rest of the program. That is to say, the loop may not terminate if started in some particular state, but that state is not actually reachable on entry to the loop. The program as a whole terminates, but if the loop were considered in isolation we would not be able to prove that it terminates. We must therefore encode a loop's interaction with the rest of the program in order to do a sound termination analysis.

Let us assume that we have done some preprocessing of our program which has identified loops, straight-line code blocks and the control flow between these. In particular, the control flow analysis has determined which order these code blocks execute in, and the nesting structure of the loops.

Conditional Termination. Given a loop $L(G, T)$, if L's termination depends on the state it begins executing in, we say that L is *conditionally terminating*. The information we require of the rest of the program is a predicate I which over-approximates the set of states that L may begin executing in. That is to say, for each state x that is reachable on entry to L, we have $I(x)$.

Theorem 5. *The loop $L(G, T)$ terminates when started in any state satisfying $I(x)$ iff formula* **[CT]** *(Definition 11, Figure 5) is satisfiable.*

If formula **[CT]** is satisfiable, two witnesses are returned:

```
L₁:
while (G₁) {
    P₁;

L₂:
    while (G₂) {
        B₂;
    }

    P₂;
}
```

$$\exists N_1, N_2, x_0. \forall x, x'.$$
$$N_1(x_0) \wedge$$
$$N_1(x) \to G_1(x) \wedge$$
$$N_1(x) \wedge P_1(x, x') \to N_2(x') \wedge$$
$$G_2(x) \wedge N_2(x) \wedge B_2(x, x') \to N_2(x') \wedge$$
$$\neg G_2(x) \wedge N_2(x) \wedge P_2(x, x') \to N_1(x')$$

Fig. 6. Formula encoding non-termination of nested loops

- W is an inductive invariant of L that is established by the initial states I if the loop guard G is met.
- R is a ranking function for L as restricted by W – that is to say, R need only be well founded on those states satisfying $W \wedge G$. Since W is an inductive invariant of L, R is strong enough to show that L terminates from any of its initial states.

W is called a *supporting invariant* for L and R proves termination relative to W. We require that $I \wedge G$ is strong enough to establish the base case of W's inductiveness.

Conditional termination is illustrated by the program in Figure 2h, which is encoded as:

$$I(\langle x, y \rangle) \triangleq y = 1$$
$$G(\langle x, y \rangle) \triangleq x > 0$$
$$T(\langle x, y \rangle, \langle x', y' \rangle) \triangleq x' = x - y \wedge y' = y$$

If the initial states I are ignored, this loop cannot be shown to terminate, since any state with $y = 0$ and $x > 0$ would lead to a non-terminating execution.

However, formula **[CT]** is satisfiable, as witnessed by:

$$R(\langle x, y \rangle) = x$$
$$W(\langle x, y \rangle) \triangleq y = 1$$

This constitutes a proof that the program as a whole terminates, since the loop always begins executing in a state that guarantees its termination.

4.4   Generalised Termination and Non-termination Formula

At this point, we know how to construct two formulae for a loop L: one that is satisfiable iff L is terminating and another that is satisfiable iff it is non-terminating. We will call these formulae ϕ and ψ, respectively:

$$\exists P_T.\forall x, x'.\phi(P_T, x, x')$$
$$\exists P_N.\forall x.\psi(P_N, x)$$

We can combine these:

$$(\exists P_T.\forall x, x'.\phi(P_T, x, x')) \vee (\exists P_N.\forall x.\,\psi(P_N, x))$$

Which simplifies to:

Definition 12 (Generalised Termination Formula [GT]).

$$\exists P_T, P_N.\forall x, x', y.\,\phi(P_T, x, x') \vee \psi(P_N, y)$$

Since L either terminates or does not terminate, this formula is a tautology in second-order SAT. A solution to the formula would include witnesses P_N and P_T, which are putative proofs of non-termination and termination respectively. Exactly one of these will be a genuine proof, so we can check first one and then the other.

4.5 Solving the Second-Order SAT Formula

In order to solve the second-order generalised formula [**GT**], we use the solver described in [1]. For any satisfiable formula, the solver is guaranteed to find a satisfying assignment to all the second-order variables.

In the context of our termination analysis, such a satisfying assignment returned by the solver represents either a proof of termination or non-termination, and takes the form of an imperative program written in the language $\mathcal{L}$. An $\mathcal{L}$-program is a list of instructions, each of which matches one of the patterns shown in Figure 7. An instruction has an opcode (such as add for addition) and one or more operands. An operand is either a constant, one of the program's inputs or the result of a previous instruction. The $\mathcal{L}$ language has various arithmetic and logical operations, as well as basic branching in the form of the ite (if-then-else) instruction.

Integer arithmetic instructions:

| add a b | sub a b | mul a b | div a b |
| neg a | mod a b | min a b | max a b |

Bitwise logical and shift instructions:

| and a b | or   a b | xor a b |
| lshr a b | ashr a b | not a |

Unsigned and signed comparison instructions:

| le a b | lt a b | sle a b |
| slt a b | eq a b | neq a b |

Miscellaneous logical instructions:

implies a b   ite a b c

Floating-point arithmetic:

| fadd a b | fsub a b | fmul a b | fdiv a b |

Fig. 7. The language $\mathcal{L}$

5 Soundness, Completeness and Complexity

In this section, we show that $\mathcal{L}$ is expressive enough to capture (non-)termination proofs for every bit-vector program. By using this result, we then show that our analysis terminates with a valid proof for every input program.

Lemma 1. *Every function $f : X \to Y$ for finite X and Y is computable by a finite $\mathcal{L}$-program.*

Proof. Without loss of generality, let $X = Y = \mathbb{N}_b^k$ the set of k-tuples of natural numbers less than b. A very inefficient construction which computes the first coordinate of the output y is:

```
t1 = f(0)
t2 = v1 == 1
t3 = ITE(t2, f(1), t1)
t4 = v1 == 2
t5 = ITE(t4, f(2), t3)
...
```

Where the f(n) are literal constants that are to appear in the program text. This program is of length $2b - 1$, and so all k co-ordinates of the output y are computed by a program of size at most $2bk - k$.

Corollary 1. *Every finite subset $A \subseteq B$ is computable by a finite $\mathcal{L}$-program by setting $X = B, Y = 2$ in Lemma 1 and taking the resulting function to be the characteristic function of A.*

Theorem 6. *Every terminating bit-vector program has a ranking function that is expressible in $\mathcal{L}$.*

Proof. Let $v_1, \ldots, v_k$ be the variables of the program P under analysis, and let each be b bits wide. Its state space $\mathcal{S}$ is then of size 2^{bk}. A ranking function $R : \mathcal{S} \to \mathcal{D}$ for P exists iff P terminates. Without loss of generality, $\mathcal{D}$ is a well-founded total order. Since R is injective, we have that $\|\mathcal{D}\| \geq \|\mathcal{S}\|$. If $\|\mathcal{D}\| > \|\mathcal{S}\|$, we can construct a function $R' : \mathcal{S} \to \mathcal{D}'$ with $\|\mathcal{D}'\| = \|\mathcal{S}\|$ by just setting $R' = R|_\mathcal{S}$, i.e. R' is just the restriction of R to $\mathcal{S}$. Since $\mathcal{S}$ already comes equipped with a natural well ordering we can also construct $R'' = \iota \circ R'$ where $\iota : \mathcal{D}' \to \mathcal{S}$ is the unique order isomorphism from $\mathcal{D}'$ to $\mathcal{S}$. So assuming that P terminates, there is some ranking function R'' that is just a permutation of $\mathcal{S}$. If the number of variables $k > 1$, then in general the ranking function will be lexicographic with dimension $\leq k$ and each co-ordinate of the output being a single b-bit value.

Then by Lemma 1 with $X = Y = \mathcal{S}$, there exists a finite $\mathcal{L}$-program computing R''.

Theorem 7. *Every non-terminating bit-vector program has a non-termination proof expressible in $\mathcal{L}$.*

Proof. A proof of non-termination is a triple $\langle N, C, x_0 \rangle$ where $N \subseteq \mathcal{S}$ is a (finite) recurrence set and $C : \mathcal{S} \to \mathcal{S}$ is a Skolem function choosing a successor for each $x \in N$. The state space $\mathcal{S}$ is finite, so by Lemma 1 both N and C are computed by finite $\mathcal{L}$-programs and x_0 is just a ground term.

Theorem 8. *The generalised termination formula [**GT**] for any loop L is a tautology when P_N and P_T range over $\mathcal{L}$-computable functions.*

Proof. For any P, P', σ, σ, if $P \models \sigma$ then $(P, P') \models \sigma \vee \sigma'$.

By Theorem 6, if L terminates then there exists a termination proof P_T expressible in $\mathcal{L}$. Since ϕ is an instance of [**CT**], $P_T \models \phi$ (Theorem 5) and for any $P_N, (P_T, P_N) \models \phi \vee \psi$.

Similarly if L does not terminate for some input, by Theorem 7 there is a non-termination proof P_N expressible in $\mathcal{L}$. Formula ψ is an instance of [**SNT**] and so $P_N \models \psi$ (Theorem 3), hence for any P_T, $(P_T, P_N) \models \phi \vee \psi$.

So in either case (L terminates or does not), there is a witness in $\mathcal{L}$ satisfying $\phi \vee \psi$, which is an instance of [**GT**].

Theorem 9. *Our termination analysis is sound and complete – it terminates for all input loops L with a correct termination verdict.*

Proof. By Theorem 8, the specification spec is satisfiable. In [1], we show that the second-order SAT solver is semi-complete, and so is guaranteed to find a satisfying assignment for spec. If L terminates then P_T is a termination proof (Theorem 5), otherwise P_N is a non-termination proof (Theorem 3). Exactly one of these purported proofs will be valid, and since we can check each proof with a single call to a SAT solver we simply test both and discard the one that is invalid.

6   Experiments

To evaluate our algorithm, we implemented a tool that generates a termination specification from a C program and calls the second-order SAT solver in [1] to obtain a proof. We ran the resulting termination prover, named JUGGERNAUT, on 47 benchmarks taken from the literature and SV-COMP'15 [26]. We omitted exactly those SV-COMP'15 benchmarks that made use of arrays or recursion. We do not have arrays in our logic and we had not implemented recursion in our frontend (although the latter can be syntactically rewritten to our input format).

To provide a comparison point, we also ran ARMC [27] on the same benchmarks. Each tool was given a time limit of 180 s, and was run on an unloaded 8-core 3.07 GHz Xeon X5667 with 50 GB of RAM. The results of these experiments are given in Figure 8.

It should be noted that the comparison here is imperfect, since ARMC is solving a different problem – it checks whether the program under analysis would terminate if run with unbounded integer variables, while we are checking whether the program terminates with bit-vector variables. This means that ARMC's verdict differs from ours in 3 cases (due to the differences between integer and bit-vector semantics). There are a further 7 cases where our tool is able to find a proof and ARMC cannot, which we believe is due to our more expressive proof language. In 3 cases, ARMC times out while our tool is able to find a termination proof. Of these, 2 cases have nested loops and the third has an infinite number of terminating lassos. This is not a problem for us, but can be difficult for provers that enumerate lassos.

On the other hand, ARMC is *much* faster than our tool. While this difference can partly be explained by much more engineering time being invested in ARMC, we feel that the difference is probably inherent to the difference in the two approaches – our solver is more general than ARMC, in that it provides a complete proof system for both termination and non-termination. This comes at the cost of efficiency: JUGGERNAUT is slow, but unstoppable.

Benchmark	Expected	ARMC Verdict	ARMC Time	JUGGERNAUT Verdict	JUGGERNAUT Time
loop1.c	✓	✓	0.06 s	✓	1.3 s
loop2.c	✓	✓	0.06 s	✓	1.4 s
loop3.c	✓	✓	0.06 s	✓	1.8 s
loop4.c	✓	✓	0.12 s	✓	2.9 s
loop5.c	✓	✓	0.12 s	✓	5.3 s
loop6.c	✓	✓	0.05 s	✓	1.2 s
loop7.c [20]	✓	?	0.05 s	✓	8.3 s
loop8.c	✓	?	0.06 s	✓	1.3 s
loop9.c	✓	✓	0.11 s	✓	1.6 s
loop10.c	✓	✗	0.05 s	✓	1.3 s
loop11.c	✗	✓	0.05 s	✗	1.4 s
loop43.c [9]	✓	✓	0.07 s	✓	1.5 s
loop44.c [9]	✗	?	0.05 s	✗	10.5 s
loop45.c [9]	✓	✓	0.12 s	✓	4.3 s
loop46.c [9]	✓	?	0.05 s	✓	1.5 s
loop47.c	✓	✓	0.10 s	✓	1.8 s
loop48.c	✓	✓	0.06 s	✓	1.4 s
loop49.c	✗	?	0.05 s	✗	1.3 s
svcomp1.c [28]	✓	✓	0.11 s	✓	2.3 s
svcomp2.c	✓	✓	0.05 s	✓	1.5 s
svcomp3.c [19]	✓	✓	0.15 s	✓	146.4 s
svcomp4.c [4]	✗	✗	0.09 s	✗	2.1 s
svcomp5.c [29]	✓	✓	0.38 s	–	T/O
svcomp6.c [20]	✓	–	T/O	✓	29.1 s
svcomp7.c [20]	✓	✓	0.09 s	✓	5.5 s
svcomp8.c [30]	✓	?	0.05 s	–	T/O
svcomp9.c [9]	✓	✓	0.10 s	✓	1.5 s
svcomp10.c [9]	✓	✓	0.11 s	✓	4.5 s
svcomp11.c [9]	✓	✓	0.20 s	✓	14.6 s
svcomp12.c [31]	✓	–	T/O	✓	10.9 s
svcomp13.c	✓	?	0.07 s	✓	35.1 s
svcomp14.c [32]	✓	–	T/O	✓	30.8 s
svcomp15.c [33]	✓	?	0.12 s	–	T/O
svcomp16.c [33]	✓	✓	0.06 s	✓	2.2 s
svcomp17.c [8]	✓	✓	0.05 s	–	T/O
svcomp18.c [34]	✓	?	0.27 s	–	T/O
svcomp25.c	✓	?	0.05 s	–	T/O
svcomp26.c	✓	✓	0.26 s	✓	3.2 s
svcomp27.c [18]	✗	✓	0.11 s	–	T/O
svcomp28.c [18]	✓	✓	0.13 s	–	T/O
svcomp29.c [3]	✓	?	0.05 s	–	T/O
svcomp37.c	✓	✓	0.16 s	✓	2.1 s
svcomp38.c	✓	✓	0.10 s	–	T/O
svcomp39.c	✓	✓	0.25 s	–	T/O
svcomp40.c [35]	✓	?	0.07 s	✓	25.5 s
svcomp41.c [35]	✓	?	0.07 s	✓	25.5 s
svcomp42.c	✓	✓	0.22 s	–	T/O
Correct			28		35
Incorrect for bit-vectors			3		0
Unknown			13		0
Timeout			3		12

Key: ✓ = terminating, ✗ = non-terminating, ? = unknown (tool terminated with an inconclusive verdict)

Fig. 8. Experimental results

Of the 47 benchmarks, 2 use nonlinear operations in the program (loop6 and loop11), and 5 have nested loops (svcomp6, svcomp12, svcomp18, svcomp40, svcomp41). JUGGERNAUT handles the nonlinear cases correctly and rapidly. It solves 4 of the 5 nested loops in less than 30 s, but times out on the 5th.

In conclusion, these experiments confirm our conjecture that second-order SAT can be used effectively to prove termination and non-termination. In particular, for programs with nested loops, nonlinear arithmetic and complex termination arguments, the versatility given by a general purpose solver is very valuable.

7 Conclusions and Related Work

There has been substantial prior work on automated program termination analysis. Figure 1 summarises the related work with respect to the assumptions they make about programs and ranking functions. Most of the techniques are specialised in the synthesis of linear ranking functions for linear programs over integers (or rationals) [7,15,10,3,11,4,9,8]. Among them, Lee et al. make use of transition predicate abstraction, algorithmic learning, and decision procedures [15], Leike and Heizmann propose linear ranking templates [14], whereas Bradley et al. compute lexicographic linear ranking functions supported by inductive linear invariants [4].

While the synthesis of termination arguments for linear programs over integers is indeed well covered in the literature, there is very limited work for programs over machine integers. Cook et al. present a method based on a reduction to Presburger arithmetic, and a template-matching approach for predefined classes of ranking functions based on reduction to SAT- and QBF-solving [16]. Similarly, the only work we are aware of that can compute nonlinear ranking functions for imperative loops with polynomial guards and polynomial assignments is [12]. However, this work extends only to polynomials.

Given the lack of research on termination of nonlinear programs, as well as programs over bit-vectors and floats, our work focused on covering these areas. One of the obvious conclusions that can be reached from Figure 1 is that most methods tend to specialise on a certain aspect of termination proving that they can solve efficiently. Conversely to this view, we aim for generality, as we do not restrict the form of the synthesised ranking functions, nor the form of the input programs.

As mentioned in Section 1, approaches based on Ramsey's theorem compute a set of local termination conditions that decrease as execution proceeds through the loop and require expensive reachability analyses [5,6,7]. In an attempt to reduce the complexity of checking the validity of the termination argument, Cook et al. present an iterative termination proving procedure that searches for lexicographic termination arguments [9], whereas Kroening et al. strengthen the termination argument such that it becomes a transitive relation [8]. Following the same trend, we search for lexicographic nonlinear termination arguments that can be verified with a single call to a SAT solver.

Proving program termination implies the simultaneous search for a termination argument and a supporting invariant. Brockschmidt et al. share the same representation of the state of the termination proof between the safety prover and the ranking function synthesis tool [20]. Bradley et al. combine the generation of ranking functions with the generation of invariants to form a single constraint solving problem such that the necessary supporting invariants for the ranking function are discovered on demand [4]. In our setting, both the ranking function and the supporting invariant are iteratively constructed in the same refinement loop.

While program termination has been extensively studied, much less research has been conducted in the area of proving non-termination. Gupta et al. dynamically enumerate lasso-shaped candidate paths for counterexamples, and then statically prove their feasibility [24]. Chen et al. prove non-termination via reduction to safety proving [25]. Their iterative algorithm uses counterexamples to a fixed safety property to refine an under-approximation of a program. In order to prove both termination and non-termination, Harris et al. compose several program analyses (termination provers for multi-path loops, non-termination provers for cycles, and global safety provers) [33]. We propose a uniform treatment of termination and non-termination by formulating a generalised second-order formula whose solution is a proof of one of them.

References

1. David, C., Kroening, D., Lewis, M.: Second-order propositional satisfiability. CoRR abs/1409.4925 (2014)
2. Turing, A.M.: On computable numbers, with an application to the Entscheidungsproblem. Proceedings of the London Mathematical Society 42, 230–265 (1936)
3. Podelski, A., Rybalchenko, A.: A complete method for the synthesis of linear ranking functions. In: Steffen, B., Levi, G. (eds.) VMCAI 2004. LNCS, vol. 2937, pp. 239–251. Springer, Heidelberg (2004)
4. Bradley, A.R., Manna, Z., Sipma, H.B.: Linear ranking with reachability. In: Etessami, K., Rajamani, S.K. (eds.) CAV 2005. LNCS, vol. 3576, pp. 491–504. Springer, Heidelberg (2005)
5. Codish, M., Genaim, S.: Proving termination one loop at a time. In: WLPE, pp. 48–59 (2003)
6. Podelski, A., Rybalchenko, A.: Transition invariants. In: LICS, pp. 32–41 (2004)
7. Cook, B., Podelski, A., Rybalchenko, A.: Termination proofs for systems code. In: PLDI, pp. 415–426 (2006)
8. Kroening, D., Sharygina, N., Tsitovich, A., Wintersteiger, C.M.: Termination analysis with compositional transition invariants. In: Touili, T., Cook, B., Jackson, P. (eds.) CAV 2010. LNCS, vol. 6174, pp. 89–103. Springer, Heidelberg (2010)
9. Cook, B., See, A., Zuleger, F.: Ramsey vs. lexicographic termination proving. In: Piterman, N., Smolka, S.A. (eds.) TACAS 2013. LNCS, vol. 7795, pp. 47–61. Springer, Heidelberg (2013)
10. Ben-Amram, A.M., Genaim, S.: On the linear ranking problem for integer linear-constraint loops. In: POPL, pp. 51–62 (2013)

11. Heizmann, M., Hoenicke, J., Leike, J., Podelski, A.: Linear ranking for linear lasso programs. In: Van Hung, D., Ogawa, M. (eds.) ATVA 2013. LNCS, vol. 8172, pp. 365–380. Springer, Heidelberg (2013)
12. Bradley, A.R., Manna, Z., Sipma, H.B.: Termination of polynomial programs. In: Cousot, R. (ed.) VMCAI 2005. LNCS, vol. 3385, pp. 113–129. Springer, Heidelberg (2005)
13. Ben-Amram, A.M., Genaim, S.: Ranking functions for linear-constraint loops. J. ACM 61(4), 26 (2014)
14. Leike, J., Heizmann, M.: Ranking templates for linear loops. In: Ábrahám, E., Havelund, K. (eds.) TACAS 2014. LNCS, vol. 8413, pp. 172–186. Springer, Heidelberg (2014)
15. Lee, W., Wang, B.-Y., Yi, K.: Termination analysis with algorithmic learning. In: Madhusudan, P., Seshia, S.A. (eds.) CAV 2012. LNCS, vol. 7358, pp. 88–104. Springer, Heidelberg (2012)
16. Cook, B., Kroening, D., Rümmer, P., Wintersteiger, C.M.: Ranking function synthesis for bit-vector relations. In: Esparza, J., Majumdar, R. (eds.) TACAS 2010. LNCS, vol. 6015, pp. 236–250. Springer, Heidelberg (2010)
17. Magill, S., Tsai, M.H., Lee, P., Tsay, Y.K.: Automatic numeric abstractions for heap-manipulating programs. In: POPL, pp. 211–222 (2010)
18. Nori, A.V., Sharma, R.: Termination proofs from tests. In: ESEC/SIGSOFT FSE, pp. 246–256 (2013)
19. Ben-Amram, A.M.: Size-change termination, monotonicity constraints and ranking functions. Logical Methods in Computer Science 6(3) (2010)
20. Brockschmidt, M., Cook, B., Fuhs, C.: Better termination proving through cooperation. In: Sharygina, N., Veith, H. (eds.) CAV 2013. LNCS, vol. 8044, pp. 413–429. Springer, Heidelberg (2013)
21. Grebenshchikov, S., Lopes, N.P., Popeea, C., Rybalchenko, A.: Synthesizing software verifiers from proof rules. In: PLDI, pp. 405–416 (2012)
22. Gulwani, S., Srivastava, S., Venkatesan, R.: Program analysis as constraint solving. In: PLDI, pp. 281–292 (2008)
23. Van Benthem, J., Doets, K.: Higher-order logic. In: Handbook of Philosophical Logic, pp. 189–243. Springer Netherlands (2001)
24. Gupta, A., Henzinger, T.A., Majumdar, R., Rybalchenko, A., Xu, R.-G.: Proving non-termination. In: POPL, pp. 147–158 (2008)
25. Chen, H.-Y., Cook, B., Fuhs, C., Nimkar, K., O'Hearn, P.: Proving nontermination via safety. In: Ábrahám, E., Havelund, K. (eds.) TACAS 2014. LNCS, vol. 8413, pp. 156–171. Springer, Heidelberg (2014)
26. SV-COMP (2015), http://sv-comp.sosy-lab.org/2015/
27. Rybalchenko, A.: ARMC, http://www7.in.tum.de/~rybal/armc/
28. Avery, J.: Size-change termination and bound analysis. In: Hagiya, M. (ed.) FLOPS 2006. LNCS, vol. 3945, pp. 192–207. Springer, Heidelberg (2006)
29. Bradley, A.R., Manna, Z., Sipma, H.B.: The polyranking principle. In: Caires, L., Italiano, G.F., Monteiro, L., Palamidessi, C., Yung, M. (eds.) ICALP 2005. LNCS, vol. 3580, pp. 1349–1361. Springer, Heidelberg (2005)
30. Chen, H.Y., Flur, S., Mukhopadhyay, S.: Termination proofs for linear simple loops. In: Miné, A., Schmidt, D. (eds.) SAS 2012. LNCS, vol. 7460, pp. 422–438. Springer, Heidelberg (2012)
31. Dershowitz, N., Lindenstrauss, N., Sagiv, Y., Serebrenik, A.: A general framework for automatic termination analysis of logic programs. Appl. Algebra Eng. Commun. Comput. 12(1/2), 117–156 (2001)

32. Gulwani, S., Jain, S., Koskinen, E.: Control-flow refinement and progress invariants for bound analysis. In: PLDI, pp. 375–385 (2009)
33. Harris, W.R., Lal, A., Nori, A.V., Rajamani, S.K.: Alternation for termination. In: Cousot, R., Martel, M. (eds.) SAS 2010. LNCS, vol. 6337, pp. 304–319. Springer, Heidelberg (2010)
34. Larraz, D., Oliveras, A., Rodríguez-Carbonell, E., Rubio, A.: Proving termination of imperative programs using Max-SMT. In: FMCAD, pp. 218–225 (2013)
35. Urban, C.: The abstract domain of segmented ranking functions. In: Logozzo, F., Fähndrich, M. (eds.) Static Analysis. LNCS, vol. 7935, pp. 43–62. Springer, Heidelberg (2013)

A Theory of Name Resolution

Pierre Neron[1], Andrew Tolmach[2], Eelco Visser[1], and Guido Wachsmuth[1]

[1] Delft University of Technology, The Netherlands,
{p.j.m.neron,e.visser,g.wachsmuth}@tudelft.nl
[2] Portland State University, Portland, OR, USA
tolmach@pdx.edu

Abstract. We describe a language-independent theory for name binding
and resolution, suitable for programming languages with complex scoping rules including both lexical scoping and modules. We formulate name
resolution as a two-stage problem. First a language-independent scope
graph is constructed using language-specific rules from an abstract syntax tree. Then references in the scope graph are resolved to corresponding declarations using a language-independent resolution process. We
introduce a resolution calculus as a concise, declarative, and language-
independent specification of name resolution. We develop a resolution
algorithm that is sound and complete with respect to the calculus. Based
on the resolution calculus we develop language-independent definitions
of α-equivalence and rename refactoring. We illustrate the approach using a small example language with modules. In addition, we show how
our approach provides a model for a range of name binding patterns in
existing languages.

1 Introduction

Naming is a pervasive concern in the design and implementation of programming
languages. Names identify *declarations* of program entities (variables, functions,
types, modules, etc.) and allow these entities to be *referenced* from other parts
of the program. Name *resolution* associates each reference to its intended decla-
ration(s), according to the semantics of the language. Name resolution underlies
most operations on languages and programs, including static checking, translation, mechanized description of semantics, and provision of editor services in
IDEs. Resolution is often complicated, because it cuts across the local inductive
structure of programs (as described by an abstract syntax tree). For example,
the name introduced by a `let` node in an ML AST may be referenced by an
arbitrarily distant child node. Languages with explicit name spaces lead to further complexity; for example, resolving a qualified reference in Java requires first
resolving the class or package name to a context, and then resolving the member
name within that context. But despite this diversity, it is intuitively clear that
the basic concepts of resolution reappear in similar form across a broad range of
lexically-scoped languages.

In practice, the name resolution rules of real programming languages are usually described using *ad hoc* and informal mechanisms. Even when a language

© Springer-Verlag Berlin Heidelberg 2015
J. Vitek (Ed.): ESOP 2015, LNCS 9032, pp. 205–231, 2015.
DOI: 10.1007/978-3-662-46669-8_9

is formalized, its resolution rules are typically encoded as part of static and dynamic judgments tailored to the particular language, rather than being presented separately using a uniform mechanism. This lack of modularity in language description is mirrored in the implementation of language tools, where the resolution rules are often encoded multiple times to serve different purposes, e.g., as the manipulation of a symbol table in a compiler, a use-to-definition display in an IDE, or a substitution function in a mechanized soundness proof. This repetition results in duplication of effort and risks inconsistencies. To see how much better this situation might be, we need only contrast it with the realm of syntax definition, where context-free grammars provide a well-established declarative formalism that underpins a wide variety of useful tools.

Formalizing Resolution. This paper describes a formalism that we believe can help play a similar role for name resolution in lexically-scoped languages. It consists of a *scope graph*, which represents the naming structure of a program, and a *resolution calculus*, which describes how to resolve references to declarations within a scope graph. The scope graph abstracts away from the details of a program AST, leaving just the information relevant to name resolution. Its nodes include name references, declarations, and "scopes," which (in a slight abuse of conventional terminology) we use to mean minimal program regions that behave uniformly with respect to name resolution. Edges in the scope graph associate references to scopes, declarations to scopes, or scopes to "parent" scopes (corresponding to lexical nesting in the original program AST). The resolution calculus specifies how to construct a path through the graph from a reference to a declaration, which corresponds to a possible resolution of the reference. Hiding of one definition by a "closer" definition is modeled by providing an ordering on resolution paths. Ambiguous references correspond naturally to multiple resolution paths starting from the same reference node; unresolved references correspond to the absence of resolution paths. To describe programs involving explicit name spaces, the scope graph also supports giving names to scopes, and can include "import" edges to make the contents of a named scope visible inside another scope. The calculus supports complex import patterns including transitive and cyclic import of scopes.

This language-independent formalism gives us clear, abstract definitions for concepts such as scope, resolution, hiding, and import. We build on these concepts to define generic notions of α-equivalence and valid renaming. We also give a practical algorithm for computing conventional static environments mapping bound identifiers to the AST locations of the corresponding declarations, which can be used to implement a deterministic, terminating resolution function that is consistent with the calculus. We expect that the formalism can be used as the basis for other language-independent tools. In particular, any tool that relies on use-to-definition information, such as an IDE offering code completion for identifiers, or a live variable analysis in a compiler, should be specifiable using scope graphs.

On the other hand, the construction of a scope graph from a given program is a language-*dependent* process. For any given language, the construction can be specified by a conventional syntax-directed definition over the language grammar; we illustrate this approach for a small language in this paper. We would also like a more generic *binding specification language* which could be used to describe how to construct the scope graph for an arbitrary object language. We do not present such a language in this paper. However, the work described here was inspired in part by our previous work on NaBL [16], a DSL that provides high-level, non-algorithmic descriptions of name binding and scoping rules suitable for use by a (relatively) naive language designer. The NaBL implementation integrated into the Spoofax Language Workbench [14] automatically generates an incremental name resolution algorithm that supports services such as code completion and static analysis. However, the NaBL language itself is defined largely by example and lacks a high-level semantic description; one might say that it works well in practice, but not in theory. Because they are language-independent, scope graphs can be used to give a formal semantics for NaBL specifications, although we defer detailed exploration of this connection to further work.

Relationship to Related Work. The study of name binding has received a great deal of attention, focused in particular on two topics. The first is how to represent (already resolved) programs in a way that makes the binding structure explicit and supports convenient program manipulation "modulo α-equivalence" [7,20,3,10,4]. Compared to this work, our system is novel in several significant respects. (i) Our representation of program binding structure is *independent* of the underlying language grammar and program AST, with the benefits described above. (ii) We support representation of ill-formed programs, in particular, programs with ambiguous or undefined references; such programs are the normal case in IDEs and other front-end tools. (iii) We support description of binding in languages with explicit name spaces, such as modules or OO classes, which are common in practice.

A second well-studied topic is binding specification languages, which are usually enriched grammar descriptions that permit simultaneous specification of language syntax and binding structure [22,8,13,23,25]. This work is essentially complementary to the design we present here.

Specific Contributions.

- *Scope Graph and Resolution Calculus*: We introduce a language-independent framework to capture the relations among *references, declarations, scopes,* and *imports* in a program. We give a declarative specification of the resolution of references to declarations by means of a calculus that defines resolution paths in a scope graph (Section 2).
- *Variants*: We illustrate the modularity of our core framework design by describing several variants that support more complex binding schemes (Section 2.5).
- *Coverage*: We show that the framework covers interesting name binding patterns in existing languages, including various flavors of let bindings, qualified names, and inheritance in Java (Section 3).

- *Scope graph construction:* We show how scope graphs can be constructed for arbitrary programs in a simple example language via straightforward syntax-directed traversal (Section 4).
- *Resolution algorithm*: We define a deterministic and terminating resolution algorithm based on the construction of binding environments, and prove that it is sound and complete with respect to the calculus (Section 5).
- *α-equivalence and renaming*: We define a language-independent characterization of α-equivalence of programs, and use it to define a notion of valid renaming (Section 6).

  The extended version of this paper [19] presents the encoding of additional name binding patterns and the details of the correctness proof of the resolution algorithm.

2   Scope Graphs and Resolution Paths

Defining name resolution directly in terms of the abstract syntax tree leads to complex scoping patterns. In unary lexical binding patterns, such as lambda abstraction, the scope of the bound variable is the subtree dominated by the binding construct. However, in name binding patterns such as the sequential `let` in ML, or the variable declarations in a block in Java, the set of abstract syntax tree locations where the bindings are visible does not necessarily form a contiguous region. Similarly, the list of declarations of formal parameters of a function is contained in a subtree of the function definition that does not dominate their use positions. Informally, we can understand these name binding patterns by a conceptual mapping from the abstract syntax tree to an underlying pattern of *scopes*. However, this mapping is not made explicit in conventional descriptions of programming languages.

We introduce the language-independent concept of a *scope graph* to capture the scoping patterns in programs. A scope graph is obtained by a language-specific mapping from the abstract syntax tree of a program. The mapping collapses all abstract syntax tree nodes that behave uniformly with respect to name resolution into a single 'scope' node in the scope graph. In this paper, we do not discuss how to specify such mappings for arbitrary languages, which is the task of a binding specification language, but we show how it can be done for a particular toy language, first by example and then systematically. We assume that it should be possible to build a scope graph in a single traversal of the abstract syntax tree. Furthermore, the mapping should be *syntactic*; *no name resolution* should be necessary to construct the mapping.

Figures 1 to 3 define the full theory. Fig. 1 defines the structure of scope graphs. Fig. 2 defines the structure of *resolution paths*, a subset of resolution paths that are *well-formed*, and a *specificity ordering* on resolution paths. Finally, Fig. 3 defines the *resolution calculus*, which consists of the definition of *edges* between scopes in the scope graph and their transitive closure, the definition of *reachable* and *visible* declarations in a scope, and the *resolution* of references to declarations. In the rest of this section we motivate and explain this theory.

<table>
<tr><td>

References and declarations

- $x_i^{\mathsf{D}}{:}S$: declaration with name x at position i and optional associated named scope S
- x_i^{R}: reference with name x at position i

Scope graph

- $\mathcal{G}$: scope graph
- $\mathcal{S}(\mathcal{G})$: scopes S in $\mathcal{G}$
- $\mathcal{D}(S)$: declarations $x_i^{\mathsf{D}}{:}S'$ in S
- $\mathcal{R}(S)$: references x_i^{R} in S
- $\mathcal{I}(S)$: imports x_i^{R} in S
- $\mathcal{P}(S)$: parent scope of S

Well-formedness properties

- $\mathcal{P}(S)$ is a partial function
- The parent relation is well-founded
- Each x_i^{R} and x_i^{D} appears in exactly one scope S

</td><td>

Resolution paths

$$s := \mathbf{D}(x_i^{\mathsf{D}}) \mid \mathbf{I}(x_i^{\mathsf{R}}, x_j^{\mathsf{D}}{:}S) \mid \mathbf{P}$$
$$p := [] \mid s \mid p \cdot p$$
(inductively generated)
$$[] \cdot p = p \cdot [] = p$$
$$(p_1 \cdot p_2) \cdot p_3 = p_1 \cdot (p_2 \cdot p_3)$$

Well-formed paths

$$WF(p) \Leftrightarrow p \in \mathbf{P}^* \cdot \mathbf{I}(\_,\_)^*$$

Specificity ordering on paths

$$\frac{}{\mathbf{D}(\_) < \mathbf{I}(\_,\_)} \quad (DI)$$

$$\frac{}{\mathbf{I}(\_,\_) < \mathbf{P}} \quad (IP)$$

$$\frac{}{\mathbf{D}(\_) < \mathbf{P}} \quad (DP)$$

$$\frac{s_1 < s_2}{s_1 \cdot p_1 < s_2 \cdot p_2} \quad (Lex1)$$

$$\frac{p_1 < p_2}{s \cdot p_1 < s \cdot p_2} \quad (Lex2)$$

</td></tr>
</table>

Fig. 1. Scope graphs

Fig. 2. Resolution paths, well-formedness predicate, and specificity ordering

---

Edges in scope graph

$$\frac{\mathcal{P}(S_1) = S_2}{\mathbb{I} \vdash \mathbf{P} : S_1 \longrightarrow S_2} \quad (P)$$

$$\frac{y_i^{\mathsf{R}} \in \mathcal{I}(S_1) \setminus \mathbb{I} \quad \mathbb{I} \vdash p : y_i^{\mathsf{R}} \longmapsto y_j^{\mathsf{D}}{:}S_2}{\mathbb{I} \vdash \mathbf{I}(y_i^{\mathsf{R}}, y_j^{\mathsf{D}}{:}S_2) : S_1 \longrightarrow S_2} \quad (I)$$

Transitive closure

$$\frac{}{\mathbb{I} \vdash [] : A \twoheadrightarrow A} \quad (N)$$

$$\frac{\mathbb{I} \vdash s : A \longrightarrow B \quad \mathbb{I} \vdash p : B \twoheadrightarrow C}{\mathbb{I} \vdash s \cdot p : A \twoheadrightarrow C} \quad (T)$$

Reachable declarations

$$\frac{x_i^{\mathsf{D}} \in \mathcal{D}(S') \quad \mathbb{I} \vdash p : S \twoheadrightarrow S' \quad WF(p)}{\mathbb{I} \vdash p \cdot \mathbf{D}(x_i^{\mathsf{D}}) : S \rightarrowtail x_i^{\mathsf{D}}} \quad (R)$$

Visible declarations

$$\frac{\mathbb{I} \vdash p : S \rightarrowtail x_i^{\mathsf{D}} \quad \forall j, p'(\mathbb{I} \vdash p' : S \rightarrowtail x_j^{\mathsf{D}} \Rightarrow \neg(p' < p))}{\mathbb{I} \vdash p : S \longmapsto x_i^{\mathsf{D}}} \quad (V)$$

Reference resolution

$$\frac{x_i^{\mathsf{R}} \in \mathcal{R}(S) \quad \{x_i^{\mathsf{R}}\} \cup \mathbb{I} \vdash p : S \longmapsto x_j^{\mathsf{D}}}{\mathbb{I} \vdash p : x_i^{\mathsf{R}} \longmapsto x_j^{\mathsf{D}}} \quad (X)$$

Fig. 3. Resolution calculus

```
program = decl*
   decl = module id { decl* } | import qid | def id = exp
    exp = qid | fun id { exp } | fix id { exp }
        | let bind* in exp | letrec bind* in exp | letpar bind* in exp
        | exp exp | exp ⊕ exp | int
    qid = id | id . qid
   bind = id = exp
```

Fig. 4. Syntax of LM

2.1   Example Language

To illustrate the scope graph framework we use the toy language LM, defined in Fig. 4, which contains a rather eclectic combination of features chosen to exhibit both simple and challenging name binding patterns. LM supports the following constructs for binding variables:

- Lambda and mu: The functional abstractions **fun** and **fix** represent lambda and mu terms, respectively; both have basic unary lexically scoped bindings.
- Let: The various flavors of let bindings (sequential **let**, **letrec**, and **letpar**) challenge the unary lexical binding model.
- Definition: A definition (**def**) declares a variable and binds it to the value of an initializing expression. The definitions in a module are not ordered (no requirement for 'def-before-use'), giving rise to mutually recursive definitions.

Most programming languages have some notion of *module* to divide a program into separate units and a notion of *imports* that make elements of one module available in another. Modules change the standard lexical scoping model, since names can be declared either in the lexical parent or in an imported module. The modules of LM support the following features:

- Qualified names: Elements of modules can be addressed by means of a qualified name using conventional dot notation.
- Imports: All declarations in an imported module are made visible without the need for qualification.
- Transitive imports: The definitions imported into an imported module are themselves visible in the importing module.
- Cyclic imports: Modules can (indirectly) mutually import each other, leading to cyclic import chains.
- Nested modules: Modules may have sub-modules, which can be accessed using dot notation or by importing the containing module.

In the remainder of this section, we use LM examples to illustrate the basic features of our framework. In Section 3 and Appendix A of [19] we explore the expressive power of the framework by applying it to a range of name binding patterns from both LM and real languages. Section 4 shows how to construct scope graphs for arbitrary LM programs.

2.2 Declarations, References, and Scopes

We now introduce and motivate the various elements of the name binding framework, gradually building up to the full system described in Figures 1 to 3. The central concepts in the framework are *declarations*, *references*, and *scopes*. A *declaration* (also known as *binding occurrence*) *introduces* a name. For example, the **def** x = e and **module** m { .. } constructs in LM introduce names of variables and modules, respectively. (A declaration may or may not also *define* the name; this distinction is unimportant for name resolution—except in the case where the declaration defines a module, as discussed in detail later.) A *reference* (also known as *applied occurrence*) is the *use* of a name that refers to a declaration with the same name. In LM, the variables in expressions and the names in import statements (e.g. the x in **import** x) are references. Each reference and declaration is unique and is distinguished not just by its name, but also by its position in the program's AST. Formally, we write x_i^R for a reference with name x at position i and x_i^D for a declaration with name x at position i.

A *scope* is an abstraction over a group of nodes in the abstract syntax tree that behave uniformly with respect to name resolution. Each program has a *scope graph* $\mathcal{G}$, whose nodes are a finite set of scopes $\mathcal{S}(\mathcal{G})$. Every program has at least one scope, the global or *root* scope. Each scope S has an associated finite set $\mathcal{D}(S)$ of declarations and finite set $\mathcal{R}(S)$ of references (at particular program positions), and each declaration and reference in a program belongs to a unique scope. A scope is the atomic grouping for name resolution: roughly speaking, each reference x_i^R in a scope resolves to a declaration of the same variable x_j^D in the scope, if one exists. Intuitively, a single scope corresponds to a group of mutually recursive definitions, e.g., a **letrec** block, the declarations in a module, or the set of top-level bindings in a program. Below we will see that edges between nodes in a scope graph determine visibility of declarations in one scope from references in another scope.

Name Resolution. We write $\mathcal{R}(\mathcal{G})$ and $\mathcal{D}(\mathcal{G})$ for the (finite) sets of all references and all declarations, respectively, in the program with scope graph $\mathcal{G}$. Name resolution is specified by a relation $\longmapsto \subseteq \mathcal{R}(\mathcal{G}) \times \mathcal{D}(\mathcal{G})$ between references and corresponding declarations in $\mathcal{G}$. In the absence of edges, this relation is very simple:

$$\frac{x_i^R \in \mathcal{R}(S) \quad x_j^D \in \mathcal{D}(S)}{x_i^R \longmapsto x_j^D} \qquad (X_0)$$

That is, a reference x_i^R resolves to a declaration x_j^D, if the scope S in which x_i^R is contained also contains x_j^D. We say that there is a *resolution path* from x_i^R to x_j^D. We will see soon that paths will grow beyond the one step relation defined by the rule above.

Scope Graph Diagrams. It can be illuminating to depict a scope graph graphically. In a scope graph diagram, a scope is depicted as a circle, a reference as a box with an arrow pointing *into* the scope that contains it, and a declaration as

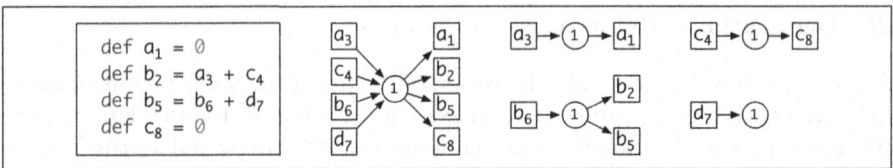

Fig. 5. Declarations and references in global scope

a box with an arrow *from* the scope that contains it. Fig. 5 shows an LM program consisting of a set of mutually-recursive global definitions; its scope graph; the resolution paths for variables a, b, and c; and an incomplete resolution path for variable d. In concrete example programs and scope diagrams we write both x_i^R and x_i^D as x_i, relying on context to distinguish references and declarations. For example, in Fig. 5, all occurrences b_i denote the *same name* b at *different positions*. In scope diagrams, the numbers in scope circles are arbitrarily chosen, and are just used to identify different scopes so that we can talk about them.

Duplicate Declarations. It is possible for a scope to contain multiple references and/or declarations with the same name. For example, scope 1 in Fig. 5 has two declarations of the variable b. While the existence of multiple references is normal, multiple declarations may give rise to multiple resolutions. For example, the b_6 reference in Fig. 5 resolves to *each* of the two declarations b_2 and b_5.

Typically, correct programs will not declare the same identifier at two different locations in the same scope, although some languages have constructs (e.g. or-patterns in OCaml [17]) that are most naturally modeled this way. But even when the existence of multiple resolutions implies an erroneous program, we want the resolution calculus to identify *all* these resolutions, since IDEs and other front-end tools need to be able to represent erroneous programs. For example, a rename refactoring should support consistent renaming of identifiers, even in the presence of ambiguities (see Section 6). The ability of our calculus to describe ambiguous resolutions distinguishes it from systems, such as nominal logic [4], that inherently require unambiguous resolution of references.

2.3   Lexical Scope

We model lexical scope by means of the *parent* relation on scopes. In a well-formed scope graph, each scope has at most one parent and the parent relation is well-founded. Formally, the partial function $\mathcal{P}(\_)$ maps a scope S to its *parent* scope $\mathcal{P}(S)$. Given a scope graph with parent relation we can define the notion of *reachable* and *visible* declarations in a scope.

Fig. 6 illustrates how the parent relation is used to model common lexical scope patterns. Lexical scoping is typically presented through nested regions in the abstract syntax tree, as illustrated by the nested boxes in Fig. 6. Expressions in inner boxes may refer to declarations in surrounding boxes, but not vice versa. Each of the scopes in the program is mapped to a scope (circle) in the scope graph. The three scopes correspond to the global scope, the scope for **fix** f_2, and the scope for **fun** n_3. The edges from scopes to scopes correspond to the parent

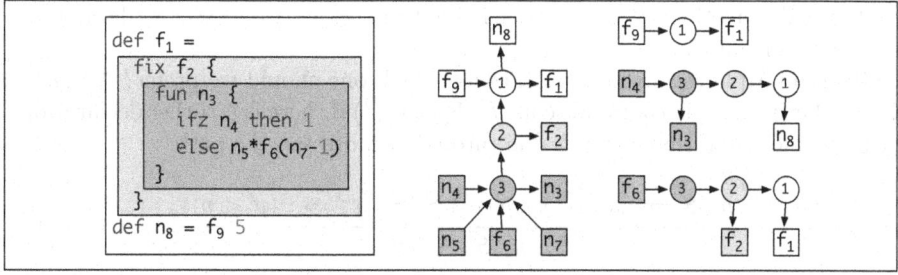

Fig. 6. Lexical scoping modeled by edges between scopes in the scope graph with example program, scope graph, and reachability paths for references

relation. The resolution paths on the right of Fig. 6 illustrate the consequences of the encoding. From reference f_6 both declarations f_1 and f_2 are *reachable*, but from reference f_9 only declaration f_1 is reachable. In languages with lexical scoping, the redeclaration of a variable inside a nested region typically *hides* the outer declaration. Thus, the duplicate declaration of variable f does not indicate a program error in this situation because only f_2 is *visible* from the scope of f_6.

Reachability. The first step towards a full resolution calculus is to take into account reachability. We redefine rule (X_0) as follows:

$$\frac{x_i^R \in \mathcal{R}(S_1) \quad p : S_1 \twoheadrightarrow S_2 \quad x_j^D \in \mathcal{D}(S_2)}{p : x_i^R \longmapsto x_j^D} \quad (X_1)$$

That is, x_i^R in scope S_1 can be resolved to x_j^D in scope S_2, if S_2 is *reachable* from S_1, i.e. if $S_1 \twoheadrightarrow S_2$. Reachability is defined in terms of the parent relation as follows:

$$\frac{\mathcal{P}(S_1) = S_2}{\mathbf{P} : S_1 \longrightarrow S_2} \qquad \frac{}{[] : A \twoheadrightarrow A} \qquad \frac{s : A \longrightarrow B \quad p : B \twoheadrightarrow C}{s \cdot p : A \twoheadrightarrow C}$$

The parent relation between scopes gives rise to a direct edge $S_1 \longrightarrow S_2$ between child and parent scope, and $A \twoheadrightarrow B$ is the reflexive, transitive closure of the direct edge relation. In order to reason about the different ways in which a reference can be resolved, we record the resolution path p. For example, in Fig. 6 reference f_6 can be resolved with path $\mathbf{P}$ to declaration f_2 and with path $\mathbf{P} \cdot \mathbf{P}$ to f_1.

Visibility. Under lexical scoping, multiple possible resolutions are not problematic, as long as the declarations reached are not declared in the same scope. A declaration is *visible* unless it is shadowed by a declaration that is 'closer by'. To formalize visibility, we first extend reachability of scopes to *reachability of declarations*:

$$\frac{x_i^D \in \mathcal{D}(S') \quad p : S \twoheadrightarrow S'}{p \cdot \mathbf{D}(x_i^D) : S \longmapsto x_i^D} \quad (R_2)$$

That is, a declaration x_i^D in S' is reachable from scope S ($S \rightarrowtail x_i^D$), if scope S' is reachable from S.

Given multiple reachable declarations, which one should we prefer? A reachable declaration x_i^D is *visible* in scope S ($S \longmapsto x_i^D$) if there is no other declaration for the same name that is reachable through a *more specific* path:

$$\frac{p : S \rightarrowtail x_i^D \quad \forall j, p'(p' : S \rightarrowtail x_j^D \Rightarrow \neg(p' < p))}{p : S \longmapsto x_i^D} \quad (V_2)$$

where the *specificity ordering* $p' < p$ on paths is defined as

$$\overline{\mathbf{D}(\_) < \mathbf{P}} \qquad \frac{s_1 < s_2}{s_1 \cdot p_1 < s_2 \cdot p_2} \qquad \frac{p_1 < p_2}{s \cdot p_1 < s \cdot p_2}$$

That is, a path with fewer parent transitions is more specific than a path with more parent transitions. This formalizes the notion that a declaration in a "nearer" scope shadows a declaration in a "farther" scope.

Finally, a reference resolves to a declaration if that declaration is visible in the scope of the reference.

$$\frac{x_i^R \in \mathcal{R}(S) \quad p : S \longmapsto x_j^D}{p : x_i^R \longmapsto x_j^D} \quad (X_2)$$

Example. In Fig. 6 the scope (labeled 3) containing reference f_6 can reach two declarations for f: $\mathbf{P} \cdot \mathbf{D}(f_2^D) : S_3 \rightarrowtail f_2^D$ and $\mathbf{P} \cdot \mathbf{P} \cdot \mathbf{D}(f_1^D) : S_3 \rightarrowtail f_1^D$. Since the first path is more specific than the second path, only f_2 is visible, i.e. $\mathbf{P} \cdot \mathbf{D}(f_2^D) : S_3 \longmapsto f_2^D$. Therefore f_6 resolves to f_2, i.e. $\mathbf{P} \cdot \mathbf{D}(f_2^D) : f_6^R \longmapsto f_2^D$.

Scopes, Revisited. Now that we have defined the notions of reachability and visibility, we can give a more precise description of the sense in which scopes "behave uniformly" with respect to resolution. For every scope S:

- Each declaration in the program is either visible at every reference in $\mathcal{R}(S)$ or not visible at any reference in $\mathcal{R}(S)$.
- For each reference in the program, either every declaration in $\mathcal{D}(S)$ is reachable from that reference, or no declaration in $\mathcal{D}(S)$ is reachable from that reference.
- Every declaration in $\mathcal{D}(S)$ is visible at every reference in $\mathcal{R}(S)$.

2.4   Imports

Introducing modules and imports complicates the name binding picture. Declarations are no longer visible only through the lexical context, but may be visible through an import as well. Furthermore, resolving a reference may require first resolving one or more imports, which may in turn require resolving further imports, and so on.

We model an *import* by means of a reference x_i^R in the set of imports $\mathcal{I}(S)$ of a scope S. (Imports are also always references and included in some $\mathcal{R}(S')$, but not

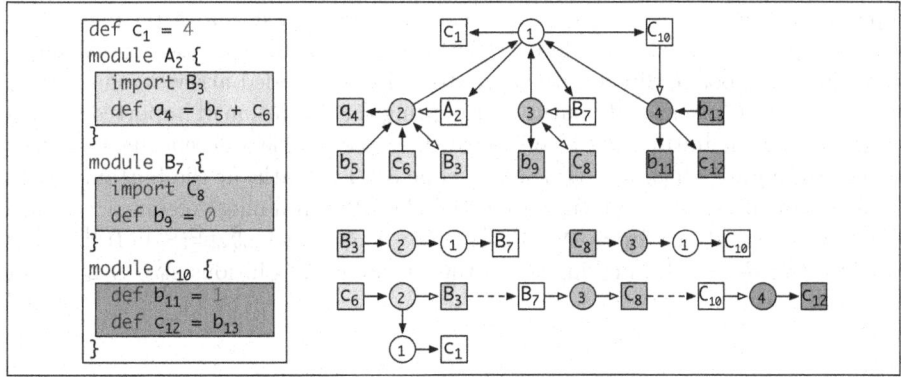

Fig. 7. Modules and imports with example program, scope graph, and reachability paths for references

necessarily in the same scope in which they are imports.) We model a *module* by associating a scope S with a declaration $x_i^D{:}S$. This associated *named scope* (i.e., named by x) represents the declarations introduced by, and encapsulated in, the module. (We write the $:S$ only in rules where it is required; where we omit it, the declaration may or may not have an associated scope.) Thus, *importing* entails resolving the import reference to a declaration and making the declarations in the scope associated with that declaration available in the importing scope.

Note that 'module' is not a built-in concept in our framework. A module is any construct that (1) is named, (2) has an associated scope that encapsulates declarations, and (3) can be imported into another scope. Of course, this can be used to model the module systems of languages such as ML. But it can be applied to constructs that are not modules at first glance. For example, a class in Java encapsulates class variables and methods, which are imported into its subclasses through the 'extends' clause. Thus, a class plays the role of module and the extends clause that of import. We discuss further applications in Section 3.

Reachability. To define name resolution in the presence of imports, we first extend the definition of reachability. We saw above that the parent relation on scopes induces an edge $S_1 \longrightarrow S_2$ between a scope S_1 and its parent scope S_2 in the scope graph. Similarly, an import induces an edge $S_1 \longrightarrow S_2$ between a scope S_1 and the scope S_2 associated with a declaration imported into S_1:

$$\frac{y_i^R \in \mathcal{I}(S_1) \quad p : y_i^R \longmapsto y_j^D{:}S_2}{\mathbf{I}(y_i^R, y_j^D{:}S_2) : S_1 \longrightarrow S_2} \tag{I_3}$$

Note the recursive invocation of the resolution relation on the name of the imported scope.

Figure 7 illustrates extensions to scope graphs and paths to describe imports. Association of a name to a scope is indicated by an open-headed arrow from the name declaration box to the scope circle. (For example, scope 2 is associated to declaration A_2.) An import into a scope is indicated by an open-headed arrow from the scope circle to the import name reference box. (For example, scope 2 imports the contents of the scope associated to the resolution of reference B_3; note that since B_3 is also a reference within scope 2, there is also an ordinary arrow in the opposite direction, leading to a double-headed arrow in the scope graph.) Edges in reachability paths representing the resolution of imported scope names to their definitions are drawn dashed. (For example, reference B_3 resolves to declaration B_7, which has associated scope 3.) The paths at the bottom right of the figure illustrate that the scope (labeled 2) containing reference c_6 can reach two declarations for c: $\mathbf{P} \cdot \mathbf{D}(c_1^D) : S_2 \rightarrowtail c_1^D$ and $\mathbf{I}(B_3^R, B_7^D : S_3) \cdot \mathbf{I}(c_8^R, c_{10}^D : S_4) \cdot \mathbf{D}(c_{12}^D) : S_2 \rightarrowtail c_{12}^D$, making use of the subsidiary resolutions $B_3^R \longmapsto B_7^D$ and $c_8^R \longmapsto c_{10}^D$.

Visibility. Imports cause new kinds of ambiguities in resolution paths, which require extension of the visibility policy.

The first issue is illustrated by Fig. 8. In the scope of reference b_{10} we can reach declaration b_7 with path $\mathbf{D}(b_7^D)$ and declaration b_4 with path $\mathbf{I}(A_6^R, A_2^D : S_A) \cdot \mathbf{D}(b_4^D)$ (where S_A is the scope named by declaration A_2). We resolve this conflict by extending the specificity order with the rule $\mathbf{D}(\_) < \mathbf{I}(\_, \_)$. That is, local declarations override imported declarations. Similarly, in the scope of reference a_8 we can reach declaration a_1 with path $\mathbf{P} \cdot \mathbf{D}(a_1^D)$ and declaration a_3 with path $\mathbf{I}(A_6^R, A_2^D : S_A) \cdot \mathbf{D}(a_3^D)$. We resolve this conflict by extending the specificity order with the rule $\mathbf{I}(\_, \_) < \mathbf{P}$. That is, resolution through imports is preferred over resolution through parents. In other words, declarations in imported modules override declarations in lexical parents.

The next issue is illustrated in Fig. 9. In the scope of reference a_8 we can reach declaration a_4 with path $\mathbf{P} \cdot \mathbf{D}(a_4^D)$ and declaration a_1 with path $\mathbf{P} \cdot \mathbf{P} \cdot \mathbf{D}(a_1^D)$. The specificity ordering guarantees that only the first of these is visible, giving the resolution we expect. However, with the rules as stated so far, there is another way to reach a_1, via the path $\mathbf{I}(B_6^R, B_2^D : S_B) \cdot \mathbf{P} \cdot \mathbf{D}(a_1^D)$. That is, we first import module B, and then go to its lexical parent, where we find the declaration. In other words, when importing a module, we import not just its declarations, but all declarations in its lexical context. This behavior seems undesirable; to our knowledge, no real languages exhibit it. To rule out such resolutions, we define a well-formedness predicate $WF(p)$ that requires paths p to be of the form $\mathbf{P}^* \cdot \mathbf{I}(\_, \_)^*$, i.e. forbidding the use of parent steps after one or more import steps. We use this predicate to restrict the reachable declarations relation by only considering scopes reachable through a well-formed path:

```
def a1 = ...
module A2 {
  def a3 = ...
  def b4 = ...
}
module C5 {
  import A6
  def b7 = a8
  def c9 = b10
}
```

Fig. 8. Parent vs Import

```
def a1 = ...
module B2 {
}
module C3 {
  def a4 = ...
  module D5 {
    import B6
    def e7 = a8
  }
}
```

Fig. 9. Parent of import

$$\frac{x_i^D \in \mathcal{D}(S') \quad p : S \twoheadrightarrow S' \quad WF(p)}{p \cdot \mathbf{D}(x_i^D) : S \rightarrowtail x_i^D} \qquad (R_3)$$

$$\cfrac{\cfrac{A_2^D{:}S_{A_2} \in \mathcal{D}(S_{A_1})}{\cfrac{A_4^R \in \mathcal{I}(S_{root}) \quad \cfrac{A_4^R \in \mathcal{R}(S_{root}) \quad A_1^D{:}S_{A_1} \in \mathcal{D}(S_{root})}{A_4^R \longmapsto A_1^D{:}S_{A_1}}}{S_{root} \longrightarrow S_{A_1} \quad (*)}}{\cfrac{S_{root} \rightarrowtail A_2^D{:}S_{A_2}}{A_4^R \in \mathcal{R}(S_{root}) \quad S_{root} \longmapsto A_2^D{:}S_{A_2}}}}{A_4^R \longmapsto A_2^D{:}S_{A_2}}$$

Fig. 10. Derivation for $A_4^R \longmapsto A_2^D{:}S_{A_2}$ in a calculus without import tracking

The complete definition of well-formed paths and specificity order on paths is given in Fig. 2. In Section 2.5 we discuss how alternative visibility policies can be defined by just changing the well-formedness predicate and specificity order.

Seen Imports. Consider the example in Fig. 11. Is declaration a_3 reachable in the scope of reference a_6? This reduces to the question whether the import of A_4 can resolve to module A_2. Surprisingly, it can, in the calculus as discussed so far, as shown by the derivation in Fig. 10 (which takes a few shortcuts). The conclusion of the derivation is that $A_4^R \longmapsto A_2^D{:}S_{A_2}$. This conclusion is obtained by *using the import at* A_4 to conclude at step (*) that $S_{root} \longrightarrow S_{A_1}$, i.e. that the body of module A_1 is reachable! In other words, the import of A_4 is used in its own resolution. Intuitively, this is nonsensical.

```
module A₁ {
  module A₂ {
    def a₃ = ...
  }
}
import A₄
def b₅ = a₆
```

Fig. 11. Self import

To rule out this kind of behavior we extend the calculus to keep track of the set of *seen imports* $\mathbb{I}$ using judgements of the form $\mathbb{I} \vdash p : x_i^R \longmapsto x_j^D$. We need to extend all rules to pass the set $\mathbb{I}$, but only the rules for resolution and import are truly affected:

$$\frac{x_i^R \in \mathcal{R}(S) \quad \{x_i^R\} \cup \mathbb{I} \vdash p : S \longmapsto x_j^D}{\mathbb{I} \vdash p : x_i^R \longmapsto x_j^D} \quad (X)$$

$$\frac{y_i^R \in \mathcal{I}(S_1) \setminus \mathbb{I} \quad \mathbb{I} \vdash p : y_i^R \longmapsto y_j^D{:}S_2}{\mathbb{I} \vdash \mathbf{I}(y_i^R, y_j^D{:}S_2) : S_1 \longrightarrow S_2} \quad (I)$$

```
module A₁ {
  module B₂ {
    def x₃ = 1
  }
}
module B₄ {
  module A₅ {
    def y₆ = 2
  }
}
module C₇ {
  import A₈
  import B₉
  def z₁₀ = x₁₁
          + y₁₂
}
```

Fig. 12. Anomalous resolution

With this final ingredient, we reach the full calculus in Fig. 3. It is not hard to see that the resolution relation is well-founded. The only recursive invocation (via the I rule) uses a strictly larger set $\mathbb{I}$ of seen imports (via the X rule); since the set $\mathcal{R}(G)$ is finite, $\mathbb{I}$ cannot grow indefinitely.

Anomalies. Although the calculus produces the desired resolutions for a wide variety of real language constructs, its behavior can be surprising on corner cases. Even with the "seen imports" mechanism, it is still possible for a single derivation to resolve a given import in two different ways, leading to unintuitive results. For example, in the program in Fig. 12, x_{11} can resolve to x_3 and y_{12} can resolve to y_6. (Derivations left as an exercise to the curious reader!) In our experience, phenomena like this occur only in the presence of mutually-recursive imports; to our knowledge, no real language has these (perhaps for good reason). We defer deeper exploration of these anomalies to future work.

2.5 Variants

The resolution calculus presented so far reflects a number of binding policy decisions. For example, we enforce imports to be transitive and local declarations to be preferred over imports. However, not every language behaves like this. We now present how other common behaviors can easily be represented with slight modifications of the calculus. Indeed, the modifications do not have to be done on the calculus itself (the $\longrightarrow$, $\twoheadrightarrow$, $\rightarrowtail$ and $\longmapsto$ relations) but can simply be encoded in the *WF* predicate and the $<$ ordering on paths.

Reachability policy. Reachability policies define how a reference can access a particular definition, i.e. what rules can be used during the resolution. We can change our reachability policy by modifying the *WF* predicate. For example, if we want to rule out transitive imports, we can change *WF* to be

$$WF(p) \Leftrightarrow p \in \mathbf{P}^* \cdot \mathbf{I}(\_, \_)?$$

where ? denotes the *at most one* operation on regular expressions. Therefore, an import can only be used once at the end of the chain of scopes.

For a language that supports both transitive and non-transitive imports, we can add a label on references corresponding to imports. If x^{R} is a reference representing a non-transitive import and x^{TR} a reference corresponding to a transitive import, then the *WF* predicate simply becomes:

$$WF(p) \Leftrightarrow p \in \mathbf{P}^* \cdot \mathbf{I}(\_^{\mathsf{TR}}, \_)^* \cdot \mathbf{I}(\_^{\mathsf{R}}, \_)?$$

Now no import can occur after the use of a non-transitive one.

Similarly, we can modify the rule to handle the *Export* declaration in Coq, which forces transitivity (a resolution can always use an exported module even after importing from a non-transitive one). Assume x^{R} is a reference representing a non-transitive import and x^{ER} a reference corresponding to an export; then we can use the following predicate:

```
module A₁ {
  def x₂ = 3
}
module B₃ {
  include A₄;
  def x₅ = 6;
  def z₆ = x₇
}
```

Fig. 13. Include

$$WF(p) \Leftrightarrow p \in \mathbf{P}^* \cdot \mathbf{I}(\_^{\mathsf{R}}, \_)? \cdot \mathbf{I}(\_^{\mathsf{ER}}, \_)^*$$

Visibility policy. We can modify the visibility policy, i.e. how resolutions shadow each other, by changing the definition of the specificity ordering. For example, we might want imports to act like textual inclusion, so the declarations in the included module have the same precedence as local declarations. This is similar to Standard ML's **include** mechanism. In the program in Fig. 13, the reference x_7 should be treated as having duplicate resolutions, to either x_5 or x_2; the

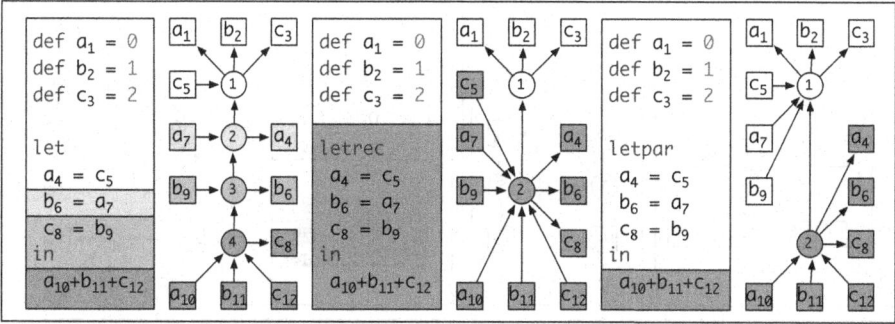

Fig. 14. Example LM programs with sequential, recursive, and parallel **let**, and their encodings as scope graphs

former should not hide the latter. To handle this situation, we can drop the rule $\mathbf{D}(\_) < \mathbf{I}(\_,\_)$ so that definitions and references will get the same precedence, and a definition will not shadow an imported definition. To handle both **include** and ordinary imports, we can once again differentiate the references, and define different ordering rules depending on the reference used in the import step.

3   Coverage

To what extent does the scope graph framework cover name binding systems that live in the world of real programming languages? It is not possible to *prove* complete coverage by the framework, in the sense of being able to encode all possible name binding systems that exist or may be designed in the future. (Indeed, given that these systems are typically implemented in compilers with algorithms in Turing-complete programming languages, the framework is likely *not* to be complete.) However, we believe that our approach handles many lexically-scoped languages. The design of the framework was informed by an investigation of a wide range of name binding patterns in existing languages, their (attempted) formalization in the NaBL name binding language [14,16], and their encoding in scope graphs. In this section, we discuss three such examples: **let** bindings, qualified names, and inheritance in Java. This should provide the reader with a good sense of how name binding patterns can be expressed using scope graphs. Appendix A of [19] provides further examples, including definition-before-use, compilation units and packages in Java, and namespaces and partial classes in C#.

Let Bindings. The several flavors of **let** bindings in languages such as ML, Haskell, and Scheme do not follow the unary lexical binding pattern in which the binding construct dominates the abstract syntax tree that makes up its scope. The LM language from Fig. 4 has three flavors of **let** bindings: sequential, recursive, and parallel **let**, each with a list of bindings and a body expression. Fig. 14 shows the encoding into scope graphs for each of the constructs and makes precise how the bindings are interpreted in each flavour. In the recursive **letrec**, the bindings are visible in all initializing expressions, so a single scope

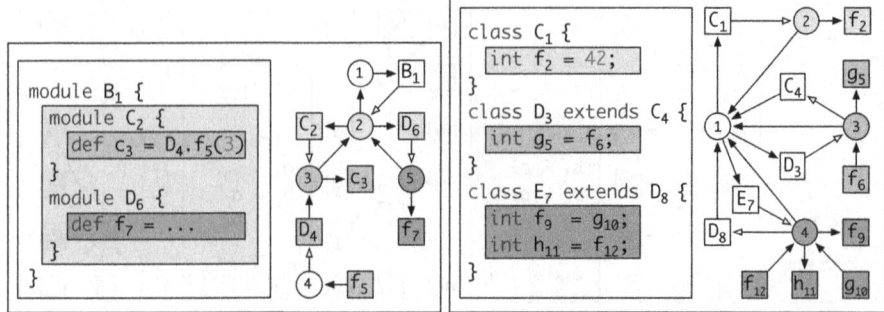

Fig. 15. Example LM program with partially-qualified name

Fig. 16. Class inheritance in Java modeled by import edges

suffices for the whole construct. In the sequential **let**, each binding is visible in the *subsequent* bindings, but not in its own initializing expression. This requires the introduction of a new scope for each binding. In the parallel **letpar**, the variables being bound are not visible in any of the initializing expressions, but only in the body. This is expressed by means of a single scope (2) in which the bindings are declared; any references in the initializing expressions are associated to the parent scope (1).

Qualified Names. Qualified names refer to declarations in named scopes outside the lexical scoping. They can be either used as simple references or as imports. For example, fully-qualified names of Java classes can be used to refer to (or import) classes from other packages. While fully-qualified names allow navigating named scopes from the root scope, partially-qualified names give access to lexical subscopes, which are otherwise hidden from lexical parent scopes.

The LM program in Fig. 15 uses a partially-qualified name D.f to access function f in submodule D. We can model this pattern using an anonymous scope (4), which is not linked to the lexical context. The relative name (f_5) is a reference in the anonymous scope. We add the qualifying scope name (D_4) as an import in the anonymous scope.

Inheritance in Java. We can model inheritance in object-oriented languages with named scopes and imports. For example, Fig. 16 shows a hierarchy of three Java classes. Class C declares a field f. Class D extends C and inherits its field f. Class E extends D, inheriting the fields of C and D. Each class name is a declaration in the same package scope (1), and associated with the scope of its class body. Inheritance is modeled with imports: a subclass body scope contains an import referring to its super class, making the declarations in the super class reachable from the body. In the example, the scope (4) representing the body of class E contains an import referring to its super class D. Using this import, g_{10} correctly resolves to g_5 . Since local declarations hide imported declarations, f_{12} also refers correctly to the local declaration f_9, which hides the transitively

$$[\![\text{ds}]\!]^{prog} \; := \; \text{let } S := \text{new}_\perp \text{ in } [\![\text{ds}]\!]_S^{recd}$$

$$[\![\text{d ds}]\!]_S^{recd} \; := \; [\![\text{d}]\!]_S^{dec}; [\![\text{ds}]\!]_S^{recd}$$

$$[\![\,]\!]_S^{recd} \; := \; ()$$

$$[\![\textbf{module } \text{x}_i\{\text{ds}\}]\!]^{dec} \; := \; \text{let } S' := \text{new}_S \text{ in } \mathcal{D}(S) \mathrel{+}= x_i^{\text{D}}{:}S'; [\![\text{ds}]\!]_{S'}^{recd}$$

$$[\![\textbf{import } \text{xs}]\!]_S^{dec} \; := \; [\![\text{xs}]\!]_S^{rqid}; [\![\text{xs}]\!]_S^{iqid}$$

$$[\![\textbf{def } \text{x}_i = \text{e}]\!]_S^{dec} \; := \; \mathcal{D}(S) \mathrel{+}= x_i^{\text{D}}; [\![\text{e}]\!]_S^{exp}$$

$$[\![\text{xs}]\!]_S^{exp} \; := \; [\![\text{xs}]\!]_S^{rqid}$$

$$[\![(\textbf{fun } | \textbf{ fix}) \; \text{x}_i\{\text{e}\}]\!]_S^{exp} \; := \; \text{let } S' := \text{new}_S \text{ in } \mathcal{D}(S') \mathrel{+}= x_i^{\text{D}}; [\![\text{e}]\!]_{S'}^{exp}$$

$$[\![\textbf{letrec } \text{bs in } \text{e}]\!]_S^{exp} \; := \; \text{let } S' := \text{new}_S \text{ in } [\![\text{bs}]\!]_{S'}^{recb}; [\![\text{e}]\!]_{S'}^{exp}$$

$$[\![\textbf{letpar } \text{bs in } \text{e}]\!]_S^{exp} \; := \; \text{let } S' := \text{new}_S \text{ in } [\![\text{bs}]\!]_{(S,S')}^{parb}; [\![\text{e}]\!]_{S'}^{exp}$$

$$[\![\textbf{let } \text{bs in } \text{e}]\!]_S^{exp} \; := \; \text{let } S' := [\![\text{bs}]\!]_S^{seqb} \text{ in } [\![\text{e}]\!]_{S'}^{exp}$$

$$[\![\text{e}_1 \; \text{e}_2]\!]_S^{exp} \; := \; [\![\text{e}_1]\!]_S^{exp}; [\![\text{e}_2]\!]_S^{exp}$$

$$[\![\text{e}_1 \oplus \text{e}_2]\!]_S^{exp} \; := \; [\![\text{e}_1]\!]_S^{exp}; [\![\text{e}_2]\!]_S^{exp}$$

$$[\![\text{n}]\!]_S^{exp} \; := \; ()$$

$$[\![\text{x}_i.\text{xs}]\!]_S^{rqid} \; := \; \mathcal{R}(S) \mathrel{+}= x_i^{\text{R}}; \text{let } S' := \text{new}_\perp \text{ in } \mathcal{I}(S') \mathrel{+}= x_i^{\text{R}}; [\![\text{xs}]\!]_{S'}^{rqid}$$

$$[\![\text{x}_i]\!]_S^{rqid} \; := \; \mathcal{R}(S) \mathrel{+}= x_i^{\text{R}}$$

$$[\![\text{x}_i.\text{xs}]\!]_S^{iqid} \; := \; [\![\text{xs}]\!]_S^{iqid}$$

$$[\![\text{x}_i]\!]_S^{iqid} \; := \; \mathcal{I}(S) \mathrel{+}= x_i^{\text{R}}$$

$$[\![\text{x}_i = \text{e}; \text{ bs}]\!]_S^{recb} \; := \; \mathcal{D}(S) \mathrel{+}= x_i^{\text{D}}; [\![\text{e}]\!]_S^{exp}; [\![\text{bs}]\!]_S^{recb}$$

$$[\![\,]\!]_S^{recb} \; := \; ()$$

$$[\![\text{x}_i = \text{e}; \text{ bs}]\!]_{(S,S')}^{parb} \; := \; \mathcal{D}(S') \mathrel{+}= x_i^{\text{D}}; [\![\text{e}]\!]_S^{exp}; [\![\text{bs}]\!]_{(S,S')}^{parb}$$

$$[\![\,]\!]_{(S,S')}^{parb} \; := \; ()$$

$$[\![\text{x}_i = \text{e}; \text{ bs}]\!]_S^{seqb} \; := \; [\![\text{e}]\!]_S^{exp}; \text{let } S' := \text{new}_S \text{ in } \mathcal{D}(S') \mathrel{+}= x_i^{\text{D}}; \text{ret}(S')$$

$$[\![\,]\!]_S^{seqb} \; := \; \text{ret}(S)$$

Fig. 17. Scope graph construction for LM via syntax-directed AST traversal

imported f_2. Note that since a scope can contain several imports, encoding multiple inheritance uses exactly the same principle.

4   Scope Graph Construction

The preceding sections have illustrated scope graph construction by means of examples corresponding to various language features. Of course, to apply our formalism in practice, one must be able to construct scope graphs systematically. Ultimately, we would like to be able to specify this process for arbitrary languages using a generic binding specification language such as NaBL [16], but that remains future work. Here we illustrate systematic scope graph construction for arbitrary programs in a *specific* language, LM (Fig. 4), via straightforward syntax-directed traversal.

Figure 17 describes the construction algorithm. For clarity of presentation, the algorithm traverses the program's concrete syntax; a real implementation would traverse the program's AST. The algorithm is presented in an *ad hoc* imperative language, explained here. The traversal is specified as a collection of (potentially) mutually recursive functions, one or more for each syntactic class of LM. Each function f is defined by a set of clauses $[\![pattern]\!]^f_{args}$. When f is invoked on a term, the clause whose *pattern* matches the term is executed. Functions may also take additional arguments *args*. Each clause body consists of a sequence of statements separated by semicolons. Functions can optionally return a value using ret(). The let statement binds a metavariable in the remainder of the clause body. An empty clause body is written ().

The algorithm is initiated by invoking $[\![\_]\!]^{prog}$ on an entire LM program. Its net effect is to produce a scope graph via a sequence of imperative operations. The construct new$_P$ creates a new scope S with parent P (or no parent if $p =\perp$) and empty sets $\mathcal{D}(S)$, $\mathcal{R}(S)$, and $\mathcal{I}(S)$. These sets are subsequently populated using the $+=$ operator, which extends a set imperatively. The program scope graph is simply the set of scopes that have been created and populated when the traversal terminates.

5 Resolution Algorithm

The calculus of Section 2 gives a precise definition of resolution. In principle, we can search for derivations in the calculus to answer questions such as "Does this variable reference resolve to this declaration?" or "Which variable declarations does this reference resolve to?" But automating this search process is not trivial, because of the need for back-tracking and because the paths in reachability derivations can have cycles (visiting the same scope more than once), and hence can grow arbitrarily long.

In this section we describe a deterministic and terminating *algorithm* for computing resolutions, which provides a practical basis for implementing tools based on scope graphs, and prove that it is sound and complete with respect to the calculus. This algorithm also connects the calculus, which talks about resolution of a single variable at a time, to more conventional descriptions of binding which use "environments" or "contexts" to describe *all* the visible or reachable declarations accessible from a program location.

For us, an *environment* is just a set of declarations x^D_i. This can be thought of as a function from identifiers to (possible empty) sets of declaration positions. (In this paper, we leave the representation of environments abstract; in practice, one would use a hash table or other dictionary data structure.) We construct an atomic environment corresponding to the declarations in each scope, and then combine atomic environments to describe the sets of reachable and visible declarations resulting from the parent and import relations. The key operator for combining environments is *shadowing*, which returns the union of the declarations in two environments restricted so that if a variable x has any declarations in the first environment, no declarations of x are included from the second environment. More formally:

Definition 1 (Shadowing). *For any environments E_1, E_2, we write:*
$$E_1 \triangleleft E_2 := E_1 \cup \{x^\mathsf{D}_i \in E_2 \mid \not\exists\, x^\mathsf{D}_{i'} \in E_1\}$$

$$Res[\mathbb{I}](x_i^R) \quad := \{x_j^D \mid \exists S \ s.t. \ x_i^R \in \mathcal{R}(S) \wedge x_j^D \in Env_V[\{x_i^R\} \cup \mathbb{I}, \emptyset](S)\}$$

$$Env_V[\mathbb{I}, \mathbb{S}](S) := Env_L[\mathbb{I}, \mathbb{S}](S) \vartriangleleft Env_P[\mathbb{I}, \mathbb{S}](S)$$

$$Env_L[\mathbb{I}, \mathbb{S}](S) := Env_D[\mathbb{I}, \mathbb{S}](S) \vartriangleleft Env_I[\mathbb{I}, \mathbb{S}](S)$$

$$Env_D[\mathbb{I}, \mathbb{S}](S) := \begin{cases} \emptyset \ \text{if} \ S \in \mathbb{S} \\ \mathcal{D}(S) \end{cases}$$

$$Env_I[\mathbb{I}, \mathbb{S}](S) := \begin{cases} \emptyset \ \text{if} \ S \in \mathbb{S} \\ \bigcup \{Env_L[\mathbb{I}, \{S\} \cup \mathbb{S}](S_y) \mid y_i^R \in \mathcal{I}(S) \setminus \mathbb{I} \wedge y_j^D{:}S_y \in Res[\mathbb{I}](y_i^R)\} \end{cases}$$

$$Env_P[\mathbb{I}, \mathbb{S}](S) := \begin{cases} \emptyset \ \text{if} \ S \in \mathbb{S} \\ Env_V[\mathbb{I}, \{S\} \cup \mathbb{S}](\mathcal{P}(S)) \end{cases}$$

Fig. 18. Resolution algorithm

Figure 18 specifies an algorithm $Res[\mathbb{I}](x_i^R)$ for resolving a reference x_i^R to a set of corresponding declarations x_j^D. Like the calculus, the algorithm avoids trying to use an import to resolve itself by maintaining a set $\mathbb{I}$ of "already seen" imports. The algorithm works by computing the full environment $Env_V[\mathbb{I}, \mathbb{S}](S)$ of declarations that are visible in the scope S containing x_i^R, and then extracting just the declarations for x. The full environment, in turn, is built from the more basic environments Env_D of immediate declarations, Env_I of imported declarations, and Env_P of lexically enclosing declarations, using the shadowing operator. The order of construction matches both the *WF* restriction from the calculus, which prevents the use of parent after an import, and the path ordering $<$, which prefers immediate declarations over imports and imports over declarations from the parent scope. (Note that the algorithm does *not* work for the variants of *WF* and $<$ described in Section 2.5.) A key difference from the calculus is that the shadowing operator is applied at each stage in environment construction, rather than applying the visibility criterion just once at the "top level" as in calculus rule *V*. This difference is a natural consequence of the fact that the algorithm computes sets of declarations rather than full derivation paths, so it does not maintain enough information to delay the visibility computation.

Termination The algorithm is terminating using the well-founded lexicographic measure $(|\mathcal{R}(\mathcal{G}) \setminus \mathbb{I}|, |\mathcal{S}(\mathcal{G}) \setminus \mathbb{S}|)$. Termination is straightforward by unfolding the calls to *Res* in Env_I and then inlining the definitions of Env_V and Env_L: this gives an equivalent algorithm in which the measure strictly decreases at every recursive call.

5.1 Correctness of Resolution Algorithm

The resolution algorithm is sound and complete with respect to the calculus.

Theorem 1. $\forall \ \mathbb{I}, x_i^R, j, (x_j^D \in Res[\mathbb{I}](x_i^R)) \iff (\exists p \ s.t. \ \mathbb{I} \vdash p : x_i^R \longmapsto x_j^D).$

We sketch the proof of this theorem here; details of the supporting lemmas and proofs are in Appendix B of [19]. To begin with, we must deal with the

Transitive closure

$$\overline{\mathbb{I}, \mathbb{S} \vdash [] : A \twoheadrightarrow A} \tag{N'}$$

$$\frac{\mathbb{I} \vdash s : A \longrightarrow B \quad B \notin \mathbb{S} \quad \mathbb{I}, \{B\} \cup \mathbb{S} \vdash p : B \twoheadrightarrow C}{\mathbb{I}, \mathbb{S} \vdash s \cdot p : A \twoheadrightarrow C} \tag{T'}$$

Reachable declarations

$$\frac{x_i^{\mathsf{D}} \in \mathcal{D}(S') \quad S \notin \mathbb{S} \quad \mathbb{I}, \{S\} \cup \mathbb{S} \vdash p : S \twoheadrightarrow S' \quad WF(p)}{\mathbb{I}, \mathbb{S} \vdash p \cdot \mathbf{D}(x_i^{\mathsf{D}}) : S \rightarrowtail x_i^{\mathsf{D}}} \tag{R'}$$

Visible declarations

$$\frac{\mathbb{I}, \mathbb{S} \vdash p : S \rightarrowtail x_i^{\mathsf{D}} \quad \forall j, p'(\mathbb{I}, \mathbb{S} \vdash p' : S \rightarrowtail x_j^{\mathsf{D}} \Rightarrow \neg(p' < p))}{\mathbb{I}, \mathbb{S} \vdash p : S \longmapsto x_i^{\mathsf{D}}} \tag{V'}$$

Reference resolution

$$\frac{x_i^{\mathsf{R}} \in \mathcal{R}(S) \quad \{x_i^{\mathsf{R}}\} \cup \mathbb{I}, \emptyset \vdash p : S \longmapsto x_j^{\mathsf{D}}}{\mathbb{I} \vdash p : x_i^{\mathsf{R}} \longmapsto x_j^{\mathsf{D}}} \tag{X'}$$

Fig. 19. "Primed" resolution calculus with "seen scopes" component

fact that the calculus can generate reachability derivations with cycles, but the algorithm does not follow cycles. In fact, *visibility* derivations cannot have cycles:

Lemma 1. *If* $\mathbb{I} \vdash p : x_i^{\mathsf{R}} \longmapsto x_j^{\mathsf{D}}$ *then* p *is cycle-free.*

We therefore begin by defining an alternative version of the calculus that prevents construction of cyclic paths. This alternative calculus consists of the original rules $(P), (I)$ from Figure 3 together with the new rules $(N'), (T'), (R'), (V'), (X')$ from Figure 19. The new rules describe transitions that include a "seen scopes" component $\mathbb{S}$ which is used to enforce acyclicity of paths. By inspection, this is the only difference between the "primed" system and original one. Thus, by Lemma 1, we have

Lemma 2. $\forall \mathbb{I}, \mathbb{S}, x_i^{\mathsf{D}}, (\exists p \text{ s.t. } \mathbb{I} \vdash p : S \longmapsto x_i^{\mathsf{D}}) \iff (\exists p \text{ s.t. } \mathbb{I}, \emptyset \vdash p : S \longmapsto x_i^{\mathsf{D}}).$

Hereinafter, we can work with the primed system.

Next we define a family of sets $\mathbb{P}$ of derivable paths in the (primed) calculus.

Definition 2 (Path Sets).

$$\mathbb{P}_D[\mathbb{I}, \mathbb{S}](S) := \{p \mid \exists\, x_i^{\mathsf{D}} \text{ s.t. } p = \mathbf{D}(x_i^{\mathsf{D}}) \wedge \mathbb{I}, \mathbb{S} \vdash p : S \rightarrowtail x_i^{\mathsf{D}}\}$$

$$\mathbb{P}_P[\mathbb{I}, \mathbb{S}](S) := \{p \mid \exists\, p'\, x_i^{\mathsf{D}} \text{ s.t. } p = \mathbf{P} \cdot p' \wedge$$
$$\mathbb{I}, \mathbb{S} \vdash p : S \rightarrowtail x_i^{\mathsf{D}} \wedge \mathbb{I}, \{S\} \cup \mathbb{S} \vdash p' : \mathcal{P}(S) \longmapsto x_i^{\mathsf{D}}\}$$

$$\mathbb{P}_I[\mathbb{I}, \mathbb{S}](S) := \{p \mid \exists\, p'\, x_i^{\mathsf{D}}\, y_j^{\mathsf{R}}\, y_{j'}^{\mathsf{D}}{:}S' \text{ s.t. } p = \mathbf{I}(y_j^{\mathsf{R}}, y_{j'}^{\mathsf{D}}{:}S') \cdot p' \wedge$$
$$\mathbb{I}, \mathbb{S} \vdash p : S \rightarrowtail x_i^{\mathsf{D}} \wedge \mathbb{I}, \{S\} \cup \mathbb{S} \vdash p' : S' \longmapsto x_i^{\mathsf{D}}\}$$

$$\mathbb{P}_L[\mathbb{I}, \mathbb{S}](S) := \{p \mid \exists\, x_i^{\mathsf{D}} \text{ s.t. } \mathbb{I}, \mathbb{S} \vdash p : S \longmapsto x_i^{\mathsf{D}} \wedge p \in \mathbf{I}(\_, \_)^* \cdot \mathbf{D}(\_)\}$$

$$\mathbb{P}_V[\mathbb{I}, \mathbb{S}](S) := \{p \mid \exists\, x_i^{\mathsf{D}} \text{ s.t. } \mathbb{I}, \mathbb{S} \vdash p : S \longmapsto x_i^{\mathsf{D}}\}$$

These sets are designed to correspond to the various classes of environments Env_C. $\mathbb{P}_D$, $\mathbb{P}_P$, and $\mathbb{P}_I$ contain all reachability derivations starting with a $\mathbf{D}(\_)$, $\mathbf{P}$, or $\mathbf{I}(\_, \_)$ respectively, with the further condition that the *tail* of each derivation is a visibility derivation (i.e. is most specific among all reachability derivations). $\mathbb{P}_V$ describes the set of all visibility derivations. ($\mathbb{P}_L$ is similar, but omits paths including $\mathbf{P}$ steps, because well-formedness prevents using these steps after an import step.) For compactness, we state the key result uniformly over all classes of sets:

Definition 3. *For any path* p, $\delta(p) := x_i^{\mathsf{D}}$ *iff* $\exists p'$ *s.t.* $p = p' \cdot \mathbf{D}(x_i^{\mathsf{D}})$ *and for any set of paths* P, $\Delta(P) := \{\delta(p) \mid p \in P\}$.

Lemma 3. *For each class* $C \in \{V, L, D, I, P\}$:
$$\forall \, \mathbb{I} \, \mathbb{S} \; S, Env_c[\mathbb{I}, \mathbb{S}](S) = \Delta(\mathbb{P}_C[\mathbb{I}, \mathbb{S}](S))$$

Proof. We first prove two auxiliary lemmas about reachability and visibility after one step:

$$\forall \, \mathbb{I} \, \mathbb{S} \; s \; p \; S \; x_i^{\mathsf{D}}, (\mathbb{I}, \mathbb{S} \vdash s \cdot p \cdot \mathbf{D}(x_i^{\mathsf{D}}) : S \rightarrowtail x_i^{\mathsf{D}} \implies \mathbb{I}, \{S\} \cup \mathbb{S} \vdash s : S \longrightarrow S' \implies$$
$$\mathbb{I}, \{S\} \cup \mathbb{S} \vdash p \cdot \mathbf{D}(x_i^{\mathsf{D}}) : S' \rightarrowtail x_i^{\mathsf{D}}) \quad (\Diamond)$$

$$\forall \, \mathbb{I} \, \mathbb{S} \; s \; p \; S \; x_i^{\mathsf{D}}, (\mathbb{I}, \mathbb{S} \vdash s \cdot p : S \longmapsto x_i^{\mathsf{D}} \implies \mathbb{I}, \{S\} \cup \mathbb{S} \vdash s : S \longrightarrow S' \implies$$
$$\mathbb{I}, \{S\} \cup \mathbb{S} \vdash p : S' \longmapsto x_i^{\mathsf{D}}) \quad (\blacklozenge)$$

Then we proceed by three nested inductions, the outer one on $\mathbb{I}$ (or, more strictly, on $|\mathcal{R}(\mathcal{G}) \setminus \mathbb{I}|$, the number of references *not* in $\mathbb{I}$), the second one on $\mathbb{S}$ (more strictly, on $|\mathcal{S}(\mathcal{G}) \setminus \mathbb{S}|$, the number of scopes *not* in $\mathbb{S}$) and the third one on the class C with the order $V > L > P, I, D$. Then we conclude using $\Diamond$ and $\blacklozenge$ and a number of other technical results. Details are in Appendix B of [19].   □

With these lemmas in hand we proceed to prove Theorem 1.

Proof. Fix $\mathbb{I}$, x_i^{R}, and j. Given S, the (unique) scope such that $x_i^{\mathsf{R}} \in \mathcal{R}(S)$:
$$x_j^{\mathsf{D}} \in Res[x_i^{\mathsf{R}}](\mathbb{I}) \Leftrightarrow x_j^{\mathsf{D}} \in Env_V[\{x_i^{\mathsf{R}}\} \cup \mathbb{I}, \emptyset](S)$$
By the V case of Lemma 3 and the definition of $\mathbb{P}_S$, this is equivalent to
$$\exists p \text{ s.t. } \{x_i^{\mathsf{R}}\} \cup \mathbb{I}, \emptyset \vdash p : S \longmapsto x_j^{\mathsf{D}}$$
which, by Lemma 2 and rule X, is equivalent to $\exists p$ s.t. $\mathbb{I} \vdash p : x_i^{\mathsf{R}} \longmapsto x_j^{\mathsf{D}}$.   □

6   α-equivalence and Renaming

The choice of a particular name for a bound identifier should not affect the meaning of a program. This notion of name irrelevance is usually referred to as α-equivalence, but definitions of α-equivalence exist only for some languages and are language-specific. In this section we show how the scope graph and resolution calculus can be used to specify α-equivalence in a language-independent way.

Free variables. A free variable is a reference that does not resolve to any declaration (x_i^R is free if $\nexists\, j, p$ s.t. $\mathbb{I} \vdash p : x_i^R \longmapsto x_j^D$); a bound variable has at least one declaration. For uniformity, we introduce for each possibly free variable x a program-independent artificial declaration $x_{\bar{x}}^D$ with an artificial position $\bar{x}$. These declarations do not belong to any scope but are reachable through a particular well-formed path $\top$, which is less specific than any other path, according to the following rules:

$$\frac{}{\mathbb{I} \vdash \top : S \longmapsto x_{\bar{x}}^D} \qquad \frac{p \neq \top}{p < \top}$$

This path representing the resolution of a free reference is shadowed by any existing path leading to a concrete declaration; therefore the resolution of bound variables is unchanged.

6.1 α-Equivalence

We now define α-equivalence using scope graphs. Except for the leaves representing identifiers, two α-equivalent programs must have the same abstract syntax tree. We write P $\simeq$ P' (pronounced "P and P' are similar") when the ASTs of P and P' are equal up to identifiers. To compare two programs we first compare their AST structures; if these are similar then we compare how identifiers behave in these programs. Since two potentially α-equivalent programs are similar, the identifiers occur at the same positions. In order to compare the identifiers' behavior, we define equivalence classes of positions of identifiers in a program: positions in the same equivalence class are declarations of, or references to, the same entity. The abstract position $\bar{x}$ identifies the equivalence class corresponding to the free variable x.

Given a program P, we write $\mathbb{P}$ for the set of positions corresponding to references and declarations and $\mathbb{PX}$ for $\mathbb{P}$ extended with the artificial positions (e.g. $\bar{x}$). We define the $\overset{P}{\sim}$ equivalence relation between elements of $\mathbb{PX}$ as the reflexive symmetric and transitive closure of the resolution relation.

Definition 4 (Position equivalence).

$$\frac{\mathbb{I} \vdash p : x_i^R \longmapsto x_{i'}^D}{i \overset{P}{\sim} i'} \qquad \frac{i' \overset{P}{\sim} i}{i \overset{P}{\sim} i'} \qquad \frac{i \overset{P}{\sim} i' \quad i' \overset{P}{\sim} i''}{i \overset{P}{\sim} i''} \qquad \frac{}{i \overset{P}{\sim} i}$$

In this equivalence relation, the class containing the abstract free variable declaration cannot contain any other declaration. So the references in a particular class are either all free or all bound.

Lemma 4 (Free variable class). *The equivalence class of a free variable does not contain any other declaration, i.e.* $\forall\, x_i^D, i \overset{P}{\sim} \bar{x} \implies i = \bar{x}$

Proof. Detailed proof is in appendix B of [19]. We first prove:
$\forall\, x_i^R, (\mathbb{I} \vdash \top : x_i^R \longmapsto x_{\bar{x}}^D) \implies \forall\, p\, i', \mathbb{I} \vdash p : x_i^R \longmapsto x_{i'}^D \implies i' = \bar{x} \wedge p = \top$
and then proceed by induction on the equivalence relation.

The equivalence classes defined by this relation contain references to or declarations of the same entity. Given this relation, we can state that two programs are α-equivalent if the identifiers at identical positions refer to the same entity, that belong to the same equivalence class:

Definition 5 (α-equivalence). *Two programs* P1 *and* P2 *are α-equivalent (denoted* P1 $\overset{\alpha}{\approx}$ P2*) when they are similar and have the same $\sim$-equivalence classes:*

$$\text{P1} \overset{\alpha}{\approx} \text{P2} \triangleq \text{P1} \simeq \text{P2} \wedge \forall\, i\ i',\ i \overset{\text{P1}}{\sim} i' \Leftrightarrow i \overset{\text{P2}}{\sim} i'$$

Remark 1. $\overset{\alpha}{\approx}$ is an equivalence relation since $\simeq$ and $\Leftrightarrow$ are equivalence relations.

Free variables. The $\overset{\text{P}}{\sim}$ equivalence classes corresponding to free variables x also contain the artificial position $\bar{x}$. Since the equivalence classes of two equivalent programs P1 and P2 have to be exactly the same, every element equivalent to $\bar{x}$ (i.e. a free reference) in P1 is also equivalent to $\bar{x}$ in P2. Therefore the free references of α-equivalent programs have to be identical.

Duplicate declarations. The definition allows us to also capture α-equivalence of programs with duplicate declarations. Assume that a reference $x_{i_1}^{\text{R}}$ resolves to two definitions $x_{i_2}^{\text{D}}$ and $x_{i_3}^{\text{D}}$; then i_1, i_2 and i_3 belong to the same equivalence class. Thus all α-equivalent programs will have the same ambiguities.

6.2   Renaming

Renaming is the substitution of a bound variable by a new variable throughout the program. It has several practical applications such as rename refactoring in an IDE, transformation to a program with unique identifiers, or as an intermediate transformation when implementing capture-avoiding substitution.

A valid renaming should respect α-equivalence classes. To formalize this idea we first define a generic transformation scheme on programs that also depends on the position of the sub-term to rewrite:

Definition 6 (Position dependent rewrite rule). *Given a program* P, *we denote by* $(t_i \rightarrow t' \mid F)$ *the transformation that replaces the occurrences of the sub-term t at positions i by t' if the condition F is true.* (T)P *denotes the application of the transformation T to the program* P.

Given this definition we can now define the renaming transformation that replaces the identifier corresponding to an entire equivalence class:

Definition 7 (Renaming). *Given a program* P *and a position i corresponding to a declaration or a reference for the name x, we denote by* $[x_i:=y]$P *the program* P*' corresponding to* P *where all the identifiers x at positions $\overset{\text{P}}{\sim}$-equivalent to i are replaced by y:*

$$[x_i := y]\text{P} \triangleq (x_{i'} \rightarrow y \mid i' \overset{\text{P}}{\sim} i)\text{P}$$

However, not every renaming is acceptable: a renaming might provoke variable captures and completely change the meaning of a program.

Definition 8 (Valid renamings). *Given a program* P, *renaming* $[x_i := y]$ *is valid only if it produces an α-equivalent program, i.e.* $[x_i := y]P \overset{\alpha}{\approx} P$

Remark 2. This definition prevents the renaming of free variables since α-equivalent programs have exactly the same free variables.

Intuitively, valid renamings are those that do not accidentally "capture" variables. Since the capture of a reference resolution also depends on the seen-import context in which this resolution occurs, a precise characterization of capture in our general setting is complex and we leave it for future work.

7   Related Work

Binding-sensitive Program Representations. There has been a great deal of work on representing program syntax in ways that take explicit note of binding structure, usually with the goal of supporting program transformation or mechanized reasoning tools that respect α-equivalence by construction. Notable techniques include de Bruijn indexing [7], Higher-Order Abstract Syntax (HOAS) [20], locally nameless representations [3], and nominal sets [10]. (Aydemir, et al. [2] give a survey in the context of mechanized reasoning.) However, most of this work has concentrated on simple lexical binding structures, such as single-argument λ-terms. Cheney [4] gives a catalog of more interesting binding patterns and suggests how nominal logic can be used to describe many of them. However, he leaves treatment of module imports as future work.

Binding Specification Languages. The *Ott* system [22] allows definition of syntax, name binding and semantics. This tool generates language definitions for theorem provers along with a notion of α-equivalence and functions such as capture-avoiding substitution that can be proven correct in the chosen proof assistant modulo α-equivalence. Avoiding capture is also the basis of hygienic macros in Scheme. Dybvig [8] gives an algorithmic description of what hygiene means. Herman and Wand [13,12] introduce static binding specifications to formalize a notion of α-equivalence that does not depend on macro expansion. Stansifer and Wand's Romeo system [23] extends these specifications to somewhat more elaborate binding forms, such as sequential **let**. *Unbound* [25] is another recent domain specific language for describing bindings that supports moderately complex binding forms. Again, none of these systems treat modules or imports.

Language Engineering. In language engineering approaches, name bindings are often realized using a random-access symbol table such that multiple analysis and transformation stages can reuse the results of a single name resolution pass [1]. Another approach is to represent the result of name resolution by means of *reference attributes*, direct pointers from the uses of a name to its definition [11]. However these representations are usually built using an implementation of a language-specific resolution algorithm. Erdweg, et al. [9] describe a system for defining capture-free transformations, assuming resolution algorithms are provided for the source and target languages. The approach represents the result of name resolution using 'name graphs' that map uses to definitions (references to declarations in our terminology) and are language independent. This notion of 'name graph' inspired our notion of 'scope graph'. The key difference is that the results of name resolution generated by the resolution calculus are *paths* that extend a use-def pair with the *language-independent evidence* for the resolution.

Semantics Engineering. Semantics engineering approaches to name binding vary from first-order representation with substitution [15], to explicit or implicit environment propagation [21,18,6], to HOAS [5]. Identifier bindings represented with environments are passed along in derivation rules, rediscovering bindings for each operation. This approach is inconvenient for more complex patterns such as mutually recursive definitions.

8 Conclusion and Future Work

We have introduced a generic, language-independent framework for describing name binding in programming languages. Its theoretical basis is the notion of a scope graph, which abstracts away from syntax, together with a calculus for deriving resolution paths in the graph. Scope graphs are expressive enough to describe a wide range of binding patterns found in real languages, in particular those involving modules or classes. We have presented a practical resolution algorithm, which is provably correct with respect to the resolution calculus. We can use the framework to define generic notions of α-equivalence and renaming.

As future work, we plan to explore and extend the theory of scope graphs, in particular to find ways to rule out anomalous examples and to give precise characterizations of variable capture and substitution. On the practical side, we will use our formalism to give a precise semantics to the NaBL DSL, and verify (using proof and/or testing) that the current NaBL implementation conforms to this semantics.

Our broader vision is that of a complete language designer's workbench that includes NaBL as the domain-specific language for name binding specification and also includes languages for type systems and dynamic semantics specifications. In this setting, we also plan to study the interaction of name resolution and types, including issues of dependent types and name disambiguation based on types. Eventually we aim to derive a complete mechanized meta-theory for the languages defined in this workbench and to prove the correspondence between static name binding and name binding in dynamics semantics as outlined in [24].

Acknowledgments. We thank the many people who reacted to our previous work on NaBL by asking "but what is its semantics?"; this paper provides our answer. We thank the anonymous reviewers for their feedback on previous versions of this paper. This research was partially funded by the NWO VICI *Language Designer's Workbench* project (639.023.206). Andrew Tolmach was partly supported by a Digiteo Chair at Laboratoire de Recherche en Informatique, Université Paris-Sud.

References

1. Aho, A.V., Sethi, R., Ullman, J.D.: Compilers: Principles, Techniques, and Tools. Addison-Wesley (1986)
2. Aydemir, B.E., Charguéraud, A., Pierce, B.C., Pollack, R., Weirich, S.: Engineering formal metatheory. In: Necula, G.C., Wadler, P. (eds.) Proceedings of the 35th ACM SIGPLAN-SIGACT Symposium on Principles of Programming Languages, POPL 2008, San Francisco, California, USA, January 7-12, pp. 3–15. ACM (2008)
3. Charguéraud, A.: The locally nameless representation. Journal of Automated Reasoning 49(3), 363–408 (2012)
4. Cheney, J.: Toward a general theory of names: binding and scope. In: Pollack, R. (ed.) ACM SIGPLAN International Conference on Functional Programming, Workshop on Mechanized Reasoning About Languages with Variable Binding, MERLIN 2005, Tallinn, Estonia, pp. 33–40. ACM (September 30, 2005)
5. Chlipala, A.J.: A verified compiler for an impure functional language. In: Hermenegildo, M.V., Palsberg, J. (eds.) Proceedings of the 37th ACM SIGPLAN-SIGACT Symposium on Principles of Programming Languages, POPL 2010, Madrid, Spain, January 17-23, pp. 93–106. ACM (2010)
6. Churchill, M., Mosses, P.D., Torrini, P.: Reusable components of semantic specifications. In: Binder, W., Ernst, E., Peternier, A., Hirschfeld, R. (eds.) 13th International Conference on Modularity, MODULARITY 2014, Lugano, Switzerland, April 22-26, pp. 145–156. ACM (2014)
7. de Bruijn, N.G.: Lambda calculus notation with nameless dummies, a tool for automatic formula manipulation, with application to the Church-Rosser theorem. Indagationes Mathematicae 34(5), 381–392 (1972)
8. Dybvig, R.K., Hieb, R., Bruggeman, C.: Syntactic abstraction in scheme. Higher-Order and Symbolic Computation 5(4), 295–326 (1992)
9. Erdweg, S., van der Storm, T., Dai, Y.: Capture-avoiding and hygienic program transformations. In: Jones, R. (ed.) ECOOP 2014. LNCS, vol. 8586, pp. 489–514. Springer, Heidelberg (2014)
10. Gabbay, M., Pitts, A.M.: A new approach to abstract syntax with variable binding. Formal Asp. Comput. 13(3-5), 341–363 (2002)
11. Hedin, G., Magnusson, E.: Jastadd–an aspect-oriented compiler construction system. Science of Computer Programming 47(1), 37–58 (2003)
12. Herman, D.: A Theory of Hygienic Macros. PhD thesis, Northeastern University, Boston, Massachusetts (May 2010)
13. Herman, D., Wand, M.: A theory of hygienic macros. In: Drossopoulou, S. (ed.) ESOP 2008. LNCS, vol. 4960, pp. 48–62. Springer, Heidelberg (2008)
14. Kats, L.C.L., Visser, E.: The Spoofax language workbench: rules for declarative specification of languages and IDEs. In: Cook, W.R., Clarke, S., Rinard, M.C. (eds.) Proceedings of the 25th Annual ACM SIGPLAN Conference on Object-Oriented Programming, Systems, Languages, and Applications, OOPSLA 2010, Reno/Tahoe, Nevada, pp. 444–463. ACM (2010)

15. Klein, C., Clements, J., Dimoulas, C., Eastlund, C., Felleisen, M., Flatt, M., McCarthy, J.A., Rafkind, J., Tobin-Hochstadt, S., Findler, R.B.: Run your research: on the effectiveness of lightweight mechanization. In: Field, J., Hicks, M. (eds.) Proceedings of the 39th ACM SIGPLAN-SIGACT Symposium on Principles of Programming Languages, POPL 2012, Philadelphia, Pennsylvania, USA, January 22-28, pp. 285–296. ACM (2012)
16. Konat, G., Kats, L., Wachsmuth, G., Visser, E.: Declarative name binding and scope rules. In: Czarnecki, K., Hedin, G. (eds.) SLE 2012. LNCS, vol. 7745, pp. 311–331. Springer, Heidelberg (2013)
17. Leroy, X., Doligez, D., Frisch, A., Garrigue, J., Rémy, D., Vouillon, J.: The OCaml system (release 4.00): Documentation and user's manual. Institut National de Recherche en Informatique et en Automatique (July 2012)
18. Mosses, P.D.: Modular structural operational semantics. Journal of Logic and Algebraic Programming 61-61, 195–228 (2004)
19. Neron, P., Tolmach, A.P., Visser, E., Wachsmuth, G.: A theory of name resolution with extended coverage and proofs. Technical Report TUD-SERG-2015-001, Software Engineering Research Group. Delft University of Technology, Extended version of this paper (January 2015)
20. Pfenning, F., Elliott, C.: Higher-order abstract syntax. In: Wexelblat, R.L. (ed.) Proceedings of the ACM SIGPLAN 1988 Conference on Programming Language Design and Implementation (PLDI), Atlanta, Georgia, USA, June 22-24, pp. 199–208. ACM (1988)
21. Pierce, B.C.: Types and Programming Languages. MIT Press, Cambridge (2002)
22. Sewell, P., Nardelli, F.Z., Owens, S., Peskine, G., Ridge, T., Sarkar, S., Strnisa, R.: Ott: Effective tool support for the working semanticist. Journal of Functional Programming 20(1), 71–122 (2010)
23. Stansifer, P., Wand, M.: Romeo: A system for more flexible binding-safe programming. In: Jeuring, J., Chakravarty, M.M.T. (eds.) Proceedings of the 19th ACM SIGPLAN International Conference on Functional Programming, Gothenburg, Sweden, September 1-3, pp. 53–65. ACM (2014)
24. Visser, E., Wachsmuth, G., Tolmach, A.P., Neron, P., Vergu, V.A., Passalaqua, A., Konat, G.D.P.: A language designer's workbench: A one-stop-shop for implementation and verification of language designs. In: Black, A.P., Krishnamurthi, S., Bruegge, B., Ruskiewicz, J.N. (eds.) Onward! 2014, Proceedings of the 2014 ACM International Symposium on New Ideas, New Paradigms, and Reflections on Programming & Software, part of SLASH 2014, Portland, OR, USA, October 20-24, pp. 95–111. ACM (2014)
25. Weirich, S., Yorgey, B.A., Sheard, T.: Binders unbound. In: Chakravarty, M.M.T., Hu, Z., Danvy, O. (eds.) Proceeding of the 16th ACM SIGPLAN International Conference on Functional Programming, ICFP 2011, Tokyo, Japan, September 19-21, pp. 333–345. ACM (2011)

A Core Calculus for XQuery 3.0

Combining Navigational and Pattern Matching Approaches

Giuseppe Castagna[1], Hyeonseung Im[2], Kim Nguyễn[3], and Véronique Benzaken[3]

[1] CNRS, PPS, Univ. Paris Diderot, Sorbonne Paris Cité, Paris, France
[2] Inria, LIG, Univ. Grenoble-Alpes, Grenoble, France
[3] LRI, Université Paris-Sud, Orsay, France

Abstract. XML processing languages can be classified according to whether they extract XML data by paths or patterns. The strengths of one category correspond to the weaknesses of the other. In this work, we propose to bridge the gap between these two classes by considering two languages, one in each class: XQuery (for path-based extraction) and CDuce (for pattern-based extraction). To this end, we extend CDuce so as it can be seen as a succinct core λ-calculus that captures XQuery 3.0. The extensions we consider essentially allow CDuce to implement XPath-like navigational expressions by pattern matching and precisely type them. The elaboration of XQuery 3.0 into the extended CDuce provides a formal semantics and a sound static type system for XQuery 3.0 programs.

1 Introduction

With the establishment of XML as a standard for data representation and exchange, a wealth of XML-oriented programming languages have emerged. They can be classified into two distinct classes according to whether they extract XML data by applying paths or patterns. The strengths of one class correspond to the weaknesses of the other. In this work, we propose to bridge the gap between these classes and to do so we consider two languages each representing a distinct class: XQuery and CDuce.

XQuery [23] is a declarative language standardized by the W3C that relies heavily on XPath [21,22] as a data extraction primitive. Interestingly, the latest version of XQuery (version 3.0, very recently released [25]) adds several functional traits: type and value case analysis and functions as first-class citizens. However, while the W3C specifies a standard for document types (XML Schema [26]), it says little about the typing of XQuery programs (the XQuery 3.0 recommendation goes as far as saying that static typing is "implementation defined" and hence optional). This is a step back from the XQuery 1.0 Formal Semantics [24] which gives sound (but sometime imprecise) typing rules for XQuery.

In contrast, CDuce [4], which is used in production but issued from academic research, is a statically-typed functional language with, in particular, higher-order functions and powerful pattern matching tailored for XML data. Its key characteristic is its type algebra, which is based on *semantic subtyping* [10] and features recursive types, type constructors (product, record, and arrow types)

© Springer-Verlag Berlin Heidelberg 2015
J. Vitek (Ed.): ESOP 2015, LNCS 9032, pp. 232–256, 2015.
DOI: 10.1007/978-3-662-46669-8_10

XQuery code

```
 1  declare function get_links($page, $print) {
 2     for $i in $page/descendant::a[not(ancestor::b)]
 3     return $print($i)
 4  }
 5  declare function pretty($link) {
 6     typeswitch($link)
 7     case $l as element(a)
 8        return switch ($l/@class)
 9           case "style1"
10              return <a href={$l/@href}><b>{$l/text()}</b></a>
11           default return $l
12     default return $link
13  }
```

CDuce code

```
14  let get_links (page: <_>_) (print: <a>_ -> <a>_) : [ <a>_ * ] =
15     match page with
16        <a>_ & x -> [ (print x) ]
17      | < (_\`b) > l -> (transform l with (i & <_>_) -> get_links i print)
18      | _ -> [ ]
19  let pretty (<a>_ -> <a>_ ; Any\<a>_ -> Any\<a>_)
20      | <a class="style1" href=h ..> l -> <a href=h>[ <b>l ]
21      | x -> x
```

Fig. 1. Document transformation in XQuery 3.0 and CDuce

and general Boolean connectives (union, intersection, and negation of types) as well as singleton types. This type algebra is particularly suited to express the types of XML documents and relies on the same foundation as the one that underpins XML Schema: regular tree languages. Moreover, the CDuce type system not only supports *ad-hoc* polymorphism (through overloading and subtyping) but also has recently been extended with parametric polymorphism [5,6].

Figure 1 highlights the key features as well as the shortcomings of both languages by defining the same two functions *get_ links* and *pretty* in each language. Firstly, *get_ links (i)* takes an XHTML document *$page* and a function *$print* as input, *(ii)* computes the sequence of all hypertext links (a-labelled elements) of the document that do not occur below a bold element (b-labelled elements), and *(iii)* applies the *print* argument to each link in the sequence, returning the sequence of the results. Secondly, *pretty* takes anything as argument and performs a case analysis. If the argument is a link whose class attribute has the value "style1", the output is a link with the same target (href attribute) and whose text is embedded in a bold element. Otherwise, the argument is unchanged.

We first look at the *get_ links* function. In XQuery, collecting every "a" element of interest is straightforward: it is done by the XPath expression at Line 2:

$$\$page/\text{descendant::a[not(ancestor::b)]}$$

In a nutshell, an XPath expression is a sequence of steps that *(i)* select sets of nodes along the specified axis (here `descendant` meaning the descendants of the root node of *$page*), *(ii)* keep only those nodes in the axis that have a particular label (here "a"), and *(iii)* further filter the results according to a Boolean condition (here `not(ancestor::b)` meaning that from a candidate "a" node, the step `ancestor::b` must return an empty result). At Lines 2–3, the `for_return` expression binds in turn each element of the result of the XPath expression to the variable *$i*, evaluates the `return` expression, and concatenates the results. Note that there is no type annotation and that this function would fail at runtime if *$page* is not an XML element or if *$print* is not a function.

In clear contrast, in the CDuce program, the interface of *get_links* is fully specified (Line 14). It is curried and takes two arguments. The first one is *page* of type `<_>_`, which denotes any XML element (`_` denotes a wildcard pattern and is a synonym of the type `Any`, the type of all values, while `<s>t` is the type of an XML element with tag of type *s* and content of type *t*). The second argument is *print* of type `<a>_` → `<a>_`, which is the type of functions that take an "a" element (whose content is anything) and return an "a" element. The final output is a value of type `[<a>_*]`, which denotes a possibly empty sequence of "a" elements (in CDuce's types, the content of a sequence is described by a regular expression on types). The implementation of *get_links* in CDuce is quite different from its XQuery counterpart: following the functional idiom, it is defined as a recursive function that traverses its input recursively and performs a case analysis through pattern matching. If the input is an "a" element (Line 16), it binds the input to the capture variable *x*, evaluates *print x*, and puts the result in a sequence (denoted by square brackets). If the input is an XML element whose tag is *not* b ("\" stands for difference, so `_\'b` matches any value different from b)[1], it captures the content of the element (a sequence) in *l* and applies itself recursively to each element of *l* using the `transform_with` construct whose behavior is the same as XQuery's `for`. Lastly, if the result is not an element (or it is a "b" element), it stops the recursion and returns the empty sequence.

For the *pretty* function (which is inspired from the example given in §3.16.2 of the XQuery 3.0 recommendation [25]), the XQuery version (Lines 5–13) first performs a "type switch", which tests whether the input *$link* has label a. If so, it extracts the value of the `class` attribute using an XPath expression (Line 8) and performs a case analysis on that value. In the case where the attribute is `"style1"`, it re-creates an "a" element (with a nested "b" element) extracting the relevant part of the input using XPath expressions. The CDuce version (Lines 19–21) behaves in the same way but collapses all the cases in a single pattern matching. If the input is an "a" element with the desired `class` attribute, it binds the contents of the `href` attribute and the element to the variables *h* and *l*, respectively (the ".." matches possible further attributes), and builds the desired output; otherwise, the input is returned unchanged. Interestingly, this function is *overloaded*. Its signature is composed of two arrow types: if the input is an "a" element, so is the output; if the input is something else than an "a" element, so

---

[1] In CDuce, one has to use 'b in conjunction with \ to denote XML tag b.

is the output (& in types and patterns stands for intersection). Note that it is safe to use the *pretty* function as the second argument of the *get_ links* function since (`<a>_→<a>_`) & (`Any\<a>_→Any\<a>_`) is a subtype of `<a>_→<a>_` (an intersection is always smaller than or equal to the types that compose it).

Here we see that the strength of one language is the weakness of the other: CDuce provides static typing, a fine-grained type algebra, and a pattern matching construct that cleanly unifies type and value case analysis. XQuery provides through XPath a declarative way to navigate a document, which is more concise and less brittle than using hand-written recursive functions (in particular, at Line 16 in the CDuce code, there is an implicit assumption that a link cannot occur below another link; the recursion stops at "a" elements).

Contributions. The main contribution of the paper is to unify the navigational and pattern matching approaches and to define a formal semantics and type system of XQuery 3.0. Specifically, we extend CDuce so as it can be seen as a succinct core λ-calculus that can express XQuery 3.0 programs as follows.

First, we allow one to navigate in CDuce values, both downward and upward. A natural way to do so in a functional setting is to use *zippers à la* Huet [18] to annotate values. Zippers denote the position in the surrounding tree of the value they annotate as well as its current path from the root. We extend CDuce not only with zipped values (*i.e.*, values annotated by zippers) but also with *zipped types*. By doing so, we show that we can navigate not only in any direction in a document but also in a *precisely typed* way, allowing one to express constraints on the path in which a value is within a document.

Second, we extend CDuce pattern matching with accumulating variables that allow us to encode *recursive* XPath axes (such as **descendant** and **ancestor**). It is well known that typing such recursive axes goes well beyond regular tree languages and that approximations in the type system are needed. Rather than giving ad-hoc built-in functions for **descendant** and **ancestor**, we define the notion of *type operators* and parameterize the CDuce type system (and dynamic semantics) with these operators. Soundness properties can then be shown in a modular way without hard-coding any specific typing rules in the language. With this addition, XPath navigation can be encoded simply in CDuce's pattern matching constructs and it is just a matter of syntactic sugar definition to endow CDuce with nice declarative navigational expressions such as those successfully used in XQuery or XSLT.

The last (but not least) step of our work is to define a "normal form" for XQuery 3.0 programs, extending both the original XQuery Core normal form of [24] and its recent adaptation to XQuery 3.0 (dubbed XQ$_H$) proposed by Benedikt and Vu [3]. In this normal form, navigational (*i.e.*, structural) expressions are well separated from data value expressions (ordering, node identity testing, *etc.*). We then provide a translation from XQuery 3.0 Core to CDuce extended with navigational patterns. The encoding provides for free an effective and efficient typechecking algorithm for XQuery 3.0 programs (described in Figure 9 of Section 5.1) as well as a formal and compact specification of their semantics. Even more interestingly, it provides a solid formal basis to start further studies on the

Pre-values	w	$::=$	$c \mid (w,w) \mid \mu f^{(t\to t;\ldots;t\to t)}(x).e$
Zippers	δ	$::=$	$\bullet \mid \mathsf{L}\,(w)_\delta \cdot \delta \mid \mathsf{R}\,(w)_\delta \cdot \delta$
Values	v	$::=$	$w \mid (v,v) \mid (w)_\delta$
Expressions	e	$::=$	$v \mid x \mid \dot{x} \mid (e,e) \mid (e)_\bullet \mid o(e,\ldots,e)$
		$\mid$	match e with $p \to e \mid p \to e$
Pre-types	u	$::=$	$b \mid c \mid u \times u \mid u \to u \mid u \vee u \mid \neg u \mid \mathbb{0}$
Zipper types	τ	$::=$	$\bullet \mid \top \mid \mathsf{L}\,(u)_\tau \cdot \tau \mid \mathsf{R}\,(u)_\tau \cdot \tau \mid \tau \vee \tau \mid \neg \tau$
Types	t	$::=$	$u \mid t \times t \mid t \to t \mid t \vee t \mid \neg t \mid (u)_\tau$
Pre-patterns	q	$::=$	$t \mid x \mid \dot{x} \mid (q,q) \mid q\mid q \mid q \& q \mid (x := c)$
Zipper patterns	φ	$::=$	$\tau \mid \mathsf{L}\,p \cdot \varphi \mid \mathsf{R}\,p \cdot \varphi \mid \varphi \mid \varphi$
Patterns	p	$::=$	$q \mid (p,p) \mid p\mid p \mid p \& p \mid (q)_\varphi$

Fig. 2. Syntax of expressions, types, and patterns

definition of XQuery 3.0 and its properties. *A minima*, it is straightforward to use this basis to add overloaded functions to XQuery (*e.g.*, to give a precise type to *pretty*). More crucially, the recent advances on polymorphism for semantic subtyping [5,6,7] can be transposed to this basis to provide a polymorphic type system and type inference algorithm both to XQuery 3.0 and to the extended ℂDuce language defined here. Polymorphic types are the missing ingredient to make higher-order functions yield their full potential and to remove any residual justification of the absence of standardization of the XQuery 3.0 type system.

Plan. Section 2 presents the core typed λ-calculus equipped with zipper-annotated values, accumulators, constructors, recursive functions, and pattern matching. Section 3 gives its semantics, type system, and the expected soundness property. Section 4 turns this core calculus into a full-fledged language using several syntactic constructs and encodings. Section 5 uses this language as a compilation target for XQuery. Lastly, Section 6 compares our work to other related approaches and concludes. Proofs and some technical definitions are given in an online appendix available at http://www.pps.univ-paris-diderot.fr/~gc/.

2 Syntax

We extend the ℂDuce language [4] with zippers *à la* Huet [18]. To ensure the well-foundedness of the definition, we stratify it, introducing first pre-values (which are standard ℂDuce values) and then values, which are pre-values possibly indexed by a zipper; we proceed similarly for types and patterns. The definition is summarized in Figure 2. Henceforth we denote by $\mathcal{V}$ the set of all values and by Ω a special value that represents runtime error and does not inhabit any type. We also denote by $\mathcal{E}$ and $\mathcal{T}$ the set of all expressions and all types, respectively.

2.1 Values and Expressions

Pre-values (ranged over by w) are the usual ℂDuce values without zipper annotations. Constants are ranged over by c and represent integers (1, 2, ...), characters ('a', 'b', ...), atoms ('nil, 'true, 'false, 'foo, ...), *etc.* A value (w, w) represents pairs of pre-values. Our calculus also features recursive functions (hence the μ binder instead of the traditional λ) with explicit, overloaded types (the set of types that index the recursion variable, forming the *interface* of the function). Values (ranged over by v) are pre-values, pairs of values, or pre-values annotated with a *zipper* (ranged over by δ). Zippers are used to record the path covered when traversing a data structure. Since the product is the only construct, we need only three kinds of zippers: the empty one (denoted by •) which intuitively denotes the starting point of our navigation, and two zippers $\mathsf{L}\,(w)_\delta \cdot \delta$ and $\mathsf{R}\,(w)_\delta \cdot \delta$ which denote respectively the path to the left and right projection of a pre-value w, which is itself reachable through δ. To ease the writing of several zipper related functions, we chose to record in the zipper the whole "stack" of values we have visited (each tagged with a left or right indication), instead of just keeping the unused component as is usual.

Example 1. Let v be the value $((1, (2, 3)))_\bullet$. Its first projection is the value $(1)_{\mathsf{L}\,((1,(2,3)))_\bullet \cdot \bullet}$ and its second projection is the value $((2, 3))_{\mathsf{R}\,((1,(2,3)))_\bullet \cdot \bullet}$, the first projection of which being $(2)_{\mathsf{L}\,((2,3))_{\mathsf{R}\,((1,(2,3)))_\bullet \cdot \bullet} \cdot \mathsf{R}\,((1,(2,3)))_\bullet \cdot \bullet}$

As one can see in this example, keeping values in the zipper (instead of pre-values) seems redundant since the same value occurs several times (see how δ is duplicated in the definition of zippers). The reason for this duplication is purely syntactic: it makes the writing of types and patterns that match such values much shorter (intuitively, to go "up" in a zipper, it is only necessary to extract the previous value while keeping it un-annotated —*i.e.*, having $\mathsf{L}\,w \cdot \delta$ in the definition instead of $\mathsf{L}\,(w)_\delta \cdot \delta$— would require a more complex treatment to reconstruct the parent). We also stress that zipped values are meant to be used only for internal representation: the programmer will be allowed to write just pre-values (not values or expressions with zippers) and be able to obtain and manipulate zippers only by applying CDuce functions and pattern matching (as defined in the rest of the paper) and never directly.

Expressions include values (as previously defined), variables (ranged over by x, y, ...), accumulators (which are a particular kind of variables, ranged over by $\dot{x}$, $\dot{y}$, ...), and pairs. An expression $(e)_\bullet$ annotates e with the empty zipper •. The pattern matching expression is standard (with a first match policy) and will be thoroughly presented in Section 3. Our calculus is parameterized by a set $\mathcal{O}$ of built-in operators ranged over by o. Before describing the use of operators and the set of operators defined in our calculus (in particular the operators for projection and function application), we introduce our type algebra.

2.2 Types

We first recall the CDuce type algebra, as defined in [10], where types are interpreted as sets of values and the subtyping relation is semantically defined by using this interpretation (*i.e.*, $[\![t]\!] = \{v \mid \vdash v : t\}$ and $s \leq t \overset{\text{def}}{\iff} [\![s]\!] \subseteq [\![t]\!]$).

Pre-types u (as defined in Figure 2) are the usual CDuce types, which are possibly infinite terms with two additional requirements:

1. (regularity) the number of distinct subterms of u is finite;
2. (contractiveness) every infinite branch of u contains an infinite number of occurrences of either product types or function types.

We use b to range over basic types (int, bool, ...). A singleton type c denotes the type that contains only the constant value c. The empty type $\mathbb{0}$ contains no value. Product and function types are standard: $u_1 \times u_2$ contains all the pairs (w_1, w_2) for $w_i \in u_i$, while $u_1 \to u_2$ contains all the (pre-)value functions that when applied to a value in u_1, if such application terminates then it returns a value in u_2. We also include type connectives for union and negation (intersections are encoded below) with their usual set-theoretic interpretation. Infiniteness of pre-types accounts for recursive types and regularity implies that pre-types are finitely representable, for instance, by recursive equations or by the explicit μ-notation. Contractiveness [2] excludes both ill-formed (*i.e.*, unguarded) recursions such as $\mu X.X$ as well as meaningless type definitions such as $\mu X.X \vee X$ or $\mu X.\neg X$ (unions and negations are finite). Finally, subtyping is defined as set-theoretic containment (u_1 is a subtype of u_2, denoted by $u_1 \leq u_2$, if all values in u_1 are also in u_2) and it is decidable in EXPTIME (see [10]).

A *zipper type* τ is a possibly infinite term that is regular as for pre-types and contractive in the sense that every infinite branch of τ must contain an infinite number of occurrences of either left or right projection. The singleton type $\bullet$ is the type of the empty zipper and $\top$ denotes the type of all zippers, while $\mathsf{L}\,(u)_\tau \cdot \tau$ (resp., $\mathsf{R}\,(u)_\tau \cdot \tau$) denotes the type of zippers that encode the left (resp., right) projection of some value of pre-type u. We use $\tau_1 \wedge \tau_2$ to denote $\neg(\neg \tau_1 \vee \neg \tau_2)$.

The type algebra of our core calculus is then defined as pre-types possibly indexed by zipper types. As for pre-types, a *type* t is a possibly infinite term that is both regular and contractive. We write $t \wedge s$ for $\neg(\neg t \vee \neg s)$, $t \setminus s$ for $t \wedge \neg s$, and $\mathbb{1}$ for $\neg \mathbb{0}$; in particular, $\mathbb{1}$ denotes the super-type of all types (it contains all values). We also define the following notations (we use $\equiv$ both for syntactic equivalence and definition of syntactic sugar):

- $\mathbb{1}_{\mathsf{prod}} \equiv \mathbb{1} \times \mathbb{1}$ the super-type of all product types
- $\mathbb{1}_{\mathsf{fun}} \equiv \mathbb{0} \to \mathbb{1}$ the super-type of all arrow types
- $\mathbb{1}_{\mathsf{basic}} \equiv \mathbb{1} \setminus (\mathbb{1}_{\mathsf{prod}} \vee \mathbb{1}_{\mathsf{fun}} \vee (\mathbb{1})_\top)$ the super-type of all basic types
- $\mathbb{1}_{\mathsf{NZ}} \equiv \mu X.(X \times X) \vee (\mathbb{1}_{\mathsf{basic}} \vee \mathbb{1}_{\mathsf{fun}})$ the type of all pre-values (*i.e.*, Not Zipped)

It is straightforward to extend the subtyping relation of pre-types (*i.e.*, the one defined in [10]) to our types: the addition of $(u)_\tau$ corresponds to the addition of a new type constructor (akin to $\to$ and $\times$) to the type algebra. Therefore, it suffices to define the interpretation of the new constructor to complete the definition of the subtyping relation (defined as containment of the interpretations). In particular, $(u)_\tau$ is interpreted as the set of all values $(w)_\delta$ such that $\vdash w : u$ and $\vdash \delta : \tau$ (both typing judgments are defined in Appendix B.1). From this we deduce that $(\mathbb{1})_\top$ (equivalently, $(\mathbb{1}_{\mathsf{NZ}})_\top$) is the type of all (pre-)values decorated with a zipper. The formal definition is more involved (see Appendix A) but the intuition is simple: a type $(u_1)_{\tau_1}$ is a subtype of $(u_2)_{\tau_2}$ if $u_1 \leq u_2$ and τ_2 is a prefix (modulo type equivalence and subtyping) of τ_1. The prefix containment translates the intuition that the more we know about the context surrounding a value, the more numerous are the situations in which it can be safely used. For instance, in XML terms, if we have a function that expects an element whose parent's first child is an integer, then we can safely apply this function to an element whose type indicates that its parent's first child has type (a subtype of) integer *and* that its grandparent is, say, tagged by a.

Finally, as for pre-types, the subtyping relation for types is decidable in EX-PTIME. This is easily shown by producing a straightforward linear encoding of zipper types and zipper values in pre-types and pre-values, respectively (the encoding is given in Definition 16 in Appendix A).

2.3   Operators and Accumulators

As previously explained, our calculus includes accumulators and is parameterized by a set $\mathcal{O}$ of operators. These have the following formal definitions:

Definition 2 (Operator). *An operator is a 4-tuple* $(o, n_o, \overset{o}{\rightsquigarrow}, \overset{o}{\rightarrow})$ *where o is the name (symbol) of the operator, n_o is its arity,* $\overset{o}{\rightsquigarrow} \subseteq \mathcal{V}^{n_o} \times \mathcal{E} \cup \{\Omega\}$ *is its reduction relation, and* $\overset{o}{\rightarrow} : \mathcal{T}^{n_o} \to \mathcal{T}$ *is its typing function.*

In other words, an operator is an applicative symbol, equipped with both a dynamic ($\rightsquigarrow$) and a static ($\rightarrow$) semantics. The reason for making $\overset{o}{\rightsquigarrow}$ a relation is to account for non-deterministic operators (*e.g.*, random choice). Note that an operator may fail, thus returning the special value Ω during evaluation.

Definition 3 (Accumulator). *An accumulator $\dot{x}$ is a variable equipped with a binary operator $\mathsf{Op}(\dot{x}) \in \mathcal{O}$ and initial value $\mathsf{Init}(\dot{x}) \in \mathcal{V}$.*

2.4   Patterns

Now that we have defined types and operators, we can define patterns. Intuitively, patterns are types with capture variables that are used either to extract subtrees from an input value or to test its "shape". As before, we first recall the definition of standard CDuce patterns (here called pre-patterns), enrich them with accumulators, and then extend the whole with zippers.

A pre-pattern q, as defined in Figure 2, is either a type constraint t, or a capture variable x, or an accumulator $\dot{x}$, or a pair (q_1, q_2), or an alternative $q_1 \,|\, q_2$, or a conjunction $q_1 \,\&\, q_2$, or a default case $(x := c)$. It is a possibly infinite term that is regular as for pre-types and contractive in the sense that every infinite branch of q must contain an infinite number of occurrences of pair patterns. Moreover, the subpatterns forming conjunctions must have distinct capture variables and those forming alternatives the same capture variables. A *zipper pattern* φ is a possibly infinite term that is both regular and contractive as for zipper types. Finally, a pattern p is a possibly infinite term with the same requirements as pre-patterns. Besides, the subpatterns q and φ forming a zipper pattern $(q)_\varphi$ must have distinct capture variables. We denote by $\mathrm{Var}(p)$ the set of capture variables occurring in p and by $\mathrm{Acc}(p)$ the set of accumulators occurring in p.

$$E ::= [\,] \mid (E, e) \mid (e, E) \mid (E)_\bullet \mid \text{match } E \text{ with } p_1 \to e_1 \mid p_2 \to e_2 \mid o(e, ..., E, ..., e)$$

$$\frac{(v_1, \ldots, v_{n_o}) \overset{o}{\rightsquigarrow} e}{o(v_1, \ldots, v_{n_o}) \rightsquigarrow e} \qquad \frac{\{\dot{x} \mapsto \mathsf{Init}(\dot{x}) \mid \dot{x} \in \mathrm{Acc}(p_1)\}; \square \vdash v/p_1 \rightsquigarrow \sigma, \gamma}{\text{match } v \text{ with } p_1 \to e_1 \mid p_2 \to e_2 \rightsquigarrow e_1[\sigma; \, \gamma]}$$

$$\frac{\{\dot{x} \mapsto \mathsf{Init}(\dot{x}) \mid \dot{x} \in \mathrm{Acc}(p_1)\}; \square \vdash v/p_1 \rightsquigarrow \Omega \quad \{\dot{x} \mapsto \mathsf{Init}(\dot{x}) \mid \dot{x} \in \mathrm{Acc}(p_2)\}; \square \vdash v/p_2 \rightsquigarrow \sigma, \gamma}{\text{match } v \text{ with } p_1 \to e_1 \mid p_2 \to e_2 \rightsquigarrow e_2[\sigma; \, \gamma]}$$

$$\frac{e \rightsquigarrow e'}{E[e] \rightsquigarrow E[e']} \qquad \qquad \frac{}{e \rightsquigarrow \Omega} \quad \left(\begin{matrix}\text{if no other rule applies}\\ \text{and } e \text{ is not a value}\end{matrix}\right)$$

Fig. 3. Operational semantics (reduction contexts and rules)

3   Semantics

In this section, the most technical one, we present the operational semantics and the type system of our calculus, and state the expected soundness properties.

3.1   Operational Semantics

We define a call-by-value, small-step operational semantics for our core calculus, using the reduction contexts and reduction rules given in Figure 3, where Ω is a special value representing a runtime error.

Of course, most of the actual semantics is hidden (the careful reader will have noticed that applications and projections are not explicitly included in the syntax of our expressions). Most of the work happens either in the semantics of operators or in the matching v/p of a value v against a pattern p. Such a matching, if it succeeds (*i.e.*, if it does not return Ω), returns two substitutions, one (ranged over by γ) from the capture variables of p to values and the other (ranged over by δ) from the accumulators to values. These two substitutions are simultaneously applied (noted $e_i[\sigma; \gamma]$) to the expression e_i of the pattern p_i that succeeds, according to a first match policy (v/p_2 is evaluated only if v/p_1 fails). Before explaining how to derive the pattern matching judgments "$\_ \vdash v/p \rightsquigarrow \_$" (in particular, the meaning of the context on the LHS of the turnstile "$\vdash$"), we introduce a minimal set of operators: application, projections, zipper erasure, and sequence building (we use sans-serif font for concrete operators). We only give their reduction relation and defer their typing relation to Section 3.2.

Function application: the operator $\mathsf{app}(\_, \_)$ implements the usual β-reduction:

$$v, v' \overset{\mathsf{app}}{\rightsquigarrow} e[v/f; v'/x] \qquad \text{if } v = \mu f^{(\cdots)}(x).e$$

and $v, v' \overset{\mathsf{app}}{\rightsquigarrow} \Omega$ if v is not a function. As customary, $e[v/x]$ denotes the capture-avoiding substitution of v for x in e, and we write $e_1\, e_2$ for $\mathsf{app}(e_1, e_2)$.

Projection: the operator $\pi_1(\_)$ (resp., $\pi_2(\_)$) implements the usual first (resp., second) projection for pairs:

$$(v_1, v_2) \overset{\pi_i}{\rightsquigarrow} v_i \qquad \text{for } i \in \{1, 2\}$$

The application of the above operators returns Ω if the input is not a pair.

Zipper erasure: given a zipper-annotated value, it is sometimes necessary to remove the zipper (*e.g.,* to embed this value into a new data structure). This is achieved by the following remove rm(_) and deep remove drm(_) operators:

$$(w)_\delta \overset{\mathsf{rm}}{\rightsquigarrow} w \qquad\qquad w \overset{\mathsf{drm}}{\rightsquigarrow} w$$

$$v \overset{\mathsf{rm}}{\rightsquigarrow} v \quad \text{if } v \not\equiv (w)_\delta \qquad (w)_\delta \overset{\mathsf{drm}}{\rightsquigarrow} w$$

$$(v_1, v_2) \overset{\mathsf{drm}}{\rightsquigarrow} (\mathsf{drm}(v_1), \mathsf{drm}(v_2))$$

The former operator only erases the top-level zipper (if any), while the latter erases all zippers occurring in its input.

Sequence building: given a sequence (encoded *à la* Lisp) and an element, we define the operators cons(_) and snoc(_) that insert an input value at the beginning and at the end of the input sequence:

$$v, v' \overset{\mathsf{cons}}{\rightsquigarrow} (v, v') \qquad v, \text{`nil} \overset{\mathsf{snoc}}{\rightsquigarrow} (v, \text{`nil})$$

$$v, (v', v'') \overset{\mathsf{snoc}}{\rightsquigarrow} (v', \mathsf{snoc}(v, v''))$$

The applications of these operators yield Ω on other inputs.

To complete our presentation of the operational semantics, it remains to describe the semantics of pattern matching. Intuitively, when matching a value v against a pattern p, subparts of p are recursively applied to corresponding subparts of v until a base case is reached (which is always the case since all values are finite). As usual, when a pattern variable is confronted with a subvalue, the binding is stored as a substitution. We supplement this usual behavior of pattern matching with two novel features. First, we add *accumulators*, that is, special variables in which results are accumulated during the recursive matching. The reason for keeping these two kinds of variables distinct is explained in Section 3.2 and is related to type inference for patterns. Second, we parameterize pattern matching by a zipper of the current value so that it can properly update the zipper when navigating the value (which should be of the pair form).

These novelties are reflected by the semantics of pattern matching, which is given by the judgment $\sigma; \delta^? \vdash v/p \rightsquigarrow \sigma', \gamma$, where v is a value, p a pattern, γ a mapping from $\mathrm{Var}(p)$ to values, and σ and σ' are mappings from accumulators to values. $\delta^?$ is an optional zipper value, which is either δ or a none value $\square$ (we consider $(v)_\square$ to be v). The judgment "returns" the result of matching the value v against the pattern p (noted v/p), that is, two substitutions: γ for capture variables and σ' for accumulators. Since the semantics is given compositionally, the matching may happen on a subpart of an "outer" matched value. Therefore, the judgment records on the LHS of the turnstile the context of the outer value explored so far: σ stores the values already accumulated during the matching, while $\delta^?$ tracks the possible zipper of the outer value (or it is $\square$ if the outer value has no zipper). The context is "initialized" in the two rules of the operational semantics of match in Figure 3, by setting each accumulator of the pattern to its initial value (function Init()) and the outer zipper to $\square$.

Judgments for pattern matching are derived by the rules given in Figure 4. The rules pat-acc, pat-pair-zip, and zpat-* are novel, as they extend pattern matching with accumulators and zippers, while the others are derived from [4,9].

$$\frac{(\vdash v : t)}{\sigma; \delta^? \vdash v/t \rightsquigarrow \sigma, \varnothing} \text{ pat-type} \qquad \frac{}{\sigma; \delta^? \vdash v/\dot{x} \rightsquigarrow \sigma[\ \mathsf{Op}(\dot{x})(v_{\delta?}, \sigma(\dot{x}))/\dot{x}\ ], \varnothing} \text{ pat-acc}$$

$$\frac{}{\sigma; \delta^? \vdash v/x \rightsquigarrow \sigma, \{x \mapsto v_{\delta?}\}} \text{ pat-var} \qquad \frac{}{\sigma; \delta^? \vdash v/(x := c) \rightsquigarrow \sigma, \{x \mapsto c\}} \text{ pat-def}$$

$$\frac{\sigma; \square \vdash v_1/p_1 \rightsquigarrow \sigma', \gamma_1 \quad \sigma'; \square \vdash v_2/p_2 \rightsquigarrow \sigma'', \gamma_2}{\sigma; \square \vdash (v_1, v_2)/(p_1, p_2) \rightsquigarrow \sigma'', \gamma_1 \oplus \gamma_2} \text{ pat-pair}$$

$$\frac{\sigma; \mathsf{L}\,(w_1, w_2)_\delta \cdot \delta \vdash w_1/p_1 \rightsquigarrow \sigma', \gamma_1 \quad \sigma'; \mathsf{R}\,(w_1, w_2)_\delta \cdot \delta \vdash w_2/p_2 \rightsquigarrow \sigma'', \gamma_2}{\sigma; \delta \vdash (w_1, w_2)/(p_1, p_2) \rightsquigarrow \sigma'', \gamma_1 \oplus \gamma_2} \text{ pat-pair-zip}$$

$$\frac{\sigma; \delta^? \vdash v/p_1 \rightsquigarrow \sigma', \gamma}{\sigma; \delta^? \vdash v/p_1 \,|\, p_2 \rightsquigarrow \sigma', \gamma} \text{ pat-or1} \qquad \frac{\sigma; \delta^? \vdash v/p_1 \rightsquigarrow \Omega \quad \sigma; \delta^? \vdash v/p_2 \rightsquigarrow \sigma', \gamma}{\sigma; \delta^? \vdash v/p_1 \,|\, p_2 \rightsquigarrow \sigma', \gamma} \text{ pat-or2}$$

$$\frac{\sigma; \delta^? \vdash v/p_1 \rightsquigarrow \sigma', \gamma_1 \quad \sigma'; \delta^? \vdash v/p_2 \rightsquigarrow \sigma'', \gamma_2}{\sigma; \delta^? \vdash v/p_1 \,\&\, p_2 \rightsquigarrow \sigma'', \gamma_1 \oplus \gamma_2} \text{ pat-and}$$

$$\frac{\sigma; \delta \vdash w/q \rightsquigarrow \sigma', \gamma_1 \quad \sigma' \vdash \delta/\varphi \rightsquigarrow \sigma'', \gamma_2}{\sigma; \square \vdash (w)_\delta/(q)_\varphi \rightsquigarrow \sigma'', \gamma_1 \oplus \gamma_2} \text{ pat-zip} \qquad \frac{(\vdash \delta : \tau)}{\sigma \vdash \delta/\tau \rightsquigarrow \sigma, \varnothing} \text{ zpat-type}$$

$$\frac{\sigma; \square \vdash (w)_\delta/p \rightsquigarrow \sigma', \gamma_1 \quad}{\sigma' \vdash \delta/\varphi \rightsquigarrow \sigma'', \gamma_2 \quad \gamma = \gamma_1 \oplus \gamma_2} \\ \frac{}{\sigma \vdash \mathsf{L}\,(w)_\delta \cdot \delta/\mathsf{L}\,p \cdot \varphi \rightsquigarrow \sigma'', \gamma} \text{ zpat-left} \qquad \frac{\sigma; \square \vdash (w)_\delta/p \rightsquigarrow \sigma', \gamma_1 \quad}{\sigma' \vdash \delta/\varphi \rightsquigarrow \sigma'', \gamma_2 \quad \gamma = \gamma_1 \oplus \gamma_2} \\ \frac{}{\sigma \vdash \mathsf{R}\,(w)_\delta \cdot \delta/\mathsf{R}\,p \cdot \varphi \rightsquigarrow \sigma'', \gamma} \text{ zpat-right}$$

$$\frac{\sigma \vdash \delta/\varphi_1 \rightsquigarrow \sigma', \gamma}{\sigma \vdash \delta/\varphi_1 \,|\, \varphi_2 \rightsquigarrow \sigma', \gamma} \text{ zpat-or1} \qquad \frac{\sigma \vdash \delta/\varphi_1 \rightsquigarrow \Omega \quad \sigma \vdash \delta/\varphi_2 \rightsquigarrow \sigma', \gamma}{\sigma \vdash \delta/\varphi_1 \,|\, \varphi_2 \rightsquigarrow \sigma', \gamma} \text{ zpat-or2}$$

$$\frac{\text{(otherwise)}}{\sigma; \delta^? \vdash v/p \rightsquigarrow \Omega} \text{ pat-error} \qquad \frac{\text{(otherwise)}}{\sigma \vdash \delta/\varphi \rightsquigarrow \Omega} \text{ zpat-error}$$

where $\gamma_1 \oplus \gamma_2 \overset{\text{def}}{=} \{x \mapsto \gamma_1(x) \mid x \in \mathsf{dom}(\gamma_1) \backslash \mathsf{dom}(\gamma_2)\}$
$\qquad\qquad \cup \ \{x \mapsto \gamma_2(x) \mid x \in \mathsf{dom}(\gamma_2) \backslash \mathsf{dom}(\gamma_1)\}$
$\qquad\qquad \cup \ \{x \mapsto (\gamma_1(x), \gamma_2(x)) \mid x \in \mathsf{dom}(\gamma_1) \cap \mathsf{dom}(\gamma_2)\}$

Fig. 4. Pattern matching

There are three base cases for matching: testing the input value against a type (rule pat-type), updating the environment σ for accumulators (rule pat-acc), or producing a substitution γ for capture variables (rules pat-var and pat-def). Matching a pattern (p_1, p_2) only succeeds if the input is a pair and the matching of each subpattern against the corresponding subvalue succeeds (rule pat-pair). Furthermore, if the value being matched was below a zipper (*i.e.*, the current zipper context is a δ and not—as in pat-pair— $\square$), we update the current zipper context (rule pat-pair-zip); notice that in this case the matched value must be a pair of pre-values since zipped values cannot be nested. An alternative pattern $p_1 \,|\, p_2$ first tries to match the pattern p_1 and if it fails, tries the pattern p_2 (rules pat-or1 and pat-or2). The matching of a conjunction pattern $p_1 \,\&\, p_2$ succeeds if and only if the matching of both patterns succeeds (rule pat-and). For a zipper constraint $(q)_\varphi$, the matching succeeds if and only if the input value is annotated by a zipper, *e.g.*, $(w)_\delta$, and both the matching of w with q and δ with φ succeed (rule pat-zip). It requires the zipper context to be $\square$ since we do not allow nested zipped values. When matching w with q, we record the zipper δ into the context so that it can be updated (in the rule pat-pair-zip) while navigating the value.

The matching of a zipper pattern φ against a zipper δ (judgments $\sigma \vdash \delta/\varphi \rightsquigarrow \sigma', \gamma$ derived by the zpat-* rules) is straightforward: it succeeds if both φ and δ are built using the same constructor (either L or R) and the componentwise matching succeeds (rules zpat-left and zpat-right). If the zipper pattern is a zipper type, the matching tests the input zipper against the zipper type (rule zpat-type), and alternative zipper patterns $\varphi_1 | \varphi_2$ follow the same first match policy as alternative patterns. If none of the rules is applicable, the matching fails (rules pat-error and zpat-error). Note that initially the environment σ contains $\mathsf{Init}(\dot{x})$ for each accumulator $\dot{x}$ in $\mathsf{Acc}(p)$ (rules for match in Figure 3).

Intuitively, γ is built when returning from the recursive descent in p, while σ is built using a *fold*-like computation. It is the typing of such fold-like computations that justifies the addition of accumulators (instead of relying on plain functions). But before presenting the type system of the language, we illustrate the behavior of pattern matching by some examples.

Example 4. Let $v \equiv (2, (\text{'true}, (3, \text{'nil)}))$, $\mathsf{Init}(\dot{x}) = \text{'nil}$, $\mathsf{Op}(\dot{x}) = \mathsf{cons}$, and $\sigma \equiv \{\dot{x} \mapsto \text{'nil}\}$. Then, we have the following matchings:

1. $\sigma; \square \vdash v/(\mathsf{int}, (x, \_)) \rightsquigarrow \varnothing, \{x \mapsto \text{'true}\}$
2. $\sigma; \square \vdash v/\mu X.((x \,\&\, \mathsf{int} \,|\, \_, X) \,|\, (x := \text{'nil})) \rightsquigarrow \varnothing, \{x \mapsto (2, (3, \text{'nil}))\}$
3. $\sigma; \square \vdash v/\mu X.((\dot{x}, X) \,|\, \text{'nil}) \rightsquigarrow \{\dot{x} \mapsto (3, (\text{'true}, (2, \text{'nil})))\}, \varnothing$

In the first case, the input v (the sequence [2 'true 3] encoded *à la* Lisp) is matched against a pattern that checks if the first element has type int (rule pat-type), binds the second element to x (rule pat-var), and ignores the rest of the list (rule pat-type, since the anonymous variable "$\_$" is just an alias for $\mathbb{1}$).

The second case is more involved since the pattern is recursively defined. Because of the first match policy of rule pat-or1, the product part of the pattern is matched recursively until the atom 'nil is reached. When that is the case, the variable x is bound to a default value 'nil. When returning from this recursive matching, since x occurs both on the left and on the right of the product (in $x \,\&\, \mathsf{int}$ and in X itself), a pair of the binding found in each part is formed (third set in the definition of $\oplus$ in Figure 4), thus yielding a mapping $\{x \mapsto (3, \text{'nil})\}$. Returning again from the recursive call, only the "$\_$" part of the pattern matches the input 'true (since it is not of type int, the intersection test fails). Therefore, the binding for this step is only the binding for the right part (second case of the definition of $\oplus$). Lastly, when reaching the top-level pair, $x \,\&\, \mathsf{int}$ matches 2 and a pair is formed from this binding and the one found in the recursive call, yielding the final binding $\{x \mapsto (2, (3, \text{'nil}))\}$.

The third case is more intuitive. The pattern just recurses the input value, calling the accumulation function for $\dot{x}$ along the way for each value against which it is confronted. Since the operator associated with $\dot{x}$ is cons (which builds a pair of its two arguments) and the initial value is 'nil, this has the effect of computing the reversal of the list.

Note the key difference between the second and third case. In both cases, the structure of the pattern (and the input) dictates the traversal, but in the second case, it also dictates *how* the binding is built (if v was a tree and not a list, the binding for x would also be a tree in the second case). In the third case, the way the binding is built is defined by the semantics of the operator and independent of the input. This allows us to reverse sequences or flatten tree structures, both of which are operations that escape the expressiveness of regular tree languages/regular patterns, but which are both necessary to encode XPath.

3.2   Type System

The main difficulty in devising the type system is to type pattern matching and, more specifically, to infer the types of the accumulators occurring in patterns.

Definition 5 (Accepted input of an operator). *The* accepted input *of an operator* $(o, n, \overset{o}{\leadsto}, \overset{o}{\to})$ *is the set* $\mathbb{I}(o)$, *defined as:*

$$\mathbb{I}(o) = \{(v_1, ..., v_n) \in \mathcal{V}^n \mid (((v_1, ..., v_n) \overset{o}{\leadsto} e) \land (e \leadsto^* v)) \Rightarrow v \neq \Omega\}$$

Definition 6 (Exact input). *An operator o has an* exact input *if and only if* $\mathbb{I}(o)$ *is (the interpretation of) a type.*

We can now state a first soundness theorem, which characterizes the set of all values that make a given pattern succeed:

Theorem 7 (Accepted types). *Let p be a pattern such that for every $\dot{x}$ in $\mathrm{Acc}(p)$, $\mathrm{Op}(\dot{x})$ has an exact input. Then, the set of all values v such that $\{\dot{x} \mapsto \mathit{Init}(\dot{x}) \mid \dot{x} \in \mathrm{Acc}(p)\}; \square \vdash v/p \not\leadsto \Omega$ is a type. We call this set the* accepted type *of p and denote it by $\langle\!\langle p \rangle\!\rangle$.*

We next define the type system for our core calculus, in the form of a judgment $\Gamma \vdash e : t$ which states that in a typing environment Γ (*i.e.*, a mapping from variables and accumulators to types) an expression e has type t. This judgment is derived by the set of rules given in Figure 10 in Appendix. Here, we show only the most important rules, namely those for accumulators and zippers:

$$\frac{}{\Gamma \vdash \dot{x} : \Gamma(\dot{x})} \qquad \frac{\vdash w : t \quad \vdash \delta : \tau \quad t \leq \mathbb{1}_{\mathsf{NZ}}}{\Gamma \vdash (w)_\delta : (t)_\tau} \qquad \frac{\vdash e : t \quad t \leq \mathbb{1}_{\mathsf{NZ}}}{\Gamma \vdash (e)_\bullet : (t)_\bullet}$$

which rely on an auxiliary judgment $\vdash \delta : \tau$ stating that a zipper δ has zipper type τ. The rule for operators is:

$$\frac{\forall i = 1..n_o, \ \Gamma \vdash e_i : t_i \quad t_1, \ldots, t_{n_o} \overset{o}{\to} t}{\Gamma \vdash o(e_1, \ldots, e_{n_o}) : t} \text{ for } o \in \mathcal{O}$$

which types operators using their associated typing function. Last but not least, the rule to type pattern matching expressions is:

$$\frac{\begin{array}{ll} t \leq \langle\!\langle p_1 \rangle\!\rangle \vee \langle\!\langle p_2 \rangle\!\rangle & \Gamma \vdash e : t \\ t_1 \equiv t \wedge \langle\!\langle p_1 \rangle\!\rangle \quad t_2 \equiv t \wedge \neg\langle\!\langle p_1 \rangle\!\rangle & \Gamma_i \equiv \square|t_i/p_i \quad \Gamma'_i \equiv \Sigma_i; \square|t_i /\!/ p_i \\ \Sigma_i \equiv \{\dot{x} \mapsto \mathit{Init}(\dot{x}) \mid \dot{x} \in \mathrm{Acc}(p_i)\} & \Gamma \cup \Gamma_i \cup \Gamma'_i \vdash e_i : t'_i \end{array}}{\Gamma \vdash \mathsf{match} \ e \ \mathsf{with} \ p_1 \to e_1 \mid p_2 \to e_2 : \bigvee_{\{i \mid t_i \not\leq 0\}} t'_i} (i = 1, 2)$$

This rule requires that the type t of the matched expression is smaller than $\wr p_1 \wr \vee \wr p_2 \wr$ (*i.e.*, the set of all values accepted by any of the two patterns), that is, that the matching is exhaustive. Then, it accounts for the first match policy by checking e_1 in an environment inferred from values produced by e and that match p_1 ($t_1 \equiv t \wedge \wr p_1 \wr$) and by checking e_2 in an environment inferred from values produced by e and that *do not* match p_1 ($t_2 \equiv t \wedge \neg \wr p_1 \wr$). If one of these branches is unused (*i.e.*, if $t_i \simeq 0$ where $\simeq$ denotes semantic equivalence, that is, $\leq \cap \geq$), then its type does not contribute to the type of the whole expression (*cf.* §4.1 of [4] to see why, in general, this must not yield an "unused case" error). Each right-hand side e_i is typed in an environment enriched with the types for capture variables (computed by $\square | t_i / p_i$) and the types for accumulators (computed by $\Sigma_i; \square | t_i /\!/ p_i$). While the latter is specific to our calculus, the former is standard except it is parameterized by a zipper type as for the semantics of pattern matching (its precise computation is described in [9] and already implemented in the CDuce compiler except the zipper-related part: see Figure 11 in Appendix for the details). As before, we write $\tau^?$ to denote an optional zipper type, *i.e.*, either τ or a none type $\square$, and consider $(t)_\square$ to be t.

To compute the types of the accumulators of a pattern p when matched against a type t, we first initialize an environment Σ by associating each accumulator $\dot{x}$ occurring in p with the singleton type for its initial value $\mathsf{Init}(\dot{x})$ ($\Sigma_i \equiv \{\dot{x} \mapsto \mathsf{Init}(\dot{x}) \mid \dot{x} \in \mathrm{Acc}(p_i)\}$). The type environment is then computed by generating a set of mutually recursive equations where the important ones are (see Figure 12 in Appendix for the complete definition):

$$\Sigma; \tau^? | t /\!/ \dot{x} \ = \Sigma[s/\dot{x}] \qquad\qquad\qquad\qquad \text{if } (t)_{\tau^?}, \Sigma(\dot{x}) \stackrel{\mathsf{Op}(\dot{x})}{\rightarrow} s$$

$$\Sigma; \tau^? | t /\!/ p_1 \mid p_2 = \Sigma; \tau^? | t /\!/ p_1 \qquad\qquad\qquad\qquad\qquad \text{if } t \leq \wr p_1 \wr$$

$$\Sigma; \tau^? | t /\!/ p_1 \mid p_2 = \Sigma; \tau^? | t /\!/ p_2 \qquad\qquad\qquad\qquad\qquad \text{if } t \leq \neg \wr p_1 \wr$$

$$\Sigma; \tau^? | t /\!/ p_1 \mid p_2 = (\Sigma; \tau^? | (t \wedge \wr p_1 \wr) /\!/ p_1) \bigsqcup (\Sigma_1; \tau^? | (t \wedge \neg \wr p_1 \wr) /\!/ p_2) \quad \text{otherwise}$$

When an accumulator $\dot{x}$ is matched against a type t, the type of the accumulator is updated in Σ, by applying the typing function of the operator associated with $\dot{x}$ to the type $(t)_{\tau^?}$ and the type computed thus far for $\dot{x}$, namely $\Sigma(\dot{x})$. The other equations recursively apply the matching on the subcomponents while updating the zipper type argument $\tau^?$ and merge the results using the "$\bigsqcup$" operation. This operation implements the fact that if an accumulator $\dot{x}$ has type t_1 in a subpart of a pattern p and type t_2 in another subpart (*i.e.*, both subparts match), then the type of $\dot{x}$ is the union $t_1 \vee t_2$.

The equations for computing the type environment for accumulators might be *not* well-founded. Both patterns and types are possibly infinite (regular) terms and therefore one has to guarantee that the set of generated equations is finite. This depends on the typing of the operators used for the accumulators. Before stating the termination condition (as well as the soundness properties of the type system), we give the typing functions for the operators we defined earlier.

Function application: it is typed by computing the minimum type satisfying the following subtyping relation: $s, t \overset{\text{app}}{\rightarrow} \min\{t' \mid s \leq t \rightarrow t'\}$, provided that $s \leq t \rightarrow \mathbb{1}$ (this min always exists and is computable: see [10]).

Projection: to type the first and second projections, we use the property that if $t \leq \mathbb{1} \times \mathbb{1}$, then t can be decomposed in a finite union of product types (we use Π_i to denote the set of the i-th projections of these types: see Lemma 19 in Appendix B for the formal definition): $t \overset{\pi_i}{\rightarrow} \bigvee_{s \in \Pi_i(t)} s$, provided that $t \leq \mathbb{1} \times \mathbb{1}$.

Zipper erasure: the top-level erasure $\overset{\text{rm}}{\rightarrow}$ simply removes the top-level zipper type annotation, while the deep erasure $\overset{\text{drm}}{\rightarrow}$ is typed by recursively removing the zipper annotations from the input type. Their precise definition can be found in Appendix B.4.

Sequence building: it is typed in the following way:

$$t_1, \text{'nil} \overset{\text{cons}}{\rightarrow} \mu X.((t_1 \times X) \vee \text{'nil})$$
$$t_1, \mu X.((t_2 \times X) \vee \text{'nil}) \overset{\text{cons}}{\rightarrow} \mu X.(((t_1 \vee t_2) \times X) \vee \text{'nil})$$

$$t_1, \text{'nil} \overset{\text{snoc}}{\rightarrow} \mu X.((t_1 \times X) \vee \text{'nil})$$
$$t_1, \mu X.((t_2 \times X) \vee \text{'nil}) \overset{\text{snoc}}{\rightarrow} \mu X.(((t_1 \vee t_2) \times X) \vee \text{'nil})$$

Notice that the output types are approximations: the operator "cons(_)" is *less* precise than returning a pair of two values since, for instance, it approximates any sequence type by an infinite one (meaning that any information on the length of the sequence is lost) and approximates the type of all the elements by a single type which is the union of all the elements (meaning that the information on the order of elements is lost). As we show next, this loss of precision is instrumental in typing accumulators and therefore pattern matching.

Example 8. Consider the matching of a pattern p against a value v of type t:

$$p \equiv \mu X.((\dot{x} \& (\text{'a} \mid \text{'b})) \mid \text{'nil} \mid (X, X))$$
$$v \equiv (\text{'a}, ((\text{'a}, (\text{'nil}, (\text{'b}, \text{'nil}))), (\text{'b}, \text{'nil})))$$
$$t \equiv \mu Y.((\text{'a} \times (Y \times (\text{'b} \times \text{'nil}))) \vee \text{'nil})$$

where $\text{Op}(\dot{x}) = \text{snoc}$ and $\text{Init}(\dot{x}) = \text{'nil}$. We have the following matching and type environment:

$$\{\dot{x} \mapsto \text{'nil}\}; \square \vdash v/p \rightsquigarrow \{\dot{x} \mapsto (\text{'a}, (\text{'a}, (\text{'b}, (\text{'b}, \text{'nil}))))\}, \varnothing$$
$$\{\dot{x} \mapsto \text{'nil}\}; \square \mid t /\!\!/ p = \{\dot{x} \mapsto \mu Z.(((\text{'a} \vee \text{'b}) \times Z) \vee \text{'nil})\}$$

Intuitively, with the usual sequence notation (precisely defined in Section 4), v is nothing but the nested sequence $[\,\text{'a}\,[\,\text{'a}\,[\,]\,\text{'b}\,]\,\text{'b}\,]$ and pattern matching just flattens the input sequence, binding $\dot{x}$ to $[\,\text{'a}\,\text{'a}\,\text{'b}\,\text{'b}\,]$. The type environment for $\dot{x}$ is computed by recursively matching each product type in t with the pattern (X, X), the singleton type 'a or 'b with $\dot{x} \& (\text{'a} \mid \text{'b})$, and 'nil with 'nil. Since the operator associated with $\dot{x}$ is snoc and the initial type is 'nil, when $\dot{x}$ is matched against 'a for the first time, its type is updated to $\mu Z.((\text{'a} \times Z) \vee \text{'nil})$. Then, when $\dot{x}$ is matched against 'b, its type is updated to the final output type which is the encoding of $[(\text{'a} \vee \text{'b})*]$. Here, the approximation in the typing function for snoc is important because the exact type of $\dot{x}$ is the union for $n \in \mathbb{N}$ of $[\text{'a}^n \, \text{'b}^n]$, that is, the sequences of 'a's followed by the same number of 'b's, which is beyond the expressivity of regular tree languages.

We conclude this section with statements for type soundness of our calculus (see Appendix C for more details).

Definition 9 (Sound operator). *An operator* $(o, n, \overset{o}{\rightsquigarrow}, \overset{o}{\rightarrow})$ *is* sound *if and only if* $\forall v_1, \ldots, v_{n_o} \in \mathcal{V}$ *such that* $\vdash v_1 : t_1, \ldots, \vdash v_{n_o} : t_{n_o}$, *if* $t_1, \ldots, t_{n_o} \overset{o}{\rightarrow} s$ *and* $v_1, \ldots, v_{n_o} \overset{o}{\rightsquigarrow} e$ *then* $\vdash e : s$.

Theorem 10 (Type preservation). *If all operators in the language are* sound, *then typing is preserved by reduction, that is, if* $e \rightsquigarrow e'$ *and* $\vdash e : t$, *then* $\vdash e' : t$. *In particular,* $e' \neq \Omega$.

Theorem 11. *The operators* app, π_1, π_2, drm, rm, cons, *and* snoc *are* sound.

4 Surface Language

In this section, we define the "surface" language, which extends our core calculus with several constructs:

- Sequence expressions, regular expression types and patterns
- Sequence concatenation and iteration
- XML types, XML document fragment expressions
- XPath-like patterns

While most of these traits are syntactic sugar or straightforward extensions, we took special care in their design so that: *(i)* they cover various aspects of XML programming and *(ii)* they are expressive enough to encode a large fragment of XQuery 3.0.

Sequences: we first add sequences to expressions

$$e ::= \quad \ldots \quad | \quad [e \cdots e]$$

where a sequence expression denotes its encoding *à la* Lisp, that is, $[e_1 \cdots e_n]$ is syntactic sugar for $(e_1, (\ldots, (e_n, \text{'nil})))$.

Regular expression types and patterns: regular expressions over types and patterns are defined as

| (Regexp. over types) | R | $::=$ | t | $|$ | $R|R$ | $|$ | $R\,R$ | $|$ | $R*$ | $|$ | ϵ |
| (Regexp. over patterns) | r | $::=$ | p | $|$ | $r|r$ | $|$ | $r\,r$ | $|$ | $r*$ | $|$ | ϵ |

with the usual syntactic sugar: $R? \equiv R|\epsilon$ and $R+ \equiv R\,R*$ (likewise for regexps on patterns). We then extend the grammar of types and patterns as follows:

$$t ::= \quad \ldots \quad | \quad [R] \qquad p ::= \quad \ldots \quad | \quad [r]$$

Regular expression types are encoded using recursive types (similarly for regular expression patterns). For instance, $[\text{int}* \text{bool}?]$ can be rewritten into the recursive type $\mu X.\text{'nil} \vee (\text{bool} \times \text{'nil}) \vee (\text{int} \times X)$.

Sequence concatenation is added to the language in the form of a binary infix operator $\_ @ \_$ defined by:

$$\begin{aligned}
\text{'nil}, v & \overset{@}{\leadsto} v \\
(v_1, v_2), v & \overset{@}{\leadsto} (v_1, v_2 @ v)
\end{aligned} \qquad [R_1], [R_2] \overset{@}{\leadsto} [R_1 R_2]$$

Note that this operator is sound but cannot be used to accumulate in patterns (since it does not guarantee the termination of type environment computation). However, it has an exact typing.

Sequence iteration is added to iterate transformations over sequences without resorting to recursive functions. This is done by a family of "transform"-like operators $\text{trs}_{p_1,p_2,e_1,e_2}(\_)$, indexed by the patterns and expressions that form the branches of the transformation (we omit trs's indexes in $\overset{\text{trs}}{\leadsto}$):

$$\begin{aligned}
\text{'nil} & \overset{\text{trs}}{\leadsto} \text{'nil} \\
(v_1, v_2) & \overset{\text{trs}}{\leadsto} (\text{match } v_1 \text{ with } p_1 \to e_1 \mid p_2 \to e_2) @ \text{trs}_{p_1,p_2,e_1,e_2}(v_2)
\end{aligned}$$

Intuitively, the construct "transform e with $p_1 \to e_1 \mid p_2 \to e_2$" iterates all the "branches" over each element of the sequence e. Each branch may return a sequence of results which is concatenated to the final result (in particular, a branch may return "'nil" to delete elements that match a particular pattern).

XML types, patterns, and document fragments: XML types (and thus patterns) can be represented as a pair of the type of the label and a sequence type representing the sequence of children, annotated by the zipper that denotes the position of document fragment of that type. We denote by $\texttt{<}t_1\texttt{>}t_{2\tau}$ the type $(t_1 \times t_2)_\tau$, where $t_1 \leq \mathbb{1}_{\text{basic}}$, $t_2 \leq [\mathbb{1}*]$, and τ is a zipper type. We simply write $\texttt{<}t_1\texttt{>}t_2$ when $\tau = \top$, that is, when we do not have (or do not require) any information on the zipper type. The invariant that XML values are always given with respect to a zipper must be maintained at the level of expressions. This is ensured by extending the syntax of expressions with the construct:

$$e ::= \ldots \mid \texttt{<}e\texttt{>}e$$

where $\texttt{<}e_1\texttt{>}e_2$ is syntactic sugar for $(e_1, \text{drm}(e_2))_\bullet$. The reason for this encoding is best understood with the following example:

Example 12. Consider the code:

```
1   match v with
2     ( <a>[ _ x _* ] )⊤ -> <b>[ x ]
3     | _ -> <c>[ ]
```

According to our definition of pattern matching, x is bound to the second XML child of v and retains its zipper (in the right-hand side, we could navigate from x up to v or even above if v is not the root). However, when x is embedded into another document fragment, the zipper must be erased so that accessing the element associated with x in the *new* value can create an appropriate zipper (with respect to its new root $\texttt{<b>}[\ldots]$).

$$\text{self}_0\{x \mid t\} \equiv \dot{x} \,\&\, t \mid\_$$

$$\text{self}\{x \mid t\} \equiv (\text{self}_0\{x \mid t\})_\top$$

$$\text{child}\{x \mid t\} \equiv (<\_>[\,(\text{self}_0\{x \mid t\})*\,] \mid\_)_\top$$

$$\text{desc-or-self}_0\{x \mid t\} \equiv \mu X.(\text{self}_0\{x \mid t\} \,\&\, <\_>[\,X*\,]) \mid\_$$

$$\text{desc-or-self}\{x \mid t\} \equiv (\text{desc-or-self}_0\{x \mid t\})_\top$$

$$\text{desc}\{x \mid t\} \equiv (<\_>[\,(\text{desc-or-self}_0\{x \mid t\})*\,] \mid\_)_\top$$

$$\text{foll-sibling}\{x \mid t\} \equiv (\_)_{\mathsf{L}\,(\_,[\,(\text{self}_0\{x \mid t\})*\,])_\top\cdot\top}$$

$$\text{parent}\{y \mid t\} \equiv (\_)_{\mathsf{L}\_\cdot\mu X.((\mathsf{R}\,(\dot{y}\,\&\,t\mid\_)_\top\cdot(\mathsf{L}\_\cdot\top\mid\bullet))\mid\mathsf{R}\_\cdot X)} \mid\_$$

$$\text{prec-sibling}\{y \mid t\} \equiv (\_)_{\mathsf{L}\_\cdot\mu X.(\mathsf{R}\,(\dot{y}\,\&\,t,\_)_\top\cdot X)\mid(\mathsf{R}\_\cdot(\mathsf{L}\_\cdot\top\mid\bullet))} \mid\_$$

$$\text{anc}\{y \mid t\} \equiv (\_)_{\mathsf{L}\_\cdot\mu X.\mu Y.((\mathsf{R}\,(\dot{y}\,\&\,t\mid\_)_\top\cdot(\mathsf{L}\_\cdot X\mid\bullet))\mid\mathsf{R}\_\cdot Y)} \mid\_$$

$$\text{anc-or-self}\{y \mid t\} \equiv (\text{self}\{y \mid t\} \,\&\, \text{anc}\{y \mid t\}) \mid\_$$

where $\text{Op}(\dot{x}) = \text{snoc}$, $\text{Init}(\dot{x}) = \text{'nil}$, $\text{Op}(\dot{y}) = \text{cons}$, and $\text{Init}(\dot{y}) = \text{'nil}$

Fig. 5. Encoding of axis patterns

XPath-like patterns are one of the main motivations for this work. The syntax of patterns is extended as follows:

(Patterns) p $::=$ $\ldots \mid axis\{x \mid t\}$

(Axes) $axis$ $::=$ self $\mid$ child $\mid$ desc $\mid$ desc-or-self $\mid$ foll-sibling
          $\mid$ parent $\mid$ anc $\mid$ anc-or-self $\mid$ prec-sibling

The semantics of $axis\{x \mid t\}$ is to capture in x all fragments of the matched document along the $axis$ that have type t. We show in Appendix D how the remaining two axes (following and preceding) as well as "multi-step" XPath expressions can be compiled into this simpler form. We encode axis patterns directly using recursive patterns and accumulators, as described in Figure 5. First, we remark that each pattern has a default branch "$\ldots \mid\_$" which implements the fact that even if a pattern fails, the value is still accepted, but the default value 'nil of the accumulator is returned. The so-called "downward" axes —self, child, desc-or-self, and desc— are straightforward. For self, the encoding checks that the matched value has type t using the auxiliary pattern self$_0$, and that the value is annotated with a zipper using the zipper type annotation $(\_)_\top$. The child axis is encoded by iterating self$_0$ on every child element of the matched value. The recursive axis desc-or-self is encoded using the auxiliary pattern desc-or-self$_0$ which matches the root of the current element (using self$_0$) and is recursively applied to each element of the sequence. Note the double recursion: vertically in the tree using a recursive binder and horizontally at a given level using a star. The non-reflexive variant desc evaluates desc-or-self$_0$ on every child element of the input.

The other axes heavily rely on the binary encoding of XML values and are better explained on an example. Consider the XML document and its binary tree representation given in Figure 6. The following siblings of a node (*e.g.*, <c>) are reachable by inspecting the first element of the zipper, which is necessarily an L one. This parent is the pair representing the sequence whose tail is the sequence of following siblings (R_3 and R_2 in the figure). Applying the self$\{x \mid t\}$ axis on each element of the tail therefore filters the following siblings that are sought (<d> and <e> in the figure). The parent axis is more involved. Consider

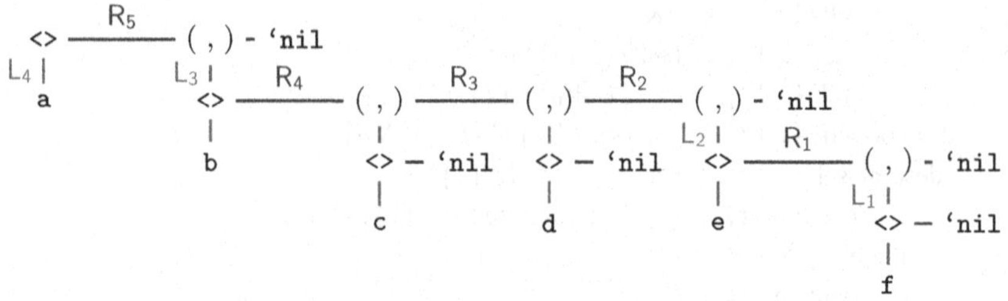

Fig. 6. A binary tree representation of an XML document
doc = <a>[<b>[<c>[ ] <d>[ ] <e>[<f>[ ]]]]

for instance node <e>. Its parent in the XML tree can be found in the zipper associated with <e>. It is the last R component of the zipper before the next L component (in the figure, the zipper of <e> starts with L_2, then contains its previous siblings reachable by R_2 and R_3, and lastly its parent reachable by R_4 (which points to node <b>). The encoding of the parent axis reproduces this walk using a recursive zipper pattern, whose base case is the last R before the next L, or the last R before the root (which has the empty zipper •). The prec-sibling axis uses a similar method and collects every node reachable by Rs and stops before the parent node (again, for node <e>, the preceding siblings are reached by R_2 and R_3). The anc axis simply iterates the parent axis recursively until there is no L zipper anymore (i.e., until the root of the document has been reached). In the example, starting from node <f>, the zippers that denote the ancestors are the ones starting with an R, just before L_2, L_3, and L_4 which is the root of the document. Lastly, anc-or-self is simply a combination of anc and self.

For space reasons, the encoding of XPath into the navigational patterns is given in Appendix D. We just stress that, with that encoding, the CDuce version of the "get_links" function of the introduction becomes as compact as in XQuery:

```
let get_links (page: <_>_) (print: <a>_ -> <a>_) : [ <a>_ * ] =
    transform page/desc::a[not(anc::b)] with x -> [ (print x) ]
```

As a final remark, one may notice that patterns of forward axes use snoc (i.e., they build the sequence of the results in order), while reverse axes use cons (thus reversing the results). The reason for this difference is to implement the semantics of XPath axis steps which return elements in document order.

5   XQuery 3.0

This section shows that our surface language can be used as a compilation target for XQuery 3.0 programs. We proceed in two steps. First, we extend the XQuery 1.0 Core fragment and XQ_H defined by Benedikt and Vu [3] to our own XQuery 3.0 Core, which we call XQ_H^+. As with its 1.0 counterpart, XQ_H^+

1. can express all navigational XQuery programs, and
2. explicitly separates navigational aspects from data value ones.

$$
\begin{aligned}
query ::= &\ () \ \mid\ c \ \mid\ <l>query</l> &\mid\ & query, query \ \mid\ x \ \mid\ x/axis\!::\!test \\
&\mid\ \text{for } x \text{ in } query \text{ return } query &\mid\ & \text{some } x \text{ in } query \text{ statisfies } query \\
&\mid\ query(query,\dots,query) &\mid\ & \text{fun } x_1 : t_1,\dots,x_n : t_n \text{ as } t.\ query
\end{aligned}
$$

| switch *query*	| typeswitch *query*
case *c* return *query*	case *t* as *x* return *query*
default return *query*	default return *query*

$$
test \quad ::= \quad \text{node()} \mid \text{text()} \mid l \qquad\qquad\qquad\qquad\qquad \text{(node test)}
$$

where t ranges over types and l ranges over element names.

Fig. 7. Syntax of $\mathsf{XQ_H^+}$

We later use the above separation in the translation to straightforwardly map navigational XPath expressions into extended CDuce pattern matching, and to encode data value operations (for which there can be no precise typing) by built-in CDuce functions.

5.1   XQuery 3.0 Core

Figure 7 shows the definition of $\mathsf{XQ_H^+}$, an *extension* of $\mathsf{XQ_H}$. To the best of our knowledge, $\mathsf{XQ_H}$ was the first work to propose a "Core" fragment of XQuery which abstracts away most of the idiosyncrasies of the actual specification while retaining essential features (*e.g.*, path navigation). $\mathsf{XQ_H^+}$ differs from $\mathsf{XQ_H}$ by the last three productions (in the yellow/gray box): it extends $\mathsf{XQ_H}$ with type and value cases (described informally in the introduction) and with *type annotations* on functions (which are only optional in the standard). It is well known (*e.g.*, see [24]) that full XPath expressions can be encoded using the XQuery fragment in Figure 7 (see Appendix E for an illustration).

Our translation of XQuery 3.0, defined in Figure 8, thus focuses on $\mathsf{XQ_H^+}$ and has following characteristics. If one considers the "typed" version of the standard, that is, XQuery programs where function declarations have an explicit signature, then the translation to our surface language *(i)* provides a formal semantics and a typechecking algorithm for XQuery and *(ii)* enjoys the soundness property that the original XQuery programs do not yield runtime errors. In the present work, we assume that the type algebra of XQuery is the one of CDuce, rather than XMLSchema. Both share regular expression types for which subtyping is implemented as the inclusion of languages, but XMLSchema also features *nominal subtyping*. The extension of CDuce types with nominal subtyping is beyond the scope of this work and is left as future work.

In XQuery, all values are sequences: the constant "42" is considered as the *singleton sequence* that contains the element "42". As a consequence, there are only "flat" sequences in XQuery and the only way to create nested data structures is to use XML constructs. The difficulty for our translation is thus twofold: *(i)* it needs to embed/extract values explicitly into/from sequences and *(ii)* it also needs to disambiguate types: an XQuery function that takes an integer as argument can also be applied to a sequence containing only one integer.

$$[\![()]\!]_{\mathsf{XC}} = \text{'nil}$$
$$[\![c]\!]_{\mathsf{XC}} = [c]$$
$$[\![<l>q</l>]\!]_{\mathsf{XC}} = [<l>[\![q]\!]_{\mathsf{XC}}]$$
$$[\![q_1, q_2]\!]_{\mathsf{XC}} = [\![q_1]\!]_{\mathsf{XC}} \,@\, [\![q_2]\!]_{\mathsf{XC}}$$
$$[\![\$x]\!]_{\mathsf{XC}} = x$$

$$\left[\!\!\left[\begin{array}{l}\text{switch } q_1 \\ \quad \text{case } c \text{ return } q_2 \\ \quad \text{default return } q_3\end{array}\right]\!\!\right]_{\mathsf{XC}} = \begin{array}{l}\text{match } [\![q_1]\!]_{\mathsf{XC}} \text{ with} \\ \quad [c] \to [\![q_2]\!]_{\mathsf{XC}} \\ \quad |\ \_ \to [\![q_3]\!]_{\mathsf{XC}}\end{array}$$

$$\left[\!\!\left[\begin{array}{l}\text{typeswitch } q_1 \\ \quad \text{case } t \text{ as } \$x \text{ return } q_2 \\ \quad \text{default return } q_3\end{array}\right]\!\!\right]_{\mathsf{XC}} = \begin{array}{l}\text{match } [\![q_1]\!]_{\mathsf{XC}} \text{ with} \\ \quad x \,\&\, \mathsf{seq}(t) \to [\![q_2]\!]_{\mathsf{XC}} \\ \quad |\ \_ \to [\![q_3]\!]_{\mathsf{XC}}\end{array}$$

$$[\![\$x/axis::test]\!]_{\mathsf{XC}} = \text{transform } x \text{ with } axis\{y \mid t(test)\} \to y$$

$$[\![\text{for } \$x \text{ in } q_1 \text{ return } q_2]\!]_{\mathsf{XC}} = \text{transform } [\![q_1]\!]_{\mathsf{XC}} \text{ with } x \to [\![q_2]\!]_{\mathsf{XC}}$$

$$[\![\text{some } \$x \text{ in } q_1 \text{ statisfies } q_2]\!]_{\mathsf{XC}} = \begin{array}{l}\text{match (transform } [\![q_1]\!]_{\mathsf{XC}} \text{ with} \\ \qquad x \to \text{match } [\![q_2]\!]_{\mathsf{XC}} \text{ with} \\ \qquad\qquad [\text{'true}] \to [\text{'dummy}] \\ \qquad\qquad |\ [\text{'false}] \to [\,] \,) \\ \text{with 'nil} \to [\text{'false}]\ |\ \_ \to [\text{'true}]\end{array}$$

$$[\![\text{fun } \$x_1 : t_1, \ldots, \$x_n : t_n \text{ as } t.\ q]\!]_{\mathsf{XC}} = \mu\,\_^{\mathsf{seq}(t_1)\times\cdots\times\mathsf{seq}(t_n)\to\mathsf{seq}(t)}(x_0).$$
$$\text{match } x_0 \text{ with } (x_1, (\ldots, x_n)) \to [\![q]\!]_{\mathsf{XC}}$$

$$[\![q(q_1,\ldots,q_n)]\!]_{\mathsf{XC}} = [\![q]\!]_{\mathsf{XC}}\,([\![q_1]\!]_{\mathsf{XC}}, (\ldots, [\![q_n]\!]_{\mathsf{XC}}))$$

where $\mathsf{seq}(t) \equiv (t \wedge [\mathbf{1}*]) \vee ([\,t \setminus [\mathbf{1}*]\,])$
and $\quad t(\mathtt{node}()) \equiv \mathbf{1},\ t(\mathtt{text}()) \equiv \mathtt{String},\ t(l) \equiv <l>1$

Fig. 8. Translation of XQ_H^+ into CDuce

The translation is defined by a function $[\![\_]\!]_{\mathsf{XC}}$ that converts an XQuery query into a CDuce expression. It is straightforward and ensures that the result of a translation $[\![q]\!]_{\mathsf{XC}}$ always has a sequence type. We assume that both languages have the same set of variables and constants. An empty sequence is translated into the atom 'nil, a constant is translated into a singleton sequence containing that constant, and similarly for XML fragments. The sequence operator is translated into concatenation. Variables do not require any special treatment. An XPath navigation step is translated into the corresponding navigational pattern, whereas "for in" loops are encoded similarly using the transform construct (in XQuery, an XPath query applied to a sequence of elements is the concatenation of the individual applications). The "switch" construct is directly translated into a "match with" construct. The "typeswitch" construct works in a similar way but special care must be taken with respect to the type t that is tested. Indeed, if t is a sequence type, then its translation returns the sequence type, but if t is something else (say int), then it must be embedded into a sequence type. Interestingly, this test can be encoded as the CDuce type $\mathsf{seq}(t)$ which keeps the part of t that is a sequence unchanged while embedding the part of t that is not a sequence (namely $t \setminus [\mathbf{1}*]$) into a sequence type (i.e., $[t \setminus [\mathbf{1}*]]$). The "some $\$x$ in q_1 statisfies q_2" expression iterates over the sequence that is the result of the translation of q_1, binding variable x in turn to each element, and evaluates (the translation of) q_2 in this context. If the evaluation of q_2 yields the singleton sequence true, then we return a dummy non-empty sequence; otherwise, we return the empty sequence. If the whole transform yields an empty sequence,

$$\cfrac{\Gamma \vdash_{\mathsf{XQ}} q : s \quad s \leq t}{\Gamma \vdash_{\mathsf{XQ}} q : t} \qquad \cfrac{\Gamma \vdash_{\mathsf{XQ}} q_1 : [s*] \quad \Gamma, x : [s] \vdash_{\mathsf{XQ}} q_2 : t \quad t \leq [\mathbb{1}*]}{\Gamma \vdash_{\mathsf{XQ}} \text{ for } \$x \text{ in } q_1 \text{ return } q_2 : t}$$

$$\cfrac{\{\dot{y} \mapsto \text{`nil'}\}; \square \,|\, s /\!\!/ axis\{y \mid \mathsf{t}(test)\} = \{\dot{y} \mapsto t\}}{\Gamma \vdash_{\mathsf{XQ}} x : [s*] \quad t \leq [\mathbb{1}*] \quad t' = \min\{t' \mid t \leq [t'*]\}}{\Gamma \vdash_{\mathsf{XQ}} \$x / axis :: test : [t'*]} \text{ typ-path}$$

$$\cfrac{\Gamma \vdash_{\mathsf{XQ}} q : t \quad \begin{cases} t \not\leq \neg[c] \Rightarrow \Gamma \vdash_{\mathsf{XQ}} q_1 : s \\ t \not\leq [c] \quad \Rightarrow \Gamma \vdash_{\mathsf{XQ}} q_2 : s \end{cases}}{\Gamma \vdash_{\mathsf{XQ}} \begin{array}{l} \text{case } c \text{ return } q_1 \quad : s \\ \text{default return } q_2 \end{array}} \qquad \cfrac{t_1 = s \wedge \mathsf{seq}(t) \quad \Gamma, x : t_1 \vdash_{\mathsf{XQ}} q_1 : t'_1}{\Gamma \vdash_{\mathsf{XQ}} q : s \quad t_2 = s \wedge \neg\mathsf{seq}(t) \quad \Gamma \vdash_{\mathsf{XQ}} q_2 : t'_2}{\Gamma \vdash_{\mathsf{XQ}} \begin{array}{l} \text{case } t \text{ as } \$x \text{ return } q_1 \quad : \bigvee_{\{i \mid t_i \not\approx 0\}} t'_i \\ \text{default return } q_2 \end{array}}$$

switch q is labeled under the left, typeswitch q under the right.

$$\cfrac{\Gamma \vdash_{\mathsf{XQ}} q_1 : [s*] \quad \Gamma, x : [s] \vdash_{\mathsf{XQ}} q_2 : [\mathsf{bool}]}{\Gamma \vdash_{\mathsf{XQ}} \text{ some } \$x \text{ in } q_1 \text{ statisfies } q_2 : [\mathsf{bool}]}$$

$$\cfrac{\Gamma, x_1 : \mathsf{seq}(t_1), \cdots, x_n : \mathsf{seq}(t_n) \vdash_{\mathsf{XQ}} q : \mathsf{seq}(t)}{\Gamma \vdash_{\mathsf{XQ}} \text{fun } \$x_1 : t_1, \ldots, \$x_n : t_n \text{ as } t. \; q \; : \; \mathsf{seq}(t_1) \times \cdots \times \mathsf{seq}(t_n) \to \mathsf{seq}(t)}$$

$$\cfrac{\Gamma \vdash_{\mathsf{XQ}} q : t_1 \times \cdots \times t_n \to t \quad \Gamma \vdash_{\mathsf{XQ}} q_i : t_i \quad (i = 1..n)}{\Gamma \vdash_{\mathsf{XQ}} q(q, \ldots, q) : t}$$

Fig. 9. Typing rules for $\mathsf{XQ_H^+}$

it means that none of the iterated elements matched satisfied the predicate q_2 and therefore the whole expression evaluates to the singleton `false`, otherwise it evaluates to the singleton `true`. Abstractions are translated into CDuce functions, and the same treatment of "sequencing" the type is applied to the types of the arguments and type of the result. Lastly, application is translated by building nested pairs with the arguments before applying the function.

Not only does this translation ensure soundness of the original XQuery 3.0 programs, it also turns CDuce into a sandbox where one can experiment various typing features that can be readily back-ported to XQuery afterwards.

5.2 Toward and beyond XQuery 3.0

We now discuss the salient features and address some shortcomings of $\mathsf{XQ_H^+}$. First and foremost, we can define a precise and sound type system directly on $\mathsf{XQ_H^+}$ as shown in Figure 9 (standard typing rules are omitted and for the complete definition, see Appendix E). While most constructs are typed straightforwardly (the typing rules are deduced from the translation of $\mathsf{XQ_H^+}$ into CDuce) it is interesting to see that the rules match those defined in XQuery Static Semantics specification [24] (with the already mentioned difference that we use CDuce types instead of XMLSchema). Two aspects however diverge from the standard. Our use of CDuce's semantic subtyping (rather than XMLSchema's nominal subtyping), and the rule typ-path where we use the formal developments of Section 3 to provide a precise typing rule for XPath navigation. Deriving the typing rules from our translation allows us to state the following theorem:

Theorem 13. *If* $\Gamma \vdash_{\mathsf{XQ}} query : t$, *then* $\Gamma \vdash [\![query]\!]_{\mathsf{XC}} : t$.

A corollary of this theorem is the soundness of the XQ_H^+ type system (since the translation of a well-typed XQ_H^+ program yields a well-typed CDuce program with the same type).

While the XQ_H^+ fragment we present here is already very expressive, it does not account for all features of XQuery. For instance, it does not feature data value comparison or sorting (*i.e.*, the order by construct of XQuery) nor does it account for built-in functions such as position(), node identifiers, and so on. However, it is known that features such as data value comparison make typechecking undecidable (see for instance [1]). We argue that the main point of this fragment is to cleanly separate structural path navigation from other data value tests for which we can add built-in operators and functions, with an hardcoded, *ad-hoc* typing rule.

Lastly, one may argue that, in practice, XQuery database engines do *not* rely on XQuery Core for evaluation but rather focus on evaluating efficiently large (multi-step, multi-predicate) XPath expressions in one go and, therefore, that normalizing XQuery programs into XQ_H^+ programs and then translating the latter into CDuce programs may seem overly naive. We show in Appendix D that XPath expressions that are purely navigational can be rewritten in a single pattern of the form: $axis\{x \mid t\}$ which can then be evaluated very efficiently (that is, without performing the unneeded extra traversals of the document that a single step approach would incur).

6 Related Work and Conclusion

Our work tackles several aspects of XML programming, the salient being: *(i)* encoding of XPath or XPath-like expressions (including reverse axes) into regular types and patterns, *(ii)* recursive tree transformation using accumulators and their typing, and *(iii)* type systems and typechecking algorithms for XQuery.

Regarding XPath and pattern matching, the work closest to ours is the implementation of paths as patterns in XTatic. XTatic [11] is an object-oriented language featuring XDuce regular expression types and patterns [16,17]. In [12], Gapeyev and Pierce alter XDuce's pattern matching semantics and encode a fragment of XPath as patterns. The main difference with our work is that they use a hard-coded all-match semantics (a variable can be bound to several subterms) to encode the accumulations of recursive axes, which are restricted by their data model to the "child" and "descendant" axes. Another attempt to use path navigation in a functional language can be found in [19] where XPath-like combinators are added to Haskell. Again, only child or descendant-like navigation is supported and typing is done in the setting of Haskell which cannot readily be applied to XML typing (results are returned as *homogeneous* sequences).

Our use of accumulators is reminiscent of Macro Tree Transducers (MTTs, [8]), that is, tree transducers (tree automata producing an output) that can also accumulate part of the input and copy it in the output. It is well known that given an input regular tree language, the type of the accumulators and results may not be regular. Exact typing may be done in the form of backward type inference, where the output type is given and a largest input type is inferred [20].

It would be interesting to use the backward approach to type our accumulators without the approximation introduced for "cons" for instance.

For what concerns XQuery and XPath, several complementary works are of interest. First, the work of Genevès *et al.* which encodes XPath and XQuery in the μ-calculus ([14,15] where zippers to manage XPath reverse axes were first introduced) supports our claim. Adding path expressions at the level of *types* is not more expensive: subtyping (or equivalently satisfiability of particular formulæ of the μ-calculus which are equivalent to regular tree languages) remains EXPTIME, even with upward paths (or in our case, zipper types). In contrast, typing path expressions and more generally XQuery programs is still a challenging topic. While the W3C's formal semantics of XQuery [24] gives a polynomial time typechecking algorithm for XQuery (in the absence of nested "let" or "for" constructs), it remains far too imprecise (in particular, reverse axes are left untyped). Recently, Genevès *et al.* [13] also studied a problem of typing reverse axes by using regular expressions of μ-calculus formulæ as types, which they call focused-tree types. Since, as our zipped types, focused-tree types can describe both the type of the current node and its context, their type system also gives a precise type for reverse axis expressions. However, while focused-tree types are more concise than zipper types, it is difficult to type construction of a new XML document, and thus their type system requires an explicit type annotation for each XML element. Furthermore, their type system does not feature arrow types. That said, it will be quite interesting to combine their approach with ours.

We are currently implementing axis patterns and XPath expressions on top of the CDuce compiler. Future work includes extensions to other XQuery constructs as well as XMLSchema, the addition of aggregate functions by associating accumulators to specific operators, the inclusion of navigational expressions in types so as to exploit the full expressivity of our zipped types (*e.g.*, to type functions that work on the ancestors of their arguments), and the application of the polymorphic type system of [5,6] to both XQuery and navigational CDuce so that for instance the function *pretty* defined in the introduction can be given the following, far more precise intersection of two arrow types:

```
(<a class="style1" href=β ..>γ -> <a href=β>[<b>γ])
   & (α\<a class="style1" href=_ ..>_ -> α\<a class="style1" href=_ ..>_ )
```

This type (where α, β, and γ denote universally quantified type variables) precisely describes, by the arrow type on the first line, the transformation of the sought links, and states, by the arrow on the second line, that in all the other cases (*i.e.*, for every type α different from the sought link) it returns the same type as the input. This must be compared with the corresponding type in Figure 1, where the types of the attribute href, of the content of the a element, and above all of any other value not matched by the first branch are not preserved.

Acknowledgments. We want to thank the reviewers of ESOP who gave detailed suggestions to improve our presentation. This work was partially supported by the ANR TYPEX project n. ANR-11-BS02-007.

References

1. Alon, N., Milo, T., Neven, F., Suciu, D., Vianu, V.: XML with data values: Type-checking revisited. In: PODS, pp. 138–149. ACM (2001)
2. Amadio, R.M., Cardelli, L.: Subtyping recursive types. ACM Trans. Program. Lang. Syst. 15(4), 575–631 (1993)
3. Benedikt, M., Vu, H.: Higher-order functions and structured datatypes. In: WebDB, pp. 43–48 (2012)
4. Benzaken, V., Castagna, G., Frisch, A.: CDuce: An XML-centric general-purpose language. In: ICFP, pp. 51–63 (2003)
5. Castagna, G., Nguyễn, K., Xu, Z., Abate, P.: Polymorphic functions with set-theoretic types. Part 2: Local type inference and type reconstruction. In: POPL, pp. 289–302 (2015)
6. Castagna, G., Nguyễn, K., Xu, Z., Im, H., Lenglet, S., Padovani, L.: Polymorphic functions with set-theoretic types. Part 1: Syntax, semantics, and evaluation. In: POPL, pp. 5–17 (2014)
7. Castagna, G., Xu, Z.: Set-theoretic foundation of parametric polymorphism and subtyping. In: ICFP, pp. 94–106 (2011)
8. Engelfriet, J., Vogler, H.: Macro tree transducers. J. Comput. Syst. Sci. 31(1), 71–146 (1985)
9. Frisch, A.: Théorie, conception et réalisation d'un langage adapté à XML. PhD thesis, Université Paris 7 Denis Diderot (2004)
10. Frisch, A., Castagna, G., Benzaken, V.: Semantic subtyping: Dealing set-theoretically with function, union, intersection, and negation types. J. ACM 55(4), 1–64 (2008)
11. Gapeyev, V., Garillot, F., Pierce, B.C.: Statically typed document transformation: An Xtatic experience. In: PLAN-X (2006)
12. Gapeyev, V., Pierce, B.C.: Paths into patterns. Technical Report MS-CIS-04-25, University of Pennsylvania (October 2004)
13. Genevès, P., Gesbert, N., Layaïda, N.: Xquery and static typing: Tackling the problem of backward axes (July 2014), http://hal.inria.fr/hal-00872426
14. Genevès, P., Layaïda, N.: Eliminating dead-code from XQuery programs. In: ICSE (2010)
15. Genevès, P., Layaïda, N., Schmitt, A.: Efficient static analysis of XML paths and types. In: PLDI (2007)
16. Hosoya, H., Pierce, B.C.: Regular expression pattern matching for XML. J. Funct. Program. 13(6), 961–1004 (2003)
17. Hosoya, H., Pierce, B.C.: XDuce: A statically typed XML processing language. ACM Trans. Internet Technol. 3(2), 117–148 (2003)
18. Huet, G.: The Zipper. J. Funct. Program. 7(5), 549–554 (1997)
19. Lämmel, R.: Scrap your boilerplate with XPath-like combinators. In: POPL (2007)
20. Maneth, S., Berlea, A., Perst, T., Seidl, H.: XML type checking with macro tree transducers. In: PODS (2005)
21. W3C: XPath 1.0 (1999), http://www.w3.org/TR/xpath,
22. W3C: XPath 2.0 (2010), http://www.w3.org/TR/xpath20,
23. W3C: XML Query (2010), http://www.w3.org/TR/xquery,
24. XQuery 1.0 and XPath 2.0 Formal Semantics, 2nd edn (2010), http://www.w3.org/TR/xquery-semantics/
25. W3C: XQuery 3.0 (2014), http://www.w3.org/TR/xquery-3.0
26. W3C: XML Schema (2009), http://www.w3.org/XML/Schema

ISOLATE: A Type System for Self-recursion

Ravi Chugh

University of Chicago
rchugh@cs.uchicago.edu

Abstract. A fundamental aspect of object-oriented languages is how recursive functions are defined. One semantic approach is to use simple record types and explicit recursion (*i.e.* `fix`) to define mutually recursive units of functionality. Another approach is to use records and recursive types to describe recursion through a "self" parameter. Many systems rely on both semantic approaches as well as combinations of universally quantified types, existentially quantified types, and mixin operators to encode patterns of method reuse, data encapsulation, and "open recursion" through self. These more complex mechanisms are needed to support many important use cases, but they often lack desirable theoretical properties, such as decidability, and can be difficult to implement, because of the equirecursive interpretation that identifies mu-types with their unfoldings. Furthermore, these systems do not apply to languages without explicit recursion (such as JavaScript, Python, and Ruby). In this paper, we present a statically typed calculus of functional objects called ISOLATE that can reason about a pattern of mixin composition without relying on an explicit fixpoint operation. To accomplish this, ISOLATE extends a standard isorecursive type system with a mechanism for checking the "mutual consistency" of a collection of functions, that is, that all of the assumptions about self are implied by the collection itself. We prove the soundness of ISOLATE via a type-preserving translation to a calculus with F-bounded polymorphism. Therefore, ISOLATE can be regarded as a stylized subset of the more expressive calculus that admits an interesting class of programs yet is easy to implement. In the future, we plan to investigate how other, more complicated forms of mixin composition (again, without explicit recursion) may be supported by lightweight type systems.

1 Introduction

Researchers have studied numerous foundational models for typed object-oriented programming in order to understand the theoretical and practical aspects of these languages. Many of these models are based on the lambda-calculus extended with combinations of explicit recursion, records, prototype delegation, references, mixins, and traits. Once the dynamic semantics of the language has been set, various type theoretic constructs are then employed in order to admit as many well-behaved programs as possible. These mechanisms include record types and recursive types [7], bounded universal quantification [9], bounded existential quantification [30], F-bounded polymorphism [6,2], and variant parametric types [20,34]. A classic survey by Bruce *et al.* [5] compares many of the core aspects of these systems.

© Springer-Verlag Berlin Heidelberg 2015
J. Vitek (Ed.): ESOP 2015, LNCS 9032, pp. 257–282, 2015.
DOI: 10.1007/978-3-662-46669-8_11

A fundamental aspect of an object calculus is how recursive functions are defined. One option is to include explicit recursion on values (*i.e.* fix) as a building block. The evaluation rule

$$\text{fix}\,(\lambda f.e) \hookrightarrow e[(\text{fix}\,\lambda f.e)/f] \quad \text{where } e = \lambda x.e'$$

repeatedly substitutes the entire expression as needed, thus realizing the recursion. Explicit recursion is straightforward to reason about: the expression fix e has type $S \to T$ as long as e has type $(S \to T) \to (S \to T)$. Similar evaluation and typechecking rules can be defined for recursive non-function values, such as records, using simple syntactic restrictions (see, for example, the notion of *statically constructive* definitions in OCaml [24]). This approach can be used to define objects of (possibly mutually) recursive functions. For example, the following simple object responds to the "message" f by multiplying increasingly large integers ad infinitum:

$$\text{o1} = \text{fix}\,(\lambda \text{this}.\{\,\text{f} = \lambda \text{n}.\text{n} * \text{this}.\text{f}\,(\text{n} + 1)\,\})$$

$$\text{o1}.\text{f}(1) \hookrightarrow^* 1 * \text{o1}.\text{f}(2) \hookrightarrow^* 1 * 2 * \text{o1}.\text{f}(3) \hookrightarrow^* \cdots$$

On the other hand, in a language without explicit recursion on values, recursive computations can be realized by passing explicit "self" (or "this") parameters through function definitions and applications. The following example demonstrates this style:

$$\text{o2} = \{\,\text{f} = \lambda\,(\text{this},\text{n}).\text{n} * \text{this}.\text{f}\,(\text{this}, \text{n} + 1)\,\}$$

$$\text{o2}.\text{f}\,(\text{o2}, 1) \hookrightarrow^* 1 * \text{o2}.\text{f}\,(\text{o2}, 2) \hookrightarrow^* 1 * 2 * \text{o2}.\text{f}\,(\text{o2}, 3) \hookrightarrow^* \cdots$$

Notice that occurrences of this, substituted by o2, do not require any dedicated evaluation rule to "tie the knot" because the record is explicitly passed as an argument through the recursive calls.

Object encodings using either of the two approaches above — explicit recursion or self-parameter passing — can be used to express many useful programming patterns of "open recursion," including: (1) the ability to define methods independently of their host objects (often referred to as *premethods*) that later get mixed into objects in a flexible way; and (2) the ability to define wrapper functions that interpose on the invocation of methods that have been previously defined.

Compared to explicit recursion, however, the self-parameter-passing approach is significantly harder for a type system to reason about, requiring a combination of equirecursive types, subtyping, and F-bounded polymorphism (where type variable bounds are allowed to be recursive). Such combinations often lack desirable theoretical properties, such as decidability [28,2], and pose implementation challenges due to the equirecursive, or "strong," interpretation of mu-types. Nevertheless, mainstream languages like Java and C# incorporate these features, but this complexity is not suitable for all language designs. For example, the popularity of dynamically typed scripting languages (such as JavaScript, Python, Ruby, and PHP) has sparked a flurry of interest in designing statically typed dialects (such as TypeScript and Hack). It is unlikely that heavyweight mechanisms like F-bounded polymorphism will be easily adopted into such language designs. Instead, it would be useful to have a lightweight type system that could reason about some of the programming patterns enabled by the more complicated systems.

Contributions. The thesis of this paper is that the two patterns of open recursion under consideration do not require the full expressive power of F-bounded polymorphism and equirecursive types. To substantiate our claim, we make two primary contributions:

- We present a statically typed calculus of functional objects called ISOLATE, which is a simple variation and extension of isorecursive types (with only a very limited form of subtyping and bounded quantification). We show that ISOLATE is able to admit an interesting class of recursive programs yet is straightforward to implement. The key feature in ISOLATE is a typing rule that treats records of premethods specially, where all assumptions about the self parameter are checked for mutual consistency.
- To establish soundness of the system, we define a type-preserving translation of well-typed ISOLATE programs into a more expressive calculus with F-bounded polymorphism. As a result, ISOLATE can be regarded as a stylized subset of the traditional, more expressive system.

In languages without `fix` and where the full expressiveness of F-bounded polymorphism is not needed, the approach in ISOLATE provides a potentially useful point in the design space for supporting recursive, object-oriented programming patterns. In future work, we plan to investigate supporting additional forms of mixin composition (beyond what ISOLATE currently supports) and applying these techniques to statically typed dialects of popular scripting languages, which often do not include a fixpoint operator.

Outline. Next, in §2, we provide background on relevant typing mechanisms and identify the kinds of programming patterns we aim to support in ISOLATE. Then, in §3, we provide an overview of how ISOLATE reasons about these patterns in a relatively lightweight way. After defining ISOLATE formally, we describe its metatheory in §4. We conclude in §5 with discussions of related and future work.

2   Background

In this section, we survey how several existing typing mechanisms can be used to define objects of mutually recursive functions, both with and without explicit recursion. We start with a simply-typed lambda-calculus and then extend it with operations for defining records, isorecursive folding and unfolding, and parametric polymorphism. We identify aspects of these systems that motivate our late typing proposal. This section is intended to provide a self-contained exposition of the relevant typing mechanisms; expert readers may wish to skip ahead to §2.3.

Core Language Features and Notation. In Figure 1, we define the expression and type languages for several systems that we will compare in this section. We assume basic familiarity with the dynamic and static semantics for all of these features [29]. In our notation, we define the language $\mathscr{L}$ to be the simply-typed lambda-calculus. We use B to range over some set of base types (int, unit, str, *etc.*) and c to range over constants (not, (*), (++), *etc.*).

Base Language $\mathscr{L}$:

$$e \; ::= \; c \; | \; \lambda x.e \; | \; x \; | \; e_1 \, e_2 \qquad\qquad T \; ::= \; B \; | \; T_1 \to T_2$$

Extensions to Expression Language (Denoted by Subscripts):

$$
\begin{aligned}
\mathscr{L}_{\texttt{fix}} \; e \; &::= \; \cdots \; | \; \texttt{fix}\,e \\
\mathscr{L}_{\{\}} \; e \; &::= \; \cdots \; | \; \{\overline{f}=\overline{e}\} \; | \; e.f \\
\mathscr{L}_{\texttt{iso}} \; e \; &::= \; \cdots \; | \; \texttt{fold}\,e \; | \; \texttt{unfold}\,e \\
\mathscr{L}_{\Lambda} \; e \; &::= \; \cdots \; | \; \Lambda A.e \; | \; e[T]
\end{aligned}
$$

Extensions to Type Language (Denoted by Superscripts):

$$
\begin{aligned}
\mathscr{L}^{\{\}} \qquad & T \; ::= \; \cdots \; | \; \{\,\overline{f}:\overline{T}\,\} \\
\mathscr{L}^{\mu} \qquad & T \; ::= \; \cdots \; | \; A \; | \; \mu A.T \\
\mathscr{L}^{\forall} \qquad & T \; ::= \; \cdots \; | \; A \; | \; \forall A.\,T \\
\mathscr{L}^{\forall A<:T} \qquad & T \; ::= \; \cdots \; | \; A \; | \; \forall A<:T.\,T' \; | \; \texttt{top} \\
\mathscr{L}^{\forall A<:T(A)} \qquad & T \; ::= \; \cdots \; | \; A \; | \; \forall A<:T(A).\,T' \; | \; \texttt{top}
\end{aligned}
$$

Comparison of Selected Language Extensions:

$$\textsc{LangFixSub} \triangleq \mathscr{L}^{\{\},<:,\forall}_{\{\},\texttt{fix},\Lambda} \qquad \textsc{LangMu} \triangleq \mathscr{L}^{\{\},\mu,\forall}_{\{\},\texttt{iso},\Lambda} \qquad \textsc{FSubRec} \triangleq \mathscr{L}^{\{\},\mu=,\forall A<:T(A)}_{\{\},\Lambda}$$

	fix	Property A	Property B	Property C	Property D	"Simplicity"
LangFixSub	Y	✓	✓	✓	✓	✓
LangMu	N	✓	✓	✗	✓	✓
IsoLate	N	✓	✓	✓	✓⁻	✓
FSubRec	N	✓	✓	✓	✓⁻	✗

Fig. 1. Core Languages of Expressions and Types

We define several extensions to the expression and type languages of $\mathscr{L}$. Our notational convention is to use subscripts to denote extensions to the language of expressions and superscripts for extensions to the language of types. In particular, we write $\mathscr{L}_{\texttt{fix}}$, $\mathscr{L}_{\{\}}$, $\mathscr{L}_{\texttt{iso}}$, and $\mathscr{L}_{\Lambda}$ to denote extensions of the base language, $\mathscr{L}$, with the usual notions of fixpoint, records, fold and unfold operators for isorecursive types, and type abstraction and application, respectively. We write $\mathscr{L}^{\{\}}$ to denote the extension of the base type system with record types, $\mathscr{L}^{\mu}$ for isorecursive types, $\mathscr{L}^{\mu=}$ for equirecursive types, $\mathscr{L}^{\forall}$ for (unbounded) universal quantification, $\mathscr{L}^{\forall A<:T}$ for bounded quantification, and $\mathscr{L}^{\forall A<:T(A)}$ for F-bounded quantification (where type bounds can recursively refer to type variables). We attach multiple subscripts or superscripts to denote multiple extensions. For example, $\mathscr{L}^{\{\}}_{\{\},\texttt{fix}}$ denotes the statically typed language with records and a fixpoint operator.

In addition to the syntactic forms defined in Figure 1, we freely use syntactic sugar for common derived forms. For example, we often write $\texttt{let}\,x=e_1\,\texttt{in}\,e_2$ instead of $(\lambda x.e_2)\,e_1$ and we often write $\texttt{let}\,f\,x\,y=e_1\,\texttt{in}\,e_2$ instead of $\texttt{let}\,f=\lambda x.\lambda y.e_1\,\texttt{in}\,e_2$.

We assume the presence of primitive if-expressions, which we write as e_1 ? e_2 : e_3. We also write val $x :: T$ as a way to ascribe an expected type to a let-bound expression.

Comparison of Systems. We define aliases in Figure 1 for three systems to which we will pay particular attention. We write LANGFIXSUB to refer to the language of records, explicit recursion, and subtyping; LANGMU to refer to the language of records and isorecursive mu-types; and FSUBREC to refer to the language of records, equirecursive mu-types, and F-bounded polymorphism. In addition, each of these systems has universal quantification, which is unbounded in LANGFIXSUB and LANGMU and F-bounded in FSUBREC. Notice that LANGMU and FSUBREC do not include explicit recursion, indicated by the absence of fix in the subscripts. The name FSUBREC is a mnemonic to describe the often-called System Fsub extended with recursive bounds and recursive types.

Next, we will compare these three languages. The table at the bottom of Figure 1 summarizes their differences along a number of dimensions that we will discuss. Properties A through D are four programming patterns that are of interest for this paper, and "Simplicity" informally refers to implementation and metatheoretic challenges that the type system presents. Then, in §3, we will explain how our IsoLate calculus identifies a new point in the design space that fits in between LANGMU and FSUBREC.

2.1   LANGFIX and LANGFIXSUB: Recursion with fix

We will start with the language LANGFIX $\stackrel{\circ}{=} \mathscr{L}^{\{\}}_{\{\},fix}$ of records and explicit recursion (without subtyping), in which mutually recursive functions can be defined as follows:

$$\text{Ticker} \stackrel{\circ}{=} \{ \text{tick:int} \rightarrow \text{str; tock:int} \rightarrow \text{str} \}$$

```
val ticker0 :: Ticker
let ticker0 = fix \this.
  let f n = n > 0 ? "tick " ++ this.tock (n)   : "" in
  let g n = n > 0 ? "tock " ++ this.tick (n-1) : "" in
  { tick = f; tock = g }

ticker0.tick 2  -- "tick tock tick tock "
```

As mentioned in §1, typechecking expressions of the form fix e is simple. We will build on this example to demonstrate several programming patterns of interest.

[Property A] Defining Premethods Separately. In the program above, all components of the mutually recursive definition appear together (*i.e.* syntactically adjacent) inside the fixpoint expression. For reasons of organization, the programmer may want to instead structure the component definitions separately and then combine them later:

```
val tick, tock :: Ticker -> int -> str

let tick this n = n > 0 ? "tick " ++ this.tock (n)   : ""
let tock this n = n > 0 ? "tock " ++ this.tick (n-1) : ""
```

```
let ticker1 = fix \this.
  let f n = tick this n in
  let g n = tock this n in
  { tick = f; tock = g }
```

```
ticker1.tick 2   -- "tick tock tick tock "
```

Furthermore, if the programmer wants to define a second quieter ticker that does not emit the string "tock", defining the component functions separately avoids the need to duplicate the implementation of tick:

```
val tock' :: Ticker -> int -> str
let tock' this n = this.tick (n-1)
```

```
let ticker2 = fix \this.
  let f n = tick this n in
  let g n = tock' this n in
  { tick = f; tock = g }
```

```
ticker2.tick 2   -- "tick tick "
```

Notice that the implementations of tick, tock, and tock' lay outside of the recursive definitions ticker1 and ticker2 and are each parameterized by a this argument. Because these three functions are not inherently tied to any record, we refer to them as *premethods*. In contrast, we say that the functions f and g in ticker1 (respectively, ticker2) are *methods* of ticker1 (respectively, ticker2) because they are tied to that particular object.[1] The method definitions are eta-expanded to ensure that the recursive definitions are syntactically well-founded (*e.g.* [24]).

[Property B] Intercepting Recursive Calls. Another benefit of defining premethods separately from their eventual host objects is that it facilitates "intercepting" recursive calls in order to customize behavior. For example, say the programmer wants to define a louder version of the ticker that emits exclamation points in between each "tick" and "tock". Notice that the following reuses the premethods tick and tock from before:

```
let ticker3 = fix \this.
  let f n = "! " ++ tick this n in
  let g n = "! " ++ tock this n in
  { tick = f; tock = g }
```

```
ticker3.tick 2   -- "! tick ! tock ! tick ! tock ! "
```

---

[1] The term premethod is sometimes used with a slightly different meaning, namely, for a lambda that closes over an implicit receiver variable rather than than explicitly taking one as a parameter. We use the term premethod to emphasize that the first (explicit) function argument is used to realize recursive binding.

[Property C] Mixing Premethods into Different Types. Having kept premethod definitions separate, the programmer may want to include them into objects with various
types, for example, an extended ticker type that contains an additional boolean field to
describe whether or not its volume is loud:

$$\texttt{TickerVol} \triangleq \{ \texttt{tick:int} \rightarrow \texttt{str; tock:int} \rightarrow \texttt{str; loud:bool} \}$$

Intuitively, the mutual requirements between the `tick` and `tock'` premethods are independent of the presence of the `loud` field, so we would like the type system to accept
the following program:

```
let quietTicker = fix \this.
  let f n = tick this n in
  let g n = tock' this n in
  { tick = f; tock = g; loud = false }
```

This program is not type-correct in LANGFIX, because the types derived for `tick` and
`tock'` pertain to `Ticker` rather than `TickerVol`. Extending LANGFIX with the usual
notion of record subtyping, resulting in a system called LANGFIXSUB, addresses the
problem, however.

In addition, LANGFIXSUB can assign the following less restrictive types to the same
premethods from before: tick :: Tock → int → str, tock :: Tick → int → str, and
tock' :: Tick → int → str. Notice that the types of the `this` arguments are described
by the following type abbreviations, rather than `Ticker`, to require only those fields
used by the definitions:

$$\texttt{Tick} \triangleq \{ \texttt{tick:int} \rightarrow \texttt{str} \} \qquad \texttt{Tock} \triangleq \{ \texttt{tock:int} \rightarrow \texttt{str} \}$$

[Property D] Abstracting Over Premethods. The last scenario that we will consider
using our running example is abstracting over premethods. For example, the following wrapper functions avoid duplicating the code to insert exclamation points in the
definition of `ticker3` from before:

```
val wrap :: all A,B,C,C'. (C->C') -> (A->B->C) -> (A->B->C')
let wrap g f x y = g (f x y)

val exclaim :: all A,B. (A -> B -> str) -> (A -> B -> str)
let exclaim = wrap _ _ _ _ (\s. "! " ++ s)

let ticker3' = fix \this.
  let f n = (exclaim _ _ tick) this n in
  let g n = (exclaim _ _ tock) this n in
  { tick = f; tock = g }

ticker3'.tick 2   -- "! tick ! tock ! tick ! tock ! "
```

The two calls to `exclaim` (and, hence, `wrap`) are made with two different premethods
as arguments. Because these premethods have different types in LANGFIXSUB, (unbounded) parametric polymorphism is required for typechecking. Note that we write
underscores where type instantiations can be easily inferred.

"Open" vs. "Closed" Objects. As we have seen throughout the previous examples, the type of a record-of-premethods differs from that of a record-of-methods. We refer to the former kind of records as *open objects* and the latter as *closed objects*. Once closed, there is no way to extract a method to be included into a closed object of a different type. The following example highlights this distinction, where the `makeLouderTicker` function takes a record of two premethods and wraps them before creating a closed `Ticker` object:

$$\text{PreTicker} \stackrel{\circ}{=} \{\text{ tick:Tock} \rightarrow \text{int} \rightarrow \text{str; tock:Tick} \rightarrow \text{int} \rightarrow \text{str }\}$$

```
val makeLouderTicker :: PreTicker -> Ticker
let makeLouderTicker openObj = fix \closedObj.
  let f n = (exclaim _ _ openObj.tick) closedObj n in
  let g n = (exclaim _ _ openObj.tock) closedObj n in
  { tick = f; tock = g }

let (ticker4, ticker5) =
  ( makeLouderTicker { tick =                tick; tock = tock }
  , makeLouderTicker { tick = exclaim _ _ tick; tock = tock } )

ticker4.tick 2   -- "! tick ! tock ! tick ! tock ! "
ticker5.tick 2   -- "! ! tick ! tock ! ! tick ! tock ! ! "
```

The first row of the table in Figure 1 summarizes that LANGFIXSUB supports the four programming scenarios outlined in the previous section. Next, we will consider how the same scenarios manifest themselves in languages *without* an explicit `fix`. Such encodings may be of theoretical interest as well as practical interest for object-oriented languages such as JavaScript, in which the semantics does not include `fix`.

2.2   LANGMU: Recursion with Mu-Types

We will consider the language LANGMU of records, isorecursive mu-types, and unbounded universal quantification. The standard rules for isorecursive types are:

$$\frac{T = \mu A.S \quad \Gamma \vdash e : S[T/A]}{\Gamma \vdash \text{fold}\, T\, e : T} \qquad \frac{T = \mu A.S \quad \Gamma \vdash e : T}{\Gamma \vdash \text{unfold}\, e : S[T/A]}$$

We define the syntactic sugar $e \,\$\, f(e') \stackrel{\circ}{=} (\text{unfold}\, e).f(e)(e')$ to abbreviate the common pattern of unfolding a recursive record, reading a method stored in one of its fields, and then calling the method with the (folded) record as the receiver (first argument) to the method.

[Properties A and B]. Defining premethods separately and interposing on recursive calls are much the same in LANGMU as they are in LANGFIXSUB. Using the type-checking rules for isorecursive types above, together with the usual rule for function application, we can write the following in LANGMU:

Ticker $\triangleq \mu A.\{$ tick:$A \to$ int $\to$ str; tock:$A \to$ int $\to$ str $\}$

```
val tick, tock, tock' :: Ticker -> int -> str
let tick this n  = n > 0 ? "tick " ++ this$tock(n)    : ""
let tock this n  = n > 0 ? "tock " ++ this$tick(n-1) : ""
let tock' this n = this$tick(n-1)

let wrap g f x y = g (f x y)
let exclaim = wrap (\s. "! " ++ s)

let (ticker1, ticker2, ticker3) =
  ( fold Ticker { tick = tick          ; tock = tock          }
  , fold Ticker { tick = tick          ; tock = tock'         }
  , fold Ticker { tick = exclaim tick ; tock = exclaim tock })

ticker1$tick(2)  -- "tick tock tick tock "
ticker2$tick(2)  -- "tick tick "
ticker3$tick(2)  -- "! tick ! tock ! tick ! tock ! "
```

[Property D]. As in LangFixSub, unbounded universal quantification in LangMu can be used to give general types to functions, such as wrap and exclaim above, that abstract over premethods.

[Property C]. Premethods in LangMu cannot be folded into *different* mu-types than the ones specified by the annotations for their receiver arguments. Consider the following example that attempts to, as before, define an extended ticker type that contains an additional boolean field:

TickerVol $\triangleq \mu A.\{$ tick:$A \to$ int $\to$ str; tock:$A \to$ int $\to$ str; loud:bool $\}$

```
let quietTicker =
  fold TickerVol { tick = tick ; tock = tock' ; loud = false }
```

The problem is that the record type

$\{$ tick,tock:Ticker $\to$ int $\to$ str; loud:bool $\}$

does not equal the unfolding of TickerVol

$\{$ tick,tock:TickerVol $\to$ int $\to$ str; loud:bool $\}$

as required by the typing rule for fold. In particular, Ticker $\neq$ TickerVol. Unlike for LangFix, simply adding subtyping to the system does not address this difficulty. In the above record type comparison, the contravariant occurrence of the mu-type would require that TickerVol be a subtype of Ticker, which seems plausible by record width subtyping. However, the standard "Amber rule"

$$\frac{\Gamma, A_1 <: A_2 \vdash T_1 <: T_2}{\Gamma \vdash \mu A_1.T_1 <: \mu A_2.T_2}$$

for subtyping on mu-types requires that the type variable vary covariantly [8,29]. As a result, TickerVol $\not<$: Ticker which means that the code snippet fails to typecheck even in LANGMU extended with subtyping.

The only recourse in LANGMU is to duplicate premethods multiple times, one for each target mu-type. The second row of the table in Figure 1 summarizes the four programming scenarios in the context of LANGMU, using an × to mark that Property C does not hold.

2.3 FSUBREC: Recursion with F-bounded Polymorphism

Adding subtyping to LANGMU is not enough, on its own, to alleviate the previous limitation, but it is when combined with a more powerful form of universal quantification. In particular, the system we need is FSUBREC, which contains (1) F-bounded polymorphism, where a bounded universally quantified type $\forall A <: S.\, T$ allows its type bound S to refer to the type variable A being constrained; and (2) equirecursive, or "strong" recursive, types where a mu-type $\mu A.\, T$ is considered definitionally equal to its unfolding $T[(\mu A.\, T)/A]$ in all contexts. That means that there are no explicit fold and unfold operations in FSUBREC as there are in languages with isorecursive, or "weak" recursive, types like LANGMU.

To see why "weak recursion is not a good match for F-bounded quantification," as described by Baldan et al. [2], consider the type instantiation rule

$$\frac{\Gamma \vdash e : \forall A <: S.\, T \quad \Gamma \vdash S' <: S[S'/A]}{\Gamma \vdash e[S'] : T[S'/A]}$$

which governs how bounded universals can be instantiated. Notice that the particular type parameter S' must be a subtype of the bound S where all (recursive) occurrences of A are replaced with S' itself. A recursive type can satisfy an equation like this only when it is considered definitionally equal to its unfolding, because the structure of the types S and S' simply do not match (in all of our examples, S' is a mu-type but S is a record type).

[Properties A and C]. Having explained the motivation for including equirecursive types in FSUBREC, we return to our example starting with premethod definitions:

$$\text{Tick}(A) \stackrel{\circ}{=} \{\, \texttt{tick}:A \to \texttt{int} \to \texttt{str} \,\}$$
$$\text{Tock}(A) \stackrel{\circ}{=} \{\, \texttt{tock}:A \to \texttt{int} \to \texttt{str} \,\}$$
$$\text{TickPre}(B,C) \stackrel{\circ}{=} (\forall A <: \text{Tick}(A).\, A \to B \to C)$$
$$\text{TockPre}(B,C) \stackrel{\circ}{=} (\forall A <: \text{Tock}(A).\, A \to B \to C)$$

```
val tick        :: TockPre (int, str)
val tock, tock' :: TickPre (int, str)

let tick this n  = n > 0 ? "tick " ++ this.tock(this)(n)   : ""
let tock this n  = n > 0 ? "tock " ++ this.tick(this)(n-1) : ""
let tock' this n = this.tick(this)(n-1)
```

There are two aspects to observe. First, the premethod types, which use F-bounded universals, require that the this parameters have only the fields used by the definitions (like in LANGFIXSUB). Second, the implementations of tick, tock, and tock' do not include unfold expressions (unlike in LANGMU), because the type system implicitly folds and unfolds equirecursive types as needed.

We can now mix the premethods into various target mu-types, such as Ticker and TickerVol as defined in LANGMU, by instantiating the F-bounded universals appropriately. In the following, we use square brackets to denote the application, or instantiation, of an expression to a particular type.

```
let (normalTicker, quietTicker) =
  ( { tick=tick[TickerVol]; tock=tock [TickerVol]; loud=false }
  , { tick=tick[TickerVol]; tock=tock'[TickerVol]; loud=false } )

normalTicker.tick(normalTicker)(2)   -- "tick tock tick tock "
quietTicker.tick(quietTicker)(2)     -- "tick tick "
```

[Property B]. Interposing on recursive calls in FSUBREC is much the same as before:

```
val ticker3 :: Ticker
let ticker3 =
  let f this n = "! " ++ tick [Ticker] this n in
  let g this n = "! " ++ tock [Ticker] this n in
  { tick = f; tock = g }

ticker3.tick 2   -- "! tick ! tock ! tick ! tock ! "
```

[Property D]. Abstracting over premethods in FSUBREC, however, comes with a caveat. Symptoms of the issue appear in the definitions of f and g in ticker3 above: notice that the tick and tock premethods are instantiated to a particular type and then wrapped. As a result, f and g are *methods* tied to the Ticker type rather than *premethods* that can work with various host object types. We can abstract the wrapper code in ticker3 as in LANGFIXSUB and LANGMU, but the fact remains that wrap and exclaim below operate on methods rather than premethods:

```
val wrap :: all A,B,C,C'. (C->C') -> (A->B->C) -> (A->B->C')
let wrap g f x y = g (f x y)
let exclaim = wrap _ _ (\s. "! " ++ s)

let loudTicker =
  { tick = exclaim _ _ (tick [TickerVol])
  ; tock = exclaim _ _ (tock [TickerVol])
  ; loud = true }

loudTicker.tick(loudTicker)(2)
  -- "! tick ! tock ! tick ! tock ! "
```

If we wanted to define a function `exclaim'` that truly abstracts over premethods (that is, which could be called with the uninstantiated `tick` and `tock` values), the type of `exclaim'` must have the form

$$\forall R.\ (\forall A <: R.\ A \to \text{int} \to \text{str}) \to (\forall A <: R.\ A \to \text{int} \to \text{str})$$

so that the type variable R could be instantiated with $\text{Tock}(A)$ or $\text{Tick}(A)$ as needed at each call-site. This kind of type cannot be expressed in FSUBREC, however, because these type instantiations need to refer to the type variable A which is not in scope.

As a partial workaround, the best one can do in FSUBREC is to define wrapper functions that work only for particular premethod types and then duplicate definitions for different types. In particular, we can specify two versions of the type signatures

$$\text{wrapTick} ::\ \forall B, C, C'.\ \text{TickPre}(B, C) \to \text{TickPre}(B, C')$$
$$\text{wrapTock} ::\ \forall B, C, C'.\ \text{TockPre}(B, C) \to \text{TockPre}(B, C')$$
$$\text{exclaimTick} ::\ \text{TickPre}(\text{int}, \text{str}) \to \text{TickPre}(\text{int}, \text{str})$$
$$\text{exclaimTock} ::\ \text{TockPre}(\text{int}, \text{str}) \to \text{TockPre}(\text{int}, \text{str})$$

and then define two versions of the wrappers as follows:

```
let wrapTick B C C' g f x y = \A. g (f[A] x y)
let wrapTock B C C' g f x y = \A. g (f[A] x y)

let exclaimTick = wrapTick _ _ _ (\s. "! " ++ s)
let exclaimTock = wrapTock _ _ _ (\s. "! " ++ s)

let loudTicker' =
  { tick = (exclaimTock tick) [TickerVol]
  ; tock = (exclaimTick tock) [TickerVol]
  ; loud = true }

loudTicker'.tick(loudTicker')(2)
  -- "! tick ! tock ! tick ! tock ! "
```

With this approach, the wrapper functions take premethods as inputs and return premethods as outputs. This code duplication is undesirable, of course, so in the FSUBREC row of the table in Figure 1 we qualify the check mark for Property D with a minus sign.

***Undecidability of* FSUBREC.** Equirecursive types and F-bounded polymorphism are powerful, indeed, which is why they are often used as the foundation for object calculi, sometimes with additional constructs like type operators and bounded existential types [5]. This power comes at a cost, however, both in theory and in practice. Subtyping for System Fsub (*i.e.* bounded quantification) is undecidable, even when type bounds are not allowed to be recursive [28], and the addition of equirecursive types poses challenges for the completeness of the system [16,2]. There exist decidable fragments of System Fsub that avoid the theoretically problematic cases without affecting many practical programming patterns. However, mainstream languages (*e.g.* Java and C#) include features beyond generics and subtyping such as variance, and the subtle interaction between these features is an active subject of research (*e.g.* [23,17]). Furthermore, an equirecursive treatment of mu-types demands more work by the type checker — treating recursive types as graphs and identifying equivalent unfoldings — than does isorecursive types, where the recursive type operator is "rigid" and, thus, easy to support [29]. As a result of these complications, we mark the "Simplicity" column for FSUBREC in Figure 1 with an ×. In settings where the full expressive power of these features is needed, then the theoretical and practical hurdles that accompany them are unavoidable. But for other settings, ideally we would have a more expressive system than LANGMU that is much simpler than FSUBREC.

3   The ISOLATE Calculus

We now present our calculus, ISOLATE, that aims to address this goal. Our design is based on two key observations about the FSUBREC encodings from the previous section: first, that the record types used to describe self parameters mention only those fields actually used by the function definitions; and second, that when creating an object out of a record of premethods, each premethod is instantiated with the mu-type that describes the resulting object.

The ISOLATE type system includes special support to handle this common programming pattern without providing the full expressive power, and associated difficulties, of FSUBREC. Therefore, as the third row of the table in Figure 1 outlines, ISOLATE satisfies the same Properties A through D as FSUBREC but fares better with respect to "Simplicity," in particular, because it is essentially as easy as LANGMU to implement.

3.1   Overview

Before presenting formal definitions, we will first work through an ISOLATE example split across Figure 2, Figure 3, and Figure 4.

Premethods. Let us first consider the premethod definitions of tick, tock, and tock' on lines 1 through 8, which bear many resemblances to the versions in FSUBREC. The special *pre-type* $(A\!:\!S) \Rightarrow T$ in ISOLATE is interpreted like the type $\forall A <\!:S.\,A \to T$ in FSUBREC. Values that are assigned pre-types are special functions called *premethods* $\varsigma x\!:\!A <\!:S.e$, which are treated like polymorphic function values $\Lambda A <\!:S.\lambda x\!:\!A.e$ in FSUBREC. A notational convention that we use in our examples is that the identifier this signifies that the enclosing function desugars to a premethod rather than an ordinary lambda. Notice that the types Tick(A) and Tock(A) describe only those fields that are referred to explicitly in the definitions. A simple rule for unfolding self parameters allows the definitions of tick, tock, and tock' to typecheck.

Closing Open Records. ISOLATE provides special support for sending messages to records of premethods. To keep subsequent definitions more streamlined, in ISOLATE we require that all premethods take two arguments (in curried style). Therefore, instead

```
1  type Tick(A)        = { tick: A -> int -> str }
2  type Tock(A)        = { tock: A -> int -> str }
3
4  val tick            :: (A:Tock(A)) => int -> str
5  val tock, tock'     :: (A:Tick(A)) => int -> str
6  let tick this n     = n > 0 ? "tick " ++ this$tock(n)   : ""
7  let tock this n     = n > 0 ? "tock " ++ this$tick(n-1) : ""
8  let tock' this n    = this$tick(n-1)
9
10 val const           :: bool -> (A:{}) => unit -> bool
11 let const b this () = b
12 let (true_, false_) = (const true, const false)
13
14 let normalTicker    = { tick = tick ; tock = tock  ; loud = false_ }
15 let quietTicker     = { tick = tick ; tock = tock' ; loud = false_ }
16
17 normalTicker # tick(2)    -- "tick tock tick tock "
18 quietTicker  # tick(2)    -- "tick tick "
```

Fig. 2. ISOLATE Example (Part 1)

of using boolean values true and false to populate a loud field of type bool, on line 12 we define premethods true_ and false_ of the following type, where the type $Bool \triangleq (A:\{\}) \Rightarrow unit \rightarrow bool$ imposes no constraints on its receivers.

Having defined the required premethods, the expressions on lines 14 through 18 show how to build and use records of premethods. The definitions of normalTicker and quietTicker create ordinary records described by the record type

$$R_0 \triangleq \text{OpenTickerVol} \triangleq \{ \text{tick}:PreTick; \text{tock}:PreTock; \text{loud}:Bool \}$$

where we make use of abbreviations $PreTick \triangleq (A:\text{Tock}(A)) \Rightarrow int \rightarrow str$ and $PreTock \triangleq (A:\text{Tick}(A)) \Rightarrow int \rightarrow str$. We refer to these two records in ISOLATE as "open" because they do not yet form a coherent "closed" object that can be used to invoke methods. The standard typing rule for fold e expressions does not apply to these open records, because the types of their receivers do not match (as was the difficulty in our LANGMU example). To use open objects, ISOLATE provides an additional expression form close e and the following typing rule:

$$\frac{\Gamma \vdash e : R \quad \text{Guar}_A(R) \supseteq \text{Rely}_A(R) \quad T = \mu A.\,\text{Coerce}(\text{Guar}_A(R))}{\Gamma \vdash \text{close}\, e : T}$$

The rule checks that an open record type R of premethods is mutually consistent (the second premise) and then freezes the type of the resulting record to be exactly what is guaranteed (the third premise). To define mutual consistency of a record R, we introduce the notions of *rely-set* and *guarantee-set* for each pre-type $(A:R_j) \Rightarrow S_j \rightarrow T_j$ stored in field f_j:

- the rely-set contains the field-type pairs (f, T) corresponding to all bindings $f : T$ in the record type R_j, where R_j may refer to the variable A, and
- the guarantee-set is the singleton set $\{\{(f_j, A \to S_j \to T_j)\}\}$, where A stands for the entire record type being checked for consistency.

The procedures Rely and Guar compute sets of field-type pairs by combining the rely-set and guarantee-set, respectively, from each premethod in R. An open record type R is consistent if $\mathsf{Guar}_A(R)$ contains all of the field-type constraints in $\mathsf{Rely}_A(R)$. The procedure Coerce converts a set of field-type pairs into a record type in the obvious way, as long as each field is mentioned in at most one pair.

Using this approach, `close normalTicker` and `close quietTicker` have type

$$\mu A.\{\ \texttt{tick}: IS(A);\ \texttt{tock}: IS(A);\ \texttt{loud}: A \to \texttt{unit} \to \texttt{bool}\ \}$$

where $IS(S) \stackrel{\circ}{=} S \to \texttt{int} \to \texttt{str}$, because the following set containment is valid:

$$\mathsf{Guar}_A(R_0) = \{(\texttt{tick}, IS(A)), (\texttt{tock}, IS(A)), (\texttt{loud}, A \to \texttt{unit} \to \texttt{bool})\}$$
$$\supseteq \mathsf{Rely}_A(R_0) = \{(\texttt{tick}, IS(A)), (\texttt{tock}, IS(A))\}$$

Notice that the use of set containment, rather than equality, in the definition of consistency allows an open record to be used even when it stores additional premethods than those required by the recursive assumptions of others. Informally, we can think of the consistency computation as a form of record width and permutation subtyping that treats constraints on self parameters specially.

Late Typing. Once open records have been closed into ordinary mu-types, typechecking method calls can proceed as in LangMu using standard typing rules for unfolding, record projection, and function application. A common pattern in IsoLate is to close an open record "late" (right before a method is invoked) rather than "early" (immediately when a record is created). We introduce the following abbreviation, used on lines 17 and 18, to facilitate this pattern (note that we could use a let-binding for the `close` expression, if needed, to avoid duplicating effects):

$$e\#f(e') \stackrel{\circ}{=} (\texttt{unfold}\,(\texttt{close}\,e)).f(\texttt{close}\,e)(e')$$

We refer to this pattern of typechecking method invocations as "late typing," hence, the name IsoLate to describe our extension of a standard isorecursive type system. The crucial difference between IsoLate and LangMu is the `close` expression, which generalizes the traditional `fold` expression while still being easy to implement. The simple set containment computation can be viewed as a way of inferring the mu-type instantiations that are required in the FsubRec encodings of our examples. As a result, IsoLate is able to make do with isorecursive types while still allowing premethods to be loosely mixed together.

Extension: Unbounded Polymorphism. The operation for closing records of premethods constitutes the main custom typing rule beyond LangMu. For convenience, our formulation also includes (unbounded) parametric polymorphism. In particular, lines

```
19  val wrap :: all A,B,C,C'. (C->C') -> (A->B->C) -> (A->B->C')
20  let wrap A B C C' g f this x = g (f this x)
21
22  type OpenTickerVol =
23    { tick: (A:Tock(A)) => int  -> str
24    ; tock: (A:Tick(A)) => int  -> str
25    ; loud: (A:{})        => unit -> bool }
26
27  type TickerVol =
28    mu A. { tick, tock : A -> int -> str ; loud : A -> unit -> bool }
29
30  val louderClose :: OpenTickerVol -> TickerVol
31  let louderClose ticker =
32    let exclaim s = "! " ++ s in
33    let o1 = close normalTicker in
34    let o2 = unfold o1 in
35      fold TickerVol
36        { tick = wrap _ _ exclaim (o2.tick)
37        ; tock = wrap _ _ exclaim (o2.tock)
38        ; loud = \_. \_. true }
39
40  louderClose(quietTicker)  $ tick(2) -- "! tick ! ! tick ! ! "
41  louderClose(normalTicker) $ tick(2) -- "! tick ! tock ! tick ! tock ! "
```

Fig. 3. ISOLATE Example (Part 2)

19 and 20 of Figure 3 show how to use parametric polymorphism to define a generic wrap function, like we saw in FSUBREC.

The rest of the example in Figure 3 demonstrates a noteworthy aspect of combining late typing with message interposition. Recall that in FSUBREC, premethods had to be instantiated with particular mu-types before wrapping (*cf.* the ticker3 definition in §2.3). Using only ordinary unbounded universal quantification, however, there is no way to instantiate a pre-type in ISOLATE. If trying to wrap record of mutually consistent premethods, the same result can be achieved by first closing the open object into a closed one described by a mu-type (line 33), unfolding it (line 34), and then using unbounded polymorphism to wrap its methods (lines 35 through 38). The louderClose function abstracts over these operations, taking an open ticker object as input and producing a closed ticker object as output. Therefore, we use unfold rather than close (signified by $ rather than #) to use the resulting objects on lines 40 and 41.

Extension: Abstracting over Premethods. The previous example demonstrates how to wrap methods using unbounded polymorphism and late typing, but as with the corresponding examples in FSUBREC, the approach does not help with truly wrapping premethods. If we wish to do so, we can extend ISOLATE with an additional rule that allows pre-types, rather than just universally quantified types, to be instantiated with type arguments. As we will discuss, this rule offers a version of the FSUBREC rule

```
42  val wrapTick :: all B,C,C'. (C -> C') ->
43                  ((A:Tick(A)) => B -> C) -> ((A:Tick(A)) => B -> C')
44  val wrapTock :: all B,C,C'. (C -> C') ->
45                  ((A:Tock(A)) => B -> C) -> ((A:Tock(A)) => B -> C')
46
47  let wrapTick B C C' g f this x = g (f[A] this x)
48  let wrapTock B C C' g f this x = g (f[A] this x)
49
50  val louder :: OpenTickerVol -> OpenTickerVol
51  let louder ticker =
52    let exclaim s = "! " ++ s in
53      { tick = wrapTock _ _ _ exclaim (ticker.tick)
54      ; tock = wrapTick _ _ _ exclaim (ticker.tock)
55      ; loud = true_ }
56
57  let (t1, t2, t3) = (louder quietTicker, louder normalTicker, louder t2)
58
59  t1 # tick(2)          -- "! tick ! ! tick ! ! "
60  t2 # tick(2)          -- "! tick ! tock ! tick ! tock ! "
61  t3 # tick(2)          -- "! ! tick ! ! tock ! ! tick ! ! tock ! ! "
```

Fig. 4. IsoLate Example (Part 3)

for type instantiations that is limited to type variables and, hence, does not require a separate subtyping relation and equirecursive treatment types in order to reason about.

The extended system allows abstracting over premethods but requires code duplication, as in FSubRec, for different pre-types. Notice that in the definitions of wrapTick and wrapTock (lines 42 through 48 of Figure 4), the extended system allows the premethod arguments f to be instantiated with the type arguments A. Making use of these wrapper functions allows us to define a louder function (lines 50 through 55) that, unlike louderClose, returns open objects. As a result, the expressions on lines 59 through 61 use the late typing form of method invocation.

Remarks. It is worth noting that the two extensions discussed, beyond the close expression, enable open object update in IsoLate. Recall that closed objects correspond to ordinary mu-types, so traditional examples of closed object update work in IsoLate as they do in LangMu and the limited fragment of FSubRec that IsoLate supports. Open objects are not a substitute for closed objects, rather, they provide support for patterns of programming with mutually recursive sets of premethods.

As we will see next, our formulation of IsoLate is designed to support the FSubRec examples from § 2.3 without offering all the power of the full system. Therefore, the third row of the table in Figure 1 summarizes that IsoLate fares as well as FSubRec with respect to the four properties of our running examples. A prototype implementation of IsoLate that typechecks the running example is available on the Web.[2]

---

[2] https://github.com/ravichugh/late-types

$$
\begin{array}{rcl}
\textbf{Expressions } e & ::= & \texttt{unit} \mid x \mid \lambda x{:}T.e \mid e_1\,e_2 \mid \Lambda A.e \mid e[T] \\
& \mid & \{\overline{f}{=}\overline{e}\} \mid e.f \mid \texttt{unfold}\,e \mid \texttt{fold}\,T\,e \\
\textit{premethod and close} & \mid & \varsigma x{:}A <{:}T.e \mid \texttt{close}\,e
\end{array}
$$

$$
\begin{array}{rcl}
\textbf{Types } R,S,T & ::= & \texttt{unit} \mid \{\,\overline{f}{:}\overline{T}\,\} \mid S \to T \mid A \mid \mu A.T \mid \forall A.\,T \\
\textit{pre-type} & \mid & (A{:}S) \Rightarrow T
\end{array}
$$

$$
\textbf{Type Environments } \Gamma \quad ::= \quad - \mid \Gamma, x{:}T \mid \Gamma, A \mid \Gamma, A <{:}T
$$

Fig. 5. ISOLATE Syntax

3.2 Syntax and Typechecking

We now present the formal definition of ISOLATE. Figure 5 defines the syntax of expressions and types, and Figure 6 defines selected typing rules; [10] provides additional definitions. We often write overbars (such as $\overline{f}{:}\overline{T}$) to denote sequences (such as $\{f_1{:}T_1;\ \ldots;\ f_n{:}T_n\,\}$).

Expressions. Expressions include the unit value, variables, lambdas, function application, type abstractions, type application, record literals, and record projection. The type abstraction and application forms are typical for a polymorphic lambda-calculus, where type arguments have no computational significance. Expressions also include isorecursive `fold` and `unfold` expressions that are semantically irrelevant, as usual. Unique to ISOLATE are the premethod expression $\varsigma x{:}A <{:}T.e$ and the `close` e expression, which triggers consistency checking in the type system but serves no computational purpose. If we consider premethods to be another form of abstraction and `close` as a more general form of `fold`, then, in the notation from earlier sections, the syntax of ISOLATE programs can be regarded as a subset of $\mathscr{L}_{\{\},\mathrm{iso},\Lambda}$, the expression language of LANGMU. The intended meaning of each expression form is standard. Instead of an operational semantics, we will define an elaboration semantics for ISOLATE in §4.

Types. Types include the unit type, record types, function types, mu-types, universally quantified types, and type variables A, B, *etc.* Custom to ISOLATE is the pre-type $(A{:}S) \Rightarrow T$ used to describe premethods, where the type A of the self parameter is bound in S (as defined by the type well-formedness rules in [10]). Type environments including bounds $A <{:}S$ for type variables that correspond to premethods and their pre-types. By convention, we use the metavariable R to describe record types.

The typechecking judgment $\Gamma \vdash e : T$ concludes that expression e has type T in an environment Γ where variables $x_1,\ldots,x_n$ have types $T_1,\ldots,T_n$, respectively. In addition to standard typechecking rules defined in [10], Figure 6 defines four custom ISOLATE rules that encode a restricted form of F-bounded polymorphism.

The T-PREMETHOD rule derives the pre-type $(A{:}S) \Rightarrow T$ for $\varsigma x{:}A <{:}S.e$ by combining the reasoning for type and value abstractions. The T-UNFOLDSELF rules allows a self parameter, which is described by bounded type variables A, to be used at its upper bound T. This allows premethod self parameters to be unfolded as if they were described by mu-types (*cf.* lines 6, 7, and 8 of Figure 2). In order to facilitate abstracting over premethods, the T-PREAPP rule allows a premethod to be instantiated with type

Type Checking (custom rules) $\boxed{\Gamma \vdash e : T}$

$$\frac{\Gamma, A <: S, x{:}A \vdash e : T}{\Gamma \vdash \varsigma x{:}A <: S.e : (A{:}S) \Rightarrow T} \text{ [T-PREMETHOD]}$$

$$\frac{A <: T \in \Gamma \quad \Gamma \vdash e : A}{\Gamma \vdash \text{unfold } e : T} \text{ [T-UNFOLDSELF]} \qquad \frac{\Gamma \vdash e : (A{:}S) \Rightarrow T \quad B <: S[B/A] \in \Gamma}{\Gamma \vdash e[B] : B \to T[B/A]} \text{ [T-PREAPP]}$$

$$\frac{\Gamma \vdash e : R \quad \text{Guar}_A(R) \supseteq \text{Rely}_A(R)}{\Gamma \vdash \text{close } e : \mu A.\text{Coerce}(\text{Guar}_A(R))} \text{ [T-CLOSE]}$$

$$\text{Guar}_A(\{\ \overline{f} : (A{:}\overline{R}) \Rightarrow \overline{S} \to \overline{T}\ \}) = \cup_i \{(f_i, A \to S_i \to T_i)\}$$
$$\text{Rely}_A(\{\ \overline{f} : (A{:}\overline{R}) \Rightarrow \overline{S} \to \overline{T}\ \}) = \cup_i \text{RelyThis}_A(R_i)$$
$$\text{RelyThis}_A(\{\ \overline{f} : A \to \overline{S} \to \overline{T}\ \}) = \cup_i \{(f_i, A \to S_i \to T_i)\}$$

Fig. 6. ISOLATE Typing

variable argument B if it has the same bound S (after substitution) as the type variable A of the premethod. The effect of these two rules is to provide some of the subtypings derived by the full subtyping relation of FSUBREC.

The premises of T-CLOSE require that (1) the types of all fields f_i bound in R have the form $(A{:}R_i) \Rightarrow S_i \to T_i$ and (2) the set $\text{Guar}_A(R)$ contains all of the field-type pairs in $\text{Rely}_A(R)$. The guarantee-set contains pairs of the form $\{(f_i, A \to S_i \to T_i)\}$, which describes the type of the record *assuming* that all of the mutual constraints on self are satisfied. The rely-set collects all of these mutual constraints by using the helper procedure RelyThis to compute the constraints from each particular self type R_i.

To understand the mechanics of this procedure, let us consider a few examples. We define three self types

$$S_0 \doteq \{\ \} \quad S_1(A) \doteq \{\ \mathtt{f}{:}A \to \mathtt{unit} \to \mathtt{int}\ \} \quad S_2(A) \doteq \{\ \mathtt{f}{:}A \to \mathtt{unit} \to \mathtt{bool}\ \}$$

that impose zero or one constraints on the receiver and three types that refer to them:

$$R_\mathtt{f} \doteq \{\ \mathtt{f}{:}(A{:}S_1(A)) \Rightarrow \mathtt{unit} \to \mathtt{int}\ \}$$
$$R_\mathtt{fg} \doteq \{\ \mathtt{f}{:}(A{:}S_1(A)) \Rightarrow \mathtt{unit} \to \mathtt{int};\ \mathtt{g}{:}(A{:}S_2(A)) \Rightarrow \mathtt{unit} \to \mathtt{bool}\ \}$$
$$R_\mathtt{fh} \doteq \{\ \mathtt{f}{:}(A{:}S_1(A)) \Rightarrow \mathtt{unit} \to \mathtt{int};\ \mathtt{h}{:}(A{:}S_0) \Rightarrow \mathtt{unit} \to \mathtt{str}\ \}$$

The first record, $R_\mathtt{f}$, is consistent because its guarantee-set matches its rely-set exactly:

$$\text{Guar}_A(R_\mathtt{f}) = \text{Rely}_A(R_\mathtt{f}) = \{(\mathtt{f}, A \to \mathtt{unit} \to \mathtt{int})\}$$

The second, $R_\mathtt{fg}$, is inconsistent because the guarantee-set does not contain the rely-set:

$$\text{Guar}_A(R_\mathtt{fg}) = \{(\mathtt{f}, A \to \mathtt{unit} \to \mathtt{int}), (\mathtt{g}, A \to \mathtt{unit} \to \mathtt{bool})\}$$
$$\not\supseteq \text{Rely}_A(R_\mathtt{fg}) = \{(\mathtt{f}, A \to \mathtt{unit} \to \mathtt{int}), (\mathtt{f}, A \to \mathtt{unit} \to \mathtt{bool})\}$$

In particular, the second constraint in the rely-set is missing from the guarantee-set. In fact, the self types S_1 and S_2 can *never* be mutually satisfied, because they require different return types for the same field. The last record, R_{fh}, is consistent because the guarantee-set is allowed to contain fields beyond those required:

$$\text{Guar}_A(R_{\text{fh}}) = \{(\text{f},A \to \text{unit} \to \text{int}),(\text{h},A \to \text{unit} \to \text{str})\}$$
$$\supseteq \text{Rely}_A(R_{\text{fh}}) = \{(\text{f},A \to \text{unit} \to \text{int})\}$$

As noted earlier, the consistency computation resembles a form of record width and permutation subtyping implemented as set containment. In §4, we will make the connection between this approach and proper subtyping in FSUBREC.

Object Update. ISOLATE can derive, for any record type R, the judgment

$$(\varsigma\text{this}:\text{Self} <: R.\text{let } y = e \text{ in this}) \ :: \ (\text{Self}:R) \ \Rightarrow \ \text{Self}$$

for a premethod that, after some well-typed expression e, returns the original self parameter. Because ISOLATE provides no analog to T-UNFOLDSELF for folding and because ISOLATE uses an isorecursive treatment of mu-types, the this variable is, in fact, the *only* expression that can be assigned the type Self. As a result, traditional (closed) object update examples require the use of mu-types, rather than pre-types, in ISOLATE.

4 Metatheory

We now show how to translate, or elaborate, ISOLATE source programs into the more expressive target language FSUBREC. Our soundness theorem establishes that well-typed programs in the source language translate to well-typed programs in the target language. The decidability of ISOLATE is evident; the primary difference beyond LANGMU is the T-CLOSE rule, which performs a straightforward computation.

4.1 The FSUBREC Calculus

We saw several examples of programming in FSUBREC in §2.3. In [10], we formally define the language and its typechecking rules. Our formulation closely follows standard presentations of equirecursive types [29] and F-bounded polymorphism [2], so we keep the discussion here brief. The language of FSUBREC expressions is standard. Notice that there are no expressions for folding and unfolding recursive types, and there is no close expression. The operational semantics can be found in the aforementioned references. The language of FSUBREC types replaces the isorecursive mu-types of ISOLATE with equirecursive mu-types, and adds bounded universal types $\forall A <: S.\, T$, where A is bound in S (in addition to T). To reason about bounded type variables, type environments Γ record assumptions $A <: S$. These assumptions are used by the subtyping rule S-TVAR that relates a type variable to its bound. The definitional equality of recursive types and their unfoldings is crucial for discharging the subtyping obligation in the second premise of the T-TAPP rule. As the soundness proof for the translation makes clear, the ISOLATE rules T-UNFOLDSELF and T-TAPP are restricted versions

Elaboration of Types $\boxed{[\![\,\Gamma\,]\!]}$ $\boxed{[\![\,T\,]\!]}$

$$[\![\,\texttt{unit}\,]\!] = \texttt{unit} \tag{5}$$

$$[\![\,\{\ldots;\,f_i\!:\!T_i;\,\ldots\,\}\,]\!] = \{\ldots;\,f_i\!:\![\![\,T_i\,]\!];\,\ldots\} \tag{6}$$

$$[\![\,-\,]\!] = - \tag{1}$$
$$[\![\,(A\!:\!S)\,\Rightarrow\,T\,]\!] = \forall A\!<\!:[\![\,S\,]\!].\,A \to [\![\,T\,]\!] \tag{7}$$

$$[\![\,\Gamma,x\!:\!T\,]\!] = [\![\,\Gamma\,]\!],x\!:\![\![\,T\,]\!] \tag{2}$$
$$[\![\,S\to T\,]\!] = [\![\,S\,]\!] \to [\![\,T\,]\!] \tag{8}$$

$$[\![\,\Gamma,A\,]\!] = [\![\,\Gamma\,]\!],A\!<\!:\texttt{top} \tag{3}$$
$$[\![\,\mu A.T\,]\!] = \mu A.[\![\,T\,]\!] \tag{9}$$

$$[\![\,\Gamma,A\!<\!:\!T\,]\!] = [\![\,\Gamma\,]\!],A\!<\!:[\![\,T\,]\!] \tag{4}$$
$$[\![\,\forall A.\,T\,]\!] = \forall A\!<\!:\texttt{top}.\,[\![\,T\,]\!] \tag{10}$$

$$[\![\,A\,]\!] = A \tag{11}$$

Fig. 7. Translation of Environments and Types

of these two FSUBREC rules. Our soundness proof does not appeal to subtyping for function, recursive, or universal types. We include the rules S-ARROW, S-AMBER, and S-KERNEL-ALL for handling these constructs anyway, however, for reference. Part of the appeal of ISOLATE is that this extra machinery need not be implemented.

4.2 Elaboration from ISOLATE to FSUBREC

The translation from ISOLATE expressions and typing derivations to FSUBREC programs is mostly straightforward. Figure 7 defines the translation of ISOLATE types and type environments recursively, where ISOLATE pre-types are translated to FSUBREC bounded universals.

We write $D :: \Gamma \vdash e : T$ to give the name D to an ISOLATE derivation of the given judgment. In Figure 8, we define a function $[\![\,D\,]\!]$ that produces an expression e' in the target language, FSUBREC. We use this translation to define the semantics for the source language, rather than specifying an operational semantics directly. Most of the translation rules are straightforward, recursively invoking translation on subderivations. Because the expression $\texttt{unfold}\,e$ (respectively, $\texttt{fold}\,T\,e$) is intended to reduce directly to e, as usual, a derivation by the T-UNFOLD (respectively, T-FOLD) rule is translated to $[\![\,D_1\,]\!]$, the translation of the derivation of e.

The key aspects of the translation relate to the custom ISOLATE rules. Premethods correspond to polymorphic functions in the target calculus (T-PREMETHOD), so applying them to type variable arguments corresponds to type instantiation (T-PREAPP). Self parameters are described by bounded type variables in the target, so unfolding them has no computational purpose (T-UNFOLDSELF). The last noteworthy aspect is how to translate T-CLOSE derivations of expressions $\texttt{close}\,e$. Motivated by the FSUBREC encodings from §2, the idea is to create a closed record of methods by instantiating all of the (universally quantified) functions in the (translated) record with the type parameter $\mu A.\,\mathsf{Coerce}(\mathsf{Guar}_A([\![\,R\,]\!]))$, a mu-type that corresponds to the (converted and translated) guarantee-set of the record. Notice that every time an open record is used in a method invocation expression $e\,\#f\,(e')$, a new closed record is created in the target program. This captures the essence of late typing in ISOLATE.

Elaboration from IsoLate to FSubRec $\boxed{[\![\,D :: \Gamma \vdash e : T\,]\!] = e'}$

$$\text{[T-Unit]} \quad \frac{}{[\![\,D :: \Gamma \vdash \text{unit} : \text{unit}\,]\!]} = \text{unit}$$

$$\text{[T-Var]} \quad \frac{x : T \in \Gamma}{[\![\,D :: \Gamma \vdash x : T\,]\!]} = x$$

$$\text{[T-Recd]} \quad \frac{\text{for } 1 \leq i \leq n,\ D_i :: \Gamma \vdash e_i : T_i}{[\![\,D :: \Gamma \vdash \{\overline{f} = \overline{e}\} : \{\,\overline{f} : \overline{T}\,\}\,]\!]} = \{ f_1 = [\![\,D_1\,]\!]; \ \cdots \ ; \ f_n = [\![\,D_n\,]\!] \}$$

$$\text{[T-Proj]} \quad \frac{D_1 :: \Gamma \vdash e : \{\,...;\ f : T;\ ...\,\}}{[\![\,D :: \Gamma \vdash e.f : T\,]\!]} = [\![\,D_1\,]\!].f$$

$$\text{[T-Fun]} \quad \frac{D_1 :: \Gamma, x : S \vdash e : T}{[\![\,D :: \Gamma \vdash \lambda x{:}S.e : S \to T\,]\!]} = \lambda x{:}S.[\![\,D_1\,]\!]$$

$$\text{[T-App]} \quad \frac{D_1 :: \Gamma \vdash e_1 : S \to T \qquad D_2 :: \Gamma \vdash e_2 : S}{[\![\,D :: \Gamma \vdash e_1\,e_2 : T\,]\!]} = [\![\,D_1\,]\!]\,[\![\,D_2\,]\!]$$

$$\text{[T-TFun]} \quad \frac{D_1 :: \Gamma, A \vdash e : T}{[\![\,D :: \Gamma \vdash \Lambda A.e : \forall A.\,T\,]\!]} = \Lambda A.[\![\,D_1\,]\!]$$

$$\text{[T-TApp]} \quad \frac{D_1 :: \Gamma \vdash e : \forall A.\,T}{[\![\,D :: \Gamma \vdash e[S] : T[S/A]\,]\!]} = [\![\,D_1\,]\!][[\![\,S\,]\!]]$$

$$\text{[T-Fold]} \quad \frac{T = \mu A.S \qquad D_1 :: \Gamma \vdash e : S[T/A]}{[\![\,D :: \Gamma \vdash \text{fold}\,T\,e : T\,]\!]} = [\![\,D_1\,]\!]$$

$$\text{[T-Unfold]} \quad \frac{T = \mu A.S \qquad D_1 :: \Gamma \vdash e : T}{[\![\,D :: \Gamma \vdash \text{unfold}\,e : S[T/A]\,]\!]} = [\![\,D_1\,]\!]$$

$$\text{[T-UnfoldSelf]} \quad \frac{A <: T \in \Gamma \qquad D_1 :: \Gamma \vdash e : A}{[\![\,D :: \Gamma \vdash \text{unfold}\,e : T\,]\!]} = [\![\,D_1\,]\!]$$

$$\text{[T-PreMethod]} \quad \frac{D_1 :: \Gamma, A <: S, x{:}A \vdash e : T}{[\![\,D :: \Gamma \vdash \varsigma x{:}A <:S.e : (A{:}S) \Rightarrow T\,]\!]} = \Lambda A.\lambda x{:}A.[\![\,D_1\,]\!]$$

$$\text{[T-PreApp]} \quad \frac{D_1 :: \Gamma \vdash e : (A{:}S) \Rightarrow T \qquad B <: S[B/A] \in \Gamma}{[\![\,D :: \Gamma \vdash e[B] : B \to T[B/A]\,]\!]} = [\![\,D_1\,]\!][B]$$

$$\text{[T-Close]} \quad \frac{\begin{array}{c} D_1 :: \Gamma \vdash e : R \qquad \text{Guar}_A(R) \supseteq \text{Rely}_A(R) \\ T = \mu A.\,\text{Coerce}(\text{Guar}_A(R)) \end{array}}{[\![\,D :: \Gamma \vdash \text{close}\,e : T\,]\!]} = \{ f_1 = ([\![\,D_1\,]\!].f_1)[[\![\,T\,]\!]]; \ ... \}$$

Fig. 8. Elaboration Semantics for IsoLate

4.3   Soundness

We now justify the correctness of our translation. Notice that because the syntactic forms of ISOLATE are so similar to those in FSUBREC, there is little value in defining an operational semantics for ISOLATE directly and then "connecting" it to FSUBREC. Instead, we use the translation to define the semantics for ISOLATE. As a result, the correctness theorem we prove needs only to state that the result of translating valid ISOLATE derivations are well-typed FSUBREC programs.

Theorem 1 (Type Soundness). *If* $D :: \Gamma \vdash e : T$, *then* $[\![\Gamma]\!] \vdash [\![D]\!] : [\![T]\!]$.

Proof. We provide the full details of the proof in [10]. Many of the cases proceed by straightforward induction. The case for T-CLOSE, which converts open records into closed records described by ordinary mu-types, is the most interesting. As discussed in §3, the key observation is that rely- and guarantee-sets can be interpreted as record types. The fact that the guarantee-set contains the rely-set can be used to argue how, with the help of definitional equality of equirecursive types, the necessary record subtypings hold via the record subtyping rule, S-RECD. As mentioned earlier, the reasoning for rules T-UNFOLDSELF and T-PREAPP appeal to T-SVAR and T-TAPP, respectively, in FSUBREC, providing a limited form of F-bounded polymorphism in ISOLATE.

5   Discussion

Our formulation of ISOLATE is a restricted version of FSUBREC that enables simple typechecking for loosely coupled premethods in a setting without explicit recursion. To conclude, we first discuss some related work that has not already been mentioned, and then we describe several ways in which future work might help to further extend the expressiveness of our system.

5.1   Related Work

Mixin and Recursive Modules. F-bounded polymorphism employs several traditional type theoretic constructs and is widely used to encode object-oriented programming patterns. Somewhat different mechanisms for combining and reusing implementations include traits, mixins, and mixin modules, which have been studied in both untyped (*e.g.* [4]) and typed (*e.g.* [14,1,19,15]) settings. Generally, these approaches distinguish expressions either as components that may get mixed in to other objects and objects which are constructed as the result of such operations. Various approaches are then used to control when it is safe to combine functionality with combinators such as sum, delete, rename, and override. Yet more expressive systems combine these approaches with full-fledged module systems and explicit recursion as found in ML (*e.g.* [33,13,32,21]).

Although all of the above approaches are more expressive than ISOLATE (which supports only sum), they rely on semantic features beyond those found in a strict lambda-calculus with records. For example, the CMS calculus [1] relies on call-by-name semantics to avoid ill-founded recursive definitions. The mixin operators of CMS can be brought to a call-by-value setting, but this requires tracking additional information (in the form of dependency graphs) in mixin signatures [19]. In contrast, ill-founded recursion is not a concern for (call-by-value) ISOLATE; because late typing is restricted to premethods in records, the function types of these fields establish that they bind syntactic values (namely, functions). Furthermore, the target of the translation in [19] (Boudol's recursive records [3]) explicitly includes letrec and uses non-standard semantics for records. In contrast, the (standard) semantics of FSUBREC does not include recursive definitions. Instead, our translation relies on F-bounded quantification to tie the knot. As a result, our formulation of mixin composition can be applied to languages without explicit recursion (such as JavaScript, Python, and Ruby). In the future, we plan to investigate whether additional mixin operators can be supported for such languages.

Row Polymorphism. Row polymorphism [35,18,31] is an alternative to record subtyping where explicit type variables bind extra fields that are not explicitly named. By ensuring disjointness between named fields in a record and those described by a type variable, row polymorphism allows functions to be mixed in to different records using record concatenation operators. It is not clear how row polymorphism on its own would help, however, in a language without fix. We might start by writing

$$\texttt{tick} :: \forall \rho_1.\mu A. \{\, \texttt{tock}\!:\! A \to \texttt{int} \to \texttt{str};\ \rho_1 \,\} \to \texttt{int} \to \texttt{str}$$

(and similarly for other premethods), but ρ_1 cannot be instantiated with a type that mentions A, as required for mutual recursion, because it is not in scope. The fact that row polymorphism is often used as an alternative to subtyping notwithstanding, simply adding F-bounded polymorphism to this example does not seem to help matters either.

Coeffects. Our special treatment of premethods can be viewed as function types that impose constraints on the context (in our case, self parameters) using a specification language besides that of the object type language. Several proof theories have been proposed to explicitly constrain the behavior of a program with respect to its context, for example, using modal logics [26] and coeffects [27]. These systems provide rather general mechanisms for defining and describing the notions of context, and they have been applied to dynamic binding structure, staged functional programming, liveness tracking, resource usage, and tracking cache requirements for dataflow programs. It would be interesting to see whether these approaches can be applied to our setting of objects and mutually recursive definitions.

Closed Recursion. Whereas we have focused on patterns of recursion for a language without fix, other researchers have studied systems with closed recursion. In particular, there have been several efforts to admit more well-behaved programs than allowed by ML-style let-polymorphism [11]. The system of polymorphic recursion [25] allows recursive calls to be instantiated nonuniformly, but the additional expressive power results in a system that is only semi-decidable. In between these two systems is Trevor Jim's proposal based on principal typings [22]. Because the notion of principal typings views the typing environment as an output of derivations, rather than an input, one can think of the environment as a set of constraints for the derived type of an expression. It could be interesting to see whether this approach can be adapted to a lambda-calculus extended with records.

5.2 Future Work and Conclusion

Our formulation is meant to emphasize that a small, syntactic variation on isorecursive mu-types can capture a set of desired usage patterns. In the future, we plan to study how additional language features — beyond the core of lambdas and records in ISOLATE — interact with late typing. Important features include reference types, existential types, type operators for supporting user-defined types and interfaces [29], and record concatenation à la Wand, Remy, Mitchell, *et al.* [18].

Similar to how mutually recursive functions can be combined through self, recursive functions can also be combined through the heap. This pattern, sometimes referred to as "backpatching" or "Landin's knot," appears in imperative languages as well as module systems for functional languages [12,13]. We are studying how to adapt the idea of late typing to the setting of lambdas and references with the goal of, as in this paper, typechecking limited patterns of mutual recursion with relatively lightweight mechanisms. Overall, because languages support various kinds of (implicit) recursion through the heap and through self parameters, we believe that late typing may be useful for typechecking common programming patterns in a relatively lightweight way.

Acknowledgements. Part of this work was done while the author was at University of California, San Diego and supported by NSF grant CNS-1223850. The author wishes to thank Cédric Fournet, Ranjit Jhala, Arjun Guha, Niki Vazou, and anonymous reviewers for many helpful comments and suggestions that have improved this paper.

References

1. Ancona, D., Zucca, E.: A Calculus of Module Systems. Journal of Functional Programming (JFP) (2002)
2. Baldan, P., Ghelli, G., Raffaetà, A.: Basic Theory of F-bounded Quantification. Information and Computation (1999)
3. Boudol, G.: The Recursive Record Semantics of Objects Revisited. Journal of Functional Programming (JFP) (2004)
4. Bracha, G.: The Programming Language Jigsaw: Mixins, Modularity and Multiple Inheritance. PhD thesis, University of Utah (1992)
5. Bruce, K.B., Cardelli, L., Pierce, B.C.: Comparing Object Encodings. Information and Computation (1999)
6. Canning, P., Cook, W., Hill, W., Olthoff, W., Mitchell, J.C.: F-bounded Polymorphism for Object-oriented Programming. In: Functional Programming Languages and Architecture (FPCA) (1989)
7. Cardelli, L.: A Semantics of Multiple Inheritance. In: Plotkin, G., MacQueen, D.B., Kahn, G. (eds.) Semantics of Data Types 1984. LNCS, vol. 173, pp. 51–67. Springer, Heidelberg (1984)
8. Cardelli, L.: Amber. In: Cousineau, G., Curien, P.-L., Robinet, B. (eds.) LITP 1985. LNCS, vol. 242, pp. 21–47. Springer, Heidelberg (1986)
9. Cardelli, L., Wegner, P.: On Understanding Types, Data Abstraction, and Polymorphism. Computing Surveys (1985)
10. Chugh, R.: IsoLate: A Type System for Self-Recursion, Extended Version (2015)
11. Damas, L., Milner, R.: Principal Type-Schemes for Functional Programs. In: Principles of Programming Languages (POPL) (1982)

12. Dreyer, D.: A Type System for Well-Founded Recursion. In: Principles of Programming Languages (POPL) (2004)
13. Dreyer, D.: A Type System for Recursive Modules. In: International Conference on Functional Programming (ICFP) (2007)
14. Duggan, D., Sourelis, C.: Mixin Modules. In: International Conference on Functional Programming, (ICFP) (1996)
15. Fisher, K., Reppy, J.: A Typed Calculus of Traits. In: Workshop on Foundations of Object-Oriented Programming (FOOL) (2004)
16. Ghelli, G.: Recursive Types Are Not Conservative Over Fsub. In: Bezem, M., Groote, J.F. (eds.) TLCA 1993. LNCS, vol. 664, pp. 146–162. Springer, Heidelberg (1993)
17. Greenman, B., Muehlboeck, F., Tate, R.: Getting F-Bounded Polymorphism Back in Shape. In: Programming Language Design and Implementation (PLDI) (2014)
18. Gunter, C.A., Mitchell, J.C. (eds.): Theoretical Aspects of Object-Oriented Programming: Types, Semantics, and Language Design. MIT Press (1994)
19. Hirschowitz, T., Leroy, X.: Mixin Modules in a Call-by-Value Setting. In: ACM Transactions on Programming Languages and Systems (TOPLAS) (2005)
20. Igarashi, A., Viroli, M.: Variant Parametric Types: A Flexible Subtyping Scheme for Generics. In: ACM Transactions on Programming Languages and Systems (TOPLAS) (2006)
21. Im, H., Nakata, K., Garrigue, J., Park, S.: A Syntactic Type System for Recursive Modules. In: Object-Oriented Programming Systems, Languages, and Applications (OOPSLA) (2011)
22. Jim, T.: What Are Principal Typings and What Are They Good For? In: Principles of Programming Languages (POPL) (1996)
23. Kennedy, A.J., Pierce, B.C.: On Decidability of Nominal Subtyping with Variance, 2006. In: FOOL-WOOD (2007)
24. Leroy, X., Doligez, D., Frisch, A., Rémy, D., Vouillon, J.: OCaml System Release 4.02: Documentation and User's Manual (2014), http://caml.inria.fr/pub/docs/manual-ocaml-4.02/
25. Mycroft, A.: Polymorphic Type Schemes and Recursive Definitions. In: Paul, M., Robinet, B. (eds.) Programming 1984. LNCS, vol. 167, pp. 217–228. Springer, Heidelberg (1984)
26. Nanevski, A., Pfenning, F., Pientka, B.: Contextual Modal Type Theory. Transactions on Computational Logic (2008)
27. Petricek, T., Orchard, D., Mycroft Coeffects, A.: Coeffects: Unified Static Analysis of Context-Dependence. In: Fomin, F.V., Freivalds, R., Kwiatkowska, M., Peleg, D. (eds.) ICALP 2013, Part II. LNCS, vol. 7966, pp. 385–397. Springer, Heidelberg (2013)
28. Benjamin, C.: Pierce. Bounded Quantification is Undecidable. In: Principles of Programming Languages (POPL) (1992)
29. Pierce, B.C.: Types and Programming Languages. MIT Press (2002)
30. Pierce, B.C., Turner, D.N.: Simple Type-Theoretic Foundations for Object-Oriented Programming. Journal of Functional Programming (JFP) (1994)
31. Rémy, D., Vouillon, J.: Objective ML: A Simple Object-Oriented Extension of ML. In: Principles of Programming Languages (POPL) (1997)
32. Rossberg, A., Dreyer, D.: Mixin' Up the ML Module System. In: ACM Transactions on Programming Languages and Systems (TOPLAS) (2013)
33. Russo, C.: Recursive Structures for Standard ML. In: International Conference on Functional Programming (ICFP) (2001)
34. Torgersen, M., Hansen, C.P., Ernst, E.: Adding Wildcards to the Java Programming Language. Journal of Object Technology (2004)
35. Wand, M.: Complete Type Inference for Simple Objects. In: Logic in Computer Science (LICS) (1987)

The Problem of Programming Language Concurrency Semantics

Mark Batty, Kayvan Memarian, Kyndylan Nienhuis, Jean Pichon-Pharabod, and Peter Sewell

University of Cambridge

Abstract. Despite decades of research, we do not have a satisfactory concurrency semantics for any general-purpose programming language that aims to support concurrent systems code. The Java Memory Model has been shown to be unsound with respect to standard compiler optimisations, while the C/C++11 model is too weak, admitting undesirable *thin-air executions*.

Our goal in this paper is to articulate this major open problem as clearly as is currently possible, showing how it arises from the combination of multiprocessor relaxed-memory behaviour and the desire to accommodate current compiler optimisations. We make several novel contributions that each shed some light on the problem, constraining the possible solutions and identifying new difficulties.

First we give a positive result, proving in HOL4 that the existing axiomatic model for C/C++11 guarantees sequentially consistent semantics for simple race-free programs that do not use low-level atomics (DRF-SC, one of the core design goals). We then describe the thin-air problem and show that it cannot be solved, without restricting current compiler optimisations, using any per-candidate-execution condition in the style of the C/C++11 model. Thin-air executions were thought to be confined to programs using relaxed atomics, but we further show that they recur when one attempts to integrate the concurrency model with more of C, mixing atomic and nonatomic accesses, and that also breaks the DRF-SC result. We then describe a semantics based on an explicit operational construction of out-of-order execution, giving the desired behaviour for thin-air examples but exposing further difficulties with accommodating existing compiler optimisations. Finally, we show that there are major difficulties integrating concurrency semantics with the C/C++ notion of undefined behaviour.

We hope thereby to stimulate and enable research on this key issue.

1   Introduction

Context. Shared-memory concurrent machines are now ubiquitous, but, despite decades of research, we still do not have a satisfactory concurrency semantics for any general-purpose programming language that aims to support concurrent systems code. The basic tension is between implementability and usability: to be efficiently implementable, such a semantics must admit the relaxed-memory

© Springer-Verlag Berlin Heidelberg 2015
J. Vitek (Ed.): ESOP 2015, LNCS 9032, pp. 283–307, 2015.
DOI: 10.1007/978-3-662-46669-8_12

behaviours that are permitted by multiprocessor architectures, and those that are introduced by compiler optimisations, but it must also provide sufficiently strong guarantees for concurrent algorithms to work correctly. It is important also for the semantics to be mathematically rigorous, as informal reasoning is particularly error-prone here, it should be as intuitive as possible, it should support testing of implementations and of concurrent algorithms, and it should support compositional reasoning.

There have been two major attempts to develop concurrency semantics for such languages, for Java and C/C++. For Java, the original language specification [20] was shown by Pugh [31] to be flawed in both directions: too strong to be implementable and too weak for some concurrent programming idioms. A new specification [25] was developed in JSR-133, and incorporated into Java 5.0, but that too has been shown to be unsound with respect to standard compiler optimisations, by Cenciarelli et al. [16] and Ševčík and Aspinall [34]. This remains unresolved.

For C and C++, an effort as part of the C++0X standardisation process led to a specification incorporated into the C++11 and C11 standards [9,2]. The basic design was outlined by Boehm and Adve [13], and Batty et al. [8] developed a formal semantics in the latter stages of the standardisation process, identifying various flaws in the draft standard and feeding back into the ratified standards and later defect reports. C/C++11 concurrency has been supported by GCC and Clang since versions 4.9 and 3.2 respectively, and the model by Batty et al. has been used for many purposes, including correctness proofs for compilation schemes to x86, by Batty et al. [8], and to IBM Power, by Batty et al. [7] and Sarkar et al. [32]; compiler testing via a theory of sound optimisations, by Morisset et al. [29]; model checking, by Norris and Demsky [30]; compositional library abstraction, by Batty et al. [6]; and program logics, by Vafeiadis and Narayan [39] and by Turon et al. [37]. Elements of the model have also been incorporated into OpenCL 2.0. The C/C++11 concurrency model is the best-developed currently in existence, but it also suffers from major problems. The model is known to admit undesirable "thin-air" executions which actual implementations are not thought to exhibit, and it has become clear that these make informal reasoning, formal compositional reasoning, and compiler optimisation very difficult [14,6,39,38]. This too is unresolved.

Without a semantics, programmers currently have to program against their folklore understanding of what the Java and C/C++ implementations provide, and research on verification, compilation, or testing for such languages is on shaky foundations.

Contributions. Our goal in this paper is to highlight and articulate this major open problem as clearly as is currently possible, explaining the difficulties with the design of concurrency semantics for shared-memory programming languages in general and for C/C++-like languages (and Java-like, albeit in less depth) languages in particular. We make several novel contributions that each shed some light on the problem, constraining the possible solutions and identifying new challenges. We begin (§2) by recalling some basic design constraints and choices, to make this paper as self-contained as possible.

Our first new contribution is a positive result: we describe a machine-checked proof, in HOL4 [21], that (for programs without loops or recursion) the model of Batty et al. satisfies one of the core design goals for C/C++11 concurrency: programs that do not use the low-level atomics of the language, and that are race-free in a sequentially consistent (SC) semantics, only exhibit sequentially consistent behaviour (§3). This *DRF-SC* property gives a relatively simple semantics for programmers using that fragment of the language.

We then consider *thin-air* reads (§4). This is a long-standing open problem in the design of the semantics for C/C++11 relaxed atomics: accesses for which races are permitted but where one does not wish to pay the cost of any barriers or other hardware instructions beyond normal reads and writes. The question is how one can define an envelope that permits current compiler optimisations and hardware behaviour, while excluding particular example executions that it is agreed should be forbidden: those with self-satisfying conditional cycles or values appearing out of thin air (this is also closely related to the difficulties with Java). Here we give an instructive negative result: the C/C++11 model is expressed in terms of candidate executions, defining which candidate executions are consistent, but we show that thin-air executions cannot be forbidden in a per-execution style by any adaptation of the C/C++11 consistency predicate that uses the same notion of candidate execution.

In §5 we identify a new problem that arises when one tries to integrate C/C++11 concurrency with semantics for more of the C language. Thin-air executions have previously been thought to be a problem only for programs using the relaxed atomics (intended only for expert use) of C/C++11, but that turns out not to be the case. The model of Batty et al. presupposes an up-front distinction between atomic and non-atomic locations, but that is not present in C, where (for example) one should be able to reuse `malloc`'d regions to store atomics and then nonatomics, or use `char` pointers to read the representation bytes of an atomic. We show that the thin-air problem essentially recurs in this setting, even in the absence of relaxed atomics, and that also breaks the DRF-SC result.

Moving away from per-candidate-execution semantics, we explore an out-of-order operational semantics construction (§6); this gives the desired behaviour for the thin-air examples of §4 but exposes further difficulties with accommodating existing compiler optimisations.

Finally we identify additional new difficulties that arise when integrating concurrency semantics with the C/C++ notion of undefined behaviour (§7). We conclude briefly in §8.

Our HOL4 proof script and the associated Lem definitions are available at www.cl.cam.ac.uk/~pes20/esop2015-supplementary-material. We introduce aspects of the C/C++11 model as required, but it is not possible to recap the whole model here; for a full description we refer to [8].

2   Background: An Introduction to the Design Space

Sequential Consistency. The most obvious shared-memory concurrency semantics is *sequential consistency (SC)*, in which, as articulated by Lamport [23], any execution has a total order over all memory writes and reads, with each read reading from the most recent write to the same location. This is attractively simple from a theoretical point of view, and it has been the underlying assumption for much research on shared-memory concurrency verification. But it does not capture the concurrency behaviour of typical current systems: multiprocessors exhibit non-SC behaviour, compilers perform optimisations that violate SC, and for C/C++-like languages the language-level memory accesses cannot reasonably be implemented as atomic machine-level accesses. We briefly summarise each of these points in turn.

Non-SC Multiprocessor Behaviour. The behaviour of Intel/AMD x86, IBM Power, and ARM multiprocessors has been clarified by a series of recent papers [35,33,32,26,4]. For x86, normal memory accesses have a Total Store Ordering (TSO) semantics, similar to SPARC TSO [1] — as if there were a FIFO write buffer (with a readback path) for each hardware thread, above a single memory. This allows the SB behaviour on the left below, but little other relaxed behaviour (in these execution diagrams x and y are shared locations, initially 0, po denotes program order, and rf denotes the reads-from relation). Power and ARM are much more relaxed, with programmer-visible out-of-order and speculative execution. For example, the MP behaviour on the right below is allowed, as the writes to different locations might be committed out-of-order, the writes might propagate out-of-order to other threads, and the reads might be satisfied out of order.

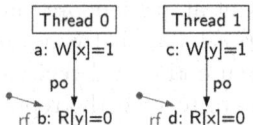
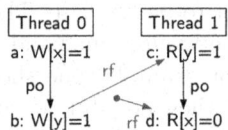

Test SB: Allowed on x86, Power, and ARM   Test MP: Allowed on Power and ARM

Moreover, Power and ARM are not multi-copy atomic: writes to different locations can propagate to multiple other threads in different orders, as in the WRC+addrs example below (pulling the a write of MP to a third thread). The address dependencies prevent local reordering, but the fact that Thread 0's write of x propagates to Thread 1 before its write of y can be committed does not guarantee that the write of x has propagated to Thread 2 before the write of y is propagated to Thread 2.

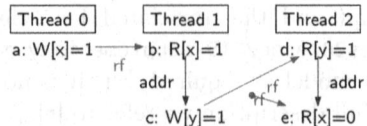

Test WRC+addrs: Allowed on Power and ARM

One can recover SC in each architecture, but at nontrivial cost: without sophisticated analysis, for x86 one needs an MFENCE barrier between shared stores and loads, while for Power, Sarkar et al. [7] prove that one needs a heavyweight sync barrier between each pair of shared memory accesses.

SC-violating Compiler Optimisations. Just as hardware optimisations can result in non-SC behaviour, compiler optimisations can too. The simplest example here is Common Subexpression Elimination (CSE): if two subexpressions are identical, e.g. perhaps just reads of the same location, typical compilers will sometimes retain the value of the first in a register for use instead of the second, effectively hoisting the second read above whatever memory accesses to other locations are in between. This is one of the ways in which the Java Memory Model is unsound with respect to (e.g.) the HotSpot implementation: the implementation does that, but the semantics (unintentionally) disallows it [16,34]. We return to other compiler optimisations in §4 and §6.

Atomicity Problems. Finally, as highlighted by Boehm [10], there is an atomicity mismatch between the language-level memory operations of C/C++-like languages and those that can be implemented reasonably in a concurrent setting. For example, C lets one access a bitfield or a byte within a larger struct, but that might have to be compiled into machine operations that also read or write some of the adjacent memory.

All this means that SC is not viable for current languages, compiler implementations, and hardware (though some authors argue that SC could be achieved at reasonable cost with modified compilers and hardware, e.g. [27,36]). It is also highly debatable whether SC is desirable: for example, McKenney argues that it does not match the intuitive programming models of those who implement high-performance concurrent algorithms, and notes that the *"Linux kernel makes heavy use of weak ordering"* [28].

TSO as a Language Semantics. As the hardware models are now tolerably well-understood, one can imagine lifting them to the programming language level, limiting compiler optimisations to those that are sound w.r.t. the hardware model. The CompCertTSO verified compiler of Ševčík et al. [41] does this for a C-like language (without bitfield accesses), and Demange et al. propose their BMM model for Java-like languages [17]. Both use TSO, which makes for simple implementation on x86 processors but would require expensive fences or sophisticated analysis on Power or ARM machines. This can be reasonable in particular circumstances, especially as x86 is very common, but it is not viable for a general-purpose language intended to support portable high-performance concurrent code.

DRF-SC or Catch Fire. The compiler optimisation and atomicity problems with SC described above are only an issue for programs in which multiple threads might be accessing the same location concurrently. Exploiting this fact,

Adve and Hill [3] and Gharachorloo et al. [19] proposed language-level models
in which programs that are free of such *data races* (in any SC execution) are
guaranteed to have only SC behaviours (DRF-SC), while other programs have
completely undefined behaviour. This model is simple to explain and to implement, and it allows a wide range of compiler optimisations (c.f. Ševčík [40] and
Morisset et al. [29]). It has two disadvantages: first, giving wholly undefined behaviour to racy programs, while perhaps acceptable for C/C++-like languages
(which already have undefined behaviour for other reasons, many of which are
not statically decidable), is not acceptable for Java-like languages, which aim to
provide memory safety guarantees for arbitrary well-typed code (that led to the
complexities of the JSR-133 Java Memory Model [25]). It also begs the question
of how one can debug code, and indeed whether there are any large programs
that are actually race-free. Second, it requires heavier synchronisation than one
wants in some concurrent algorithms.

The C/C++11 Model. The C/C++11 model [13,8,2,9] aims to support DRF-
SC for simple programs (those using only locks and *SC atomics*), but also provides a range of *low-level atomics* that provide less synchronisation but without
the cost of restoring full SC: *release/acquire* write/read pairs for message-passing
synchronisation, *relaxed atomics* that should be implementable just with single
machine-level loads and stores, and *release/consume* pairs to expose some dependency preservation guarantees of the hardware to make them available in the
language. As we shall see, the semantics of all these remains problematic.

3   DRF-Sc: Sequential Consistency for Race-Free Programs

The design of the C/C++11 model could not simply adopt DRF-SC/catch-fire
as its definition, due to the need to provide low-level atomics, but it aimed to provide a DRF-SC *property* (for programs that do not use those) as a consequence
of its actual definition. We now report on a proof that, for the first time, establishes DRF-SC for the full C/C++11 concurrency model: for programs that do
not use low-level atomics and that are race-free in an SC semantics (and subject
to conditions detailed below), the full model permits only SC executions. The
proof is mechanised in HOL4 and is included in the supplementary material (ap-
prox. 23k lines of proof script, including additional model equivalence results).
For a more complete account of the proof, see Batty's thesis [5]. Recalling that
the prose ISO standards for C++11 and C11 [9,2] and the mathematical formalisation of the model by Batty et al. [8] correspond closely, this is effectively
a mechanised proof of a key metatheoretic property of a mainstream language
definition.

There have been two previous results along these lines, but both were preliminary: Boehm and Adve [13] give a hand proof for a preliminary model that
omits many features, while Batty et al. [7, Thm. 5] give a hand proof based on
an earlier version of their formal model that uses that model's notion of races
for the SC semantics. This is a major simplification: the point of a DRF-SC

theorem is to let programmers in the DRF fragment reason solely in terms of an SC semantics, but that result required users to grapple with the full model complexity to understand whether their program contained races. In contrast, the result we present here uses the straightforward SC notion of race based on identifying two conflicting adjacent actions. The mechanisation of the current proof adds assurance, particularly desirable for a fundamental result about an industry-standard model of this intricacy.

To state DRF-SC, we first define a memory model for C/C++ executions, the *total model*, that is manifestly sequentially consistent. We start with a graph over memory accesses, called a *pre-execution* [8], that captures the syntactic structure of the program with a relation for parent-to-child thread ordering and another (*sequenced-before*) for program order. The total model and C/C++11 differ in the relations added to the pre-execution to form their candidate-executions: C/C++11 represents the dynamic behaviour of memory with many partial orders (modification order, lock order and SC order), whereas the total model has only a single total order over all memory accesses in the pre-execution. Reads must read from the immediately preceding write to the same location in the total order, and two accesses race if they access the same location, at least one is a write, they are not both atomic, and they are adjacent in the total order.

The theorem requires that the program ensures that atomic initialisation happens before all atomic accesses for each location. To simplify the proof, we also restrict its statement to programs that satisfy a strong finiteness condition: there must be a finite bound on the size of the pre-executions allowed by the threadwise semantics (this lets us use a simple form of induction). This means it does not apply to programs with recursion or loops. However, intuitively those are orthogonal to the concurrency semantics; we do not know of any reason why including them might affect the truth of the theorem.

Theorem 1. *For programs whose pre-executions (i) use only mutex, non-atomic and SC-atomic accesses, (ii) have atomic initialisations ordered by sequenced-before and parent-to-child thread synchronisation before all atomic accesses at the same location, and (iii) are bounded in size by some N, either both the C/C++11 model and the total model give undefined behaviour, or the sets of consistent executions in each, projected down to the pre-execution and the reads-from relation, are equal.*

PROOF OUTLINE The proof first involves several steps of simplifying the C/C++11 model for programs that do not use low-level atomics. The remaining proof can be split into one part for race-free programs and another for racy ones. For race-free programs there are two cases.

Given a consistent execution in the C/C++11 model, we must construct a consistent execution in the total model with the same pre-execution and reads-from relation. The union of happens-before and SC order is acyclic, so we extend this to a total order and show that that is consistent according to the total model. In the other direction, given a consistent execution in the total model, we project partial relations from the total relation that serve as modification order, SC order and lock order in a C/C++11 candidate execution, and then show that it is consistent.

Given a racy execution in one model, e.g. the execution in the total model on the left below, we construct a (potentially different) racy execution in the other, e.g. the C/C++11 execution on the right. As one might expect, given a race in the C/C++11 model, constructing a consistent racy execution in the total model is quite involved, and this execution might be very different to its progenitor. Perhaps surprisingly, the other direction is similar: a direct translation, with identical read values, of a consistent execution in the total model is not necessarily consistent in C/C++11. Take the execution on the left below: reads-from would violate the C/C++11 non-atomic reads-from condition that requires the write to happen before the read, so we have to construct a different execution with a race.

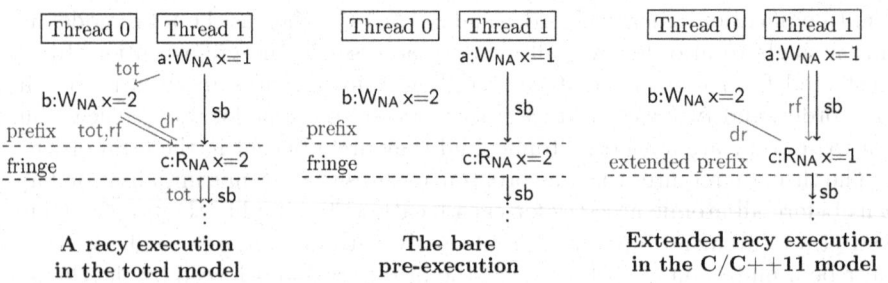

**A racy execution
in the total model**   **The bare
pre-execution**   **Extended racy execution
in the C/C++11 model**

To build the execution, we rely on several definitions and an assumption about the *thread-local semantics*: the part of the semantics that enumerates the pre-executions of a particular program. We illustrate these on the example executions above. We define a *prefix* as a part of an execution where every sequenced-before or thread-synchronisation predecessor of any action within the part is also included: e.g. all nodes above the "prefix" lines in the executions above. The *fringe actions* of a prefix are all actions that are not in the prefix, but are immediate sequenced-before or thread-synchronisation successors of an action in the prefix, e.g. precisely c in the left and central executions above. The central diagram above is just the pre-execution of the consistent execution on the left, and hence is allowed by the thread-local semantics. We must assume that the thread-local semantics is *receptive*: for any read or lock in the fringe of a prefix of a pre-execution, allowed by the thread-local semantics, e.g. c in the centre above, and for every other value or lock outcome, there exists a pre-execution with the same prefix, but where the fringe action is changed accordingly, e.g. c in the underlying pre-execution of the right-hand diagram.

Given a racy execution in the total model, we find the first race according to the total relation, e.g. b and c above left, and take the prefix made up of all strict predecessors of the later action (c) with respect to the total order. The prefix is consistent and race free, so we can translate it to a consistent prefix in the C/C++11 memory model with the same set of fringe actions. We extend this to a consistent prefix containing the second racy action, appealing to receptiveness to change its value if necessary for consistency, producing the execution on the right above, and we show that there is a race in the extended prefix, again between b and c. This is all inside an induction on the size of the prefix: we show that for each larger finite prefix size, n, either there exists a racy consistent execution, or a racy consistent prefix with at least n actions. Finally, we appeal to the boundedness of executions to establish that there is a racy consistent execution of the program in the C/C++11 memory model.

Given a racy execution in the C/C++11 model, the steps involved in the proof are similar, but finding the first race differs. For each race in the execution, we identify the set containing the racy actions and all of their happens-before predecessors. The execution is finite, so the set of all such sets is finite, and the subset relation is acyclic over them, so we can find a subset-minimal set made up of a pair of racing actions and their happens-before predecessors. We identify one of the racy actions and the happens-before predecessors of both as a race-free prefix. This prefix is consistent, so we can translate it to a consistent prefix in the total model. We then add the previously-racy fringe action to the prefix, and establish that it is consistent and racy, appealing to receptiveness, if necessary for consistency. In a similar fashion to the previous case, we complete the consistent racy prefix to get a consistent racy execution in the total model.

4   The Thin-Air Problem has No Per-candidate-execution Solution

The question of "thin-air" reads is a longstanding issue in the design of memory models for C and C++, specifically for C/C++11 relaxed atomics: accesses for which races are permitted but which should be implemented with normal load and store instructions, without the cost of additional barriers or synchronisation instructions. Related questions arise in the semantics of C as used in the Linux Kernel (for ACCESS_ONCE accesses), and for normal accesses in Java [25].

The C++11 standard [9] included text intended to forbid thin-air executions (29.3p9), and it says explicitly (29.3p10) that that text forbids the LB+data example below, but the text was already recognised as flawed: a non-normative note in the standard (29.3p11) observed that *"The requirements do allow [the LB+ctrldata+ctrl-single example below]. However, implementations should not allow such behavior."*. Batty et al. identified further problems [8, §4], and their formal model does not attempt to capture that text or to exclude thin-air executions in any other way. The current proposal [12] for C++14 acknowledges difficulties with the C++11 version and proposes a deliberately vague placeholder as an interim replacement: *"Implementations should ensure that no "out-of-thin-air" values are computed that circularly depend on their own computation."*.

There is not a precise definition of what it means for a read to be "out of thin air" (if there were, the problem would be solved, as the semantics could simply exclude those). Rather, there are some example executions for which there is a consensus that the language should forbid them, and that current hardware and compiler optimisations do not exhibit. This is a high-level-language specification problem: there is no suggestion that thin-air executions occur in practice with current compilers and hardware; the problem is rather how to exclude them without preventing desired compiler optimisations.

In this section, we describe the thin-air problem via a series of examples, and we show that thin-air executions cannot be forbidden without restricting current compiler optimisations by any per-candidate-execution condition using the C/C++11 notion of candidate executions.

For each example we identify a particular execution by specifying the values read, and discuss whether it should be allowed by the semantics or not.

Example LB (language must allow)

```
r1=load_rlx(x) //reads 42
store_rlx(y,42)
─────────────────────────
r2=load_rlx(y) //reads 42
store_rlx(x,42)
```

$a{:}R_{RLX}\ x{=}42 \qquad c{:}R_{RLX}\ y{=}42$
$\quad sb\downarrow \qquad\qquad\qquad\qquad \downarrow sb$
$\qquad\qquad\qquad rf\ \ rf$
$b{:}W_{RLX}\ y{=}42 \qquad d{:}W_{RLX}\ x{=}42$

Here r1 and r2 are thread-local variables (which do not have memory actions in the model), while x and y are shared variables; initially all are 0. This execution (the dual of the first example of §2) is permitted by the ARM and IBM POWER architectures (presuming the code is compiled in the obvious way into machine load and store instructions): the actions of the each thread are to manifestly different addresses and so can be done out of order; it is moreover experimentally observable on current ARM multiprocessors [33]. Hence, the language semantics must allow it for relaxed atomics.

Example LB+datas (language can and should forbid)

```
r1=load_rlx(x) //reads 42
store_rlx(y,r1)
─────────────────────────
r2=load_rlx(y) //reads 42
store_rlx(x,r2)
```

$a{:}R_{RLX}\ x{=}42 \qquad c{:}R_{RLX}\ y{=}42$
$\ sb,dd\downarrow \qquad\qquad\qquad\qquad \downarrow sb,dd$
$\qquad\qquad\qquad rf\ \ rf$
$b{:}W_{RLX}\ y{=}42 \qquad d{:}W_{RLX}\ x{=}42$

There are two paradigmatic kinds of thin-air execution, the *thin-air read value* executions like this one, in which a value (here 42) "appears out of thin air", and the *self-satisfying conditional* example we discuss below. This example is architecturally forbidden on current hardware (x86, ARM, and IBM POWER), we do not expect future hardware to adopt the load-value prediction that would be required to make it observable, and to the best of our knowledge it cannot be exhibited by any reasonable current compiler optimisation combined with current hardware. Hence, the language semantics could forbid it.

Moreover, it is clearly desirable to forbid it, to make the language semantics as intuitive as possible. Boehm and Demsky [14] give examples where programming with relaxed atomics that permit thin-air values would be problematic, and in languages that aim to preserve implementation invariants at some types (such as that all pointer values point to allocated memory) it would be essential.

As for *how* it might be forbidden, the example suggests that one might simply forbid candidate executions with cycles in the union of the reads-from and dependency relations (the model has a data dependency relation shown as *dd* above). But the next two examples show that a combination of hardware behaviour and compiler optimisations make that infeasible.

Example LB+ctrldata+po (language must allow)

```
r1=load_rlx(x) //reads 42
if (r1 == 42)
    store_rlx(y,r1)
r2=load_rlx(y) //reads 42
store_rlx(x,42)
```

a:R_{RLX} x=42   c:R_{RLX} y=42

sb,dd,cd ↓   ↓ sb

b:W_{RLX} y=42   d:W_{RLX} x=42

rf   rf

This is architecturally allowed on ARM and Power (for the same reason as LB), and likewise observable on ARM, hence the language must allow it.

Example LB+ctrldata+ctrl-double (language must allow)

```
r1=load_rlx(x) //reads 42
if (r1 == 42)
    store_rlx(y,r1)
r2=load_rlx(y) //reads 42
if (r2 == 42)
    store_rlx(x,42)
else
    store_rlx(x,42)
```

a:R_{RLX} x=42   c:R_{RLX} y=42

sb,dd,cd ↓   ↓ sb,cd

b:W_{RLX} y=42   d:W_{RLX} x=42

rf   rf

This is forbidden on hardware if compiled naively, as the architectures respect read-to-write control dependencies, but in practice compilers will collapse conditionals like that of the second thread, removing the control dependencies from the read of y to the writes of x and making the code identical to the previous example. As that example is allowed and observable on hardware (and we presume that it would be impractical to outlaw such optimisation for C or C++), the language must also allow this execution. But this execution has a cycle in the union of reads-from and dependency, so we cannot simply exclude all those.

Then one might hope for some other adaptation of the C/C++11 model, but the following example shows at least that there is no per-candidate-execution solution.

Example LB+ctrldata+ctrl-single (language can and should forbid)

```
r1=load_rlx(x) //reads 42
if (r1 == 42)
    store_rlx(y,r1)
r2=load_rlx(y) //reads 42
if (r2 == 42)
    store_rlx(x,42)
```

a:R_{RLX} x=42   c:R_{RLX} y=42

sb,dd,cd ↓   ↓ sb,cd

b:W_{RLX} y=42   d:W_{RLX} x=42

rf   rf

This is the paradigmatic "self-satisfying conditional" example. It is forbidden on hardware if compiled naively (both ARM and POWER architectures prevent speculative writes becoming visible to other threads), and applying reasonable thread-local compiler optimisation does not change that. Hence, the language could forbid it. Moreover, it is problematic for informal and formal compositional reasoning [14,6,39], so the language should forbid it.

But the candidate execution that we want to forbid here is identical to the execution of the previous example that we have to allow. This immediately gives:

Theorem 2. *No adaptation of the C/C++11 per-candidate-execution definition that uses the same notion of candidate execution can give the desired behaviour for both of these examples.*

The basic point here is that compiler optimisations (such as the collapse of the LB+ctrldata+ctrl-double conditional) are operating over a representation of the *program*, covering all its executions, while the C/C++11 definition of candidate execution and consistency for those considers each candidate *execution* independently (it ignores the set of all executions); it is not able to capture the fact that the conditional is unnecessary because the two candidate executions corresponding to taking the two branches are equivalent. We develop this observation in §6.

Restricting optimisation involving relaxed atomics?. One might think that it would be feasible to restrict just compiler optimisations involving relaxed atomics, e.g. requiring that the compiler should respect all dependencies between relaxed atomic operations, while permitting more optimisation elsewhere. But (as observed by Boehm [11]) dependencies can be via functions in other compilation units that only involve non-atomic accesses, e.g. as in the version of LB+ctrldata+ctrl-double below, where the second thread's conditional is factored out into a function f() that does not involve atomics and that is in a different compilation unit. When compiling f() the compiler cannot tell whether it might be used in a dependency chain between atomic accesses, and so it would have to preserve all such dependencies. The cost of that is unknown, and worth investigating experimentally, but we suspect it to be unacceptable.

```
// in one compilation unit
void f(int ra, int*rb) {
  if (ra==42)
    *rb = 42;
  else
    *rb = 42; }
```

```
// in another compilation unit
r1=load_rlx(x) //reads 42      r2=load_rlx(y) //reads 42
if (r1 == 42)                  f(r2,&r3)
  store_rlx(y,r1)              store_rlx(x,r3)
```

In practice, GCC (checked with 4.6.3 on x86) does optimise away the control dependency in f(), at 01, 02, or 03.

Preventing load-store reordering. If one relaxes the requirement that relaxed atomics must be implementable with simple machine accesses, one might restrict all shared-variable load-to-store reordering, as proposed by Boehm and Demsky [12,14], adding barriers and somewhat restricting compiler optimisation. The cost has not yet been quantitatively assessed. For C/C++ it might be viable due to the small number of relaxed atomics (though if practitioners resorted to in-line assembly instead, that would defeat the purpose). But for normal Java accesses on ARM or Power, the cost seems likely to be prohibitive.

5 Integrating Non-atomics and Atomics Leads Back to Thin Air

We now show that the thin-air problem is not confined to relaxed atomics. The C++11 standard prose refers to *"atomic objects"* as if they are quite different from non-atomic objects, and the mathematical model of Batty et al. [8] for the C++11 and C11 concurrency primitives followed suit by imposing a simple type discipline: a *location kind* map in each candidate execution partitioned locations into atomic, nonatomic, and mutex locations. The definition of consistent execution permitted atomic accesses only at atomic locations, and the only nonatomic accesses allowed at atomic locations were atomic initialisations[1].

However, when one considers generalising that semantics for the concurrency primitives to cover more of C, it becomes clear that an up-front location-kind distinction is unrealistic, for several reasons:

1. In C it is permitted to reuse a region of allocated storage (e.g. from malloc) at a new type, simply by overwriting the bytes of memory with a new value. Restricting that to prevent strong updates from atomic to nonatomic (or v.v.) would not give a usable language.
2. In C one can inspect the representation bytes of a value by casting a pointer to (char *), or by type-punning via a union.
3. In C one can copy a value by copying its representation bytes, e.g. using memcpy. This could perhaps be deemed illegal for structures containing atomic values (indeed, it would have to be if atomic values had to be registered somewhere in the implementation), but it would be preferable, and in keeping with the rest of the language, to permit it.
4. In C11 one can construct atomic versions of structure and union types (with _Atomic(*type-name*) or the _Atomic qualifier), but their members can be accessed only via a non-atomic object which is assigned to or from the atomic object, not directly [2, 6.5.2.3p5].

---

[1] It is desirable to have nonatomic initialisations so that they do not require fences, but then to obtain a DRF-SC result initialisation had to be limited to be happens-before all other accesses, and without reinitialisation.

Hence, contrary to [8], we have to allow mixtures of atomic and nonatomic accesses at the same location, at least where the nonatomic accesses do not race with each other or with any atomic accesses.

But what should the semantics be for these? The standard text does not directly address these mixtures, but for the entirely nonatomic and entirely atomic cases it and the formal model [8] are clear:

- for the non-atomic case, the definition of consistent execution requires, in *consistent_non_atomic_rf*, the read to read from the most recent happens-before-visible write to the same location; while
- for the atomic case, the analogous *consistent_atomic_rf* lets the read read from any write that is not after it in happens-before (subject to the other predicates of the model).

Neither of these predicates are suitable to govern mixtures of atomic and non-atomic accesses, as the following two examples show. Our first example program uses memcpy to mix atomic and non-atomic accesses at the same location. The C/C++11 memory model as it stands suggests that the mixed accesses would be governed by *consistent_atomic_rf*, because the location has an atomic type. However, this breaks DRF-SC: the example program is race-free in every SC execution, but it has racy executions in the C/C++11 memory model:

```
// parent thread
size_t s = sizeof(atomic_int)
atomic_int x = 0
atomic_int y = 0
atomic_int a = 1

int r1 = load_sc(x)
if(r1 != 0)
    memcpy(&y,&a,s)
int r2 = load_sc(y)
if (r2 != 0)
    memcpy(&x,&a,s)
```

In the execution above, each atomic load reads from the non-atomic write implicit in the memcpy of the other thread. The execution is consistent and has data races. Breaking DRF-SC makes *consistent_atomic_rf* unsuitable to govern non-atomic reads from atomic writes. By swapping the atomics and non-atomics in the example, we see that it is also not suitable to govern atomic reads from non-atomic writes.

Our second example establishes that we also cannot use the *consistent_non_atomic_rf* predicate for mixtures. In the program below, there is a reading thread that spins until it sees the other thread's writes of z and y, and then reads from x twice: once with acquire memory order and once with consume. After the loop, there are two memcpy's of location x:

```
// parent thread
size_t s = sizeof(atomic_int)
atomic_int n=0, x=0, y=0, z=0
```

```
storerlx(x,1)
storerel(z,1)
storerlx(x,2)
storerel(y,&x)
do { r1 = loadacq(z)
     r2 = loadcon(y)}
while (r1==0 || r2==0)
memcpy(&n,r2,s)
memcpy(&n,&x,s)
```

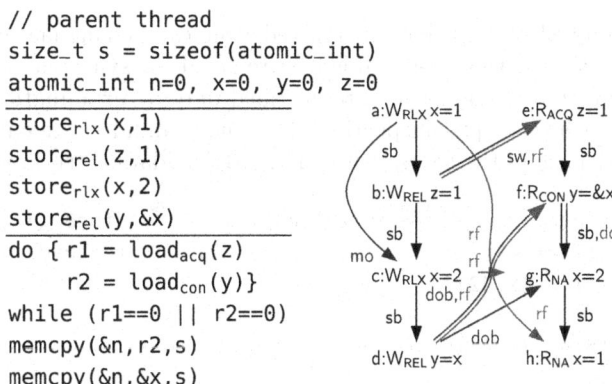

In the candidate execution on the right above, the loop exits (we elide the implicit write of the memcpy's, and the initialisation writes). The first memcpy happens after all atomic writes of x, but before the write implicit in the second memcpy, so according to *consistent_non_atomic_rf*, it must read write c. The second memcpy reads a pointer provided by the consume read, creating a dependency and forcing it to read a, but this execution, shown above, contains a CoRR coherence violation between accesses a, c, g and h, making the execution inconsistent, so the only behaviour that the model allows of this program is spinning on the conditional of the loop (similar executions arise if we swap atomics with non-atomics and vice versa), when in fact the program contains a race. Using *consistent_non_atomic_rf* for the mixtures cuts out executions we need to allow: it can make reasonable executions of race-free programs inconsistent and remove racy executions from racy programs, making them race-free and well-defined.

Vafeiadis et al. provide another alternative semantics for non-atomic reads [38]: modification order and coherence are extended to cover all locations (including non-atomics), atomic reads use the existing condition for reads at atomic locations, and the condition on non-atomic reads is replaced with a requirement that a new relation, the union of happens-before and rf edges to or from non-atomic accesses, is acyclic. This semantics provides the desired behaviour in the examples above, but, as noted by Vafeiadis et al., it forbids compiler optimisations from reordering loads followed by stores. Morisset et al. observe that this sort of reordering results from loop invariant code motion [29], an optimisation performed by both GCC and LLVM [18,24], so this attractive semantics comes with the unacceptable cost of forbidding routine compiler optimisations over blocks of non-atomic code.

We have seen that using *consistent_nonatomic_rf* to govern the behaviour of non-atomic reads at locations accessed atomically removes too many behaviours; we cannot use *consistent_atomic_rf* to govern such reads either (that would break DRF-SC); and the suggestion of Vafeiadis et al. comes at too high a cost. It is not clear what the semantics of non-atomic reads should be in C11.

6   An Out-of-order Operational Construction

The examples of §4 showed that, for relaxed atomics, the language semantics has to admit reorderings that are enabled by removals of syntactic control dependencies, where those removals can be justified only by examination of multiple control-flow paths (not just inspection of a single candidate execution). For example, consider again the second thread of LB+ctrldata+ctrl-double:

$$
\begin{array}{lll}
\texttt{r2=load}_\texttt{rlx}\texttt{(x)} & \xrightarrow{\text{compiler}} & \texttt{r2=load}_\texttt{rlx}\texttt{(y)} \\
\texttt{if (r2 == 42)} & & \texttt{store}_\texttt{rlx}\texttt{(x,42)} \\
\quad\texttt{store}_\texttt{rlx}\texttt{(x,42)} & & \\
\texttt{else} & & \\
\quad\texttt{store}_\texttt{rlx}\texttt{(x,42)} & &
\end{array}
\xrightarrow{\text{h/w}}
\begin{array}{l}
\texttt{store}_\texttt{rlx}\texttt{(x,42)} \\
\texttt{r2=load}_\texttt{rlx}\texttt{(y)}
\end{array}
$$

The key fact here is that the store$_\texttt{rlx}$(x,42) is possible on all control-flow paths of this thread, and a sufficiently "smart" compiler can detect that and then remove the control dependency from the read of y. In this section we generalise this observation: we give a semantics for relaxed and nonatomic accesses (and locks and fences) that correctly accounts for all the thin-air examples of §4 in an interesting and reasonably clean way. But those examples only involve reorderings; in §6.2 we use this semantics to highlight difficulties with other common optimisations.

6.1   The Semantics for Reorderings

We start from a standard labelled transition system (LTS) semantics for each thread in isolation, describing its interactions with memory by transitions labelled a:Rx=v and b:Wx=v for a read or write of value v at location x. This thread-local base semantics does not constrain the values read from memory in any way; it simply has a transition for each possible read value. For example, looking at some of the threads from the §4 tests, we have:

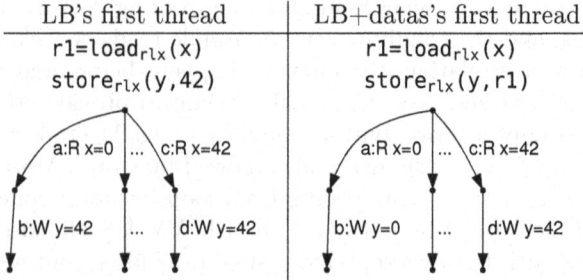

LB's first thread	LB+datas's first thread
r1=load$_\texttt{rlx}$(x)	r1=load$_\texttt{rlx}$(x)
store$_\texttt{rlx}$(y,42)	store$_\texttt{rlx}$(y,r1)

In LB's first thread, there is a write of 42 to y in all branches of the LTS, and we will allow the thread to write 42 before reading, letting both threads read 42. On the other hand, in LB+datas's first thread, it is not the case that a write of 42 is available in all branches, so it will have to do the read first, preventing LB+datas from exhibiting out-of-thin-air behaviour.

We capture this by constructing a derived *out-of-order* labelled transition system for each thread. Its states are copies of the entire base in-order LTS with some edges *ticked*. The initial state is the base LTS with no edge ticked. For example, part of the out-of-order LTS for LB's first thread is shown below. From now on, we only show the branches for some interesting values; in reality there is one branch per possible value, as we assume the base LTS is receptive.

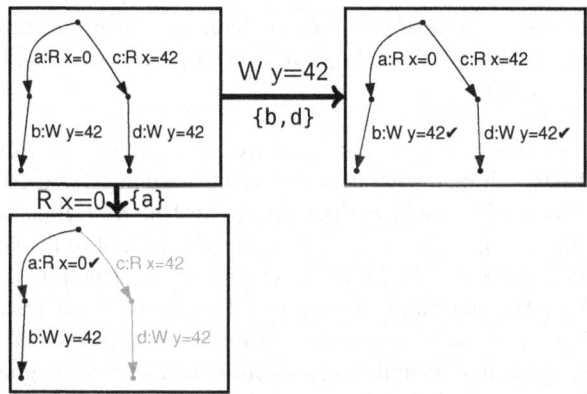

The transitions are labelled with the same memory actions as the base semantics; each transition of the derived LTS corresponds to ticking a set of base transitions. But the base transitions can be performed out-of-order, when they are not blocked (as defined below) in any branch by coherence or fences. Specifically: a set of edges can be ticked iff it forms a *frontier*, that is, (1) it is non-empty, (2) the edges are not ticked, (3) the edges have the same memory action label, (4) each non-discarded path either has a single edge in the frontier, or becomes discarded by this ticking, and (5) no edge is blocked (see below). Here an edge is *discarded* if it has a ticked sibling, and a path is discarded if it contains a discarded edge.

For example, the horizontal transition above is justified by the frontier on the left below consisting of all the W y=42 edges (b, d, and all the similar edges in elided paths), while the vertical transition is justified by the frontier on the right below consisting just of a (and there is a similar transition, not shown, for each base transition with a different read value).

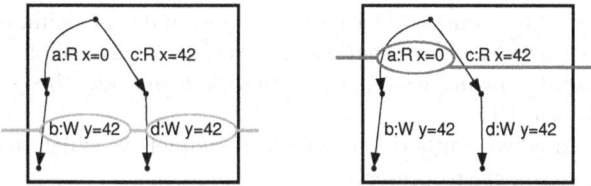

An edge is blocked by another if its action cannot be reordered before the other's. To maintain coherence (the fact that execution respects a per-location total order over writes to each location, consistent with program order, as guaranteed by standard hardware and by C11 relaxed atomics), actions to the same location cannot be reordered. Fences cannot be reordered before or after actions, so that all the actions before the fence have to be ticked before the fence can be ticked, and all the actions before the fence and the fence itself have to be ticked before actions after the fence can be ticked. Unlock and lock actions cannot be reordered before and after actions, respectively, but can in some cases be reordered the other way around, to allow for roach motel reordering.

Handling nonatomics. Non-atomic accesses can be executed out-of-order, like relaxed accesses, but in addition, they can also cause races, which the semantics has to be able to detect.

Non-multi-copy-atomic memory. For two-thread examples, one can combine the derived LTS of each thread with an underlying sequentially consistent shared memory (and that is what we have done for the testing described below). But in general the language semantics must also admit the lack of multi-copy atomicity permitted by the Power and ARM architectures, as described in §2. This can be handled by taking the parallel composition of the thread subsystems given by the derived LTSs with a storage subsystem following that of Sarkar et al. [33], which provides a generic non-multi-copy-atomic memory by keeping track of (a) the coherence commitments made among write events, and (b) the lists of writes and barriers propagated to each thread. The storage and thread subsystems are then synchronised on write requests, read requests and responses, etc.

This semantics gives the desired behaviour for each of the thin-air examples of §4: it is liberal enough to allow the reordering (introduced by compiler or hardware) that gives rise to the "must be allowed" examples, and restrictive enough to prevent the "should be forbidden" examples, ruling out thin-air executions basically by executing along a totally ordered trace of the derived LTS, with reads reading from previous writes in that trace. We have a precise Lem definition of the out-of-order semantics, and have built a tool that lets one explore the semantics of small examples, based on OCaml code generated from the Lem and integrated with an underlying SC memory. It has several good features:

- It is operational and relatively concrete, which makes it easier to understand than (say) the C11 axiomatic memory model.
- The construction is independent from the language syntax and thread-local operational semantics, which is highly desirable for tackling a complex language like C. This contrasts with explicit-speculation calculi, e.g. [15,22].
- For entirely thread-local computation, as thread-local variables do not create memory events, optimisations are already factored into the computation of the thread-local LTS.
- It does not involve syntactic notions of dependency, which are difficult for optimising compilers to preserve.

However, this semantics does not allow behaviour that is introduced by many other common compiler optimisations. Looking at these other optimisations highlights some subtle issues that any semantics for a C-like language will have to tackle.

6.2   Optimisations Beyond Reordering

In contrast to hardware semantics, there is (to date) no good characterisation of the envelope of all compiler optimisations normally performed in practice. The syntactic optimisations that are performed by compilers are numerous (GCC and Clang each have of the order of 100 passes) and they have unclear effects and interactions. Ševčík [40] and Morisset et al. [29] consider some abstract classes of optimisations, but these are only thread-local. In this section we give a preliminary discussion of some optimisations that go beyond reordering, in the context of the out-of-order semantics.

Elimination of subsumed memory actions. Many common compiler optimisations, like constant propagation and common subexpression elimination (CSE), can be explained in terms of eliminations of individual memory accesses [40]: read after read, read after write, write after read, and overwritten write elimination, which consist in conflating actions when the effect of one subsumes that of the others. For example, in the following program, the second read of x can be merged into the first as a very simple instance of CSE (by a read after read elimination); then, both branches of the conditional write 1 to x, so this write can be executed out-of-order, so there is an execution where both r1 and r3 are 1.

```
r1=load_rlx(x)       r3=load_rlx(y)
if (r1 == 1)         store_rlx(x,r3)
    r2=load_rlx(x)
    store_rlx(y,r2)
else
    store_rlx(y,1)
```

We conjecture that the notion of frontier can be relaxed to deal with these, e.g. with extended frontiers as below. We interleave optimisations (extended frontiers) with execution (ticking) on purpose to account for adaptive optimisations. When compilers perform this kind of optimisation, they effectively identify extended frontiers, and collapse them into elementary frontiers, but work on finite foldings of the LTSs, like SSA.

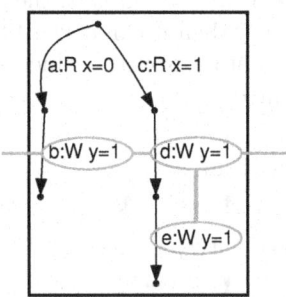

These optimisations need information about multiple paths, but only in a limited way: they only need the existence of particular actions (in a non-blocked path context) in each path. However, this is not the case for all optimisations, as we show next.

Irrelevant read elimination. Intuitively, irrelevant read elimination consists in removing a read action when its result does not affect the thread's behaviour: for example, if the branches of a read have identical subtrees, it is certainly irrelevant. But in general a read is irrelevant if its subtrees are in some sense semantically equivalent, where equivalence is up to optimisations, including re-ordering, eliminations, and irrelevant read elimination. For example, in the following program, the read of x is irrelevant only up to reordering of the writes to y and z, overwritten write elimination of the first write to z in the else branch, and irrelevant read elimination of the read of w. This suggests a recursive construction of the memory model, but it is not clear at what level: thread-local read-irrelevance, whole-program read irrelevance, etc.

```
r1=load_rlx(x)
if (r1 == 1) {
    store_rlx(y,1)
    store_rlx(z,1)
    r2=load_rlx(w)
} else {
    store_rlx(z,42)
    store_rlx(z,1)
    store_rlx(y,1)
}
```

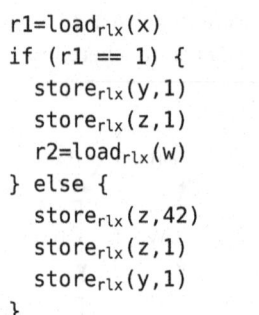
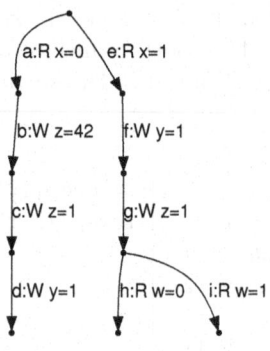

Inter-thread optimisations. The previous optimisations were all thread-local. Inter-thread optimisations (alias analysis, pointer analysis, ...) turn out to be even more challenging. The out-of-order construction makes no assumption about what values can be read, and thread-local LTSs thus have a branch for every value of each read. Identifying a value restriction amounts to discarding some "impossible" branches of the LTS. This can create more valid frontiers, and hence permit more out-of-order behaviour. For example, in the LTS below, if, by looking at all the writes to x by all the threads, the compiler determines that x can only contain values 0 and 1, then it can discard the branch where the value 2 is read, which makes {b, d} into a frontier, which allows the write to y to be executed before the read from x:

```
r1=load_rlx(x)
if (r1 == 2)
    store_rlx(y,0)
else
    store_rlx(y,1)
```

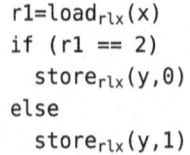
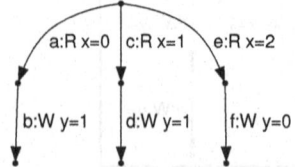
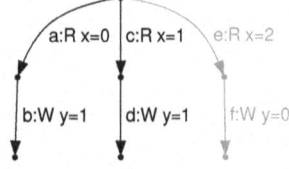

Moreover, some optimisations restrict behaviour, which creates more opportunities for inter-thread analyses, so inter-thread optimisations cannot be separated to an initial phase, but have to be intertwined with the other optimisations. This again suggests a recursive construction of the memory model. For example, in the following program, the second read of x can be merged into the first (by read after read elimination); value-range analysis can then remove the conditional, which allows additional behaviour: r1 and r3 can be 42.

```
r1=load_rlx(x)       r3=load_rlx(y)
r2=load_rlx(x)       store_rlx(x,r3)
if (r1 == r2)
    store_rlx(y,42)
else
    store_rlx(y,43)
```

The additional behaviour introduced by the analysis can invalidate it, or enable more optimisations that can invalidate it, so the semantics cannot be defined by a naive fixpoint.

Thread-local and shared variables. Finally, the out-of-order semantics is defined over a calculus that has a syntactic distinction between thread-local variables and potentially-shared variables. This distinction is important, as the semantics does not need to consider interference on thread-local variables, and thread-local optimisations on them are built into the base LTS construction, and can be much more aggressive. For example, in the following program, if x is determined to be thread-local, then constant propagation (in our framework, read after write elimination) can be done across the synchronisation.

```
x = 7
unlock(l)
...
lock(l)
r1 = x
```

However, C does not have such a distinction, and whether a variable behaves thread-locally depends on the dynamic behaviour of the program, which in turn depends on which variables behave thread-locally.

7   Concurrency and Undefined Behaviour

For our final contribution, we observe that there is a fundamental mismatch between the concurrency models of C/C++11 and the treatment of undefined behaviour in their preexisting specifications.

The C and C++ standards impose many constraints on programs by attributing *undefined behaviour* to programs that exhibit them (for C these are collected in [2, J.2]). Some of these are static properties (e.g. programs should define a main function) but many are dynamic, e.g. there should be no division by zero or out-of-bounds array access (OOBAA). For programs with undefined behaviour, the standard does not say that execution fails or behaves arbitrarily at that point. Instead, the compiler is completely unconstrained in the code it produces for the whole program [2, §3.4.3#1]:

> NOTE Possible undefined behavior ranges from ignoring the situation completely with unpredictable results, to behaving during translation or program execution in a documented manner characteristic of the environment (with or without the issuance of a diagnostic message), to terminating a translation or execution (with the issuance of a diagnostic message).

This is important because optimisations can involve significant code motion. For example, in an execution in which x=0, the following reaches a division-by-zero after the puts, both in the sequential execution model of the standard and in a non-optimising implementation. But an optimising compiler that does loop-invariant code motion might well hoist the 1/x before the loop, reaching the division-by-zero error before the puts. That code motion is made legal in general by giving this program entirely undefined behaviour.

```
for(int i=0; i<5; i++) {
    puts("foo\n");
    ret += i + 1/x;
}
```

Integrating the concurrency model into the language changes things. There are new sources of undefined behaviour: any program with a data race has undefined behaviour, which (for example) licenses the conventional implementation of bitfield operations mentioned in §2. But the overall form of the semantics also changes: instead of that simple sequential execution model (used to discover the division-by-zero on a reachable path) the definition calculates the set of candidate complete executions (essentially graphs like the examples shown in §4 and §5) that satisfy the *consistency predicate* of the concurrency model; if none of those contains a data race, then they are the allowable behaviour of the program (otherwise the program is undefined). There is a tension between this global completed-execution structure and the implicit use of the sequential execution model to discover the earlier forms of undefined behaviour.

For example, the C standard says that out-of-bounds array access is undefined behaviour [2, §6.5.6#8 (for an access from one-past an array)]. In the sequential setting (or indeed in an SC concurrent setting) there is a clear notion of execution prefix, and to identify such an undefined behaviour one only has to consider such a prefix leading up to it. But in the concurrency model, LB-like tests show that parts of a candidate complete execution that follow (in program order) the offending access might influence whether it is performed; we cannot restrict attention to simple prefixes. Consider the following example, where x and y are atomic integers initialised to 0, and a is an integer array with two elements:

$$r1 = \text{load}_{rlx}(x)$$
$$r3 = a[r1]$$
$$\text{store}_{rlx}(y,2)$$

$$r2 = \text{load}_{rlx}(y)$$
$$\text{store}_{rlx}(x,r2)$$

In any sequentially consistent execution of the program, the first thread loads 0 from x, and there is no OOBAA. But with the intended implementation of relaxed atomics above the ARM or Power architectures, there can be an execution where the second thread loads the store of 2 to y then writes to x, and the first thread loads 2 from x and then performs an OOBAA[2]. As a consequence, the language must provide this program with undefined behaviour.

But to identify this undefined behaviour, we need to consider executions that go past it in program order, and that means we need to choose some semantics for the out-of-bounds array access, and the other sources of undefined behaviour, to provide a context for the subsequent execution. This leads to a great many questions about the semantics of constructs that might introduce undefined behaviour. Taking out-of-bounds array access as an example, what should the semantics of an out-of-bounds load be, what if control flow is decided by the result of the load, what if the access is a store, or if the access loads or stores a function pointer? In each of these cases, it is unclear what the semantics should be. The point of undefined behaviour in the C and C++ semantics is to cover cases where the language semantics cannot easily reflect what an implementation might do, so one would prefer not to have to answer such questions.

8   Conclusion

The C/C++11 concurrency model remains the state of the art for the semantics of a general-purpose shared-memory concurrent programming languages; it is, to the best of our knowledge, sound with respect to the compiler optimisation behaviour of implementations [29] (in contrast to the JMM [16,34]), it is provably compilable to relaxed hardware models [8,7,32], and our work here establishes a machine-checked DRF-SC theorem. But the thin-air problem shows that it allows too many behaviours, and we have seen here that that cannot be solved in a simple per-candidate-execution way, that the problem is not specific to relaxed atomics, that, while an operational solution for those examples is possible, it brings other difficulties, and that there are further problems with undefined behaviour.

Disturbingly, 40+ years after the first relaxed-memory hardware was introduced (the IBM 370/158MP), the field still does not have a credible proposal for the concurrency semantics of any general-purpose high-level language that includes high-performance shared-memory concurrency primitives. This is a major open problem for programming language semantics.

---

[2] Note that this is not a thin-air execution, just a normal LB shape, with the reads and writes to x and y related by program order on the first thread and a data dependency on the second, extended just by using the read value of the first thread in an array access.

Acknowledgements. We would like to thank Hans Boehm, Paul McKenney, Jaroslav Ševčík, Ali Sezgin, Viktor Vafeiadis, and Francesco Zappa Nardelli for discussions about parts of this work. We acknowledge funding from EPSRC grants EP/H005633 (Leadership Fellowship, Sewell) and EP/K008528 (REMS Programme Grant), and a Gates Cambridge Scholarship (Nienhuis).

References

1. The SPARC architecture manual, v. 9, http://www.sparc.org/technical-documents/, http://www.dev
2. Programming Languages — C (2011), ISO/IEC 9899:2011, http://www.open-std. org/jtc1/sc22/wg14/docs/n1539.pdf
3. Adve, S.V., Hill, M.D.: Weak ordering — a new definition. In: ISCA (1990)
4. Alglave, J., Maranget, L., Tautschnig, M.: Herding cats: Modelling, simulation, testing, and data mining for weak memory. ACM TOPLAS, 36(2) (2014)
5. Batty, M.: The C11 and C++11 concurrency model. PhD thesis, University of Cambridge (2014), http://www.cl.cam.ac.uk/~mjb220/battythesis.pdf
6. Batty, M., Dodds, M., Gotsman, A.: Library abstraction for C/C++ concurrency. In: Proc. POPL (2013)
7. Batty, M., Memarian, K., Owens, S., Sarkar, S., Sewell, P.: Clarifying and compiling C/C++ concurrency: from C++11 to POWER. In: Proc. POPL (2012)
8. Batty, M., Owens, S., Sarkar, S., Sewell, P., Weber, T.: Mathematizing C++ concurrency. In: Proc. POPL (2011)
9. Becker, P. (ed.): Programming Languages — C++ (2011), ISO/IEC 14882:2011, http://www.open-std.org/jtc1/sc22/wg21/docs/papers/2011/n3242.pdf
10. Boehm, H.-J.: Threads cannot be implemented as a library. In: Proc. PLDI (2005)
11. Boehm, H.-J.: Memory model rationales (March 2007), http://open-std.org/jtc1/ sc22/wg21/docs/papers/2007/n2176.html
12. Boehm, H.-J.: N3786: Prohibiting "out of thin air" results in C++14 (September 2013), http://www.open-std.org/jtc1/sc22/wg21/docs/papers/2013/n3786.htm
13. Boehm, H.-J., Adve, S.V.: Foundations of the C++ concurrency memory model. In: Proc. PLDI (2008)
14. Boehm, H.-J., Demsky, B.: Outlawing ghosts: Avoiding out-of-thin-air results. In: Proc. MSPC (2014)
15. Boudol, G., Petri, G.: A theory of speculative computation. In: Gordon, A.D. (ed.) ESOP 2010. LNCS, vol. 6012, pp. 165–184. Springer, Heidelberg (2010)
16. Cenciarelli, P., Knapp, A., Sibilio, E.: The Java memory model: Operationally, denotationally, axiomatically. In: De Nicola, R. (ed.) ESOP 2007. LNCS, vol. 4421, pp. 331–346. Springer, Heidelberg (2007)
17. Demange, D., Laporte, V., Zhao, L., Jagannathan, S., Pichardie, D., Vitek, J.: Plan B: A buffered memory model for Java. In: POPL (2013)
18. Free Software Foundation, Inc., RTL Passes — GNU Compiler Collection (GCC) Internals (October 2014), https://gcc.gnu.org/onlinedocs/gccint/RTL-passes. html.
19. Gharachorloo, K., Adve, S.V., Gupta, A., Hennessy, J.L., Hill, M.D.: Programming for different memory consistency models. Journal of Parallel and Distributed Computing 15, 399–407 (1992)
20. Gosling, J., Joy, B., Steele, G.: The Java Language Specification (1996)
21. The HOL 4 system, http://hol.sourceforge.net/

22. Jagadeesan, R., Pitcher, C., Riely, J.: Generative operational semantics for relaxed memory models. In: Gordon, A.D. (ed.) ESOP 2010. LNCS, vol. 6012, pp. 307–326. Springer, Heidelberg (2010)
23. Lamport, L.: How to make a multiprocessor computer that correctly executes multiprocess programs. IEEE Trans. Comput. C-28(9), 690–691 (1979)
24. LLVM Project. LLVM's Analysis and Transform Passes — LLVM 3.6 documentation (October 2014), http://llvm.org/docs/Passes.html
25. Manson, J., Pugh, W., Adve, S.V.: The Java memory model. In: POPL (2005)
26. Maranget, L., Sarkar, S., Sewell, P.: A tutorial introduction to the ARM and POWER relaxed memory models (October 2012), http://www.cl.cam.ac.uk/~pes20/ppc-supplemental/test7.pdf
27. Marino, D., Singh, A., Millstein, T., Musuvathi, M., Narayanasamy, S.: A case for an SC-preserving compiler. In: PLDI (2011)
28. McKenney, P.: Reordering and verification at the linux kernel reorder workshop in vienna summer of logic. In: Invited talk at REORDER Workshop, Vienna Summer of Logic (July 2014), http://www2.rdrop.com/users/paulmck/scalability/paper/LinuxRCUVerif.2014.07.17a.pdf
29. Morisset, F.R., Pawan, P., Nardelli, Z.: Compiler testing via a theory of sound optimisations in the C11/C++11 memory model. In: Proc. PLDI (2013)
30. Norris, B., Demsky, B.: CDSchecker: Checking concurrent data structures written with C/C++ atomics. In: Proc. OOPSLA (2013)
31. Pugh, W.: Fixing the Java memory model. In: Proc. ACM 1999 Conference on Java Grande (1999)
32. Sarkar, S., Memarian, K., Owens, S., Batty, M., Sewell, P., Maranget, L., Alglave, J., Williams, D.: Synchronising C/C++ and POWER. In: Proc. PLDI (2012)
33. Sarkar, S., Sewell, P., Alglave, J., Maranget, L., Williams, D.: Understanding POWER multiprocessors. In: Proc. PLDI (2011)
34. Ševčík, J., Aspinall, D.: On Validity of Program Transformations in the Java Memory Model. In: Vitek, J. (ed.) ECOOP 2008. LNCS, vol. 5142, pp. 27–51. Springer, Heidelberg (2008)
35. Sewell, P., Sarkar, S., Owens, S., Zappa Nardelli, F., Myreen, M.O.: x86-TSO: A rigorous and usable programmer's model for x86 multiprocessors. C. ACM 53(7), 89–97 (2010), (Research Highlights)
36. Singh, A., Narayanasamy, S., Marino, D., Millstein, T., Musuvathi, M.: End-to-end sequential consistency. In: Proc. ISCA (2012)
37. Turon, A., Vafeiadis, V., Dreyer, D.: GPS: Navigating weak memory with ghosts, protocols, and separation. In: Proc. OOPSLA (2014)
38. Vafeiadis, V., Balabonski, T., Chakraborty, S., Morisset, R., Zappa Nardelli, F.: Common compiler optimisations are invalid in the C11 memory model and what we can do about it. In: Proc. POPL (2015)
39. Vafeiadis, V., Narayan, C.: Relaxed separation logic: A program logic for C11 concurrency. In: Proc. OOPSLA (2013)
40. Ševčík, J.: Safe optimisations for shared-memory concurrent programs. In: PLDI (2011)
41. Ševčík, J., Vafeiadis, V., Zappa Nardelli, F., Jagannathan, S., Sewell, P.: CompCertTSO: A verified compiler for relaxed-memory concurrency. J. ACM 60, 22:1–22:50 (2013)

The Best of Both Worlds:
Trading Efficiency and Optimality
in Fence Insertion for TSO*

Parosh Aziz Abdulla, Mohamed Faouzi Atig, and Tuan-Phong Ngo

Department of Information and Technology, Uppsala University, Sweden
{parosh,mohamed_faouzi.atig,tuan-phong.ngo}@it.uu.se

Abstract. We present a method for automatic fence insertion in concurrent programs running under weak memory models that provides the best known trade-off between efficiency and optimality. On the one hand, the method can efficiently handle complex aspects of program behaviors such as unbounded buffers and large numbers of processes. On the other hand, it is able to find small sets of fences needed for ensuring correctness of the program. To this end, we propose a novel notion of correctness, called *persistence*, that compares the behavior of the program under the weak memory semantics with that under the classical interleaving (SC) semantics. We instantiate our framework for the Total Store Ordering (TSO) memory model, and give an algorithm that reduces the fence insertion problem under TSO to the reachability problem for programs running under SC. Furthermore, we provide an abstraction scheme that substantially increases scalability to large numbers of processes. Based on our method, we have implemented a tool and run it successfully on a wide range benchmarks.

1 Introduction

Most modern processor architectures implement weak (relaxed) memory models for performance reasons [2,15]. However, this comes at a price since a program may exhibit behaviors that deviate substantially from its behaviors under the usual *Sequentially Consistent (SC)* semantics. The standard way to eliminate the undesired behaviors is to insert memory *fence* instructions that typically prevent reordering of instructions issued before and after the fence. The most common model corresponds to TSO (for Total Store Ordering) that is adopted by Sun's SPARC multiprocessors and x86 multiprocessors [25,26].

Challenge. An important problem in concurrent programming is to find sets of fences that ensure program correctness without compromising efficiency. Manual fence placement is time-consuming and error-prone due to complex behaviors introduced by weak memory models. The challenge then is to develop methods

---

* An open source tool with all the experimental data are publicly available at
  https://github.com/PhongNgo/persistence

© Springer-Verlag Berlin Heidelberg 2015
J. Vitek (Ed.): ESOP 2015, LNCS 9032, pp. 308–332, 2015.
DOI: 10.1007/978-3-662-46669-8_13

for *automatic* fence placement. A fence insertion algorithm requires an underlying verification algorithm that checks the correctness of the program for a given set of fences. This is necessary in order to be able to decide whether the current set of fences is sufficient, or whether additional fences are needed to achieve correctness. Designing such an algorithm is hard since we face a crucial trade-off between two criteria, namely:

- *Efficiency.* The algorithm needs to be able to carry out efficient analysis of complex program behaviors that arise due to intricate reorderings of program events. This complexity is for instance reflected by the fact that standard operational definitions for weak memory models [25,26] use *unbounded store-buffer* semantics, thus giving rise to an infinite state space even in the case where the original program is finite-state. Furthermore, since we are dealing with concurrent programs, the algorithm should scale well when increasing the number of processes and the number of variables.

- *Optimality.* The algorithm should derive sets of fences that are as close to optimal as possible. More precisely, we are required to avoid *under-fencing*, i.e., inserting too few fences since this would result in unsound program behaviors; and avoid *over-fencing*, i.e., inserting too many fences since this would result in a degradation of program performance (see e.g., [3,14,9,13], for descriptions of the high cost of fences on CPU-intensive concurrent programs).

In this context, identifying "good correctness properties" is crucial since a given property represents a particular choice in the trade-off between efficiency and optimality. For instance, at one extreme, we may require that the program is *data race free* (DRF) under SC (e.g. [24,22]). However, this will cause over-fencing, and hence failing the optimality criterion (see §2). In fact, some data races are in reality not harmful. For example, two *racy* implementations of a work-stealing queue [23,19] perform well under TSO without requiring fences. At the other extreme, we may consider *SC properties* such as safety and liveness properties. This would result in smaller sets of fences than in the previous case, but the verification problem becomes significantly harder (a *non-primitive recursive* lower-bound) or even undecidable [6,7], thus failing the efficiency criterion. Between these two extremes, the works in [11,5,9] consider the *robustness* (called also stability) property, i.e., checking whether a program generates the same set of traces *à la* Shasa and Snir [27] under weak memory and SC semantics. Robustness represents a correctness criterion between DRF property and SC properties since it is a weaker condition than the former and hence it would generate smaller sets of fences, while it could be more efficient than the latter since its verification problem belongs to a lower complexity class for finite-state programs under TSO (PSPACE-COMPLETE [10]). Robustness can be checked through a reduction to the reachability problem for a set of target programs under SC [9]. However, checking robustness causes state space explosion (see an explanation in the related work, §2), and furthermore, robustness may insert unnecessary fences as demonstrated by our experimental results.

Contribution. In this paper, we present a tool for automatic fence insertion in concurrent programs running under TSO that gives a good trade-off between efficiency and optimality. To this end, we make the following contributions.

• *Persistence:* We introduce a novel notion of correctness, called *persistence*, that as demonstrated by our experimental data provides a good trade-off between efficiency and optimality. Persistence considers the traces of a program and extracts two parameters, namely (i) *program order:* the order in which instructions are executed within the same process; and (ii) *store order:* the order in which different write operations hit the shared memory. The program is deemed to be persistent if it generates the same program and store orders under the TSO and SC semantics. If a program is persistent then it reaches identical sets of configurations (a configuration is a global state of the program) under TSO and SC. In particular, if the program is correct wrt. a given safety property under SC, then it will also be correct wrt. the same property under TSO.

• *Pattern:* We present an algorithm that automatically reduces the problem of checking persistence to the problem checking whether a given program exhibits a certain behavior pattern under SC . Despite the high complexity of the proof, the definition of the pattern is extremely simple. Crucially, from the efficiency point of view: (i) we need only to perform one reachability analysis query on a single target program, and (ii) there is no explosion in the size of the target program, since it contains the same number of processes, and only two extra variables compared to the original program (regardless of the number of processes).

• *Fence Insertion:* We present an algorithm that produces a *minimal* set of fences needed to ensure the program is persistent. The set is minimal in the sense that removing any fence in the set makes the program non-persistent. The algorithm is *counter-example-guided*, using counter-examples that are produced by the persistence checking algorithm. The fact that we reduce checking persistence to reachability analysis of SC programs allows using existing tools for program verification (we use SPIN [16] in the current implementation of our tool).

• *Abstraction:* We present a general abstraction framework that is compatible with the notion of persistence, in the sense that persistence of the abstract program implies persistence of the concrete program. We instantiate the framework by defining an abstraction function that allows to reduce the number of variables, thus significantly limiting the state space explosion problem.

• *Tool:* We have implemented an open-source and publicly available tool, called Persist, that we use to evaluate our framework on a wide range of benchmarks. Persist uses SPIN as a backend tool for checking reachability queries for programs under SC. Since SPIN runs on finite-state programs, our experiments are carried out only on such programs. We do an extensive comparison with state-of-the-art tools, such as Trencher [9], Memorax [1], Remmex [20] , and Musketeer [3]. Our data shows that persistence indeed provides a good trade-off between efficiency and optimality.

2   Related Work

Fig. 1 shows the relevant correctness criteria ordered according to their strength. In this paper, we consider the *Persistence* condition (PER). The strongest condition is Data Race Freedom (DRF) [24,22] where the program is declared incorrect in case it contains a trace with a data race. The main drawback of this approach is over-fencing. In view of this, more precise techniques, based on weaker conditions, have been developed to uncover real violations.

In [24], *triangular race freedom* (TRF) is introduced where a program is considered to be correct if the traces of the program under TSO and SC agree on (i) program order, (ii) store order, and (iii) the source relation (in some works, called the read-from relation). Condition (iii) records the write operation from which a given read operation fetches its value. The main limitation of TRF approach is that it does not come with a method for checking program correctness w.r.t. TRF. Our approach is a weakening of TRF in the sense that we have removed the source relation, and therefore using TRF will cause over-fencing compared to our method. Observe that the pattern for checking persistence (despite the high complexity of the proof) is similar to the pattern for checking TRF. Hence, we can remove the read-from relation from TRF without paying a huge cost.

Another weakening of the TRF condition is *Robustness* (ROB) (known also as stability) [27,11,12,5,9], where the store order condition is replaced by a weaker condition, namely *variable store order* (sometimes called *coherence* order). The latter considers the order of memory updates performed on each variable individually. In [9], a tool (called Trencher) is provided for *exact* checking of the robustness criterion. Our approach offers two advantages over this approach: (i) *Efficiency*: In [9], the robustness problem for TSO is reduced to the reachability problem for a set of target programs under the SC semantics.

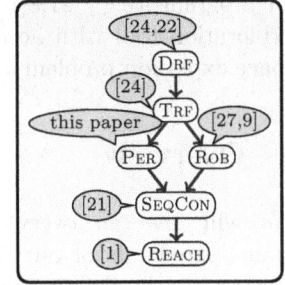

Fig. 1. Correctness criteria

However, the number of reachability queries issued, i.e., the size of the set of target programs, is quadratic in the size of the original program (this number is given by the number of pairs of instructions that can be reordered). Furthermore, the reduction triples the number of variables in each target program compared to the original program. In contrast, we reduce the persistence problem to a single reachability query for a program under SC which contains only 2 additional variables compared the original program (regardless of its number of variables or processes). This means that checking robustness is much more sensitive to the state space explosion problem than checking persistence (which is also visible in our experimental results, where we use the same backend tool SPIN). (ii) *Optimality*: Although the approaches are incomparable in general from the point of view of optimality (robustness-based analysis may insert fewer or more fences than persistence-based analysis), the absence of the source relation implies, in almost all examples, that we insert at most the same number of fences. In fact, in several examples, we insert a strict subset of the set of fences inserted by Trencher. Finally, no abstraction techniques are known for checking robustness.

Tools have been developed for *approximate* analysis of robustness (e.g., [11,12,5,3]). For instance, Musketeer [3] is based on static detection of critical cycles (that may violate robustness) in the control-flow graph of the program. The tool scales well to large programs but may cause over-fencing. In all examples we consider, we insert a subset of the set of fences inserted by Musketeer.

Liu et al. [21] consider even weaker conditions. One of them is *Sequential Consistence* (SEQCON) (called *state-based robustness* in [9]), i.e., checking whether there are any states of the program that are reachable under the weak memory semantics, but not under SC. The weakest condition is considered in [1], where the method checks REACH, i.e., whether a given state of the program is reachable under the weak memory semantics. The latter approach is used to implement Memorax which is a sound and complete tool for correcting finite-state programs under TSO wrt. safety properties. In contrast to our approach, checking the conditions SEQCON and REACH have non-primitive recursive complexities even for finite-state programs (as shown in [6,1]). This is reflected in our experimentation by the number of cases in which Memorax runs out of time/memory.

Several approximate tools have been developed for checking SC properties for programs (e.g., [21,8,17,20,4,18]). For instance, Remmex [20] is a state-space exploration tool with acceleration techniques. Remmex suffers from the state-space explosion problem as shown by our experimental data (see §11).

3   Overview

We will give an overview of the main ingredients of our framework, illustrating the definitions and algorithms through a simple toy example. We present our model for describing concurrent programs, and then introduce the notions of *runs* and *traces* using them to define the notion of *persistence*. A non-persistent program is said to be *fragile*. We solve the problem of checking whether a

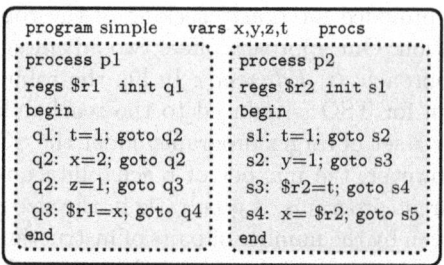

Fig. 2. A simple running example

given program is persistent (or fragile) in two steps. First, we show that any fragile program has a run containing a certain *fragility pattern*. Second, we reduce the problem of checking the existence of runs with fragility patterns in a given program $\mathcal{P}$ to the *reachability problem* under SC for a target programs that we derive automatically from $\mathcal{P}$. Then, we present our counter-example guided fence insertion procedure. Finally, we introduce our abstraction technique.

Model. Fig. 2 shows an example of a toy program, named *simple*, consisting of two concurrent processes (threads), called p_1 and p_2. Communication between processes is performed through four shared variables t, x, y, and z to which the processes can read and write. The processes have one register each, namely $\$r_1$ and $\$r_2$. The registers and shared variables are allowed to range over infinite domains (here they range over set of integers).

The behavior of a process is defined by a list of assembly-like instructions, each consisting of a label and a statement such as a read or write statement (see §5 for the full list of statements). For instance, at q_1, process p_1 performs a write statement in which it assigns the value 1 to the shared variable t. Notice that there are

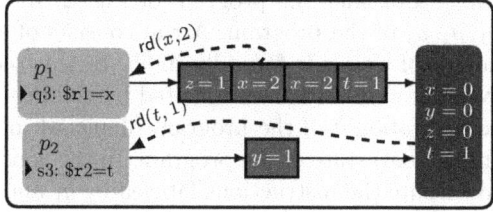

Fig. 3. Store buffers and the shared memory of a program under TSO

two instructions in p_1 labeled with q_2. This means that, once p_1 has executed the instruction labeled with q_1, it may non-deterministically choose to move to any of the two instructions labeled with q_2. This allows to encode conditional branching, iteration, and non-determinism, all using the same light syntax.

To define the runs of the program, we will use the operational semantics for TSO given in [25,26]. Conceptually, the model adds a FIFO buffer, called a *store buffer*, between each process and the main memory (cf. Fig. 3). The buffer is used to store the write operations performed by the process. A process executing a write instruction inserts it into its store buffer and immediately continues executing subsequent instructions. Memory updates are performed by non-deterministically choosing a process and by executing the oldest write operation in its buffer (the right-most element in the buffer). If a process p performs a read operation on a variable x then there are two possible scenarios. More precisely, if the buffer contains some write operations on x, then the read value must correspond to the value of the most recent such a write operation (the one that lies closest to the entry of the buffer). Otherwise, the value is fetched from the memory. Essentially, this means

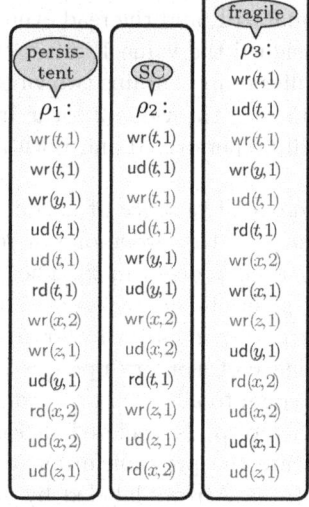

Fig. 4. Runs from configuration d of Fig. 9

that a read operation on a variable x may *overtake* a sequence of write operations stored in its own buffer provided that all these operations concern variables that are different from x. For example, in the given configuration of Fig. 3, p_1 can read the value 2 from x, and p_2 can read the value 1 from t. A fence means that the buffer of the process must be flushed before the program can continue beyond the fence. The store buffers of processes are *unbounded* since there is *a priori* no limit on the number of write operations that can be issued by a process before a memory update occurs. For instance, in the program of Fig. 2 the loop labeled by q_2 in process p_1 may generate an unbounded number of write operations, and hence create an unbounded number of elements in the buffer of p_1.

Runs. Consider the program of Fig. 2. In Fig. 4 we depict three typical runs ρ_1, ρ_2, ρ_3 of the program. A run consists of a sequence of *events*. We define the notion of an event formally in §5. In this section, it is sufficient to think of an event as an instruction executed by the process. The runs start from the initial configuration d of the program (depicted in Fig. 9). A configuration represents the (global) state of the program. In the configuration d, the processes are about to execute the instructions labeled by q_1 and s_1, and the values of all the shared variables and registers are equal to 0, while the store buffers are empty.

We describe ρ_1 below. To simplify the presentation, we identify an event with the operation that the process performs. For instance, we write the first step of process p_1 in ρ_1 as $\mathsf{wr}(t, 1)$ since p_1 will perform a write operation in which it assigns 1 to x. First, three write events are issued by the processes, namely $\mathsf{wr}(t, 1)$ by p_1, and $\mathsf{wr}(t, 1)$, $\mathsf{wr}(y, 1)$ by p_2. The new values are stored inside the corresponding buffers. Next, the two update events $\mathsf{ud}(t, 1)$ and $\mathsf{ud}(t, 1)$ are performed by the processes after which the value of t in the memory will be equal to 1, and the read event $\mathsf{rd}(t, 1)$ can be performed by p_2, where $\$r_2$ will be assigned the value 1. After the next three events $\mathsf{wr}(x, 2), \mathsf{wr}(z, 1), \mathsf{ud}(y, 1)$, the buffer of p_1 contains two write events $\mathsf{wr}(x, 2)$ and $\mathsf{wr}(z, 1)$, which means that p_1 can read the value 2 for x from the buffer. Finally, the last two update events will be performed and both buffers will now be empty.

Traces. The *trace* of a run π, records (i) the *program order*, i.e., the order of the read and write events executed by each process in π. The program order of ρ_1 in Fig. 4 is given by $\mathsf{wr}(t, 1)\mathsf{wr}(x, 2)\mathsf{wr}(z, 1)\mathsf{rd}(x, 2)$ for p_1 and by $\mathsf{wr}(t, 1)\mathsf{wr}(y, 1)\mathsf{rd}(t, 1)$ for p_2. (ii) the *store order* is the sequence of memory updates performed during the run. This is equal to $\mathsf{ud}(t, 1)\mathsf{ud}(t, 1)\mathsf{ud}(y, 1)\mathsf{ud}(x, 2)\mathsf{ud}(z, 1)$ for ρ_1. The trace of ρ_1 is depicted in Fig. 5. Arrows labeled by *po* indicate the program order, e.g., the arrow from $\mathsf{wr}(y, 1)$ to $\mathsf{rd}(t, 1)$. Arrows labeled by *so* indicate the store order. For instance, the arrow from $\mathsf{wr}(y, 1)$ to $\mathsf{wr}(x, 2)$ means that the event $\mathsf{ud}(y, 1)$ corresponding to $\mathsf{wr}(y, 1)$ occurs before the event $\mathsf{ud}(x, 2)$ corresponding to $\mathsf{wr}(x, 2)$.

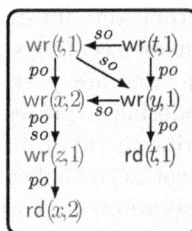

Fig. 5. The trace of ρ_1 and ρ_2 in Fig. 4

Persistence. A run is *Sequentially Consistent (SC)* if every write event is immediately followed by the corresponding update. Intuitively, this means that write events are atomic. An example is the run ρ_2 in Fig. 4. A run π is *persistent* if there is an SC run π' such that the traces of π and π' are identical, otherwise we say that π is *fragile*. For instance, in Fig. 4, the run ρ_1 is persistent since its trace is identical to that of ρ_2 (namely the trace shown in Fig. 5). We argue that the run ρ_3 in Fig. 4 is fragile. If not, there is an SC run ρ, with a trace identical to the one in Fig. 6. From the equality of store orders of ρ and ρ_3 and the fact that the update events, $\mathsf{ud}(x, 2)$, $\mathsf{ud}(x, 1)$, and $\mathsf{ud}(z, 1)$, occur in that order in ρ_3, it follows that the corresponding write events, $\mathsf{wr}(x, 2)$, $\mathsf{wr}(x, 1)$, and $\mathsf{wr}(z, 1)$, will occur in the same order in ρ. From the equality of program orders it follows that $\mathsf{rd}(x, 2)$ will occur after $\mathsf{wr}(z, 1)$ (and hence also after $\mathsf{wr}(x, 1)$) in ρ. Then ρ contains the sequence $\mathsf{wr}(x, 2)\mathsf{wr}(x, 1)\mathsf{wr}(z, 1)\mathsf{rd}(x, 2)$ which is not possible.

A program is *persistent* if all runs from its initial configuration are persistent, otherwise it is *fragile*. In the *persistence problem*, we check whether a given program is persistent or fragile. Notice that a persistent program reaches the same set of configurations under TSO and SC, and hence it satisfies the same safety properties under SC and TSO (see §6, paragraph on Safety Properties).

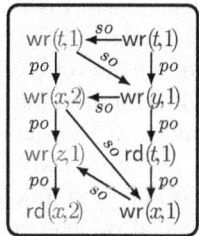

Fig. 6. The trace of ρ_3 in Fig. 4

Patterns. We reduce the persistence problem to the problem of checking the existence of a certain type of runs, called runs of type ⊛ (see Theorem 1). A run of of type ⊛ is defined with respect to one of the processes, called the *pivot* of the run (the other processes are called *fringe* processes). Fig. 7 shows an example of such a run for the program of Fig. 2, ρ_*, where the pivot is p_1 and the (only) fringe process is p_2. A run π of type ⊛ is the concatenation $\pi_1 \cdot \pi_2 \cdot \pi_3 \cdot \pi_4$ of four parts. In addition to being SC, π satisfies a number of conditions. We do not place any constraints on the first part π_1 (except that is it SC). The second part π_2, consists of two events performed by the pivot, namely a write event e_1 followed by the matching update event u_1. In our example, p_1 writes the value 1 to the variable z in π_2. The third part π_3 (which is empty in our example) consists only of events performed by the pivot, although it is not allowed to perform any write, update, atomic-read-write, or fence events. The fourth part π_4, consists of two events performed by the fringe process, namely a write event e_2 followed by the matching update event u_2. Furthermore, the variable updated here should be different from the variable updated in π_2. In our example, p_2 writes the value 1 to the variable x (which is different from z). Finally, there should be a read event e of the pivot (the event corresponding to the instruction labeled q_3 in our example), called the *complementary event* of π, such that the following properties hold: (i) e should be enabled after $\pi_1 \cdot \pi_2 \cdot \pi_3$. In our example, the event labeled q_3 is enabled after $\pi_1 \cdot \mathsf{wr}(z, 1)\mathsf{ud}(z, 1)$. (ii) In e, the process reads a value from the same variable as the one updated in π_4. In our example, this variable is x. (iii) The value read during e (i.e., the value of the read variable in the memory after $\pi_1 \cdot \pi_2 \cdot \pi_3$) should be different from the value assigned to it in π_4. In our example, the value of x in the memory after $\pi_1 \cdot \mathsf{wr}(z, 1)\mathsf{ud}(z, 1)$ is equal to 2 (which is different form the value 1 assigned to x in π_4).

The proof of existence of runs of type ⊛ is highly non-trivial. However, once done, it allows to define a surprisingly simple pattern for detecting fragile runs. More precisely, we can derive the fragile run $\pi_1 \cdot e_1 \cdot \pi_3 \cdot e \cdot e_2 \cdot u_2 \cdot u_1$, which we call the *witness*. In the above example, the witness is given by $\pi_1 \cdot \mathsf{wr}(z, 1)\mathsf{rd}(x, 2)\mathsf{wr}(x, 1)\mathsf{ud}(x, 1)\mathsf{ud}(z, 1)$.

316   P.A. Abdulla, M.F. Atig, and T.-P. Ngo

Pattern Detection. Given a program $\mathcal{P}$ we generate a new *target* program $\mathcal{Q}$ such that $\mathcal{P}$ contains a witness iff $\mathcal{Q}$ can reach a given set of states under SC. The program $\mathcal{Q}$ will run in three phases 0, 1, 2, and find the existence of a witness (as described above). Recall that such a witness is derived from a type ⊛ run $\pi = \pi_1 \cdot \pi_2 \cdot \pi_3 \cdot \pi_4$ which is defined wrt. pivot. In phase 0 (see Fig. 10), $\mathcal{Q}$ simulates π_1, and hence all the processes will simulate their moves in $\mathcal{P}$. At the end of phase 0 (see Fig. 11), one of the processes will non-deterministically decide to play the role of the pivot. At the same time, it records the variable on which it performs a write event in π_2 (in the above example, this variable is z). In our construction we do this by assigning a special value $\mathfrak{c}_1$ to the variable z. In phase 1 (see Fig. 12), $\mathcal{Q}$ simulates π_3 in which only the pivot is active. At the end of phase 1 (see Fig. 13), the pivot ensures the existence of the complementary event. This is done by (i) choosing

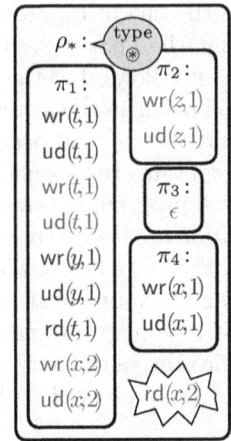

Fig. 7. Run of type ⊛ from d of Fig. 9

an enabled read event, (ii) ensuring that the involved variable is different from the one that was updated during π_2 (at the end of phase 0). This can be done by checking that its value is different from $\mathfrak{c}_1$. (iii) It records the read event of the complementary event by copying the value of the variable x to a new variable new and announcing the variable x to the fringe processes by assigning to it a new value $\mathfrak{c}_2$. Finally, in phase 2 (see Fig. 14), a fringe process verifies the existence of π_4, by finding the (only) variable whose value is $\mathfrak{c}_2$ and check that it can indeed assign to it a different value from the one that was assigned by the complementary event (by comparing with the value stored in new).

An important aspect of our scheme is that $\mathcal{Q}$ contains only two additional variables compared to $\mathcal{P}$, namely the variable new, and a variable with a small domain that we use to record the current simulation phase. This holds regardless of the number of variables in $\mathcal{P}$. Thus, the verification problem for the target program is as efficient as that for the source program (when run under SC).

Fence Insertion. A naive way to find the minimal set of fences is to simply try out all combinations. Obviously, such an algorithm would not work in practice due to the large number of possible combinations. Instead, we use counter-examples analysis, where we use witnesses provided by our persistence detection algorithm. A witness of the form shown above is fragile since complementary event overtakes the event e_1. Therefore, inserting a fence somewhere along $e_1 \cdot \pi_3 \cdot e_2 \cdot u_2 \cdot u_1$ is both necessary and sufficient to disable the witness. This allows to derive a set of fences by repeatedly calling the pattern detection algorithm, each time inserting a new fence, until the program becomes persistent. Notice that when the program becomes persistent the pattern detection algorithm declares that no witnesses exists any more. The derived set is optimal since each of the inserted is necessary to eliminate a witness, and hence removing any of them would make the program fragile. In the program of Fig. 2, our algorithm would put a fence after the second instruction labeled by q_2 in p_1, making the program persistent.

Abstraction. We develop an abstraction framework, called *observation abstraction*, that exploits the fact that persistence is not sensitive to the *source* relation. A process needs only to know the value of a variable in its own buffer or in the memory. It does not need to figure out the process (or instruction) that produced this value. The idea is to reason about the memory view of each process p: Each process can only observe the memory changes due to its own instructions on shared variables or the changes caused by the other processes over the set variables that p can read from. We abstract the rest of the processes, in a way that p has at least the same sequences of memory views as in the original program $\mathcal{P}$. This ensures that the persistence of the abstract program implies persistence of $\mathcal{P}$. We instantiate our framework by defining an abstraction function, called *Flattening* (see Fig. 15), that can be used for efficient checking of persistence.

4 Preliminaries

We use $\mathbb{N}$ to denote the set of natural numbers. For sets A and B, we use $f : A \mapsto B$ to denote that f is a function that maps A to B. For $a \in A$ and $b \in B$, we use $f[a \hookleftarrow b]$ to denote the function f' where $f'(a) = b$ and $f'(a') = f(a')$ for all $a' \neq a$. Let A be a finite set. We use $|A|$ to denote its size. We use A^* (resp. A^+) to denote the set of words (resp. non-empty words) over A; and ϵ to denote the empty word. Consider a word $w = a_1 a_2 \cdots a_n \in A^*$. We define $|w| := n$ and $\mathsf{last}(w) := a_n$. For $i : 1 \leq i \leq n$, we define $w[i] := a_i$. For $a \in A$, we write $a \in w$ if a appears in w, i.e., $a = w[i]$ for some $i : 1 \leq i \leq |w|$. For $B \subseteq A$, we define $w \odot B := a_{i_1} a_{i_2} \cdots a_{i_m}$ to be the maximal subword of w such that $a_{i_j} \in B$ for all $j : 1 \leq j \leq m$, i.e., we keep the elements that belong to B; and define $w \otimes B := i_1 i_2 \cdots i_m$, i.e., it gives the sequence of indices of the elements that belong to B.

5 Concurrent Programs

We define the syntax and the semantics we use for concurrent processes communicating through shared memory. Moreover, we define SC computations as a subclass of TSO ones. Finally, we introduce the reachability problem.

Syntax. The syntax of a program is given in Fig. 8. In the following, we assume a program with name $\mathcal{P}$, a set of shared variables $\mathbb{X}$, and a set of processes ProcSet. Each process $p \in$ ProcSet has a list of registers $\mathbb{R}_p$ and

```
⟨prog⟩ ::= program ⟨progid⟩
              vars ⟨var⟩*  procs ⟨proc⟩*
⟨proc⟩ ::= process ⟨procid⟩ regs ⟨reg⟩*
              init ⟨label⟩ begin ⟨inst⟩* end
⟨inst⟩ ::= ⟨label⟩ : ⟨stmt⟩ ; goto ⟨label⟩
⟨stmt⟩ ::= ⟨var⟩ = ⟨exp⟩ | ⟨reg⟩ = ⟨var⟩
           | ⟨reg⟩ = ⟨exp⟩ | fence
           | arw( ⟨var⟩ , ⟨exp⟩ , ⟨exp⟩ )
           | skip | assume ⟨exp⟩
⟨exp⟩ ::= ⟨fun⟩ ( ⟨reg⟩* )
```

Fig. 8. Syntax of a concurrent program

a list of assembly-like instructions $\mathbb{I}_p$. Each process starts from a statement with initial label $init_p$. We let $\mathbb{Q}_p$ be the set of labels that occur in the instructions of process p, and assume w.l.o.g. that the sets of labels and also the sets of registers of the different processes are disjoint. We define the sets $\mathbb{R} := \cup_{p \in \mathsf{ProcSet}} \mathbb{R}_p$, $\mathbb{I} := \cup_{p \in \mathsf{ProcSet}} \mathbb{I}_p$, and $\mathbb{Q} := \cup_{p \in \mathsf{ProcSet}} \mathbb{Q}_p$. Sometimes, we represent a program $\mathcal{P}$ by a tuple $\langle \mathbb{X}, \mathsf{ProcSet}, \mathbb{Q}, \mathbb{I}, init \rangle$, where $\mathbb{X}$ is the set of instructions, $\mathsf{ProcSet}$ is the set of processes, $\mathbb{Q}$ is the set of labels, $\mathbb{I}$ is the set of instructions, and $init : \mathsf{ProcSet} \mapsto \mathbb{Q}$ is mapping such that $init(p) = init_p$ for all $p \in \mathsf{ProcSet}$.

The registers and shared variables range over some (potentially) infinite domain V. Here, we assume w.l.o.g. that V is the set of integers. An instruction ins is of the form $\boxed{q_1 : \mathsf{s}; \mathsf{goto}\ q_2}$, where q_1, q_2 are labels, and s is a *statement*. After the program has executed the statement s it jumps to an instruction labeled with q_2. If several instructions are labeled with q_2, the process chooses one of them non-deterministically. If none exists, the process terminates. We assume that a program comes with a set $\mathbb{F}$ of functions. Our method is not dependent on the particular set of functions that occur in the programs, and therefore we will not specify the set precisely in the grammar. The set may include all the standard functions, such as addition, subtraction, multiplication, division, etc. An *expression* $\mathfrak{e}$ is either a constant (a member of V), register $\$r$, or of the form $f(\mathfrak{e}_1, \ldots, \mathfrak{e}_n)$ where $f \in \mathbb{F}$ is a function and $\mathfrak{e}_1, \ldots, \mathfrak{e}_n$ are expressions. A statement s is one of the following forms: (i) wr (write statement): $x = \mathfrak{e}$ writes the value of the expression $\mathfrak{e}$ to the shared variable x. This value will be stored in the buffer of the process. (ii) rd (read statement): $\$r = x$ reads the value of the shared variable x (either from the buffer of the process or from the memory), and stores it in the register $\$r$. (iii) arw (atomic-read-write statement): $arw(x, \mathfrak{e}_1, \mathfrak{e}_2)$ checks atomically whether the value of the shared variable x is equal to the value of $\mathfrak{e}_1$; if true it assigns the value of $\mathfrak{e}_2$ to x, otherwise the execution of the instruction is blocked. (iv) fn (fence statement): flushes the buffer of the process. (v) skip statement: is the empty statement. (vi) asgn (assign statement): $\$r = \mathfrak{e}$ assigns the value of the expression $\mathfrak{e}$ to the register $\$r$. (vii) asm (assume statement): assume $\mathfrak{e}$ checks whether $\mathfrak{e}$ evaluates to *true*. If not, the execution of the instruction is blocked. We use *source* (ins), *stmt* (ins), and *target* (ins), to denote q_1, s, and q_2 respectively. We classify instructions according to the forms of their statements. First, for a process p, and $i \in \{\mathsf{wr}, \mathsf{rd}, \mathsf{arw}, \mathsf{fn}, \mathsf{skip}, \mathsf{asgn}, \mathsf{asm}\}$, we define $\mathbb{I}_p^i$ to be the set of instructions in $\mathbb{I}_p$ with an i statement. For instance, $\mathbb{I}_p^{\mathsf{wr}}$ consists of the instructions ins $\in \mathbb{I}_p$ with write statements, i.e., *stmt* (ins) is of the form $x = \mathfrak{e}$ for some x and $\mathfrak{e}$. Furthermore, for $i \in \{\mathsf{wr}, \mathsf{rd}, \mathsf{arw}\}$, and a variable $x \in \mathbb{X}$ we define $\mathbb{I}_p^{i,x}$ to be the set of instructions in $\mathbb{I}_p^i$ that operate on the variable x. For instance, $\mathbb{I}_p^{\mathsf{wr},x}$ consists of the instructions ins $\in \mathbb{I}_p$ such that *stmt* (ins) is of the form $x = \mathfrak{e}$ for some $\mathfrak{e}$. In the program of Fig. 2, the instruction labeled with q_1 is a member of $\mathbb{I}_{p_1}^{\mathsf{wr},t}$.

Configurations. To define configurations, we introduce the following concepts. A *label definition* $\overline{q} : \mathsf{ProcSet} \mapsto \mathbb{Q}$ is a function such that $\overline{q}(p) \in \mathbb{Q}_p$ for each $p \in \mathsf{ProcSet}$. Intuitively, for a process $p \in \mathsf{ProcSet}$, $\overline{q}(p)$ gives the label of the instruction that p will execute in its next step. A *register*

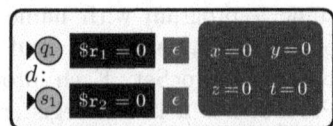

Fig. 9. Initial configuration d

state is a function $\overline{r} : \mathbb{R} \mapsto V$. For a register $\$r \in \mathbb{R}$, the value of $\overline{r}(\$r)$ is the content of $\$r$. A *buffer state* is a function $\overline{b} : \mathsf{ProcSet} \mapsto (\mathbb{X} \times V)^*$. The value of $\overline{b}(p)$ is the content of the buffer belonging to p. This buffer contains a sequence of write operations, where each write operation is defined by a pair, namely a variable x and a value v that is assigned to x. In our model, messages will be appended to the buffer from the left, and fetched from the right. A *memory state* is a function $mem : \mathbb{X} \mapsto V$ that defines the value of each variable in the memory. A *configuration* c is a tuple $\langle \overline{q}, \overline{r}, \overline{b}, mem \rangle$ where $\overline{q}$ is a label definition, $\overline{r}$ is a register state, $\overline{b}$ is a buffer state, and mem is a memory state. We use $\mathsf{LabelOf}\,(c)$, $\mathsf{BuffersOf}\,(c)$, $\mathsf{RegsOf}\,(c)$, and $\mathsf{MemoryOf}\,(c)$ to denote $\overline{q}$, $\overline{r}$, $\overline{b}$, and mem respectively. We define $\mathsf{EmptyBuffer}$ to be the buffer state such that $\mathsf{EmptyBuffer}(p) = \epsilon$ for all processes $p \in \mathsf{ProcSet}$. We say that c is *plain* if $\mathsf{BuffersOf}\,(c) = \mathsf{EmptyBuffer}$. The *initial configuration* c_{init} is defined by $\langle \overline{q}_{init}, \overline{r}_{init}, \mathsf{EmptyBuffer}, mem_{init} \rangle$ where $\overline{r}_{init}(\$r) = 0$ for all $\$r \in \mathbb{R}$, $\overline{q}_{init}(p) = init_p$ for all $p \in \mathsf{ProcSet}$, and $mem_{init}(x) = 0$ for all $x \in \mathbb{X}$. In other words, to start with, each process is in the initial label, all buffers are empty, and all registers and shared variables have value 0. We use C to denote the set of all configurations. For a configuration c and an expression $\mathfrak{e}$, we define the *evaluation* $c(\mathfrak{e})$ of $\mathfrak{e}$ in c inductively by $c(\$r) := \mathsf{RegsOf}\,(c)(\$r)$, and $c(f(\mathfrak{e}_1, \ldots, \mathfrak{e}_n)) := f(c(\mathfrak{e}_1), \ldots, c(\mathfrak{e}_n))$ (we regard a constant expression as a function with zero arguments). Fig. 9 illustrates an initial configuration d of the program in Fig. 2.

Events. We introduce the notion of *events* that describe two aspects of process behavior, namely the internal actions of a process and its interaction with the memory. The former consists of fence, skip, assign, and assume events, while the latter consists of write, read, and atomic-read-write events. For these three types of events we will also record the variable involved together with its value. Formally, we define the set of events Δ as follows. Let $p \in \mathsf{ProcSet}$ be a process. For $i \in \{\mathsf{fn}, \mathsf{skip}, \mathsf{asgn}, \mathsf{asm}\}$, we define the set $\Delta^i_p := \mathbb{I}^i_p$, i.e., for internal events, we define the set simply to be the set of instructions. For a variable $x \in \mathbb{X}$, a value $v \in V$, and $i \in \{\mathsf{wr}, \mathsf{rd}, \mathsf{arw}\}$, we define $\Delta^{i,x,v}_p := \mathbb{I}^{i,x}_p \times \{v\}$, i.e., the event is a pair including the instruction (whose statement operates on x) and the used value v. We define the sets $\Delta^{i,x}_p := \cup_{v \in V} \Delta^{i,x,v}_p$, $\Delta^i_p := \cup_{x \in \mathbb{X}} \Delta^{i,x}_p$, $\Delta_p := \cup_{i \in \{\mathsf{wr}, \mathsf{rd}, \mathsf{arw}, \mathsf{fn}, \mathsf{skip}, \mathsf{asgn}, \mathsf{asm}\}} \Delta^i_p$, and $\Delta := \cup_{p \in \mathsf{ProcSet}} \Delta_p$. For an event $e \in \Delta$ of the form $\mathfrak{ins}$ or the form $\langle \mathfrak{ins}, i \rangle$, we define $source\,(e) := source\,(\mathfrak{ins})$, $stmt\,(e) := stmt\,(\mathfrak{ins})$, and $target\,(e) := target\,(\mathfrak{ins})$. Together with the above sets of events (that are induced by the instructions of the processes), we define an additional type of events, namely update events as follows. For a process $p \in \mathsf{ProcSet}$, a variable $x \in \mathbb{X}$ and value $v \in V$, we define an event $\mathsf{ud}_p(x, v)$. We will use this event in the semantics to update the memory using the oldest message (x, v) in the buffer of process p. We define the sets $\Delta^{\mathsf{ud},x,v}_p := \{\mathsf{ud}_p(x, v)\}$, $\Delta^{\mathsf{ud},x}_p := \{\mathsf{ud}_p(x, v) \mid v \in V\}$, $\Delta^{\mathsf{ud}}_p := \cup_{x \in \mathbb{X}} \Delta^{\mathsf{ud},x}_p$, and $\Delta^{\mathsf{ud}} := \cup_{p \in \mathsf{ProcSet}} \Delta^{\mathsf{ud}}_p$. Furthermore, we define $\Delta^{\bullet}_p := \Delta_p \cup \Delta^{\mathsf{ud}}_p$ and $\Delta^{\bullet} := \Delta \cup \Delta^{\mathsf{ud}}$. In Fig. 2, let $\mathfrak{ins}_1$ (resp. $\mathfrak{ins}_2$) be the instruction labeled by q_1 (resp. s_3) in p_1 (resp. p_2). Then, $\langle \mathfrak{ins}_1, 1 \rangle \in \Delta^{\mathsf{wr},t,1}_{p_1}$, and $\langle \mathfrak{ins}_2, 1 \rangle \in \Delta^{\mathsf{rd},t,1}_{p_2}$. Furthermore, $\mathsf{ud}_{p_1}(x, 2) \in \Delta^{\mathsf{ud},x,2}_{p_1}$.

Transition Relation. We define the transition relation $\longrightarrow \subseteq C \times \Delta^\bullet \times C$ as follows. For configurations $c = \langle \bar{q}, \bar{r}, \bar{b}, mem \rangle$, $c' = \langle \bar{q}', \bar{r}', \bar{b}', mem' \rangle$, $p \in$ ProcSet, and $e \in \Delta_p^\bullet$, we write $c \xrightarrow{e} c'$ to denote that one of the following conditions holds:

- skip: $e \in \Delta_p^{\text{skip}}$, $\bar{q}(p) = source\,(e)$, $\bar{q}' = \bar{q}\,[p \hookleftarrow target\,(e)]$, $\bar{r}' = \bar{r}$, $\bar{b}' = \bar{b}$, and $mem' = mem$. The process jumps to a statement with the target label, while the register, buffer, and memory contents remain unchanged.
- write: $e \in \Delta_p^{\text{wr},x,v}$, $stmt\,(e)$ is of the form $x = \mathfrak{e}$ with $c(\mathfrak{e}) = v$, $\bar{q}(p) = source\,(e)$, $\bar{q}' = \bar{q}\,[p \hookleftarrow target\,(e)]$, $\bar{r}' = \bar{r}$, $\bar{b}' = \bar{b}\,[p \hookleftarrow (x,v) \cdot \bar{b}(p)]$, and $mem' = mem$.
- update: $e \in \Delta_p^{\text{ud},x,v}$, $\bar{q}' = \bar{q}$, $\bar{r}' = \bar{r}$, $\bar{b} = \bar{b}'\,\Big[p \hookleftarrow \bar{b}'(p) \cdot (x,v)\Big]$, and $mem' = mem\,[x \hookleftarrow v]$.
- read: $e \in \Delta_p^{\text{rd},x,v}$, $stmt\,(e)$ is of the form $\$r = x$, $\bar{q}(p) = source\,(e)$, $\bar{q}' = \bar{q}\,[p \hookleftarrow target\,(e)]$, $\bar{r}' = \bar{r}\,[\$r \hookleftarrow v]$, $\bar{b}' = \bar{b}$, $mem' = mem$, and one of the following conditions holds:
  - read-own-write: There is an $i : 1 \leq i \leq |\bar{b}(p)|$ s.t. $\bar{b}(p)[i] = (x,v)$, and there are no $1 \leq j < i$ and $v' \in V$ s.t. $\bar{b}(p)[j] = (x,v')$. If there is a write on x in the buffer of p then we consider the most recent of such a write (the left-most one in the buffer). This operation should assign v to x.
  - read-memory: $(x,v') \notin \bar{b}(p)$ for all $v' \in V$ and $mem(x) = v$. If there is no write operation on x in the buffer of p then the value v of x is fetched from the memory.
- fence: $e \in \Delta_p^{\text{fn}}$, $\bar{q}(p) = source\,(e)$, $\bar{q}' = \bar{q}\,[p \hookleftarrow target\,(e)]$, $\bar{r}' = \bar{r}$, $\bar{b}(p) = \epsilon$, $\bar{b}' = \bar{b}$, and $mem' = mem$. A fence operation may be performed by a process only if its buffer is empty.
- arw: $e \in \Delta_p^{\text{arw},x,v}$, $stmt\,(e)$ is of the form $\text{arw}(x, \mathfrak{e}_1, \mathfrak{e}_2)$ with $c(\mathfrak{e}_2) = v$, $\bar{q}(p) = source\,(e)$, $\bar{q}' = \bar{q}\,[p \hookleftarrow target\,(e)]$, $\bar{r}' = \bar{r}$, $\bar{b}(p) = \epsilon$, $\bar{b}' = \bar{b}$, $mem(x) = c(\mathfrak{e}_1)$, and $mem' = mem\,[x \hookleftarrow v]$. The arw operation is performed by a process only if its buffer is empty. The operation checks whether the value of x is equal to the evaluation of $\mathfrak{e}_1$ in c. In such a case, it changes that value to v.
- assume: $e \in \Delta_p^{\text{asm}}$, $stmt\,(e)$ is of the form asm $\mathfrak{e}$ with $c(\mathfrak{e}) = true$, $\bar{q}(p) = source\,(e)$, $\bar{q}' = \bar{q}\,[p \hookleftarrow target\,(e)]$, $\bar{r}' = \bar{r}$, $\bar{b}' = \bar{b}$, and $mem' = mem$. The instruction can be performed only if $\mathfrak{e}$ evaluates to true in c.
- assign: $e \in \Delta_p^{\text{asgn}}$, $stmt\,(e)$ is of the form $\$r = \mathfrak{e}$, $\bar{q}(p) = source\,(e)$, $\bar{q}' = \bar{q}\,[p \hookleftarrow target\,(e)]$, $\bar{r}' = \bar{r}\,[\$r \hookleftarrow c(\mathfrak{e})]$, $\bar{b}' = \bar{b}$, and $mem' = mem$. The content of $\$r$ is updated to the value of $\mathfrak{e}$.

A event $e \in \Delta^\bullet$ is said to be *enabled* from a configuration c if there is a configuration c' with $c \xrightarrow{e} c'$. If e is enabled from c, then we use $e(c)$ to denote the unique c' such that $c \xrightarrow{e} c'$. We define $\longrightarrow := \cup_{e \in \Delta^\bullet} \xrightarrow{e}$, and use $\xrightarrow{*}$ to denote the reflexive transitive closure of $\longrightarrow$.

Runs. A *run* π in $\mathcal{P}$ is a sequence of events $e_1 e_2 \cdots e_n \in (\Delta^\bullet)^*$. We generalize the notion of enabledness to runs, and say that π is *enabled* from a configuration c

(or simply π is a run *from* c) if $c_0 \xrightarrow{e_1} c_1 \xrightarrow{e_2} \cdots \xrightarrow{e_n} c_n$ for some configurations $c_0, c_1, \ldots, c_n$ with $c_0 = c$. In such a case, we define $\pi(c) := c_n$ (notice that c_n is unique given the configuration c and the run π). We write $c \xrightarrow{\pi} c'$ to denote that $\pi(c) = c'$. Notice that, for configurations c and c', we have that $c \xrightarrow{*} c'$ iff $c \xrightarrow{\pi} c'$ for some run π. For a run π, we define the relation $\mathsf{match}(\pi) \subseteq \mathbb{N} \times \mathbb{N}$ such that $\mathsf{match}(\pi)(j, k)$ $(1 \leq j < k \leq |\pi|)$ holds if and only if there is a process $p \in \mathsf{ProcSet}$ and $\ell \in \mathbb{N}$ where $\left(\pi \otimes \Delta_p^{\mathsf{wr}}\right)[\ell] = j$, and $\left(\pi \otimes \Delta_p^{\mathsf{ud}}\right)[\ell] = k$. In other words e_j is the ℓth write operation performed by p, and e_k is the (matching) ℓ^{th} update operation performed by p. We use Π to denote the set of all runs, and use $\Pi(c)$ to denote the set of runs from c. For a set $\Pi' \subseteq \Pi$, we define $\Pi'(c) := \Pi(c) \cap \Pi'$, i.e., it is the subset of runs in Π' that are enabled from c. For a set of runs $\Pi' \subseteq \Pi$ and a set of events $\Delta' \subseteq \Delta$, we use $\Pi'(\Delta') := \Pi' \cap (\Delta')^*$, i.e., it is the subset of π' that uses only events from Δ'. For instance, $\Pi\left(\Delta_p^{\mathsf{rd}}\right)$ is the set of all runs consisting only of read events performed by p, $\Pi\left(\Delta_p^{\mathsf{rd}}\right)(c)$ is the subset of the latter enabled from c, and $\Pi\left(\neg\Delta_p^{\mathsf{rd}}\right)$ is the set of all runs that do not contain read events performed by p. We say that π is *complete* if $|\pi \odot \Delta^{\mathsf{wr}}| = |\pi \odot \Delta^{\mathsf{ud}}|$, i.e., the numbers of write and update events in π are equal. We use $\Pi^{\mathsf{Complete}}(c)$ to denote the set of runs from c that are complete. Notice that if c is plain and $\pi \in \Pi^{\mathsf{Complete}}(c)$ then $\pi(c)$ is plain.

Fig. 4 depicts different runs from the configuration d of Fig. 9 in the program *simple* of Fig. 2. All these runs are complete. To simplify the presentation of the runs in Fig. 4, we represent an event in $\Delta^{\mathsf{wr},x,v}$ by the triple $\mathsf{wr}(x, v)$, and an event in $\Delta^{\mathsf{rd},x,v}$ by the triple $\mathsf{rd}(x, v)$. This does not introduce any ambiguity, since each instruction in the program *simple* has a unique statement. For instance, the event $\mathsf{wr}(t, 1)$ corresponds to the process p_1 performing the event labeled by q_1.

SC Semantics. We will define *SC* runs as special cases of *TSO* runs. For a process $p \in \mathsf{ProcSet}$, a run $\pi \in \Pi$ is said to be *Sequentially Consistent* (or *SC* for short) wrt. p if whenever $\pi[j] \in \Delta_p^{\mathsf{wr},x,v}$ then $1 \leq j < |\pi|$ and $\pi[j+1] \in \Delta_p^{\mathsf{ud},x,v}$. In other words, any write operation of the process p should be immediately followed by the matching update operation from the same process. We use Π_p^{SC} to denote the set of runs that are SC wrt. p. We say that π is *SC* if it is SC wrt. all processes $p \in \mathsf{ProcSet}$. We use Π^{SC} to denote the set of SC runs. We say that π is *singly TSO* wrt. $p \in \mathsf{ProcSet}$ if $\pi \in \Pi_r^{\mathsf{SC}}$ for all processes $r \in \mathsf{ProcSet} - \{p\}$, i.e., π is SC wrt. all processes except (possibly) p. We use $\Pi_p^{\mathsf{SinglyTSO}}$ to denote the set of runs that are singly TSO wrt. p. In Fig. 4, the run ρ_2 is SC, while ρ_1 is not singly TSO wrt. p_1 or p_2 (it is not SC wrt. p_1 or p_2).

Reachability Problem. An instance of the *reachability problem* is defined by a program and a finite set of label definitions Final. The question is whether there is a configuration c such that $c_{init} \xrightarrow{*} c$ for some configuration c with $\mathsf{LabelOf}(c) \in \mathsf{Final}$. Recall that $c_{init} \xrightarrow{*} c$ is equivalent to whether $c_{init} \xrightarrow{\pi} c$ for some run $\pi \in \Pi(c_{init})$. In the *SC reachability problem*, we restrict the program to SC runs, and ask whether there is a configuration c s.t. $c_{init} \xrightarrow{\pi} c$ for some c with $\mathsf{LabelOf}(c) \in \mathsf{Final}$ and $\pi \in \Pi^{\mathsf{SC}}(c_{init})$.

6 Persistence

We formulate the persistence problem by first introducing our notion of traces, and then comparing the set of traces of the program under TSO and SC. We explain the relation between persistence and correctness wrt. safety properties.

Traces. For a run π, we define the *program order* $\mathsf{ProgOrder}(\pi) : \mathsf{ProcSet} \mapsto \Delta^*$ by $\mathsf{ProgOrder}(\pi)(p) := \pi \odot \left(\Delta_p^{\mathsf{wr}} \cup \Delta_p^{\mathsf{rd}} \cup \Delta_p^{\mathsf{arw}} \right)$ for each $p \in \mathsf{ProcSet}$. In other words it extracts, for each process p, the sequence of write, read, and atomic-read-write events performed by the process. We define the *store order* by $\mathsf{StoreOrder}(\pi) := \pi \odot \left(\Delta^{\mathsf{ud}} \cup \Delta^{\mathsf{arw}} \right)$, i.e., it extracts the sequence of update and atomic-read-write events of all processes from π. Observe that, in the store order definition, we keep the two events that modify the memory. We define $\mathsf{Trace}(\pi)$ as $\langle \mathsf{ProgOrder}(\pi), \mathsf{StoreOrder}(\pi) \rangle$. Fig. 5 depicts a trace.

Persistence. Consider a plain configuration c and a complete run $\pi \in \Pi^{\mathsf{Complete}}(c)$. We say that π is *persistent* from c if there is an SC run $\pi' \in \Pi^{\mathsf{SC}}(c)$ with $\mathsf{Trace}(\pi) = \mathsf{Trace}(\pi')$; otherwise we say that π is *fragile* from c. We use $\Pi^{\mathsf{Persistent}}(c)$ and $\Pi^{\mathsf{Fragile}}(c)$ to denote the set of persistent and fragile runs from c respectively. A plain configuration c is said to be *persistent* if each complete run from c is persistent from c (i.e., $\Pi^{\mathsf{Persistent}}(c) \subseteq \Pi^{\mathsf{Complete}}(c)$); otherwise it is called *fragile*. An instance of *the persistence* problem defined on a program $\mathcal{P}$ asks whether the initial configuration c_{init} is persistent or not. In Fig. 4, the run ρ_1 is persistent from the configuration d (of Fig. 9), while ρ_3 is fragile from d.

Safety Properties. We can show that if a program is persistent then it is *strongly persistent*. This means that, for any plain configuration c and run π, it is the case that $c_{init} \xrightarrow{\pi} c$ iff $c_{init} \xrightarrow{\pi'} c$ from some $\pi' \in \Pi^{\mathsf{SC}}$. In other words, c is reachable from c_{init} under TSO iff it is reachable under SC. It is well-known that checking safety properties can be expressed as reachability of sets of (plain) configurations. This implies that a persistent program satisfies the same safety properties under SC and TSO.

7 Fragility Pattern

We perform the first step in solving the persistence problem. We show that the persistence problem can be reduced to searching for runs of a special form More precisely, for a given plain configuration c, there is a fragile run from c iff there is another run from c that will follow a certain *fragility pattern*.

Fix a program $\mathcal{P} = \langle \mathbb{X}, \mathsf{ProcSet}, \mathbb{Q}, \mathbb{I}, init \rangle$. For a plain configuration c, a process $p \in \mathsf{ProcSet}$, a variable $x \in \mathbb{X}$, and a value $v \in V$, we define $\Pi_{p,x,v}^{\circledast}(c)$ to be the set of runs π such that $\pi = \pi_1 \cdot \pi_2 \cdot \pi_3 \cdot \pi_4$, and the following conditions are satisfied: (i) $\pi \in \Pi^{\mathsf{SC}}(c)$. (ii) $\pi_2 = e_1 \cdot u_1$, where $e_1 \in \Delta_p^{\mathsf{wr},y}$ and $u_1 \in \Delta_p^{\mathsf{ud},y}$ for some $y \neq x$. (iii) $\pi_3 \in \Pi \left(\Delta_p - \left(\Delta_p^{\mathsf{wr}} \cup \Delta_p^{\mathsf{ud}} \cup \Delta_p^{\mathsf{arw}} \cup \Delta_p^{\mathsf{fn}} \right) \right)$, i.e., π_3 consists only of events performed by p excluding write, update, atomic-read-write, and fence events. (iv) $\pi_4 = e_2 \cdot u_2$, where $e_2 \in \Delta_r^{\mathsf{wr},x,v'}$ and $u_2 \in \Delta_r^{\mathsf{ud},x,v'}$ for some $r \neq p$ and $v' \neq v$. (v) There is an event $e \in \Delta_p^{\mathsf{rd},x,v}$, called the *complementary event* of π, such that $\pi_1 \cdot \pi_2 \cdot \pi_3 \cdot e \in \Pi(c)$, i.e., if we replace π_4 by e, then it results in a run from c. We call the process p the *pivot* of π, and call the rest of the processes the *fringe processes* of π. In §3, we motivate why $\rho_* \in \Pi_{p_1,x,2}^{\circledast}(d)$. We define $\Pi^{\circledast}(c) := \cup_{p \in \mathsf{ProcSet}} \cup_{x \in \mathbb{X}} \cup_{v \in V} \Pi_p^{\circledast}(c)$, and call $\Pi^{\circledast}(c)$ the set of runs of *type* $\circledast$ from c. We can show:

Theorem 1. $\Pi^{\circledast}(c) = \emptyset$ *iff* $\Pi^{\mathsf{Fragile}}(c) = \emptyset$.

In other words, to check whether c is fragile, we need only to check whether there is a run of type $\circledast$ from c. Although a run π of type $\circledast$ is SC, its existence together with the complementary event e show the fragility of c. More precisely, we define the *witness* run $\mathsf{Witness}(\pi) := \pi_1 \cdot e_1 \cdot \pi_3 \cdot e \cdot e_2 \cdot u_2 \cdot u_1$, and observe that $\mathsf{Witness}(\pi) \in \Pi^{\mathsf{Fragile}}(c)$. The run $\mathsf{Witness}(\pi)$ is TSO since e overtakes e_1. However, it is not persistent since there is no SC run with the same program and store order. In §3, we give the witness run corresponding to ρ_*.

Theorem 1 holds for any program respecting the syntax of §5. In particular, the program may contain any number of variables, and the variables may range over unbounded data domains (see §3, Model).

8   Pattern Detection

We perform the second step in solving the persistence problem. We translate the persistence problem for programs running under TSO to the SC reachability problem. We exploit Theorem 1 which shows that the persistence problem is reducible to the problem of checking whether the initial configuration has a type $\circledast$ run. Consider an instance of the persistence problem, defined by a program $\mathcal{P} = \langle \mathbb{X}, \mathsf{ProcSet}, \mathbb{Q}, \mathbb{I}, init \rangle$ and its initial configuration c_{init}. We will translate the problem of whether $\mathcal{P}$ has a run in $\Pi^{\circledast}(c_{init})$ to an instance of the SC reachability problem defined on a new program $\mathcal{Q} = \langle \mathbb{X}^{\mathcal{Q}}, \mathsf{ProcSet}^{\mathcal{Q}}, \mathbb{Q}^{\mathcal{Q}}, \mathbb{I}^{\mathcal{Q}}, init^{\mathcal{Q}} \rangle$.

We define $\mathsf{ProcSet}^{\mathcal{Q}} := \mathsf{ProcSet}$, and $init^{\mathcal{Q}} := init$, i.e., $\mathcal{Q}$ uses the same set of processes as $\mathcal{P}$ and the processes start from identical initial labels. Each process p in $\mathcal{Q}$ will simulate the corresponding process in $\mathcal{P}$. Recall that a run $\pi \in \Pi_{p,x,v}^{\circledast}(c)$ is of the form $\pi = \pi_1 \cdot e_1 \cdot u_1 \cdot \pi_3 \cdot e_2 \cdot u_2$, and it is defined in terms of a pivot p, and a complementary event e. We know that π induces a witness (fragile run) $\mathsf{Witness}(\pi) = \pi_1 \cdot e_1 \cdot \pi_3 \cdot e \cdot e_2 \cdot u_2 \cdot u_1$. A run of $\mathcal{Q}$ is divided into three phases: $0, 1, 2$, where each phase will accomplish a particular task in the simulation of $\mathsf{Witness}(\pi)$. To carry out the simulation we add two new constants, $\mathfrak{c}_1$ and $\mathfrak{c}_2$ to the domain V, and add two new variables, ph with domain $\{-1, 0, 1, 2\}$, and new with domain V to the set of variables. In other words, we define $\mathbb{X}^{\mathcal{Q}} := \mathbb{X} \cup \{\mathsf{ph}, \mathsf{new}\}$. Furthermore, we define $\mathbb{Q}^{\mathcal{Q}} := \mathbb{Q} \cup \{q' \mid q \in \mathbb{Q}\} \cup \{\mathsf{final}_p \mid p \in \mathsf{ProcSet}\} \cup \mathbb{Q}^{\mathsf{tmp}}$, and define $\mathsf{Final} := \cup_{p \in \mathsf{ProcSet}} \{c \mid \mathsf{LabelOf}(c)(p) = \mathsf{final}_p\}$. Intuitively, for each label q in $\mathcal{P}$, we create a copy q' (that will be used to simulate the moves of the pivot).

Furthermore, we provide each process with an additional label that marks its "accepting state". Finally, $\mathbb{Q}^{\mathrm{tmp}}$ contains a set of "temporary labels", each of which is either of the form $q.i$ or $q'.i$, where $q \in \mathbb{Q}$, and $i \in \mathbb{N}$. To simplify the definition of $\mathcal{Q}$, we will extend the set of expressions (see §5) by allowing the use of shared variables in expressions. Since the program $\mathcal{Q}$ runs under the SC semantics, the evaluation $c(x)$ of a shared variable x in a configuration c is straightforward (it is given by $\mathtt{MemoryOf}\,(c)\,(x)$).

Phase 0. Phase 0 corresponds to simulating π_1. During this phase, the processes will run the same code as in $\mathcal{P}$. A process needs to make sure that the program is currently running in phase 0. This is accomplished as follows. For each process $p \in$ ProcSet, and each instruction $\boxed{\texttt{q1: s; goto q2}}$ in $\mathbb{I}_p$,

```
q1: arw(ph,0,-1); goto q1.1
q1.1: s; goto q1.2
q1.2: ph=0; goto q2
```

Fig. 10. Phase 0

we add the instructions of Fig. 10 to $\mathbb{I}_p^{\mathcal{Q}}$. First, the process changes the phase to -1 (thus blocking the moves of all other processes). In the next step, the process simulates the given instruction, after which it puts the phase back to 0 (thus unblocking the rest of the processes).

At any point where a process p is about to execute a write event (on a variable y), it may decide to declare the event to be e_1. In this case, p will play the role of the pivot for the rest of the run, while deeming the other processes to become fringe

```
q1: arw(ph,0,-1); goto q1.3
q1.3: y=c1; goto q1.4
q1.4: ph=1; goto q'2
```

Fig. 11. Change to phase 1

processes. This signals the end of phase 0, and the phase is changed to 1. At the same time, we assign the value c_1 to y. The value of y will not be changed in the rest of the run, and y will be the only variable whose value is equal to c_1. Notice that, by definition of a type $\circledast$ run, y will not be used again in the rest of the run and therefore it is safe to change its value to c_1. To carry out the above steps, we add, for each write instruction $\mathsf{ins} \in \mathbb{I}_p$, of the form $\boxed{\texttt{q1: y=c; goto q2}}$ the instructions shown in Fig. 11 to $\mathbb{I}_p^{\mathcal{Q}}$.

Phase 1. The purpose of phase 1 is to simulate the π_3. Since π_3 consists only of events performed by the pivot, only this process is active during phase 1. Furthermore, the pivot only performs read, skip, assignment, and assume events.

```
q'1: assume ph=1; goto q'1.1
q'1.1: s; goto q'2
```

Fig. 12. Phase 1

For each instruction in $\mathbb{I}_p^{\mathrm{rd}} \cup \mathbb{I}_p^{\mathrm{skip}} \cup \mathbb{I}_p^{\mathrm{asgn}} \cup \mathbb{I}_p^{\mathrm{asm}}$, of the form $\boxed{\texttt{q1: s; goto q2}}$, we add the instructions of Fig. 12 to $\mathbb{I}_p^{\mathcal{Q}}$. Notice that, since only p is active during this phase, we need not lock the variable ph.

At any point where the pivot is about to execute a read event, it may decide to verify the existence of the complementary event e. Recall that $e \in \Delta_p^{\mathrm{rd},x,v}$, where $x \neq y$. For each read instruction $\mathsf{ins} \in \mathbb{I}_p^{\mathrm{rd},x}$, of the form $\boxed{\texttt{q1: \$r=x; goto q2}}$, we add the instructions shown in Fig. 13 to $\mathbb{I}_p^{\mathcal{Q}}$. The instruction at q'1.2 checks that the read variable is different from y. Recall that

y is the only variable whose value is c_1. At q'1.3, we store the current value of x in new (this will be used in phase 2). At q'1.4, the value c_2 is stored in x. Notice that x is now the only variable whose value is c_2. By this assignment, the pivot has declared (i) that it was about to perform a read on x; and (ii) that the value it was about to read is stored in new. At this point, the pivot changes the phase to 2. This is the last instruction executed by the pivot. Observe that the process does not execute

```
q'1: assume ph=1; goto q'1.2
q'1.2: assume !(x=c₁); goto q'1.3
q'1.3: new=x; goto q'1.4
q'1.4: x=c₂; goto q'1.5
q'1.5: ph=2; goto q'2
```

Fig. 13. Change to phase 2

the instruction ins itself, but it marks its existence through the instructions of Fig. 13.

Phase 2. In phase 2, only the fringe processes are active. The purpose of this phase is to verify the existence of the event e_2. This event is performed by a fringe process and it should write a value different from c_2 to x. A process that is about to

```
q1: assume ph=2; goto q1.5
q1.5: assume z=c₂; goto q1.6
q1.6: assume !(new=c); goto finalₚ
```

Fig. 14. Phase 2

execute a write event, may verify that this event corresponds to e_2. The process recognizes the variable x, since x is the only variable carrying the value c_2. For each write instruction of the form $\boxed{\text{q1: z=c; goto q2}}$ in $\mathbb{I}_p^{wr}$, we add the instructions shown in Fig. 14 to $\mathbb{I}_p^{\mathcal{Q}}$. The test $z = c_2$ ensures that z and x are identical. The test $!(\text{new} = c)$ ensures that the current instruction assigns a value different from the value of x during the complementary event. In such a case, a witness has been found and the process moves to the accepting label final_p.

Remarks. Notice that $\mathcal{Q}$ contains only two additional variables, namely new and ph, compared to $\mathcal{P}$; and that we increase the variable domain V by two elements, namely c_1 and c_2.

9   Fence Insertion

In this section we describe our fence insertion procedure that finds a minimal set of fences sufficient for making the program persistent. The algorithm builds a set of fences successively using fragile runs generated by the pattern detection algorithm. First, we define the fence insertion operation, and then show how to use a type ⊛ run generated by the pattern detection algorithm of §8 to derive a set of fences such that the insertion of at least one element of the set is necessary in order to eliminate the run from the behavior of the program. Based on that, we introduce the fence insertion algorithm.

Fence Insertion. We define the operation of inserting a fence in a program. Intuitively, we identify the instruction after which we insert the fence. For a program $\mathcal{P} = \langle \mathbb{X}, \text{ProcSet}, \mathbb{Q}, \mathbb{I}, \textit{init} \rangle$ and an instruction $f \in \mathbb{I}$, we use $\mathcal{P} \oplus f$ to denote the program we get by inserting a fence instruction just after $\mathfrak{f}$ in $\mathcal{P}$. Formally, let $\mathfrak{f} \in \mathbb{I}$ be of the form $\boxed{q_1 : \mathfrak{s}; \texttt{goto } q_2}$. Then, $\mathcal{P} \oplus \mathfrak{f}$ is the program we get by replacing $\mathfrak{f}$ by the following two instructions (where $q' \notin \mathbb{Q}$ is a unique new label): $\boxed{q_1 : \mathfrak{s}; \texttt{goto } q'}$, and $\boxed{q' : \texttt{fn}; \texttt{goto } q_2}$ (recall from §5 that fn is the fence statement). For a set $F = \{\mathfrak{f}_1, \ldots \mathfrak{f}_n\} \subseteq \mathbb{I}$, we define $\mathcal{P} \oplus F := \mathcal{P} \oplus \mathfrak{f}_1 \cdots \oplus \mathfrak{f}_n$. We say F is *minimal* wrt. $\mathcal{P}$ if (i) $\mathcal{P} \oplus F$ is persistent, and (ii) $\mathcal{P} \oplus (F \setminus \{\mathfrak{f}\})$ is fragile for all $\mathfrak{f} \in F$. That is, removing any fence from F makes $\mathcal{P}$ fragile.

Fence Inference. Let $p \in \mathsf{ProcSet}$, and let π be a type ⊛ run generated by applying an arbitrary SC reachability analysis algorithm on the program $\mathcal{Q}$ defined in §8. Recall from §7, that from π, we can derive a new run $\mathsf{Witness}(\pi)$ of the form $\pi_1 \cdot e_1 \cdot \pi_3 \cdot e \cdot e_2 \cdot u_2 \cdot u_1$, then $\mathsf{Witness}(\pi) \in \Pi^{\mathsf{Fragile}}(c)$, where e overtakes e_1. Let π_3 be of the form $e'_1 e'_2 \cdots e'_n$, and define $\mathsf{NewFences}(\pi) := \{e_1, e'_1, e'_2, \ldots, e'_n\}$. Intuitively, inserting a fence after one of the instructions in $\mathsf{NewFences}(\pi)$ is both necessary and sufficient to prevent e from overtaking e_1, and hence eliminating the run π from the behavior of the program.

Algorithm. We present our fence insertion algorithm (Algorithm 1). It takes a concurrent program $\mathcal{P}$ and returns a minimal set of fences that is sufficient to make $\mathcal{P}$ persistent. The algorithm uses a set of sets of fences, namely $\mathcal{W}$ for sets of fences that have been *partially* constructed (but not yet large enough to make the program persistent). During each iteration, a set F is picked and removed from $\mathcal{W}$. We use the construction of §8, together with an SC reachability analysis algorithm, to check whether the set F is sufficient to make the program persistent. If *yes*, we return F as a possible set of minimal fences. If *no*, we compute the set of fences N such that inserting a member of N is sufficient and necessary to eliminate the generated type ⊛ run π. For each $\mathfrak{f} \in N$ we add $F' = F \cup \{\mathfrak{f}\}$ back to $\mathcal{W}$ unless there is already a subset of F' in the set $\mathcal{W}$.

Algorithm 1. Fence Insertion.

input : A concurrent program $\mathcal{P}$
output: A minimal set of fences

1 $\mathcal{W} \leftarrow \{\emptyset\}$;
2 **while** *true* **do**
3 $\quad$ Pick and remove a set F from $\mathcal{W}$;
4 $\quad$ **if** $\exists \pi : \pi \in \Pi_p^\circledast (c_{init})$ *in* $\mathcal{P} \oplus F$ **then**
5 $\quad\quad$ N $\leftarrow$ NewFences(π);
6 $\quad\quad$ **foreach** $\mathfrak{f} \in$ N **do**
7 $\quad\quad\quad$ $F' \leftarrow F \cup \{\mathfrak{f}\}$;
8 $\quad\quad\quad$ **if**
$\quad\quad\quad$ $\exists F'' \in \mathcal{W} : F'' \subseteq F'$
$\quad\quad\quad$ **then discard** F' **else**
$\quad\quad\quad$ $\mathcal{W} \leftarrow \mathcal{W} \cup \{F'\}$
9 $\quad$ **else**
10 $\quad\quad$ **return** F;

Theorem 2. *If each call to the pattern detection algorithm (line 4) returns, then Algorithm 1 terminates and returns a minimal set of fences wrt.* $\mathcal{P}$.

In particular, Theorem 2 implies that Algorithm 1 terminates when $\mathcal{P}$ is a finite-state program.

In the program of Fig. 2, our algorithm would insert a single fence, replacing the instruction $\boxed{\texttt{q2. z=1; goto q3}}$ in p_1 by the instructions $\boxed{\texttt{q2. z=1; goto q5}}$ and $\boxed{\texttt{q5. fn; goto q3}}$.

10   Observation Abstraction

In this section, we present a general abstraction framework, called *observation abstraction*, that is compatible with the notion of persistence (*compatibility* means that persistence of the abstract program implies persistence of the concrete program). The abstraction considers a process p and captures the sequences of events that can be observed from p. We instantiate observation abstraction by defining an abstraction function, namely *Flattening*, whose efficiency is demonstrated in the experimental results (see §11). Let us fix a program $\mathcal{P} = \langle \mathbb{X}, \mathsf{ProcSet}, \mathbb{Q}, \mathbb{I}, init \rangle$ and a process $p \in \mathsf{ProcSet}$. Let $R := \mathsf{ProcSet} \setminus \{p\}$.

Notation. We define $\mathsf{ReadFrom}_p := \left\{ x \in \mathbb{X} \mid \left(\Delta_p^{\mathsf{rd},x} \cup \Delta_p^{\mathsf{arw},x} \right) \neq \emptyset \right\}$, i.e., it is the set of shared variables on which p may perform read or atomic-read-write events. We define a new type of events, namely $\mathsf{MemEvent}(x, v)$, where $x \in \mathbb{X}$ and $v \in V$, that we use to abstract update and atomic-read-write events. The event records changes in the state of the memory (changing the value of x to v), while hiding the identity of the process performing the event. For a variable $x \in \mathbb{X}$, a value $v \in V$, an event $e \in \Delta^{\mathsf{arw},x,v} \cup \Delta^{\mathsf{ud},x,v}$, and a process $p \in \mathsf{ProcSet}$, we will write $[e]_p$ to describe "how much of e can be observed by p". Formally: (i) $[e]_p := e$ if $e \in \Delta_p$, i.e., we keep the event if it is performed by p (p can observe its own events). (ii) $[e]_p := \mathsf{MemEvent}(x, v)$ if $e \in \Delta_r^{\mathsf{arw},x,v} \cup \Delta_r^{\mathsf{ud},x,v}$, for some $r \neq p$ and $x \in \mathsf{ReadFrom}_p$. If the event is performed by another process on a variable in $\mathsf{ReadFrom}_p$ then p can observe the change in memory although it cannot see the process making the change. (iii) $[e]_p := \epsilon$ if $e \in \Delta_r^{\mathsf{arw},x} \cup \Delta_r^{\mathsf{ud},x}$, for some $r \neq p$ and $x \notin \mathsf{ReadFrom}_p$. The event is not observed by p in case it is performed by another process on a variable not in $\mathsf{ReadFrom}_p$. For a run $\pi = e_1 e_2 \cdots e_n \in \left(\Delta^{\mathsf{arw},x} \cup \Delta^{\mathsf{ud},x} \right)^*$, we define $[\pi]_p := [e_1]_p [e_2]_p \cdots [e_n]_p$.

Framework. In addition to (the concrete program) $\mathcal{P}$ with initial configuration c_{init}, we consider an abstract program $\mathcal{A}_p = \left\langle \mathbb{X}^{\mathcal{A}_p}, \mathsf{ProcSet}^{\mathcal{A}_p}, \mathbb{Q}^{\mathcal{A}_p}, \mathbb{I}^{\mathcal{A}_p}, init^{\mathcal{A}_p} \right\rangle$, with initial configuration $c_{init}^{\mathcal{A}_p}$. We assume that the following conditions hold: (i) $\mathbb{X}^{\mathcal{A}_p} = \mathbb{X}$, i.e., $\mathcal{P}$ and $\mathcal{A}_p$ operate on the same set $\mathbb{X}$ of shared variables. (ii) The process p is in $\mathsf{ProcSet}^{\mathcal{A}_p}$. (iii) $\mathbb{I}_p = \mathbb{I}_p^{\mathcal{A}_p}$, i.e., the process p executes the same code in $\mathcal{P}$ and $\mathcal{A}_p$. We say that $\mathcal{A}_p$ is an observation abstraction of $\mathcal{P}$ wrt. p, denoted $\mathcal{P} \sqsubseteq_p \mathcal{A}_p$, if for every run $\pi \in \Pi_p(c_{init})$ in $\mathcal{P}$, there is a run $\pi' \in \Pi_p\left(c_{init}^{\mathcal{A}_p} \right)$ in $\mathcal{A}_p$ such that the following two conditions are satisfied: (i) $\mathsf{ProgOrder}(\pi)(p) = \mathsf{ProgOrder}(\pi')(p)$. (ii) $[\mathsf{StoreOrder}(\pi)]_p = [\mathsf{StoreOrder}(\pi')]_p$. In other words, the programs $\mathcal{P}$ and $\mathcal{A}_p$ agree on the "parts of runs" that are observable by p: on the one hand, p observes all events it performs itself; on the other hand, it observes modifications of the memory performed by other processes provided that they concern variables from which it can read (in the latter case, the actual identity of the process performing the event is not relevant).

The next theorem shows that persistence can be established by analyzing the abstract programs.

Theorem 3. *For every $p \in$ ProcSet, let $\mathcal{A}_p$ be an abstract program such that $\mathcal{P} \sqsubseteq_p \mathcal{A}_p$. If $\mathcal{A}_p$ is persistent for all $p \in$ ProcSet then $\mathcal{P}$ is persistent.*

Flattening. We define a family of functions that build observation abstractions $\mathcal{A}_p$ of $\mathcal{P}$ wrt. p for $k \in \mathbb{N}$. Flattening keeps all processes in $\mathcal{P}$ applying abstraction individually on each process. Although the number of processes remains the same, the abstraction simplifies the behavior of each process, again limiting the state explosion problem. The precision of the abstraction increases with increasing the value of k. More precisely, we keep the behavior of a process r during its first k steps after which we replace each statement, except write and atomic-read-write statements on variables in ReadFrom$_p$, by the empty statement. Flattening abstraction is precise in the sense that it does not add extra fences (compared to the set of fences that would be added to the concrete program). In our experiments, no additional fences are added when applying this abstraction.

```
process p1
regs   init s
begin
  q1.  t=1; goto q2
  q2.  skip; goto q2
  q2.  skip; goto q3
  q3.  skip; goto q4
end
```

Fig. 15. Flattening abstraction of the program of Fig. 2 wrt. p_2

11   Experimental Results

Tool. We have implemented our techniques from §8-§10 in an open-source tool called Persist. The syntax of the input language of Persist is defined by the grammar of Fig. 8 in §5. Persist uses SPIN [16] as backend model-checker to solve the SC reachability queries for programs under SC. Since SPIN is a finite-state model checker, all the programs in our experiments are finite-state. We compare our method with state-of-the-art tools[1]: Trencher [9] (a sound and complete tool for robustness analysis of finite-state programs under TSO, that uses SPIN as backend tool), Memorax [1] (a sound and complete tool for the correctness analysis of finite-state programs under TSO wrt. safety properties), Remmex [20] (a tool based on state-space exploration with acceleration, for correctness analysis of programs under TSO wrt. to safety properties), and Musketeer [3] (a static analysis tool for correctness analysis of programs under weak memory models wrt. robustness). We perform the comparisons based on two aspects, namely the *number of fences* (and their placement) and the *running time*. The experiments are performed on an Intel x86-32 Core2 2.4 Ghz machine and 4GB of RAM. To insert the fences, Memorax and Remmex require a safety property as an additional input which is not always given; while Persist, Trencher, and Musketeer are fully automatic. Notice that both persistence and robustness guarantee SEQCON.

In the following, we present two sets of results. The first set concerns the comparison of Persist (without the abstraction) with the other tools (see Table 1). The second set shows the scalability of Persist (with/without abstraction) compared to Musketeer and Trencher when increasing the number of processes (see

---

[1] Except Dfence [21] which requires a special manual specification encoding for each example.

Fig. 16). In all experiments, we set up the time out to 2400 seconds, and $k = 2$ for abstraction. Our examples are from [21,9,1,20,3].

Performance of Persist without abstraction. The results are given in Table 1. Below, we summarise the main observations: (i) Persist manages to return for all the benchmarks. Trencher and Musketeer fail to answer for 8 out of 35 examples within 2400 seconds. Memorax (Remmex) manages to return for only 4 (9) out of 30 examples within 2400 seconds. For the comparison among tools, we compute the average of the ratio of the running times over examples where the tools terminate. On average Persist is 51 times faster than Trencher and 2.48 times faster than Musketeer. (ii) Persist returns 54 fences while Musketeer returns 119 fences for all the examples where Musketeer terminates within 2400 seconds (and thus, Persist inserts 54% less fences in total). Furthermore, Persist returns 44 fences while Trencher returns 62 fences for all the examples

Table 1. Experimental results. Per, Tre, Mus, Mem, and Rem stand for Persist, Trencher, Musketeer, Memorax, and Remmex, respectively. The columns #P, #F, and #T give the number of processes, number of fences, and running time in seconds, respectively. If a tool runs out of time (resp. memory), we put "TO" (resp. "OM") in the #T column, and • in #F column. We use "−" when a tool is not tested due to a missing specification.

Program	#P	Per		Tre		Mus		Mem		Rem	
		#F	#T	#F	#T	#F	#T	#F	#T	#F	#T
SimDekker	4	4	4	4	163	8	1	•	OM	•	OM
Dekker	4	8	14	•	TO	16	783	•	OM	•	OM
Peterson	4	4	223	•	TO	•	TO	•	OM	•	OM
LamBak	3	6	104	•	TO	18	110	•	OM	6	372
LamBak	4	8	286	•	TO	•	TO	•	OM	•	OM
Dijkstra	4	8	30	•	TO	•	TO	•	OM	•	OM
Dc-Lock	6	0	7	0	139	0	1	•	TO	0	5
SpinLock	2	0	1	0	1	0	1	0	1	0	1
TSpinLock	2	0	1	0	1	0	1	•	OM	0	4
InlinePgsql	8	0	8	0	426	•	TO	•	OM	0	42
Burns	5	9	119	•	TO	•	TO	•	OM	•	TO
Szymaski	2	8	3	8	642	11	1	3	1	3	4
Szymaski	4	16	88	•	TO	•	TO	•	OM	•	TO
LamFast2	2	8	6	4	48	15	1	4	136	4	6
LamFast2	3	12	10	•	TO	•	TO	•	OM	6	108
CLH Lock	4	4	176	4	674	•	TO	•	OM	•	OM
Parker	6	1	76	1	425	3	1	•	OM	•	OM
Pgsql	6	7	26	7	260	7	1	•	OM	•	TO
AltenatingBit	2	4	2	4	5	5	1	0	4	0	3
IncSequence	6	0	21	0	194	0	1	•	OM	•	TO
TaskSchedule	10	0	5	0	418	0	1	•	TO	•	OM
NBW_W_WR	2	0	18	1	595	6	1	•	OM	•	TO
NBW_W_5R	6	0	3	0	514	0	1	•	OM	•	TO
SeqLock	4	0	63	0	364	0	1	•	OM	•	TO
write+r	5	0	2	4	7	4	1	•	OM	•	TO
r+detour	5	0	1	3	3	3	1	•	OM	•	TO
r+detours	5	0	1	3	1	3	1	•	OM	•	TO
write+r+co	6	0	7	4	38	4	1	•	OM	•	TO
sb+detours	6	2	3	5	3	5	1	•	OM	•	TO
sb+detour+co	6	0	1	4	3	4	1	•	OM	•	TO
Cilk WSQ	2	2	3	2	107	3	1	−	−	−	−
CL WSQ	2	1	2	1	796	1	1	−	−	−	−
FIFO iWSQ	2	1	2	1	354	1	1	−	−	−	−
LIFO iWSQ	2	1	7	1	558	1	1	−	−	−	−
Anchor iWSQ	2	1	2	1	20	1	1	−	−	−	−

where Trencher terminates within 2400 seconds (and thus, Persist inserts 30% less fences in total).

Scalability wrt. the number of processes. We compare the scalability of Trencher, Musketeer, Persist, and our abstraction Flattening) while increasing the number of processes in several examples. In all these examples, no additional fences are added due to flattening abstraction (compared to the case of Persist without abstraction). The results are given in Fig. 16. We observe that: (i) Persist without

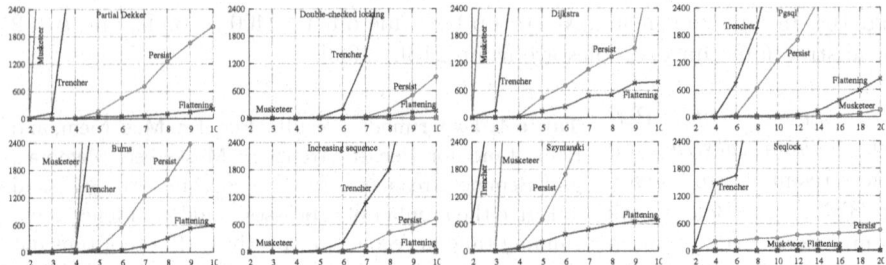

Fig. 16. Persist with/without abstraction compared with Trencher and Musketeer. The x axis is number of processes, the y axis is running time in seconds.

abstraction always scales better than Trencher for all the examples. Furthermore, Persist without abstraction scales better than Musketeer in 4 out of 8 examples (Partial Dekker, Burns, Dijkstra, Szymanski). (ii) Persist with the abstraction always performs better than Persist without the abstractions while inserting no extra fences. (iii) Persist with the abstraction outperforms Musketeer in 4 out of 8 examples (Partial Dekker, Burns, Dijkstra, Szymanski) while having comparable running time for the rest.

In all the reported results (Table 1 and Fig. 16), except *Lamport Fast 2 with two processes (LamFast2)*, the set of fences returned by Persist is a subset of the ones returned by Trencher and Musketeer. Hence, Persist presents a good trade-off between efficiency and optimality.

12   Conclusion, Discussion, and Future Work

We have presented a framework for automatic fence insertion under TSO that provides an excellent trade-off between efficiency and optimality. We have implemented our framework in a tool and evaluated it on a wide range of benchmarks. The correctness criteria of Fig. 1, namely Data Race Freedom (DRF), Triangular Race Freedom (TRF), Robustness (ROB), Persistence (PER), Sequential Consistency (SEQCON), and state Reachability (REACH) can be seen as "stability conditions", in the sense that they measure how stable the behaviors of the program is

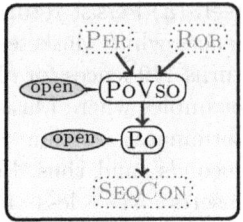

Fig. 17. Relevant correctness criteria

under TSO compared to SC. Only a small number of stability conditions are relevant, since each condition corresponds to relaxing one of three parameters: program order, source, and (variable) store order. A stability condition should imply that the program under SC and TSO have the same reachable (control) states, and that they satisfy the same safety properties (see §1, Persistence; and §6, Safety Properties). We believe that our work is an important step in this investigation. There are several open and hard questions to consider in future work (see Fig. 17). This includes studying two remaining important stability conditions, namely POVSO where a program is considered to be correct if the traces of the program under TSO and SC agree on (i) program order, and (ii) variable store (coherence) order. PoVso gives an entirely different stability condition compared to persistence (in fact, PoVso is a weakening of both robustness and persistence). Checking PoVso (e.g., through finding appropriate patterns) is an important (and difficult) open problem. Notice that, if persistence and PoVso were equivalent, then robustness would be stronger than persistence which is not the case (they are incomparable). Another open problem is checking the condition Po, a weakening of PoVso, where the program is considered if its TSO and SC traces need only to agree on program order. Finally, it is important to develop frameworks that allow checking the different stability conditions for other weak memory models, as done in [4].

Acknowledgment. The authors would like to thank Jade Alglave, Ahmed Bouajjani, Roland Meyer, Madan Musuvathi and Viktor Vafeiadis for helpful discussions. We thank Madan Musuvathi again for pointing us to some benchmarks. We also thank Carl Leonardsson, Jade Alglave, Egor Derevenetc, Daniel Kroening, Alexander Linden, Roland Meyer, and Martin Vechev for giving us access to their tools and their benchmarks.

References

1. Abdulla, P.A., Atig, M.F., Chen, Y.-F., Leonardsson, C., Rezine, A.: Counterexample guided fence insertion under TSO. In: Flanagan, C., König, B. (eds.) TACAS 2012. LNCS, vol. 7214, pp. 204–219. Springer, Heidelberg (2012)
2. Adve, S.V., Gharachorloo, K.: Shared memory consistency models: A tutorial. Computer 29(12), 66–76 (1996)
3. Alglave, J., Kroening, D., Nimal, V., Poetzl, D.: Don't sit on the fence. In: Biere, A., Bloem, R. (eds.) CAV 2014. LNCS, vol. 8559, pp. 508–524. Springer, Heidelberg (2014)
4. Alglave, J., Kroening, D., Nimal, V., Tautschnig, M.: Software verification for weak memory via program transformation. In: Felleisen, M., Gardner, P. (eds.) ESOP 2013. LNCS, vol. 7792, pp. 512–532. Springer, Heidelberg (2013)
5. Alglave, J., Maranget, L.: Stability in weak memory models. In: Gopalakrishnan, G., Qadeer, S. (eds.) CAV 2011. LNCS, vol. 6806, pp. 50–66. Springer, Heidelberg (2011)
6. Atig, M.F., Bouajjani, A., Burckhardt, S., Musuvathi, M.: On the verification problem for weak memory models. In: Hermenegildo, M.V., Palsberg, J. (eds.) POPL 2010, pp. 7–18. ACM (2010)
7. Atig, M.F., Bouajjani, A., Burckhardt, S., Musuvathi, M.: What's decidable about weak memory models? In: Seidl, H. (ed.) ESOP 2012. LNCS, vol. 7211, pp. 26–46. Springer, Heidelberg (2012)
8. Atig, M.F., Bouajjani, A., Parlato, G.: Getting rid of store-buffers in TSO analysis. In: Gopalakrishnan, G., Qadeer, S. (eds.) CAV 2011. LNCS, vol. 6806, pp. 99–115. Springer, Heidelberg (2011)
9. Bouajjani, A., Derevenetc, E., Meyer, R.: Checking and enforcing robustness against TSO. In: Felleisen, M., Gardner, P. (eds.) ESOP 2013. LNCS, vol. 7792, pp. 533 553. Springer, Heidelberg (2013)

10. Bouajjani, A., Meyer, R., Möhlmann, E.: Deciding robustness against total store ordering. In: Aceto, L., Henzinger, M., Sgall, J. (eds.) ICALP 2011, Part II. LNCS, vol. 6756, pp. 428–440. Springer, Heidelberg (2011)
11. Burckhardt, S., Musuvathi, M.: Effective program verification for relaxed memory models. In: Gupta, A., Malik, S. (eds.) CAV 2008. LNCS, vol. 5123, pp. 107–120. Springer, Heidelberg (2008)
12. Burnim, J., Sen, K., Stergiou, C.: Sound and complete monitoring of sequential consistency for relaxed memory models. In: Abdulla, P.A., Leino, K.R.M. (eds.) TACAS 2011. LNCS, vol. 6605, pp. 11–25. Springer, Heidelberg (2011)
13. Duan, Y., Muzahid, A., Torrellas, J.: Weefence: toward making fences free in TSO. In: Mendelson, A. (ed.) ISCA 2013, pp. 213–224. ACM (2013)
14. Fraser, K.: Practical lock-freedom. Tech. Rep. UCAM-CL-TR-579, University of Cambridge, Computer Laboratory (2004)
15. Gharachorloo, K., Gupta, A., Hennessy, J.: Performance evaluation of memory consistency models for shared-memory multiprocessors. SIGPLAN Not. 26(4), 245–257 (1991)
16. Holzmann, G.: SPIN Model Checker, the: Primer and Reference Manual, 1st edn. Addison-Wesley Professional (2003)
17. Kuperstein, M., Vechev, M.T., Yahav, E.: Automatic inference of memory fences. In: Bloem, R., Sharygina, N. (eds.) FMCAD 2010, pp. 111–119. IEEE (2010)
18. Kuperstein, M., Vechev, M.T., Yahav, E.: Partial-coherence abstractions for relaxed memory models. In: Hall, M.W., Padua, D.A. (eds.) PLDI 2011, pp. 187–198. ACM (2011)
19. Leijen, D., Schulte, W., Burckhardt, S.: The design of a task parallel library. In: Arora, S., Leavens, G.T. (eds.) OOPSLA 2009, pp. 227–242. ACM (2009)
20. Linden, A., Wolper, P.: A verification-based approach to memory fence insertion in relaxed memory systems. In: Groce, A., Musuvathi, M. (eds.) SPIN Workshops 2011. LNCS, vol. 6823, pp. 144–160. Springer, Heidelberg (2011)
21. Liu, F., Nedev, N., Prisadnikov, N., Vechev, M.T., Yahav, E.: Dynamic synthesis for relaxed memory models. In: Vitek, J., Lin, H., Tip, F. (eds.) PLDI 2012, pp. 429–440. ACM (2012)
22. Marino, D.L.: Simplified Semantics and Debugging of Concurrent Programs via Targeted Race Detection. Ph.D. thesis, University of California at Los Angeles, Los Angeles, CA, USA (2011)
23. Michael, M.M., Vechev, M.T., Saraswat, V.A.: Idempotent work stealing. In: Reed, D.A., Sarkar, V. (eds.) PPOPP 2009, pp. 45–54. ACM (2009)
24. Owens, S.: Reasoning about the implementation of concurrency abstractions on x86-TSO. In: D'Hondt, T. (ed.) ECOOP 2010. LNCS, vol. 6183, pp. 478–503. Springer, Heidelberg (2010)
25. Owens, S., Sarkar, S., Sewell, P.: A better x86 memory model: x86-TSO. In: Berghofer, S., Nipkow, T., Urban, C., Wenzel, M. (eds.) TPHOLs 2009. LNCS, vol. 5674, pp. 391–407. Springer, Heidelberg (2009)
26. Sewell, P., Sarkar, S., Owens, S., Nardelli, F.Z., Myreen, M.O.: X86-TSO: A rigorous and usable programmer's model for x86 multiprocessors. Commun. ACM 53(7), 89–97 (2010)
27. Shasha, D., Snir, M.: Efficient and correct execution of parallel programs that share memory. ACM Trans. Program. Lang. Syst. 10(2), 282–312 (1988)

Specifying and Verifying Concurrent Algorithms with Histories and Subjectivity

Ilya Sergey, Aleksandar Nanevski, and Anindya Banerjee

IMDEA Software Institute, Spain
{ilya.sergey,aleks.nanevski,anindya.banerjee}@imdea.org

Abstract. We present a lightweight approach to Hoare-style specifications for fine-grained concurrency, based on a notion of *time-stamped histories* that abstractly capture atomic changes in the program state. Our key observation is that histories form a *partial commutative monoid*, a structure fundamental for representation of concurrent resources. This insight provides us with a unifying mechanism that allows us to treat histories just like heaps in separation logic. For example, both are subject to the same assertion logic and inference rules (*e.g.*, the frame rule). Moreover, the notion of ownership transfer, which usually applies to heaps, has an equivalent in histories. It can be used to formally represent helping—an important design pattern for concurrent algorithms whereby one thread can execute code on behalf of another. Specifications in terms of histories naturally abstract away the internal interference, so that sophisticated fine-grained algorithms can be given the same specifications as their simplified coarse-grained counterparts, making them equally convenient for client-side reasoning. We illustrate our approach on a number of examples and validate all of them in Coq.

1 Introduction

For sequential programs and data structures, Hoare-style specifications (or specs) in the form of pre- and postconditions are a declarative way to express a program's behavior. For example, an abstract specification of stack operations can be given as follows:

$$\{ s \mapsto xs \} \; \mathsf{push}(x) \; \{ s \mapsto x :: xs \}$$

$$\{ s \mapsto xs \} \quad \mathsf{pop}() \quad \left\{ \begin{array}{l} \mathsf{res} = \mathsf{None} \wedge xs = \mathsf{nil} \wedge s \mapsto \mathsf{nil} \; \vee \\ \mathsf{res} = \mathsf{Some}\, x \wedge \exists xs',\, xs = x :: xs' \wedge s \mapsto xs' \end{array} \right\} \tag{1}$$

where s is an "abstract pointer" to the data structure's logical contents, and the logical variable xs is universally quantified over the spec. The result res of pop is either $\mathsf{Some}\, x$, if x was on the top of the stack, or None if the stack was empty. The spec (1) is usually accepted as canonical for stacks: it hides the details of method implementation, but exposes what is important about the method behavior, so that a verification of a stack *client* does not need to explore the implementations of push and pop.

The situation is much more complicated in the case of concurrent data structures. In the concurrent setting, the spec (1) is of little use for implementations with server-side locking, as the interference of the threads executing concurrently may invalidate the assertions about the stack. For example, a call to pop may encounter an empty stack, and decide to return None, but by the time it returns, the stack may be filled by the other threads, thus invalidating the postcondition of pop in (1). To soundly reason

© Springer-Verlag Berlin Heidelberg 2015
J. Vitek (Ed.): ESOP 2015, LNCS 9032, pp. 333–358, 2015.
DOI: 10.1007/978-3-662-46669-8_14

about concurrent data structures, one has to devise specs that are *stable* (*i.e.*, invariant under interference), but this may require trade-offs with respect to the specifications' expressivity and precision for the client's needs.

Reasoning about concurrent data structures is further complicated by the fact that their implementations are often *fine-grained*. Striving for better performance, they avoid explicit locking, and implement sophisticated synchronization patterns that deliberately rely on interference. For reasoning purposes, however, it is desirable that the clients can perceive such fine-grained implementations as if they were *coarse-grained*; that is, as if the effects of their methods take place *atomically*, at singular points in time. The standard correctness criteria of *linearizability* [16] establishes that a fine-grained data structure implementation *contextually refines* a coarse-grained one [10]. One can make use of a refined, fine-grained, implementation for efficiency in programming, but then soundly replace it with a more abstract coarse-grained implementation to simplify the reasoning about clients.

Semantically, one program linearizes to another if the *histories* of the first program (*i.e.*, the sequence of actions it executed) can be transformed, in a suitable sense, into the histories of the second. Thus, histories are an essential ingredient in specifying fine-grained concurrent data structures. However, while a number of logical methods exist for establishing the linearizability relation between two programs, for a class of data structures [7, 20, 24, 33], in general, it is a non-trivial property to prove and use. First, in a setting that employs Hoare-style reasoning, showing that a fine-grained structure refines a coarse-grained one is not an end in itself. One still needs to ascribe a stable spec to the coarse-grained version [20, 31]. Second, the standard notion of linearizability does not directly account for modern programming features, such as ownership transfer of state between threads, pointer aliasing, and higher-order procedures. Theoretical extensions required to support these features are a subject of active ongoing research [4, 13]. Finally, being a relation on *two* programs, deriving linearizability by means of logical inference inherently requires a *relational program logic* [20, 31], even though the spec one is ultimately interested in may be expressed using a Hoare triple that operates over a *single* program.

In this paper, we propose a novel method to specify and verify fine-grained programs by directly reasoning about histories in the specs of an elementary Hoare logic. We propose using *timestamped* histories, which carry information about the atomic changes in the abstract state of the program, indexed by discrete timestamps, and tracking the history of a program as a form of auxiliary state.

While using histories in Hoare-style specs is a simple and natural idea, and has been used before [1, 11, 12], in our paper it comes with two additional novel observations.

First, timestamped histories are technically very similar to heaps, as both satisfy the algebraic properties of a *partial commutative monoid* (PCM). A PCM is a set $\mathbb{U}$ with an associative and commutative *join* operation $\bullet$ and unit element $\mathbb{1}$. Both heaps and histories (considered as sets of actions with distinct timestamps, correspondingly) form a PCM with disjoint union and empty heap/history as the unit. Also, a singleton history $t \mapsto a$ is similar to the singleton heap $x \mapsto v$ containing only the pointer x with value v. We emphasize the connection by using the same notation for both. The common PCM structure makes it possible to reuse for histories the ideas and results developed for heaps in the work on separation logic [3]. In particular, in this paper, we make both heaps and histories subject to the same assertion logic, the same rules of inference (*e.g.*, the frame rule), and thus the same style of *local reasoning*. Moreover, concepts such as ownership transfer, well-studied for heaps, apply to histories as well. For example, in Section 5, we use ownership transfer on histories to formalize the important design pattern of *helping* [14], whereby a concurrent thread may execute a task on behalf of other threads. That helping corresponds to a kind of ownership transfer (though not on histories, but on auxiliary commands) has been noticed before [20, 32]. However, commands do not form a PCM, while histories do—a fact that makes our development simple and uniform.

Second, we argue that precise history-based specs have to differentiate between the actions that have been performed by the specified thread, from the actions that have been performed by the thread's concurrent environment. Thus, our specs will range over *two* different history-typed variables, capturing the timestamped actions of the specified thread (*self*) and its environment (*other*), respectively. This split between self and other will provide us with a novel and very direct way of relating the functional behavior of a program to the interference of its environment, leading to specs that have a similar canonical "feel" in the concurrent setting, as the specs (1) have in the sequential one.

The self/other dichotomy required of histories is a special case of the more general specification pattern of *subjectivity*, observed in the recent related work on Subjective and Fine-grained Concurrent Separation Logic (FCSL) [19, 22]. That work generalized Concurrent Separation Logic (CSL) [23] to apply not only to heaps, but to any abstract notion of state (real or auxiliary) satisfying the PCM properties. We thus reuse FCSL [22] *off-the-shelf*, and instantiate it with histories, *without any additions to the logic or its meta-theory*. Surprisingly, the FCSL style of auxiliary state is sufficient to enable expressive history-based proofs of realistic fine-grained algorithms, including those with helping.

Specifications with histories also allow the clients of a fine-grained data structure to pretend, for the sake of simplifying their own reasoning, that they are using a coarse-grained version of the data structure. In this sense, we consider a program *logically* atomic (irrespective of the physical granularity of its implementation), if its specification is a singleton history $t \mapsto a$, containing only an abstract action a time-stamped with t. This spec provides an abstraction that the effect a of the program takes place at a singular point in time t, as if the program were coarse-grained, thus providing a uniform way to reason about coarse- and fine-grained programs.[1]

We show how a number of well-known algorithms can be proved logically atomic, and illustrate how the specs with histories facilitate client-side reasoning. We consider an atomic pair snapshot data structure [20, 26] (Section 2), a Treiber stack [30] along with its clients (Section 4), and Hendler *et al.*'s flat combining algorithm [14], a non-trivial example employing first-class functions and helping (Section 5). All our proofs, including the theory of histories, have been checked mechanically in Coq, and the sources are available online [27].

---

[1] An orthogonal aspect of granularity abstraction is the ability of a logical framework to express synchronized changes to auxiliary state that is spread across several shared data structures. We do not consider such abstraction in this paper, but elaborate in Section 6 on how to extend FCSL to support it, as well as on related approaches [5, 17, 28, 29].

2   Overview: Specifying Snapshots with Histories

In this section, we illustrate history-based specifications by applying them to the fine-grained *atomic pair snapshot* data structure [20, 26]. This data structure contains a pair of pointers, x and y, pointing to tuples (c_x, v_x) and (c_y, v_y), respectively. The components c_x and c_y of type A represent the accessible contents of x and y, that may be read and updated by the client. The components v_x and v_y are nats, encoding "version numbers" for x and y. They are internal to the structure and not directly accessible by the client.

The data structure interface exports three methods: readPair, writeX, and writeY. readPair is the main method, and the focus of the section. It returns the *snapshot* of the data structure, *i.e.*, the accessible contents of x and y as they appear together at the moment of the call. However, while x and y are being read by readPair, other threads may change them, by invoking writeX or writeY. Thus, a naïve implementation of readPair which first reads x, then y, and returns the pair (c_x, c_y) does not guarantee that c_x and c_y ever appeared

Fig. 1. Atomic pair snapshot
1   `readPair(): A × A {`
2   `(cx, vx) <- readX();`
3   `(cy, _) <- readY();`
4   `(_, tx) <- readX();`
5   **`if`**` vx == tx`
6   **`then return`**` (cx, cy);`
7   **`else`**` readPair();}`

together in the structure. One may have readPair first lock x and y to ensure exclusive access, but here we consider a fine-grained implementation which relies on the version numbers to ensure that readPair returns a valid snapshot.

The idea is that writeX(cx) (and symmetrically, writeY(cy)), changes the logical contents of x to cx, while incrementing the internal version number, *simultaneously*. Since the operation involves changes to the contents of a single pointer, in this paper we assume that it can be performed atomically (*e.g.*, by some kind of read-modify-write operation [15, §5.6]). We also assume atomic operations readX and readY for reading from x and y respectively. Then the implementation of readPair (Figure 1) reads from x and y in succession, but makes a check (line 5) to compare the version numbers for x obtained before and after the read of y. If x's version has changed, the procedure restarts.

We want to specify and prove that such an implementation of readPair is correct; that is, if it returns a pair (c_x, c_y), then c_x and c_y occurred simultaneously in the structure. To do so, we use histories as auxiliary state of every method of the structure. Histories, ranged over by τ, are finite maps from the natural numbers to pairs of elements of some type S; *i.e.*, hist $S \triangleq$ nat $\rightharpoonup S \times S$.[2] The natural numbers represent the moments in time, and the pairs represent the change of state. Thus, a singleton history $t \mapsto (s_1, s_2)$ encodes an atomic change from abstract state s_1 to abstract state s_2 at the time moment t. We will only consider *continuous* histories, for which $t \mapsto (s_1, s_2)$ and $t+1 \mapsto (s_3, s_4)$ implies $s_2 = s_3$. We use the following abbreviations to work with histories:

$$
\begin{aligned}
\tau[t] &\triangleq s, \text{ such that } \exists s', \tau(t) = (s', s) \\
\tau \le t &\triangleq \forall t' \in \mathrm{dom}(\tau), t' \le t \\
\tau \sqsubseteq \tau' &\triangleq \tau \text{ is a subset of } \tau'
\end{aligned}
\tag{2}
$$

Similarly to heaps, histories form a PCM under the operation $\uplus$ of disjoint union, with the empty history as the unit. The type S can be chosen arbitrarily, depending on the

---

[2] Other sets for time-stamping are possible besides nat, as will be mentioned in Section 6.

application, to capture whichever logical aspects of the actual physical state are of interest. For the snapshot structure, we take $S = A \times A \times \mathsf{nat}$. That is, the entries in the histories for pair snapshot will be of the form

$$t \mapsto (\langle c_x, c_y, v_x \rangle, \langle c'_x, c'_y, v'_x \rangle). \tag{3}$$

The entry encodes that at time moment t, the contents of x, y, and the version of x have changed from (c_x, c_y, v_x) to (c'_x, c'_y, v'_x). We ignore v_y, as it does not factor in the implementation of readPair (even though it is present for the sake of symmetry).

All the threads working over the pair snapshot structure respect a protocol on histories consisting of the following three properties. We explain in Section 3 how these are formally specified and enforced, but for now simply assume them. They will be important in the proof outline for readPair.

(*i*) Whenever a thread modifies x or y (*e.g.*, by calling writeX or writeY), its history gets augmented by an entry such as (3), where the timestamp t is chosen afresh. Thus, histories only grow, and only by adding valid snapshots (*i.e.*, snapshots corresponding to values of x and y, *simultaneously* present in the data structure).

(*ii*) Whenever the contents of x is changed in a history, its version number changes too. In contrapositive form, if $\tau[t_1] = \langle c_1, -, v \rangle$ and $\tau[t_2] = \langle c_2, -, v \rangle$, then $c_1 = c_2$.

(*iii*) Version numbers in a history grow monotonically. That is, if $\tau[t_1] = \langle -, -, v_1 \rangle$ and $\tau[t_2] = \langle -, -, v_2 \rangle$ and $t_1 \le t_2$, then $v_1 \le v_2$.

Specification. We now describe an FCSL spec for readPair and explain how it captures that its result is a valid snapshot of x and y.

$$\{\exists \tau_0. \ell \xmapsto{s} \mathsf{empty} \wedge \ell \xmapsto{o} \tau_0 \wedge \tau \sqsubseteq \tau_0\}$$
$$\mathsf{readPair}() \tag{4}$$
$$\left\{ \exists \tau_0\, t. \ell \xmapsto{s} \mathsf{empty} \wedge \ell \xmapsto{o} \tau_0 \wedge \tau \sqsubseteq \tau_0 \wedge \tau \le t \wedge \tau_0[t] = \langle \mathsf{res.1}, \mathsf{res.2}, - \rangle \right\}$$

First, note the label ℓ, which serves as an "abstract pointer" that differentiates the instance of the pair snapshot structure from any other structure that may exist in the program. In particular, ℓ identifies the histories of concern to readPair. Each thread keeps track of two such histories: the self-history, describing the operations that the thread itself has executed, and the other-history for the operations executed by all the other threads combined. They are captured by the assertions $\ell \xmapsto{s} \tau$ and $\ell \xmapsto{o} \tau$, respectively.

Thus, the precondition in (4) requires that readPair starts with the empty self-history, *i.e.*, the calling thread has not performed any updates to x or y. We show in Section 3 that the frame rule can be used to relax the requirement, so that readPair can be invoked by threads with an arbitrary self history. The precondition allows an arbitrary initial other-history τ_0. As τ_0 is bound locally in the precondition, we use the logical variable τ and a conjunct $\tau \sqsubseteq \tau_0$ to propagate the information about τ_0 into the postcondition. Because τ and τ_0 are related by inclusion, the precondition is stable under growth of τ_0 due to interfering threads, according to (*i*).

The postcondition states that readPair does not perform any changes to x and y; it is a *pure* method, thus its self-history remains empty. The main novelty of the specification is that the postcondition directly relates the result of readPair to the interference of the environment, *i.e.*, to the value of τ_0. This is in contrast to the extant logics, which do not keep track of the *other* component, and hence cannot specify readPair as directly.

In particular, the postcondition says that $\tau_0[t] = \langle \text{res}.1, \text{res}.2, -\rangle$, *i.e.*, that the components of the returned pair res appear in the environment history. Since according to the property (*i*) above, the histories only store valid snapshots, the resulting pair must be a valid snapshot too. In other words, readPair behaves as if it read x and y atomically, at time t. Moreover, $\tau \le t$, *i.e.*, the read occurred after readPair was invoked.

The specification pattern whereby a logical variable τ names the initial history of the environment is very common, so we streamline it by introducing the following notation.

$$\ell \hookrightarrow (\tau_s, \tau_0, \tau) \mathrel{\widehat{=}} \ell \overset{s}{\mapsto} \tau_s \wedge \ell \overset{o}{\mapsto} \tau_0 \wedge \tau \sqsubseteq \tau_s \cup \tau_0 \tag{5}$$

Proof outline. Figure 2 contains the proof outline for readPair, which we discuss next. The relation $\tau \sqsubseteq \tau_0$ is folded into the definition of $\ell \hookrightarrow (\text{empty}, \tau_0, \tau)$. Lines 1 and 3 abbreviate the precondition in (4). The readX method has the following spec:

$$\left\{ \ell \hookrightarrow (\text{empty}, -, \tau) \right\} \text{readX}() \left\{ \exists \tau_0\, t.\, \ell \hookrightarrow (\text{empty}, \tau_0, \tau) \wedge \tau \le t \wedge \tau_0[t] = \langle \text{res}.1, -, \text{res}.2 \rangle \right\}$$

Since the "initial" other-history is bounded by τ in the precondition, and the "final" τ_0 may only grow, we require $\tau \le t$ in the postcondition to ensure that we will not get a value from the history, which has "expired" *before* the call to readX. Thus in line 5 of the proof, we infer the existence of the history τ_1 and time stamp $t_1 \ge \tau$, such that the cx and vx appear in τ_1 at the time t_1. Similarly, readY has the spec:

$$\left\{ \ell \hookrightarrow (\text{empty}, -, \tau) \right\} \text{readY}() \left\{ \exists \tau_0\, t.\, \ell \hookrightarrow (\text{empty}, \tau_0, \tau) \wedge \tau \le t \wedge \tau_0[t] = \langle -, \text{res}.1, - \rangle \right\}$$

To obtain line 7, instantiate τ with τ_1 in the spec of readY. This derives the existence of τ_2, t_2, c and v, such that $\ell \hookrightarrow (\text{empty}, \tau_2, \tau_1)$, $\tau_1 \le t_2$, and $\tau_2[t_2] = \langle c, \text{cy}, v \rangle$. Because $t_1 \in \text{dom}(\tau_1)$, it must be that $t_1 \le t_2$. Moreover, because $\tau \sqsubseteq \tau_1 \sqsubseteq \tau_2$, we further obtain $\ell \hookrightarrow (\text{empty}, \tau_2, \tau)$, and $\tau \le t_2$, and lifting from line 5, $\tau_2[t_1] = \langle \text{cx}, -, \text{vx} \rangle$. Because t_1, t_2 appear in the same history τ_2, with versions vx and v, respectively, by property (*iii*), vx $\le v$. Similarly, instantiating τ in the spec of readX with τ_2, and invoking (*iii*), derives line 9 of the proof outline, and in particular vx $\le v \le$ tx.

Fig. 2. Proof outline for readPair

```
1  { ℓ ↪ (empty, −, τ) }
2  readPair() : A × A {
3  { ℓ ↪ (empty, −, τ) }
4  (cx, vx) <- readX();
5  { ℓ ↪ (empty, τ₁, τ) ∧ τ ≤ t₁ ∧ τ₁[t₁] = ⟨cx, −, vx⟩ }
6  (cy, _) <- readY();
7  { ℓ ↪ (empty, τ₂, τ) ∧ τ ≤ t₁ ≤ t₂ ∧ vx ≤ v ∧
      τ₂[t₁] = ⟨cx, −, vx⟩ ∧ τ₂[t₂] = ⟨c, cy, v⟩ }
8  (_, tx) <- readX();
9  { ℓ ↪ (empty, τ₃, τ) ∧ τ ≤ t₁ ≤ t₂ ≤ t₃ ∧ vx ≤ v ≤ tx ∧
      τ₃[t₁] = ⟨cx, −, vx⟩ ∧ τ₃[t₂] = ⟨c, cy, v⟩ ∧ τ₃[t₃] = ⟨−, −, tx⟩ }
10 if vx == tx
11    { ℓ ↪ (empty, τ₃, τ) ∧ τ ≤ t₂ ∧ cx = c ∧ τ₃[t₂] = ⟨cx, cy, v⟩ }
12    then return (cx, cy);
13    { ∃τ₀ t. ℓ ↪ (empty, τ₀, τ) ∧ τ ≤ t ∧ τ₀[t] = ⟨res.1, res.2, −⟩ }
14 else readPair();}
15 { ∃τ₀ t. ℓ ↪ (empty, τ₀, τ) ∧ τ ≤ t ∧ τ₀[t] = ⟨res.1, res.2, −⟩ }
```

From this property, if vx $=$ tx in the conditional on line 10, it must be that vx $= v$, and thus by (*ii*), cx $= c$. Substituting c by cx in line 9 gives us $\tau_3[t_2] = \langle \text{cx}, \text{cy}, v \rangle$, which, after (cx, cy) are returned in res, obtains the postcondition of readPair. Otherwise, if

$vx \neq tx$ in the conditional 10, we perform the recursive call to readPair. The precondition for the call is $\ell \hookrightarrow (\text{empty}, -, \tau)$, which is clearly met in line 9, so the postcondition immediately follows.

Monolithic histories. We compare the spec (4) with an alternative spec where the history is not split into self/other portions, but is kept monolithically as a *joint* (or shared) state. We use the predicate $\ell \overset{j}{\mapsto} \tau$ to specify such state:

$$\{\exists \tau_0. \ell \overset{j}{\mapsto} \tau_0 \wedge \tau \sqsubseteq \tau_0\} \; \text{readPair}() \; \left\{ \begin{array}{l} \exists \tau_0 \; t. \ell \overset{j}{\mapsto} \tau_0 \wedge \tau \sqsubseteq \tau_0 \wedge \\ \tau \leq t \wedge \tau_0[t] = \langle \text{res.1}, \text{res.2}, - \rangle \end{array} \right\} \qquad (6)$$

Note that the spec (6) imposes no restrictions on the growth of τ_0 (unlike (4) which keeps the self history empty). Thus, (6) is weaker than (4), as it allows more behaviors. In particular, it can be ascribed to any program which, in addition to calling readPair, also modifies x and y. This substantiates our claim from Section 1 that the self/other dichotomy is required to prevent history-based specs from losing precision. We provide further evidence for this claim in Section 4, where we show that subjective specs for *stacks* generalize the sequential canonical ones (1). The latter can be derived from the former by restricting τ_0 to be the empty history. Such a restriction is not possible if the history is kept monolithic.

3   Background: A Review of FCSL

In this section we review the relevant aspects of the previous work on Fine-grained Concurrent Separation Logic (FCSL) [22]. We explain FCSL by showing how it can be specialized to our novel contribution of specifying concurrent objects by means of histories. FCSL has been previously implemented as a shallow embedding in Coq; thus our assertions will freely use Coq's higher-order logic and datatype definition mechanism.

FCSL is a Hoare logic, generalizing CSL, hence its assertions are predicates on state. But unlike in CSL where state is a heap, in FCSL state may consist of a number of labeled components (sometimes dubbed as "regions" or "islands" in the literature [6, 28, 31]), each of which may represent state by a different type. If the type used by some label is non-heap, then that label encodes auxiliary state, used for logical specification, but erased at run time. For example, histories are an auxiliary state identified by the label ℓ in the atomic snapshot example. If we had a program which used two different atomic snapshot structures, we may label these by ℓ_1 and ℓ_2, etc.

3.1   Subjectivity

The state recorded in labels is further divided across another orthogonal axis – ownership. Each label identifies three different chunks of state: self, joint and other portion. The self portion is private to the specified thread, and cannot be accessed by the other threads. Dually, other is private to the environment threads, and cannot be accessed by the one being specified. Finally, the joint section is shared and can be accessed by everyone. The self and other portions of any given label have to belong to a common PCM (the joint portion, though, is not required to be a PCM element, as it is not a subject of a split between threads, as we will see below), and are often combined together by means of the • operation of that PCM. Of course, different labels can use different PCMs, and, therefore, the points-to assertions are implicitly parametrized with a PCM type.

The FCSL assertions reflect the division across these axes. We have already illustrated the assertions $\ell \overset{s}{\mapsto} v$, $\ell \overset{j}{\mapsto} v$ and $\ell \overset{o}{\mapsto} v$, which identify the self/joint/other component stored in the label ℓ of the state. These three basic assertions, constraining only one state component correspondingly (and leaving the two other unconstrained), can be, therefore, combined by the usual propositional connectives, such as $\wedge$ and $\vee$, as we have already shown in Section 2. FCSL further provides two connectives that generalize the *separating conjunction* $*$ from separation logic, along the two axes of state splitting. We next illustrate the *subjective separating conjunction* $\circledast$, and defer the discussion of the *resource separating conjunction* $*$ until additional technical material has been introduced. The formal definitions of all the connectives can be found in [27, Appendix A]. The subjective conjunction $\circledast$ models the division of state between concurrent threads upon forking and joining. In particular, the parallel composition rule of FCSL is:

$$\frac{\{p_1\} c_1 \{q_1\}@\mathcal{U} \qquad \{p_2\} c_2 \{q_2\}@\mathcal{U}}{\{p_1 \circledast p_2\} c_1 \parallel c_2 \{q_1 \circledast q_2\}@\mathcal{U}} \tag{7}$$

Ignoring $\mathcal{U}$ and the result types of c_1 and c_2 for now, we describe how $\circledast$ works. In this rule, it splits the pre-state of $c_1 \parallel c_2$ into two parts, satisfying p_1 and p_2 respectively. The parts contain the same labels, and equal joint portions, but the self and other portions are recombined to match the thread-relative views of c_1 and c_2. Concretely, in the case of one label ℓ, with a PCM $\mathbb{U}$ and values $a, b, c \in \mathbb{U}$, we have the following implication.

$$\ell \overset{s}{\mapsto} a \bullet b \wedge \ell \overset{o}{\mapsto} c \implies (\ell \overset{s}{\mapsto} a \wedge \ell \overset{o}{\mapsto} b \bullet c) \circledast (\ell \overset{s}{\mapsto} b \wedge \ell \overset{o}{\mapsto} a \bullet c) \tag{8}$$

Thus, if before the fork, the self-state of the parent thread contained $a \bullet b$, and the other-state contained c, then after the fork, the children will have self-states a and b, and the other-states $b \bullet c$ and $a \bullet c$, respectively. In the opposite direction:

$$(\ell \overset{s}{\mapsto} a \wedge \ell \overset{o}{\mapsto} c_1) \circledast (\ell \overset{s}{\mapsto} b \wedge \ell \overset{o}{\mapsto} c_2) \implies$$
$$\exists c. c_1 = b \bullet c \wedge c_2 = a \bullet c \wedge \ell \overset{s}{\mapsto} a \bullet b \wedge \ell \overset{o}{\mapsto} c \tag{9}$$

That is, if the state can be subjectively split between two child threads so that their other-views are c_1, c_2 (with self-views a, b), then there exists a common c—the other-view of the parent thread—such that $c_1 = b \bullet c$ and $c_2 = a \bullet c$. In this sense, the rule for parallel composition models the important effect that upon a split, c_1 becomes an environment thread for c_2, and vice-versa.

There are a few further equations that illustrate the interaction between the different assertions. First, every label contains all three of the self/joint/other components. Thus:

$$\ell \overset{s}{\mapsto} a \iff \ell \overset{s}{\mapsto} a \wedge \ell \overset{j}{\mapsto} - \wedge \ell \overset{o}{\mapsto} - \tag{10}$$

and similarly for $\ell \overset{j}{\mapsto} a$ and $\ell \overset{o}{\mapsto} a$. Also:

$$\ell \overset{s}{\mapsto} a \bullet b \iff \ell \overset{s}{\mapsto} a \circledast \ell \overset{s}{\mapsto} b \tag{11}$$

which is provable from (8), (9) and (10).

FCSL also provides a *frame rule*, obtained as a special case of parallel composition when c_2 is the idle thread, and $p_2 = q_2 = r$ is a stable predicate, as usual in fine-grained logics [6, 8, 33].

$$\frac{\{p\} c \{q\}@\mathcal{U}}{\{p \circledast r\} c \{q \circledast r\}@\mathcal{U}} \quad r \text{ stable under } \mathcal{U} \tag{12}$$

We illustrate the frame rule by deriving from the `readPair` spec (4) a relaxed spec which allows `readPair` to apply when the calling thread has non-trivial self history τ_S:

$$\{ \ell \hookrightarrow (\tau_S, -, \tau) \} \; \texttt{readPair}() \; \left\{ \begin{array}{l} \exists \tau_0 \, t. \, \ell \hookrightarrow (\tau_S, \tau_0, \tau) \wedge \tau \leq t \wedge \\ (\tau_S \cup \tau_0)[t] = \langle \mathrm{res.1}, \mathrm{res.2}, - \rangle \end{array} \right\} \tag{13}$$

Note that (13), when compared to (4), changes the self component from empty to τ_S, but also $\tau_0[t]$ changes into $(\tau_S \cup \tau_0)[t]$. The latter accounts for the possibility that the returned snapshot may have been recorded in τ_S as a consequence of the thread itself changing x or y, immediately before invoking `readPair`.

The spec (13) derives from (4) by framing with the predicate $r = \ell \overset{s}{\mapsto} \tau_S$. r is trivially stable, as it describes self-state, which is inaccessible to the interfering threads. We only show how to weaken the framed postcondition of (4) to the postcondition in (13); the preconditions can be strengthened similarly. Abbreviating $\tau \sqsubseteq \tau_0 \wedge \tau \leq t \wedge \tau_0[t] = \langle \mathrm{res.1}, \mathrm{res.2}, - \rangle$ by $P(\tau_0)$, which is a label-free (*i.e.*, pure) assertion, and thus commutes with $\circledast$, we get:

$$(\ell \overset{s}{\mapsto} \mathrm{empty} \wedge \ell \overset{o}{\mapsto} \tau_0 \wedge P(\tau_0)) \circledast (\ell \overset{s}{\mapsto} \tau_S) \implies \text{by (10) and } P\text{-pure}$$
$$(\ell \overset{s}{\mapsto} \mathrm{empty} \wedge \ell \overset{o}{\mapsto} \tau_0) \circledast (\ell \overset{s}{\mapsto} \tau_S \wedge \ell \overset{o}{\mapsto} -) \wedge P(\tau_0) \implies \text{by (9)}$$
$$\exists \tau_0'. \tau_0 = \tau_S \cup \tau_0' \wedge \ell \overset{s}{\mapsto} \tau_S \wedge \ell \overset{o}{\mapsto} \tau_0' \wedge P(\tau_0) \implies \text{by substituting } \tau_0$$
$$\exists \tau_0'. \ell \hookrightarrow (\tau_S, \tau_0', \tau) \wedge \tau \leq t \wedge (\tau_S \cup \tau_0')[t] = \langle \mathrm{res.1}, \mathrm{res.2}, - \rangle.$$

Intuitively, in (13) the frame history τ_S is "subtracted" from the other-history τ_0 of (4), and moved to the self-history, illustrating one important difference between the frame rule of FCSL and that of CSL. In FCSL, the frame is always subtracted from the other component, whereas in CSL it simply materializes out of nowhere. On the flip side, CSL does not consider the other component, and cannot easily express a spec such as (4).

3.2 Concurroids

We now turn to the component $\mathcal{U}$ of the FCSL specs, which is called *concurroid*. Concurroids are responsible for enforcing the invariants on the evolution of the state. For example, the properties (*i*)–(*iii*) in Section 2 will be enforced by defining an appropriate concurroid to govern the pair-snapshot structure. Thus, concurroids formally represent concurrent data structures, over which the programs operate.

A concurroid is (a form of) a state transition system (STS). It is a quadruple $\mathcal{U} = (L, W, I, E)$ where: (1) L is a set of labels, identifying different data structures; (2) W is a set of admissible states (alternatively, an FCSL assertion); (3) I is the set of *internal transitions* on W; (4) E is a set of pairs (α, ρ), where α is a *heap-acquiring* and ρ is a *heap-releasing* transition, collectively called *external* transitions. The internal transitions are relations on states, describing how a state of the STS evolves in one atomic step. The external transitions serve for transfer of state ownership. The concurroids thus bound the moves of the concurrent programs that operate on a data structure, and therefore represent a structured form of rely/guarantee transitions from Rely/Guarantee logics [8, 9, 18, 33, 34]. We next illustrate concurroids by example.

Pair-snapshot concurroid. Given a label ℓ, pointers x, y, and the type A of the accessible contents of x and y, the concurroid for the pair-snapshot structure is $S = (\{\ell\}, W_S, \{wr_x, wr_y, \mathrm{id}\}, \emptyset)$. The set of states W_S is described below. We assume that τ_S, τ_0 are histories, $c_x, c_y : A$ and $v_x, v_y : \mathrm{nat}$, and are implicitly existentially quantified.

$$W_S \;\widehat{=}\; \ell \overset{s}{\mapsto} \tau_S \wedge \ell \overset{j}{\mapsto} (x \mapsto (c_x, v_x) \cup y \mapsto (c_y, v_y)) \wedge \ell \overset{o}{\mapsto} \tau_O \wedge$$
$$\tau_S, \tau_O \text{ satisfy } (ii) - (iii), \tau_S \cup \tau_O \text{ is continuous, and}$$
$$\text{if } t = \mathsf{last}(\tau_S \cup \tau_O), \text{ then } (\tau_S \cup \tau_O)[t] = (c_x, c_y, v_x)$$

A state in W_S consists of the auxiliary part, which are histories in the self and other components, and concrete part, which is a joint heap, storing pointers x and y, with accessible contents c_x, c_y, and version numbers v_x, v_y, respectively.[3] It requires several additional properties of the auxiliary histories. First, the combined history $\tau_S \cup \tau_O$ is continuous; that is, adjacent timestamps have matching states. Second, the last time-stamp in $\tau_S \cup \tau_O$ correctly reflects what is stored in x and y. Finally, W_S also bakes in the properties $(ii) - (iii)$ required in the proof outline of readPair, so the specification (4) and its proof were, in fact, carried out in the concurroid context @S, which was omitted.

The internal transitions wr_x and wr_y synchronize the changes to x and y with histories. The transitions operate only on self and joint portions of the state, and the other-portion, τ_O, is fixed (cf. notation (10)). That is, the transitions essentially define the concurroid's Guarantee. In both transitions, $t_{\mathrm{fresh}}^{\tau_S \cup \tau_O}$ is the smallest timestamp unused by τ_S and τ_O.

$$wr_x \;\widehat{=}\; \ell \overset{j}{\mapsto} (x \mapsto (c_x, v_x) \cup y \mapsto (c_y, v_y)) \wedge \ell \overset{s}{\mapsto} \tau_S \quad\rightsquigarrow$$
$$\ell \overset{j}{\mapsto} (x \mapsto (c'_x, v_x + 1) \cup y \mapsto (c_y, v_y)) \wedge \ell \overset{s}{\mapsto} \tau_S \cup t_{\mathrm{fresh}}^{\tau_S \cup \tau_O} \mapsto (\langle c_x, c_y, v_x\rangle, \langle c'_x, c_y, v_x + 1\rangle)$$
$$wr_y \;\widehat{=}\; \ell \overset{j}{\mapsto} (x \mapsto (c_x, v_x) \cup y \mapsto (c_y, v_y)) \wedge \ell \overset{s}{\mapsto} \tau_S \quad\rightsquigarrow$$
$$\ell \overset{j}{\mapsto} (x \mapsto (c_x, v_x) \cup y \mapsto (c'_y, v_y + 1)) \wedge \ell \overset{s}{\mapsto} \tau_S \cup t_{\mathrm{fresh}}^{\tau_S \cup \tau_O} \mapsto (\langle c_x, c_y, v_x\rangle, \langle c_x, c'_y, v_x\rangle)$$

The first conjunct after $\rightsquigarrow$ in wr_x (and wr_y is similar) allows that the version number of x can only increase by 1 in an atomic step. The second conjunct shows that simultaneously with the change of x, the snapshot of the changed state is committed to the self-history of the invoking thread. Together, wr_x and wr_y ensure that histories only grow, and only by adding valid snapshots; i.e., precisely the property (i) from Section 2.

$\mathcal{U}$ also contains the identity transition id, whose presence enables programs that do not modify the state at all. In the pair-snapshot example, these are the readX and readY actions, and the readPair method. The pair-snapshot example does not involve ownership transfer, so S has no external transitions, but these will be important in the forthcoming examples.

Entanglement and private heaps. Larger concurroids may be constructed out of smaller ones. A particularly common construction is *entanglement* [22]. Given concurroids $\mathcal{U}$ and $\mathcal{V}$, the entanglement $\mathcal{U} \bowtie \mathcal{V}$ is a concurroid whose state space is the Cartesian product $W_{\mathcal{U}} \times W_{\mathcal{V}}$, and the transitions allow the $\mathcal{U}$ portion to perform a $\mathcal{U}$ transition, while the $\mathcal{V}$ portion remains idle, and vice-versa. Additionally, $\mathcal{U}$ and $\mathcal{V}$ portions can communicate to *transfer a heap* between themselves, by having one take a heap-acquiring, and the other *simultaneously* taking a heap-releasing transition.

The most common is the entanglement with the concurroid $\mathcal{P}$ of *private heaps* [27, Appendix B]. Entangling with $\mathcal{P}$ lets the concurroids temporarily move heaps to a private section, via the communication discussed above, where threads may then perform the customary operations of reading, writing, allocating, and deallocating pointers,

---

[3] Notice the overloading of the $\mapsto$ notation for singleton heaps and histories.

without interference.[4] $\mathcal{P}$ comes with a dedicated label pv. As an illustration, the following assertion may describe one possible state in the state space of the entanglement $\mathcal{P} \bowtie \mathcal{S}$ with the snapshot concurroid.

$$\mathsf{pv} \overset{s}{\mapsto} (z \mapsto 0) * \ell \overset{j}{\mapsto} (x \mapsto (c_x, v_x) \cup y \mapsto (c_y, v_y))$$

The $\ell \overset{j}{\mapsto} -$ portion describes the part of the state coming from $\mathcal{S}$, which is joint, containing pointers x and y, as explained before. The $\mathsf{pv} \overset{s}{\mapsto} (z \mapsto 0)$ describes the part of the state coming from $\mathcal{P}$. In this case, it contains a heap with a single pointer z. The heap is private, *i.e.*, owned by the self thread, so z cannot be modified by other threads. Notice that the assertions about pv and ℓ are separated by the resource separating conjunction $*$, which splits the state into portions with disjoint labels and heaps. In this particular case, it signifies that the labels pv and ℓ are distinct, as are the pointers z, x and y.

3.3 Extending and Hiding Concurroids

Concurroids represent concurrent data structures; thus it is important to be able to introduce and eliminate them. FCSL provides two programming constructors (both no-ops operationally), and corresponding inference rules for that purpose. For completeness, we introduce them here, but postpone the illustration until Section 4.

The injection rule shows that if a program is proved correct with respect to a smaller concurroid $\mathcal{U}$, then it can be extended to $\mathcal{U} \bowtie \mathcal{V}$, without invalidating the proof.

$$\frac{\{p\} c \{q\} @ \mathcal{U}}{\{p * r\} [c] \{q * r\} @ \mathcal{U} \bowtie \mathcal{V}} \quad r \subseteq W_{\mathcal{V}} \text{ stable under } \mathcal{V} \tag{14}$$

This is a form of framing rule, along the axis of adding new resources. The operator $*$ splits the state into portions with disjoint labels, and the side-condition that $r \subseteq W_{\mathcal{V}}$ forces r to remove the labels of the concurroid $\mathcal{V}$, so that c is verified *wrt.* the labels of $\mathcal{U}$. The program constructor $[-]$ is a coercion from $\mathcal{U}$ to $\mathcal{U} \bowtie \mathcal{V}$.

Hiding is the ability to introduce a concurroid $\mathcal{V}$, *i.e.*, install it in a private heap, for the scope of a thread c. The children forked by c can interfere on $\mathcal{V}$'s state, respecting $\mathcal{V}$'s transitions, but $\mathcal{V}$ is hidden from the environment of c, To the environment, $\mathcal{V}$'s state changes look like changes of the private heap of c. Upon termination of c, $\mathcal{V}$ is deinstalled.

$$\frac{\{\mathsf{pv} \overset{s}{\mapsto} h * p\} c \{\mathsf{pv} \overset{s}{\mapsto} h' * q\} @ (\mathcal{P} \bowtie \mathcal{U}) \bowtie \mathcal{V}}{\{\Psi \, g \, h * (\Phi(g) \twoheadrightarrow p)\} \, \mathsf{hide}_{\Phi,g} \, c \, \{\exists g'. \Psi \, g' \, h' * (\Phi(g') \twoheadrightarrow q)\} @ \mathcal{P} \bowtie \mathcal{U}}$$
$$\text{where } \Psi \, g \, h = \exists k{:}\mathsf{heap}. \, \mathsf{pv} \overset{s}{\mapsto} h \cup k \wedge \Phi(g) \text{ erases to } k \tag{15}$$

Since installing $\mathcal{V}$ consumes a chunk of private heap, the rule requires the overall concurroid to support private heaps, *i.e.*, to be an entanglement of $\mathcal{P}$ with an arbitrary $\mathcal{U}$. In programs, we use the coercion hide c to indicate the change from $(\mathcal{P} \bowtie \mathcal{U}) \bowtie \mathcal{V}$ to $\mathcal{P} \bowtie \mathcal{U}$. If $\mathcal{U}$ is of no interest, one can take it to be the empty concurroid $\mathcal{E}$, which is a right unit for $\bowtie$ [27, Appendix B.4].

---

[4] Our Coq proofs actually use two different concurroids, one for reading/writing, another for allocation/deallocation, which we entangle to provide all four operations. For simplicity, here we assume a monolithic implementation.

The annotation Φ is a predicate; it describes an invariant that holds within the scope of hide, parametrized by an argument. It is subject to a number of conditions [27, Appendix D.3]. g is the initial argument, so $\Phi(g)$ holds in the initial state into which $\mathcal{V}$ is placed upon installation. The rule guarantees that the ending state of c satisfies $\exists g'.\ \Phi(g')$. The surrounding connectives $*$ and $-\!\!*$ merely mediate between $\mathcal{U}$, $\mathcal{V}$, and the erasure of $\mathcal{V}$ to heaps. We explain the precondition, and the postcondition is similar.

In the precondition, $*$ separates private heaps from $\mathcal{U}$, and Ψ requires that every state in $\Phi(g)$ obtains the same private heap when the auxiliary fields are erased. $-\!\!*$ is inherited from separation logic. $\Phi(g) -\!\!* p$ says that if the initial state (which is in $W_{\mathcal{U}}$) is extended with a state from $\Phi(g)$ (which is in $W_{\mathcal{V}}$), then the result is a state satisfying p. In other words, if a state satisfying $\Phi(g)$ is installed in the initial state of c, while its heap footprint is removed from the private heaps, then c's precondition is satisfied.

4   Treiber Stack and Its Client

In this section we illustrate how histories can be used to specify and verify the fine-grained data structure of Treiber stack [30]. We also show how the specs can be used by clients, where they provide an abstraction that facilitates client reasoning as if the structure were coarse-grained.

The Treiber stack works as follows. Physically, the stack is kept as a singly-linked list in the heap, with a sentinel pointer snt pointing to the stack top p1. The call to push(e) allocates a node p that is supposed to go to the top of stack, and attempts to link the node into the stack, by changing the sentinel to p. Clearly, this operation should not succeed if some interfering thread has in the meantime changed the top by pushing or popping elements. Thus push applies a CAS read-modify-write operation [15], which atomically reads snt, compares its contents with p1, and if the two are equal (*i.e.*, if the stack's top has not changed), writes p into snt, thus en-linking the new top. Otherwise, push is restarted. pop() behaves similarly. It reads the first node p, pointed to by snt, and obtains its value e and pointer p1 to the next node. Then it tries to de-link p, by changing the sentinel to p1 using a CAS to identify interference. Note that pop does not deallocate the de-linked node p (this is

Fig. 3. Treiber stack methods

```
1 push(e : A): Unit {
2   p <- alloc();
3   fix loop() {
4     p1 <- readSentinel();
5     write(p, (e, p1));
6     ok <- tryPush(p1, p);
7     if ok then return ();
8     else loop();}();
9 }
```

```
1 pop(): option A {
2   p <- readSentinel();
3   if p == null
4   then return None;
5   else {
6     (e,p1) <- readNode(p);
7     ok      <- tryPop(p,p1);
8     if ok then return Some e;
9     else pop();}}
```

enforced by the design of the appropriate concurroid as we will soon see), which thus remains in the data structure as garbage. This is by design, to prevent the ABA problem [15, §10]: if p is deallocated, then some other push may allocate it again, and place it back on top of the stack. A procedure that observed p on top of the stack, but has not performed its CAS yet may thus be fooled as follows. Its CAS may encounter p on top of the stack, and proceed as if the stack had not changed, producing invalid results.

The described code of the Treiber stack operations is given in Figure 3, where we used descriptive names for the atomic operations. Instead of CAS, we used tryPush and tryPop, and instead of pointer read, we used readSentinel and readNode. The reason for the descriptive names is that the atomic operations in FCSL operate not only on concrete heap pointers, but on auxiliary state as well. In the particular case of Treiber, the auxiliary state will be histories, which tryPush and tryPop change in different ways, even though they both operationally perform a CAS. Similarly, readSentinel and readNode deduce different facts about the histories, even though they both simply read from a pointer. We elide here any further discussion on how the atomic operations are specified and verified in FCSL (it can be found in [22] and [27, Appendix C]). Instead, whenever needed, we simply state the Hoare specs for the atomics and proceed to use them in proof outlines, as if the atomics were ordinary procedures. Of course, our Coq files contain proofs that all such Hoare triples are valid.

Treiber concurroid. Given a label tb, the sentinel pointer *snt*, and the type A of the stack elements, the state space of the Treiber concurroid $\mathcal{T}$ is described as follows. Its auxiliary self/other components are histories τ_S and τ_O that store mathematical sequences l corresponding to the logical contents of the stack at various timestamps. The joint component contains a heap h_s storing a sentinel *snt* pointing to a linked list, a heap h implementing the list, and a garbage section *grb* of de-linked nodes.

$$W_{\mathcal{T}} \,\widehat{=}\, \exists \tau_S\, \tau_O\, h_s.\, \text{tb} \overset{s}{\mapsto} \tau_S \wedge \text{tb} \overset{o}{\mapsto} \tau_O \wedge \text{tb} \overset{j}{\mapsto} h_s \wedge I\,(\tau_S \cup \tau_O)\,h_s$$

$$I\,\tau\,h_s \,\widehat{=}\, \exists p\, h\, grb\, l.\, h_s = (snt \mapsto p) \cup h \cup grb \wedge \text{list}(p, l, h) \wedge \qquad (16)$$
$$\text{complete}(\tau) \wedge \text{continuous}(\tau) \wedge \text{stacklike}(\tau) \wedge \tau[\text{last}(\tau)] = l$$

The auxiliary predicates are:

$$\text{list}(p, l, h) \,\widehat{=}\, p = \text{null} \wedge l = \text{nil} \wedge h = \text{empty}\, \vee$$
$$\exists e\, p'\, l'\, h'.\, l = e :: l' \wedge h = p \mapsto (e, p') \cup h' \wedge \text{list}(p', l', h')$$
$$\text{complete}(\tau) \,\widehat{=}\, \exists l_0.\, \tau(0) = (l_0, l_0) \wedge \forall t.\, t < |\text{dom}(\tau)| \Rightarrow t \in \text{dom}(\tau)$$
$$\text{stacklike}(\tau) \,\widehat{=}\, \forall t \in \text{dom}(\tau).\, t > 0 \Rightarrow \exists l\, e.\, \tau(t) = (l, e :: l) \vee \tau(t) = (e :: l, e)$$

In particular: (1) the overall history $\tau_S \cup \tau_O$ is complete, *i.e.* no gaps exist between timestamps (this property was irrelevant for the pair snapshot structure, but essential for stacks to ensure the absence of the ABA-problem); (2) aside from the initialization in timestamp 0, the history only stores events corresponding to pushing or popping, and (3) the last recorded state in the history captures the current contents of the stack. For simplicity, we disable reasoning about the structure's inherent memory leak by not relating histories to *grb* in (16).

The transitions of $\mathcal{T}$ allow for popping and pushing only.

$$pop \,\widehat{=}\, \text{tb} \overset{j}{\mapsto} snt \mapsto p \cup h \cup grb \wedge \text{tb} \overset{s}{\mapsto} \tau_S \wedge h = (p \mapsto (e, p') \cup h') \wedge \text{list}(p, (e :: l), h) \rightsquigarrow$$
$$\text{tb} \overset{j}{\mapsto} snt \mapsto p' \cup h' \cup (p \mapsto (e, p') \cup grb) \wedge \text{tb} \overset{s}{\mapsto} \tau_S \cup t^{\tau_S \cup \tau_O}_{\text{fresh}} \mapsto (e :: l, l)$$

$$push_{p',e,p} \,\widehat{=}\, \text{tb} \overset{j}{\mapsto} snt \mapsto p \cup h \cup grb \wedge \text{tb} \overset{s}{\mapsto} \tau_S \wedge \text{list}(p, l, h) \qquad\qquad \rightsquigarrow$$
$$\text{tb} \overset{j}{\mapsto} snt \mapsto p' \cup (p' \mapsto (e, p) \cup h) \cup grb \wedge \text{tb} \overset{s}{\mapsto} \tau_S \cup t^{\tau_S \cup \tau_O}_{\text{fresh}} \mapsto (l, e :: l)$$

In *pop*, the sentinel pointer is swapped from used-to-be head p to its next one, p', whereas $(p \mapsto -)$ logically joins the garbage. The transition *push* describes how a heap of the shape $p' \mapsto (e, p)$, describing the node to be pushed, is acquired and placed at the top of the stack. It is an external transition, which means it only fires when entangled with a concurroid from which the heap $p' \mapsto (e, p)$ can be taken away. In our case, that will be the concurroid $\mathcal{P}$ for private state. Indeed, both transitions preserve the state invariant I (16). Importantly, $\mathcal{T}$ does not have a release transition; once a memory chunk is in the joint state, it never leaves, capturing that $\mathcal{T}$ does not allow deallocation.

Method specs. We give the following history-based specs.

$$\left\{ \begin{array}{l} \text{pv} \overset{s}{\mapsto} \text{empty} * \\ \text{tb} \hookrightarrow (\text{empty}, -, \tau) \end{array} \right\} \text{push}(e) \left\{ \begin{array}{l} \exists t\, l.\, \text{pv} \overset{s}{\mapsto} \text{empty} * \\ \text{tb} \hookrightarrow (t \mapsto (l, e :: l), -, \tau) \wedge \tau < t \end{array} \right\} @\mathcal{P} \bowtie \mathcal{T}$$

$$\left\{ \text{tb} \hookrightarrow (\text{empty}, -, \tau) \right\} \tag{17}$$
$$\text{pop}()$$
$$\left\{ \begin{array}{l} \exists e\, t\, l.\, \text{res} = \text{Some } e \wedge \text{tb} \hookrightarrow (t \mapsto (e :: l, l), -, \tau) \wedge \tau < t \vee \\ \exists \tau_0\, t.\, \text{res} = \text{None} \wedge \text{tb} \hookrightarrow (\text{empty}, \tau_0, \tau) \wedge \tau_0[t] = \text{nil} \end{array} \right\} @\mathcal{T}$$

A call to push runs with empty private heap and history, thus by framing, it can run with any private heap and history. After termination, the self history is incremented by a singleton exposing that a push event has been executed at a time stamp t; $\tau < t$ indicates that the push event appeared strictly after the events preceding the call. The spec for pop is slightly more complicated as pop checks for stack emptiness, but ultimately proceeds in the similar manner. push works over the entangled concurroid $\mathcal{P} \bowtie \mathcal{T}$, as it needs to allocate memory; pop works over $\mathcal{T}$ only, as it does not deallocate.

Fig. 4. A proof outline of Treiber's push method

```
1  { pv ⤳ empty * tb ↪ (empty, -, τ) }
2  p <- [alloc()];
3  { pv ⤳ p ↦ - * tb ↪ (empty, -, τ) }
4  fix loop() {
5  { pv ⤳ p ↦ - * tb ↪ (empty, -, τ) }
6  p1 <- [readSentinel()];
7  { pv ⤳ p ↦ - * tb ↪ (empty, -, τ) }
8  [write(p, (e, p1))];
9  { pv ⤳ p ↦ (e, p1) * tb ↪ (empty, -, τ) }
10 ok <- tryPush(p1, p);
11 { ok ∧ ∃t l. pv ⤳ empty * tb ↪ (t ↦ (l, e :: l), -, τ) ∧ τ < t ∨
        ¬ok ∧ pv ⤳ p ↦ (e, p1) * tb ↪ (empty, -, τ) }
12 if ok then return ();
13 { ∃t l. pv ⤳ empty * tb ↪ (t ↦ (l, e :: l), -, τ) ∧ τ < t }
14 else
15 { pv ⤳ p ↦ - * tb ↪ (empty, -, τ) }
16 loop();}();
17 { ∃t l. pv ⤳ empty * tb ↪ (t ↦ (l, e :: l), -, τ) ∧ τ < t }
```

Verification of push and pop implementations relies on the specifications of the atomic actions alloc and write, which are specific to the $\mathcal{P}$ concurroid.

$$\{ \text{pv} \overset{s}{\mapsto} \text{empty} \} \quad \text{alloc}() \quad \{ \text{pv} \overset{s}{\mapsto} \text{res} \mapsto - \} @\mathcal{P}$$
$$\{ \text{pv} \overset{s}{\mapsto} x \mapsto - \} \text{write}(x, e) \{ \text{pv} \overset{s}{\mapsto} x \mapsto e \} @\mathcal{P} \tag{18}$$

In Figure 4, we present the proof outline for push (the proof for pop can be found in the Coq files). It is mostly self-explanatory, so we only point out a few technicalities. The actions alloc and write have to be explicitly injected into $\mathcal{P} \bowtie \mathcal{T}$, by means of the coercion $[-]$, introduced in Section 3. Similarly for readSentinel, whose concurroid is $\mathcal{T}$. Somewhat surprisingly, the call to readSentinel in line 6 is irrelevant for the (partial) correctness of tryPush; thus, line 7 does not say anything about p1.[5] The proof rule for

---

[5] Though, taking a random p1 here will affect liveness, as push will keep looping until it finds the chosen p1 at the top of the stack.

fix allows assuming the spec of a procedure in the proof of the body, and is presented in [27, Appendix D]. The tryPush action appears in the proof outline with its precise specification; that is, line 9 contains its precondition, and 11 contains the postcondition, describing that a successful outcome of tryPush removed a heap from $\mathcal{P}$, moved it to the joint heap of $\mathcal{T}$, and updated the history, following the *push* transition.

Recovering sequential specifications. We next show that the subjective spec (17) is a generalization of the canonical sequential spec (1). In particular, if there is no interference from other threads, (17) can be reduced to (1). The mechanism for achieving the reduction relies on the self/other dichotomy, thus substantiating our point that the dichotomy is important for precise reasoning with histories.

To this end, we use the hide constructor from Section 3. It introduces a concurroid in a delimited scope, and prohibits
the environment threads from interfering
on it. The heap for the introduced concurroid is appropriated from the private
heap. In the case of push, we will appropriate a heap storing the sentinel and the
linked list of the stack, install the $\mathcal{T}$ concurroid over this heap, perform push with
interference disabled, then return the heap
back to private heaps. We will derive the
following specification, which is essentially an elaborated version of (1), modulo
the memory leak inherent to Treiber stack
(hence *grb* in the postcondition).

Fig. 5. Proof of sequential spec for push.

1 $\{\exists p\, h.\ \mathsf{pv} \overset{s}{\mapsto} (snt \mapsto p \cup h) \wedge \mathsf{list}(p, l, h)\}$
2 $\{\Psi\ \mathsf{empty}\ \mathsf{empty} * (\Phi(\mathsf{empty}) \twoheadrightarrow \mathsf{tb} \hookrightarrow (0 \mapsto (l, l), -, -))\}$
3 $\mathsf{hide}_{\Phi,\mathsf{empty}}\ \{$
4 $\{\ \mathsf{pv} \overset{s}{\mapsto} \mathsf{empty} * \mathsf{tb} \hookrightarrow (0 \mapsto (l, l), -, -)\ \}$
5 $\mathsf{push}(e);$
6 $\left\{ \begin{array}{l} \exists l'.\, \mathsf{pv} \overset{s}{\mapsto} \mathsf{empty} * \\ \mathsf{tb} \hookrightarrow (0 \mapsto (l, l) \cup t \mapsto (l', e :: l'), -, -) \end{array} \right\}$ }
7 $\left\{ \begin{array}{l} \exists \tau.\ \Psi\ \tau\ \mathsf{empty}\ * \\ (\Phi(\tau) \twoheadrightarrow \exists t\, l'.\, \mathsf{tb} \hookrightarrow (0 \mapsto (l, l) \cup t \mapsto (l', e :: l'), -, -)) \end{array} \right\}$
8 $\left\{ \begin{array}{l} \exists t\, l'\, \tau.\, \tau = 0 \mapsto (l, l) \cup t \mapsto (l', e :: l') \wedge \\ \mathsf{complete}(\tau) \wedge \mathsf{continuous}(\tau) \wedge \Psi\ \tau\ \mathsf{empty} \end{array} \right\}$
9 $\{\exists \tau.\ \tau = 0 \mapsto (l, l) \cup 1 \mapsto (l, e :: l) \wedge \Psi\ \tau\ \mathsf{empty}\}$
10 $\{\exists p'\, h.\ \mathsf{pv} \overset{s}{\mapsto} (snt \mapsto p' \cup h \cup -) \wedge \mathsf{list}(p', e :: l, h)\}$

$$\{\ \exists p\, h.\ \mathsf{pv} \overset{s}{\mapsto} (snt \mapsto p \cup h) \wedge \mathsf{list}(p, l, h)\ \}$$

$$\mathsf{hide}_{\Phi,\mathsf{empty}}\ \{\ \mathsf{push}(e);\ \} \tag{19}$$

$$\{\ \exists p\, h\, grb.\ \mathsf{pv} \overset{s}{\mapsto} (snt \mapsto p \cup h \cup grb) \wedge \mathsf{list}(p, e :: l, h)\ \}@\mathcal{P}$$

The self/other dichotomy affords explicit access to other-owned histories, so that we can define the following predicate Φ stating that other-histories remain empty within the scope of hide.

$$\Phi(\tau) \ \widehat{=}\ \exists l.\ \mathsf{tb} \overset{s}{\mapsto} ((0 \mapsto (l, l)) \cup \tau) \wedge \mathsf{tb} \overset{o}{\mapsto} \mathsf{empty} \wedge W_{\mathcal{T}} \tag{20}$$

Inside hide, the stack is initialized (the history contains the singleton $0 \mapsto (l, l)$), there is no interference (tb $\overset{o}{\mapsto}$ empty), and the state is a valid one for $\mathcal{T}$ (*i.e.*, it is captured by the definition (16)).

One can prove that if the histories are erased from any state in $\Phi(\tau)$, the remaining concrete heap consists of *snt* and the stack. Moreover, the contents of the stack is the last entry of τ (or l if τ is empty). In other words, using Ψ (15), defined in Section 3:

$$\Psi\ \tau\ \mathsf{empty} \iff \exists p\, h.\ \mathsf{pv} \overset{s}{\mapsto} (snt \mapsto p \cup h \cup -) \wedge \mathsf{list}(p, l', h) \tag{21}$$

where $l' = \tau[\mathsf{last}(\tau)]$ (or $l' = l$ if τ is empty).

Fig. 6. A parallel stack-based producer/consumer program

```
1 produce(n: nat, i: nat) {    1 consume(n: nat, i: nat) {    1 exchange(n: nat): Unit {
2  if i == n                   2  if i == n                   2   hideΦ,empty {
3  then return ();             3  then return ();             3     produce(n, 0); || consume(n, 0);
4  else {                      4  else {                      4   }
5    e <- ap[i];               5    r <- poptb();              5 }
6    pushtb(e);                6    if r == Some e
7    produce(i + 1);           7    then {
8  }                           8      ac[i] := e;
9 }                            9      consume(i + 1);}
                              10    else consume(i);}}
```

The derivation is in Figure 5, and we comment on the main points. In line 2, the right conjunct uses the property inherent in Ψ, that Φ(empty) erases to the heap storing l. Thus, this is the l that appears in the consequent of $\twoheadrightarrow$. In line 7, the second conjunct implies that the history τ, whose existence obtains from the rule for hiding (15), must be the self-history returned by push. Hence, it is equal to $0 \mapsto (l, l) \cup t \mapsto (l', e :: l')$ for some t and l'. But, we also know that τ must be complete (no gaps between timestamps) and continuous. Hence $t = 1$ and $l' = l$ in line 9, which derives the postcondition by (21).

A stack client. We next illustrate how the specs (17) are exploited by the *concurrent* clients of Treiber stack to abstract from the fine-grained nature of Treiber's implementation. The example code in Figure 6 presents two procedures, produce and consume, that communicate via a common Treiber stack tb. produce pushes onto the stack the elements of its array ap in order, whereas consume pops from the stack, to fill its array ac. Both arrays are of equal size n. The procedure exchange runs produce and consume concurrently. We will prove that after exchange terminates, ap has been copied to ac, modulo element permutation. The inference will only use the specs (17) but not the code of stack methods, thus obtaining a coarse-grained view of effects provided by histories.

We use several auxiliary predicates. First, $\mathsf{Arr}_n(a, l, h)$ defines an array of size n as a sequence of consecutive pointers in the heap h, starting from pointer a, and storing elements of the list l:

$$\mathsf{Arr}_n(a, l, h) \;\hat{=}\; |l| = n \wedge h = \bigcup_{i<n}(a + i) \mapsto l(i) \tag{22}$$

Next, the predicates Pushed and Popped extract the lists of pushed and popped elements from a stack history τ.

$$\begin{aligned}
\mathsf{Pushed}(\tau, l) &\;\hat{=}\; l =_{/\mathrm{mset}} \{\!\{e \mid \exists t\, l.\, t \mapsto (l, e :: l) \in \tau \vee 0 \mapsto (l, l) \in \tau \wedge e \in l\}\!\} \\
\mathsf{Popped}(\tau, l) &\;\hat{=}\; l =_{/\mathrm{mset}} \{\!\{e \mid \exists t\, l.\, t \mapsto (e :: l, l) \in \tau\}\!\}
\end{aligned} \tag{23}$$

The notation $\{\!\{-\}\!\}$ stands for multisets, and $=_{/\mathrm{mset}}$ is multiset equality, which we conflate with list equality modulo permutation. We can now ascribe the following specs to produce and consume:

$$\begin{aligned}
&\big\{\mathsf{Pr}(h_p, l_{<i}) \wedge \mathsf{Arr}_n(\mathsf{ap}, l, h_p)\big\}\ \mathbf{produce}(n, i)\ \big\{\mathsf{Pr}(h_p, l) \wedge \mathsf{Arr}_n(\mathsf{ap}, l, h_p)\big\} \\
&\big\{\exists h_c\, l.\, \mathsf{Cn}(h_c, l_{<i}) \wedge \mathsf{Arr}_n(\mathsf{ac}, l, h_c)\big\}\ \mathbf{consume}(n, i)\ \big\{\exists h_c\, l.\, \mathsf{Cn}(h_c, l) \wedge \mathsf{Arr}_n(\mathsf{ac}, l, h_c)\big\}
\end{aligned} \tag{24}$$

both over the $\mathcal{P} \bowtie \mathcal{T}$ concurroid. Pr and Cn are defined as follows:

$$\begin{aligned}
\mathsf{Pr}(h_p, l) &\;\hat{=}\; \mathsf{pv} \xmapsto{s} h_p * \mathsf{tb} \xmapsto{s} \tau_s \wedge \mathsf{Pushed}(\tau_s, l) \wedge \mathsf{Popped}(\tau_s, \mathsf{nil}) \\
\mathsf{Cn}(h_c, l) &\;\hat{=}\; \mathsf{pv} \xmapsto{s} h_c * \mathsf{tb} \xmapsto{s} \tau_s \wedge \mathsf{Pushed}(\tau_s, \mathsf{nil}) \wedge \mathsf{Popped}(\tau_s, l),
\end{aligned}$$

so they essentially describe the producer/consumer loop invariants; $l_{<i}$ is a prefix of l for elements with indices less than i. The specs (24) show that produce pushes all the elements from ap, and consume fills ac with elements of some sequence of the length n. The proofs of both specs (available in our Coq development) derive easily from (17) after these are framed to allow running in arbitrary initial self heap and history.

The interesting part of the example is proving exchange, where we compose produce and consume in parallel, and then use hiding to infer that the ap and ac arrays in the end contain the same elements, modulo permutation. The proof outline is in Figure 7, and it relies on the following important lemmas about histories.

Lemma 1. $\mathsf{Pushed}(\tau_1, l_1) \wedge \mathsf{Popped}(\tau_1, \mathsf{nil}) \wedge \mathsf{Popped}(\tau_2, l_2) \wedge \mathsf{Pushed}(\tau_2, \mathsf{nil}) \implies \mathsf{Pushed}(\tau_1 \cup \tau_2, l_1) \wedge \mathsf{Popped}(\tau_1 \cup \tau_2, l_2)$.

Lemma 2. *If* $\mathsf{complete}(\tau)$ *and* $\mathsf{stacklike}(\tau)$ *then* $\mathsf{Pushed}(\tau, l_1) \wedge \mathsf{Popped}(\tau, l_2) \wedge |l_1| = |l_2| \implies l_1 =_{/\mathsf{mset}} l_2$.

The proof outline in Figure 7 starts in the concurroid $\mathcal{P}$, which extends to $\mathcal{P} \bowtie \mathcal{T}$ in the scope of hide. The invariant Φ of hide is the one we already used, defined in (20). It introduces a Treiber stack structure with an initial history $0 \mapsto (\mathsf{nil}, \mathsf{nil})$. Also, the heaplet $snt \mapsto \mathsf{null}$ with the sentinel pointer has been donated to the state space of the Treiber stack, so it is removed from the private heap. Next, the self-heap and history are split via $\circledast$; the parts are given to produce and consume, respectively, according to the parallel composition rule (7). Next, we reason out of specifications (24) for producer/consumer and combine the subjective views back via $\circledast$ upon

Fig. 7. Proof outline for producer/consumer

$$\left\{ \mathsf{pv} \xmapsto{s} h_p \cup h_c \cup snt \mapsto \mathsf{null} \wedge \mathsf{Arr}_n(\mathsf{ap}, l, h_p) \wedge \mathsf{Arr}_n(\mathsf{ac}, -, h_c) \right\}$$

$$\mathsf{hide}_{\Phi,\mathsf{empty}} \{$$

$$\left\{ \mathsf{pv} \xmapsto{s} h_p \cup h_c \wedge \mathsf{Arr}_n(\mathsf{ap}, l, h_p) \wedge \mathsf{Arr}_n(\mathsf{ac}, -, h_c) * \atop \mathsf{tb} \xmapsto{s} 0 \mapsto (\mathsf{nil}, \mathsf{nil}) \wedge \mathsf{tb} \xmapsto{s} \mathsf{empty} \right\}$$

$$\left\{ \left(\mathsf{pv} \xmapsto{s} h_p \wedge \mathsf{Arr}_n(\mathsf{ap}, l, h_p) \atop * \, \mathsf{tb} \xmapsto{s} 0 \mapsto (\mathsf{nil}, \mathsf{nil}) \right) \circledast \left(\mathsf{pv} \xmapsto{s} h_c \wedge \mathsf{Arr}_n(\mathsf{ac}, -, h_c) \atop * \, \mathsf{tb} \xmapsto{s} \mathsf{empty} \right) \right\}$$

$$\begin{array}{c|c} \left\{ \mathsf{Pr}(h_p, l_{<0}) \wedge \mathsf{Arr}_n(\mathsf{ap}, l, h_p) \right\} & \left\{ \exists l'. \; \mathsf{Cn}(h_c, l'_{<0}) \wedge \mathsf{Arr}_n(\mathsf{ac}, l', h'_c) \right\} \\ \mathsf{produce}(n, 0); & \mathsf{consume}(n, 0); \\ \left\{ \mathsf{Pr}(h_p, l) \wedge \mathsf{Arr}_n(\mathsf{ap}, l, h_p) \right\} & \left\{ \exists h'_c \; l'. \; \mathsf{Cn}(h_c, l') \wedge \mathsf{Arr}_n(\mathsf{ac}, l', h'_c) \right\} \end{array}$$

$$\left\{ \left(\mathsf{Pr}(h_p, l) \wedge \mathsf{Arr}_n(\mathsf{ap}, l, h_p) \right) \circledast \left(\exists h'_c \; l'. \; \mathsf{Cn}(h_c, l') \wedge \mathsf{Arr}_n(\mathsf{ac}, l', h'_c) \right) \right\}$$

$$\left\{ \exists h'_c \; l'. \mathsf{pv} \xmapsto{s} h_p \cup h_c \wedge \mathsf{Arr}_n(\mathsf{ap}, l, h_p) \wedge \mathsf{Arr}_n(\mathsf{ac}, l', h'_c) \atop * \, \exists \tau_s. \mathsf{tb} \xmapsto{s} \tau_s \wedge \mathsf{Pushed}(\tau_s, l) \wedge \mathsf{Popped}(\tau_s, l') \wedge \mathsf{tb} \xmapsto{s} \mathsf{empty} \right\} \}$$

$$\left\{ \exists h'_c \; l'. \mathsf{pv} \xmapsto{s} h_p \cup h_c \cup (snt \mapsto -) \cup - \wedge \atop \mathsf{Arr}_n(\mathsf{ap}, l, h_p) \wedge \mathsf{Arr}_n(\mathsf{ac}, l', h'_c) \wedge l =_{/\mathsf{mset}} l' \right\}$$

joining of the parallel threads: we thus derive that the contents of ap and ac, are l and l' respectively. By unfolding the definitions of Pr and Cn, and using Lemma 1, we derive $\mathsf{Pushed}(\tau_s, l) \wedge \mathsf{Popped}(\tau_s, l')$, where τ_s is the combined history of produce and consume. Finally, τ_s is complete and stack-like (since other-history is provably empty thanks to hiding). Moreover, both l and l' have size n, as ensured by the assertion Arr_n constraining both of them. Thus, in the last assertion, we can use Lemma 2 to obtain the desired equality of l and l' modulo permutation. Note also that the sentinel pointer is returned back to the private heap, along with the garbage heap (existentially abstracted by the − placeholder).

5 Flat Combining

This section shows how PCMs in general, and histories in particular, can formalize the concurrent algorithm design pattern of helping, whereby one concurrent thread may execute code on behalf of another. We use Hendler *et al.*'s flat combining algorithm as an example [14]. Unlike other proofs of this algorithm [4, 31], we do not require any additional logical infrastructure aside from ordinary auxiliary state, represented by a PCM [19, 22]. We verify the algorithm *wrt.* a generic PCM, and then instantiate with the PCM of histories. Thus, our proof is usable even in examples where the specs do not rely on histories.

The flat combiner structure (FC) generalizes a coarse-grained lock [22,23,25]. In the case of a lock, threads acquire exclusive access to the shared resource protected by the lock, *in succession*. With the flat combiner, threads register the work that they want to perform over the shared resource. The lock-acquiring thread (aka. the *combiner*) then executes all the registered work, so the other threads do not need to compete for the lock anymore. This reduces the contention on the lock, and improves performance.

The higher-order `flatCombine` procedure (Figure 8) works as follows.[6] It takes as input a *sequential* function f and argument x, and registers the invoking thread for help with executing $f\ x$ over the shared resource. It does so by storing Req f x into the shared *publication* array, at index tid (line 2), where tid is the id of the invoking thread. It next enters the main loop (line 3) and tries to acquire the lock to the shared heap (line 4). The acquiring thread becomes a combiner (line 5); it traverses the publication array, where the global variable n bounds the number of threads, checking for help requests (lines 6–11). For each request found (which can arrive even while the combiner holds the lock), the combiner executes the appro-

Fig. 8. Flat combining algorithm.

```
1  flatCombine(f: A → B, x: A): B {
2    reqHelp(tid, f, x);
3    fix loop() {
4      locked <- tryLock();
5      if locked then {
6        for i∈{0, ..., n−1} {
7          req <- readReq(i);
8          if req == Req fi xi then {
9            w <- fi(xi);
10           doHelp(i, w);
11         }}
12       unlock();}
13     rc <- tryCollect(tid);
14     if rc == Some w
15     then return w;
16     else loop();}();}
```

priate function with the provided arguments (line 9) over the shared heap. It informs the requesting thread i of the result w, by writing Resp w into the slot i of the publication array (line 10). After the traversal, the combiner releases the lock (line 12). Finally, the thread (combiner or otherwise), checks the publication array to see if it has been helped (line 13). If so, it extracts the result w from its slot in the publication array, and fills the slot with Init (all line 13). The result of the help, if one exists, is returned in line 15. Otherwise, the thread loops for help again.

To supply the intuition behind the spec for FC, we first review how ordinary locks work with auxiliary state, in the subjective setting of FCSL [22]. In CSL [23], and the Owicki-Gries method [25], a lock comes with a resource invariant I that restricts the heap of the shared resource. Such restriction implicitly assumes a presence of "hard-coded" auxiliary state, describing the contents of the corresponding shared heap (the

---

[6] For simplicity, we consider a modified version of the original algorithm. In particular, (a) we use an array rather than a priority queue for registration of help requests, and (b) we do not expunge help requests that have not been served for sufficiently long time.

explicit parametrization over the auxiliary state, which we make use of here, is explained in the introduction of [19]). When the lock is not taken, the shared heap satisfies I. When the lock is taken, the heap is in the exclusive possession of the acquiring thread, which can invalidate I, but has to restore it before releasing the lock. The subjective setting is similar, except the values of the auxiliary state are drawn from a PCM $\mathbb{U}$, and specs keep track of two values g_S and g_O, describing how much the thread (*self*) and its environment (*other*) have contributed to the resource, respectively. When the lock is free, the heap of the shared resource satisfies $I(g_S \bullet g_O)$. When the lock is released by a thread, the thread may update its g_S by some value g_Δ, reflecting that its contribution to the resource changed. Thus, if before locking, the resource satisfied $I(g_S \bullet g_O)$, after unlocking it will satisfy $I(g_S \bullet g_\Delta \bullet g_O)$, as shown by examples in Section 3 of [22].

The setup of the flat combiner is similar, but in addition to g_S and g_O, FC also keeps an array g_p storing a $\mathbb{U}$-value for each thread. The entry $g_p[i]$ signifies how much the thread i has been helped by the combiner. If $g_p[i] = g_\Delta$ is non-unit, i can collect the help by joining g_Δ to its own g_S, and setting $g_p[i]$ to the unit $\mathbb{1}$ of $\mathbb{U}$, after which it can ask for help again. Thus, the overall relation between the auxiliary state and the resource heap, when the lock is free, is captured by the invariant $I\left(\bigodot_{i=1}^{n} g_p[i] \bullet g_S \bullet g_O\right)$.

5.1 Flat Combiner State and Transitions

The states of the FC concurroid $\mathcal{F}$ are described by the assertion:

$$W_\mathcal{F} \cong \mathsf{fc} \xmapsto{s} (\mathsf{t}_S, \mathsf{m}_S, \mathsf{g}_S) \wedge \mathsf{fc} \xmapsto{o} (\mathsf{t}_O, \mathsf{m}_O, \mathsf{g}_O) \wedge \mathsf{fc} \xmapsto{j} \langle lk \mapsto b \uplus h_p \uplus h_r, \mathsf{g}_p \rangle \wedge \exists l_p.\, \mathsf{Arr}_n(a_p, l_p, h_p)$$

The auxiliary state in the self/other components consists of the following. t_S and t_O are sets of thread ids, which form a PCM under disjoint union.[7] m_S and m_O are elements of the *mutual exclusion* set $O = \{\overline{\mathsf{Own}}, \mathsf{Own}\}$ [19, 22] and record whether the lock lk is owned by the thread, or the environment. O is a PCM under the operation defined as $x \bullet \overline{\mathsf{Own}} = \overline{\mathsf{Own}} \bullet x = x$, with $\mathsf{Own} \bullet \mathsf{Own}$ undefined. The unit element is $\overline{\mathsf{Own}}$, and the undefinedness of $\mathsf{Own} \bullet \mathsf{Own}$ means that two threads cannot simultaneously own the lock. g_S and g_O are elements of a generic PCM $\mathbb{U}$, as described above. The self/other triples form a PCM with component-wise lifted joins and units.

The joint component of $\mathcal{F}$ contains a concrete heap, and the auxiliary array g_p. The concrete heap keeps the pointer $lk \mapsto b$, which stands for the lock, with the boolean b representing the lock status. It also stores the publication array with the origin pointer a_p into the heaplet h_p (see notation (22)). The array stores elements of type $\mathsf{Stat} \cong \mathsf{Init} \mid \mathsf{Req}\ f\ x \mid \mathsf{Resp}\ w$, as already apparent from Figure 8. We abuse the notation and refer to the array represented by h_p as a_p. The heap h_r is the resource protected by the FC lock. Upon locking it moves to the exclusive ownership of the combiner.

We further assume the following properties of $W_\mathcal{F}$:

(*i*) for any *tid*, if $\mathsf{g}_p[tid] \neq \mathbb{1}$, then $a_p[tid] = \mathsf{Resp}\ w$ for some w;

(*ii*) if b is true then $h_r = $ empty and $\mathsf{m}_S \bullet \mathsf{m}_O = \mathsf{Own}$; otherwise $\mathsf{m}_S \bullet \mathsf{m}_O = \overline{\mathsf{Own}}$ and $I\left(\bigodot_{i=1}^{n} \mathsf{g}_p[i] \bullet \mathsf{g}_S \bullet \mathsf{g}_O\right) h_r$.

Property (*i*) ensures that the auxiliary array g_p holds a pending contribution in a cell *tid* only if the corresponding entry in the publication array a_p points to the response

---

[7] One thread may hold many thread id's, which it distributes between its children upon forking.

with some (uncollected) result. Property *(ii)* formally relates the auxiliary state to the resource heap h_r, as already described.

Flat combinator concurroid's external transitions intuitively correspond to locking and unlocking the heap h_r, thus moving it from the joint to private state, and vice-versa. We do not present them formally, as they are similar to the transitions in CSL [22]. The internal transitions *req*, *help* and *coll* synchronously change the contents of a_p and g_p for a particular thread id i (one at a time) as the following diagram illustrates.

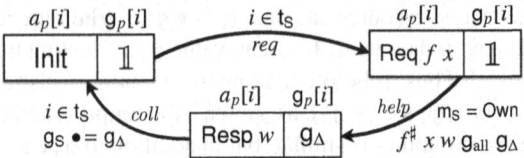

The transition *req* can be taken only by a thread holding the thread id i; it changes the value of $a_p[i]$ from Init to Req f x for some f and x. The transition *help* can be performed by any thread that owns the lock (not necessarily the one with the id i); it replaces the contents of $a_p[i]$ and $g_p[i]$ with an appropriate result w and an auxiliary delta g_Δ, respectively. The two are valid *wrt.* the input x and the cumulative auxiliary g_{all}, as ensured by the constraint $f^\sharp$. Finally, *coll* is invoked by the thread with id i; it flushes the contents of $g_p[i]$, into the self-contribution g_S and puts Init into $a_p[i]$.

5.2 Flat Combiner Specification

We now provide a spec for flatCombine in terms of the concurroid $\mathcal{F}$. We assume $f : A \to B$, $x : A$, and f comes with the following spec.[8]

$$\{ \exists h.\ \text{pv} \overset{s}{\mapsto} h \land I\ g\ h \}\ f(x)\ \{ \exists h'\ g_\Delta.\ \text{pv} \overset{s}{\mapsto} h' \land I\ (g \bullet g_\Delta)\ h' \land f^\sharp\ x\ \text{res}\ g\ g_\Delta \}@\mathcal{P} \quad (25)$$

The spec allows the input heap h to change to h'. The resource invariant I has to be preserved, up to a change of the auxiliary state, from g to g $\bullet$ g_Δ. $f^\sharp$ is a client-supplied predicate which specifies f. We call it *validity predicate*; it is functional with respect to g_Δ, and relates the input value v, the result value res, the initial auxiliary state g and the "auxiliary delta" g_Δ resulting from the invocation of f. For instance, if f were a sequential push operation on stacks, with g and g_Δ being set to histories τ and τ_Δ, we might choose the following validity predicate:

$$\text{push}^\sharp\ x\ \text{res}\ \tau\ \tau_\Delta \cong \text{res} = () \land \tau_\Delta = t^\tau_{\text{fresh}} \mapsto (l, x :: l), \quad (26)$$

where $l = \tau[\text{last}(\tau)]$. That is, $\text{push}^\sharp$ fixes the result of push to be unit and its effect to be the singleton history describing the action of pushing.

For the flatCombine spec, we need two auxiliary predicates. NoReq indicates that the thread *tid* does not request help. $\cdot \hookrightarrow (\cdot)$, generalizes (5) from histories to PCM $\mathbb{U}$.

$$\text{NoReq}(tid) \cong \text{fc} \overset{s}{\mapsto} (\{tid\}, \text{Own}, -) \land a_p[tid] = \text{Init}$$

$$\text{fc} \hookrightarrow (g_S, g_O, g) \cong \text{fc} \overset{s}{\mapsto} (-, -, g_S) \land \text{fc} \overset{o}{\mapsto} (-, -, g_O) \land g \sqsubseteq \bigodot_{i=1}^n g_p[i] \bullet g_S \bullet g_O \quad (27)$$

---

[8] Thus, we do not require f to be sequential (*i.e.*, in addition to just manipulating the privately-owned state, f can also allocate new concurroids via hiding, and fork children threads), but every sequential function can be given a spec in $\mathcal{P}$.

Here, the partial order $\sqsubseteq$ on PCM elements is defined as $g_1 \sqsubseteq g_2 \hat{=} \exists g, g_2 = g_1 \bullet g$. It generalizes the relation $\sqsubseteq$ from histories to the PCM $\mathbb{U}$, and in the specs captures that the value g_1 was "current" before g_2.

The spec for flatCombine is given *wrt.* a specific thread id *tid*.

$$\{pv \xmapsto{s} empty * fc \hookrightarrow (\mathbb{1}, -, g) \wedge \mathsf{NoReq}(tid)\}$$
$$\mathtt{flatCombine}(f, x) : B \tag{28}$$
$$\left\{\exists g' \; g_\Delta. \; pv \xmapsto{s} empty * fc \hookrightarrow (g_\Delta, -, g') \wedge \mathsf{NoReq}(tid) \wedge g \sqsubseteq g' \wedge f^\sharp \; x \; res \; g' \; g_\Delta\right\} @ \mathcal{P} \rtimes \mathcal{F}$$

A call to flatCombine starts and ends in a state in which the thread *tid* does not request the help (NoReq), and in which g names the sum total of the contributions. It does not change the privately-owned heap, but increases self-contribution by amount of an auxiliary delta g_Δ. The mediating value g' is a sum-total of the contributions at the moment when the thread received help; thus, $f^\sharp \; x \; res \; g' \; g_\Delta$. As g' is current sometime after the initial g, the spec postulates $g \sqsubseteq g'$. Due to space limitations, we omit a detailed discussion on verification of the spec (28) of the flat combiner (it can be found in [27, Appendix E] or in the accompanying Coq files).

To strengthen the analogy with coarse-grained CSL-style locks, let us note that if one were to implement a procedure coarseGrainedCombine$(f, x) = \{$lock(); $f(x)$; unlock()$\}$, its specification would be the same as (28), modulo the NoReq conjunct and the join with all $g_p[i]$ components in (27), which would not be present in the coarse-grained case, as they are artefacts of the helping machinery.[9]

5.3   Instantiating the Flat Combiner for Stacks

To illustrate that the abstract spec for the flat combiner follows the expected intuition, we consider an instance where g_s, g_o, g_p are histories, and f is the sequential push method for stacks, satisfying the generic sequential spec (25) with the validity predicate push$^\sharp$ defined by (26) and the stack invariant (16). So by instantiating (28), after some simplification, we obtain:

$$\left\{pv \xmapsto{s} empty * fc \hookrightarrow (empty, -, \tau) \wedge \mathsf{NoReq}(tid)\right\}$$
$$\mathtt{flatCombine}(push, e) : \mathsf{Unit} \tag{29}$$
$$\left\{\exists t \; l. \; pv \xmapsto{s} empty * fc \hookrightarrow (t \mapsto (l, e :: l), -, \tau) \wedge \tau < t \wedge \mathsf{NoReq}(tid)\right\}$$

Note that (29) is very similar to the spec (17) for Treiber push; the only difference, again, is in the FC-specific components such as thread id's, the NoReq predicate, and the lock status views used in the definition of NoReq. Thus, the spec (28) is adequate. A similar derivation can be done for an FC-specification of pop.

6   Related and Future Work

Histories are a recurring idea in the semantics of shared-memory concurrency, in one form or another. For example, the classical Brookes' semantics [2] uses *traces* to give a model for CSL. Traces are similar to histories, but do not contain time stamps. The explicit time-stamping makes it straightforward to define a merge (*i.e.*, join) for histories,

---

[9] To provide truly *the same* specs, we need abstract predicates to hide these artefacts. As abstract predicates are easily available in Coq, we omit the further discussion.

and endows them with PCM structure. While Brookes uses traces in the semantics, we use histories in the specs.

Temporal reasoning about shared-memory concurrent programs has also been employed before. For example, O'Hearn *et al.* [24] advocate *hindsight lemmas* to directly and elegantly capture the intuition about linearizability of a class of concurrent data structures. In this paper, we put histories to use in ordinary Hoare-style specs. This avoids the relational reasoning about permuting traces of *two* programs, as required by linearizability, but is strong enough to provide Hoare logic specs that are expressive, and capable of abstracting granularity. In our experience, deriving history-based specs very much resembles reasoning by hindsight (*e.g.*, verifying `locate` [24] and `readPair`).

HLRG by Fu *et al.* is a Hoare logic for concurrency that admits history-based assertions [11]. However, their histories are hard-coded into the logic. In contrast, our histories are just a specific PCM, that one can use to instantiate the general framework of FCSL. This affords greater flexibility: if history-based specifications are not needed (*e.g.*, the incrementation example [22]), they do not have to be used. HLRG defines separating conjunction ∗ over histories as follows: conjoined histories must have equal length, and their corresponding entry heaps are merged via disjoint union. In contrast, our histories are not required to have heaps in the codomain. One can choose an arbitrary datatype to capture what is important for an example at hand.

Bell *et al.* use a variant of concurrent separation logic augmented with a monoid of *sets of histories* to reason about programs with asynchronous communication via channels [1]. Their logic is tailored for producer/consumer pattern (similar to the example we have considered in Section 4), and it features dedicated produce/consume predicates PHist and CHist defined for a particular channel and a set of histories. However, without time-stamping, Bell *et al.*'s sets of histories do not enjoy the unifomity with heaps, hence, they are a subject of a series of dedicated inference rules.

Gotsman *et al.* use temporal reasoning to verify several concurrent memory reclamation algorithms using the notion of *grace period* [12]. Their logic extends RGSep [34] with a very specific notion of histories, which live in the shared state. In contrast, we use histories not as shared, but as private auxiliary state, following the self/other dichotomy. This enables us to directly reuse the frame rule and other logical infrastructure from the separation logic FCSL, without any extensions.

Several recent approaches, such as Turon *et al.*'s CaReSL [31] (which also verifies the flat combiner), and the logic of Liang and Feng (L&F) [20] support granularity abstraction by unifying Hoare-style reasoning with linearizability and contextual refinement. In contrast, in this paper, we argue that a form of granularity abstraction achieved by these works can already be obtained *without* relying on linearizability. Instead, by using histories, one obtains Hoare-style specs which hide the fine-grained nature of the underlying programs. This can be done in a simple Hoare logic (and we reuse FCSL off-the-shelf), whereas CaReSL and L&F require significant additional logical infrastructure [21, 32], as linearizability is a stronger property than our specs. One example of the additional infrastructure has to do with helping (*e.g.*, in the flat combiner), where these logics consider the refined effectful commands as resources, and make them subject to ownership transfer [31]. While on the surface there is a similarity between commands-as-resources and histories-as-resources, there are also significant differences. Commands-as-resources are about executing specification-level programs (and an effectful abstract program, once executed, cannot be "re-executed", since it has reached a value), while histories are about what has transpired. Unlike commands-as-resources, histories also contain information about the order in which something happened in the form of timestamps, thus enabling temporal reasoning by hindsight [24]. Histories have a PCM structure, whereas commands-as-resources do not. Hence, histories in FCSL are subject to the same set of inference rules as heaps, in contrast to commands-as-resources which requires a number of dedicated inference rules.

Many of our history-based proofs are very close in spirit to proofs of linearizability (*e.g.*, the proofs of Treiber stack in Section 4 compared to the proofs in L&F [20]), since adding an entry to a self-history can be seen as linearizing an effectful operation. However, we obtain some simplification in the proofs of pure methods such as readPair. In particular, L&F and related logics require *prophecy variables* [26] (or, equivalently, *speculations* [20, 32]) in their proofs of readPair, but we do not. We do expect, however, that prophecy variables will be required in examples where the shape of the event to be inserted into the history cannot be fully determined at the moment when it logically takes place (*e.g.*, Harris *et al.*'s MCAS [33]). We plan to address such examples in the future work, by choosing another history-based PCM; that of branching-time histories, in contrast to the linear-time ones used here.

In this work, we argued for the abstraction of granularity via the singleton histories of the form $t \mapsto (s_1, s_2)$, which describe the atomic changes in the abstract state, although other ways are possible to express what it means for a program to behave "like an atomic one" in a setting of a Hoare-style logic.

In particular, a different approach to express atomicity abstraction is suggested by da Rocha Pinto *et al.*'s logic TaDA [5] (a successor of the Concurrent Abstract Predicates framework (CAP) [6]) using the notion of an "atomic Hoare triple" of the form $\langle p \rangle\, c\, \langle q \rangle$, where the precondition p is required to be stable, whereas q is not. TaDA proposes a *make_atomic* command and a number of related inference rules, which allow one to specify *synchronized changes* of auxiliary resources across *several* shared regions. The changes themselves do not have to be physically atomic; it is sufficient that they appear atomic from the point of view of specs. TaDA's assertions range over *atomic tracking* resources, similar to the operations-as-resources [20, 31]. Unlike histories, these resources do not have the PCM structure, and thus require special treatment in TaDA's metatheory. The atomic tracking resources are not subject of ownership transfer, which is why TaDA currently does not support reasoning about helping.

Yet another view of atomicity abstraction and canonical concurrent specifications, which also bypasses linearizability, is advocated by Svendsen *et al.* in a series of papers on Higher-Order and Impredicative Concurrent Abstract Predicates [28, 29]. Both HO-CAP and iCAP leverage the idea, originated by Jacobs and Piessens [17], of parametrizing specs of concurrent data types by a user-provided auxiliary code. Such auxiliary code can be seen as a callback, which, when invoked at some point during the execution of a specified method, changes the values of auxiliary resources in several regions simultaneously. Thus, when proving a parametrized spec, one should locate a right moment to invoke the provided auxiliary code, so its precondition would be ensured and the postcondition handled properly, a reasoning similar to locating a linearization point.

The use of the first-class auxiliary code can introduce circularity in the domain underlying the logic—the issue tackled in HOCAP by means of indirection via "region types" and resolved in iCAP by providing a (non-elementary) model in the topos of trees. One difference between iCAP and TaDA is that *make_atomic* in TaDA presents a more *localized* view of atomicity, whereas the specs in iCAP have to predict the uses of the data structure, and provide hooks for callbacks. The hooks lead to somewhat indirect specs, and propagate client-side information into the reasoning about the structure.

We have not considered either of these two ways of exploiting abstract atomicity in the current paper, but plan to add *make_atomic* to FCSL in the future work. The challenge will be to generalize *make_atomic* to work with different notions of histories (*e.g.*, branching-time histories may be useful, as mentioned above). We believe that the PCM approach (together with subjectivity), neither of which is exploited by TaDA and iCAP, will be beneficial in that respect. In particular, we plan to use PCMs to generalize the notion of logical atomicity afforded by histories, that we explored in this paper. Given a PCM $\mathbb{U}$, the element $x \in \mathbb{U}$ is *prime* if it cannot be represented as $x = x_1 \bullet x_2$, for non-unit x_1, x_2. For example, in the PCM of heaps, the prime elements are the singleton heaps. In the PCM of natural numbers with multiplication, the prime elements are the prime numbers. In the PCM of histories, the prime elements are the singleton histories $t \mapsto a$. A program can be considered logically atomic if it augments the self-owned portion of its state by a prime element, or by a unit. According to this definition, all the examples presented in this paper are atomic. We expect it should be possible to soundly apply *make_atomic* to programs that are atomic in this logical sense.

7   Conclusion

In this work we proposed using specifications over auxiliary state in the form of histories as means of providing general and expressive specifications for fine-grained concurrent data structures in a separation style logic.

Histories satisfy the algebraic properties of PCMs, and thus can directly reuse the underlying infrastructure from an employed separation logic, such as its assertion logic and frame rule, enabling a separation logic style of local reasoning about histories that has usually been reserved for heaps. Moreover, as we illustrated with the formalization of the flat combiner Section 5, the concept of ownership transfer from separation logic, when specialized to the PCM of histories, captures the design pattern of helping.

In addition to the flat combiner, we have verified a number of benchmark fine-grained structures, such as the pair snapshot structure, and the Treiber stack. The novelty of the specs and the proofs is that they all rely in an essential way on the subjective dichotomy between self and other auxiliary state, in order to directly relate the result of a program execution with the interference of other threads. Such explicit dichotomy provides for what we consider very concise proofs, as demonstrated by our implementation in Coq.

Acknowledgements. We thank the anonymous ESOP 2015 reviewers for their feedback. This research was partially supported by Ramon y Cajal grant RYC-2010-0743.

References

1. Bell, C.J., Appel, A.W., Walker, D.: Concurrent separation logic for pipelined parallelization. In: Cousot, R., Martel, M. (eds.) SAS 2010. LNCS, vol. 6337, pp. 151–166. Springer, Heidelberg (2010)
2. Brookes, S.: A semantics for concurrent separation logic. Th. Comp. Sci. 375(1-3) (2007)
3. Calcagno, C., O'Hearn, P.W., Yang, H.: Local action and abstract separation logic. In: LICS (2007)
4. Cerone, A., Gotsman, A., Yang, H.: Parameterised Linearisability. In: Esparza, J., Fraigniaud, P., Husfeldt, T., Koutsoupias, E. (eds.) ICALP 2014, Part II. LNCS, vol. 8573, pp. 98–109. Springer, Heidelberg (2014)
5. da Rocha Pinto, P., Dinsdale-Young, T., Gardner, P.: TaDA: A Logic for Time and Data Abstraction. In: Jones, R. (ed.) ECOOP 2014. LNCS, vol. 8586, pp. 207–231. Springer, Heidelberg (2014)
6. Dinsdale-Young, T., Dodds, M., Gardner, P., Parkinson, M.J., Vafeiadis, V.: Concurrent Abstract Predicates. In: D'Hondt, T. (ed.) ECOOP 2010. LNCS, vol. 6183, pp. 504–528. Springer, Heidelberg (2010)
7. Elmas, T., Qadeer, S., Sezgin, A., Subasi, O., Tasiran, S.: Simplifying linearizability proofs with reduction and abstraction. In: Esparza, J., Majumdar, R. (eds.) TACAS 2010. LNCS, vol. 6015, pp. 296–311. Springer, Heidelberg (2010)
8. Feng, X.: Local rely-guarantee reasoning. In: POPL (2009)
9. Feng, X., Ferreira, R., Shao, Z.: On the relationship between concurrent separation logic and assume-guarantee reasoning. In: De Nicola, R. (ed.) ESOP 2007. LNCS, vol. 4421, pp. 173–188. Springer, Heidelberg (2007)
10. Filipovic, I., O'Hearn, P.W., Rinetzky, N., Yang, H.: Abstraction for concurrent objects. Theor. Comput. Sci. 411(51-52) (2010)
11. Fu, M., Li, Y., Feng, X., Shao, Z., Zhang, Y.: Reasoning about optimistic concurrency using a program logic for history. In: Gastin, P., Laroussinie, F. (eds.) CONCUR 2010. LNCS, vol. 6269, pp. 388–402. Springer, Heidelberg (2010)
12. Gotsman, A., Rinetzky, N., Yang, H.: Verifying concurrent memory reclamation algorithms with grace. In: Felleisen, M., Gardner, P. (eds.) ESOP 2013. LNCS, vol. 7792, pp. 249–269. Springer, Heidelberg (2013)
13. Gotsman, A., Yang, H.: Linearizability with Ownership Transfer. In: Koutny, M., Ulidowski, I. (eds.) CONCUR 2012. LNCS, vol. 7454, pp. 256–271. Springer, Heidelberg (2012)
14. Hendler, D., Incze, I., Shavit, N., Tzafrir, M.: Flat combining and the synchronization-parallelism tradeoff. In: SPAA (2010)
15. Herlihy, M., Shavit, N.: The art of multiprocessor programming. M. Kaufmann (2008)
16. Herlihy, M., Wing, J.M.: Linearizability: A correctness condition for concurrent objects. ACM Trans. Prog. Lang. Syst. 12(3) (1990)
17. Jacobs, B., Piessens, F.: Expressive modular fine-grained concurrency specification. In: POPL (2011)
18. Jones, C.B.: Specification and design of (parallel) programs. In: IFIP Congress (1983)
19. Ley-Wild, R., Nanevski, A.: Subjective auxiliary state for coarse-grained concurrency. In: POPL (2013)
20. Liang, H., Feng, X.: Modular verification of linearizability with non-fixed linearization points. In: PLDI (2013)
21. Liang, H., Feng, X., Fu, M.: A rely-guarantee-based simulation for verifying concurrent program transformations. In: POPL (2012)
22. Nanevski, A., Ley-Wild, R., Sergey, I., Delbianco, G.A.: Communicating State Transition Systems for Fine-Grained Concurrent Resources. In: Shao, Z. (ed.) ESOP 2014. LNCS, vol. 8410, pp. 290–310. Springer, Heidelberg (2014)

23. O'Hearn, P.W.: Resources, concurrency, and local reasoning. Th. Comp. Sci. 375(1-3) (2007)
24. O'Hearn, P.W., Rinetzky, N., Vechev, M.T., Yahav, E., Yorsh, G.: Verifying linearizability with hindsight. In: PODC (2010)
25. Owicki, S.S., Gries, D.: Verifying properties of parallel programs: An axiomatic approach. Commun. ACM 19(5) (1976)
26. Qadeer, S., Sezgin, A., Tasiran, S.: Back and forth: Prophecy variables for static verification of concurrent programs. Technical Report MSR-TR-2009-142 (2009)
27. Sergey, I., Nanevski, A., Banerjee, A.: Specifying and verifying concurrent algorithms with histories and subjectivity. Extended Version and Supporting Material, http://ilyasergey.net/projects/histories
28. Svendsen, K., Birkedal, L.: Impredicative Concurrent Abstract Predicates. In: Shao, Z. (ed.) ESOP 2014. LNCS, vol. 8410, pp. 149–168. Springer, Heidelberg (2014)
29. Svendsen, K., Birkedal, L., Parkinson, M.: Modular reasoning about separation of concurrent data structures. In: Felleisen, M., Gardner, P. (eds.) ESOP 2013. LNCS, vol. 7792, pp. 169–188. Springer, Heidelberg (2013)
30. Treiber, R.K.: Systems programming: coping with parallelism. Technical Report RJ 5118, IBM Almaden Research Center (1986)
31. Turon, A., Dreyer, D., Birkedal, L.: Unifying refinement and Hoare-style reasoning in a logic for higher-order concurrency. In: ICFP (2013)
32. Turon, A.J., Thamsborg, J., Ahmed, A., Birkedal, L., Dreyer, D.: Logical relations for fine-grained concurrency. In: POPL (2013)
33. Vafeiadis, V.: Modular fine-grained concurrency verification. PhD thesis, University of Cambridge (2007)
34. Vafeiadis, V., Parkinson, M.: A Marriage of Rely/Guarantee and Separation Logic. In: Caires, L., Vasconcelos, V.T. (eds.) CONCUR 2007. LNCS, vol. 4703, pp. 256–271. Springer, Heidelberg (2007)

Witnessing (Co)datatypes

Jasmin Christian Blanchette[1,2], Andrei Popescu[3], and Dmitriy Traytel[4]

[1] Inria Nancy & LORIA, Villers-lès-Nancy, France
[2] Max-Planck-Institut für Informatik, Saarbrücken, Germany
[3] Department of Computer Science, School of Science and Technology,
Middlesex University, UK
[4] Fakultät für Informatik, Technische Universität München, Germany

Abstract. Datatypes and codatatypes are useful for specifying and reasoning about (possibly infinite) computational processes. The Isabelle/HOL proof assistant has recently been extended with a definitional package that supports both. We describe a complete procedure for deriving nonemptiness witnesses in the general mutually recursive, nested case—nonemptiness being a proviso for introducing types in higher-order logic.

1 Introduction

Proof assistants, or interactive theorem provers, are becoming increasingly popular as vehicles for formalizing the metatheory of logical systems and programming languages. Such developments often involve datatypes and codatatypes in various constellations. For example, Lochbihler's formalization of the Java memory model represents possibly infinite executions using a codatatype [26]. Codatatypes are also useful for capturing lazy data structures, such as Haskell's lists.

A popular and expanding family of proof assistants, heavily used in software and hardware verification, are those based on higher-order logic (HOL)—examples include HOL4 [37], HOL Light [16], HOL Zero [3], Isabelle/HOL [30], and ProofPower–HOL [4]. They are traditionally built on top of a trusted inference kernel through which all theorems are generated. Various definitional packages reduce high-level specifications to primitive inferences; characteristic theorems are derived rather than postulated. This reduces the amount of code that must be trusted. We recently extended Isabelle/HOL with a definitional package for mutually recursive, nested (co)datatypes [8, 39]. While some proof assistants support codatatypes (notably, Agda, Coq, Matita, and PVS), Isabelle is the first to provide a *definitional* implementation.

In this paper, we focus on a fundamental problem posed by any HOL development that extends the type infrastructure: proofs of, or "witnesses" for, the nonemptiness of newly introduced types. Besides its importance to formal logic engineering, the problem also enjoys theoretical relevance, since it essentially amounts to the decision problem for the nonemptiness of open-ended, mutual, nested (co)datatypes. Furthermore, our modular witness generation algorithm is relevant outside the proof assistant world, in areas such as program synthesis [15].

Our starting point is the nonemptiness requirement on HOL types. This is a well-known design decision connected to the presence of Hilbert choice in HOL [13, 31]. In

© Springer-Verlag Berlin Heidelberg 2015
J. Vitek (Ed.): ESOP 2015, LNCS 9032, pp. 359–382, 2015.
DOI: 10.1007/978-3-662-46669-8_15

all HOL-based provers, the following inductive specification of "finite streams" must be rejected because it would lead to an empty datatype:

datatype α fstream $=$ FSCons α (α fstream)

While checking nonemptiness appears to be an easy reachability test, nested recursion complicates the picture, as shown by this attempt to define infinitely branching trees with finite branches by nested recursion via a codatatype of (infinite) streams:

codatatype α stream $=$ SCons α (α stream)
datatype α tree $=$ Node α ((α tree) stream)

The second definition should fail: To get a witness for α tree, we would need a witness for (α tree) stream, and vice versa. Replacing streams with finite lists should make the definition acceptable, because the empty list stops the recursion. Even though final coalgebras are never empty (except in trivial cases), here the datatype provides a better witness (the empty list) than the codatatype (which requires an α tree to build an (α tree) stream). Mutual, nested datatype specifications can be arbitrarily complex:

datatype (α,β) tree $=$ Leaf β | Branch $((\alpha + (\alpha,\beta)$ tree) stream)
codatatype (α,β) ltree $=$ LNode β $((\alpha + (\alpha,\beta)$ ltree) stream)
datatype $t_1 = T_{11} (((t_1, t_2)$ ltree) stream) | $T_{12} (t_1 \times (t_2 + t_3)$ stream)
and $t_2 = T_2 ((t_1 \times t_2)$ list) and $t_3 = T_3 ((t_1, (t_3, t_3)$ tree) tree)

The definitions are legitimate, but the last group should be rejected if t_2 is replaced by t_3 in the constructor T_{11}.

What makes the problem interesting is our open-endedness assumption: The type constructors handled by the (co)datatype package are not syntactically predetermined. In particular, they are not restricted to polynomial functors—the user can register new type constructors in the package database after establishing a few semantic properties.

Our solution exploits the package's abstract, functorial view of types. Each (co)datatype, and more generally each functor (type constructor) that participates in a definition, carries its own witnesses together with soundness proofs. Operations such as functorial composition, initial algebra, and final coalgebra derive their witnesses from those of the operands. Each computational step performed by the package is certified in HOL. The solution is complete: Given precise information about the functors participating in a definition, all nonempty datatypes are identified as such.

We start by recalling the package's abstract layer, which is based on category theory (Section 2). Then we look at a concrete instance: a variation of context-free grammars acting on finite sets and their associated possibly infinite derivation trees (Section 3). The example supplies precious building blocks to the nonemptiness proofs (Section 4). It also displays some unique characteristics of the package, such as support for nested recursion through nonfree types. Other features and user conveniences are described elsewhere [8, 11]. The formalization covering the results presented here is publicly available [9]. It employs similar notations to this text but presents more details. The implementation is part of Isabelle [30] (Section 5).

Conventions. We work informally in a mathematical universe $\mathscr{S}$ of sets but adopt many conventions from higher-order logic and functional programming. Function application is normally written in prefix form without parentheses (e.g., $f \, x \, y$). Sets are ranged over by capital Roman letters $(A, B, \ldots)$ and Greek letters $(\alpha, \beta, \ldots)$. For n-ary functions, we often prefer the curried form $f : \alpha_1 \to \cdots \to \alpha_n \to \beta$ to the tuple form $f : \alpha_1 \times \cdots \times \alpha_n \to \beta$ but occasionally pass tuples to curried functions. Polymorphic operators are regarded as families of higher-order constants indexed by sets.

Operators on sets are normally written in postfix form: α set is the powerset of α, consisting of sets of elements of α; α fset is the set of finite sets over α. Given $f : \alpha \to \beta$, $A \subseteq \alpha$, and $B \subseteq \beta$, image $f \, A$, or $f \cdot A$, is the image of A through f, and $f^- B$ is the inverse image of B through f. The set unit contains a single element (), and $[n] = \{1, \ldots, n\}$. Prefix and postfix operators bind more tightly than infixes, so that $\alpha \times \beta$ set is read as $\alpha \times (\beta \text{ set})$ and $f \cdot g \, x$ as $f \cdot (g \, x)$.

The notation $\bar{a}_n$, or simply $\bar{a}$, denotes the tuple $(a_1, \ldots, a_n)$. Given $\bar{a}_m$ and $\bar{b}_n$, $(\bar{a}, \bar{b})$ denotes the flat tuple $(a_1, \ldots, a_m, b_1, \ldots, b_n)$. Given n m-ary functions $f_1, \ldots, f_n$, the notation $\bar{f} \, \bar{a}$ stands for $(f_1 \, \bar{a}, \ldots, f_n \, \bar{a})$, and similarly $\bar{\alpha} \, \mathsf{F} = (\bar{\alpha} \, \mathsf{F}_1, \ldots, \bar{\alpha} \, \mathsf{F}_n)$. Depending on the context, $\bar{\alpha}_n \, \mathsf{F}$ either denotes the application of F to $\bar{\alpha}$ or merely indicates that F is an n-ary set operator.

2   The Category Theory behind the Package

User-specified (co)datatypes and their characteristic theorems are derived from underlying constructions adapted from category theory. The central concept is that of bounded natural functors, a well-behaved class of functors with additional structure.

2.1   Functors and Functor Operations

We consider operators F on sets, which we call *set constructors*. We are interested in set constructors that are *functors* on the category of sets and functions, i.e., that are equipped with an action on morphisms commuting with identities and composition. This action is a polymorphic constant $\mathsf{Fmap} : (\alpha_1 \to \beta_1) \to \cdots \to (\alpha_n \to \beta_n) \to \bar{\alpha} \, \mathsf{F} \to \bar{\beta} \, \mathsf{F}$ that satisfies $\mathsf{Fmap} \, \overline{\mathsf{id}} = \mathsf{id}$ and $\mathsf{Fmap} \, (g_1 \circ f_1) \ldots (g_n \circ f_n) = \mathsf{Fmap} \, \bar{g} \circ \mathsf{Fmap} \, \bar{f}$. Formally, functors are pairs $(\mathsf{F}, \mathsf{Fmap})$. Basic instances are presented below.

Identity functor $(\mathsf{ID}, \mathsf{id})$.   The identity maps any set and any function to itself.

(n, α)-*Constant functor* $(\mathsf{C}_{n,\alpha}, \mathsf{Cmap}_{n,\alpha})$.   The (n, α)-constant functor $(\mathsf{C}_{n,\alpha}, \mathsf{Cmap}_{n,\alpha})$ is the n-ary functor consisting of the set constructor $\bar{\beta} \, \mathsf{C}_{n,\alpha} = \alpha$ and the action $\mathsf{Cmap}_{n,\alpha} \, f_1 \ldots f_n = \mathsf{id}$. We write C_α for $\mathsf{C}_{1,\alpha}$.

Sum functor $(+, \oplus)$.   The sum $\alpha_1 + \alpha_2$ consists of a copy $\mathsf{Inl} \, a_1$ of each element $a_1 : \alpha_1$ and a copy $\mathsf{Inr} \, a_2$ of each element $a_2 : \alpha_2$. Given $f_1 : \alpha_1 \to \beta_1$ and $f_2 : \alpha_2 \to \beta_2$, let $f_1 \oplus f_2 : \alpha_1 + \alpha_2 \to \beta_1 + \beta_2$ be the function sending $\mathsf{Inl} \, a_1$ to $\mathsf{Inl} \, (f_1 \, a_1)$ and $\mathsf{Inr} \, a_2$ to $\mathsf{Inr} \, (f_2 \, a_2)$.

Product functor $(\times, \otimes)$.   Let $\mathsf{fst} : \alpha_1 \times \alpha_2 \to \alpha_1$ and $\mathsf{snd} : \alpha_1 \times \alpha_2 \to \alpha_2$ denote the two projection functions from pairs. Given $f_1 : \alpha \to \beta_1$ and $f_2 : \alpha \to \beta_2$, let $\langle f_1, f_2 \rangle :$

$\alpha \to \beta_1 \times \beta_2$ be the function $\lambda a. (f_1\ a, f_2\ a)$. Given $f_1 : \alpha_1 \to \beta_1$ and $f_2 : \alpha_2 \to \beta_2$, let $f_1 \otimes f_2 : \alpha_1 \times \alpha_2 \to \beta_1 \times \beta_2$ be $\langle f_1 \circ \mathsf{fst},\ f_2 \circ \mathsf{snd} \rangle$.

α-Function space functor (func_α, comp_α).   Given a set α, let $\beta\ \mathsf{func}_\alpha = \alpha \to \beta$. For all $g : \beta \to \gamma$, let $\mathsf{comp}_\alpha\ g : \beta\ \mathsf{func}_\alpha \to \gamma\ \mathsf{func}_\alpha$ be $\mathsf{comp}_\alpha\ g\ f = g \circ f$.

Powerset functor (set, image).   For all $f : \alpha \to \beta$, the function $\mathsf{image}\ f : \alpha\ \mathsf{set} \to \beta\ \mathsf{set}$ sends each subset A of α to the image of A through the function $f : \alpha \to \beta$.

Bounded k-powerset functor (set_k, image).   Given an infinite cardinal k, for all sets α, the set $\alpha\ \mathsf{set}_k$ carves out from $\alpha\ \mathsf{set}$ only those sets of cardinality less than k. The finite powerset functor fset corresponds to $\mathsf{set}_{\aleph_0}$.

Functors can be composed to form complex functors. Composition requires the functors F_j to take the same type arguments $\overline{\alpha}$ in the same order. The operations of permutation and lifting, together with the identity and (n, α)-constant functors, make it possible to compose functors freely. Let Func_n be the collection of n-ary functions.

Composition.   Given $\overline{\alpha}\ F_j$ for $j \in [n]$ and $\overline{\beta}_n\ G$, the *functor composition* $G \circ \overline{F}$ is defined as $(\overline{\alpha}\ \overline{F})\ G$ on objects and similarly on morphisms.

Permutation.   Given $F \in \mathsf{Func}_n$ and $i, j \in [n]$ with $i < j$, the (i,j)-*permutation* of F, written $F^{(i,j)} \in \mathsf{Func}_n$, is defined on objects as $\overline{\alpha}\ F^{(i,j)} = (\alpha_1, \dots, \alpha_{i-1}, \alpha_j, \alpha_{i+1}, \dots, \alpha_{j-1}, \alpha_i, \alpha_{j+1}, \dots, \alpha_n)\ F$ and similarly on morphisms.

Lifting.   Given $F \in \mathsf{Func}_n$, the *lifting* of F, written $F\!\uparrow\ \in \mathsf{Func}_{n+1}$, is defined on objects as $(\overline{\alpha}_n, \alpha_{n+1})\ F\!\uparrow\ = \overline{\alpha}_n\ F$ and similarly on morphisms. In other words, $F\!\uparrow$ is obtained from F by adding a superfluous argument α_{n+1}.

Datatypes are defined by taking the initial algebra of a set of functors and codatatypes by taking the final coalgebra. Both operations are partial.

Initial algebra.   Given n $(m+n)$-ary functors $(\overline{\alpha}_m, \overline{\beta}_n)\ F_j$, their *(mutual) initial algebra* consists of n m-ary functors $\overline{\alpha}\ \mathsf{IF}_j$ that satisfy the isomorphism $\overline{\alpha}\ \mathsf{IF}_j \cong (\overline{\alpha}, \overline{\alpha}\ \overline{\mathsf{IF}})\ F_j$ minimally (i.e., as the least fixpoint). The variables $\overline{\alpha}$ are the passive parameters, and $\overline{\beta}$ are the fixpoint variables. The functors IF_j are characterized by

- n polymorphic *folding bijections* (constructors) $\mathsf{ctor}_j : (\overline{\alpha}, \overline{\alpha}\ \overline{\mathsf{IF}})\ F_j \to \overline{\alpha}\ \mathsf{IF}_j$ and
- n polymorphic *iterators* $\mathsf{fold}_j : \left(\prod_{k \in [n]} (\overline{\alpha}, \overline{\beta})\ F_k \to \beta_k \right) \to \overline{\alpha}\ \mathsf{IF}_j \to \beta_j$

and subject to the following properties (for all $j \in [n]$):

- Iteration equations: $\mathsf{fold}_j\ \overline{s} \circ \mathsf{ctor}_j = s_j \circ \mathsf{Fmap}\ \overline{\mathsf{id}}\ (\overline{\mathsf{fold}\ \overline{s}})$.
- Unique characterization of iterators: Given $\overline{\beta}$ and $\overline{s}$, the only functions $f_j : \overline{\alpha}\ \mathsf{IF}_j \to \beta_j$ satisfying $f_j \circ \mathsf{ctor}_j = s_j \circ \mathsf{Fmap}\ \overline{\mathsf{id}}\ \overline{f}$ are $\mathsf{fold}_j\ \overline{s}$.

The functorial actions IFmap_j for IF_j are defined by iteration in the standard way.

Final coalgebra.   The final coalgebra operation is categorically dual to initial algebra. Given n $(m+n)$-ary functors $(\overline{\alpha}_m, \overline{\beta}_n)\ F_j$, their *(mutual) final coalgebra* consists of n m-ary functors $\overline{\alpha}\ \mathsf{JF}_j$ that satisfy the isomorphism $\overline{\alpha}\ \mathsf{JF}_j \cong (\overline{\alpha}, \overline{\alpha}\ \overline{\mathsf{JF}})\ F_j$ maximally (i.e., as the greatest fixpoint). The functors JF_j are characterized by

- n polymorphic *unfolding bijections* (destructors) $\mathrm{dtor}_j : \overline{\alpha}\ \mathrm{JF}_j \to (\overline{\alpha}, \overline{\alpha}\ \overline{\mathrm{JF}})\ \mathrm{F}_j$ and
- n polymorphic *coiterators* $\mathrm{unfold}_j : \left(\prod_{k \in [n]} \beta_k \to (\overline{\alpha}, \overline{\beta})\ \mathrm{F}_k \right) \to \beta_j \to \overline{\alpha}\ \mathrm{JF}_j$

and subject to the following properties:

- Coiteration equations: $\mathrm{dtor}_j \circ \mathrm{unfold}_j\ \overline{s} = \mathrm{Fmap}\ \overline{\mathrm{id}}\ (\overline{\mathrm{unfold}\ \overline{s}}) \circ s_j$.
- Unique characterization of coiterators: Given $\overline{\beta}$ and $\overline{s}$, the only functions $f_j : \beta_j \to \overline{\alpha}\ \mathrm{JF}_j$ satisfying $\mathrm{dtor}_j \circ f_j = \mathrm{Fmap}\ \overline{\mathrm{id}}\ \overline{f} \circ s_j$ are $\mathrm{unfold}_j\ \overline{s}$.

The functorial actions JFmap_j for JF_j are defined by coiteration in the standard way.

2.2 Bounded Natural Functors

The (co)datatype package is based on a class $\mathscr{B}$ of functors, called *bounded natural functors (BNFs)*. The particular axioms defining $\mathscr{B}$ are described in previous papers [8, 39]. The class $\mathscr{B}$ contains all the basic functors except for unbounded powerset and is closed under the operations described in Section 2.1.

Unlike the (co)datatype specification mechanisms of other proof assistants, in our package the involved types are not syntactically predetermined by a fixed grammar. $\mathscr{B}$ includes the class of polynomial functors but is additionally open-ended in the sense that users can register further functors as members of $\mathscr{B}$.

Besides closure under functor operations, another important question for theorem proving is how to state induction and coinduction abstractly, irrespective of the shape of the functor. We know how to state induction on lists, or trees, but how about initial algebras of arbitrary functors?

The answer we propose enriches the structure of functors $\overline{\alpha}_n\ \mathrm{F}$ with additional data: For each $i \in [n]$, BNFs must provide a natural transformation $\mathrm{Fset}^i : \overline{\alpha}\ \mathrm{F} \to \alpha_i$ set that gives, for $x \in \overline{\alpha}\ \mathrm{F}$, the set of α_i-atoms that take part in x. For example, if $(\alpha_1, \alpha_2)\ \mathrm{F} = \alpha_1 \times \alpha_2$, then $\mathrm{Fset}^1\ (a_1, a_2) = \{a_1\}$ and $\mathrm{Fset}^2\ (a_1, a_2) = \{a_2\}$; if $\alpha\ \mathrm{F} = \alpha$ list (the list functor, obtained as minimal solution to $\beta \cong \mathrm{unit} + \alpha \times \beta$), then $\mathrm{Fset}\ (= \mathrm{Fset}^1)$ applied to a list x gives all the elements appearing in x.[1] The abstract (co)induction principles can be massaged to account for multiple curried constructors (Appendices B and C).

Given $j \in [n]$, the elements of $\mathrm{Fset}_j^{m+k}\ x$ (for $k \in [n]$) are the recursive components of $\mathrm{ctor}_j\ x$. (Notice that subscripts select functors F_j in the tuple $\overline{\mathrm{F}}$, whereas superscripts select Fset operators for different arguments of F_j.) The explicit modeling of the recursive components makes it possible to state induction and coinduction abstractly for arbitrary BNFs (Appendix A).

Briefly, the registration process is as follows. The user provides a type constructor F and its associated BNF structure (in the form of polymorphic HOL constants), including the Fmap functorial action on objects. Then the user establishes the BNF properties (e.g., that $(\mathrm{F}, \mathrm{Fmap})$ is indeed a functor). After this, the new BNF is integrated and can appear nested in future (co)datatype definitions. Following this procedure, Isabelle users

---

[1] This Fset has similarities with Pierce's notion of support from his account of (co)inductive types [33] and with Abel and Altenkirch's urelement relation from their framework for strong normalization [1]. A distinguishing feature of our notion is the consideration of categorical structure [39].

have already introduced the BNF α bag of finite bags (multisets) over α and the BNF α pmf of probability mass functions with domain α. Other nonstandard BNFs can be produced by using the quotient package [22,23] and the nonfree datatype package [36].

As an example, the type constructor α bag is registered as a BNF by the following command:

bnf α bag
  map: bmap : $(\alpha \to \beta) \to \alpha$ bag $\to \beta$ bag
  sets: bset : α bag $\to \alpha$ set
  bd: $\aleph_0$: (nat × nat) set
  wits: $\{\#\} : \alpha$ bag
  rel: brel : $(\alpha \to \beta \to$ bool$) \to \alpha$ bag $\to \beta$ bag $\to$ bool

The command provides the necessary infrastructure that makes α bag a BNF, consisting of various previously introduced constants (whose definitions are not shown here):

- the functorial action (bmap);
- the natural transformation (bset);
- a cardinal bound represented as minimal well-order relations [10] (here, that of natural numbers, $\aleph_0$);
- a witness term (the empty bag $\{\#\}$);
- a custom relator (brel).

The user is then requested to discharge the BNF assumptions [8, Section 2]:

$$\text{bmap id} = \text{id} \qquad \text{bmap } (f \circ g) = \text{bmap } f \circ \text{bmap } g \qquad \frac{\forall x.\ x \in \text{bset } xs \Rightarrow f\, x = g\, x}{\text{bmap } f\, xs = \text{bmap } g\, xs}$$

$$|\text{bset } xs| \leq_o \aleph_0 \qquad \text{bset} \circ \text{bmap } f = \text{image } f \circ \text{bset}$$

$$\text{brel } R\, x\, y \iff \exists z.\ \text{bset } z \subseteq \{(x, y) \mid R\, x\, y\} \wedge \text{bmap fst } z = x \wedge \text{bmap snd } z = y$$

$$\text{brel } R \circ\circ \text{brel } S \sqsubseteq \text{brel } (R \circ\circ S)$$

(The operator $\leq_o$ is a well-order on ordinals [10], $\sqsubseteq$ denotes implication lifted to binary predicates, and $\circ\circ$ denotes the relational composition of binary predicates.) In addition, the user is invited to discharge the nonemptiness witness property bset $\{\#\} = \{\}$.

3   Coinductive Derivation Trees

Before turning to the nonemptiness witnesses, we first study a concrete codatatype definable with our package. It consists of derivation trees for a context-free grammar, where we perform the following changes to the usual setting: Trees are possibly infinite and the generated words are not lists, but finite sets. The Isabelle formalization of this example [9] lays at the heart of the results presented in the next section. Indeed, this particular codatatype will provide the infrastructure for tracking nonemptiness of arbitrary (co)datatypes.

We take a few liberties with Isabelle notations to lighten the presentation; in particular, until Section 4, we always ignore the distinction between sets and types.

Definition of Derivation Trees.   We fix a set T of *terminals* and a set N of *nonterminals*. The command

codatatype dtree = Node (root: N) (cont: (T + dtree) fset)

introduces a constructor Node : $N \rightarrow (T + \text{dtree})$ fset $\rightarrow$ dtree and two selectors root : dtree $\rightarrow$ N, cont : dtree $\rightarrow$ $(T + \text{dtree})$ fset. A tree has the form Node $n\ as$, where n is a nonterminal (the tree's *root*) and as is a finite set of terminals and trees (its *continuation*). The codatatype keyword indicates that this tree formation rule may be applied an infinite number of times.

Given the above definition of dtree, the package first composes the input BNF to the final coalgebra operation $\text{pre\_dtree} = (\times) \circ (C_N, \text{fset} \circ ((+) \circ (C_T, ID)))$ from the constants N and T, identity, sum, product, and finite set. In the sequel, we prefer the more readable notation α pre_dtree $= N \times (T + \alpha)$ fset. Then it constructs the final coalgebra dtree $(= \text{JF})$ from pre_dtree $(= F)$.

The unfolding bijection dtor : dtree $\rightarrow$ dtree pre_dtree is decomposed in two selectors: root $=$ fst $\circ$ dtor and cont $=$ snd $\circ$ dtor. The constructor Node is defined as the inverse of the unfolding bijection. The basic properties of constructors and selectors (e.g., injectivity, distinctness) are derived from those of sums and products.

After some massaging that involves splitting according to the indicated destructors, the abstract coiterator from Section 2.2 leaves the stage to the dtree coiterator unfold : $(\beta \rightarrow N) \rightarrow (\beta \rightarrow (T + \beta) \text{ fset}) \rightarrow \beta \rightarrow \text{dtree}$ characterized as follows: For all sets β, functions $r : \beta \rightarrow N$, $c : \beta \rightarrow (T + \beta)$ fset, and elements $b \in \beta$,

$$\text{root (unfold } r\ c\ b) = r\ b \qquad \text{cont (unfold } r\ c\ b) = (\text{id} \oplus \text{unfold } r\ c) \bullet c\ b$$

Intuitively, the coiteration contract reads as follows: Given a set β, to define a function $f : \beta \rightarrow \text{dtree}$ we must indicate how to build a tree for each $b \in \beta$. The root is given by r, and its continuation is given corecursively by c. Formally, $f = \text{unfold } r\ c$.

A Variation of Context-Free Grammars. We consider a variation of context-free grammars, acting on finite sets instead of sequences. We assume that the previously fixed sets T and N, of terminals and nonterminals, are finite and that we are given a set of *productions* $P \subseteq N \times (T + N)$ fset. The triple $Gr = (T, N, P)$ forms a *(set) grammar*, which is fixed for the rest of this section. Both finite and infinite derivation trees are of interest. The codatatype dtree constitutes a suitable universe for defining well-formed trees as a coinductive predicate.

Fixpoint (or Knaster–Tarski) (co)induction is provided in Isabelle/HOL by a separate package [32]. Fixpoint induction relies on the minimality of a predicate (the least fixpoint); dually, fixpoint coinduction relies on maximality (the greatest fixpoint). It is well known that datatypes interact well with definitions by fixpoint induction. For codatatypes, both fixpoint induction and fixpoint coinduction play an important role—the former to express safety properties, the latter to express liveness.

Well-formed derivation trees for Gr are defined coinductively as the greatest predicate wf : dtree $\rightarrow$ bool such that, for all $t \in$ dtree,

$$\text{wf } t \iff (\text{root } t, (\text{id} \oplus \text{root}) \bullet \text{cont } t) \in P \wedge \text{root is injective on } \text{Inr}^- (\text{cont } t) \wedge$$
$$\forall t' \in \text{Inr}^- (\text{cont } t). \text{ wf } t'$$

Each nonterminal node of a well-formed derivation tree t represents a production. This is achieved by three conditions: (1) the root of t forms a production together with the terminals constituting its successor leaves and the roots of its immediate subtrees; (2) no two immediate subtrees of t have the same root; (3) properties 1 and 2 also hold for the immediate subtrees of t. The definition's coinductive nature ensures that these properties hold for arbitrarily deep subtrees of t, even if t has infinite depth.

In contrast to well-formedness, the notions of subtree, interior (the set of nonterminals appearing in a tree), and frontier (the set of terminals appearing in a tree) require inductive definitions. The *subtree* relation subtr : dtree $\rightarrow$ dtree $\rightarrow$ bool is defined inductively as the least predicate satisfying the rules

> subtr $t\,t$
> subtr $t\,t'' \wedge$ Inr $t'' \in$ cont $t' \Rightarrow$ subtr $t\,t'$

We write Subtr t for the set of subtrees of t. The *interior* Itr : dtree $\rightarrow$ N set is defined inductively by the rules

> root $t \in$ Itr t
> Inr $t_1 \in$ cont $t \wedge n \in$ Itr $t_1 \Rightarrow n \in$ Itr t

The *frontier* Fr : dtree $\rightarrow$ N set is defined inductively by

> Inl $t \in$ cont $t \Rightarrow t \in$ Fr t
> Inr $t_1 \in$ cont $t \wedge t \in$ Fr $t_1 \Rightarrow t \in$ Fr t

The language generated by the grammar Gr from a nonterminal $n \in$ N (via possibly infinite derivation trees) is defined as $\mathscr{L}_{\mathsf{Gr}}(n) = \{\mathsf{Fr}\,t \mid \mathsf{wf}\,t \wedge \mathsf{root}\,t = n\}$.

Regular Derivation Trees. A derivation tree is *regular* if each subtree is uniquely determined by its root. Formally, we define regular t as the existence of a function $f : \mathsf{N} \rightarrow$ Subtr t such that $\forall t' \in$ Subtr t. f (root t') $= t'$. The regular language of a nonterminal is defined as $\mathscr{L}_{\mathsf{Gr}}^{\mathsf{r}}(n) = \{\mathsf{Fr}\,t \mid \mathsf{wf}\,t \wedge \mathsf{root}\,t = n \wedge \mathsf{regular}\,t\}$.

Given a possibly nonregular derivation tree t_0, a *regular cut* of t_0 is a regular tree rcut t_0 such that Fr (rcut t_0) $\subseteq$ Fr t_0. Here is one way to perform the cut:

1. Choose a subtree of t_0 for each interior node $n \in$ Itr t_0 via a function pick : Itr $t_0 \rightarrow$ Subtr t_0 with $\forall n \in$ Itr t_0. root (pick n) $= n$.

2. Traverse t_0 and substitute pick n for each subtree with root n. Perform this substitution hereditarily, i.e., also in the emerging subtree pick n.

This substitution task is elegantly achieved by the corecursive function H : Itr $t_0 \rightarrow$ dtree defined as unfold $r\,c$, where $r :$ Itr $t_0 \rightarrow$ N and $c :$ Itr $t_0 \rightarrow$ (T $+$ Itr t_0) fset are specified as follows: $r\,n = n$ and $c\,n = (\mathsf{id} \oplus \mathsf{root}) \bullet \mathsf{cont}$ (pick n). The function H is therefore characterized by the corecursive equations root (H n) $= n$ and cont (H n) $= (\mathsf{id} \oplus (\mathsf{H} \circ \mathsf{root})) \bullet$ cont (pick n). It is not hard to prove the following by fixpoint coinduction:

Lemma 1. For all $n \in$ Itr t_0, H n is regular and Fr (H n) $\subseteq$ Fr t_0. Moreover, H n is well-formed provided t_0 is well-formed.

Proof. H n is regular by construction: If a subtree of it has root n', then it is equal to H n'. The frontier inclusion Fr (H n) $\subseteq$ Fr t_0 follows by routine fixpoint induction on the definition of Fr (since at each node $n' \in$ Itr (H n) we only have the immediate leaves of pick n', which is a subtree of Fr t_0). Finally, assume that t_0 is well-formed. Then the well-formedness of H n follows by routine fixpoint coinduction on the definition of wf (since, again, at each $n' \in$ Itr (H n) we have the production of pick n'). □

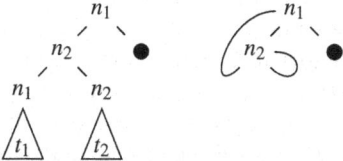

Fig. 1. A derivation tree (left) and a minimal regular cut (right)

We define rcut t_0 to be H (root t_0). Figure 1 shows a derivation tree and a minimal regular cut. The bullets denote terminals, and t_1 and t_2 are arbitrary trees with roots n_1 and n_2. The loop indicates an infinite tree that is its own subtree.

4   Computing Nonemptiness Witnesses

In the previous two sections, we referred to the codatatype dtree and other collections of elements as *sets*, ignoring an important aspect of HOL. While for most purposes sets and types can be identified in an abstract treatment of the logic, empty types are not allowed. The main primitive way to define custom types in HOL is to specify from an existing type α a nonempty subset $A : \alpha$ set that is isomorphic to the desired type. Hence, to register a collection of elements as a HOL type (and take advantage of the associated convenience, notably static type checking), it is necessary to prove it nonempty.

Datatype definitions are an instance of the above scenario, with the additional requirement that nonemptiness should be discharged automatically. When producing the relevant nonemptiness proofs, the package must take into consideration arbitrary combinations of basic and user-defined BNFs, datatypes, and codatatypes.

A first idea would be to follow the traditional approach of HOL datatype packages [6, 14]: Unfold all the definitions of the involved nested datatypes, inlining them as additional components of the mutual definition, until only sums of products remain, and then perform a reachability analysis. However, this approach is problematic in our framework. Due to open-endedness, there is no fixed set of basic types. Delving into nested types requires reproving nonemptiness facts, which scales poorly. Moreover, it is not clear how to unfold datatypes nested in codatatypes or vice versa.

By relying on all specifications being eventually reducible to the fixed situation of sums of products, the traditional approach needs to consider nonemptiness only at the point of a datatype definition. Here, we look for a prophylactic solution instead, trying to prepare the BNFs for future nonemptiness checks involving them. To this end, we ask: Given a mutual datatype definition involving several n-ary BNFs, what is the relevant information we need to know about their nonemptiness *without knowing what they look like* (hence, with no option to delve into them)? To answer this, we use a generalization of pointed types [20,25], by maintaining witnesses that assert conditional nonemptiness for combinations of arguments. We introduce the solution by examples.

4.1   Introductory Examples

We start with the simple cases of products and sums. For $\alpha \times \beta$, the proof is as follows: Assuming $\alpha \neq \emptyset$ and $\beta \neq \emptyset$, we construct the witness $(a, b) \in \alpha \times \beta$ for some $a \in \alpha$ and

$b \in \beta$. For $\alpha + \beta$, two proofs are possible: Assuming $\alpha \neq \emptyset$, we can construct $\mathsf{Inl}\ a$ for some $a \in \alpha$; alternatively, assuming $\beta \neq \emptyset$, we can construct $\mathsf{Inr}\ b$ for some $b \in \beta$.

With each BNF $\overline{\alpha}$ F, we associate a set of witnesses, each of the form $\mathsf{Fwit} : \alpha_{i_1} \to \cdots \to \alpha_{i_k} \to \overline{\alpha}$ F for a subset $\{i_1, \ldots, i_k\} \subseteq [n]$. From a witness, we can construct a set-theoretic proof by following its signature, in the spirit of the Curry–Howard correspondence. Accordingly, $\mathsf{Inr} : \beta \to \alpha + \beta$ can be read as the following contract: Given a proof that β is nonempty, Inr yields a proof that $\alpha + \beta$ is nonempty.

When BNFs are composed, so are their witnesses. The two possible witnesses for the list-defining functor (α, β) pre_list $= \mathsf{unit} + \alpha \times \beta$ are $\mathsf{wit\_pre\_list}_1 = \mathsf{Inl}\ ()$ and $\mathsf{wit\_pre\_list}_2\ a\ b = \mathsf{Inr}\ (a, b)$. The first witness subsumes the second one, because it unconditionally shows the collection nonempty, regardless of the potential emptiness of α and β. From this witness, we obtain the witness list_ctor $\mathsf{wit\_pre\_list}_1$ (i.e., Nil).

Because they can store infinite objects, codatatype set constructors are never empty provided their arguments are nonempty. Compare the following:

datatype α fstream $= \mathsf{FSCons}\ \alpha\ (\alpha$ fstream$)$

codatatype α stream $= \mathsf{SCons}\ \alpha\ (\alpha$ stream$)$

The datatype definition fails because the optimal witness has a circular signature: $\alpha \to \alpha$ fstream $\to \alpha$ fstream. In contrast, the codatatype definition succeeds and produces the witness $(\lambda a.\, \mu s.\, \mathsf{SCons}\ a\ s) : \alpha \to \alpha$ stream, namely the (unique) stream s such that $s = \mathsf{SCons}\ a\ s$ for a given $a \in \alpha$. This stream is easy to define by coiteration.

Let us now turn to a pair of examples involving nesting:

datatype (α, β) tree $= \mathsf{Leaf}\ \beta \mid \mathsf{Branch}\ ((\alpha + (\alpha, \beta)$ tree$)$ stream$)$

codatatype (α, β) ltree $= \mathsf{LNode}\ \beta\ ((\alpha + (\alpha, \beta)$ ltree$)$ stream$)$

In the tree definition, the two constructors hide a sum BNF, giving us some flexibility. For the Leaf constructor, all we need is a witness $b \in \beta$, from which we construct $\mathsf{Leaf}\ b$. For Branch, we can choose the left-hand side of the nested $+$, completely avoiding the recursive right-hand side: From a witness $a \in \alpha$, we construct $\mathsf{Branch}\ (\mu s.\, \mathsf{SCons}\ (\mathsf{Inl}\ a)\ s)$.

For the ltree functor, the two arguments to LNode are hiding a product, so the ltree-defining functor is (α, β, γ) pre_ltree $= \beta \times (\alpha + \gamma)$ stream with γ representing the corecursive component. Composition yields two witnesses for pre_ltree:

$$\mathsf{wit\_pre\_ltree}_1\ a\ b = (b,\ \mu s.\, \mathsf{SCons}\ (\mathsf{Inl}\ a)\ s)$$
$$\mathsf{wit\_pre\_ltree}_2\ b\ c = (b,\ \mu s.\, \mathsf{SCons}\ (\mathsf{Inr}\ c)\ s)$$

These can serve to build infinitely many witnesses for ltree. Fig. 2 enumerates the possible combinations, starting with $\mathsf{wit\_pre\_ltree}_1$. This witness requires only the noncorecursive components α and β to be nonempty, and hence immediately yields a witness $\mathsf{wit\_ltree}_1 : \alpha \to \beta \to (\alpha, \beta)$ ltree (by applying the constructor LNode). The second witness $\mathsf{wit\_pre\_ltree}_2$ requires both β and the corecursive component γ to be nonempty; it effectively "consumes" another ltree witness through γ. The consumed witness can again be either $\mathsf{wit\_pre\_ltree}_1$ or $\mathsf{wit\_pre\_ltree}_2$, and so on. At the limit, $\mathsf{wit\_pre\_ltree}_2$ is used infinitely often. The corresponding witness $\mathsf{wit\_ltree}_2 : \beta \to (\alpha, \beta)$ ltree can be defined by coiteration as $\lambda b.\, \mu t.\, \mathsf{wit\_pre\_ltree}_2\ b\ t$. It subsumes $\mathsf{wit\_ltree}_1$ and all the

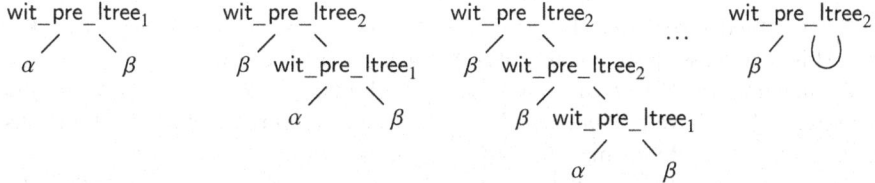

Fig. 2. Derivation trees for ltree witnesses

other finite witnesses. But had ltree been defined as a datatype instead of a codatatype, wit_ltree$_1$ would have been its optimal witness.

4.2 A General Solution

The nonemptiness problem for an n-ary set constructor F and a set of indices $I \subseteq [n]$ can be stated as follows: Is $\bar{\alpha}$ F $\neq \emptyset$ whenever $\forall i \in I.\ \alpha_i \neq \emptyset$, for all sets $\bar{\alpha}_n$? We call F *I-witnessed* if the answer is yes. Thus, set sum $(+)$ is $\{1\}$-, $\{2\}$-, and $\{1,2\}$-witnessed; set product $(\times)$ is $\{1,2\}$-witnessed; and α list is $\emptyset$- and $\{1\}$-witnessed. This leads to the following notion of soundness: Given an n-ary functor F, a set $\mathscr{I} \subseteq [n]$ set is *(witness-)sound* for F if F is I-witnessed for all $I \in \mathscr{I}$.

The next question is: When is such a set $\mathscr{I}$ also complete, in that it covers all witnesses? Clearly, if $I_1 \subseteq I_2$, then I_1-witnesshood implies I_2-witnesshood. Therefore, we are interested in retaining the witnesses completely only up to inclusion of sets of indices. A set $\mathscr{I} \subseteq [n]$ set is *(witness-)complete* for F if for all $J \subseteq [n]$ such that F is J-witnessed, there exists $I \in \mathscr{I}$ such that $I \subseteq J$; *(witness-)perfect* for F if it is both sound and complete.

Here are perfect sets $\mathscr{I}_F$ for basic BNFs:

- Identity: $\mathscr{I}_{\alpha\, \mathsf{ID}} = \{\{\alpha\}\}$
- Constant: $\mathscr{I}_{\mathsf{C}_{n,\alpha}} = \{\emptyset\}$ $(\alpha \neq \emptyset)$
- Sum: $\mathscr{I}_{\alpha+\beta} = \{\{\alpha\}, \{\beta\}\}$
- Product: $\mathscr{I}_{\alpha\times\beta} = \{\{\alpha,\beta\}\}$
- Function space: $\mathscr{I}_{\beta\,\mathsf{func}_\alpha} = \{\{\beta\}\}$ $(\alpha \neq \emptyset)$
- Bounded k-powerset: $\mathscr{I}_{\alpha\,\mathsf{set}_k} = \{\emptyset\}$

Parameters α_j are identified with their indices j to improve readability.

Perfect sets must be maintained across BNF operations. Let us start with composition, permutation, and lifting.

Theorem 1. *Let* $H = G \circ \bar{F}_n$, *where* $G \in \mathsf{Func}_n$ *has a perfect set* $\mathscr{J}$ *and each* $F_j \in \mathsf{Func}_m$ *has a perfect set* $\mathscr{I}_j$. *Then* $\{\bigcup_{j\in J} I_j \mid J \in \mathscr{J} \wedge (I_j)_{j\in J} \in \prod_{j\in J} \mathscr{I}_j\}$ *is a perfect set for* H.

Proof sketch. Let $\mathscr{K} = \{\bigcup_{j\in J} I_j \mid J \in \mathscr{J} \wedge (I_j)_j \in \prod_{j\in J} \mathscr{I}_j\}$. We first prove that $\mathscr{K}$ is sound for H. Let $K \in \mathscr{K}$ and $\bar{\alpha}_m$ be such that $\forall i \in K.\ \alpha_i \neq \emptyset$. By the definition of $\mathscr{K}$, we obtain $J \in \mathscr{J}$ and $(I_j)_{j\in J}$ such that (1) $K = \bigcup_{j\in J} I_j$ and (2) $\forall j \in J.\ I_j \in \mathscr{I}_j$. Using (1), we have $\forall j \in J.\ \forall i \in I_j.\ \alpha_i \neq \emptyset$. Hence, since each $\mathscr{I}_j$ is sound for F_j, $\forall j \in J.\ \bar{\alpha}\, F_j \neq \emptyset$. Finally, since $\mathscr{J}$ is sound for G, we obtain $\bar{\alpha}\, \bar{F}\, G \neq \emptyset$, i.e., $\bar{\alpha}\, H \neq \emptyset$.

We now prove that $\mathcal{K}$ is complete for H. Let $K \subseteq [m]$ be such that H is K-witnessed. Let $\overline{\beta}_n$ be defined as $\beta_j = $ unit if $j \in K$ and $\emptyset$ otherwise, and let $J = \{j \in [n] \mid \overline{\beta} \, \mathsf{F}_j \neq \emptyset\}$. Since H is K-witnessed, we obtain that $\overline{\beta} \, \mathsf{H} \neq \emptyset$, i.e., (3) $\overline{\beta} \, \overline{\mathsf{F}} \, \mathsf{G} \neq \emptyset$.

We show that (4) G is J-witnessed. Let $\overline{\gamma}_n$ such that $\forall j \in J. \, \gamma_j \neq \emptyset$. Thanks to the definition of J, we have $\forall j \in [n]. \, \mathsf{F}_j \neq \emptyset \Rightarrow \gamma_j \neq \emptyset$, and therefore we obtain the functions $(f_j : \overline{\beta} \, \mathsf{F}_j \to \gamma_j)_{i \in [n]}$. With $\mathsf{Gmap} \, \overline{f} : \overline{\beta} \, \overline{\mathsf{F}} \, \mathsf{G} \to \overline{\gamma} \, \mathsf{G}$, by (3) we obtain $\overline{\gamma} \, \mathsf{G} \neq \emptyset$.

From (4), since J is complete for G, we obtain $J_1 \in \mathcal{J}$ such that $J_1 \subseteq J$. Let $j \in J_1$. By the definition of J, we have $\overline{\beta} \, \mathsf{F}_j \neq \emptyset$, making $\overline{\beta} \, \mathsf{F}_j$ K-witnessed (by definition of $\overline{\beta}$); hence, since $\mathcal{J}_j$ is F_j-complete, we obtain $I_j \in \mathcal{J}_j$ such that $I_j \subseteq K$. Then $K_1 = \bigcup_{j \in J_1} I_j$ belongs to $\mathcal{K}$ and is included in K. $\square$

Theorem 2. Let $\mathcal{J} \subseteq [n]$ set be a perfect set for F. Then $\mathcal{J}$ and $\mathcal{J}^{(i,j)}$ are perfect sets for $\mathsf{F}\!\uparrow$ and $\mathsf{F}^{(i,j)}$, respectively, where $\mathcal{J}^{(i,j)}$ is $\mathcal{J}$ with i and j exchanged in each of its elements.

Theorems 1 and 2 hold not only for functors but also for plain set constructors (with a further cardinality-monotonicity assumption needed for the completeness part of Theorem 1). The most interesting cases are the genuinely functorial ones of initial algebras and final coalgebras. Witnesses for initial algebras and final coalgebras are essentially obtained by repeated compositions of the witnesses of the involved BNFs and the folding bijections, inductively in one case and coinductively in the other. The derivation trees from Section 3 turn out to be perfectly suited for recording the combinatorics of these compositions, so that both soundness and completeness follow easily.

For the rest of this subsection, we fix n $(m+n)$-ary functors $\overline{\beta} \, \mathsf{F}_j$ and assume each F_j has a perfect set $\mathcal{K}_j$. We start by constructing a (set) grammar $\mathsf{Gr} = (\mathsf{T}, \mathsf{N}, \mathsf{P})$ with $\mathsf{T} = [m]$, $\mathsf{N} = [n]$, and $\mathsf{P} = \{(j, \mathsf{cp}(K)) \mid K \in \mathcal{K}_j\}$, where, for each $K \subseteq [m+n]$, $\mathsf{cp}(K)$ is its copy to $[m] + [n]$ defined as $\mathsf{Inl} \cdot ([m] \cap K) \cup \mathsf{Inr} \cdot \{k \in [n] \mid m+k \in K\}$.

The intuition is as follows. A mutual datatype definition introduces n isomorphisms $\overline{\alpha} \, \mathsf{IF}_j \cong (\overline{\alpha}, \overline{\alpha} \, \mathsf{IF}_j, \ldots, \overline{\alpha} \, \mathsf{IF}_n) \, \mathsf{F}_j$. We are looking for conditions that guarantee nonemptiness of the functors IF_j. To this end, we traverse these isomorphisms from left to right, reducing nonemptiness of $\overline{\alpha} \, \mathsf{IF}_j$ to that of $(\overline{\alpha}, \overline{\alpha} \, \mathsf{IF}_1, \ldots, \overline{\alpha} \, \mathsf{IF}_n) \, \mathsf{F}_j$. Nonemptiness of the latter can be reduced to nonemptiness of some $\alpha_{i_1}, \ldots, \alpha_{i_p}$ and some $\overline{\alpha} \, \mathsf{IF}_{j_1}, \ldots, \overline{\alpha} \, \mathsf{IF}_{j_q}$, via a witness for F_j of the form $\{i_1, \ldots, i_p, m+j_1, \ldots, m+j_q\}$. This yields a grammar production $j \to \{\mathsf{Inl} \, i_1, \ldots, \mathsf{Inl} \, i_p, \mathsf{Inr} \, j_1, \ldots, \mathsf{Inr} \, j_q\}$, where the i_k's are terminals and the j_l's are, like j, nonterminals. The ultimate goal is to reduce the nonemptiness of $\overline{\alpha} \, \mathsf{IF}_j$ to that of components of $\overline{\alpha}$ alone, i.e., to terminals. This precisely corresponds to derivations in the grammar of terminal sets. It should be intuitively clear that by considering finite derivations, we obtain sound witnesses for IF_j. We actually prove more: For initial algebras, finite derivations are also witness-complete; for final coalgebras (substituting $\overline{\mathsf{JF}}$ for $\overline{\mathsf{IF}}$), accepting infinite derivations is sound and also required for completeness.

Theorem 3. Assume that the final coalgebra of $\overline{\mathsf{F}}$ exists and consists of n m-ary functors $\overline{\alpha}_m \, \mathsf{JF}_j$ (cf. Section 2.1). Then $\mathcal{L}^{\mathrm{r}}_{\mathsf{Gr}}(j)$ is a perfect set for JF_j, for $j \in [n]$.

To prove soundness, we define a nonemptiness witness to $\overline{\alpha} \, \mathsf{JF}_j$ corecursively (by abstract $\overline{\mathsf{JF}}$-corecursion). Showing completeness is more interesting: We define a function to dtree corecursively (by concrete tree corecursion), obtaining a derivation tree, from which we then cut a regular derivation tree by exploiting Lemma 1.

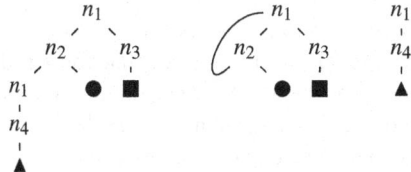

Fig. 3. A finite derivation tree (left), a regular cut (middle), and a finite regular cut (right)

Proof sketch. Let $j_0 \in [n]$. We first show that $\mathscr{L}^r_{Gr}(j_0)$ is sound. Let t_0 be a well-formed regular derivation tree with root j_0. We must prove that F_{j_0} is $Fr\ t_0$-witnessed. For this, we fix $\overline{\alpha}_m$ such that $\forall i \in Fr\ t_0.\ \alpha_i \neq \emptyset$, and aim to show that $\overline{\alpha}\ JF_{j_0} \neq \emptyset$.

For each $j \in Itr\ t_0$, let t_j be the corresponding subtree of t_0. (It is well-defined, since t_0 is regular.) Note that $t_0 = t_{j_0}$. For each K such that $(j, cp(K)) \in P$, since $K \in \mathscr{K}_j$ and $\mathscr{K}_j$ is sound for F_j, we obtain a K-witness for F_j, i.e., a function $w_{j,K} : (\gamma_k)_{k \in K} \to \overline{\gamma}\ F_j$ (polymorphic in $\overline{\gamma}$).

Let $\overline{\beta}_n$ be defined as $\beta_j =$ unit if $j \in Itr\ t_0$ and $\emptyset$ otherwise. We build a coalgebra structure on $\overline{\beta}$, $(s_j : \beta_j \to (\overline{\alpha}, \overline{\beta})\ F_j)_{j \in [n]}$, as follows: If $j \notin Itr\ t_0$, then s_j is the unique function from $\emptyset$. Otherwise, let $s_j\ () = w_{j,K}\ (a_i)_{i \in K \cap [m]}\ ()^{|K \cap [m+1,m+n]|}$, where $cp(K)$ is the right-hand side of the top production of t_j, i.e., $(id \oplus root) \bullet cont\ t_j$. For each $j \in Itr\ t_0$, unfold$_j\ \overline{s}$: unit $\to \overline{\alpha}\ JF_j$ ensures the nonemptiness of $\overline{\alpha}\ JF_j$. In particular, $\overline{\alpha}\ JF_{j_0} \neq \emptyset$.

We now show that $\mathscr{L}^r_{Gr}(j_0)$ is complete. Let $I \subseteq [m]$ such that JF_{j_0} is I-witnessed. We must find $I_1 \in \mathscr{L}^r_{Gr}(j_0)$ such that $I_1 \subseteq I$. Let $\overline{\alpha}_m$ be defined as $\alpha_i =$ unit if $i \in I$ and $\emptyset$ otherwise. Let $J = \{j \mid \overline{\alpha}\ F_j \neq \emptyset\}$. We define $c : J \to ([m] + J)$ fset by $c\ j = cp(K_j)$, where K_j is such that $(j, cp(K_j)) \in P$ and $K_j \subseteq I \cup \{m+j \mid j \in J\}$.

Now let $g : J \to$ dtree be unfold id c. Thus, for all $j \in J$, root $(g\ j) = j$ and cont $(g\ j) = (id \oplus g) \bullet c\ j = Inl \bullet (K_j \cap I) \cup Inr \bullet \{g\ j \mid m+j \in K_j\}$. Taking $t_0 = g\ j_0$ and using Lemma 1, we obtain the regular well-formed tree t_1 such that $Fr\ t_1 \subseteq Fr\ t_0 \subseteq I$. Hence $Fr\ t_1$ is the desired index set I_1. □

The above completeness proof provides an example of self-application of codatatypes: A specific codatatype, of infinite derivation trees, arises in the metatheory of general codatatypes. And this may well be unavoidable: While for soundness the regular trees are replaceable by some equivalent (finite) inductive items, it is not clear how completeness could be proved without first considering arbitrary infinite derivation trees and then cutting them down to regular trees.

An analogous result holds for initial algebras. For each $i \in \mathsf{N}$, let $\mathscr{L}^{rf}_{Gr}(i)$ be the language generated by i by means of regular finite derivation trees for grammar Gr. Since N is finite, these can be described more directly as trees for which every nonterminal path has no repetitions.

In the following proofs, we exploit an embedding of datatypes as finite codatatypes. Using this embedding, we can transfer the recursive definition and structural induction principles from $\overline{IF}$ to finite elements of $\overline{JF}$, and in particular from a datatype fdtree of finite trees (Appendix C) to finite trees in dtree.

The regular cut of a tree works well with respect to the metatheory of codatatypes, but for datatypes it has the disadvantage that it may produce infinite trees out of finite ones, as depicted in Fig. 3 (left and middle). We need a slightly different concept for datatypes: the finite regular cut (right). Let t_0 be a finite derivation tree. We define the function fpick : ltr $t_0 \to$ Subtr t_0 similarly to pick from Section 3, but making sure that the choice of the subtrees fpick n is minimal, in that fpick n does not have n in the interior of a proper subtree (and hence does not have any proper subtree of root n). Such a choice is possible owing to the finiteness of t_0. We define the finite regular cut of t_0, rfcut t_0, analogously to rcut t_0, using fpick instead of pick.

Lemma 2. Assume t_0 is a finite derivation tree. Then:

(1) The statement of Lemma 1 holds if we replace rcut by rfcut.
(2) rfcut t_0 is finite.

Proof. (1) Similar to the proof of Lemma 1. (2) By routine induction on t_0.   □

Theorem 4. Assume that the initial algebra of $\overline{\mathsf{F}}$ exists and consists of n m-ary functors $\overline{\alpha}_m$ IF_j (cf. Section 2.1). Then $\mathscr{L}_{\mathrm{Gr}}^{\mathrm{rf}}(j)$ is a perfect set for IF_j, for $j \in [n]$.

Proof. Let $j_0 \in [n]$. We first show that $\mathscr{L}_{\mathrm{Gr}}^{\mathrm{rf}}(j_0)$ is sound. Let t_0 be a well-formed finite regular derivation tree with root j_0. We must prove that F_{j_0} is Fr t_0-witnessed. For this, we fix $\overline{\alpha}_m$ such that $\forall i \in$ Fr t_0. $\alpha_i \neq \emptyset$, and aim to show that $\overline{\alpha}$ $\mathsf{IF}_{j_0} \neq \emptyset$.

For each $j \in$ ltr t_0, let t_j be the corresponding subtree of t_0. (It is well-defined, since t_0 is regular.) Note that $t_0 = t_{j_0}$. For each K such that $(j, \mathsf{cp}(K)) \in \mathsf{P}$, since $K \in \mathscr{K}_j$ and $\mathscr{K}_j$ is sound for F_j, we obtain a K-witness for F_j, i.e., a function $w_{j,K} : (\alpha_k)_{k \in K} \to \overline{\alpha} \, \mathsf{F}_j$.

We verify the following fact by induction on the finite derivation tree t: If $\exists j \in$ ltr t_0. $t = t_j$, then $\overline{\alpha}$ $\mathsf{IF}_j \neq \emptyset$. The induction step goes as follows: Assume $t = t_j$ has the form Node j as, and let J be the set of all roots of the immediate subtrees of t, namely, root • $(\mathsf{Inr}^-(\mathsf{cont}\ t))$. By the induction hypothesis, $\overline{\alpha}$ $\mathsf{IF}_{j'} \neq \emptyset$ (say, $b_{j'} \in \overline{\alpha}$ $\mathsf{IF}_{j'}$) for all $j' \in J$. Then $w_{j,K} \, (a_i)_{i \in |\mathsf{In}|^- t} \, (b_{j'})_{j' \in J} \in \overline{\alpha}$ IF_j, making $\overline{\alpha}$ IF_j nonempty. In particular, $\overline{\alpha}$ $\mathsf{JF}_{j_0} \neq \emptyset$.

We now show that $\mathscr{L}_{\mathrm{Gr}}^{\mathrm{rf}}(j_0)$ is complete. Let $I \subseteq [m]$ such that IF_{j_0} is I-witnessed. We must find $I_1 \in \mathscr{L}_{\mathrm{Gr}}^{\mathrm{rf}}(j_0)$ such that $I_1 \subseteq I$. Let $\overline{\alpha}_m$ be defined as $\alpha_i =$ unit if $i \in I$ and $\emptyset$ otherwise. We verify, by structural $\overline{\mathsf{IF}}$-induction on b, that for all $j \in [n]$ and $b \in \overline{\alpha}$ IF_j, there exists a finite well-formed derivation tree t such that root $t = j$ and Fr $t \subseteq I$. For the inductive step, assume ctor$_j$ $x \in \overline{\alpha}$ IF_j, where $x \in (\overline{\alpha}, \overline{\alpha} \, \overline{\mathsf{IF}})$ F_j. By the induction hypotheses, we obtain the finite well-formed derivation trees $\overline{t}_n$ such that root $t_j = j$ and Fr $t_j \subseteq I$ for all $j \in [n]$. Let $J = \{j' \in [n] \mid \overline{\alpha} \, \mathsf{IF}_{j'} \neq \emptyset\}$. Then F_j is $(I \cup J)$-witnessed, hence by the F_j-completeness of $\mathscr{K}_j$ we obtain $K \in \mathscr{K}_j$ such that $K \subseteq I \cup \{m + j' \mid j' \in J\}$. We take t to have j as root, $I \cap K$ as leaves and $(t_{j'})_{j' \in J}$ as immediate subtrees; namely, $t = $ Node j $(|\mathsf{In}| \cdot I \cup \mathsf{Inr} \cdot \{t_{j'} \mid j' \in J\})$.

Let t_0 be a tree as above corresponding to j_0 (since $\overline{\alpha}$ $\mathsf{IF}_{j_0} \neq \emptyset$). Then, by Lemma 2, $t_1 =$ rcut t_0 is a well-formed finite derivation tree such that Fr $t_1 \subseteq$ Fr $t_0 \subseteq I$. Thus, taking $I_1 =$ Fr t_1, we obtain $I_1 \in \mathscr{L}_{\mathrm{Gr}}^{\mathrm{rf}}(j_0)$ and $I_1 \subseteq I$.   □

Let us see how Theorems 1 to 4 can be combined in establishing or refuting non-emptiness for some of our motivating examples from Sections 1 and 4.1.

- $\mathscr{I}_{(\alpha,\beta)\ \mathsf{pre\_list}} = \{\emptyset\}$ by Theorem 1; $\mathscr{I}_{\alpha\ \mathsf{list}} = \{\emptyset\}$ by Theorem 4
- $\mathscr{I}_{(\alpha,\beta)\ \mathsf{pre\_fstream}} = \{\{\alpha,\beta\}\}$; $\mathscr{I}_{\alpha\ \mathsf{fstream}} = \emptyset$ by Theorem 4 (i.e., α fstream is empty)
- $\mathscr{I}_{(\alpha,\beta)\ \mathsf{pre\_stream}} = \{\{\alpha,\beta\}\}$; $\mathscr{I}_{\alpha\ \mathsf{stream}} = \{\{\alpha\}\}$ by Theorem 3
- $\mathscr{I}_{(\alpha,\beta,\gamma)\ \mathsf{pre\_ltree}} = \{\{\alpha,\beta\},\{\beta,\gamma\}\}$ by Theorem 1;
 $\mathscr{I}_{(\alpha,\beta)\ \mathsf{ltree}} = \{\{\beta\}\}$ by Theorem 3
- $\mathscr{I}_{(\alpha,\beta,\gamma)\ \mathsf{pre\_t_1}} = \{\{\beta\},\{\alpha,\gamma\}\}$, $\mathscr{I}_{(\alpha,\beta,\gamma)\ \mathsf{pre\_t_2}} = \{\emptyset\}$, and
 $\mathscr{I}_{(\alpha,\beta,\gamma)\ \mathsf{pre\_t_3}} = \{\{\alpha\},\{\gamma\}\}$ by Theorem 1; $\mathscr{I}_{t_i} = \{\emptyset\}$ by Theorem 4

Since we have maintained perfect sets throughout all the BNF operations, we obtain the following central result.

Theorem 5. Any BNF built from other BNFs endowed with perfect sets of witnesses (in particular all basic BNFs discussed in this paper) by repeated applications of the composition, initial algebra, and final coalgebra operations has a perfect set defined as indicated in Theorems 1 to 4.

Corollary 1. The nonemptiness problem is decidable for arbitrarily nested, mutual (co)datatypes.

Consequently, a procedure implementing Theorems 1 to 4 will preserve enough nonemptiness witnesses to ensure that all specifications describing nonempty datatypes are accepted. The next subsection presents such a procedure.

4.3   Computational Aspects

Theorem 3 reduces the computation of perfect sets for final coalgebras to that of $\mathscr{L}_{\mathsf{Gr}}^{\mathsf{r}}(n)$. The use of infinite regular trees in the definition of $\mathscr{L}_{\mathsf{Gr}}^{\mathsf{r}}(n)$ allows a simple proof of soundness, and the only natural proof of completeness we could think of, relating the coinductive nature of arbitrary mutual codatatypes with that of infinite trees. However, from a computational point of view, the use of infinite trees is excessive.

In fact, $\mathscr{L}_{\mathsf{Gr}}(n)$ and $\mathscr{L}_{\mathsf{Gr}}^{\mathsf{f}}(n)$, the nonregular versions of the generated languages, are computable by fixpoint iteration on finite sets. It is not hard to show that $\mathscr{L}_{\mathsf{Gr}}$ and $\mathscr{L}_{\mathsf{Gr}}^{\mathsf{f}}$ are the greatest and least solutions of the following fixpoint equation, involving the variable $X : \mathsf{N} \to ((\mathsf{T} + \mathsf{N})\ \mathsf{set})\ \mathsf{set}$, where the order is componentwise inclusion:

$$X\,n = \left\{ \mathsf{Inl}^-\,ss \cup \bigcup\nolimits_{n' \in \mathsf{Inr}^-\,ss} K_{n'} \;\middle|\; (n,ss) \in \mathsf{P} \wedge K \in \prod\nolimits_{n' \in \mathsf{Inr}^-\,ss} X\,n' \right\}$$

The equation simply states the expected closure under the grammar productions, familiar from formal language theory. But since the "words" are finite sets and not lists, a fixpoint is reached after at most card N iterations.

However, it is easier to settle this computational aspect by working with the regular versions $\mathscr{L}_{\mathsf{Gr}}^{\mathsf{r}}(n)$ and $\mathscr{L}_{\mathsf{Gr}}^{\mathsf{rf}}(n)$, whose structure nicely exhibits boundedness. Namely, we prove for these languages a bounded version of the above fixpoint equation, featuring a decumulator that witnesses the finite convergence of the computation.

First, we relativize the notion of frontier to that of "frontier through ns," Fr $ns\ t$, containing the leaves of t accessible by paths of nonterminals from $ns \subseteq \mathsf{N}$. We also define the corresponding *ns*-restricted regularly generated language $\mathscr{L}^{r}_{Gr}$ *ns n*. Thus, what used to be denoted by Fr *t* and $\mathscr{L}^{r}_{Gr}$ *n* now becomes Fr N *t* and $\mathscr{L}^{r}_{Gr}$ N *n*.

In what follows, by "word" we mean "finite set of terminals." We can think of a generated word as being more precise than another provided the former is a subword (subset) of the latter. This leads us to defining, for languages (sets of words), the notions of word-inclusion subsumption,[2] $\leq$, by $L \leq L'$ iff $\forall w \in L. \exists w' \in L'. w' \subseteq w$, and equivalence, $\equiv$, by $L \equiv L'$ iff $L \leq L'$ and $L' \leq L$. It is easy to see that any set $\equiv$-equivalent to a perfect set is again perfect. Note also that Lemma 1 implies $\mathscr{L}^{r}_{Gr}(n) \equiv \mathscr{L}_{Gr}(n)$, which qualifies regular trees as a generated-language optimization of arbitrary trees.

We compute $\mathscr{L}^{r}_{Gr}$ *ns n* up to word-inclusion equivalence $\equiv$ by recursively applying available productions whose source nonterminals are in *ns*, removing each time from *ns* the expanded nonterminal. Thus, if *n* is in *ns*, $\mathscr{L}^{r}_{Gr}$ *ns n* calls $\mathscr{L}^{r}_{Gr}$ *ns' n'* recursively with $ns' = ns \setminus \{n'\}$ for each nonterminal *n'* in the chosen production from *n*, and so on, until the current node is no longer in the decumulator *ns*:

Theorem 6. For all $ns \subseteq N$ and $n \in N$, $\mathscr{L}^{r}_{Gr}$ *ns n* $\equiv$

$$\begin{cases} \{\emptyset\} & \text{if } n \notin ns \\ \{\ln^{-} ss \cup \bigcup_{n' \in \ln r^{-} ss} K_{n'} \mid (n, ss) \in P \wedge K \in \prod_{n' \in \ln r^{-} ss} \mathscr{L}^{r}_{Gr} (ns \setminus \{n\}) \, n'\} & \text{otherwise} \end{cases}$$

Proof sketch. $\mathscr{L}^{r}_{Gr}$ *ns n* $\subseteq \{\emptyset\}$, since Fr *ns t* $= \emptyset$ for all *t* such that root *t* $= n$. It remains to show that $\emptyset \in \mathscr{L}^{r}_{Gr}$ *ns t*, i.e., to find a derivation tree with root *n*. Using the assumption that there are no unused nonterminals, we can build a "default derivation tree" deftr *n* for each *n* as follows. We pick, for each *n*, a set $S \, n \in (T + N)$ fset such that $(n, S\,n) \in P$. Then we define deftr : N $\rightarrow$ dtree corecursively as deftr $=$ unfold id *S*, i.e., such that root (deftr *n*) $= n$ and cont (deftr *n*) $= (\text{id} \oplus \text{deftr}) \bullet S\,n$. It is easy to prove by fixpoint coinduction that deftr *n* is a derivation tree for each *n*.

Now assume $n \notin ns$, and let $ns' = ns \setminus \{n\}$. For the left-to-right direction, we prove more than $\leq$, namely, actual inclusion between $\mathscr{L}^{r}_{Gr}$ *ns n* and the righthand side. Assume *t* is a well-formed regular derivation tree of root *n*. We must find $ss \in (T + N)$ fset and $U : \ln r^{-} ss \rightarrow$ dtree such that, for all $n' \in \ln r^{-} ss$, *U n'* is a well-formed regular derivation tree of root *n'* and Fr *ns t* $= \ln^{-} ss \cup \bigcup_{n' \in \ln r^{-} ss}$ Fr *ns'* (*U n'*). Clearly, *ss* should be the right-hand side of the top production of *t*. As for *U*, the immediate subtrees of *t* would appear to be suitable candidates; however, these do not work, since our goal is to have Fr *ns t* covered by ($\ln^{-} ss$ in conjunction with) Fr *ns'* (*U n'*), while the immediate subtrees only guarantee this property with respect to Fr *ns* (*U n'*), i.e., allowing paths to go through *n* as well. A correct solution is again offered by a corecursive definition: We build the tree t_0 from *t* by substituting hereditarily each subtree with root *n* by *t*. Formally, we take $t_0 = $ unfold *r c*, where *r t'* $=$ root *t'* and *c t'* $=$ cont *t* if root *t'* $= n$ and *c t'* $=$ cont *t'* otherwise. It is easy to prove that t_0, like *t*, is a regular derivation tree. Thus, we can define *U* to give, for any *n'*, the corresponding immediate subtree of t_0.

To prove the right-to-left direction, let $ss \in (T+N)$ fset and $K \in \prod_{n' \in \ln r^{-} ss} \mathscr{L}^{r}_{Gr}$ *ns' n'* such that $ts = \ln^{-} ss \cup \bigcup_{n' \in \ln r^{-} ss} K_{n'}$. Unfolding the definition of $\mathscr{L}^{r}_{Gr}$, we obtain *U* : $\ln r^{-} ss \rightarrow$ dtree such that, for all $n' \in \ln r^{-} ss$, *U n'* is a regular derivation tree of root *n'* such that $K_{n'} \in$ Fr *ns'* (*U n'*). Then the tree of immediate leafs $\ln^{-} ss$ and immediate

---

[2] This is in effect the Smyth preorder extension [38] of the subword relation.

subtrees $\{U\,n'\mid n'\in \mathsf{Inr}^-\,ss\}$, namely, Node $n\,((\mathrm{id}\oplus U)\bullet ss)$, is the desired regular derivation tree whose frontier is included ts.   □

Theorem 6 provides an alternative, recursive definition of $\mathscr{L}^{\mathrm{r}}_{\mathsf{Gr}}\,ns\,n$. The definition terminates because the argument ns is finite and decreases strictly in the recursive case. This shows that the height of the recursive call stack is bounded by the number of non-terminals, which corresponds to the number of simultaneously introduced codatatypes.

Here is how the above recursion operates on the ltree example. We have $\mathsf{T}=\{\alpha,\beta\}$, $\mathsf{N}=\{\gamma\}$, and $\mathsf{P}=\{p_1,p_2\}$, where $p_1=(\gamma,\{\mathsf{Inl}\,\alpha,\mathsf{Inl}\,\beta\})$ and $p_2=(\gamma,\{\mathsf{Inl}\,\beta,\mathsf{Inr}\,\gamma\})$. Note that

- $\mathsf{Inl}^-\,ss=\{\alpha,\beta\}$ and $\mathsf{Inr}^-\,ss=\emptyset$ for $(n,ss)=p_1$
- $\mathsf{Inl}^-\,ss=\{\beta\}$ and $\mathsf{Inr}^-\,ss=\{\gamma\}$ for $(n,ss)=p_2$

The computation has one single recursive call, yielding

$$
\begin{aligned}
\mathscr{L}^{\mathrm{r}}_{\mathsf{Gr}}\,\gamma &= \mathscr{L}^{\mathrm{r}}_{\mathsf{Gr}}\,\{\gamma\}\,\gamma\\
&\equiv \{\{\alpha,\beta\}\cup\emptyset\}\cup\{\{\beta\}\cup\bigcup_{n'\in\{\gamma\}}K_{n'}\mid K\in\prod_{n'\in\{\gamma\}}\mathscr{L}^{\mathrm{r}}_{\mathsf{Gr}}\,\emptyset\,n'\}\\
&= \{\{\alpha,\beta\}\}\cup\{\{\beta\}\cup K_\gamma\mid K_\gamma\in\mathscr{L}^{\mathrm{r}}_{\mathsf{Gr}}\,\emptyset\,\gamma\}\\
&= \{\{\alpha,\beta\}\}\cup\{\{\beta\}\cup\emptyset\}\\
&= \{\{\alpha,\beta\},\{\beta\}\}\\
&\equiv \{\{\beta\}\}
\end{aligned}
$$

For datatypes, the computation of $\mathscr{L}^{\mathrm{rf}}_{\mathsf{Gr}}$ is achieved analogously to Theorem 6, defining $\mathscr{L}^{\mathrm{rf}}_{\mathsf{Gr}}\,ns\,n$ as a generalization of $\mathscr{L}^{\mathrm{rf}}_{\mathsf{Gr}}\,n$.

In what follows, nl ranges over lists of nonterminals and the centered dot operator $(\cdot)$ denotes list concatenation. If n is a nonterminal, n also denotes the n-singleton list. The predicate path $nl\,t$, stating that nl is a path in t (starting from the root), is defined inductively as follows:

path $(\mathrm{root}\,t)\,t$
$\mathsf{Inr}\,t'\in\mathrm{cont}\,t\wedge\mathrm{path}\,nl\,t'\Rightarrow\mathrm{path}\,((\mathrm{root}\,t)\cdot nl)\,t'$

Lemma 3. Let t be a finite regular derivation tree. Then t has no paths that contain repetitions.

Proof. Assume, by absurdity, that a path nl in t contains repetitions, i.e., has the form $nl_1\cdot n\cdot nl_2\cdot n$, and let t_1 and t_2 be the subtrees corresponding to the paths $nl_1\cdot n$ and nl, respectively. Then t_2 is a proper subtree of t_1; on the other hand, by the regularity of t, we have $t_1=t_2$, which is impossible since t_1 and t_2 are finite.   □

Theorem 7. The statement of Theorem 6 still holds if we substitute $\mathscr{L}^{\mathrm{rf}}_{\mathsf{Gr}}$ for $\mathscr{L}^{\mathrm{r}}_{\mathsf{Gr}}$ and $\emptyset$ for $\{\emptyset\}$.

Proof. By Lemma 3 and the properties of regular cuts, we have (1) $\mathscr{L}^{\mathrm{rf}}_{\mathsf{Gr}}\,ns'\,n\equiv\mathscr{L}^{\mathrm{pf}}_{\mathsf{Gr}}\,ns'\,n$, where $\mathscr{L}^{\mathrm{pf}}_{\mathsf{Gr}}\,ns'\,n$ is the language defined like $\mathscr{L}^{\mathrm{rf}}_{\mathsf{Gr}}\,ns'\,n$ but replacing "regular" with "having no paths that contain repetitions." Moreover, it is easy to see that (2) the desired facts hold if we replace $\mathscr{L}^{\mathrm{rf}}_{\mathsf{Gr}}\,ns'\,n$ with $\mathscr{L}^{\mathrm{pf}}_{\mathsf{Gr}}\,ns'\,n$ and $\equiv$ with equality. The result follows from (1) and (2).   □

5   Implementation in Isabelle

The package maintains nonemptiness information for producing nonemptiness proofs arising when defining datatypes. The equations from Theorems 6 and 7 involve only executable operations over finite sets of numbers, sums, and products. Since the descriptions of Theorems 1 and 2 are also executable, the implementation task emerges clearly: Store a perfect set with each basic BNF, and have each BNF operation compute witnesses from those of its operands.

However, as it stands, I-witnesshood cannot be expressed in HOL because types are always nonempty: How can we state that (α, β) tree $\neq \emptyset$ conditionally on $\alpha \neq \emptyset$ or $\beta \neq \emptyset$, in the context of α and β being assumed nonempty in the first place? The solution is to work not with operators $\overline{\alpha}\,\mathsf{F}$ on HOL types directly but rather with their *internalization* to sets, expressed as a polymorphic function $\mathsf{Fin} : \alpha_1 \text{ set} \to \cdots \to \alpha_n \text{ set} \to (\overline{\alpha}\ \mathsf{F})$ set defined as $\mathsf{Fin}\,\overline{A} = \{x \mid \forall i \in [n].\ \mathsf{Fset}^i\ x \subseteq A_i\}$. I-witnesshood is then expressible as $(\forall i \in I.\ A_i \neq \emptyset) \Rightarrow \mathsf{Fin}\,\overline{A} \neq \emptyset$.

For each n-ary BNF F, the package stores a set of sets $\mathscr{I}$ of numbers in $[n]$ (the perfect set) and, for each set $I \in \mathscr{I}$, a polymorphic constant $w_I : (\alpha_i)_{i \in I} \to \overline{\alpha}\ \mathsf{F}$ and an equivalent formulation of I-witnesshood: $\forall i \in I.\ \mathsf{Fset}^i\ (w_I\ (a_j)_{j \in I}) \neq \emptyset$.

Due to the logic's restricted expressiveness, we cannot prove the theorems presented in this paper in their most general form for arbitrary functors and have the package instantiate them for specific functors. Instead, the package proves the theorems dynamically for the specific functors involved in the datatype definitions. Only the soundness part of the theorems is needed. Completeness is desirable, because in its absence some legitimate definitions would be rejected. To paraphrase Krauss and Nipkow [24], completeness belongs to the realm of metatheory and is not required to obtain actual nonemptiness proofs—it merely lets you sleep better.

A HOL definitional package bears the burden of computing terms and certifying the computation, i.e., ensuring that certain terms are theorems. The combinatorial computation of witnessing sets of indices described in Theorems 6 and 7 would be expensive if performed through Isabelle, that is, by executing the equations stated in these theorems as term rewriting in the logic. Instead, we perform the computation outside the logic, employing a Standard ML datatype aimed at efficiently representing the finite and the regular derivation trees inhabiting the Isabelle type dtree from Section 3:

```
datatype wit_tree = Wit_Leaf of int
                  | Wit_Node of (int ∗ int ∗ int list) ∗ wit_tree list
```

Here, Wit_Node $((i, j, is), ts)$ stores the root nonterminal i, a numeric identifier of the used production j, and the continuation consisting of the terminals is and the further nonterminal expanded trees ts. Moreover, Wit_Leaf i stores, in the case of regular infinite trees, the nonterminal where a regularity loop occurs, i.e., such that it has a previous occurrence on the path to the root.

From this tree datatype, we produce witnesses represented as Isabelle constants of appropriate types (the w_I's described above), by essentially mimicking the (co)recursive definitions employed in the proofs of the soundness parts of Theorems 3 and 4. We certify the witnesses by producing the relevant Isabelle proof goals and discharging them by mirroring the corresponding (co)inductive arguments from the aforementioned proofs. In summary: The witnesses are computed outside the logic, but they are verified by Isabelle's kernel. After introducing a BNF, redundant witnesses are silently removed.

The development devoted to the production and certification of witnesses amounts to about 1000 lines of Standard ML [9].

6 Related Work

Coinductive (or coalgebraic) datatypes have become popular in recent years in the study of infinite behaviors and nonterminating computation. Whereas inductive datatypes are well studied and widely available in most programming languages and proof assistants, coinductive types are still not mainstream, and their integration into existing systems poses many challenges.

In the context of theorem proving, much research has been done in the past few years on how to add coinductive types or improve support of coinductive proofs, notably in Agda [2], CIRC [27], and Coq [7, 29]. The work described in this paper is in line with this research. The results are applicable to other proof assistants from the HOL family.

In HOL-based systems, other definitional packages must also prove nonemptiness of newly defined types, but typically the proofs are easy. For example, Homeier's quotient package for HOL4 [19] exploits the observation that quotients of nonempty sets are nonempty, and Huffman's (co)recursive domain package for Isabelle/HOLCF [21] can rely on a minimal element $\perp$. For the traditional datatype packages introduced by Melham [28], and implemented in Isabelle/HOL by Berghofer and Wenzel [6], proving nonemptiness is nontrivial, but by reducing nested definitions to mutual definitions, they could employ a standard reachability analysis [6, § 4.1]. To our knowledge, the completeness of the analysis has not been proved (or even formulated) for these.

Obviously, our overall approach to (co)datatypes is heavily inspired by category-theory developments [5, 12, 17, 18, 35]—this is discussed in detail in a previous paper [39], which puts forward a program for integrating insight from category theory in proof assistants based on higher-order logic, to achieve better structure and functionality. A similar program is pursed on a larger scale in the context of homotopy type theory [40], targeting proof assistants based on type theory, notably Agda and Coq. Our nonemptiness witness maintenance is similar to the preservation of enriched types along various constructions—for example, initial algebras and final coalgebras of pointed functors are also pointed [20]. However, existing analysis techniques are only concerned with soundness (not completeness) results.

7 Conclusion

We presented a complete solution to the nonemptiness problem for open-ended, mutual, nested codatatypes. This problem arose in the context of Isabelle's new (co)datatype package and has broad practical applicability in terms of the popularity of HOL-based provers. The problem and its solution also enjoy an elegant metatheory, which itself is best expressed in terms of codatatypes. Our solution, like the rest of the definitional package, is part of the latest edition of Isabelle.

Acknowledgment. Tobias Nipkow made this work possible. Andreas Lochbihler suggested many improvements, notably concerning the format of the concrete coinduction principles. Brian Huffman suggested major conceptual simplifications to the package. Florian Haftmann, Christian Urban, and Makarius Wenzel guided us through the jungle of package writing. Stefan Milius and Lutz Schröder found an elegant proof to eliminate one of the BNF cardinality assumptions. Andreas Abel and Martin Hofmann pointed out relevant work. Mark Summerfield and several anonymous reviewers commented on various versions of this paper.

Blanchette was supported by the Deutsche Forschungsgemeinschaft (DFG) project Hardening the Hammer (grant Ni 491/14-1). Popescu was supported by the project Security Type Systems and Deduction (grant Ni 491/13-2) as part of the DFG program Reliably Secure Software Systems (RS³, Priority Program 1496). Traytel was supported by the DFG program Program and Model Analysis (PUMA, doctorate program 1480). The authors are listed in alphabetical order.

References

[1] Abel, A., Altenkirch, T.: A predicative strong normalisation proof for a λ-calculus with interleaving inductive types. In: Coquand, T., Nordström, B., Dybjer, P., Smith, J. (eds.) TYPES 1999. LNCS, vol. 1956, pp. 21–40. Springer, Heidelberg (2000)

[2] Abel, A., Pientka, B., Thibodeau, D., Setzer, A.: Copatterns: Programming infinite structures by observations. In: Giacobazzi, R., Cousot, R. (eds.) POPL 2013, pp. 27–38. ACM (2013)

[3] Adams, M.: Introducing HOL Zero (extended abstract). In: Fukuda, K., van der Hoeven, J., Joswig, M., Takayama, N. (eds.) ICMS 2010. LNCS, vol. 6327, pp. 142–143. Springer, Heidelberg (2010)

[4] Arthan, R.D.: Some mathematical case studies in ProofPower–HOL. In: Slind, K. (ed.) TPHOLs 2004 (Emerging Trends). pp. 1–16. School of Computing. University of Utah (2004)

[5] Barr, M.: Terminal coalgebras in well-founded set theory. Theor. Comput. Sci. 114(2), 299–315 (1993)

[6] Berghofer, S., Wenzel, M.: Inductive datatypes in HOL—lessons learned in formal-logic engineering. In: Bertot, Y., Dowek, G., Hirschowitz, A., Paulin, C., Théry, L. (eds.) TPHOLs 1999. LNCS, vol. 1690, pp. 19–36. Springer, Heidelberg (1999)

[7] Bertot, Y.: Filters on coinductive streams, an application to Eratosthenes' sieve. In: Urzyczyn, P. (ed.) TLCA 2005. LNCS, vol. 3461, pp. 102–115. Springer, Heidelberg (2005)

[8] Blanchette, J.C., Hölzl, J., Lochbihler, A., Panny, L., Popescu, A., Traytel, D.: Truly modular (co)datatypes for Isabelle/HOL. In: Klein, G., Gamboa, R. (eds.) ITP 2014. LNCS, vol. 8558, pp. 93–110. Springer, Heidelberg (2014)

[9] Blanchette, J.C., Popescu, A., Traytel, D.: Supplementary material associated with this paper, https://github.com/dtraytel/Witnessing-Codatatypes

[10] Blanchette, J.C., Popescu, A., Traytel, D.: Cardinals in Isabelle/HOL. In: Klein, G., Gamboa, R. (eds.) ITP 2014. LNCS, vol. 8558, pp. 111–127. Springer, Heidelberg (2014)

[11] Blanchette, J.C., Popescu, A., Traytel, D.: Unified classical logic completeness. In: Demri, S., Kapur, D., Weidenbach, C. (eds.) IJCAR 2014. LNCS, vol. 8562, pp. 46–60. Springer, Heidelberg (2014)

[12] Ghani, N., Johann, P., Fumex, C.: Generic fibrational induction. Log. Meth. Comput. Sci. 8(2:12), 1–27 (2012)

[13] Gordon, M.J.C., Melham, T.F. (eds.): Introduction to HOL: A Theorem Proving Environment for Higher Order Logic. Cambridge University Press (1993)

[14] Gunter, E.L.: Why we can't have SML-style datatype declarations in HOL. In: Claesen, L.J.M., Gordon, M.J.C. (eds.) TPHOLs 1992. IFIP Transactions, vol. A-20, pp. 561–568. North-Holland/Elsevier (1993)

[15] Gvero, T., Kuncak, V., Piskac, R.: Interactive synthesis of code snippets. In: Gopalakrishnan, G., Qadeer, S. (eds.) CAV 2011. LNCS, vol. 6806, pp. 418–423. Springer, Heidelberg (2011)

[16] Harrison, J.: HOL Light: A tutorial introduction. In: Srivas, M., Camilleri, A. (eds.) FM-CAD 1996. LNCS, vol. 1166, pp. 265–269. Springer, Heidelberg (1996)

[17] Hasegawa, R.: Two applications of analytic functors. Theor. Comput. Sci. 272(1–2), 113–175 (2002)

[18] Hermida, C., Jacobs, B.: Structural induction and coinduction in a fibrational setting. Inf. Comput. 145(2), 107–152 (1998)

[19] Homeier, P.V.: A design structure for higher order quotients. In: Hurd, J., Melham, T. (eds.) TPHOLs 2005. LNCS, vol. 3603, pp. 130–146. Springer, Heidelberg (2005)

[20] Howard, B.T.: Inductive, coinductive, and pointed types. In: Harper, R., Wexelblat, R.L. (eds.) ICFP 1996, pp. 102–109. ACM Press, New York (1996)

[21] Huffman, B.: A purely definitional universal domain. In: Berghofer, S., Nipkow, T., Urban, C., Wenzel, M. (eds.) TPHOLs 2009. LNCS, vol. 5674, pp. 260–275. Springer, Heidelberg (2009)

[22] Huffman, B., Kunčar, O.: Lifting and Transfer: A modular design for quotients in Isabelle/HOL. In: Gonthier, G., Norrish, M. (eds.) CPP 2013. LNCS, vol. 8307, pp. 131–146. Springer, Heidelberg (2013)

[23] Kaliszyk, C., Urban, C.: Quotients revisited for Isabelle/HOL. In: Chu, W.C., Wong, W.E., Palakal, M.J., Hung, C.-C. (eds.) SAC 2011, pp. 1639–1644. ACM (2011)

[24] Krauss, A., Nipkow, T.: Proof pearl: Regular expression equivalence and relation algebra. J. Autom. Reasoning 49(1), 95–106 (2012)

[25] Lenisa, M., Power, J., Watanabe, H.: Distributivity for endofunctors, pointed and co-pointed endofunctors, monads and comonads. Electr. Notes Theor. Comput. Sci. 33, 230–260 (2000)

[26] Lochbihler, A.: Java and the Java memory model—A unified, machine-checked formalisation. In: Seidl, H. (ed.) ESOP 2012. LNCS, vol. 7211, pp. 497–517. Springer, Heidelberg (2012)

[27] Lucanu, D., Goriac, E.-I., Caltais, G., Roşu, G.: CIRC: A behavioral verification tool based on circular coinduction. In: Kurz, A., Lenisa, M., Tarlecki, A. (eds.) CALCO 2009. LNCS, vol. 5728, pp. 433–442. Springer, Heidelberg (2009)

[28] Melham, T.F.: Automating recursive type definitions in higher order logic. In: Birtwistle, G., Subrahmanyam, P.A. (eds.) Current Trends in Hardware Verification and Automated Theorem Proving, pp. 341–386. Springer, Heidelberg (1989)

[29] Nakata, K., Uustalu, T., Bezem, M.: A proof pearl with the fan theorem and bar induction—Walking through infinite trees with mixed induction and coinduction. In: Yang, H. (ed.) APLAS 2011. LNCS, vol. 7078, pp. 353–368. Springer, Heidelberg (2011)

[30] Nipkow, T., Paulson, L.C., Wenzel, M.: Isabelle/HOL. LNCS, vol. 2283. Springer, Heidelberg (2002)

[31] Paulson, L.C.: A formulation of the simple theory of types (for Isabelle). In: Martin-Löf, P., Mints, G. (eds.) COLOG 1988. LNCS, vol. 417, pp. 246–274. Springer, Heidelberg (1990)

[32] Paulson, L.C.: A fixedpoint approach to (co)inductive and (co)datatype definitions. In: Plotkin, G.D., Stirling, C., Tofte, M. (eds.) Proof, Language, and Interaction—Essays in Honour of Robin Milner, pp. 187–212. MIT Press (2000)

[33] Pierce, B.C.: Types and Programming Languages. MIT Press (2002)

[34] Rutten, J.J.M.M.: Relators and metric bisimulations. Electr. Notes Theor. Comput. Sci. 11, 252–258 (1998)

[35] Rutten, J.J.M.M.: Universal coalgebra: A theory of systems. Theor. Comput. Sci. 249, 3–80 (2000)

[36] Schropp, A., Popescu, A.: Nonfree datatypes in Isabelle/HOL—Animating a many-sorted metatheory. In: Gonthier, G., Norrish, M. (eds.) CPP 2013. LNCS, vol. 8307, pp. 114–130. Springer, Heidelberg (2013)

[37] Slind, K., Norrish, M.: A brief overview of HOL4. In: Mohamed, O.A., Muñoz, C., Tahar, S. (eds.) TPHOLs 2008. LNCS, vol. 5170, pp. 28–32. Springer, Heidelberg (2008)

[38] Smyth, M.B.: Power domains. J. Comput. Syst. Sci. 16(1), 23–36 (1978)

[39] Traytel, D., Popescu, A., Blanchette, J.C.: Foundational, compositional (co)datatypes for higher-order logic—Category theory applied to theorem proving. In: LICS 2012, pp. 596–605. IEEE (2012)

[40] Univalent Foundations Program: Homotopy Type Theory—Univalent Foundations of Mathematics. Institute for Advanced Study (2013), http://homotopytypetheory.org/book/

A   Abstract (Co)induction

Using the atomic infrastructure described in Section 2.2, the induction principle can be expressed abstractly for the mutual initial algebra $\overline{\mathsf{IF}}$ of functors $\overline{\mathsf{F}}$ as follows for sets $\overline{\alpha}$ and predicates $\varphi_j : \overline{\alpha}\,\mathsf{IF}_j \to \mathsf{bool}$:

$$\frac{\bigwedge_{j=1}^n \forall x \in (\overline{\alpha}, \overline{\alpha}\,\overline{\mathsf{IF}})\,\mathsf{F}_j.\,(\bigwedge_{k=1}^n \forall b \in \mathsf{Fset}_j^{m+k} x.\,\varphi_k\,b) \Rightarrow \varphi_j\,(\mathsf{ctor}_j\,x)}{\bigwedge_{j=1}^n \forall b \in \overline{\alpha}\,\mathsf{IF}_j.\,\varphi_j\,b}$$

For lists, this instantiates to

$$\frac{\forall x \in \mathsf{unit} + \alpha \times \alpha\,\mathsf{list}.\,(\forall b \in \mathsf{Fset}^2 x.\,\varphi\,b) \Rightarrow \varphi\,(\mathsf{ctor}\,x)}{\forall b \in \alpha\,\mathsf{list}.\,\varphi\,b}$$

which, by taking $\mathsf{Nil} = \mathsf{ctor}\,(\mathsf{Inl}\,())$ and $\mathsf{Cons}\,a\,b = \mathsf{ctor}\,(\mathsf{Inr}\,(a, b))$, can be recast into the familiar rule

$$\frac{\varphi\,\mathsf{Nil} \qquad \forall a \in \alpha.\,\forall b \in \alpha\,\mathsf{list}.\,\varphi\,b \Rightarrow \varphi\,(\mathsf{Cons}\,a\,b)}{\forall b \in \alpha\,\mathsf{list}.\,\varphi\,b}$$

Moving to coinduction, we need a further well-known assumption: that our functors preserve weak pullbacks, or, equivalently, that they induce relators [34]. For a functor $\overline{\alpha}_n\,\mathsf{F}$, we lift its action $\mathsf{Fmap} : (\alpha_1 \to \beta_1) \to \cdots \to (\alpha_n \to \beta_n) \to \overline{\alpha}\,\mathsf{F} \to \overline{\beta}\,\mathsf{F}$ on functions to an action $\mathsf{Frel} : (\alpha_1 \to \beta_1 \to \mathsf{bool}) \to \cdots \to (\alpha_n \to \beta_n \to \mathsf{bool}) \to (\overline{\alpha}\,\mathsf{F} \to \overline{\beta}\,\mathsf{F} \to \mathsf{bool})$, the *relator*, defined as follows:

$$\mathsf{Frel}\,\overline{\varphi}\,x\,y \Leftrightarrow \exists z.\,\mathsf{Fmap}\,\overline{\mathsf{fst}}\,z = x \wedge \mathsf{Fmap}\,\overline{\mathsf{snd}}\,z = y \wedge$$
$$\bigwedge_{i=1}^n \forall (a, b) \in \mathsf{Fset}^i\,z.\,\varphi_i\,a\,b$$

Structural coinduction can also be expressed abstractly, for the mutual final coalgebra $\overline{\mathsf{JF}}$ of functors $\overline{\mathsf{F}}$:

$$\frac{\bigwedge_{j=1}^{n} \forall a\, b \in (\overline{\alpha}, \overline{\alpha}\, \overline{\mathsf{JF}})\, \mathsf{F}_j.\ \theta_j\, a\, b \Rightarrow \mathsf{Frel}_j\, (=)^m\, \overline{\theta}\, (\mathsf{dtor}_j\, a)\, (\mathsf{dtor}_j\, b)}{\bigwedge_{j=1}^{n} \forall a\, b.\ \theta_j\, a\, b \Rightarrow a = b}$$

for sets $\overline{\alpha}_n$ and binary predicates $\theta_j \in \overline{\alpha}\, \mathsf{JF}_j \to \overline{\alpha}\, \mathsf{JF}_j \to \mathsf{bool}$. The rule is parameterized by predicates $\theta_j : \overline{\alpha}\, \mathsf{JF}_j \to \overline{\alpha}\, \mathsf{JF}_j \to \mathsf{bool}$ required by the antecedent to form an $\overline{\mathsf{F}}$-bisimulation. The principle effectively states that equality is the largest $\overline{\mathsf{F}}$-bisimulation [35].

B   Concrete Coiteration and Coinduction

Coiteration.   The abstract coiteration principle described in Section 2.1 relies on a co-iterator $\mathsf{unfold} : (\beta \to \beta\ \mathsf{pre\_dtree}) \to \beta \to \mathsf{dtree}$ such that $\mathsf{dtor} \circ \mathsf{unfold}\, s = \mathsf{map\_pre\_dtree}\ (\mathsf{unfold}\, s) \circ s$. Writing s as $\langle r, c \rangle$ for $r : \beta \to \mathsf{N}$ and $c : \beta \to (\mathsf{T} + \alpha)\ \mathsf{fset}$ and recasting the equation in pointful form yields $\mathsf{dtor}\ (\mathsf{unfold}\ \langle r, c \rangle\ b) = \mathsf{map\_pre\_dtree}\ (\mathsf{unfold}\ s)\ (r\, b, c\, b)$ This can be further improved by unfolding the definition of $\mathsf{map\_pre\_}$ dtree, expressing dtor as $\langle \mathsf{root}, \mathsf{cont} \rangle$, and splitting the result into a pair of equations: $\mathsf{root}\ (\mathsf{unfold}\ \langle r, c \rangle\ b) = r\, b$ and $\mathsf{cont}\ (\mathsf{unfold}\ \langle r, c \rangle\ b) = (\mathsf{id} \oplus \mathsf{unfold}\ \langle r, c \rangle) \bullet c\, b$. The coiteration rule of Section 2.1 emerges by replacing unfold with the curried unfold' : $(\beta \to \mathsf{N}) \to (\beta \to (\mathsf{T} + \beta)\ \mathsf{fset}) \to \beta \to \mathsf{dtree}$ defined as $\mathsf{unfold}'\, r\, c = \mathsf{unfold}\ \langle r, c \rangle$.

Coinduction.   The abstract coinduction principle of Appendix A is customized into the following concrete coinduction for dtree:

$$\frac{\forall t_1\, t_2.\ \theta\, t_1\, t_2 \Rightarrow \mathsf{root}\, t_1 = \mathsf{root}\, t_2 \wedge \mathsf{fset\_rel}\ (\mathsf{sum\_rel}\ (=)\ \theta)\ (\mathsf{cont}\, t_1)\ (\mathsf{cont}\, t_2)}{\theta\, t_1\, t_2 \Rightarrow t_1 = t_2}$$

where the predicate $\mathsf{fset\_rel}\ (\mathsf{sum\_rel}\ (=)\ \theta)$ is an instance of the abstract Frel: It gives the componentwise extension of θ to $(\mathsf{T} + \mathsf{dtree})\ \mathsf{fset}$. Unfolding the characteristic theorems for $\mathsf{fset\_rel}$ and $\mathsf{sum\_rel}$ yields the antecedent

$$\begin{aligned}
\forall t_1\, t_2.\ \theta\, t_1\, t_2 &\Rightarrow \mathsf{root}\, t_1 = \mathsf{root}\, t_2\ \wedge \\
&\quad \mathsf{Inl}^-\,(\mathsf{cont}\, t_1) = \mathsf{Inl}^-\,(\mathsf{cont}\, t_2)\ \wedge \\
&\quad \forall t_1' \in \mathsf{Inr}^-\,(\mathsf{cont}\, t_1).\ \exists t_2' \in \mathsf{Inr}^-\,(\mathsf{cont}\, t_2).\ \theta\, t_1'\, t_2'\ \wedge \\
&\quad \forall t_2' \in \mathsf{Inr}^-\,(\mathsf{cont}\, t_2).\ \exists t_1' \in \mathsf{Inr}^-\,(\mathsf{cont}\, t_1).\ \theta\, t_1'\, t_2'
\end{aligned}$$

where $\mathsf{Inl}^-\,(\mathsf{cont}\, t)$ is the set of t's successor leaves and $\mathsf{Inr}^-\,(\mathsf{cont}\, t)$ is the set of its immediate subtrees. Informally: If two trees are in relation θ, then they have the same root and the same successor leaves and for each immediate subtree of one, there exists an immediate subtree of the other in relation θ with it.

C   Concrete Iteration and Induction

Finite trees can be defined by

$$\mathsf{datatype}\ \mathsf{fdtree} = \mathsf{FNode}\ (\mathsf{froot} : \mathsf{N})\ (\mathsf{fcont} : (\mathsf{T} + \mathsf{dtree})\ \mathsf{fset})$$

This produces the operations FNode, froot, and fcont, with the same constructor–selector properties as Node, root and cont from the codatatype dtree introduced in Section 3. The differences concern (co)induction and (co)recursion.

Iteration. The general principle described in Section 2.1 employs in the unary case an iterator fold of (polymorphic) type $(\beta \text{ pre\_fdtree} \to \beta) \to \text{fdtree} \to \beta$, for which it yields $\forall s : \beta \text{ pre\_fdtree} \to \beta.$ fold $s \circ \text{ctor} = s \circ \text{map\_pre\_fdtree}$ (fold s), that is,

$$\forall s : \beta \text{ pre\_fdtree} \to \beta. \forall k. \text{ fold } s \text{ (ctor } k) = s \text{ (map\_pre\_fdtree (fold } s) \text{ } k)$$

The fdtree-defining BNF coincides with the dtree-defining BNF: β pre_fdtree $= \text{N} \times (\text{T} + \beta)$ fset and map_pre_fdtree $f = \text{id} \otimes (\text{image (id} \oplus f))$.

The above characterization needs some customization. Using the FNode instead of ctor and unfolding the definition of map_pre_fdtree, we obtain $\forall s : \text{N} \times (\text{T} + \beta)$ fset $\to \beta. \forall n \text{ } as.$ fold s (FNode $n \text{ } as) = s$ (map_pre_fdtree (fold s) (n, as)). By unfolding the definition of map_pre_fdtree, we obtain

$$\forall s : \text{N} \times (\text{T} + \beta) \text{ fset} \to \beta. \forall n \text{ } as. \text{ fold } s \text{ (FNode } n \text{ } as) = s \text{ } (n, (\text{id} \oplus \text{fold } s) \bullet as)$$

Finally, replacing fold with its more convenient curried version fold$'$: $(\text{N} \to (\text{T} + \beta) \text{ fset} \to \beta) \to \text{fdtree} \to \beta$ defined as fold$'$ $s = \text{fold } (\lambda(n, as). \text{ } s \text{ } n \text{ } as)$, we obtain the following customized iteration principle, where we write fold instead of fold$'$: For all sets β, functions $s : \text{N} \to (\text{T} + \beta) \text{ fset} \to \beta$ and elements $n \in \text{N}$ and $as \in (\text{T} + \text{fdtree}) \text{ fset}$, it holds that fold s (FNode $n \text{ } as) = s \text{ } n \text{ } ((\text{id} \oplus \text{fold } s) \bullet as)$.

Induction. The induction principle from Section A yields for $\varphi : \alpha \text{ fdtree} \to \text{bool}$

$$\frac{\forall k \in \alpha \text{ pre\_fdtree}. \text{ } (\forall t \in \text{Fset } k. \text{ } \varphi \text{ } t) \Rightarrow \varphi \text{ (ctor } k)}{\forall t \in \alpha \text{ fdtree}. \text{ } \varphi \text{ } t}$$

i.e., using the curried variation FNode of dtor,

$$\frac{\forall n \text{ } as. \text{ } (\forall t \in \text{Fset } (n, as). \text{ } \varphi \text{ } t) \Rightarrow \varphi \text{ (FNode } n \text{ } as)}{\forall t \in \alpha \text{ fdtree}. \text{ } \varphi \text{ } t}$$

Unfolding the definition of Fset, namely, Fset $(n, as) = \text{Inr}^- as$, we obtain the end-product customized induction for finite trees:

$$\frac{\forall n \text{ } as. \text{ } (\forall t \in \text{Inr}^- as. \text{ } \varphi \text{ } t) \Rightarrow \varphi \text{ (FNode } n \text{ } as)}{\forall t \in \alpha \text{ fdtree}. \text{ } \varphi \text{ } t}$$

Making Random Judgments: Automatically Generating Well-Typed Terms from the Definition of a Type-System

Burke Fetscher[1], Koen Claessen[2], Michał Pałka[2], John Hughes[2],
and Robert Bruce Findler[1]

[1] Northwestern University
[2] Chalmers University of Technology

Abstract. This paper presents a generic method for randomly generating well-typed expressions. It starts from a specification of a typing judgment in PLT Redex and uses a specialized solver that employs randomness to find many different valid derivations of the judgment form.

Our motivation for building these random terms is to more effectively falsify conjectures as part of the tool-support for semantics models specified in Redex. Accordingly, we evaluate the generator against the other available methods for Redex, as well as the best available custom well-typed term generator. Our results show that our new generator is much more effective than generation techniques that do not explicitly take types into account and is competitive with generation techniques that do, even though they are specialized to particular type-systems and ours is not.

1 Introduction

Redex (Felleisen et al. 2010) employs property-based testing to help semantics engineers uncover bugs in their models. Semantics engineers write down properties that should hold of their models (e.g., type soundness) and Redex can randomly generate example expressions in an attempt to falsify those properties. Until recently, Redex used a naive generation strategy: it simply randomly picks productions from the grammar of the language to build a term and then checks to see if that falsifies the property of interest. For untyped models, or when the model author writes a "fixing" function that makes expressions more likely to type-check (e.g., by writing a post-processing function that binds free variables), this naive technique is effective (Klein 2009; Klein et al. 2012; Klein et al. 2013). With typed models, however, such randomly generated terms rarely type check and so the testing process spends most of its time rejecting ill-typed terms instead of actually testing the model.

To make testing more effective, we built a solver that randomly generates solutions to problems involving a subset of first-order logic with equality and inequality constraints, and we use that to transform a Redex specification of a type-system into a random generator of well-typed terms.

We evaluate our generator on a benchmark suite of buggy Redex models and show that it is far more effective than the naive approach and less effective than the fixing function approach, but still competitive. We also evaluate our generator against the best known, hand-tuned generator for random well-typed terms (Pałka et al. 2011). This

© Springer-Verlag Berlin Heidelberg 2015
J. Vitek (Ed.): ESOP 2015, LNCS 9032, pp. 383–405, 2015.
DOI: 10.1007/978-3-662-46669-8_16

generator handles only a language closely matched to the GHC Haskell compiler intermediate language, but is better than our generic generator, overall. We compared the two generators by searching for counterexamples to two properties using a buggy version of GHC. A straightforward translation into Redex using our generator is able to find one bug infrequently, and to investigate the difficulties we refined that translation into a non-polymorphic model that was much more effective, demonstrating how polymorphism can be a difficult issue to tackle with random testing. We carefully explore why and discuss the issues in section 4.

Section 2 works through the generation process for a specific model in order to explain our method. Section 3 gives a small, formal model of our generator. Section 4 explains the evaluation of our generator. Section 5 discusses related work and section 6 concludes.

2   Example: Generating a Well-Typed Term

This section gives an overview of our method for generating well-typed terms by working through the generation of an example term. We will build a derivation satisfying the judgment form definition in figure 1, a typing judgment for simply-typed lambda calculus with a single base type of natural numbers. We begin with a goal pattern, which we will want the conclusion of the generated derivation to match.

Our goal pattern will be the following:

$$\bullet \vdash e^0 : \tau^0$$

stating that we would like to generate an expression with arbitrary type in the empty type environment. We then randomly select one of the type rules. This time, the generator selects the abstraction rule, which requires us to specialize the values of e^0 and τ^0 in order to agree with the form of the rule's conclusion. To do that, we first generate a new set of variables to replace the ones in the abstraction rule, and then unify our conclusion with the specialized rule. We put a super-script 1 on these variables to indicate that they were introduced in the first step of the derivation building process, giving us this partial derivation.

$$e ::= (e\ e)\ |\ (\lambda\ (x\ \tau)\ e)\ |\ x\ |\ n$$
$$\tau ::= (\tau \to \tau)\ |\ num$$
$$\Gamma ::= (x\ \tau\ \Gamma)\ |\ \bullet$$

$$\Gamma \vdash n : num$$

$$\frac{(x\ \tau_x\ \Gamma) \vdash e : \tau_e}{\Gamma \vdash (\lambda\ (x\ \tau_x)\ e) : (\tau_x \to \tau_e)}$$

$$\frac{\tau = \mathsf{lookup}\ [\![\Gamma, x]\!]}{\Gamma \vdash x : \tau}$$

$$\frac{\Gamma \vdash e_1 : (\tau_2 \to \tau) \quad \Gamma \vdash e_2 : \tau_2}{\Gamma \vdash (e_1\ e_2) : \tau}$$

Fig. 1. Grammar and type system for the simply-typed lambda calculus used in the example derivation

$$\frac{(x^l\,\tau_x^l\,\bullet)\;\vdash\;e^l\;:\;\tau^l}{\bullet\;\vdash\;(\lambda\,(x^l\,\tau_x^l)\,e^l)\;:\;(\tau_x^l\to\tau^l)}$$

The abstraction rule has added a new premise we must now satisfy, so we follow the same process with the premise. If the generator selects the abstraction rule again and then the application rule, we arrive at the following partial derivation, where the superscripts on the variables indicate the step where they were generated:

$$\frac{\dfrac{(x^2\,\tau_x^2\,(x^l\,\tau_x^l\,\bullet))\;\vdash\;e_1^3\;:\;(\tau_2^3\to\tau^2)\quad(x^2\,\tau_x^2\,(x^l\,\tau_x^l\,\bullet))\;\vdash\;e_2^3\;:\;\tau_2^3}{(x^2\,\tau_x^2\,(x^l\,\tau_x^l\,\bullet))\;\vdash\;(e_1^3\,e_2^3)\;:\;\tau^2}}{\dfrac{(x^l\,\tau_x^l\,\bullet)\;\vdash\;(\lambda\,(x^2\,\tau_x^2)\,(e_1^3\,e_2^3))\;:\;(\tau_x^2\to\tau^2)}{\bullet\;\vdash\;(\lambda\,(x^l\,\tau_x^l)\,(\lambda\,(x^2\,\tau_x^2)\,(e_1^3\,e_2^3)))\;:\;(\tau_x^l\to(\tau_x^2\to\tau^2))}}$$

Application has two premises, so there are now two unfinished branches of the derivation. Working on the left side first, suppose the generator chooses the variable rule:

$$\frac{\dfrac{\mathsf{lookup}\,[\![(x^2\,\tau_x^2\,(x^l\,\tau_x^l\,\bullet)),x^4]\!]\;=\;(\tau_2^3\to\tau^2)}{(x^2\,\tau_x^2\,(x^l\,\tau_x^l\,\bullet))\;\vdash\;x^4\;:\;(\tau_2^3\to\tau^2)\quad(x^2\,\tau_x^2\,(x^l\,\tau_x^l\,\bullet))\;\vdash\;e_2^3\;:\;\tau_2^3}}{\dfrac{(x^2\,\tau_x^2\,(x^l\,\tau_x^l\,\bullet))\;\vdash\;(x^4\,e_2^3)\;:\;\tau^2}{\dfrac{(x^l\,\tau_x^l\,\bullet)\;\vdash\;(\lambda\,(x^2\,\tau_x^2)\,(x^4\,e_2^3))\;:\;(\tau_x^2\to\tau^2)}{\bullet\;\vdash\;(\lambda\,(x^l\,\tau_x^l)\,(\lambda\,(x^2\,\tau_x^2)\,(x^4\,e_2^3)))\;:\;(\tau_x^l\to(\tau_x^2\to\tau^2))}}}$$

To continue, we need to use the lookup metafunction, whose definition is shown on the left-hand side of figure 2. Unlike judgment forms, however, Redex metafunction clauses are ordered, meaning that as soon as one of the left-hand sides matches an input, the corresponding right-hand side is used for the result. Accordingly, we cannot freely choose a clause of a metafunction without considering the previous clauses. Internally, our method treats a metafunction as a judgment form, however, adding premises to reflect the ordering.

$$\mathsf{lookup}\,[\![(x\,\tau\,\Gamma),x]\!]\;=\;\tau$$
$$\mathsf{lookup}\,[\![(x_l\,\tau\,\Gamma),x_2]\!]\;=\;\mathsf{lookup}\,[\![\Gamma,x_2]\!]$$
$$\mathsf{lookup}\,[\![\bullet,x]\!]\qquad\;=\;\#f$$

$$\frac{}{\mathsf{lookup}\,[\![(x\,\tau\,\Gamma),x]\!]\;=\;\tau}$$

$$\frac{}{\mathsf{lookup}\,[\![\bullet,x]\!]\;=\;\#f}$$

$$\frac{x_l\neq x_2\quad\mathsf{lookup}\,[\![\Gamma,x_2]\!]\;=\;\tau}{\mathsf{lookup}\,[\![(x_l\,\tau_x\,\Gamma),x_2]\!]\;=\;\tau}$$

Fig. 2. Lookup as a metafunction (left), and the corresponding judgment form (right)

For the lookup function, we can use the judgment form shown on the right of figure 2. The only additional premise appears in the bottom rule and ensures that we only recur with the tail of the environment when the head does not contain the variable we're looking for. The general process is more complex than lookup suggests and we return to this issue in section 3.1.

If we now choose that last rule, we have this partial derivation:

$$\cfrac{\cfrac{x^2 \neq x^4 \quad \text{lookup}\,[\![(x^1\ \tau_x^1\ \bullet),x^4]\!]\ =\ (\tau_2^3 \to \tau^2)}{\text{lookup}\,[\![(x^2\ \tau_x^2\ (x^1\ \tau_x^1\ \bullet)),x^4]\!]\ =\ (\tau_2^3 \to \tau^2)}}{\cfrac{\cfrac{(x^2\ \tau_x^2\ (x^1\ \tau_x^1\ \bullet)) \vdash x^4\ :\ (\tau_2^3 \to \tau^2) \qquad (x^2\ \tau_x^2\ (x^1\ \tau_x^1\ \bullet)) \vdash e_2^3\ :\ \tau_2^3}{(x^2\ \tau_x^2\ (x^1\ \tau_x^1\ \bullet)) \vdash (x^4\ e_2^3)\ :\ \tau^2}}{\cfrac{(x^1\ \tau_x^1\ \bullet) \vdash (\lambda\ (x^2\ \tau_x^2)\ (x^4\ e_2^3))\ :\ (\tau_x^2 \to \tau^2)}{\bullet \vdash (\lambda\ (x^1\ \tau_x^1)\ (\lambda\ (x^2\ \tau_x^2)\ (x^4\ e_2^3)))\ :\ (\tau_x^1 \to (\tau_x^2 \to \tau^2))}}}$$

The generator now chooses lookup's first clause, which has no premises, thus completing the left branch.

$$\cfrac{\cfrac{x^2 \neq x^1 \quad \text{lookup}\,[\![(x^1\ (\tau_2^3 \to \tau^2)\ \bullet),x^1]\!]\ =\ (\tau_2^3 \to \tau^2)}{\text{lookup}\,[\![(x^2\ \tau_x^2\ (x^1\ (\tau_2^3 \to \tau^2)\ \bullet)),x^1]\!]\ =\ (\tau_2^3 \to \tau^2)}}{\cfrac{\cfrac{(x^2\ \tau_x^2\ (x^1\ (\tau_2^3 \to \tau^2)\ \bullet)) \vdash x^1\ :\ (\tau_2^3 \to \tau^2) \qquad (x^2\ \tau_x^2\ (x^1\ (\tau_2^3 \to \tau^2)\ \bullet)) \vdash e_2^3\ :\ \tau_2^3}{(x^2\ \tau_x^2\ (x^1\ (\tau_2^3 \to \tau^2)\ \bullet)) \vdash (x^1\ e_2^3)\ :\ \tau^2}}{\cfrac{(x^1\ (\tau_2^3 \to \tau^2)\ \bullet) \vdash (\lambda\ (x^2\ \tau_x^2)\ (x^1\ e_2^3))\ :\ (\tau_x^2 \to \tau^2)}{\bullet \vdash (\lambda\ (x^1\ (\tau_2^3 \to \tau^2))\ (\lambda\ (x^2\ \tau_x^2)\ (x^1\ e_2^3)))\ :\ ((\tau_2^3 \to \tau^2) \to (\tau_x^2 \to \tau^2))}}}$$

Because pattern variables can appear in two different premises (for example the application rule's τ_2 appears in both premises), choices in one part of the tree affect the valid choices in other parts of the tree. In our example, we cannot satisfy the right branch of the derivation with the same choices we made on the left, since that would require $\tau_2^3 = (\tau_2^3 \to \tau^2)$.

This time, however, the generator picks the variable rule and then picks the first clause of the lookup, resulting in the complete derivation:

$$\cfrac{\cfrac{\vdots \qquad \qquad \cfrac{}{\text{lookup}\,[\![(x^2\ \tau_x^2\ (x^1\ (\tau_x^2 \to \tau^2)\ \bullet)),x^2]\!]\ =\ \tau_x^2}}{\cfrac{(x^2\ \tau_x^2\ (x^1\ (\tau_x^2 \to \tau^2)\ \bullet)) \vdash x^1\ :\ (\tau_x^2 \to \tau^2) \qquad (x^2\ \tau_x^2\ (x^1\ (\tau_x^2 \to \tau^2)\ \bullet)) \vdash x^2\ :\ \tau_x^2}{(x^2\ \tau_x^2\ (x^1\ (\tau_x^2 \to \tau^2)\ \bullet)) \vdash (x^1\ x^2)\ :\ \tau^2}}}{\cfrac{(x^1\ (\tau_x^2 \to \tau^2)\ \bullet) \vdash (\lambda\ (x^2\ \tau_x^2)\ (x^1\ x^2))\ :\ (\tau_x^2 \to \tau^2)}{\bullet \vdash (\lambda\ (x^1\ (\tau_x^2 \to \tau^2))\ (\lambda\ (x^2\ \tau_x^2)\ (x^1\ x^2)))\ :\ ((\tau_x^2 \to \tau^2) \to (\tau_x^2 \to \tau^2))}}$$

To finish the construction of a random well-typed term, we choose random values for the remaining, unconstrained variables, e.g.:

- ⊢ (λ (f (num → num)) (λ (a num) (f a))) : ((num → num) → (num → num))

We must be careful to obey the constraint that x^1 and x^2 are different, which was introduced earlier during the derivation, as otherwise we might not get a well-typed term. For example, (λ (f (num → num)) (λ (f num) (f f))) is not well-typed but is an otherwise valid instantiation of the non-terminals.

3 Derivation Generation in Detail

This section describes a formal model[1] of the derivation generator. The centerpiece of the model is a relation that rewrites programs consisting of metafunctions and judgment forms into the set of possible derivations that they can generate. Our implementation has a structure similar to the model, except that it uses randomness and heuristics to select just one of the possible derivations that the rewriting relation can produce. Our model is based on Jaffar et al. (1998)'s constraint logic programming semantics.

$$
\begin{array}{lll}
P ::= (D \ ...) & & \\
D ::= (r \ ...) & C ::= (\wedge\ (\wedge\ (x = p)\ ...) & p ::= (\text{lst } p \ ...) \\
r ::= ((d\,p) \leftarrow a \ ...) & \quad\ (\wedge\ \delta\ ...)) & \quad |\ m \\
a ::= (d\,p)\ |\ \delta & e ::= (p = p) & \quad |\ x \\
d ::= \textit{Identifier} & \delta ::= (\forall\ (x\ ...)\ (\vee\ (p \neq p)\ ...)) & m ::= \textit{Constant} \\
& & x ::= \textit{Variable}
\end{array}
$$

| Programs | Formulas | Patterns |

Fig. 3. The syntax of the derivation generator model

The grammar in figure 3 describes the language of the model. A program P consists of definitions D, which are sets of inference rules $((d\,p) \leftarrow a\ ...)$, here written horizontally with the conclusion on the left and premises on the right. (Note that ellipses are used in a precise manner to indicate repetition of the immediately previous expression, in this case a, following Scheme tradition. They do not indicate elided text.) Definitions can express both judgment forms and metafunctions. They are a strict generalization of judgment forms, and metafunctions are compiled into them via a process we discuss in section 3.1.

The conclusion of each rule has the form $(d\,p)$, where d is an identifier naming the definition and p is a pattern. The premises a may consist of literal goals $(d\,p)$ or disequational constraints δ. We dive into the operational meaning behind disequational constraints later in this section, but as their form in figure 3 suggests, they are a disjunction of negated equations, in which the variables listed following $\forall$ are universally

---

[1] The corresponding Redex model is available from this paper's website (listed after the conclusion), including a runnable simple example that may prove helpful when reading this section.

$$(P \vdash ((d \; p_g) \; a \; ...) \parallel C_1) \qquad\qquad\qquad \text{[reduce]}$$
$$\longrightarrow (P \vdash (a_f \; ... \; a \; ...) \parallel C_2)$$
$$\text{where } (D_0 \; ... \; (r_0 \; ... \; ((d \; p_r) \leftarrow a_r \; ...) \; r_1 \; ...) \; D_1 \; ...) = P,$$
$$((d \; p_f) \leftarrow a_f \; ...) = \text{freshen} \, [\![((d \; p_r) \leftarrow a_r \; ...)]\!],$$
$$C_2 = \text{solve} \, [\![(p_f = p_g), C_1]\!]$$

$$(P \vdash (\delta_g \; a \; ...) \parallel C_1) \qquad\qquad\qquad \text{[new constraint]}$$
$$\longrightarrow (P \vdash (a \; ...) \parallel C_2)$$
$$\text{where } C_2 = \text{dissolve} \, [\![\delta_g, C_1]\!]$$

Fig. 4. Reduction rules describing generation of the complete tree of derivations

quantified. The remaining variables in a disequation are implicitly existentially quantified, as are the variables in equations.

The reduction relation shown in figure 4 generates the complete tree of derivations for the program P with an initial goal of the form $(d \; p)$, where d is the identifier of some definition in P and p is a pattern that matches the conclusion of all of the generated derivations. The relation is defined using two rules: [reduce] and [new constraint]. The states that the relation acts on are of the form $(P \vdash (a \; ...) \parallel C)$, where $(a \; ...)$ represents a stack of goals, which can either be incomplete derivations of the form $(d \; p)$, indicating a goal that must be satisfied to complete the derivation, or disequational constraints that must be satisfied. A constraint store C is a set of simplified equations and disequations that are guaranteed to be satisfiable. The notion of equality we use here is purely syntactic; two ground terms are equal to each other only if they are identical.

Each step of the rewriting relation looks at the first entry in the goal stack and rewrites to another state based on its contents. In general, some reduction sequences are ultimately doomed, but may still reduce for a while before the constraint store becomes inconsistent. In our implementation, discovery of such doomed reduction sequences causes backtracking. Reduction sequences that lead to valid derivations always end with a state of the form $(P \vdash () \parallel C)$, and the derivation itself can be read off of the reduction sequence that reaches that state.

When a goal of the form $(d \; p)$ is the first element of the goal stack (as is the root case, when the initial goal is the sole element), then the [reduce] rule applies. For every rule of the form $((d \; p_r) \leftarrow a_r \; ...)$ in the program such that the definition's id d agrees with the goal's, a reduction step can occur. The reduction step first freshens the variables in the rule, asks the solver to combine the equation $(p_f = p_g)$ with the current constraint store, and reduces to a new state with the new constraint store and a new goal state. If the solver fails, then the reduction rule doesn't apply (because solve returns $\perp$ instead of a C_2). The new goal stack has all of the previously pending goals as well as the new ones introduced by the premises of the rule.

The [new constraint] rule covers the case where a disequational constraint δ is the first element in the goal stack. In that case, the disequational solver is called with the current constraint store and the disequation. If it returns a new constraint store, then the disequation is consistent and the new constraint store is used.

The remainder of this section fills in the details in this model and discusses the correspondence between the model and the implementation in more detail. Metafunctions are added via a procedure generalizing the process used for lookup in section 2, which we explain in section 3.1. Section 3.2 describes how our solver handles equations and disequations. Section 3.3 discusses the heuristics in our implementation and section 3.4 describes how our implementation scales up to support features in Redex that are not covered in this model.

3.1   Compiling Metafunctions

The primary difference between a metafunction, as written in Redex, and a set of $((d\ p) \leftarrow a\ ...)$ clauses from figure 3 is sensitivity to the ordering of clauses. Specifically, when the second clause in a metafunction fires, then the pattern in the first clause must not match, in contrast to the rules in the model, which fire regardless of their relative order. Accordingly, the compilation process that translates metafunctions into the model must insert disequational constraints to capture the ordering of the cases.

As an example, consider the metafunction definition of g on the left and some example applications on the right:

$$g[\![(\text{lst}\ p_1\ p_2)]\!] = 2 \qquad g[\![(\text{lst}\ 1\ 2)]\!] = 2$$
$$g[\![p]\!] \qquad\quad = 1 \qquad g[\![(\text{lst}\ 1\ 2\ 3)]\!] = 1$$

The first clause matches any two-element list, and the second clause matches any pattern at all. Since the clauses apply in order, an application where the argument is a two-element list will reduce to 2 and an argument of any other form will reduce to 1. To generate conclusions of the judgment corresponding to the second clause, we have to be careful not to generate anything that matches the first.

Applying the same idea as lookup in section 2, we reach this incorrect translation:

$$\frac{}{g[\![(\text{lst}\ p_1\ p_2)]\!]\ =\ 2} \qquad \frac{(\text{lst}\ p_1\ p_2)\ \neq\ p}{g[\![p]\!]\ =\ 1}$$

This is wrong because it would let us derive $g[\![(\text{list}\ 1\ 2)]\!] = 1$, using 3 for p_1 and 4 for p_2 in the premise of the right-hand rule. The problem is that we need to disallow all possible instantiations of p_1 and p_2, but the variables can be filled in with just specific values to satisfy the premise.

The correct translation, then, universally quantifies the variables p_1 and p_2:

$$\frac{}{g[\![(\text{lst}\ p_1\ p_2)]\!]\ =\ 2} \qquad \frac{(\forall (p_1\ p_2)\ (\text{lst}\ p_1\ p_2)\ \neq\ p)}{g[\![p]\!]\ =\ 1}$$

Thus, when we choose the second rule, we know that the argument will never be able to match the first clause.

In general, when compiling a metafunction clause, we add a disequational constraint for each previous clause in the metafunction definition. Each disequality is between the left-hand side patterns of one of the previous clauses and the left-hand side of the current clause, and it is quantified over all variables in the previous clause's left-hand side.

3.2   The Constraint Solver

The constraint solver maintains a set of equations and disequations that captures invariants of the current derivation that it is building. These constraints are called the constraint store and are kept in the canonical form C, as shown in figure 3, with the additional constraint that the equational portion of the store can be considered an idempotent substitution. That is, it always equates variables with with ps and, no variable on the left-hand side of an equality also appears in any right-hand side. Whenever a new constraint is added, consistency is checked again and the new set is simplified to maintain the canonical form.

Figure 5 shows solve, the entry point to the solver for new equational constraints. It accepts an equation and a constraint store and either returns a new constraint store that is equivalent to the conjunction of the constraint store and the equation or ⊥, indicating that adding e is inconsistent with the constraint store. In its body, it first applies the equational portion of the constraint store as a substitution to the equation. Second, it performs syntactic unification (Baader and Snyder 2001) of the resulting equation with the equations from the original store to build a new equational portion of the constraint. Third, it calls check, which simplifies the disequational constraints and checks their consistency. Finally, if all that succeeds, check returns a constraint store that combines the results of unify and check. If either unify or check fails, then solve returns ⊥.

Figure 6 shows dissolve, the disequational counterpart to solve. It applies the equational part of the constraint store as a substitution to the new disequation and then calls disunify. It disunify returns ⊤, then the disequation was already guaranteed in the current constraint store and thus does not need to be recorded. If disunify returns ⊥ then the disequation is inconsistent with the current constraint store and thus dissolve itself returns ⊥. In the final situation, disunify returns a new disequation, in which case dissolve adds that to the resulting constraint store.

The disunify function exploits unification and a few cleanup steps to determine if the input disequation is satisfiable. In addition, disunify is always called with a disequation that has had the equational portion of the constraint store applied to it (as a substitution).

The key trick in this function is to observe that since a disequation is always a disjunction of inequalities, its negation is a conjuction of equalities and is thus suitable as an input to unification. The first case in disunify covers the case where unification fails. In this situation we know that the disequation must have already been guaranteed to be false in constraint store (since the equational portion of the constraint store was applied as a substitution before calling disunify). Accordingly, disunify can simply return ⊤ to indicate that the disequation was redundant.

Ignoring the call to param-elim in the second case of disunify for a moment, consider the case where unify returns an empty conjunct. This means that unify's argument is guaranteed to be true and thus the given disequation is guaranteed to be false.

solve : $e\ C \to C$ or $\bot$

solve $[\![e_{new}, (\land\ (\land\ (x = p)\ ...)\ (\land\ \delta\ ...))]\!] =$

$$\begin{cases} (\land\ (\land\ (x_2 = p_2)\ ...) & \text{if } (\land\ (x_2 = p_2)\ ...) = \text{unify}\,[\![(e_{new}\{x \to p,\ ...\}), (\land\ (x = p)\ ...)]\!], \\ \quad (\land\ \delta_2\ ...)) & (\land\ \delta_2\ ...) = \text{check}\,[\![(\land\ \delta\{x_2 \to p_2,\ ...\}\ ...)]\!] \\ \bot & \text{otherwise} \end{cases}$$

unify : $(e\ ...)\ (\land\ (x = p)\ ...) \to (\land\ (x = p)\ ...)$ or $\bot$

unify $[\![((p = p)\ e\ ...), (\land\ e_s\ ...)]\!]$ $= \text{unify}\,[\![(e\ ...), (\land\ e_s\ ...)]\!]$

unify $[\![(((\text{lst}\ p_1\ ..._1) = (\text{lst}\ p_2\ ..._1))\ e\ ...), (\land\ e_s\ ...)]\!]$ $= \text{unify}\,[\![((p_1 = p_2)\ ...\ e\ ...), (\land\ e_s\ ...)]\!]$
where $|(p_{1..})| = |(p_2\ ...)|$

unify $[\![((x = p)\ e\ ...), (\land\ e_s\ ...)]\!]$ $= \bot$
where occurs? $[\![x, p]\!]$, $x \neq p$

unify $[\![((x = p)\ e\ ...), (\land\ e_s\ ...)]\!]$ $= \text{unify}\,[\![(e\{x \to p\}\ ...),$
                                                                                                    $(\land\ (x = p)\ e_s\{x \to p\}\ ...)]\!]$

unify $[\![((p = x)\ e\ ...), (\land\ e_s\ ...)]\!]$ $= \text{unify}\,[\![((x = p)\ e\ ...), (\land\ e_s\ ...)]\!]$

unify $[\![(), (\land\ e\ ...)]\!]$ $= (\land\ e\ ...)$

unify $[\![(e\ ...), (\land\ e_s\ ...)]\!]$ $= \bot$

Fig. 5. The Solver for Equations

dissolve : $\delta\ C \to C$ or $\bot$

dissolve $[\![\delta_{new}, (\land\ (\land\ (x = p)\ ...)\ (\land\ \delta\ ...))]\!] =$

$$\begin{cases} (\land\ (\land\ (x = p)\ ...)\ (\land\ \delta\ ...)) & \text{if } \top = \text{disunify}\,[\![\delta_{new}\{x \to p,\ ...\}]\!] \\ \bot & \text{if } \bot = \text{disunify}\,[\![\delta_{new}\{x \to p,\ ...\}]\!] \\ (\land\ (\land\ (x = p)\ ...)\ (\land\ \delta_0\ \delta\ ...)) & \text{if } \delta_0 = \text{disunify}\,[\![\delta_{new}\{x \to p,\ ...\}]\!] \end{cases}$$

disunify : $\delta \to \delta$ or $\top$ or $\bot$

disunify $[\![(\forall\ (x\ ...)\ (\lor\ (p_1 \neq p_2)\ ...))]\!] =$

$$\begin{cases} \top & \text{if } \bot = \text{unify}\,[\![((p_1 = p_2)\ ...), (\land)]\!] \\ \bot & \text{if } (\land) = \text{param-elim}\,[\![\text{unify}\,[\![((p_1 = p_2)\ ...), (\land)]\!], \\ & \qquad\qquad (x\ ...)]\!] \\ (\forall\ (x\ ...) & \text{if } (\land\ (x_p = p)\ ...) = \text{param-elim}\,[\![\text{unify}\,[\![((p_1 = p_2)\ ...), (\land)]\!], \\ \quad (\lor\ (x_p \neq p)\ ...)) & \qquad\qquad (x\ ...)]\!] \end{cases}$$

Fig. 6. The Solver for Disequations

param-elim : $(\wedge\ e\ ...)\ (x\ ...) \to (\wedge\ e\ ...)$ or $\perp$

param-elim $[\![(\wedge\ (x_0 = p_0)\ ...\ (x = p)\ (x_1 = p_1)\ ...),\ (x_2\ ...\ x\ x_3\ ...)]\!] =$
    param-elim $[\![(\wedge\ (x_0 = p_0)\ ...\ (x_1 = p_1)\ ...),\ (x_2\ ...\ x\ x_3\ ...)]\!]$

param-elim $[\![(\wedge\ (x_0 = p_0)\ ...\ (x_1 = x)\ (x_2 = p_2)\ ...),\ (x_4\ ...\ x\ x_5\ ...)]\!] =$
    param-elim $[\![(\wedge\ (x_0 = p_0)\ ...\ (x_3 = p_3)\ ...),\ (x_4\ ...\ x\ x_5\ ...)]\!]$

where $x \notin (p_0\ ...),\ ((x_3 = p_3)\ ...) =$ elim-x $[\![x,\ ((x_1 = x)\ (x_2 = p_2)\ ...)]\!]$

param-elim $[\![(\wedge\ e\ ...),\ (x\ ...)]\!] = (\wedge\ e\ ...)$

check : $(\wedge\ \delta\ ...) \to (\wedge\ \delta\ ...)$ or $\perp$

check $[\![(\wedge\ \delta_1\ ...\ (\forall\ (x_a\ ...)\ (\vee\ ((\mathsf{lst}\ p_l\ ...) \neq p_r)\ ...))\ \delta_2\ ...)]\!] =$

$\begin{cases} \text{check}\,[\![(\wedge\ \delta_1\ ...\ \delta_s\ \delta_2\ ...)]\!] & \text{if } \delta_s = \text{disunify}\,[\![(\forall\ (x_a\ ...)\ (\vee\ ((\mathsf{lst}\ p_l\ ...) \neq p_r)\ ...))]\!] \\ \text{check}\,[\![(\wedge\ \delta_1\ ...\ \delta_2\ ...)]\!] & \text{if } \top = \text{disunify}\,[\![(\forall\ (x_a\ ...)\ (\vee\ ((\mathsf{lst}\ p_l\ ...) \neq p_r)\ ...))]\!] \\ \perp & \text{if } \perp = \text{disunify}\,[\![(\forall\ (x_a\ ...)\ (\vee\ ((\mathsf{lst}\ p_l\ ...) \neq p_r)\ ...))]\!] \end{cases}$

check $[\![(\wedge\ \delta\ ...)]\!] = (\wedge\ \delta\ ...)$

Fig. 7. Metafunctions used to process disequational constraints

In this case, we have failed to generate a valid derivation because one of the negated disequations must be false (in terms of the original Redex program, this means that we attempted to use some later case in a metafunction with an input that would have satisfied an earlier case) and so disunify must return $\perp$.

But there is a subtle point here. Imagine that unify returns only a single clause of the form $(x = p)$ where x is one of the universally quantified variables. We know that in that case, the corresponding disequation $(\forall\ (x)\ (x \neq p))$ is guaranteed to be false because every pattern admits at least one concrete term. This is where param-elim comes in. It cleans up the result of unify by eliminating all clauses that, when negated and placed back under the quantifier would be guaranteed false, so the reasoning in the previous paragraph holds and the second case of disunify behaves properly.

The last case in disunify covers the situation where unify composed with param-elim returns a non-empty substitution. In this case, we do not yet know if the disequation is true or false, so we collect the substitution that unify returned back into a disequation and return it, to be saved in the constraint store.

This brings us to param-elim, in figure 7. Its first argument is a unifier, as produced by a call to unify to handle a disequation, and the second argument is the universally quantified variables from the original disequation. Its goal is to clean up the unifier by removing redundant and useless clauses.

There are two ways in which clauses can be false. In addition to clauses of the form $(x = p)$ where x is one of the universally quantified variables, it may also be the case that we have a clause of the form $(x_1 = x)$ and, as before, x is one of the universally quantified variables. This clause also must be dropped, according to the same reasoning (since $=$ is symmetric). But, since variables on the right hand side of an equation may also appear elsewhere, some care must be taken here to avoid losing transitive inequalities. The function elim-x (not shown) handles this situation, constructing a new set of clauses without x but, in the case that we also have $(x_2 = x)$, adds back the equation $(x_1 = x_2)$. For the full definition of elim-x and a proof that it works correctly, we refer the reader to the first author's masters dissertation (Fetscher 2014).

Finally, we return to check, shown in figure 7, which is passed the updated disequations after a new equation has been added in solve (see figure 5). It verifies the disequations and maintains their canonical form, once the new substitution has been applied. It does this by applying disunify to any non-canonical disequations.

3.3   Search Heuristics

To pick a single derivation from the set of candidates, our implementation must make explicit choices when there are differing states that a single reduction state reduces to. Such choices happen only in the [reduce] rule, and only because there may be multiple different clauses, $((d\ p) \leftarrow a\ ...)$, that could be used to generate the next reduction state.

To make these choices, our implementation collects all of the candidate cases for the next definition to explore. It then randomly permutes the candidate rules and chooses the first one of the permuted rules, using it as the next piece of the derivation. It then continues to search for a complete derivation. That process may fail, in which case the implementation backtracks to this choice and picks the next rule in the permuted list. If none of the choices leads to a successful derivation, then this attempt is failure and the implementation either backtracks to an earlier such choice, or fails altogether.

There are two refinements that the implementation applies to this basic strategy. First, the search process has a depth bound that it uses to control which production to choose. Each choice of a rule increments the depth bound and when the partial derivation exceeds the depth bound, then the search process no longer randomly permutes the candidates. Instead, it simply sorts them by the number of premises they have, preferring rules with fewer premises in an attempt to finish the derivation off quickly.

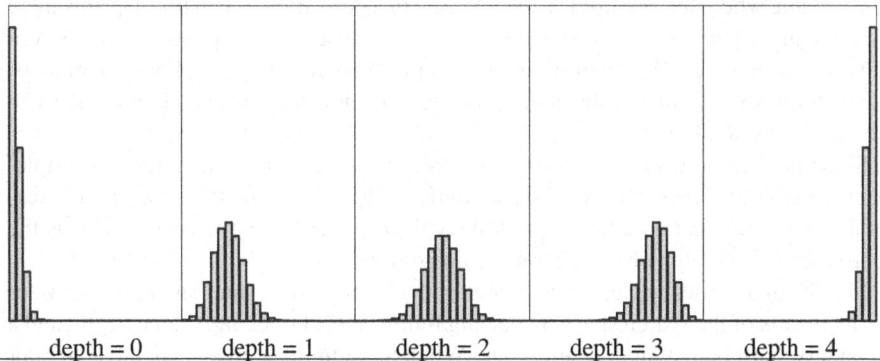

| depth = 0 | depth = 1 | depth = 2 | depth = 3 | depth = 4 |

Fig. 8. Density functions of the distributions used for the depth-dependent rule ordering, where the depth limit is 4 and there are 4 rules

The second refinement is the choice of how to randomly permute the list of candidate rules, and the generator uses two strategies. The first strategy is to just select from the possible permutations uniformly at random. The second strategy is to take into account how many premises each rule has and to prefer rules with more premises near the beginning of the construction of the derivation and rules with fewer premises as the search gets closer to the depth bound. To do this, the implementation sorts all of the possible permutations in a lexicographic order based on the number of premises of each choice. Then, it samples from a binomial distribution whose size matches the number of permutations and has probability proportional to the ratio of the current depth and the maximum depth. The sample determines which permutation to use.

More concretely, imagine that the depth bound was 4 and there are also 4 rules available. Accordingly, there are 24 different ways to order the premises. The graphs in figure 8 show the probability of choosing each permutation at each depth. Each graph has one x-coordinate for each different permutation and the height of each bar is the chance of choosing that permutation. The permutations along the x-axis are ordered lexicographically based on the number of premises that each rule has (so permutations that put rules with more premises near the beginning of the list are on the left and permutations that put rules with more premises near the end of the list are on the right). As the graph shows, rules with more premises are usually tried first at depth 0 and rules with fewer premises are usually tried first as the depth reaches the depth bound.

These two permutation strategies are complementary, each with its own drawbacks. Consider using the first strategy that gives all rule ordering equal probability with the rules shown in figure 1. At the initial step of our derivation, we have a 1 in 4 chance of choosing the type rule for numbers, so one quarter of all expressions generated will just be a number. This bias towards numbers also occurs when trying to satisfy premises of the other, more recursive clauses, so the distribution is skewed toward smaller derivations, which contradicts commonly held wisdom that bug finding is more effective when using larger terms. The other strategy avoids this problem, biasing the generation towards rules with more premises early on in the search and thus tending to produce larger terms. Unfortunately, our experience testing Redex program suggests that it is not uncommon for there to be rules with large number of premises that are completely unsatisfiable when they are used as the first rule in a derivation (when this happens there are typically a few other, simpler rules that must be used first to populate an environment or a store before the interesting and complex rule can succeed). For such models, using all rules with equal probability still is less than ideal, but is overall more likely to produce terms at all.

Since neither strategy for ordering rules is always better than the other, our implementation decides between the two randomly at the beginning of the search process for a single term, and uses the same strategy throughout that entire search. This is the approach the generator we evaluate in section 4 uses.

Finally, in all cases we terminate searches that appear to be stuck in unproductive or doomed parts of the search space by placing limits on backtracking, search depth, and a secondary, hard bound on derivation size. When these limits are violated, the generator simply abandons the current search and reports failure.

3.4 A Richer Pattern Language

The model we present in section 3 uses a much simpler pattern language than Redex itself. The portion of Redex's internal pattern language supported by the generator[2] is shown in figure 9. We now discuss briefly the interesting differences between this language and the language of our model and how we support them in Redex's implementation.

$$
\begin{array}{ll}
p ::= (\text{nt } s) & b ::= \text{any} \\
\quad | \ (\text{name } s \ p) & \quad | \ \text{number} \\
\quad | \ (\text{mismatch-name } s \ p) & \quad | \ \text{string} \\
\quad | \ (\text{list } p \ ...) & \quad | \ \text{natural} \\
\quad | \ b & \quad | \ \text{integer} \\
\quad | \ v & \quad | \ \text{real} \\
\quad | \ c & \quad | \ \text{boolean} \\
v ::= \text{variable} & s ::= symbol \\
\quad | \ (\text{variable-except } s \ ...) & c ::= constant \\
\quad | \ (\text{variable-prefix } s) \\
\quad | \ \text{variable-not-otherwise-mentioned}
\end{array}
$$

Fig. 9. The subset of Redex's pattern language supported by the generator. Racket symbols are indicated by s, and c represents any Racket constant.

Named patterns of the form (name s p) correspond to variables x in the simplified version of the pattern language from figure 3, except that the variable s is paired with a pattern p. From the matcher's perspective, this form is intended to match a term with the pattern p and then bind the matched term to the name s. The generator pre-processes all patterns with a first pass that extracts the attached pattern p and attempts to update the current constraint store with the equation ($s = p$), after which s can be treated as a logic variable.

The b and v non-terminals are built-in patterns that match subsets of Racket values. The productions of b are straightforward; integer, for example, matches any Racket integer, and any matches any Racket s-expression. From the perspective of the unifier, integer is a term that may be unified with any integer, the result of which is the integer itself. The value of the term in the current substitution is then updated. Unification of built-in patterns produces the expected results; for example unifying real and natural produces natural, whereas unifying real and string fails.

The productions of v match Racket symbols in varying and commonly useful ways; variable-not-otherwise-mentioned, for example, matches any symbol that is not used as a literal elsewhere in the language. These are handled similarly to the patterns of the b non-terminal within the unifier.

Patterns of the from (mismatch-name s p) match the pattern p with the constraint that two occurrences of the same name s may never match equal terms. These are straightforward: whenever a unification with a mismatch takes place, disequations are

---

[2] The generator is not able to handle parts of the pattern language that deal with evaluation contexts or "repeat" patterns (ellipses).

added between the pattern in question and other patterns that have been unified with the same mismatch pattern.

Patterns of the form (nt s) refer to a user-specified grammar, and match a term if it can be parsed as one of the productions of the non-terminal s of the grammar. It is less obvious how such non-terminal patterns should be dealt with in the unifier. To unify two such patterns, the intersection of two non-terminals should be computed, which reduces to the problem of computing the intersection of tree automata, for which there is no efficient algorithm (Comon et al. 2007). Instead a conservative check is used at the time of unification. When unifying a non-terminal with another pattern, we attempt to unify the pattern with each production of the non-terminal, replacing any embedded non-terminal references with the pattern any. We require that at least one of the unifications succeeds. Because this is not a complete check for pattern intersection, we save the names of the non-terminals as extra information embedded in the constraint store until the entire generation process is complete. Then, once we generate a concrete term, we check to see if any of the non-terminals would have been violated (using a matching algorithm). This means that we can get failures at this stage of generation, but it tends not to happen very often for practical Redex models.[3]

4   Evaluating the Generator

We evaluate the generator in two ways. First, we compare its effectiveness against the standard Redex generator on Redex's benchmark suite. Second, we compare it against the best known hand-tuned typed term generator.

4.1   The Redex Benchmark

Our first effort at evaluating the effectiveness of the derivation generator compares it to the existing random expression generator included with Redex (Klein and Findler 2009), which we term the "ad hoc" generation strategy in what follows. This generator is based on the method of recursively unfolding non-terminals in a grammar.

To compare the two generators, we used the Redex Benchmark (Findler et al. 2014), a suite of buggy models developed specifically to evaluate methods of automated testing for Redex. Models included in the benchmark define a soundness property and come in a number of different versions, each of which introduces a single bug that can violate the soundness property into the model. Most models are of programming languages and most soundness properties are type-soundness. For each version of each model, we define one soundness property and two generators, one using the method explained in this paper and one using Redex's ad hoc generation strategy. For a single test run, we pair a generator with its soundness property and repeatedly generate test cases using the generator, testing them with the soundness property, and tracking the intervals between instances where the test case causes the soundness property to fail, exposing the bug.

---

[3] To be more precise, on the Redex benchmark (see section 4.1) such failures occur on all "delim-cont" models $2.9\pm1.1\%$ of the time, on all "poly-stlc" models $3.3\pm0.3\%$ of the time, on the "rvm-6" model $8.6\pm2.9\%$ of the time, and are not observed on the other models.

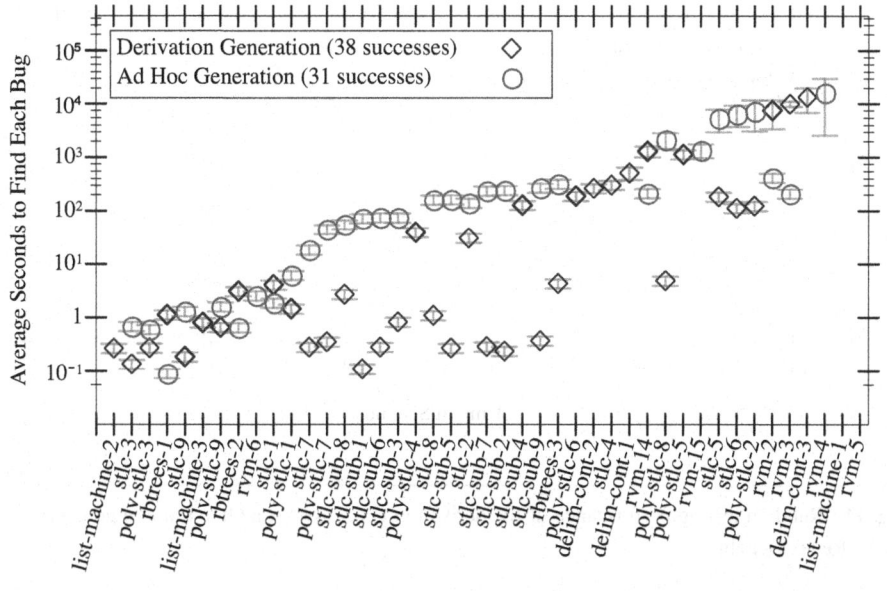

Fig. 10. Performance results by individual bug on the Redex Benchmark

For this study, each run continued for either 24 hours[4] or until the uncertainty in the average interval between such counterexamples became acceptably small.

This study used 6 different models, each of which has between 3 and 9 different bugs introduced into it, for a total of 40 different bugs. The models in the benchmark come from a number of different sources, some synthesized based on our experience for the benchmark, and some drawn from outside sources or pre-existing efforts in Redex. The latter are based on Appel et al. (2012)'s list machine benchmark, the model of contracts for delimited continuations developed by Takikawa et al. (2013), and the model of the Racket virtual machine from Klein et al. (2013). Detailed descriptions of all the models and bugs in the benchmark can be found in Findler et al. (2014).

Figure 10 summarizes the results of the comparison on a per-bug basis. The y-axis is time in seconds, and for each bug we plot the average time it took each generator to find a counterexample. The bugs are arranged along the x-axis, sorted by the average time for both generators to find the bug. The error bars represent 95% confidence intervals in the average, and in all cases the errors are small enough to clearly differentiate the averages. The two blank columns on the right are bugs that neither generator was able to find. The vertical scale is logarithmic, and the average time ranges from a tenth of a second to several hours, an extremely wide range in the rarity of counterexamples.

To depict more clearly the relative testing effectiveness of the two generation methods, we plot our data slightly differently in figure 11. Here we show time in seconds

---

[4] With one exception: we ran the derivation generator on "rvm-3" for a total of 32 days of processor time to reduce its uncertainty.

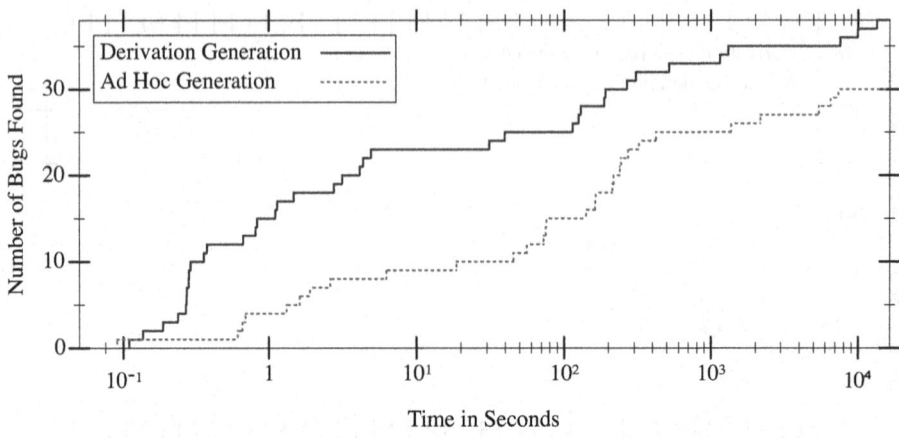

Fig. 11. Random testing performance of the derivation generator vs. ad hoc random generation on the Redex Benchmark

on the x-axis (the y-axis from figure 10, again on a log scale), and the total number of bugs found for each point in time on the y-axis. This plot makes it clear that the derivation generator is much more effective, finding more bugs more quickly at almost every time scale. In fact, an order of magnitude or more on the time scale separates the two generators for almost the entire plot.

While the derivation generator is more effective when it is used, it cannot be used with every Redex model, unlike the ad hoc generator. There are three broad categories of models to which it may not apply. First, the language may not have a type system, or the type system's implementation might use constructs that the generator fundamentally cannot handle (like escaping to Racket code to run arbitrary computation). Second, the generator currently cannot handle ellipses (aka repetition or Kleene star); we hope to someday figure out how to generalize our solver to support those patterns, however. And finally, some judgment forms thwart our termination heuristics. Specifically, our heuristics make the assumptions that the cost of completing the derivation is proportional to the size of the goal stack, and that terminal nodes in the search space are uniformly distributed. Typically these are safe assumptions, but not always; the benchmark's "let-poly" model, for example, is a CPS-transformed type judgment, embedding the search's continuation in the model, and breaking the first assumption.

4.2 Testing GHC: A Comparison with a Specialized Generator

We also compared the derivation generator we developed for Redex to a specialized generator of typed terms. This generator was designed to be used for differential testing of GHC, and generates terms for a specific variant of the lambda calculus with polymorphic constants, chosen to be close to the compiler's intermediate language. The generator is implemented using Quickcheck (Claessen and Hughes 2000), and is able to leverage its extensive support for writing random test case generators. Writing a generator for well-typed terms in this context required significant effort, essentially implementing a function from types to terms in Quickcheck. The effort yielded significant benefit, however, as implementing the entire generator from the ground up provided many opportunities for specialized optimizations, such as variations of type rules that are more likely to succeed, or varying the frequency with which different constants are chosen. Pałka (2012) discusses the details.

Generator	Terms/Ctrex.	Gen. Time (s)	Check Time (s)	Time/Ctrex. (s)
Property 1				
Hand-written (size: 50)	25K	0.007	0.009	413.79
Hand-written (size: 70)	16K	0.009	0.01	293.06
Hand-written (size: 90)	12K	0.011	0.01	260.65
Redex poly (depth: 6)	∞	0.361	0.008	∞
Redex poly (depth: 7)	∞	0.522	0.009	∞
Redex poly (depth: 8)*	4000K	0.63	0.008	2549K
Redex non-poly (depth: 6)*	500K	0.038	0.008	23K
Redex non-poly (depth: 7)	668	0.082	0.01	61.33
Redex non-poly (depth: 8)	320	0.076	0.01	27.29
Property 2				
Hand-written (size: 50)	100K	0.005	0.007	1K
Hand-written (size: 70)	125K	0.007	0.008	2K
Hand-written (size: 90)	83K	0.009	0.009	2K
Redex poly (depth: 6)	∞	0.306	0.005	∞
Redex poly (depth: 7)	∞	0.447	0.005	∞
Redex poly (depth: 8)	∞	0.588	0.005	∞
Redex non-poly (depth: 6)	∞	0.059	0.005	∞
Redex non-poly (depth: 7)	∞	0.17	0.01	∞
Redex non-poly (depth: 8)	∞	0.142	0.008	∞
Redex non-poly (depth: 10)*	4000K	0.196	0.01	823K

Fig. 12. Comparison of the derivation generator and a hand-written typed term generator. ∞ indicates runs where no counterexamples were found. Runs marked with * found only one counterexample, which gives low confidence to their figures.

Implementing this language in Redex was easy: we were able to port the formal description in Pałka (2012) directly into Redex with little difficulty. Once a type system is defined in Redex we can use the derivation generator immediately to generate well-typed terms. Such an automatically derived generator is likely to make some performance tradeoffs versus a specialized one, and this comparison gave us an excellent opportunity to investigate those.

We compared the generators by testing two of the properties used in Pałka (2012), and using same baseline version of the GHC (7.3.20111013) that was used there. **Property 1** checks whether turning on optimization influences the strictness of the compiled Haskell code. The property fails if the compiled function is less strict with optimization turned on. **Property 2** observes the order of evaluation, and fails if optimized code has a different order of evaluation compared to unoptimized code.

Hand-written (size: 90)   Redex poly (depth: 8)   Redex non-poly (depth: 8)

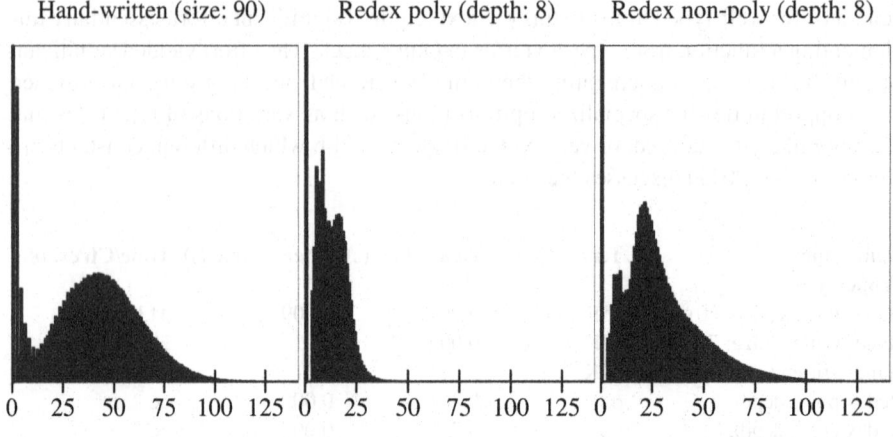

Fig. 13. Histograms of the sizes (number of internal nodes) of terms produced by the different runs. The vertical scale of each plot is one twentieth of the total number of terms in that run.

Counterexamples from the first property demonstrate erroneous behavior of the compiler, as the strictness of Haskell expressions should not be influenced by optimization. In contrast, changing the order of evaluation is allowed for a Haskell compiler to some extent, so counterexamples from the second property usually demonstrate interesting cases of the compiler behavior, rather than bugs.

Figure 12 summarizes the results of our comparison of the two generators. Each row represents a run of one of the generators, with a few varying parameters. We refer to Pałka (2012)'s generator as "hand-written." It takes a size parameter, which we varied over 50, 70, and 90 for each property. "Redex poly" is our initial implementation of this system in the Redex, the direct translation of the language from Pałka (2012). The Redex generator takes a depth parameter, which we vary over 6,7,8, and, in one case, 10. The depths are chosen so that both generators target terms of similar size.[5] (Figure 13 compares generated terms at targets of size 90 and depth 8). "Redex non-poly" is a modified version of our initial implementation, the details of which we discuss below. The columns show approximately how many tries it took to find a counterexample, the average time to generate a term, the average time to check a term, and finally the average time per counterexample over the entire run. Note that the goal type of terms used to test the two properties differs, which may affect generation time for otherwise identical generators.

A generator based on our initial Redex implementation was able to find counterexamples for only one of the properties, and did so and at significantly slower rate than the hand-written generator. The hand-written generator performed best when targeting a size of 90, the largest, on both properties. Likewise, Redex was only able to find coun-

---

[5] Although we are able to generate terms of larger depth, the runtime increases quickly with the depth. One possible explanation is that well-typed terms become very sparse as term size increases. Grygiel and Lescanne (2013) show how scarce well-typed terms are even for simple types. In our experience polymorphism exacerbates this problem.

terexamples when targeting the largest depth on property one. There, the hand-written generator was able to find a counterexample every 12K terms, about once every 260 seconds. The Redex generator both found counterexamples much less frequently, at one in 4000K, and generated terms several orders of magnitude more slowly. Property two was more difficult for the hand-written generator, and our first try in Redex was unable to find any counterexamples there.

Comparing the test cases from both generators, we found that Redex was producing significantly smaller terms than the hand-written generator. The left two histograms in figure 13 compare the size distributions, which show that most of the terms made by the hand-written generator are larger than almost all of the terms that Redex produced (most of which are clumped below a size of 25). The majority of counterexamples we were able to produce with the hand-written generator fell in this larger range.

Digging deeper, we found that Redex's generator was backtracking an excessive amount. This directly affects the speed at which terms are generated, and it also causes the generator to fail more often because the search limits discussed in section 3.3 are exceeded. Finally, it skews the distribution toward smaller terms because these failures become more likely as the size of the search space expands. We hypothesized that the backtracking was caused by making doomed choices when instantiating polymorphic types and only discovering that much later in the search, causing it to get stuck in expensive backtracking cycles. The hand-written generator avoids such problems by encoding model-specific knowledge in heuristics.

To test this hypothesis, we built a new Redex model identical to the first except with a pre-instantiated set of constants, removing polymorphism. We picked the 40 most common instantiations from a set of counterexamples to both models generated by the hand-written generator. Runs based on this model are referred to as "Redex non-poly" in both figure 12 and figure 13.

As figure 13 shows, we get a much better size distribution with the non-polymorphic model, comparable to the hand-written generator's distribution. A look at the second column of figure 12 shows that this model produces terms much faster than the first try in Redex, though still slower than the hand-written generator. This model's counterexample rate is especially interesting. For property one, it ranges from one in 500K terms at depth 6 to, astonishingly, one in 320 at depth 8, providing more evidence that larger terms make better test cases. This success rate is also much better than that of the hand-written generator, and in fact, it was this model that was most effective on property 1, finding a counterexample approximately every 30 seconds, significantly faster than the hand-written generator. Thus, it is interesting that it did much worse on property 2, only finding a counterexample once every 4000K terms, and at very large time intervals. We don't presently know how to explain this discrepancy.

Overall, our conclusion is that our generator is not competitive with the hand-tuned generator when it has to cope with polymorphism. Polymorphism, in turn, is problematic because it requires the generator to make parallel choices that must match up, but where the generator does not discover that those choices must match until much later in the derivation. Because the choice point is far from the place where the constraint is discovered, the generator spends much of its time backtracking. The improvement in generation speed for the Redex generator when removing polymorphism provides evidence for our explanation of what makes generating these terms difficult. The ease with which we were able to implement this language in Redex, and as a result, conduct this experiment, speaks to the value of a general-purpose generator.

5 Related Work

We first address work which our constraint solver draws on, and then related work in the field of random testing.

5.1 Disequations

Colmerauer (1984) is the first to introduce a method of solving disequational constraints of the type we use, but his work handles only existentially quantified variables. Like him, we too use the unification algorithm to simplify disequations.

Comon and Lescanne (1989) address the more general problem of solving all first order logical formulas where equality is the only predicate, which they term "equational problems," of which our constraints are a subset. They present a set of rules as rewrites on such formulas to transform them into solved forms. We believe our solver is essentially a way of factoring a stand-alone unifier out of their rules.

Byrd (2009) notes that a related form of disequality constraints has been available in many Prolog implementations and constraint programming systems since Prolog II. Notably, miniKanren (Byrd 2009) and cKanren (Alvis et al. 2011) implement them in a way similar to us, using unification as a subroutine. However, as far as we know, none of these systems supports the universally quantified constraints we require.

We are currently investigating extending our solver to handle Redex's repeat patterns. In this area, we note Kutsia (2002)'s work on sequence unification, which handles patterns similar to Redex's.

5.2 Random Testing

The most closely related work to ours is Claessen et al. (2014)'s typed term generator. Their work addresses specifically the problem of generating well-formed lambda terms based an implementation of a type-checker (in Haskell). They measured their approach against property 1 from section 4.2 and it performs better than Redex's 'poly' generator, but they are working from a lower-level specification of the type system than we are. Also, their approach observes the order of evaluation of the predicate, and prunes the search space based on that; it does not use constraint solving.

Quickcheck (Claessen and Hughes 2000) is a widely-used library for random testing in Haskell. It provides combinators supporting the definition of testable properties, random generators, and analysis of results. Although Quickcheck's approach is much more general than the one taken here, it has been used to implement a random generator for well-typed terms robust enough to find bugs in GHC (Pałka 2012). This generator provides a good contrast to the approach of this work, as it was implemented by hand, albeit with the assistance of a powerful test framework. Significant effort was spent on adjusting the distribution of terms and optimization, even adjusting the type system in clever ways. Our approach, on the other hand, is to provide a straightforward way to implement a test generator. The relationship to Pałka's work is discussed in more detail in section 4.2.

Random program generation for testing purposes is not a new idea and goes back at least to Hanford (1970), who details the development and application of the "syntax machine", a generator of random program expressions. The tool was intended for testing compilers, a common target for this type of random generation. Other uses of random testing for compiler testing throughout the years are discussed in Bourjarwah and Saleh (1997)'s survey.

In the area of random testing for compilers, of special note is Csmith (Yang et al. 2011), a highly effective tool for generating C programs for compiler testing. Csmith generates C programs that avoid undefined or unspecified behavior. These programs are then used for differential testing, where the output of a given program is compared across several compilers and levels of optimization, so that if the results differ, at least one of test targets must contain a bug. Csmith represents a significant development effort at 40,000+ lines of C++ and the programs it generates are finely tuned to be effective at finding bugs based on several years of experience. This approach has been effective, finding over 300 bugs in mainstream compilers as of 2011.

Efficient random generation of program terms has seen some interesting advances in previous years, much of which focuses on enumerations. Feat (Duregard et al. 2012), or "Functional Enumeration of Algebraic Types," is a Haskell library that exhaustively enumerates a datatype's possible values. The enumeration is made very efficient by memoising cardinality metadata, which makes it practical to access values that have very large indexes. The enumeration also weights all terms equally, so a random sample of values can in some sense be said to have a more uniform distribution. Feat was used to test Template Haskell by generating AST values, and compared favorably with Smallcheck in terms of its ability to generate terms above a certain size. (QuickCheck was excluded from this particular case study because it was "very difficult" to write a QuickCheck generator for "mutual recursive datatypes of this size", the size being around 80 constructors. This provides some insight into the effort involved in writing the generator described in Pałka (2012).)

Another, more specialized, approach to enumerations was taken by Grygiel and Lescanne (2013). Their work addresses specifically the problem of enumerating well-formed lambda terms. (Terms where all variables are bound.) They present a variety of combinatorial results on lambda terms, notably some about the extreme scarcity of simply-typable terms among closed terms. As a by-product they get an efficient generator for closed lambda terms. To generate typed terms their approach is simply to filter the closed terms with a typechecker. This approach is somewhat inefficient (as one would expect due to the rarity of typed terms) but it does provide a uniform distribution.

Instead of enumerating terms, Kennedy and Vytiniotis (2012) develop a bit-coding scheme where every string of bits either corresponds to a term or is the prefix of some term that does. Their approach is quite general and can be used to encode many different types. They are able to encode a lambda calculi with polymorphically-typed constants and discuss its possible extension to even more challenging languages such as System-

404   B. Fetscher et al.

F. This method cannot be used for random generation because only bit-strings that have a prefix-closure property correspond to well-formed terms.

6 Conclusion

As this paper demonstrates, random test-case generation is an effective tool for finding bugs in formal models. Even better, this work demonstrates how to build a generic random generator that is competitive with hand-tuned generators. We believe that employing more such lightweight techniques for debugging formal models can help the research community more effectively communicate research results, both with each other and with the wider world. Eliminating bugs from our models makes our results more approachable, as it means that our papers are less likely to contain frustrating obstacles that discourage newcomers.

Acknowledgments. Thanks to Casey Klein for help getting this project started and for an initial prototype implementation, to Asumu Takikawa for his help with the delimited continuations model, and to Larry Henschen for his help with earlier versions of this work. Thanks to Spencer Florence for helpful discussions and comments on the writing. Thanks to Hai Zhou, Li Li, Yuankai Chen, and Peng Kang for graciously sharing their compute servers with us. Thanks to the Ministry of Science and Technology of the R.O.C. for their support (under Contract MOST 103-2811-E-002-015) when Findler visited the CSIE department at National Taiwan University. Thanks also to the NSF for their support of this work.

This paper is available online at:

```
http://users.eecs.northwestern.edu/~baf111/random-judgments/
```

along with Redex models for all of the definitions in the paper and the raw data used to generate all of the plots.

References

1. Claire, E., Alvis, J.J., Willcock, K.M., Carter, W.E.: Byrd, and Daniel P. Friedman. cKanren: miniKanren with Constraints. In: Proc. Scheme and Functional Programming (2011)
2. Appel, A.W., Dockins, R., Leroy, X.: A list-machine benchmark for mechanized metatheory. Journal of Automated Reasoning 49(3), 453–491 (2012)
3. Baader, F., Snyder, W.: Unification Theory. In: Handbook of Automated Reasoning, vol. 1, pp. 445–532 (2001)
4. Bourjarwah, A.S., Saleh, K.: Compiler test case generation methods: a survey and assessment. Information & Software Technology 39(9), 617–625 (1997)
5. Byrd, W.E.: Relational Programming in miniKanren: Techniques, Applications, and Implementations. PhD dissertation, Indiana University (2009)
6. Claessen, K., Duregard, J., Palka, M.H.: Generating Constrained Random Data with Uniform Distribution. In: Proc. Intl. Symp. Functional and Logic Programming, pp. 18–34 (2014)
7. Claessen, K., Hughes, J.: QuickCheck: A Lightweight Tool for Random Testing of Haskell Programs. In: Proc. ACM Intl. Conf. Functional Programming, pp. 268–279 (2000)

8. Colmerauer, A.: Equations and Inequations on Finite and Infinite Trees. In: Proc. Intl. Conf. Fifth Generation Computing Systems, pp. 85–99 (1984)
9. Comon, H., Dauchet, M., Gilleron, R., Loding, C., Jacquemard, F., Lugiez, D., Tison, S., Tommasi, M.: Tree Automata Techniques and Applications (2007), http://www.grappa.univ-lille3.fr/tata
10. Comon, H., Lescanne, P.: Equational Problems and Disunification. Journal of Symbolic Computation 7, 371–425 (1989)
11. Duregard, J., Jansson, P., Wang, M.: Feat: Functional Enumeration of Algebraic Types. In: Proc. ACM SIGPLAN Haskell Wksp, pp. 61–72 (2012)
12. Felleisen, M., Findler, R.B., Flatt, M.: Semantics Engineering with PLT Redex. MIT Press (2010)
13. Fetscher, B.: The Random Generation of Well-Typed Terms. Northwestern University, NU-EECS-14-05 (2014)
14. Findler, R.B., Klein, C., Fetscher, B.: The Redex Reference (2014), http://docs.racket-lang.org/redex
15. Grygiel, K., Lescanne, P.: Counting and generating lambda terms. J. Functional Programming 23(5), 594–628 (2013)
16. Hanford, K.V.: Automatic Generation of Test Cases. IBM Systems Journal 9(4), 244–257 (1970)
17. Jaffar, J., Maher, M.J., Marriott, K., Stuckey, P.J.: The Semantics of Constraint Logic Programming. Journal of Logic Programming 37(1-3), 1–46 (1998)
18. Kennedy, A.J., Vytiniotis, D.: Every bit counts: The binary representation of typed data and programs. J. Functional Programming 22, 529–573 (2012)
19. Klein, C.: Experience with Randomized Testing in Programming Language Metatheory. MS dissertation, Northwestern University (2009)
20. Klein, C., Clements, J., Dimoulas, C., Eastlund, C., Felleisen, M., Flatt, M., McCarthy, J.A., Rafkind, J., Tobin-Hochstadt, S., Findler, R.B.: Run Your Research: On the Effectiveness of Lightweight Mechanization. In: Proc. ACM Symp. Principles of Programming Languages (2012)
21. Klein, C., Findler, R.B.: Randomized Testing in PLT Redex. In: Proc. Scheme and Functional Programming, pp. 26–36 (2009)
22. Klein, C., Findler, R.B., Flatt, M.: The Racket virtual machine and randomized testing. In: Higher-Order and Symbolic Computation (2013)
23. Kutsia, T.: Unification with Sequence Symbols and Flexible Arity Symbols and Its Extension with Pattern-Terms. In: Proc. Intl. Conf. Artificial Intelligence, Automated Reasoning, and Symbolic Computation, pp. 290–304 (2002)
24. Pałka, M.H.: Testing an Optimising Compiler by Generating Random Lambda Terms. Licentiate dissertation, Chalmers University of Technology, Göteborg (2012)
25. Pałka, M.H., Claessen, K., Russo, A., Hughes, J.: Testing an Optimising Compiler by Generating Random Lambda Terms. In: Proc. International Workshop on Automation of Software Test (2011)
26. Takikawa, A., Strickland, T.S., Tobin-Hochstadt, S.: Constraining Delimited Control with Contracts. In: Proc. Euro. Symp. Programming, pp. 229–248 (2013)
27. Yang, X., Chen, Y., Eide, E., Regehr, J.: Finding and Understanding Bugs in C Compilers. In: Proc. ACM Conf. Programming Language Design and Implementation, pp. 283–294 (2011)

Refinement Types for Incremental Computational Complexity

Ezgi Çiçek[1], Deepak Garg[1], and Umut Acar[2]

[1] Max Planck Institute for Software Systems
[2] Carnegie Mellon University

Abstract. With recent advances, programs can be compiled to efficiently respond to incremental input changes. However, there is no language-level support for reasoning about the time complexity of incremental updates. Motivated by this gap, we present CostIt, a higher-order functional language with a lightweight refinement type system for proving asymptotic bounds on incremental computation time. Type refinements specify which parts of inputs and outputs may change, as well as dynamic stability, a measure of time required to propagate changes to a program's execution trace, given modified inputs. We prove our type system sound using a new step-indexed cost semantics for change propagation and demonstrate the precision and generality of our technique through examples.

1 Introduction

Many applications operate on data that change over time: compilers respond to source code changes by recompiling as necessary, robots interact with the physical world as it naturally changes over time, and scientific simulations compute with objects whose properties change over time. Although it is possible to develop such applications using ad hoc algorithms and data structures to handle changing data, such algorithms can be challenging to design even for problems that are simple in the batch/static setting where changes to data are not allowed. A striking example is the two-dimensional convex hull problem, whose dynamic (incremental) version required decades of more research [30,7] than its static version (e.g., [17]). The field of incremental computation aims at deriving software that can respond automatically and efficiently to changing data. Earlier work investigated techniques based on static dependency graphs [14,38] and memoization [32,21]. More recent work on self-adjusting computation introduced dynamic dependency graphs [3] and a way to integrate them with a form of memoization [2,4]. Several flavors of self-adjusting computation have been implemented in programming languages such as C [19], Haskell [8], Java [34] and Standard ML [27,10].

However, in all prior work on incremental computation, the programmer must reason about the time complexity of incremental execution, which we call *dynamic stability*, by direct analysis of the cost semantics of programs [26]. While this analytical technique makes efficient design possible, dynamic stability is

© Springer-Verlag Berlin Heidelberg 2015
J. Vitek (Ed.): ESOP 2015, LNCS 9032, pp. 406–431, 2015.
DOI: 10.1007/978-3-662-46669-8_17

a difficult property to establish because it requires reasoning about execution traces, which can be viewed as graphs of computations and their (run-time) data and control dependencies.

Therefore, we are interested in designing static techniques to help a programmer reason about the dynamic stability of programs. As a first step in this direction, we equip a higher-order functional programming language with a refinement type system for establishing the dynamic stability of programs. Our type system, called CostIt, soundly approximates the dynamic stability of a program as an effect. CostIt builds on index refinement types [37] and type annotations to track which program values may change after an update and which may not [11]. To improve precision, we add subtyping rules motivated by co-monadic types [28]. Together, these give enough expressive power to perform non-trivial, asymptotically tight analysis of dynamic stability of programs.

We provide an overview of CostIt's design, highlighting some challenges and our solutions. First, dynamic stability is a function of changes to a program's inputs and, hence, analysis of dynamic stability requires knowing which of its free variables and, more generally, which of its subexpressions' result values may change after an update. To differentiate changeable and unchangeable values statically, we rely on refinement type annotations from Chen et al.'s work on implicit self-adjusting computation [11]:[1] $(\tau)^{\mathbb{S}}$ ascribes values of type τ which cannot change whereas $(\tau)^{\mathbb{C}}$ ascribes all values of type τ. Second, the dynamic stability of a program is often a function of the length of an input list or the number of elements of the list that may change. To track such attributes of inputs in the type system, we add standard index refinement types in the style of Xi and Pfenning's DML [37] or Gaboardi et al.'s DFuzz [16].

Centrally, our type system treats dynamic stability as an effect [29]. Expression typing has the form $e :_\kappa \tau$, where κ is an upper bound on the cost of propagating changes through any trace of e. Similarly, if changes to any trace of a function can be propagated in time at most κ, we give the function a type of the form $\tau_1 \xrightarrow{\kappa} \tau_2$. The cost κ may depend on refinement parameters (e.g., list lengths) that are shared with τ_1. For example, the usual higher-order list mapping function $\mathtt{map} : (\tau_1 \to \tau_2) \to \mathtt{list}\,\tau_1 \to \mathtt{list}\,\tau_2$ can be given the following refined type: $(\tau_1 \xrightarrow{\kappa} \tau_2) \xrightarrow{0} \forall n, \alpha.\ \mathtt{list}\,[n]^\alpha\,\tau_1 \xrightarrow{\alpha \cdot \kappa} \mathtt{list}\,[n]^\alpha\,\tau_2$. Roughly, the type says that if each application of the mapping function can be updated in time κ and at most α elements of the mapped list change, then the entire map can be updated in time $\alpha \cdot \kappa$. (This refined type is approximate; the exact type is shown later.)

Change propagation has the inherent property that if the inputs to a computation do not change, then propagation on the trace of the computation is bypassed and, hence, incurs zero cost. Often, this property must be taken into account in reasoning about dynamic stability. A key insight in CostIt is that this property corresponds to a *co-monadic* reasoning principle in the type system: If all free variables of an expression have types of the form $(\cdot)^{\mathbb{S}}$, then that

---

[1] Nearly identical annotations are also used for other purposes, e.g., binding-time analysis [29] and information flow analysis [31].

expression's dynamic stability is 0 and its result type can also be annotated $(\cdot)^{\mathbb{S}}$, irrespective of what or how the expression computes. Thus, $(\tau)^{\mathbb{S}}$ can be treated like the co-monadic type $\Box\tau$ [28]. A novelty in CostIt is that whether a type's label is $(\cdot)^{\mathbb{S}}$ or $(\cdot)^{\mathbb{C}}$ may depend on index refinements (this flexibility is essential for inductive proofs of dynamic stability in many of our examples). Hence, co-monadic rules are represented in an expanded subtyping relation, which, as usual, takes index refinements into account.

We prove that any dynamic stability derived in our type system is an upper bound on the actual cost of trace update (i.e., that our type system is sound). To do this, we develop an abstract cost semantics for trace update. The cost semantics is formalized using a novel syntactic class called *bi-expression*, which simultaneously represents the original expression and the modified expression, and indicates (syntactically) where the two differ. We interpret types using a step-indexed logical relation over bi-expressions (i.e. relationally) with a stipulated change propagation semantics. Bi-expressions are motivated by largely unrelated work in analysis of security protocols [6].

In summary, we make the following contributions. 1) We develop the first type system for establishing dynamic stability of programs. 2) We combine lightweight dependent types, immutability annotations and co-monadic reasoning principles to facilitate static proofs of dynamic stability. 3) We prove the type system sound relative to a new cost semantics for change propagation. 4) We demonstrate the precision and generality of our technique on several examples. An online appendix, available from the authors' homepages, includes parametric polymorphism, many additional examples, higher-order sorts that are needed to type some of the additional examples, proofs of theorems and several inference rules that are omitted from this paper.

Scope. This paper focuses on laying the foundations of type-based analysis of dynamic stability. The issue of implementing CostIt's type system is beyond the scope of this paper. We comment on an ongoing implementation in Section 7.

2   Types for Dynamic Stability by Example

Dynamic stability. Suppose a program e has been executed with some input v and, subsequently, we want to re-run the program with a slightly different input v'. Dynamic stability measures the amount of time needed for the second execution, given the entire trace of the first execution. The advantage of using the first trace for the second execution is that the runtime can reuse parts of the first trace that are not affected by changes to the input; for parts that are affected, it can selectively *propagate changes* [2]. This can be considerably faster than a from-scratch evaluation. Consider the program $(1+(2+\ldots+10))+x$. Suppose the input x is 0 in the first run and 1 in the second. A naive evaluation of the second run requires 10 additions. However, if a trace of the first execution is available, then the runtime can reuse the result of the first 9 of these additions, which involve unchanged constants. Assuming that an addition takes exactly 1 unit of time (and, for simplicity, that no other runtime operation incurs a cost), the cost of the re-run or the dynamic stability of this program would be 1 unit of time. Abstractly, dynamic stability is a property of two executions of a program and is dependent on a specification of the language's change propagation semantics. For instance, our conclusion that $(1 + (2 + \ldots + 10)) + x$ has dynamic stability 1 assumes that change propagation directly reuses the result of the first 9 additions during the second run. If change propagation is naive, the program might be re-run in its entirety, resulting in a dynamic stability of 10, not 1.

Change propagation. We assume a simple, standard change propagation semantics. We formalize the semantics in Section 4, but explain it here intuitively. During the first execution of a program expression, we record the expression's execution trace. The trace is a tree, a reification of the big-step derivation of the expression's execution. For the second execution, we allow updates to some of the values embedded in the expression (some of the trace's leaves). Change propagation recomputes the result of the modified expression by propagating changes upward through the trace, starting at the modified leaves. Pointers to modified leaves are an input to change propagation and finding them incurs zero cost. Primitive functions (like $+$, $-$, etc.) on the trace whose arguments change are recomputed, but large parts of the trace may *not* be recomputed from scratch, which makes change propagation asymptotically faster than from-scratch evaluation in many cases. The maximum amount of work done in change propagation of an expression's trace (given assumptions on allowed changes to the expression's leaves) is called the expression's dynamic stability. CostIt helps establish this dynamic stability statically.

If the shape of the execution trace of an updated expression is different from the shape of the trace of the original expression (i.e., if the control flow of the execution changes), then change propagation must, in general, construct some parts of the new trace by evaluating subexpressions from scratch. Analysis of dynamic stability in such cases requires also an analysis of worse-case execution time complexity. In this paper, we disallow (through our type system) control flow dependence on data that may change. This simplifying choice mirrors prior work like DFuzz [16] and still allows us to type several interesting programs like sorting and matrix algorithms. In Section 7, we comment on a CostIt extension that can handle control flow changes.

During change propagation, only re-execution of primitive functions incurs a non-zero cost. Although this may sound counter-intuitive, prior work has shown that by storing values in modifiable reference cells and updating them in-place during change propagation, the cost for structural operations like pairing, projection and list consing can be avoided during change propagation [2,11]. The details of such implementations are not important here; readers only need to be aware that our change propagation incurs a cost only for re-executing primitive functions of the language.

Type system overview. We build on a λ-calculus with lists. The simple types of our language are $\texttt{real}, \texttt{unit}, \tau_1 \times \tau_2, \texttt{list } \tau$ and $\tau_1 \to \tau_2$. Since the dynamic stability of an expression depends on sizes of input lists as well as knowledge of which of its free variables (inputs) may change, we add type refinements. First, we refine the type list τ to $\text{list}\,[n]^\alpha\,\tau$, which specifies lists of length exactly n, of which *at most* α elements are allowed to change before the second execution. Technically, n and α are natural numbers in an index domain, over which types may quantify. Second, any type τ may be refined to $(\tau)^\mu$ where μ belongs to an index sort with two values, $\mathbb{S}$ and $\mathbb{C}$. $(\tau)^\mathbb{S}$ specifies those values of type τ that will not change in the second execution ($\mathbb{S}$ is read "stable"). $(\tau)^\mathbb{C}$ specifies all values of type τ ($\mathbb{C}$ is read "potentially changeable"). τ and $(\tau)^\mathbb{C}$ are subtypes of each other. Our typing judgment takes the form $\Gamma \vdash e :_\kappa \tau$. Here, κ is an upper bound on the dynamic stability of the expression e. (For simplicity, we omit several contexts from the typing judgment in this section.)

Example 1 (Warm-up). Assume that computing a primitive operation like addition from scratch costs 1 unit of time. Consider the expression $x + 1$ with one input x. This expression can be typed in at least two ways: $x : (\text{real})^\mathbb{S} \vdash x + 1 :_0$ $(\text{real})^\mathbb{S}$ and $x : (\text{real})^\mathbb{C} \vdash x + 1 :_1 (\text{real})^\mathbb{C}$. When $x : (\text{real})^\mathbb{S}$, x cannot change. So change propagation bypasses the expression $x + 1$ and its cost is 0. Moreover, the value of $x+1$ does not change. This justifies the first typing judgment. When $x : (\text{real})^\mathbb{C}$, change propagation may incur a cost of 1 to recompute the addition in $x + 1$ and the value of $x + 1$ may change. This justifies the second judgment.

Example 2 (List map). The CostIt type $\tau_1 \xrightarrow{\kappa} \tau_2$ specifies a function whose body has a change propagation cost upper-bounded by κ and whose type is $\tau_1 \to \tau_2$. For instance, based on Example 1, the function $\lambda x.(x + 1)$ can be given either of the types $(\text{real})^\mathbb{S} \xrightarrow{0} (\text{real})^\mathbb{S}$ and $(\text{real})^\mathbb{C} \xrightarrow{1} (\text{real})^\mathbb{C}$. Consider the standard list map function of simple type $(\tau_1 \to \tau_2) \to \text{list}\,\tau_1 \to \text{list}\,\tau_2$.

$$\text{fix}\,\text{map}(f).\,\lambda l.\,\text{case}_\text{L}\,l\,\text{of}\;\;\text{nil}\;\to\text{nil}\;\mid\;\text{cons}(h,\,tl)\;\to\;\text{cons}(f\,h,\,\text{map}\,f\,tl)$$

Suppose that the mapping function f has dynamic stability κ, i.e., its type is $\tau_1 \xrightarrow{\kappa} \tau_2$ and that the list l has type $\text{list}\,[n]^\alpha\,\tau_1$ (exactly n elements of which at most α may change). What can we say about the type of the result and the dynamic stability of map? If we know that f *will not change*, then change propagation will reapply f at most α times (because at most α list elements will change), so the total cost can be bounded by $\alpha \cdot \kappa$. Moreover, at most α elements of the output will change, so we can give map the following type.[2]

$$\text{map} : (\tau_1 \xrightarrow{\kappa} \tau_2)^\mathbb{S} \to \forall n.\,\forall\alpha.\,\text{list}\,[n]^\alpha\,\tau_1 \xrightarrow{\alpha\cdot\kappa} \text{list}\,[n]^\alpha\,\tau_2 \qquad (1)$$

If f may change, then change propagation may have to remap every element of the list and all elements in the output may change. This yields a dynamic stability of $n \cdot \kappa$ and the following type.

$$\text{map} : (\tau_1 \xrightarrow{\kappa} \tau_2)^\mathbb{C} \to \forall n.\,\forall\alpha.\,\text{list}\,[n]^\alpha\,\tau_1 \xrightarrow{n\cdot\kappa} \text{list}\,[n]^n\,\tau_2 \qquad (2)$$

---

[2] If κ is omitted from $\tau_1 \xrightarrow{\kappa} \tau_2$, then it is treated as 0. Our expressions (Section 3) have explicit annotations for introducing and eliminating universal and existential quantifiers ($\Lambda.\,e, e[], \text{pack}\,e, \text{unpack}\,e_1\,\text{as}\,x\,\text{in}\,e_2$). We omit those annotations from our examples for better readability.

We explain how the type in (1) is derived as it highlights our co-monadic reasoning principle. The interesting part of the typing is establishing the change propagation cost of the $\mathtt{cons}(h, tl)$ branch in the definition of $\mathtt{map}$. We are trying to bound this cost by $\alpha \cdot \kappa$. We know from l's type that at most α elements in $\mathtt{cons}(h, tl)$ will change in the second run. However, we do not know whether h is one of those elements. So, our case analysis rule (Section 3, Figure 4) has *two* *premises for the* $\mathtt{cons}$ *branch* (a total of three premises, including the premise for $\mathtt{nil}$). In the first of these two premises, we assume that h may change, so $h : \tau_1$ and $tl : \mathtt{list}\,[n-1]^{\alpha-1}\,\tau_2$. In the second premise, we assume that h cannot change, so $h : (\tau_1)^{\mathbb{S}}$ and $tl : \mathtt{list}\,[n-1]^{\alpha}\,\tau_2$. Analysis of the first premise is straightforward: $(f\ h)$ incurs cost κ (from f's type $(\tau_1 \xrightarrow{\kappa} \tau_2)^{\mathbb{S}}$) and, inductively, $(\mathtt{map}\ f\ tl)$ incurs cost $(\alpha-1)\cdot\kappa$, for a total cost $\kappa + (\alpha-1)\cdot\kappa = \alpha\cdot\kappa$. Analysis of the second premise requires nonstandard reasoning. Here, $tl : \mathtt{list}\,[n-1]^{\alpha}\,\tau_2$, so the inductive cost of $(\mathtt{map}\ f\ tl)$ is already $\alpha\cdot\kappa$. Hence, we must show that $(f\ h)$ has 0 change propagation cost. For this, we rely on our co-monadic reasoning principle: If all of an expression's free variables have types of the form $(\cdot)^{\mathbb{S}}$ (i.e., their substitutions will not change), then the expression's change propagation cost is 0. Since we know that $f : (\tau_1 \xrightarrow{\kappa} \tau_2)^{\mathbb{S}}$ and $h : (\tau_1)^{\mathbb{S}}$, we can immediately conclude that $(f\ h)$ has 0 change propagation cost.

The same reasoning cannot be applied to the second premise in type (2), where $f : (\tau_1 \xrightarrow{\kappa} \tau_2)^{\mathbb{C}}$. Instead, we can show only that $(f\ h)$ incurs cost κ. This results in a dynamic stability of $n \cdot \kappa$. Note that in both the types above, the dynamic stability depends on attributes of the input list (α and n, respectively). This demonstrates the importance of index refinements in CostIt.

Example 3 (Balanced list fold). Standard list fold operations ($\mathtt{foldl}$ and $\mathtt{foldr}$) can be typed easily in CostIt but are uninteresting for incremental computation because they have linear traces and, hence, have $O(n)$ dynamic stability even for single element changes to the input list (n is the list's length). A more interesting operation is what we call the balanced fold. Given an *associative and commutative* binary function f of simple type $\tau \times \tau \to \tau$, a list of simple type ($\mathtt{list}\ \tau$) can be folded by splitting it into two nearly equal sized lists, folding the sublists recursively and then applying f to the two results. This results in a balanced tree-like trace, whose depth is $\lceil \log_2(n) \rceil$. A single change to the list causes $\lceil \log_2(n) \rceil$ recomputations of f. So, if f has dynamic stability κ, the dynamic stability with one change to the list is $O(\kappa \cdot \log_2(n))$. More generally, it can be shown that if α changes are allowed to the list, then the dynamic stability is $O(\kappa \cdot (\alpha + \alpha \cdot \log_2(n/\alpha)))$. This simplifies to $O(\kappa \cdot n)$ when $\alpha = n$ (entire list may change) and $O(\kappa \cdot \log_2(n))$ when $\alpha = 1$. In the following we implement such a balanced fold operation, $\mathtt{bfold}$, and derive its dynamic stability in CostIt.

Our first ingredient is the function $\mathtt{bsplit}$, which splits a list of length n into two lists of lengths $\lceil \frac{n}{2} \rceil$ and $\lfloor \frac{n}{2} \rfloor$. This function is completely standard. Its CostIt type, although easily established, is somewhat interesting because it uses an existential quantifier to split the allowed number of changes α into the two split lists. The dynamic stability of $\mathtt{bsplit}$ is 0 because $\mathtt{bsplit}$ uses no primitive functions (*cf.* discussion earlier in this section).

$\text{bsplit} : \forall n. \, \forall \alpha. \, \text{list} \, [n]^{\alpha} \, \tau \xrightarrow{0} \exists \beta. \, (\text{list} \, [\lceil \frac{n}{2} \rceil]^{\beta} \, \tau \times \text{list} \, [\lfloor \frac{n}{2} \rfloor]^{\alpha - \beta} \, \tau)$
$\text{fix bsplit}(l). \, \text{case}_L \, l \, \text{of}$
$\quad \text{nil} \, \rightarrow (\text{nil}, \text{nil})$
$\quad | \, \text{cons}(h_1, \, tl_1) \, \rightarrow \, \text{case}_L \, tl_1 \, \text{of nil} \, \rightarrow ([h_1], \text{nil})$
$\qquad\qquad\qquad\qquad\qquad\qquad | \, \text{cons}(h_2, \, tl_2) \, \rightarrow \, \text{let} \, (z_1, z_2) \, = \, \text{bsplit} \, tl_2 \, \text{in}$
$\qquad\qquad\qquad\qquad\qquad\qquad\qquad (\text{cons}(h_1, z_1), \text{cons}(h_2, z_2))$

Using bsplit we define the balanced fold function, bfold. The function applies only to non-empty lists (reflected in its type later), so the nil case is omitted.

$\quad\quad \text{fix bfold}(f). \, \lambda l. \, \text{case}_L \, l \, \text{of}$
$\quad\quad\quad\quad \text{nil} \, \rightarrow \dots$
$\quad\quad\quad | \, \text{cons}(h_1, \, tl_1) \, \rightarrow \, \text{case}_L \, tl_1 \, \text{of}$
$\quad\quad\quad\quad\quad\quad\quad\quad \text{nil} \, \rightarrow h_1$
$\quad\quad\quad\quad\quad\quad | \, \text{cons}(\_, \, \_) \, \rightarrow \, \text{let} \, (z_1, z_2) \, = \, (\text{bsplit} \, l) \, \text{in}$
$\quad\quad\quad\quad\quad\quad\quad\quad\quad f \, (\text{bfold} \, f \, z_1, \text{bfold} \, f \, z_2)$

We first derive a type for bfold informally, and then show how the type is established in CostIt. Assume that the argument l has type $\text{list} \, [n]^{\alpha} \, \tau$. We count how many times change propagation may have to reapply f in updating bfold's trace, which is a nearly balanced tree of height $H = \lceil \log_2(n) \rceil$. Counting levels from the deepest leaves upward (leaves have level 0), the number of applications of f at level k in the trace is at most 2^{H-k}. If α leaves change, at most α of these applications must be recomputed. Consequently, the maximum number of recomputations of f at level k is $\min(\alpha, 2^{H-k})$. If the dynamic stability of f is κ, the dynamic stability of bfold is $P(n, \alpha, \kappa) = \sum_{k=0}^{\lceil \log_2(n) \rceil} \kappa \cdot \min(\alpha, 2^{\lceil \log_2(n) \rceil - k})$. So, in principle, we should be able to give bfold the following type.

$$\text{bfold} : (\tau \times \tau \xrightarrow{\kappa} \tau)^{\mathbb{S}} \rightarrow \forall n > 0. \, \forall \alpha. \, \text{list} \, [n]^{\alpha} \, \tau \xrightarrow{P(n, \alpha, \kappa)} \tau$$

The expression $P(n, \alpha, \kappa)$ may look complex, but it is in $O(\kappa \cdot (\alpha + \alpha \cdot \log_2(n/\alpha)))$. (To prove this, split the summation in $P(n, \alpha, \kappa)$ into two: one for $k \leq \lceil \log_2(n) \rceil - \lceil \log_2(\alpha) \rceil$ and the other for $k > \lceil \log_2(n) \rceil - \lceil \log_2(\alpha) \rceil$. Our appendix has the details.) Although the type above is correct, we will see soon that in typing the recursive calls in bfold, we need to know that bfold's type is annotated $(\cdot)^{\mathbb{S}}$. Hence, the actual type we assign to bfold is stronger.

$$\text{bfold} : ((\tau \times \tau \xrightarrow{\kappa} \tau)^{\mathbb{S}} \rightarrow \forall n > 0. \, \forall \alpha. \, \text{list} \, [n]^{\alpha} \, \tau \xrightarrow{P(n, \alpha, \kappa)} \tau)^{\mathbb{S}} \qquad (3)$$

We explain how bfold's type is established in CostIt. The interesting case starts where bsplit is invoked. From the type of bsplit, we know that variables z_1 and z_2 in the body of bfold have types $\text{list} \, [\lceil \frac{n}{2} \rceil]^{\beta} \, \tau$ and $\text{list} \, [\lfloor \frac{n}{2} \rfloor]^{\alpha - \beta} \, \tau$, respectively for some β. Inductively, the change propagation costs of $(\text{bfold} \, f \, z_1)$ and $(\text{bfold} \, f \, z_2)$ are $P(\lceil \frac{n}{2} \rceil, \beta, \kappa)$ and $P(\lfloor \frac{n}{2} \rfloor, \alpha - \beta, \kappa)$, respectively. Hence, the change propagation cost of the whole body of bfold is $\kappa + P(\lceil \frac{n}{2} \rceil, \beta, \kappa) +$

$P(\lfloor \frac{n}{2} \rfloor, \alpha - \beta, \kappa)$. The additional κ accounts for the only application of f in the body of $\texttt{bfold}$ (non-primitive operations have zero cost and $\texttt{bsplit}$ also has zero cost). Hence, to complete the typing, we must establish the following inequality.

$$\kappa + P(\lceil \frac{n}{2} \rceil, \beta, \kappa) + P(\lfloor \frac{n}{2} \rfloor, \alpha - \beta, \kappa) \leq P(n, \alpha, \kappa) \qquad (4)$$

This is an easily established arithmetic tautology (our online appendix has a proof), *except* when $\alpha \doteq 0$. When $\alpha \doteq 0$, the right side of the inequality is 0 but we don't necessarily have $\kappa \leq 0$. So, in order to proceed, we consider the cases $\alpha \doteq 0$ and $\alpha > 0$ separately. This requires a typing rule for case analysis on the index domain, which poses no theoretical difficulty. The $\alpha > 0$ case succeeds as described above. For $\alpha \doteq 0$, we use our co-monadic reasoning principle. With $\alpha \doteq 0$, the types of z_1 and z_2 are equivalent (formally, via subtyping) to $\texttt{list} \left[\lceil \frac{n}{2} \rceil \right]^0 \tau$ and $\texttt{list} \left[\lfloor \frac{n}{2} \rfloor \right]^0 \tau$, respectively. Since, no elements in these lists can change, we use another subtyping rule to promote the types to $(\texttt{list} \left[\lceil \frac{n}{2} \rceil \right]^0 \tau)^{\mathbb{S}}$ and $(\texttt{list} \left[\lfloor \frac{n}{2} \rfloor \right]^0 \tau)^{\mathbb{S}}$, respectively. At this point, the type of every variable occurring in the expression f ($\texttt{bfold } f \ z_1, \texttt{bfold } f \ z_2$), including the variable $\texttt{bfold}$, has annotation $(\cdot)^{\mathbb{S}}$. By our co-monadic reasoning principle, the change propagation cost of this expression and, hence, the body of $\texttt{bfold}$, must be 0, which is trivially no more than $P(n, \alpha, \kappa)$. This completes our argument.

Observe that the inference of the annotation $(\cdot)^{\mathbb{S}}$ on the types of z_1 and z_2 is conditional on the constraint $\alpha \doteq 0$. Subtyping, which is aware of constraints, plays an essential role in determining these annotations and in making our co-monadic reasoning principle useful. Also, the fact that we have to consider the cases $\alpha \doteq 0$ and $\alpha > 0$ separately is not as surprising as it may seem. The case $\alpha \doteq 0$ corresponds to a sub-trace whose leaves have not changed. Since change propagation is a bottom-up procedure, it will bypass this sub-trace completely, incurring no cost. This is exactly what our analysis for $\alpha \doteq 0$ establishes.

Using the type (3) of $\texttt{bfold}$, we can show that for $f : (\tau \times \tau \xrightarrow{\kappa} \tau)^{\mathbb{S}}$ and $l : \texttt{list}[n]^\alpha \tau$, the dynamic stability of ($\texttt{bfold } f \ l$) is in $O(\log_2(n))$ when $\alpha \in O(1)$ and in $O(n)$ when $\alpha \in O(n)$, assuming κ constant. This dynamic stability is asymptotically tight.

Example 4 (Merge sort). The analysis of Example 3 generalizes to other divide-and-conquer algorithms. We illustrate this generalization using merge sort as a second example; our appendix describes a generic template for establishing the dynamic stability of divide-and-conquer algorithms. Abstractly, the trace of merge sort on a list of length n is a tree of height $\lceil \log_2(n) \rceil$, where each node receives a list (a sublist of the original list) as input, partitions the list into two nearly equal length sublists, recursively sorts the sublists and then merges the sorted sublists. During change propagation, cost is incurred at a node only in merging the sorted sublists. In the worst case, this cost is $O(m)$, where m is the length of the list being sorted at that node because merging is a linear-time operation. Counting levels from the deepest leaves upward to the root, at level k, $m \leq 2^k$. If a single element of the list changes, change propagation might re-merge at each node on the path from this changed element to the root. Hence, the cost is upper-bounded by $1 + 2 + 4 + \ldots + 2^{\lceil \log_2(n) \rceil} \in O(n)$. If all elements of the list may change, the change propagation cost is $O(n \cdot \log_2(n))$. More generally, as we prove below, if α elements of the list change, then change propagation cost is bounded by $O(n \cdot (1 + \log_2(\alpha)))$. Importantly, this calculation does not require an analysis of the change propagation cost of the merge function: A completely pessimistic assumption that all merges on any path from a changed element to the root must be re-executed from scratch yields these bounds. Accordingly, we assume that we have a merge function with the most pessimistic bounds. Using this function, we can define the merge sort function, msort.

$$\text{merge} : (\forall n, m, \alpha, \beta.\ (\text{list}\,[n]^{\alpha}\ \text{real} \times \text{list}\,[m]^{\beta}\ \text{real})$$
$$\xrightarrow{n+m} \text{list}\,[n+m]^{n+m}\ \text{real})^{\mathbb{S}}$$

```
fix msort(l). case_L l of
  nil  → nil
| cons(h₁, tl₁) →  case_L tl₁ of
                     nil → cons(h₁, nil)
                   | cons(_, _) → let (z₁, z₂) = (bsplit l) in
                                     merge (msort z₁, msort z₂)
```

Almost exactly as for bfold, msort can be given the following type:

$$\text{msort} : (\forall n.\ \forall \alpha.\ \text{list}\,[n]^{\alpha}\ \tau \xrightarrow{Q(n,\alpha)} \text{list}\,[n]^{n}\ \tau)^{\mathbb{S}}$$

where for $Q(n,\alpha) = \displaystyle\sum_{k=0}^{\lceil \log_2(n) \rceil} 2^k \cdot \min(\alpha, 2^{\lceil \log_2(n) \rceil - k})$. $Q(n,\alpha)$ is in $O(n \cdot (1 + \log_2(\alpha)))$. Using this type for msort, we can show that for $l : \text{list}\,[n]^{\alpha}\ \tau$, (msort l) has dynamic stability in $O(n)$ for $\alpha \in O(1)$ and in $O(n \cdot \log_2(n))$ for $\alpha \in O(n)$. This dynamic stability is asymptotically tight.

Note that the syntactic cumbersomeness of the expressions $P(n, \alpha, \kappa)$ (Example 3, bfold) and $Q(n, \alpha)$ (Example 4, msort) is inherent to the dynamic stability of the two algorithms. It is not an artifact of CostIt. We tried to find simpler expressions that would support inductive proofs. For bfold, the simpler form $P(n, \alpha, \kappa) = \kappa \cdot (\alpha - 1 + \alpha \cdot (\lceil \log_2(n) \rceil - \log_2(\alpha)))$ for $\alpha > 0$ can be used, but the constraint corresponding to (4) is more difficult to establish and requires real analysis. We do not know of a useful simpler form for $Q(n, \alpha)$.

Other examples. Our appendix contains several other examples. We briefly list some of these with their asymptotic CostIt-established dynamic stability for single element changes in parenthesis: list append (1), list pair zip (1), matrix transpose (1), dot product $(\log_2(n))$ and matrix multiplication $(n \cdot \log_2(n))$. The matrix examples demonstrate that CostIt can establish asymptotically tight bounds on dynamic stability even when the latter depends on the sizes of nested inner lists.

We note that the dynamic stability proved using CostIt is asymptotically tight for all the examples in this section and our appendix. Nonetheless, like other type systems, CostIt abstracts over concrete program values and, hence, we cannot expect CostIt's analysis to be asymptotically tight on all programs.

Types	τ	$::=$	$\texttt{real} \mid \tau_1 \times \tau_2 \mid \texttt{list}\,[n]^\alpha\,\tau \mid \tau_1 \xrightarrow{\kappa} \tau_2 \mid \forall i \stackrel{\kappa}{::} S.\,\tau \mid \exists i.\,\tau \mid$
			$\texttt{unit} \mid C \to \tau \mid C \wedge \tau \mid (\tau)^\mu$
Sorts	S	$::=$	$\mathbb{N} \mid \mathbb{R}^+ \mid \mathbb{V}$
Index terms	$I, \mu, \kappa,$	$::=$	$i \mid \mathbb{S} \mid \mathbb{C} \mid 0 \mid I+1 \mid I_1 + I_2 \mid I_1 - I_2 \mid \frac{I_1}{I_2} \mid I_1 \cdot I_2 \mid$
	n, α		$\lceil I \rceil \mid \lfloor I \rfloor \mid \log_2(I) \mid I_1^{I_2} \mid \min(I_1, I_2) \mid \max(I_1, I_2) \mid \sum\limits_{k=I_1}^{I_n} I$
Constraints	C	$::=$	$I_1 \doteq I_2 \mid I_1 < I_2 \mid \neg C$
Constraint env.	Φ	$::=$	$\emptyset \mid C \mid \Phi_1 \wedge \Phi_2$
Sort env.	Δ	$::=$	$\emptyset \mid \Delta, i :: S$
Type env.	Γ	$::=$	$\emptyset \mid \Gamma, x : \tau$
Primitive env.	Υ	$::=$	$\emptyset \mid \Upsilon, \zeta : \forall \overline{t_i}.\,\tau_1 \xrightarrow{\kappa} \tau_2$

Fig. 1. Syntax of types

3   Syntax and Type System

This section describes CostIt's language, types and type system. Section 4 defines CostIt's dynamic semantics. CostIt is a refinement type system on a call-by-value λ-calculus with lists, similar to DFuzz [16]. The syntax of CostIt's types and type refinements is listed in Figure 1.

Index terms and constraints. CostIt's types are refined by index terms, denoted $I, \mu, \kappa, n, \alpha$, etc. Index terms are sorted as follows: (a) natural numbers, $\mathbb{N}$, which are used to specify list lengths and number of changes allowed in a list, (b) non-negative real numbers, $\mathbb{R}^+$, that show up in logarithmic expressions in change propagation costs, and (c) the two-valued sort *variation*, $\mathbb{V} = \{\mathbb{S}, \mathbb{C}\}$, used as a type refinement to specify whether a value may change or not from the first to the second execution. The syntax of index terms includes various arithmetic operators, with their usual meanings. Most operators are overloaded for the sorts $\mathbb{R}^+$ and $\mathbb{N}$ and there is an implicit coercion from $\mathbb{N}$ to $\mathbb{R}^+$. A standard sorting judgment $\Delta \vdash I :: S$ assigns sort S to index term I. The sort environment Δ, assigns sorts to index variables, i, t. We use different letters for index terms in different roles: n for list lengths, α for the number of allowed changes in a list, μ for terms of sort $\mathbb{V}$, κ for change propagation costs and I for generic terms.

Propositions over index terms are called constraints, denoted C. For our examples, we only need comparison and negation. Constraints are collected in a context called the constraint environment, denoted Φ. As usual, logical entailment over constraints is defined by the black-box judgment $\Delta; \Phi \models C$, which is assumed to embody the usual algebraic laws of arithmetic. Constraints are also subject to standard syntactic sorting rules, which we omit.

Types. CostIt types refine types of the simply typed λ-calculus. The list type $\texttt{list}\,[n]^\alpha\,\tau$ contains two index refinements — n and α — which specify, respectively, the precise length of the list and the maximum number of elements of the list that may be updated before change propagation. The cost annotation κ

Values v $::=$ $\mathtt{r} \mid (v_1, v_2) \mid \mathtt{nil} \mid \mathtt{cons}(v_1, v_2) \mid \mathtt{fix}\ f(x).e \mid \Lambda.e \mid \mathtt{pack}\ v \mid ()$

Expressions $e, f ::=$ $x \mid \mathtt{r} \mid (e_1, e_2) \mid \mathtt{fst}\ e \mid \mathtt{snd}\ e \mid \mathtt{nil} \mid \mathtt{cons}(e_1, e_2) \mid$
                  $\mathtt{case_L}\ e\ \mathtt{of\ nil}\ \rightarrow\ e_1 \mid \mathtt{cons}(h, tl)\ \rightarrow\ e_2 \mid$
                  $\mathtt{fix}\ f(x).e \mid e_1\ e_2 \mid \Lambda.e \mid e[] \mid \mathtt{pack}\ e \mid \mathtt{unpack}\ e\ \mathtt{as}\ x\ \mathtt{in}\ e' \mid$
                  $\mathtt{let}\ x\ =\ e_1\ \mathtt{in}\ e_2 \mid () \mid \zeta\ e$

Fig. 2. Syntax of expressions and values

in function types $\tau_1 \xrightarrow{\kappa} \tau_2$ and universally quantified types $\forall i \overset{\kappa}{::} S$. τ is an upper bound on the change propagation cost of closures contained in the type. The type $C \rightarrow \tau$ reads "τ if constraint C is true, else every expression". Any type τ may be annotated with a variation term μ, written $(\tau)^\mu$. $(\tau)^S$ specifies values of type τ that cannot change (in our relational interpretation, $(\tau)^S$ is the diagonal relation on τ). $(\tau)^C$ is equivalent (via subtyping) to τ. There is one representative, unrefined base type $\mathtt{real}$. Other refined and unrefined base types can be added, as in our appendix. We note that it is not obvious how refinements may be extended to algebraic datatypes beyond lists, because needed refinements vary by application. In the case of lists, the refinements length and the number of allowed changes suffice for many applications, so we adopt them. CostIt supports standard type quantification (parametric polymorphism). Type quantification does not interact with our technical development in any significant way, so we defer its details to the appendix.

Expressions. Figure 2 shows the grammar of CostIt values and expressions. The syntax is mostly standard. $\mathtt{r}$ denotes constants of type $\mathtt{real}$. ζ denotes a primitive function and $\zeta\ e$ is application of the function to e. Primitive functions have a special role in our dynamic semantics because only they incur a non-zero cost during change propagation. The construct $\mathtt{case_L}$ is case analysis on lists.

Our expressions do not mention index terms or index variables. For instance, the introduction and elimination forms for the universal quantifier are $\Lambda.e$ and $e[]$ instead of the more common and more elaborate forms $\Lambda i.e$ and $e\ [I]$. Index terms are absent from expressions for a reason. As explained in Section 2, the list case analysis rule has two premises for the $\mathtt{cons}(\cdot, \cdot)$ branch. Often, universally quantified terms must be instantiated differently in the two premises, which means that if index terms were included in expressions, we would have to write *two expressions* for the $\mathtt{cons}(\cdot, \cdot)$ branch. This would be cumbersome at best, so we do not include index terms in expressions. If necessary, the two separate fully annotated expressions can be created by elaboration after type-checking.

Subtyping. Like all other index refinement type systems, CostIt relies heavily on subtyping. Selected rules of our subtyping judgment $\Delta; \Phi \vdash \tau_1 \sqsubseteq \tau_2$ are shown in Figure 3. The judgment $\tau_1 \sqsubseteq \tau_2$ means that τ_1 is a subtype of τ_2 and $\tau_1 \equiv \tau_2$ is shorthand for ($\tau_1 \sqsubseteq \tau_2$ and $\tau_2 \sqsubseteq \tau_1$). The rule $\rightarrow 2$ defines standard subtyping for function types, covariant in the result and contravariant in the argument. Additionally, function subtyping is covariant in the cost κ, because

$$\boxed{\Delta; \Phi \models \tau_1 \sqsubseteq \tau_2} \quad \tau_1 \text{ is a subtype of } \tau_2$$

$$\frac{}{\Delta; \Phi \models (\tau_1 \xrightarrow{\kappa} \tau_2)^\mu \sqsubseteq (\tau_1)^\mu \xrightarrow{\kappa} (\tau_2)^\mu} \to 1$$

$$\frac{\Delta; \Phi \models \tau_1' \sqsubseteq \tau_1 \quad \Delta; \Phi \models \tau_2 \sqsubseteq \tau_2' \quad \Delta; \Phi \models \kappa \le \kappa'}{\Delta; \Phi \models \tau_1 \xrightarrow{\kappa} \tau_2 \sqsubseteq \tau_1' \xrightarrow{\kappa'} \tau_2'} \to 2$$

$$\frac{}{\Delta; \Phi \models (\tau_1 \times \tau_2)^\mu \equiv (\tau_1)^\mu \times (\tau_2)^\mu} \times 1$$

$$\frac{}{\Delta; \Phi \models (\text{list } [n]^\alpha \ \tau)^\mu \equiv \text{list } [n]^\alpha \ (\tau)^\mu} 11$$

$$\frac{\Delta; \Phi \models \mu \doteq \mathbb{S}}{\Delta; \Phi \models (\text{list } [n]^\alpha \ \tau)^\mu \equiv \text{list } [n]^0 \ \tau} 12$$

$$\frac{\Delta; \Phi \models n \doteq n' \quad \Delta; \Phi \models \alpha \le \alpha' \quad \Delta; \Phi \models \tau \sqsubseteq \tau'}{\Delta; \Phi \models \text{list } [n]^\alpha \ \tau \sqsubseteq \text{list } [n']^{\alpha'} \ \tau'} 14$$

$$\frac{}{\Delta; \Phi \models (\forall t \overset{\kappa}{::} S. \ \tau)^\mu \equiv \forall t \overset{\kappa}{::} S. \ (\tau)^\mu} \forall 2$$

$$\frac{}{\Delta; \Phi \models (\tau)^\mu \sqsubseteq \tau} \mathbf{T} \qquad \frac{}{\Delta; \Phi \models \tau \sqsubseteq (\tau)^\mathbb{C}} \mathbf{I}$$

Fig. 3. Selected subtyping rules

κ is an upper bound on the dynamic stability. Rule **14** makes list subtyping invariant in the list size n and covariant in the number α of elements allowed to change (because the former is exact but the latter is an upper-bound).

The remaining subtyping rules shown in Figure 3 mention variation annotations $(\tau)^\mu$. These rules are best understood separately for the cases $\mu \doteq \mathbb{S}$ and $\mu \doteq \mathbb{C}$. Rules **T** and **I** imply that $(\tau)^\mathbb{C} \equiv \tau$ (expressions are allowed to change unless specified, so the annotation $(\cdot)^\mathbb{C}$ provides no additional information). Given this observation, the remaining rules state obvious identities for the case $\mu \doteq \mathbb{C}$.

We describe the rules for the case $\mu \doteq \mathbb{S}$. As expected, $(\tau)^\mathbb{S} \sqsubseteq \tau$ (rule **T**), but the converse is not true in general. Rule $\to 1$ says that $(\tau_1 \xrightarrow{\kappa} \tau_2)^\mathbb{S} \sqsubseteq (\tau_1)^\mathbb{S} \xrightarrow{\kappa} (\tau_2)^\mathbb{S}$. This can be read as follows: If a function will not change and it is given an argument that will not change, then the result will not change. The converse is not true: If given a non-changing argument, a function's result will not change, this does not imply that the function itself will not change (e.g., some dead code in the function may change). Rule **12** implies that $(\text{list } [n]^\alpha \ \tau)^\mathbb{S} \equiv \text{list } [n]^0 \ \tau$. This equivalence is justified as follows: The annotation 0 in the type on the right forbids changes to any elements of the list and its length is fixed at n, so the list cannot change. This rule is critical to typing Examples 3 and 4 of Section 2.

Readers familiar with co-monadic types or constructive modal logic will notice that our subtyping rules for $(\tau)^\mathbb{S}$ mirror rules for a co-monad $\Box\tau$: $\Box\tau \sqsubseteq \tau$ (but not the converse), $\Box(\tau_1 \to \tau_2) \sqsubseteq (\Box\tau_1 \to \Box\tau_2)$ and $\Box(\tau_1 \times \tau_2) \equiv (\Box\tau_1 \times \Box\tau_2)$.

Typing rules. Our typing judgment has the form $\Delta; \Phi; \Gamma \vdash e :_\kappa \tau$. Here, κ is an upper bound on the dynamic stability of e. It is treated as an effect. Important typing rules are shown in Figure 4. Technically, all rules include a fourth context Υ that specifies the types of primitive functions ζ, but this context does not change in the rules, so we exclude it from the presentation. The rules follow some general principles. First, if an expression contains subexpressions, then the change propagation costs (κ's) of subexpressions are added to obtain the change

$\boxed{\Delta; \Phi; \Gamma \vdash e :_\kappa \tau}$ expression e has type τ and dynamic stability at most κ

$$\frac{}{\Delta; \Phi; \Gamma, x : \tau \vdash x :_0 \tau} \text{ var} \qquad \frac{}{\Delta; \Phi; \Gamma \vdash \mathbf{r} :_0 (\mathbf{real})^{\mathbb{S}}} \text{ real}$$

$$\frac{}{\Delta; \Phi; \Gamma \vdash \mathbf{nil} :_0 \mathbf{list}\,[0]^0\,\tau} \text{ nil}$$

$$\frac{\Delta; \Phi; \Gamma \vdash e_1 :_{\kappa_1} (\tau)^{\mathbb{S}} \qquad \Delta; \Phi; \Gamma \vdash e_2 :_{\kappa_2} \mathbf{list}\,[n]^\alpha\,\tau}{\Delta; \Phi; \Gamma \vdash \mathbf{cons}(e_1, e_2) :_{\kappa_1+\kappa_2} \mathbf{list}\,[n+1]^\alpha\,\tau} \text{ cons1}$$

$$\frac{\Delta; \Phi; \Gamma \vdash e_1 :_{\kappa_1} \tau \qquad \Delta; \Phi; \Gamma \vdash e_2 :_{\kappa_2} \mathbf{list}\,[n]^{\alpha-1}\,\tau \qquad \Delta; \Phi \models \alpha > 0}{\Delta; \Phi; \Gamma \vdash \mathbf{cons}(e_1, e_2) :_{\kappa_1+\kappa_2} \mathbf{list}\,[n+1]^\alpha\,\tau} \text{ cons2}$$

$$\frac{\begin{array}{c} \Delta; \Phi; \Gamma \vdash e :_\kappa \mathbf{list}\,[n]^\alpha\,\tau \qquad \Delta; \Phi \wedge n \doteq 0; \Gamma \vdash e_1 :_{\kappa'} \tau' \\ i :: \iota, \Delta; \Phi \wedge n \doteq i+1; h : (\tau)^{\mathbb{S}}, tl : \mathbf{list}\,[i]^\alpha\,\tau, \Gamma \vdash e_2 :_{\kappa'} \tau' \\ i :: \iota, \beta :: \iota, \Delta; \Phi \wedge n \doteq i+1 \wedge \alpha \doteq \beta+1; h : (\tau)^{\mathbb{C}}, tl : \mathbf{list}\,[i]^\beta\,\tau, \Gamma \vdash e_2 :_{\kappa'} \tau' \end{array}}{\Delta; \Phi; \Gamma \vdash \mathbf{case_L}\ e\ \mathbf{of\ nil}\ \rightarrow\ e_1\ |\ \mathbf{cons}(h, tl)\ \rightarrow\ e_2 :_{\kappa+\kappa'} \tau'} \text{ caseL}$$

$$\frac{\Delta; \Phi; x : \tau_1, f : \tau_1 \xrightarrow{\kappa} \tau_2, \Gamma \vdash e :_\kappa \tau_2}{\Delta; \Phi; \Gamma \vdash \mathbf{fix}\ f(x).\,e :_0 \tau_1 \xrightarrow{\kappa} \tau_2} \text{ fix1} \qquad \frac{\begin{array}{c} \Delta; \Phi; \Gamma \vdash e_1 :_{\kappa_1} \tau_1 \xrightarrow{\kappa} \tau_2 \\ \Delta; \Phi; \Gamma \vdash e_2 :_{\kappa_2} \tau_1 \end{array}}{\Delta; \Phi; \Gamma \vdash e_1\ e_2 :_{(\kappa_1+\kappa_2+\kappa)} \tau_2} \text{ app}$$

$$\frac{t :: S, \Delta; \Phi; \Gamma \vdash e :_\kappa \tau}{\Delta; \Phi; \Gamma \vdash \Lambda.\,e :_0 \forall t \overset{\kappa}{::} S.\ \tau} \forall\text{I} \qquad \frac{\Delta; \Phi; \Gamma \vdash e :_\kappa \forall t \overset{\kappa'}{::} S.\ \tau \qquad \Delta \vdash I :: S}{\Delta; \Phi; \Gamma \vdash e[] :_{\kappa+\kappa'\{I/t\}} \tau\{I/t\}} \forall\text{E}$$

$$\frac{\Delta; \Phi; \Gamma \vdash e :_\kappa \tau \qquad \Delta; \Phi \models \tau \sqsubseteq \tau' \qquad \Delta; \Phi \models \kappa \leq \kappa'}{\Delta; \Phi; \Gamma \vdash e :_{\kappa'} \tau'} \sqsubseteq$$

$$\frac{\Upsilon(\zeta) = \zeta : \forall \overline{t_i {::} S_i}.\ \tau_1 \xrightarrow{\kappa} \tau_2 \qquad \Delta \vdash \overline{I_i} :: \overline{S_i} \qquad \Delta; \Phi; \Gamma \vdash e :_{\kappa_e} \tau_1[\overline{I_i/t_i}]}{\Delta; \Phi; \Gamma \vdash \zeta\ e :_{\kappa_e + \kappa[\overline{I_i/t_i}]} \tau_2[\overline{I_i/t_i}]} \text{ primApp}$$

$$\frac{\Delta; \Phi; \Gamma \vdash e :_\kappa \tau \qquad \forall y \in \Gamma.\ \Delta; \Phi \models \Gamma(y) \sqsubseteq (\Gamma(y))^{\mathbb{S}}}{\Delta; \Phi; \Gamma, \Gamma' \vdash e :_0 (\tau)^{\mathbb{S}}} \text{ nochange}$$

$$\frac{\Delta; \Phi; x : \tau_1, f : (\tau_1 \xrightarrow{\kappa} \tau_2)^{\mathbb{S}}, \Gamma \vdash e :_\kappa \tau_2 \qquad \forall y \in \Gamma.\ \Delta; \Phi \models \Gamma(y) \sqsubseteq (\Gamma(y))^{\mathbb{S}}}{\Delta; \Phi; \Gamma, \Gamma' \vdash \mathbf{fix}\ f(x).\,e :_0 (\tau_1 \xrightarrow{\kappa} \tau_2)^{\mathbb{S}}} \text{ fix2}$$

Fig. 4. Selected typing rules. The context Υ carrying types of primitive functions is omitted from all rules.

propagation cost of the expression. This is akin to accumulation of effects in a type and effect system. Second, values incur 0 change propagation cost because they are either updated before change propagation starts or by earlier steps of change propagation (which account for the cost of their update).

Variables represent values, so they have $\kappa = 0$ (rule **var**). All primitive constants like r can be given the type annotation $(\cdot)^S$ as in the rule **real**. (Modifiable constants can be modeled as variables with types without the $(\cdot)^S$ annotation and given two different substitutions in the two runs. This is standard in relational semantics and should be clear in Section 5.) This also applies to the empty list nil, but in its typing (rule **nil**) we do not explicitly write the annotation $(\cdot)^S$ because this annotation can be established through the subtyping rule **l2**. The term cons(e_1, e_2) can be typed at list $[n + 1]^\alpha \tau$ using one of two rules (**cons1** and **cons2**) depending on whether e_1 may change or not. If e_1 cannot change (it has type $(\tau)^S$), then e_2 is allowed α changes (rule **cons1**). If e_1 may change, then e_2 is allowed $\alpha - 1$ changes (rule **cons2**). The elimination rule for a list expression $e :$ list $[n]^\alpha \tau$ has three premises for the case branches (rule **caseL**). The first of these premises applies when e evaluates to nil. In this premise, we assume that the size of the list n and the number of allowed changes α are both 0. The remaining two premises correspond to the two typing rules for cons. In one premise, we assume that the head of the list (variable h) cannot change, so it has type $(\tau)^S$ and the tail may have α changes. In the other premise, we assume that the head may change, so it has type τ, but the tail may have only $\alpha - 1$ changes ($\alpha - 1$ is denoted by a new index variable β in the rule).

Rules **fix1** and **app** type recursive functions and function applications, respectively. A function is a value, so $\kappa = 0$ in rule **fix1**. In rule **app**, we add the function's change propagation cost κ to the cost of the application, as expected. Rule $\sqsubseteq$ allows weakening an ascribed type to any supertype and also allows weakening the change propagation cost upper-bound κ. Rule **primApp** types primitive function applications. This rule eliminates both $\forall$ and $\rightarrow$ from the type of the primitive function. $\overline{I_i}$ denotes a vector of index terms.

The rule **nochange** embodies our co-monadic reasoning principle. It says: If $e :_\kappa \tau$ in some context Γ (first premise) and the type of every variable in type Γ is a *subtype* of the same type annotated with $(\cdot)^S$ (second premise), then we can also give e the type $(\tau)^S$ and change propagation cost 0. In other words, if an expression depends only on unchanging variables, then its result cannot change and no change propagation is required. This rule is a strict generalization of the introduction rule for the type $\Box \tau$ in co-monadic type systems like [28]: If $e : \tau$ and all of e's free variables have types of the form $\Box \tau'$, then $e : \Box \tau$. The generalization here is that whether or not a variable in context has annotation $(\cdot)^S$ can depend on the constraints in Φ (via subtyping). We showed an application of this general rule in Example 3 of Section 2. Finally, we need an additional rule to type some recursive functions with annotation $(\cdot)^S$ (an example is the function bfold of Section 2). This rule, **fix2**, has the same condition on the function's free variables as the rule **nochange**. In typing the body of the recursive function, **fix2** allows us to assume that the function itself has a type annotated $(\cdot)^{\mathbb{S}}$. This rule cannot be derived using the rules **fix1** and **nochange**.

4   Dynamic Semantics

We define a tracing evaluation semantics and a cost-counting change propagation semantics for our language (Sections 4.1 and 4.2, respectively). We then prove our type system sound relative to the change propagation semantics (Section 5).

4.1   Evaluation Semantics and Traces

Our big-step, call-by-value evaluation judgment has the form $e \Downarrow v, T$ where e is the evaluated program, value v is the result of evaluating e and T is a reification of the big-step derivation tree, called a trace. The trace is used for change propagation after e has been modified. Traces have the following syntax.

$$
\begin{array}{rcl}
\text{Traces} \quad T & ::= & \mathbf{r} \mid () \mid (T_1, T_2) \mid \mathbf{fst}\ T \mid \mathbf{snd}\ T \mid \mathbf{nil} \mid \mathbf{cons}(T_1, T_2) \mid \\
& & \mathbf{case}_{\mathbf{nil}}(T, T') \mid \mathbf{case}_{\mathbf{cons}}(T, T') \mid \mathbf{fix}\ f(x).e \mid \mathbf{app}(T_1, T_2, T_r) \mid \\
& & \Lambda.e \mid \mathbf{iApp}(T, T_r) \mid \mathbf{pack}\ T \mid \mathbf{unpack}\ T\ \mathbf{as}\ x\ \mathbf{in}\ T' \mid \\
& & \mathbf{let}\ x = T_1\ \mathbf{in}\ T_2 \mid \mathbf{primApp}(T, v_r, \zeta)
\end{array}
$$

This syntax has one constructor for every evaluation rule and is largely self-explanatory. The trace of a value is the value itself. The trace of a primitive function application $\zeta\ e$ has the form $\mathbf{primApp}(T, v_r, \zeta)$, where T is the trace of e and v_r is the result of the application. Recording v_r is important: During change propagation, if the argument to the primitive function has not changed, then we simply reuse v_r, without re-computing the primitive function.

Selected evaluation rules are shown in Figure 5. The rules are self-explanatory, given the description of traces above. In the rule **primapp**, $\widehat{\zeta}$ denotes the semantic interpretation of the primitive ζ. For every value v, $\widehat{\zeta}(v)$ is a pair (c_r, v_r), where v_r is the result of evaluating the primitive ζ with argument v and c_r is the cost of this primitive evaluation.

4.2   Cost-counting Change Propagation Semantics

Change propagation takes as input the trace of an expression and a modified expression, and computes the trace of the modified expression by propagating changes through the original trace. This begs two questions: First, what kinds of expression modifications we allow and, second, how do we specify the modifications. The answer to the first question is that changes *stem* from replacing primitive values of a base type like **real** with other primitive values of the same type. Because our language contains closures and lists, changes lift to higher types, e.g., if a function receives the function $\lambda x.(x+1)$ as argument in the original execution, it may receive $\lambda x.(x + 2)$ after modification. However, it is not possible to receive $\lambda x.(x+1)$ as argument in the original execution and $\lambda x.(x+x)$

$\boxed{e \Downarrow v, T}$ Expression e evaluates to value v with trace T

$$\frac{}{\mathbf{r} \Downarrow \mathbf{r}, \mathbf{r}} \ \mathbf{r} \qquad \frac{e_1 \Downarrow v_1, T_1 \qquad e_2 \Downarrow v_2, T_2}{\mathtt{cons}(e_1, e_2) \Downarrow \mathtt{cons}(v_1, v_2), \mathtt{cons}(T_1, T_2)} \ \mathbf{cons}$$

$$\frac{}{\mathtt{fix}\, f(x).\, e \Downarrow \mathtt{fix}\, f(x).\, e, \mathtt{fix}\, f(x).\, e} \ \mathbf{fix} \qquad \frac{e \Downarrow v, T \qquad \widehat{\zeta}(v) = (c_r, v_r)}{\zeta\, e \Downarrow v_r, \mathtt{primApp}(T, v_r, \zeta)} \ \mathbf{primapp}$$

$$\frac{e_1 \Downarrow \mathtt{fix}\, f(x).\, e, T_1 \qquad e_2 \Downarrow v_2, T_2 \qquad e[v_2/x, (\mathtt{fix}\, f(x).\, e)/f] \Downarrow v_r, T_r}{e_1\, e_2 \Downarrow v_r, \mathtt{app}(T_1, T_2, T_r)} \ \mathbf{app}$$

Fig. 5. Selected evaluation rules

Bi-values $\quad \mathbf{w} ::= \quad \mathtt{keep}(\mathbf{r}) \mid \mathtt{repl}(\mathbf{r}, \mathbf{r}') \mid (\mathbf{w}_1, \mathbf{w}_2) \mid \mathtt{nil} \mid \mathtt{cons}(\mathbf{w}_1, \mathbf{w}_2) \mid$
$\qquad\qquad\qquad \mathtt{fix}\, f(x).\mathit{ee} \mid \Lambda.\mathit{ee} \mid \mathtt{pack}\, \mathbf{w} \mid ()$

Bi-expr. $\quad\; \mathit{ee} ::= \quad x \mid \mathtt{keep}(\mathbf{r}) \mid \mathtt{repl}(\mathbf{r}, \mathbf{r}') \mid (\mathit{ee}_1, \mathit{ee}_2) \mid \mathtt{fst}\, \mathit{ee} \mid \mathtt{snd}\, \mathit{ee} \mid$
$\qquad\qquad\qquad \mathtt{nil} \mid \mathtt{fix}\, f(x).\mathit{ee} \mid \mathit{ee}_1\, \mathit{ee}_2 \mid \Lambda.\mathit{ee} \mid \mathit{ee}[] \mid \mathtt{pack}\, \mathit{ee} \mid$
$\qquad\qquad\qquad \mathtt{unpack}\, \mathit{ee}\, \mathtt{as}\, x\, \mathtt{in}\, \mathit{ee}' \mid \mathtt{let}\, x = \mathit{ee}_1\, \mathtt{in}\, \mathit{ee}_2 \mid \zeta\, \mathit{ee} \mid () \mid$
$\qquad\qquad\qquad \mathtt{cons}(\mathit{ee}_1, \mathit{ee}_2) \mid (\mathtt{case}_{\mathsf{L}}\, \mathit{ee}\, \mathtt{of}\, \mathtt{nil} \rightarrow \mathit{ee}_1 \mid \mathtt{cons}(h, tl) \rightarrow \mathit{ee}_2)$

Fig. 6. Syntax of bi-values and bi-expressions

after modification. Similarly, the list $[1, 2, 3]$ may be modified to $[2, 2, 4]$, but because the refined list type mentions a statically determined length, it is not possible to modify the length of a list.

To *specify* expression changes and to prove soundness of our type system, we find it convenient to define a new syntactic category called a *bi-expression*, denoted ee. A bi-expression represents two nearly identical expressions (the original and the modified) that differ only in some primitive constants. The functions $\mathrm{L}(\mathit{ee})$ and $\mathrm{R}(\mathit{ee})$ project out the left and right (original and modified) expressions from ee. The syntax of bi-expressions, shown in Figure 6, is identical to that of expressions, except that instead of primitive constants $\mathbf{r}$, we have the forms $\mathtt{keep}(\mathbf{r})$ and $\mathtt{repl}(\mathbf{r}, \mathbf{r}')$. Roughly, $\mathtt{keep}(\mathbf{r})$ means that the original constant $\mathbf{r}$ has not been modified, whereas $\mathtt{repl}(\mathbf{r}, \mathbf{r}')$ means that the original constant $\mathbf{r}$ has been replaced by the constant $\mathbf{r}'$. Analogous to bi-expressions, we define bi-values, denoted $\mathbf{w}$, that represent pairs of values differing only in primitive constants. As an example, if the original value $\mathtt{fix}\, f(x).\, (x + 1)$ is modified to $\mathtt{fix}\, f(x).\, (x+2)$, then the two values can be represented together as the bi-value $\mathtt{fix}\, f(x).\, (x + \mathtt{repl}(1, 2))$. The left and right projections of a bi-expression/bi-value are defined as the homomorphic lifting of the following definitions.

$$\mathrm{L}(\mathtt{keep}(\mathbf{r})) = \mathbf{r} \ \Big\| \ \mathrm{R}(\mathtt{keep}(\mathbf{r})) = \mathbf{r}$$
$$\mathrm{L}(\mathtt{repl}(\mathbf{r}, \mathbf{r}')) = \mathbf{r} \ \Big\| \ \mathrm{R}(\mathtt{repl}(\mathbf{r}, \mathbf{r}')) = \mathbf{r}'$$

Bi-values and bi-expressions are typed as shown in Figure 7. The judgment $\Delta; \Phi; \Gamma \vdash \mathbf{w} \gg \tau$ means that the bi-value $\mathbf{w}$ represents two (related) values

$\boxed{\Delta; \Phi; \Gamma \vdash \mathbf{w} \gg \tau \text{ and } \Delta; \Phi; \Gamma \vdash \mathbf{ee} \gg_\kappa \tau}$ Bi-value and bi-expression typing

$$\frac{}{\Delta; \Phi; \Gamma \vdash \mathtt{keep}(\mathbf{r}) \gg (\mathtt{real})^{\mathbb{S}}} \text{keep} \qquad \frac{}{\Delta; \Phi; \Gamma \vdash \mathtt{repl}(\mathbf{r}, \mathbf{r}') \gg (\mathtt{real})^{\mathbb{C}}} \text{repl}$$

$$\frac{\Delta; \Phi; x : \tau_1, f : \tau_1 \xrightarrow{\kappa} \tau_2, \Gamma \vdash \mathbf{ee} \gg_\kappa \tau_2}{\Delta; \Phi; \Gamma \vdash \mathtt{fix}\, f(x).\, \mathbf{ee} \gg \tau_1 \xrightarrow{\kappa} \tau_2} \text{fix1}$$

$$\frac{\Delta; \Phi; \Gamma \vdash \mathbf{w} \gg \tau \quad \forall z \in \Gamma.\ \Delta; \Phi \models \Gamma(z) \sqsubseteq (\Gamma(z))^{\mathbb{S}} \quad \mathtt{stable}(\mathbf{w})}{\Delta; \Phi; \Gamma, \Gamma' \vdash \mathbf{w} \gg (\tau)^{\mathbb{S}}} \text{nochange}$$

$$\frac{\Delta; \Phi; x : \tau_1, f : (\tau_1 \xrightarrow{\kappa} \tau_2)^{\mathbb{S}}, \Gamma \vdash \mathbf{ee} \gg_\kappa \tau_2}{\forall z \in \Gamma.\ \Delta; \Phi \models \Gamma(z) \sqsubseteq (\Gamma(z))^{\mathbb{S}} \quad \mathtt{stable}(\mathbf{ee})}{\Delta; \Phi; \Gamma, \Gamma' \vdash \mathtt{fix}\, f(x).\, \mathbf{ee} \gg (\tau_1 \xrightarrow{\kappa} \tau_2)^{\mathbb{S}}} \text{fix2} \qquad \frac{\Delta; \Phi; \Gamma \vdash \mathbf{w} \gg \tau \quad \Delta; \Phi \models \tau \sqsubseteq \tau'}{\Delta; \Phi; \Gamma \vdash \mathbf{w} \gg \tau'} \sqsubseteq$$

$$\frac{\Delta; \Phi; \Gamma \vdash \mathbf{w}_i \gg \tau_i \quad \Delta; \Phi; \overline{x_i : \tau_i}, \Gamma \vdash e :_\kappa \tau}{\Delta; \Phi; \Gamma \vdash \ulcorner e \urcorner [\overline{\mathbf{w}_i / x_i}] \gg_\kappa \tau} \text{exp}$$

Fig. 7. Selected typing rules for bi-values and bi-expressions

of type τ. Its rules mirror those of value typing, mostly. The bi-value $\mathtt{keep}(\mathbf{r})$ has the type $(\mathtt{real})^{\mathbb{S}}$, whereas the bi-value $\mathtt{repl}(\mathbf{r}, \mathbf{r}')$ has the type $(\mathtt{real})^{\mathbb{C}}$, reflecting the difference between the refinements $(\cdot)^{\mathbb{S}}$ and $(\cdot)^{\mathbb{C}}$. Rules **fix2** and **nochange** are analogous to their homonyms from expression typing and introduce the annotation $(\cdot)^{\mathbb{S}}$. In these rules, we have to additionally check that the bi-value being typed contains no syntactic occurrences of $\mathtt{repl}(\cdot, \cdot)$ because the annotation $(\cdot)^{\mathbb{S}}$ means absence of syntactic change. This is formalized by the proposition $\mathtt{stable}(\mathbf{ee})$, which means that $\mathbf{ee}$ has no occurrences of $\mathtt{repl}(\cdot, \cdot)$.

The judgment $\Delta; \Phi; \Gamma \vdash \mathbf{ee} \gg_\kappa \tau$ means that $\mathbf{ee}$ represents two related expressions of type τ and that the trace of any one of those expressions can be change propagated for the other expression, incurring cost at most κ. This judgment is defined by only one rule, **exp**, that relies on the typing judgments for expressions and bi-values. Let $\ulcorner e \urcorner$ denote the bi-expression obtained by replacing all occurrences of $\mathbf{r}$ in e with $\mathtt{keep}(\mathbf{r})$. It is easy to see that every bi-expression $\mathbf{ee}$ can be written as $\ulcorner e \urcorner [\overline{\mathbf{w}_i / x_i}]$ for some expression e and some sequence of bi-values $\overline{\mathbf{w}_i}$. The rule **exp** types $\mathbf{ee}$ by typing e (using the expression typing rules) and $\overline{\mathbf{w}_i}$ (using the bi-value typing rules). Setting up bi-expression typing this way is primarily for technical convenience in proving the soundness of our type system. An equivalent type system is obtained by mirroring all the expression typing rules for bi-expressions.

Change Propagation. We formalize change propagation abstractly by the judgment $\langle T, \mathbf{ee} \rangle \curvearrowright \mathbf{w}', T', c'$, which has inputs T and $\mathbf{ee}$ and outputs $\mathbf{w}'$, T' and c'. The input T must be the trace of the original expression $\mathrm{L}(\mathbf{ee})$. The output $\mathbf{w}'$ represents two values, $\mathrm{L}(\mathbf{w}')$ and $\mathrm{R}(\mathbf{w}')$, which are the results of evaluating the original and modified expressions, respectively. The output T' is the trace of the

$$\boxed{\langle T, \text{\ae}\rangle \curvearrowright \mathbf{w}', T', c'} \quad \text{Change propagation with cost-counting}$$

$$\frac{\text{stable}(\text{\ae})}{\langle \text{primApp}(T, v_r, \zeta), \zeta\ \text{\ae}\rangle \curvearrowright \ulcorner v_r \urcorner, \text{primApp}(T, v_r, \zeta), 0}\ \textbf{r-prim-s}$$

$$\frac{\neg\text{stable}(\text{\ae}) \qquad \langle T, \text{\ae}\rangle \curvearrowright \mathbf{w}', T', c' \qquad \widehat{\zeta}(\text{R}(\mathbf{w}')) = (c_r', v_r')}{\langle \text{primApp}(T, v_r, \zeta), \zeta\ \text{\ae}\rangle \curvearrowright \text{merge}(v_r, v_r'), \text{primApp}(T', v_r', \zeta),\ c' + c_r'}\ \textbf{r-prim}$$

$$\frac{}{\langle r, \text{keep}(\_)\rangle \curvearrowright \text{keep}(r), r, 0}\ \textbf{r-keep} \qquad\qquad \frac{}{\langle r, \text{repl}(\_, r')\rangle \curvearrowright \text{repl}(r, r'), r', 0}\ \textbf{r-repl}$$

$$\frac{\langle T_1, \text{\ae}_1\rangle \curvearrowright \mathbf{w}_1', T_1', c_1' \qquad \langle T_2, \text{\ae}_2\rangle \curvearrowright \mathbf{w}_2', T_2', c_2'}{\langle \text{cons}(T_1, T_2), \text{cons}(\text{\ae}_1, \text{\ae}_2)\rangle \curvearrowright\ \text{cons}(\mathbf{w}_1', \mathbf{w}_2'), \text{cons}(T_1', T_2'), c_1' + c_2'}\ \textbf{r-cons}$$

$$\frac{\langle T, \text{\ae}\rangle \curvearrowright \text{nil}, T', c' \qquad \langle T_1, \text{\ae}_1\rangle \curvearrowright \mathbf{w}_1', T_1', c_1'}{\langle \text{case}_{\text{nil}}(T, T_1), \text{case}_L\ \text{\ae}\ \text{of nil}\ \to\ \text{\ae}_1 \mid \text{cons}(h, tl)\ \to\ \text{\ae}_2\rangle \curvearrowright\ \mathbf{w}_1', \text{case}_{\text{nil}}(T', T_1'), c' + c_1'}\ \textbf{r-case-nil}$$

$$\frac{\langle T, \text{\ae}\rangle \curvearrowright \text{cons}(\mathbf{w}_h, \mathbf{w}_{tl}), T', c' \qquad \langle T_2, \text{\ae}_2[\mathbf{w}_h/h, \mathbf{w}_{tl}/tl]\rangle \curvearrowright \mathbf{w}_2', T_2', c_2'}{\langle \text{case}_{\text{cons}}(T, T_1), \text{case}_L\ \text{\ae}\ \text{of nil}\ \to\ \text{\ae}_1 \mid \text{cons}(h, tl)\ \to\ \text{\ae}_2\rangle \curvearrowright\ \mathbf{w}_2', \text{case}_{\text{cons}}(T', T_2'), c' + c_2'}\ \textbf{r-case-cons}$$

Fig. 8. Selected Replay Rules

modified expression. Most importantly, c' is the total cost incurred in change propagation. The output $\mathbf{w}'$ is an artifact of our formalization and important for an inductive proof of our soundness theorem. Actual implementations of change propagation never construct it and, hence, we do not count any cost for constructing or analyzing it during change propagation. As part of our soundness theorem, we show that $\curvearrowright$ is a total function on well-typed programs.

Rules defining the judgment $\curvearrowright$ case analyze the input trace T. Representative rules are shown in Figure 8. To change propagate the trace $\text{primApp}(T, v_r, \zeta)$ for the primitive function application bi-expression $\zeta\ \text{\ae}$, we case analyze whether the original expression in $\text{\ae}$ changed or not. If $\text{stable}(\text{\ae})$, then the argument to ζ has not changed ($\text{stable}(\text{\ae})$ implies $L(\text{\ae}) = R(\text{\ae})$). So we simply reuse the result v_r stored in the original trace. The output bi-value is $\ulcorner v_r \urcorner$ (which represents v_r paired with itself). The output trace is the same as the input trace and the cost is 0. This is summarized in the rule **r-prim-s**. If, on the other hand, $\neg\text{stable}(\text{\ae})$ (rule **r-prim**), then the argument to ζ has changed, so we change propagate through the argument (second premise) and reapply the primitive function ζ to the updated argument (third premise). The bi-value in the output is obtained by *merging* the original result v_r with the new result v_r'. Merge is defined as follows: If $L(\mathbf{w}) = v_r$ and $R(\mathbf{w}) = v_r'$, then $\text{merge}(v_r, v_r') = \mathbf{w}$. In general, merge is a partial function. But, if the primitive function's interpretation lies in the semantic interpretation of its type (semantic interpretations are defined in the next section), then the merge must be defined. The cost of change propagation is the sum of the cost c' of change propagating the argument of ζ and the cost c_r' of evaluating ζ on the new argument. This rule is the only source of non-zero costs during change propagation. All other rules either incur zero cost, or simply aggregate costs from the premises.

The trace of a primitive constant r is change propagated using rules **r-keep** and **r-repl**. If the constant has not changed (rule **r-keep**) then the trace does not change and no cost is incurred. If the constant has changed, the resulting trace is the new value of the constant (rule **r-repl**). Even in this case, no cost is incurred, because in an implementation of change propagation, the trace and the expression can share a pointer to the constant so the update to the expression (which happens before change propagation starts) implicitly updates the trace [11]. At constructors like cons, change propagation simply recurses on argument sub-traces and adds the costs (rule **cons**). Elimination forms like case$_L$ are handled similarly. Because control flow changes are forbidden, the original trace determines the branch of the case analysis to which changes must be propagated (rules **r-case-nil** and **r-case-cons**).

Implementation. The relation $\curvearrowright$ formalizes change propagation and its cost *abstractly.* An obvious question is whether change propagation can be *implemented* with the costs stipulated by $\curvearrowright$. The answer is affirmative. Prior work on libraries and compilers for self-adjusting computation already shows how to implement change propagation with these costs using imperative traces, leaf-to-root traversals and in-place update of values [1,10]. Since values are updated in-place, no cost is incurred for structural operations like pairing, projection, consing, etc; cost is incurred only for re-evaluating primitive functions on paths starting in updated leaves, exactly as in the judgment $\curvearrowright$. To double-check, we implemented most of our examples on an existing library, AFL [1], and observed exactly the costs stipulated by $\curvearrowright$. Due to lack of space, we omit the experimental results.

5   Soundness

We prove our type system sound in two ways: (a) Trace propagation is total and produces correct results on typed expressions, and (b) The cost of change propagation (determined by $\curvearrowright$) on a typed expression is no more than the cost κ estimated in the expression's typing judgment. We combine these two statements together in the following theorem. This theorem considers an expression e with one free variable x, which receives two potentially different substitutions (the two projections of a bi-value w) in the original and modified execution. A more general theorem with any number of free variables (and, hence, any number of independent changes) holds as well, but we skip it here to improve readability.

Theorem 1 (Type soundness). *Suppose that (a) $x : \tau \vdash e :_\kappa \tau'$; (b) $\vdash w \gg \tau$; and (c) $e[L(w)/x] \Downarrow v', T$. Then the following hold for some T', w' and c: (1) $\langle T, \ulcorner e \urcorner [w/x] \rangle \curvearrowright w', T', c$; (2) $e[R(w)/x] \Downarrow R(w'), T'$; and (3) $c \le \kappa$.*

In words, the theorem says that if expression e types with dynamic stability κ and we execute e with an initial substitution $L(\mathbf{w})/x$ to obtain a trace T, then we can successfully change propagate T with a new substitution $R(\mathbf{w})/x$ in e (statement 1) to obtain the correct new output and trace (statement 2) with cost c of change propagation no more than the statically estimated dynamic stability κ (statement 3). Briefly, (1) states totality of change propagation for typed programs, (2) states its functional correctness, and (3) shows that our type system estimates dynamic stability conservatively. Note that this theorem models changes to expressions as different substitutions to the expression's free variable. Syntactic constants in e cannot change, which explains why we can type constants with annotation $(\cdot)^{\mathbb{S}}$ in Figure 4.

$$\boxed{\;[\![\tau]\!]_v \subseteq \text{Step index} \times \text{Bi-values and } [\![\tau]\!]_\varepsilon^\kappa \subseteq \text{Step index} \times \text{Bi-expressions}\;}$$

$$[\![(\tau)^{\mathbb{S}}]\!]_v = \{(m, \mathbf{w}) \mid (m, \mathbf{w}) \in [\![\tau]\!]_v \wedge \text{stable}(\mathbf{w})\}$$

$$[\![(\tau)^{\mathbb{C}}]\!]_v = [\![\tau]\!]_v$$

$$[\![\text{real}]\!]_v = \{(m, \text{keep}(\mathbf{r})) \mid \top\} \cup \{(m, \text{repl}(\mathbf{r}, \mathbf{r}')) \mid \top\}$$

$$[\![\text{list}\,[0]^\alpha\,\tau]\!]_v = \{(m, \text{nil}) \mid \top\}$$

$$[\![\text{list}\,[n{+}1]^\alpha\,\tau]\!]_v = \{(m, \text{cons}(\mathbf{w}_1, \mathbf{w}_2)) \mid ((m, \mathbf{w}_1) \in [\![(\tau)^{\mathbb{S}}]\!]_v \wedge (m, \mathbf{w}_2) \in [\![\text{list}\,[n]^\alpha\,\tau]\!]_v)$$
$$\vee\, ((m, \mathbf{w}_1) \in [\![\tau]\!]_v \wedge (m, \mathbf{w}_2) \in [\![\text{list}\,[n]^{\alpha-1}\,\tau]\!]_v \wedge \alpha > 0)\}$$

$$[\![\tau_1 \xrightarrow{\kappa} \tau_2]\!]_v = \{(m, \text{fix}\,f(x).\mathbf{e}\mathbf{e}) \mid$$
$$\forall j < n,\ \forall \mathbf{w}\ (j, \mathbf{w}) \in [\![\tau_1]\!]_v \Rightarrow (j, \mathbf{e}\mathbf{e}[\text{fix}\,f(x).\mathbf{e}\mathbf{e}/f][\mathbf{w}/x]) \in [\![\tau_2]\!]_\varepsilon^\kappa\}$$

$$[\![\forall t \overset{\kappa}{::} S.\ \tau]\!]_v = \{(m, \Lambda.\mathbf{e}\mathbf{e}) \mid \forall I\ I :: S\ \ (m, \mathbf{e}\mathbf{e}) \in [\![\tau[I/t]]\!]_\varepsilon^{\kappa[I/t]}\}$$

$$[\![\exists t.\ \tau]\!]_v = \{(m, \text{pack}\,\mathbf{w}) \mid \exists I.I :: S\ \wedge (m, \mathbf{w}) \in [\![\tau[I/t]]\!]_v\}$$

$$[\![\tau_1 \times \tau_2]\!]_v = \{(m, (\mathbf{w}_1, \mathbf{w}_2)) \mid (m, \mathbf{w}_1) \in [\![\tau_1]\!]_v \wedge (m, \mathbf{w}_2) \in [\![\tau_2]\!]_v\}$$

$$[\![\tau]\!]_\varepsilon^\kappa = \{(m, \mathbf{e}\mathbf{e}) \mid \forall j < n.\ L(\mathbf{e}\mathbf{e}) \Downarrow v, T\ \wedge\ j = |T|\ \Rightarrow \exists\, v', T', \mathbf{e}\mathbf{e}', c' :$$
$$\qquad 1.\ \langle T, \mathbf{e}\mathbf{e}\rangle \curvearrowright \mathbf{w}', T', c'$$
$$\qquad 2.\ c' \leq \kappa$$
$$\qquad 3.\ (m - j, \mathbf{w}') \in [\![\tau]\!]_v$$
$$\qquad 4.\ R(\mathbf{e}\mathbf{e}) \Downarrow v', T'$$
$$\qquad 5.\ v' = R(\mathbf{w}') \wedge\ v = L(\mathbf{w}')\}$$

$$\mathcal{D}[\![\cdot]\!], \mathcal{G}[\![\cdot]\!] = \{\emptyset\}$$

$$\mathcal{D}[\![\Delta, t :: S]\!] = \{\sigma[t \mapsto I] \mid \sigma \in \mathcal{D}[\![\Delta]\!] \wedge I :: S\}$$

$$\mathcal{G}[\![\Gamma, x : \tau]\!] = \{(m, \theta[x \mapsto \mathbf{w}]) \mid (m, \theta) \in \mathcal{G}[\![\Gamma]\!] \wedge (m, \mathbf{w}) \in [\![\tau]\!]_v\}$$

Fig. 9. Step-indexed interpretation of selected types

To prove this theorem, we build a *relational* model of types interpreted as sets of bi-values and bi-expressions. To handle recursive functions, we step-index our model [5]. The index counts trace size in our model. Trace size is proportional to the number of steps in complete reductions of small-step semantics. The size $|T|$ of a trace T is defined as follows: Primitive constants and functions have size 0 and each trace constructor adds 1 to the size.

For every closed type τ we define a value interpretation $[\![\tau]\!]_v$ and an expression interpretation $[\![\tau]\!]_\varepsilon^\kappa$. The value interpretation $[\![\tau]\!]_v$ is a set of pairs of the form $(m, \mathbf{w})$, where m is a step index. The expression interpretation $[\![\tau]\!]_\varepsilon^\kappa$ is a set of pairs of the form $(m, \textbf{æ})$, where change propagating the trace of $L(\textbf{æ})$ with $\textbf{æ}$ costs no more than κ if the size of that trace is less than m. The two interpretations of types, shown in Figure 9, are defined simultaneously by induction on τ. In the definition of the value interpretation of the list type $\texttt{list}\,[n]^\alpha\,\tau$, we subinduct on n. Our definitions are unsurprising but we mention a few salient points. First, $[\![(\tau)^C]\!]_v = [\![\tau]\!]_v$ and $[\![(\tau)^S]\!]_v \subseteq [\![\tau]\!]_v$. Moreover, $(m, \mathbf{w}) \in [\![(\tau)^S]\!]_v$ implies $\texttt{stable}(\mathbf{w})$, as expected. The value interpretation of $\texttt{list}\,[n+1]^\alpha\,\tau$ has two clauses corresponding to the two typing rules for $\texttt{cons}$. Most importantly, the expression interpretation $[\![\tau]\!]_\varepsilon^\kappa$ captures enough invariants about change propagation to enable us to prove the soundness theorem above. Figure 9 also shows the definitions of semantic substitutions σ and θ for the contexts Δ and Γ, respectively. As usual, the substitution for each variable in Γ must lie in the value interpretation of the variable's type.

We prove the following fundamental theorem for our type interpretations. The theorem consists of three statements for three different syntactic classes: expressions, bi-values and bi-expressions (in that order). The statement for expressions is established by an induction on expression typing, with a subinduction on step-indices for recursive functions. The other two statements follow by simultaneous induction on bi-value and bi-expression typing. The theorem relies on the assumption that the interpretation of every primitive function lies in the interpretation of the function's type. The formal statement of this assumption and the proof of the theorem are in our online appendix. Type soundness, Theorem 1, is an immediate corollary of the first two statements of this theorem.

Theorem 2 (Fundamental Theorem). *1. If $\Delta; \Phi; \Gamma \vdash e :_\kappa \tau$ and $\sigma \in \mathcal{D}[\![\Delta]\!]$ and $(m, \theta) \in \mathcal{G}[\![\sigma\Gamma]\!]$ and $\models \sigma\Phi$, then $(m, \theta^\ulcorner e^\urcorner) \in [\![\sigma\tau]\!]_\varepsilon^{\sigma\kappa}$.*

2. If $\Delta; \Phi; \Gamma \vdash \mathbf{w} \gg \tau$ and $\sigma \in \mathcal{D}[\![\Delta]\!]$ and $(m, \theta) \in \mathcal{G}[\![\sigma\Gamma]\!]$ and $\models \sigma\Phi$, then $(m, \theta\mathbf{w}) \in [\![\sigma\tau]\!]_v$.

3. If $\Delta, \Phi, \Gamma \vdash \textbf{æ} \gg_\kappa \tau$ and $\sigma \in \mathcal{D}[\![\Delta]\!]$ and $(m, \theta) \in \mathcal{G}[\![\sigma\Gamma]\!]$ and $\models \sigma\Phi$, then $(m, \theta(\textbf{æ})) \in [\![\sigma\tau]\!]_\varepsilon^{\sigma\kappa}$.

6   Related Work

Incremental and self-adjusting computation. Incremental computation has been studied extensively in the last three decades (reduction in the lambda calculus [15], graph algorithms [24], attribute grammars [14], programming languages [8] etc.). While most work focuses on efficient data-structures and memoization techniques for incremental computation, recent work develops type-directed techniques for automatic incrementalization of batch programs [10]. Ley-Wild *et al.* propose a cost semantics for program execution and bound the change propagation time of self-adjusting programs using a metric of trace distances [26]. Their analysis only yields that change propagation is no slower than from-scratch evaluation, asymptotically. Although they are able to prove tight bounds for some benchmark programs, this analysis requires comparing trace distances by hand for each change. Unlike our work, no existing approach provides a general, static technique for establishing tight asymptotic dynamic stability.

Chen *et al.* [11] use variation annotations similar to CostIt's, but do not address the problem of estimating dynamic stability. Instead, they focus on compiling a higher-order functional language to AFL, a language with change propagation semantics. Their translation is facilitated by types annotated $(\cdot)^{\mathbb{S}}$ and $(\cdot)^{\mathbb{C}}$, which CostIt uses for a different purpose. In turn, Chen *et al.* borrow these type annotations from Simonet and Pottier's work on type inference for information flow analysis [31].

In contrast to our co-monadic interpretation of $(\tau)^{\mathbb{S}}$ and identification of $(\tau)^{\mathbb{C}}$ with τ, a significant amount of prior work on implementation of incremental programs equates $(\tau)^{\mathbb{S}}$ to τ and gives a monadic interpretation to the type $(\tau)^{\mathbb{C}}$ [1,8,11]. Although a deeper study of the connection between these two approaches is necessary, the choice so far seems to be motivated by the task at hand. For executing programs, it is natural to confine changes (and change propagation) to a monad, whereas for reasoning about dynamic stability it is often necessary to conclude by looking at an expression's inputs that the expression's result cannot change, which is easier in our co-monadic interpretation.

Continuity and program sensitivity. Also closely related to our work in concept, but not in the end-goal, is work on analysis of program continuity. There, the goal is to prove that the outputs of two runs of a program are closely related if the inputs are. Program continuity does not account for dynamic stability. Our type system also proves a limited form of program continuity, as an intermediate step in establishing dynamic stability. Reed and Pierce present a linear type system called Fuzz for proving continuity [33], as an intermediate step in verifying differential privacy properties. Gaboardi *et al.* extend Fuzz with lightweight dependent types in a type system called DFuzz [16]. DFuzz's syntax and use of lightweight dependent types influenced our work significantly. A technical difference from DFuzz (and Fuzz) is that our types capture where two values differ whereas in DFuzz, the "distance" between related values is not explicit in the type, but only in the relational model. As a result, our type system does not need linearity, which DFuzz does. Unlike CostIt and DFuzz, Chaudhuri *et al.*'s static analysis can prove program continuity even with control flow changes as long as perturbations to the input result in branches that are close to each other [9].

Static computation of resource bounds/complexity analysis. The programming languages community is rife with work on static computation of resource bounds, particularly worse-case execution time complexity, using different techniques such as abstract interpretation [18,35], linear dependent types [13], amortized resource analysis [22] and sized types [12,25,36]. A common denominator of these techniques is that they all reason about a single execution of a program. In contrast, our focus — dynamic stability — is a two-trace property. It requires a relational model of execution which accounts for change propagation, as well as a relational model of types to track what parts of values can change across the executions, both of which we develop in this paper.

We mention some type-theoretic approaches to inferring and verifying resource usage bounds in programs. Dal Lago *et al.* present a complete time complexity analysis for PCF [13]. They use linear types to statically limit the number of times a function may be applied by the context. This allows reasoning about the time complexity of recursive functions precisely. We could adopt a similar approach in our work, although we have not found this necessary so far. Hoffmann *et al.* [23,22] infer polynomial-shaped bounds on resource usage of RAML (Resource Aware ML) programs. A significant advantage of their technique is automation. A similar analysis for dynamic stability may be possible although the compatibility of logarithmic functions (which are necessary to state the dynamic stability of interesting programs) with Hoffmann *et al.*'s approach remains an open problem.

We use sized types [25] for lists. Sized types are often used in termination checking and analysis of heap and stack space [35]. Our types are precise on list lengths, unlike conventional uses where the size in the type is an upper-bound. For the number of allowed changes, our types specify upper-bounds.

7   Conclusion and Future Work

Existing work on incremental computation has been very successful at improving efficiency of incremental runs of a program, but does not consider the equally important question of developing static tools to analyze dynamic stability. Our work, CostIt, takes a first step in this direction by equipping a higher-order functional language with a type system to analyze dynamic stability of programs. We find that index refinements, immutability annotations, co-monadic reasoning and constraint-aware subtyping are useful in analyzing dynamic stability. Our type system is sound relative to a cost semantics for change propagation. We demonstrate the expressiveness and precision of CostIt on several examples.

Our ongoing work builds on the content of this paper in three ways. First, we are working on a prototype implementation of CostIt using bidirectional type-checking. We reduce type-checking and type inference to constraint satisfiability as in Dependent ML [37]. There is no new conceptual difficulty, but the constraint domain is largely intractable, as demonstrated by the occurrence of logarithmic and exponential functions in Examples 3 and 4. Consequently, we are exploring the possibility of using a combination of automatic and semi-automatic constraint solving (Dal Lago *et al.* use a similar approach in the context of worse-case execution time complexity analysis [13]).

Second, in work done after the review of this paper, we have extended CostIt's type system, relational model and soundness theorem to cover situations where program control flow may change with input changes. This is a nontrivial extension, beyond the scope of this paper. Briefly, we extend the type system with a standard worse-case execution time complexity analysis for branches which might execute from scratch during change propagation. The resulting type system is a significant refinement of the pure fragment of Pottier and Simonet's (simple) information flow type system for ML [31] (in contrast, the work in this paper corresponds to the special case where Pottier and Simonet's program counter or pc is always "low" or unchanging).

Finally, motivated by recent work on demand-driven incremental computation [20], we are planning to work on a version of CostIt for lazy evaluation semantics.

Acknowledgments. The research of Umut Acar is partially supported by the European Research Council under grant number ERC-2012-StG-308246 and by the National Science Foundation under grant numbers CCF-1320563 and CCF-1408940.

References

1. Acar, U., Blelloch, G., Blume, M., Harper, R., Tangwongsan, K.: A library for self-adjusting computation. Elec. Notes in Theor. Comp. Sci. 148(2), 127–154 (2006)
2. Acar, U.A., Blelloch, G.E., Blume, M., Harper, R., Tangwongsan, K.: An experimental analysis of self-adjusting computation. ACM Trans. Program. Lang. Syst. 3, 3:1–3:53 (2009)
3. Acar, U.A., Blelloch, G.E., Harper, R.: Adaptive functional programming. ACM Trans. Program. Lang. Syst. 28(6), 990–1034 (2006)
4. Acar, U.A., Blume, M., Donham, J.: A consistent semantics of self-adjusting computation. The Journal of Functional Programming (2013)
5. Ahmed, A.: Step-indexed syntactic logical relations for recursive and quantified types. In: Sestoft, P. (ed.) ESOP 2006. LNCS, vol. 3924, pp. 69–83. Springer, Heidelberg (2006)
6. Blanchet, B., Abadi, M., Fournet, C.: Automated verification of selected equivalences for security protocols. The Journal of Logic and Algebraic Programming 75(1), 3–51 (2008)
7. Brodal, G.S., Jacob, R.: Dynamic planar convex hull. In: Proceedings of the 43rd Annual IEEE Symposium on Foundations of Computer Science, pp. 617–626 (2002)
8. Carlsson, M.: Monads for incremental computing. In: Proceedings of the 7th International Conference on Functional Programming, ICFP 2002, pp. 26–35. ACM (2002)
9. Chaudhuri, S., Gulwani, S., Lublinerman, R.: Continuity and robustness of programs. Communications of the ACM 55(8), 107–115 (2012)
10. Chen, Y., Dunfield, J., Acar, U.A.: Type-directed automatic incrementalization. In: Proceedings of the 33rd Conference on Programming Language Design and Implementation, PLDI 2012, pp. 299–310. ACM (2012)
11. Chen, Y., Dunfield, J., Hammer, M.A., Acar, U.A.: Implicit self-adjusting computation for purely functional programs. In: International Conference on Functional Programming, ICFP 2011, pp. 129–141 (2011)
12. Chin, W.N., Khoo, S.C.: Calculating sized types. In: Proceedings of the Workshop on Partial Evaluation and Semantics-based Program Manipulation, PEPM 2000, pp. 62–72. ACM (1999)

13. Dal Lago, U., Petit, B.: The geometry of types. In: Proceedings of the 40th Annual Symposium on Principles of Programming Languages, POPL 2013, pp. 167–178. ACM (2013)
14. Demers, A., Reps, T., Teitelbaum, T.: Incremental evaluation for attribute grammars with application to syntax-directed editors. In: Proceedings of the 8th Symposium on Principles of Programming Languages, POPL 1981, pp. 105–116. ACM (1981)
15. Field, J.: Incremental Reduction in the Lambda Calculus and Related Reduction Systems. Ph.D. thesis, Department of Computer Science, Cornell University (1991)
16. Gaboardi, M., Haeberlen, A., Hsu, J., Narayan, A., Pierce, B.C.: Linear dependent types for differential privacy. In: Proceedings of the 40th Annual Symposium on Principles of Programming Languages, POPL 2013, pp. 357–370. ACM (2013)
17. Graham, R.L.: An efficient algorithm for determining the convex hull of a finite planar set. Information Processing Letters 1(4), 132–133 (1972)
18. Gulwani, S., Mehra, K.K., Chilimbi, T.: Speed: Precise and efficient static estimation of program computational complexity. In: Proceedings of the 36th Annual Symposium on Principles of Programming Languages, POPL 2009, pp. 127–139. ACM (2009)
19. Hammer, M.A., Acar, U.A., Chen, Y.: Ceal: A C-based language for self-adjusting computation. In: Proceedings of the 2009 Conference on Programming Language Design and Implementation, PLDI 2009, pp. 25–37. ACM (2009)
20. Hammer, M.A., Phang, K.Y., Hicks, M., Foster, J.S.: Adapton: Composable, demand-driven incremental computation. In: Proceedings of the 35th Conference on Programming Language Design and Implementation, PLDI 2014, pp. 156–166. ACM (2014)
21. Heydon, A., Levin, R., Yu, Y.: Caching function calls using precise dependencies. In: Proceedings of the Conference on Programming Language Design and Implementation, PLDI 2000, pp. 311–320. ACM (2000)
22. Hoffmann, J., Aehlig, K., Hofmann, M.: Multivariate amortized resource analysis. In: Proceedings of the 38th Annual Symposium on Principles of Programming Languages, POPL 2011, pp. 357–370. ACM (2011)
23. Hoffmann, J., Hofmann, M.: Amortized resource analysis with polynomial potential: A static inference of polynomial bounds for functional programs. In: Gordon, A.D. (ed.) ESOP 2010. LNCS, vol. 6012, pp. 287–306. Springer, Heidelberg (2010)
24. Holm, J., de Lichtenberg, K.: Top-trees and dynamic graph algorithms. Tech. Rep. DIKU-TR-98/17, Department of Computer Science, University of Copenhagen (1998)
25. Hughes, J., Pareto, L.: Recursion and dynamic data-structures in bounded space: Towards embedded ML programming. In: Proceedings of the Fourth International Conference on Functional Programming, ICFP 1999, pp. 70–81. ACM (1999)
26. Ley-Wild, R., Acar, U.A., Fluet, M.: A cost semantics for self-adjusting computation. In: Proceedings of the 36th Annual Symposium on Principles of Programming Languages, POPL 2009, pp. 186–199. ACM (2009)
27. Ley-Wild, R., Fluet, M., Acar, U.A.: Compiling self-adjusting programs with continuations. In: Proceedings of the 13th International Conference on Functional Programming, ICFP 2008, pp. 321–334. ACM (2008)
28. Nanevski, A., Pfenning, F.: Staged computation with names and necessity. J. Funct. Program. 15(6), 893–939 (2005)
29. Nielson, F., Riis Nielson, H.: Type and effect systems. In: Olderog, E.-R., Steffen, B. (eds.) Correct System Design. LNCS, vol. 1710, pp. 114–136. Springer, Heidelberg (1999)

30. Overmars, M.H., van Leeuwen, J.: Maintenance of configurations in the plane. Journal of Computer and System Sciences 23, 166–204 (1981)
31. Pottier, F., Simonet, V.: Information flow inference for ML. ACM Trans. Program. Lang. Syst. 25(1), 117–158 (2003)
32. Pugh, W., Teitelbaum, T.: Incremental computation via function caching. In: Proceedings of the 16th Annual ACM Symposium on Principles of Programming Languages, POPL 1989, pp. 315–328. ACM (1989)
33. Reed, J., Pierce, B.C.: Distance makes the types grow stronger: A calculus for differential privacy. In: Proceedings of the 15th International Conference on Functional Programming, ICFP 2010, pp. 157–168. ACM (2010)
34. Shankar, A., Bodík, R.: Ditto: Automatic incrementalization of data structure invariant checks (in Java). In: Proceedings of the Conference on Programming Language Design and Implementation, PLDI 2007, pp. 310–319. ACM (2007)
35. Vasconcelos, P.: Space cost analysis using sized types. Ph.D. thesis, School of Computer Science, University of St Andrews (2008)
36. Vasconcelos, P.B., Hammond, K.: Inferring cost equations for recursive, polymorphic and higher-order functional programs. In: Trinder, P., Michaelson, G.J., Peña, R. (eds.) IFL 2003. LNCS, vol. 3145, pp. 86–101. Springer, Heidelberg (2005)
37. Xi, H., Pfenning, F.: Dependent types in practical programming. In: Proceedings of the 26th Symposium on Principles of Programming Languages, POPL 1999, pp. 214–227. ACM (1999)
38. Yellin, D., Strom, R.: Inc: A language for incremental computations. In: Proceedings of the Conference on Programming Language Design and Implementation, PLDI 1988, pp. 115–124. ACM (1988)

Monotonic References for Efficient Gradual Typing

Jeremy G. Siek[1], Michael M. Vitousek[1], Matteo Cimini[1],
Sam Tobin-Hochstadt[1], and Ronald Garcia[2]

[1] Indiana University Bloomington
jsiek@indiana.edu
[2] University of British Columbia
rxg@cs.ubc.ca

Abstract. Gradual typing enables both static and dynamic typing in
the same program and makes it convenient to migrate code regions between the two typing disciplines. One goal of gradual typing is to provide all the benefits of static typing, such as efficiency, in statically-typed
regions. However, this goal is elusive: the standard approach to mutable references imposes run-time overhead in statically-typed regions and
alternative approaches are too conservative, either statically or at run-
time. In this paper we present a new semantics called *monotonic refer-*
ences which imposes none of the run-time overhead of dynamic typing in
statically typed regions. With this design, casting a reference may cause
a heap cell to become more statically typed (but not less). Retaining
type safety is challenging with strong updates to the heap. Nevertheless,
we have a mechanized proof of type safety. Further, we present blame
tracking for monotonic references and prove a blame theorem.

1 Introduction

Static and dynamic type systems have well-known strengths and weaknesses.
Static type systems provide machine-checked documentation, catch bugs early,
and enable efficient code. Dynamic type systems provide the flexibility often
needed during prototyping and enable powerful features such as reflection. Over
the years, many languages blurred the boundary between static and dynamic typing, such as type hints in Lisp and the addition of a dynamic type to otherwise
statically typed languages (Abadi et al., 1989). But the seamless and sound integration of static and dynamic typing remained problematic until two pieces fell
into place: the gradual type system of Siek and Taha (2006) and the blame theorems of Tobin-Hochstadt and Felleisen (2006) and Wadler and Findler (2009).

However, there are challenges regarding the efficiency of gradual typing. One
issue concerns mutable references in statically-typed regions of code. Consider
the following statically-typed function f that dereferences its parameter x.

$$\texttt{let } f = \lambda x{:}\textsf{Ref Int. } !x \textsf{ in}$$
$$f(\textsf{ref } 4);$$
$$f(\textsf{ref } (4 \textsf{ as } \star))$$

© Springer-Verlag Berlin Heidelberg 2015
J. Vitek (Ed.): ESOP 2015, LNCS 9032, pp. 432–456, 2015.
DOI: 10.1007/978-3-662-46669-8_18

In the first application of f, a normal reference to an integer flows into f. For the second application, we allocate a reference of type Ref $\star$ ($\star$ is the dynamic type) then implicitly cast it to Ref Int before applying f. According to the semantics of Herman et al. (2007), this cast wraps the reference in a proxy which performs dynamic checks on reads and writes. Thus code generated for the dereference in the body of f must inspect the reference to find out whether it is a normal reference or a proxied reference, and in the proxied case, apply a coercion.

Before discussing solutions to this problem, we recall the *gradual guarantee* of Boyland (2014) and Siek et al. (2015), an important property of the standard semantics for mutable references, and of gradual typing in general. The gradual guarantee promises that removing type annotations, or changing type annotations to be less precise, does not affect the behavior of a program: it should still type check and the result should be the same modulo proxies. (Adding or making type annotations more precise, on the other hand can sometimes induce static type errors and runtime cast errors.) Consider the statically-typed program on the left that allocates a reference to an integer and then dereferences it from within a function. In the code on the right, we change the annotation on h from Ref Int to Ref $\star$, but the program still type checks and the result remains 42.

$$\begin{array}{ll} \texttt{let } r = \texttt{ref } 42 \texttt{ in} & \texttt{let } r = \texttt{ref } 42 \texttt{ in} \\ \texttt{let } f = \lambda h{:}\texttt{Ref Int}.\,!h \implies \texttt{let } f = \lambda h{:}\texttt{Ref } \star.\,!h \\ \texttt{in } f(r) & \texttt{in } f(r) \end{array}$$

Wrigstad et al. (2010) address the efficiency problem by introducing a distinction between *like types* and *concrete types*. Concrete types are the usual types of a statically-typed language and incur zero run-time overhead, but dynamically-typed values cannot flow into concrete types. Like types, on the other hand, may refer to dynamically-typed values but incur run-time overhead. The distinction between like types and concrete types achieves the efficiency goals, but the restrictions in their type system mean that removing concrete type annotations, as in the above example, can trigger a static type error.

In this paper we investigate this run-time overhead problem in the context of the gradually-typed lambda calculus with mutable references. We propose a semantics, *monotonic references*, that enables the compilation of statically-typed regions to machine code that is free of any of the indirection or run-time checking associated with dynamic typing, like boxing or bit tags. Monotonic references allow dynamically-typed values to flow into code with (concrete) static types. When a reference flows through a cast, the cast may coerce its underlying heap cell to become more statically typed. In general, this means that values in the heap may evolve monotonically with respect to the precision relation (Section 2). The idea for monotonic references came out of our work on implementing and evaluating gradual typing for Python (Vitousek et al., 2014).

Monotonic references preserve a global invariant that a value in the heap is at least as precise as any reference that points to it. Thus, a static reference always points to a value of the same type, so there is no overhead associated with reading or writing through the reference: the reads and writes may be implemented as machine loads and stores. By a *static* reference we mean that there are no occurrences of the dynamic type $\star$ in the pointed-to type of the reference, such as `Ref Int` and `Ref (Int × Bool)`. Reads and writes to references that are not static, such as `Ref ⋆` and `Ref (⋆ × Bool)`, still require casts: the dynamic regions of code have to pay their own way. The intermediate representation that we compile to contains different instructions for fast, static loads and stores versus non-static loads and stores that require casts.

Swamy et al. (2014) and Rastogi et al. (2014) integrate static and dynamic typing in the context of TypeScript with the TS* and Safe TypeScript languages. Both use a notion of monotonicity in the heap, but with respect to subtyping, treating $\star$ as a universal supertype, instead of with respect to the precision relation. Because these languages compile to JavaScript, they inherit the overhead of dynamic typing, whereas with monotonic references, the overhead of dynamic typing occurs only in dynamically-typed code. In the example above, making the type annotation on h less precise causes TS* to halt the program with a cast error at the implicit cast from `Ref Int` to `Ref ⋆`. TS* does not allow casts from one mutable reference type to a different one because its references are invariant with respect to subtyping. Thus, TS* does not satisfy the gradual guarantee.

In gradually-typed languages with higher-order features such as first-class functions and objects, blame tracking plays an important role in providing meaningful error messages when casts fail. Blame tracking enables fine-grained guarantees, via a blame theorem, regarding which regions of the code are statically type safe. In this paper we present blame tracking for monotonic references and prove a blame theorem. Our design uses the labeled types of Siek and Wadler (2010) as run-time type information (RTTI), together with three new operations on labeled types: a bidirectional cast operator that captures the dual read/write nature of mutable references, a merge operator that models how casts on separate aliases to the same heap cell interact over time, and an operator that casts heap cells between labeled types.

To summarize, this paper presents a new semantics for gradually-typed mutable references that delivers guaranteed efficiency for the statically-typed parts of a program, maintains type safety, and provides blame tracking, while continuing to enable fine-grained migration between static and dynamic code. This paper makes the following technical contributions:

1. We define the semantics of monotonic references (Sections 3 and 5).
2. We discuss our proof of type safety, mechanized in Isabelle (Section 4).
3. We augment monotonic references with blame tracking and prove the blame-subtyping theorem (Section 6).

We review the gradually-typed lambda calculus with references in Section 2 and discuss the run-time overhead associated with mutable references. We address an implementation concern regarding strong updates in Section 7. The paper concludes in Section 9.

2   Background and Problem Statement

Figure 1 reviews the syntax and static semantics of the gradually-typed lambda calculus with references. The primary difference between gradual typing and static typing is that uses of type equality are replaced with *consistency* (aka. compatibility), also defined in Figure 1. The consistency relation enables implicit casts to and from ⋆. (In contrast, an object-oriented language only allows implicit casts to the top `Object` type.) This consistency relation is a congruence, even for reference types (Herman et al., 2007), which differs from the original treatment of references as invariant (Siek and Taha, 2006). The more flexible treatment of references enables the passing of references between more and less dynamically typed regions of code, but is also the source of the difficulties that we solve in this paper. The precision relation, which says whether one type is more or less dynamic than another, is also defined in Figure 1, and is closely related to consistency. Two types are consistent when there exists a greatest lower bound with respect to the precision relation. This relation is also known as naïve subtyping (Wadler and Findler, 2009).

All of the types, except for ⋆, classify unboxed values. So, for example, `Int` is the type for native integers (e.g. 64-bit integers). The auxiliary relations *fun, pair*, and *ref*, defined in Figure 1, implement pattern matching on types, enabling a more concise presentation of the typing rules compared to prior presentations of gradual type systems. Labels ℓ represent source code locations that are captured during parsing.

The dynamic semantics of the gradually-typed lambda calculus is defined by a type-directed translation to the coercion calculus (Henglein, 1994), using the standard semantics for mutable references due to Herman et al. (2007).

Each use of consistency between types T_1 and T_2 in the type system, and each use of one of the auxiliary relations, becomes an explicit cast from T_1 to T_2. The coercion calculus expresses casts in terms of combinators that say how to cast from one type to another. Figure 2 gives the compilation of casts into coercions, written $(T \Rightarrow^\ell T) = c$. The compilation of gradually-typed terms into the coercion-based calculus is otherwise straightforward, so we give just the function application rule as an example:

$$\frac{\Gamma \vdash e_1 \rightsquigarrow e_1' : T_1 \qquad \Gamma \vdash e_2 \rightsquigarrow e_2' : T_2 \\ fun(T_1, T_{11}, T_{12}) \qquad T_2 \sim T_{11} \\ (T_1 \Rightarrow^\ell T_{11} \rightarrow T_{12}) = c_1 \qquad (T_2 \Rightarrow^\ell T_{11}) = c_2}{\Gamma \vdash (e_1\ e_2)^\ell \rightsquigarrow e_1'\langle c_1\rangle\ e_2'\langle c_2\rangle : T_{12}}$$

Figures 3 and 4 define the coercion-based calculus. We highlight the parts of the definition related to references, as they are of particular interest here. We review the coercion calculus in the context of discussing the run-time overhead problem in the next subsection. For an introduction to the coercion calculus, we refer to Henglein (1994).

Syntax

$$\begin{array}{lll}
\text{Base types} & B & ::= \text{Int} \mid \text{Bool} \\
\text{Types} & T & ::= B \mid T \to T \mid T \times T \mid \text{Ref}\, T \mid \star \\
\text{Labels} & \ell & \\
\text{Operators} & op & ::= \text{plus} \mid \text{minus} \mid \text{is} \mid \cdots \\
\text{Expressions} & e & ::= k \mid op^\ell(\vec{e}) \mid x \mid \lambda x{:}T.\,e \mid (e\,e)^\ell \mid e\,\text{as}^\ell\, T \mid \\
& & \quad (e,e) \mid \text{fst}^\ell e \mid \text{snd}^\ell e \mid \text{ref}\, e \mid !^\ell e \mid e :=^\ell e
\end{array}$$

$$\lambda x.\,e \equiv \lambda x{:}\star.\,e$$

Consistency $\boxed{T \sim T}$

$$\frac{}{\star \sim T} \qquad \frac{}{T \sim \star} \qquad \frac{}{B \sim B} \qquad \frac{T_1 \sim T_2}{\text{Ref}\, T_1 \sim \text{Ref}\, T_2}$$

$$\frac{T_1 \sim T_3 \quad T_2 \sim T_4}{T_1 \to T_2 \sim T_3 \to T_4} \qquad \frac{T_1 \sim T_3 \quad T_2 \sim T_4}{T_1 \times T_2 \sim T_3 \times T_4}$$

Precision $\boxed{T \sqsubseteq T}$

$$T \sqsubseteq \star \quad B \sqsubseteq B \quad \frac{T_1 \sqsubseteq T_2}{\text{Ref}\, T_1 \sqsubseteq \text{Ref}\, T_2}$$

$$\frac{T_1 \sqsubseteq T_3 \quad T_2 \sqsubseteq T_4}{T_1 \to T_2 \sqsubseteq T_3 \to T_4} \qquad \frac{T_1 \sqsubseteq T_3 \quad T_2 \sqsubseteq T_4}{T_1 \times T_2 \sqsubseteq T_3 \times T_4}$$

Expression typing $\boxed{\Gamma \vdash e : T}$

$$\frac{k : B}{\Gamma \vdash k : B} \qquad \frac{\Gamma \vdash \vec{e} : \vec{T} \quad op : \vec{B} \to B \quad \vec{T} \sim \vec{B}}{\Gamma \vdash op^\ell(\vec{e}) : B} \qquad \frac{\Gamma \vdash e : T_1 \quad T_1 \sim T_2}{\Gamma \vdash e\,\text{as}^\ell\, T_2 : T_2}$$

$$\frac{\Gamma(x) = T}{\Gamma \vdash x : T} \qquad \frac{\Gamma(x \mapsto T_1) \vdash e : T_2}{\Gamma \vdash \lambda x{:}T_1.\,e : T_1 \to T_2} \qquad \frac{\Gamma \vdash e_1 : T_1 \quad \Gamma \vdash e_2 : T_2 \quad fun(T_1, T_{11}, T_{12}) \quad T_2 \sim T_{11}}{\Gamma \vdash (e_1\,e_2)^\ell : T_{12}}$$

$$\frac{\Gamma \vdash e_1 : T_1 \quad \Gamma \vdash e_2 : T_2}{\Gamma \vdash (e_1, e_2) : T_1 \times T_2} \qquad \frac{\Gamma \vdash e : T \quad pair(T, T_1, T_2)}{\Gamma \vdash \text{fst}^\ell e : T_1} \qquad \frac{\Gamma \vdash e : T \quad pair(T, T_1, T_2)}{\Gamma \vdash \text{snd}^\ell e : T_2}$$

$$\frac{\Gamma \vdash e : T}{\Gamma \vdash \text{ref}\, e : \text{Ref}\, T} \qquad \frac{\Gamma \vdash e : T \quad ref(T, T')}{\Gamma \vdash !^\ell e : T'} \qquad \frac{\Gamma \vdash e_1 : T_1 \quad \Gamma \vdash e_2 : T_2 \quad ref(T_1, T_1') \quad T_2 \sim T_1'}{\Gamma \vdash e_1 :=^\ell e_2 : T_1}$$

Type matching

$$\frac{}{fun(T_{11} \to T_{12}, T_{11}, T_{12})} \qquad \frac{}{fun(\star, \star, \star)}$$

$$\frac{}{pair(T_{11} \times T_{12}, T_{11}, T_{12})} \qquad \frac{}{pair(\star, \star, \star)}$$

$$\frac{}{ref(\text{Ref}\, T, T)} \qquad \frac{}{ref(\star, \star)}$$

Fig. 1. Gradually-typed λ calculus with mutable references

$$\boxed{(T \Rightarrow^\ell T) = c}$$

$$(B \Rightarrow^\ell B) = \iota \qquad (I \Rightarrow^\ell \star) = I!$$
$$(\star \Rightarrow^\ell \star) = \iota \qquad (\star \Rightarrow^\ell I) = I?^\ell$$

$$(T_1 \to T_2) \Rightarrow^\ell (T_1' \to T_2') = (T_1' \Rightarrow^\ell T_1) \to (T_2 \Rightarrow^\ell T_2')$$
$$(T_1 \times T_2) \Rightarrow^\ell (T_1' \times T_2') = (T_1 \Rightarrow^\ell T_1') \times (T_2 \Rightarrow^\ell T_2')$$
$$\mathtt{Ref}\, T \Rightarrow^\ell \mathtt{Ref}\, T' = \mathtt{Ref}\, (T \Rightarrow^\ell T')\, (T' \Rightarrow^\ell T)$$

Fig. 2. Compile casts to coercions

Expressions	e	$::= k \mid op(\vec{e}) \mid x \mid \lambda x.e \mid e\, e \mid (e,e) \mid \mathtt{fst}\, e \mid \mathtt{snd}\, e \mid$
		$\mathtt{ref}\, e \mid !e \mid e := e \mid e\langle c\rangle \mid \mathtt{blame}\, \ell$
Injectibles	I	$::= B \mid T \to T \mid T \times T \mid \mathtt{Ref}\, T$
Coercions	c	$::= \iota \mid I?^\ell \mid I! \mid c \to c \mid c \times c \mid c\,;c \mid \mathtt{Ref}\, c\,c$
Values	v	$::= k \mid \lambda x.e \mid (v,v) \mid v\langle I!\rangle \mid a \mid v\langle\mathtt{Ref}\, c\,c\rangle$
Heap	μ	$::= \emptyset \mid \mu(a \mapsto v)$
Heap Typing	Σ	$::= \emptyset \mid \Sigma(a \mapsto T)$
Frames	F	$::= op(\vec{v},\square,\vec{e}) \mid \square\, e \mid v\, \square \mid (\square,e) \mid (v,\square) \mid \mathtt{fst}\, \square \mid \mathtt{snd}\, \square \mid$
		$\mathtt{ref}\, \square \mid !\square \mid \square := e \mid v := \square \mid \square\langle c\rangle$

Fig. 3. Syntax for the coercion-based calculus with mutable references

2.1 Run-Time Overhead in Fully-Static Code

Recall the example in Section 1 in which the dereference of a statically-typed reference must first check whether the reference is proxied or not.

```
let f = λx:Ref Int. !x in
f(ref 4);
let r = ref (4 as ⋆) in f(r)
```

The overhead can be seen in the dynamic semantics (Figure 4), where there are two reduction rules for dereferencing: (DEREF) and (DEREFCAST), and two reduction rules for updating references: (UPDATE) and (UPDATECAST). Another way to look at this problem is that there are two canonical forms of type Ref Int, a plain address a and also a value wrapped in a reference coercion, $v\langle\mathtt{Ref}\, c_1\, c_2\rangle$, so operations on values of this type need to dispatch on which form occurs at runtime. To eliminate this overhead we need a design with only a single canonical form for values of reference type.

The run-time overhead for references affects every read and write to the heap and is particularly detrimental in tight loops over arrays. When adding support for contracts to mutable data structures in Racket, Strickland et al. (2012, Figure 9) measured this overhead at approximately 25% for fully-typed code on a bubble-sort microbenchmark.

Coercion typing $\boxed{c : T \Rightarrow T}$

$$\frac{}{\iota : T \Rightarrow T} \qquad \frac{c_1 : T_3 \Rightarrow T_1 \quad c_2 : T_2 \Rightarrow T_4}{c_1 \rightarrow c_2 : (T_1 \rightarrow T_2) \Rightarrow (T_3 \rightarrow T_4)}$$

$$\frac{}{I?^\ell : \star \Rightarrow I} \qquad \frac{c_1 : T_1 \Rightarrow T_3 \quad c_2 : T_2 \Rightarrow T_4}{c_1 \times c_2 : (T_1 \times T_2) \Rightarrow (T_3 \times T_4)}$$

$$\frac{}{I! : I \Rightarrow \star} \qquad \frac{c_1 : T_1 \Rightarrow T_2 \quad c_2 : T_2 \Rightarrow T_3}{c_1 \,;\, c_2 : T_1 \Rightarrow T_3}$$

$$\frac{c_1 : T_1 \Rightarrow T_2 \quad c_2 : T_2 \Rightarrow T_1}{\mathbf{Ref}\, c_1\, c_2 : \mathbf{Ref}\, T_1 \Rightarrow \mathbf{Ref}\, T_2}$$

Expression typing $\boxed{\Gamma; \Sigma \vdash e : T}$

$$\cdots \qquad \frac{\Sigma(a) = T}{\Gamma; \Sigma \vdash a : T} \qquad \frac{\Gamma; \Sigma \vdash e : T_1 \quad c : T_1 \Rightarrow T_2}{\Gamma; \Sigma \vdash e\langle c \rangle : T_2}$$

Reduction rules for functions, primitives, and pairs $\boxed{e \longrightarrow e}$

$$(\lambda x.\, e)\, v \longrightarrow [x := v]e \qquad \mathbf{fst}\,(v_1, v_2) \longrightarrow v_1$$
$$op(\vec{k}) \quad \longrightarrow \delta(op, \vec{k}) \qquad \mathbf{snd}\,(v_1, v_2) \longrightarrow v_2$$

Cast reduction rules $\boxed{e \longrightarrow_c e}$

$$v\langle \iota \rangle \longrightarrow_c v$$
$$v\langle I_1! \rangle \langle I_2?^\ell \rangle \longrightarrow_c v\langle I_1 \Rightarrow^\ell I_2 \rangle \quad \text{if } I_1 \sim I_2$$
$$v\langle I_1! \rangle \langle I_2?^\ell \rangle \longrightarrow_c \mathbf{blame}\, \ell \quad \text{if } I_1 \not\sim I_2$$
$$v\langle c_1 \rightarrow c_2 \rangle \longrightarrow_c \lambda x.\, v\, (x\langle c_1 \rangle)\langle c_2 \rangle$$
$$(v_1, v_2)\langle c_1 \times c_2 \rangle \longrightarrow_c (v_1\langle c_1 \rangle, v_2\langle c_2 \rangle)$$
$$v\langle c_1 \,;\, c_2 \rangle \longrightarrow_c v\langle c_1 \rangle\langle c_2 \rangle$$

Reference reduction rules $\boxed{e, \mu \longrightarrow_r e, \mu}$

$$\mathbf{ref}\, v, \mu \longrightarrow_r a, \mu(a \mapsto v) \qquad \text{if } a \notin dom(\mu) \qquad \text{(ALLOCREF)}$$
$$!a, \mu \longrightarrow_r \mu(a), \mu \qquad\qquad\qquad\qquad\qquad \text{(DEREF)}$$
$$!(v\langle \mathbf{Ref}\, c_1\, c_2 \rangle), \mu \longrightarrow_r (!v)\langle c_1 \rangle, \mu \qquad\quad\; \text{(DEREFCAST)}$$
$$a := v, \mu \longrightarrow_r a, \mu(a \mapsto v) \qquad\qquad\qquad \text{(UPDATE)}$$
$$v_1\langle \mathbf{Ref}\, c_1\, c_2 \rangle := v_2, \mu \longrightarrow_r v_1 := v_2\langle c_2 \rangle, \mu \qquad \text{(UPDATECAST)}$$

State reduction rules

$$\frac{e \longrightarrow e'}{e, \mu \longrightarrow e', \mu} \qquad \frac{e \longrightarrow_c e'}{e, \mu \longrightarrow e', \mu} \qquad \frac{e, \mu \longrightarrow_r e', \mu'}{e, \mu \longrightarrow e', \mu'}$$

$$\frac{e, \mu \longrightarrow e', \mu'}{F[e], \mu \longrightarrow F[e'], \mu'} \qquad \frac{}{F[\mathbf{blame}\, \ell], \mu \longrightarrow \mathbf{blame}\, \ell, \mu}$$

Fig. 4. Coercion-based calculus with mutable references

2.2   Non-determinism in Multi-threaded Code

This standard semantics for mutable references produces an error only if type inconsistency is witnessed by some read or write to a particular reference, so in a non-deterministic multi-threaded program, whether a check will fail at run-time is difficult to predict.

The contract system in Racket implements the standard semantics (Flatt and PLT, 2014). For example, the following program sometimes fails and blames b1, sometimes fails and blames b2, and sometimes succeeds, as explained below.

```
#lang racket
(define b (box #f))
(define/contract b1 (box/c integer?) b)
(define/contract b2 (box/c string?)  b)

(thread (lambda ()
          (for ([i 2])
            (set-box! b1 5)
            (sleep 0.000000001)
            (add1 (unbox b1)))))
(thread (lambda ()
          (for ([i 2])
            (set-box! b2 "hello")
            (sleep 0.000000001)
            (string-append "world" (unbox b2)))))
```

The program creates a single heap cell b, and accesses it through two distinct proxies, b1 and b2, each with its own dynamic check. When the two threads do not interleave, the program succeeds, but if the second thread changes b2 to contain a string between the set-box! and unbox calls for b1, the system halts, blaming one of the parties.

In contrast, if box/c implemented monotonic references, then an error would *deterministically* occur when define/contract is used for the second time.

3   Monotonic References Without Blame

Figures 5 and 6 define the syntax and semantics of our new coercion calculus with monotonic references, but without blame. Figure 8 defines the compilation of casts to monotonic coercions, also without blame. The addition of blame adds considerable complexity, so we postpone its treatment to Section 5. Typical of gradually-typed languages, there is a value form for values that have been boxed and injected to $\star$, which is $v\langle I!\rangle$. The I plays the role of a tag that records the type of v. The values at all other types are unboxed, as they would be in a statically-typed language.

With monotonic references, only one kind of value has reference type: normal addresses. When a cast is applied to a reference, instead of wrapping the reference with a cast, we cast the underlying value on the heap. To make sure that the new type of the value is consistent with all the outstanding references, we require that a cast only make the type of the value more precise (Figure 1). Otherwise the cast results in a run-time error. Thus, we maintain the heap invariant that the type of each reference in the program is less or equally precise as the type of the value on the heap that it points to, as captured in the typing rule (WTREF).

One might wonder why our heap invariant uses the precision relation instead of subtyping. Could we obtain the same efficiency goals using subtyping instead? Consider the following program in which a function of type $\star{\rightarrow}\mathtt{Int}$ is referenced from the static type $\mathtt{Int}{\rightarrow}\mathtt{Int}$. (We have $\star \rightarrow \mathtt{Int} <: \mathtt{Int} \rightarrow \mathtt{Int}$.)

$$
\begin{aligned}
&\mathtt{let}\ r_1 = \mathtt{ref}\ (\lambda x : \star.\ x\ \mathtt{as\ Int})\ \mathtt{in} \\
&\mathtt{let}\ r_2 = (r_1\ \mathtt{as\ Ref}\ (\mathtt{Int} \rightarrow \mathtt{Int}))\ \mathtt{in} \\
&!r_2\ 42
\end{aligned}
$$

The dereference of r_2 should not require overhead, but we have a function of type $\star{\rightarrow}\mathtt{Int}$ that is to be applied to an integer, and the conversion from $\mathtt{Int}$ to $\star$ requires boxing in our setting. Thus, the dereference of r_2 is not simply a load instruction, but it must handle the casting from $\star{\rightarrow}\mathtt{Int}$ to $\mathtt{Int}{\rightarrow}\mathtt{Int}$. (Other systems, such as Reticulated Python and TS*, box all values. In these systems, upcasts on dereferences are unnecessary, but instead overhead is incurred in nearly every operation.) In general, given a reference of type $\mathtt{Ref}\ T_2$, even when T_2 is a static type, there are many types T_1 such that $T_1 <: T_2$ and $T_1 \neq T_2$.

The syntax of the monotonic calculus differs from the standard calculus in that there are two kinds of dereference and update expressions. Programmers need not worry about choosing which of the two dereference or update expressions to use because this choice is type-directed and therefore is handled during compilation from the source language to the coercion calculus. We reserve the forms $!e$ and $e_1 := e_2$ for situations in which the reference type is fully static. In these situations we know that the value in the heap has the same type as the reference. Thus, if a reference has a fully static type, such as $\mathtt{Ref\ Int}$, the corresponding value on the heap must be an actual integer (and not an injection to $\star$), so we need only one reduction rule for dereferencing a fully-static reference (DEREFM), and one rule for updating a fully-static reference (UPDM).

For expressions of reference type that are not fully-static, we introduce the syntactic forms $!e@T$ and $e_1 := e_2@T$ for dereference and update, respectively. The type annotation T records the compile-time type of e, that is, e has type $\mathtt{Ref}\ T$. For example, T could be $\star$, $\star \times \star$, or $\star \times \mathtt{Int}$. Because the value on the heap might be more precise than T, a cast is needed to mediate between T and the run-time type of the heap cell.

The reduction rule (DYNDEREFM) casts from the addresses' run-time type, which we store next to the heap cell, to the compile-time type T. We write $\mu(a)_{\mathsf{rtti}}$ for the run-time type information for reference a and we write $\mu(a)_{\mathsf{val}}$ for the value in the heap cell. The reduction rule (DYNUPDM) casts the incoming value v from T to the address's run-time type, so the new content of the cell

Expressions	e	$::=$	$.. \mid \mathbf{ref}_T\, e \mid !e@T \mid e := e@T \mid \mathbf{error}$
Coercions	c	$::=$	$\iota \mid I? \mid I! \mid c{\rightarrow}c \mid c \times c \mid c\,;c \mid \mathbf{Ref}\,T$
Values	v	$::=$	$k \mid \lambda x.\,e \mid (v,v) \mid v\langle I!\rangle \mid a$
Casted Values	cv	$::=$	$v \mid v\langle c\rangle \mid (cv,cv)$
Heap	μ	$::=$	$\emptyset \mid \mu(a \mapsto v : T)$
Evolving Heap	ν	$::=$	$\emptyset \mid \nu(a \mapsto cv : T)$
Frames	F	$::=$	$.. \mid !\square@T \mid \square := e@T \mid v := \square@T$

Fig. 5. Syntax for monotonic references without blame

is $cv = v\langle T \Rightarrow \mu(a)_{\mathsf{rtti}}\rangle$. This cv is not a value yet, so storing it in the heap is unusual. In earlier versions of the semantics we tried to reduce cv to a value before storing it in the heap, but there are complications that force this design, which we discuss later in this section . To summarize our treatment of dereference and update, we present efficient semantics for the fully-static dereference and update but have slightly increased the overhead for dynamic dereferences and updates. This is a price we are willing to pay to have dynamic typing "pay its own way".

The crux of the monotonic semantics is in the reduction rules that apply a reference coercion to an address: (CASTREF1), (CASTREF2), and (CASTREF3). In (CASTREF1) we have an address that maps to cv of type T_1 and we cast cv so that it is no more dynamic than (i.e. at least as static as) both the target type T_2 and all of the existing references to the cell. To accomplish this, we take the greatest lower bound $T_3 = T_1 \sqcap T_2$ (Figure 7) to be the new type of the cell, so the new contents is $cv' = cv\langle T_1 \Rightarrow T_3\rangle$. There are two side conditions on (CASTREF1): $T_1 \sqcap T_2$ must be defined and $T_3 \neq T_1$. If $T_1 \sqcap T_2$ is undefined, or equivalently, if $T_1 \not\sim T_2$, we instead signal an error, as handled by (CASTREF3). If $T_3 = T_1$, then there is no need to cast cv, which is handled by (CASTREF2).

The last coercion reduction rule (PURECAST) imports the reduction rules from the standard semantics (Figure 4) though here we ignore blame, i.e., replace $\mathbf{blame}\,\ell$ with $\mathbf{error}$, $I_2?^\ell$ with $I_2?$, and $I_1 \Rightarrow^\ell I_2$ with $I_1 \Rightarrow I_2$.

The meet function defined in Figure 7 computes the greatest lower bound with respect to the precision relation.

To motivate our organization of the heap, we present two examples that demonstrate why we store run-time type information and casted values, not just values, on the heap.

Cycles and termination. The first complication is that there can be cycles in the heap and we need to make sure that when we apply a cast to an address in a cycle, the cast terminates. Consider the following example in which we create a pair whose second element is a reference back to itself.

```
let r₁ = ref (42, 0 as ⋆) in
r₁ := (42, r₁ as ⋆);
let r₂ = r₁ as Ref (Int × Ref ⋆)in
fst !r₂
```

Once the cycle is established, we cast r_1 from type $\mathbf{Ref}\,(\mathbf{Int} \times \star)$ to $\mathbf{Ref}\,(\mathbf{Int} \times \mathbf{Ref}\,\star)$. The presence of the nested $\mathbf{Ref}\,\star$ in the target type means that the cast

Expression typing $\qquad\qquad\qquad\qquad\qquad\qquad\qquad\qquad\boxed{\Gamma; \Sigma \vdash e : T}$

$$\frac{\Gamma; \Sigma \vdash e : \mathbf{Ref}\, T \quad static\ T}{\Gamma; \Sigma \vdash\, !e : T} \qquad \frac{\Gamma; \Sigma \vdash e_1 : \mathbf{Ref}\, T \quad \Gamma; \Sigma \vdash e_2 : T \quad static\ T}{\Gamma; \Sigma \vdash e_1 := e_2 : \mathbf{Ref}\, T} \qquad \frac{\Gamma; \Sigma \vdash e : \mathbf{Ref}\, T}{\Gamma; \Sigma \vdash\, !e@T : T}$$

$$\frac{\Gamma; \Sigma \vdash e_1 : \mathbf{Ref}\, T \quad \Gamma; \Sigma \vdash e_2 : T}{\Gamma; \Sigma \vdash e_1 := e_2@T : \mathbf{Ref}\, T} \quad \cdots \quad \frac{\Sigma(a) \sqsubseteq T_2}{\Gamma; \Sigma \vdash a : T_2} \qquad \text{(WTREF)}$$

Cast reduction rules $\qquad\qquad\qquad\qquad\qquad\qquad\qquad\qquad\boxed{e, \nu \longrightarrow_{cr} e, \nu}$

$$\frac{e \longrightarrow_c e'}{e, \nu \longrightarrow_{cr} e', \nu} \qquad\qquad \text{(PURECAST)}$$

$$\frac{\nu(a) = cv : T_1 \quad T_3 = T_1 \sqcap T_2}{T_3 \neq T_1 \quad cv' = cv\langle T_1 \Rightarrow T_3\rangle}{a\langle \mathbf{Ref}\, T_2\rangle, \nu \longrightarrow_{cr} a, \nu(a \mapsto cv' : T_3)} \qquad \text{(CASTREF1)}$$

$$\frac{\nu(a) = cv : T_1 \quad T_1 = T_1 \sqcap T_2}{a\langle \mathbf{Ref}\, T_2\rangle, \nu \longrightarrow_{cr} a, \nu} \qquad \text{(CASTREF2)}$$

$$\frac{\nu(a) = cv : T_1 \quad T_1 \not\sim T_2}{a\langle \mathbf{Ref}\, T_2\rangle, \nu \longrightarrow_{cr} \mathbf{error}, \nu} \qquad \text{(CASTREF3)}$$

Program reduction rules $\qquad\qquad\qquad\qquad\qquad\qquad\qquad\boxed{e, \mu \longrightarrow_e e, \nu}$

$$e, \mu \longrightarrow_e e', \mu \qquad \text{if } e \longrightarrow e'$$

$$\mathbf{ref}_T\, v, \mu \longrightarrow_e a, \mu(a \mapsto v : T) \qquad \text{if } a \notin dom(\mu)$$

$$!a, \mu \longrightarrow_e \mu(a)_{\mathsf{val}}, \mu \qquad\qquad\qquad\qquad \text{(DEREFM)}$$

$$!a@T, \mu \longrightarrow_e \mu(a)_{\mathsf{val}}\langle \mu(a)_{\mathsf{rtti}} \Rightarrow T\rangle, \mu \qquad \text{(DYNDEREFM)}$$

$$a := v, \mu \longrightarrow_e a, \mu(a \mapsto v : \mu(a)_{\mathsf{rtti}}) \qquad\quad \text{(UPDM)}$$

$$a := v@T, \mu \longrightarrow_e a, \mu(a \mapsto cv : \mu(a)_{\mathsf{rtti}}) \qquad \text{(DYNUPDM)}$$

$$\text{where } cv = v\langle T \Rightarrow \mu(a)_{\mathsf{rtti}}\rangle$$

For $X \in \{cr, e\}$:

$$\frac{e, \nu \longrightarrow_X e', \nu'}{F[e], \nu \longrightarrow_X F[e'], \nu'} \qquad F[\mathbf{error}], \nu \longrightarrow_X \mathbf{error}, \nu$$

State reduction rules $\qquad\qquad\qquad\qquad\qquad\qquad\qquad\qquad\boxed{e, \nu \longrightarrow e, \nu}$

$$\frac{e, \mu \longrightarrow_X e', \nu \quad X \in \{cr, e\}}{e, \mu \longrightarrow e', \nu} \qquad \frac{\nu(a) = cv : T \quad cv, \nu \longrightarrow_{cr} cv', \nu'}{\nu'(a)_{\mathsf{rtti}} = T}{e, \nu \longrightarrow e, \nu'(a \mapsto cv' : T)} \qquad \text{(HCAST)}$$

$$\frac{\nu(a) = cv : T \quad cv, \nu \longrightarrow_{cr} \mathbf{error}, \nu'}{e, \nu \longrightarrow \mathbf{error}, \nu'} \qquad \frac{\nu(a) = cv : T \quad cv, \nu \longrightarrow_{cr} cv', \nu'}{\nu'(a)_{\mathsf{rtti}} \neq T}{e, \nu \longrightarrow e, \nu'}$$

$$\text{(HDROP)}$$

Fig. 6. Monotonic references without blame

$$\boxed{T \sqcap T = T}$$

$$\star \sqcap T = T$$
$$T \sqcap \star = T$$
$$B \sqcap B = B$$

$$(T_1 \times T_2) \sqcap (T_3 \times T_4) = (T_1 \sqcap T_3) \times (T_2 \sqcap T_4)$$
$$(T_1 \to T_2) \sqcap (T_3 \to T_4) = (T_1 \sqcap T_3) \to (T_2 \sqcap T_4)$$

Fig. 7. The meet function (greatest lower bound)

on r_1 will trigger another cast on r_1. The correct result of this program is 42 but a naïve dynamic semantics would diverge. Our semantics avoids divergence by checking whether the new run-time type is equal to the old run-time type; in such cases the heap cell is left unchanged (see rule (CASTREF2)).

Casted values in the heap. Consider the following example in which we create a triple of type $\star \times \star \times \star$ whose third element is a reference back to itself.

```
let r₀ = ref (42 as ⋆, 7 as ⋆, 0 as ⋆)in
r₀ := (42 as ⋆, 7 as ⋆, r₀ as ⋆);
let r₁ = r₀ as Ref (Int × ⋆ × Ref (Int × Int × ⋆))in
fst (fst !r₁)
```

Suppose a_0 is the address created in the allocation on the first line. On line three we cast a_0 in such a way that we trigger two casts on a_0. Consider the action of these casts on just the first two elements of the triple, we have:

$$\star \times \star \Rightarrow \mathtt{Int} \times \star \Rightarrow \mathtt{Int} \times \mathtt{Int}$$

The second cast occurs while the first is still in progress. Now, suppose we delayed updating the heap cell until we finished reducing to a value. At the moment when we apply the second cast, we would still have the original value, of type $\star \times \star \times \star$, in the heap. This is problematic because our next step would be to apply a cast from $\mathtt{Int} \times \star \Rightarrow \mathtt{Int} \times \mathtt{Int}$ to this value, but the value's type and the source type of the cast don't match! In fact, in this example the result would be incorrect; we would get $42\langle\mathtt{Int!}\rangle$ instead of 42.

There are several solutions to this problem, and they all require storing more information on the heap or as a separate map. Here we take the most straightforward approach of immediately updating the heap with casted values, that is, with values that are in the process of being cast.

We walk through the execution of the above example, explaining our rules for reducing casted values in the heap and showing snapshots of the heap. We use the following abbreviations.

$$T_0 = \star \times \star \times \star$$
$$T_1 = \mathtt{Int} \times \star \times \mathtt{Ref}\, T_2$$
$$T_2 = \mathtt{Int} \times \mathtt{Int} \times \star$$
$$c = \mathtt{Int?} \times \iota \times (\mathtt{Ref}\, T_2)?$$

The first line of the program allocates a triple.

$$a_0 \mapsto (42\langle \text{Int}!\rangle, 7\langle \text{Int}!\rangle, 0\langle \text{Int}!\rangle) : T_0$$

The second line sets the third element to be a reference to itself.

$$a_0 \mapsto (42\langle \text{Int}!\rangle, 7\langle \text{Int}!\rangle, a_0\langle (\text{Ref } T_0)!\rangle) : T_0$$

The third line casts the reference to $\text{Ref } T_1$ via (CASTREF1).

$$a_0 \mapsto (42\langle \text{Int}!\rangle, 7\langle \text{Int}!\rangle, a_0\langle (\text{Ref } T_0)!\rangle)\langle c\rangle : T_1$$

We have a casted value in the heap that needs to be reduced. We apply (HCAST) and (PURECAST) to get

$$a_0 \mapsto (42, 7\langle \text{Int}!\rangle, a_0\langle \text{Ref } T_2\rangle) : T_1$$

We cast address a_0 again, this time to $T_1 \sqcap T_2$, via rule (HDROP) and (CASTREF1).

$$a_0 \mapsto (42, 7\langle \text{Int}!\rangle, a_0)\langle \iota \times \text{Int?} \times \text{Ref } T_2\rangle : \text{Int} \times \text{Int} \times \text{Ref } T_2$$

A few reductions via (HCAST) and (PURECAST) give us

$$a_0 \mapsto (42, 7, a_0\langle \text{Ref } T_2\rangle) : \text{Int} \times \text{Int} \times \text{Ref } T_2$$

The final cast applied to a_0 is a no-op because the run-time type is already more precise than T_2. So we reduce via (HCAST) and (CASTREF2) to:

$$a_0 \mapsto (42, 7, a_0) : \text{Int} \times \text{Int} \times \text{Ref } T_2$$

Even though we allow casted values on the heap, we require the normalization of all such casts before returning to the execution of the program. We distinguish between normal heaps of values, μ, and evolving heaps, ν, that may contain both values and casted values. Normal heaps are a subset of the evolving heaps.

Encoding permissive references. The monotonic discipline and its run-time invariant-enforcement seems to restrict how developers can formulate their programs. It is natural to ask whether monotonic references are compatible with the flexibility that is expected in dynamic languages. In this section we show that the monotonic discipline admits permissive references through a syntactic discipline that can be conveniently provided to programmers.

Consider the following program that uses an allocated reference cell at two incompatible types, Int and Bool.

```
let x = ref (4 as ⋆) in
let y = (x as Ref Int) in
let z = (x as Ref Bool) in
!y;
z := true;
!z
```

Under the standard reference semantics, this program runs without incident, but under monotonic references it fails in the cast to Ref Bool. We can regain this flexibility under monotonic references via a disciplined use of $\star$ typed reference cells. Consider the following rewrite of this program:

```
let x = ref (4 as ⋆) in
let y = x in                // treat y like Ref Int
let z = x in                // treat z like Ref Bool
(!y) as Int;
(z := (true) as ⋆) as Bool;
(!z) as Bool
```

In this encoding, all references have type Ref $\star$, and typing is enforced only at dereferences and updates, using ascriptions. This program runs successfully under the monotonic semantics, but it would be tedious and error prone to insert these ascriptions by hand.

Luckily there is no need: we codify this permissive reference discipline by introducing a surface language that makes this convenient. We extend the expressions with *permissive references* $\widetilde{\mathsf{ref}}\ e$, and the types with a corresponding type $\widetilde{\mathsf{Ref}}\ T$. Consistency is extended so that permissive references have the same consistency properties as monotonic references, but permissive references are not consistent with monotonic references.

Finally we introduce a type-directed transformation $\Gamma \vdash e : T \leadsto e$ that translates permissive references to monotonic references. The interesting cases are presented below.

$$\frac{x : \widetilde{\mathsf{Ref}}\ T \in \Gamma}{\Gamma \vdash x : \widetilde{\mathsf{Ref}}\ T \leadsto x} \qquad \frac{\Gamma \vdash e : T \leadsto e'}{\Gamma \vdash \widetilde{\mathsf{ref}}\ e : \widetilde{\mathsf{Ref}}\ T \leadsto \mathsf{ref}\ (e'\ \text{as}\ \star)}$$

$$\frac{\Gamma \vdash e : \widetilde{\mathsf{Ref}}\ T \leadsto e'}{\Gamma \vdash !e : T \leadsto (!e')\ \text{as}\ T} \qquad \frac{\Gamma \vdash e_1 : \widetilde{\mathsf{Ref}}\ T_1 \leadsto e_1' \quad \Gamma \vdash e_2 : T_2 \leadsto e_2'}{\Gamma \vdash e_1 := e_2 : T_1 \leadsto (e_1' := (e_2'\ \text{as}\ \star))\ \text{as}\ T_1}$$

Note that the static semantics for permissive references enforces type consistency at assignments, even though the assigned value is ultimately cast to $\star$. Furthermore, reference values translate to themselves, so object identity is preserved. However cast overhead is introduced at each dereference and update, so permissive references pay their own way with respect to performance.

If we revisit the initial example in this section and replace ref with $\widetilde{\mathsf{ref}}$ and Ref with $\widetilde{\mathsf{Ref}}$, then this judgment translates the first program above into the second.

Proposition 1 (Translation). *If $\Gamma \vdash e : T \leadsto e'$ then $|\Gamma| \vdash e' : |T|$, Where $|\cdot|$ is the compatible extension of the equation $|\widetilde{\mathsf{Ref}}\ T| = \mathsf{Ref}\ \star$.*

This syntactic extension gives programmers access to both permissive references and monotonic references as desired.

$$\boxed{(T \Rightarrow T) = c}$$

$$(B \Rightarrow B) = \iota \qquad (I \Rightarrow \star) = I!$$
$$(\star \Rightarrow \star) = \iota \qquad (\star \Rightarrow I) = I?$$

$$(T_1 {\rightarrow} T_2) \Rightarrow (T_1' {\rightarrow} T_2') = (T_1' \Rightarrow T_1) {\rightarrow} (T_2 \Rightarrow T_2')$$
$$(T_1 \times T_2) \Rightarrow (T_1' \times T_2') = (T_1 \Rightarrow T_1') \times (T_2 \Rightarrow T_2')$$
$$\textbf{Ref } T \Rightarrow \textbf{Ref } T' = \textbf{Ref } T'$$

Fig. 8. Compile casts to monotonic coercions (without blame)

Permissive references are a useful abstraction for the programmer and provide strong guarantees. However, such guarantees are provided only as long as permissive references do not flow into monotonic references. Consider the program above (with permissive references) where the following code comes after the let statements.

$$\texttt{let } w_1 = (x \texttt{ as } \star) \texttt{ in}$$
$$\texttt{let } w_2 = (w_1 \texttt{ as Ref Bool}) \texttt{ in}$$
$$w_2 := \texttt{true};$$

The program places us in a same situation as the original program that the monotonic semantics could not run without error. This example shows an important syntactic discipline for programmers that want to employ the monotonic paradigm for gradual references: *permissive references should not flow into monotonic references.*

4   Type Safety for Monotonic References

We present the high-points of the type safety proof here. The full proof is mechanized in Isabelle 2013 and available on arxiv (Siek and Vitousek, 2013). The semantics in the mechanized version differs from the semantics presented here in that it uses an abstract machine instead of a reduction semantics, as we found the mechanized proof easier to carry out on an abstract machine while the reduction semantics is more approachable.

We begin by lifting the precision relation to heap typings.

Definition 1 (Precision relation on heap typings). $\Sigma' \sqsubseteq \Sigma$ *iff* $dom(\Sigma') = dom(\Sigma)$ *and* $\Sigma(a) = T$ *implies* $\Sigma'(a) = T'$ *where* $T' \sqsubseteq T$.

Our first lemma below is important: expression typing is preserved when moving to a more precise heap typing.

Lemma 1 (Strengthening wrt. the heap typing). *If* $\Gamma; \Sigma \vdash e : T$ *and* $\Sigma' \sqsubseteq \Sigma$, *then* $\Gamma; \Sigma' \vdash e : T$.

Proof (Proof sketch). The interesting case is for addresses. We have

$$\frac{\Sigma(a) \sqsubseteq T}{\Gamma; \Sigma \vdash a : T}$$

From $\Sigma' \sqsubseteq \Sigma$ and transitivity of $\sqsubseteq$, we have $\Sigma'(a) \sqsubseteq T$. Therefore $\Gamma; \Sigma' \vdash a : T$.

The definition of well-typed heaps is standard.

Definition 2 (Well-typed heaps). *A heap ν is well-typed with respect to heap typing Σ, written $\Sigma \vdash \nu$, iff $\forall a\, T$. $\Sigma(a) = T$ implies $\nu(a) = cv : T$ and $\emptyset; \Sigma \vdash cv : T$ for some cv.*

From the strengthening lemma, we have the following corollary.

Corollary 1 (Monotonic heap update). *If $\Sigma \vdash \nu$ and $\Sigma(a) = T$ and $T' \sqsubseteq T$ and $\emptyset; \Sigma \vdash cv : T'$, then $\Sigma(a \mapsto T') \vdash \nu(a \mapsto cv : T')$.*

Proof (sketch). Let $\Sigma' = \Sigma(a \mapsto T')$. From $T' \sqsubseteq T$ we have $\Sigma' \sqsubseteq \Sigma$, so by Lemma 1 we have $\emptyset; \Sigma' \vdash cv : T'$ and $\Sigma' \vdash \nu$. Thus, $\Sigma(a \mapsto T') \vdash \nu(a \mapsto cv : T')$.

Lemma 2 (Progress and Preservation). *Suppose $\emptyset; \Sigma \vdash e : T$ and $\Sigma \vdash \nu$. Exactly one of the following holds:*

1. *(a) e is a value, or*
   (b) $e = \mathtt{error}$, or
   (c) $e, \nu \longrightarrow e', \nu'$ for some e' and ν'.
2. *for all e', ν', if $e, \nu \longrightarrow e', \nu'$ then $\emptyset; \Sigma' \vdash e' : T$ and $\Sigma' \vdash \nu'$ and $\Sigma' \sqsubseteq \Sigma$ for some Σ'.*

Theorem 1 (Type Safety). *Suppose $\emptyset; \Sigma \vdash e : T$ and $\Sigma \vdash \nu$. Exactly one of the following holds:*

1. *$e, \nu \longrightarrow^* v, \nu'$ and $\emptyset; \Sigma' \vdash v : T$ for some Σ', or*
2. *$e, \nu \longrightarrow^* \mathtt{error}, \nu'$, or*
3. *e diverges.*

Proof. If e diverges we immediately conclude the proof. Otherwise, suppose e does not diverge. Then $e, \nu \longrightarrow^* e', \nu'$ and e' cannot reduce. We proceed by induction on the length $e, \nu \longrightarrow^* e', \nu'$, and use Lemma 2 to conclude.

5   Monotonic References with Blame

We turn to the challenge of designing blame tracking for monotonic references, presenting several examples that motivate and provide intuitions for the design. The later part of this section presents the dynamic semantics of monotonic references with blame tracking.

Consider the following example in which we allocate a reference of dynamic type and then, separately, cast from **Ref $\star$** to **Ref Int** and to **Ref Bool**.

$$B <: B \qquad\qquad T <: \star \qquad\qquad \text{Ref } T <: \text{Ref } T$$

$$\dfrac{T_1' <: T_1 \quad T_2 <: T_2'}{T_1 \rightarrow T_2 <: T_1' \rightarrow T_2'} \qquad\qquad \dfrac{T_1 <: T_1' \quad T_2 <: T_2'}{T_1 \times T_2 <: T_1' \times T_2'}$$

Fig. 9. Subtyping relation

$$
\begin{aligned}
&\text{let } r_0 = \text{ref } (42 \text{ as}^{\ell_1} \star) \text{ in}\\
&\text{let } r_1 = r_0 \text{ as}^{\ell_2} \text{ Ref Int in}\\
&\text{let } r_2 = r_0 \text{ as}^{\ell_3} \text{ Ref Bool in}\\
&!r_2
\end{aligned}
$$

With monotonic references, the cast at ℓ_3 triggers an error, because Int and Bool are inconsistent. But what blame labels should the error message include? Is it only the fault of ℓ_3? Not really; because ℓ_3 would not cause an error if it were not for the cast at ℓ_2. The casts at ℓ_2 and ℓ_3 disagree with each other regarding the type of the heap cell, so we blame both. The result of this program is blame $\{\ell_2, \ell_3\}$.

Next consider an example in which we allocate a reference at type Ref Int, cast it to Ref $\star$, and then attempt to write a Boolean.

$$
\begin{aligned}
&\text{let } r_0 = \text{ref } 42 \text{ in}\\
&\text{let } r_1 = r_0 \text{ as}^{\ell_1} \text{ Ref } \star \text{ in}\\
&r_1 := ^{\ell_3} (\text{true as}^{\ell_2} \star)
\end{aligned}
$$

The update on the third line triggers an error, and we have three possible locations to blame: ℓ_1, ℓ_2, and ℓ_3. The cast at ℓ_2 is from Bool to $\star$, which is harmless. There is no cast at ℓ_3, we are just writing a value of type $\star$ to a reference of type Ref $\star$. The real culprit here is ℓ_1, which casts from Ref Int to Ref $\star$, thereby opening up the potential for the later cast error. Naïvely, this looks like an upcast, but a proper treatment of subtyping for references makes references invariant. So we have Ref Int $\not<:$ Ref $\star$ and the result of this program is blame $\{\ell_1\}$. Figure 9 presents the subtyping relation.

We consider a pair of examples below that differ only on the fourth line. We allocate a reference to a pair at type Ref $(\star \times \star)$ then cast it to Ref $(\text{Int} \times \star)$ and to Ref $(\star \times \text{Int})$. In the first example, we update through the original reference, writing a Boolean and integer, whereas in the second example we write an integer and a Boolean. Here is the first example:

$$
\begin{aligned}
&\text{let } r_0 = \text{ref } (1 \text{ as}^{\ell_1} \star, 2 \text{ as}^{\ell_2} \star) \text{in}\\
&\text{let } r_1 = r_0 \text{ as}^{\ell_3} \text{ Ref } (\text{Int} \times \star) \text{in}\\
&\text{let } r_2 = r_0 \text{ as}^{\ell_4} \text{ Ref } (\star \times \text{Int}) \text{in}\\
&r_0 := (\text{true as}^{\ell_5} \star, 2 \text{ as}^{\ell_6} \star);\\
&\text{fst } !r_0
\end{aligned}
$$

and here is the second example, just showing the fourth line:

$$\ldots$$
$$r_0 := (1 \text{ as}^{\ell_7} \star, \text{true as}^{\ell_8} \star);$$
$$\ldots$$

The first example should produce $\text{blame}\,\{\ell_3\}$ while the second example should produce $\text{blame}\,\{\ell_4\}$, but the challenge is how can we associate multiple blame labels with the same heap cell?

We take inspiration from Siek and Wadler (2010) and use *labeled types* for our run-time type information. With a labeled type, each type constructor within the type can be labeled with a type. Figure 10 gives the syntax of labeled types and operations on them, which we shall explain later in this section. In the above examples, the run-time type information for the heap cell evolves as follows:

$$(\star \times^{\emptyset} \star) \Rightarrow (\text{Int}^{\ell_3} \times^{\emptyset} \star) \Rightarrow (\text{Int}^{\ell_3} \times^{\emptyset} \text{Int}^{\ell_4})$$

In the first example, when we write true into the first element of the pair, the cast to Int fails and blames ℓ_3, as desired. In the second example, when we write true into the second element, the cast to Int fails and blames ℓ_4, as desired.

Our next example brings up a somewhat ambiguous situation. We allocate a reference at type $\text{Ref}\,\star$, cast it to Ref Int twice, then write a Boolean.

$$\begin{aligned}
&\text{let } r_0 = \text{ref}\,(42 \text{ as}^{\ell_1} \star)\text{in} \\
&\text{let } r_1 = r_0 \text{ as}^{\ell_2} \text{ Ref Int in} \\
&\text{let } r_2 = r_0 \text{ as}^{\ell_3} \text{ Ref Int in} \\
&r_0 := (\text{true as}^{\ell_4} \star)
\end{aligned}$$

Should we blame ℓ_2 or ℓ_3? In some sense, they are both just as guilty and the ideal would be to blame them both. On the other hand, maintaining potentially large sets of blame labels would induce some space overhead. Our design instead blames the first cast with respect to execution order, in this case ℓ_2.

For our final example, we adapt the above example to have a function in the heap cell so that we can consider the behavior to the left of the arrow.

$$\begin{aligned}
&\text{let } r_0 = \text{ref}\,(\lambda x{:}\,\star\,.\,\text{true})\text{in} \\
&\text{let } r_1 = r_0 \text{ as}^{\ell_1} \text{ Ref}\,(\text{Int} \rightarrow \text{Bool})\text{in} \\
&\text{let } r_2 = r_0 \text{ as}^{\ell_2} \text{ Ref}\,(\text{Int} \rightarrow \text{Bool})\text{in} \\
&r_0 := \lambda x{:}\text{Int}.\,\text{zero?}(x); \\
&!r_0\,(\text{true as}^{\ell_3} \star)
\end{aligned}$$

The run-time type information for the heap cell evolves in the following way:

$$(\star \rightarrow^{\emptyset} \text{Bool}^{\emptyset}) \Rightarrow (\text{Int}^{\ell_1} \rightarrow^{\emptyset} \text{Bool}^{\emptyset}) \Rightarrow (\text{Int}^{\ell_1} \rightarrow^{\emptyset} \text{Bool}^{\emptyset})$$

The function application on the last line of the example triggers a cast error, with the blame going to ℓ_1, again because we wish to blame the first cast with respect to execution order. However, to obtain this semantics some care must be taken. On the second cast, we merge the labeled type for the second cast with the current run-time type information:

$$(\text{Int}^{\ell_1} \rightarrow^{\emptyset} \text{Bool}^{\emptyset}) \vartriangle (\text{Int}^{\ell_2} \rightarrow^{\emptyset} \text{Bool}^{\emptyset})$$

If we were to use the composition function from Siek and Wadler (2010), the result would be $\text{Int}^{\ell_2} \to^{\emptyset} \text{Bool}^{\emptyset}$ because that composition function is contravariant for function parameters. Here we instead want to be covariant on function parameters, so the result is $\text{Int}^{\ell_1} \to^{\emptyset} \text{Bool}^{\emptyset}$. We define a new function for merging labeled types, $\triangle$, in Figure 10.

5.1 Semantics of Monotonic References with Blame

Armed with the intuitions from the above examples, we discuss the semantics of monotonic references with blame, defined in Figures 12 and 13. The semantics is largely similar to the semantics without blame except that the run-time type information is represented as labeled types and we replace the functions, such as meet ($\sqcap$) that operate on types, with functions such as merge ($\triangle$) that operate on labeled types.

Proposition 2 (Meet is the erasure of merge)
If $|P_1| \sim |P_2|$, then $|P_1 \triangle P_2| = |P_1| \sqcap |P_2|$.
If $|P_1| \not\sim |P_2|$, then $P_1 \triangle P_2 = \perp^L$ for some L.

As discussed with the example above, the definition of $P_1 \triangle P_2$ takes into account that P_1 is temporally prior to P_2 and should therefore take precedence with respect to blame responsibility. We use the auxiliary function $p \triangle q$ to choose between two optional labels, returning the first if it is present and the second otherwise.

When we cast a reference via rule (CASTR1B), we need to update the heap cell from labeled type P_1 to P_3. We accomplish this with a new operator $P_1 \Rightarrow P_3$ that produces a coercion. The most interesting line of its definition is for reference types. There we use a different operator, $P \Leftrightarrow Q$, that produces a labeled type and captures the bidirectional read/write nature of mutable references.

The definitions of $\triangle$, $\Rightarrow$, and $\Leftrightarrow$ need to percolate errors, which we write as $\perp^L$ where L is a set of blame labels. We use "smart" constructors $\hat{\to}$, $\hat{\times}$, and $\hat{\text{Ref}}$ that return $\perp^L$ if either argument is $\perp^L$ (with precedent to the left if both arguments are errors), but otherwise act like the underlying constructor.

In the rule for allocation, we initialize the RTTI to $T^{\emptyset}$. (Figure 11 defines converting a type to a labeled type.) In the rule for a dynamic dereference, (DYNDRFMB), we cast from the reference's run-time labeled type to T by promoting T to the labeled type $T^{\emptyset}$ and then applying the $\Rightarrow$ function to cast between labeled types, so we have $\mu(a)_{\text{rtti}} \Rightarrow T^{\emptyset}$. Suppose that $\mu(a)_{\text{rtti}}$ is Ref Int^{ℓ} and T is $\text{Ref} \star$. Then the coercion we apply during the dereference is $\text{Int}^{\ell}!$; so our injection coercions contain labeled types. The rule for dynamic update, (DYNUPDMB), is dual: we perform the cast $T^{\emptyset} \Rightarrow \mu(a)_{\text{rtti}}$.

Because our injection and projection coercions contain labeled types, the (COLLAPSE) rule becomes

$$v\langle P_1! \rangle \langle P_2? \rangle \longrightarrow_c v\langle P_1 \Rightarrow P_2 \rangle \qquad \text{if } |P_1| \sim |P_2|$$

We make similar changes to the (CONFLICT) rule.

Optional labels p, q ::= $\emptyset \mid \{\ell\}$
Label sets L ::= $\emptyset \mid \{\ell\} \mid \{\ell_1, \ell_2\}$
Labeled types P, Q ::= $B^p \mid P \to^p P \mid P \times^p P \mid \mathbf{Ref}\,^p P \mid \star$

Erase labels $\boxed{|P| = T}$

$$|B^p| = B \quad |P \to^p Q| = |P| \to |Q| \quad |P \times^p Q| = |P| \times |Q| \quad |\mathbf{Ref}\,^p P| = \mathbf{Ref}\,|P| \quad |\star| = \star$$

Top label $\boxed{lab(P) = L}$

$$lab(B^p) = p \quad lab(P \to^p Q) = p \quad lab(P \times^p Q) = p \quad lab(\mathbf{Ref}\,^p P) = p \quad lab(\star) = \emptyset$$

Merge optional labels $\boxed{p \bigtriangleup p = p}$

$$\{\ell\} \bigtriangleup q = \{\ell\} \qquad \emptyset \bigtriangleup q = q$$

Merge labeled types $\boxed{P \bigtriangleup P = P \text{ or } \bot^L}$

$$B^p \bigtriangleup B^q = B^{p \bigtriangleup q}$$
$$P \bigtriangleup \star = P \qquad \star \bigtriangleup Q = Q$$
$$(P \to^p P') \bigtriangleup (Q \to^q Q') = (P \bigtriangleup Q) \dot{\to}^{p \bigtriangleup q} (P' \bigtriangleup Q')$$
$$(P \times^p P') \bigtriangleup (Q \times^q Q') = (P \bigtriangleup Q) \hat{\times}^{p \bigtriangleup q} (P' \bigtriangleup Q')$$
$$\mathbf{Ref}\,^p P \bigtriangleup \mathbf{Ref}\,^q Q = \hat{\mathbf{Ref}}\,^{p \bigtriangleup q} (P \bigtriangleup Q)$$
$$P \bigtriangleup Q = \bot^{lab(P) \cup lab(Q)} \qquad \text{otherwise}$$

Bidirectional cast between labeled types $\boxed{P \Leftrightarrow P = P \text{ or } \bot^L}$

$$B^p \Leftrightarrow B^q = B^\emptyset$$
$$P \Leftrightarrow \star = P \qquad \star \Leftrightarrow Q = Q$$
$$(P \to^p P') \Leftrightarrow (Q \to^q Q') = (P \Leftrightarrow Q) \dot{\to}^\emptyset (P' \Leftrightarrow Q')$$
$$(P \times^p P') \Leftrightarrow (Q \times^q Q') = (P \Leftrightarrow Q) \hat{\times}^\emptyset (P' \Leftrightarrow Q')$$
$$\mathbf{Ref}\,^p P \Leftrightarrow \mathbf{Ref}\,^q Q = \hat{\mathbf{Ref}}\,^\emptyset (P \Leftrightarrow Q)$$
$$P \Leftrightarrow Q = \bot^{lab(P) \cup lab(Q)} \qquad \text{otherwise}$$

Cast between labeled types $\boxed{P \Rightarrow P = c \text{ or } \bot^L}$

$$B^p \Rightarrow B^q = \iota \qquad \star \Rightarrow \star = \iota$$
$$P \Rightarrow \star = P! \qquad \star \Rightarrow Q = Q?$$
$$(P \to^p P') \Rightarrow (Q \to^q Q') = (Q \Rightarrow P) \dot{\to} (P' \Rightarrow Q')$$
$$(P \times^p P') \Rightarrow (Q \times^q Q') = (P \Rightarrow Q) \hat{\times} (P' \Rightarrow Q')$$
$$\mathbf{Ref}\,^p P \Rightarrow \mathbf{Ref}\,^q Q = \hat{\mathbf{Ref}}\,(P \Leftrightarrow Q)$$
$$P \Rightarrow Q = \bot^{lab(P) \cup lab(Q)} \qquad \text{otherwise}$$

Fig. 10. Labeled types and their operations

$$\boxed{(T \Rightarrow^\ell T) = c}$$

$$(B \Rightarrow^\ell B) = \iota \qquad (T \Rightarrow^\ell \star) = T^\emptyset!$$
$$(\star \Rightarrow^\ell \star) = \iota \qquad (\star \Rightarrow^\ell T) = T^\ell?$$

$$(T_1 \to T_2) \Rightarrow^\ell (T_1' \to T_2') = (T_1' \Rightarrow^\ell T_1) \to (T_2 \Rightarrow^\ell T_2')$$
$$(T_1 \times T_2) \Rightarrow^\ell (T_1' \times T_2') = (T_1 \Rightarrow^\ell T_1') \times (T_2 \Rightarrow^\ell T_2')$$
$$\text{Ref } T_1 \Rightarrow^\ell \text{Ref } T_2 = \text{Ref } (T_1^\ell \Leftrightarrow T_2^\ell)$$

Add labels to a type $\qquad\qquad\qquad\qquad\qquad\qquad \boxed{T^\ell = P}$

$$B^\ell = B^\ell \quad (T_1 \to T_2)^\ell = T_1^\ell \to^\ell T_2^\ell \quad (T_1 \times T_2)^\ell = T_1^\ell \times^\ell T_2^\ell$$

$$(\text{Ref } T)^\ell = \text{Ref } {}^\ell T^\ell \quad \star^\ell = \star$$

Fig. 11. Compile casts to monotonic coercions (with blame)

Expressions	$e ::= \cdots \mid \textbf{blame } L$
Coercions	$c ::= \iota \mid P? \mid P! \mid c \to c \mid c \times c \mid c \mathbin{;} c \mid \text{Ref } P$
Values	$v ::= k \mid \lambda x.e \mid (v,v) \mid v\langle P! \rangle \mid a$
Heap	$\mu ::= \emptyset \mid \mu(a \mapsto v : P)$
Evolving Heap	$\nu ::= \emptyset \mid \nu(a \mapsto cv : P)$

Fig. 12. Syntax for monotonic references with blame

Figure 11 defines the compilation of casts to monotonic coercions. Compared to the compilation without blame (Figure 8), there are three differences. The first two concern injection and projection coercions: instead of only having a blame label on projections we have labeled types inside both injections and projections, as noted above. In the compilation of a cast labeled ℓ, we generate a labeled type for the injection from T by adding the empty label to T, and for the projection to T by adding ℓ to T. The third difference is in the formation of the reference coercion. Instead of simply taking the target type, we use the bidirectional operator $\Leftrightarrow$. Recall the second example of this section in which we blamed the cast from Ref Int to Ref $\star$. By using $\Leftrightarrow$, the resulting coercion is Ref Int$^{\ell_1}$ instead of Ref $\star$.

6   The Blame-Subtyping Theorem

The blame-subtyping theorem pin-points the source of cast errors in gradually-typed programs. The blame-subtyping theorem states that if a program results in a cast error, blame L, then the blame labels in L identify the location of implicit casts that did not respect subtyping. That is, the blame labels that occur in a safe implicit cast, $T_1 \Rightarrow T_2$ where $T_1 <: T_2$, can never be blamed.

We prove the blame-subtyping theorem via a preservation-style proof in which we preserve the e safe ℓ predicate (Wadler and Findler, 2009). This proof is

Coercion typing $\boxed{c : T \Rightarrow T}$

$$\frac{}{P? : \star \Rightarrow |P|} \qquad \frac{}{P! : |P| \Rightarrow \star} \qquad \cdots$$

Pure cast reduction rules $\boxed{e \longrightarrow_c e}$

$$\cdots \quad v\langle P_1!\rangle\langle P_2?\rangle \longrightarrow_c v\langle P_1 \Rightarrow P_2\rangle \quad \text{if } |P_1| \sim |P_2| \tag{COLLAPSE}$$

$$v\langle P_1!\rangle\langle P_2?\rangle \longrightarrow_c \textbf{blame } L \quad \text{if } P_1 \Rightarrow P_2 = \bot^L \tag{CONFLICT}$$

Cast reduction rules $\boxed{e, \nu \longrightarrow_{cr} e, \nu}$

$$\frac{e \longrightarrow_c e'}{e, \nu \longrightarrow_{cr} e', \nu} \tag{PCASTB}$$

$$\frac{\nu(a) = cv : P_1 \qquad P_3 = P_1 \vartriangle P_2 \\ |P_3| \neq |P_1| \quad cv' = cv\langle P_1 \Rightarrow P_3\rangle}{a\langle \textbf{Ref } P_2\rangle, \nu \longrightarrow_{cr} a, \nu(a \mapsto cv' : P_3)} \tag{CASTR1B}$$

$$\frac{\nu(a) = cv : P_1 \qquad P_1 = P_1 \vartriangle P_2}{a\langle \textbf{Ref } P_2\rangle, \nu \longrightarrow_{cr} a, \nu} \tag{CASTR2B}$$

$$\frac{\nu(a) = cv : P_1 \qquad P_1 \vartriangle P_2 = \bot^L}{a\langle \textbf{Ref } P_2\rangle, \nu \longrightarrow_{cr} \textbf{blame } L, \nu} \tag{CASTR3B}$$

Program reduction rules $\boxed{e, \mu \longrightarrow_e e, \mu}$

$$\textbf{ref}_T\, v, \mu \longrightarrow_e a, \mu(a \mapsto v : T^\emptyset) \qquad \text{if } a \notin dom(\mu)$$

$$!a, \mu \longrightarrow_e \mu(a)_{\mathsf{val}}, \mu \tag{DEREFMB}$$

$$!a@T, \mu \longrightarrow_e \mu(a)_{\mathsf{val}}\langle \mu(a)_{\mathsf{rtti}} \Rightarrow T^\emptyset\rangle, \mu \tag{DYNDRFMB}$$

$$a := v, \mu \longrightarrow_e a, \mu(a \mapsto v : \mu(a)_{\mathsf{rtti}}) \tag{UPDMB}$$

$$a := v@T, \mu \longrightarrow_e a, \mu(a \mapsto cv : \mu(a)_{\mathsf{rtti}}) \tag{DYNUPDMB}$$

$$\text{where } cv = v\langle T^\emptyset \Rightarrow \mu(a)_{\mathsf{rtti}}\rangle$$

For $X \in \{cr, e\}$:

$$\frac{e, \nu \longrightarrow_X e', \nu'}{F[e], \nu \longrightarrow_X F[e'], \nu'} \qquad \frac{}{F[\textbf{blame } L], \nu \longrightarrow_X \textbf{blame } L, \nu}$$

State reduction rules $\boxed{e, \nu \longrightarrow e, \nu}$

$$\frac{e, \mu \longrightarrow_X e', \nu \quad X \in \{cr, e\}}{e, \mu \longrightarrow e', \nu} \qquad \frac{\nu(a) = cv : P \quad cv, \nu \longrightarrow_{cr} \textbf{blame } L, \nu'}{e, \nu \longrightarrow \textbf{blame } L, \nu'}$$

$$\frac{\nu(a) = cv : P \quad cv, \nu \longrightarrow_{cr} cv', \nu' \quad |\nu'(a)_{\mathsf{rtti}}| = |P|}{e, \nu \longrightarrow e, \nu'(a \mapsto cv' : P)}$$

$$\frac{\nu(a) = cv : P \quad cv, \nu \longrightarrow_{cr} cv', \nu' \quad |\nu'(a)_{\mathsf{rtti}}| \neq |P|}{e, \nu \longrightarrow e, \nu'}$$

Fig. 13. Monotonic references with blame

conducted on the coercion calculus, so to relate the result back to the gradually-typed λ-calculus, we need a theorem concerning the relationship between subtyping and coercion blame safety, Theorem 2. Recall that subtyping is defined in Figure 9 and compilation to coercions is defined in Figure 11.

Theorem 2 (Blame-Subtyping Theorem for coercion calculus). *For all T_1, T_2, and ℓ, it holds that $T_1 <: T_2$ iff $(T_1 \Rightarrow^\ell T_2)$ safe ℓ.*

Lemma 3 (Preservation of blame safety)
For all e, e', ν, ν', and ℓ, if e, ν safe ℓ and $e, \nu \longrightarrow e', \nu'$ then e', ν' safe ℓ.

We now move away from the coercion calculus and prove these important results on the gradually typed λ calculus with references. This latter language is indeed the one that programmers are expected to use. The following definitions will help to recast the results into the setting of the gradually typed language.

Definition 3 (Casts for a label in an expression). *Let e be an expression and ℓ a label, we say that e contains the cast $T_1 \Rightarrow T_2$ for ℓ whenever, in the derivation of $\Gamma \vdash e \rightsquigarrow e' : T$, there is the creation of a coercion via $T_1 \Rightarrow^\ell T_2$.*

Definition 4 (Blame safety for gradually-typed expressions). *A gradually-typed expression e is safe for ℓ if all the casts contained in e labeled ℓ respect subtyping.*

We now have all the ingredients to state and prove one of the main contributions of the paper, i.e. the Blame-Subtyping Theorem for the gradually-typed λ calculus with references.

Lemma 4 (Translation preserves blame safety). *If e safe ℓ and $\Gamma \vdash e \rightsquigarrow e' : T$, then e' safe ℓ.*

Proof. The proof is a straightforward induction on $\Gamma \vdash e \rightsquigarrow e' : T$.

Theorem 3 (Blame-Subtyping Theorem). *For all e, e', T_1, T_2, ℓ, if $\emptyset \vdash e \rightsquigarrow e' : T$, e safe ℓ, and $e', \emptyset \longrightarrow$ blame L, ν, then $\ell \notin L$.*

Proof. From the assumptions we have e' safe ℓ by Lemma 4. Then we conclude by applying the Blame-Subtyping Theorem for the coercion calculus.

7   Implementation Concerns w.r.t. Strong Updates

The monotonic semantics for references performs in-place updates to the heap with values of different type. In languages where values have uniform size, like many functional and object-oriented languages, this does not pose a problem. However, for languages where values may have different sizes, in-place updates pose a problem. This issue can be addressed using an approach inspired by garbage collection techniques. When the semantics is to update a cell with a larger value than the current one, the implementation allocates a new piece of memory and places a forwarding pointer in the old location. When reading and writing through dynamic references, the implementation must check for and follow the forwarding pointers. However, when reading and writing through fully-static references, the implementation does not need to consider forwarding pointers because fully-static heap cells never move. Then during a garbage collection, the implementation can collapse sequences of forwarding pointers to reduce overhead in subsequent execution.

8 Related Work

Here we mention related work that is not discussed in the introduction or elsewhere in the paper.

The casts and coercions studied in this paper bear many similarities with contracts (Findler and Felleisen, 2002). Racket (Flatt and PLT, 2014) provides contracts for mutable values in the form of impersonators (Strickland et al., 2012), which, for our purposes, can be viewed as implementing the standard semantics of Herman et al. (2007), as we saw in Section 2.

Fähndrich and Leino (2003) introduce a technique similar to monotonic references with their monotonic typestate. In this design, objects may flow from less restrictive to more restrictive typestates, but not vice versa. Unlike monotonic references, which require runtime checks due to the existence of dynamically-typed regions of code, their system enforces monotonicity statically.

Gradual typing was added to $C^\sharp$ with the addition of the **dynamic** type. Bierman et al. (2010) define a formal model of $C^\sharp$, named $FC_4^\sharp$, and present an operational semantics. The semantics is similar to that of Swamy et al. (2014) in that they use an RTTI-based approach and subtype checks to implement casts.

9 Conclusion

We have presented a new design for gradually-typed mutable references, called monotonic references, the first to incur zero-overhead for reference accesses in statically typed code while maintaining the full expressiveness of a gradual type system. We defined a dynamic semantics for monotonic references and presented a mechanized proof of type safety. Further, we defined blame tracking based on using labeled types in the run-time type information and proved the blame-subtyping theorem.

References

Abadi, M., Cardelli, L., Pierce, B., Plotkin, G.: Dynamic typing in a statically-typed language. In: Symposium on Principles of programming languages (1989)

Bierman, G., Meijer, E., Torgersen, M.: Adding dynamic types to C#. In: European Conference on Object-Oriented Programming (2010)

Boyland, J.T.: The problem of structural type tests in a gradual-typed language. In: Foundations of Object Oriented Languages, FOOL 2014, pp. 675–681. ACM (2014)

Fähndrich, M., Leino, K.R.M.: Heap monotonic typestate. In: International Workshop on Alias Confinement and Ownership (2003)

Findler, R.B., Felleisen, M.: Contracts for higher-order functions. In: International Conference on Functional Programming, ICFP, pp. 48–59 (2002)

Flatt, M.: The Racket reference 6.0. Technical report, PLT Inc (2014), http://docs.racket-lang.org/reference/index.html

Henglein, F.: Dynamic typing: syntax and proof theory. Science of Computer Programming 22(3), 197–230 (1994)

Herman, D., Tomb, A., Flanagan, C.: Space-efficient gradual typing. In: Trends in Functional Prog (TFP), p. XXVIII (April 2007)

Rastogi, A., Swamy, N., Fournet, C., Bierman, G., Vekris, P.: Safe & efficient gradual typing for TypeScript. Technical Report MSR-TR-2014-99 (2014)

Siek, J.G., Taha, W.: Gradual typing for functional languages. In: Scheme and Functional Programming Workshop, pp. 81–92 (September 2006)

Siek, J.G., Vitousek, M.M.: Monotonic references for gradual typing. In: Computing Research Repository (2013), http://arxiv.org/abs/1312.0694

Siek, J.G., Wadler, P.: Threesomes, with and without blame. In: Symposium on Principles of Programming Languages, POPL, pp. 365–376 (January 2010)

Siek, J.G., Vitousek, M.M., Cimini, M., Boyland, J.T.: Refined criteria for gradual typing. Under review for publication at SNAPL 2015 (2015)

Strickland, T.S., Tobin-Hochstadt, S., Findler, R.B., Flatt, M.: Chaperones and impersonators: run-time support for reasonable interposition. In: OOPSLA (2012)

Swamy, N., Fournet, C., Rastogi, A., Bhargavan, K., Chen, J., Strub, P.-Y., Bierman, G.: Gradual typing embedded securely in JavaScript. In: Symposium on Principles of Programming Languages, POPL (January 2014)

Tobin-Hochstadt, S., Felleisen, M.: Interlanguage migration: From scripts to programs. In: Dynamic Languages Symposium (2006)

Vitousek, M.M., Siek, J.G., Kent, A., Baker, J.: Design and evaluation of gradual typing for Python. In: Dynamic Languages Symposium (2014)

Wadler, P., Findler, R.B.: Well-typed programs can't be blamed. In: European Symposium on Programming, ESOP, pp. 1–16 (March 2009)

Wrigstad, T., Nardelli, F.Z., Lebresne, S., Östlund, J., Vitek, J.: Integrating typed and untyped code in a scripting language. In: Symposium on Principles of Programming Languages, POPL, pp. 377–388 (2010)

Inter-procedural Two-Variable
Herbrand Equalities

Stefan Schulze Frielinghaus, Michael Petter, and Helmut Seidl

Technische Universität München, Boltzmannstrasse 3, 85748 Garching, Germany
{schulzef,petter,seidl}@in.tum.de
http://www2.in.tum.de/~{schulzef,petter,seidl}

Abstract. We prove that all valid Herbrand equalities can be inter-procedurally inferred for programs where all assignments are taken into account whose right-hand sides depend on at most one variable. The analysis is based on procedure summaries representing the weakest pre-conditions for finitely many generic post-conditions with template variables. In order to arrive at effective representations for all occurring weakest pre-conditions, we show for almost all values possibly computed at run-time, that they can be uniquely factorized into tree patterns and a terminating ground term. Moreover, we introduce an approximate notion of subsumption which is effectively decidable and ensures that finite conjunctions of equalities may not grow infinitely. Based on these technical results, we realize an effective fixpoint iteration to infer all inter-procedurally valid Herbrand equalities for these programs.

How can we infer that an equality such as $\mathbf{x} \doteq \mathbf{y}$ holds at some program point, if the operators by which the program variables $\mathbf{x}$ and $\mathbf{y}$ are computed, do not satisfy obvious algebraic laws? This is the case, e.g., when either very high-level operations such as sqrt, or very low-level operations such as bit-shift are involved or, generally, for floating-point calculations. Still, the equality $\mathbf{x} \doteq \mathbf{y}$ can be inferred, if $\mathbf{x}$ and $\mathbf{y}$ are computed by means of *syntactically* identical terms of operator applications. The equality then is called *Herbrand* equality. The problem of inferring valid Herbrand equalities dates back to [1] where it was introduced as the famous *value numbering* problem. Since quite a while, algorithms are known which, in absence of procedures, infer *all* valid Herbrand equalities [11,21]. These algorithms can even be tuned to run in polynomial time, if only invariants of polynomial size are of interest [7]. Surprisingly little is known about Herbrand equalities if recursive procedure calls are allowed. In [17] it has been observed that the intra-procedural techniques can be extended to programs with local variables and *functions* – but without global variables. The ideas there are strong enough to generally infer all Herbrand *constants* in programs with procedures and both local and global variables, i.e., invariants of the form $\mathbf{x} \doteq t$ where t is ground. Another tractable case of invariants is obtained if only assignments are taken into account whose right-hand sides have at most *one occurrence* of a variable [18]. Thus, assignment $\mathbf{x} = f(\mathbf{y}, a);$ is considered while assignments such as $\mathbf{x} = f(\mathbf{y}, \mathbf{y});$ or $\mathbf{x} = f(\mathbf{y}, \mathbf{z});$ are approximated with $\mathbf{x} = ?;$,

© Springer-Verlag Berlin Heidelberg 2015
J. Vitek (Ed.): ESOP 2015, LNCS 9032, pp. 457–482, 2015.
DOI: 10.1007/978-3-662-46669-8_19

i.e., by an assignment of an unknown value to $\mathbf{x}$. The idea is to encode ground terms as numbers. Then Herbrand equalities can be represented as polynomial equalities with a fixed number of variables and of bounded degree. Accordingly, techniques from linear algebra are sufficient to infer all valid Herbrand equalities for such programs. As a special case, Petter's class of programs from [18] subsumes those programs where only *unary* operators are involved. Such programs have been considered by [8]. Interestingly, the latter paper arrives at decidability by a completely different line of argument, namely, by exploiting properties of the free monoid generated from the unary operators. Another avenue to decidability is to restrict the control structure of programs to be analyzed. In [5], the restricted class of *Sloopy* Programs is introduced where the format of loop as well as recursion is drastically restricted. For this class an algorithm is not only provided to decide arbitrary equalities between variables but also disequalities.

On the other hand, when only affine numerical expressions as well as affine program invariants are of concern, the set of valid invariants at a program point form a *vector space* which can be effectively represented. This observation is exploited in [14] to apply methods from linear algebra to infer all valid affine program invariants. These methods later have been adapted to the case where values of variables are not from a field, but where integers will overflow at some power of 2, i.e., are taken from a modular ring. Note that in the latter structure, some number different from 0 may be a zero divisor and thus does not have a multiplicative inverse [15]. For some applications, an analysis of *general* equalities is not necessary. In applications such as coalescing of registers [16] or detection of local variables in low-level code [4], it suffices to infer equalities involving two variables only. In the affine case, algorithms for inferring all two-variable equalities can be constructed which have better complexities as the corresponding algorithms for general equalities [4].

The question whether or not *all* inter-procedurally valid Herbrand equalities can be inferred, is still open. Here, we consider the case of Herbrand equalities containing two variables only. These are equalities such as $\mathbf{x} \doteq f(g(\mathbf{y}), \mathbf{y}, a)$, i.e., right-hand sides of equalities may contain only a single variable, but this multiple times. Accordingly, in programs only assignments are taken into account whose right-hand sides contain (arbitrarily many) occurrences of at most one variable. Our main result is that under this provision, *all* inter-procedurally valid two-variable Herbrand equalities can be inferred.

Our novel analysis is based on calculating weakest pre-conditions for all occurring post-conditions. Since there may be infinitely many potential post-conditions for a called procedure, we rely on *generic* post-conditions to obtain finite representations of procedure summaries. In a generic post-condition *second-order* variables are used as place-holders for yet unknown relationships between program variables. In the generic post-condition

$$A(\mathbf{x}) \doteq B(\mathbf{y})$$

the second-order variables A and B take as values terms with (possibly multiple occurrences of) *holes* (which we call *templates*). To realize our algorithm for inferring all inter-procedurally valid two-variable equalities, we thus require

- a method to finitely represent all occurring conjunctions of equalities,
- a method for proving that one conjunction subsumes another conjunction, i.e., a method to detect when the greatest fixpoint computation has terminated;
- a guarantee that the fixpoint ever will be reached.

Note here that the equalities occurring during the weakest pre-condition computation of a generic post-condition may contain occurrences of second-order variables. Thus, subsumption between conjunctions of equalities is subtly related to second-order unification [6]. Second-order unification asks whether a conjunction of equalities possibly containing second-order variables is satisfiable. Since long, it is known that generally, second-order unification is undecidable. Undecidability of second-order unification even holds if only a single unary second-order variable is involved [12]. In contrast, the problem of *context* unification, i.e., the variant of second-order unification where second-order variables range over terms with single occurrences of holes only, has recently been proven to be decidable [10]. It is worth mentioning that neither of the two cases directly applies to our application, since we consider unary second-order variables (as context unification) but let variables range over terms with one or multiple occurrences of holes (differently from context unification). To the best of our knowledge, decidability of satisfiability is still open for our case.

In this paper, we will not solve the satisfiability problem for the given unification problem. Instead, we introduce two novel ideas to circumvent this problem and still infer all inter-procedurally valid two-variable Herbrand equalities. First, we introduce a notion of *approximate* subsumption. This means that our algorithm does not allow to prove implications between all conjunctions of equalities — but at least sufficiently many so that accumulation of *infinite* conjunctions is ruled out. Second, we note that subsumption is not required for arbitrary valuations of program variables. Instead it suffices to consider values which may possibly be constructed by the program at run-time. For programs where every right-hand side of assignments contain occurrences of single variables only, we observe that the ground terms possibly occurring at run-time, have a specific structure, which allows for a *unique factorization* of these terms into irreducible templates — at least, if these ground terms are sufficiently *large*. Our factorization result applied to these kind of values, enables us to make use of the monoidal methods of [8]. This approach, which works for sufficiently large terms, then is complemented with a dedicated treatment of finitely many exceptional cases. By that, we ultimately succeed to construct an effective approximative subsumption algorithm which allows us to restrict the number of equalities in occurring conjunctions and to determine all valid two-variable Herbrand equalities.

In order to arrive at our key result, namely an algorithm to infer all valid inter-procedural two-variable Herbrand equalities, we thus build on the following two novel technical constructions:

- a method to uniquely factorize the kind of values possibly occuring at run-time (except finitely many) of a given program;
- a notion of approximative subsumption which is decidable and still guarantees that every occurring conjunction of equalities is effectively equivalent to a finite conjunction.

Subsequently, we sketch how not only all two-variable equalities, but *all* interprocedurally valid Herbrand equalities can be inferred, if only all right-hand sides in assignments each contain occurrences of at most one variable.

Our paper is organized as follows. Section 1 briefly introduces our programming model. Section 2 presents our basic **WP** based approach of inferring all valid program invariants. In Section 3, we provide general background on the cancellation and factorization properties of terms and prove a first compactness result for equalities with template variables but no occurrences of program variables. In Section 4 we then provide an algorithm for inferring all two-variable equalities — at least, for programs which are *initialization-restricted* (see Section 4 for a precise definition of this restriction). Technically, this restriction implies that all occurring terms can be uniquely factorized into irreducible terms. In order to arrive at an algorithm for programs which are not initialization-restricted, we complement this approach in Section 5 with a dedicated treatment of values where a unique factorization is not possible. Finally, Section 6 indicates how our methods can be extended to general Herbrand equalities.

1   Programs

For the purpose of this paper, we consider imperative programs which consist of a finite set P of procedures such as:

```
0:   Herbrand x, y;
1:   main() {                    6:   p() {
2:       x = a;                   7:       if (*) {
3:       y = a;                   8:           x = f(x, x);
4:       p();                     9:           p();
5:   }                           10:           y = f(y, y);
                                 11:       }
                                 12:   }
```

Instead of operating on the syntax of programs, we prefer to represent each procedure by a (non-deterministic) control flow graph. Figure 1 shows, e.g., the control flow graphs for the given example program. Formally, the control flow graph for a procedure p consists of:

- A finite set N_p of program points where $s_p, r_p \in N_p$ represent the start and return point of the procedure p;
- A finite set E_p of edges (u, s, v) where $u, v \in N_p$ are program points and s denotes a basic statement.

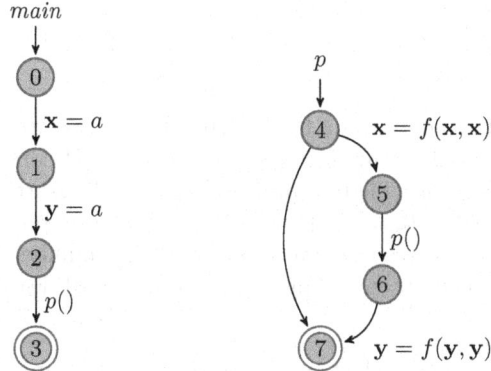

Fig. 1. The corresponding CFGs for the example program

For simplicity, we proceed in the style of Sharir/Pnueli in [20] and consider pa-
rameterless procedures which operate on global variables only. In the following,
X denotes the finite set of program variables. As *values*, we consider uninterpreted operator expressions only. Thus, values are constructed from atomic values by means of (uninterpreted) operator applications. Let Ω denote a signature
containing a non-empty set of atomic values Ω_0 and sets $\Omega_k, k > 0$, of constructors of rank k. Then $\mathcal{T}_\Omega$ denotes the set of all possible (ground) terms over Ω,
and $\mathcal{T}_\Omega(\mathbf{X})$ the set of all possible terms over Ω and (possibly) occurrences of program variables from **X**. In general, we will omit brackets around the argument
of unary symbols. Thus, we may, e.g., write $h\mathbf{x}$ instead of $h(\mathbf{x})$.

As basic statements, we only consider assignments and procedure calls. An
assignment $\mathbf{x} = ?$ non-deterministically assigns *any* value to the program variable
x, whereas an assignment $\mathbf{x} = t$ assigns the value constructed according to
the right-hand side term $t \in \mathcal{T}_\Omega(\mathbf{X})$. A procedure call is of the form $p()$ for a
procedure name p.

In this paper, we only consider assignments whose right-hand sides contain
occurrences of at most one variable. The assignments occurring in the example
program from Figure 1 have this property. Note that this program does not fall
into Petter's class, since the right-hand sides of assignments contain more than
one occurrence of a variable. In *general* programs with arbitrary assignments,
the assignments with right-hand sides not conforming to the given restriction
may, e.g., be abstracted by the non-deterministic assignment of *any* value.

2   Computing Weakest Pre-conditions

In order to prove a given assertion or infer all valid invariants, we would like
to calculate weakest pre-conditions, to determine for every program point the
assumptions to be met for the queried assertion to hold at the given program
point. Since the program model makes use of non-deterministic branching, we
may assume w.l.o.g. that every program point is *reachable*. In particular, this

implies that no procedure is definitely non-terminating, i.e., that for every procedure p, there is at least one execution path from the start point of p reaching the end point of p.

Example 1. Consider the program from Figure 1. At program exit, the invariant $\mathbf{x} \doteq \mathbf{y}$ holds. In a proof of this fact by means of a **WP** computation, weakest pre-conditions must be provided for procedure p and all assertions $\mathbf{x} \doteq t_k$, $k \geq 0$, where $t_0 = \mathbf{y}$ and for $k > 0$, $t_k = f(t_{k-1}, t_{k-1})$. This set of post-conditions is not only infinite, but also makes use of an ever increasing number of variable occurrences. Thus, an immediate encoding, e.g., into bounded degree polynomials as in [18] is not obvious.   □

In order to summarize the effect of a procedure for multiple but similar post-conditions, we tabulate the weakest pre-conditions for *generic* post-conditions only. Generic post-conditions are assertions which contain *template variables* which later may be instantiated differently in different contexts for arriving post-conditions. This idea has been applied, e.g., for affine equalities [14,16,4], for polynomial equalities [13,18], or for Herbrand equalities with unary operators [8]. The generic post-conditions which are of interest here, are of the forms

$$A\mathbf{x} \doteq C \quad \text{or} \quad A\mathbf{x} \doteq B\mathbf{y}$$

where $\mathbf{x}, \mathbf{y}$ are program variables, the *ground* template variable C is meant to receive a constant value, and the template variables A, B take *templates* as values, i.e., terms over the ranked alphabet Ω and having at least one occurrence of the (fresh) place holder variable $\bullet$. Computing weakest pre-conditions operates on assertions where an assertion is a (possibly infinite) conjunction of equalities. The equalities occurring during weakest pre-condition calculations are of the forms:

$$As \doteq C \quad \text{or} \quad As \doteq Bt$$

where s, t are terms possibly containing a program variable, i.e., $s, t \in \mathcal{T}_\Omega(\mathbf{X})$.

Consider a mapping σ which assigns *appropriate values* to the program variables from $\mathbf{X}$ as well as to the (non-ground or ground) template variables A, B, C. This means that σ assigns ground terms to the variables in $\mathbf{X} \cup \{C\}$ and templates to A, B. Such a mapping is called *variable assignment*. The variable assignment σ *satisfies* the equality $s \doteq t$ ($\sigma \models (s \doteq t)$ for short) iff $\sigma^*(s) = \sigma^*(t)$ where σ^* is the natural tree homomorphism corresponding to σ, which is the identity on all operators in Ω. The homomorphism σ^* maps, e.g., the application At of the template variable A to the term t into $\sigma(A)[\sigma^*(t)/\bullet]$, i.e., the *substitution* of the term $\sigma^*(t)$ into the occurrences of the dedicated variable $\bullet$ in the template $\sigma(A)$. Substitution into the dedicated variable $\bullet$ is an associative binary operation where the neutral element is the template consisting of $\bullet$ alone. In the following, we denote this operation by juxtaposition.

Consider, e.g., an assignment σ with $\sigma(A) = h(\bullet, \bullet)$, and $\sigma(B) = \bullet$, and $\sigma(\mathbf{x}) = a$. Then

$$\sigma^*(A\mathbf{x}) = h(\bullet, \bullet)\, a = h(a, a) = \bullet\, h(a, a) = \sigma^*(Bh(\mathbf{x}, a))$$

holds. Therefore, σ satisfies the equality $A\mathbf{x} \doteq Bh(\mathbf{x}, a)$. In the following, we will no longer distinguish between σ and σ^*.

The variable assignment σ satisfies the conjunction ϕ of equalities ($\sigma \models \phi$ for short), iff $\sigma \models e$ for all equalities $e \in \phi$.

In our application, it will be convenient not to consider arbitrary variable assignments, but only those which map program variables to *reasonable* values as shown in the following. For a subset $T \subseteq \mathcal{T}_\Omega$ of ground terms, we call a variable assignment σ a T-*assignment*, if σ maps program variables $\mathbf{x}$ to values $\sigma(\mathbf{x}) \in T$ only.

The conjunction ϕ then is called T-*satisfiable* if there is some T-assignment σ with $\sigma \models \phi$. Otherwise, it is T-*unsatisfiable*. Conjunctions ϕ, ϕ' are T-*equivalent* if for every T-assignment σ, $\sigma \models \phi$ iff $\sigma \models \phi'$. Obviously, an empty conjunction is satisfied by every variable assignment and therefore equal to $\top$ (true), while all T-unsatisfiable conjunctions are T-equivalent. As usual, these are denoted by $\bot$ (false). Finally, a conjunction ϕ' is T-*subsumed* by a conjunction ϕ, if ϕ is T-equivalent to $\phi \wedge \phi'$.

If the set T by which we have relativized the notions of satisfiability, equivalence and subsumption equals the full set $\mathcal{T}_\Omega$, we may also drop the prefixing with T. In particular, we have for any T that satisfiability, equivalence and subsumption imply T-satisfiability, T-equivalence and T-subsumption, while the reverse implication may not necessarily hold.

In the following, we recall the ingredients of weakest pre-condition computation for assignments as well as for procedure calls as provided, e.g. in [9] or [2]. The weakest pre-condition of ϕ w.r.t. assignments are given by:

$$[\![\mathbf{x} = t]\!]^\top \, \phi = \phi[t/\mathbf{x}]$$
$$[\![\mathbf{x} = ?]\!]^\top \, \phi = \forall \mathbf{x}. \, \phi$$

Thus, the weakest pre-condition for an assignment $\mathbf{x} = t$ is given by substitution of the term t into all occurrences of the variable $\mathbf{x}$ in the post-conditions, while the weakest pre-condition for a non-deterministic assignment $\mathbf{x} = ?$ of any value is given by universal quantification. For Herbrand equalities, universal quantification can be computed as follows. Recall that universal quantification commutes with conjunction. Therefore, it suffices to consider single equalities e. If $\mathbf{x}$ does not occur in e, then $\forall \mathbf{x}. \, e$ is equivalent to e. If $\mathbf{x}$ occurs only on one side of e, then $\forall \mathbf{x}. \, e = \bot$. Now assume that $\mathbf{x}$ occurs on both sides of e. If e is of the form $s\mathbf{x} \doteq t\mathbf{x}$ for templates s, t (no template variables), then either $s = t$ and hence e as well as $\forall \mathbf{x}. \, e$ is equivalent to $\top$, or $s \neq t$, in which case $\forall \mathbf{x}. \, e$ equals $\bot$. If e is of the form $As\mathbf{x} \doteq Bt\mathbf{x}$ for templates s, t, then $\forall \mathbf{x}. \, e$ is equivalent to $As \doteq Bt$.

Every transformation f which is specified for generic post-conditions to conjunctions of pre-conditions, can be uniquely extended to a transformation $\bar{f}$ of *arbitrary* post-conditions by

$$\bar{f}(\bigwedge E) \quad = \quad \bigwedge_{e \in E} \bar{f}(e)$$

where the transformation $\bar{f}$ for an arbitrary equality e is defined as follows:

$$\bar{f}(s \doteq t) = \begin{cases} f(Ax \doteq By)[s'/A, t'/B] & \text{if } s = s'\mathbf{x}, t = t'\mathbf{y} \\ f(Ax \doteq C)[s'/A, t/C] & \text{if } s = s'\mathbf{x}, t \text{ ground} \\ f(Ax \doteq C)[s/C, t'/A] & \text{if } t = t'\mathbf{x}, s \text{ ground} \\ s \doteq t & \text{otherwise} \end{cases}$$

Subsequently, the extended function $\bar{f}$ is denoted by f as well. The procedure summaries are then characterized by the constraint system $\mathbf{S}$:

$$\begin{array}{lll} [\![r_p]\!]^\mathsf{T} & \Longrightarrow \mathsf{Id} & \text{for each procedure } p \\ [\![u]\!]^\mathsf{T} & \Longrightarrow [\![s_p]\!]^\mathsf{T} \circ [\![v]\!]^\mathsf{T} & \text{for each } (u, p(), v) \in E \\ [\![u]\!]^\mathsf{T} & \Longrightarrow [\![s]\!]^\mathsf{T} \circ [\![v]\!]^\mathsf{T} & \text{for each } (u, s, v) \in E, \\ & & s \text{ assignment} \end{array}$$

where $\circ$ means the composition of the weakest pre-condition transformers and Id is the identity transformer. Thus, accumulation of weakest pre-conditions for a generic post-condition e at procedure exit r_p with e and then propagates its pre-conditions backward to the start point of p by applying the transformations corresponding to the traversed edges. Here, the subsumption relation $\Longrightarrow$ as defined for conjunction of equalities, has silently been raised to the function level. Thus, $f \Longrightarrow g$ if $f(e)$ subsumes $g(e)$ for all generic post-conditions e.

W.r.t. the ordering $\sqsubseteq$ given by $\Longrightarrow$, the **WP** transformer of procedure p then is obtained as the value for the variable corresponding to the start point s_p in the *greatest* solution to the constraint system $\mathbf{S}$.

The **WP** transformers for all program points are characterized by the greatest solution of the constraint system $\mathbf{R}$:

$$\begin{array}{lll} [s_{main}]^\mathsf{T} & \Longrightarrow \mathsf{Id} & \\ [s_p]^\mathsf{T} & \Longrightarrow [u]^\mathsf{T} & \text{for each } (u, p(), \_) \in E \\ [v]^\mathsf{T} & \Longrightarrow [u]^\mathsf{T} \circ [\![s_p]\!]^\mathsf{T} & \text{for each } (u, p(), v) \in E \\ [v]^\mathsf{T} & \Longrightarrow [u]^\mathsf{T} \circ [\![s]\!]^\mathsf{T} & \text{for each } (u, s, v) \in E, \\ & & s \text{ assignment} \end{array}$$

The value for $[v]^\mathsf{T}$ for program point v is meant to transform every assertion at program point v, into the corresponding weakest pre-condition at the start point of the program. Note that the constraint system for characterizing these functions makes use of the weakest pre-condition transformers of procedures as characterized by the constraint system $\mathbf{S}$.

Assume that we are somehow given the greatest solution of the constraint system $\mathbf{R}$ where $[v]^\mathsf{T}$ is the corresponding transformation for program point v. In order to determine all one- or two-variable equalities which are valid when reaching the program point v, we conceptually proceed as follows:

One-variable Equality. For a program variable $\mathbf{x}$, let ψ denote the *universal closure* of $[v]^\mathsf{T}(A\mathbf{x} \doteq C)$. If $\psi = \bot$, then program variable $\mathbf{x}$ does not receive a constant value at program point v. Otherwise ψ is equivalent to an equality $As \doteq C$ where s is ground, i.e., $\mathbf{x} \doteq s$ is an invariant at v.

Two-variable Equality. For distinct program variables $\mathbf{x}$ and $\mathbf{y}$, let ψ denote the universal closure of $[v]^{\top}(A\mathbf{x} \doteq B\mathbf{y})$. If $\psi = \bot$, then no equality between $\mathbf{x}$ and $\mathbf{y}$ holds. Otherwise, ψ equals a conjunction of equalities $As_i \doteq Bt_i$, $i \in I$, for some index set I where for each $i \in I$, s_i, t_i are both ground. Then $r_1\mathbf{x} \doteq r_2\mathbf{y}$ is an invariant at v iff $r_1s_i \doteq r_2t_i$ for all i, i.e., any assignment σ with $\sigma(A) = r_1, \sigma(B) = r_2$ satisfies the conjunction.

Here, the *universal closure* of a conjunction ϕ is given by $\forall \mathbf{x}_1 \ldots \forall \mathbf{x}_n.\phi$, if the set of program variables equals $\mathbf{X} = \{\mathbf{x}_1, \ldots, \mathbf{x}_n\}$.

Example 2. Consider the main procedure of the program in Section 1, as defined by the control flow graph in Figure 1. The **WP** transformer $[3]^{\top}$ for the endpoint 3 of the main program is given by:

$$[3]^{\top} = [\mathbf{x} = a]^{\top} \circ [\mathbf{y} = a]^{\top} \circ [4]^{\top}$$

where 4 is the entry point of the procedure p. Assume that

$$[4]^{\top}(A\mathbf{x} \doteq B\mathbf{y}) = (A\mathbf{x} \doteq B\mathbf{y}) \wedge (Af(\mathbf{x}, \mathbf{x}) \doteq Bf(\mathbf{y}, \mathbf{y}))$$

holds. For the program variables $\mathbf{x}, \mathbf{y}$, we therefore obtain:

$$\begin{aligned} [3]^{\top}(A\mathbf{x} \doteq B\mathbf{y}) &= (A\mathbf{x} \doteq B\mathbf{y})[a/\mathbf{y}][a/\mathbf{x}] \wedge (Af(\mathbf{x}, \mathbf{x}) \doteq Bf(\mathbf{y}, \mathbf{y}))[a/\mathbf{y}][a/\mathbf{x}] \\ &= (Aa \doteq Ba) \wedge (Af(a, a) \doteq Bf(a, a)) \end{aligned}$$

This assertion does not contain occurrences of the program variables $\mathbf{x}, \mathbf{y}$. Therefore, it is preserved by universal quantification over program variables. Since $A = B = \bullet$ is a solution, $\mathbf{x} \doteq \mathbf{y}$ holds whenever program point 3 is reached. $\square$

In order to turn these definitions into an effective analysis algorithm, several obstacles must be overcome. So, it is not clear how general subsumption, as required in our characterization of the **WP** transformers, can be decided in presence of template variables. We observe, however, that instead of general subsumption, it suffices to rely on T-subsumption only — for a well-chosen subset $T \subseteq \mathcal{T}_\Omega$. Note that the smaller the set T is, the coarser is the subsumption relation. In particular for $T = \emptyset$, all conjunctions are T-equivalent. Since every assertion expresses a property of reaching program states, it suffices for our application to choose T as a superset of all run-time values of program variables.

The following wish list collects properties which enable us to construct an effective inter-procedural analysis of all two-variable Herbrand equalities:

T-**Compactness.** Every occurring conjunction ϕ is T-subsumed by a conjunction of a *finite* subset of equalities in ϕ.

Effectiveness of Subsumption. T-subsumption for *finite* conjunctions can be effectively decided.

Solvability of Ground Equalities. The set of solutions of finite systems of equalities with template variables only, i.e., *without* occurrences of program variables can be explicitly computed.

By the first assumption, a standard fixpoint iteration for the constraint systems **S** and **R** will terminate after finitely many iterations (up to T-equivalence). By the second assumption, termination can effectively be detected, while the third assumption guarantees that for every program point and every program variable (pair of program variables) the set of all valid invariants can be extracted out of the greatest solution of **R**. In total, we arrive at an effective algorithm for inferring all valid two-variable equalities.

The assumption on decidability of T-subsumption can be further relaxed. Instead, we provide an *approximate* notion of T-subsumption which is decidable. Our approximate T-subsumption implies T-subsumption. Moreover, it is still strong enough to guarantee that every occurring conjunction of equalities is approximately T-subsumed by a finite subset of the equalities. Notions for approximate T-subsumption will be introduced in Sections 4 and 5.

In the upcoming section, we recall basic properties of the set of terms, possibly containing the variable •. These properties will allow us to deal with conjunctions of equalities where template variables are applied to ground terms only, i.e., the case of ground equalities.

3   Factorization of Terms

Let $\mathcal{T}_\Omega(\bullet)$ denote the set of terms constructed from the symbols in Ω, possibly together with the dedicated variable •. In [3], Engelfriet presents the following cancellation and factorization properties for terms in $\mathcal{T}_\Omega(\bullet)$:

Bottom Cancellation
   Assume that $t_1 \neq t_1'$. Then $s_1 t_1 = s_2 t_1$ and $s_1 t_1' = s_2 t_1'$ implies $s_1 = s_2$.
Top Cancellation
   Assume • occurs in s. Then $s t_1 = s t_2$ implies $t_1 = t_2$.
Factorization
   Assume $t_i \neq t_i'$ for $i = 1, 2$. Then $s_1 t_1 = s_2 t_2$ and $s_1 t_1' = s_2 t_2'$ implies that $s_1 r_1 = s_2 r_2$ for some r_1, r_2 each containing • where at least one of the r_i equals •. In that case (by top cancellation), we furthermore have that both $r_2 t_1 = r_1 t_2$ and $r_2 t_1' = r_1 t_2'$.

Using these cancellation properties, we obtain a complete method for dealing with equalities *without* occurrences of program variables.

For one-variable equalities alone, we have the following results concerning subsumption and compactness:

Theorem 1

1. *A single equality $As \doteq C$ for some ground term s has exactly one solution where $A = \bullet$.*
2. *Consider the conjunction $As_1 \doteq C \wedge As_2 \doteq C$ for terms $s_1 \neq s_2$ containing the same variable **x**. If the conjunction is satisfiable, then the value of **x** is uniquely determined.*

Proof. We only prove the second assertion. The conjunction $As_1 \doteq C \wedge As_2 \doteq C$ is equivalent to the conjunction $As_1 \doteq C \wedge s_1 \doteq s_2$. The most general unifier of s_1, s_2 maps $\mathbf{x}$ to a ground subterm of s_1, s_2 if the conjunction is satisfiable.   □

As a consequence, we obtain:

Corollary 1. *Consider finite conjunctions of equalities of the form $As \doteq C$.*

1. *Subsumption for these is decidable.*
2. *Every satisfiable conjunction is equivalent to a conjunction of at most $n + 1$ equalities where n is the number of program variables.*

Since the weakest pre-condition of a generic one-variable equality consists of equalities of the form $As \doteq C$ only, Corollary 1 suffices to infer all inter-procedurally valid one-variable equalities. In the following, we therefore concentrate on the two-variable case where the weakest pre-condition consists of conjunctions of equalities of the form $As \doteq Bt$. First, we observe:

Theorem 2

1. *A single equality $As \doteq Bt$ for ground terms s, t has only finitely many solutions $A = r_1, B = r_2$ with templates r_1, r_2 of which at least one equals •.*
2. *Consider the conjunction $As_1 \doteq Bt_1 \wedge As_2 \doteq Bt_2$ for ground terms $s_1 \neq s_2$ and $t_1 \neq t_2$. Then it has either no solution or there are templates r_1, r_2 of which at least one equals • such that the conjunction is equivalent to $Ar_1 \doteq Br_2$. In the latter case, $A = r_2$, $B = r_1$ is the single solution where at least one of the templates equals •.*
3. *Consider the (finite) conjunction $\bigwedge_{i=1}^{k}(As_i \doteq Bt_i)$ for ground terms s_i, t_i. Then the set of all solutions where either the template for A or for B equals •, can be effectively computed.*

Proof For a proof of the first statement, w.l.o.g. assume that s is at least as large as t. Then for size reasons, $r_1 = \bullet$. This means that $s = r_2 t$ must hold. If t is not a subterm of s, there is no solution at all. Otherwise, i.e., if s contains occurrences of t, then every solution r_2 is obtained from s by replacing a non-empty set of occurrences of t with •.

Now consider the second statement. If the pair of equalities is satisfiable then by factorization, there are templates r_1, r_2 of which at least one equals • such that $Ar_1 \doteq Br_2$ holds. Since at the same time $r_2 s_i \doteq r_1 t_i$ holds, the equality $Ar_1 \doteq Br_2$ is equivalent to the conjunction. Moreover, there is exactly one solution $A = r_1', B = r_2'$ where at least one of the templates r_i' equals •, namely, $r_1' = r_2, r_2' = r_1$.

Finally, consider the third statement. If $k = 1$, the assertion follows from statement 1. Therefore now let $k > 1$. First assume that for some i, j, $s_i \neq s_j$ and $t_i \neq t_j$. Then by statement 2, the conjunction is unsatisfiable or there is exactly one pair r_1, r_2 of templates one of which equals •, such that $A = r_1, B = r_2$ is a solution of the conjunction $As_i \doteq Bt_i \wedge As_j \doteq Bt_j$. If in the latter case, $r_1 s_l \doteq r_2 t_l$ for all l, we have obtained a single solution. Otherwise, the conjunction is unsatisfiable. Now assume that no such i, j exists. Then either the conjunction is unsatisfiable or all equalities are syntactically equal.   □

Example 3 Consider the two equalities:

$$Af(a, gb, gb) \doteq Bgb \qquad Af(a, gc, gb) \doteq Bgc$$

Then $A = \bullet$ and $B = f(a, \bullet, gb)$ is the only solution for A, B where at least one of the templates equals $\bullet$.   □

Applying the arguments which we used to prove Theorem 2, we obtain:

Corollary 2. *Consider a conjunction $\bigwedge_{i=1}^{n} As_i \doteq Bt_i$ with ground terms s_i, t_i.*

1. *If it is satisfiable, it is equivalent to the conjunction of at most two conjuncts.*
2. *If it is unsatisfiable, there are at most three conjuncts whose conjunction is unsatisfiable.*

By Theorem 2, the assumption **solvability of ground equalities** from Section 2 is met. Thus, it remains to solve the constraint systems **S** and **R**, i.e., to construct an approximate T-subsumption relation which is both effective and guarantees that every conjunction is approximately T-subsumed by the conjunction of a finite subset of equalities. In order to construct such a relation, we require stronger insights into the structure of templates and their compositions. Let $\mathcal{C}_\Omega$ denote the subset of all terms in $\mathcal{T}_\Omega(\bullet)$ which contain at least one occurrence of $\bullet$, i.e., $\mathcal{C}_\Omega = \mathcal{T}_\Omega(\bullet) \setminus \mathcal{T}_\Omega$. The terms in $\mathcal{C}_\Omega$ have also been called *templates*. The set $\mathcal{C}_\Omega$, equipped with substitution, is a *free monoid* with neutral element $\bullet$. This monoid is *infinitely* generated from the irreducible elements in $\mathcal{C}_\Omega$. As usual, we call an element t *irreducible* if t cannot be non-trivially decomposed into a product, i.e., $t = uv$ implies that $t = u$ with $v = \bullet$ or $t = v$ with $u = \bullet$.

While templates can be uniquely factored, this is no longer the case for ground terms, i.e., terms without variable occurrences.

Example 4. Consider the ground term $t = h(f(h(1), h(1)))$, together with the templates $s_1 = h(f(\bullet, h(1)))$, $s_2 = h(f(h(1), \bullet))$ and $s_3 = h(f(\bullet, \bullet))$. All these three templates are distinct. Still,

$$t = s_1 \, h(\bullet) \, 1 = s_2 \, h(\bullet) \, 1 = s_3 \, h(\bullet) \, 1 \qquad\qquad □$$

Thus, unique factorization of arbitrary ground terms cannot be hoped for. Still, we observe that unique factorization can be obtained — at least up to any fixed finite set of ground terms. Let G denote a finite set of ground terms which is closed by subterms.

Let M_G denote the sub-monoid of all templates $m \in \mathcal{C}_\Omega$ whose ground subterms all are contained in G. Then we have:

Theorem 3. *Assume that $S \subseteq \mathcal{T}_\Omega$ which is closed by subterms. If $G \subseteq S$, then every ground term $t \in \mathcal{T}_\Omega \setminus S$, can be uniquely factored into $t = mx$ such that*

(A) $m \in M_G$ and $x \notin S$;
(B) x is minimal with property *(A)*.

Example 5. Consider the term

$$t = f(h(f(2, h(1))), h(f(2, h(1))))$$

and assume that the set G of forbidden ground subterms is given by $G = \{h(1), 1\}$ and $S = G$. Then t can be decomposed into:

$$f(\bullet, \bullet) \ h(\bullet) \ f(\bullet, h(1)) \ 2$$

If on the other hand, $S = G = \{2\}$, we obtain the decomposition:

$$f(\bullet, \bullet) \ h(\bullet) \ f(2, \bullet) \ h(\bullet) \ 1$$

If finally, S and G are empty, the term x of Theorem 3 is the minimal subterm such that the occurrences of x contains all ground leaves of t. This means that $x = f(2, h(1))$, and we obtain the decomposition:

$$f(\bullet, \bullet) \ h(\bullet) \ f(2, h(1)) \qquad\qquad \square$$

The unique decomposition of the ground term t claimed by Theorem 3, is constructed as follows. Let X denote the set of minimal subterms x' of t such that $x' \notin G$. Then we construct the least subterm $x \notin S$ of t such that all occurrences of subterms $x' \in X$ in t are contained in some occurrence of x. This subterm is uniquely determined. Then define m as the term obtained from t by replacing all occurrences of x with $\bullet$. This term m is also uniquely determined with $t = mx$. Moreover by construction, all ground subterms of m are contained in G.

Example 6. Consider the program from example 1. In this program, no non-ground right-hand side contains ground subterms. Accordingly, the set G is empty. Since the only ground right-hand side equals the atom a, the decomposition Theorem 3 allows to uniquely decompose all run-time values of this program into right-hand sides of assignments. $\qquad\qquad \square$

Theorem 3 allows to extend the monoidal techniques of Gulwani et al. [8] for unary operators to programs where all run-time values can be uniquely factorized into right-hand sides. This extension is given in Section 4. For completeness reasons, we also present simplified versions of the algorithms for monoidal equalities from [8] in Appendix A. The general case where unique factorization of all run-time values can no longer be guaranteed, subsequently is presented in Section 5.

4   Initialization-Restricted Programs

Assume that R is the set of ground right-hand sides of assignments, and G is the set of ground subterms of non-ground right-hand sides of assignments of our program. Then generally, each value x possibly constructed at run-time by the program is of the form $x = x'r$ where $r \in R$ and $x' \in M_G$. This means that for pre-conditions ϕ possibly occurring in a **WP** calculation for a program invariant, we are only interested in variable assignments σ which map each program variable **x** to a possible run-time value for **x**, i.e., to a value from the set $M_G R$. Henceforth, we therefore no longer consider general satisfiability, equivalence and subsumption, but only T-satisfiability, T-equivalence and T-subsumption for $T = M_G R$. This restriction is crucial for the generalization of the monoidal techniques from [8]. In the following, we first consider the sub-class of programs p where set R of ground right-hand sides of p satisfies the two properties:

1. $R \cap G = \emptyset$.
2. The elements in R are mutually incomparable ground terms, i.e., for $r_1, r_2 \in R$, r_1 is a subterm of r_2 iff $r_1 = r_2$.

The program p then is called *initialization-restricted* (or IR for short).

Example 7. Assume that the non-ground right-hand sides of assignments of the program are $f(\mathbf{x}, h(1))$ and $f(2, h(\mathbf{y}))$. Then the set G is given by $G = \{1, h(1), 2\}$. A suitable set R of ground right-hand sides might be, e.g., $R = \{0, a\}$.   □

Our condition here is not as restrictive as it might seem. Programs where each variable is initialized by a non-deterministic assignment, are all IR. The same holds true for programs where all non-ground right-hand sides of assignments do not contain ground terms, and variables are initialized with atoms only. The latter property is met by our example 1. By suitably massaging variable initializations, it also comprises all programs using monadic operators only (as in [8]).

We distinguish between two-variable equalities of the following formats:

$$
\begin{array}{lll}
[F_{\mathbf{x},\mathbf{y}}] & As\mathbf{x} \doteq Bt\mathbf{y} & \text{where } s, t \in M_G \\
[F_{\cdot,\mathbf{x}}] & As \doteq Bt\mathbf{x} & \text{where } s \in T \text{ and } t \in M_G \\
[F_{\mathbf{x},\cdot}] & At\mathbf{x} \doteq Bs & \text{where } s \in T \text{ and } t \in M_G
\end{array}
$$

For each format separately, we observe:

Theorem 4

T-subsumption. *For finite sets E, E' of two-variable equalities of the same format it is decidable whether $\bigwedge E$ T-subsumes $\bigwedge E'$ or not.*

T-compactness. *Every T-satisfiable conjunction of a set E of two-variable equalities of the same format is T-subsumed by a conjunction of a subset of at most three equalities in E.*

For a proof see Appendix B. It relies on the unique factorization property together with the monoidal techniques from Section A. Since T-subsumption is decidable, at least for equalities of the same format, we define an approximate T-subsumption relation $\bigwedge E \implies^\sharp \bigwedge E'$ for conjunctions of equalities as follows. Let E_F and E'_F denote the subsets of equalities of the same format F in E and E', respectively. Then $\bigwedge E \implies^\sharp \bigwedge E'$ holds iff $\bigwedge E_F$ T-subsumes $\bigwedge E'_F$ for all formats F. Hence, by Theorem 4, we obtain:

Corollary 3. *Assume that n is the number of program variables.*

Approximate T-subsumption. *For finite sets E, E' of two-variable equalities, it is decidable whether $\bigwedge E$ approximately T-subsumes $\bigwedge E'$ or not.*

Approximate T-compactness. *Every T-satisfiable conjunction of a set E of two-variable equalities is approximately T-subsumed by a conjunction of a subset of at most $\mathcal{O}(n^2)$ equalities in E.*

Overall, we therefore conclude for IR programs:

Theorem 5. *Assume that p is an IR program. Then for every program point u, the set of all two-variable equalities can be determined that are valid when reaching program point u.*

Proof. By Corollary 3, the greatest solutions of the constraint systems **S** and **R** can be effectively computed. Let $[u]^\mathsf{T}$, u program point, denote the greatest solution of the system **R**. Then the set of valid equalities $s\mathbf{x} \doteq t\mathbf{y}$ between program variables $\mathbf{x}, \mathbf{y}$ is given by the set of solutions to a system of ground equalities which are obtained by universal quantification over all program variables of the conjunction of equalities $[u]^\mathsf{T}(A\mathbf{x} \doteq B\mathbf{y})$. By Theorem 2, a representation of the set of solutions for the template variables A, B in this conjunction can be explicitly computed. Likewise, the set of valid equalities $x \doteq t$ for program variable $\mathbf{x}$ and ground term t can be extracted from the universal quantification over all program variables of the conjunction of equalities $[u]^\mathsf{T}(A\mathbf{x} \doteq C)$. The resulting conjunction may either equal $\bot$ (no constant value for $\mathbf{x}$) or contain only the variable C. Consequently, the possible constant value for $\mathbf{x}$ and program point u can also be effectively computed. This completes the proof. □

Example 8. According to our constructions in Section 2 and Theorem 2, the set of all inter-procedurally valid assertions can be obtained from the greatest solutions to the constraint systems **S** and **R**. Consider, e.g., the constraint system **R** for the recursive procedure p from Section 1, as defined by the control flow graph of Figure 1. If Round-Robin iteration is applied to calculate the transformers $[u]^\mathsf{T}$ for the program points $u = 4, 5, 6, 7$, we obtain for the generic post-condition $A\mathbf{x} \doteq B\mathbf{y}$ the result depicted by Table 1 where in the ith column, we have only displayed pre-conditions which have additionally been attained in the ith iteration for the program points $7, 6, 5$ and 4, respectively. For convenience, we have displayed the terms in equalities according to their unique factorizations. For program point 4, the two equalities after the second iteration, imply:

$$Af(\bullet, \bullet)A^- \doteq Bf(\bullet, \bullet)B^-$$

The second equality for program point 4 together with this identity imply that

$$Af(\bullet, \bullet)A^- Af(\bullet, \bullet)\mathbf{x} \doteq Bf(\bullet, \bullet)B^- Bf(\bullet, \bullet)\mathbf{y}$$

from which the third equality for program point 4 as provided by the third iteration follows. Thus, Round-Robin fixpoint iteration reaches the greatest fixpoint after the third iteration.   □

Table 1. Round-Robin iteration for the procedure p from Figure 1

	1	2	3
7	$A\mathbf{x} \doteq B\mathbf{y}$		
6	$A\mathbf{x} \doteq Bf(\bullet, \bullet)\mathbf{y}$		
5	$\top$	$A\mathbf{x} \qquad \doteq Bf(\bullet, \bullet)\mathbf{y}$	$Af(\bullet, \bullet)\mathbf{x} \qquad \doteq Bf(\bullet, \bullet)f(\bullet, \bullet)\mathbf{y}$
4	$A\mathbf{x} \doteq B\mathbf{y}$	$Af(\bullet, \bullet)\mathbf{x} \doteq Bf(\bullet, \bullet)\mathbf{y}$	$Af(\bullet, \bullet)f(\bullet, \bullet)\mathbf{x} \doteq Bf(\bullet, \bullet)f(\bullet, \bullet)\mathbf{y}$

5   Unrestricted Programs

Our analysis of IR programs relied on the fact that all run-time values of program variables can be uniquely factorized. This was made possible since in IR programs the "bottom end" of values can be uniquely identified by means of the ground right-hand sides from R. In general, though, ground right-hand sides could very well also occur as subterms of other right-hand sides in the program. In this case, we can no longer assume that R serves as such a handy set of end marker terms. At first sight, therefore, the monoidal method seems no longer applicable. A second look, however, reveals that the monoidal method essentially fails only, where program variables take *small* values. Again, let R and G denote the set of all ground right-hand sides and the set of all ground subterms of non-ground right-hand sides of assignments in the program, respectively. We call a term $t \in M_G R$ *small* if it is a ground subterm of a right-hand side of an assignment. Let us denote the (finite) set of all small terms by S. The terms in $M_G R$ which are not small, are called *large*. Let $\bar{R}$ be the set of *minimal* elements in $M_G R$ which are large, i.e., not contained in S. Then by Theorem 3, every large term t, i.e., every term $t \in L$ can be uniquely factored such that $t = mr$ where $m \in M_G$ and $r \in \bar{R}$. For small terms, i.e., for terms in S, on the other hand, we cannot hope for unique factorizations. Since there are finitely many small terms only, we take care of small terms by two means:

- We restrict the formats $[F_{\mathbf{x},\cdot}]$ and $[F_{\cdot,\mathbf{x}}]$ from the last section to the case where the occurring ground terms are large and introduce dedicated subformats $[F_{\mathbf{x},s}]$ and $[F_{s,\mathbf{x}}]$ for each small term s in the equalities.
- For T-subsumption, we single out the case of subsumption w.r.t. assignments of large terms only and treat subsumption w.r.t. assignments assigning small terms separately.

Thus, we now consider the following formats of two-variable equalities:

$$
\begin{array}{lll}
[F_{\mathbf{x},\mathbf{y}}] & A s \mathbf{x} \doteq B t \mathbf{y} & \text{where } s, t \in M_G \\
[F_{.,\mathbf{x}}] & A s \doteq B t \mathbf{x} & \text{where } s \in L \text{ and } t \in M_G \\
[F_{s,\mathbf{x}}] & A s \doteq B t \mathbf{x} & \text{where } s \in S \text{ and } t \in M_G \\
[F_{\mathbf{x},.}] & A t \mathbf{x} \doteq B s & \text{where } s \in L \text{ and } t \in M_G \\
[F_{\mathbf{x},s}] & A t \mathbf{x} \doteq B s & \text{where } s \in S \text{ and } t \in M_G
\end{array}
$$

In the following, let us call a substitution σ of program variables *small*, if for every program variable $\mathbf{x}$, $\sigma(\mathbf{x})$ either equals $\mathbf{x}$ or is a small ground term. The notions of satisfiability, equivalence and subsumption restricted to the set T can be inferred by means of the corresponding notions restricted to the set L of large terms only. We have:

- A conjunction ϕ of equalities is T-satisfiable iff there is a small substitution σ such that $\sigma(\phi)$ is L-satisfiable.
- A conjunction ϕ T-subsumes an equality e, iff for every small substitution σ, $\sigma(\phi)$ L-subsumes $\sigma(e)$.

According to this observation, it seems plausible to consider the analogue of Theorem 4 for L-subsumption and L-compactness only. We obtain:

Theorem 6

L-subsumption. For finite sets E, E' of two-variable equalities of the same format it is decidable whether $\bigwedge E$ L-subsumes $\bigwedge E'$ or not.

L-compactness. Every L-satisfiable conjunction of a set E of two-variable equalities of the same format is L-subsumed by a conjunction of a subset of at most three equalities in E.

Proof For equalities of the formats $[F_{\mathbf{x},\mathbf{y}}], [F_{\mathbf{x},.}], [F_{.,\mathbf{x}}]$ the proofs are analogous to the corresponding proofs for Theorem 4 where the set T is replaced with the set $L = M_G \bar{R}$, i.e., instead of the set R we rely on the set $\bar{R}$. Therefore now consider equalities of the format $[F_{s,\mathbf{x}}]$ for a small term $s \in S$.

W.l.o.g., let $A s \doteq B t \mathbf{x}$ and $A s \doteq B t' \mathbf{x}$ be two equalities of this format. If $t \neq t'$, then their conjunction is either contradictory, or $t\mathbf{x}, t'\mathbf{x}$ have a ground unifier which maps $\mathbf{x}$ to a value from G — in contradiction to the assumption that $\mathbf{x}$ takes values from L only.

Therefore, each conjunction of a set E of equalities of the format $[F_{s,\mathbf{x}}]$ either is L-equivalent to $\bot$ or to a single equality in E, and the assertion of the theorem follows. The same argument also applies for the format $[F_{\mathbf{x},s}]$.   □

Given that L-subsumption is decidable, at least for equalities of the same format, and that also L-compactness holds, we define an approximate T-subsumption relation $\bigwedge E \implies^\sharp \bigwedge E'$ as follows. Let E_F and E'_F denote the subsets of equalities of format F, in E and E', respectively. Then $\bigwedge E \implies^\sharp \bigwedge E'$ holds iff for all small substitutions σ, $\bigwedge \sigma(E_F)$ L-subsumes $\bigwedge \sigma(E'_F)$ for all formats F. As a consequence of Theorem 6, we obtain:

Theorem 7. *Assume that n is the number of program variables and m is the cardinality of the set S of small terms.*

Approximate T-subsumption. *For finite sets E, E' of two-variable equalities, it is decidable whether $\bigwedge E$ approximately T-subsumes $\bigwedge E'$ or not.*

Approximate T-compactness. *Every T-satisfiable conjunction of a set E of two-variable equalities is approximately T-subsumed by a conjunction of a subset of at most $\mathcal{O}(n^2 \cdot m^2)$ equalities in E.*

A proof is provided in the long version of this paper [19]. Due to Theorem 7, representations of the greatest solutions of the constraint systems **S** and **R** can be effectively computed. By that, we arrive at our main result:

Theorem 8. *Assume that all right-hand sides of assignments in an arbitrary program contain at most one variable. Then all valid inter-procedurally two-variable Herbrand equalities can be inferred.*

The proof is analogous to the proof of Theorem 5 — only that Theorem 7 is used instead of Corollary 3.

Example 9. Consider a variant of the program from Section 1 where the non-ground assignments are given by:

$$\mathbf{x} = f(\mathbf{x}, a, \mathbf{x}) \qquad \text{and} \qquad \mathbf{y} = f(\mathbf{y}, a, \mathbf{y})$$

The set of small terms then is given by $S = \{a\}$, while the set of smallest large terms is given by $\bar{R} = \{f(a, a, a)\}$.

Now consider the constraint system **R** for the recursive procedure p as defined by the control flow graph of Figure 1 with the modified assignments. Let us concentrate on the start point 4 of p. Round-Robin iteration for the transformer $[\![4]\!]^\mathsf{T}$ for the generic post-condition $A\mathbf{x} \doteq B\mathbf{y}$, successively will produce the equalities depicted by Table 2, where in the ith column, we again only have displayed pre-conditions which have additionally been attained in the ith iteration for the program points $7, 6, 5$ and 4, respectively. For program point 4, we can argue as in Example 8 in order to verify that the first two equalities L-subsume the third one. Therefore, it remains to consider the given iteration for any small assignment to the program variables $\mathbf{x}, \mathbf{y}$.

If $\mathbf{x} = \mathbf{y} = a$, then $A = B$ must hold and the third equality is implied. If $\mathbf{x} = a$, but $\mathbf{y}$ is bound to large terms, then the first equality is of the format

Table 2. Round-Robin iteration of Example 9

	1	2	3
7	$A\mathbf{x} \doteq B\mathbf{y}$		
6	$A\mathbf{x} \doteq Bf(\mathbf{y}, a, \mathbf{y})$		
5	$\top$	$A\mathbf{x} \doteq Bf(\mathbf{y}, a, \mathbf{y})$ $Af(\mathbf{x}, a, \mathbf{x})$	$\doteq Bf(f(\mathbf{y}, a, \mathbf{y}), a, f(\mathbf{y}, a, \mathbf{y}))$
4	$A\mathbf{x} \doteq B\mathbf{y}$	$Af(\mathbf{x}, a, \mathbf{x}) \doteq Bf(\mathbf{y}, a, \mathbf{y})$ $Af(f(\mathbf{x}, a, \mathbf{x}), a, f(\mathbf{x}, a, \mathbf{x}))$	$\doteq Bf(f(\mathbf{y}, a, \mathbf{y}), a, f(\mathbf{y}, a, \mathbf{y}))$

$[F_{a,\mathbf{y}}]$ while the subsequent equalities are of the format $[F_{\cdot,\mathbf{y}}]$. Accordingly, the first equality must be kept separately. For the second and third equalities the techniques from Theorem 6 again allow to derive the monoidal equality:

$$Af(\bullet, a, \bullet)A^- \doteq Bf(\bullet, a, \bullet)B^-$$

implying that the equality provided in the fourth iteration will be subsumed. A similar argument applies to the case where $\mathbf{y} = a$ while $\mathbf{x}$ is bound to large values only. Thus, Round-Robin fixpoint iteration reaches the greatest fixpoint after the fourth iteration.   □

6   Multi-variable Equalities

In this section, we extend our methods to arbitrary equalities such as

$$\mathbf{x} \doteq f(g\mathbf{y}, \mathbf{z})$$

where, w.l.o.g., the left-hand side is a plain program variable while the right-hand side is a term possibly containing occurrences of more than one variable. Still, we consider programs where each right-hand side contains occurrences of at most one variable only. Here, we indicate how for any program point v and any given candidate Herbrand equality $\mathbf{x} \doteq s$, we verify whether or not the equality is valid whenever v is reached. There are only constantly many candidate equalities of this form, namely, all equalities which hold for a variable assignment σ_v computed by a single run of the program reaching v. Since such a single run can be effectively computed before-hand, we conclude:

Theorem 9. *Assume that all right-hand sides of assignments in an arbitrary program contain at most one variable. Then all inter-procedurally valid Herbrand equalities can be inferred.*

Now consider the single Herbrand equality $\mathbf{x} \doteq s$, where s contains occurrences of the program variables $\mathbf{y}_1, \ldots, \mathbf{y}_r$. Then we construct new generic post-conditions as follows. First, we consider all substitutions σ which map each variable $\mathbf{y}_i$ in s either to a fresh template variable C_i or an expression $A_i \mathbf{y}'_i$ for a fresh template variable A_i and any program variable $\mathbf{y}'_i$. Then the new generic post-conditions are of the form $\mathbf{x}' \doteq s'$ where $\mathbf{x}'$ is any program variable, and s' is a subterm of $s\sigma$. Note that this set may be large but is still finite. In a practical implementation, we may, however, tabulate for each procedure the weakest pre-conditions only for those post-conditions which are really required. Since we envision that for realistic programs, only few of these equalities for each procedure will be necessary to prove the queried assertion e_t at target point u_t, the potential exponential blow-up will still be not an obstacle.

Example 10. Assume the equality we are interested in is $\mathbf{x} \doteq f(g\mathbf{y}, \mathbf{z})$, then, e.g.,

$$\mathbf{x} \doteq f(gA_1\mathbf{y}, A_2\mathbf{z}) \qquad \mathbf{y} \doteq f(gA_1\mathbf{x}, A_2\mathbf{z})$$

are new generic post-conditions to be considered, as well as

$$\mathbf{z} \doteq f(gC, A\mathbf{y}) \qquad \mathbf{y} \doteq f(gA\mathbf{z}, C) \qquad\qquad\qquad □$$

Starting from a new generic post-condition $\mathbf{x} \doteq p$, repeatedly computing weakest pre-conditions w.r.t. assignments may result in conjunctions of equalities which can be simplified to one of the following forms:

- $s \doteq C_i$ or $s \doteq A_i t_i$ where s and t_i contain occurrences of at most one program variable each;
- $\mathbf{y} \doteq p'$, i.e., the left-hand side is a plain program variable, and the right-hand side p' is obtained from a subterm of p by substituting each occurrence of a program variable $\mathbf{y}_i$ with some term t_i containing occurrences of at most one program variable each.

Example 11. Consider, e.g., the generic post-condition $\mathbf{x} \doteq f(gA_1\mathbf{y}, A_2\mathbf{z})$. Then

$$\llbracket \mathbf{x} = f(\mathbf{x}, h\mathbf{x}) \rrbracket^\mathsf{T}(\mathbf{x} \doteq f(gA_1\mathbf{y}, A_2\mathbf{z})) = f(\mathbf{x}, h\mathbf{x}) \doteq f(gA_1\mathbf{y}, A_2\mathbf{z})$$
$$= (\mathbf{x} \doteq gA_1\mathbf{y}) \wedge (h\mathbf{x} \doteq A_2\mathbf{z})$$

which means that we equivalently obtain two two-variable equalities. Likewise, for an assignment to one of the program variables on the right, we have:

$$\llbracket \mathbf{y} = f(b, \mathbf{y}) \rrbracket^\mathsf{T}(\mathbf{x} \doteq f(gA_1\mathbf{y}, A_2\mathbf{z})) = \mathbf{x} \doteq f(gA_1 f(b, \mathbf{y}), A_2\mathbf{z})$$

which is an equality of the form described in the second item.   □

The equalities from the first item contain at most program variable on each side. They can be dealt with in the same way as we did for plain two-variable equalities. They are even somewhat simpler, in that only one template variable occurs (instead of two). The equalities of the second item, on the other hand, we may group into equalities which agree in the variable on the left as well as in the constructor applications outside the template variables A_i. Of each such group it suffices to keep exactly one equality. Any conjunction with another equality from the same group will allow us to simplify the second equality to a conjunction of equalities with at most one program variable on each side.

Example 12. Assume that we are given the conjunction of the two equalities:

$$\mathbf{x} \doteq f(gA_1\mathbf{y}, A_2\mathbf{z}) \qquad \mathbf{x} \doteq f(gA_3h\mathbf{y}, A_4g\mathbf{z})$$

This conjunction is equivalent to the first equality together with:

$$f(gA_1\mathbf{y}, A_2\mathbf{z}) \doteq f(gA_3h\mathbf{y}, A_4g\mathbf{z})$$

The latter equality, now, is equivalent to the conjunction of:

$$A_1\mathbf{y} \doteq A_3h\mathbf{y} \qquad A_2\mathbf{z} \doteq A_4g\mathbf{z}$$

which is a finite conjunction of two-variable equalities.   □

Thus, in the course of **WP** computation for any of the new generic post-conditions, we obtain conjunctions which (up to finitely many exceptions) consists of two-variable equalities only, to which we can apply our methods from Section 5. In summary, we thus find that it can be effectively checked whether or not a general Herbrand equality is inter-procedurally valid at a given program point v.

7 Conclusion

We provided an analysis which infers all inter-procedurally valid Herbrand equalities for programs where all assignments are taken into account whose right-hand sides depend on at most one variable. The novel analysis is based on three main ideas. First, we restricted general satisfiability, subsumption and equivalence to satisfiability, subsumption and equivalence w.r.t. a set of values subsuming all possible run-time values of a given program. Together with our factorization theorem, this allowed us to apply the monoidal methods from [8] to effectively infer all inter-procedurally valid two-variable Herbrand equalities, at least for programs, which we called *initialization-restricted*. In the second step, we abandoned this restriction by introducing the extra distinction between *large* values (which can be uniquely factored) and *small* ones (of which there are only finitely many). Finally, we showed how general Herbrand equalities could be handled. For convenience, we presented the construction for programs with global variables only. The techniques, however, can be extended to programs with both local and global variables as provided in the long version of this paper [19]. In addition we show in the long version of this paper that an implementation of the analysis can be provided which runs in polynomial time.

References

1. Cocke, J., Schwartz, J.T.: Programming Languages and Their Compilers: Preliminary Notes. Courant Institute of Mathematical Sciences, New York University (1970)
2. Cousot, P.: Methods and logics for proving programs. In: van Leeuwen, J. (ed.) Handbook of Theoretical Computer Science, Formal Models and Semantics, ch. 15, pp. 843–993. Elsevier Science Publishers, Amsterdam (1990)
3. Engelfriet, J.: Some open questions and recent results on tree transducers and tree languages. In: Book, R. (ed.) Formal Language Theory: Perspectives and Open Problems, pp. 241–286. Academic Press (1980)
4. Flexeder, A., Müller-Olm, M., Petter, M., Seidl, H.: Fast interprocedural linear two-variable equalities. ACM Trans. Program. Lang. Syst. 33(6), 21:1–21:33 (2011)
5. Godoy, G., Tiwari, A.: Invariant checking for programs with procedure calls. In: Palsberg, J., Su, Z. (eds.) SAS 2009. LNCS, vol. 5673, pp. 326–342. Springer, Heidelberg (2009)
6. Goldfarb, W.D.: The undecidability of the second-order unification problem. Theoretical Computer Science 13(2), 225–230 (1981)
7. Gulwani, S., Necula, G.C.: A polynomial-time algorithm for global value numbering. In: Giacobazzi, R. (ed.) SAS 2004. LNCS, vol. 3148, pp. 212–227. Springer, Heidelberg (2004)
8. Gulwani, S., Tiwari, A.: Computing procedure summaries for interprocedural analysis. In: De Nicola, R. (ed.) ESOP 2007. LNCS, vol. 4421, pp. 253–267. Springer, Heidelberg (2007)
9. Hoare, C.A.R.: An axiomatic basis for computer programming. Communications of the ACM 12(10), 576–580 (1969)

10. Jeż, A.: Context unification is in PSPACE. In: Esparza, J., Fraigniaud, P., Husfeldt, T., Koutsoupias, E. (eds.) ICALP 2014, Part II. LNCS, vol. 8573, pp. 244–255. Springer, Heidelberg (2014)
11. Kildall, G.A.: A unified approach to global program optimization. In: 1st Annual ACM SIGACT-SIGPLAN Symposium on Principles of Programming Languages (POPL), pp. 194–206. ACM (1973)
12. Levy, J., Veanes, M.: On the undecidability of second-order unification. Information and Computation 159(1-2), 125–150 (2000)
13. Müller-Olm, M., Petter, M., Seidl, H.: Interprocedurally analyzing polynomial identities. In: Durand, B., Thomas, W. (eds.) STACS 2006. LNCS, vol. 3884, pp. 50–67. Springer, Heidelberg (2006)
14. Müller-Olm, M., Seidl, H.: Precise interprocedural analysis through linear algebra. In: Jones, N.D., Leroy, X. (eds.) 31st Annual ACM SIGPLAN-SIGACT Symposium on Principles of Programming Languages (POPL), pp. 330–341. ACM (January 2004)
15. Müller-Olm, M., Seidl, H.: Analysis of modular arithmetic. ACM Trans. Program. Lang. Syst. 29, 29:1–29:27 (2007)
16. Müller-Olm, M., Seidl, H.: Upper adjoints for fast inter-procedural variable equalities. In: Drossopoulou, S. (ed.) ESOP 2008. LNCS, vol. 4960, pp. 178–192. Springer, Heidelberg (2008)
17. Müller-Olm, M., Seidl, H., Steffen, B.: Interprocedural herbrand equalities. In: Sagiv, M. (ed.) ESOP 2005. LNCS, vol. 3444, pp. 31–45. Springer, Heidelberg (2005)
18. Petter, M.: Interprocedural Polynomial Invariants. PhD thesis, Institut für Informatik, Technische Universität München (September 2010)
19. Schulze Frielinghaus, S., Petter, M., Seidl, H.: Inter-procedural two-variable herbrand equalities. arXiv e-prints (2014), http://arxiv.org/abs/1410.4416
20. Sharir, M., Pnueli, A.: Two approaches to interprocedural data flow analysis. In: Muchnick, S.S., Jones, N.D. (eds.) Program Flow Analysis: Theory and Application, pp. 189–233. Prentice-Hall (1981)
21. Steffen, B., Knoop, J., Rüthing, O.: The value flow graph: A program representation for optimal program transformations. In: Jones, N.D. (ed.) ESOP 1990. LNCS, vol. 432, pp. 389–405. Springer, Heidelberg (1990)

A   Equalities over a Free Monoid

Consider a free monoid M_Σ with set of generators Σ. As usual, the neutral element of M_Σ is denoted by ϵ. Let F_Σ be the corresponding free group. F_Σ can be considered as the free monoid generated from $\Sigma \cup \Sigma^-$ (where $\Sigma^- = \{a^- \mid a \in \Sigma\}$ is the set of formal inverses of elements in Σ with $\Sigma \cap \Sigma^- = \emptyset$) modulo exhaustive application of the cancellation rules $a \cdot a^- = a^- \cdot a = \epsilon$ for all $a \in \Sigma$. In particular, the neutral element of F_Σ is given by ϵ, and the inverse g^{-1} of an element $g = a_1 \ldots a_k, a_i \in \Sigma \cup \Sigma^-$, is given by $g^{-1} = a_k^{-1} \ldots a_1^{-1}$ where $x^{-1} = x^-$ and $(x^-)^{-1} = x$ for $x \in \Sigma$.

For every $w \in M_{\Sigma \cup \Sigma^-}$, the *balance* $|w|$ is the difference between the number of occurrences of positive and negative letters in w, respectively. Thus, $|aba^-b^-c| = 1$ and $|a^-b| = 0$. Note that the balance stays invariant under application of the cancellation rules. Also, $|uv| = |u| + |v|$ and $|u^{-1}| = -|u|$. Accordingly, the balance $|\cdot| : F_\Sigma \to \mathbb{Z}$ is a group homomorphism. Furthermore, we call w *non-negative* if $|w'| \geq 0$ for all prefixes w' of w. This property is also preserved by cancellation and concatenation but not by inverses. Instead, we have:

Lemma 1. *If both u and v are non-negative, and $|u| \geq |v|$ then also uv^{-1} is non-negative.*

Proof. Consider a prefix x of uv^{-1}. If x is a prefix of u, $|x| \geq 0$ since u is non-negative. Otherwise, $x = uv'^{-1}$ for some suffix v' of v. Then $|v'| \leq |v|$, since v is non-negative. Therefore, $|uv'^{-1}| = |u| - |v'| \geq |u| - |v| \geq 0$. $\qquad\square$

We consider equations of the form:

$$AuA^{-1} = Bu'B^{-1} \tag{1}$$

where A, B are variables which take values in M_Σ and u, u' are maximally canceled. If the equation is satisfiable, then necessarily $|u| = |u'|$ holds. Assume from now on that u, u' are maximally canceled, and $|u| = |u'|$. Furthermore, we assume that $|u| \geq 0$ and u, u' are both non-negative. We then have:

Lemma 2. *If $|u| = |u'| = 0$, then the equation (1) either is trivial, is equivalent to an equation $As = B$ or an equation $A = Bs$ for some $s \in M_\Sigma$ or is contradictory.*

Proof. Assume $u = \epsilon$. Then $B = Bu'$. Thus either $u' = \epsilon$ and the equation is trivial, or $u' \neq \epsilon$ and the equation is contradictory.

Therefore, assume that $u \neq \epsilon \neq u'$. Then u and u' must be of the form $u = xyz^{-1}$, $u' = x'y'z'^{-1}$ for maximal $x, x', z, z' \in M_\Sigma$, i.e., y, y' each are either equal to ϵ or of the form $a^- wb$ for some $a, b \in \Sigma$. Then all x, x', z, z' are different from ϵ. Then equation (1) is equivalent to:

$$Ax = Bx' \qquad \wedge \qquad y = y' \qquad \wedge \qquad Az = Bz'$$

By bottom cancellation, these three equations either are equivalent to one fixed relation between $As = B$ or $A = Bs$ for some $s \in M_\Sigma$, or to a contradiction. $\qquad\square$

Example 13. Consider the equation

$$Affg^{-1}f^{-1}A^{-1} \doteq Bfg^{-1}B^{-1}$$

which is, according to Lemma 2, equivalent to

$$Aff \doteq Bf \qquad \wedge \qquad \epsilon \doteq \epsilon \qquad \wedge \qquad Afg \doteq Bg$$

By bottom cancellation, we conclude that the conjunction is equivalent to a solved equation $Af \doteq B$. $\qquad\square$

Now assume that there is another equation:

$$AvA^{-1} = Bv'B^{-1} \tag{2}$$

with non-negative v, v' where $|v| = |v'|$.

Theorem 10. *The two equations* (1) *and* (2) *are effectively equivalent either to one solved equation, or to a single equation of the form* (1) *or are contradictory.*

Proof. We perform an induction on the sum of balances $|u|+|v|$. W.l.o.g., assume that $|u| \geq |v|$. If $|v| = 0$, then the assertion follows from Lemma 2. Therefore, assume that $|v| > 0$, and $r \geq 1$ is the maximal number such that $|v^r| = r \cdot |v| \leq |u|$. Then we construct the elements uv^{-r} and $u'v'^{-r}$, which are both non-negative by Lemma 1. Let w, w' be obtained from uv^{-r} and $u'v'^{-r}$ by exhaustively applying the cancellation rules. By construction, these are non-negative as well. Then we consider the equation:

$$AwA^{-1} = Bw'B^{-1} \tag{3}$$

which is implied by the two equations (1) and (2).

If $w = \epsilon$, then either $w' = \epsilon$ holds and the equation (3) is trivial, or $w' \neq \epsilon$ and equation (3) is contradictory. In the first case, the equation (2) is implied by equation (1), while in the second case the two given equations (1) and (2) are contradictory. The same argument applies when $w' = \epsilon$ with the roles of A, B exchanged. Therefore now assume that $w \neq \epsilon \neq w'$. Otherwise, the pair of equations (1) and (2) is equivalent to the pair of equations (2) and (3), where the sum of balances $|w| + |v| \leq |w| + r \cdot |v| = |u| < |u| + |v|$ has decreased. For these, the claim follows by inductive hypothesis. $\square$

In [8] a similar argument is presented. The argument there together with the resulting algorithm has been significantly simplified by introducing the extra notion of *non-negativity*.

B   Proof of Theorem 4

In order to prove the theorem we show that every T-satisfiable conjunction of equalities of the same format is effectively T-subsumed by a conjunction of at most three equalities. Furthermore, the proof indicates that, given three equalities, it can be effectively decided whether or not a fourth equality is T-subsumed or not. We consider one case of the assertion of the theorem after the other.

Same Variable on Both Sides. Consider the two distinct equalities

$$As_1\mathbf{x} \doteq Bt_1\mathbf{x} \qquad As_2\mathbf{x} \doteq Bt_2\mathbf{x}$$

where $s_i, t_i \in M_G$, and assume that the conjunction of them is T-satisfiable. We claim that then $s_1\mathbf{x} \neq s_2\mathbf{x}$ and $t_1\mathbf{x} \neq t_2\mathbf{x}$. For that, we convince ourselves first that $s_1 \neq s_2$ and $t_1 \neq t_2$ must hold. Then for a contradiction, assume that $s_1\mathbf{x} \doteq s_2\mathbf{x}$. Since $s_1 \neq s_2$, their unifier must map $\mathbf{x}$ to a ground term of s_1 and s_2. These ground terms are all contained in G, whereas we only consider values for $\mathbf{x}$ in $M_G R$, which is disjoint from G. A similar argument also shows that $t_1\mathbf{x} \neq t_2\mathbf{x}$ holds. Thus by factorization, $Ar_1 \doteq Br_2$ must hold for some $r_1, r_2 \in M_G$ of which at least one equals $\bullet$. Due to unique factorization, we then may cancel $\mathbf{x}$ on both sides, resulting in the equalities $As_1 \doteq Bt_1$ and $As_2 \doteq Bt_2$. These can be simplified to one equality $Ar_1 \doteq Br_2$ for some $r_1, r_2 \in M_G$ where $r_i = \bullet$ for at least one i. Hence, the second equality is T-subsumed by the first one.

One-Sided Single Variable. Consider the three distinct equalities

$$As_1 \doteq Bt_1\mathbf{x} \qquad As_2 \doteq Bt_2\mathbf{x} \qquad As_3 \doteq Bt_3\mathbf{x}$$

where $s_i \in M_G R$ and $t_i \in M_G$, and assume that the conjunction of them is T-satisfiable. Again, we argue that all s_i must be distinct as well as all $t_i\mathbf{x}$. Then again by factorization, $Ar_1 \doteq Br_2$ for some templates r_1, r_2 of which at least one equals $\bullet$. By unique factorization, $s_1 = s_1' r$ for some $s_1' \in M_G$ and $r \in R$. Therefore, again by unique factorization, the value for $\mathbf{x}$ also must terminate in the term r, i.e., is of the form $\mathbf{x} = x'r$ for some $x' \in M_G$. Accordingly, also s_2, s_3 can be factored as $s_i = s_i' r$ for suitable $s_i' \in M_G$. Canceling out the ground terms r, we obtain the monoid equalities:

$$As_1' \doteq Bt_1 x' \qquad As_2' \doteq Bt_2 x' \qquad As_3' \doteq Bt_3 x'$$

Assume w.l.o.g., that the balance of s_1 is less or equal to the balances of s_2 and s_3. Then the conjunction of the three equalities is T-equivalent to:

$$As_1' \doteq Bt_1 x' \qquad As_2' s_1'^{-1} A^{-1} \doteq Bt_2 t_1^{-1} B^{-1} \qquad As_3' s_1'^{-1} A^{-1} \doteq Bt_3 t_1^{-1} B^{-1}$$

where $s_2' s_1'^{-1}, t_2 t_1^{-1}, s_3' s_1'^{-1}, t_3 t_1^{-1}$ all are non-negative. According to Theorem 10, the two last equalities are either T-equivalent to each other, which means that the initial conjunction is T-equivalent to the conjunction of the two equalities

$$As_1 \doteq Bt_1\mathbf{x} \qquad As_2 \doteq Bt_2\mathbf{x}$$

and the assertion follows. Otherwise, they are T-equivalent to an equality $Ar_1 \doteq Br_2$ for templates r_1, r_2 of which at least one equals $\bullet$. A fourth equality is then either T-subsumed or falsifies the conjunction of equalities. A similar argument applies to equalities of the form $At_i\mathbf{x} \doteq Bs_i$.

Different Variables on Both Sides. Consider the three distinct equalities

$$As_1\mathbf{x} \doteq Bt_1\mathbf{y} \qquad As_2\mathbf{x} \doteq Bt_2\mathbf{y} \qquad As_3\mathbf{x} \doteq Bt_3\mathbf{y}$$

for distinct program variables $\mathbf{x}, \mathbf{y}$ where $s_i, t_i \in M_G$, and assume that the conjunction of them is T-satisfiable. As before, we argue that $s_i\mathbf{x} \neq s_j\mathbf{x}, t_i\mathbf{y} \neq t_j\mathbf{y}$ for all $i \neq j$ must hold. Then by factorization, A is a prefix of B or vice versa. But then, due to unique factorization, also As_1 is a prefix of Bt_1 or vice versa. This means that there are $\mathbf{u}, \mathbf{v} \in M_G$ of which one equals $\bullet$ such that $As_1\mathbf{u} \doteq Bt_1\mathbf{v}$, which (by top cancellation) implies that $\mathbf{vx} = \mathbf{uy}$ holds. From that, we conclude that $As_i\mathbf{u} \doteq Bt_i\mathbf{v}$ for all i. Assume again w.l.o.g. that the balance of s_1 is less or equal to the balances of s_2 and s_3. We then proceed as in the last case to obtain the T-equivalent three equalities:

$$As_1\mathbf{u} \doteq Bt_1\mathbf{v} \qquad As_2 s_1^{-1} A^{-1} \doteq Bt_2 t_1^{-1} B^{-1} \qquad As_3 s_1^{-1} A^{-1} \doteq Bt_3 t_1^{-1} B^{-1}$$

where $s_2s_1^{-1}, t_2t_1^{-1}, s_3s_1^{-1}, t_3t_1^{-1}$ all are non-negative. According to Theorem 10, the latter two equalities again are T-equivalent to an equality $Ar_1 \doteq Br_2$ for templates r_1, r_2 of which at least one equals $\bullet$, or are T-equivalent to each other, and the assertion of the theorem follows. This completes the proof. $\square$

Desynchronized Multi-State
Abstractions for Open Programs in Dynamic Languages

Arlen Cox[1,2], Bor-Yuh Evan Chang[1], and Xavier Rival[2]

[1] University of Colorado Boulder
{arlen.cox,evan.chang}@colorado.edu
[2] INRIA/CNRS/ENS Paris
xavier.rival@ens.fr

Abstract. Dynamic language library developers face a challenging problem: ensuring that their libraries will behave correctly for a wide variety of client programs without having access to those client programs. This problem stems from the common use of two defining features for dynamic languages: callbacks into client code and complex manipulation of attribute names within objects. To remedy this problem, we introduce two state-spanning abstractions. To analyze callbacks, the first abstraction desynchronizes a heap, allowing partitions of the heap that may be affected by a callback to an unknown function to be frozen in the state prior to the call. To analyze object attribute manipulation, building upon an abstraction for dynamic language heaps, the second abstraction tracks attribute name/value pairs across the execution of a library. We implement these abstractions and use them to verify modular specifications of class-, trait-, and mixin-implementing libraries.

1 Introduction

"Don't Repeat Yourself!" This DRY mantra leads JavaScript developers to minimize the code that they write and thus minimize the number of places bugs can occur. As a result, there is a proliferation of generic libraries and code reuse in the JavaScript community. Unfortunately, even though library authors would like to know that their libraries work correctly with any client, current verification techniques cannot verify this because they do not also follow the DRY mantra – they require reverifying libraries along with each and every client [15, 17–19, 27]. This paper brings the DRY mantra to automatic dynamic language verification by modularly verifying libraries *without the presence of client code*.

While there are many kinds of libraries for many dynamic languages, this paper focuses on meta-feature libraries for JavaScript. *Meta-feature libraries* add functionality that is commonly built-in to languages, such as mixins, traits, classes, and memoization. These features are not first-class features of the JavaScript language, but they aid software engineering, so nearly every program includes them in some form or another. For example, the ubiquitous jQuery, Prototype, and MooTools libraries all include implementations of mixins. Similarly, MooTools, Prototype, and the Microsoft Ajax Library include class implementations. What makes these libraries unique is their use of open

© Springer-Verlag Berlin Heidelberg 2015
J. Vitek (Ed.): ESOP 2015, LNCS 9032, pp. 483–509, 2015.
DOI: 10.1007/978-3-662-46669-8_20

```
1   var Class = function(cfg) { //make class
2     var copy = function(src,exc) {...};
3     var attrs = copy(cfg,{});
4     var init = cfg.init;
5     return function(args) { //make instance
6       var result = copy(attrs,{init:null});
7       init(result, args);
8       return result;
9     };
10  }
```

Fig. 1. Class implements a simple version of classes. The class is essentially allocated by the call to copy on line 3. The instance is allocated by the call to copy on line 6. Line 7 calls the initialization function on the instance.

object manipulation, functions, and encapsulation to implement language features as libraries.

For example, while JavaScript does not contain classes, a simple version of classes can be implemented with the few lines shown in Figure 1. These few lines implement classes by constructing a class instantiation function (highlighted) that is responsible for creating new instances of the class. This class instantiation function is derived from a configuration object cfg that describes not only a template for the instance object, but an initialization function init. The init function is run on each newly created instance, completing the initialization of the new object using arguments passed to the instantiation function. Note that because JavaScript allows the attributes of objects to be mutated (i.e. objects are *open*), it is necessary to copy the configuration object twice to create an instance. The first copy (underlined) creates a backup that ensures that if the configuration object is mutated, already constructed classes are not mutated as well. The second copy (line 6) creates the instance object.

A key challenge of verifying library implementations is that developers specify libraries in terms of input/output behaviors. If a particular kind of input is given, a particular, but related kind of output is given. For example, in Class, the object generated by instantiating a class is related to the cfg object that was passed in to Class. This means inputs to a library must be treated as unknowns that can be related to the outputs of that library — even when the inputs are *unknown functions* or objects with *unknown attribute name/value relationships*.

The core problem with unknown functions (such as init) as input to a library is that they may be called by the library. If they are, they may have wide-reaching effects on the state of the program. However, developers are not stymied by these function calls when reading code because the effects are usually well contained by the surrounding code. Developers use conventions such as copying into local, non-escaping variables (like attrs and init) to ensure that certain parts of the program's state cannot be affected by calls to unknown functions. Therefore, when developers are reasoning about this code, they optimistically assume that when a call to one of these unknown functions occurs, there are two parts of the program memory: (1) the part unaffected by the call, which may be freely accessed and modified after the call and (2) the part affected by the call, which, over the remainder of the function is solely described as "the result of calling the function on whatever that part was before the call." In this paper, we observe that analyses that are designed for such library code can optimistically split the heap into two parts, where the analysis can proceed on the unaffected part and the affected part can be saved along with the function that affected it until that function is known.

Furthermore, existing analyses have problems with input objects that have unknown attribute name/value relationships. Most analyses represent containers by partitioning them. However only using partitions, it is not possible to represent the fact that attribute/value pairs are often preserved. For example, when cfg is copied to attrs, it is clear that every attribute/value pair is copied and therefore, all attribute/value pairs are preserved as-is across this computation. As Halbwachs and Péron [14] discovered for arrays, it is beneficial to capture relations between individual attributes and values and to share those relations between multiple containers. However, these relations can be generalized beyond arrays to any container and can be extended to relate partitions across multiple states. This allows proving that attrs is equal to what cfg was at the beginning of the class creation.

To verify modular specifications of JavaScript libraries, even when client code is absent, and thus enabling reuse of specifications, and improving library reliability, we make the following contributions:

- To abstract open objects and containers with unknown attribute name/value relationships, we introduce *attribute/value trackers* that extend existing container and open-object abstractions with the ability to perform fully precise partitioning when attributes and values are copied. Trackers represent a form of parametric polymorphism for attribute/value relationships that can be applied across multiple abstract heaps to relate unknown input objects to unknown output objects.
- For the analysis of a call to an unknown function, we introduce *desynchronized separation*, which splits off a region of the heap by representing it as an old analysis state along with the code required to synchronize that portion of the state with the rest of the analysis. This creates a form of assume-guarantee reasoning that mimics the programmer intuition for simple, well-contained callbacks, while enabling automatic analysis.
- We extend the heap with open objects abstraction (HOO) with attribute/value trackers and desynchronized separation and evaluate these additions to HOO by automatically verifying specifications written for JavaScript meta-feature libraries. We analyze the core functionality of libraries that implement mixins, traits, classes, and memoization. By utilizing HOO along with both desynchronization and attribute/value trackers, we are able to fully precisely analyze these library cores, even without any knowledge of specific attribute names used in input objects or code for client-supplied callbacks.

2   Overview

In this section we demonstrate the power of attribute/value trackers and desynchronized separation applied to HOO (the Heap with Open Objects Abstraction [8]) by showing

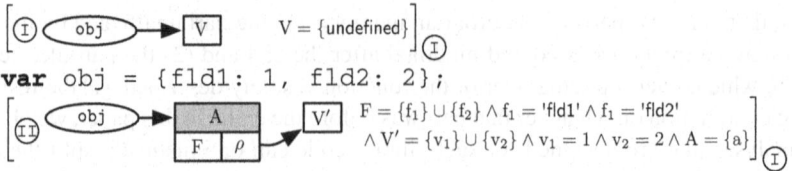

var obj = {fld1: 1, fld2: 2};

Fig. 2. The HOO abstract domain represents a heap of open objects using a combination of a heap graph and pure side constraints

key parts of the analysis of instantiating a class created by the Class library introduced in Figure 1. First we show how attribute/value trackers enhance open-object and container abstractions with the analysis of the copy function used by Class. Then, we show how desynchronization allows analysis of calls to unknown functions.

2.1 Preliminaries

Before we explain attribute/value trackers and desynchronized separation, we introduce the basics of the HOO abstraction. HOO is a separation-logic-based abstraction for dynamic language heaps that supports reasoning about open objects, which behave like containers mapping strings representing attribute names to values. HOO supports the basic requirements for both attribute/value trackers and desynchronization. It partitions open objects by the attributes (as most container abstractions do) and it supports partitioning the heap (as all separation-logic-based abstractions do). What makes HOO unique is its use of a set abstraction to relate partitions to one another. However, this functionality is not strictly required to make use of trackers and desynchronized separation.

Because we are concerned with input/output relationships, Figure 2 shows a simple program annotated with two-state HOO invariants. The first invariant ① shows the initial heap containing the variable obj pointing to the value undefined. The input heap is indicated with its program point in the lower right hand corner of an invariant. In the case of ①, the input heap is the same as the heap shown in brackets at ①. The current heap, relative to that input heap is shown in the brackets along with a constraint on the logic variables used in both heaps. This constraint is represented and manipulated by an abstract domain for sets.

This program creates a new object pointed to by obj that has two attributes: 'fld1' corresponds to the value 1 and 'fld2' corresponds to the value 2. The abstract state ② highlights the important parts of the abstraction. The heap part in brackets shows an abstract object that is represented as a table. The shaded top row is the set symbol A for the base address of objects. If this is not a singleton set, the object is a summary. On the right, A is constrained to be a singleton set of addresses and thus it is not a summary object. Below the shaded top row are rows each describing a partition of attribute names for that object. Here we have decided to represent these two attribute names 'fld1' and 'fld2' using a single partition that conflates the two attribute names. This partition is represented with the set symbol F, where it is equated to the union of two singleton sets with attribute names f_i. Additionally, this partition has been assigned the attribute/value tracker ρ, which can keep track of specific attribute/value pairs from the beginning of

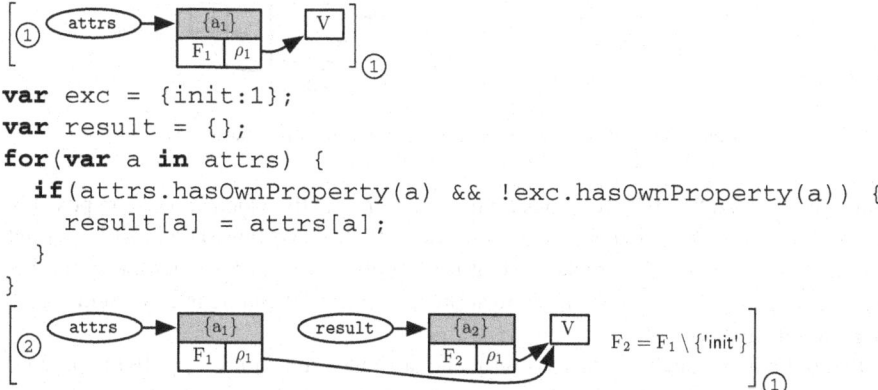

```
var exc = {init:1};
var result = {};
for(var a in attrs) {
  if(attrs.hasOwnProperty(a) && !exc.hasOwnProperty(a)) {
    result[a] = attrs[a];
  }
}
```

Fig. 3. Analysis of class instantiation uses attribute/value trackers to maintain precision when attributes are copied

the function to the end, as will be demonstrated in next section. Finally, the partition points to a set of values that is made up of individual values v_i. Note that this is not the most precise abstraction because the two attributes have been summarized into a single partition. An alternative abstraction would construct a separate partition for each known attribute name.

In this paper we will often use a shorthand notation where instead of showing a set symbol such as A in the heap, we will show instead a singleton set in brackets, such as $\{a\}$. This is equivalent to having a set symbol and then constraining that set symbol to be equal to the singleton set. This is useful for improving the readability of the notation, but formally all symbols in the heap are set symbols.

2.2 Attribute/Value Trackers

At the start of the analysis of the class instantiation function (the highlighted part of Figure 1), the first code that the analysis encounters is the call to the copy function. Figure 3 shows the body of the copy function after it has been inlined into the context of the class instantiation. This function iteratively copies one open object attrs to another open object result by first checking if the attribute name that is being copied is in the exclusion object exc. Accompanying the copy function are pre/postconditions that show a portion of heap that is relevant to this function.

An abstraction such as HOO does a nice job of incrementally inferring the relationship that forms between the result object and the attrs object. While, as a two-state abstraction, HOO can relate initial objects to final objects, it still conflates all of the attributes and values that may have been in that object into a single partition. This means that while HOO can prove that the result object has a subset of the attributes of the initial attrs object, it cannot prove by itself that the attribute/value relationship was maintained for everything that was copied. This is where attribute/value trackers come in.

An attribute/value tracker is an *uninterpreted symbol* for some relationship between attributes and values. When a tracker is applied to a particular partition and corresponding

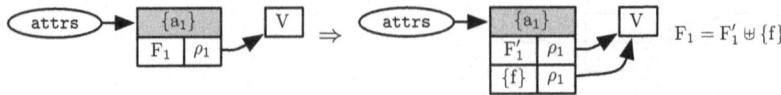

Fig. 4. Materialization maintains the attribute/value tracker ρ_1

values, it uses that "global" relationship to constrain exactly which values can possibly correspond to which attributes that are described by the partition. The most important aspect of an attribute/value tracker is that it is "global" in the sense that the symbol is shared between the two-states of the invariant. A tracker's meaning is consistent across these two abstract heaps.

Throughout this analysis, there is only one attribute/value tracker ρ_1. In the precondition the attribute/value tracker ρ_1 can be automatically added, as at that point the true relationship between attributes and values is unknown. But in the postcondition, the fact that ρ_1 is used for two partitions means not only that attrs and result have the same attribute and value relationship after the loop, but that the relationship is the same one that existed before the loop.

Critically, once a tracker is associated with a partition, that tracker can be reused with any other partition that is a subset of that initial, associated partition. Here, we see that the same tracker ρ_1 is used in the F_2 partition of the object at address a_2. Even though the F_2 partition is a subset of F_1 used in the object at address a_1, the same tracker can be used. As a result, this constraint says that the result object is *exactly* the same as the attrs object except that the 'init' attribute has been removed if it was present.

Materialization with Attribute/Value Trackers: In the loop body, before the object pointed to by attrs can be read, the single attribute that will be read must be materialized in that object. This ensures strong updates occur. An example materialization is shown in Figure 4. On the left is the object at address a_1 before materialization and on the right is the same object after materialization. Here, we assume that the particular attribute is represented by the symbol f, and while f is not explicitly constrained, it is known that f is one of the attributes from F_1.

What is special about attribute/value trackers is that rather than requiring a new description of the partition when a materialization occurs; here they can be duplicated. On the right the tracker ρ_1 occurs in both partitions F_1' and $\{f\}$. This is because the tracker only restricts the values that correspond to those in the partition. Since the partition has been refined, the same restriction can be applied to both new partitions.

Transfer of Attribute/Value Trackers: As part of analyzing code like copy above, there is a transfer of an attribute/value pair from one object to another. This transfer maintains the relationship between attributes and values. When transfer occurs, the attribute/value tracker can be transfered along with the attribute and value. Therefore, even if the particular attribute and particular value cannot be identified from their sets, the tracker maintains whatever the original relationship was and allows it to be transfered to other objects.

Here this property of trackers ensures that ρ_1 is transfered from the attrs object to the results object. Since the transfer occurs whenever the attribute/value pair is copied, the tracker can be unconditionally copied. However, because the resulting partition F_2 is restricted, this simply limits the scope of where the tracker can be applied.

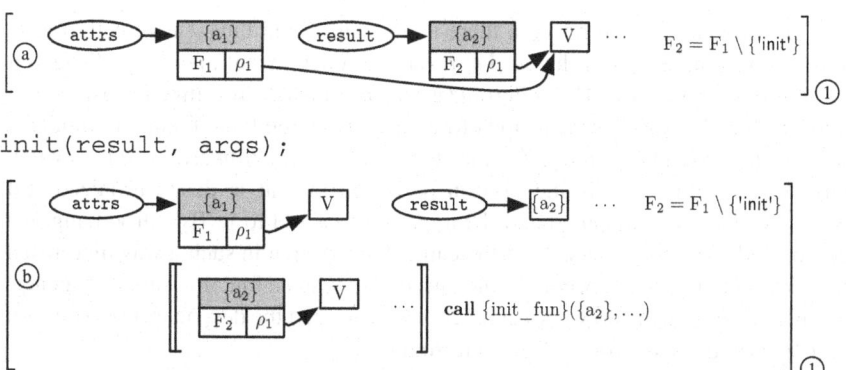

```
init(result, args);
```

Fig. 5. Desynchronized terms are introduced by function calls to unresolvable functions

While we have not demonstrated the use of summaries generated by this analysis, attribute/value trackers are critical to this application. With the use of attribute/value trackers, a general precondition can be specialized for a particular calling context, essentially a form of parametric polymorphism. That is, partitions can be more finely specified corresponding to the actual objects passed into library functions. This very same partitioning can be applied to postcondition, allowing precision that was made available well after the analysis was completed to be preserved by the analysis.

2.3 Desynchronized Separation

When analysis reaches the call to the client-supplied initializer that is shown in Figure 5, there is a problem. The actual function that is called is an input to the class library and as a result it is unknown to the analysis. However, despite the fact that this function is unknown, developers might optimistically reason about what this class library does as follows: attrs is protected by lexical scoping, so it should not change, and result is initialized by the copy and then the return value is whatever result is after running the client-supplied initializer init on it. Desynchronized separation is a means for capturing this kind of optimistic reasoning in a sound manner using a form of assume-guarantee reasoning.

Immediately before the call to the initializer, there are two objects shown: a_1 is the attrs object, which is the backup copy of the cfg object that was passed in to Class and a_2 is the result object that is the class instance that is currently in the process of being constructed. The relationship between F_1 and F_2 carries over from the copy as before. Other parts of the heap are not shown, as they are not necessary for explaining desynchronized separation.

When the analysis reaches the call to init, desynchronized separation optimistically splits the heap into two separate parts: (1) the part that shall not be used by the client-supplied initializer and (2) the part that shall be used by the client-supplied initializer. In our algorithm, we make this split based on reachability: optimistically *assuming* the post-call code in the caller does not use anything reachable from the arguments to the call. Thus for the unused portion, there is no change and thus it is directly represented in the post-state ⓑ. For the used portion, the function call may have changed it and thus it is desynchronized. Desynchronization represents the resulting heap as a term that stores the used portion of the heap before the call and the function that is applied.

The desynchronization process introduces a *desynchronized term*, written $[\![H]\!]$ call $V(\ldots)$, where H is the portion of the heap that is desynchronized and call $V(\ldots)$ is the function called and the arguments passed to it. By introducing this desynchronized term, the post-state of the call can be written in such a way that, when the client-supplied initializer becomes known, such as when a function summary generated by HOO is reused, the now known function can convert the desynchronized portion back into a normal, "synchronized" heap formula.

In ⓑ we can see that the heap has been split so that a_2 has been desynchronized. Because it may have been modified by the call, it is "locked" in a state from before the function call. That is, we *guarantee* in the analysis that the post-call code does not access desynchronized sub-heaps by ensuring the analysis gets stuck (raises a warning) if accessing desynchronized memory. Desynchronization is different from simply separating the two parts of the heap because the desynchronized region represents the portion of the heap that results from calling the desynchronized function on the desynchronized part of the heap. In this way it soundly abstracts calls to unknown functions by explicitly representing the precondition to the call and implicitly representing the postcondition of the call.

A significant part of implementing desynchronized separation is the operation used to split the heap into the desynchronized and non-desynchronized parts. In this paper we outline a simple means of splitting the heap based on reachability as in [25] that exploits the fact that JavaScript developers, by convention, protect regions of the heap using closures for encapsulation. Here, attrs is protected in such a way. Consequently, the heap split that is automatically inferred leaves a_1 and all local variables outside the desynchronized term and places a_2 inside the desynchronized term. With this split, it is possible to verify that attrs is unmodified by class instantiation, which means that classes are immutable and it is possible to verify that the object built in the class is the one returned by the class after calling the client-supplied initializer on it starting from elements copied from attrs.

3 Abstracting Callbacks and Objects with Multi-State Abstraction

In this section we define attribute/value trackers and desynchronization as an extension to the heap with open objects (HOO) abstract domain [8]. First we present attribute/value trackers and how they are added to HOO. Then we present desynchronized separation, also adding it to HOO.

Throughout these sections we utilize the following symbols in the definitions.

$$\overline{\text{Address}} \subseteq \overline{\text{Value}} \qquad d \subseteq \overline{\text{Attribute}}$$
$$\overline{\text{Attribute}} \subseteq \overline{\text{Value}} \qquad o \in \overline{\text{Object}} = \overline{\text{Attribute}} \rightharpoonup \overline{\text{Value}}$$
$$v \in \overline{\text{Value}} \qquad \sigma \in \overline{\text{State}} = \overline{\text{Address}} \rightharpoonup \overline{\text{Object}}$$
$$f \in \overline{\text{Attribute}}$$

$\overline{\text{Address}}$ is the set of all concrete addresses, $\overline{\text{Attribute}}$ is the set of all concrete attributes (strings), and $\overline{\text{Value}}$ is the set of all values including addresses and attributes. $\overline{\text{Object}}$ is the set of partial functions from attributes to values, where unmapped attributes are not attributes in the object. Similarly concrete states are a partial function from addresses to objects. Individual concrete values v, attributes f, and object domains d are used in defining semantics.

3.1 Attribute/Value Trackers on HOO

Attribute/value trackers extend an existing domain for containers that supports strong updates. Attribute/value trackers significantly increase the precision of the existing container domains by precisely keeping track of the relationship between individual attributes and individual values, even when the container has summarized many attributes and values into a single partition. An *attribute/value tracker* is an uninterpreted partial function ρ that is optionally added to each container partition in an existing abstract domain for containers.

HOO is a separation-logic-based approximation of a heap that is restricted by an abstraction for sets of values. This abstraction for sets restricts relationships between symbols each representing a set:

$$\{a\}, \{f\}, \{v\}, A, F, V \in \overline{\text{Symbol}}$$

where A represents a set of addresses, F represents a set of attributes, and V represents a set of values. The $\{a\}$, $\{f\}$, and $\{v\}$ sets are the respective singleton forms.

Definition 1 (Attribute/Value Trackers with HOO). *The heap with open objects abstract domain, when extended with attribute/value trackers, is represented with the following logical syntax:*

$$\widehat{\text{Heap}} \ni H ::= H_1 * H_2 \mid A \cdot \langle O \rangle \mid \text{EMP} \mid \text{TRUE}$$
$$\widehat{\text{Object}} \ni O ::= O_1; O_2 \mid F : \rho \mapsto V \mid F : - \mapsto V \mid \text{NONE}$$
$$\widehat{\text{Domain}} \ni D ::= D_1 \vee D_2 \mid [H_2]_{H_1} \mid P$$

An abstract state D is either a disjunction of abstract states, or a triple $[H_2]_{H_1} \mid P$ representing an initial heap H_1 and a current heap H_2 restricted by a domain instance P for sets. The domain responsible for representing P is a parameter to this abstraction and unspecified. An individual heap H is a standard separation logic heap consisting of two disjoint parts combined with separating conjunction, a set of objects $A \cdot \langle O \rangle$ at addresses described by A with structure O, or the empty EMP or unknown TRUE heap. Objects are a form of container, which is represented by a number of disjoint partitions of the attributes. A single partition is represented as either $F : \rho \mapsto V$ or $F : - \mapsto V$ depending on whether the attribute/value tracker ρ is present or not. Partitions are joined together into objects using another form of separating conjunction ; whose unit is the empty object NONE.

Figure 6 shows that an instance of HOO concretizes to a set of pairs of concrete states along with a valuation. The σ_0 state represents a starting state for a library function and

$$\gamma : \widehat{\text{Object}} \to \mathcal{P}\left(\overline{\text{Valuation}} \times \overline{\text{TrackerMap}} \times \overline{\text{Object}} \times \mathcal{P}\left(\overline{\text{Attribute}}\right)\right)$$

$$\gamma(O_1;O_2) = \left\{ \eta,\mu,o,d \;\middle|\; \begin{array}{l} \exists o_1, o_2, d_1, d_2.\ (\eta,\mu,o_1,d_1) \in \gamma(O_1) \wedge (\eta,\mu,o_2,d_2) \in \gamma(O_2) \\ \wedge\, o = o_1 \cup o_2 \wedge d = d_1 \uplus d_2 \wedge \text{Dom}\,(o_1) \cap \text{Dom}\,(o_2) = \emptyset \end{array} \right\}$$

$$\gamma(F : \rho \mapsto V) = \left\{ \eta,\mu,o,d \;\middle|\; \begin{array}{l} d = \eta(F) \wedge \forall f \in \eta(F). \\ o(f) \in \eta(V) \wedge \mu(\rho)(f) = o(f) \end{array} \right\}$$

$$\gamma(F : - \mapsto V) = \left\{ \eta,\mu,o,d \;\middle|\; d = \eta(F) \wedge \forall f \in \eta(F).\ o(f) \in \eta(V) \right\}$$

$$\gamma(\textsc{None}) = \{ \eta,\mu,[],\emptyset \}$$

$$\gamma : \widehat{\text{Heap}} \to \mathcal{P}\left(\overline{\text{Valuation}} \times \overline{\text{TrackerMap}} \times \overline{\text{State}}\right)$$

$$\gamma(H_1 * H_2) = \left\{ \eta,\mu,\sigma \;\middle|\; \begin{array}{l} \exists \sigma_1, \sigma_2.\ (\eta,\mu,\sigma_1) \in \gamma(H_1) \wedge (\eta,\mu,\sigma_2) \in \gamma(H_2) \\ \wedge\, \sigma = \sigma_1 \cup \sigma_2 \wedge \text{Dom}\,(\sigma_1) \cap \text{Dom}\,(\sigma_2) = \emptyset \end{array} \right\}$$

$$\gamma(A \cdot \langle O \rangle) = \left\{ \eta,\mu,\sigma \;\middle|\; \begin{array}{l} \forall a \in \eta(A).\ \exists o, d. \\ \sigma(a) = o \wedge (\eta,\mu,o,d) \in \gamma(O) \wedge \text{Dom}\,(o) = d \end{array} \right\}$$

$$\gamma(\textsc{Emp}) = \{ \eta,\mu,[] \}$$

$$\gamma(\textsc{True}) = \overline{\text{Valuation}} \times \overline{\text{TrackerMap}} \times \overline{\text{State}}$$

$$\gamma : \widehat{\text{Domain}} \to \mathcal{P}\left(\overline{\text{Valuation}} \times \overline{\text{State}} \times \overline{\text{State}}\right)$$

$$\gamma(D_1 \vee D_2) = \left\{ \eta,\sigma_1,\sigma_2 \;\middle|\; (\eta,\sigma_1,\sigma_2) \in \gamma(D_1) \vee (\eta,\sigma_1,\sigma_2) \in \gamma(D_2) \right\}$$

$$\gamma([H_2]_{H_1} \,|\, P) = \left\{ \eta,\sigma_1,\sigma_2 \;\middle|\; \begin{array}{l} \exists \mu.\ (\eta,\mu,\sigma_1) \in \gamma(H_1) \\ \wedge\, (\eta,\mu,\sigma_2) \in \gamma(H_2) \wedge \eta \in \gamma(P) \end{array} \right\}$$

Fig. 6. Concretization of HOO abstract states along with attribute/value trackers

the σ_1 state represents the current state relative to σ_0. The valuation maps each symbol that occurs in the heap formula, including those representing sets of addresses, attributes and values to a set of concrete addresses, attributes, or values:

$$\eta : \overline{\text{Valuation}} = \overline{\text{Symbol}} \to \mathcal{P}\left(\overline{\text{Value}}\right)$$

The valuation ensures that symbols map to consistent values throughout a concretization, even if the symbol is used multiple times. The concretization of P produces a set of these valuations as must be defined by the abstraction for sets. The concretization for any instance of the abstraction for sets must have the following type.

$$P \in \widehat{\text{Sets}} \qquad \gamma_P : \widehat{\text{Sets}} \to \mathcal{P}\left(\overline{\text{Valuation}}\right)$$

For the concretization of heaps and objects, there is an additional value that is returned besides the valuation η and the state σ. The attribute/value tracker map μ binds trackers to their corresponding partial functions:

$$\rho \in \overline{\text{TrackSym}}$$

$$\mu \in \overline{\text{TrackerMap}} = \overline{\text{TrackSym}} \rightharpoonup \overline{\text{Attribute}} \rightharpoonup \overline{\text{Value}}$$

An element $\mu \in \overline{\text{TrackerMap}}$ maps a tracker symbol to a partial function from attributes to values. The domain of that function is fixed when the tracker is introduced (Section 4.2).

Example 1 (Attribute/Value Trackers with HOO). In the following state, there are two abstract heaps and a single pure domain instance.

$$[\{a\} \cdot \langle F' : \rho \mapsto \{v\}\rangle]_{\{a\} \cdot \langle F : \rho \mapsto \{v\}\rangle} | F' \subseteq F$$

This constrains the relationship between the pre-state and the current state so that they both refer to the same object because they use the same symbol $\{a\}$ and the number of attributes has been possibly reduced: some attributes may have been deleted. All other attributes remain the same and no attributes can have been observably added (added and then later removed is acceptable).

Additionally, the attribute/value tracker ensures that the partition F' is exactly the same as F except for the elements that are removed.

3.2 Desynchronized Separation

Desynchronized separation is an extension to a separation logic that adds a desynchronized term to the logical formulas. It is useful for representing different parts of the heap from different times during an analysis. As a result, it allows a meaningful representation of the heap after a call to an unknown function has been made.

Example 2 (Desynchronization). To demonstrate the power of desynchronization, Figure 7 shows the process of desynchronization pictorially. The program being considered has four separate regions of memory A, B, C, and D that are entirely self contained (no pointers between regions) and the program is about to evaluate three function calls whose bodies are unknown in sequence: $\text{fun}_1(D)$; $\text{fun}_2(B)$; $\text{fun}_3(C, D)$. Figures 7 (a), (b), and (c) show the state of desynchronization after each of these calls. Initially, at time 1, all memory is synchronized and represented at time 1.

When analyzing the call to $\text{fun}_1(D)$, the body is unknown and thus the analysis cannot continue. However, because the function can only affect the memory region D, it

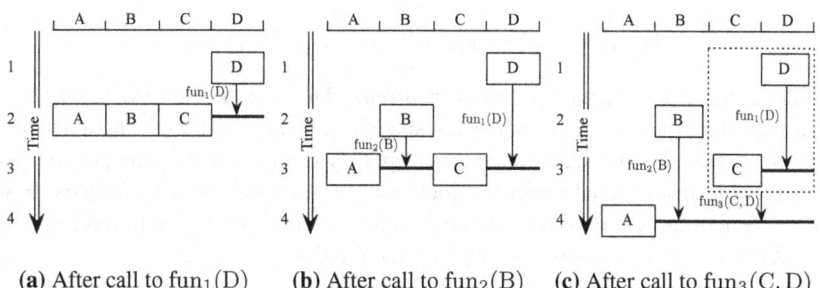

(a) After call to $\text{fun}_1(D)$ (b) After call to $\text{fun}_2(B)$ (c) After call to $\text{fun}_3(C, D)$

Fig. 7. Three separate desynchronizations after calling three successive functions on four regions of memory. In (c), A is the current analysis state where as regions B, C, and D have all been desynchronized. The D region has been desynchronized twice.

is possible to proceed if we desynchronize the heap. The result of the desynchronization is shown in Figure 7a. Regions A, B, and C are allowed to proceed on to time 2, but region D stays locked at time 1 and becomes inaccessible. This inaccessibility is critical because any of that memory in region D may have been mutated by the call to fun_1, and without any knowledge of what fun_1 did, it is impossible to say what the effect of accessing such memory would be.

Even though D has been desynchronized, we can still know a lot about the region after the function has been evaluated. Specifically, we can save which function was supposed to be evaluated, thus we know not only the state of the program before the function call, but we know the function call. With this information, if the function body were provided later, we could resynchronize D with A, B, and C by applying the analysis to that function body starting from D.

In Figure 7b we show the result after the call to $fun_2(B)$. The only accessible region is B and thus it is desynchronized from the A and C regions. Because D is still inaccessible, it just becomes farther in time from being synchronized, but it is no more challenging to resynchronize it. Because B and D are completely distinct regions, there is no affect on B (or A or C) when resynchronizing D and thus even though B and D were desynchronized at different times, the resynchronization is no different.

Finally, in Figure 7c we show the result after the call to $fun_3(C, D)$. Because it is possible that the result of region D is accessed here, the same region must be desynchronized again. We show this nested desynchronization in the dashed box. Both C and D are desynchronized from A, which D is also now desynchronized from C.

To resynchronize everything after Figure 7c, the three functions must be evaluated. However, the order in which the functions are evaluated is irrelevant. Evaluating $fun_1(D)$ first would resynchronize D with C (but not with A). Evaluating $fun_2(B)$ first would resynchronize B with A. Evaluating $fun_3(C, D)$ first would resynchronize C with A and would allow D to be resynchronized with A by only evaluating $fun_1(D)$.

Definition 2 (Desynchronized Separation). *Desynchronized separation extends the logic presented in HOO with a desynchronizing term, an extra kind of heap H that represents a desynchronized portion of the heap along with the function to call and the arguments to pass to resynchronize that portion of the heap with the surrounding heap. The heap H now has the following grammar:*

$$\widehat{\text{Heap}} \ni H ::= [\![H]\!] \text{ call } V_f(V_1, \dots, V_n) \mid \dots$$

To define the concretization of a desynchronized term, concrete values must be extended with functions. We do not give any specific semantics to these functions, but we do assume that while they can mutate the heap, they can only mutate the portion of the heap reachable from global variables, local variables or any closed variables. Essentially, the functions adhere to the standard framing conditions of separation logic [24]. The evaluation of a function is described by the relation

$$\langle\sigma\rangle\text{call } v(v_1, \dots, v_n)\langle\sigma'\rangle$$

which evaluates a call to the function v starting from state σ, passing arguments v_1 to v_n and results in state σ'. Note that we assume all variables have been resolved to values before evaluating this function and thus no environment is necessary to express this computation. This minimizes the reachable heap, which may reduce the footprint of the desynchronized term.

The concretization of HOO with desynchronization is defined as an extension to the concretization of HOO. Because the signature of the function is not required to change, we only define the concretization of the new desynchronized terms:

$$\gamma(\llbracket H \rrbracket \, \mathbf{call} \, V_f(V_1,\ldots,V_n)) \overset{\text{def}}{=} \left\{ \eta,\mu,\sigma \; \middle| \; \begin{array}{l} (\eta,\mu,\sigma_o) \in \gamma(H) \wedge v \in \eta(V_f) \\ \wedge \, (v_1,\ldots,v_n) \in \eta(V_1) \times \ldots \times \eta(V_n) \\ \wedge \, \langle \sigma_o \rangle \mathbf{call} \, v(v_1,\ldots,v_n) \langle \sigma \rangle \end{array} \right\}$$

The γ function concretizes the embedded heap H to a pre-state σ_o and its corresponding valuation. Then for each possible concrete value of the function and each argument, the state σ is the result of evaluating that function on those arguments starting from σ_o. Of course, what makes it possible to reason about applying a function to a portion of the heap is separating conjunction. This dictates that the portion of the heap σ_o was disjoint from the rest of the heap when the desynchronization was created and thus, after this call to a possibly unknown function, σ must be disjoint from the rest of the heap as well.

4 Analysis Using Multi-State Abstraction

In this section we formalize analysis using HOO with desynchronized separation and attribute/value trackers. Because most of the JavaScript language has little effect on desynchronization or attribute value trackers, we focus on the analysis of two core commands. Other commands are either critical to HOO (loops and branches) and documented in [8] or are not critical to any of these analyses. The two core commands are:

$$c ::= \mathbf{call} \, x(y_1,\ldots,y_n) \mid x_1[x_2] := x_3[x_4]$$

The first command is a call to a function, where the function has been closure converted. We assume the corresponding closure and the global object are passed as arguments. The second command is responsible for copying an attribute/value pair from one object to another (handling missing attributes appropriately).

Analysis using HOO is standard abstract interpretation [7]. It infers invariants for each point in the program. Because HOO is a heap abstraction, each command in the language mutates the heap graph, but does not mutate the pure set abstraction P. Destructive updates are achieved by swinging pointers to fresh symbols and constraining those fresh symbols in P.

HOO's inclusion checking, join, and widening algorithms involve an object matching procedure where variables are matched, then objects pointed to by those variables are correspondingly matched. Within each of those objects, partitions are matched. This matching process proceeds summarizing objects from the same allocation site until all objects are matched (and summarized).

Inclusion checking: When performing an inclusion check such as the following, there are two kinds of mappping. The address mapping $M : \overline{\text{Symbol}} \to \overline{\text{Symbol}}$ maps each object symbol from the left-hand side to an object symbol from the right-hand

496   A. Cox, B.-Y. Evan Chang, and X. Rival

side. Whereas the attribute mapping $J : \mathcal{P}\left(\mathcal{P}\left(\overline{\text{Symbol}}\right) \times \overline{\text{Symbol}}\right)$ is a set of sets of attribute partitions from the left-hand side and the corresponding attribute partition from the right-hand side.

$$H \mid P \sqsubseteq_M^J H' \mid P'$$

For each matched partition $(\bar{\text{F}}, \text{F}) \in J$ if each $\text{F}_i \in \bar{\text{F}}$ is included in F, the inclusion check can hold. Otherwise it fails. Similarly, for each $\text{A}_1 \mapsto \text{A}_2 \in M$, if A_1 is included in A_2, the inclusion check can hold.

Join and widening: When performing join or widening the underlying operation is similar. The objects must be matched. The difference between the two algorithms is that the widening algorithm makes use of the underlying widening algorithm for pure operations and may produce different matchings in order to ensure analysis convergence. There are three kinds of matching for the following join: (1) $M_1 : \overline{\text{Symbol}} \to \overline{\text{Symbol}}$ is a mapping from the left-hand side to the result object; (2) $M_2 : \overline{\text{Symbol}} \to \overline{\text{Symbol}}$ is a mapping from the right-hand side to the result object; and (3) $J : \mathcal{P}\left(\mathcal{P}\left(\overline{\text{Symbol}}\right) \times \mathcal{P}\left(\overline{\text{Symbol}}\right) \times \overline{\text{Symbol}}\right)$, which is a set of mappings where each mapping contains a set of attribute partitions from the left-hand side and a set of attribute partitions from the right-hand side to a single partition in the join result.

$$H \mid P \sqcup_{M_1,M_2}^J H' \mid P' = H'' \mid P''$$

For each matched partition $(\bar{\text{F}}_1, \bar{\text{F}}_2, \text{F}') \in J$, F' must over-approximate $\bigcup \bar{\text{F}}_1$ and $\bigcup \bar{\text{F}}_2$. Similarly, for each pair $\text{A}_i \mapsto \text{A}'$ in M_1 and M_2, A' in the output of the join must over-approximate A_i in the appropriate input to the join. The algorithm for the join is detailed in [8].

4.1 Desynchronized Separation

Desynchronized terms can be introduced at any function call. They are automatically derived by evaluating all of the arguments to symbols, possibly eliminating already existing desynchronized terms to do so. Once this has been completed, a special function reach is used to determine the desynchronized region.

$$\text{reach} : \mathcal{P}\left(\overline{\text{Symbol}}\right) \times \widehat{\text{Heap}} \to \widehat{\text{Heap}}_u \times \widehat{\text{Heap}}_r$$

The function reach returns a partitioning (H_u, H_r) of the passed heap. The partition H_r is the part possibly reachable from the arguments of the function, including the global object and any closed variables. The partition H_u is the part unreachable from the arguments of the function. With reach, a frame H_u is inferred. The introduction of desynchronization is given with a transfer function judgment and relies on an abstract environment $\hat{E}$ to map variables to abstract addresses and then relates a pre abstract state D_1 to a post abstract state D_2 via a command c:

$$\hat{E} \vdash [D_1] \ c \ [D_2]$$

DESYNC-INTRO

$$\hat{E}(x) = \mathrm{V}_f$$
$$\hat{E}(y_1) = \mathrm{V}_1 \quad \cdots \quad \hat{E}(y_n) = \mathrm{V}_n \qquad \mathrm{reach}(\{\mathrm{V}_f, \mathrm{V}_1, \ldots, \mathrm{V}_n\}, H) = (H_u, H_r)$$
$$\frac{H' = H_u * [\![H_r]\!] \ \mathbf{call} \ \mathrm{V}_f(\mathrm{V}_1, \ldots, \mathrm{V}_n)}{\hat{E} \vdash [H \,|\, P] \ \mathbf{call} \ x(y_1, \ldots, y_n) \ [H' \,|\, P]}$$

The splitting of the heap into the function frame H_u and the function footprint H_r is heuristic. For the analysis to be successful on resynchronization, the footprint H_r should over-approximate all memory that could be accessed by any function to which this call could resolve. But then this desynchronized memory H_r is no longer accessible in the analysis of the code after this call (i.e., when accessed, a warning will be raised). Thus, it may be that with an imprecise reach(), the analysis cannot proceed either on the code after the call or on resynchronization. In our implementation, we define reach to yield H_r as the entire reachable heap from the arguments [25], so we allow any function be used for later resynchronization.

Example 3 (Desynchronization introduction). In Figure 5 there is a call to the client-supplied initialization function. This is a function that originated outside the class library and thus is necessarily undefined. When this call occurs, we introduce a desynchronized term representing the effects of this constructor. We use an "arrow-following" reach() function that determines that one (shown) object is reachable from the arguments and thus in H_r at ⓐ: $\{a_2\}$. This leaves the objects pointed to by attrs in H_u. The resulting introduced desynchronized term is shown in ⓑ.

In other abstract domain operations such as transfer functions, join, widening, or inclusion checking on a domain constructed with desynchronized separation, desynchronized terms must be treated as unknown, but separate portions of the heap. As a consequence desynchronized memory is inaccessible as part of transfer functions and any transfer function that must access it may not proceed:

DESYNC-FRAME

$$\frac{\hat{E} \vdash [H \,|\, P] \ c \ [H' \,|\, P]}{\hat{E} \vdash [H * [\![H_d]\!] \ \mathbf{call} \ \mathrm{V}_f(\mathrm{V}_1, \ldots, \mathrm{V}_n) \,|\, P] \ c \ [H' * [\![H_d]\!] \ \mathbf{call} \ \mathrm{V}_f(\mathrm{V}_1, \ldots, \mathrm{V}_n) \,|\, P]}$$

This DESYNC-FRAME rule is a special case of the separation logic frame rule that frames out the desynchronized part of memory and applies the transfer function to the remainder of memory. If this is not well defined because memory in the result of the desynchronized term must be accessed, either a different definition of reach() should be used or the code must be changed to ensure that the needed memory is not in a desynchronized region.

Similar rules apply for join, widening, and inclusion checking. Desynchronized regions can be joined or widened if they syntactically match, producing the same

desynchronized region. Otherwise, without employing a variety of precondition generalization, a join or widening can only be completed if the logic supports TRUE, in which
case all precision for this region is lost. Similarly for inclusion checking, only if there
is a syntactic match does it return true for desynchronized regions.

Introduction heuristics and elimination: For the purposes of analyzing JavaScript libraries, we use a simple introduction heuristic for desynchronized terms: if a function
call can be resolved to a known function, a desynchronized term should not be introduced. This policy has the effect that desynchronized terms only represent unknown
functions and thus we do not want to eliminate these terms from the heap. In fact, they
nicely represent the callback behavior that occurs in the library in the library's inferred
postcondition.

However, there are circumstances where such a simple heuristic may be non-optimal,
and it may be desirable to introduce desynchronized terms even when the code for a
called function is available. For example, sufficiently surjective functions [28] are functions where after a number of recursions the effect of continued recursion does not matter. In these situations desynchronization can represent the behavior of the unbounded
number of recursive calls without actually evaluating all of those calls. Another situation where desynchronization can benefit is in speeding up the analysis when known
functions may take too long to analyze but where they do not affect the result in any
meaningful way. In these situations, the postcondition includes a desynchronized term
that refers to the known function, but the result of that function has not been evaluated.

If desynchronized terms are introduced anywhere, it may be necessary that due to
access of desynchronized memory, the term that describes that memory has to be eliminated. This can be done if, for example, the synchronizing function's code is available.
The resynchronization process takes advantage of the separation logic frame rule by running the analysis on the synchronizing function starting from the desynchronized term:

DESYNC-ELIM
$$\frac{\cdot \vdash [H_d \mid P] \; \textbf{call} \; V_f(V_1,\ldots,V_n) \; [H_d' \mid P] \qquad \hat{E} \vdash [H * H_d' \mid P] \; c \; [H' \mid P]}{\hat{E} \vdash [H * [\![H_d \mid P]\!] \, \textbf{call} \; V_f(V_1,\ldots,V_n)] \; c \; [H' \mid P]}$$

With such an elimination rule it is possible to eagerly introduce desynchronized terms
on every function call and then lazily eliminate them as portions of the heap are needed.

When employing such an elimination rule, it is possible to consider the variety of
ways in which the $\cdot \vdash [H_d \mid P]$ **call** $V_f(\ldots)$ $[H_d' \mid P]$ judgment could be satisfied.
One way is if each function in V_f can be resolved to known code. In this case the
analyzer can be run on each resolvent and a disjunction of postconditions considered.
Alternatively, the formula H could carry the information to satisfy this judgment in the
form of a nested Hoare triple [26].

Example 4 (Desynchronization elimination). A region of the heap can be resynchronized by eliminating a desynchronized term:

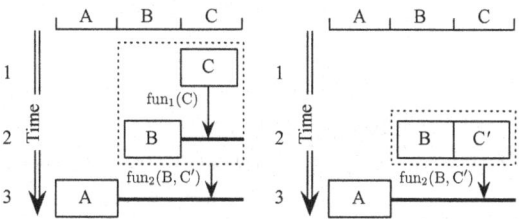

Here, the region C is resynchronized with B by analyzing the call to $\text{fun}_1(\text{C})$ starting from the memory state C resulting in memory state C'. Note that this resynchronization does not require analyzing $\text{fun}_2(\text{B}, \text{C}')$. This combined region can stay desynchronized if none of the desynchronized memory is required to proceed with the analysis.

Theorem 1 (Soundness of desynchronization introduction). *Desynchronization introduction is sound because the following property holds: for all* $E, \hat{E}, \sigma, \sigma', H, H', P.$
$E \vdash \langle\sigma\rangle\text{call } x(y_1, \ldots, y_n)\langle\sigma'\rangle$ *and* $\hat{E} \vdash [H \mid P] \quad \text{call } x(y_1, \ldots, y_n) \quad [H' \mid P]$ *and* $(\eta, \sigma) \in \gamma(H \mid P)$ *implies that there exists* η' *such that* $(\eta', \sigma') \in \gamma(H' \mid P).$

4.2 Attribute/Value Trackers

The primary benefit of attribute/value trackers occurs when they can be preserved from one abstract state to the next. To do so requires extending HOO transfer functions for the multi-state abstractions. The extension is trivial by appending the abstract heap from the precondition to each state in the transfer functions:

$$\frac{\hat{E} \vdash [H \mid P] \ c \ [H' \mid P]}{\hat{E} \vdash [[H]_{H_1} \mid P] \ c \ [[H']_{H_1} \mid P]}$$

To utilize attribute/value trackers, they must be introduced and managed appropriately. The goal is to reuse the same tracker whenever it is possible to do so and to only introduce fresh trackers when it is otherwise impossible. A key aspect of trackers is that the domain of a tracker is determined by the corresponding attribute set F at the point of introduction and thus the same tracker can be applied to any attribute set F' such that $F' \subseteq F$ if the values also match appropriately.

There are three key steps in managing this behavior of attribute/value trackers. First, materialization is responsible for splitting a singleton set off of a summary. In doing so, trackers can be preserved, even when the partition tied to a particular tracker is split. Second, trackers can be transfered along with attributes and values when an attribute/-value pair is copied from one object to another. Finally, trackers can be introduced when not otherwise available.

Materializing with attribute/value trackers: Since JavaScript does not have operations that allow many attributes and values to be copied or manipulated at once, a key operation for maintaining precision with attribute/value trackers is preserving them when splitting summarized objects/attributes/values so that there is a single attribute/value pair from a single object to be copied to another object. This operation is *materialization* and is described in Figure 8 in three parts.

$$\boxed{D_1 \Rightarrow D_2}$$

MAT-VALUE
$$\frac{v \text{ is fresh} \qquad P' = P \wedge \{v\} \subseteq V}{[H_2 * \{a\} \cdot \langle O; \{f\} : \rho \mapsto V \rangle]_{H_1} \,\vert\, P \Rightarrow [H_2 * \{a\} \cdot \langle O; \{f\} : \rho \mapsto \{v\} \rangle]_{H_1} \,\vert\, P'}$$

MAT-ATTR
$$\frac{F' \text{ is fresh} \qquad P' = P \wedge \{f\} \uplus F' = F \qquad P'' = P \wedge \{f\} \cap F = \emptyset}{\begin{array}{l} [H_2 * \{a\} \cdot \langle O; F : \rho \mapsto V \rangle]_{H_1} \,\vert\, P \Rightarrow \\ \qquad [H_2 * \{a\} \cdot \langle O; F' : \rho \mapsto V; \{f\} : \rho \mapsto V \rangle]_{H_1} \,\vert\, P' \vee [H_2 * \{a\} \cdot \langle O; F : \rho \mapsto V \rangle]_{H_1} \,\vert\, P'' \end{array}}$$

MAT-ADDR
$$\frac{A' \text{ is fresh} \qquad P' = P \wedge \{a\} \uplus A' = A \qquad P'' = P \wedge \{a\} \cap A = \emptyset}{[H_2 * A \cdot \langle O \rangle]_{H_1} \,\vert\, P \Rightarrow [H_2 * \{a\} \cdot \langle O \rangle * A' \cdot \langle O \rangle]_{H_1} \,\vert\, P' \vee [H_2 * A \cdot \langle O \rangle]_{H_1} \,\vert\, P''}$$

Fig. 8. Materialization of all of the parts of objects never produces fresh attribute/value trackers. It reuses existing trackers.

Each materialization rule is of the form $D_1 \Rightarrow D_2$ and thus intended to be used with the rule of consequence from Hoare logic [16] to allow a future rule to be applied. For example, rules for assignment (next section) can only be applied to singleton object addresses, singleton attributes, and often singleton values. By applying materialization correctly, an abstract heap element that consists of summary object addresses, summary attributes, and summary values can be converted to the appropriate singleton form without loss of precision, assuming a precise pure domain.

The first rule for materialization MAT-VALUE materializes a single value from a summary value, assuming that the object address, and attribute are already materialized. Because the object address and attribute are singletons, it must be that there is a singleton value $\{v\}$ and thus it can be materialized from the summary V. Doing so produces the additional constraint that $\{v\}$ is a subset of V. Because the materialized value $\{v\}$ is a fresh variable, this added constraint does not affect soundness.

The second rule for materialization is the primary rule for materializing attribute/value trackers. The MAT-ATTR rule splits an attribute set F into two attribute sets F' and $\{f\}$. There are two possible outcomes of this split. Either $\{f\}$ was already a subset of F, in which case the materialization can proceed, or $\{f\}$ is disjoint from F, in which case there is no materialization. In the case that the materialization proceeds, when the set F is split into two, both new partitions can be assigned the same tracker as was present in the original partition. This is because such a split does not require an extension of the domain of the tracker.

This second rule is applied whenever an object is being read. The attribute that is being read must be materialized from each partition of the object that may contain the attribute in question. Therefore, the read operation must consider a case where the attribute is in each partition of the object. The resulting pure constraints of MAT-ATTR may thus produce conflicts, causing such cases to be dropped.

A-OVERWRITE-DISTINCT

$$\hat{E}(x_1) = a_1 \qquad \hat{E}(y) = f \qquad \hat{E}(x_2) = a_2$$

$$\hat{E} \vdash \dfrac{\left[[H_1 * \{a_1\} \cdot \langle O_1; \{f\} : \rho_1 \mapsto V_1 \rangle * \{a_2\} \cdot \langle O_2; \{f\} : \rho_2 \mapsto \{v_2\} \rangle]_{H_0} \dashv P \right]}{x_1[y] := x_2[y]}$$
$$\left[[H_1 * \{a_1\} \cdot \langle O_1; \{f\} : \rho_2 \mapsto \{v_2\} \rangle * \{a_2\} \cdot \langle O_2; \{f\} : \rho_2 \mapsto \{v_2\} \rangle]_{H_0} \dashv P \right]$$

Fig. 9. Example abstract transfer function for assignment where the attribute/value tracker ρ_2 is transfered from the object at a_2 to the object at a_1

The third rule for materialization MAT-ADDR also manipulates attribute/value trackers, but less directly than the previous rule. This rule materializes a particular address $\{a\}$ from a summary of addresses A. Like the previous rule, if $\{a\}$ is a subset of A, the summary can be split. When this split occurs, the whole object definition is duplicated. Consequently each tracker is also duplicated. In the event that the materialization cannot occur, this constraint is added to indicate in the future that such an attempt was already considered.

Example 5 (Materializing a summary). Consider the heap abstraction $[A \cdot \langle F : \rho \mapsto V \rangle]_{H_1} \dashv \{a\} \subseteq A \wedge \{f\} \subseteq F$. If the analysis needs to read from a[f], this must be materialized. To achieve the following heap abstraction first the MAT-ADDR rule is applied, then the MAT-ATTR rule is applied to the result, then the rule MAT-VALUE is applied:

$$\left[\begin{array}{l} A' \cdot \langle F : \rho \mapsto V \rangle * \\ \{a\} \cdot \langle F' : \rho \mapsto V; \{f\} : \rho \mapsto \{v\} \rangle \end{array} \right]_{H_1} \dashv \begin{array}{l} \{a\} \uplus A' = A \\ \wedge \{f\} \uplus F' = F \\ \wedge \{v\} \subseteq V \end{array}$$

Transfering attribute/value trackers: Attribute/value trackers are transfered from one object to another by assignment. For simplicity, we assume here that all assignments between objects are transformed into the form of a simultaneous read from an object and a write to another object. When the attribute being read and written matches so that an attribute/value pair is being copied, there is an opportunity to transfer that attribute/value pair from one object to the other. When this transfer happens, the attribute/value tracker can be transfered as well.

Figure 9 shows one of the transfer functions that enables an attribute/value tracker transfer. The A-OVERWRITE-DISTINCT rule uses the abstract environment $\hat{E}$ to map variables onto addresses and then if the same attribute exists in two distinct objects the transfer occurs, in this case replacing ρ_1 with ρ_2.

Introducing attribute/value trackers: Attribute/value trackers should be introduced at chosen program points where the first of the paired states is selected. For example, when constructing an initial abstract state, it would be reasonable to express it as $[H]_H \dashv P$ where the two described heaps are identical. In this instance, fresh attribute/value trackers should be introduced for each partition in H. This establishes the initial relationship between the initial abstract state and the current abstract state and then any attribute/-value trackers that are preserved strengthen the relationship between the two states.

Additionally, attribute/value trackers can be introduced at other times. The benefits of doing so are less significant as freshly introduced trackers cannot relate objects from one time to another, but instead are limited to relating multiple objects in the same time. However as trackers are incomparable unless they are equal, freely introducing fresh trackers will prevent inclusion checking from succeeding and prevent the analysis from terminating. In the current implementation, we avoid this problem by only introducing absent trackers – after the precondition.

Other domain operations: Other domain operations such as join, widening, and inclusion check are largely the same as with HOO. Attribute/value trackers form a partition-by-partition lattice where any tracker $\rho \sqsubseteq -$. Join, widening, and inclusion follow from this: identical trackers can be matched and maintained through join and widening. Differing trackers must be replaced with –.

Theorem 2 (Soundness of tracker materialization). *Tracker materialization is sound because the following property holds:*
For all $D, D', \eta, \sigma_1, \sigma_2$. $D \Rightarrow D'$ and $(\eta, \sigma_1, \sigma_2) \in \gamma(D)$ implies that $(\eta, \sigma_1, \sigma_2) \in \gamma(D')$.

Theorem 3 (Soundness of transfer functions). *Transfer functions including desynchronization introduction, elimination, framing, and attribute/value tracker transfer are sound because the following property holds:*
For all $D, D', \sigma, \sigma', \sigma_0, \eta$. $E \vdash \langle \sigma \rangle c \langle \sigma' \rangle$ and $\hat{E} \vdash [D] \ c \ [D']$ and $(\eta, \sigma_0, \sigma) \in \gamma(D)$ implies that there exists a η' such that $(\eta', \sigma_0, \sigma') \in \gamma(D')$

5   Empirical Evaluation

In this section, we evaluate the use of desynchronized separation and attribute/value trackers on JavaScript meta-feature libraries – libraries that add language features to JavaScript through the use of object manipulation and callbacks. To do so, we test two hypotheses: (1) Does desynchronization provide the necessary precision for analyzing libraries that call unknown functions. (2) Do attribute/value trackers provide necessary precision for analyzing libraries that manipulate objects with unknown attribute/value relationships.

To evaluate these hypotheses, we identified several classes of meta-feature libraries that are available in JavaScript: classes, traits[1], mixins[2], and memoization[3]. From each of these candidates, we selected a small, but complex core (Table 1a) and annotated that functionality with preconditions. These preconditions indicate aliasing in the heap as well as give names to sets of attributes. Then, on each library, we compared expected postconditions against those generated by the JSAna analyzer for JavaScript, which is based on HOO with desynchronized separation and attribute value trackers.

---

[1] Extracted   from   `http://soft.vub.ac.be/~tvcutsem/traitsjs/`
[2] Extracted   from   `http://prototypejs.org/`   [3] Extracted   from
`https://developers.google.com/closure/library/`

Table 1. Results of running HOO with desynchronized separation and attribute/value trackers on JavaScript meta-feature libraries

(a) Test Library Code: Stmts is the number of statements in the program after preprocessing and lowering. Vars is the peak number of pure symbols used in the analysis. JP is the number of join points.

Test	Stmts	Vars	JP	Time (s)
Mixin	33	52	1	0.16
Traits	131	111	1	7.20
Memo	149	179	0	0.24
Class	128	118	1	8.13

(b) Properties: HOO is a property proven solely by HOO. D is HOO with desynchronized separation. T is HOO with attribute/value trackers. D+T is HOO with both enhancements.

Test	Property	HOO	D	T	D+T
Traits	Conflict managed	✓	✓	✓	✓
Memo	In table	✓	✓	✓	✓
Class	Constructor Call	✗	✓	✗	✓
Memo	Call saved	✗	✓	✗	✓
Mixin	Object extended	✗	✗	✓	✓
Traits	Object extended	✗	✗	✓	✓
Class	Resulting Object	✗	✗	✗	✓

The results of these experiments are shown in Table 1b. The first two properties are able to be proven solely with HOO. In the Traits example, which combines two objects into one, when the same attribute is present in both source objects, a single, global conflict value is used in the place of either source value. Because it is a single value, partitioning is sufficient to distinguish it. Similarly, in Memo, while Memo makes a call to an unknown function, if the precondition indicates that the call has already been memoized, that function call never happens and thus HOO's object-level reasoning, given a sufficiently precise set domain, is fully precise.

The second two properties actually require analyzing calls to unknown functions. In Class, this is the call to the initializer, and in Memo, this is the call to the memoized function. In both cases, the reachability analysis identifies suitable heap regions to allow the analysis to be fully precise. By comparison HOO, without desynchronization, cannot handle these calls and thus cannot prove the desired property.

The two object extended properties reason about the precise extension of objects that occurs in mixins and traits. In Mixin, an existing object has a number of attributes and corresponding values that may be overwritten by adding attributes and values from another object into it. Similarly, the Traits adds attribute and values from two different objects. Maintaining exact relationships between attributes and values is impossible without the use of attribute/value trackers, which allow the inferred postconditions for these analyses to be fully precise.

The last property, which checks that the instance created by the class is correct requires both attribute/value trackers and desynchronization to be precise. Because it uses both object manipulations and calls to the initializer, this indicates that these two additions are complementary and necessary for analyzing meta-feature libraries in JavaScript.

While it is not a goal to highly optimize for performance at this time, the results suggest that the analysis time is dependent on the number of pure symbols (Vars) and the number of join points (JP). When the number of variables increases (as long as there are join points), the overall analysis slows down. As in [8], nearly all of the cost can be attributed to the exponential set domain, which is implemented using binary decision diagrams. On top of this, the overhead of adding desynchronization and attribute/value trackers is negligible.

6 Discussion

In this section, we discuss the features and limitations of the analysis by considering two of the benchmarks in more detail. Additionally, we give some perspective on situations where the analysis loses precision.

6.1 Case Study: Class

The class benchmark is similar to the function Class presented in the introduction and the overview. Here we examine the similarities between the theory presented in the overview and what occurs in practice. We use program points from the overview for reference to the code used in the benchmark (which is complicated by more complete JavaScript support).

The analysis of the copy function proceeded exactly as shown in the overview. On each iteration of the analysis, a tracker was duplicated via materialization. That tracker was transfered to the result object. Consequently, the postcondition ② of copy was fully precise.

The desynchronization also works as expected. Critically, reachability identifies that a_1 and the local variable result are both *outside* the desynchronized region. This means that these things are unmodified by the call to the client-supplied initializer. Consequently, the resulting postcondition shows that the result object is the object created by the constructor and that constructor always produces exactly the same object attributes *and values* prior to the call to the client-supplied initializer regardless of how many times it is called.

6.2 Case Study: Memoization

The Memo benchmark transforms a function into a memoized version of that function. To accomplish this, it first translates the arguments array into a unique identifier by calling a uid() function passing it the entire arguments object. Then it determines if that unique identifier is already in the memoization table. If so, it returns the value from the table. Otherwise it calls the function to be memoized, f, passing it arguments (via JavaScript's apply functionality) and then memoizing the result.

Each of the function calls is challenging. The uid() function is essentially a hash function. It is responsible for converting data of any type into a unique string suitable for use in indexing into an object. Because hash functions are typically hard to analyze and this is a hash function that hashes to strings, this function presents a problem for analysis. Even if we had the code for it, it would be undesirable to analyze it.

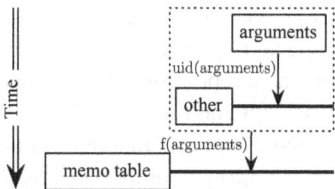

Fig. 10. Desynchronization phases of the memoization example

The second function call is also challenging because it is a callback into client-supplied code. The behavior of the function could be anything. It could have side-effects or it could be pure. Its only restriction is from JavaScript being memory safe (it cannot create pointers to previously unreachable parts of the heap).

Both of these problems are addressed by desynchronization as shown in Figure 10. Figure 10 shows as representation of the postcondition of the library function. In it we can see that not only was the callback to the client-supplied function f() desynchronized, but the call to uid() was desynchronized. Additionally, because the arguments object may have been modified by the uid() function, it is necessary to nest the desynchonizations to represent the result.

Nested desynchronization allows continuation-like behavior to be analyzed over parts of the program. Here the arguments object was possibly modified by the uid function before being possibly modified by the callback. The benefit of this nested structure is even if there is a sequence of functions that all touch the same memory, analysis can proceed by nesting all of these individual functions.

6.3 Boundaries of Analysis and Future Improvements

While our results suggest that both desynchronization and attribute/value trackers can be effective on JavaScript code, there are limitations to the precision. The most significant limitation is that attribute/value trackers are dropped when direct copies are not used. In particular, complex, nested copies are not currently supported by these trackers. For example, the following code wraps each value inside a newly allocated object.

```
result[a] = {value: attrs[a]};
```

Without the ability to reason about intermediary objects, full precision cannot be maintained and such abstractions fall back to what HOO can do. However, this behavior does not appear to occur in most libraries and thus may not be a significant issue. Adding support for this particular case is another form of tracker, but the inference of such trackers remains challenging.

While we find that reachability is a suitable heuristic for the analysis of many libraries, it may be overly pessimistic. In certain situations developers intentionally make portions of libraries globally mutable, but mutation is still not the common case.

7 Related Work

JavaScript specification and analysis: JuS [12] is an abduction-based inference tool for JavaScript targeted at the prototype and the scope chain. It is based on a detailed model of JavaScript semantics [2, 11] and thus automation is limited to resolving variable lookup through a prototype and scope chain. DJS [5, 6], which is a specification language and a dependent refinement type system for JavaScript, by comparison is more restricted in its support of JavaScript and thus offers more automation in that straight-line code can be reliably analyzed (loops and functions require annotations). The work presented in this paper automates discovery of loop invariants and callback summaries. This is significantly more automation than is provided by existing systems without sacrificing language features.

TAJS [17, 18], WALA [27], JSAI [15, 19], and SAFE [1, 21] are whole-program JavaScript analyses. Unlike the above systems, they require no annotations at all and are highly automated. However, they are ill suited to analyzing partial programs as is the case when verifying libraries. Because whole-program analysis has extensive context information, including object attributes and function bodies, there is less complexity involved in handling first-class functions (the function body can usually be resolved) or open objects (the attribute names are often fully known) and thus the abstractions used by these analyzers are incomparable to those that we employ.

The idea of attribute/value trackers comes from correlation tracking [27], which is implemented in both WALA and TAJS. Correlation tracking uses context sensitivity to exactly determine the constant attribute/symbolic value pairs needed for loops. Attribute/value trackers generalize this to symbolic attribute/symbolic value pairs that are each elements of summaries.

Higher-order separation logic and contracts: Desynchronized separation is closely tied to the concept of nested Hoare triples [26] and higher-order separation logic [20]. However, there are several key differences.

The goal of desynchronized separation is fundamentally different from that of nested Hoare triples. Unlike desynchronized separation, nested Hoare triples are intended to be used in program logics and not for automated inference. While there are efforts to automate some amount of reasoning [4], current techniques require significant annotation overhead and perform no inference, only inclusion checking.

The other significant difference is that nested Hoare triples strive for complete generality. A desynchronized term carries the following correspondence with nested Hoare triples:

$$[\![H_1]\!] \textbf{ call } V_f(V_1, \ldots V_n) * H_o \quad \Rightarrow \quad \exists H_2. \ [H_1] \textbf{ call } V_f(V_1, \ldots V_n) \ [H_2] \wedge H_2 * H_o$$

where an equivalence holds if an appropriate H_2 is chosen. The additional heap H_o is here to illustrate the key differentiating factor. A nested Hoare triple is a pure part of a formula that describes a value whereas a desynchronized term describes a heap that results from calling a function. The $* H_o$ illustrates which parts of the description are heap and which are pure.

The process of inference using desynchronization is significantly simpler than using nested Hoare tripes. This is due to the fact that desynchronization is less expressive than nested Hoare triples. There are fewer existentially quantified variables, and there is no need to treat portions of the heap that are simply passed through the unknown function call as separate portions of the heap that are manipulated. As a result, it is possible to

(1) easily adapt existing separation-logic-based analyses to certain higher-order tasks and (2) easily perform necessary heap splits during the analysis because there are two possible ways the heap can be split.

The key idea of nested Hoare triples is also similar to static contract checking for higher order languages [23, 29], which requires a pure specification of any callback's behavior up front. It is also similar to [22], except that it relies on separation logic and is applied to a stronger heap abstraction.

The goal of desynchronized separation is to not require a specification for callbacks at all, if the developer is judicious with built-in language protection mechanisms. In the event that memory is insufficiently protected, or the reachability analysis is too coarse, our analysis could be extended with nested Hoare triple specifications. In such a scenario, the nested Hoare triple is essentially the same as a resolvable function call. However, it is possible to imagine a simpler specification where only a footprint for the unknown function is specified. In this case, desynchronization would be required, but it would be applied to specified footprint (instead of using heap reachability to determine the split).

Container analysis: A significant part of HOO resembles an analysis for containers. Keeping track of object attribute names and values is similar to what is required for reasoning about mapping containers. The analysis in [10] also uses uninterpreted functions. However, the purpose of their uninterpreted functions is not to keep track of unknown attribute/value relationships, but instead to handle the sparsity problem of containers. Instead, they use uninterpreted functions to map a elements of a key/attribute type to a natural number that is the array index containing the value. The value arrays are then represented and manipulated using fluid updates [9].

Uninterpreted functions: There are several analyses [3, 13] that use uninterpreted functions to combine multiple abstract domains. While this work is also used for object and heap abstractions, the purpose of uninterpreted functions is different from attribute/value trackers. The uninterpreted functions in [3, 13] are used to transfer information between multiple abstract domains, whereas attribute/value trackers disambiguate individual symbolic elements of summaries across an analysis.

8   Conclusion

In this paper, we presented two multi-state abstractions that build upon abstract domains for heaps like HOO. Desynchronized separation gives a means for automatically reasoning about callbacks to unknown functions, while attribute/value trackers improve upon the partitioning of object attributes performed by HOO by maintaining consistent relationships between symbolic attribute names and symbolic values that are both members of summaries. Collectively these multi-state abstractions enable precise analysis of several core routines in JavaScript libraries.

Acknowledgements. Thank you to Anders Møller, our anonymous reviewers, and members of CUPLV and Antique for the helpful reviews and feedback. This material is based upon work supported in part by a Chateaubriand Fellowship, by the National Science Foundation under Grant Numbers CCF-1055066 and CCF-1218208, and by the European Research Council under the FP7 grant agreement 278673 (Project MemCAD).

References

[1] Bae, S.G., Cho, H., Lim, I., Ryu, S.: SAFE$_{WAPI}$: Web API misuse detector for web applications. In: FSE 2014 (2014)

[2] Bodin, M., Charguéraud, A., Filaretti, D., Gardner, P., Maffeis, S., Naudziuniene, D., Schmitt, A., Smith, G.: A trusted mechanised JavaScript specification. In: POPL, pp. 87–100 (2014)

[3] Bor-Yuh Evan Chang and K. Rustan M. Leino. Abstract interpretation with alien expressions and heap structures. In: VMCAI, pp. 147–163 (2005)

[4] Charlton, N., Horsfall, B., Reus, B.: Crowfoot: A verifier for higher-order store programs. In: VMCAI, pp. 136–151 (2012)

[5] Chugh, R., Herman, D., Jhala, R.: Dependent types for JavaScript. In: OOPSLA, pp. 587–606 (2012a)

[6] Chugh, R., Rondon, P.M., Jhala, R.: Nested refinements: a logic for duck typing. In: POPL, pp. 231–244 (2012b)

[7] Cousot, P., Cousot, R.: Abstract interpretation: A unified lattice model for static analysis of programs by construction or approximation of fixpoints. In: POPL, pp. 238–252 (1977)

[8] Cox, A., Chang, B.-Y.E., Rival, X.: Automatic analysis of open objects in dynamic language programs. In: Müller-Olm, M., Seidl, H. (eds.) Static Analysis. LNCS, vol. 8723, pp. 134–150. Springer, Heidelberg (2014)

[9] Dillig, I., Dillig, T., Aiken, A.: Fluid Updates: Beyond Strong vs. Weak Updates. In: Gordon, A.D. (ed.) ESOP 2010. LNCS, vol. 6012, pp. 246–266. Springer, Heidelberg (2010)

[10] Dillig, I., Dillig, T., Aiken, A.: Precise reasoning for programs using containers. In: POPL, pp. 187–200 (2011)

[11] Gardner, P., Maffeis, S., Smith, G.D.: Towards a program logic for JavaScript. In: POPL, pp. 31–44 (2012)

[12] Gardner, P., Naudziuniene, D., Smith, G.: JuS: Squeezing the sense out of JavaScript programs. In: JSTools (2013)

[13] Gulwani, S., Tiwari, A.: Combining abstract interpreters. In: PLDI, pp. 376–386 (2006)

[14] Halbwachs, N., Péron, M.: Discovering properties about arrays in simple programs. In: PLDI, pp. 339–348 (2008)

[15] Hardekopf, B., Wiedermann, B., Churchill, B.R., Kashyap, V.: Widening for control-flow. In: VMCAI, pp. 472–491 (2014)

[16] Hoare, C.A.R.: An axiomatic basis for computer programming. Commun. ACM 12(10), 576–580 (1969)

[17] Jensen, S.H., Møller, A., Thiemann, P.: Type analysis for JavaScript. In: SAS, pp. 238–255 (2009)

[18] Jensen, S.H., Møller, A., Thiemann, P.: Interprocedural analysis with lazy propagation. In: SAS, pp. 320–339 (2010)

[19] Kashyap, V., Sarracino, J., Wagner, J., Wiedermann, B., Hardekopf, B.: Type refinement for static analysis of JavaScript. In: DLS, pp. 17–26 (2013)

[20] Krishnawami, N.R.: Verifying Higher-Order Imperative Programs with HIgher-Order Separation Logic. PhD thesis, Carnegie Mellon University (2011)

[21] Lee, H., Won, S., Jin, J., Cho, J., Ryu, S.: SAFE: Formal specification and implementation of a scalable analysis framework for ECMAScript. In: FOOL 2012 (2012)

[22] Madhavan, R., Ramalingam, G., Vaswani, K.: Modular heap analysis for higher-order programs. In: SAS, pp. 370–387 (2012)

[23] Nguyen, P.C., Tobin-Hochstadt, S., Van Horn, D.: Soft contract verification. In: ICFP (2014)

[24] Reynolds, J.C.: Separation logic: A logic for shared mutable data structures. In: LICS, pp. 55–74 (2002)

[25] Rinetzky, N., Sagiv, M., Yahav, E.: Interprocedural shape analysis for cutpoint-free programs. In: SAS, pp. 284–302 (2005)
[26] Schwinghammer, J., Birkedal, L., Reus, B., Yang, H.: Nested Hoare triples and frame rules for higher-order store. Logical Methods in Computer Science 7(3) (2011)
[27] Sridharan, M., Dolby, J., Chandra, S., Schäfer, M., Tip, F.: Correlation Tracking for Points-To Analysis of JavaScript. In: Noble, J. (ed.) ECOOP 2012. LNCS, vol. 7313, pp. 435–458. Springer, Heidelberg (2012)
[28] Suter, P., Köksal, A.S., Kuncak, V.: Satisfiability modulo recursive programs. In: SAS, pp. 298–315 (2011)
[29] Xu, D.N., Jones, S.L.P., Claessen, K.: Static contract checking for Haskell. In: POPL, pp. 41–52 (2009)

Fine-Grained Detection of Privilege Escalation Attacks on Browser Extensions

Stefano Calzavara[1], Michele Bugliesi[1], Silvia Crafa[2], and Enrico Steffinlongo[1]

[1] Università Ca' Foscari Venezia
[2] University of Padova

Abstract. Even though their architecture relies on robust security principles, it is well-known that poor programming practices may expose browser extensions to serious security flaws, leading to privilege escalations by untrusted web pages or compromised extension components. We propose a formal security analysis of browser extensions in terms of a fine-grained characterization of the privileges that an active opponent may escalate through the message passing interface and we discuss to which extent current programming practices take this threat into account. Our theory builds on a formal language that embodies the essential features of JavaScript, together with few additional constructs dealing with the security aspects specific to the browser extension architecture. We then present a flow logic specification estimating the safety of browser extensions modelled in our language against the threats of privilege escalation and we prove its soundness. Finally, we show the feasibility of our approach by means of CHEN, a prototype static analyser for Google Chrome extensions based on our flow logic specification.

1 Introduction

Browser extensions customize and enhance the functionalities of standard web browsers by intercepting and reacting to a number of events triggered by navigation, page rendering or updates to specific browser data structures. While many extensions are simple and just installed to customize the navigation experience, other extensions serve security-critical tasks and have access to powerful APIs, providing access to the download manager, the cookie jar, or the navigation history of the user. Hence, the security of the web browser (and the assets stored therein) ultimately hinges on the security of the installed browser extensions. Just like browsers, extensions typically interact with untrusted and potentially malicious web pages: thus, all modern browser extension architectures rely on robust security principles, such as *privilege separation* [31].

Browser Extension Architecture. Privilege separated software architectures require programmers to structure their code in separated modules, running with different privileges. In the realm of browser extensions, privilege separation is implemented by structuring the extension in two different types of components: a privileged *background page*, which has access to the browser APIs and runs

J. Vitek (Ed.): ESOP 2015, LNCS 9032, pp. 510–534, 2015.
DOI: 10.1007/978-3-662-46669-8_21

isolated from web pages; and a set of unprivileged *content scripts*, which are injected into specific web pages, interact with them and are at a higher risk of attacks [4,10]. The permissions available to the background page are defined at installation time in a manifest file, to limit the dangers connected to the compromise of the background page. Content scripts interacting with different web pages are isolated one from each other by the same-origin policy of the browser, while process isolation protects the background page. The message passing interface available to extensions only allows the exchange of serialized JSON objects[1] between different components, hence pointers cannot cross trust boundaries.

Language Support for Privilege Separation. We are interested here in understanding to which extent current browser extension development frameworks, such as the Google Chrome extension APIs, naturally support privilege separation and comply with the underlying security architecture. Worryingly, we notice that in these frameworks a single privileged module typically offers a unified entry point to security-sensitive functionalities to all the other extension components, even though not all the components need to access the same functionalities and different trust relationships exist between different components.

To make matters worse, current programming patterns adopted in browser extensions do not safeguard the programmer against *compromised* components, even though the underlying privilege separated architecture was designed with compromise in mind. Compromise adds another layer of complexity to security-aware extension development, since corrupted extension components may get access to surprisingly powerful privileges.

1.1 Motivating Example

We illustrate our argument with a simple, but realistic example, inspired by one of the many cookie managers available in the Chrome Web Store (e.g., `EditThisCookie`). Consider an extension which allows users to add, delete or modify any cookie stored in the browser through an intuitive user interface. Additionally, it allows web pages to specify a set of security policies for the cookies they register: these client-side security policies are enforced by the extension and can be used to significantly strengthen web authentication [6,7].

The extension is composed of three components: two content scripts C and O, and a background page B. The background page is given the `cookies` permission, which grants it access to the browser cookie jar. The content script O is injected in the `options.html` page packaged with the extension and it provides facilities for cookie editing; when the user is done with his changes, O sends B a message and instructs it to update the cookie jar. The content script C, instead, is injected in the DOM of any HTTPS web page P opened by the browser: it is essentially a proxy, which forwards to B the security policies specified by P using the message passing interface. The messages sent by P are extended by C with an additional information: the website which specified the security policy.

---

[1] http://json.org

A possible run involving the described components is the following, where the last message is triggered by a user click:

$$P \to C : \{\text{tag: "policy", spec: "read-only"}\}$$
$$C \to B : \{\text{tag: "policy", site: "paypal.com", spec: "read-only"}\}$$
$$O \to B : \{\text{tag: "upd", ck: \{dom: "a.com", name: "res", value: "1440x900"\}}\}$$

Using the Google Chrome extension API, the components are programmed in JavaScript, typically by registering appropriate listeners for incoming messages. For instance, the content script C can be programmed as follows:

```
1  window.addEventListener("message", function(event) {
2    /* Accept only internal messages */
3    if (event.source != window) { return; }
4    /* Get the payload of the message */
5    var obj = event.data;
6    /* Extend the message with the site and forward it */
7    obj.site = window.location.hostname;
8    chrome.runtime.sendMessage (obj);
9  }, false);
```

Web pages can communicate with C by using the `window.postMessage` method available in JavaScript, thus opting-in to custom client-side protection.

The background page B, instead, is typically programmed as follows:

```
1   chrome.runtime.onMessage.addListener (
2     function (msg, sender, sendResp) {
3       /* Handle the reception of new policies */
4       if (msg.tag == "policy") {
5         /* Store a new (valid) policy for the site */
6         if (is_valid (msg.spec))
7            localStorage.setItem (msg.site, msg.spec);
8         else console.log ("Invalid policy");
9       }
10      /* Handle requests for cookie updates */
11      else if (msg.tag == "upd") {
12         chrome.cookies.set (msg.ck);
13      }
14      else console.log ("Invalid message");
15  });
```

This tag-based coding style featuring a single entry point to the background page is very popular, since it is easy to grasp and allows for fast prototyping, but it also fools programmers into underestimating the attack surface against the extensions they write. In this example, a malicious web page can compromise the integrity of the cookie jar by exploiting the poorly programmed content script C through the following method invocation:

```
window.postMessage ({tag: "upd", ck: {dom: "google.com",
                     name: "SID", value: "aQe73ajs..."}});
```

This allows the web page to carry out dangerous attacks, like session fixation or login CSRF on arbitrary websites [7]. The issue can be rectified by including a *sanitization* in the code of C and by ensuring that only messages with the `"policy"` tag are delivered to the background page.

The revised code is more robust than the original one and it safeguards the extension against the threats posed by malicious (or compromised) web pages. Unfortunately, it does not yet protect the background page against a compromised content script: if an attacker is able to exploit a code injection vulnerability in C, he may force the content script into deviating from the intended communication protocol. Specifically, an attacker with scripting capabilities in C may forge arbitrary messages to the background page and taint the cookie jar.

A much more robust solution then consists in introducing two distinct communication ports for C and O, and dedicating these ports to the reception of the two different message types (see Section 5). This is relatively easy to do in this simple example, but, in general, decoupling the functionalities available to the background page to shield it against privilege escalation is complex, since n different content scripts or extensions may require access to m different, possibly overlapping sets of privileged functionalities.

1.2 Contributions

Our contributions can be summarized as follows:

1. we model browser extensions in a formal language that embodies the essential features of JavaScript, together with a few additional constructs dealing with the security aspects specific to the browser extension architecture;
2. we formalize a fine-grained characterization of the privileges which can be escalated by an active opponent through the message passing interface, assuming the compromise of some untrusted extension components;
3. we propose a flow logic specification estimating the safety of browser extensions against the threats of privilege escalation and we prove its soundness, despite the best efforts of an active opponent. We show how the static analysis works on the example above and supports its secure refactoring;
4. we present CHEN (CHrome Extension aNalyser), a prototype tool that implements our flow logic specification, providing an automated security analysis of existing Google Chrome extensions. The tool opens the way to an automatic security-oriented refactoring of existing extensions. We show CHEN at work on ShareMeNot [30], a real extension for Google Chrome, and we discuss how the tool spots potentially dangerous programming practices.

2   Related Work

Browser Extension Security. Carlini *et al.* performed a security evaluation of the Google Chrome extension architecture by means of a manual review of 100 popular extensions [10]. Liu *et al.* further analysed the Google Chrome extension architecture, highlighting that it is inadequate to provide protection against malicious extensions [21]. Guha *et al.* [15] proposed a methodology to write provably secure browser extensions, based on refinement typing; the approach requires extensions to be coded in Fine, a dependently-typed ML dialect. Karim *et al.* developed Beacon, a static detector of capability leaks for Firefox extensions [20]. A capability leak happens when a component exports a pointer to a privileged piece of code. These leaks violate the desired modularity of Firefox extensions, but they cannot be directly exploited by content scripts, since the message passing interface prevents the exchange of pointers. Finally, information flow control frameworks have been proposed for browser extensions [13,3].

Privilege Escalation Attacks. Privilege escalation attacks have been extensively studied in the context of Android applications, starting with [12,29]. Fragkaki *et al.* formalized protection against privilege escalation in Android applications as a noninterference property, which is then enforced by a dynamic reference monitor [14]. Bugliesi *et al.* presented a stronger security notion and discussed a static type system for Android applications, which provably enforces protection against privilege escalation [8]. The present paper generalizes both these proposals, by providing a fine-grained view of the privileges leaked to an arbitrarily powerful opponent. Akhawe *et al.* [2] pointed out severe limitations in how privilege separation is implemented in browser extension architectures. Their work has been very inspiring for the present paper, which provides a formal counterpart to many interesting observations contained therein. For instance, [2] defines *bundling* as the collection of disjoint functionalities inside a single module running with the union of the privileges required by each functionality. Our formal notion of privilege leak captures the real dangers of permission bundling.

Formal Analysis of JavaScript. Maffeis *et al.* formalized the first detailed operational semantics for JavaScript [22] and used it to verify the (in)security of restricted JavaScript subsets [23]. Jensen *et al.* proposed an abstract interpretation framework for JavaScript in the realm of type analysis [18]. Guha *et al.* defined λ_{JS} as a relatively small core calculus based on a few well-understood constructs, where the numerous quirks of JavaScript can be encoded with a reasonable effort [16]. The adequacy of the semantics has been assessed by extensive automatic testing. The calculus has been used to support static analyses to detect type errors in JavaScript [17] and to verify the correctness of JavaScript sandboxing [28]. We also develop our flow analysis on top of λ_{JS}, extending it to reason about browser extension security. An alternate solution would have been to base our work on S5 [27]. This approach would have allowed to analyse browser extensions using ECMA5-specific features, but at the cost of significantly complicating the formal development.

3   Modelling Browser Extensions

Our language embodies the essential features of JavaScript, formalized as in λ_{JS} [16], up to a number of changes needed to deal with the security aspects specific to the browser extension architecture. In our model, several expressions run in parallel with different permissions and are isolated from each other: communication is based on asynchronous message exchanges.

3.1 Syntax

We assume disjoint sets of channel names $\mathcal{N}$ (a, b, m, n) and variables $\mathcal{V}$ (x, y, z). We let r range over a set of references $\mathcal{R}$, and we assume a lattice of permissions $(\mathcal{P}, \sqsubseteq)$, letting ρ range over $\mathcal{P}$. The syntax of the language is given below:

Constants	$c ::= num \mid str \mid bool \mid$ **unit** $\mid$ **undefined**,	
Values	$v ::= n \mid x \mid c \mid r_\ell \mid \lambda x.e \mid \{\overrightarrow{str_i : v_i}\}$	
Expressions	$e ::= v \mid$ **let** $x = e$ **in** $e \mid e\,e \mid op(\overrightarrow{e_i}) \mid$ **while** $(e) \{ e \}$	
	$\mid$ **if** $(e) \{ e \}$ **else** $\{ e \} \mid e; e \mid e[e] \mid e[e] = e$	
	$\mid$ **delete** $e[e] \mid$ **ref**$_\ell\, e \mid$ **deref** $e \mid e := e$	
	$\mid \bar{e}\langle e \rhd \rho \rangle \mid$ **exercise**(ρ)	
Systems	$s ::= \mu; h; i$	*Memories* $\quad \mu ::= \emptyset \mid \mu, r_\ell \overset{\rho}{\mapsto} v$
Handlers	$h ::= \emptyset \mid h, a(x \lhd \rho : \rho').e$	*Instances* $\quad i ::= \emptyset \mid i, a\{\!\!\{e\}\!\!\}_\rho$

All the value forms are standard, we just note that references r_ℓ bear a label ℓ, taken from a set of labels $\mathcal{L}$. Labels identify the program point where references are created: this is needed for the static analysis and plays no role in the semantics. As usual, the lambda abstraction $\lambda x.e$ binds x in e.

As to expressions, the first three lines correspond to standard constructs inherited from λ_{JS}, including function applications, basic control-flow operators, and the usual operations on records (field selection, field update/creation, field deletion) and references (allocation, dereference and update). As anticipated, reference allocation comes with an annotation ℓ. We leave unspecified the precise set of primitive operations op. The expression **let** $x = e$ **in** e' binds x in e'.

The last line of the productions includes the new constructs added to λ_{JS}. The expression $\bar{a}\langle v \rhd \rho \rangle$ sends the value v on channel a. In order for the sender to protect the message, the expression specifies that the value can be received by any *handler* with at least permission ρ that is listening on a. The expression **exercise**(ρ) exercises the privilege ρ. This construct uniformly abstracts any security-sensitive operation, such as the call to a privileged API, which requires the permission ρ to successfully complete the task.

We let h range over multisets of *handlers* of the form $a(x \lhd \rho : \rho').e$. The handler $a(x \lhd \rho : \rho').e$ listens for messages on the channel a. When a value v is sent over a, a new *instance* of the handler is spawned to run the expression e with permission ρ', with the bound variable x replaced by v. The handler protects its body against untrusted senders by specifying that only instances with permission ρ can be granted access. Intuitively, the body of a handler corresponds to the function passed as a parameter to the `addListener` method of `chrome.runtime.onMessage`. Different handlers can listen on the same channel: in this case, only one handler is non-deterministically dispatched. We often refer to a handler with the name of the channel where it is registered.

Table 1. Small-step operational semantics of systems ($s \xrightarrow{\alpha} s'$)

(R-SYNC)

$$\frac{h = h',\, b(x \triangleleft \rho_s : \rho_b).e \qquad \rho_s \sqsubseteq \rho_a \qquad \rho_r \sqsubseteq \rho_b \qquad v \text{ serializable}}{\mu; h; a\{\!|E\langle \overline{b}\langle v \triangleright \rho_r\rangle\rangle|\!\}_{\rho_a} \xrightarrow{\langle a:\rho_a, b:\rho_b\rangle} \mu; h; a\{\!|E\langle\mathbf{unit}\rangle|\!\}_{\rho_a}, b\{\!|e[v/x]|\!\}_{\rho_b}}$$

(R-SET)

$$\mu; h; i, i'' \xrightarrow{\alpha} \mu'; h'; i', i''$$

(R-EXERCISE)

$$\frac{\rho \sqsubseteq \rho_a}{\mu; h; a\{\!|E\langle\mathbf{exercise}(\rho)\rangle|\!\}_{\rho_a} \xrightarrow{a:\rho_a \gg \rho} \mu; h; a\{\!|E\langle\mathbf{unit}\rangle|\!\}_{\rho_a}}$$

(R-INTERNAL)

$$\frac{\mu; e \hookrightarrow_\rho \mu'; e'}{\mu; h; a\{\!|e|\!\}_\rho \xrightarrow{\cdot} \mu'; h; a\{\!|e'|\!\}_\rho}$$

$$\begin{aligned}
E ::=\ &\bullet \mid \mathbf{let}\ x = E\ \mathbf{in}\ e \mid E\,e \mid v\,E \mid op(\overrightarrow{v_i}, E, \overrightarrow{e_j}) \mid \mathbf{if}\ (E)\ \{e\}\ \mathbf{else}\ \{e\} \\
&\mid E[e] \mid v[E] \mid E[e] = e \mid v[E] = e \mid v[v] = E \mid E; e \mid \overline{E}\langle e \triangleright \rho\rangle \mid \overline{v}\langle E \triangleright \rho\rangle \\
&\mid \mathbf{delete}\ E[e] \mid \mathbf{delete}\ v[E] \mid \mathbf{ref}_\ell\ E \mid \mathbf{deref}\ E \mid E := e \mid v := E.
\end{aligned}$$

We let i range over multisets of running *instances* of the form $a\{\!|e|\!\}_\rho$. The instance $a\{\!|e|\!\}_\rho$ is a running expression e, which is granted permission ρ. The instance is annotated with the channel name a corresponding to the handler which spawned it.

We let μ range on *memories*, i.e., sets of bindings of the form $r_\ell \xmapsto{\rho} v$. A memory is a partial map from (labelled) references to values. The annotation ρ on the arrow records the permission of the instance that created the reference, and at the same time tracks the permissions required to have read/write access on the reference. Given a memory μ, we let $dom(\mu) = \{r \mid r_\ell \xmapsto{\rho} v \in \mu\}$.

Finally, a *system* is defined as a triple $s = \mu; h; i$. Intuitively, a system evolves by letting running instances (i) communicate through the memory μ when they are granted exactly the same permissions, (*ii*) spawn new instances by sending messages to handlers in h, and (*iii*) perform internal computations.

3.2 Semantics

The small-step operational semantics of the calculus is defined in terms of a labelled reduction relation between systems $s \xrightarrow{\alpha} s'$. Labels play no role in the semantics of systems: they are just used to track useful information that is needed in the proofs. The syntax of labels α is defined as follows:

$$\alpha ::= \cdot \mid a:\rho_a \gg \rho \mid \langle a:\rho_a, b:\rho_b\rangle.$$

The label $a:\rho_a \gg \rho$ records the exercise of the privilege ρ by an instance a running with permissions ρ_a. The send label $\langle a:\rho_a, b:\rho_b\rangle$ records that an instance a with permissions ρ_a is sending a message to a handler b with permissions ρ_b. Finally, the empty label $\cdot$ tracks no information. We denote traces by $\overrightarrow{\alpha}$ and we write $\xRightarrow{\overrightarrow{\alpha}}$ for the reflexive-transitive closure of $\xrightarrow{\alpha}$. Table 1 collects the reduction rules for systems and the definition of evaluation contexts. We write $E\langle e\rangle$ when the hole $\bullet$ in E is filled with the expression e.

Table 2. Small-step operational semantics of expressions $(\mu; e \hookrightarrow_\rho \mu'; e')$

(JS-Expr)

$$\dfrac{e_1 \hookrightarrow e_2}{\mu; e_1 \hookrightarrow_\rho \mu; e_2}$$

(JS-Ref)

$$\dfrac{r \notin dom(\mu) \quad \mu' = \mu, r_\ell \overset{\rho}{\mapsto} v}{\mu; \mathbf{ref}_\ell\ v \hookrightarrow_\rho \mu'; r_\ell}$$

(JS-Deref)

$$\dfrac{\mu = \mu', r_\ell \overset{\rho}{\mapsto} v}{\mu; \mathbf{deref}\ r_\ell \hookrightarrow_\rho \mu; v}$$

(JS-SetRef)

$$\dfrac{\mu = \mu', r_\ell \overset{\rho}{\mapsto} v'}{\mu; r_\ell := v \hookrightarrow_\rho \mu', r_\ell \overset{\rho}{\mapsto} v; v}$$

(JS-Context)

$$\dfrac{\mu; e_1 \hookrightarrow_\rho \mu'; e_2}{\mu; E\langle e_1\rangle \hookrightarrow_\rho \mu'; E\langle e_2\rangle}$$

Rule (R-Sync) implements a security cross-check between the sender a and the receiver b: by specifying a permission ρ_r on the send expression, the instance a requires the handler b to have at least ρ_r, while by specifying a permission ρ_s in its definition, the handler b requires the instance a to have at least ρ_s. If the security check succeeds, a new instance of b is created and the sent value v is substituted to the bound variable x in the body of the handler. Communication is restricted to *serializable* values, according to the following definition.

Definition 1 (Serializable Value). *A value v is* serializable *iff either (1) v is a name n or a constant c; or (2) $v = \{\overrightarrow{str_i : v_i}\}$ and each v_i is serializable.*

This restriction is consistent with the browser extension security architecture, which prevents the exchange of pointers between different components [10].

Rule (R-Exercise) reduces the expression **exercise**(ρ). Reduction takes place only when the expression runs in an instance a which is granted permission $\rho_a \sqsupseteq \rho$. Rule (R-Set) allows for reducing any of the parallel instances running in a system, while rule (R-Internal) performs an internal reduction step based on the auxiliary transition relation $\mu; e \hookrightarrow_\rho \mu'; e'$, annotated with the permission ρ granted to the instance. The internal reduction relation is defined in Table 2; it relies on a basic reduction $e \hookrightarrow e'$, which is directly inherited from λ_{JS} and lifted to the internal reduction by rule (JS-Expr). The definition of the basic reduction is standard and given in the full version [9].

A reference is allocated by means of rule (JS-Ref). According to this rule, two references may have the same label (e.g., when reference allocation occurs inside a program loop) but each reference is guaranteed to have a distinct name. Since read/write operations on memory ultimately depend on the reference name, this ensures that labels on references do not play any role at runtime.

Finally, rules (JS-SetRef) and (JS-Deref) deal with reference update and dereference. Observe that, according to these rules, both read and write access to memory requires *exactly* the permission ρ annotated on the reference. In other words, instances with different privileges cannot communicate through the memory. This corresponds to the heap separation policy implemented in modern browser extension architectures.

3.3   Privilege Leak

We now define the notion of *privilege leak*, which dictates an upper bound to the privileges which can be escalated by an opponent when interacting with the system. We start by defining when a system exercises a given permission.

Definition 2 (Exercise). *Given a system s, we say that s exercises ρ iff there exist s' and $\vec{\alpha}$ such that $s \overset{\vec{\alpha}}{\Rightarrow} s'$ and $a{:}\rho_a \gg \rho \in \{\vec{\alpha}\}$.*

In our threat model, an opponent can mount an attack against the system by registering new handlers, which may intercept messages sent to trusted components, and/or by spawning new instances, which may tamper with the system by writing in shared memory cells and by using the message passing interface.

Formally, an opponent is defined as a pair (h, i), with an upper bound ρ for the permissions granted to h and i. For technical reasons, we assume that the set of variables $\mathcal{V}$ is partitioned into the sets $\mathcal{V}_t$ and $\mathcal{V}_u$ (trusted and untrusted variables). We stipulate that all the variables occurring in the system are drawn from $\mathcal{V}_t$, while all the variables occurring in the opponent code belong to $\mathcal{V}_u$.

Definition 3 (Opponent). *A ρ-opponent is a closed pair (h, i) where*

- *for any handler $a(x \triangleleft \rho : \rho').e \in h$, we have $\rho' \sqsubseteq \rho$;*
- *for any instance $a\{\!|e|\!\}_{\rho'} \in i$, we have $\rho' \sqsubseteq \rho$;*
- *for any $x \in vars(h) \cup vars(i)$, we have $x \in \mathcal{V}_u$.*

Definition 4 (Privilege Leak). *A (initial) system $s = \mu; h; \emptyset$ leaks ρ against ρ' (with $\rho \not\sqsubseteq \rho'$) iff, for any ρ'-opponent (h_o, i_o), the system $s' = \mu; h, h_o; i_o$ exercises at most ρ.*

Our security property is given over *initial* systems, that is systems with no running instances, since we are interested in understanding the interplay between the exercised permissions and the communication interface exposed by the handlers in the system. Intuitively, a system s is "more secure" than another system s' if it leaks fewer privileges than s' against any possible ρ.

3.4   Encoding the Example

To illustrate, we encode in our formal language the example in Section 1.1. Consider the system $s = \mu; h_c, h_o, h_b; \emptyset$, where the handlers h_c, h_o and h_b encode the two content scripts and the background page. The memory μ encodes the private memory of the background page, and it is used to store library functions. We grant the background page two different permissions: MemB to access the references under its control and Cookies to access the cookie jar.

Let B = MemB $\sqcup$ Cookies, we let $\mu = lib_\ell \overset{B}{\mapsto} obj$, where:

$$obj = \{ \text{``set''} : \lambda x.\textbf{exercise}(\text{Cookies}); \textit{set/update the cookie } x,$$
$$\text{``is\_valid''} : \lambda x.\textit{check validity of policy } x,$$
$$\text{``store''} : \lambda x.\lambda y.\textbf{exercise}(\text{MemB}); \textit{bind policy } y \textit{ to site } x,$$
$$\text{``log''} : \lambda x.\textit{print message } x\}$$

We omit the internal logic of the functions, we just observe that we put in place the exercise expressions corresponding to the usage of the required privileges. The definition of the handler h_b modelling the background page is given below, where C and O are the permissions granted to the two content scripts in order to let them contact B through the message passing interface.

$$h_b \triangleq b(x \triangleleft \mathsf{C} \sqcap \mathsf{O} : \mathsf{B}).$$
$$\mathbf{let}\ mylib = \mathbf{deref}\ lib_\ell\ \mathbf{in}$$
$$\mathbf{if}\ (x[\text{``}tag\text{''}] == \text{``}policy\text{''})\ \{$$
$$\quad \mathbf{if}\ (mylib[\text{``}is\_valid\text{''}]\,(x[\text{``}spec\text{''}]))\ \{$$
$$\quad\quad (mylib[\text{``}store\text{''}]\,(x[\text{``}site\text{''}]))\,(x[\text{``}spec\text{''}])$$
$$\quad \}$$
$$\quad \mathbf{else}\ \{\ mylib[\text{``}log\text{''}]\ \text{``}invalid\ policy\text{''}\ \}$$
$$\}$$
$$\mathbf{else}\ \{$$
$$\quad \mathbf{if}\ (x[\text{``}tag\text{''}] == \text{``}upd\text{''})\ \{\ (mylib[\text{``}set\text{''}])\,(x[\text{``}ck\text{''}])\ \}$$
$$\quad \mathbf{else}\ \{\ mylib[\text{``}log\text{''}]\ \text{``}invalid\ message\text{''}\ \}$$
$$\}$$

The handler can be accessed by both C and O, as modelled by the guard $\mathsf{C} \sqcap \mathsf{O}$.

A simplified encoding of the content scripts, corresponding to the handlers h_c and h_o respectively, is given below. This simple encoding will be enough to explain the most important aspects of the flow analysis in Section 4.3.

$$h_c \triangleq c(y \triangleleft \mathsf{P} : \mathsf{C}).\mathbf{let}\ y' = (y[\text{``}site\text{''}] = \ldots)\ \mathbf{in}\ \overline{b}\langle y' \triangleright \mathsf{B}\rangle$$
$$h_o \triangleq o(z \triangleleft \top : \mathsf{O}).\mathbf{let}\ z' = \{\ \text{``}tag\text{''} : \text{``}upd\text{''},\ \text{``}ck\text{''} : \ldots\}\ \mathbf{in}\ \overline{b}\langle z' \triangleright \mathsf{B}\rangle$$

The only notable point here is that h_o is protected with permission $\top$, since it is injected in the trusted options page of the extension, while h_c is protected with permission P, modelling access to the `window.postMessage` method used to communicate with C from a web page. As a consequence, any P-opponent has the ability to activate h_c through the message passing interface.

Based on the encoding, we estimate the robustness against privilege escalation attacks. It turns out that the system s leaks B against P, since a P-opponent can force h_c into forwarding an arbitrary (up to the choice of the *"site"* field) message to h_b, hence all the privileges available to h_b may be escalated.

Assume then that h_c is replaced by a new handler h'_c, defined as follows:

$$h'_c \triangleq c(y \triangleleft \mathsf{P} : \mathsf{C}).\ \mathbf{let}\ y_{new} = \{\ \text{``}tag\text{''} : \text{``}policy\text{''},\ \text{``}site\text{''} : \ldots\}\ \mathbf{in}$$
$$\mathbf{let}\ y' = (y_{new}[\text{``}spec\text{''}] = y[\text{``}spec\text{''}])\ \mathbf{in}\ \overline{b}\langle y' \triangleright \mathsf{B}\rangle$$

The new system $s_{tag} = \mu; h'_c, h_o, h_b; \emptyset$ leaks MemB against P, since a P-opponent can only communicate with h_b through the proxy h'_c, which ensures that only messages tagged with *"policy"* are delivered to the background page and the integrity of the cookie jar is preserved. However, s_{tag} leaks B against C, since a C-opponent can send arbitrary messages to h_b and thus escalate all the available privileges.

3.5   Fixing the Example

The key observation here is that there is no good reason to let C and O share the same entry point to B, since they request distinct functionalities. We can then split the logic of h_b into two different handlers: h_{b_1} protected by permission C, and h_{b_2} protected by permission O.

$b_1(x \triangleleft \mathsf{C} : \mathsf{B})$.

   let $mylib = $ **deref** lib_ℓ **in**

   if $(x[\,\text{"}tag\text{"}] == \text{"}policy\text{"})$ { ... }

   else $\{mylib[\,\text{"}log\text{"}]\,\text{"}invalid\ policy\text{"}\}$

$b_2(x \triangleleft \mathsf{O} : \mathsf{B})$.

   let $mylib = $ **deref** lib_ℓ **in**

   if $(x[\,\text{"}tag\text{"}] == \text{"}upd\text{"})$ { ... }

   else $\{mylib[\,\text{"}log\text{"}]\,\text{"}invalid\ message\text{"}\}$

Clearly, the code of h_c and h_o must also be changed to communicate on the new channels b_1 and b_2 respectively: call these new handlers $\hat{h}_c$ and $\hat{h}_o$. Now the handler h_{b_1} is only accessible by $\hat{h}_c$, while the handler h_{b_2} can only be accessed by $\hat{h}_o$, hence, if O is not compromised, the integrity of the cookie jar is preserved.

Unfortunately, the current extension architecture does not support a fine-grained assignment of permissions to different portions of the background page [2], hence we are forced to violate the principle of least privilege and assign to both h_{b_1} and h_{b_2} the full set of permissions $\mathsf{B} = \mathsf{MemB} \sqcup \mathsf{Cookies}$ available to the original h_b, even though h_{b_1} and h_{b_2} only require a subset of these permissions. Still, the system $s_{chan} = \mu; \hat{h}_c, \hat{h}_o, h_{b_1}, h_{b_2}; \emptyset$ only leaks MemB against C.

Notice that this refactoring can be performed on existing Google Chrome extensions by using the `chrome.runtime.connect` API for the dynamic creation of communication ports towards the background page.

4   Security Analysis: Flow Logic

To precisely reason about privilege escalation, it is crucial to statically capture the interplay between the format of the exchanged messages and the exercised privileges: we then resort to the flow logic framework [24]. The main judgement of our flow analysis is $\mathcal{E} \Vdash s$ **despite** ρ, meaning that the environment $\mathcal{E}$ represents an acceptable analysis estimate for s, even when s interacts with a ρ-opponent. This implies that any ρ-opponent will at most escalate privileges up to an upper bound which can be immediately computed from $\mathcal{E}$ (see Theorem 1).

4.1   Analysis Specification

Abstract Values. We let $\hat{V}$ stand for the set of abstract values $\hat{v}$, defined as sets of abstract pre-values (we often omit brackets around singletons):

$$
\begin{aligned}
\textit{Abstract pre-values} \quad & \hat{u} ::= n \mid \hat{c} \mid \ell \mid \lambda x^\rho \mid \langle\!\langle \overrightarrow{str_i : \hat{v}_i} \rangle\!\rangle_{\mathcal{E},\rho} \\
\textit{Abstract values} \quad & \hat{v} ::= \{\hat{u}_1, \ldots, \hat{u}_n\}.
\end{aligned}
$$

Channel names n are abstracted into themselves. The abstract pre-value $\hat{c}$ stands for the abstraction of the constant c. We dispense from listing all the abstract pre-values corresponding to the constants of our calculus, but we assume that they include at least **true**, **false**, **unit** and **undefined**.

A reference r_ℓ is abstracted into the label ℓ. A function $\lambda x.e$ is abstracted into the simpler representation λx^ρ, keeping track of the privileges ρ exercised by the expression e. The abstract pre-value $\langle\!\overrightarrow{str_i : v_i}\rangle\!\rangle_{\mathcal{E},\rho}$ is the abstract representation of the concrete record $\{\overrightarrow{str_i : v_i}\}$ in the environment $\mathcal{E}$, assuming that the record is created in a context with permission ρ. We do not fix any apriori abstract representation for records, e.g., both field-sensitive and field-insensitive representations are admissible.

We associate to each concrete operation op an abstract counterpart $\widehat{op}$ on abstract values. We also assume three abstract operations $\widehat{get}$, $\widehat{set}$ and $\widehat{del}$, mirroring the standard get field, set field and delete field operations on records. Finally, we assume that abstract values are ordered by a pre-order $\sqsubseteq$ containing set inclusion, with the intuition that smaller abstract values are more precise (we overload the symbol used to order permissions, to keep the notation lighter). All the abstract operations and the abstract value pre-order can be chosen arbitrarily, as long as they satisfy some relatively mild and well-established conditions needed in the proofs. For instance, we require abstract operations to be monotonic and to soundly over-approximate their concrete counterparts (see the full version [9] for details).

Abstract Environments. The judgements of the analysis are specified relative to an abstract environment $\mathcal{E} = \hat{\Upsilon}; \hat{\Phi}; \hat{\Gamma}; \hat{\mu}$, consisting of the following four components, where $\Lambda = \{\lambda x \mid x \in \mathcal{V}\}$ is used to store the abstract return value for lambdas:

Abstract variable environment	$\hat{\Gamma} : \mathcal{V} \cup \Lambda \to \hat{V}$
Abstract memory	$\hat{\mu} : \mathcal{L} \times \mathcal{P} \to \hat{V}$
Abstract stack	$\hat{\Upsilon} : \mathcal{N} \times \mathcal{P} \to \mathcal{P} \times \mathcal{P}$
Abstract network	$\hat{\Phi} : \mathcal{N} \times \mathcal{P} \to \hat{V}.$

Abstract variable environments are standard: they associate abstract values to variables and to functions, corresponding to the abstraction of their return value. Abstract memories are also standard: they associate abstract values to labels denoting references. Specifically, if $\hat{\mu}(\ell, \rho) = \hat{v}$, then $\hat{v}$ is a sound abstraction of any value stored in a reference labelled with ℓ and protected with permission ρ.

Abstract stacks are novel and are central to the privilege escalation analysis. This part of the environment is used to keep track of the permissions required to get access to each handler and the privileges which are exercised (also *transitively*, i.e., by communicating with other components) by the handlers themselves. Specifically, if $\hat{\Upsilon}(a, \rho_a) = (\rho_s, \rho_e)$, then the handler a with permission ρ_a can be accessed by any component with permission ρ_s and it will be able to exercise privileges up to ρ_e, possibly by calling other handlers in the system.

Finally, abstract networks are adapted from flow logic specifications for process calculi [26] and they are used to keep track of the messages sent to the handlers in the system. For instance, if we have $\hat{\Phi}(a, \rho_a) = \hat{v}$, then $\hat{v}$ is a sound

Table 3. Flow analysis for values

(PV-Name)	(PV-Var)	(PV-Cons)	(PV-Ref)
$n \in \hat{v}$	$\mathcal{E}_{\hat{\Gamma}}(x) \sqsubseteq \hat{v}$	$\{\hat{c}\} \sqsubseteq \hat{v}$	$\ell \in \hat{v}$
$\mathcal{E} \Vdash_\rho n \rightsquigarrow \hat{v}$	$\mathcal{E} \Vdash_\rho x \rightsquigarrow \hat{v}$	$\mathcal{E} \Vdash_\rho c \rightsquigarrow \hat{v}$	$\mathcal{E} \Vdash_\rho r_\ell \rightsquigarrow \hat{v}$

$$\frac{\lambda x^{\rho_e} \in \hat{v} \qquad \mathcal{E} \Vdash_\rho e : \hat{v}' \gg \rho' \qquad \hat{v}' \sqsubseteq \mathcal{E}_{\hat{\Gamma}}(\lambda x) \qquad \rho' \sqsubseteq \rho_e}{\mathcal{E} \Vdash_\rho \lambda x.e \rightsquigarrow \hat{v}} \text{(PV-Fun)}$$

$$\frac{\{\overrightarrow{(\!|str_i : \hat{v}_i|\!)_{\mathcal{E},\rho}}\} \sqsubseteq \hat{v}}{\mathcal{E} \Vdash_\rho \{\overrightarrow{str_i : \hat{v}_i}\} \rightsquigarrow \hat{v}} \text{(PV-Rec)}$$

abstraction of any message received by the handler a with permission ρ_a. Given an abstract environment $\mathcal{E}$, we denote by $\mathcal{E}_{\hat{\Gamma}}, \mathcal{E}_{\hat{\mu}}, \mathcal{E}_{\hat{\Upsilon}}, \mathcal{E}_{\hat{\Phi}}$ its four components.

Flow Analysis for Values and Expressions. The flow analysis for values and expressions consists of two mutually inductive judgements: $\mathcal{E} \Vdash_\rho v \rightsquigarrow \hat{v}$ and $\mathcal{E} \Vdash_\rho e : \hat{v} \gg \rho'$. The first judgement means that, assuming permission ρ, the concrete value v is mapped to the abstract value $\hat{v}$ in the abstract environment $\mathcal{E}$. The judgement $\mathcal{E} \Vdash_\rho e : \hat{v} \gg \rho'$ means that in the context of a handler (or an instance) with permission ρ, and under the abstract environment $\mathcal{E}$, the expression e may evaluate to a value abstracted by $\hat{v}$ and exercise at most ρ'.

The rules to derive $\mathcal{E} \Vdash_\rho v \rightsquigarrow \hat{v}$ are collected in Table 3. Most of these rules are straightforward. The only rule worth commenting on here is (PV-Fun), which can be explained as follows: to abstract $\lambda x.e$ into $\hat{v}$, we first analyse the function body e to compute an approximation $\hat{v}'$ of the value it may evaluate to and an upper bound ρ' for the exercised privileges. Then, we check that $\lambda x^{\rho_e} \in \hat{v}$ for some $\rho_e \sqsupseteq \rho'$, i.e., we ensure that the exercised privileges are over-approximated in $\hat{v}$. Finally, we check that $\hat{v}' \sqsubseteq \mathcal{E}_{\hat{\Gamma}}(\lambda x)$, i.e., we guarantee that the abstract variable environment correctly over-approximates the return value of the function.

The analysis rules for expressions are collected in Table 4. We comment on some representative rules below. Rule (PE-Let) can be explained as follows: to analyse **let** $x = e_1$ **in** e_2, we first analyse e_1 to compute an approximation $\hat{v}_1$ of the value it may evaluate to and an upper bound ρ_1 for the exercised privileges. We then ensure that the abstract variable environment $\mathcal{E}_{\hat{\Gamma}}(x)$ contains an over-approximation of $\hat{v}_1$ for the bound variable x, and we analyse e_2 to approximate its value as $\hat{v}_2$ and the exercised privileges as ρ_2. The analysis is acceptable if the abstract value $\hat{v}$ given to the let expression is an over-approximation of $\hat{v}_2$ and the estimated exercised privileges ρ are an upper bound for $\rho_1 \sqcup \rho_2$.

Rule (PE-App) deals with function applications: it states that, to analyse $e_1 e_2$, we first analyse the e_i's to compute the approximations $\hat{v}_i$ of the value they may evaluate to and the upper bounds ρ_i for the exercised privileges. We then focus on each λx^{ρ_e} contained in $\hat{v}_1$ and we check that: (1) the abstract variable environment binds x to an over-approximation of the abstraction of the actual argument of the function, (2) the abstract value $\hat{v}$ given to the application

Table 4. Flow analysis for expressions

(PE-VAL)

$$\frac{\mathcal{E} \Vdash_{\rho_s} v \rightsquigarrow \hat{v}}{\mathcal{E} \Vdash_{\rho_s} v : \hat{v} \gg \rho}$$

(PE-LET)

$$\frac{\mathcal{E} \Vdash_{\rho_s} e_1 : \hat{v}_1 \sqsubseteq \mathcal{E}_{\hat{\Gamma}}(x) \gg \rho_1 \sqsubseteq \rho \qquad \mathcal{E} \Vdash_{\rho_s} e_2 : \hat{v}_2 \sqsubseteq \hat{v} \gg \rho_2 \sqsubseteq \rho}{\mathcal{E} \Vdash_{\rho_s} \text{let } x = e_1 \text{ in } e_2 : \hat{v} \gg \rho}$$

(PE-APP)

$$\frac{\mathcal{E} \Vdash_{\rho_s} e_1 : \hat{v}_1 \gg \rho_1 \sqsubseteq \rho \qquad \mathcal{E} \Vdash_{\rho_s} e_2 : \hat{v}_2 \gg \rho_2 \sqsubseteq \rho \qquad \forall \lambda x^{\rho_e} \in \hat{v}_1 . \hat{v}_2 \sqsubseteq \mathcal{E}_{\hat{\Gamma}}(x) \wedge \mathcal{E}_{\hat{\Gamma}}(\lambda x) \sqsubseteq \hat{v} \wedge \rho_e \sqsubseteq \rho}{\mathcal{E} \Vdash_{\rho_s} e_1 \, e_2 : \hat{v} \gg \rho}$$

(PE-SEQ)

$$\frac{\mathcal{E} \Vdash_{\rho_s} e_1 : \hat{v}_1 \gg \rho_1 \sqsubseteq \rho \qquad \mathcal{E} \Vdash_{\rho_s} e_2 : \hat{v}_2 \sqsubseteq \hat{v} \gg \rho_2 \sqsubseteq \rho}{\mathcal{E} \Vdash_{\rho_s} e_1 ; e_2 : \hat{v} \gg \rho}$$

(PE-OP)

$$\frac{\forall i. \, \mathcal{E} \Vdash_{\rho_s} e_i : \hat{v}_i \gg \rho_i \sqsubseteq \rho \qquad \widehat{op}(\vec{\hat{v}_i}) \sqsubseteq \hat{v}}{\mathcal{E} \Vdash_{\rho_s} op(\vec{e_i}) : \hat{v} \gg \rho}$$

(PE-COND)

$$\frac{\mathcal{E} \Vdash_{\rho_s} e_0 : \hat{v}_0 \gg \rho_0 \sqsubseteq \rho \qquad \textbf{true} \in \hat{v}_0 \Rightarrow \mathcal{E} \Vdash_{\rho_s} e_1 : \hat{v}_1 \sqsubseteq \hat{v} \gg \rho_1 \sqsubseteq \rho \qquad \textbf{false} \in \hat{v}_0 \Rightarrow \mathcal{E} \Vdash_{\rho_s} e_2 : \hat{v}_2 \sqsubseteq \hat{v} \gg \rho_2 \sqsubseteq \rho}{\mathcal{E} \Vdash_{\rho_s} \textbf{if } (e_0) \, \{ \, e_1 \, \} \textbf{ else } \{ \, e_2 \, \} : \hat{v} \gg \rho}$$

(PE-WHILE)

$$\frac{\mathcal{E} \Vdash_{\rho_s} e_1 : \hat{v}_1 \gg \rho_1 \sqsubseteq \rho \qquad \textbf{true} \in \hat{v}_1 \Rightarrow \mathcal{E} \Vdash_{\rho_s} e_2 : \hat{v}_2 \gg \rho_2 \sqsubseteq \rho \qquad \textbf{false} \in \hat{v}_1 \Rightarrow \textbf{undefined} \in \hat{v}}{\mathcal{E} \Vdash_{\rho_s} \textbf{while } (e_1) \, \{ \, e_2 \, \} : \hat{v} \gg \rho}$$

(PE-GETFIELD)

$$\frac{\mathcal{E} \Vdash_{\rho_s} e_1 : \hat{v}_1 \gg \rho_1 \sqsubseteq \rho \qquad \mathcal{E} \Vdash_{\rho_s} e_2 : \hat{v}_2 \gg \rho_2 \sqsubseteq \rho \qquad \widehat{get}(\hat{v}_1, \hat{v}_2) \sqsubseteq \hat{v}}{\mathcal{E} \Vdash_{\rho_s} e_1[e_2] : \hat{v} \gg \rho}$$

(PE-SETFIELD)

$$\frac{\mathcal{E} \Vdash_{\rho_s} e_0 : \hat{v}_0 \gg \rho_0 \sqsubseteq \rho \qquad \mathcal{E} \Vdash_{\rho_s} e_1 : \hat{v}_1 \gg \rho_1 \sqsubseteq \rho \qquad \mathcal{E} \Vdash_{\rho_s} e_2 : \hat{v}_2 \gg \rho_2 \sqsubseteq \rho \qquad \widehat{set}(\hat{v}_0, \hat{v}_1, \hat{v}_2) \sqsubseteq \hat{v}}{\mathcal{E} \Vdash_{\rho_s} e_0[e_1] = e_2 : \hat{v} \gg \rho}$$

(PE-DELFIELD)

$$\frac{\mathcal{E} \Vdash_{\rho_s} e_1 : \hat{v}_1 \gg \rho_1 \sqsubseteq \rho \qquad \mathcal{E} \Vdash_{\rho_s} e_2 : \hat{v}_2 \gg \rho_2 \sqsubseteq \rho \qquad \widehat{del}(\hat{v}_1, \hat{v}_2) \sqsubseteq \hat{v}}{\mathcal{E} \Vdash_{\rho_s} \textbf{delete } e_1[e_2] : \hat{v} \gg \rho}$$

(PE-REF)

$$\frac{\mathcal{E} \Vdash_{\rho_s} e : \hat{v}' \gg \rho' \sqsubseteq \rho \qquad \hat{v}' \sqsubseteq \mathcal{E}_{\hat{\mu}}(\ell, \rho_s) \qquad \ell \in \hat{v}}{\mathcal{E} \Vdash_{\rho_s} \textbf{ref}_\ell \, e : \hat{v} \gg \rho}$$

(PE-DEREF)

$$\frac{\mathcal{E} \Vdash_{\rho_s} e : \hat{v}' \gg \rho' \sqsubseteq \rho \qquad \forall \ell \in \hat{v}' . \mathcal{E}_{\hat{\mu}}(\ell, \rho_s) \sqsubseteq \hat{v}}{\mathcal{E} \Vdash_{\rho_s} \textbf{deref } e : \hat{v} \gg \rho}$$

(PE-SETREF)

$$\frac{\mathcal{E} \Vdash_{\rho_s} e_1 : \hat{v}_1 \gg \rho_1 \sqsubseteq \rho \qquad \mathcal{E} \Vdash_{\rho_s} e_2 : \hat{v}_2 \sqsubseteq \hat{v} \gg \rho_2 \sqsubseteq \rho \qquad \forall \ell \in \hat{v}_1 . \hat{v}_2 \sqsubseteq \mathcal{E}_{\hat{\mu}}(\ell, \rho_s)}{\mathcal{E} \Vdash_{\rho_s} e_1 := e_2 : \hat{v} \gg \rho}$$

(PE-SEND)

$$\frac{\mathcal{E} \Vdash_{\rho_s} e_1 : \hat{v}_1 \gg \rho_1 \sqsubseteq \rho' \qquad \mathcal{E} \Vdash_{\rho_s} e_2 : \hat{v}_2 \gg \rho_2 \sqsubseteq \rho' \qquad \forall m \in \hat{v}_1 . \forall \rho_m \sqsupseteq \rho . \mathcal{E}_{\hat{\Gamma}}(m, \rho_m) = (\rho_r, \rho_e) \wedge \rho_r \sqsubseteq \rho_s \Rightarrow \rho_e \sqsubseteq \rho' \wedge \hat{v}_2 \sqsubseteq \mathcal{E}_{\hat{\Phi}}(m, \rho_m) \wedge \textbf{unit} \in \hat{v}}{\mathcal{E} \Vdash_{\rho_s} \overline{e_1} \langle e_2 \triangleright \rho \rangle : \hat{v} \gg \rho'}$$

(PE-EXERCISE)

$$\frac{\rho \sqsubseteq \rho_s \Rightarrow \rho \sqsubseteq \rho' \wedge \textbf{unit} \in \hat{v}}{\mathcal{E} \Vdash_{\rho_s} \textbf{exercise}(\rho) : \hat{v} \gg \rho'}$$

is an over-approximation of the abstract return value of the function $\mathcal{E}_{\hat{F}}(\lambda x)$, and (3) the exercised privileges $\rho_1 \sqcup \rho_2 \sqcup \rho_e$ are bounded above by the privileges ρ assigned to the application.

The rules in the central portion of the table should be relatively easy to understand. Notice that the rules for control flow operators, i.e., (PE-COND) and (PE-WHILE), allow for excluding from the static analysis some program branches which are never reached at runtime. The rules for references use the information ρ_s annotated on the turnstile, corresponding to the privileges granted to the handler/instance that is accessing the reference. These rules ensure that any value stored in a reference is correctly over-approximated by the abstract memory; and dually, that any value retrieved from a reference is abstracted with an over-approximation of the content of the abstract memory. This ensures that any value which is first stored in a reference and then retrieved from it is over-approximated correctly by the flow logic.

Rule (PE-SEND) first analyses e_1 and e_2 to compute the approximations of the recipient ($\hat{v}_1$) and the sent message ($\hat{v}_2$). Then, the last premise enforces two invariants: (1) the privileges ρ_e escalated by communicating with other handlers in the system are bounded above by the privileges ρ' assigned to the send expression, and (2) the abstraction of the sent message $\hat{v}_2$ is over-approximated by the information in the abstract network for each possible recipient. We also check that **unit** is included in the abstract value assigned to the expression, accordingly to the operational semantics of the send construct. Finally, rule (PE-EXERCISE) ensures that, whenever an instance with permission ρ_s exercises $\rho \sqsubseteq \rho_s$, then ρ is bounded above by the privileges ρ' assigned to the expression.

Flow Analysis for Systems. Finally, we extend the flow analysis to systems by defining the main judgement $\mathcal{E} \Vdash s$ **despite** ρ, which follows from similar judgements for memories, handlers and instances. The definition is given in Table 5.

In the rules for memories we just need to ensure (cf. rule (PM-SINGLE)) that, whenever a value v is stored in a reference r_ℓ protected with permission ρ_r, then v can be abstracted to some $\hat{v}$ over-approximated by the abstract memory entry $\mathcal{E}_{\hat{\mu}}(\ell, \rho_r)$. As for instances, rule (PI-SINGLE) computes an approximation of the privileges ρ_e exercised by the running expression. Then, if the instance is granted permission $\rho_a \not\sqsubseteq \rho$, i.e., if it is not compromised, we check that the abstract stack correctly approximates with ρ_e the privileges exercised by the instance body. This check is not enforced for instances that might be under the control of the opponent, according to the idea that any opponent must be accepted by a sufficiently weak abstract environment. This is needed to prove an *opponent acceptability* result (Lemma 2), which allows for a convenient soundness proof technique for the analysis [1,5].

Handlers are accepted by rule (PH-SINGLE), which states that, to analyse $a(x \vartriangleleft \rho_s : \rho_a).e$ despite ρ-opponents, we first lookup the abstract stack $\hat{\Upsilon}$: let $\hat{\Upsilon}(a, \rho_a) = (\rho'_s, \rho'_e)$. If we are not analysing a (possibly) compromised handler, i.e., if $\rho_a \not\sqsubseteq \rho$, we ensure that the permission ρ'_s in the abstract stack matches the permission ρ_s guarding access to the handler. We then lookup the abstract network $\hat{\Phi}$: if $\hat{\Phi}(a, \rho_a) = \emptyset$, no instance of the system will ever communicate

Table 5. Flow analysis for systems

(PM-EMPTY)
$$\mathcal{E} \Vdash \emptyset \text{ despite } \rho$$

(PM-SINGLE)
$$\frac{\mathcal{E} \Vdash_{\rho_r} v \rightsquigarrow \hat{v} \qquad \hat{v} \sqsubseteq \mathcal{E}_{\hat{\mu}}(\ell, \rho_r)}{\mathcal{E} \Vdash r_\ell \xrightarrow{\rho_r} v \text{ despite } \rho}$$

(PM-MANY)
$$\frac{\mathcal{E} \Vdash \mu_1 \text{ despite } \rho \qquad \mathcal{E} \Vdash \mu_2 \text{ despite } \rho}{\mathcal{E} \Vdash \mu_1, \mu_2 \text{ despite } \rho}$$

(PH-EMPTY)
$$\mathcal{E} \Vdash \emptyset \text{ despite } \rho$$

(PH-MANY)
$$\frac{\mathcal{E} \Vdash h \text{ despite } \rho \qquad \mathcal{E} \Vdash h' \text{ despite } \rho}{\mathcal{E} \Vdash h, h' \text{ despite } \rho}$$

(PH-SINGLE)
$$\frac{\mathcal{E}_{\hat{\Phi}}(a, \rho_a) \neq \emptyset \;\Rightarrow\; \mathcal{E}_{\hat{\Gamma}}(x) \sqsupseteq \mathcal{E}_{\hat{\Phi}}(a, \rho_a) \wedge \mathcal{E} \Vdash_{\rho_a} e : \hat{v} \gg \rho_e \wedge (\rho_a \not\sqsubseteq \rho \Rightarrow \rho'_e = \rho_e)}{\mathcal{E} \Vdash a(x \triangleleft \rho_s : \rho_a).e \text{ despite } \rho}$$
with the numerator side-conditions $\mathcal{E}_{\hat{\Gamma}}(a, \rho_a) = (\rho'_s, \rho'_e) \quad \rho_a \not\sqsubseteq \rho \Rightarrow \rho'_s = \rho_s$

(PI-EMPTY)
$$\mathcal{E} \Vdash \emptyset \text{ despite } \rho$$

(PI-SINGLE)
$$\frac{\mathcal{E} \Vdash_{\rho_a} e : \hat{v} \gg \rho_e \qquad \rho_a \not\sqsubseteq \rho \Rightarrow \exists \rho_s. \mathcal{E}_{\hat{\Gamma}}(a, \rho_a) = (\rho_s, \rho_e)}{\mathcal{E} \Vdash a\{|e|\}_{\rho_a} \text{ despite } \rho}$$

(PI-MANY)
$$\frac{\mathcal{E} \Vdash i \text{ despite } \rho \qquad \mathcal{E} \Vdash i' \text{ despite } \rho}{\mathcal{E} \Vdash i, i' \text{ despite } \rho}$$

(PS-SYS)
$$\frac{\mathcal{E} \Vdash \mu \text{ despite } \rho \quad \mathcal{E} \Vdash h \text{ despite } \rho \qquad \mathcal{E} \Vdash i \text{ despite } \rho \quad \mathcal{E} \text{ is } \rho\text{-conservative}}{\mathcal{E} \Vdash \mu; h; i \text{ despite } \rho}$$

with the handler and we can skip the analysis of its body. Otherwise, we ensure that the abstract variable environment maps the bound variable x to an over-approximation of the incoming message, abstracted by $\hat{\Phi}(a, \rho_a)$, and we analyse the body of the handler, to detect the exercised privileges ρ_e. If we are not analysing the opponent, we further ensure that ρ_e matches the permissions ρ'_e annotated in the abstract stack, i.e., we guarantee that the abstract stack contains reliable information.

Finally, rule (PS-SYS) states that a system $s = \mu; h; i$ is acceptable for $\mathcal{E}$ only whenever μ, h and i are all acceptable for $\mathcal{E}$, and $\mathcal{E}$ is a ρ-*conservative* abstract environment. This notion corresponds to the informal idea of "sufficiently weak abstract environment" needed to prove the opponent acceptability result. In order to define ρ-conservativeness, we first define the notion of *static leak* for an abstract environment.

Definition 5 (Static Leak). *We define the* static leak *of $\mathcal{E}$ against ρ as:* $SLeak_\rho(\mathcal{E}) = \bigsqcup_{\rho_e \in L} \rho_e$, *where* $L = \{\rho_e \mid \exists a, \rho_a, \rho_s. \mathcal{E}_{\hat{\Gamma}}(a, \rho_a) = (\rho_s, \rho_e) \wedge \rho_s \sqsubseteq \rho\}$.

Intuitively, $SLeak_\rho(\mathcal{E})$ is the upper bound of all the permissions ρ_e that can be (transitively) exercised by any handler that can be called by a ρ-opponent. We then define the set $\mathcal{V}_\rho(\mathcal{E})$ of the opponent-controlled variables as:

$$\mathcal{V}_\rho(\mathcal{E}) = \mathcal{V}_u \cup \{x \mid \exists \rho_e, \ell, \rho_r \sqsubseteq \rho. \lambda x^{\rho_e} \in \mathcal{E}_{\hat{\mu}}(\ell, \rho_r)\}.$$

The set contains all the variables $\mathcal{V}_u$ occurring in the opponent code, together with all the variables bound in lambda abstractions stored in references under the control of the opponent. All these variables can be instantiated at runtime with values chosen by the opponent. We use this set of variables also to define a sound abstraction of any value which can be generated by/flow to the opponent.

Definition 6 (Canonical Disclosed Abstract Value). *Given an abstract environment $\mathcal{E}$ and a permission ρ, the canonical disclosed abstract value is defined as:* $\hat{v}_\rho(\mathcal{E}) = \{\hat{u} \mid vars(\hat{u}) \subseteq \mathcal{V}_\rho(\mathcal{E})\}$.

The canonical disclosed abstract value is a canonical representation of any abstract value under the control of a ρ-opponent in a system accepted by $\mathcal{E}$. It is the set of all the pre-values which contain only opponent-controlled variables.

Based on the notions above, we define ρ-conservativeness.

Definition 7 (ρ-Conservative Abstract Environment). *An abstract environment $\mathcal{E}$ is ρ-conservative if and only if all the following conditions hold true:*

1. $\forall n \in \mathcal{N}, \forall \rho' \sqsubseteq \rho. \mathcal{E}_{\hat{\Upsilon}}(n, \rho') = (\bot, SLeak_\rho(\mathcal{E}))$;
2. $\forall n \in \mathcal{N}, \forall \rho' \sqsubseteq \rho. \mathcal{E}_{\hat{\Phi}}(n, \rho') = \hat{v}_\rho(\mathcal{E})$;
3. $\forall n \in \mathcal{N}, \forall \rho_n, \rho_s, \rho_e. \mathcal{E}_{\hat{\Upsilon}}(n, \rho_n) = (\rho_s, \rho_e) \wedge \rho_s \sqsubseteq \rho \Rightarrow \mathcal{E}_{\hat{\Phi}}(n, \rho_n) = \hat{v}_\rho(\mathcal{E})$;
4. $\forall \ell \in \mathcal{L}, \forall \rho' \sqsubseteq \rho. \mathcal{E}_{\hat{\mu}}(\ell, \rho') = \hat{v}_\rho(\mathcal{E})$;
5. $\forall x \in \mathcal{V}_\rho(\mathcal{E}). \mathcal{E}_{\hat{\Gamma}}(x) = \mathcal{E}_{\hat{\Gamma}}(\lambda x) = \hat{v}_\rho(\mathcal{E})$.

In words, an abstract environment is ρ-conservative whenever: (1) any handler that can be under the control of the opponent is in fact assumed to be accessible by the opponent and to escalate up to the static leak; (2) any handler that can be under the control of the opponent, or (3) that can be contacted by the opponent, is assumed to receive the canonical disclosed abstract value $\hat{v}_\rho(\mathcal{E})$; (4) any reference possibly under the control of the opponent is assumed to contain $\hat{v}_\rho(\mathcal{E})$; and (5) the argument of any function which can be called by the opponent is assumed to contain the canonical disclosed abstract value $\hat{v}_\rho(\mathcal{E})$ and similarly these functions are assumed to return $\hat{v}_\rho(\mathcal{E})$.

4.2 Formal Results

Our main formal result defines an upper bound for the privileges which can be escalated by the opponent in a system accepted by the flow analysis. Complete proofs are in the full version [9]; here, we start proving the soundness of the flow logic specification by means of a subject reduction result, which ensures that the acceptability of the analysis is preserved upon reduction.

Lemma 1 (Subject Reduction). *If $\mathcal{E} \Vdash s$ despite ρ and $s \xrightarrow{\alpha} s'$, then $\mathcal{E} \Vdash s'$ despite ρ.*

The next lemma states that any ρ-opponent is accepted by a ρ-conservative abstract environment. Intuitively, the combination of this result with subject reduction ensures that the acceptability of the analysis is preserved at runtime, even when the analysed system interacts with the opponent.

Lemma 2 (Opponent Acceptability). *If (h,i) is a ρ-opponent and $\mathcal{E}$ is ρ-conservative, then $\mathcal{E} \Vdash h$ despite ρ and $\mathcal{E} \Vdash i$ despite ρ.*

Moreover, proving the safety theorem requires to explicitly track the call chains carried out by the system reduction, to collect the privileges transitively exercised by system components. The next lemma then relies on the following definition of call chain to prove that the abstract stack contains a static approximation of the privileges which are exercised by each system component either directly or by communicating with other components.

Definition 8 (Call Chain). *A call chain $(\overrightarrow{\alpha}, a{:}\rho_a \gg \rho')$ is a trace of length $(n+1)$ such that:*

1. *the trace $\overrightarrow{\alpha} = \langle a_1{:}\rho_{a_1}, b_1{:}\rho_{b_1} \rangle, \ldots, \langle a_n{:}\rho_{a_n}, b_n{:}\rho_{b_n} \rangle$ is a sequence of send labels where the sender occurring in each label is the receiver occurring in the previous label, i.e., $\forall i \in [1, n-1]. \, a_{i+1} = b_i \wedge \rho_{a_{i+1}} = \rho_{b_i}$, and*
2. *the component exercising the privilege ρ' at the end of the call chain corresponds to the last receiver, i.e., $b_n = a \wedge \rho_{b_n} = \rho_a$.*

A trace $\overrightarrow{\beta}$ includes a call chain $\overrightarrow{\alpha}$ iff $\overrightarrow{\alpha}$ is a sub-trace of $\overrightarrow{\beta}$.

According to the intuition given above, proving the soundness of the abstract stack amounts to showing that, given a call chain leading to the exercise of some privilege ρ' not available to the opponent, the abstract stack $\mathcal{E}_{\hat{\Upsilon}}$ approximates the privileges exercised by any component involved in the chain with a permission greater than or equal to ρ'. The proof uses the subject reduction result.

Lemma 3 (Soundness of the Abstract Stack). *If $\mathcal{E} \Vdash s$ despite ρ and $s \stackrel{\overrightarrow{\beta}}{\Rightarrow} s'$ for a trace $\overrightarrow{\beta}$ including the call chain $(\overrightarrow{\alpha}, a{:}\rho_a \gg \rho')$ for some $\rho' \not\sqsubseteq \rho$, then for each label $\alpha_j = \langle a_j{:}\rho_{a_j}, b_j{:}\rho_{b_j} \rangle \in \{\overrightarrow{\alpha}\}$ we have $\mathcal{E}_{\hat{\Upsilon}}(b_j, \rho_{b_j}) = (\rho_{s_{b_j}}, \rho_{e_{b_j}})$ with $\rho' \sqsubseteq \rho_{e_{b_j}}$ and $\mathcal{E}_{\hat{\Upsilon}}(a_j, \rho_{a_j}) = (\rho_{s_{a_j}}, \rho_{e_{a_j}})$ with $\rho' \sqsubseteq \rho_{e_{a_j}}$.*

Theorem 1 (Flow Safety). *Let $s = \mu; h; \emptyset$. If $\mathcal{E} \Vdash s$ despite ρ, then s leaks $SLeak_\rho(\mathcal{E})$ against ρ.*

Proof. By contradiction. Let $\hat{s}$ be the system obtained by composing s with a ρ-opponent and assume that $\hat{s}$ eventually reaches a state s' such that s' exercises privileges ρ_{bad}, with $\rho_{bad} \not\sqsubseteq \rho$ and $\rho_{bad} \not\sqsubseteq SLeak_\rho(\mathcal{E})$.

By inverting rule (PS-SYS) on the hypothesis $\mathcal{E} \Vdash s$ despite ρ, we have that $\mathcal{E}$ is ρ-conservative. Using Lemma 2 (Opponent Acceptability), we show that $\mathcal{E} \Vdash \hat{s}$ despite ρ. Given that $\rho_{bad} \not\sqsubseteq \rho$, the privileges ρ_{bad} cannot be directly exercised by the opponent, hence there must exist a call chain leading to ρ_{bad} from $\hat{s}$. Let a_i range over the components in the call chain and ρ_i range over their corresponding permissions. Consider now the first sender a_1 in the call chain: given that the original system s does not have running instances, it turns out that a_1 must be the opponent, hence $\rho_1 \sqsubseteq \rho$. Since $\mathcal{E}$ is ρ-conservative and $\rho_1 \sqsubseteq \rho$, we have $\mathcal{E}_{\hat{\gamma}}(a_1, \rho_1) = (\bot, SLeak_\rho(\mathcal{E}))$. By Lemma 3 (Soundness of the Abstract Stack), for each component a_i with permissions ρ_i occurring in the call chain we must have $\mathcal{E}_{\hat{\gamma}}(a_i, \rho_i) = (\rho_{s_i}, \rho_{e_i})$ for some ρ_{s_i} and some $\rho_{e_i} \sqsupseteq \rho_{bad}$. But then we get $\rho_{bad} \sqsubseteq SLeak_\rho(\mathcal{E})$, which is contradictory.

4.3   Analysing the Example

We now show the analysis at work on our running example in its three variants, namely the systems s, s_{tag} and s_{chan} introduced in Section 3. We assume that the abstract domain for strings includes all the string literals syntactically occurring in the program code, plus the distinguished symbol * to represent all the other strings (or any string which we cannot statically reconstruct). We let $\widehat{str}$ range over elements of this abstract domain and we assume that $\widehat{str} \sqsubseteq *$ for any $\widehat{str}$. As to records, we choose the field-sensitive representation $\langle\!\langle \widehat{str}_i : \hat{v}_i \rangle\!\rangle$ where both the field names and contents are inductively abstracted. In the following we mostly focus on the intuitions behind the analysis: additional details, including the formal definitions of the expected abstract record operations and the abstract value pre-order, are given in the full version [9].

The Original System. We start by studying the robustness of the original system s against a P-opponent, i.e., an opponent with the only ability to dispatch the content script C attached to untrusted web pages. We have that $\mathcal{E} \Vdash s$ **despite** P, where $\mathcal{E} = \hat{\Gamma}; \hat{\mu}; \hat{\gamma}; \hat{\Phi}$ satisfies the following assumptions:

$$\hat{\Phi}(c, \mathsf{C}) = \hat{v}_\mathsf{P}(\mathcal{E}) \qquad \hat{\Phi}(o, \mathsf{O}) = \emptyset \qquad \hat{\Phi}(b, \mathsf{B}) = \{\!\!\{\langle\!\langle \text{``site''} : \hat{v}_\mathsf{P}(\mathcal{E}), * : \hat{v}_\mathsf{P}(\mathcal{E}) \rangle\!\rangle\}\!\!\}$$
$$\hat{\gamma}(c, \mathsf{C}) = (\mathsf{P}, \mathsf{B}) \qquad \hat{\gamma}(o, \mathsf{O}) = (\top, \bot) \qquad \hat{\gamma}(b, \mathsf{B}) = (\mathsf{C} \sqcap \mathsf{O}, \mathsf{B})$$

Since C can be accessed by the opponent, the value of $\hat{\Phi}(c, \mathsf{C})$ must be equal to $\hat{v}_\mathsf{P}(\mathcal{E})$ to ensure the P-conservativeness of $\mathcal{E}$. Conversely, O can never be accessed by the opponent or by any other component in the system, hence $\hat{\Phi}(o, \mathsf{O}) = \emptyset$. By rule (PH-SINGLE), this implies that there is no need to analyse the body of O, which allows for ignoring the format of the messages sent by O: this explains why the value of $\hat{\Phi}(b, \mathsf{B})$ includes just one element, corresponding to the message sent by C. Indeed, observe that $\widehat{set}(\hat{v}_\mathsf{P}(\mathcal{E}), \text{``site''}, str) \sqsubseteq \{\!\!\{\langle\!\langle \text{``site''} : \hat{v}_\mathsf{P}(\mathcal{E}), * : \hat{v}_\mathsf{P}(\mathcal{E}) \rangle\!\rangle\}\!\!\}$ for any str to accept the send expression in the body of C.

Now observe that $\{ \text{``policy''}, \text{``upd''}\} \sqsubseteq \widehat{get}(\langle\!\langle \text{``site''} : \hat{v}_\mathsf{P}(\mathcal{E}), * : \hat{v}_\mathsf{P}(\mathcal{E}) \rangle\!\rangle, \text{``tag''})$, hence both branches of the conditional in the body of B are reachable and the conditional expression may exercise B; we then let $\hat{\gamma}(b, \mathsf{B}) = (\mathsf{C} \sqcap \mathsf{O}, \mathsf{B})$ by rule (PH-SINGLE). Given that C communicates with B, the privileges exercised by C must be greater or equal than B by rule (PE-SEND), and propagated into $\hat{\gamma}(c, \mathsf{C})$ by rule (PH-SINGLE). Since $SLeak_\mathsf{P}(\mathcal{E}) = \mathsf{B}$, we know that the system s leaks B against P by Theorem 1.

The System with Tags. Let us focus now on the system s_{tag} and a P-opponent. We have that $\mathcal{E} \Vdash s_{tag}$ **despite** P, where $\mathcal{E} = \hat{\Gamma}; \hat{\mu}; \hat{\Upsilon}; \hat{\Phi}$ is such that:

$$\hat{\Phi}(c, \mathsf{C}) = \hat{v}_{\mathsf{P}}(\mathcal{E}) \qquad \hat{\Phi}(o, \mathsf{O}) = \emptyset$$

$$\hat{\Phi}(b, \mathsf{B}) = \{\langle \text{``tag''} : \text{``policy''}, \text{``site''} : *, \text{``spec''} : \hat{v}_{\mathsf{P}}(\mathcal{E})\rangle\}$$

$$\hat{\Upsilon}(c, \mathsf{C}) = (\mathsf{P}, \mathsf{MemB}) \qquad \hat{\Upsilon}(o, \mathsf{O}) = (\top, \bot) \qquad \hat{\Upsilon}(b, \mathsf{B}) = (\mathsf{C} \sqcap \mathsf{O}, \mathsf{MemB})$$

Based on this information, rule (PE-COND) allows for analysing only the program branch of B corresponding to the processing of a message with tag *"policy"*, which only exercises the privilege MemB: this motivates the precise choice of $\hat{\Upsilon}(b, \mathsf{B})$. Since $SLeak_{\mathsf{P}}(\mathcal{E}) = \mathsf{MemB}$, the system leaks MemB against P.

Assume now an opponent with permission C, then we have $\mathcal{E}' \Vdash s_{tag}$ **despite** C, where $\mathcal{E}' = \hat{\Gamma}'; \hat{\mu}'; \hat{\Upsilon}'; \hat{\Phi}'$ is such that:

$$\hat{\Phi}'(c, \mathsf{C}) = \hat{v}_{\mathsf{C}}(\mathcal{E}') \qquad \hat{\Phi}'(o, \mathsf{O}) = \emptyset \qquad \hat{\Phi}'(b, \mathsf{B}) = \hat{v}_{\mathsf{C}}(\mathcal{E}')$$

$$\hat{\Upsilon}'(c, \mathsf{C}) = (\bot, \mathsf{B}) \qquad \hat{\Upsilon}'(o, \mathsf{O}) = (\top, \bot) \qquad \hat{\Upsilon}'(b, \mathsf{B}) = (\mathsf{C} \sqcap \mathsf{O}, \mathsf{B})$$

With respect to the previous scenario, the abstract network entry for B contains $\hat{v}_{\mathsf{C}}(\mathcal{E}')$, abstracting all the values which may be generated by a C-opponent: this is needed for C-conservativeness. The consequence is that all the program branches of B are reachable, hence B may exercise its full set of privileges B. Since $SLeak_{\mathsf{C}}(\mathcal{E}') = \mathsf{B}$, the system leaks B against C by Theorem 1.

The System with Channels. We are able to prove $\mathcal{E} \Vdash s_{chan}$ **despite** C for an abstract environment $\mathcal{E} = \hat{\Gamma}; \hat{\mu}; \hat{\Upsilon}; \hat{\Phi}$ such that:

$$\hat{\Phi}(c, \mathsf{C}) = \hat{v}_{\mathsf{C}}(\mathcal{E}) \qquad \hat{\Phi}(o, \mathsf{O}) = \emptyset \qquad \hat{\Phi}(b_1, \mathsf{B}) = \hat{v}_{\mathsf{C}}(\mathcal{E}) \qquad \hat{\Phi}(b_2, \mathsf{B}) = \emptyset$$

$$\hat{\Upsilon}(c, \mathsf{C}) = (\bot, \mathsf{MemB}) \quad \hat{\Upsilon}(o, \mathsf{O}) = (\top, \bot) \quad \hat{\Upsilon}(b_1, \mathsf{B}) = (\mathsf{C}, \mathsf{MemB}) \quad \hat{\Upsilon}(b_2, \mathsf{B}) = (\mathsf{O}, \bot)$$

For the new abstract environment $\mathcal{E}$ we have $SLeak_{\mathsf{C}}(\mathcal{E}) = \mathsf{MemB}$, which ensures that the new system only leaks MemB against C. Since the privilege Cookies cannot be escalated by a compromised C anymore, there is no way to corrupt the cookie jar without compromising the background page B itself (or the options page O). Interestingly, this is a formal characterization of the dangers connected to the development of *bundled* browser extensions in a realistic setting [2].

5   Implementation: CHEN

CHEN is a prototype Google Chrome extension analyser written in F#. Given a Chrome extension, CHEN translates it into a corresponding system in our formalism and computes an acceptable flow analysis estimate by constraint solving. CHEN can be used by programmers to evaluate the robustness of their extensions against privilege escalation attacks and to support their security refactoring.

5.1 Flow Logic Implementation

Implementing the flow logic specification amounts to defining an algorithm that, given a system s and a permission ρ characterizing the power of the opponent, computes an abstract environment $\mathcal{E}$ such that $\mathcal{E} \Vdash s$ **despite** ρ. Following a standard approach [25], we first define a verbose variant of the flow logic, which associates an analysis estimate to each sub-expression of s, and then we devise a constraint-based formulation of the analysis. Any solution of the constraints is an abstract environment $\mathcal{E}$ which accepts s.

We initially implemented in CHEN a simple worklist algorithm for constraint solving. However, consistently with what has been reported by Jensen *et al.* in the context of JavaScript analysis [19], we observed that this solution does not scale, taking hours to perform the analysis even on small examples. Therefore, in our implementation we use a variant of the worklist algorithm where most of the constraint generation is performed *on demand* during the solving process. Even though this approach does not allow us to reuse existing solvers, it leads to a dramatic improvement in the performances of the analysis.

The current prototype implements a context-insensitive analysis, which is enough to capture the privileges escalated by the content scripts, provided that some specific library functions introduced by the desugaring process from JavaScript to λ_{JS} (see below) are inlined. The choice of the abstract pre-values for constants is standard: in the current implementation, we represent numbers with their sign and we approximate strings with finite prefixes [11]. The representation of records is field-sensitive, but we collapse into a single label * all the entries bound to approximate labels (string prefixes). As to the ordering, we consider a standard pre-order $\sqsubseteq_p$ on abstract pre-values, and we lift it to abstract values using a lower powerset construction, i.e., we let $\hat{v} \sqsubseteq \hat{v}'$ if and only if $\forall \hat{u} \in \hat{v}. \exists \hat{u}' \in \hat{v}'. \hat{u} \sqsubseteq_p \hat{u}'$.

5.2 Using CHEN to Assess Google Chrome Extensions

Given an extension, CHEN takes as an input a sequence of *component* names, along with the JavaScript files corresponding to their implementation. Components represent isolation domains, in that different components must be able to communicate only using the message passing interface. Different content scripts which may injected in the same web page should be put inside the same component, since Google Chrome does not separate their heaps. The background page should be put in a separate component, since it runs in an isolated process[2].

From JavaScript to the Model. Let c be a component name and $f_1, \ldots, f_n$ the corresponding JavaScript files: our tool concatenates $f_1, \ldots, f_n$ into a single file f, which is desugared into a closed λ_{JS} expression using an existing tool [16]. The adequacy of the translation from JavaScript to λ_{JS} has been assessed by extensive automatic testing, hence safety guarantees for JavaScript programs can be provided just by analysing their λ_{JS} translation; see [16] for further details.

---

[2] An appropriate mapping of JavaScript files to components can be derived from the manifest file of the extension, but the current prototype does not support this feature.

The obtained λ_{JS} expression is then transformed into a set of handlers: more precisely, for any function $\lambda x.e'$ passed as an argument to the addListener method of chrome.runtime.onMessage, we introduce a new handler on a channel with the same name of the component, whose body is obtained by closing e' with the introduction of all the bindings defined before the registration of the listener. For each component we introduce a unique permission for memory access, granted to each handler in the component; handlers corresponding to the background page are also given the permissions specified in the manifest of the extension. Any invocation of chrome.runtime.sendMessage in the definition of a content script is translated to a send expression over a channel with the name of the component corresponding to the background page.

Notice that CHEN exploits an existing tool to translate JavaScript to λ_{JS}, but our target language has two new constructs: message sending and privilege exercise. In JavaScript, both operations correspond to function calls to the Chrome extension API, hence, to introduce the syntactic forms corresponding to them in the translation to our formalism, we extend the JavaScript code to redefine the functions of interest in the Chrome API with *stubs*. For instance, chrome.cookies.set is redefined to a function including the special tag "#Cookies#", which is preserved when desugaring JavaScript to λ_{JS}: we then post-process the λ_{JS} expression to replace this tag with **exercise**(Cookies).

Running the Analysis. The tool supports two analyses. The option -compromise instructs CHEN to analyse the privileges which may be escalated by an opponent assuming the full compromise of an arbitrary content script, i.e., it estimates the safety of the system despite the permission that protects the background page. If the background page requests some permission ρ intended for internal use, but ρ is available to some content script according to the results of the analysis, then the developer is recommended to review the communication interface.

Alternatively, the option -target n allows to get an approximation of the privileges available to the content scripts in the component n in absence of compromise. We model absence of compromise by considering a $\bot$-opponent as the threat model, since this opponent cannot directly communicate with the background page: if the option -target n is specified, CHEN transforms the system by protecting with permission $\bot$ all the handlers included in n, and computes a permission ρ such that the system is ρ-safe despite $\bot$. This allows to estimate which privileges are enabled by messages sent from n, so as to identify potential room for a security refactoring, as we discuss below.

Both the analyses additionally support the option -flag p, which allows to define a dummy permission p assigned to the background page. The programmer may then annotate specific program points with the tag "#p#, corresponding to the exercise of this dummy permission; by checking the presence of the flag among the escalated privileges, CHEN can be used to implement an opponent-aware reachability analysis on the extension code.

Supporting a Security Refactoring. To exemplify, we analyse with CHEN our motivating example. By first specifying the option -target O, the tool detects that the options page O is only accessing the privilege Cookies as part of its standard functionalities, even though the background page B is given the permissions MemB $\sqcup$ Cookies. To support least privilege, the developer is thus recommended to introduce a distinct communication port for B. Notably, the permission gap arises from the presence in the code of B of program branches which are never triggered by messages sent by O in absence of compromise: in principle, CHEN could then automatically introduce the new port, replicate the code from the handler of the background page, and improve its security against compromise by eliminating the dead branches, even though the current prototype does not implement this feature.

Then, by using the option -target C, the tool outputs that the privilege MemB $\sqcup$ Cookies can be escalated by the content script C. Hence, no automated refactoring is possible, but the output of the analysis is still helpful for a careful developer, who realizes that C should not be able to access the Cookies privilege. Based on the output of the analysis, the developer may opt for a manual reviewing and refactoring of the extension.

Current Limitations. Being a proof-of-concept implementation, the current version of CHEN lacks a full coverage of the Chrome extension APIs. Moreover, CHEN cannot analyse extensions which use ports to communicate: in our model, ports are just channels and do not pose any significant problem to the analysis. Unfortunately, the current Chrome API makes it difficult to support the analysis of extensions using ports, since the underlying programming patterns make massive usage of callbacks. Based on our experience and a preliminary investigation, however, ports are not widely used in practice, hence many extensions can still be analysed by CHEN.

5.3   Case Study: ShareMeNot

ShareMeNot [30] is a popular privacy-enhancing extension developed at the University of Washington. The extension looks for social sharing buttons in the web pages and replaces them with dummy buttons: only when the user clicks one of these buttons, its original version is loaded and the cookies registered by the corresponding social networks are sent. This means that the social network can track the user only when the user is willing to share something.

ShareMeNot consists of four components: a content script, a background page, an option page and a popup, for a total of approximately 1,500 lines of JavaScript code. The background page offers a unique entry point to all the other extension components and handles seven different message types. Interestingly, one of these messages allows to unblock all the trackers in an arbitrary tab, by invoking the unblockAllTrackersOnTab function: this message should only be sent by the popup page. We then put a flag in the body of the function and we performed the analysis of ShareMeNot with the -compromise option, observing that the flag is reachable: hence, a compromised content script could entirely deactivate the extension. The analysis took around 150 seconds on a standard commercial machine.

We then ran the analysis with the -target C option, where C is the name of the component including only the content script, and we observed that the flag was not reachable. This means that C does not need to access the function `unblockAllTrackersOnTab` as part of its standard functionalities, hence the code should be refactored to comply with the principle of the least privilege and prevent a potential security risk. The analysis took around 210 seconds on the same machine.

6 Conclusions

We presented a core calculus to reason about browser extensions security and we proposed a flow analysis aimed at detecting which privileges may be leaked to an opponent which compromises some (arbitrarily chosen) untrusted extension components. The analysis has been proved sound and it has been implemented in CHEN, a prototype static analyser for Google Chrome extensions. We discussed how CHEN can assist developers in writing more robust extensions.

As future work, we plan to further engineer CHEN, to make it support more sophisticated communication patterns used in Google Chrome extensions. We ultimately plan to evolve CHEN into a compiler, which automatically refactors the extension code to make it more secure, by unbundling functionalities based on their exercised permissions. Based on a preliminary investigation, this will require a non-trivial programming effort.

Acknowledgements. We would like to thank Arjun Guha for insightful discussions about the λ_{JS} semantics. Alvise Spanò provided useful F# libraries and advices for the development of CHEN. This work was partially supported by the MIUR projects ADAPT and CINA, and by the University of Padova under the PRAT project BECOM.

References

1. Abadi, M.: Secrecy by typing in security protocols. J. ACM 46, 749–786 (1999)
2. Akhawe, D., Saxena, P., Song, D.: Privilege separation in HTML5 applications. In: USENIX Security Symposium, pp. 429–444 (2012)
3. Bandhakavi, S., Tiku, N., Pittman, W., King, S.T., Madhusudan, P., Winslett, M.: Vetting browser extensions for security vulnerabilities with VEX. Communications of the ACM 54(9), 91–99 (2011)
4. Barth, A., Porter Felt, A., Saxena, P., Boodman, A.: Protecting browsers from extension vulnerabilities. In: NDSS (2010)
5. Bodei, C., Buchholtz, M., Degano, P., Nielson, F., Nielson, H.R.: Static validation of security protocols. Journal of Computer Security 13(3), 347–390 (2005)
6. Bugliesi, M., Calzavara, S., Focardi, R., Khan, W.: Automatic and robust client-side protection for cookie-based sessions. In: Jürjens, J., Piessens, F., Bielova, N. (eds.) ESSoS. LNCS, vol. 8364, pp. 161–178. Springer, Heidelberg (2014)
7. Bugliesi, M., Calzavara, S., Focardi, R., Khan, W., Tempesta, M.: Provably sound browser-based enforcement of web session integrity. In: CSF, pp. 366–380 (2014)
8. Bugliesi, M., Calzavara, S., Spanò, A.: Lintent: Towards security type-checking of android applications. In: Beyer, D., Boreale, M. (eds.) FORTE 2013 and FMOODS 2013. LNCS, vol. 7892, pp. 289–304. Springer, Heidelberg (2013)

9. Calzavara, S., Bugliesi, M., Crafa, S., Steffinlongo, E.: Fine-grained detection of privilege escalation attacks on browser extensions (full version), http://www.dais.unive.it/textasciitildecalzavara/papers/esop15-full.pdf

10. Carlini, N., Porter Felt, A., Wagner, D.: An evaluation of the Google Chrome extension security architecture. In: USENIX Security Symposium, pp. 97–111 (2012)

11. Costantini, G., Ferrara, P., Cortesi, A.: Static analysis of string values. In: ICFEM, pp. 505–521 (2011)

12. Davi, L., Dmitrienko, A., Sadeghi, A.-R., Winandy, M.: Privilege escalation attacks on android. In: Burmester, M., Tsudik, G., Magliveras, S., Ilić, I. (eds.) ISC 2010. LNCS, vol. 6531, pp. 346–360. Springer, Heidelberg (2011)

13. Dhawan, M., Ganapathy, V.: Analyzing information flow in JavaScript-based browser extensions. In: ACSAC, pp. 382–391 (2009)

14. Fragkaki, E., Bauer, L., Jia, L., Swasey, D.: Modeling and enhancing Android's permission system. In: ESORICS, pp. 1–18 (2012)

15. Guha, A., Fredrikson, M., Livshits, B., Swamy, N.: Verified security for browser extensions. In: 32nd IEEE Symposium on Security and Privacy, pp. 115–130 (2011)

16. Guha, A., Saftoiu, C., Krishnamurthi, S.: The essence of javaScript. In: D'Hondt, T. (ed.) ECOOP 2010. LNCS, vol. 6183, pp. 126–150. Springer, Heidelberg (2010)

17. Guha, A., Saftoiu, C., Krishnamurthi, S.: Typing local control and state using flow analysis. In: Barthe, G. (ed.) ESOP 2011. LNCS, vol. 6602, pp. 256–275. Springer, Heidelberg (2011)

18. Jensen, S.H., Møller, A., Thiemann, P.: Type analysis for javaScript. In: Palsberg, J., Su, Z. (eds.) SAS 2009. LNCS, vol. 5673, pp. 238–255. Springer, Heidelberg (2009)

19. Jensen, S.H., Møller, A., Thiemann, P.: Interprocedural analysis with lazy propagation. In: Cousot, R., Martel, M. (eds.) SAS 2010. LNCS, vol. 6337, pp. 320–339. Springer, Heidelberg (2010)

20. Karim, R., Dhawan, M., Ganapathy, V., Shan, C.-c.: An analysis of the mozilla jetpack extension framework. In: Noble, J. (ed.) ECOOP 2012. LNCS, vol. 7313, pp. 333–355. Springer, Heidelberg (2012)

21. Liu, L., Zhang, X., Yan, G., Chen, S.: Chrome extensions: Threat analysis and countermeasures. In: NDSS (2012)

22. Maffeis, S., Mitchell, J.C., Taly, A.: An operational semantics for JavaScript. In: APLAS, pp. 307–325 (2008)

23. Maffeis, S., Taly, A.: Language-based isolation of untrusted JavaScript. In: CSF, pp. 77–91 (2009)

24. Nielson, F., Nielson, H.R.: Flow logic and operational semantics. Electronic Notes on Theoretical Computer Science 10, 150–169 (1997)

25. Nielson, F., Nielson, H.R., Hankin, C.: Principles of program analysis. Springer (1999)

26. Nielson, H.R., Nielson, F., Pilegaard, H.: Flow logic for process calculi. ACM Computing Surveys 44(1), 1–39 (2012)

27. Politz, J.G., Carroll, M.J., Lerner, B.S., Pombrio, J., Krishnamurthi, S.: A tested semantics for getters, setters, and eval in javascript. In: DLS, pp. 1–16 (2012)

28. Politz, J.G., Eliopoulos, S.A., Guha, A., Krishnamurthi, S.: Adsafety: Type-based verification of JavaScript sandboxing. In: USENIX Security Symposium (2011)

29. Porter Felt, A., Wang, H.J., Moshchuk, A., Hanna, S., Chin, E.: Permission redelegation: Attacks and defenses. In: USENIX Security Symposium (2011)

30. Roesner, F., Kohno, T., Wetherall, D.: Detecting and defending against third-party tracking on the web. In: NSDI, pp. 155–168 (2012)

31. Saltzer, J.H., Schroeder, M.D.: The protection of information in computer systems. Proceedings of the IEEE 63(9), 1278–1308 (1975)

Analysis of Asynchronous Programs
with Event-Based Synchronization[*]

Michael Emmi[1], Pierre Ganty[1], Rupak Majumdar[2], and Fernando Rosa-Velardo[3]

[1] IMDEA Software Institute, Madrid, Spain
{michael.emmi,pierre.ganty}@imdea.org
[2] MPI-SWS, Kaiserslautern, Germany
rupak@mpi-sws.org
[3] Universidad Complutense de Madrid, Spain
fernandorosa@sip.ucm.es

Abstract. Asynchronous event-driven programming has become a central model for building responsive and efficient software systems, from low-level kernel modules, device drivers, and embedded systems, to consumer application on platforms such as .Net, Android, iOS, as well as in the web browser. Being fundamentally concurrent, such systems are vulnerable to subtle and elusive programming errors which, in principle, could be systematically discovered with automated techniques such as model checking. However, current development of such automated techniques are based on formal models which make great simplifications in the name of analysis decidability: they ignore event-based synchronization, and they assume concurrent tasks execute serially. These simplifications can ultimately lead to false positives, in reporting errors which are infeasible considering event-based synchronization, as well as false negatives, overlooking errors which arise due to interaction between concurrent tasks.

In this work, we propose a formal model of asynchronous event-driven programs which goes a long way in bridging the semantic gap between programs and existing models, in particular by allowing the dynamic creation of concurrent tasks, events, task buffers, and threads, and capturing precisely the interaction between these quantities. We demonstrate that (1) the analogous program analysis problems based on our new model remain decidable, and (2) that our new model is strictly more expressive than the existing Petri net based models. Our proof relies on a class of high-level Petri nets called *Data Nets*, whose tokens carry names taken from an infinite and linearly ordered domain. This result represents a significant expansion to the decidability frontier for concurrent program analyses.

1 Introduction

The asynchronous event-driven programming model has emerged as a common approach to building responsive and efficient software. Rather than assigning each computing *task* to a dedicated *thread* which becomes blocked as the task polls for some condition, the system maintains lightweight sets of *events* on which tasks are pending, *buffers* of tasks whose events have been triggered, and worker threads to execute

[*] This work is supported in part by the N-GREENS Software project (Ref. S2013/ICE-2731), STRONGSOFT (TIN2012-39391-C04-04), and AMAROUT-II (EU-FP7-COFUND-291803).

© Springer-Verlag Berlin Heidelberg 2015
J. Vitek (Ed.): ESOP 2015, LNCS 9032, pp. 535–559, 2015.
DOI: 10.1007/978-3-662-46669-8_22

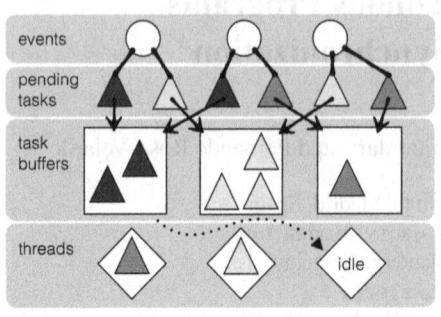

Fig. 1. In asynchronous event-driven programs, pending tasks (drawn as triangles) are moved to their designated task buffers (drawn as boxes) once their designated events (drawn as circles) are triggered. Threads (drawn as diamonds) execute buffered tasks to completion, such that no two tasks from the same buffer (drawn with the same color) execute in parallel.

$$
\begin{aligned}
s ::= \ &s;\ s\ |\ \textbf{skip} \\
&|\ \textbf{assume}\ e\ |\ \textbf{assert}\ e \\
&|\ x := e \\
&|\ \textbf{if}\ e\ \textbf{then}\ s\ \textbf{else}\ s \\
&|\ \textbf{while}\ e\ \textbf{do}\ s \\
&|\ \textbf{call}\ x := p\ e\ |\ \textbf{return}\ e \\
&|\ x := (\textbf{rep})\ \textbf{task}\ p\ e\ Y\ z \\
&|\ y := \textbf{new event} \\
&|\ y := \textbf{event} \\
&|\ z := \textbf{new buffer} \\
&|\ z := \textbf{buffer} \\
&|\ \textbf{cancel}\ x\ |\ \textbf{wait}\ Y\ |\ \textbf{sync}\ y
\end{aligned}
$$

Fig. 2. The grammar of program statements. Here $x, y, z \in$ Vars $\cup \{\bot\}$ range over program variables or $\bot$, $Y \subseteq$ Vars over sets of program variables, e over expressions, and p over procedure names. We assume $\bot$ does not appear on the left-hand side of assignment.

buffered tasks; Figure 1 illustrates the architecture. Besides the possibility of using events for synchronization between tasks, the task buffers themselves provide another means of orchestration: the system can allow tasks from distinct buffers to execute in parallel while ensuring that tasks from the same buffer execute serially. Delegating the management of events, tasks, buffers, and threads to the system generally increases efficiency.

High-performance asynchronous event-driven programs require subtle management of concurrent tasks [1], increasing the possibility of anomalous behavior due to unforeseen task schedules. One important research direction is the development of static program analyses for these programs. Indeed, such analyses have been developed, based on the formal models of *multi-set pushdown systems* (MPDS) [2] and Petri nets [3]. Led by decidability considerations, MPDSs (and Petri nets) model systems with, effectively, just one thread and task buffer, and without events. However, multiple threads, buffers, and events are fundamental to the use of many, if not most, systems; consider the most basic libevent API [4] function

    event_new(buff,fd,flags,proc,arg),

which returns a new task destined for buffer buff pending an event on the file descriptor fd, or the most basic libdispatch API [5] functions

    dispatch_source_create(ty,fd,flags,buff),

which returns a new event whose attached tasks are destined to buffer buff, and

    dispatch_source_set_handler(evt,proc),

which attaches to event `evt` a task to execute procedure `proc`. Accordingly, MPDS-based analyses may report false positives due to abstraction of event-based synchronization, as well as false negatives due to the restriction to executions in which all tasks are executed serially, by one single thread.

We propose a formal model of event-driven asynchronous programs which captures event-based synchronization and task-buffer partitioning precisely, and yet retains a basis for decidable program analyses. Besides task creation, our model allows event creation and task-buffer creation as primitive operations. Each newly-created task has a set of events on which it is *pending*, and a task buffer in which it resides once activated by the triggering of any such event. An *active* task can be dispatched to run on any idle thread until it completes its execution, possibly blocking its thread by *waiting* for some other set of events. Alternatively, in the spirit of asynchronous event-driven programming, tasks can also create more tasks to continue their work once certain events are triggered, rather than blocking their execution threads. Our model also permits tasks to trigger events and to cancel previously-created tasks. This allows our model to capture many intricate aspects of existing languages and libraries supporting asynchronous programming. We show that these features make our model *strictly* more expressive[1] than usual models for asynchronous programs [2,3].

Our main result is that decidability of safety verification is retained despite such heightened modeling precision. While verification expectedly becomes undecidable once tasks can execute concurrently and call recursive procedures [6], or when task buffers are FIFO-ordered [7], even if data is abstracted into finite domains, we demonstrate that with only non-recursive procedures and unordered buffers, the decidability boundary can be stretched unexpectedly far to include unbounded dynamic creation of tasks *and* events.

We prove our result by a reduction to the coverability (or control-state reachability) problem of *Data nets*, an extension of *Petri nets* [8]. In standard Petri nets each place may contain an arbitrary number of *identical* (black) tokens. These tokens may be consumed from the so-called preconditions of transitions and created in postconditions. In the paradigm of *colored Petri nets*, tokens are distinguishable (colored), and their color is relevant for the semantics of the net. *Data Nets* [9] are a class of colored Petri nets in which tokens carry names taken from an infinite, linearly-ordered, and dense domain. In particular, only the relative order among these names is relevant for the semantics of nets. This reduction essentially works by modeling program tasks as tokens in a Data Net, whose event identifiers are encoded by names, and whose task identifiers, buffer identifiers, and local variable valuations are all encoded by the places of the Data Net. Even if Data Nets can represent unboundedly many names in a configuration, they cannot represent unboundedly many *sets* of names. Therefore, the challenge in the simulation is to represent unboundedly-many tasks that are pending on multiple events, even if we cannot directly represent such sets. We use the linear order to specify the order in which events can take place in the future.

The main contributions and outline of this work are:

§2 A formal model of event-driven asynchronous programs which captures event-based synchronization and task-buffer partitioning.

---

[1] With respect to the control-state reachability problem.

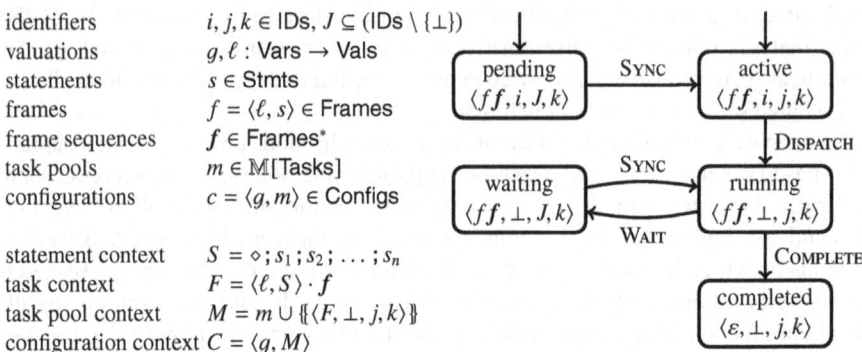

identifiers	$i, j, k \in \mathsf{IDs}, J \subseteq (\mathsf{IDs} \setminus \{\bot\})$
valuations	$g, \ell : \mathsf{Vars} \to \mathsf{Vals}$
statements	$s \in \mathsf{Stmts}$
frames	$f = \langle \ell, s \rangle \in \mathsf{Frames}$
frame sequences	$\boldsymbol{f} \in \mathsf{Frames}^*$
task pools	$m \in \mathbb{M}[\mathsf{Tasks}]$
configurations	$c = \langle g, m \rangle \in \mathsf{Configs}$
statement context	$S = \diamond; s_1; s_2; \dots; s_n$
task context	$F = \langle \ell, S \rangle \cdot \boldsymbol{f}$
task pool context	$M = m \cup \{\!\{\langle F, \bot, j, k \rangle\}\!\}$
configuration context	$C = \langle g, M \rangle$

Fig. 3. Syntactic conventions and the task-state transition diagram. Note that $i \neq \bot$ and $J \neq \emptyset$.

§4 The decidability of the control-state reachability problem for this formal model.
§5 A hardness result for the control-state reachability problem.

2 Asynchronous Programs with Event-Based Synchronization

We fix the sets Procs of procedure names, Vars of variables, Vals of values, and IDs $\subseteq$ Vals of resource identifiers, such that $\bot \in$ IDs and **true, false** $\in$ Vals. A *procedure* $p \in$ Procs is a sequence of parameter and local variable declarations, along with a statement s_p representing the body of p. A *program* P is a sequence of global variable and procedure declarations; we write Stmts to denote the finite set of statements included in the grammar of Figure 2, restricted to those which appear[2] in P. While the syntax of program expressions is mostly unimportant for our concurrency-centric considerations, we do suppose that program expressions are statically typed with distinct types for task, buffer, and event identifiers, and the type of $\bot$ is polymorphic, e.g., as null in Java. We also include the nullary nondeterministic choice operator $\star$, which can evaluate to any value of (Vals $\setminus$ IDs), in order to model programs in which data values have been abstracted [10].

Intuitively, the $y :=$ **new event** and $z :=$ **new buffer** statements store a fresh identifier in variable y, resp., z, which is later used to refer to an event or buffer. The $x :=$ (**rep**) **task** p e Y z statement stores in x the identifier of a new (repeating) task to execute procedure p with argument e, which is to be placed in the buffer identified by z once (resp., each time) any of the events identified by Y are triggered; the **cancel** x statement discards any task(s) identified by x. The **sync** y statement triggers the event identified by y. The **wait** Y statement suspends its executing thread until any of the events identified by Y are triggered. We suppose that local variables do not store identifiers; otherwise, as Appendix B demonstrates, the program analysis problems we consider in this work would become undecidable.

A *(procedure) frame* $f = \langle \ell, s \rangle \in$ Frames is a (finite) valuation $\ell :$ Vars $\to$ Vals to the procedure local variables, along with a statement $s \in$ Stmts describing the entire

---
[2] See Appendix A for the precise meaning of *appear* in P.

body of a procedure that remains to be executed, initially set to s_p. A *task* $\langle f, i, j, k \rangle$ or $\langle f, i, J, k \rangle$ is a procedure frame sequence $f \in$ Frames*, along with a repeating or non-repeating task identifier $i \in$ IDs, an event identifier $j \in$ IDs or non-empty set of event identifiers $J \subseteq$ (IDs \ {$\bot$}), and a buffer identifier $k \in$ IDs. The set of tasks is denoted Tasks. A task $\langle \_, i, \_, \_ \rangle$ is *repeating* when i is a repeating task identifier. A task $\langle f, \_, \_, \_ \rangle$ is *completed* when $f = \varepsilon$.[3] A non-completed task $\langle \_, i, J, \_ \rangle$ is *pending* when $i \neq \bot$, and *waiting* when $i = \bot$. A non-completed task $\langle \_, i, j, \_ \rangle$ is *active* when $i \neq \bot$, and *running* when $i = \bot$. Note that we only write capital J for waiting and pending tasks, and lowercase $j \in$ IDs for active and running tasks. A *task pool* is a finite-support[4] multiset $m \in \mathbb{M}$[Tasks]. A *configuration* $c = \langle g, m \rangle \in$ Configs is a valuation g : Vars $\rightarrow$ Vals to the global variables, along with a task pool m. Figure 3 summarizes our syntactic conventions, and the transitions between the various task states.

The (nondeterministic) evaluation $e(g, \ell) \subseteq$ Vals is the set of values to which the program expression e can evaluate, given the variable valuations g and ℓ. While we are mostly agnostic to the meaning of program expressions, we assume that $\star(g, \ell) =$ (Vals \ IDs), and that evaluation of each IDs-typed expression e is deterministic, i.e., $e(g, \ell) = \{i\}$ for some $i \in$ IDs; accordingly, we treat $e(g, \ell)$ as an element of IDs rather than as a subset of IDs.

To reduce clutter in the operational program semantics, we introduce a notion of context. A *statement context S* is a term derived from the grammar $S ::= \diamond \mid S ; s$, where $s \in$ Stmts. We write $S[s]$ for the statement obtained by substituting a statement s for the unique occurrence of $\diamond$ in S. Intuitively, a context filled with s, e.g., $S[s]$, indicates that s is the next statement to execute in the statement sequence $S[s]$. Similarly, a *task context* $F = \langle \ell, S \rangle \cdot f$ is a frame sequence in which the first frame's statement is replaced with a statement context, and we write $F[s]$ to denote the frame sequence $\langle \ell, S[s] \rangle \cdot f$. A *task pool context* $M = m \cup \{\!\{\langle F, \bot, j, k \rangle\}\!\}$ is a task pool in which the frame sequence of one *running* task is replaced by a context, and we write $M[s]$ to denote the task pool $m \cup \{\!\{\langle F[s], \bot, j, k \rangle\}\!\}$. A *configuration context* $C = \langle g, M \rangle$ is a configuration in which the task pool is replaced by a task pool context, and we write $C[s]$ to denote the configuration $\langle g, M[s] \rangle$. We overload expression evaluation to contexts, writing $e(g, F)$, $e(g, M)$, or $e(C)$ for the evaluation $e(g, \ell)$ using the global valuation g and the local valuation ℓ of the selected task's first frame.

Figure 4 defines the program transition relation $\rightarrow$ as a set of operational steps on configurations. The NEW rule simply stores a freshly-allocated identifier, which can be subsequently used to identify events or buffers. The DISPATCH rule makes some active task running, so long the number of running and waiting tasks does not exceed the number $T \in (\mathbb{N} \cup \{\omega\})$ of threads,[5] nor does the number of running tasks from any given buffer k exceed the buffer's concurrency limit $T_k \in (\mathbb{N} \cup \{\omega\})$. While our semantics is indifferent, the most common use case sets $T_k = 1$. The COMPLETE rule turns a running task with no statements but **skip** to execute into a completed task. The TASK rule adds a newly-created task to the task pool, while the CANCEL rule removes any task identified

---

[3] We write "$\_$" to denote irrelevant entities, and "ε" for the empty sequence.

[4] A multiset m has *finite support* when $m(x) > 0$ for only finitely many $x \in$ dom(m).

[5] While we consider for simplicity a fixed number T of threads, we claim our theoretical results continue to hold if the number of threads can be changed dynamically, e.g., by the program.

NEW
$$\frac{i \text{ is fresh} \qquad g_2 = g_1(y \mapsto i)}{\langle g_1, M[y := \textbf{new } \_]\rangle \to \langle g_2, M[\textbf{skip}]\rangle}$$

DISPATCH
$$\frac{i \neq \bot \qquad |\{\!|\langle f\!f, i', \_, \_\rangle \in m : i' = \bot\}\!| < T}{|\{\!|\langle f\!f, i', \_, k\rangle \in m : i' = \bot \text{ and } k' = k\}\!| < T_k} \qquad \text{COMPLETE} \qquad \frac{f = \langle \_, \textbf{skip}\rangle}{\langle g, m \cup \{\!|\langle f, \bot, j, k\rangle\}\!|\rangle \to \langle g, m \cup \{\!|\langle \varepsilon, \bot, j, k\rangle\}\!|\rangle}$$

$$\langle g, m \cup \{\!|\langle f, i, j, k\rangle\}\!|\rangle \to \langle g, m \cup \{\!|\langle f, \bot, j, k\rangle\}\!|\rangle$$

TASK
$$\frac{i \text{ is a fresh (repeating) task ID} \qquad g_2 = g_1(x \mapsto i)}{\ell \in e(g_1, M) \qquad m = \{\!|\langle\langle \ell, s_p\rangle, i, Y(g_1, M), z(g_1, M)\rangle\rangle\}\!|}{\langle g_1, M[x := (\textbf{rep}) \textbf{ task } p \, e \, Y \, z]\rangle \to \langle g_2, M[\textbf{skip}] \cup m\rangle}$$

CANCEL
$$\frac{g_2 = g_1(x \mapsto \bot)}{m = \{\!|\langle \_, i, \_, \_\rangle \in M : i = x(g_1, M) \neq \bot\}\!|}{\langle g_1, M[\textbf{cancel } x]\rangle \to \langle g_2, M[\textbf{skip}] \setminus m\rangle}$$

WAIT
$$m_1 = \{\!|\langle F[\textbf{wait } Y], \bot, \_, k\rangle\}\!|$$
$$m_2 = \{\!|\langle F[\textbf{skip}], \bot, Y(g, M), k\rangle\}\!|$$

$$\frac{}{\langle g, m \cup m_1\rangle \to \langle g, m \cup m_2\rangle}$$

SYNC
$$j = y(g, M_1) \qquad m \in \{\!|\langle \_, \_, J, \_\rangle \in M_1 : j \in J\}\!|$$
$$m' = \{\!|\langle \_, i, \_, \_\rangle \in m : i \text{ is not a repeating task ID}\}\!|$$
$$M_2 = (M_1 \setminus m') \cup \{\!|\langle f, i, j, k\rangle : \langle f, i, \_, k\rangle \in m\}\!|$$

$$\frac{}{\langle g, M_1[\textbf{sync } y]\rangle \to \langle g, M_2[\textbf{skip}]\rangle}$$

CURRENT-EVENT
$$m_1 = \{\!|\langle F[y := \textbf{event}], \bot, j, k\rangle\}\!|$$
$$m_2 = \{\!|\langle F[\textbf{skip}], \bot, j, k\rangle\}\!| \qquad g_2 = g_1(y \mapsto j)$$

$$\frac{}{\langle g_1, m \cup m_1\rangle \to \langle g_2, m \cup m_2\rangle}$$

CURRENT-BUFFER
$$m_1 = \{\!|\langle F[z := \textbf{buffer}], \bot, j, k\rangle\}\!|$$
$$m_2 = \{\!|\langle F[\textbf{skip}], \bot, j, k\rangle\}\!| \qquad g_2 = g_1(z \mapsto k)$$

$$\frac{}{\langle g_1, m \cup m_1\rangle \to \langle g_2, m \cup m_2\rangle}$$

Fig. 4. Rules defining the program transition relation $\to$. The variables $i, j, k \in$ IDs range over task, event, and buffer identifiers respectively, and $J \subseteq (\text{IDs} \setminus \{\bot\})$ over non-empty event-identifier sets. We write $Y(g, M)$ in the TASK and WAIT rules for the set $\{y(g, M) \in \text{IDs} : y \in Y\} \setminus \{\bot\}$ of identifiers referred to by Y, interpreting $Y(g, M)$ as $\bot$ when $Y(g, M) = \emptyset$. We denote irrelevant values with "$\_$" and write $\{\!| \cdot \}\!|$, $\cup$, and $\setminus$ to denote the multiset constructor, union, and difference operators.

by x. The WAIT rule makes waiting one running task. The SYNC rule makes active or running all pending or waiting tasks whose event sets contain the event identified by y. The CURRENT-BUFFER and CURRENT-EVENT rules allow to recover and store in a variable the buffer to which the task belongs and the event that activated the task, respectively. The transition rules for the usual sequential program statements are standard, and are included in Appendix A; those rules modify the state of exactly one *running* task at a time.

The *initial configuration* $c_0 = \langle g_0, m_0\rangle$ of a program P is the valuation g_0 mapping each global variable to $\bot$, along with the task pool m_0 containing a single running task $\langle\langle \ell_0, s_{\text{main}}\rangle, \bot, \bot, \bot\rangle$ such that ℓ_0 maps each variable of the main procedure to $\bot$. An *execution of P to c_j* is a configuration sequence $c_0 c_1 \ldots c_j$ such that $c_i \to c_{i+1}$ for $0 \leq i < j$. The *control-state reachability problem* asks, given a procedure p of a program P, whether p can be executed in some execution of P, i.e. whether there is a reachable configuration $\langle g, m\rangle$ with $\langle\langle \_, s_p\rangle f, \bot, \_, \_\rangle \in m$ for some $f \in \text{Frames}^*$. Typical safety

```
        var x: int                          var x: int
        var b1, b2: buffer id               var e: event id
        var t1, t2: task id                 var t1, t2: task id

        proc main                           proc main
s₁:     x := 1000;               s₁:        e := new event;
s₂:     b1 := new buffer;        s₂:        t1 := task p1() {} ⊥;
s₃:     b2 := new buffer;        s₃:        t2 := rep task p2() {e} ⊥
s₄:     t1 := task p(1000) {} b1;           end
s₅:     t2 := task p(1) {} b2
        end                                 proc p1()
                                 s₄:        x := 1;
        proc p(var y: int)       s₅:        sync e
          var z: int                        end
s₆:     z := x - y;
s₇:     x := z                              proc p2()
        end                      s₆:        assert x > 0
                                            end
```

Fig. 5. A program which executes two tasks
from different buffers to subtract a value
from global variable x   **Fig. 6.** A program using synchronization

verification questions (e.g., program assertions, mutual exclusion properties) can be
reduced to this problem.

Example 1. Consider an execution from the initial configuration

$$\left\langle \begin{pmatrix} x \mapsto \bot \\ b1 \mapsto \bot, b2 \mapsto \bot \\ t1 \mapsto \bot, t2 \mapsto \bot \end{pmatrix}, \{\!\!\{ \langle \langle \emptyset, s_1 ; s_2 ; s_3 ; s_4 ; s_5 \rangle, \bot, \bot, \bot \rangle \}\!\!\} \right\rangle$$

of the program listed in Figure 5. By applying the ASSIGN and NEW rules to the first
three program statements s_1–s_3, and advancing past reduced **skip** statements via the
SKIP rule, we arrive to the configuration

$$\left\langle \begin{pmatrix} x \mapsto 1000 \\ b1 \mapsto b_1, b2 \mapsto b_2 \\ t1 \mapsto \bot, t2 \mapsto \bot \end{pmatrix}, \{\!\!\{ \langle \langle \emptyset, s_4 ; s_5 \rangle, \bot, \bot, \bot \rangle \}\!\!\} \right\rangle$$

in which buffer identifiers b_1 and b_2 have been created and stored in variables b1 and
b2. Now applying the TASK rule to statement s_4 creates a fresh task identified by t_1 to
execute p(1000) from buffer b_1

$$\left\langle \begin{pmatrix} x \mapsto 1000 \\ b1 \mapsto b_1, b2 \mapsto b_2 \\ t1 \mapsto t_1, t2 \mapsto \bot \end{pmatrix}, \left\{\!\!\left\{ \begin{matrix} \langle \langle \emptyset, s_5 \rangle, \bot, \bot, \bot \rangle \\ \langle \langle [y \mapsto 1000, z \mapsto \bot], s_6 ; s_7 \rangle, t_1, \bot, b_1 \rangle \end{matrix} \right\}\!\!\right\} \right\rangle$$

and subsequently applying the TASK rule to statement s_5 creates a fresh task identified by t_2 to execute $p(1)$ from buffer b_2

$$\left\langle \begin{pmatrix} x \mapsto 1000 \\ b1 \mapsto b_1, b2 \mapsto b_2 \\ t1 \mapsto t_1, t2 \mapsto t_2 \end{pmatrix}, \left\{\!\!\left\{ \begin{array}{l} \langle \varepsilon, \bot, \bot, \bot \rangle \\ \langle\langle [y \mapsto 1000, z \mapsto \bot], s_6 ; s_7 \rangle, t_1, \bot, b_1 \rangle \\ \langle\langle [y \mapsto 1, z \mapsto \bot], s_6 ; s_7 \rangle, t_2, \bot, b_2 \rangle \end{array} \right\}\!\!\right\} \right\rangle$$

resulting in the completion of the initially-running task, by subsequently applying the COMPLETE rule. Supposing that Task t_2 executes statement s_6 before t_1 executes, via the DISPATCH and ASSIGN rules, we arrive at the configuration

$$\left\langle \begin{pmatrix} x \mapsto 1000 \\ b1 \mapsto b_1, b2 \mapsto b_2 \\ t1 \mapsto t_1, t2 \mapsto t_2 \end{pmatrix}, \left\{\!\!\left\{ \begin{array}{l} \langle \varepsilon, \bot, \bot, \bot \rangle \\ \langle\langle [y \mapsto 1000, z \mapsto \bot], s_6 ; s_7 \rangle, t_1, \bot, b_1 \rangle \\ \langle\langle [y \mapsto 1, z \mapsto 999], s_7 \rangle, \bot, \bot, b_2 \rangle \end{array} \right\}\!\!\right\} \right\rangle .$$

Since Task t_1 executes from a different buffer than t_2, it may execute to completion before t_2 makes another move, by applying the DISPATCH, ASSIGN, and COMPLETE rules, arriving at the configuration

$$\left\langle \begin{pmatrix} x \mapsto 0 \\ b1 \mapsto b_1, b2 \mapsto b_2 \\ t1 \mapsto t_1, t2 \mapsto t_2 \end{pmatrix}, \left\{\!\!\left\{ \begin{array}{l} \langle \varepsilon, \bot, \bot, \bot \rangle \\ \langle \varepsilon, \bot, \bot, b_1 \rangle \\ \langle\langle [y \mapsto 1, z \mapsto 999], s_7 \rangle, \bot, \bot, b_2 \rangle \end{array} \right\}\!\!\right\} \right\rangle$$

in which t_1 has subtracted 1000 from x, yet t_2 has yet to complete its subtraction of 1 from 1000. Allowing t_1 to execute to completion, via the ASSIGN and COMPLETE rules, we arrive at the configuration

$$\left\langle \begin{pmatrix} x \mapsto 999 \\ b1 \mapsto b_1, b2 \mapsto b_2 \\ t1 \mapsto t_1, t2 \mapsto t_2 \end{pmatrix}, \left\{\!\!\left\{ \begin{array}{l} \langle \varepsilon, \bot, \bot, \bot \rangle \\ \langle \varepsilon, \bot, \bot, b_1 \rangle \\ \langle \varepsilon, \bot, \bot, b_2 \rangle \end{array} \right\}\!\!\right\} \right\rangle$$

in which both subtraction tasks have completed, yet only the second's effects are accounted for in the global state. Note that this global state would not be admitted were both tasks to execute serially.

Example 2. Consider an execution from the initial configuration

$$\left\langle \begin{pmatrix} x \mapsto \bot, e \mapsto \bot \\ t1 \mapsto \bot, t2 \mapsto \bot \end{pmatrix}, \left\{\!\!\left\{ \langle\langle \emptyset, s_1 ; s_2 ; s_3 \rangle, \bot, \bot, \bot \rangle \right\}\!\!\right\} \right\rangle$$

of the program listed in Figure 6. By applying the NEW and TASK rules for the first three program statements s_1–s_3 we arrive to the configuration

$$\left\langle \begin{pmatrix} x \mapsto \bot, e \mapsto e_1 \\ t1 \mapsto t_1, t2 \mapsto t_2 \end{pmatrix}, \left\{\!\!\left\{ \begin{array}{l} \langle \varepsilon, \bot, \bot, \bot \rangle \\ \langle\langle \emptyset, s_4 ; s_5 \rangle, t_1, \bot, \bot \rangle \\ \langle\langle \emptyset, s_6 \rangle, t_2, \{e_1\}, \bot \rangle \end{array} \right\}\!\!\right\} \right\rangle$$

in which Task t_1 is active and t_2 is pending. Note that neither would have been able to execute before the initial task had completed, since all three tasks are executed from the same buffer, identified by $\bot$. Now applying DISPATCH, which is the only enabled transition, and executing t_1 to completion, we arrive to the configuration

$$\left\langle \left(\begin{array}{l} x \mapsto 1, e \mapsto e_1 \\ t1 \mapsto t_1, t2 \mapsto t_2 \end{array} \right), \left[\begin{array}{l} \langle \varepsilon, \bot, \bot, \bot \rangle \\ \langle \varepsilon, \bot, \bot, \bot \rangle \\ \langle \langle \emptyset, s_6 \rangle, t_2, \{e_1\}, \bot \rangle \\ \langle \langle \emptyset, s_6 \rangle, t_2, e_1, \bot \rangle \end{array} \right] \right\rangle$$

in which event e_1 has been triggered, and variable x set to 1. Since t_2 is a repeating task, a pending copy of it remains in the task pool in addition to the newly-activated copy. Now the DISPATCH rule can apply to the active copy, which can be executed to completion, succeeding the assertion of statement s_6, resulting in the configuration

$$\left\langle \left(\begin{array}{l} x \mapsto 1, e \mapsto e_1 \\ t1 \mapsto t_1, t2 \mapsto t_2 \end{array} \right), \left[\begin{array}{l} \langle \varepsilon, \bot, \bot, \bot \rangle \\ \langle \varepsilon, \bot, \bot, \bot \rangle \\ \langle \langle \emptyset, s_6 \rangle, t_2, \{e_1\}, \bot \rangle \\ \langle \varepsilon, \bot, e_1, \bot \rangle \end{array} \right] \right\rangle .$$

Had we abstracted the event-based synchronization from our formal model, it would have been possible to violate the assertion of statement s_6, which is not possible in the program as it is written.

While our general model of asynchronous programs gives semantics to arbitrary programs with infinite data domains, our decidability arguments in the following sections rely on the following key assumptions/restrictions:

1. The set (Vals \ IDs) of non-identifier values is finite.
2. Procedures are not recursive: two frames of the same procedure cannot appear on the same procedure stack.
3. Identifiers are not stored in local variables (previously mentioned).
4. Buffer identifiers and repeating task identifiers (other than $\bot$) are stored in global variables from the time they are created.

Assumption 1 is a standard assumption in model checking [11] and data-flow analysis [12], which is obtained, e.g., by predicate abstraction [10]; allowing general-purpose infinite data domains quickly leads to undecidability. Assumption 2 avoids a well-established undecidable class, in which concurrent threads with unbounded procedure stacks (only 2 threads are required) can synchronize (e.g., through global variables) [6]. Assumption 4 together with 1, essentially limits the number of task buffers and repeating tasks (prohibiting unbounded dynamic creation), since their identifiers are always stored among a finite number of global variables. While Appendix B shows Assumption 3 cannot be relaxed, we do not know whether Assumption 4 can be weakened. We believe that it is reasonable to assume that programs keep references to their repeating tasks, in order to eventually cancel them, and do not create unboundedly-many task buffers. Note that these assumptions do not preclude an unbounded number of dynamically-created events correlating tasks in the task pool.

Though we model repeating tasks explicitly in order to describe their semantics precisely, and to stipulate Assumption 4, they can be simulated with regular tasks. Essentially, each **sync** action triggering events on which a repeating task is waiting is augmented to create active copies of triggered repeating tasks. The required bookkeeping is possible since the number of repeating task identifiers is bounded via Assumption 4 by the number of global variables: for each repeating task i, additional global variables are added to store the identifiers of i's buffer, and events on which i waits.

3 Data Nets

In this section we present basic facts about *Petri Data Nets (PDN)* [9], which extend the classical model of Petri nets [8] with identity-carrying tokens. Despite the fact that PDNs are strictly-more expressive than Petri nets, their coverability problem remains decidable [9]. In Section 4, we show a reduction from the control-state reachability problem of asynchronous programs to the coverability problem on PDNs.

We denote the null tuple $(0, \ldots, 0) \in \mathbb{N}^k$ (for any k) as $\mathbf{0}$, and for $x = (a_1, \ldots, a_k)$ we write $x(i) = a_i$. We denote as $(\mathbb{N}^k)^*$ set of finite words over $\mathbb{N}^k$. For a word $w = x_1 \cdots x_n$ we write $|w| = n$ and $w(i) = x_i$. Formally, a *Petri Data Net* is a Petri net where each token carries an *identity* from a linearly ordered and dense domain $\mathbb{D}$.

Petri Data Nets. A k-dimensional *Petri Data Net (PDN)* is a tuple $N = (\mathsf{P}, \mathsf{T}, F, H)$, where:

- $\mathsf{P} = \{p_1, \ldots, p_k\}$ is a finite set of *places*,
- T is a finite set of *transitions*, disjoint from P,
- for every $t \in \mathsf{T}$, F_t and H_t are finite sequences in $(\mathbb{N}^k)^*$ with $|F_t| = |H_t| = n$ (for some n specific to t), and we say t has arity n.

Every transition t is endowed with two sequences F_t and H_t of the same length of (possibly null) vectors; $F_t(i)$ specifies the tokens carrying the i-th identity that are consumed and analogously, $H_t(i)$ specifies the tokens of that identity that are produced, when transition t is taken.

Markings. A marking m of a *PDN* can be seen as a mapping m that maps every place p to a multiset of identities. This will be the intuition that will guide our graphical notations. However, in the formal exposition, we use a different representation of markings, guided by the two following observations:

1. A marking m only has finitely many tokens, carrying some identities $d_1 < \cdots < d_n$. For each i, we can gather all the tokens carrying the name d_i in m, thus obtaining assuming k places a non-null vector $v_i \in \mathbb{N}^k$ (the j-th component of v_i standing for the number of tokens in the j-th place carrying the name d_i). Therefore, m can be written as $(d_1, v_1) \cdots (d_n, v_n)$.
2. The concrete identities d_i are irrelevant, and only their relative *order* is useful with respect to the semantics of *PDN*. Thus, m can be safely abstracted as the sequence $v_1 \cdots v_n$ in $(\mathbb{N}^{|\mathsf{P}|} \setminus \mathbf{0})^*$.

Formally, a *marking* of a k-dimensional *PDN* is a word in $(\mathbb{N}^k \setminus \mathbf{0})^*$. We say a marking $m = x_1 \cdots x_n$ marks p_i if $x_j(i) > 0$ for some $j \in \{1, \ldots, n\}$.

Prior to formally defining the transition relation, we start with some intuition. Consider a marking $m \in (\mathbb{N}^k \setminus \mathbf{0})^*$. In order to fire a transition t with arity n, the net nondeterministically selects n identities, consumes some tokens with these identities as specified

Fig. 7. Firing of a *PDN* transition (assuming $a < c < b$)

by F_t, and produces new tokens with the identities specified by H_t. In order to deal with identities that are not present in m, or identities that are removed due to the firing of t, we introduce/remove null vectors where needed. We say $m' \in (\mathbb{N}^k)^*$ is a **0**-*extension* of a marking m (or m is the **0**-*contraction* of m') if m is obtained by removing every tuple **0** from m'.

Transition Relation of *PDN*. Let m, m' be two markings and $t \in \mathsf{T}$ with arity n. We say t can be *fired* in m, reaching m' if:

1. there exists a **0**-extension $u_0 x_1 u_1 \cdots u_{n-1} x_n u_n$ of m with $u_i \in (\mathbb{N}^k)^*$ for $i \in \{0, \ldots, n\}$ and $x_i \in \mathbb{N}^k$ for $i \in \{1, \ldots, n\}$,
2. $x_i \geq F_t(i)$ for $i \in \{1, \ldots, n\}$,
3. and taking $y_i = (x_i - F_t(i)) + H_t(i)$, m' is the **0**-contraction of $u_0 y_1 u_1 \cdots u_{n-1} y_n u_n$.

We write $m \rightarrow m'$ if m' can be reached from m by firing some transition $t \in \mathsf{T}$, and denote $\rightarrow^*$ the reflexive and transitive closure of $\rightarrow$. We assume an initial marking m_0, and say a marking m is reachable if $m_0 \rightarrow^* m$.

We rely on the often-used graphical depiction for *PDN* and use pictures of Petri nets where arcs connected to a transition t are labelled with bags of variables that must be instantiated by ordered identities. The number of these variables is exactly the arity of t and the ordering of the corresponding identities is carried by the transition.

Using these graphical conventions, Figure 7 depicts a *PDN* with a single transition t given by $F_t = (1, 0, 0)(0, 0, 0)(0, 1, 0)$ and $H_t = (0, 0, 0)(0, 0, 1)(0, 0, 0)$ (where places are ordered by their index). The marking shown in the left of the figure is given by the word $m = (1, 0, 0)(0, 1, 0)$, where the first tuple represents the identity a (that appears only in p_1) and the second tuple represents the identity b (that appears only in p_2). Since the transition has arity 3, we need to have three tuples in our marking, for which we can add a **0**-tuple, thus obtaining the **0**-expansion $m'' = (1, 0, 0)(0, 0, 0)(0, 1, 0)$. Notice that $m''(i) \geq F_t(i)$ for $i = 1, 2, 3$. After subtracting F_t and adding H_t we obtain $m''' = (0, 0, 0)(0, 0, 1)(0, 0, 0)$, which is **0**-contracted to the marking $m' = (0, 0, 1)$ shown in the right of Figure 7.

For brevity and readability, we allow variables that are not totally ordered, which stands for a choice among all possible linearizations. Also, we will allow the labelling of some arcs by an expression of the form $\min\{x_1, \ldots, x_n\}$, which is replaced in each linearization by the minimum variable. For instance, we can simulate a transition t in which two unrelated variables x and y appear, by a non-deterministic choice between three transitions, the first one assuming $x < y$ (and replacing $\min\{x, y\}$ by x), the second one assuming $y < x$ (replacing $\min\{x, y\}$ by y) and the last one assuming $x = y$ (with y and $\min\{x, y\}$ replaced by x). Analogously, a transition with variables x and y so that $x \leq y$, can be simulated by two transitions, assuming $x < y$ and $x = y$, respectively.

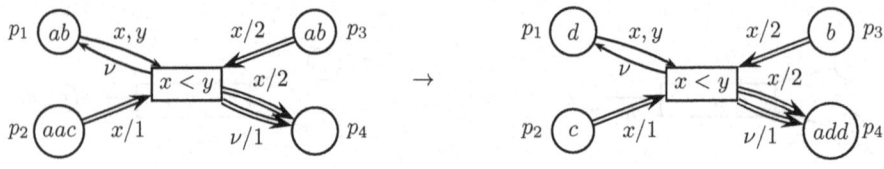

Fig. 8. Firing of a *PDN* transition *t* with extra features (with $a < b$, and c and d unrestricted)

Additionally, we shall use the following features in Section 4. Abdulla *et al.* [13] prove that each of these features, not present in the basic *PDN* model, can be simulated by the basic model.

Fresh Name Creation. In the *PDN* definition a new name *may* be created whenever some $F_t(i)$ is the null tuple, but freshness is not guaranteed. Abdulla *et al.* [13] give a construction that guarantees some name is created fresh. In pictures we will represent fresh name creation by labeling some postarc of a transition *t* by a special variable v, that can only be instantiated to names not appearing in the marking that enables *t*.

Transfers and Renamings. In our simulation we will need to transfer every token carrying a given name from one place to another. We represent transfers in our pictures by having double arcs labelled by some variable. If a transition has several transfers, we will distinguish them in pictures by numbering them. Moreover, if the variables in the prearc and the postarc of a transfer do not coincide, then the names in the precondition are renamed accordingly in the postcondition.

Figure 8 shows an example that illustrates all the extra features we will need in our simulation. In the firing of *t* the variables x and y are instantiated to a and b, respectively (we are assuming $a < b$). After the firing of the transition (*i*) a and b are removed from p_1, (*ii*) a fresh name d is put in p_1, (*iii*) every a-token is transfered from p_3 to p_4 (transfer arc labelled by $x/2$), and (*iv*) every a-token in p_2 is transfered to p_4, and renamed to the same fresh name d (transfer labelled by $x/1$ and $v/1$). Notice that the transition specifies a partial order $x < y$ over the set of variables $\{x, y, v\}$, so that the position of the fresh name in the order is left unspecified.

We define the *control-state reachability problem*, that given a *PDN* N, an initial marking m_0, and a subset S of places of N, asks whether some marking reachable from m_0 in N marks some place $p \in S$. This problem is proved to be decidable in [9].

In Sect. 5, we show a reduction from a subclass of Data Nets in which names are unordered, and in which transfers are disallowed. This class is referred to as *v-PN* in the literature [14]. The control-state reachability problem for *v-PN* is known to be Ackermann-hard [14].

4   Simulation

We now demonstrate a reduction from the control-state reachability problem for asynchronous programs to the control-state reachability problem of Petri Data Nets. Our reduction models program tasks as tokens in a *PDN*, whose event identifiers are encoded by identities, and whose task identifiers, buffer identifiers, and local variable valuations are all encoded by the places of the *PDN*. We demonstrate that the runs of the *PDN* N_P constructed from a program P simulate the executions of P, transition for transition, such that a given procedure p of P can be executed if and only if one of N_P's reachable markings marks a place which can only be marked when some task executing p is dispatched. For presentational simplicity, we give our simulation result assuming the only program variables have buffer- or event-identifier type, respectively denoted Vars_b and Vars_e. This assumption implies that

- together with Assumption 3, there are no local variables, hence the valuation component from Frames always equals $\emptyset$;
- because there are no local variables, there are no local assignments. Thus, the statement **call** $x := p\ e$ is simply replaced by **call** p and $x := \mathbf{task}\ p\ e\ Y\ z$ is replaced by $x := \mathbf{task}\ p\ Y\ z$;
- because there are no variables storing task identifiers $x := \mathbf{task}\ p\ Y\ z$ is further simplified to $\mathbf{task}\ p\ Y\ z$; further, there are no **cancel** statements.

We start by describing the encoding of the global valuations for the variables Vars_b and Vars_e. Notice that, because of Assumption 4, the set of buffers identifiers is limited to $|\mathsf{Vars}_b| + 1$ distinct values. Define IDs_B to be the set buffer identifiers given by $\{\bot\} \cup \{1, \ldots, |\mathsf{Vars}_b|\}$.

The *PDN* N_P simulating the program P has a set of places given by

$$\{p_b(z, k) \mid z \in \mathsf{Vars}_b, k \in \mathsf{IDs}_B\}$$
$$\cup \{p_e(y) \mid y \in \mathsf{Vars}_e\}$$
$$\cup \{p_t(f, q, k) \mid f \in \mathsf{Frames}^{\leq F}, q \in \{\mathsf{A}, \mathsf{P}, \mathsf{W}, \mathsf{R}\}, k \in \mathsf{IDs}_B\}$$
$$\cup \{p_{\mathrm{tlim}}\} \cup \{p_{\mathrm{blim}}(k) \mid k \in \mathsf{IDs}_B\} \cup \{p_a\}$$

Intuitively, a token in $p_b(z, k)$ means that variable z stores the buffer identifier k. A token in $p_e(y)$ with identity j means that variable y stores the event identifier j. This difference of encoding for event typed variable and buffer typed variable stems from the fact that the number of buffer identifiers is bounded across all executions of P (Assumption 4), whereas the number of event identifiers is not. For each variable $z \in \mathsf{Vars}_b$, the simulation will enforce that if $p_b(z, k)$ and $p_b(z, k')$ are marked then $k = k'$. Intuitively, this is because a buffer typed variable stores exactly one buffer identifier. Also no variables $p_e(y)$ contains more than 1 token. The places $p_t(f, q, k)$ are meant to encode the task buffer. In concordance with Assumption 1, we assume a finite set $\mathsf{Frames}^{\leq F} \stackrel{\mathrm{def}}{=} \{f \in \mathsf{Frames}^* : |f| \leq F\}$ of bounded frame sequences for some $F \in \mathbb{N}$, and we assume the procedure frame sequence of each task belongs to $\mathsf{Frames}^{\leq F}$. Recall that (*i*) it follows from the previous developments that a *frame* f consists of an empty valuation and a statement $s \in \mathsf{Stmts}$ describing the entire body of a procedure p that remains to be executed; and that (*ii*) the set Stmts is finite as explained in §2.

A token in $p_t(f, q, k)$ with identity j encodes a task given by

$$\begin{cases} \langle f, \bot, j, k \rangle & \text{if } q = \mathsf{R}(unning) \\ \langle f, \bot, \{j\}, k \rangle & \text{if } q = \mathsf{W}(aiting) \\ \langle f, \neg, j, k \rangle & \text{if } q = \mathsf{A}(ctive) \\ \langle f, \neg, \{j\}, k \rangle & \text{if } q = \mathsf{P}(ending) \end{cases}$$

Observe that, since no variable can store a task identifier, they become irrelevant. This is why in the third and fourth case, the component identifying a task is left unspecified (but different from $\bot$).

We remark that the set of places of N_P can be effectively built, since all the finite sets appearing in the definition are not only finite, but they can be statically obtained. This is clearly the case for Vars_b, Vars_e and IDs_B, which can be obtained by a simple inspection of the program P, but also for $\mathsf{Frames}^{\leq F}$ because of Assumptions 1 to 3.

The initial configuration of P defines the initial marking of N_P, in particular all the variables are initialized to $\bot$.

The encoding of pending tasks, using places $p_t(f, q, k)$, disallows more than one event in their event set. The single event case (i.e., the case in which tasks are pending on a single event) is a special case of the general case, in which tasks are pending on multiple events. Next, we will see how we overcome what seems to be a loss of generality. Since this is the most delicate point in our simulation of programs using *PDN*, let us start by explaining some intuitions (formal developments will follow).

In the simulation using *PDN*, the event set of a task is limited to single events because tokens in *PDN* carry only a single identity.[6] So our goal is to simulate a program where tasks are pending on multiple events using, instead, tasks that are pending on a single event.

We first observe that although tasks are pending on multiple events, only one of them activates the task. At the task creation time, it is not possible to know which event from the set Y will activate the task. However, at creation time, one could guess the event which will activate the task. The non-determinism which is inherent in the model covers all possible such guesses of the event that will eventually activate the task.

However, this is not correct because guesses have to be consistent with the order in which events are triggered along the computation. To see this, consider the following scenario: a task t is created as pending on events $\{e_1, e_2\}$, and every computation triggers first e_2 and then e_1. In this case, the guess which associates to t the single event e_1 wrongly yields a computation with no counterpart in the original program, that in which e_1 and not e_2 activates t.

This example shows that these arbitrary guesses yield a loss of precision by introducing new behaviors. When creating a task, the event that has to be chosen is determined by the ordering in which events are triggered in the future of the computation.

Instead of guessing at task creation time which event will activate the task, we will guess at *event creation time* when this event will be triggered. Formally, this order in the

---

[6] Similar models in which multiple values can be carried by tokens are undecidable [15], and only become decidable when extra semantic restrictions are added [16], which are too restrictive for us.

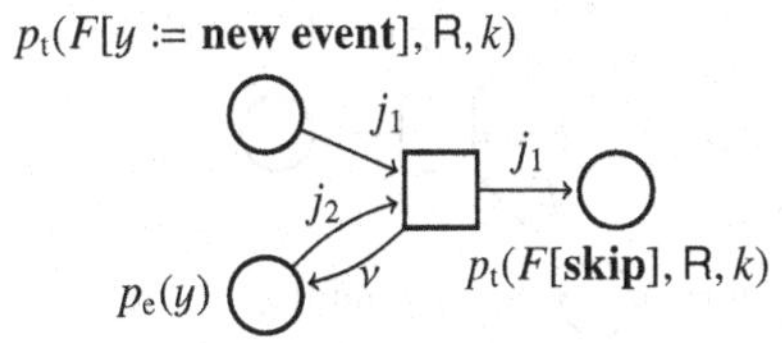

Fig. 9. Widget simulating event creation

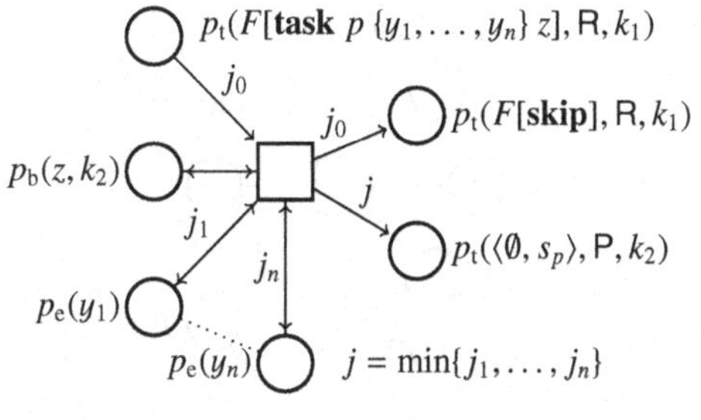

Fig. 10. Widget simulating task creation

triggering of events is given by the linear order on identities in the *PDN*. When creating a task pending on multiple events, the event chosen to activate the task is thus given by the minimal one with respect to that linear order.

Back to the example, since first e_2 and then e_1 are triggered, we have that $e_2 < e_1$ in the *PDN*. Hence, when the task t is created, only $e_2 = \min\{e_1, e_2\}$ can be chosen as the single event associated with t.

Summing up, event identities will be ordered in the simulation according to the (guessed) order in which they are triggered. Then, the simulation needs to guarantee that this order is correct, i.e., that the current computation is consistent with that order. For that purpose, we use a place p_a that holds a token whose identity is the next event that can be triggered, thus separating past events from future events. Also, this token will be used to guarantee that past events (those below the identity in p_a in the linear order) are no longer used. In our example, if the computation guesses $e_1 < e_2$, so that e_1 should first be triggered, and a sync over e_2 is attempted, then the simulation blocks.

Finally, p_{tlim} is a budget place for the number of threads. Its content corresponds to the number of threads available to execute a task. It initially contains $T - 1$ tokens (the -1 accounts for the thread running `main`). The simulation of DISPATCH will remove a token from p_{tlim} while the simulation of COMPLETE will add a token into it. Optionally, the budget can vary along the execution. Moreover, for each $k \in \{\bot\} \cup \{1, \ldots, |\text{Vars}_b|\}$ $p_{\text{blim}}(k)$ is a budget place (one per buffer). As above $p_{\text{blim}}(k)$ accounts for the remaining concurrency limit for buffer k. Initially, it is set to the value T_j by putting T_j tokens at buffer creation time, and it is modified as for p_{tlim}.

Let us next see the simulation of each type of instructions. At the end of the section, we will discuss how to encode our programs when we relax the assumptions on variables so that we allow variables to store finite data and task identifiers.

New Event. An instruction $y := $ **new event** is simply simulated by replacing the name in $p_e(y)$ by a fresh name. The newly created name must be inserted at an arbitrary position in the linear order of identities of the *PDN*, so that the variable v must be left unrestricted.

The widget of Figure 9 simulates event creation. For every place $p_t(f, \mathsf{R}, k)$ where $f = F[y := $ **new event**], there is a transition tr such that:

- tr moves a token from $p_t(f, \mathsf{R}, k)$ to $p_t(f', \mathsf{R}, k)$ where $f' = F[\mathbf{skip}]$. It does so preserving the identity, j_1, carried by the token.
- tr replaces the token in $p_e(y)$ with one whose identity is fresh.

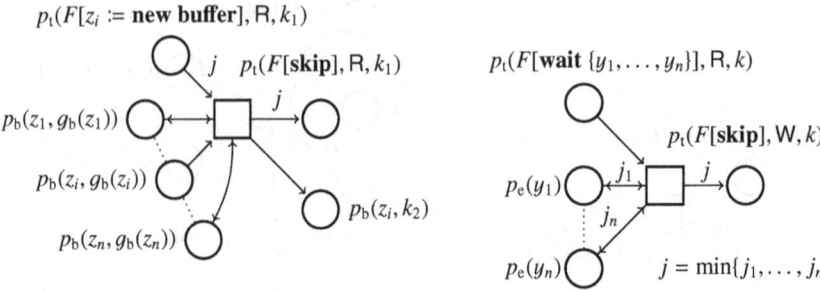

Fig. 11. Widget simulating task buffer creation   **Fig. 12.** Widget simulating waiting

New Task. We first treat the case in which the set Y is not empty (the other case is much simpler). The instruction **task** p $\{y_1,\ldots,y_n\}$ z selects the minimum event in places $p_e(y_1),\ldots,p_e(y_n)$ and puts this name in the place representing the initial state of p (in the buffer given by z), so that the new task is pending on the first event to be triggered.

The widget of Figure 10 simulates task creation. For every place $p_t(f,\mathsf{R},k_1)$ where $f = F[\textbf{task } p\ Y\ z]$, $Y = \{y_1,\ldots,y_n\}$, for every $k_2 \in \mathsf{IDs}_B$, there is a transition tr such that:

- tr moves a token from $p_t(f,\mathsf{R},k_1)$ to $p_t(f',\mathsf{R},k_1)$ where $f' = F[\textbf{skip}]$. It does so preserving the identity, j_0, carried by the token.
- tr reads n identities stored in the places $p_e(y_1)$ to $p_e(y_n)$.
- tr tests whether the place $p_b(z,k_2)$ is marked.
- tr adds a token with identity j into $p_t(\langle\emptyset, s_p\rangle,\mathsf{P},k_2)$ where j is given by the minimum among the identities stored in the places $\{p_e(y_1),\ldots,p_e(y_n)\}$.

For the case $Y = \emptyset$, it is enough to test the place $p_b(z,k_2)$, move the token with identity j_0 and add a token with constant identity $\bot$ to $p_t(\langle\emptyset, s_p\rangle,\mathsf{A},k_2)$ instead (the task is already active).

New buffer. For the simulation of the creation of buffers, we mostly have to deal with the places of the form $p_b(z,k)$ that contain the valuation of buffer variables.

The widget of Figure 11 simulates task buffer creation. For every place $p_t(f,\mathsf{R},k)$ where $f = F[z_i := \textbf{new buffer}]$, buffer identifier $k_2 \neq \bot$ and valuation g_b of the buffer typed variables whose range excludes k_2 there is a transition tr such that:

- tr moves a token from $p_t(f,\mathsf{R},k_1)$ to $p_t(f',\mathsf{R},k_1)$ where $f' = F[\textbf{skip}]$. It does so preserving the identity, j, carried by the token.
- tr tests that no buffer variables stores k_2. It does so by checking whether $p_b(z,g_b(z))$ is marked for every buffer typed variable z.
- tr moves the token from $p_b(z_i,g_b(z_i))$ to $p_b(z_i,k_2)$.

Sync. The simulation of a **sync** y is perhaps the most involved. On the one hand, we need to guarantee that the triggered event is legal according to the guessed linear order of events. For that purpose, the identity j_2 in $p_e(y)$ must coincide with that in p_a. Also, all the tokens carrying j_2 in each place of the form $p_t(f,P/W,k)$ must be transferred to

$p_1, \ldots, p_n$ an enumeration of $\{p_t(f, P, k), p_t(f, W, k), f \in \mathsf{Frames}, k \in \mathsf{IDs}_B\}$
and $\hat{p} = p_t(f, A/R, k)$ where $p = p_t(f, P/W, k)$

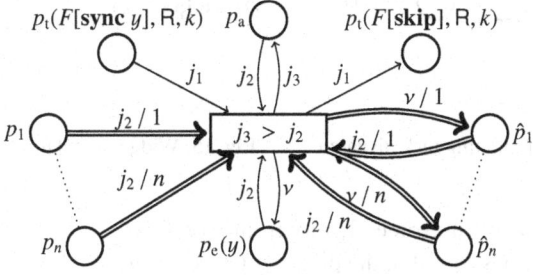

Fig. 13. Widget simulating synchronization

the corresponding place $p_t(f, A/R, k)$, because every task that is pending (waiting) on that name becomes active (running). Then an identity greater than j_2, say j_3, is chosen so as to correspond to the next event to be triggered, by replacing, in p_a, j_2 by j_3. Finally, in order to be able to repeat a **sync** over j_2 we should move j_2 in the linear order to a later position. This is not possible (we cannot change the linear order), but we can replace every j_2 token by a fresh one, which is created at an arbitrary position.

The widget of Figure 13 simulates synchronization. Formally, for every place $p_t(f, \mathsf{R}, k)$ where $f = F[\textbf{sync } y]$, there is a transition tr such that:

- tr moves a token from $p_t(f, \mathsf{R}, k)$ to $p_t(f', \mathsf{R}, k)$ where $f' = F[\textbf{skip}]$. It does so preserving the identity, j_1, carried by the token.
- tr tests that the identity of the token in p_a coincides with that of $p_e(y)$.
- tr transfers, for each place $p_t(f_1, \mathsf{P}, k_1)$ all the tokens whose identity coincide with j_2, the identity of the token in place $p_e(y)$, into $p_t(f_1, \mathsf{A}, k_1)$.
- tr transfers similarly from places $p_t(f_1, \mathsf{W}, k_1)$ into $p_t(f_1, \mathsf{R}, k_1)$.
- tr replace the identity j_2 of the token from p_a with an identity $j_3 > j_2$.
- tr replaces the identity j_2 of the token from $p_e(y)$ with a new, fresh name, v.
- tr also renames all the tokens in the *PDN* carrying the old name j_2 with this new fresh name v.

Let us remark that when choosing the next event to be triggered (by instantiating j_3 to a name greater than the instance of j_2) we may choose an event which is different from the one which immediately follows the instance of j_2 in the linear order.[7] For example, if $e_1 < e_2 < e_3$ is the current order of events and p_a contains e_1, the firing of tr may replace e_1 by e_3 in p_a. This means that in that execution e_2 can no longer be triggered, which only implies a loss of behaviour that preserves control-state reachability.

Wait. The simulation of this instruction is similar to the creation of tasks. The widget of Figure 12 simulates waiting. For every place $p_t(f, \mathsf{R}, k)$ where $f = F[\textbf{wait } Y]$, $Y = \{y_1, \ldots, y_n\}$, there is a transition tr such that:

---

[7] If we allow that operation in PDN the model becomes Turing-complete.

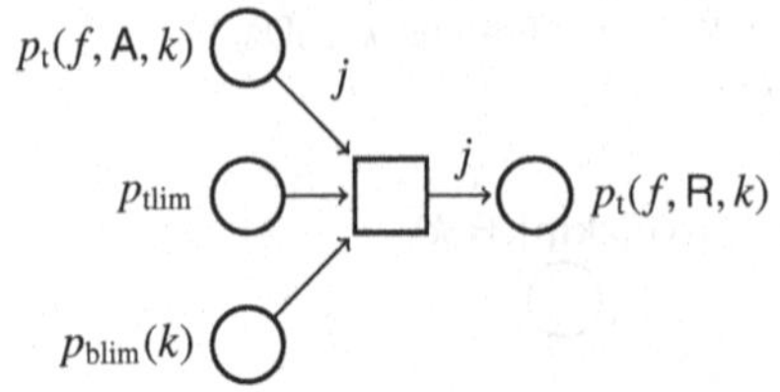

Fig. 14. Widget simulating task dispatch 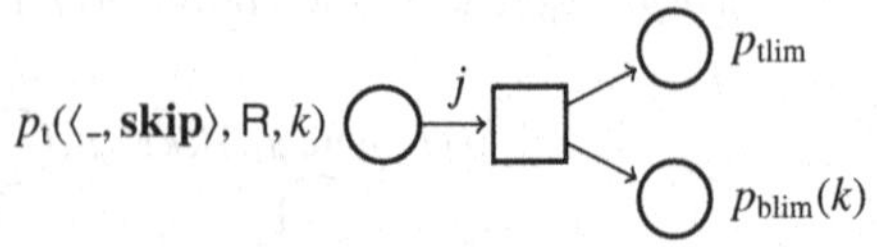   **Fig. 15.** Widget simulating task completion

- tr removes a token with some identity from $p_t(f, R, k)$.
- tr reads n identities stored in the places $p_e(y_1)$ to $p_e(y_n)$.
- tr adds a token with identity j to $p_t(f', W, k)$ where $f' = F[\textbf{skip}]$ and j is given by the minimum among the identities stored in the places $\{p_e(y_1), \ldots, p_e(y_n)\}$.

Dispatch. The widget of Figure 14 simulates task dispatch. For every place $p_t(f, A, k)$, there is a transition tr that: moves a token from $p_t(f, A, k)$ to $p_t(f, R, k)$ while preserving its identity j, removes one token from p_{tlim}, the budget place for threads, and also from $p_{\text{blim}}(k)$, the budget place for concurrency limit of buffer k.

Complete. The widget of Figure 15 simulates task completion. For every place $p_t(f, R, k)$ where $f = \langle \emptyset, \textbf{skip} \rangle$, there is a transition tr that removes a token form $p_t(f, R, k)$, and adds one token into p_{tlim} and one into $p_{\text{blim}}(k)$.

Current-Buffer. For every place $p_t(f, R, k)$ where $f = F[z := \textbf{buffer}]$ and $k_1 \in \textsf{IDs}_B$, there is a transition tr that: moves a token from $p_t(f, R, k)$ to $p_t(f', R, k)$ where $f' = F[\textbf{skip}]$. It does so preserving the identity carried by the token. Transition tr also removes a token from $p_b(z, k_1)$ and adds one into $p_b(z, k)$.

Current-Event. For every place $p_t(f, R, k)$ where $f = F[y := \textbf{event}]$, there is a transition tr that moves a token from $p_t(f, R, k)$ to $p_t(f', R, k)$ where $f' = F[\textbf{skip}]$. It does so preserving the identity j carried by the token. It also replaces the identity of the token from $p_e(y)$ with j.

At this point, it should be easy for the reader to define the widgets for the remaining statements. They present no particular difficulty.

Variables Over Finite Data Domain. At the beginning of this section we made the simplifying assumptions that there were no other variables than buffer or event typed variables. Let us now explain, how we simulate variables ranging over a finite set Vals of values. They are simulated similarly to the buffer typed variables, namely there is a place in the *PDN* for each pair variable-value. The presence of a token in such a place means the variable currently holds the value. The rest is tedious but follows easily.

Variables Over Task Identifiers. Variables storing task identifiers are trickier to simulate. This is because, although we have a fixed number of them, they store values from the unbounded domain of task identifers. This situation is similar to that of variables storing event identifiers with an important difference: there is no comparable mecanism like the $x := \textbf{event}$ statement that allows to recover an event identifer stored in no variables. It follows that, for task identifiers, once no variable stores it, its precise value does not matter anymore. Rather than the actual identifiers, our encoding stores symbolic identifiers. If we call Vars_t the set of task typed variables then define IDs_T to be the set $\{\bot, \top\} \cup \{id_1, id_2, \ldots, id_{|\mathsf{Vars}_t|}\}$ of symbolic identifiers. The encoding keeps track of the symbolic identifers currently in use. Since this information is finite and can be statically obtained, it is easily encoded in the *PDN*. The execution of a $x :=$ **task** $p\ e\ Y\ z$ statement requires a symbolic identifer, not currently in use, to be stored in x. Two situations can occur: either there is a symbolic identifer among $\{id_1, id_2, \ldots, id_{|\mathsf{Vars}_t|}\}$ not in use; or not. However, in the latter case, the symbolic identifier, call it id_j, already stored in x can be re-used as long as all the pending or active tasks with identifier id_j are updated with the symbolic identifier $\top$. Intuitively, a task t with symbolic identifier $\top$ means no program variables storing a task identifier refers to t, hence t cannot be canceled. Furthermore, the set $\{p_t(f, q, k) \mid f \in \mathsf{Frames}^{\leq F}, q \in \{\mathsf{A}, \mathsf{P}, \mathsf{W}, \mathsf{R}\}, k \in \mathsf{IDs}_B\}$ of places encoding the task pool is extended with the information about symbolic identifiers. Hence each place $p_t(f, q, k)$ is replaced by the set $\{p_t(f, i, q, k) \mid i \in \mathsf{IDs}_T\}$. This enables the precise simulation of the **cancel** x statement.

The preceding encoding into *PDN* and the decidability of coverability for *PDN* implies the following main result.

Theorem 3. *The control-state reachability problem is decidable for asynchronous event-driven programs satisfying Assumptions 1 through 4.*

5   Hardness of Control State Reachability

We now show that the control state reachability problem for asynchronous programs is Ackermann-hard by showing that asynchronous programs can simulate ν-*PN*, for which the control state reachability problem is Ackermann-hard [14]. Notice that the usual models of asynchronous programs [2,3] are equivalent to Petri nets and therefore have an EXPSPACE-complete coverability problem. Our model is strictly more expressive than these models.

Theorem 4. *The control state reachability problem for asynchronous programs satisfying Assumptions 1 through 4 is Ackermann-hard, even if no tasks are pending on multiple events.*

We simulate a ν-*PN* as follows. For each place p, we define a different procedure `place-p` (see Figure 17), so that a task executing `place-p` simulates a token in place p. Furthermore, the procedure `main` schedules the firing of transitions. This scheduler uses boolean global variables `remove-p` and `done-p` for each place p, in order to synchronize with the rest of the tasks. Furthermore, we consider two global event variables, `current` and `aux`, for name matching (whenever a variable labels more than one prearc). Figure 18 shows the general scheme of the scheduler, with an infinite loop that non-deterministically selects the transition that is fired next. Whenever the scheduler decides to fire a transition that is not enabled, the system blocks (notice that this preserves control-state reachability). Instead of showing how it orchestrates the firing of an arbitrary transition, Figure 18 shows the simulation of the transition in Fig. 16 that (*i*) removes two tokens carrying the same name from p_1 and p_2, (*ii*) puts a token with that same name in p_3, and *(iii)* puts a fresh token in p_4. For that purpose it first signals that a token from p_1 should be removed (by setting `remove-p1 := true`) and blocks until some task executing `place-p1` is done (instruction `assume done-p1`). Notice that this task is done only after putting its own event name in the global variable `aux`, and then it is completed. Then the scheduler does the same with p_2, but it also checks that both events coincide. Then it creates a new task executing `place-p3`, activated by that same event, and finally it creates a fresh event that is used to create and activate the task executing `place-p4`. The general case follows these ideas, but is more tedious.

We are assuming in the reduction that the number of threads and the concurrency limit of the only buffer $\bot$ is 2 ($T = T_{\bot}$=2), since the task executing `main` must be executed in parallel at each time with at most *one* task of the form `place-p`. The simulation of a transition completes and creates the tasks that represent the tokens involved in the firing of a transition. Therefore, every task of the form `place-p` can be active (not running), except one that has to be completed, which must be dispatched, only to be immediately completed. Notice that if a task that cannot be completed (one with its `remove-p` set to `false`) is dispatched, then the system blocks with that task blocked on `assume remove-p` and `main` blocked on `assume done-p`.

This simulation preserves control-state reachability, that is, a place p in the *v-PN* can be marked iff some reachable configuration contains a pending task executing `place-p`.

Fig. 16. Simple *v-PN* transition

```
var remove-p1, done-p1: bool
...
var current, aux: event id
proc place-p()
    assume remove-p;
    remove-p := false;
    aux := event;
    done-p := true
end
```

Fig. 17. Global variables and procedure modelling a place

```
proc main
    while true do
        if ⋆ then //transition t1
            remove-p1 := true;
            assume done-p1;
            done-p1 := false;
            current := aux;
            remove-p2 := true;
            assume done-p2 && current = aux;
            done-p2 := false;
            _ := task place-p3() {current} ⊥;
            sync current;
            current := new event;
            _ := task place-p4() {current} ⊥;
            sync current
        else if ⋆ then //transition t2
        ...
end
```

Fig. 18. Scheduler

6 Related Work

Existing decidable models of asynchronous event-driven programs are based on *multiset pushdown systems* (MPDS) [2], a model with unbounded task creation and recursion, yet without events, without multiple task buffers, without multiple threads, and without task cancellation. Sen and Viswanathan showed that control-state reachability in MPDS is decidable [2], and Ganty and Majumdar showed that control-state reachability is equivalent to Petri net coverability [3]. Chadha and Viswanathan, Cai and Ogawa generalize these results to show that coverability for well-structured transition systems with one unbounded procedure stack remains decidable, so long as decreasing transitions only occur from a bounded set of configurations [17,18]. Others have shown decidable extensions to MPDS, e.g., with task cancellation [3], or task priorities [19]. While the *Actor Communicating Systems* (ACS) of D'Osualdo et al. do capture multiple task buffers and threads [20] (albeit without recursion, which is modeled in MPDS), they do not model events, and their task buffers are addressed only imprecisely: rather than sending a task to a particular task buffer, tasks are sent to an arbitrary member of a set of equivalent task buffers. Consequently, control-state reachability in ACS is also equivalent to Petri net coverability, and thus to control-state reachability in MPDS. Similarly, while Geeraerts et al. consider multithreaded asynchronous programs with *FIFO-ordered* task buffers, they show decidability for the case of "concurrent queues," in which arbitrarily-many tasks from the same buffer can run concurrently [21]. Effectively, the order in which tasks are added to buffers becomes irrelevant, and their control-state reachability problem is, again, equivalent to Petri net coverability. In contrast, Theorem 4 rules out the possibility of a polynomial time reduction to Petri nets.

Departing from MPDS, Babic and Rakamaric consider an expressive decidable class of asynchronous systems which leverages the decidable properties of visibly pushdown languages [22], equipping visibly pushdown processes with ordered, visibly pushdown, task/message buffers [23]. Kochems and Ong consider a relaxation of MPDS which allows concurrent *and* recursive tasks at the expense of a stack-shape restriction [24]. Bouajjani and Emmi consider decidable subclasses of a hierarchical generalization of MPDS, without communication between concurrent threads [25]. While Atig et al. and Emmi et al. consider MPDSs with multiple buffers and threads [26,27], their algorithms are only under-approximate, analyzing only up to a context-bound [28]. None of these works model events, nor propose sound and complete analysis algorithms in the presence of dynamic creation of threads, events, and buffers.

References

1. Adya, A., Howell, J., Theimer, M., Bolosky, W.J., Douceur, J.R.: Cooperative task management without manual stack management. In: USENIX ATC, pp. 289–302. USENIX (2002)
2. Sen, K., Viswanathan, M.: Model checking multithreaded programs with asynchronous atomic methods. In: Ball, T., Jones, R.B. (eds.) CAV 2006. LNCS, vol. 4144, pp. 300–314. Springer, Heidelberg (2006)
3. Ganty, P., Majumdar, R.: Algorithmic verification of asynchronous programs. ACM Trans. Program. Lang. Syst. 34(1), 6 (2012)
4. Mathewson, N., Provos, N.: libevent: an event notification library, http://libevent.org
5. The GCD team: libdispatch, https://libdispatch.macosforge.org
6. Ramalingam, G.: Context-sensitive synchronization-sensitive analysis is undecidable. ACM Trans. Program. Lang. Syst. 22(2), 416–430 (2000)
7. Brand, D., Zafiropulo, P.: On communicating finite-state machines. J. ACM 30(2), 323–342 (1983)

8. Reisig, W.: Place/transition systems. In: Brauer, W., Reisig, W., Rozenberg, G. (eds.) APN 1986. LNCS, vol. 254, pp. 117–141. Springer, Heidelberg (1987)
9. Lazic, R., Newcomb, T., Ouaknine, J., Roscoe, A.W., Worrell, J.: Nets with tokens which carry data. Fundam. Inform. 88(3), 251–274 (2008)
10. Graf, S., Saïdi, H.: Construction of abstract state graphs with PVS. In: Grumberg, O. (ed.) CAV 1997. LNCS, vol. 1254, pp. 72–83. Springer, Heidelberg (1997)
11. Clarke, E.M., Emerson, E.A.: Design and synthesis of synchronization skeletons using branching-time temporal logic. In: Kozen, D. (ed.) Logic of Programs 1981. LNCS, vol. 131, pp. 52–71. Springer, Heidelberg (1982)
12. Reps, T.W., Horwitz, S., Sagiv, S.: Precise interprocedural dataflow analysis via graph reachability. In: POPL 1995: Proc. 22th ACM SIGPLAN-SIGACT Symposium on Principles of Programming Languages, pp. 49–61. ACM (1995)
13. Abdulla, P.A., Delzanno, G., Van Begin, L.: A classification of the expressive power of well-structured transition systems. Inf. Comput. 209(3), 248–279 (2011)
14. Rosa-Velardo, F., de Frutos-Escrig, D.: Decidability and complexity of Petri nets with unordered data. Theor. Comput. Sci. 412(34), 4439–4451 (2011)
15. Rosa-Velardo, F., de Frutos-Escrig, D.: Decidability problems in Petri nets with names and replication. Fundam. Inform. 105(3), 291–317 (2010)
16. Meyer, R.: On boundedness in depth in the pi-calculus. In: Fifth IFIP International Conference On Theoretical Computer Science - TCS 2008, IFIP 20th World Computer Congress, TC 1, Foundations of Computer Science, Milano, Italy, September 7-10, 2008. IFIP, vol. 273, pp. 477–489. Springer, Heidelberg (2008)
17. Chadha, R., Viswanathan, M.: Decidability results for well-structured transition systems with auxiliary storage. In: Caires, L., Vasconcelos, V.T. (eds.) CONCUR 2007. LNCS, vol. 4703, pp. 136–150. Springer, Heidelberg (2007)
18. Cai, X., Ogawa, M.: Well-structured pushdown systems. In: D'Argenio, P.R., Melgratti, H. (eds.) CONCUR 2013 – Concurrency Theory. LNCS, vol. 8052, pp. 121–136. Springer, Heidelberg (2013)
19. Atig, M.F., Bouajjani, A., Touili, T.: Analyzing asynchronous programs with preemption. In: FSTTCS 2008: Proc. IARCS Annual Conference on Foundations of Software Technology and Theoretical Computer Science. LIPIcs, vol. 2, pp. 37–48. Schloss Dagstuhl - Leibniz-Zentrum fuer Informatik (2008)
20. D'Osualdo, E., Kochems, J., Ong, C.-H.L.: Automatic verification of erlang-style concurrency. In: Logozzo, F., Fähndrich, M. (eds.) Static Analysis. LNCS, vol. 7935, pp. 454–476. Springer, Heidelberg (2013)
21. Geeraerts, G., Heußner, A., Raskin, J.-F.: Queue-dispatch asynchronous systems. CoRR abs/1201.4871 (2012)
22. Alur, R., Madhusudan, P.: Visibly pushdown languages. In: STOC 2004: Proc. 36th Annual ACM Symposium on Theory of Computing, pp. 202–211. ACM (2004)
23. Babić, D., Rakamarić, Z.: Asynchronously communicating visibly pushdown systems. In: Beyer, D., Boreale, M. (eds.) FORTE 2013 and FMOODS 2013. LNCS, vol. 7892, pp. 225–241. Springer, Heidelberg (2013)
24. Kochems, J., Ong, C.-H.L.: Safety verification of asynchronous pushdown systems with shaped stacks. In: D'Argenio, P.R., Melgratti, H. (eds.) CONCUR 2013 – Concurrency Theory. LNCS, vol. 8052, pp. 288–302. Springer, Heidelberg (2013), http://dx.doi.org/10.1007/978-3-642-40184-8_21
25. Bouajjani, A., Emmi, M.: Analysis of recursively parallel programs. In: POPL 2012: Proc. 39th ACM SIGPLAN-SIGACT Symposium on Principles of Programming Languages, pp. 203–214. ACM (2012)
26. Atig, M.F., Bouajjani, A., Qadeer, S.: Context-bounded analysis for concurrent programs with dynamic creation of threads. Logical Methods in Computer Science 7(4) (2011)

27. Emmi, M., Lal, A., Qadeer, S.: Asynchronous programs with prioritized task-buffers. In: SIG-SOFT FSE 2012: Proc. 20th ACM SIGSOFT Symposium on the Foundations of Software Engineering, p. 48. ACM (2012)

28. Qadeer, S., Rehof, J.: Context-bounded model checking of concurrent software. In: Halbwachs, N., Zuck, L.D. (eds.) TACAS 2005. LNCS, vol. 3440, pp. 93–107. Springer, Heidelberg (2005)

A   Sequential Program Semantics

The finite set Stmts of statements appearing in a program P, with finite sets Procs of procedures and Vals of values, is defined formally by the inference rules of Figure 19. These rules are defined with respect to the notion of *context* of Section 2, which we define here to avoid circularity: a *context* S is a term derived from the grammar $S ::= \diamond \mid S ; s$. We write $S[s]$ for the object obtained by substituting s for the unique occurrence of $\diamond$ in S. Intuitively, a context filled with s, e.g., $S[s]$, indicates that s is the first object in a sequence separated by ";".

The transition rules for the sequential program statements complementing those of Figure 4 in Section 2 are listed in Figure 20.

B   Storing Unboundedly-Many Identifiers

The ability to store an unbounded number of identifiers, e.g., using the local variables of unboundedly-many running tasks, makes coverability undecidable. In essence, those stored identifiers allow point-to-point communication between arbitrarily many running tasks. While we give an undecidability proof for when *event* identifier storage is unbounded, very similar proofs are carried out using unbounded *buffer* or *task* identifier storage as well. Furthermore, while we assume unlimited threads and buffer concurrency ($T = T_k = \omega$) in what follows, a similar construction is possible with only one thread and one running task per buffer ($T = T_k = 1$), by continuously creating new tasks. Note that we assume an alternate program semantics, in which the CURRENT-EVENT rule of Figure 4 allows local-variable storage.

$$\frac{p \in \text{Procs}}{s_p \in \text{Stmts}} \qquad \frac{S[\_] \in \text{Stmts}}{S[\textbf{skip}] \in \text{Stmts}} \qquad \frac{S[\textbf{skip}; s] \in \text{Stmts}}{S[s] \in \text{Stmts}} \qquad \frac{S[\textbf{if } e \textbf{ then } s_1 \textbf{ else } s_2] \in \text{Stmts}}{S[s_1] \in \text{Stmts}}$$

$$\frac{S[\textbf{if } e \textbf{ then } s_1 \textbf{ else } s_2] \in \text{Stmts}}{S[s_2] \in \text{Stmts}} \qquad \frac{S[\textbf{while } e \textbf{ do } s] \in \text{Stmts}}{S[s; \textbf{while } e \textbf{ do } s] \in \text{Stmts}}$$

$$\frac{v \in \text{Vals} \qquad S[\textbf{call } x := \_] \in \text{Stmts}}{S[x := v] \in \text{Stmts}}$$

Fig. 19. Rules defining the finite set of statements appearing in a program with procedures Procs.

$$\text{Skip} \quad \frac{}{C[\mathbf{skip};s] \to C[s]}$$

$$\text{Assume} \quad \frac{\mathbf{true} \in e(C)}{C[\mathbf{assume}\ e] \to C[\mathbf{skip}]}$$

$$\text{If-Then} \quad \frac{\mathbf{true} \in e(C)}{C[\mathbf{if}\ e\ \mathbf{then}\ s_1\ \mathbf{else}\ s_2] \to C[s_1]}$$

$$\text{If-Else} \quad \frac{\mathbf{false} \in e(C)}{C[\mathbf{if}\ e\ \mathbf{then}\ s_1\ \mathbf{else}\ s_2] \to C[s_2]}$$

$$\text{Loop-Do} \quad \frac{\mathbf{true} \in e(C)}{C[\mathbf{while}\ e\ \mathbf{do}\ s] \to C[s;\mathbf{while}\ e\ \mathbf{do}\ s]}$$

$$\text{Loop-End} \quad \frac{\mathbf{false} \in e(C)}{C[\mathbf{while}\ e\ \mathbf{do}\ s] \to C[\mathbf{skip}]}$$

$$\text{Assign-Global} \quad \frac{x \in \mathrm{dom}(g) \quad v \in e(g,M)}{\langle g, M[x := e]\rangle \to \langle g(x \mapsto v), M[\mathbf{skip}]\rangle}$$

$$\text{Assign-Local} \quad \frac{f_1 = \langle \ell, S[x := e]\rangle \quad x \in \mathrm{dom}(\ell) \quad v \in e(g,\ell) \quad f_2 = \langle \ell(x \mapsto v), S[\mathbf{skip}]\rangle}{\langle g, m \cup \{\!\{\langle f_1 f, \perp, j, k\rangle\}\!\}\rangle \to \langle g, m \cup \{\!\{\langle f_2 f, \perp, j, k\rangle\}\!\}\rangle}$$

$$\text{Call} \quad \frac{f_1 = F[\mathbf{call}\ x := p\ e] \quad \ell \in e(g,F) \quad f_2 = \langle \ell, s_p\rangle}{\langle g, m \cup \{\!\{\langle f_1, \perp, j, k\rangle\}\!\}\rangle \to \langle g, m \cup \{\!\{\langle f_2 f_1, \perp, j, k\rangle\}\!\}\rangle}$$

$$\text{Return} \quad \frac{f_1 = \langle \ell_1, S_1[\mathbf{return}\ e]\rangle \quad v \in e(g,\ell_1) \quad f_2 = F[\mathbf{call}\ x := \_] \quad f_2' = F[x := v]}{\langle g, m \cup \{\!\{\langle f_1 f_2, \perp, j, k\rangle\}\!\}\rangle \to \langle g, m \cup \{\!\{\langle f_2', \perp, j, k\rangle\}\!\}\rangle}$$

Fig. 20. The semantics for sequential program statements. In the Call rule, we suppose that the valuation ℓ is obtained by assigning the call arguments of e to the parameters of procedure p.

Theorem 5. *The coverability problem for event-driven asynchronous programs (with unbounded identifier storage) is undecidable.*

Proof. By reduction from the language emptiness problem for Turing machines: given a Turing machine $\mathcal{M}$, we construct the program P_M of Figure 22 which simulates P_M according to Figure 21: for each tape cell we have one running task which executes the procedure `cell`; we assume no limit on the maximum number of threads, $T = \omega$, nor on the running tasks per buffer, $T_k = \omega$. The initial condition dictates that one active task executing `cell` begins with `state` set to the initial state $q_0 \in Q$ of P_M. By answering whether the procedure `reached` (whose code is irrelevant and therefore not given) can be executed in some execution of P_M, we thus answer whether q_f is reachable in $\mathcal{M}$.
□

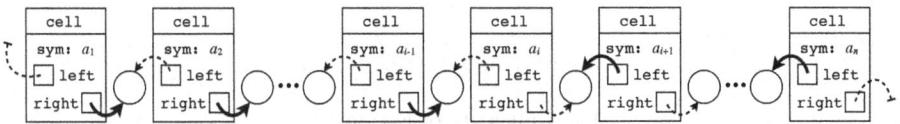

Fig. 21. Encoding a Turing machine tape. Several waiting `cell` tasks, drawn as rectangles, maintain two event variables, `left` and `right`. Arrows depict references to event identifiers, which are drawn as circles; bold arrows originate from tasks waiting on a given event, while dashed arrows denote an otherwise-stored event identifier. The only non-waiting `cell` task is the one pointed to by the tape head; cells to the left (resp., right) wait for their `right` (resp., `left`) event to be signaled.

```
1  proc cell ()
2      var symbol: Σ
3      var left, right: IDs
4
5      // initialize this cell,
6      // and its left neighbor
7      symbol := ⋆;
8      right := event;
9      if ⋆ then
10         left := new event;
11         _ := task cell() {left} ⊥;
12         sync left;
13     wait {left};
14     else left := ⊥
15
16     while true do
17         // test reachability
18         if state = qf then
19             call _ := reached();
20         // choose a transition
21         ...
22         // make the transition
23         ...
24     return
```

```
25  // TM-state stored in
26  // a global variable
27  var state: Q
28
29  // code to choose
30  // an enabled transition
31  let q1,q2: Q, a,b: Σ, d: {L,R} in
32  assume TX(q1,a,d,q2,b);
33  assume state = q1;
34  assume symbol = a;
35
36  // code to make the
37  // chosen transition
38  state := q2;
39  symbol := b;
40  if d = L then
41      sync left;
42      wait {left};
43  else
44      sync right;
45      wait {right};
```

Fig. 22. The program P_M simulating a Turing machine $M = \langle Q, \Sigma, q_0, q_f, \delta \rangle$ with states Q and alphabet Σ. The predicate $TX(q_1, a, d, q_2, b)$ holds for $\langle q_1, a, d, q_2, b \rangle \in \delta$. The assignment on Line 8 (described in Appendix A) assigns the event identifier on which the current task was activated. Note that here we assume a few trivial syntactic extensions, e.g., **let .. in**, for clarity; they are easily encoded into the rigid syntax of Section 2.

A Semantics for Propositions as Sessions

Sam Lindley and J. Garrett Morris

The University of Edinburgh
{Sam.Lindley,Garrett.Morris}@ed.ac.uk

Abstract. Session types provide a static guarantee that concurrent programs respect communication protocols. Recently, Caires, Pfenning, and Toninho, and Wadler, have developed a correspondence between propositions of linear logic and session typed π-calculus processes. We relate the cut-elimination semantics of this approach to an operational semantics for session-typed concurrency in a functional language. We begin by presenting a variant of Wadler's session-typed core functional language, GV. We give a small-step operational semantics for GV. We develop a suitable notion of deadlock, based on existing approaches for capturing deadlock in π-calculus, and show that all well-typed GV programs are deadlock-free, deterministic, and terminating. We relate GV to linear logic by giving translations between GV and CP, a process calculus with a type system and semantics based on classical linear logic. We prove that both directions of our translation preserve reduction; previous translations from GV to CP, in contrast, failed to preserve β-reduction. Furthermore, to demonstrate the modularity of our approach, we define two extensions of GV which preserve deadlock-freedom, determinism, and termination.

1 Introduction

From massively distributed programs running across entire data centres, to hand-held apps reliant on remote services for functionality, concurrency has become a critical aspect of modern programs, and thus a central problem in program correctness. Assuring correct concurrent behaviour requires reasoning not just about the types of data communicated, but the order in which the communication takes place. For example, the messages between an SMTP client and server are all strings, but a client that sends the recipient's address before the sender's address is in violation of the protocol despite sending the correct type of data.

Session types, originally proposed by Honda [13], provide a mechanism to reason about the state of channel-based communication. The type of a channel captures the expected behaviour of a process communicating on that channel. For example, we might express a simplified session type for an SMTP client as:

$$!FromAddress.!ToAddress.!Message.\mathsf{end}$$

where $!T.S$ is the type of a channel that sends a value of type T, then continues with behaviour specified by S. An important feature of session types is duality: the session type of an SMTP server is the dual of the session type of the client:

$$?FromAddress.?ToAddress.?Message.\mathsf{end}$$

© Springer-Verlag Berlin Heidelberg 2015
J. Vitek (Ed.): ESOP 2015, LNCS 9032, pp. 560–584, 2015.
DOI: 10.1007/978-3-662-46669-8_23

where $?T.S$ is the type of a channel that sends a value of type T, then continues with behaviour specified by S. Honda originally defined session types for process calculi; recent work [10, 25] has investigated the use of session types for concurrency in functional languages.

Session type systems are necessarily substructural—if processes can freely discard or duplicate channels, then the type system cannot guarantee that observable messages on channels match their expected types. Recent work seeks to establish a correspondence between session types and linear logic, an archetypal substructural logic for reasoning about state. Caires and Pfenning [5] develop a correspondence between cut elimination in intuitionistic linear logic and process reduction in a session-typed process calculus. Wadler [26] adapts their approach to classical linear logic, emphasising the role of duality in typing; the semantics of his system is given directly by the cut elimination rules of classical linear logic. He gives a type-preserving translation from a simple functional calculus (GV), inspired by Gay and Vasconcelos [10], to a process calculus (CP), inspired by Caires and Pfenning [5]. However, he gives no semantics for GV other than by translation to CP.

In this paper, we develop a session-typed functional core calculus, also called GV. (Our language shares most of the distinctive features of Wadler's, although it differs in some details.) We present a small-step operational semantics for GV, factored into functional and concurrent portions following the approach of Gay and Vasconcelos [10]. The functional portion of our semantics differs from standard presentations of call-by-value reduction only in that we adopt a weak form of explicit substitution to enable a direct correspondence with cut reduction. The concurrent portion of our semantics includes the typical reductions and equivalences of π-calculus-like process calculi.

Ultimately, our goal is to build and reason about functional programming languages extended with session types. Thus GV is a natural fit. Indeed, we are currently implementing an asynchronous variant of GV as part of the Links web programming language [8]. Developing a direct semantics for GV provides a number of benefits over relying on the translation to CP.

- It provides a simple semantic characterisation of deadlock. Unlike Wadler's proof of deadlock freedom, ours does not depend on normalisation, and thus extends to non-terminating processes.
- The proof technique itself is modular: as illustrated by the extensions, the same technique can be applied to practical (sometimes non-logical) extensions of the language.
- Compared to cut-elimination in CP, the GV semantics is much closer to something one might actually want to implement in practice, as witnessed by our Links implementation.

We believe in modularity, and so re-use as much of the standard linear lambda calculus machinery as possible, while limiting non-standard extensions. The paper proceeds as follows.

- We define a core linearly-typed functional language, GV, by extending linear lambda calculus with session-typed communication primitives (§2.1).

Session types $S ::= \; !T.S \mid ?T.S \mid \mathsf{end}_! \mid \mathsf{end}_? \mid S^\sharp$

Types $T, U ::= \; S \mid \mathbf{1} \mid T \times U \mid \mathbf{0} \mid T + U \mid T \multimap U$

Terms $L, M, N ::= \; x \mid K\,M \mid \lambda x.M \mid M\,N$
$\qquad\qquad\qquad \mid (M, N) \mid \mathsf{let}\;(x, y) = M \;\mathsf{in}\; N$
$\qquad\qquad\qquad \mid \mathsf{inl}\,M \mid \mathsf{inr}\,M \mid \mathsf{case}\;M\;\{\mathsf{inl}\,x \mapsto N; \mathsf{inr}\,x \mapsto N\}$
$\qquad\qquad\qquad \mid () \mid \mathsf{let}\;() = M \;\mathsf{in}\; N \mid \mathsf{absurd}\;M$

Constants $K ::= \; \mathsf{send} \mid \mathsf{receive} \mid \mathsf{fork} \mid \mathsf{wait} \mid \mathsf{link}$

Fig. 1. Syntax of GV Terms and Types

We present an (untyped) synchronous operational semantics for GV (§2.2). We characterise deadlock and normal forms; we show that typed terms are deadlock-free, that closed typed terms evaluate to normal forms (§2.3), and that evaluation is deterministic and terminating (§2.4).

- We connect GV to the interpretation of session types as linear logic propositions, by establishing a correspondence between the semantics of GV and that of CP. We begin by introducing CP (§3.1). We show that we can simulate CP reduction in GV (§3.2), and GV reduction in CP (§3.3). (As π-calculus-like process calculi provide substitution only for names, not entire process expressions, the latter depends crucially on the use of weak explicit substitutions in the semantics of GV lambda abstractions.)
- We consider two extensions of GV: one which has a single self-dual type for closed channels, harmonising the treatment of closed channels with that of other session-typed calculi (§4.1), and another which adds unlimited types and replicated behaviour (§4.2). We show that these extensions preserve the essential meta theoretic properties of the core language.

We conclude by discussing related (§5) and future (§6) work.

2   A Session-Typed Functional Language

2.1   Syntax and Typing

Figure 1 gives the syntax of GV types and terms. The types T include nullary ($\mathbf{0}$) and binary ($T + U$) linear sums, nullary ($\mathbf{1}$) and binary ($T \times U$) linear products, and linear implication ($T \multimap U$). We frequently write $M; N$ as the elimination form of $\mathbf{1}$ in place of the more verbose $\mathsf{let}\;() = M\;\mathsf{in}\;N$. Session types S include input ($?T.S$), output ($!T.S$), and closed channels ($\mathsf{end}_!$, $\mathsf{end}_?$). We also include a type $S^\sharp$ of channels; values of channel type cannot be used directly in terms, but will appear in the typing of thread configurations. The terms are the standard λ-calculus terms, augmented with constructs for pairs and sums. Figure 2 gives both typing rules and type schemas for the constants. Note that core GV judgements are linear, i.e., not subject to weakening or contraction.

Concurrency. Concurrent behaviour is provided by the constants. Communication is provided by send and receive. For example (assuming an extension of our core language with numbers and arithmetic operators), a program M that

Typing rules

$$\frac{T \neq S^\sharp}{x : T \vdash x : T} \qquad\qquad \frac{K : T \multimap U \quad \Gamma \vdash M : T}{\Gamma \vdash K\,M : U}$$

$$\frac{\Gamma, x : T \vdash M : U}{\Gamma \vdash \lambda x.M : T \multimap U} \qquad\qquad \frac{\Gamma \vdash M : T \multimap U \quad \Gamma' \vdash N : T}{\Gamma, \Gamma' \vdash M\,N : U}$$

$$\frac{\Gamma \vdash M : T \quad \Gamma' \vdash N : U}{\Gamma, \Gamma' \vdash (M, N) : T \times U} \qquad \frac{\Gamma \vdash M : T \times T' \quad \Gamma', x : T, y : T' \vdash N : U}{\Gamma, \Gamma' \vdash \mathsf{let}\ (x, y) = M\ \mathsf{in}\ N : U}$$

$$\frac{\Gamma \vdash M : T}{\Gamma \vdash \mathsf{inl}\ M : T + U} \qquad \frac{\Gamma \vdash M : T + T' \quad \Gamma', x : T \vdash N : U \quad \Gamma', x : T' \vdash N' : U}{\Gamma, \Gamma' \vdash \mathsf{case}\ M\ \{\mathsf{inl}\ x \mapsto N ; \mathsf{inr}\ x \mapsto N'\} : U}$$

$$\frac{}{\vdash () : 1} \qquad \frac{\Gamma \vdash M : 1 \quad \Gamma' \vdash N : T}{\Gamma, \Gamma' \vdash \mathsf{let}\ () = M\ \mathsf{in}\ N : T} \qquad \frac{\Gamma \vdash M : 0}{\Gamma, \Gamma' \vdash \mathsf{absurd}\ M : T}$$

Type schemas for constants

$$\mathsf{send} : T \times {!T.S} \multimap S \qquad \mathsf{receive} : {?T.S} \multimap T \times S \qquad \mathsf{fork} : (S \multimap \mathsf{end}_!) \multimap \overline{S}$$

$$\mathsf{wait} : \mathsf{end}_? \multimap 1 \qquad\qquad \mathsf{link} : S \times \overline{S} \multimap \mathsf{end}_!$$

Duality

$$\overline{!T.S} = {?T.\overline{S}} \qquad \overline{?T.S} = {!T.\overline{S}} \qquad \overline{\mathsf{end}_!} = \mathsf{end}_? \qquad \overline{\mathsf{end}_?} = \mathsf{end}_!$$

Fig. 2. GV Typing Rules

receives a pair of numbers along channel z and then sends their sum along the same channel can be expressed as

$$M \triangleq \mathsf{let}\ ((x, y), z) = \mathsf{receive}\ z\ \mathsf{in}\ \mathsf{send}\ (x + y, z)$$

(where the interpretation of nested patterns by sequences of bindings is standard). Channels are treated linearly in GV. Thus, receive returns not only the received value (the pair of x and y) but also a new copy of the channel used for receiving z; similarly, send returns a copy of the channel used for sending. Thus, the term above is well-typed in the context $z : {?(Int \times Int).!Int.S}$, and evaluates to a channel of type S. Session initiation is provided by fork. If f is a function from a channel of type S to a closed channel (of type $\mathsf{end}_!$), then fork f forks a new thread in which f is applied to a fresh channel of type S, and returns a channel of type $\overline{S}$ in order to communicate with the thread. For example, the term fork $(\lambda z.M)$ returns a channel of type $!(Int \times Int).?Int.\mathsf{end}_?$. Given a thread created by fork f, the channel returned from f is closed by fork, whereas the other end of the channel must be closed by calling wait. A client of the process M can be defined as follows:

$$N \triangleq \mathsf{let}\ z = \mathsf{send}\ ((6, 7), z)\ \mathsf{in}\ \mathsf{let}\ (x, z) = \mathsf{receive}\ z\ \mathsf{in}\ \mathsf{wait}\ z; x$$

The combined process let $x = $ fork $(\lambda z.M)$ in N evaluates to 13. The expression link (x, y) forwards messages sent on x to be received on y and vice versa. We choose to include it as a primitive as it corresponds to the axiom rule of linear logic, which is standard in logical accounts of session types.

Choice. In addition to input and output, typical session type systems also provide session types representing internal $(S_1 \oplus S_2)$ and external $(S_1 \mathbin{\&} S_2)$ choice (also known as selection and branching, respectively). For example, we might write a process that can either sum two numbers or negate one:

$$\text{offer } z \ \{ \ \text{inl } z \mapsto \text{let } ((x, y), z) = \text{receive } z \text{ in send } (x + y, z) \\ \text{inr } z \mapsto \text{let } (x, z) = \text{receive } z \text{ in send } (-x, z) \ \}$$

This term initially requires $z : (?(Int \times Int).!Int.S) \mathbin{\&} (?Int.!Int.S)$. A client of this process begins by choosing which branch of the session to take; for example, we can extend the preceding example as follows:

$$\text{let } z = \text{select inl } z \text{ in let } z = \text{send } ((6, 7), z) \text{ in let } (x, z) = \text{receive } z \text{ in wait } z; x$$

While we would expect a surface language to include selection and branching, we omit them from our core calculus. Instead, we show that they are macro-expressible using the linear sum type. The intuition is that selection is implemented by sending a suitably tagged process, while branching is implemented by a term-level branch on a received value. Concretely, we define the types by:

$$S_1 \oplus S_2 \triangleq !(\overline{S_1} + \overline{S_2}).\text{end}_! \qquad S_1 \mathbin{\&} S_2 \triangleq ?(S_1 + S_2).\text{end}_?$$

Note that we have the expected duality relationship: $\overline{S_1 \oplus S_2} = \overline{S_1} \mathbin{\&} \overline{S_2}$. We implement select and offer as follows (where ℓ ranges over $\{\text{inl}, \text{inr}\}$):

$$\text{select } \ell \, M \triangleq \text{fork}(\lambda x.\text{send } (\ell \, x, M))$$
$$\text{offer } M \ \{\text{inl } x \mapsto P; \text{inr } x \mapsto Q\} \triangleq \text{let } (x, y) = \text{receive } M \text{ in} \\ \text{wait } y; \text{case } x \ \{\text{inl } x \mapsto P; \text{inr } x \mapsto Q\}$$

Correspondingly, nullary choice and selection are encoded using the $\mathbf{0}$ type:

$$\oplus\{\} \triangleq !\mathbf{0}.\text{end}_! \qquad \mathbin{\&}\{\} \triangleq ?\mathbf{0}.\text{end}_?$$

$$\text{offer } M \ \{\} \triangleq \text{let } (x, y) = \text{receive } M \text{ in wait } y; \text{absurd}\{\}$$

2.2   Semantics

Following Gay and Vasconcelos [10], we factor the semantics of GV into a (deterministic) reduction relation on terms (called $\longrightarrow_V$) and a (non-deterministic) reduction on configurations of processes (called $\longrightarrow$). Figure 3 gives the syntax of values, configurations, and evaluation and configuration contexts.

Terms. To preserve a close connection between the semantics of our term language and cut-reduction in linear logic, we define term reduction using weak

Values	$V, W ::= x \mid \lambda^\sigma x.M$
	$\mid \ () \mid (V, W) \mid \mathsf{inl}\ V \mid \mathsf{inr}\ V$
Substitutions	$\sigma ::= \{ V_1/x_1, \ldots, V_n/x_n \}$
	where the x_i are pairwise distinct
Evaluation contexts	$E ::= [\,] \mid E\ M \mid V\ E \mid K\ E \mid \mathsf{let}\ () = E\ \mathsf{in}\ M$
	$\mid (E, M) \mid (V, E) \mid \mathsf{let}\ (x, y) = E\ \mathsf{in}\ M$
	$\mid\ \mathsf{inl}\ E \mid \mathsf{inr}\ E \mid \mathsf{case}\ E\ \{\mathsf{inl}\ x \mapsto N; \mathsf{inr}\ x \mapsto N'\}$
	$F ::= \phi E$
Configurations	$C, D ::= \phi M \mid C \parallel C' \mid (\nu x) C$
Configuration contexts	$G ::= [\,] \mid G \parallel C \mid (\nu x) G$
Flags	$\phi ::= \circ \mid \bullet$

Fig. 3. Syntax of Values, Configurations, and Contexts

explicit substitutions [18]. In this approach, we intercept substitutions at λ-terms rather than immediately applying them to the body of the term. Thus, our language of terms includes closures $\lambda^\sigma x.M$, where σ provides the intercepted substitution. We extend the typing judgement to include closures, as follows:

$$\frac{\Gamma, x : T \vdash M\sigma : U \quad dom(\sigma) = (fv(M) \setminus \{x\})}{\Gamma \vdash \lambda^\sigma x.M : T \multimap U}$$

The free variables of a closure $\lambda^\sigma x.M$ are the free variables of the range of σ, not the free variables of M. The capture avoiding substitution $M\sigma$ of σ applied to M is defined as usual on the free variables of M. Note that the side condition on the domain of σ is preserved under substitution. We implicitly treat plain lambda abstractions $\lambda x.M$ as closures $\lambda^\sigma x.M$, where σ is a renaming substitution restricted to the free variables of M less $\{x\}$; concretely:

$$\lambda x.M \triangleq \lambda^\sigma x.(M\sigma')$$
where $fv(M) \setminus \{x\} = \{x_1, \ldots, x_n\}$ $y_1, \ldots, y_n$ are fresh variables
$$\sigma = \{x_1/y_1, \ldots, x_n/y_n\} \qquad \sigma' = \{y_1/x_1, \ldots, y_n/x_n\}$$

We lift the typing judgement on terms pointwise to substitutions:

$$\frac{\Gamma_1 \vdash \sigma(x_1) : \Delta(x_1) \cdots \Gamma_k \vdash \sigma(x_k) : \Delta(x_k) \quad dom(\sigma) = dom(\Delta)}{\Gamma_1, \ldots, \Gamma_k \vdash \sigma : \Delta}$$

Configurations. The grammar of configurations includes the usual π-calculus forms for composition and name restriction. However, because functional computations return values (which may, in turn, contain channels), we distinguish between the "main" thread $\bullet M$ (which returns a value) and the threads $\circ M$ created by fork (which do not).

Reduction. Reduction rules for terms and configurations, and equivalences for configurations, are given in Figure 4. Term reduction ($\longrightarrow_V$) implements

---

Term reduction

$$(\lambda^\sigma x.M)\,V \longrightarrow_{\text{v}} M(\{V/x\} \uplus \sigma)$$
$$();M \longrightarrow_{\text{v}} M$$
$$\text{let } (x,y) = (V,V') \text{ in } M \longrightarrow_{\text{v}} M\{V/x, V'/y\}$$
$$\text{case (inl } V)\,\{\text{inl } x \mapsto N; \text{inr } x \mapsto N'\} \longrightarrow_{\text{v}} N\{V/x\}$$
$$E[M] \longrightarrow_{\text{v}} E[M'] \qquad \text{if } M \longrightarrow_{\text{v}} M'$$

Configuration equivalence

$$F[\text{link}\,(x,y)] \equiv F[\text{link}\,(y,x)] \qquad C \parallel D \equiv D \parallel C \qquad C \parallel (D \parallel E) \equiv (C \parallel D) \parallel E$$

$$C \parallel (\nu x)D \equiv (\nu x)(C \parallel D) \text{ if } x \notin \mathit{fv}(C) \qquad\qquad G[C] \equiv G[D] \text{ if } C \equiv D$$

Configuration reduction

SEND

$$\overline{(\nu x)(F[\text{send}\,(V,x)] \parallel F'[\text{receive}\,x]) \longrightarrow (\nu x)(F[x] \parallel F'[(V,x)])}$$

LIFT
$$\frac{C \longrightarrow C'}{G[C] \longrightarrow G[C']}$$

FORK
$$\frac{x \text{ is a fresh channel name}}{F[\text{fork}\,(\lambda^\sigma y.M)] \longrightarrow (\nu x)(F[x] \parallel M(\{x/y\} \uplus \sigma))}$$

WAIT
$$\overline{(\nu x)(F[\text{wait}\,x] \parallel \phi x) \longrightarrow F[()]}$$

LINK
$$\frac{x \in \mathit{fv}(M)}{(\nu x)(F[\text{link}\,(x,y)] \parallel F'[M]) \longrightarrow (\nu x)(F[x] \parallel F'[\text{wait}\,x; M\{y/x\}])}$$

LIFTV
$$\frac{M \longrightarrow_{\text{v}} M'}{G[M] \longrightarrow G[M']}$$

Fig. 4. Reduction Rules and Equivalences for Terms and Configurations

call-by-value, left-to-right evaluation. Configuration equivalence ($\equiv$) is standard. Communication is provided by SEND and session initiation by FORK. Rule WAIT combines synchronisation of closed channels with garbage collection of the associated name restriction. Rule LINK is complicated by the need to produce a channel of type end$_!$; the inserted wait synchronises with the produced channel.

Relation Notation. We write $R\,R'$ for sequential composition and $R \cup R'$ for union of R and R'. We write R^+ for transitive closure and $R^\star$ for the reflexive, transitive closure of R.

Configuration Typing. Our syntax of configurations permits various forms of deadlocked configurations. For example, if we define the terms M and N by

$$M \triangleq \text{let } (z,y) = \text{receive } y \text{ in} \qquad\qquad N \triangleq \text{let } (z,x) = \text{receive } x \text{ in}$$
$$\text{let } x = \text{send}\,(z,x) \text{ in } M' \qquad\qquad\qquad \text{let } y = \text{send}\,(z,y) \text{ in } N'$$

given suitable terms M' and N', then it is apparent that configurations such as $(\nu xy)M$, $(\nu xy)(M \parallel M)$ and $(\nu xy)(M \parallel N)$ cannot reduce further, even though M and N can be individually well-typed. To exclude such cases, we provide a type discipline for configurations (Figure 5). It is based on type systems for linear π-calculus [17], but with two important differences.

Configuration typing

$$\frac{\Gamma \vdash M : T \quad T \neq \mathsf{end}_!}{\Gamma \vdash^\bullet \bullet M} \qquad \frac{\Gamma \vdash M : \mathsf{end}_!}{\Gamma \vdash^\circ \circ M} \qquad \frac{\Gamma, x : S^\sharp \vdash^\phi C}{\Gamma \vdash^\phi (\nu x) C}$$

$$\frac{\Gamma, x : S \vdash^\phi C \quad \Gamma', x : \overline{S} \vdash^{\phi'} C'}{\Gamma, \Gamma', x : S^\sharp \vdash^{\phi+\phi'} C \parallel C'}$$

Combination of flags

$$\circ + \circ = \circ \qquad \circ + \bullet = \bullet \qquad \bullet + \circ = \bullet \qquad \bullet + \bullet \text{ undefined}$$

Fig. 5. Configuration Typing

- First, we seek to assure that there is at most one main thread. This constraint is enforced by the flags ($\bullet$ and $\circ$) on the derivations: a derivation $\Gamma \vdash^\bullet C$ indicates that configuration C contains the main thread, while $\Gamma \vdash^\circ C$ indicates that C does not contain the main thread. We write $\Gamma \vdash C$ to abbreviate $\exists \phi. \Gamma \vdash^\phi C$, that is, C may include a main thread.
- Second, we require that exactly one channel is shared at each composition of processes. This is more restrictive than standard type systems for linear π-calculus, which allow an arbitrary number of channels (including none) to be shared at a composition of processes.

Notice that the above stuck examples are ill-typed in this system: $(\nu x y) M$ because y must have a type $S^\sharp$ in M; $(\nu x y)(M \parallel M)$ because there is no type $S^\sharp$ such that both S and $\overline{S}$ are of the form $?T.S'$, as required by receive; and, $(\nu x y)(M \parallel N)$ because both x and y must be shared between M and N, but the typing rule for composition only allows one channel to be shared.

Now we can show that reduction preserves typing. We begin with terms.

Lemma 1. *If $\Gamma \vdash M : T$ and $M \longrightarrow_V M'$, then $\Gamma \vdash M' : T$*

The proof is by induction on M; the cases are all standard. We now extend this result to configurations.

Theorem 2. *If $\Gamma \vdash C$ and $C \longrightarrow C'$ then $\Gamma \vdash C'$.*

The proof is by induction on the derivation of $C \longrightarrow C'$.

Typing and Configuration Equivalence. Alas, our notion of typing is not preserved by configuration equivalence. For example, assume that $\Gamma \vdash (\nu x y)(C \parallel (D \parallel E))$, where $x \in fv(C), y \in fv(D)$, and $x, y \in fv(E)$. We have that $C \parallel (D \parallel E) \equiv (C \parallel D) \parallel E$, but $\Gamma \nvdash (\nu x y)((C \parallel D) \parallel E)$, as both x and y must be shared between the processes $C \parallel D$ and E. However, we can show that starting from a well-typed configuration, we need never rely on an ill-typed equivalent configuration to expose possible reductions.

Theorem 3. *If* $\Gamma \vdash C$, $C \equiv C'$ *and* $C' \longrightarrow D'$, *then there exists* D *such that* $D \equiv D'$, *and* $\Gamma \vdash D$.

Proof. Observe that if $\Gamma \vdash C$, then for any pair of terms M_1, M_2 appearing in C, there are environments Γ_1, Γ_2 and types T_1, T_2 such that $\Gamma_1 \vdash M_1 : T_1, \Gamma_2 \vdash M_2 : T_2$, and (because of the typing rule for composition) Γ_1 and Γ_2 share at most one variable. By examination of the reduction rules, we can conclude that there are well-typed C_0, D_0 such that $C' = G[C_0]$, $C_0 \longrightarrow D_0$ and $D' = G[D_0]$. The result then follows by structural induction on C, examining the possible equivalences in each case.   □

We extend Theorem 3 to sequences of reductions, defining $\Longrightarrow$ as $(\equiv\longrightarrow\equiv)^\star$.

Corollary 4. *If* $\Gamma \vdash C$ *and* $C \Longrightarrow D$, *then there exists* D' *such that* $D \equiv D'$, *and* $\Gamma \vdash D'$.

2.3   Deadlock and Its Absence

In the previous section, we saw examples of deadlocked terms which were rejected by our type system. We now present a general account of deadlock: we characterise deadlocked configurations, and show that well-typed configurations do not evaluate to deadlocked configurations.

We begin by observing that many examples of stuck configurations are already excluded by existing session-typing disciplines: in particular, those configurations in which either too many or too few threads attempt to synchronise on a given channel, or those with inconsistent use of channels. The cases of interest to us are those in which the threads individually obey the session-typing discipline, but the order of synchronisation in the threads creates deadlock. We say that a thread M is blocked on a channel x, written $\mathsf{blocked}(x, M)$, if M has evaluated to some context surrounding a communication primitive applied to x:

$$\mathsf{blocked}(x, M) \stackrel{\text{def}}{\Longleftrightarrow} \exists N.\, M = E[\mathsf{send}\,(N, x)] \vee M = E[\mathsf{receive}\,x] \vee M = E[\mathsf{wait}\,x]$$

In such a case, M can only reduce further in composition with another thread blocked on x, and any communication on other channels in M will be delayed until a communication on x has occurred. We abstract over the property that y depends on x in M, abbreviated $\mathsf{depends}(x, y, M)$; in other words, M is blocked on x, but has y as one of its (other) free variables. We extend this notion of dependency from single threads to configurations of threads, with the observation that in a larger configuration intermediate channels may participate in the dependency.

$$\mathsf{depends}(x, y, E[M]) \stackrel{\text{def}}{\Longleftrightarrow} \mathsf{blocked}(x, M) \wedge y \in fv(E[M])$$
$$\mathsf{depends}(x, y, C) \stackrel{\text{def}}{\Longleftrightarrow} (C \equiv G[M] \wedge \mathsf{depends}(x, y, M)) \vee (C \equiv G[D \parallel D']$$
$$\wedge\,(\exists z.\mathsf{depends}(x, z, D) \wedge \mathsf{depends}(z, y, D')))$$

We now define deadlocked configurations as those with cyclic dependencies:

$$\mathsf{deadlocked}(C) \stackrel{\text{def}}{\Longleftrightarrow} C \equiv G[D \parallel D'] \wedge \exists x, y.\mathsf{depends}(x, y, D) \wedge \mathsf{depends}(y, x, D')$$

Because the definition of dependency permits intermediate channels, this definition encompasses cycles involving an arbitrary number of channels. We say that a configuration C is deadlock free if, for all D such that $C \Longrightarrow D$, $\neg\mathsf{deadlocked}(D)$. Observe that if $C \equiv D$, $\mathsf{deadlocked}(C) \Longleftrightarrow \mathsf{deadlocked}(D)$.

At this point, we can observe that in any deadlocked configuration there must be a composition of configurations that shares more than one channel. This is precisely the situation that is excluded by our configuration type system.

Lemma 5. *If $\Gamma \vdash C$, and $C = G[D \parallel D']$, then $fv(D) \cap fv(D') = \{x\}$ for some variable x.*

Proof. By structural induction on the derivation of $\Gamma \vdash C$; the only interesting case is for parallel composition, where the desired result is assured by the partitioning of the environment. ☐

To extend this observation to deadlock freedom, we must take equivalence into account. While it is true that equivalence need not preserve typing, there are no equivalence rules that affect the free variables of individual threads. Thus, cycles of dependent channels are preserved by equivalence.

Lemma 6. *If $\Gamma \vdash C$ then $\neg\mathsf{deadlocked}(C)$.*

Proof. By contradiction. Suppose $\mathsf{deadlocked}(C)$, then by expanding the definition of $\mathsf{deadlocked}$ we know that there must exist variables $x_1, \ldots, x_n$ and processes $M_1, \ldots, M_n$ in C such that:

$$\mathsf{depends}(x_1, x_2, M_1) \land \mathsf{depends}(x_2, x_3, M_2) \land \cdots \land \mathsf{depends}(x_n, x_1, M_n)$$

Either $n = 1$, which violates linearity, or configuration C must partition the cycle. However, any cut of the cycle is crossed by at least two channels, so C must be ill-typed by Lemma 5. ☐

Finally, we can combine the previous result with preservation of typing to show that well-typed terms never evaluate to deadlocked configurations.

Theorem 7. *If $\Gamma \vdash M : T$, then $\bullet M$ is deadlock-free.*

Proof. If $\Gamma \vdash M : T$, then $\Gamma \vdash \bullet M$, and so $\neg\mathsf{deadlocked}(\bullet M)$ and, for any D such that $\bullet M \Longrightarrow D$, we know that there is a well-typed $D' \equiv D$, and so $\neg\mathsf{deadlocked}(D)$. ☐

Progress and Canonical Forms. We conclude this section by describing a canonical form for configurations, and characterising the stuck terms resulting from the evaluation of well-typed terms. One might hope that evaluation of a well-typed term would always produce a value; however, this is complicated because terms may return channels. For a simple example, consider the term:

$$\bullet \mathsf{fork}\,(\lambda x.\mathsf{let}\,(y, x) = \mathsf{receive}\,x\,\mathsf{in}\,\mathsf{send}\,(y, x))$$

This term spawns a thread (which simply echoes once), and then returns the resulting channel; thus, the result of evaluation is a configuration equivalent to:

$$(\nu x)(\bullet\, x \parallel \circ\, \mathsf{let}\ (y, x) = \mathsf{receive}\ x\ \mathsf{in}\ \mathsf{send}\ (y, x))$$

Clearly, no more evaluation is possible, even though the configuration still contains blocked threads. However, it turns out that we can show that evaluation of terms that do not return channels must always produce a value (Corollary 12).

Definition 8. *A process C is in canonical form if there is some sequence of variables $x_1, \ldots, x_{n-1}$ and terms $M_1, \ldots, M_n$ such that:*

$$C = (\nu x_1)(\circ\, M_1 \parallel (\nu x_2)(\circ\, M_2 \parallel \cdots \parallel (\nu x_{n-1})(\circ\, M_{n-1} \parallel \phi M_n) \ldots))$$

Note that canonical forms need not be unique. For example, consider the configuration $\vdash (\nu xy)(C \parallel D \parallel E)$ where $x \in fv(C)$, $y \in fv(D)$, and $x, y \in fv(E)$. Both $(\nu x)(C \parallel (\nu y)(D \parallel E))$ and $(\nu y)(D \parallel (\nu x)(C \parallel E))$ are canonical forms of the original configuration. We can show that any well-typed term must be equivalent to a term in canonical form; again, the key insight is that captured by Lemma 5: if any two sub-configurations share at most one channel, then we can order the threads by the channels they share.

Lemma 9. *If $\Gamma \vdash C$, then there is some $C' \equiv C$ such that $\Gamma \vdash C'$ and C' is in canonical form.*

The proof is by induction on the count of bound variables.

We can now state some progress results. We begin with open configurations: each thread must be blocked on either a free variable or a ν-bound variable.

Theorem 10. *Let $\Gamma \vdash C$, $C \not\longrightarrow$ and let $C' = (\nu x_1)(\circ\, M_1 \parallel (\nu x_2)(\circ\, M_2 \parallel \cdots \parallel (\nu x_{n-1})(\circ\, M_{n-1} \parallel \phi M_n) \ldots))$ be a canonical form of C. Then:*

1. *For $1 \le i \le n-1$ either $\mathsf{blocked}(x_j, M_i)$ where $j \le i$ or $\mathsf{blocked}(y, M_i)$ for some $y \in dom(\Gamma)$; and,*
2. *Either M_n is a value or $\mathsf{blocked}(y, M_n)$ for some $y \in \{x_i \mid 1 \le i \le n-1\} \cup dom(\Gamma)$.*

Proof. By induction on the derivation of $\Gamma \vdash C'$, using the definition of $\longrightarrow$. $\square$

We can strengthen the result significantly when we move to configurations without free variables. To see why, consider just the first two threads of a configuration $(\nu x_1)(M_1 \parallel (\nu x_2)(M_2 \parallel \ldots))$. As there are no free variables, thread M_1 can only be blocked on x_1. Now, from the previous result, thread M_2 can be blocked on either x_1 or x_2. But, were it blocked on x_1, it could reduce with thread M_1; we can conclude it is blocked on x_2. Generalising this observation gives the following progress result.

Theorem 11. *Let $\vdash C$, $C \not\longrightarrow$ and let $C' = (\nu x_1)(\circ\, M_1 \parallel (\nu x_2)(\circ\, M_2 \parallel \cdots \parallel (\nu x_{n-1})(\circ\, M_{n-1} \parallel \phi M_n) \ldots))$ be a canonical form of C. Then:*

1. *For* $1 \leq i \leq n-1$, blocked(x_i, M_i); *and,*
2. M_n *is a value.*

Proof. By induction on the derivation of $\vdash C'$, relying on Theorem 10.   □

Finally, observe that some subset of the variables $x_1, \ldots, x_n$ must appear in the result V. Therefore, if the original expression returns a value that does not contain any channels, it will evaluate to a configuration with no blocked threads (i.e., a single value).

Corollary 12. *Let* $\vdash C$, $C \not\longrightarrow$ *and* C' *be a canonical form of* C *such that the value returned by* C' *contains no channels, then* $C' = \phi V$ *for some value* V.

2.4 Determinism and Termination

It is straightforward to show that GV is deterministic. In fact, GV enjoys a strong form of determinism, called the *diamond property* [2].

Theorem 13. *If* $\Gamma \vdash C$, $C \equiv\longrightarrow\equiv D_1$, *and* $C \equiv\longrightarrow\equiv D_2$, *then either* $D_1 \equiv D_2$ *or there exists* D_3 *such that* $D_1 \equiv\longrightarrow\equiv D_3$, *and* $D_2 \equiv\longrightarrow\equiv D_3$.

Proof. First, observe that $\longrightarrow_V$ is deterministic, and furthermore configuration reductions always treat $\longrightarrow_V$ redexes linearly. This means we need only consider the interaction between different configuration reductions. Linear typing ensures that two configuration reductions cannot overlap. Furthermore, each configuration reduction is linear in the existing redexes, so we can straightforwardly perform the reductions in either order.   □

It is not hard to see that the system remains deterministic if we extend the functional part of GV with any well-typed confluent reduction rules at all.

Theorem 14 (Strong normalisation). *If* $\Gamma \vdash C$, *then there are no infinite* $\equiv\longrightarrow\equiv$ *reduction sequences beginning from* C.

To prove strong normalisation for core GV, one can use an elementary argument based on linearity. When we add replication (§4.2) and other features, a logical relations argument along the lines of that of Perez et al. [21] suffices. Weak normalisation (the existence of a finite reduction sequence to an irreducible configuration) also follows as a direct corollary of Theorem 23 and the cut-elimination theorem for classical linear logic.

3 Classical Linear Logic

3.1 The Process Calculus CP

Figure 6 gives the syntax and typing rules for the multiplicative-additive fragment of CP; we let Δ range over typing environments. CP types and duality are the standard propositions and duality function of classical linear logic, while

---

Syntax

$$\text{Types } A, B ::= A \otimes B \mid A \,\bindnasrepma\, B \mid 1 \mid \bot \mid A \oplus B \mid A \,\&\, B \mid 0 \mid \top$$
$$\text{Terms } P, Q ::= x \leftrightarrow y \mid \nu y\,(P \mid Q) \mid x(y).P \mid x[y].(P \mid Q)$$
$$\mid x[in_i].P \mid \text{case } x\,\{P; Q\} \mid x().P \mid x[].0 \mid \text{case } x\,\{\}$$

Duality

$$(A \otimes B)^\bot = A^\bot \,\bindnasrepma\, B^\bot \qquad 1^\bot = \bot \qquad (A \oplus B)^\bot = A^\bot \,\&\, B^\bot \qquad \top^\bot = 0$$
$$(A \,\bindnasrepma\, B)^\bot = A^\bot \otimes B^\bot \qquad \bot^\bot = 1 \qquad (A \,\&\, B)^\bot = A^\bot \oplus B^\bot \qquad 0^\bot = \top$$

Typing

$$\frac{}{x \leftrightarrow w \vdash x : A, w : A^\bot} \qquad \frac{P \vdash \Delta, y : A \quad Q \vdash \Delta', y : A^\bot}{\nu y\,(P \mid Q) \vdash \Delta, \Delta'} \qquad \frac{}{x[].0 \vdash x : 1}$$

$$\frac{P \vdash \Delta, y : A, x : B}{x(y).P \vdash \Delta, x : A \,\bindnasrepma\, B} \qquad \frac{P \vdash \Delta, y : A \quad Q \vdash \Delta', x : B}{x[y].(P \mid Q) \vdash \Delta, \Delta', x : A \otimes B} \qquad \frac{P \vdash \Delta}{x().P \vdash \Delta, x : \bot}$$

$$\frac{P \vdash \Delta, x : A_i}{x[in_i].P \vdash \Delta, x : A_1 \oplus A_2} \qquad \frac{P \vdash \Delta, x : A \quad Q \vdash \Delta, x : B}{\text{case } x\,\{P; Q\} \vdash \Delta, x : A \,\&\, B} \qquad \frac{}{\text{case } x\,\{\} \vdash \Delta, x : \top}$$

Fig. 6. CP Syntax and Typing

the terms are based on a subset of the π-calculus. The types $\&$ and $\oplus$ are interpreted as external and internal choice; the types $\bindnasrepma$ and $\otimes$ are interpreted as input and output, while their units $\bot$ and 1 are interpreted as nullary input and output. Note that CP's typing rules implicitly rebind identifiers: for example, in the hypothesis of the rule for $\bindnasrepma$, x identifies a proof of B, while in the conclusion it identifies a proof of $A \,\bindnasrepma\, B$.

CP includes two rules that are logically derivable: the axiom rule, which is interpreted as channel forwarding, and the cut rule, which is interpreted as process composition. Two of CP's terms differ from standard π-calculus terms. The first is composition—rather than having distinct name restriction and composition operators, CP provides one combined operator. This syntactically captures the restriction that composed processes must share exactly one channel. The second is output: the CP term $x[y].(P \mid Q)$ includes output, composition, and name restriction (the name y designates a new channel, bound in P).

A Simpler Send. The CP send rule is appealing because if one erases the terms it is exactly the classical linear logic rule for tensor. However, this correspondence comes at a price. Operationally, the process $x[y].(P \mid Q)$ does three things: it introduces a fresh variable y, it sends y to a freshly spawned process P, and in parallel it continues as process Q. This complicates both the reduction semantics of CP (as the cut reduction of $\otimes$ against $\bindnasrepma$ must account for all three behaviours) and the equivalence of CP and GV (where the behaviour of send is simpler).

Structural congruence

$$x \leftrightarrow w \equiv w \leftrightarrow x$$
$$\nu y \, (P \mid Q) \equiv \nu y \, (Q \mid P)$$
$$\nu y \, (P \mid \nu z \, (Q \mid R)) \equiv \nu z \, (\nu y \, (P \mid Q) \mid R), \quad \text{if } y \notin fv(R)$$
$$\nu x \, (P_1 \mid Q) \equiv \nu x \, (P_2 \mid Q), \quad \text{if } P_1 \equiv P_2$$

Primary cut reduction rules

$$\nu x \, (w \leftrightarrow x \mid P) \longrightarrow_C P[w/x]$$
$$\nu x \, (x[y].(P \mid Q) \mid x(y).R) \longrightarrow_C \nu x \, (Q \mid \nu y \, (P \mid R))$$
$$\nu x \, (x[].0 \mid x().P) \longrightarrow_C P$$
$$\nu x \, (x[in_i].P \mid \text{case } x \, \{Q_1; Q_2\}) \longrightarrow_C \nu x \, (P \mid Q_i)$$
$$\nu x \, (P_1 \mid Q) \longrightarrow_C \nu x \, (P_2 \mid Q), \quad \text{if } P_1 \longrightarrow_C P_2$$

Commuting conversions

$$\nu z \, (x[y].(P \mid Q) \mid R) \longrightarrow_{CC} x[y].(\nu z \, (P \mid R) \mid Q), \quad \text{if } z \notin fv(Q)$$
$$\nu z \, (x[y].(P \mid Q) \mid R) \longrightarrow_{CC} x[y].(P \mid \nu z \, (Q \mid R)), \quad \text{if } z \notin fv(P)$$
$$\nu z \, (x(y).P \mid Q) \longrightarrow_{CC} x(y).\nu z \, (P \mid Q)$$
$$\nu z \, (x().P \mid Q) \longrightarrow_{CC} x().\nu z \, (P \mid Q)$$
$$\nu z \, (x[in_i].P \mid Q) \longrightarrow_{CC} x[in_i].\nu z \, (P \mid Q)$$
$$\nu z \, (\text{case } x \, \{P; Q\} \mid R) \longrightarrow_{CC} \text{case } x \, \{\nu z \, (P \mid R); \nu z \, (Q \mid R)\}$$
$$\nu z \, (\text{case } x \, \{\} \mid Q) \longrightarrow_{CC} \text{case } x \, \{\}$$

Fig. 7. CP Congruences and Cut Reduction

Following Boreale [4], we can give an alternative formulation of send, avoiding the additional name restriction and composition, as follows:

$$\frac{P \vdash \Delta, x : B, y : A}{x\langle y\rangle.P \vdash \Delta, x : A \otimes B, y : A^\perp}$$

where $x\langle y\rangle.P$ is defined as $x[z].(y \leftrightarrow z \mid P)$. In particular, note that

$$\nu x \, (x\langle y\rangle.P \mid x(z).Q) = \nu x \, (x[z].(y \leftrightarrow z \mid P) \mid x(z).Q)$$
$$\longrightarrow_C \nu z \, (y \leftrightarrow z \mid \nu x \, (P \mid Q))$$
$$\longrightarrow_C \nu x \, (P \mid Q\{y/z\})$$

as we would expect for synchronising a send and a receive. Similarly, we note that any process $x[y].(P \mid Q)$ can also be expressed as a process $\nu y \, (P \mid x\langle y\rangle.Q)$, which reduces to the original by one application of the commuting conversions. However, the two formulations are not quite identical. Let us consider the possible reductions of the two terms. Notice that in $x[y].(P \mid Q)$, both P and Q are blocked on x; however, the same is not true for $\nu y \, (P \mid x\langle y\rangle.Q)$; the latter permits reductions in P before synchronising on x.

Cut Elimination. The semantics of CP terms are given by cut reduction, as shown in Figure 7. We write $fv(P)$ for the free names of process P. Terms are identified up to structural congruence $\equiv$ (as name restriction and composition are

On types

$$(\!|A \otimes B|\!) = !\overline{(\!|A|\!)}.(\!|B|\!) \quad (\!|1|\!) = \text{end}_! \quad (\!|A \oplus B|\!) = (\!|A|\!) \oplus (\!|B|\!) \quad (\!|0|\!) = \oplus\{\}$$
$$(\!|A \mathbin{\bindnasrepma} B|\!) = ?(\!|A|\!).(\!|B|\!) \quad (\!|\perp|\!) = \text{end}_? \quad (\!|A \mathbin{\&} B|\!) = (\!|A|\!) \mathbin{\&} (\!|B|\!) \quad (\!|\top|\!) = \mathbin{\&}\{\}$$

On terms

$$(\!|\nu x\,(P \mid Q)|\!) = \text{let } x = \text{fork } (\lambda x.(\!|P|\!)) \text{ in } (\!|Q|\!)$$
$$(\!|x \leftrightarrow y|\!) = \text{link } (x, y)$$
$$(\!|x[y].(P \mid Q)|\!) = \text{let } x = \text{send } (\text{fork } (\lambda y.(\!|P|\!)), x) \text{ in } (\!|Q|\!)$$
$$(\!|x(y).P|\!) = \text{let } (y, x) = \text{receive } x \text{ in } (\!|P|\!)$$
$$(\!|x[].0|\!) = x$$
$$(\!|x().P|\!) = \text{let } () = \text{wait } x \text{ in } (\!|P|\!)$$
$$(\!|x[\ell].P|\!) = \text{let } x = \text{select } l\ x \text{ in } (\!|P|\!)$$
$$(\!|\text{case } x\ \{P; Q\}|\!) = \text{offer } x\ \{\text{inl } x \mapsto (\!|P|\!); \text{inr } x \mapsto (\!|Q|\!)\}$$
$$(\!|\text{case } x\ \{\}|\!) = \text{let } (y, x) = \text{receive } x \text{ in absurd } y$$

$$\mathcal{C}(\!|\nu x\,(P \mid Q)|\!) = (\nu x)(\mathcal{C}(\!|P|\!) \parallel \mathcal{C}(\!|Q|\!))$$
$$\mathcal{C}(\!|P|\!) = \circ\,(\!|P|\!), \quad P \text{ is not a cut}$$

Fig. 8. Translation of CP Terms into GV

combined into one form, composition is not always associative). We write $\longrightarrow_C$ for the cut reduction relation, $\longrightarrow_{CC}$ for the commuting conversion relation, and $\longrightarrow$ for $\longrightarrow_C \cup \longrightarrow_{CC}$. The majority of the cut reduction rules correspond closely to synchronous reductions in π-calculus—for example, the reduction of $\&$ against $\oplus$ corresponds to the synchronisation of an internal and external choice. The rule for reduction of $\mathbin{\bindnasrepma}$ against $\otimes$ is more complex than synchronisation of input and output in GV, as it must also manipulate the implicit name restriction and composition in CP's output term. We write $\Longrightarrow$ for $(\equiv\longrightarrow\equiv)^+$, $\Longrightarrow_C$ for $(\equiv\longrightarrow_C\equiv)^+$, and $\Longrightarrow_{CC}$ for $\Longrightarrow_C \longrightarrow^\star_{CC}$.

Just as cut elimination in logic ensures that any proof may be transformed into an equivalent cut-free proof, the reduction rules of CP transform any term into a term blocked only on external communication—that is to say, if $P \vdash \Delta$, then $P \Longrightarrow_{CC} P'$ where $P' \neq \nu x\,(Q \mid Q')$ for any x, Q, Q'. The final commuting conversions play a central role in this transformation, moving any remaining internal communication after an external communication. However, note that the commuting conversions do not correspond to computational steps (i.e., any reduction rule in π-calculus).

3.2   Translation from CP to GV

In this section, we show that GV can simulate CP. Figure 8 gives the translation of CP into GV; typing environments are translated by the pointwise extension of the translation on types. We rely on our encoding of choice in GV (§2.1).

In translating CP terms to GV terms, the key observation is that CP terms contain their continuations; for example, the translation of input includes both a call to receive and the translation of the continuation. Additionally, the rebinding that is implicit in CP syntax is made explicit in GV. The translation $\mathcal{C}(\!|-|\!)$

Session types

$$[\![!T.S]\!] = [\![T]\!]^{\perp} \otimes [\![S]\!] \qquad [\![?T.S]\!] = [\![T]\!] \,\wp\, [\![S]\!] \qquad [\![\mathsf{end}_!]\!] = 1 \qquad [\![\mathsf{end}_?]\!] = \perp$$

Functional types

$$[\![T]\!] = [\![T]\!]^{\perp}, \quad \text{if } T \text{ is not a session type}$$

$$[\![\mathbf{0}]\!] = 0 \qquad\qquad\qquad [\![\mathbf{1}]\!] = 1$$
$$[\![T + U]\!] = [\![T]\!] \oplus [\![U]\!] \qquad [\![T \times U]\!] = [\![T]\!] \otimes [\![U]\!]$$
$$[\![T \multimap U]\!] = [\![T]\!]^{\perp} \,\wp\, [\![U]\!]$$
$$[\![S]\!] = [\![S]\!]$$

Fig. 9. Translation of GV Types into CP

translates top-level cuts to GV configurations; cuts that appear under prefixes are translated to applications of fork. As CP processes do not have return values, the translation of a CP process contains no main thread.

It is straightforward to see that the translation preserves typing; note that the channels in the CP typing environment become free variables in its translation.

Theorem 15. *If* $P \vdash \Delta$ *then* $(\!|\Delta|\!) \vdash^{\circ} \mathcal{C}(\!|P|\!)$.

Structural congruence in CP is a subset of the structural congruence relation for GV configurations; thus the translation trivially preserves congruence.

Theorem 16. *If* $P \equiv Q$, *then* $\mathcal{C}(\!|P|\!) \equiv \mathcal{C}(\!|Q|\!)$.

Finally, observe that the translation of any prefixed CP term is a GV thread of either the form $F[K\ M]$ for $K \in \{\mathsf{send}, \mathsf{receive}, \mathsf{wait}\}$ or is $\circ x$ for some variable x. Thus, we can see that any cut reduction immediately possible for a process P is similarly possible for $(\!|P|\!)$. Following such a reduction, several additional GV reductions may be necessary to expose the next possible communication, such as substituting the received values into the continuation in the case of the translation of input, or spawning new threads in the translation of composition.

Theorem 17. *If* $P \vdash \Delta$ *and* $P \longrightarrow_C Q$, *then* $\mathcal{C}(\!|P|\!) \longrightarrow^+ \mathcal{C}(\!|Q|\!)$.

Proof. By induction on P; the cases are all straightforward.

The commuting conversions in CP do not expose additional reductions, but are only necessary to assure that the result of evaluation does not have a cut at the top level. Our characterisation of deadlock freedom in GV has no such requirement, so we have no need for corresponding steps in GV.

3.3 Translation from GV to CP

In this section, we show that CP can simulate GV. Figure 9 gives the translation on types and Figure 10 gives the translation on terms, substitutions, and configurations; we translate type environments pointwise on types.

The translation on session types is homomorphic except for output, where the output type is dualised. This accounts for the discrepancy between $\overline{!T.S} = ?T.S$

Session terms

$$\llbracket \text{fork } M \rrbracket z = \nu w \ (w \leftrightarrow z \mid \nu x \ (\llbracket M \rrbracket x \mid \nu y \ (x\langle w\rangle.x \leftrightarrow y \mid y[])))$$
$$\llbracket \text{link } (M, N) \rrbracket z = \nu v \ (v \leftrightarrow z \mid \nu w \ (v \leftrightarrow w \mid \nu x \ (\llbracket M \rrbracket x \mid \nu y \ (\llbracket N \rrbracket y \mid w().x \leftrightarrow y))))$$
$$\llbracket \text{send } (M, N) \rrbracket z = \nu x \ (\llbracket N \rrbracket x \mid \nu y \ (\llbracket M \rrbracket y \mid x\langle y\rangle.x \leftrightarrow z))$$
$$\llbracket \text{receive } M \rrbracket z = \nu y \ (\llbracket M \rrbracket y \mid y(x).\nu w \ (w \leftrightarrow y \mid z\langle x\rangle.w \leftrightarrow z))$$
$$\llbracket \text{wait } M \rrbracket z = \nu y \ (y \leftrightarrow z \mid \llbracket M \rrbracket y)$$

Functional terms

$$\llbracket x \rrbracket z = x \leftrightarrow z$$
$$\llbracket \lambda^\sigma x.M \rrbracket z = \llbracket \sigma \rrbracket(z(x).\llbracket M \rrbracket z)$$
$$\llbracket L \ M \rrbracket z = \nu x \ (\llbracket M \rrbracket x \mid \nu y \ (\llbracket L \rrbracket y \mid y\langle x\rangle.y \leftrightarrow z))$$
$$\llbracket () \rrbracket z = z[]$$
$$\llbracket \text{let } () = M \text{ in } N \rrbracket z = \nu y \ (\llbracket M \rrbracket y \mid y().\llbracket N \rrbracket z)$$
$$\llbracket (M, N) \rrbracket z = \nu x \ (\llbracket M \rrbracket x \mid \nu y \ (\llbracket N \rrbracket y \mid z\langle x\rangle.y \leftrightarrow z))$$
$$\llbracket \text{let } (x, y) = M \text{ in } N \rrbracket z = \nu y \ (\llbracket M \rrbracket y \mid y(x).\llbracket N \rrbracket z)$$
$$\llbracket \text{inl } M \rrbracket z = \nu x \ (\llbracket M \rrbracket x \mid z[in_1].x \leftrightarrow z)$$
$$\llbracket \text{inr } M \rrbracket z = \nu x \ (\llbracket M \rrbracket x \mid z[in_2].x \leftrightarrow z)$$
$$\llbracket \text{case } L \ \{\text{inl } x \mapsto M ; \text{inr } x \mapsto N\} \rrbracket z = \nu x \ (\llbracket L \rrbracket x \mid \text{case } x \ \{\llbracket M \rrbracket z; \llbracket N \rrbracket z\})$$
$$\llbracket \text{absurd } L \rrbracket z = \nu x \ (\llbracket L \rrbracket x \mid \text{case } x \ \{\})$$

Substitutions

$$\llbracket \{V_i/x_i\} \rrbracket(P) = \hat\nu(x_i \mapsto \llbracket V_i \rrbracket x_i)_i[P]$$

$$\hat\nu(x_i \mapsto P_i)_i[P] \triangleq \nu x_1 \ (P_1 \mid \ldots \nu x_n \ (P_n \mid P) \ldots)$$

Configurations

$$\llbracket \circ M \rrbracket z = \nu y \ (\llbracket M \rrbracket y \mid y[])$$
$$\llbracket \bullet M \rrbracket z = \llbracket M \rrbracket z$$
$$\llbracket (\nu x) C \rrbracket z = \llbracket C \rrbracket z$$
$$\llbracket C \parallel_x C' \rrbracket z = \nu x \ (\llbracket C \rrbracket z \mid \llbracket C' \rrbracket z)$$

Fig. 10. Translation of GV Terms, Substitutions, and Configurations into CP

and $(A \otimes B)^\perp = A^\perp \mathbin{⅋} B^\perp$. Following our previous work [19], the translation on functional types is factored through an auxiliary translation $\lceil - \rceil$. The intuition is that the translation $\llbracket T \rrbracket$ of a functional type T is the type of its *interface*, whereas $\lceil T \rceil$ is the type of its *implementation*.

As CP processes do not have return values, the translation $\llbracket M \rrbracket z$ of a term M of type T includes the additional argument $z : \llbracket T \rrbracket^\perp$, which is a channel for simulating the return value. The translation on session terms is somewhat complicated by the need to include apparently trivial axiom cuts (highlighted in grey). These are needed to align with the translation of values, which permit further reduction inside the value constructors. The output in the translation of a fork arises from the need to apply the argument to a freshly generated channel (notice that application is simulated by an output). Linking is simulated by a link ($\leftrightarrow$) guarded by a nullary input which matches the nullary output of the output channel. Sending is simulated by output as one might expect. Receiving is simulated by input composed with sending the result to the return channel. Waiting is simulated by simply connecting the result to the return channel.

Variables are linked to the return channel. Closures are simulated by input, subject to an appropriate substitution, and application by output. Unit values are simulated by empty output to the return channel. Pairs are simulated by evaluating both components in parallel, transmitting the first along the return channel, and linking the second to the continuation of the return channel. Injections are simulated by injections. Each elimination form (other than application) guards the continuation with a suitable prefix, delaying reduction of the continuation until a value has been computed to pass to it. Substitutions are translated to right-nested sequences of cuts.

The translation of configurations is quite direct. We write $C \parallel_x C'$ to indicate that the variable x is shared by C and C'; in a well-typed GV configuration, there will always be exactly one such variable, so the translation is unambiguous.

Our translation differs from both Wadler's [26] and our previous one [19], neither of which simulate even plain β-reduction. This is because the obvious translation to CP cannot simulate substitution under a lambda abstraction, motivating our use of closures / weak explicit substitution. Indeed, others have taken advantage of full explicit substitutions in order simulate small-step semantics of λ-calculi in the full π-calculus [24].

Another departure from the previous translations to CP is that, despite the call-by-value semantics of GV, our translation is more in the spirit of call-by-name. For instance, in the translation of an application $L\,M$, the evaluation of L and M can happen in parallel, and β-reduction can occur before M has reduced to a value. The previous translations hide the evaluation of M behind the prefix $y\langle x\rangle$, which means that reduction of M can get stuck in the case that L is a free variable. Short of performing a CPS transformation on the translation, our new approach seems necessary in order to ensure that $[\![-]\!]$ preserves reduction.

It is straightforward to show that the translation preserves typing.

Theorem 18.

1. If $\Gamma \vdash M : T$, then $[\![M]\!]z \vdash [\![\Gamma]\!], z : [\![T]\!]^{\perp}$.
2. If $\Gamma \vdash C$, then $\exists T.[\![C]\!]z \vdash [\![\Gamma]\!], z : [\![T]\!]^{\perp}$.

Proof. By induction on derivations.   □

We now show that reduction in GV is preserved by reduction in CP. First, we observe that structural equivalence is preserved.

Theorem 19. *If $\Gamma \vdash C$, $\Gamma \vdash D$, and $C \equiv D$, then $[\![C]\!]z \equiv [\![D]\!]z$.*

Proof. By induction on the derivation of $\Gamma \vdash C$.   □

As the translations on terms and configurations are compositional, we can mechanically lift them to translations on evaluation contexts and configuration contexts such that the following lemma holds by construction. Each translation of a context takes two arguments: a function that describes the CP term to plug into the hole, and an output channel.

Lemma 20. *For $X \in \{E, F, G\}$, $[\![X[M]]\!]z = [\![X]\!][\![M]\!]z$*

We will make implicit use of Lemma 20 throughout our proofs. We write $x \mapsto P$ for a function that maps a name x to a process P that depends on x.

We now show that substitution commutes with $[\![-]\!]$.

Lemma 21. *If $\Gamma \vdash M : T$, $\Gamma \vdash \sigma : \Delta$, and $z \notin dom(\sigma)$, then $[\![\sigma]\!]([\![M]\!]z) \Longrightarrow$ $[\![M\sigma]\!]z$.*

Proof. By induction on the structure of M. Here we show the cases for variables and closures.

- Case x. By linearity there exists V such that $\sigma = \{V/x\}$.

$$[\![\sigma]\!]([\![x]\!]z) = \nu x \,([\![V]\!]x \mid x \leftrightarrow z) \longrightarrow [\![V]\!]z = [\![x\sigma]\!]z$$

- Case $\lambda^{\sigma'} x.M$.

$$
\begin{aligned}
&[\![\sigma]\!]([\![\lambda^{\sigma'} x.M]\!]) \\
=\quad &(\sigma' = \{V_i/x_i\}_i) \\
&[\![\sigma]\!](\hat{\nu}(x_i \mapsto ([\![V_i]\!]x_i))_i[z(x).[\![M]\!]z]) \\
=\quad &(\sigma = \sigma_1 \uplus \cdots \uplus \sigma_n \text{ where } dom(\sigma_i) = fv(V_i)) \\
&[\![\sigma_1]\!](\ldots [\![\sigma_n]\!](\hat{\nu}(x_i \mapsto [\![V_i]\!]x_i)_i[z(x).[\![M]\!]z])) \\
=\quad &(\text{structural equivalence}) \\
&\hat{\nu}(x_i \mapsto [\![\sigma_i]\!]([\![V_i]\!]x_i))_i[z(x).[\![M]\!]z] \\
\Longrightarrow\quad &(\text{IH}) \\
&\hat{\nu}(x_i \mapsto [\![V_i\sigma_i]\!]x_i)_i[z(x).[\![M]\!]z] \\
=\quad &(V_i\sigma_i = V\sigma) \\
&\hat{\nu}(x_i \mapsto [\![V_i\sigma]\!]x_i)_i[z(x).[\![M]\!]z] \\
=\quad &(\text{definition of } [\![-]\!]) \\
&[\![\lambda^{\sigma'}\sigma x.M]\!] \\
=\quad &(\text{definition of substitution}) \\
&[\![\lambda^{\sigma'} x.M\sigma]\!]
\end{aligned}
$$

Each of the remaining non-binding form cases follows straightforwardly using the induction hypothesis. Each of the remaining binding form cases requires a commuting conversion to push the appropriate substitution through a prefix. □

Using the substitution lemma, we prove that $[\![-]\!]$ preserves reduction on terms.

Theorem 22. *If $\Gamma \vdash M$, and $M \longrightarrow_V N$, then $[\![M]\!]z \Longrightarrow [\![N]\!]v$.*

Proof. By induction on the derivation of $M \longrightarrow_V N$. Here we show the case of β-reduction.

Syntax

$$\text{Session types} \quad S ::= \; !T.S \mid ?T.S \mid \text{end} \mid S^\sharp$$
$$\text{Constants} \quad K ::= \text{send} \mid \text{receive} \mid \text{fork} \mid \text{close} \mid \text{link}$$

Changes to duality

$$\overline{\text{end}} = \text{end}$$

Changes to type schemas for constants

$$\text{fork} : (S \multimap 1) \multimap \overline{S} \qquad \text{close} : \text{end} \multimap 1 \qquad \text{link} : S \times \overline{S} \multimap 1$$

Fig. 11. Syntax and Typing Rules for Combined Closed Channels

– Case $(\lambda^\sigma x.M)\, V \longrightarrow_V M(\{V/x\} \cup \sigma)$.

$$\llbracket (\lambda^\sigma x.M)\, V \rrbracket z$$
$$= \quad (\text{definition of } \llbracket - \rrbracket)$$
$$\nu w\, (\llbracket V \rrbracket w \mid \nu y\, (\llbracket \sigma \rrbracket (y(x).\llbracket M \rrbracket y) \mid y[x](w \leftrightarrow x \mid y \leftrightarrow z)))$$
$$\Longrightarrow_C \quad (\text{cut send against receive})$$
$$\nu w\, (\llbracket V \rrbracket w \mid \nu y\, (y \leftrightarrow z \mid \nu x\, (w \leftrightarrow x \mid \llbracket \sigma \rrbracket (\llbracket M \rrbracket y))))$$
$$\Longrightarrow_C \quad (\text{cut links and } \alpha \text{ rename})$$
$$\nu x\, (\llbracket V \rrbracket x \mid \llbracket \sigma \rrbracket (\llbracket M \rrbracket z))$$
$$\Longrightarrow \quad (\text{by Lemma 21})$$
$$\llbracket M(\{V/x\} \uplus \sigma) \rrbracket$$

The remaining base cases are similarly direct. The inductive case for reduction inside an evaluation context follows straightforwardly by observing that the translation of an evaluation context never places its argument inside a prefix. □

Finally, we prove that $\llbracket - \rrbracket$ preserves reduction on configurations.

Theorem 23. *If* $\Gamma \vdash C$, $\Gamma \vdash D$, *and* $C \longrightarrow D$, *then* $\llbracket C \rrbracket z \Longrightarrow \llbracket D \rrbracket z$.

Proof. By induction on the derivation of $C \longrightarrow D$. The inductive cases follow straightforwardly from the compositionality of the definitions and Theorem 22.

□

4 Extending GV

In this section, we consider two variants of our core calculus: the first adopts a single self-dual type for closed channels; the second adds unlimited types.

4.1 Unifying end₁ and end?

We begin by defining a language, based on GV, but combining the types end₁ and end? of closed channels. Figure 11 gives the alterations to the syntax and typing rules. The dual session types end₁ and end? are replaced by a single, self-dual type end; a new constant, close is provided to eliminate channels of type end. (In many existing systems, channels of type end are treated as unlimited, subject to weakening, rather than requiring an explicit close. We have left close

Extended configuration equivalence

$$C \parallel \circ () \equiv C$$

Extended reduction rules (all other reduction rules apply as in GV)

CLOSE

$$\overline{(\nu x)(F[\text{close } x] \parallel F'[\text{close } x]) \longrightarrow F[()] \parallel F'[()]}$$

LINK

$$\overline{(\nu x)(F[\text{link } (x,y)] \parallel C) \longrightarrow F[()] \parallel C\{y/x\}}$$

Fig. 12. Updated Configuration Evaluation Rules

explicit to simplify the presentation.) The type schemas for fork and link have been simplified, as we no longer need to build elimination of closed channels into fork. Figure 12 gives the updated evaluation rules for the extended language. In addition to a new rule for close (replacing the one for wait), the rule for link can be simplified (as it can now return a unit value instead of a closed channel).

Our modified language is, perhaps surprisingly, strictly more expressive than GV. Consider the following term:

$$\text{let } w = \text{fork } (\lambda w.\text{close } w; M) \text{ in close } w; N$$

Initially, the forked thread and its parent share channel w. After both threads close w, there can be no further communication between the threads; in contrast, in core GV, there must always be a final synchronisation with wait. To account for the increase in expressivity, we must extend the existing configuration typing rules (Figure 5) with a rule for composition in which no channels are shared:

$$\frac{\Gamma \vdash^\phi C \quad \Gamma' \vdash^{\phi'} C'}{\Gamma, \Gamma' \vdash^{\phi+\phi'} C \parallel C'}$$

Despite the additional expressivity of the modified calculus, we might hope that our results on deadlock freedom and progress (Theorems 7 and 10) would apply to this calculus as well. For the modified calculus, we must adapt Lemma 5:

Lemma 5A. *If $\Gamma \vdash C$ and $C = G[D \parallel D']$, then $fv(D) \cap fv(D')$ is either empty or the singleton set $\{x\}$ for some variable x.*

Clearly, this change does not allow the introduction of cyclic dependencies. Thus, the adaptation of the deadlock freedom and progress results to the modified calculus is entirely mechanical. It is straightforward to show that the other theorems of (§2.2) still hold in the presence of a single self-dual type for closed channels.

The additional expressivity does mean that we cannot define a translation from the modified calculus to CP. We believe that we could do so were CP extended with terms corresponding to the mix rules:

$$\overline{0 \vdash} \qquad \frac{P \vdash \Delta \quad Q \vdash \Delta'}{P \mid Q \vdash \Delta, \Delta'}$$

Syntactic extensions

Types	$T ::= \Box T \mid \ldots$	
Terms	$M, N ::= \text{let } !x = M \text{ in } N \mid !M \mid \ldots$	
Values	$V ::= !^{\sigma} E \mid \ldots$	
Evaluation contexts	$E ::= \text{let } !x = E \text{ in } M \mid \ldots$	

Typing rules

$$\frac{\Gamma \vdash M : T \quad \Box \Gamma}{\Gamma \vdash !M : \Box T} \qquad \frac{\Gamma \vdash M : \Box T \quad \Gamma', x : T \vdash N : U}{\Gamma \vdash \text{let } !x = M \text{ in } N : U}$$

$$\frac{\Gamma \vdash M : T}{\Gamma, x : \Box U \vdash M : T} \qquad \frac{\Gamma, x : \Box T, x' : \Box T \vdash M : U}{\Gamma, x : \Box T \vdash M\{x/x'\} : T}$$

Reduction

$$\text{let } !x = !^{\sigma} M \text{ in } N \longrightarrow_{\text{V}} N\{(M\sigma)/x\}$$

Fig. 13. GV Extensions for Unlimited Types

4.2 Unlimited Types

So far, we have treated only linear types. In this section, we consider one standard approach to extending the term language to include unlimited types.

Figure 13 gives the extension of GV. We begin by adding a new class of types, $\Box T$, representing unlimited types. (The typical notation for such types in linear logic, $!T$, clashes with the notation for output in session types.) We add terms to construct and deconstruct values of type $\Box T$; $\Box \Gamma$ denotes that every type in Γ must be of the form $\Box U$ for some type U. Values of type $\Box T$ can be weakened (discarded) and contracted (duplicated). We extend the language of values with unlimited values $!^{\sigma} E$; note that, as an unlimited value behaves similarly to a closure, we must introduce an explicit substitution. As in the treatment of λ-terms, we extend the typing relation to take account of the substitution

$$\frac{\Gamma \vdash M\sigma : T \quad dom(\sigma) = fv(M) \quad \Box \Gamma}{\Gamma \vdash !^{\sigma} M : \Box T}$$

and treat a term $!M$ as an abbreviation as follows:

$$!M \triangleq !^{\sigma}(M\sigma')$$
where $fv(M) = \{x_1, \ldots, x_n\}$ $y_1, \ldots, y_n$ are fresh variables
$\sigma = \{x_1/y_1, \ldots, x_n/y_n\}$ $\sigma' = \{y_1/x_1, \ldots, y_n/x_n\}$

The reduction rule for $\Box T$ values is unsurprising—however, unlike in the other reductions, x may be used non-linearly in M. As the concurrent semantics is unchanged from the base calculus, the extension of deadlock freedom and progress to this calculus is mechanical. Similarly, it is not difficult to show that the other theorems of (§2.2) still hold in the presence of either or both extensions.

The only non-trivial feature is the need for a logical relations argument in order to prove strong normalisation in the presence of unlimited types.

It is straightforward to extend CP with replication (following Wadler [26]) and correspondingly adapt the translations between CP and GV. The details are omitted due to lack of space.

5   Related Work

Session Types and Functional Languages. Session types were originally proposed by Honda [13], and later extended by Takeuchi et al. [22] and by Honda et al. [14]. Honda's system relies on a substructural type system (in which channels cannot be duplicated or discarded) and adopts the syntax & and ⊕ for choice; however, he does not draw a connection between his type system and the connectives of linear logic, and his system includes a single, self-dual closed channel. Vasconcelos et al. [25] develop a language that integrates session-typed communication primitives and a functional language. Gay and Vasconcelos [10] extend the approach to describe asynchronous communication with statically-bounded buffers. Their approach provides a more flexible mechanism of session initiation, distinct from their construct for thread creation, and they do not consider deadlock. Kobayashi [15] describes an embedding of session-typed π-calculus in polyadic linear π-calculus, relying on multi-argument send and receive to capture the state of a communication and variant types to capture choice; Dardha et al. [9] extend his approach to subtyping and polymorphism.

Linear Logic and Session Types. When he originally described linear logic, Girard [12] suggested that it would be suited to reasoning about concurrency. Abramsky [1] and Bellin and Scott [3] give embeddings of linear logic proofs in π-calculus, and show that cut reduction is simulated by π-calculus reduction. Their work is not intended to provide a type system for π-calculus: there are many processes which are not the image of some proof.

Caires and Pfenning [5] present a session type system for π-calculus that exactly corresponds to the proof system for the dual intuitionistic linear logic, and show that (up to congruence) cut reductions corresponds to process reductions or process equivalences. Toninho et al. [23] consider embeddings of the λ-calculus into session-typed π-calculus; their focus is on expressing the concurrency inherent in λ-calculus terms, rather than simulating standard reduction. Wadler [26] adapts the approach of Caires and Pfenning to classical (rather than intuitionistic) linear logic, and gives a translation from GV (his functional language) to CP (his process calculus). He does not give a direct semantics for GV. In previous work [19], we give a type-preserving translation from CP to GV.

Deadlock Freedom and Progress. There have been several approaches to guarantee deadlock freedom in π-calculus. Kobayashi [16] and Padovani [20] extend type systems for linear π-calculus with priority information, capturing the order in which channels are used. Giachino et al. [11] give a type system that expresses dependencies directly in the types of CCS terms. These systems permit more programs than ours, at the cost of significantly more complex type systems; they also do not enjoy the close correspondence with linear logic (or other well-known logical systems).

Carbone and Debois [7] give a graphical characterisation of session-typed processes; this allows them to directly identify cycles in channel dependencies. They show that all possible interactions eventually take place in cycle-free processes. Carbone et al. [6] show similar results for well-typed processes under Kobayashi's type system for deadlock freedom; their approach accommodates processes with open channels by defining a type-directed closure of a process, and showing that open processes progress only if their typed closures progress.

6 Conclusion and Future Work

We have presented a small-step operational semantics for GV, a session-typed functional core language. We have proved that it is deadlock-free, deterministic, and terminating, and have established simulations both ways between our semantics for GV and cut-reduction in a process calculus based on linear logic. Furthermore, we have shown that GV provides a promising basis for future modular language development by illustrating two extensions to GV, both of which preserve deadlock-freedom, determinism, and termination.

We identify two important directions for future work: recursion and asynchronous communication. Recursion is essential both for channels (to capture repeating behaviour, such as adding recipients to a mail message) and for functional programming. Adding unchecked recursion to GV would clearly compromise termination and introduce the possibility of livelock; we hope that adapting approaches used for fixed points in linear logic might mitigate this issue. Asynchronous communication naturally lends itself to practical implementation. We hope to develop the approach of Gay and Vasconcelos [10] and show a correspondence between synchronous and asynchronous semantics for GV.

Acknowledgements. Thanks to Philip Wadler and the anonymous reviewers. This work was funded by EPSRC grant number EP/K034413/1.

References

[1] Abramsky, S.: Proofs as processes. Theor. Comput. Sci. 135(1), 5–9 (1992)
[2] Barendregt, H.P.: The Lambda Calculus Its Syntax and Semantics, vol. 103. North-Holland (1984)
[3] Bellin, G., Scott, P.J.: On the π-Calculus and linear logic. Theoretical Computer Science 135(1), 11–65 (1994)
[4] Boreale, M.: On the expressiveness of internal mobility. In: Sassone, V., Montanari, U. (eds.) CONCUR 1996. LNCS, vol. 1119, Springer, Heidelberg (1996)
[5] Caires, L., Pfenning, F.: Session types as intuitionistic linear propositions. In: Gastin, P., Laroussinie, F. (eds.) CONCUR 2010. LNCS, vol. 6269, pp. 222–236. Springer, Heidelberg (2010)

[6] Carbone, M., Dardha, O., Montesi, F.: Progress as compositional lock-freedom. In: COORDINATION 2014, Springer (2014)

[7] Carbone, M., Debois, S.: A graphical approach to progress for structured communication in web services. In: ICE (2010)

[8] Cooper, E., Lindley, S., Yallop, J.: Links: Web programming without tiers. In: de Boer, F.S., Bonsangue, M.M., Graf, S., de Roever, W.-P. (eds.) FMCO 2006. LNCS, vol. 4709, pp. 266–296. Springer, Heidelberg (2007)

[9] Dardha, O., Giachino, E., Sangiorgi, D.: Session types revisited. In: PPDP. ACM (2012)

[10] Gay, S.J., Vasconcelos, V.T.: Linear type theory for asynchronous session types. Journal of Functional Programming 20(01), 19–50 (2010)

[11] Giachino, E., Kobayashi, N., Laneve, C.: Deadlock analysis of unbounded process networks. In: Baldan, P., Gorla, D. (eds.) CONCUR 2014. LNCS, vol. 8704, pp. 63–77. Springer, Heidelberg (2014)

[12] Girard, J.-Y.: Linear logic. Theoretical Computer Science 50(1), 1–101 (1987)

[13] Honda, K.: Types for dyadic interaction. In: Best, E. (ed.) CONCUR 1993. LNCS, vol. 715, Springer, Heidelberg (1993)

[14] Honda, K., Vasconcelos, V.T., Kubo, M.: Language primitives and type discipline for structured communication-based programming. In: Hankin, C. (ed.) ESOP 1998. LNCS, vol. 1381, p. 122. Springer, Heidelberg (1998)

[15] Kobayashi, N.: Type systems for concurrent programs. In: Aichernig, B.K. (ed.) Formal Methods at the Crossroads. From Panacea to Foundational Support. LNCS, vol. 2757, pp. 439–453. Springer, Heidelberg (2003)

[16] Kobayashi, N.: A new type system for deadlock-free processes. In: Baier, C., Hermanns, H. (eds.) CONCUR 2006. LNCS, vol. 4137, pp. 233–247. Springer, Heidelberg (2006)

[17] Kobayashi, N., Pierce, B.C., Turner, D.N.: Linearity and the π-calculus. In: POPL. ACM (1996)

[18] Lévy, J.-J., Maranget, L.: Explicit substitutions and programming languages. In: Pandu Rangan, C., Raman, V., Sarukkai, S. (eds.) FST TCS 1999. LNCS, vol. 1738, p. 181. Springer, Heidelberg (1999)

[19] Lindley, S., Morris, J.G.: Sessions as propositions. In: PLACES (2014)

[20] Padovani, L.: Deadlock and lock freedom in the linear π-calculus. In: LICS, ACM (2014)

[21] Pérez, J.A., Caires, L., Pfenning, F., Toninho, B.: Linear logical relations and observational equivalences for session-based concurrency. Inf. Comput. 239, 254–302 (2014)

[22] Takeuchi, K., Honda, K., Kubo, M.: An interaction-based language and its typing system. In: Halatsis, C., Philokyprou, G., Maritsas, D., Theodoridis, S. (eds.) PARLE 1994. LNCS, vol. 817, Springer, Heidelberg (1994)

[23] Toninho, B., Caires, L., Pfenning, F.: Functions as session-typed processes. In: Birkedal, L. (ed.) FOSSACS 2012. LNCS, vol. 7213, pp. 346–360. Springer, Heidelberg (2012)

[24] van Bakel, S., Vigliotti, M.G.: A logical interpretation of the λ-calculus into the π-calculus, preserving spine reduction and types. In: Bravetti, M., Zavattaro, G. (eds.) CONCUR 2009. LNCS, vol. 5710, pp. 84–98. Springer, Heidelberg (2009)

[25] Vasconcelos, V.T., Gay, S.J., Ravara, A.: Type checking a multithreaded functional language with session types. Theor. Comput. Sci. 368(1-2), 64–87 (2006)

[26] Wadler, P.: Propositions as sessions. J. Funct. Program. 24(2-3), 384–418 (2014)

Composite Replicated Data Types*

Alexey Gotsman[1] and Hongseok Yang[2]

[1] IMDEA Software Institute
[2] University of Oxford

Abstract. Modern large-scale distributed systems often rely on eventually consistent replicated stores, which achieve scalability in exchange for providing weak semantic guarantees. To compensate for this weakness, researchers have proposed various abstractions for programming on eventual consistency, such as replicated data types for resolving conflicting updates at different replicas and weak forms of transactions for maintaining relationships among objects. However, the subtle semantics of these abstractions makes using them correctly far from trivial.

To address this challenge, we propose composite replicated data types, which formalise a common way of organising applications on top of eventually consistent stores. Similarly to an abstract data type, a composite data type encapsulates objects of replicated data types and operations used to access them, implemented using transactions. We develop a method for reasoning about programs with composite data types that reflects their modularity: the method allows abstracting away the internals of composite data type implementations when reasoning about their clients. We express the method as a denotational semantics for a programming language with composite data types. We demonstrate the effectiveness of our semantics by applying it to verify subtle data type examples and prove that it is sound and complete with respect to a standard non-compositional semantics.

1 Introduction

Background. To achieve availability and scalability, many modern networked systems use *replicated stores*, which maintain multiple *replicas* of shared data. Clients can access the data at any of the replicas, and these replicas communicate changes to each other using message passing. For example, large-scale Internet services use data replicas in geographically distinct locations, and applications for mobile devices keep replicas locally as well as in the cloud to support offline use. Ideally, we would like replicated stores to provide *strong consistency*, i.e., to behave as if a single centralised replica handles all operations. However, achieving this ideal usually requires synchronisation among replicas, which slows down the store and even makes it unavailable if network connections between replicas fail [14, 3]. For this reason, modern replicated stores often provide weaker guarantees, described by the umbrella term of *eventual consistency* [5].

---

* We thank Giovanni Bernardi, Sebastian Burckhardt and Andrea Cerone for many interesting discussions and thoughtful comments about our results and the paper. Our work was supported by the EU FET project ADVENT and EPSRC.

J. Vitek (Ed.): ESOP 2015, LNCS 9032, pp. 585–609, 2015.
DOI: 10.1007/978-3-662-46669-8_24

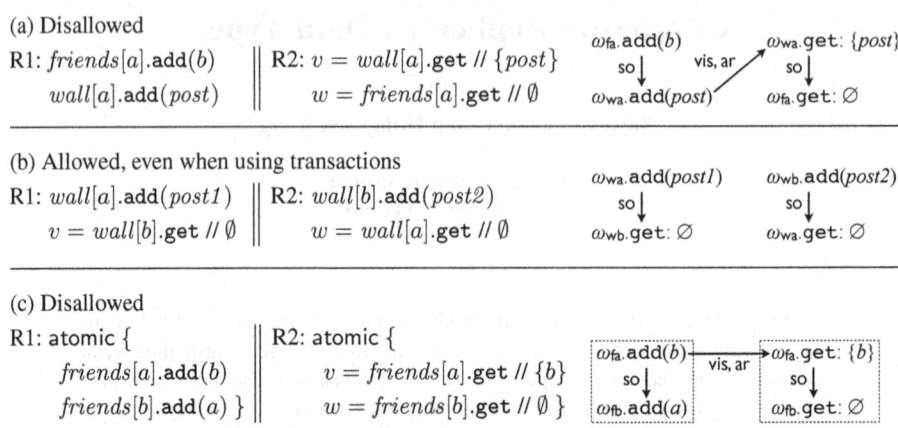

(a) Disallowed

R1: $friends[a].\text{add}(b)$ ‖ R2: $v = wall[a].\text{get}$ // $\{post\}$

    $wall[a].\text{add}(post)$ ‖ $w = friends[a].\text{get}$ // $\emptyset$

$\omega_{fa}.\text{add}(b)$ $\omega_{wa}.\text{get}$: $\{post\}$
so↓   vis, ar   so↓
$\omega_{wa}.\text{add}(post)$ $\omega_{fa}.\text{get}$: $\emptyset$

(b) Allowed, even when using transactions

R1: $wall[a].\text{add}(post1)$ ‖ R2: $wall[b].\text{add}(post2)$

    $v = wall[b].\text{get}$ // $\emptyset$ ‖ $w = wall[a].\text{get}$ // $\emptyset$

$\omega_{wa}.\text{add}(post1)$ $\omega_{wb}.\text{add}(post2)$
so↓   so↓
$\omega_{wb}.\text{get}$: $\emptyset$ $\omega_{wa}.\text{get}$: $\emptyset$

(c) Disallowed

R1: atomic { ‖ R2: atomic {

    $friends[a].\text{add}(b)$ ‖ $v = friends[a].\text{get}$ // $\{b\}$

    $friends[b].\text{add}(a)$ } ‖ $w = friends[b].\text{get}$ // $\emptyset$ }

$\omega_{fa}.\text{add}(b)$ $\omega_{fa}.\text{get}$: $\{b\}$
so↓   vis, ar   so↓
$\omega_{fb}.\text{add}(a)$ $\omega_{fb}.\text{get}$: $\emptyset$

Fig. 1. Anomalies illustrating the semantics of causal consistency and causally consistent transactions. The outcomes of operations are shown in comments. The variables v and w are local to clients. The structures shown on the right are explained in §3.2.

Eventually consistent stores adopt an architecture where a replica performs an operation requested by a client locally without any synchronisation with others and immediately returns to the client; the effects of the operation are propagated to other replicas only *eventually*. As a result, different replicas may find out about an operation at different points in time. This leads to *anomalies*, one of which is illustrated by the outcome in Figure 1(a). The program shown there consists of two clients operating on set objects $friends[a]$ and $wall[a]$, which represent information about a user a in a social network application. The first client, connected to replica 1, makes b a friend of a's and then posts b's message on a's wall. After each of these operations, replica 1 might send a message with an update to replica 2. If the messages carrying the additions of b to $friends[a]$ and $post$ to $wall[a]$ arrive at replica 2 out of order, the second client can see b's $post$, but does not know that b has become a's friend. This outcome cannot be produced by any interleaving of the operations shown in Figure 1(a) and, hence, is not strongly consistent.

The *consistency model* of a replicated store restricts the anomalies that it exhibits. In this paper, we consider the popular model of *causal consistency* [18], a variant of eventual consistency that strikes a reasonable balance between programmability and efficiency. A causally consistent store disallows the anomaly in Figure 1(a), because it respects causal dependencies between operations: if the programmer sees b's $post$ to a's wall, she is also guaranteed to see all events that led to this posting, such as the addition of b to the set of a's friends. Causal consistency is weaker than strong consistency; in particular, it allows reading stale data. This is illustrated by the outcome in Figure 1(b), which cannot be produced by any interleaving of the operations shown. In a causally consistent store it may be produced because each message about an addition sent by the replica performing it may be slow to get to the other replica.

Due to such subtle semantics, writing correct applications on top of eventually consistent stores is very difficult. In fact, finding a good programming model for eventual consistency is considered one of the major research challenges in the systems community [5]. We build on two programming abstractions proposed by researchers to address this challenge, which we now describe.

One difficulty of programming for eventually consistent stores is that their clients can concurrently issue conflicting operations on the same data item at different replicas. For example, spouses sharing a shopping cart in an online store can add and concurrently remove the same item. To deal with these situations, eventually consistent stores provide *replicated data types* [23] that implement *objects*, such as registers, counters or sets, with various strategies for resolving conflicting updates to them. The strategies can be as simple as establishing a total order on all operations using timestamps and letting the last writer win, but can also be much more subtle. For example, a set data type, which can be used to implement a shopping cart, can process concurrent operations trying to add and concurrently remove the same element so that ultimately the element ends up in the set.

Another programming abstraction that eventually consistent stores are starting to provide is *transactions*, which make it easier to maintain relationships among different objects. In this paper we focus on *causally consistent transactions*, implemented (with slight variations) by a number of stores [26, 18, 19, 24, 17, 2, 4]. When a causally consistent transaction performs several updates at a replica, we are guaranteed that these will be delivered to every other replica together. For example, consider the execution in Figure 1(c), where at replica 1 two users befriend each other by adding their identifiers to set objects in the array *friends*. If we did not use transactions, the outcome shown would be allowed by causal consistency, as replica 2 might receive the addition of b to $friends[a]$, but not that of a to $friends[b]$. This would break the expected invariant that the friendship relation encoded by *friends* is symmetric. Causally consistent transactions disallow this anomaly, but nevertheless provide weaker guarantees than the classical serialisable ACID transactions. The latter guarantee that operations done within a transaction can be viewed as taking effect instantaneously at all replicas. With causally consistent transactions, even though each separate replica sees updates done by a transaction together, different replicas may see them at different times. For example, the outcome in Figure 1(b) could occur even if we executed the pair of commands at each replica in a transaction, again because of delays in message delivery.

A typical way of using replicated data types and transactions for writing applications on top of an eventually consistent store is to keep the application data as a set of objects of replicated data types, and update them using transactions over these objects [26, 24, 17, 2]. Then replicated data types ensure sensible conflict resolution, and transactions ensure the maintenance of relationships among objects. However, due to the subtle semantics of these abstractions, reasoning about the behaviour of applications organised in this way is far from trivial. For example, it is often difficult to trace how the choice of conflict-resolution policies on separate objects affects the policy for the whole application: as we show in §5, a wrong choice can lead to violations of integrity invariants across objects, resulting in undesirable behaviour.

Contributions. To address this challenge, we propose a new programming concept of a *composite replicated data type* that formalises the above way of organising applications using eventually consistent stores. Similarly to a class or an abstract data type, a composite replicated data type encapsulates *constituent objects* of replicated data types and *composite operations* used to access them, each implemented using a transaction. For example, a composite data type representing the friendship relation in a social network may consist of a number of set objects storing the friends of each user, with transactions used to keep the relation symmetric. Composite data types can also capture the modular structure of applications, since we can construct complex data types from simpler ones in a nested fashion.

We further propose a method for reasoning about programs with composite data types that reflects their modularity: the method allows one to abstract from the internals of composite data type implementations when reasoning about the clients of these data types. Technically, we express our reasoning method as a denotational semantics for a programming language that allows defining composite data types (§4). As any denotational semantics, ours is compositional and is thus able to give a denotation to every composite data type separately. This denotation abstracts from the internal data type structure using what we term *granularity abstraction*: it does not record *fine-grained* events describing operations on the constituent objects that are performed by composite operations, but represents every invocation of a composite operation by a single *coarse-grained* event. Thereby, the denotation allows us to pretend that the composite data type represents a single monolithic object, no different from an object of a *primitive* data type implemented natively by the store. The denotation then describes the data type behaviour using a mechanism recently proposed for specifying primitive replicated data types [11]. The granularity abstraction achieved by this *coarse-grained* denotational semantics is similar (but not identical, as we discuss in §7) to atomicity abstraction, which has been extensively investigated in the context of shared-memory concurrency [13, 25].

Our coarse-grained semantics enables modular reasoning about programs with composite replicated data types. Namely, it allows us to prove a property of a program by: *(i)* computing the denotations of the composite data types used in it; and *(ii)* proving that the program satisfies the property assuming that it uses *primitive* replicated data types with the specifications equal to the denotations of the composite ones. We thus never have to reason about composite data type implementations and their clients together.

Since we use an existing specification mechanism [11] to represent a composite data type denotation, our technical contribution lies in identifying *which* specification to pick. We show that the choice we make is correct by proving that our coarse-grained semantics is sound with respect to a *fine-grained semantics* of the programming language (§6), which records the internal execution of composite operations and follows the standard way of defining language semantics on weak consistency models [11]. We also establish that the coarse-grained semantics is complete with respect to the fine-grained one: we do not lose precision by reasoning with denotations of composite data types instead of their implementations. The soundness and completeness results also imply that our coarse-grained denotational semantics is *adequate*, i.e., can be used for proving the observational equivalence of two composite data type implementations.

We demonstrate the usefulness of the coarse-grained semantics by applying the composite data type denotation it defines to specify and verify small but subtle data types, such as a social graph (§5). In particular, we show how our semantics lets one

Primitive data types $B \in \mathsf{PrimType}$ Data-type variables $\alpha, \beta \in \mathsf{DVar}$

Ordinary variables $v, w \in \mathsf{Var} = \{v_{\mathsf{in}}, v_{\mathsf{out}}, \ldots\}$ Object variables $x, y \in \mathsf{OVar}$

Data-type contexts $\Gamma ::= \alpha_1 : O_1, \ldots, \alpha_k : O_k$ Ordinary contexts $\Sigma ::= v_1, \ldots, v_k$

Object contexts $\Delta ::= x_1 : O_1, \ldots, x_k : O_k$

$D ::= \mathsf{let}\ \{x_j = \mathsf{new}\ T_j\}_{j=1..m}\ \mathsf{in}\ \{o = \mathsf{atomic}\ \{C_o\}\}_{o \in O}$ $T ::= B \mid D \mid \alpha$

$G ::= v \mid G + G \mid G \wedge G \mid G \vee G \mid \neg G \mid \ldots$

$C ::= \mathsf{var}\ v.\, C \mid v = x.o(G) \mid v = G \mid C; C \mid \mathsf{if}\ G\ \mathsf{then}\ C\ \mathsf{else}\ C \mid \mathsf{while}\ G\ \mathsf{do}\ C \mid \mathsf{atomic}\ \{C\}$

$P ::= C_1 \parallel \ldots \parallel C_n \mid \mathsf{let}\ \alpha = T\ \mathsf{in}\ P \mid \mathsf{let}\ x = \mathsf{new}\ T\ \mathsf{in}\ P$

$$\boxed{\Delta \mid \Sigma \vdash C} \quad \dfrac{\mathsf{FV}(G) \cup \{v\} \subseteq (\Sigma - \{v_{\mathsf{in}}, v_{\mathsf{out}}\})}{\Delta, x : \{o\} \cup O \mid \Sigma \vdash v = x.o(G)} \quad \dfrac{\Delta \mid \Sigma, v \vdash C}{\Delta \mid \Sigma \vdash \mathsf{var}\ v.\, C} \quad \dfrac{\Delta \mid \Sigma \vdash C}{\Delta \mid \Sigma \vdash \mathsf{atomic}\ \{C\}}$$

$$\boxed{\Gamma \vdash T : O} \quad \dfrac{}{\Gamma \vdash B : \mathsf{sig}(B)} \quad \dfrac{\Gamma \vdash T_j : O_j\ \text{for all}\ j = 1..m}{x_1 : O_1, \ldots, x_m : O_m \mid v_{\mathsf{in}}, v_{\mathsf{out}} \vdash C_o\ \text{for all}\ o \in O}$$
$$\dfrac{}{\Gamma, \alpha : O \vdash \alpha : O} \quad \dfrac{}{\Gamma \vdash \mathsf{let}\ \{x_j = \mathsf{new}\ T_j\}_{j=1..m}\ \mathsf{in}\ \{o = \mathsf{atomic}\ \{C_o\}\}_{o \in O} : O}$$

$$\boxed{\Gamma \mid \Delta \vdash P} \quad \dfrac{\Gamma \vdash T : O \qquad \Gamma \mid \Delta, x : O \vdash P}{\Gamma \mid \Delta \vdash \mathsf{let}\ x = \mathsf{new}\ T\ \mathsf{in}\ P} \quad \dfrac{\Delta \mid \emptyset \vdash C_j\ \text{for all}\ j = 1..n}{\Gamma \mid \Delta \vdash C_1 \parallel \ldots \parallel C_n}$$

Fig. 2. Programming language and sample typing rules

understand the consequences of different design decisions in the implementation of a composite data type on its behaviour.

2   Programming Language and Composite Replicated Data Types

Store Data Model. We consider a replicated store organised as a collection of ***primitive objects***. Clients interact with the store by invoking ***operations*** on objects from a set Op, ranged over by o. Every object in the store belongs to one of the ***primitive replicated data types*** $B \in \mathsf{PrimType}$, implemented by the store natively. The ***signature*** $\mathsf{sig}(B) \subseteq \mathsf{Op}$ determines the set of operations allowed on objects of the type B. As we explain in §3, the data type also determines the semantics of the operations and, in particular, the conflict-resolution policies implemented by them. For uniformity of definitions, we assume that each operation takes a single parameter and returns a single value from a set of ***values*** Val, whose elements are ranged over by a, b, c, d. We assume that Val includes at least Booleans and integers, their sets and tuples thereof. We use a special value $\bot \in \mathsf{Val}$ to model operations that take no parameter or return no value. For example, primitive data types can include sets with operations `add`, `remove`, `contains` and `get` (the latter returning the set contents).

Composite Replicated Data Types. We develop our results for a language of client programs interacting with the replicated store, whose syntax we show in Figure 2. We consider only programs well-typed according to the rules also shown in the figure. The interface to the store provided by the language is typical of existing implementations [26, 2]. It allows programs to declare objects of primitive replicated data types, residing in the store, invoke operations on them, and combine these into transactions. Crucially, the language also allows declaring **composite replicated data types** from the given primitive ones and **composite objects** of these types. These composite objects do not actually reside in the store, but serve as client-side anchors for compositions of primitive objects. A declaration D of a composite data type includes several **constituent objects** of specified types T_j, which can be primitive types, composite data type declarations or **data-type variables** $\alpha \in$ DVar, bound to either. The constituent objects are bound to distinct **object variables** x_j, $j = 1..m$ from a set OVar. The declaration D also defines a set of **composite operations** O (the type's **signature**), with each $o \in O$ implemented by a **command** C_o executed as a **transaction** accessing the objects x_j. We emphasise the use of transactions by wrapping C_o into an atomic block. Since a store implementation executes a transaction at a replica without synchronising with other replicas, transactions never abort.

The syntax of commands includes the form var $v.\,C$ for declaring **ordinary variables** $v, w \in$ Var, to be used by C, which store values from Val and are initialised to $\perp$. Commands C_o in composite data type declarations D can additionally access two distinguished ordinary variables v_{in} and v_{out} (never declared explicitly), used to pass parameters and return values of operations: the parameter gets assigned to v_{in} at the beginning of the execution of C_o and the return value is read from v_{out} at the end. The command $v = x.o(G)$ executes the operation o on the object bound to the variable x with parameter G and assigns the result to v.[1]

Our type system enforces that commands only invoke operations on objects consistent with the signatures of their types and that all variables be used within the correct scope; in particular, constituent objects of composite types can only be accessed by their composite operations. For simplicity, we do not adopt a similar type discipline for values and treat all expressions as untyped. Finally, for convenience of future definitions, the typing rule for $v = x.o(G)$ requires that v_{in} and v_{out} do not appear in v or G.

Example: Social Graph.   Figure 3 gives our running example of a composite data type soc, which maintains friendship relationships and requests between accounts in a toy social network application. To concentrate on core issues of composite data type correctness, we consider a language that does not allow creating unboundedly many objects; hence, we assume a fixed number of accounts N. Using syntactic sugar, the constituent objects are grouped into arrays *friends* and *requesters* and have the type RWset of sets with a particular conflict-resolution policy (defined in §3.1). We use these sets to store account identifiers: *friends* [a] gives the set of a's friends, and *requesters* [a] the set of accounts with pending friendship requests to a. The implementation maintains the expected integrity invariants that the friendship relation is symmetric and the friend and requester sets of any account are disjoint:

$$\forall a, b.\ \mathit{friends}[a].\texttt{contains}(b) \Leftrightarrow \mathit{friends}[b].\texttt{contains}(a); \tag{1}$$

$$\forall a.\ \mathit{friends}[a].\texttt{get} \cap \mathit{requesters}[a].\texttt{get} = \emptyset. \tag{2}$$

---

[1] Since the object bound to x may itself be composite, this may result in atomic blocks being nested. Their semantics is the same as the one obtained by discarding all blocks except the top-level one. In particular, the atomic blocks that we include into the syntax of commands have no effect inside operations of composite data types.

D_{soc} = let { $friends$ = new RWset$[N]$; $requesters$ = new RWset$[N]$ } in {
    request$(from, to)$ = atomic {
        if $(friends[to].\text{contains}(from) \lor requesters[to].\text{contains}(from))$ then v_{out} = false
        else { $requesters[to].\text{add}(from); v_{\text{out}}$ = true } };
    accept$(from, to)$ = atomic {
        if $(\neg requesters[to].\text{contains}(from))$ then v_{out} = false
        else { $requesters[to].\text{remove}(from); requesters[from].\text{remove}(to);$
                $friends[to].\text{add}(from); friends[from].\text{add}(to); v_{\text{out}}$ = true } };
    reject$(from, to)$ = atomic {
        if $(\neg requesters[to].\text{contains}(from))$ then v_{out} = false
        else { $requesters[to].\text{remove}(from); requesters[from].\text{remove}(to); v_{\text{out}}$ = true } };
    breakup$(from, to)$ = atomic {
        if $(\neg friends[to].\text{contains}(from))$ then v_{out} = false
        else { $friends[to].\text{remove}(from); friends[from].\text{remove}(to); v_{\text{out}}$ = true } };
    get(id) = atomic {v_{out} = $(friends[id].\text{get}, requesters[id].\text{get})$ } } }

Fig. 3. A social graph data type soc

The composite operations allow issuing a friendship request, accepting or rejecting it, breaking up and getting the information about a given account. For readability, we use some further syntactic sugar in the operations. Thus, we replace v_{in} with more descriptive names, recorded after the operation name and, in the case when the parameter is meant to be a tuple, introduce separate names for its components. Thus, $from$ and to desugar to fst(v_{in}) and snd(v_{in}). We also allow invoking operations on objects inside expressions and omit unimportant parameters to operations.

The code of the composite operations is mostly as expected. For example, request adds the user sending the request to the requester set of the user being asked, after checking, e.g., that the former is not already a friend of the latter. However, this simplicity is deceptive: when reasoning about the behaviour of the data type, we need to consider the possibility of operations being issued concurrently at different replicas. For example, what happens if two users concurrently issue friendship requests to each other? What if two users managing the same institutional account take conflicting decisions, such as concurrently accepting and rejecting a request? As we argue in §5, it is nontrivial to implement the data type so that the behaviour in the above situations be acceptable. Using the results in this paper, we can specify the desired social graph behaviour and prove that the composite data type in Figure 3 satisfies such a specification. Our specification abstracts from the internal structure of the data type, thereby allowing us to view it as no different from the primitive set data types it is constructed from. This facilitates reasoning about programs using the data type, which we describe next.

Programs. A *program* P consists of a series of data type and object variable declarations followed by a *client*. The latter consists of several commands $C_1, \ldots, C_n$, each representing a user *session* accessing the store concurrently with others; a session is thus an analogue of a thread in shared-memory languages. An implementation would connect each session to one of the store replicas (as in examples in Figure 1), but this is transparent on the language level. Data type variables declared in P are used to specify the types of objects declared afterwards, and object variables are used inside sessions C_j, as per the typing rules. Sessions can thus invoke operations on a number of objects of primitive or composite types. By default, every such operation is executed within a dedicated transaction. However, like in composite data type implementations, we allow sessions to group multiple operations into transactions using atomic blocks included into the syntax of commands. We consider data types T and programs P up to the standard alpha-equivalence, adjusted so that v_{in} and v_{out} are not renamed.

Technical Restriction. To simplify definitions, we assume that commands inside atomic blocks always terminate and, thus, so do all operations of composite data types. We formalise this restriction when presenting the semantics of the language in §4. It can be lifted at the expense of complicating the presentation. Note that the sessions C_j do not have to terminate, thereby allowing us to model the reactive nature of store clients.

3   Replicated Store Semantics

A replicated store holds objects of primitive replicated data types and implements operations on these objects. The language of §2 allows us to write programs that interact with the store by invoking the operations while grouping primitive objects into composite ones to achieve modularity. The main contribution of this paper is a denotational semantics of the language that allows the reasoning about a program to reflect this modularity. But before presenting it (in §4), we need to define the semantics of the store itself: which values can operations on primitive objects return in an execution of the store? This is determined by the consistency model of causally consistent transactions [26, 18, 19, 24, 17, 12, 4], which we informally described in §1. To formalise it, we use a variant of the framework proposed by Burckhardt et al. [11, 12, 10], which defines the store semantics declaratively, without referring to implementation-level concepts such as replicas or messages. The framework models store executions using structures on events and relations in the style of weak memory models and allows us to define the semantics of the store in two stages. We first specify the semantics of single operations on primitive objects using *replicated data type specifications* (§3.1), which are certain functions on events and relations. We then specify allowed executions of the store, including multiple operations on different objects, by constraining the events and relations using *consistency axioms* (§3.2).

A correspondence between the declarative store specification and operational models closer to implementations was established elsewhere [11, 12, 10]. Although we do not present an operational model in this paper, we often explain various features of the store specification framework by referring to the implementation-level concepts they are meant to model.

The granularity abstraction of the denotational semantics we define in §4 allows us to pretend that a composite data type is a primitive one. Hence, when defining the semantics, we reuse the replicated data type specifications introduced here to specify the behaviour of a composite data type, such as the one in Figure 3, while abstracting from the internals of its implementation.

3.1 Semantics of Primitive Replicated Data Types

In a strongly consistent system, there is a total order on all operations on an object, and each operation takes into account the effects of all operations preceding it in this order. In an eventually consistent system, the result of an operation o is determined in a more complex way:

1. The result of o depends on the set of operations information about which has been delivered to the replica performing o—those *visible* to o. For example, in Figure 1(a) the operation $friends[a]$.get returns $\emptyset$ because the message about $friends[a]$.add(b) has not yet been delivered to the replica performing the get.
2. The result of o may also depend on additional information used to order some events. For example, we may decide to order concurrent updates to an object using timestamps, as is the case when we use the last-writer-wins conflicts resolution policy mentioned in §1.

Hence, we specify the semantics of a replicated data type by a function F that computes the return value of an operation o given its *operation context*, which includes all we need to know about the store execution to determine the value: the set of events visible to o, together with a pair of relations on them that specify the above relationships.

   Assume a countably-infinite set Event of *events*, representing operations issued to the store. A relation is a **strict partial order** if it is transitive and irreflexive. A **total order** is a strict partial order such that for every two distinct elements e and f, the order relates e to f or f to e. We call a pair $p \in$ Op $\times$ Val $=$ AOp of an operation o together with its parameter a an **applied operation**, written as $o(a)$.

DEFINITION 1 *An **operation context** is a tuple* $N = (p, E, \text{aop}, \text{vis}, \text{ar})$*, where* $p \in$ AOp*, E is a finite subset of* Event*, aop* $: E \to$ AOp*, and vis **(visibility)** and ar **(arbitration)** are strict partial orders on E such that* vis $\subseteq$ ar*.*
   We call the tuple $M = (E, \text{aop}, \text{vis}, \text{ar})$ *a **partial operation context**.*

We write Ctxt for the set of all operation contexts and denote components of N and similar structures as in $N.E$. For a relation R we write $(e, f) \in R$ and $e \xrightarrow{R} f$ interchangeably. Informally, the orders vis and ar record the relationships between events in E motivated by the above points 1 and 2, respectively. In implementation terms, the requirement vis $\subseteq$ ar guarantees that timestamps are consistent with message delivery: if e is visible to f, then e has a lower timestamp than f. We define where vis and ar come from formally in §3.2; for now we just assume that they are given and define *replicated data type specifications* as certain functions of operation contexts including them.

DEFINITION 2 *A **replicated data type specification** is a partial function $F :$ Ctxt $\rightharpoonup$ Val that returns the same value on isomorphic operation contexts and preserves it on arbitration extensions. Formally, let us order operation contexts by the pre-order $\sqsubseteq$:*

$$(p, E, \text{aop}, \text{vis}, \text{ar}) \sqsubseteq (p', E', \text{aop}', \text{vis}', \text{ar}') \iff$$

$$p = p' \wedge \exists \pi \in E \to_{\text{bijective}} E'. \ \pi(\text{aop}) = \text{aop}' \wedge \pi(\text{vis}) = \text{vis}' \wedge \pi(\text{ar}) \subseteq \text{ar}',$$

where we use the expected lifting of π to relations. Then we require

$$\forall N, N' \in \text{Ctxt}. \ N \sqsubseteq N' \wedge N \in \text{dom}(F) \implies N' \in \text{dom}(F) \wedge F(N) = F(N'). \quad (3)$$

A. Gotsman and H. Yang

Let Spec be the set of data type specifications F and assume a fixed F_B for every primitive type $B \in \mathrm{PrimType}$ provided by the store. The requirement (3) states that, once arbitration gives all the information that is needed in addition to visibility to determine the outcome of an operation, arbitrating more events does not change this outcome.

Replicated Sets. We illustrate the above definitions by specifying replicated set data types with different conflict-resolution policies. The semantics of a replicated set is straightforward when it is **_add-only_**, i.e., its signature is $\{\mathrm{add}, \mathrm{contains}, \mathrm{get}\}$. An element a is in the set if there is an $\mathrm{add}(a)$ event in the context, or informally, if the replica performing $\mathrm{contains}(a)$ has received a message about the addition of a:

$$F_{\mathrm{AOset}}(\mathrm{contains}(a), E, \mathrm{aop}, \mathrm{vis}, \mathrm{ar}) \ = \ (\exists e \in E. \, \mathrm{aop}(e) = \mathrm{add}(a)).$$

We define the result to be $\bot$ for add operations and define the result of get as expected.[2]

Things become more subtle if we allow removing elements, since we need to define the outcome of concurrent operations adding and removing the same element, as in the context $N = (\mathrm{contains}(42), \{e, f\}, \mathrm{aop}, \mathrm{vis}, \mathrm{ar})$, where $\mathrm{aop}(e) = \mathrm{add}(42)$ and $\mathrm{aop}(f) = \mathrm{remove}(42)$. There are several possible ways of resolving this conflict [8]: in **_add-wins sets_** (AWset) adds always win against concurrent removes (so that the element ends up in the set), **_remove-wins sets_** (RWset) act vice versa, and **_last-writer-wins sets_** (LWWset) apply operations in the order of their timestamps. We specify the result of contains in these cases using the vis and ar orders in the operation context:

$$F_{\mathrm{AWset}}(\mathrm{contains}(a), E, \mathrm{aop}, \mathrm{vis}, \mathrm{ar}) =$$
$$\exists e \in E. \, \mathrm{aop}(e) = \mathrm{add}(a) \wedge (\forall f \in E. \, \mathrm{aop}(f) = \mathrm{remove}(a) \implies \neg(e \xrightarrow{\mathrm{vis}} f));$$
$$F_{\mathrm{RWset}}(\mathrm{contains}(a), E, \mathrm{aop}, \mathrm{vis}, \mathrm{ar}) =$$
$$\exists e \in E. \, \mathrm{aop}(e) = \mathrm{add}(a) \wedge (\forall f \in E. \, \mathrm{aop}(f) = \mathrm{remove}(a) \implies f \xrightarrow{\mathrm{vis}} e);$$
$$F_{\mathrm{LWWset}}(\mathrm{contains}(a), E, \mathrm{aop}, \mathrm{vis}, \mathrm{ar}) =$$
$$\exists e \in E. \, \mathrm{aop}(e) = \mathrm{add}(a) \wedge (\forall f \in E. \, \mathrm{aop}(f) = \mathrm{remove}(a) \implies f \xrightarrow{\mathrm{ar}} e),$$
$$\text{if ar is total on } \{e \in E \mid \mathrm{aop}(e) \in \{\mathrm{add}(\_), \mathrm{remove}(\_)\}\};$$
$$F_{\mathrm{LWWset}}(\mathrm{contains}(a), E, \mathrm{aop}, \mathrm{vis}, \mathrm{ar}) = \text{undefined, otherwise.}$$

Thus, the add-wins semantics is formalised by mandating that remove operations cancel only the add operations that are visible to them; the remove-wins semantics additionally mandates that they cancel concurrent add operations, but not those that follow them in visibility. On the above context N, the operation $\mathrm{contains}(42)$ returns true iff: $\neg(e \xrightarrow{\mathrm{vis}} f)$ for AWset; $f \xrightarrow{\mathrm{vis}} e$ for RWset; and $f \xrightarrow{\mathrm{ar}} e$ for LWWset. As we show in §5, using a remove-wins set for _requesters_ in Figure 3 is crucial for preserving the integrity invariant (2); _friends_ could well be add-wins, which would lead to different, but also sensible, data type behaviour.

3.2 Whole-Store Semantics

We define the semantics of a causally consistent store by the set of its _histories_, which are certain structures on events recording all client-store interactions that can be pro-

---

[2] F_{AOset} is undefined on contexts with operations other than those from the signature. The type system of our language ensures that such contexts do not arise in its semantics.

duced during a run of the store; these include operations invoked on all objects and their return values. The store has no control over the operations occurring in histories, since these are chosen by the client; hence, the semantics only constrains return values. Replicated data type specifications define return values of operations in terms of visibility and arbitration, but where do these orders come from? As we explained in §3.1, intuitively, they are determined by the way messages are delivered and timestamps assigned in a run of a store implementation. Since this highly non-deterministic, in general, visibility and arbitration orders are arbitrary, but not entirely. A causally consistent store provides to its clients a guarantee that these orders in the contexts of different operations in the same run are related in certain ways, and this guarantee disallows anomalies such as the one in Figure 1(a).

We formalise the guarantee using the notion of an *execution*, which extends a history with visibility and arbitration orders on its events. A history is allowed by the store semantics if there is a way to extend it to an execution such that: *(i)* the return values of operations in the execution are obtained by applying replicated data type specifications to contexts extracted from it; and *(ii)* the execution satisfies certain *consistency axioms*, which constrain visibility and arbitration and, therefore, operation contexts.

Histories, Executions and the Satisfaction of Data Type Specifications. We identify objects (primitive or composite) by elements of the set Obj, ranged over by ω. A strict partial order R is **prefix-finite** if $\{ f \mid (f, e) \in R \}$ is finite for every e.

DEFINITION 3 *A **history** is a tuple* $H = (E, \mathsf{label}, \mathsf{so}, \sim)$, *where:*

- $E \subseteq \mathsf{Event}$.
- label : $E \to \mathsf{Obj} \times \mathsf{AOp} \times \mathsf{Val}$ *describes the events in* E: *if* $\mathsf{label}(e) = (\omega, p, a)$, *then the event* e *describes the applied operation* p *on the object* ω *returning the value* a.
- so $\subseteq E \times E$ *is a **session order**, ordering events in the same session according to the order in which they were submitted to the store. We require that* so *be prefix-finite and be the union of finitely many total orders defined on disjoint subsets of* E, *which correspond to events in different sessions.*
- $\sim \; \subseteq E \times E$ *is an equivalence relation grouping events in the same transaction. Since all transactions terminate (§2), we require that every equivalence class of* $\sim$ *be a finite set. Since every transaction is performed by a single session, we require that any two distinct events by the same transaction be related by* so *one way or another:*

$$\forall e, f.\, e \sim f \wedge e \neq f \implies e \xrightarrow{\mathsf{so}} f \vee f \xrightarrow{\mathsf{so}} e.$$

We also require that a transaction be contiguous in so:

$$\forall e, f, g.\, e \xrightarrow{\mathsf{so}} f \xrightarrow{\mathsf{so}} g \wedge e \sim g \implies e \sim f \sim g.$$

*An **execution** is a triple* $X = (H, \mathsf{vis}, \mathsf{ar})$ *of a history* H *and prefix-finite strict partial orders* vis *and* ar *on* $H.E$, *such that* $\mathsf{vis} \cup \mathsf{ar} \subseteq \{ (e, f) \mid H.\mathsf{obj}(e) = H.\mathsf{obj}(f) \}$ *and* $\mathsf{vis} \subseteq \mathsf{ar}$.

We denote the sets of all histories and executions by Hist and Exec. We write $H.\mathsf{obj}(e)$, $H.\mathsf{aop}(e)$ and $H.\mathsf{rval}(e)$ for the components of $H.\mathsf{label}(e)$ and shorten, e.g., $X.H.\mathsf{so}$

to X.so. Note that the set $H.E$ can be infinite, which models infinite runs. Figure 1(a) graphically represents an execution corresponding to the causality violation anomaly explained in §1. The relation $\sim$ is an identity in this case, and the objects in this and other executions in Figure 1 are add-only sets (AOset, §3.1).

Given an execution X, we extract the operation context of an event $e \in X.E$ by selecting all events visible to it according to X.vis:

$$\text{ctxt}(X, e) = (X.\text{aop}(e), E, (X.\text{aop})|_E, (X.\text{vis})|_E, (X.\text{ar})|_E), \tag{4}$$

where $E = (X.\text{vis})^{-1}(e)$ and $\cdot|_E$ is the restriction to events in E. Then, given a function $\mathbb{F} : \text{Obj} \rightharpoonup \text{Spec}$ that associates data type specifications with some objects, we say that an execution X satisfies $\mathbb{F}$ if the return value of every event in X is computed on its context according to the specification that $\mathbb{F}$ gives for the accessed object.

DEFINITION 4 *An execution X **satisfies** $\mathbb{F}$, written $X \models \mathbb{F}$, if*

$$\forall e \in X.E. \, (X.\text{obj}(e) \in \text{dom}(\mathbb{F}) \implies X.\text{rval}(e) = \mathbb{F}(X.\text{obj}(e))(\text{ctxt}(X, e))).$$

Since a context does not include return values, the above equation determines them uniquely for the events e satisfying the premise. For example, in the execution in Figure 1(a) the context of the get from ω_{fa} is empty. Hence, to satisfy $\mathbb{F} = (\lambda \omega. \, F_{\text{AOset}})$, the get returns $\emptyset$. If we had a vis edge from the add(b) to the get, then the latter would have to return $\{b\}$.

Consistency Axioms.   We now formulate additional constraints that executions have to satisfy. They restrict the anomalies allowed by the consistency model we consider and, in particular, rule out the execution in Figure 1(a).

To define the semantics of transactions we use the following operation. For a relation R on a set of events E and an equivalence relation $\sim$ on E (meant to group events in the same transaction), we define the *factoring* $R/\sim$ of R over $\sim$ as follows:

$$R/\sim \; = \; R \cup ((\sim; R; \sim) - (\sim)), \tag{5}$$

where ; composes relations. Thus, $R/\sim$ includes all edges from R and those obtained from such edges by relating any actions coming from the same transactions as their endpoints, excluding the case when the endpoints themselves are from the same transaction. We also let sameobj$(X)(e, f) \iff X.\text{obj}(e) = X.\text{obj}(f)$.

DEFINITION 5 *An execution $X = ((E, \text{label}, \text{so}, \sim), \text{vis}, \text{ar})$ is **causally consistent** if it satisfies the following **consistency axioms**:*

CAUSALVIS. $((\text{so} \cup \text{vis})/\sim)^+ \cap \text{sameobj}(X) \subseteq \text{vis};$
CAUSALAR. $(\text{so} \cup \text{ar})/\sim$ *is acyclic;*
EVENTUAL. $\forall e \in E. \, |\{f \in E \mid \text{sameobj}(X)(e, f) \land \neg(e \xrightarrow{\text{vis}} f)\}| < \infty.$

We write $X \models_{\text{CC}} \mathbb{F}$ if $X \models \mathbb{F}$ and X is causally consistent.

The axioms follow the informal description of the consistency model we gave in §1. We explain them below; however, their details are not crucial for understanding the rest of the paper. Before explaining the axioms, we note that Definitions 4 and 5 allow us to define the semantics of a store with object specifications given by $\mathbb{F} : \mathrm{Obj} \rightharpoonup \mathrm{Spec}$ as the set of histories that can be extended to a causally consistent execution satisfying $\mathbb{F}$:

$$\mathrm{HistCC}(\mathbb{F}) = \{H \mid \exists \mathsf{vis}, \mathsf{ar}. \, (H, \mathsf{vis}, \mathsf{ar}) \models_{\mathrm{CC}} \mathbb{F}\}. \tag{6}$$

To prove that a particular store implementation satisfies this specification, for every history H the implementation produces we have to come up with vis and ar that satisfy the constraint in (6); this is usually done by constructing them from message delivery and timestamps in the run of the implementation producing H. Here we rely on previous correctness proofs of store implementations [11, 12, 10] and use the above declarative specification of the store semantics without fixing the store implementation.

Causal Consistency. The axioms CAUSALVIS and CAUSALAR in Definition 5 ensure that visibility and arbitration respect causality between operations. CAUSALVIS guarantees that an event sees all events on the same object that causally affect it, i.e., those preceding it in a chain of session order and visibility edges (ignore the use of factoring over $\sim$ for now). Thus, CAUSALVIS disallows the execution in Figure 1(a). CAUSALAR similarly requires that arbitration be consistent with session order on all objects (recall that $X.\mathsf{vis} \subseteq X.\mathsf{ar}$). EVENTUAL formalises the liveness property that every replica eventually sees every update: it ensures that an event cannot be invisible to infinitely many other events on the same object.

Transactions. The use of factoring over the $\sim$ relation in CAUSALVIS formalises the guarantee provided by causally consistent transactions that we noted in §1: updates done by a transaction get delivered to replicas together. According to CAUSALVIS, a causal dependency established between two actions of different transactions results in a dependency also being established between any other actions in the two transactions. Thus, CAUSALVIS disallows the execution in Figure 1(c), where the dashed rectangles group events into transactions. The axioms allow the execution in Figure 1(b) even when the operations by the same session are done within a transaction—an outcome that would not be allowed with serialisable transactions.

4   Coarse-Grained Language Semantics

We now describe our main contribution—a coarse-grained denotational semantics of programs in the language of §2 that enables modular reasoning. We establish a correspondence between this semantics and the reference fine-grained semantics in §6.

4.1   Session-Local Semantics of Commands

The semantics of the replicated store defined by (6) in §3 describes the store behaviour under any client and thus produces histories with all possible sets of client operations. However, a particular command C in the language of §2 generates only histories with certain sequences of operations. Thus, our first step is to define a ***session-local*** semantics that, for each (sequential) command C, gives the set of histories that C can possibly generate. This semantics takes into account only the structure of the command C and

$$\langle \Delta \mid \Sigma \vdash C \rangle \; : \; (\mathrm{dom}(\Delta) \to_{\mathrm{inj}} \mathrm{Obj}) \times \mathsf{LState}(\Sigma) \to \mathcal{P}((\mathsf{FHist} \times \mathsf{LState}(\Sigma)) \cup \mathsf{IHist})$$

$$\langle v = G \rangle (obj, \sigma) = \{ (H_{\mathrm{emp}}, \sigma[v \mapsto [\![G]\!]\sigma]) \mid H_{\mathrm{emp}} = (\emptyset, [\,], \emptyset, \emptyset) \}$$

$$\langle v = x.o(G) \rangle (obj, \sigma) = \{ (H_e, \sigma[v \mapsto a]) \mid e \in \mathsf{Event} \wedge a \in \mathsf{Val}$$
$$\wedge\, H_e = (\{e\}, [e \mapsto (obj(x), o([\![G]\!]\sigma), a)], \emptyset, \{(e, e)\}) \}$$

$$\langle \mathsf{atomic}\,\{C\} \rangle (obj, \sigma) = \{ ((E, \mathsf{label}, \mathsf{so}, E \times E), \sigma') \mid ((E, \mathsf{label}, \mathsf{so}, \sim), \sigma') \in \langle C \rangle (obj, \sigma) \}$$

Fig. 4. Key clauses of the session-local semantics of commands. Here FHist and IHist are respectively sets of histories with finite and infinite event sets; $\sigma[v \mapsto a]$ denotes the function that has the same value as σ everywhere except v, where it has the value a; and $[\,]$ is a nowhere-defined function. We assume a standard semantics of expressions $[\![G]\!] : \mathsf{LState}(\Sigma) \to \mathsf{Val}$.

operations on local variables; the return values of operations executed on objects in the store are chosen arbitrarily. Later (§4.3), we intersect the set of histories produced by the session-local semantics with (6) to take the store semantics into account.

To track the values of local variables Σ in the session-local semantics of a command $\Delta \mid \Sigma \vdash C$ (Figure 2), we use **local states** $\sigma \in \mathsf{LState}(\Sigma) = \Sigma \to \mathsf{Val}$. The semantics interprets commands by the function $\langle \Delta \mid \Sigma \vdash C \rangle$ in Figure 4. Its first parameter obj determines the identities of objects bound to object variables in Δ. Given an initial local state σ as the other parameter, $\langle \Delta \mid \Sigma \vdash C \rangle$ returns the set of histories produced by C when run from σ, together with final local states when applicable. The semantics is mostly standard and therefore we give only key clauses; see [1, §A] for the remaining ones. Recall that, to simplify our formalism, we require every transaction to terminate (§2). To formalise this assumption, the clause for atomic filters out infinite histories.

4.2   Composite Data Type Semantics

The distinguishing feature of our coarse-grained semantics is its support for granularity abstraction: the denotation of a composite data type abstracts from its internal structure. Technically, this means that composite data types are interpreted in terms of replicated data type specifications, which we originally used for describing the meaning of primitive data types (§3.1). Thus, type variable environments Γ and data types $\Gamma \vdash T : O$ (Figure 2) are interpreted over the following domains:

$$[\![\Gamma]\!] = \mathrm{dom}(\Gamma) \to \mathsf{Spec}; \qquad [\![\Gamma \vdash T : O]\!] \in [\![\Gamma]\!] \to \mathsf{Spec}.$$

We use *type* to range over elements of $[\![\Gamma]\!]$. Two cases in the definition of $[\![\Gamma \vdash T : O]\!]$ are simple. We interpret a primitive data type $B \in \mathsf{PrimType}$ as the corresponding data type specification F_B, which is provided as part of the store specification (§3.1): $[\![B]\!]type = F_B$. We define the denotation of a type variable α by looking it up in the environment *type*: $[\![\alpha]\!]type = type(\alpha)$.

The remaining and most interesting case is the interpretation $[\![\Gamma \vdash D : O]\!]$ of a composite data type

$$D \;=\; \mathsf{let}\,\{x_j = \mathsf{new}\,T_j\}_{j=1..m} \;\mathsf{in}\; \{o = \mathsf{atomic}\,\{C_o\}\}_{o \in O}. \tag{7}$$

For *type* $\in [\![\Gamma]\!]$, the data type specification $F = [\![\Gamma \vdash D : O]\!]type$ returns a value given a context consisting of **coarse-grained** events that represent composite operations on

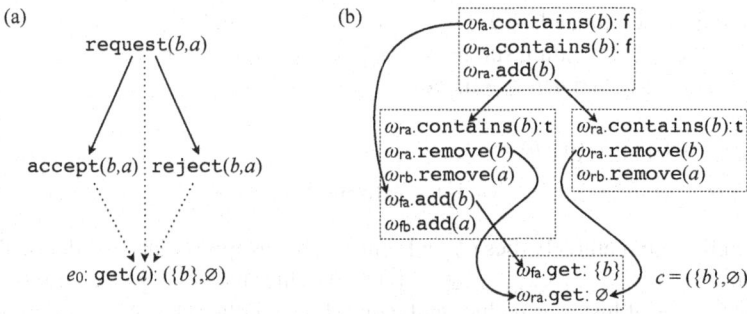

Fig. 5. (a) A context N of coarse-grained events for the social graph data type soc in Figure 3, with an event e_0 added to represent the operation $N.p$. Solid edges denote both visibility and arbitration (equal, since the data type does not use arbitration). The dashed edges show the additional edges in vis′ and ar′ introduced in Definition 7. (b) An execution X belonging to the concretisation of N. The objects $\omega_{\text{fa}}, \omega_{\text{fb}}, \omega_{\text{ra}}, \omega_{\text{rb}}$ correspond to the variables $friends[a]$, $friends[b]$, $requesters[a]$, $requesters[b]$ of type RWset. Solid edges denote both visibility and arbitration. We have omitted the session order inside transactions, the visibility and arbitration edges it induces and the transitive consequences of the edges shown. Dashed rectangles group events into transactions. The function β maps events in X to the horizontally aligned events in N.

an object of type D (e.g., the one in Figure 5(a)). This achieves granularity abstraction, because, once a denotation of this form is computed, it can be used to determine the return value of a composite operation without knowing the operations on the constituent objects x_j that were done by the implementations C_o of the composite operations in its context (e.g., the ones in Figure 3). We call events describing the operations on x_j *fine-grained*.

Informally, our approach to defining the denotation F of D is to determine the value that F has to return on a context N of coarse-grained events by "running" the implementations C_o of the composite operations invoked in N. This produces an execution X over fine-grained events that describes how C_o acts on the constituent objects x_j—a *concretisation* of N. The execution X has to be causally consistent and satisfy the data type specifications for the objects x_j. We then define $F(N)$ to be the return value that the implementation of the composite operation $N.p$ gives in X. However, concretising N into X is easier said than done: while the history part of X is determined by the session-local semantics of the implementations C_o (§4.1), determining the visibility and arbitration orders so that the resulting denotation be sound (in the sense described in §6) is nontrivial and represents our main insight.

To define the denotation of (7) formally, we first gather all histories that an implementation C_o of a composite operation can produce in the session-local semantics $\langle \cdot \rangle$ into a *summary*: given an applied composite operation and a return value, a summary defines the set of histories that its implementation produces when returning the value.

DEFINITION 6 *A **summary** ρ is a partial map $\rho : \text{AOp} \times \text{Val} \rightharpoonup \mathcal{P}(\text{FHist})$ such that for every $(p, a) \in \text{dom}(\rho)$, $\rho(p, a)$ is closed under the renaming of events, and for every $H \in \rho(p, a)$, $H.\text{so}$ is a total order on $H.E$ and $H.\sim = H.E \times H.E$.*

For a family of commands $\{\Delta \mid v_{\text{in}}, v_{\text{out}} \vdash C_o\}_{o \in O}$ and $obj : \text{dom}(\Delta) \rightarrow_{\text{inj}} \text{Obj}$, we define the corresponding summary $[\![\{C_o\}_{o \in O}]\!](obj) : \text{AOp} \times \text{Val} \rightharpoonup \mathcal{P}(\text{FHist})$ as follows: for $o' \in O$ and $a, b \in \text{Val}$, we let

$$[\![\{C_o\}_{o \in O}]\!](obj)(o'(a), b) =$$
$$\{H \mid (H, [v_{\text{in}} \mapsto \_, v_{\text{out}} \mapsto b]) \in \langle \text{atomic } \{C_{o'}\}\rangle(obj, [v_{\text{in}} \mapsto a, v_{\text{out}} \mapsto \perp])\}.$$

For example, the method bodies C_o in Figure 3 and an appropriate obj define the summary $\rho_{\text{soc}} = [\![\{C_o\}_{o \in \{\text{request},\text{accept},\dots\}}]\!](obj)$. This maps the get operation in Figure 5(a) to a set of histories including the one shown to the right of it in Figure 5(b).

We now define the executions X that may result from "running" the implementations of composite operations in a coarse-grained context N given by a summary ρ. The definition below pairs these executions X with the value c returned in them by the implementation of $N.p$, since this is what we are ultimately interested in. We first state the formal definition, and then explain it in detail. We write id for the identity relation.

DEFINITION 7 *A pair $(X, c) \in \text{Exec} \times \text{Val}$ is a **concretisation** of a context N with respect to a summary $\rho : \text{AOp} \times \text{Val} \rightharpoonup \mathcal{P}(\text{FHist})$ if for some event $e_0 \notin N.E$ and function $\beta : X.E \rightarrow N.E \uplus \{e_0\}$ we have*

$$(\forall f \in (N.E).\ (X.H)|_{\beta^{-1}(f)} \in \rho(N.\text{aop}(f), \_)) \wedge ((X.H)|_{\beta^{-1}(e_0)} \in \rho(N.p, c)); \quad (8)$$
$$\beta(X.\text{so}) \subseteq \text{id}; \quad (9)$$
$$\beta(X.\text{vis}) - \text{id} \subseteq \text{vis}'; \quad (10)$$
$$\beta^{-1}(\text{vis}') \cap \text{sameobj}(X) \subseteq X.\text{vis}; \quad (11)$$
$$\beta(X.\text{ar}) - \text{id} \subseteq \text{ar}', \quad (12)$$

where $\text{vis}' = N.\text{vis} \cup \{(f, e_0) \mid f \in N.E\}$ and $\text{ar}' = N.\text{ar} \cup \{(f, e_0) \mid f \in N.E\}$. We write $\gamma(N, \rho)$ for the set of all concretisations of N with respect to ρ.

For example, the pair of the execution and the value in Figure 5(b) belongs to $\gamma(N, \rho_{\text{soc}})$ for N in Figure 5(a). When X concretises N with respect to ρ, the history $X.H$ is a result of expanding every composite operation in N into a history of its implementation according to ρ. The function β maps every event in $X.E$ to the event from N it came from, with an event e_0 added to $N.E$ to represent the operation $N.p$; this is formalised by (8). The condition (9) further requires that the implementation of every composite operation be executed in a dedicated session. As it happens, it is enough to consider concretisations of this form to define the denotation.

The conditions (10)–(12) represent the main insight of our definition of the denotation: they tell us how to select the visibility and arbitration orders in X given those in N. They are best understood by appealing to the intuition about how an implementation of the store operates. Recall that, from this perspective, visibility captures message delivery: an event is visible to another event if and only if the information about the former has been delivered to the replica of the latter (§3.1). Also, in implementations of causally consistent transactions, updates done by a transaction are delivered to every replica together (§1). Since composite operations execute inside transactions, the visibility order in N can thus be intuitively thought of as specifying the delivery of groups of updates made by them: we have an edge $e' \xrightarrow{\text{vis}'} f'$ between coarse-grained events e' and f' in N (e.g., request and accept in Figure 5(a)) if and only if the updates performed by the transaction denoted by e' have been delivered to the replica of f'. Now consider fine-grained events $e, f \in X.E$ on the same constituent object describing updates made inside the transactions of e' and f', so that $\beta(e) = e'$ and $\beta(f) = f'$ (e.g., $\omega_{\text{ra}}.\text{add}(b)$ and $\omega_{\text{ra}}.\text{contains}(b)$ in Figure 5(b)). Then we can have $e \xrightarrow{X.\text{vis}} f$ if and only if $e' \xrightarrow{\text{vis}'} f'$. This is formalised by (10) and (11).

To explain (12), recall that arbitration captures the order of timestamps assigned to events by the store implementation. Also, in implementations the timestamps of all updates done by a transaction are contiguous in this order. Thus, arbitration in N can be thought of as specifying the timestamp order on the level of whole transactions corresponding to the composite operations in N. Then (12) states that the order of timestamps of fine-grained events in X is consistent with that over transactions these events come from.

To define the denotation, we need to consider only those executions concretising N that are causally consistent and satisfy data type specifications. Hence, for $\mathbb{F} : \text{Obj} \rightharpoonup \text{Spec}$ we let

$$\gamma(N, \rho, \mathbb{F}) = \{(X, c) \in \gamma(N, \rho) \mid X \models_{\text{CC}} \mathbb{F}\}.$$

For example, the execution in Figure 5(b) belongs to $\gamma(N, \rho_{\text{soc}}, \mathbb{F})$ for N in Figure 5(a) and $\mathbb{F} = (\lambda\omega. F_{\text{RWset}})$. As the following theorem shows, the constraints (8)–(12) are so tight that the set of concretisations defined in this way never contains two different return values; this holds even if we allow choosing object identities differently.

THEOREM 8 *Given a family* $\{\Delta \mid v_{\text{in}}, v_{\text{out}} \vdash C_o\}_{o \in O}$, *we have:*

$$\forall N. \forall obj_1, obj_2 \in [\text{dom}(\Delta) \rightarrow_{\text{inj}} \text{Obj}].$$
$$\forall \mathbb{F}_1 \in [\text{range}(obj_1) \rightarrow \text{Spec}]. \forall \mathbb{F}_2 \in [\text{range}(obj_2) \rightarrow \text{Spec}].$$
$$(\forall x \in \text{dom}(\Delta). \mathbb{F}_1(obj_1(x)) = \mathbb{F}_2(obj_2(x))) \implies$$
$$\forall (X_1, c_1) \in \gamma(N, [\![\{C_o\}_{o \in O}]\!](obj_1), \mathbb{F}_1).$$
$$\forall (X_2, c_2) \in \gamma(N, [\![\{C_o\}_{o \in O}]\!](obj_2), \mathbb{F}_2). c_1 = c_2.$$

This allows us to define the denotation of (7) according to the outline we gave before.

DEFINITION 9 *For (7) we let* $[\![\Gamma \vdash D]\!]type = F$, *where* $F : \text{Ctxt} \rightharpoonup \text{Val}$ *is defined as follows: for* $N \in \text{Ctxt}$ *and* $c \in \text{Val}$, *if*

$$\exists obj \in [\{x_j \mid j = 1..m\} \rightarrow_{\text{inj}} \text{Obj}]. \exists \mathbb{F} \in [\text{range}(obj) \rightarrow \text{Spec}].$$
$$(\forall j = 1..m. \mathbb{F}(obj(x_j)) = [\![T_j]\!]type) \wedge (\_, c) \in \gamma(N, [\![\{C_o\}_{o \in O}]\!](obj), \mathbb{F}),$$

then $F(N) = c$; *otherwise* $F(N)$ *is undefined.*

The existence and uniqueness of F in the definition follow from Theorem 8. It is easy to check that F defined above satisfies all the properties required in Definition 2 and, hence, $F \in \text{Spec}$. According to the above definition, the denotation of the data type in Figure 3 has to give $(\{b\}, \emptyset)$ on the context in Figure 5(a).

$$\llbracket \Gamma \mid \Delta \vdash P \rrbracket \ : \ \llbracket \Gamma \rrbracket \to \prod_{obj \in [\mathrm{dom}(\Delta) \to_{\mathrm{inj}} \mathrm{Obj}]} ((\mathrm{range}(obj) \rightharpoonup \mathsf{Spec}) \to \mathcal{P}(\mathsf{Hist}))$$

$$\llbracket \mathsf{let}\ \alpha = T\ \mathsf{in}\ P \rrbracket(type, obj, \mathbb{F}) = \llbracket P \rrbracket(type[\alpha \mapsto \llbracket T \rrbracket type], obj, \mathbb{F})$$

$$\llbracket \mathsf{let}\ x = \mathsf{new}\ T\ \mathsf{in}\ P \rrbracket(type, obj, \mathbb{F}) = \bigcup \{ \llbracket P \rrbracket(type, obj[x \mapsto \omega], \mathbb{F}[\omega \mapsto \llbracket T \rrbracket type]) \mid$$
$$\omega \notin \mathrm{range}(obj) \}$$

$$\llbracket C_1 \parallel \ldots \parallel C_n \rrbracket(type, obj, \mathbb{F}) = \mathsf{HistCC}(\mathbb{F}) \cap \big\{ \biguplus_{j=1}^{n} H_j \mid \forall j = 1..n.$$
$$(H_j, \_) \in \langle C_j \rangle(obj, []) \ \vee\ H_j \in \langle C_j \rangle(obj, []) \big\}$$

Fig. 6. Semantics of $\Gamma \mid \Delta \vdash P$. Here $H \uplus H' = (H.E \uplus H'.E,\ H.\mathrm{label} \uplus H'.\mathrm{label},\ H.\mathrm{so} \cup H'.\mathrm{so},\ H.\sim \cup H'.\sim)$; undefined if so is $H.E \uplus H'.E$.

4.3   Program Semantics

Having defined the denotations of composite data types, we give the semantics to a program in the language of §2 by instantiating (6) with an $\mathbb{F}$ computed from these denotations and by intersecting the result with the set of histories that can be produced by the program according to the session-local semantics of its sessions (§4.1). A program $\Gamma \mid \Delta \vdash P$ is interpreted with respect to environments $type$, obj and $\mathbb{F}$, which give the semantics of data type variables in Γ, the identities of objects in Δ and the specifications associated with these objects (Figure 6). A data type variable declaration extends the $type$ environment with the specification of the data type computed from its declaration as described in §4.2. An object variable declaration extends obj with a fresh object and $\mathbb{F}$ with the specification corresponding to its type. A client is interpreted by combining all histories its sessions produce in the session-local semantics with respect to obj and intersecting the result with (6). Note that we originally defined the store semantics (6) under the assumption that all replicated data types are primitive. Here we are able to reuse the definition because our denotations of composite data types have the same form as those of primitive ones.

Using the Semantics.   Our denotational semantics enables modular reasoning about programs with composite replicated data types. Namely, it allows us to check if a program P can produce a given history H by: *(i)* computing the denotations $\mathbb{F}$ of the composite data types used in P; and *(ii)* checking if the client of P can produce H assuming it uses *primitive* data types with the specifications $\mathbb{F}$. Due to the granularity abstraction in our denotation, it represents every invocation of a composite operation by a single event and thereby abstracts from its internal structure. In particular, different composite data type implementations can have the same denotation describing the data type behaviour. As a consequence, in *(ii)* we can pretend that composite data types are primitive and thus do not have to reason about the behaviour of their implementations and the client together. For example, we can determine how a program using the social graph data type behaves in the situation shown in Figure 5(a) using the result the data type denotation gives on this context, without considering how its implementation behaves (cf. Figure 5(b)). We get the same benefits when reasoning about a complex composite data type D constructed from simpler composite data types T_j as in (7): we can first compute the denotations of T_j and then use the results in reasoning about D.

   In practice, we do not compute the denotation of a composite data type D using Definition 9 directly. Instead, we typically invent a specification F that describes the desired behaviour of D, and then prove that F is equal to the denotation of D, i.e., that D is *correct* with respect to F. Definition 9 and, in particular, constraints (8)–(12), give a *proof method* for establishing this. The next section illustrates this on an example.

5 Example: Social Graph

We have applied the composite data type denotation in §4 to specify and prove the correctness of three composite data types: *(i)* the social graph data type in Figure 3; *(ii)* a shopping cart data type implemented using an add-wins set, which resolves conflicts between concurrent changes to the quantity of the same product; *(iii)* a data type that uses transactions to simultaneously update several objects that resolve conflicts using the last-writer-wins policy (cf. LWWset from §3.1). The latter example uses arbitration in a nontrivial way. Due to space constraints, we focus here on the social graph data type and defer the others to [1, §D].

Below we give a specification F_{soc} to the social graph data type, which we have proved to be the denotation of its implementation D_{soc} in Figure 3. The proof is done by considering an arbitrary context N and its concretisation (X, c) according to Definition 7 and showing that $F_{\text{soc}}(N) = c$. The constraints (8)–(12) make the required reasoning mostly mechanical and therefore we defer the easy proof to [1, §D] and only illustrate the correspondence between D_{soc} and F_{soc} on examples.

The function F_{soc} is defined recursively using the following operation that selects a subcontext of a given event in a context, analogously to the ctxt operation on executions (4) from §3.2. For a partial context M and an event $e \in M.E$, we let

$$\text{ctxt}(M, e) = (M.\text{aop}(e), E, (M.\text{aop})|_E, (M.\text{vis})|_E, (M.\text{ar})|_E),$$

where $E = (M.\text{vis})^{-1}(e)$. Then

$F_{\text{soc}}(\texttt{get}(a), M) =$
$(\{b \mid \exists e \in (M.E).\,(M.\text{aop}(e) = \texttt{accept}((b,a) \mid (a,b))) \wedge F_{\text{soc}}(\text{ctxt}(M, e)) \wedge$
$\quad (\forall f \in (M.E).\,(M.\text{aop}(f) \in \texttt{breakup}((b,a) \mid (a,b))) \wedge F_{\text{soc}}(\text{ctxt}(M, f))$
$$\implies f \xrightarrow{\text{vis}} e)\},$$
$\quad \{b \mid \exists e \in (M.E).\,(M.\text{aop}(e) = \texttt{request}(b,a)) \wedge F_{\text{soc}}(\text{ctxt}(M, e)) \wedge$
$\quad\quad (\forall f \in (M.E).\,(M.\text{aop}(f) \in (\texttt{accept} \mid \texttt{reject})((b,a) \mid (a,b))) \wedge F_{\text{soc}}(\text{ctxt}(M, f))$
$$\implies f \xrightarrow{\text{vis}} e)\});$$
$F_{\text{soc}}(\texttt{accept}(b, a), M) = (b \in \text{snd}(F_{\text{soc}}(\texttt{get}(a), M))).$

The results of request, reject and breakup are defined similarly to accept. For brevity, we use the notation $(G_1 \mid G_2)$ above to denote the set arising from picking either G_1 or G_2 as the subexpression of the expression where it occurs. Even though the definition looks complicated, its conceptual idea is simple and has a temporal flavour. Our definition takes into account that: after breaking up, users can become friends again; and sometimes data type operations are unsuccessful, in which case they return false. According to the two components of $F_{\text{soc}}(\texttt{get}(a), M)$:

1. a's friends are the accounts b with a successful accept operation between a and b such that any successful breakup between them was in its past, as formalised by visibility. We determine whether an operation was successful by calling F_{soc} recursively on its subcontext.

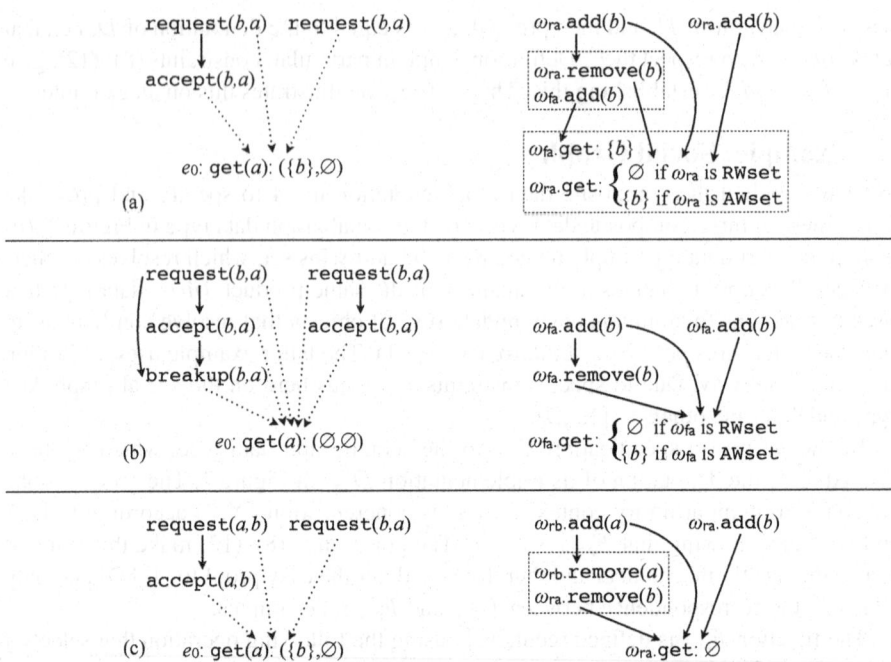

Fig. 7. (Left) Coarse-grained contexts of the social graph data type together with the result that F_{soc} gives on them. (Right) Relevant events of the fine-grained executions of the implementation in Figure 3 resulting from concretising the contexts according to Definition 7. We use the same conventions as in Figure 5.

2. a's requesters are the accounts b with a successful $\texttt{request}(b, a)$ operation such that any successful $\texttt{accept}$ or $\texttt{reject}$ between a and b was in its past.

This specifies the behaviour of the data type while abstracting from its implementation, thereby enabling modular reasoning about programs using it (§4.3).

Our specification F_{soc} can be used to analyse the behaviour of the implementation in Figure 3. By a simple unrolling of the definition of F_{soc}, it is easy to check that the two sets returned by $F_{\text{soc}}(\texttt{get}(a), M)$ are disjoint and, hence, the invariant (2) in §2 holds; (1) can be checked similarly. Also, since F_{soc} returns $(\{b\}, \emptyset)$ on the context in Figure 5(a), when the same friendship request is concurrently accepted and rejected, the accept wins. Different behaviour could also be reasonable; the decision ultimately depends on application requirements.

We now illustrate the correspondence between D_{soc} and F_{soc} on examples and, on the way, show that our coarse-grained semantics lets one understand how the choice of conflict-resolution policies on constituent objects affects the policy of the composite data type. First, we argue that making *requesters* remove-wins in Figure 3 is crucial for preserving the integrity invariant (2) and satisfying F_{soc}. Indeed, consider the scenario shown in Figure 7(a). Here two users managing the same account b concurrently issue friendship requests to a, which initially sees only one of them. If *requesters* were add-wins, the $\texttt{accept}$ by a would affect only the request that it sees. The remaining request would eventually propagate to all replicas in the system, and the calls to get in the implementation would thus return b as being both a friend and a requester of a's, violating (2). The remove-wins policy of *requesters* ensures that, when a user accepts or rejects a request, this also removes all identical requests issued concurrently.

If we made *friends* add-wins, this would make the data type behave differently, but sensibly, as illustrated in Figure 7(b). Here we again have two concurrently issued requests from b to a. The account a may also be managed by multiple users, which concurrently accept the requests they happen to see. One of the users then immediately breaks up with a. Since *friends* are remove-wins, this cancels the addition of b to $friends[a]$ (i.e., ω_{fa}) resulting from the concurrent accept by the other user; thus, b ends up not being a's friend, as prescribed by F_{soc}. Making *friends* add-wins would result in the reverse outcome, and F_{soc} would have to change accordingly. Thus, the conflict-resolution policy on *friends* determines the way conflicts between accept and breakup are resolved.

Finally, if users a and b issue friendship requests to each other concurrently, a decision such as an accept taken on one of them will also affect the other, as illustrated in Figure 7(c). To handle this situation without violating (2), accept removes not only the request it is resolving, but also the symmetric one.

6 Fine-Grained Language Semantics, Soundness and Completeness

To justify that the coarse-grained semantics from §4 is sensible, we relate it to a *fine-grained semantics* that follows the standard way of defining language semantics on weak consistency models [11]. Unlike the coarse-grained semantics, the fine-grained one is defined non-compositionally: it considers only certain *complete* programs and defines the denotation of a program as a whole, without separately defining denotations of composite data types in it. This denotation is computed using histories that record all operations on all primitive objects comprising the composite data types in the program; hence, the name fine-grained. The semantics includes those histories that can be produced by the program in the session-local semantics (§4.1) and are allowed by the semantics of the store managing the primitive objects the program uses (§3).

We state the correspondence between the coarse-grained and fine-grained semantics as an equivalence of the *externally-observable behaviour* of a program in the two semantics. Let us fix a variable $x_{io} \in$ OVar and an object io $\in$ Obj used to interpret x_{io}. A program P is **complete** if $\emptyset \mid x_{io} : \{o_{io}\} \vdash P$. The operation o_{io} on x_{io} models a combined user input-output action, rather than an operation on the store, and the externally-observable behaviour of a complete program P is given by operations on x_{io} it performs. Formally, for a history H let observ(H) be its projection to events on io: $\{e \in H \mid H.\text{obj}(e) = \text{io}\}$. We lift observ to sets of histories pointwise. Then we define the set of externally-observable behaviours of a complete program P in the coarse-grained semantics of §4 as $[\![P]\!]_{CG} = \text{observ}([\![P]\!]([\,], [x_{io} : \text{io}], [\,]))$. Note that our semantics does not restrict the values returned by o_{io}, thus accepting any input.

To define the fine-grained semantics of a complete program P, we flatten P by inlining composite data type definitions using a series of reductions $\longrightarrow$ on programs (defined shortly). Applying the reductions exhaustively yields programs with only objects of primitive data types, which have the following normal form:

$$\bar{P} ::= C_1 \| \dots \| C_n \mid \text{let } x = \text{new } B \text{ in } \bar{P}$$

Given a complete program P, consider the unique $\bar{P}$ such that $P \longrightarrow^* \bar{P}$ and $\bar{P} \nrightarrow \_$. Then we define the denotation of P in the fine-grained semantics by the set of externally-observable behaviours that $\bar{P}$ produces when interacting with a causally consistent store managing the primitive objects it uses. To formalise this, we reuse the definition of the coarse-grained semantics and define the denotation of P in the fine-grained semantics as $[\![P]\!]_{\text{FG}} = [\![\bar{P}]\!]_{\text{CG}}$. Since $\bar{P}$ contains only primitive data types, this does not use the composite data type denotation of §4.2.

We now define the reduction $\longrightarrow$. Let Comm be the set of commands C in Figure 2. We use an operator $subst$ that takes a mapping $S : \text{OVar} \times \text{Op} \rightharpoonup \text{Comm}$ and a command C or a program P, and replaces invocations of object operations in C or P according to S. The key clauses defining $subst$ are as follows:

$$subst(S, v = x.o(G)) \;=\; \text{if } ((x, o) \notin \text{dom}(S)) \text{ then } (v = x.o(G))$$
$$\text{else } (\text{atomic } \{\text{var } v_1. \text{ var } v_2.\, v_1 = G; (S(x, o)[v_1/v_{\text{in}}, v_2/v_{\text{out}}]); v = v_2\})$$
$$subst(S, \text{let } x = \text{new } T \text{ in } P) \;=\; \text{let } x = \text{new } T \text{ in } subst(S|_{\neg x}, P)$$
$$subst(S, \text{let } \alpha = T \text{ in } P) \;=\; \text{let } \alpha = T \text{ in } subst(S, P)$$
$$subst(S, C_1 \| \dots \| C_n) \;=\; subst(S, C_1) \| \dots \| subst(S, C_n)$$

Here v_1, v_2 are fresh ordinary variables, and $S|_{\neg x}$ denotes S with its domain restricted to $(\text{OVar} \setminus \{x\}) \times \text{Op}$. Applying $subst$ to an assignment command does not change the command, and applying it to all others results in recursive applications of $subst$ to their subexpressions. Then the relation $\longrightarrow$ is defined as follows:

$$\mathcal{P} ::= [-] \mid \text{let } x = \text{new } T \text{ in } \mathcal{P} \mid \text{let } \alpha = T \text{ in } \mathcal{P}$$
$$\mathcal{P}[\text{let } \alpha = T \text{ in } P] \;\longrightarrow\; \mathcal{P}[P[T/\alpha]]$$
$$\mathcal{P}[\text{let } x = \text{new } (\text{let } \{x_j = \text{new } T_j\}_{j=1..m} \text{ in } \{o = \text{atomic } \{C_o\}\}_{o \in O}) \text{ in } P]$$
$$\longrightarrow \mathcal{P}[\text{let } \{x_j = \text{new } T_j\}_{j=1..m} \text{ in } subst(\{(x, o) \mapsto C_o \mid o \in O\}, P)],$$

where x_j do not occur in P. The first reduction rule replaces data-type variables by their definitions, and the second defines the semantics of composite operations via inlining.

Our central technical result is that the coarse-grained semantics of §4 is sound and complete with respect to the fine-grained semantics presented here: the sets of externally-observable behaviours of programs in the two semantics coincide.

THEOREM 10 *For every complete program P we have $[\![P]\!]_{\text{FG}} = [\![P]\!]_{\text{CG}}$.*

We give a (highly nontrivial) proof in [1, §C]. The theorem allows us to reason about programs using the coarse-grained semantics, which enables granularity abstraction and modular reasoning (§4.3). It also implies that our denotational semantics is *adequate*, i.e., can be used to prove the observational equivalence of two data type implementations D_1 and D_2: if $[\![D_1]\!] = [\![D_2]\!]$, then $[\![\mathcal{C}[D_1]]\!]_{\text{FG}} = [\![\mathcal{C}[D_2]]\!]_{\text{FG}}$ for all contexts $\mathcal{C}$ of the form $\mathcal{P}[\text{let } \alpha = [-] \text{ in } P]$. Note that both soundness and completeness are needed to imply this property.

7   Related Work

One of the classical questions of data abstraction is: how can we define the semantics of a data type implementation that abstracts away the implementation details, including a particular choice of data representation? Our results can be viewed as revisiting this question, which has so far been investigated in the context of sequential [15] and shared-memory concurrent [13, 25] programs, in the emerging domain of eventually consistent distributed systems. Most of the work on data abstraction for concurrency has considered a strongly consistent setting [13, 25]. Thus, it usually aimed to achieve *atomicity abstraction*, which allows one to pretend that a composite command takes effect atomically throughout the system. Here we consider data abstraction in the more challenging setting of weak consistency and achieve a weaker and more subtle guarantee of *granularity abstraction*: although our coarse-grained semantics represents composite operations by single events, these events are still subject to anomalies of causal consistency, with different replicas being able to see the events at different times.

We are aware of only a few previous data abstraction results for weak consistency [16, 9, 6]. The most closely related is the one for the C/C++ memory model by Batty et al. [6]. Like the consistency model we consider, the C/C++ model is defined axiomatically, which leads to some similarities in the general approach followed in [6] and in this paper. However, other features of the settings considered are different. First, we consider arbitrary replicated data types, whereas, as any model of a shared-memory language, the C/C++ one considers only registers with the last-writer-wins conflict-resolution policy. Second, the artefacts related during abstraction in [6] and in this paper are different. Instead of composite replicated data types, [6] considers *libraries*, which encapsulate last-writer-wins registers and operations accessing them implemented by arbitrary code without using transactions. A specification of a library is then just another library, but with operations implemented using atomic blocks reminiscent of our transactions. Hence, a single invocation of an operation of a specification library is still represented by multiple events and therefore [6] does not support granularity abstraction to the extent achieved here. Our work can roughly be viewed as starting where [6] left off, with composite constructions whose operations are implemented using transactions, and specifying their behaviour more declaratively with replicated data type specifications over contexts of coarse-grained events. It is thus possible that our approach can be adapted to give more declarative specifications to C/C++ libraries.

Researchers and developers have often implemented complex objects with domain-specific conflict resolution policies inside replicated stores [22], which requires dealing with low-level details, such as message exchange between replicas. Burckhardt et al. [11] also proposed a method for proving the correctness of such replicated data type implementations with respect to specifications of Definition 2. Our results show that, using causally consistent transactions, complex domain-specific objects can often be implemented as composite replicated data types, using a high-level programming model to compose replicated objects and their conflict-resolution policies. Furthermore, due to the granularity abstraction we established, the resulting objects can be viewed as no different from those implemented inside the store. The higher-level programming model we consider makes our proof method significantly different from that of Burckhardt et al.

Partial orders, such as event structures [20] and Mazurkiewicz traces [21], have been used to define semantics of concurrent or distributed programs by explicitly expressing the dependency relationships among events such programs generate. Our results extend this line of semantics research by considering new kinds of relations among events, describing computations of eventually consistent replicated stores, and studying how consistency axioms on these relations interact with the granularity abstraction for composite replicated data types.

8 Conclusion

In this paper we have proposed the concept of composite replicated data types, which formalises a common way of organising applications on top of eventually consistent stores. We have presented a coarse-grained denotational semantics for these data types that supports granularity abstraction: the semantics allows us to abstract from the internals of a composite data type implementation and pretend that it represents a single monolithic object, which simplifies reasoning about client programs. We have also shown that our semantics is sound and complete with respect to a standard non-compositional semantics.

One important derivative of our semantics is a mechanism for specifying composite data types where we regard all operations of these data types as atomic, and describe their return values for executions that consist of such atomic operations. As our soundness and completeness results show, this mechanism is powerful enough to capture all essential aspects of composite replicated data types. Using a nontrivial example, we have illustrated how the denotation of a data type in our semantics specifies its behaviour in tricky situations and thereby lets one understand the consequences of different design decisions in its implementation.

As we explained in §1, developing correct programs on top of eventually consistent stores is a challenging yet unavoidable task. Our results mark the first step towards providing developers with methods and tools for specifying and verifying programs in this new programming environment and expanding the rich theories of programming languages, such as data abstraction, to this environment. Even though our results were developed for a particular popular variant of eventual consistency—causally consistent transactions—we hope that in the future the results can be generalised to other consistency models with similar formalisations [10, 4]. Another natural future direction is to use our coarse-grained semantics to propose a logic for reasoning about composite data types symbolically.

References

[1] Extended version of this paper. Available from the submission system
[2] Microsoft TouchDevelop, https://www.touchdevelop.com/
[3] Abadi, D.: Consistency tradeoffs in modern distributed database system design: CAP is only part of the story. IEEE Computer (2012)
[4] Bailis, P., Davidson, A., Fekete, A., Ghodsi, A., Hellerstein, J.M., Stoica, I.: Highly Available Transactions: virtues and limitations. In: VLDB (2014)
[5] Bailis, P., Ghodsi, A.: Eventual consistency today: Limitations, extensions, and beyond. CACM 56(5) (2013)

[6] Batty, M., Dodds, M., Gotsman, A.: Library abstraction for C/C++ concurrency. In: POPL (2013)

[7] Batty, M., Owens, S., Sarkar, S., Sewell, P., Weber, T.: Mathematizing C++ concurrency. In: POPL (2011)

[8] Bieniusa, A., Zawirski, M., Preguiça, N.M., Shapiro, M., Baquero, C., Balegas, V., Duarte, S.: Semantics of eventually consistent replicated sets. In: DISC (2012)

[9] Burckhardt, S., Gotsman, A., Musuvathi, M., Yang, H.: Concurrent library correctness on the TSO memory model. In: Seidl, H. (ed.) Programming Languages and Systems. LNCS, vol. 7211, pp. 87–107. Springer, Heidelberg (2012)

[10] Burckhardt, S., Gotsman, A., Yang, H.: Understanding eventual consistency. Technical Report MSR-TR-2013-39, Microsoft (2013)

[11] Burckhardt, S., Gotsman, A., Yang, H., Zawirski, M.: Replicated data types: specification, verification, optimality. In: POPL (2014)

[12] Burckhardt, S., Leijen, D., Fähndrich, M., Sagiv, M.: Eventually consistent transactions. In: Seidl, H. (ed.) Programming Languages and Systems. LNCS, vol. 7211, pp. 67–86. Springer, Heidelberg (2012)

[13] Filipovic, I., O'Hearn, P.W., Rinetzky, N., Yang, H.: Abstraction for concurrent objects. Theor. Comput. Sci. 411(51-52) (2010)

[14] Gilbert, S., Lynch, N.: Brewer's conjecture and the feasibility of consistent, available, partition-tolerant web services. SIGACT News 33(2) (2002)

[15] Hoare, C.A.R.: Proof of correctness of data representations. Acta. Inf. 1 (1972)

[16] Jagadeesan, R., Petri, G., Pitcher, C., Riely, J.: Quarantining weakness. In: Felleisen, M., Gardner, P. (eds.) ESOP 2013. LNCS, vol. 7792, pp. 492–511. Springer, Heidelberg (2013)

[17] Li, C., Porto, D., Clement, A., Rodrigues, R., Preguiça, N., Gehrke, J.: Making geo-replicated systems fast if possible, consistent when necessary. In: OSDI (2012)

[18] Lloyd, W., Freedman, M.J., Kaminsky, M., Andersen, D.G.: Don't settle for eventual: scalable causal consistency for wide-area storage with COPS. In: SOSP (2011)

[19] Lloyd, W., Freedman, M.J., Kaminsky, M., Andersen, D.G.: Stronger semantics for low-latency geo-replicated storage. In: NSDI (2013)

[20] Nielsen, M., Plotkin, G.D., Winskel, G.: Petri nets, event structures and domains. In: Semantics of Concurrent Computation (1979)

[21] Nielsen, M., Sassone, V., Winskel, G.: Relationships between models of concurrency. In: REX School/Symposium (1993)

[22] M. Shapiro, N. Preguiça, C. Baquero, and M. Zawirski. A comprehensive study of Convergent and Commutative Replicated Data Types. Technical Report 7506, INRIA (2011)

[23] Shapiro, M., Preguiça, N., Baquero, C., Zawirski, M.: Conflict-free replicated data types. In: Défago, X., Petit, F., Villain, V. (eds.) SSS 2011. LNCS, vol. 6976, pp. 386–400. Springer, Heidelberg (2011)

[24] Sovran, Y., Power, R., Aguilera, M.K., Li, J.: Transactional storage for geo-replicated systems. In: SOSP (2011)

[25] Turon, A., Dreyer, D., Birkedal, L.: Unifying refinement and Hoare-style reasoning in a logic for higher-order concurrency. In: ICFP (2013)

[26] Zawirski, M., Bieniusa, A., Balegas, V., Duarte, S., Baquero, C., Shapiro, M., Preguiça, N.: SwiftCloud: Fault-tolerant geo-replication integrated all the way to the client machine. Technical Report 8347, INRIA (2013)

Relaxed Stratification: A New Approach to Practical Complete Predicate Refinement[*]

Tachio Terauchi[1] and Hiroshi Unno[2]

[1] JAIST
terauchi@jaist.ac.jp
[2] University of Tsukuba
uhiro@cs.tsukuba.ac.jp

Abstract. In counterexample-guided abstraction refinement, a predicate refinement scheme is said to be *complete* for a given theory if it is guaranteed to eventually find predicates sufficient to prove the given property, when such exist. However, existing complete methods require deciding if a proof of the counterexample's spuriousness exists in some finite language of predicates. Such an *exact* finite-language-restricted predicate search is quite hard for many theories used in practice and incurs a heavy overhead. In this paper, we address the issue by showing that the language restriction can be relaxed so that the refinement process is restricted to infer proofs from some finite language $L_{base} \cup L_{ext}$ but is only required to return a proof when the counterexample's spuriousness can be proved in L_{base}. Then, we show how a proof-based refinement algorithm can be made to satisfy the relaxed requirement and be complete by restricting only the theory-level reasoning in SMT to emit L_{base}-restricted partial interpolants (while such an approach has been proposed previously, we show for the first time that it can be done for languages that are not closed under conjunctions and disjunctions). We also present a technique that uses a property of counterexample patterns to further improve the efficiency of the refinement algorithm while still satisfying the requirement. We have experimented with a prototype implementation of the new refinement algorithm, and show that it is able to achieve complete refinement with only a small overhead.

1 Introduction

Predicate abstraction with counterexample-guided abstraction refinement (CEGAR) is a promising approach to automated verification of safety (i.e., reachability) properties (see, e.g., [6] for a survey). Briefly, the CEGAR approach works as follows. Let $\mathcal{T}$ be a first-order logic (FOL) theory. The verifier picks some finite set of predicates from $\mathcal{T}$ as the initial *candidate predicate set*, and iterates the following two processes until convergence (here, we use the term "predicate" for an arbitrary formula, and not limited to just atomic predicates).

---

[*] This work was supported by MEXT Kakenhi 23220001, 26330082, 25280023, and 25730035.

© Springer-Verlag Berlin Heidelberg 2015
J. Vitek (Ed.): ESOP 2015, LNCS 9032, pp. 610–633, 2015.
DOI: 10.1007/978-3-662-46669-8_25

(1) The *abstraction process* checks if the current candidates form a sufficient proof of the program's safety (i.e., an inductive invariant – sometimes called "safe" inductive invariant). If so, then the program is proved safe and the iteration halts. Otherwise, the process generates a *counterexample* as an evidence that the current candidates are insufficient, and (2) is invoked.

(2) The *refinement process* analyzes the given counterexample. If the counterexample cannot be proved spurious by predicates from $\mathcal{T}$ (i.e., "the counterexample is real"), then the iteration halts and the program is detected to be unsafe. Otherwise, the predicates inferred as a proof of the counterexample's spuriousness are added to the candidates, and we repeat from (1).

Note that the verifier halts either when sufficient predicates are inferred to prove the program safe, or a real counterexample is discovered.

For an unsafe program, the state-of-the-art CEGAR-based verifiers are usually able to eventually discover a real counterexample and converge, by exploring the state space in a fair manner (if somewhat slowly for ones requiring large counterexamples). By contrast, when a program is safe and the underlying theory $\mathcal{T}$ is sufficient for proving the safety, most verifiers have no guarantee of convergence and can diverge by having the refinement process indefinitely produce incorrect candidate proofs.

For example, consider the C-like program shown in Figure 1. Here, ndet() returns a non-deterministic integer. The goal is to verify that the assertion failure is unreachable, that is, $a = b \Rightarrow y = x$ whenever line 10 is reached.

Suppose we start the verification process with the candidate set comprising the boolean closure of the predicates $z = 0$, $a = b$ and $y = x$. A possible counterexample is a path that passes through the first loop (lines 4-6) once, reaching line 7 with the abstract state $(a = b \Rightarrow y \neq x) \wedge z \neq 0$, and then passes through the second loop (lines 7-9) once, reaching line 10 with the abstract state $z = 0$, which does not

```
1:   void main(void) {
2:     int a = ndet();int b = ndet();
3:     int x = a;int y = b;int z = 0;
4:     while (ndet()) {
5:       y++;z++;
6:     }
7:     while (z != 0) {
8:       y--;z--;
9:     }
10:    if (a=b && y!=x) { assert false; }
11:  }
```

Fig. 1. A program on which CEGAR may diverge

imply $a = b \Rightarrow y = x$. A possible proof of the counterexample's spuriousness (i.e., proof that the path is actually safe) is the predicate $a = b \Rightarrow y = x + z$. The predicate turns out to be an inductive invariant for the program, and the verification process halts in the next iteration.

Unfortunately, the refinement process is not guaranteed to infer such a predicate but may choose any predicates that can prove the counterexample's safety. For instance, another possibility is the predicate $a = b \Rightarrow y = x + 1$. Adding this to the candidate set is sufficient for proving the safety of the counterexample but not that of the program, and the abstraction process in the subsequent iteration would return yet another counterexample. For example, it may return the counterexample that passes the first loop twice, reaching line 7 with the abstract state $a = b \Rightarrow y \neq x + 1$, and then passes through the second loop twice, reaching line 10 with the abstract state $z = 0$ again. Then, the refinement process may choose the predicate $a = b \Rightarrow y = x + 2$ to prove the spuriousness of this new counterexample, which is still insufficient to prove the whole program correct. The abstract-and-refine iteration may repeat indefinitely in this manner, adding to the candidates the predicate $a = b \Rightarrow y = x + i$ in each i-th run of the refinement process. A refinement process is said to be *complete* (w.r.t. $\mathcal{T}$) if the CEGAR process is guaranteed to converge and eventually discover a proof of the program's safety, when one exists in $\mathcal{T}$.

Previous works [7,11] have proposed to achieve complete refinement in CEGAR by stratifying $\mathcal{T}$ into an infinite sequence of predicate languages $L^0 \subseteq L^1 \subseteq \ldots L^k \ldots$ such that $\mathcal{T} = \bigcup_{k \in \omega} L^k$, and requiring each i-th run of the refinement to only infer predicates from the stratum $L^{lvl(i)}$ where $lvl(i)$ is the stratum level at the i-th CEGAR iteration. By requiring each L^k to be finite[1] and raising the stratum level just when the refinement process reports that no proof exists for the given counterexample in the current stratum, the approach guarantees completeness. However, the approach requires the refinement process to exactly decide if there is a proof of the given counterexample in the current stratum. Indeed, completeness would be lost if the refinement process was allowed to report that the current stratum does not have a proof when it actually does. For many theories used in practice, such as the theory of linear real arithmetic, such an exact finite-language-restricted proof search incurs heavy overhead and is prohibitive (see Section 4, Section 5, and the extended report [18] for analysis and discussion).

The first contribution of this paper is the observation that exact finite language restricted proof search is actually unnecessary for completeness. Instead, we show that the following more relaxed scheme is sufficient: in each i-th run of the refinement process, we restrict the returned proof to some finite language of predicates $L_{base}^{lvl(i)} \cup L_{ext}^{lvl(i)}$ (*base* and *extension*) such that the refinement process may report that no proof exists only when no proof exists in $L_{base}^{lvl(i)}$. There are no further restrictions on the refinement process, and so, the refinement process may return a proof that is not in L_{base} (but in $L_{base} \cup L_{ext}$) even if a proof exists in L_{base}, or may report that no proof exists even if a proof exists in $L_{base} \cup L_{ext}$ (but not in L_{base}). We show that this relaxed approach still ensures completeness when the stratum level is raised just when the refinement process reports that there is no proof in the current stratum, as before, and L_{base} grows to eventually cover $\mathcal{T}$ (i.e., $L_{base}^0 \subseteq L_{base}^1 \subseteq \ldots L_{base}^k \ldots$ such that $\mathcal{T} = \bigcup_{k \in \omega} L_{base}^k$). We formalize this observation in a refinement algorithm scheme called *relaxed stratification* (contra the *exact stratification* approach described above) and prove that it is indeed complete.

---

[1] The term "finite predicate language" is used synonymously with "finite set of predicates".

As the second contribution, we present a concrete refinement algorithm that implements the relaxed scheme. The algorithm is a modification of the proof-based refinement [5] in which the theory-level reasoning is restricted so that partial (tree-)interpolants at that level is restricted to L_{base}.[2] We also present a technique that uses a certain property of the counterexample patterns to further improve the efficiency of the algorithm while still satisfying the requirement of the scheme. We formalize the refinement algorithm as a constraint solver for recursion-free Horn-clause constraints [3,12,13,21] which has gained popularity as the standard format for describing refinement algorithms. We have implemented a prototype of the refinement algorithm, and we show empirically that it is able to achieve complete predicate refinement with low overhead.

In summary, the paper's contributions are as follows:

- A new scheme for practical complete predicate refinement called *relaxed stratification* and the proof of its completeness (Section 2).
- A new predicate refinement algorithm as concrete instance of the relaxed stratification scheme (Section 3).
- Experiments with a prototype implementation of the refinement algorithm (Section 4).

The rest of the paper is organized as follows. Section 2 formally defines the relaxed stratification scheme and proves its completeness. Section 3 presents the concrete refinement algorithm implementing the scheme. Section 4 presents experimental results with the prototype implementation of the refinement algorithm. We discuss related work in Section 5 and conclude the paper in Section 6. Supplementary material contains the extended report with proofs and extra materials omitted from the main body of the paper, and the benchmarks used in the experiments [18].

2   The Relaxed Stratification Scheme

Let $\mathcal{T}$ be a FOL theory. For a formula θ in the signature of $\mathcal{T}$ (a $\mathcal{T}$-formula), we write $fvs(\theta)$ for the free variables in θ. A *predicate* in $\mathcal{T}$ is of the form $\lambda x_1, \ldots, x_n.\theta$ where θ is a $\mathcal{T}$-formula such that $fvs(\theta) \subseteq \{x_1, \ldots, x_n\}$. For readability, we often omit the explicit λ abstraction and treat a formula θ as the predicate $\lambda \bar{x}.\theta$ where $\{\bar{x}\} = fvs(\theta)$. We overload $\mathcal{T}$ for the set of predicates in $\mathcal{T}$.

2.1   Assumptions on the Abstraction Process

Relaxed stratification only concerns the refinement process part of CEGAR. We show that the scheme is quite general and can be used in a wide range of CEGAR-based verifiers. To this end, we delineate the conditions that the

---

[2] While this approach has already been suggested in [7], they require the restricting language to be closed under conjunctions and disjunctions (see Section 5 for further discussion).

abstraction process part needs to satisfy. As we shall show below, the conditions are quite weak and satisfied by virtually any CEGAR-based verifier.

We assume that the abstraction process Abs takes as input a program and a finite set of predicates in $\mathcal{T}$ (the set of *candidate proofs*). For a program M and a finite set of predicates $F \subseteq \mathcal{T}$, we require that $\mathsf{Abs}(M, F)$ either returns safe, indicating that M has been proved safe using the predicates from F, or returns a *counterexample*. For generality, we assume that a counterexample is also simply a program so that, for a counterexample $\{M$, we write $\mathsf{Abs}(M, F)$ = safe when F is sufficient for the abstraction process to prove the spuriousness of M (in practice, a counterexample is not an arbitrary program, but, e.g., an unwound program slice of the input program, and concrete instances of the relaxed stratification scheme take advantage of the counterexample structure – cf. Section 3). We sometimes say that F *refutes* the counterexample M when $\mathsf{Abs}(M, F)$ = safe.

We require Abs to be monotonic on the candidates, that is, if $\mathsf{Abs}(M, F)$ = safe and $F \subseteq F'$ then $\mathsf{Abs}(M, F')$ = safe (i.e., having more predicates can only increase Abs's ability to prove). We also require that if $\mathsf{Abs}(M, F) = \mathsf{cex}(M')$ then $\mathsf{Abs}(M', F) \neq$ safe, that is, the returned counterexample is actually a counterexample and cannot be refuted by the given predicates. Finally, we require that if $\mathsf{Abs}(M, F)$ = safe and $\mathsf{Abs}(M, F') = \mathsf{cex}(M')$ then $\mathsf{Abs}(M', F)$ = safe, that is, if a set of predicates is a proof for program's safety then it is also a proof for any counterexample of the program. We say that Abs is *sound* when it only proves safe programs safe, that is, $\mathsf{Abs}(M, F)$ = safe only if M is safe.[3]

We note that the assumptions on the abstraction process are quite liberal and do not demand, for example, the process uses the given set of predicates by decomposing them into atomic predicates and taking their boolean closure, taking the cartesian closure, or using them directly as loop invariants. In Example 1 below, we describe an example abstraction process that uses the predicates directly.

Example 1. Let $\mathcal{T}$ be the quantifier-free theory of linear real arithmetic. Let Abs be the abstraction process that, given a program (or counterexample) M and the set of predicates $F \subseteq \mathcal{T}$, checks if there exists an assignment from each loop-head location in M to a predicate in F that forms an inductive invariant of M. Recall the example from Section 1. Let M_{ex} be the program shown in Figure 1. Then, the map ρ such that $\rho(\mathsf{L4}) = \rho(\mathsf{L7}) = \theta_{ex1}$ where $\theta_{ex1} \equiv a = b \Rightarrow y = x + z$ is an inductive invariant of M_{ex}, and therefore, $\mathsf{Abs}(M_{ex}, \{\theta_{ex1}\})$ = safe.

When given an insufficient set of predicates as the candidates, Abs returns a counterexample. For instance, as discussed in Section 1, a possible counterexample of M_{ex} is M_{exa}, shown in Figure 2, that passes through each loop once to reach line 10. (The semantics of assume (*b*) is to safely halt if *b* is false, and proceed otherwise.) Here, a1–a4 label the entry points of the unwound loops. Viewing them as one-iteration loops where invariants are asserted, it can be seen that ρ such that $\rho(\ell) = \theta_{ex1}$ for each $\ell \in \{$a1–a4$\}$ is an inductive invariant of M_{exa}, and therefore $\mathsf{Abs}(M_{exa}, \{\theta_{ex1}\})$ = safe. However, as discussed in

---

[3] Soundness of Abs is not required for completeness of relaxed stratification.

```
int a=ndet();int b=ndet();          int a=ndet();int b=ndet();
int x=a;int y=b;int z=0;            int x=a;int y=b;int z=0;
a1:assume (ndet());                 b1:assume (ndet());
   y++;z++;                            y++;z++;
a2:assume (ndet());                 b2:assume (ndet());
a3:assume (z != 0);                    y++;z++;
   y--;z--;                         b3:assume (ndet());
a4:assume (z == 0);                 b4:assume (z != 0);
   assume (a=b && y!=x);               y--;z--;
   assert false;                    b5:assume (z != 0);
                                       y--;z--;
                                    b6:assume (z == 0);
                                       assume (a=b && y!=x);
                                       assert false;
```

$$M_{exa} \qquad\qquad\qquad M_{exb}$$

Fig. 2. Counterexamples of the program from Figure 1

Section 1, asserting θ_0 at a1, a4, and θ_1 at a2, a3 where $\theta_0 \equiv a = b \Rightarrow y = x$ and $\theta_1 \equiv a = b \Rightarrow y = x + 1$ also constitutes a sufficient loop invariant of M_{exa}. Therefore, we also have $\mathsf{Abs}(M_{exa}, \{\theta_0, \theta_1\}) = \mathsf{safe}$.

Similarly, M_{exb} shown in Figure 2 is a counterexample that passes through each loop twice to reach line 10. By reasoning similar to the above, we have $\mathsf{Abs}(M_{exb}, \{\theta_{ex1}\}) = \mathsf{Abs}(M_{exb}, \{\theta_0, \theta_1, \theta_2\}) = \mathsf{safe}$ where $\theta_2 \equiv a = b \Rightarrow y = x + 2$. ▲

2.2 The Relaxed Stratification Scheme

We are now ready to formalize the relaxed stratification scheme. The core of the scheme is the relaxed finite-language-restricted refinement process RlxRef that takes as input a counterexample and a *restricting predicate language* (L_{base}, L_{ext}), and returns either unsafe indicating that the counterexample is real, a set of predicates $F \subseteq L_{base} \cup L_{ext}$ that proves the safety of the counterexample, or noproof indicating that it could not find a proof for the counterexample within the given restriction.

We prepare strata of restricting predicate languages:

$$(L_{base}^0, L_{ext}^0),\ (L_{base}^1, L_{ext}^1),\ \dots\ (L_{base}^k, L_{ext}^k),\ \dots$$

We require each restricting predicate language to be finite, and the base-part to eventually cover $\mathcal{T}$. Formally, we impose the following condition on the restricting predicate languages: 1.) for each $k \in \omega$, $L_{base}^k \cup L_{ext}^k$ is a finite subset of $\mathcal{T}$, 2.) for each $k \in \omega$, $L_{base}^k \subseteq L_{base}^{k+1}$, and 3.) $\mathcal{T} = \bigcup_{k\in\omega} L_{base}^k$.

Figure 3 shows the overview of the relaxed stratification verification process. The verification procedure RlxCegar takes as input the program M to be verified, and first initializes the candidate predicate set *Cands* to $\varnothing$ (line 2) and the restricting language stratum k to 0 (line 3). Then, it repeats the abstract-and-refine loop (lines 4-10) until convergence. The loop first calls $\mathsf{Abs}(M, Cands)$ to

```
01:  RlxCegar(M) =
02:    Cands := ∅;
03:    k := 0;
04:    while true do
05:      match Abs(M,Cands) with
06:        safe → return safe
07:        | cex(M') → match RlxRef(M',L^k_base,L^k_ext) with
08:                      unsafe → return unsafe
09:                      | prf(F) → Cands := Cands ∪ F
10:                      | noproof → k := k + 1
```

Fig. 3. The relaxed stratification verification process

check if M can be proved safe with the current candidates. If so, then we exit
the verification process, returning safe (line 6). Otherwise, a counterexample
M' is obtained, and we call RlxRef on M' and the current restricting language
(L^k_{base}, L^k_{ext}) (line 7). If RlxRef returns unsafe, then the counterexample is real
and we exit the verification process, returning unsafe (line 8). Otherwise, RlxRef
either returns a set of predicates that refutes the counterexample (line 9), or
returns noproof indicating that it has failed to find a proof for the counterexample
in the current language stratum (line 10). In the former case, the returned set
of predicates are added to $Cands$, and in the latter case, the language stratum
is raised to the next level.

We require RlxRef to only report unsafe on a real counterexample, that is,
$RlxRef(M, L_{base}, L_{ext})$ = unsafe only if $\forall F \subseteq \mathcal{T}.Abs(M, F) \neq$ safe, and we require the returned proof to be actually a proof of the counterexample, that is
$RlxRef(M, L_{base}, L_{ext})$ = prf(F) only if Abs(M, F) = safe (these conditions are
not particular to relaxed stratification and usually assumed for any refinement
process in CEGAR). In addition, we require RlxRef to only infer proofs from the
given restricting predicate language and be able to return some proof if the given
counterexample is refutable just in the base part of the language. Formally, we
impose the following additional conditions on RlxRef:

  – If $RlxRef(M, L_{base}, L_{ext})$ = prf(F) then $F \subseteq L_{base} \cup L_{ext}$; and
  – If $\exists F \subseteq L_{base}.Abs(M, F)$ = safe, then $RlxRef(M, L_{base}, L_{ext})$ = prf(F') for
    some F'.

We state and prove the completeness of the relaxed stratification scheme.

Theorem 1 (Completeness). *If $\exists F \subseteq \mathcal{T}.Abs(M, F)$ = safe, then RlxCegar(M)
terminates and returns safe.*

We remind that safety verification is undecidable in general, and our "completeness" only states that the verification terminates under the promise that a proof
of the program's safety exists in $\mathcal{T}$.[4]

---

[4] This notion of completeness is the same as the one from previous works [7,11].

Also, assuming that Abs is sound (cf. Section 2.1), it is easy to see that RlxCegar is also sound in that it only proves safe programs safe (in fact, this holds independently of the behavior of RlxRef).

Theorem 2 (Soundness). *If Abs is sound, then RlxCegar(M) returns safe only if M is safe.*

Example 2. We show how the relaxed stratification scheme would ensure the convergence of a verifier on the program M_{ex} from Example 1. Suppose that we run RlxCegar(M_{ex}), and for contradiction, it diverges by generating the following infinite series of refinements discussed in Section 1:

$$a = b \Rightarrow y = x, \quad a = b \Rightarrow y = x + 1, \quad \ldots \quad a = b \Rightarrow y = x + i, \quad \ldots$$

By the definition of RlxCegar, it must be the case that the restricting predicate language at the i-th CEGAR iteration is $(L_{base}^{lvl(i)}, L_{ext}^{lvl(i)})$ such that $y = x + i \in L_{base}^{lvl(i)} \cup L_{ext}^{lvl(i)}$ where $lvl(i)$ is the restricting language stratum level in the i-th iteration. Because each $L_{base}^{lvl(i)} \cup L_{ext}^{lvl(i)}$ is finite, the stratum level of the language must have been raised infinitely many times. Therefore, $a = b \Rightarrow y = x + z \in L_{base}^{j}$ for some j because $\mathcal{T} = \bigcup_{k \in \omega} L_{base}^{k}$.

But, as argued in Example 1, $a = b \Rightarrow y = x + z$ is a sufficient proof of M_{ex}'s safety, and therefore also that of its counterexamples. Therefore, by the fact that RlxRef refutes any counterexample refutable in the base part of the given restricting language, for any counterexample M' of M_{ex}, RlxRef($M', L_{base}^{j}, L_{ext}^{j}$) = prf($F'$) for some $F' \subseteq L_{base}^{j} \cup L_{ext}^{j}$. Then, because $L_{base}^{j} \cup L_{ext}^{j}$ is finite, RlxCegar must have eventually inferred a sufficient set of predicates that constitutes a proof of M_{ex}'s safety without further raising the language stratum.   ▲

3   Concrete Refinement Algorithm Instances

We show how to implement the relaxed finite-language-restricted refinement process RlxRef. In fact, we describe a technique that takes as module an exact L_{base}-restricted refinement algorithm and turn it into a relaxed $(L_{base}, L_{base}^{\wedge\vee})$-restricted refinement algorithm. (We write $L^{\wedge\vee}$ for the closure of L under conjunctions and disjunctions.) We focus on the case where the given counterexample is spurious.[5]

Following the recent trend [17,3,8,2,1,13,12,21], we formalize the refinement algorithm as a constraint solver for recursion-free Horn-clause constraints. Specifically, we present a relaxed (L_{base}, L_{ext})-restricted constraint solver that takes as module an exact L_{base}-restricted constraint solver (cf. Section 3.1 for the definition of exact/relaxed finite-language-restricted constraint solvers). We review Horn-clause constraints in Section 3.1, and describe the constraint solver, that we call RlxSolveA, in Section 3.2.

---

[5] Detecting if the counterexample real and returning unsafe if so can be handled via usual unrestricted refinement (cf. Section 4).

We also present a technique that takes as module a relaxed (L_{base}, L_{ext})-restricted constraint solver, an unrestricted constraint solver $\mathcal{AU}$, and a positive integer parameter ℓ, and turn them into a relaxed $(L_{base}, \mathsf{LB}(L_{base} \cup L_{ext}, \mathcal{AU}, \ell))$-restricted constraint solver where $\mathsf{LB}(L, \mathcal{AU}, \ell)$ is a certain finite language of predicates determined by L, $\mathcal{AU}$, and ℓ. We formalize the technique as the constraint solver RlxSolveB, described in Section 3.3. The technique applies the relaxed finite-language-restricted constraint solver provided as the module to only a small subset of the constraint solving problem, and can be used to improve the efficiency of the given relaxed finite-language-restricted constraint solver.

We remind that the exact finite-language-restricted proof search is an inherently expensive process (cf. Section 5, Section 4, and the extended report [18]), and the key idea in these constraint solvers is to use the expensive exact finite-language-restricted proof search process (given as a module) only on small subparts of the problem. This is made possible thanks to the relaxed requirement on the language restriction where the refinement process is not required to exactly decide the existence of a restricted solution for the whole problem. Informally, the trick is to choose the subproblems just large enough to guarantee that if a subproblem is not L_{base} solvable then neither is the whole and that there can only be finitely many solutions for the whole obtainable from L_{base}-restricted solutions for the subproblems.

3.1   Horn Clause Constraints

For concreteness, in what follows, we assume that the underlying theory $\mathcal{T}$ is the quantifier-free theory of linear real arithmetic (QFLRA). However, the techniques presented in Sections 3.2 and 3.3 can be applied to any quantifier-free theory.

A *formula* θ in the signature of QFLRA comprises *atomic predicate* p of the form $a_1 x_1 + a_2 x_2 \cdots + a_n x_n \le a_{n+1}$ where $a_1, \ldots, a_{n+1} \in \mathbb{Z}$, and is closed under the usual boolean operations $\neg$, $\wedge$, $\vee$, and $\Rightarrow$. As usual, we let $\neg$ bind the tightest and $\Rightarrow$ the weakest. A *literal* l is either an atomic predicate or its negation. A *clause* C is a disjunction of literals. A conjunctive normal form (CNF) is a conjunction of clauses. We often use a set to represent a clause or a CNF so that $\{l_1, \ldots, l_n\}$ represents $l_1 \vee \cdots \vee l_n$ and $\{C_1, \ldots, C_n\}$ represents $C_1 \wedge \cdots \wedge C_n$. We write $\bot$ for contradiction and $\top$ for tautology. We write $\vDash \theta$ when θ is valid in $\mathcal{T}$.

Horn Clauses and Horn-Clause Constraints. A *predicate variable application* is of the form $P(\bar{x})$ where P is a *predicate variable* of arity $|\bar{x}|$. A *Horn clause* hc is of the form $\theta \wedge B_1 \wedge \cdots \wedge B_n \rightarrow H$ where θ is a formula in $\mathcal{T}$, each B_i is a predicate variable application, and H is a predicate variable application or $\bot$. We call H the *head* of the Horn clause, and $\theta \wedge B_1 \wedge \cdots \wedge B_n$ the *body*. A Horn clause whose head is $\bot$ is called a *root clause*.

A *Horn-clause constraint set* (HCCS) $\mathcal{H}$ is a finite set of Horn clauses. We write $pvs(\mathcal{H})$ for the predicate variables in $\mathcal{H}$. We write $leaves(\mathcal{H})$ for the set of predicate variables in $\mathcal{H}$ that do not occur as a head in $\mathcal{H}$. We define $\rightsquigarrow_{\mathcal{H}}$ to be the relation $\{(P, Q) \mid \theta \wedge \ldots P(\bar{x}) \ldots \rightarrow Q(\bar{y}) \in \mathcal{H}\}$. We say that a Horn clause

$$\theta_{p1} \rightarrow P(\bar{x})$$
$$\theta_{p2} \wedge P(\bar{x}) \rightarrow P(\bar{x}')$$
$$P(\bar{x}) \rightarrow Q(\bar{x})$$
$$\theta_{p3} \wedge Q(\bar{x}) \rightarrow Q(\bar{x}')$$
$$\theta_{p4} \wedge Q(\bar{x}) \rightarrow \bot$$

$$\theta_{p1} \rightarrow P_1(\bar{x})$$
$$\theta_{p2} \wedge P_1(\bar{x}) \rightarrow P_2(\bar{x}')$$
$$P_2(\bar{x}) \rightarrow Q_1(\bar{x})$$
$$\theta_{p3} \wedge Q_1(\bar{x}) \rightarrow Q_2(\bar{x}')$$
$$\theta_{p4} \wedge Q_2(\bar{x}) \rightarrow \bot$$

$$\theta_{p1} \rightarrow P_1(\bar{x})$$
$$\theta_{p2} \wedge P_1(\bar{x}) \rightarrow P_2(\bar{x}')$$
$$\theta_{p2} \wedge P_2(\bar{x}) \rightarrow P_3(\bar{x}')$$
$$P_3(\bar{x}) \rightarrow Q_1(\bar{x})$$
$$\theta_{p3} \wedge Q_1(\bar{x}) \rightarrow Q_2(\bar{x}')$$
$$\theta_{p3} \wedge Q_2(\bar{x}) \rightarrow Q_3(\bar{x}')$$
$$\theta_{p4} \wedge Q_3(\bar{x}) \rightarrow \bot$$

$\mathcal{H}_{ex}$ $\mathcal{H}_{exa}$ $\mathcal{H}_{exb}$

Fig. 4. HCCS examples

$\theta \wedge B_1 \wedge \cdots \wedge B_n \rightarrow H$ is *conjunctive* if θ is a conjunction of literals. We say that an HCCS $\mathcal{H}$ is conjunctive if each $hc \in \mathcal{H}$ is conjunctive.

We say that $\mathcal{H}$ is *recursion-free* if $\leadsto_{\mathcal{H}}$ is acyclic. We say that a recursion-free HCCS $\mathcal{H}$ is *tree-like* [12,13] if 1.) there is exactly one root clause in $\mathcal{H}$ and every $P \in pvs(\mathcal{H})$ can reach a predicate variable occurring in the body of the root clause via $\leadsto_{\mathcal{H}}^{*}$; and 2.) for any $P \in pvs(\mathcal{H})$, at most one $hc \in \mathcal{H}$ contains P in its body, at most one $hc \in \mathcal{H}$ contains P as its head, and no $hc \in \mathcal{H}$ has multiple occurrences of P. For a tree-like HCCS $\mathcal{H}$, we define the *depth* of $\mathcal{H}$, $depth(\mathcal{H})$, to be the length of the longest $\leadsto_{\mathcal{H}}$ path. For η a mapping from predicate variables to predicate variables, we write $\eta(hc)$ for the Horn clause hc with each predicate variable application $P(\bar{x})$ replaced by $\eta(P)(\bar{x})$. We write $\eta(\mathcal{H})$ for $\{\eta(hc) \mid hc \in \mathcal{H}\}$. We say that a tree-like HCCS $\mathcal{H}'$ is an *unwound instance* of a (possibly recursive) HCCS $\mathcal{H}$ if there exists a mapping η from $pvs(\mathcal{H}')$ to $pvs(\mathcal{H})$ such that $\eta(\mathcal{H}') \subseteq \mathcal{H}$.

Constraint Solutions and Restricted Constraint Solvers. For σ a mapping from predicate variables to predicates in $\mathcal{T}$, we write $\sigma(hc)$ for hc with each predicate variable application $P(\bar{x})$ replaced by $\theta[\bar{x}/\bar{y}]$ where $\sigma(P) = \lambda\bar{y}.\theta$. We say that the map σ from $pvs(\mathcal{H})$ to predicates in $\mathcal{T}$ is a *solution* of $\mathcal{H}$, written $\sigma \vDash \mathcal{H}$, if for each $hc \in \mathcal{H}$, $\vDash \sigma(hc)$, interpreting $\rightarrow$ as $\Rightarrow$. We define $ran(\sigma)$, the *range* of σ, to be the set of predicates $\{\sigma(P) \mid P \in dom(\sigma)\}$.

We focus on constraint solving algorithms for tree-like HCCSs (they can be extended to arbitrary recursion-free HCCSs by adopting the technique from [13]). We say that an algorithm is an *unrestricted constraint solver* if given a tree-like HCCS $\mathcal{H}$, it returns a solution of $\mathcal{H}$ or decides that $\mathcal{H}$ has no solution. We say that an algorithm is an *exact L-restricted constraint solver* if given a tree-like HCCS $\mathcal{H}$, it decides if there is a solution σ of $\mathcal{H}$ such that $ran(\sigma) \subseteq L$ and returns such a solution if so. We say that an algorithm is a *relaxed (L_{base}, L_{ext})-restricted constraint solver* if given a tree-like HCCS $\mathcal{H}$, it either returns a solution σ of $\mathcal{H}$ such that $ran(\sigma) \subseteq L_{base} \cup L_{ext}$ or returns noproof indicating that it has failed to find a solution, with the requirement that it returns some solution (whose range is in $L_{base} \cup L_{ext}$) if there exists a solution σ' of $\mathcal{H}$ such that $ran(\sigma') \subseteq L_{base}$.

Example 3. Consider the HCCS $\mathcal{H}_{ex}$ shown in Figure 4. Here, $\bar{x} = a, b, x, y, z$, $\bar{x}' = a', b', x', y', z'$, and

$$\theta_{p1} \equiv x = a \wedge y = b \wedge z = 0$$
$$\theta_{p2} \equiv z' = z + 1 \wedge y' = y + 1 \wedge x' = x \wedge a' = a \wedge b' = b$$
$$\theta_{p3} \equiv z \neq 0 \wedge z' = z - 1 \wedge y' = y - 1 \wedge x' = x \wedge a' = a \wedge b' = b$$
$$\theta_{p4} \equiv z = 0 \wedge a = b \wedge x \neq y$$

$\mathcal{H}_{ex}$ is not tree-like (in fact, $\leadsto_S$ is cyclic). Figure 4 shows HCCSs $\mathcal{H}_{exa}$ and $\mathcal{H}_{exb}$ that are tree-like. In addition, they are unwound instances of $\mathcal{H}_{ex}$ because $\eta_a(\mathcal{H}_{exa}) \subseteq \mathcal{H}_{ex}$ and $\eta_b(\mathcal{H}_{exb}) \subseteq \mathcal{H}_{ex}$ where $\eta_a = \{P_1 \mapsto P, P_2 \mapsto P, Q_1 \mapsto Q, Q_2 \mapsto Q\}$ and $\eta_b = \{P_1 \mapsto P, P_2 \mapsto P, P_3 \mapsto P, Q_1 \mapsto Q, Q_2 \mapsto Q, Q_3 \mapsto Q\}$.

Recall the predicates $\theta_{ext1}, \theta_0, \theta_1, \theta_2$ from Example 1. Let the maps $\sigma_{a_1}, \sigma_{a_2}$, σ_{b_1}, and σ_{b_2} be defined as below.

$$\sigma_{a_1} = \{P \mapsto \theta_{ext1} \mid P \in pvs(\mathcal{H}_{exa})\} \qquad \sigma_{b_1} = \{P \mapsto \theta_{ext1} \mid P \in pvs(\mathcal{H}_{exb})\}$$

$$\sigma_{a_2} = \{P \mapsto \theta_0 \mid P \in \{P_1, Q_2\}\} \qquad \sigma_{b_2} = \{P \mapsto \theta_0 \mid P \in \{P_1, Q_3\}\}$$
$$\cup \{P \mapsto \theta_1 \mid P \in \{P_2, Q_1\}\} \qquad \cup \{P \mapsto \theta_1 \mid P \in \{P_2, Q_2\}\}$$
$$\qquad\qquad\qquad\qquad\qquad\qquad\quad\; \cup \{P \mapsto \theta_2 \mid P \in \{P_3, Q_1\}\}$$

Then, σ_{a_1} and σ_{a_2} are solutions of $\mathcal{H}_{exa}$, and σ_{b_1} and σ_{b_2} are solutions of $\mathcal{H}_{exb}$. ▲

Relating Refinement Process to Constraint Solving. We relate constraint solving to refinement process. Roughly, the relationship says that, for any counterexample, there is a corresponding tree-like HCCS such that the range of its solutions are the proofs of the counterexample's spuriousness. We further assume that such a tree-like HCCS is always an unwound instance of some fixed "generator" HCCS determined by the given program.

We formalize the relationship. Let M be a program. We assume that there exists an HCCS $\mathcal{H}_{gen(M)}$ such that for any counterexample M' of M (i.e., $\mathsf{Abs}(M, F) = \mathsf{cex}(M')$ for some F), there exists an unwound instance $\mathcal{H}_{M'}$ of $\mathcal{H}_{gen(M)}$ that satisfies the following:

- if $\sigma \models \mathcal{H}_{M'}$ then $\mathsf{Abs}(M', ran(\sigma)) = \mathsf{safe}$ (i.e., the range of a solution of $\mathcal{H}_{M'}$ is a proof of M''s spuriousness); and
- if $\mathsf{Abs}(M', F) = \mathsf{safe}$ then $\exists \sigma. ran(\sigma) \subseteq F \wedge \sigma \models \mathcal{H}_{M'}$ (i.e., if M' can be refuted by F, then there is a solution for $\mathcal{H}_{M'}$ whose range is in F).

Hence, the task of implementing a relaxed language restricted refinement process RlxRef for the restricting language (L_{base}, L_{ext}) is now reduced to implementing a relaxed (L_{base}, L_{ext})-restricted constraint solver.

We remark that the relationship stated above is quite general and many CEGAR-based verifiers [17,8,2,1,13,12,21] use the relationship to implement the refinement process as a constraint solver for tree-like HCCSs. For example, refuting a counterexample in a typical CEGAR-based verification of sequential imperative programs is equivalent to solving a tree-like HCCS of the form below where $\bar{x}$ are the variables in the program, and each θ_i is a formula on $\bar{x}$ and $\bar{x}'$ that expresses the semantics of symbolically executing the corresponding segment (e.g., basic block) in the path:

$$\theta_1 \dashrightarrow P_1(\bar{x}) \qquad\qquad \theta_i \wedge P_i(\bar{x}) \dashrightarrow P_{i+1}(\bar{x}')$$
$$\theta_2 \wedge P_1(\bar{x}) \dashrightarrow P_2(\bar{x}') \qquad\qquad \vdots$$
$$\vdots \qquad\qquad\qquad P_n(\bar{x}) \dashrightarrow \bot$$

In such a verification, the generator HCCS $\mathcal{H}_{gen(M)}$ can be described as follows. Let $\bar{x}$ be the variables in the program. For each node a in the program's control flow graph (CFG), we associate a predicate variable P_a of arity $|\bar{x}|$. For each edge from node a to node b in the CFG, we add to $\mathcal{H}_{gen(M)}$ the Horn clause $\theta_{ab} \wedge P_a(\bar{x}) \dashrightarrow P_b(\bar{x}')$ where θ_{ab} is a formula on $\bar{x}$ and $\bar{x}'$ expressing the effect of symbolically executing the CFG path from a to b (with $\bar{x}$ representing the current and $\bar{x}'$ representing the post state). For the entry node a, we add the Horn clause $\theta_{init} \dashrightarrow P_a(\bar{x})$ where θ_{init} is a formula on $\bar{x}$ expressing the program's initial state. Finally, for each error node a (i.e., `assert false` statement), we add the Horn clause $P_a(\bar{x}) \dashrightarrow \bot$.

Example 4. Recall the program M_{ex} from Example 1. The corresponding generator HCCS $\mathcal{H}_{gen(M_{ex})}$ is $\mathcal{H}_{ex}$ from Example 3. Roughly, the predicate variable P in the HCCS represents the program states at the time when the first loop is entered, and Q represents the states when the second loop is entered.

Recall the counterexamples M_{exa} and M_{exb} from Example 1, and the tree-like HCCSs $\mathcal{H}_{exa}$ and $\mathcal{H}_{exb}$ from Example 3. $\mathcal{H}_{exa}$ corresponds to M_{exa} and $\mathcal{H}_{exb}$ corresponds to M_{exb}. Indeed, as shown in Example 1, $\{\theta_{ex1}\}$ (resp. $\{\theta_0, \theta_1\}$) is a proof of M_{exa}, and $\mathcal{H}_{exa}$ has the corresponding solution σ_{a_1} (resp. σ_{a_2}) from Example 3. Similarly, the solutions σ_{b_1} and σ_{b_2} of $\mathcal{H}_{exb}$ and the proofs $\{\theta_{ex1}\}$ and $\{\theta_0, \theta_1, \theta_2\}$ of M_{exb} correspond. ▲

3.2   The Constraint Solver RlxSolveA

RlxSolveA is a relaxed $(L_{base}, L_{base}{}^{\wedge\vee})$-restricted constraint solver. It is parameterized by an exact L_{base}-restricted constraint solver that it takes as module. Let us fix the exact solver, $\mathcal{AE}_{L_{base}}$, and write RlxSolveA$[\mathcal{AE}_{L_{base}}]$ for RlxSolveA parameterized by the exact solver. Note that $L_{base}{}^{\wedge\vee}$ is finite for a finite L_{base}.

We briefly overview the construction of RlxSolveA. First, we leverage the equivalence of solving tree-like HCCS and tree interpolation [13] to reduce the problem to tree interpolation. Then, we adopt the standard proof-based interpolation technique that obtains interpolants from resolution proofs generated via SMT solving [10], except that we modify the SMT solver to use the exact L_{base}-restricted solver $\mathcal{AE}_{L_{base}}$ for the theory solver so as to infer L_{base}-restricted (partial) interpolants at the theory level of the resolution proof. As we shall show, this guarantees that if the SMT solver fails to prove, then no L_{base} solution exists, and conversely, any inferred solution is guaranteed to be in $L_{base}{}^{\wedge\vee}$.

We describe the approach more formally. First, we review *tree interpolation*. The tree interpolation problem takes as input (V, E, Θ) where (V, E) is a finite directed tree with the node set V and $(v, v') \in E$ denoting that the node v is a direct child of the node v', and the map Θ labels each node $v \in V$ with the $\mathcal{T}$-formula $\Theta(v)$. The goal is to find a map I from V to $\mathcal{T}$-formulas, called a *tree interpolant* of (V, E, Θ), that satisfies the following.

- $I(v_{rt}) = \bot$ for the root node v_{rt};
- for each $v \in V$, $\vDash \Theta(v) \wedge \bigwedge_{(v',v) \in E} I(v') \Rightarrow I(v)$; and
- for each $v \in V$, $fvs(I(v)) \subseteq (\bigcup_{(v',v) \in E^*} fvs(\Theta(v'))) \cap (\bigcup_{(v',v) \notin E^*} fvs(\Theta(v')))$.

We reduce constraint solving for a tree-like HCCS to tree interpolation as follows.[6] Let $\mathcal{H}$ be the input tree-like HCCS. We transform $\mathcal{H}$ to an equivalent HCCS that satisfies: 1.) for each predicate variable $P \in pvs(\mathcal{H})$, there exists a vector of fresh variables $\bar{x}_P$ such that P only occurs in the form $P(\bar{x}_P)$, and 2.) the only sharing of variables among Horn clauses are $\bar{x}_P$'s between two Horn clauses both containing P. Then, the transformed $\mathcal{H}$ is reduced to the tree interpolation problem $(V_{\mathcal{H}}, E_{\mathcal{H}}, \Theta_{\mathcal{H}})$ where

- $V_{\mathcal{H}} = pvs(\mathcal{H}) \cup \{v_{rt}\}$ where $v_{rt} \notin pvs(\mathcal{H})$;
- $E_{\mathcal{H}} = \leadsto_{\mathcal{H}} \cup \{(P, v_{rt}) \mid \theta \wedge \ldots P(\bar{x}_P) \ldots \dashrightarrow \bot \in \mathcal{H}\}$;
- For each P, $\Theta_{\mathcal{H}}(P) = \theta_P$ if $\theta_P \wedge \bigwedge_i B_i \rightarrow P(\bar{x}_P) \in \mathcal{H}$ and otherwise $\Theta_{\mathcal{H}}(P) = \bot$; and
- $\Theta_{\mathcal{H}}(v_{rt}) = \theta_{rt}$ where $\theta_{rt} \wedge \bigwedge_i B_i \dashrightarrow \bot \in \mathcal{H}$.

The theorem below follows from the construction, and shows the one-to-one correspondence between the tree interpolants of $(V_{\mathcal{H}}, E_{\mathcal{H}}, \Theta_{\mathcal{H}})$ and the solutions of $\mathcal{H}$.

Theorem 3 ([13]). *Let $\mathcal{H}$ be a tree-like HCCS. Let σ and I be such that $I(v_{rt}) = \bot$ and for each $P \in pvs(S)$, $\sigma(P) = \lambda \bar{x}_P . I(P)$. Then, $\sigma \vDash \mathcal{H}$ if and only if I is a tree interpolant of $(V_{\mathcal{H}}, E_{\mathcal{H}}, \Theta_{\mathcal{H}})$.*

Example 5. Recall the tree-like HCCS $\mathcal{H}_{exa}$ from Example 3. The corresponding tree interpolation problem (V, E, Θ) is shown below where each $\bar{x}_{P_1}$, $\bar{x}_{P_2}$, $\bar{x}_{Q_1}$, $\bar{x}_{Q_2}$ is a quintuple of fresh variables.

$$V = \{v_{rt}, P_1, P_2, Q_1, Q_2\}$$
$$E = \{(P_1, P_2), (P_2, Q_1), (Q_1, Q_2), (Q_2, v_{rt})\}$$
$$\Theta(P_1) = \theta_{p1}[\bar{x}_{P_1}/\bar{x}] \qquad \Theta(P_2) = \theta_{p2}[\bar{x}_{P_1}/\bar{x}][\bar{x}_{P_2}/\bar{x}']$$
$$\Theta(Q_1) = \bar{x}_{P_2} = \bar{x}_{Q_1} \qquad \Theta(Q_2) = \theta_{p3}[\bar{x}_{Q_1}/\bar{x}][\bar{x}_{Q_2}/\bar{x}']$$
$$\Theta(v_{rt}) = \theta_{p4}[\bar{x}_{Q_2}/\bar{x}]$$

▲

Now, the relaxed $(L_{base}, L_{base}^{\wedge\vee})$-restricted constraint solving problem is reduced to relaxed $(L_{base}, L_{base}^{\wedge\vee})$-restricted tree interpolation. That is, we would like to find tree interpolants restricted to $L_{base} \cup L_{base}^{\wedge\vee}$ (i.e., $L_{base}^{\wedge\vee}$), with the guarantee to return one if there exists a L_{base}-restricted tree interpolant.

---

[6] The reduction is adopted from [13].

$$\text{THY} \quad \frac{\mathcal{AE}_{L_{base}}(\mathcal{H}_{thy(C,V,E,\Theta)}) = \sigma \quad I(v_{rt}) = \bot \quad \forall P.\lambda\bar{x}_P.I(P) = \sigma(P)}{(V,E,\Theta) \vdash_{itp} C : I}$$

$$\text{HYP} \quad \frac{C \in \Theta(v) \quad \forall v'.I(v') = \begin{cases} C\!\uparrow_{v'} & \text{if } (v,v') \in E^* \\ \top & \text{otherwise} \end{cases}}{(V,E,\Theta) \vdash_{itp} C : I}$$

$$\text{RES} \quad \frac{\begin{array}{l}(V,E,\Theta) \vdash_{itp} p \vee C_1 : I_1 \\ (V,E,\Theta) \vdash_{itp} \neg p \vee C_2 : I_2\end{array} \quad \forall v.I_3(v) = \begin{cases} I_1(v) \wedge I_2(v) & \text{if } p \in outs(v) \\ I_1(v) \vee I_2(v) & \text{otherwise}\end{cases}}{(V,E,\Theta) \vdash_{itp} C_1 \vee C_2 : I_3}$$

Fig. 5. The tree interpolation rules

Next, we describe the process of relaxed $(L_{base}, L_{base}{}^{\wedge\vee})$-restricted tree interpolation. In what follows, we assume familiarity with lazy SMT and the proof-based technique for obtaining interpolants from resolution proofs [16,10]. Let (V,E,Θ) be the tree interpolation instance to be solved. In an ordinary proof-based tree interpolation, one looks for tree interpolants by having the SMT solver check the unsatisfiability of $\bigwedge_{v \in V} \Theta(v)$ and analyzing the output resolution proof to compute the interpolant. However, this decides the existence of, and infers, a tree interpolant from the entire $\mathcal{T}$, and is unsuitable for our task (i.e., this results in an unrestricted constraint solver).

Instead, we modify the SMT solver so that its theory-level reasoning is delegated to the exact L_{base}-restricted constraint solver $\mathcal{AE}_{L_{base}}$. More specifically, when the SMT solver builds a model of possible (propositional) satisfying assignment $\neg C$, instead of passing the model to a theory solver as in ordinary SMT, we build a "fragment" HCCS $\mathcal{H}_{thy(C,V,E,\Theta)}$ that just contains the part of the tree interpolation problem touched by the literals in C. Formally,

$$\mathcal{H}_{thy(C,V,E,\Theta)} = \{\neg C\!\downarrow_v \wedge \bigwedge_{(P,v) \in E} P(\bar{x}_P) \rightarrow H_v \mid v \in V\}$$

where $C\!\downarrow_v$ is the set of literals of C over atomic predicates occurring in $\Theta(v)$, and $H_v = \bot$ if $v = v_{rt}$ and $H_v = Q(\bar{x}_Q)$ if v is a predicate variable Q. We pass $\mathcal{H}_{thy(C,V,E,\Theta)}$ to $\mathcal{AE}_{L_{base}}$ to decide if it has a L_{base} restricted solution, and if so, we set the obtained solution as the partial tree interpolant for the theory-level resolution proof nodes where C occurs as the theory lemma. Otherwise, we can safely reject that the whole problem as having no L_{base}-restricted tree interpolant and return noproof. To generate the tree interpolant for the whole, we adopt the proof-based approach that builds the tree interpolant in a bottom up manner following the rules shown in Figure 5.[7] Here, $outs(v)$ is the set of atomic predicates occurring outside of the subtree rooted at v (i.e., $outs(v) =$

---

[7] We assume that each $\Theta(v)$ is CNF (if not, they can be transformed so via the Tseitin transformation [19]).

$\{p \mid p$ occurs in $\Theta(v')$ where $(v',v) \notin E^*\})$, and $C\uparrow_v$ is the set of literals of C over the atomic predicates occurring outside of the subtree rooted at v (i.e., $C\uparrow_v = \{p \in C \mid p \in outs(v)\} \cup \{\neg p \in C \mid p \in outs(v)\})$. The rules HYP for clauses in the input tree and RES for resolution steps extend the analogous rules from the standard proof-based interpolation [10] to tree interpolation. As described above, THY uses the L_{base}-restricted solution computed by the exact solver for the partial tree interpolant. As we show in Theorems 4 and 5 below, this achieves the desired relaxed $(L_{base}, L_{base}{}^{\wedge\vee})$-restricted tree interpolation.

First, we show that any tree interpolant obtained by the method is restricted to $L_{base}{}^{\wedge\vee}$.

Theorem 4. *Let $\mathcal{H}$ be a tree-like HCCS and (V, E, Θ) be the corresponding tree interpolation problem. Suppose $(V, E, \Theta) \vdash_{itp} \bot : I$. Then, I is a tree interpolant of (V, E, Θ), and for all $P \in pvs(\mathcal{H})$, $\lambda\bar{x}_P.I(P) \in L_{base}{}^{\wedge\vee}$.*

Next, we prove that if there is a L_{base}-restricted tree interpolant for the given tree interpolation instance, then the method infers some tree interpolant (and by Theorem 4 above, such a tree interpolant will be restricted to $L_{base}{}^{\wedge\vee}$).

Theorem 5. *Let $\mathcal{H}$ be a tree-like HCCS and (V, E, Θ) be the corresponding tree interpolation problem. Suppose there is a tree interpolant I of (V, E, Θ) such that for all $P \in pvs(\mathcal{H})$, $\lambda\bar{x}_P.I(P) \in L_{base}$. Then, $(V, E, \Theta) \vdash_{itp} \bot : I'$ for some I'.*

We note that $\mathcal{H}_{thy(C,V,E,\Theta)}$ is always a conjunctive HCCS and is often much smaller than the input HCCS. Therefore, the expensive exact L_{base}-restricted constraint solver $\mathcal{AE}_{L_{base}}$ is only applied to small conjunctive HCCSs, thereby making its job easier. Also, we note that, while we have presented RlxSolveA to be parameterized by an exact L_{base}-restricted constraint solver passed as module, the algorithm actually works even if a relaxed (L_{base}, L_{ext})-restricted constraint solver is used in place of the exact L_{base}-restricted constraint solver.[8] Therefore, RlxSolveA can actually be parameterized by RlxSolveA itself, but the solvers must be "primed" by some exact solver (e.g., RlxSolveA[RlxSolveA[$\mathcal{AE}_{L_{base}}$]]).

Minimizing Theory Lemmas. When $\neg C$ is given as a possible propositional model by the SMT solver, we use the exact finite-language-restricted constraint solver $\mathcal{AE}_{L_{base}}$ to find a solution for $\mathcal{H}_{thy(C,V,E,\Theta)}$. But, using C directly as the theory lemma after $\mathcal{AE}_{L_{base}}$ finds a solution could result in the SMT solver producing many propositional models and lead to bad performance. (This is analogous to using C directly as the theory lemma in an ordinary lazy SMT solving when the theory solver finds $\neg C$ unsatisfiable.) Instead, we let $\mathcal{AE}_{L_{base}}$ return the subset of the literals of C that it used to find the solution, and use it to obtain a smaller theory lemma. (We refer to the extended report [18] for more detail.)

---

[8] More precisely, it becomes a relaxed $(L_{base}, L_{ext}{}^{\wedge\vee})$-restricted constraint solver when passed a relaxed (L_{base}, L_{ext})-restricted constraint solver.

3.3   The Constraint Solver RlxSolveB

RlxSolveB is a relaxed $(L_{base}, \mathsf{LB}(L_{base} \cup L_{ext}, \mathcal{AU}, \ell))$-restricted constraint solver which takes as module a relaxed (L_{base}, L_{ext})-restricted constraint solver, an unrestricted constraint solver $\mathcal{AU}$, and a positive integer parameter ℓ. $\mathsf{LB}(L, \mathcal{AU}, \ell)$ is a finite language of predicates determined by L, $\mathcal{AU}$, and ℓ.

We informally describe RlxSolveB. We select some fraction of predicate variables in a certain "fair" manner based on the parameter ℓ and use the relaxed (L_{base}, L_{ext})-restricted solver provided as a module to look for solutions to just the selected predicate variables. After restricted solutions are obtained for the selected predicate variables, we use $\mathcal{AU}$ to look for unrestricted solutions to the remaining predicate variables, and return the combined solution as the solution for the input HCCS. Note that this technique reduces the number of predicate variables that the given relaxed (L_{base}, L_{ext})-restricted solver needs to solve for, and therefore can be used to improve the performance of a relaxed (L_{base}, L_{ext})-restricted solver. The key observation we use here is that HCCSs solved in a refinement process are all unwound instances of a fixed "generator" HCCS (i.e., $\mathcal{H}_{gen(M)}$). As we shall show next, the observation can be used to guarantee that the result is a relaxed finite-language-restricted solver, when the predicate variable selection is done in a certain proper way.

We describe the constraint solving algorithm in detail. In what follows, we extend the definition of a tree-like HCCS (cf. Section 3.1) so that the root clause can be a Horn clause whose head is of the form $P(\bar{x})$ where P does not occur anywhere else in the HCCS. For such an HCCS $\mathcal{H}$, we say that P is the *root* of $\mathcal{H}$ and write $root(\mathcal{H}) = P$ ($root(\mathcal{H}) = \bot$ for $\mathcal{H}$ with a $\bot$-head root clause).

```
01:  RlxSolveB[AR_(L_base, L_ext), AU, ℓ](H) =
02:    let Y = partition(ℓ, H) in
03:    let A = ⋃_{H'∈Y}{root(H')} ∖ {⊥} in
04:    let H_A = rewrite(H, A) in
05:    match AR_(L_base, L_ext)(H_A) with
06:      noproof → return noproof
07:      | sol(σ_A) → return sol(σ_A ∪ ⋃_{H'∈Y} AU(σ_A(H')))
```

Fig. 6. The overview of RlxSolveB

Let RlxSolveB be parameterized by the relaxed (L_{base}, L_{ext})-restricted constraint solver $AR_{(L_{base}, L_{ext})}$, the unrestricted constraint solver $\mathcal{AU}$, and the positive integer ℓ. Figure 6 shows the overview of RlxSolveB. RlxSolveB first partitions the input HCCS $\mathcal{H}$ into a set of tree-like HCCSs of depth at most ℓ in a top-down manner (line 2). Formally, partition is defined as follows.

$$\mathsf{partition}(\ell, \mathcal{H}) =$$
$$\textbf{if } depth(\mathcal{H}) \leq \ell \textbf{ then } \{\mathcal{H}\}$$
$$\textbf{else let } \mathcal{H}', X = subtrees(\ell, \mathcal{H}) \textbf{ in}$$
$$\{\mathcal{H}'\} \cup \bigcup_{\mathcal{H} \in X} \mathsf{partition}(\ell, \mathcal{H})$$

Here, $subtrees(\ell, \mathcal{H})$ returns the pair $(\mathcal{H}', X)$ such that 1.) $\mathcal{H}'$ is the largest tree-like subset of $\mathcal{H}$ containing the root clause of $\mathcal{H}$ and $depth(\mathcal{H}') = \ell$, and 2.) X is the set of subtrees of $\mathcal{H}$ rooted at each leaf of $\mathcal{H}'$. It is easy to see that Y partitions $\mathcal{H}$ (i.e., $\mathcal{H} = \bigcup Y$ and $\forall \mathcal{H}_1, \mathcal{H}_2 \in Y . \mathcal{H}_1 \neq \mathcal{H}_2 \Rightarrow \mathcal{H}_1 \cap \mathcal{H}_2 = \varnothing$), and that each $\mathcal{H}' \in Y$ is a tree-like HCCS of depth at most ℓ. In fact, the partition is the coarsest of such partitions, where the only HCCSs in the partition having depth less than ℓ are the ones whose leaf predicate variables do not appear anywhere else in the partition.

By construction, only the root predicate variables are shared by different HCCSs in Y (more precisely, a root of one HCCS appears as a leaf in another HCCS). RlxSolveB selects these shared predicate variables to be the ones to infer restricted solutions (line 3). It can be seen that the fraction of the selected predicate variables, that is $|A|/|pvs(\mathcal{H})|$, is inversely proportional to the size of a depth ℓ tree-like subset of $\mathcal{H}$, and decreases rapidly as ℓ is increased.

To infer restricted solutions to A, RlxSolveB constructs the HCCS $\mathcal{H}_A$ such that $pvs(\mathcal{H}_A) = A$, and solutions of $\mathcal{H}_A$ correspond exactly to the solutions of $\mathcal{H}$ restricted to A (i.e., $\sigma \vDash \mathcal{H}_A$ if and only if $\exists \sigma'.\sigma' \upharpoonright_A = \sigma \wedge \sigma' \vDash \mathcal{H}$). This is done by the operation $\mathsf{rewrite}(\mathcal{H}, A)$ (line4), defined to be the application of the following rewriting relation $\twoheadrightarrow$ to $\mathcal{H}$ until convergence.

$$\mathcal{H}' \cup \{\Phi_1 \twoheadrightarrow P(\bar{x}), \Phi_2 \wedge P(\bar{y}) \twoheadrightarrow H\} \twoheadrightarrow \mathcal{H}' \cup \{\Phi_2 \wedge \Phi_1[\bar{y}/\bar{x}] \twoheadrightarrow H\}$$

Here, $P \in pvs(\mathcal{H}) \setminus A$, and Φ_i's range over Horn clause bodies. (We assume that each Horn clause in $\mathcal{H}$ is over disjoint variables. Otherwise, we transform $\mathcal{H}$ into such a form by variable renaming.)

Then, RlxSolveB calls $\mathcal{AR}_{(L_{base}, L_{ext})}$ to find a $L_{base} \cup L_{ext}$-restricted solution for $\mathcal{H}_A$ (line 5). If $\mathcal{AR}_{(L_{base}, L_{ext})}$ returns $\mathsf{noproof}$ then no L_{base} solution exists for $\mathcal{H}_A$ by the property of $\mathcal{AR}_{(L_{base}, L_{ext})}$, and by the construction above, it can be shown that no L_{base} solution exists for the input HCCS $\mathcal{H}$ either, and we safely return $\mathsf{noproof}$ (line 6) (see Theorem 7 for the proof). Otherwise, we obtain a solution σ_A for $\mathcal{H}_A$, and RlxSolveB calls $\mathcal{AU}$ on each element of the partition with the solution σ_A substituted (i.e., $\mathcal{AU}(\sigma_A(\mathcal{H}'))$ for each $\mathcal{H}' \in Y$).[9] This gives solutions for the remaining predicate variables in $\mathcal{H}$, and we return the union of σ_A and these solutions as the final solution (line 7).

We argue that the produced solution is indeed a solution of $\mathcal{H}$. Let $Y = \{\mathcal{H}_1, \ldots, \mathcal{H}_n\}$. Note that the only predicate variables shared by different elements in Y are A. Therefore, $\sigma_A(\mathcal{H}_1), \ldots, \sigma_A(\mathcal{H}_n)$, and $\mathcal{H}_A$ are over disjoint predicate variables, and their solutions are over disjoint domains. Then, because Y partitions $\mathcal{H}$, it follows that $\sigma_A \cup \bigcup_{\mathcal{H}' \in Y} \mathcal{AU}(\sigma_A(\mathcal{H}'))$ is a solution of $\mathcal{H}$.

To show that RlxSolveB is a relaxed finite-language-restricted constraint solver, it remains to show that the obtained solution is restricted to a finite language of predicates. We show that it is restricted to $LB(L_{base} \cup L_{ext}, \mathcal{AU}, \ell)$ which is

---

[9] Here, we extend the notion of substitution so that the result is tree-like: for $P \in dom(\sigma_A)$, $\sigma_A(\Phi \twoheadrightarrow P(\bar{x})) = \neg \sigma_A(P)[\bar{x}/\tilde{\nu}(P)] \wedge \sigma_A(\Phi) \twoheadrightarrow \bot$.

defined as follows. Let L be a finite language. We define $\mathsf{LB}(L, \mathcal{AU}, \ell)$ as follows.

$$\mathsf{LB}(L, \mathcal{AU}, \ell) = L \cup \bigcup_{\mathcal{H}' \in X} ran(\mathcal{AU}(\mathcal{H}'))$$
where
$$X = \{\sigma(\mathcal{H}') \mid \mathcal{H}' \in unwds(\ell, \mathcal{H}_{gen(M)}) \text{ and } \sigma \succeq_L \mathcal{H}'\}$$

Here, M is the program being verified (i.e., the input to the top-level procedure RlxCegar), $unwds(\ell, \mathcal{H})$ is the set of unwound instances of $\mathcal{H}$ of depth at most ℓ, and $\sigma \succeq_L \mathcal{H}$ if and only if σ is a map from the leaves and the root of $\mathcal{H}$ to the predicates in L (i.e., $dom(\sigma) = leaves(\mathcal{H}) \cup \{root(\mathcal{H})\} \setminus \{\bot\}$ and $ran(\mathcal{H}) \subseteq L$). Note that $\mathcal{H}_{gen(M)}$ is a constant for the entire run of RlxCegar(M), and $unwds(\ell, \mathcal{H})$ is finite for any $\mathcal{H}$ and ℓ. Therefore, $\mathsf{LB}(L, \mathcal{AU}, \ell)$ is a finite language that is determined by L, $\mathcal{AU}$ and ℓ.

We formally prove that RlxSolveB is indeed a relaxed $(L_{base}, \mathsf{LB}(L_{base} \cup L_{ext}, \mathcal{AU}, \ell))$-restricted constraint solver. First, we prove that the solution returned is indeed a solution of the input HCCS and that it is restricted to $\mathsf{LB}(L_{base} \cup L_{ext}, \mathcal{AU}, \ell)$.

Theorem 6. *Suppose RlxSolveB$[\mathcal{AR}_{(L_{base}, L_{ext})}, \mathcal{AU}, \ell](\mathcal{H})$ returns sol(σ). Then, $\sigma \models \mathcal{H}$ and $ran(\sigma) \subseteq \mathsf{LB}(L_{base} \cup L_{ext}, \mathcal{AU}, \ell)$.*

Next, we show that some solution is returned if there exists a L_{base}-restricted solution to the given HCCS (and by Theorem 6, such a solution is restricted to $\mathsf{LB}(L_{base} \cup L_{ext}, \mathcal{AU}, \ell)$).

Theorem 7. *Suppose that there exists σ such that $\sigma \models \mathcal{H}$ and $ran(\sigma) \subseteq L_{base}$. Then, RlxSolveB$[\mathcal{AR}_{(L_{base}, L_{ext})}, \mathcal{AU}, \ell](\mathcal{H})$ infers some σ' such that $\sigma' \models \mathcal{H}$.*

Example 6. Recall the HCCS $\mathcal{H}_{exa}$ from Example 3. Running RlxSolveB on $\mathcal{H}_{exa}$ with $\ell = 2$, we have $Y = \mathsf{partition}(2, \mathcal{H}_{exa}) = \{\mathcal{H}_1, \mathcal{H}_2, \mathcal{H}_3\}$ where

$$\mathcal{H}_1 = \{\theta_{p3} \wedge Q_1(\bar{x}) \dashrightarrow Q_2(\bar{x}'), \theta_{p4} \wedge Q_2(\bar{x}) \dashrightarrow \bot\}$$
$$\mathcal{H}_2 = \{\theta_{p2} \wedge P_1(\bar{x}) \dashrightarrow P_2(\bar{x}'), P_2(\bar{x}) \dashrightarrow Q_1(\bar{x})\}$$
$$\mathcal{H}_3 = \{\theta_{p1} \dashrightarrow P_1(\bar{x})\}$$

Then, the shared predicate variables that are selected to be restricted are $A = \{root(\mathcal{H}_1), root(\mathcal{H}_2), root(\mathcal{H}_3)\} \setminus \{\bot\} = \{P_1, Q_1\}$. And, $\mathcal{H}_A = \mathsf{rewrite}(\mathcal{H}_{exa}, A)$ is as shown below.

$$\mathcal{H}_A = \{\theta_{p3} \wedge Q_1(\bar{x}) \wedge \theta_{p4}[\bar{x}', \bar{x}''/\bar{x}, \bar{x}'] \dashrightarrow \bot, \theta_{p2} \wedge P_1(\bar{x}) \dashrightarrow Q_1(\bar{x}'), \theta_{p1} \dashrightarrow P_1(\bar{x})\}$$

where $\bar{x}''$ is a quintuple of fresh variables. RlxSolveB then calls $\mathcal{AR}_{(L_{base}, L_{ext})}$ on $\mathcal{H}_A$ to obtain a restricted solution for A. Suppose the returned solution is $\sigma_A = \{P \mapsto a = b \Rightarrow y = z + x \mid P \in \{P_1, Q_1\}\}$. Then, σ_A is applied to each element of the partition and we obtain $\sigma_A(\mathcal{H}_1)$, $\sigma_A(\mathcal{H}_2)$, and $\sigma_A(\mathcal{H}_3)$ shown below.

$$\sigma_A(\mathcal{H}_1) = \{\theta_{p3} \wedge (a = b \Rightarrow y = x + z) \dashrightarrow Q_2(\bar{x}'), \theta_{p4} \wedge Q_2(\bar{x}) \dashrightarrow \bot\}$$
$$\sigma_A(\mathcal{H}_2) = \{\theta_{p2} \wedge (a = b \Rightarrow y = x + z) \dashrightarrow P_2(\bar{x}'), a = b \wedge y \neq x + z \wedge P_2(\bar{x}) \dashrightarrow \bot\}$$
$$\sigma_A(\mathcal{H}_3) = \{a = b \wedge y \neq x + z \wedge \theta_{p1} \dashrightarrow \bot\}$$

$\mathcal{AU}$ is called on $\sigma_A(\mathcal{H}_1)$, $\sigma_A(\mathcal{H}_2)$, and $\sigma_A(\mathcal{H}_3)$ to infer unrestricted solutions to the remaining predicate variables. Suppose we have obtained the solutions $\sigma_1 = \{Q_2 \mapsto a = b \Rightarrow z \neq -1 \wedge y = x + z\}$, $\sigma_2 = \{P_2 \mapsto a = b \Rightarrow y = x + z\}$, and $\sigma_3 = \varnothing$ for $\sigma_A(\mathcal{H}_1)$, $\sigma_A(\mathcal{H}_2)$, and $\sigma_A(\mathcal{H}_3)$ respectively. Finally, RlxSolveB returns the combined map, $\sigma_A \cup \sigma_1 \cup \sigma_2 \cup \sigma_3$, as the solution inferred for the input HCCS $\mathcal{H}_{exa}$.   ▲

4 Implementation and Experiments

We have implemented the new refinement algorithms RlxSolveA and RlxSolveB described in Section 3. The refinement algorithms require an exact finite-language-restricted constraint solver and an unrestricted constraint solver to be provided as modules. An unrestricted constraint solver finds unrestricted solutions to the given tree-like HCCS. This is the ordinary constraint solving for tree-like HCCSs which is a well-studied problem, and we use the existing technique that iteratively solve the constraints one predicate variable at a time by using interpolation as a blackbox process (see [20,4,17] for details).[10]

Exact Finite-Language-Restricted Constraint Solver. For the exact solver, we use a simple approach in which the finite predicate languages are represented by predicate templates containing unknowns of bounded range. We use an SMT solver[11] to find an assignment to the unknowns within the bound that makes the templates into an actual solution. Below, we informally describe the process by an example, and defer the detailed description to the extended report [18].

Example 7. Recall the program M_{ex} from Example 1. Let L_{base} be the finite language of predicates consisting of conjunctions of at most two atomic predicates whose numeric constants are bounded in the range $\{-1, 0, 1\}$. We represent the language by the *bounded predicate template* shown below

$$\lambda a, b, x, y, z.\ c_1 a + c_2 b + c_3 x + c_4 y + c_5 z + c_6 \leq 0 \wedge$$
$$c_7 a + c_8 b + c_9 x + c_{10} y + c_{11} z + c_{12} \leq 0$$

where c_i's are *unknown constants* each associated with the *bound* $\{-1, 0, 1\}$. Bounded predicate templates can concisely represent a finite language of predicates.[12]

Let ξ be a bounded predicate template. To check if the given HCCS $\mathcal{H}$ has a solution in the language represented by ξ, we make a *solution template* σ_ξ that maps each predicate variable in $\mathcal{H}$ to a copy of ξ with fresh unknowns. Then, we check if there exists an assignment to the unknowns within the bounds that makes the solution template into an actual solution of $\mathcal{H}$, that is, we look for

---

[10] The implementation uses MathSAT 5 (http://mathsat.fbk.eu/) for the backend interpolation process.

[11] The implementation uses Z3 (http://z3.codeplex.com/).

[12] Note that such a language is generally not closed under conjunctions or disjunctions.

A New Approach to Practical Complete Predicate Refinement   629

assignments to the unknowns within the bounds that satisfy $\forall hc \in \sigma_\xi(\mathcal{H}). \models hc$.
For QFLRA, the latter can be done by applying the Motzkin's transposition
theorem [15] to reduce the problem to the satisfiability problem for quantifier-
free non-linear real arithmetic, and using an SMT solver to solve the resulting
problem.   ▲

We note that the exact finite-language-restricted constraint solving is a highly
expensive process and using it directly solve the whole HCCS is prohibitive.
Indeed, as we show in the experiments, the exact solver fails to scale even on
relatively small constraint sets (see also the discussion in Section 5 and the
complexity theoretic analysis in the extended report [18]).

Experiment Setup. We have experimented with the new refinement algorithms
by using them in the refinement process of MoCHi [8]. MoCHi is a state-of-the-art
software model checker for higher-order functional programs based on predicate
abstraction, CEGAR, and higher-order-recursion-scheme (HORS) model checking. MoCHi verifies assertion safety of OCaml programs. A verifier for functional
programs such as MoCHi is suited for experimenting with the new refinement
algorithm because Horn-clause constraints generated in such a verifier often contain non-trivial tree-like structure. (Intuitively, this is because the constraints
express the flow of data in the program, and data often flow in a complex way
in a functional program, e.g., passed to and returned from recursive functions,
captured in closures, etc.)

The new refinement algorithms RlxSolveA and RlxSolveB are parametric. For
this experiment, we parameterize them as follows to obtain a single refinement
algorithm:

$$\text{RlxSolveB}[\text{RlxSolveA}[\mathcal{AE}_{L_{base}}], \mathcal{AU}, 4]$$

Here, the exact L_{base}-restricted constraint solver $\mathcal{AE}_{L_{base}}$ and the unrestricted
solver $\mathcal{AU}$ are the ones described above. That is, we use RlxSolveB parameterized to use as modules the relaxed $(L_{base}, (L_{base})^{\wedge\vee})$-restricted constraint solver
RlxSolveA (itself parameterized by the exact L_{base}-restricted constraint solver
$\mathcal{AE}_{L_{base}}$) and the unrestricted constraint solver $\mathcal{AU}$, and with the parameter
$\ell = 4$. The strata of restricting predicate languages are built "dynamically" as
the CEGAR iteration progresses, by starting from a small fixed (L_{base}^0, L_{ext}^0)
and enlarging the current (L_{base}, L_{ext}) whenever the refinement algorithm returns noproof by using the unrestricted refinement process.[13]

We compare the new refinement algorithm with two other refinement methods: 1.) the ordinary (incomplete) unrestricted predicate search, and 2.) exact
finite-language-restricted predicate search. The unrestricted predicate search algorithm is $\mathcal{AU}$, and the exact finite-language restricted predicate search algorithm is $\mathcal{AE}$. For $\mathcal{AE}$, we give (the L_{base} part of) the same restricting predicate
language given to the new algorithm when solving the corresponding HCCS.

---

[13] Formally, this is done by having a non-decreasing preorder of restricting predicate
languages where the limit of any ω-chain is $\mathcal{T}$, and when noproof is returned, the
language raised to the least one containing the predicate inferred by $\mathcal{AU}$.

We have ran the three refinement algorithms on 318 HCCSs generated by running MoCHi on 139 programs, measuring the time spent in each run of the refinement process. The benchmark programs are mostly taken from the previous work on MoCHi [8,14,22,9]. To obtain the benchmark HCCS set, we ran MoCHi on each benchmark program with the new refinement algorithm until completion or timeout and recorded the HCCS given as the input to each run of the refinement process. We also compare the overall verification speed of MoCHi when using the three refinement algorithms. This is done by running MoCHi with each of the refinement algorithm on the 139 benchmark programs. We have run the experiments on a machine with 2.69 GHz i7-4600U processor with 16 GB of RAM, with the time limit of 100 seconds. The benchmark programs, the benchmark HCCSs and the experiment results data are available online [18].

Experiment Aim and Hypothesis. Because of the overhead from computing restricted proofs, we expect the individual refinement runs to be slower with the new refinement algorithm compared to an ordinary incomplete approach which only does unrestricted refinement, but faster than the more naïve complete approach that directly applies the exact finite-language-restricted proof search to the entire refinement problem. The main purpose of the experiment is to test this hypothesis. We also compare the overall verification speeds, but we do not expect a significant improvement on this aspect because of the inherent complexity of the verification problem. (For any sound and QFLRA-complete verifier, one can always find a program on which the verifier takes arbitrarily long time.)

Experiment Results and Analysis. Figure 7 shows the plots comparing the the run times of the new refinement algorithm (New Algorithm), the unrestricted refinement algorithm (Unrestricted), and the exact finite-language-restricted refinement algorithm (Exact) on each of the 318 benchmark HCCSs. As we have expected, the unrestricted refinement algorithm is the fastest of the three. The new algorithm performs quite competitively, however, and shows that it is able to achieve completeness with only a low overhead. Also, the plots show that the exact finite-language-restricted refinement algorithm is significantly slower, timing out on many instances that the other two algorithms were able to solve quickly.

Figure 8 shows the plots comparing the run times of the overall verification process on each of the 139 benchmark programs for each refinement algorithm. The plots show that there is no clear winner in this comparison and none of the three outperformed the others on all benchmarks (while the unrestricted refinement edged out in the number of instances solved within the time limit, it also timed out on some instances the complete methods were able to solve). This is due to the inherent undecidability of the program verification problem, and the fact that the speed of overall verification depends heavily on subtle heuristic choices made by MoCHi. Such issues are largely outside of the scope of this paper, but they give interesting insights into what would be the good

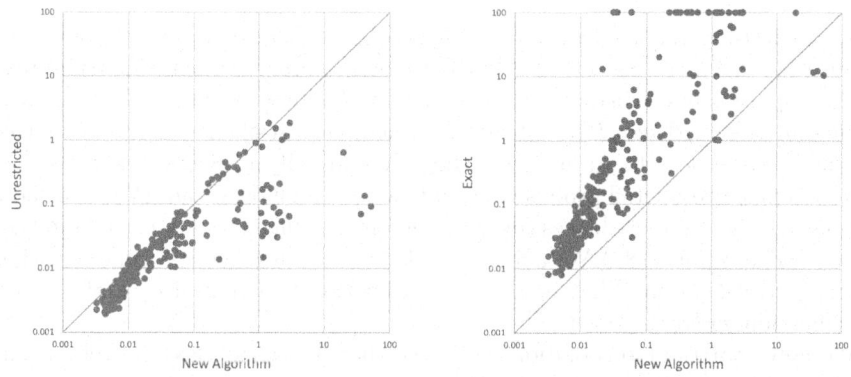

Fig. 7. Run time comparison of the refinement algorithms on benchmarks HCCSs

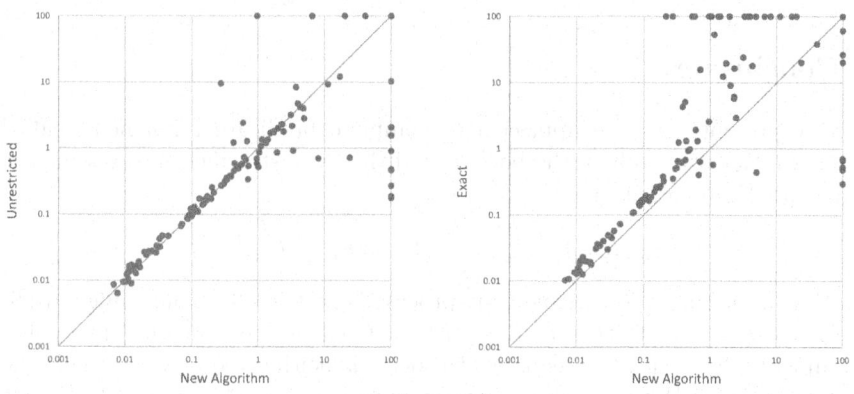

Fig. 8. Run time comparison of the refinement algorithms on benchmarks programs

heuristics to use with the new refinement algorithm. For instance, an interesting behavior we have observed is that the stratified approaches (New Algorithm and Exact) sometimes infer more useless predicates than the ordinary unrestricted refinement because a stratified approach needs to add predicates to raise the language stratum till it reaches the level where a proof of the given program exists. Because MoCHi does eager predicate abstraction, its performance degrades exponentially in the number of predicates that are added to the candidate predicate set. This seems to have had a large negative impact on the stratified approaches. A possible way to address the issue maybe is to have MoCHi take a more lazy approach to predicate abstraction, or allow the language strata "coarseness" to be dynamically adjustable so that we can immediately jump to a large predicate language when it seems beneficial.

5   Related Work

Previous work [7,11] has considered an *exact* stratification approach which requires the refinement process to exactly decide if a proof of the given counterexample's spuriousness exists in the current finite language stratum. As remarked before, an issue with exact stratification is the high cost of exact finite-language-restricted proof search. As we have shown empirically in Section 4, the exact finite-language-restricted proof search suffers from high overhead. (We also show complexity theoretic evidences for the inherent hardness of the exact search in the extended report [18].) We note that relaxed stratification is a generalization of exact stratification. That is, exact stratification is a special case of relaxed stratification where $L_{ext} = \varnothing$.

We note that the interpolation technique that limits the theory-level reasoning to only emit restricted partial interpolants (cf. Section 3.2) has also been proposed in [7]. But, they target exact stratification and therefore requires the restricting language to be closed under conjunctions and disjunctions (so that $L^{\wedge\vee} = L$), which substantially reduces the applicability of the technique.[14]

6   Conclusion

We have presented a new approach to complete predicate refinement, called relaxed stratification, where the background theory is stratified into a sequence of finite predicate languages

$$(L^0_{base}, L^0_{ext}), (L^1_{base}, L^1_{ext}), \dots (L^k_{base}, L^k_{ext}), \dots$$

such that each run of the refinement process is restricted to only infer predicates from the current stratum $L_{base} \cup L_{ext}$. Contrary to previous approaches to complete refinement, the refinement process is neither required to decide the existence of a proof for the given counterexample in $L_{base} \cup L_{ext}$ nor in L_{base}, but is only required to return some proof if one exists in L_{base}. We have proved that the approach is complete despite the relaxed requirement, assuming that the strata of L_{base}'s grow to eventually cover the predicates of the underlying theory. We have shown that the relaxed requirement can be used to build practical refinement algorithms that have low overhead and the completeness guarantee.

References

1. Grebenshchikov, S., Lopes, N.P., Popeea, C., Rybalchenko, A.: Synthesizing software verifiers from proof rules. In: Vitek, J., Lin, H., Tip, F. (eds.) PLDI, pp. 405–416. ACM (2012)

---

[14] Contrary to [7], in QFLRA, it is insufficient to only look for an atomic interpolant (i.e., a separating hyperplane) when interpolating between even just conjunctions of literals (i.e., polytopes) under a finite-language restriction. For example, consider interpolating between $y \le 1 \wedge 2x + y \le -3 \wedge x \le -1$ and $y + x \ge 1$ under the restriction that interpolants' constants are in $\{-1, 0, 1\}$.

2. Gupta, A., Popeea, C., Rybalchenko, A.: Predicate abstraction and refinement for verifying multi-threaded programs. In: Ball, T., Sagiv, M. (eds.) POPL, pp. 331–344. ACM (2011)
3. Gupta, A., Popeea, C., Rybalchenko, A.: Solving recursion-free horn clauses over LI+UIF. In: Yang, H. (ed.) APLAS 2011. LNCS, vol. 7078, pp. 188–203. Springer, Heidelberg (2011)
4. Heizmann, M., Hoenicke, J., Podelski, A.: Nested interpolants. In: Hermenegildo, M.V., Palsberg, J. (eds.) POPL, pp. 471–482. ACM (2010)
5. Henzinger, T.A., Jhala, R., Majumdar, R., McMillan, K.L.: Abstractions from proofs. In: Jones, N.D., Leroy, X. (eds.) POPL, pp. 232–244. ACM (2004)
6. Jhala, R., Majumdar, R.: Software model checking. ACM Computing Surveys 41(4) (2009)
7. Jhala, R., McMillan, K.L.: A practical and complete approach to predicate refinement. In: Hermanns, H., Palsberg, J. (eds.) TACAS 2006. LNCS, vol. 3920, pp. 459–473. Springer, Heidelberg (2006)
8. Kobayashi, N., Sato, R., Unno, H.: Predicate abstraction and CEGAR for higher-order model checking. In: Hall, M.W., Padua, D.A. (eds.) PLDI, pp. 222–233. ACM (2011)
9. Kuwahara, T., Terauchi, T., Unno, H., Kobayashi, N.: Automatic termination verification for higher-order functional programs. In: Shao, Z. (ed.) ESOP 2014 (ETAPS). LNCS, vol. 8410, pp. 392–411. Springer, Heidelberg (2014)
10. McMillan, K.L.: An interpolating theorem prover. Theoretical Computer Science 345(1), 101–121 (2005)
11. McMillan, K.L.: Quantified invariant generation using an interpolating saturation prover. In: Ramakrishnan, C.R., Rehof, J. (eds.) TACAS 2008. LNCS, vol. 4963, pp. 413–427. Springer, Heidelberg (2008)
12. Rümmer, P., Hojjat, H., Kuncak, V.: Classifying and solving horn clauses for verification. In: Cohen, E., Rybalchenko, A. (eds.) VSTTE 2013. LNCS, vol. 8164, pp. 1–21. Springer, Heidelberg (2014)
13. Rümmer, P., Hojjat, H., Kuncak, V.: Disjunctive interpolants for horn-clause verification. In: Sharygina, N., Veith, H. (eds.) CAV 2013. LNCS, vol. 8044, pp. 347–363. Springer, Heidelberg (2013)
14. Sato, R., Unno, H., Kobayashi, N.: Towards a scalable software model checker for higher-order programs. In: Albert, E., Mu, S. (eds.) PEPM, pp. 53–62. ACM (2013)
15. Schrijver, A.: Theory of linear and integer programming. Wiley (1998)
16. Sebastiani, R.: Lazy satisability modulo theories. JSAT 3(3-4), 141–224 (2007)
17. Terauchi, T.: Dependent types from counterexamples. In: Hermenegildo, M.V., Palsberg, J. (eds.) POPL, pp. 119–130. ACM (2010)
18. Terauchi, T., Unno, H.: Relaxed stratification: A new approach to practical complete predicate refinement (2015), http://www.jaist.ac.jp/~terauchi
19. Tseitin, G.S.: On the complexity of derivation in propositional calculus. Studies in Constructive Mathematics and Mathematical Logic 2(115-125), 10–13 (1968)
20. Unno, H., Kobayashi, N.: Dependent type inference with interpolants. In: Porto, A., López-Fraguas, F.J. (eds.) PPDP, pp. 277–288. ACM (2009)
21. Unno, H., Terauchi, T.: Inferring simple solutions to recursion-free horn clauses via sampling. In: TACAS (2015) (to appear)
22. Unno, H., Terauchi, T., Kobayashi, N.: Automating relatively complete verification of higher-order functional programs. In: Giacobbazzi, R., Cousot, R. (eds.) POPL, pp. 75–86. ACM (2013)

Spatial Interpolants

Aws Albargouthi[1], Josh Berdine[2], Byron Cook[3], and Zachary Kincaid[4]

[1] University of Wisconsin-Madison
[2] Microsoft Research
[3] University College London
[4] University of Toronto

Abstract. We propose SPLINTER, a new technique for proving properties of heap-manipulating programs that marries (1) a new *separation
logic–based* analysis for heap reasoning with (2) an *interpolation-based*
technique for refining heap-shape invariants with data invariants. SPLIN-
TER is *property directed, precise,* and produces counterexample traces
when a property does not hold. Using the novel notion of *spatial interpolants modulo theories,* SPLINTER can infer complex invariants over
general recursive predicates, e.g., of the form *all elements in a linked list
are even* or a *binary tree is sorted.* Furthermore, we treat interpolation
as a black box, which gives us the freedom to encode data manipulation
in any suitable theory for a given program (e.g., bit vectors, arrays, or
linear arithmetic), so that our technique immediately benefits from any
future advances in SMT solving and interpolation.

1 Introduction

Since the problem of determining whether a program satisfies a given property
is undecidable, every verification algorithm must make some compromise. There
are two classical schools of program verification, which differ in the compromise
they make: the *static analysis* school gives up refutation soundness (i.e., may
report *false positives*); and the *software model checking* school gives up the guarantee of termination. In the world of integer program verification, both schools
are well explored and enjoy cross-fertilization of ideas: each has its own strengths
and uses in different contexts. In the world of heap-manipulating programs, the
static analysis school is well-attended [36,15,13,11], while the software model
checking school has remained essentially vacant. This paper initiates a program
to rectify this situation, by proposing one of the first path-based software model
checking algorithms for proving combined shape-and-data properties.

The algorithm we propose, SPLINTER, marries two celebrated program verification ideas: McMillan's *lazy abstraction with interpolants* (IMPACT) algorithm for
software model checking [26], and *separation logic,* a program logic for reasoning
about shape properties [33]. SPLINTER (like IMPACT) is based on a path-sampling
methodology: given a program P and safety property φ, SPLINTER constructs a
proof that P is memory safe and satisfies φ by sampling a finite number of paths
through the control-flow graph of P, proving them safe, and then assembling proofs
for each sample path into a proof for the whole program. The key technical advance
which enables SPLINTER is an algorithm for *spatial interpolation,* which is used to

ⓒ Springer-Verlag Berlin Heidelberg 2015
J. Vitek (Ed.): ESOP 2015, LNCS 9032, pp. 634–660, 2015.
DOI: 10.1007/978-3-662-46669-8_26

construct proofs in *separation logic* for the sample traces (serving the same function as *Craig interpolation* for first-order logic in IMPACT).

SPLINTER is able to prove properties requiring integrated heap and data (e.g., integer) reasoning by strengthening separation logic proofs with *data refinements* produced by classical Craig interpolation, using a technique we call *spatial interpolation modulo theories*. Data refinements are *not tied to a specific logical theory*, giving us a rather generic algorithm and freedom to choose an appropriate theory to encode a program's data.

Fig. 1 summarizes the high-level operation of our algorithm. Given a program with no heap manipulation, SPLINTER only computes theory interpolants and behaves exactly like IMPACT, and thus one can thus view SPLINTER as a proper extension of IMPACT to heap manipulating programs. At the other extreme, given a program with no data manipulation, SPLINTER is a new shape analysis that uses path-based relaxation to construct memory safety proofs in separation logic.

There is a great deal of work in the static analysis school on shape analysis and on combined shape-and-data analysis, which we will discuss further in Sec. 8. We do not claim superiority over these techniques (which have had the benefit of 20 years of active development). SPLINTER, as the first member of the software model checking school, is not *better*; however, it *is* fundamentally *different*. Nonetheless, we will mention two of the features of SPLINTER (not enjoyed by any previous verification algorithm for shape-and-data properties) that make our approach worthy of exploration: path-based refinement and property-direction.

- *Path-based refinement*: This supports a progress guarantee by tightly correlating program exploration with refinement, and by avoiding imprecision due to lossy join and widening operations employed by abstract domains. SPLINTER does not report false positives, and produces counterexamples for violated properties. This comes, as usual, at the price of potential divergence.
- *Property-direction*: Rather than seeking the strongest invariant possible, we compute one that is *just strong enough* to prove that a desired property holds. Property direction enables scalable reasoning in rich program logics like the one described in this paper, which combines separation logic with first-order data refinements.

We have implemented an instantiation of our generic technique in the T2 verification tool [38], and used it to prove correctness of a number of programs, partly drawn from open source software, requiring combined data and heap invariants. Our results indicate the usability and promise of our approach.

Contributions. We summarize our contributions as follows:

1. A generic property-directed algorithm for verifying and falsifying safety of programs with heap and data manipulation.
2. A precise and expressive separation logic analysis for computing memory safety proofs of program paths using a novel technique we term *spatial interpolation*.
3. A novel interpolation-based technique for strengthening separation logic proofs with data refinements.
4. An implementation and an evaluation of our technique for a fragment of separation logic with linked lists enriched with linear arithmetic refinements.

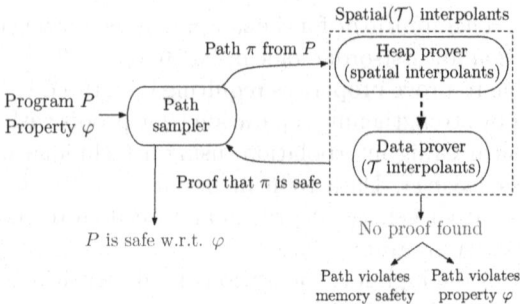

Fig. 1. Overview of SPLINTER verification algorithm

The extended version [2] of this paper contains additional details and material.

2   Overview

In this section, we demonstrate the operation of SPLINTER (Fig. 1) on the simple linked list example shown in Fig. 2. We assume that integers are unbounded (i.e., integer values are drawn from $\mathbb{Z}$ rather than machine integers) and that there is a **struct** called **node** denoting a linked list node, with a next pointer N and an integer (data) element D. The function **nondet()** returns a nondeterministic integer value. This program starts by building a linked list in the loop on location 2. The loop terminates if the initial value of i is ≥ 0, in which case a linked list of size i is constructed, where data elements D of list nodes range from 1 to i. Then, the loop at location 3 iterates through the linked list asserting that the data element of each node in the list is ≥ 0. Our goal is to prove that the assertion at location 4 is never violated.

```
1:  int i = nondet();
    node* x = null;
2:  while (i != 0)
        node* tmp = malloc
            (node);
        tmp->N = x;
        tmp->D = i;
        x = tmp;
        i--;
3:  while (x != null)
4:      assert(x->D >= 0);
        x = x->N;
```

Fig. 2. Illustrative Example

Sample a Program Path. To start, we need a path π through the program to the assertion at location 4. Suppose we start by sampling the path 1,2,2,3,4, that is, the path that goes through the first loop once, and enters the second loop arriving at the assertion. This path is illustrated in Fig. 3 (where 2a indicates the second occurrence of location 2). Our goal is to construct a Hoare-style proof of this path: an annotation of each location along the path with a formula describing reachable states, such that location 4 is annotated with a formula implying that x->D >= 0. This goal is accomplished in two phases. First, we use *spatial interpolation* to compute a memory safety proof for the path π (Fig. 3(b)). Second, we use *theory refinement* to strengthen the memory safety proof and establish that the path satisfies the post-condition x->D >= 0 (Fig. 3(c)).

Compute Spatial Interpolants. The first step in constructing the proof is to find *spatial interpolants*: a sequence of separation logic formulas *approximating* the shape of the heap at each program location, and forming a Hoare-style memory safety proof of the path. Our spatial interpolation procedure is a two step process that first symbolically executes the path in a forward pass and then derives a weaker proof using a backward pass. The backward pass can be thought of as an under-approximate weakest precondition computation, which uses the symbolic heap from the forward pass to guide the under-approximation.

We start by showing the *symbolic heaps* in Fig. 3(a), which are the result of the forward pass obtained by symbolically executing *only* heap statements along this program path (i.e., the strongest postcondition along the path). The separation logic annotations in Fig. 3 follow standard notation (e.g., [15]), where a formula is of the form $\Pi : \Sigma$, where Π is a Boolean first-order formula over heap variables (pointers) as well as data variables (e.g., $x =$ null or $i > 0$), and Σ is a *spatial conjunction* of *heaplets* (c.g., emp, denoting the empty heap, or $Z(x, y)$, a recursive predicate, e.g., that denotes a linked list between x and y). For the purposes of this example, we assume a recursive predicate $\mathsf{ls}(x, y)$ that describes linked lists. In our example, the symbolic heap at location **2a** is $true : x \mapsto [d', \text{null}]$, where the heap consists of a node, pointed to by variable x, with null in the N field and the (implicitly existentially quantified) variable d' in the D field (since so far we are only interested in heap shape and not data).

The symbolic heaps determine a memory safety proof of the path, but it is too strong and would likely not generalize to other paths. The goal of spatial interpolation is to find a sequence of annotations that are weaker than the symbolic heaps, but that still prove memory safety of the path. A sequence of spatial interpolants is shown in Fig. 3(b). Note that all spatial interpolants are implicitly spatially conjoined with true; for clarity, we avoid explicitly conjoining formulas with true in the figure. For example, location 2 is annotated with $true : \mathsf{ls}(x, \text{null}) * true$, indicating that there is a list on the heap, as well as other potential objects not required to show memory safety. We compute spatial interpolants by going backwards along the path and asking questions of the form: *how much can we weaken the symbolic heap while still maintaining memory safety?* We will describe how to answer such questions in Section 4.

Refine with Theory Interpolants. Spatial interpolants give us a memory safety proof as an approximate heap shape at each location. Our goal now is to strengthen these heap shapes with data refinements, in order to prove that the assertion at the end of the path is not violated. To do so, we generate a system of Horn clause constraints from the path in some first-order theory admitting interpolation (e.g., linear arithmetic). These Horn clauses carefully encode the path's data manipulation along with the spatial interpolants, which tell us heap shape at each location along the path. A solution of this constraint system, which can be solved using off-the-shelf interpolant generation techniques (e.g., [27,35]), is a *refinement* (strengthening) of the memory safety proof.

In this example, we encode program operations over integers in the theory of linear integer arithmetic, and use Craig interpolants to solve the system of constraints. A solution of this system is a set of linear arithmetic formulas that refine our spatial interpolants and, as a result, imply the assertion we want to

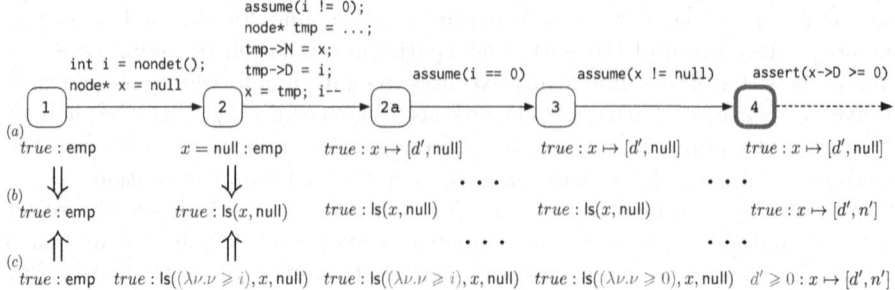

Fig. 3. Path through program in Fig. 2, annotated with (a) results of forward symbolic execution, (b) spatial interpolants, and (c) spatial($\mathcal{T}$) interpolants, where $\mathcal{T}$ is linear integer arithmetic. Arrows $\Rightarrow$ indicate implication (entailment) direction.

prove holds. One possible solution is shown in Fig. 3(c). For example, location 2a is now labeled with $true : \mathsf{ls}((\lambda\nu.\nu \geqslant i), x, \mathsf{null})$, where the green parts of the formula are those added by refinement. Specifically, after refinement, we know that *all* elements in the list from x to null after the first loop have data values greater than or equal to i, as indicated by the predicate $(\lambda\nu.\nu \geqslant i)$. (In Section 3, we formalize recursive predicates with data refinements.)

Location 4 is now annotated with $d' \geqslant 0 : x \mapsto [d', n'] * true$, which implies that x->D >= 0, thus proving that the path satisfies the assertion.

From Proofs of Paths to Proofs of Programs. We go from proofs of paths to whole program proofs implicitly by building an *abstract reachability tree* as in IMPACT [26]. To give a flavour for how this works, consider that the assertions at 2 and 2a are identical: this implies that this assertion is an inductive invariant at line 2. Since this assertion also happens to be strong enough to prove safety of the program, we need not sample any longer unrollings of the first loop. However, since we have not established the inductiveness of the assertion at 3, the proof is not yet complete and more traces need to be explored (in fact, exploring one more trace will do: consider the trace that unrolls the second loop once and shows that the second time 3 is visited can also be labeled with $true : \mathsf{ls}((\lambda\nu.\nu \geqslant 0), x, \mathsf{null}))$.

Since our high-level algorithm is virtually the same as IMPACT [26], we will not describe it further in the paper. For the remainder of this paper, we will concentrate on the novel contribution of our algorithm: computing spatial interpolants with theory refinements for program paths.

3   Preliminaries

3.1   Separation Logic

We define RSep, a fragment of separation logic formulas featuring points-to predicates and general recursive predicates refined by theory propositions.

Fig. 4 defines the syntax of RSep formulas. In comparison with the standard list fragment used in separation logic analyses (e.g., [4,14,28]), the differentiating features of RSep are: (1) General *recursive predicates*, for describing unbounded

$$
\begin{array}{ll}
x, y \in \mathsf{HVar} & \text{(Heap variables)} \\
a, b \in \mathsf{DVar} & \text{(Data variables)} \\
A \in \mathsf{DTerm} & \text{(Data terms)} \\
\varphi \in \mathsf{DFormula} & \text{(Data formulas)} \\
Z \in \mathsf{RPred} & \text{(Rec. predicates)} \\
\theta \in \mathsf{Refinement} ::= \lambda \vec{a}.\varphi \\
X \subseteq \mathsf{Var} \qquad ::= x \mid a
\end{array}
\qquad
\begin{array}{l}
E, F \in \mathsf{HTerm} ::= \mathsf{null} \mid x \\
\quad \mathit{\mathbb{E}} ::= A \mid E \\
\Pi \in \mathsf{Pure} ::= \mathit{true} \mid E = E \mid E \neq E \mid \\
\qquad \varphi \mid \Pi \wedge \Pi \\
H \in \mathsf{Heaplet} ::= \mathsf{true} \mid \mathsf{emp} \mid E \mapsto [\vec{A}, \vec{E}] \mid Z(\vec{\theta}, \vec{E}) \\
\Sigma \in \mathsf{Spatial} ::= H \mid H * \Sigma \\
P \in \mathsf{RSep} ::= (\exists X.\ \Pi : \Sigma)
\end{array}
$$

Fig. 4. Syntax of RSep formulas

pointer structures like lists, trees, etc. (2) Recursive predicates are augmented with a vector of *refinements*, which are used to constrain the data values appearing on the data structure defined by the predicate, detailed below. (3) Each heap cell (points-to predicate), $E \mapsto [\vec{A}, \vec{E}]$, is a *record* consisting of *data* fields (a vector $\vec{A}$ of DTerm) followed by *heap* fields (a vector $\vec{E}$ of HTerm). (Notationally, we will use d_i to refer to the ith element of the vector $\vec{d}$, and $\vec{d}[t/d_i]$ to refer to the vector $\vec{d}$ with the ith element modified to t.) (4) Pure formulas contain heap and first-order data constraints.

Our definition is (implicitly) parameterized by a first-order theory $\mathcal{T}$. DVar denotes the set of theory variables, which we assume to be disjoint from HVar (the set of heap variables). DTerm and DFormula denote the sets of theory terms and formulas, and we assume that heap variables do not appear in theory terms.

For an RSep formula P, $\mathsf{Var}(P)$ denotes its free (data and heap) variables. We treat a Spatial formula Σ as a multiset of heaplets, and consider formulas to be equal when they are equal as multisets. For RSep formulas $P = (\exists X_P.\ \Pi_P : \Sigma_P)$ and $Q = (\exists X_Q.\ \Pi_Q : \Sigma_Q)$, we write $P * Q$ to denote the RSep formula

$$
P * Q = (\exists X_P \cup X_Q.\ \Pi_P \wedge \Pi_Q : \Sigma_P * \Sigma_Q)
$$

assuming that X_P is disjoint from $\mathsf{Var}(Q)$ and X_Q is disjoint from $\mathsf{Var}(P)$ (if not, then X_P and X_Q are first suitably renamed). For a set of variables X, we write $(\exists X.\ P)$ to denote the RSep formula

$$
(\exists X.\ P) = (\exists X \cup X_P.\ \Pi_P : \Sigma_P)
$$

Recursive Predicates. Each recursive predicate $Z \in \mathsf{RPred}$ is associated with a definition that describes how the predicate is unfolded. Before we formalize these definitions, we will give some examples.

The definition of the list segment predicate from Sec. 2 is:

$$
\begin{aligned}
\mathsf{ls}(R, x, y) \equiv\ & (x = y : \mathsf{emp}) \vee \\
& (\exists d, n'.\ x \neq y \wedge R(d) : x \mapsto [d, n'] * \mathsf{ls}(R, n', y))
\end{aligned}
$$

In the above, R is a *refinement variable*, which may be instantiated to a concrete refinement $\theta \in \mathsf{Refinement}$. For example, $\mathsf{ls}((\lambda a.a \geqslant 0), x, y)$ indicates that there is a list from x to y where every element of the list is at least 0.

A refined binary tree predicate is a more complicated example:

$$bt(Q, L, R, x) = (x = \mathsf{null} : \mathsf{emp})$$
$$\vee \; (\exists d, l, r. \; Q(d) : x \mapsto [d, l, r]$$
$$* \; bt((\lambda a.Q(a) \wedge L(d, a)), L, R, l)$$
$$* \; bt((\lambda a.Q(a) \wedge R(d, a)), L, R, r))$$

This predicate has three refinement variables: a unary refinement Q (which must be satisfied by every node in the tree), a binary refinement L (which is a relation that must hold between every node and its descendants to the left), and a binary refinement R (which is a relation that must hold between every node and its descendants to the right). For example,

$$bt((\lambda a.true), (\lambda a, b.a \geqslant b), (\lambda a, b.a \leqslant b), x)$$

indicates that x is the root of a *binary search tree*, and

$$bt((\lambda a.a \geqslant 0), (\lambda a, b.a \leqslant b), (\lambda a, b.a \leqslant b), x)$$

indicates that x is the root of a *binary min-heap* with non-negative elements.

To formalize these definitions, we first define *refinement terms* and *refined formulas*: a refinement term τ is either (1) a refinement variable R or (2) an abstraction $(\lambda a_1, \ldots, a_n.\Phi)$, where Φ is a refined formula. A *refined formula* is a conjunction where each conjunct is either a data formula (DFormula) or the application $\tau(\vec{A})$ of a refinement term to a vector of data terms (DTerm).

A *predicate definition* has the form

$$Z(\vec{R}, \vec{x}) \equiv (\exists X_1. \; \Pi_1 \wedge \Phi_1 : \Sigma_1) \vee \cdots \vee (\exists X_n. \; \Pi_n \wedge \Phi_n : \Sigma_n)$$

where $\vec{R}$ is a vector of refinement variables, $\vec{x}$ is a vector of heap variables, and where refinement terms may appear as refinements in the spatial formulas Σ_i. We refer to the disjuncts of the above formula as the *cases* for Z, and define $cases(Z(\vec{R}, \vec{x}))$ to be the set of cases of Z. $\vec{R}$ and $\vec{x}$ are bound in $cases(Z(\vec{R}, \vec{x}))$, and we will assume that predicate definitions are closed, that is, for each case of Z, the free refinement variables belong to $\vec{R}$, the free heap variables belong to $\vec{x}$, and there are no free data variables. We also assume that they are well-typed in the sense that each refinement term τ is associated with an arity, and whenever $\tau(\vec{A})$ appears in a definition, the length of $\vec{A}$ is the arity of τ.

Semantics. The semantics of our logic, defined by a satisfaction relation $s, h \models Q$, is essentially standard. Each predicate $Z \in \mathsf{RPred}$ is defined to be the least solution[1] to the following equivalence:

$$s, h \models Z(\vec{\theta}, \vec{E}) \iff \exists P \in cases(Z(\vec{R}, \vec{x})). \; s, h \models P[\vec{\theta}/\vec{R}, \vec{E}/\vec{x}]$$

Note that when substituting a λ-abstraction for a refinement variable, we implicitly β-reduce resulting applications. For example, $R(b)[(\lambda a.a \geqslant 0)/R] = b \geqslant 0$.

Semantic entailment is denoted by $P \models Q$, and provable entailment by $P \vdash Q$. When referring to a proof that $P \vdash Q$, we will mean a sequent calculus proof.

---

[1] Our definition does not preclude ill-founded predicates; such predicates are simply unsatisfiable, and do not affect the technical development in the rest of the paper.

3.2   Programs

A program $\mathcal{P}$ is a tuple $\langle V, E, v_i, v_e \rangle$, where
- V is a set of control locations, with a distinguished *entry* node $v_i \in V$ and *error* (exit) node $v_e \in V$, and
- $E \subseteq V \times V$ is a set of directed edges, where each $e \in E$ is associated with a program command e^c.

We impose the restriction that all nodes $V \setminus \{v_i\}$ are reachable from v_i via E, and all nodes can reach v_e. The syntax for program commands appears below. Note that the allocation command creates a record with n data fields, $D_1, \ldots, D_n$, and m heap fields, $N_1, \ldots, N_m$. To access the ith data field of a record pointed to by x, we use x->D_i (and similarly for heap fields). We assume that programs are well-typed, but not necessarily memory safe.

Assignment: x := ⅇ	**Assumption:** assume(Π)	**Allocation:** x := new(n, m)
Heap store: x->N_i := E	**Data store:** x->D_i := A	**Disposal:** free(x)
Heap load: y := x->N_i	**Data load:** y := x->D_i	

As is standard, we compile assert commands to reachability of v_e.

4   Spatial Interpolants

In this section, we first define the notion of spatial path interpolants, which serve as memory safety proofs of program paths. We then describe a technique for computing spatial path interpolants. This algorithm has two phases: the first is a (forwards) *symbolic execution* phase, which computes the strongest memory safety proof for a path; the second is a (backwards) *interpolation* phase, which weakens the proof so that it is more likely to generalize.

Spatial path interpolants are bounded from below by the strongest memory safety proof, and (implicitly) from above by the weakest memory safety proof. Prior to considering the generation of inductive invariants using spatial path interpolants, consider what could be done with only one of the bounds, in general, with either a path-based approach or an iterative fixed-point computation. Without the upper bound, an interpolant or invariant could be computed using a standard forward transformer and widening. But this suffers from the usual problem of potentially widening too aggressively to prove the remainder of the path, necessitating the design of analyses which widen conservatively at the price of computing unnecessarily strong proofs. The upper bound neatly captures the information that must be preserved for the future execution to be proved safe. On the other hand, without the lower bound, an interpolant or invariant could be computed using a backward transformer (and lower widening). But this suffers from the usual problem that backward transformers in shape analysis explode, due to issues such as not knowing the aliasing relationship in the pre-state. The lower bound neatly captures such information, heavily reducing the potential for explosion. These advantages come at the price of operating over full paths from entry to error. Compared to a forwards iterative analysis, operating over full paths has the advantage of having information about the execution's past and future when weakening at each point along the path. A forwards iterative

$\mathrm{exec}(\mathtt{x} := \mathtt{new}(k, l),\ (\exists X.\ \Pi : \Sigma)) = (\exists X \cup \{x', \vec{d}, \vec{n}\}.\ (\Pi : \Sigma)[x'/x] * x \mapsto [\vec{d}, \vec{n}])$
$$\text{where } x', \vec{d}, \vec{n} \text{ are fresh, } \vec{d} = (d_1, \ldots, d_k), \text{ and } \vec{n} = (n_1, \ldots, n_l).$$

$\mathrm{exec}(\mathtt{free}(\mathtt{x}),\ (\exists X.\ \Pi : \Sigma * z \mapsto [\vec{d}, \vec{n}]) = (\exists X.\ \Pi \wedge \Pi^{\neq} : \Sigma)$
$$\text{where } \Pi : \Sigma * z \mapsto [\vec{d}, \vec{n}] \vdash x = z \text{ and } \Pi^{\neq} \text{ is the}$$
$$\text{conjunction of all disequalities } x \neq y \text{ s.t } y \mapsto [\_,\_] \in \Sigma.$$

$\mathrm{exec}(\mathtt{x} := \mathtt{E},\ (\exists X.\ \Pi : \Sigma)) = (\exists X \cup \{x'\}.\ (x = E[x'/x]) * (\Pi : \Sigma)[x'/x])$
$$\text{where } x' \text{ is fresh.}$$

$\mathrm{exec}(\mathtt{assume}(\Pi'),\ (\exists X.\ \Pi : \Sigma)) = (\exists X.\ \Pi \wedge \underline{\Pi'} : \Sigma)\ .$

$\mathrm{exec}(\mathtt{x\text{->}N}_i := \mathtt{E},\ (\exists X.\ \Pi : \Sigma * z \mapsto [\vec{d}, \vec{n}])) = (\exists X.\ \Pi : \Sigma * x \mapsto [\vec{d}, \vec{n}[E/n_i]])$
$$\text{where } i \leqslant |\vec{n}| \text{ and } \Pi : \Sigma * z \mapsto [\vec{d}, \vec{n}] \vdash x = z\ .$$

$\mathrm{exec}(\mathtt{y} := \mathtt{x\text{->}N}_i,\ (\exists X.\ \Pi : \Sigma * z \mapsto [\vec{d}, \vec{n}])) =$
$$(\exists X \cup \{y'\}.\ (y = n_i[y'/y]) * (\Pi : \Sigma * z \mapsto [\vec{d}, \vec{n}])[y'/y])$$
$$\text{where } i \leqslant |\vec{n}| \text{ and } \Pi : \Sigma * z \mapsto [\vec{d}, \vec{n}] \vdash x = z, \text{ and } y' \text{ is fresh.}$$

Fig. 5. Symbolic execution for heap statements. Data statements are treated as skips.

analysis, on the other hand, trades the information about the future for information about many past executions through the use of join or widening operations.

The development in this section is purely spatial: we do not make use of data variables or refinements in recursive predicates. Our algorithm is thus of independent interest, outside of its context in this paper. We use Sep to refer to the fragment of RSep in which the only data formula (appearing in pure assertions and in refinements) is *true* (this fragment is equivalent to classical separation logic). An RSep formula P, in particular including those in recursive predicate definitions, determines a Sep formula $\underline{P}$ obtained by replacing all refinements (both variables and λ-abstractions) with $(\lambda \vec{a}.true)$ and all DFormulas in the pure part of P with *true*. Since recursive predicates, refinements, and DFormulas appear only positively, $\underline{P}$ is no stronger than any refinement of P. Since all refinements in Sep are trivial, we will omit them from the syntax (e.g., we will write $Z(\vec{E})$ rather than $Z((\lambda \vec{a}.true), \vec{E}))$.

4.1 Definition

We define a *symbolic heap* to be a Sep formula where the spatial part is a *-conjunction of points-to heaplets and the pure part is a conjunction of pointer (dis)equalities. Given a command c and a symbolic heap S, we use $\mathrm{exec}(c, S)$ to denote the symbolic heap that results from symbolically executing c starting in S (the definition of exec is essentially standard [4], and is shown in Fig. 5).

Given a program path $\pi = e_1, \ldots, e_n$, we obtain its strongest memory safety proof by symbolically executing π starting from the empty heap emp. We call this sequence of symbolic heaps the symbolic execution sequence of π, and say that a path π is *memory-feasible* if every formula in its symbolic execution sequence is consistent. The following proposition justifies calling this sequence the strongest memory safety proof.

Proposition 1. *For a path* π, *if the symbolic execution sequence for* π *is defined, then* π *is memory safe. If* π *is memory safe and memory-feasible, then its symbolic execution sequence is defined.*

Recall that our strategy for proving program correctness is based on sampling and proving the correctness of several program paths (*á la* IMPACT [26]). The problem with *strongest* memory safety proofs is that they do not generalize well (i.e., do not generate inductive invariants).

One solution to this problem is to take advantage of property direction. Given a desired postcondition P and a (memory-safe and -feasible) path π, the goal is to come up with a proof that is weaker than π's symbolic execution sequence, but still strong enough to show that P holds after executing π. Coming up with such "weak" proofs is how traditional path interpolation is used in IMPACT. In light of this, we define *spatial path interpolants* as follows:

Definition 1 (Spatial path interpolant). *Let* $\pi = e_1, \ldots, e_n$ *be a program path with symbolic execution sequence* $S_0, \ldots, S_n$, *and let* P *be a* Sep *formula (such that* $S_n \models P$). *A spatial path interpolant for* π *is a sequence* $I_0, \ldots, I_n$ *of* Sep *formulas such that*
 - *for each* $i \in [0, n]$, $S_i \models I_i$;
 - *for each* $i \in [1, n]$, $\{I_{i-1}\}\, e_i^c\, \{I_i\}$ *is a valid triple in separation logic; and*
 - $I_n \models P$.

Our algorithm for computing spatial path interpolants is a backwards propagation algorithm that employs a *spatial interpolation* procedure at each backwards step. Spatial interpolants for a single command are defined as:

Definition 2 (Spatial interpolant). *Given* Sep *formulas* S *and* I' *and a command* c *such that* $\mathsf{exec}(c, S) \models I'$, *a spatial interpolant (for* S, c, *and* I') *is a* Sep *formula* I *such that* $S \models I$ *and* $\{I\}\, c\, \{I'\}$ *is valid.*

Before describing the spatial interpolation algorithm, we briefly describe how spatial interpolation is used to compute path interpolants. Let us use $\mathsf{itp}(S, c, I)$ to denote a spatial interpolant for S, c, I, as defined above. Let $\pi = e_1, \ldots, e_n$ be a program path and let P be a Sep formula. First, symbolically execute π to compute a sequence $S_0, \ldots, S_n$. Suppose that $S_n \vdash P$. Then we compute a sequence $I_0, \ldots, I_n$ by taking $I_n = P$ and (for $k < n$) $I_k = \mathsf{itp}(S_k, e_{k+1}^c, I_{k+1})$. The sequence $I_0, \ldots, I_n$ is clearly a spatial path interpolant.

4.2 Bounded Abduction

Our algorithm for spatial interpolation is based on an abduction procedure. Abduction refers to the inference of explanatory hypotheses from observations (in contrast to deduction, which derives conclusions from given hypotheses). The variant of abduction we employ in this paper, which we call *bounded abduction*, is simultaneously a form of abductive and deductive reasoning. Seen as a variant of abduction, bounded abduction adds a constraint that the abduced hypothesis be at least weak enough to be derivable from a given hypothesis. Seen as a variant of deduction, bounded abduction adds a constraint that the deduced conclusion be at least strong enough to imply some desired conclusion. Formally, we define bounded abduction as follows:

Definition 3 (Bounded abduction). *Let* L, M, R *be Sep formulas, and let* X *be a set of variables. A solution to the* bounded abduction problem

$$L \vdash (\exists X. \ M * [\ ]) \vdash R$$

is a Sep formula A *such that* $L \models (\exists X. \ M * A) \models R$.

Note how, in contrast to bi-abduction [11] where a solution is a pair of formulas, one constrained from above and one from below, a solution to bounded abduction problems is a single formula that is simultaneously constrained from above and below. The fixed lower and upper bounds in our formulation of abduction give considerable guidance to solvers, in contrast to bi-abduction, where the bounds are part of the solution.

Sec. 6 presents our bounded abduction algorithm. For the remainder of this section, we will treat bounded abduction as a black box, and use $L \vdash (\exists X. \ M * [A]) \vdash R$ to denote that A is a solution to the bounded abduction problem.

4.3 Computing Spatial Interpolants

We now proceed to describe our algorithm for spatial interpolation. Given a command c and Sep formulas S and I' such that $\mathsf{exec}(c, S) \vdash I'$, this algorithm must compute a Sep formula $\mathsf{itp}(S, c, I')$ that satisfies the conditions of Definition 2. Several examples illustrating this procedure are given in Fig. 3.

This algorithm is defined by cases based on the command c. We present the cases for the spatial commands; the corresponding data commands are similar.

Allocate. Suppose c is $\mathtt{x \ := \ new}(n, m)$. We take $\mathsf{itp}(S, c, I') = (\exists x. \ A)$, where A is obtained as a solution to $\mathsf{exec}(c, S) \vdash (\exists \vec{a}, \vec{z}. \ x \mapsto [\vec{a}, \vec{z}] * [A]) \vdash I'$, and $\vec{a}$ and $\vec{z}$ are vectors of fresh variables of length n and m, respectively.

Deallocate. Suppose c is $\mathtt{free(x)}$. We take $\mathsf{itp}(S, c, I') = (\exists \vec{a}, \vec{z}. \ I' * x \mapsto [\vec{a}, \vec{z}])$, where $\vec{a}$ and $\vec{z}$ are vectors of fresh variables whose lengths are determined by the unique heap cell which is allocated to x in S.

Assignment. Suppose c is $\mathtt{x \ := \ E}$. We take $\mathsf{itp}(S, c, I') = I'[E/x]$.

Store. Suppose c is $\mathtt{x\text{-}>N}_i \ \mathtt{:= \ E}$. We take $\mathsf{itp}(S, c, I') = (\exists \vec{a}, \vec{z}. \ A * x \mapsto [\vec{a}, \vec{z}])$, where A is obtained as a solution to $\mathsf{exec}(c, S) \vdash (\exists \vec{a}, \vec{z}. \ x \mapsto [\vec{a}, \vec{z}[E/z_i]] * [A]) \vdash I'$ and where $\vec{a}$ and $\vec{z}$ are vectors of fresh variables whose lengths are determined by the unique heap cell which is allocated to x in S.

Example 1. Suppose that S is $t \mapsto [4, y, \mathsf{null}] * x \mapsto [2, \mathsf{null}, \mathsf{null}]$ where the cells have one data and two pointer fields, c is $\mathtt{t\text{-}>N_0 \ := \ x}$, and I' is $\mathsf{bt}(t)$. Then we can compute $\mathsf{exec}(c, S) = t \mapsto [4, x, \mathsf{null}] * x \mapsto [2, \mathsf{null}, \mathsf{null}]$, and then solve the bounded abduction problem

$$\mathsf{exec}(c, S) \vdash (\exists a, z_1. \ t \mapsto [a, x, z_1] * [\ ]) \vdash I' \ .$$

One possible solution is $A = \mathsf{bt}(x) * \mathsf{bt}(z_1)$, which yields

$$\mathsf{itp}(S, c, I') = (\exists a, z_0, z_1. \ t \mapsto [a, z_0, z_1] * \mathsf{bt}(z_1) * \mathsf{bt}(x)) \ . \qquad \lrcorner$$

Load. Suppose c is $y := x\text{->}N_i$. Suppose that $\vec{a}$ and $\vec{z}$ are vectors of fresh variables of lengths $|\vec{A}|$ and $|\vec{E}|$ where S is of the form $\Pi : \Sigma * w \mapsto [\vec{A}, \vec{E}]$ and $\Pi : \Sigma * w \mapsto [\vec{A}, \vec{E}] \vdash x = w$ (this is the condition under which $\mathsf{exec}(\mathsf{c}, S)$ is defined, see Fig. 5). Let y' be a fresh variable, and define $\overline{S} = (y = z_i[y'/y]) * (\Pi : \Sigma * w \mapsto [\vec{a}, \vec{z}])[y'/y]$. Note that $\overline{S} \vdash (\exists y'.\ \overline{S}) \equiv \mathsf{exec}(\mathsf{c}, S) \vdash I'$.

We take $\mathsf{itp}(S, \mathsf{c}, I') = (\exists \vec{a}, \vec{z}.\ A[z_i/y, y/y'] * x \mapsto [\vec{a}, \vec{z}])$ where A is obtained as a solution to $\overline{S} \vdash (\exists \vec{a}, \vec{z}.\ x[y'/y] \mapsto [\vec{a}, \vec{z}] * [A]) \vdash I'$.

Example 2. Suppose that S is $y = t : y \mapsto [1, \mathsf{null}, x] * x \mapsto [5, \mathsf{null}, \mathsf{null}]$, c is y := $y\text{->}N_1$, and I' is $y \neq \mathsf{null} : \mathsf{bt}(t)$. Then $\overline{S}$ is

$$y = x \wedge y' = t : y' \mapsto [1, \mathsf{null}, x] * x \mapsto [5, \mathsf{null}, \mathsf{null}]$$

We can then solve the bounded abduction problem

$$\overline{S} \vdash (\exists a, z_0, z_1.\ y' \mapsto [a, z_0, z_1] * [\ ]) \vdash I'$$

A possible solution is $y \neq \mathsf{null} \wedge y' = t : \mathsf{bt}(z_0) * \mathsf{bt}(z_1)$, yielding
$\mathsf{itp}(S, \mathsf{c}, I') = (\exists a, z_0, z_1.z_1 \neq \mathsf{null} \wedge y = t : \mathsf{bt}(z_0) * \mathsf{bt}(z_1) * y \mapsto [a, z_0, z_1])$.   ⌐

Assumptions. The interpolation rules defined up to this point cannot introduce recursive predicates, in the sense that if I' is a *-conjunction of points-to predicates then so is $\mathsf{itp}(S, \mathsf{c}, I')$.[2] A *-conjunction of points-to predicates is *exact* in the sense that it gives the full layout of some part of the heap. The power of recursive predicates lies in their ability to be *abstract* rather than exact, and describe only the shape of the heap rather than its exact layout. It is a special circumstance that $\{P\}$ c $\{I'\}$ holds when I' is exact in this sense and P is not: intuitively, it means that by executing c we somehow gain information about the program state, which is precisely the case for **assume** commands.

For an example of how spatial interpolation can introduce a recursive predicate at an **assume** command, consider the problem of computing an interpolant

$$\mathsf{itp}(S, \mathsf{assume}(x \neq \mathsf{null}), (\exists a, z.\ x \mapsto [a, z] * \mathsf{true}))$$

where $S \equiv x \mapsto [d, y] * y \mapsto [d', \mathsf{null}]$: a desirable interpolant may be $\mathsf{ls}(x, \mathsf{null}) * \mathsf{true}$. The disequality introduced by the assumption ensures that one of the *cases* of the recursive predicate $\mathsf{ls}(x, \mathsf{null})$ (where the list from x to null is empty) is impossible, which implies that the other case (where x is allocated) must hold.

Towards this end, we now define an auxiliary function intro which we will use to introduce recursive predicates for the **assume** interpolation rules. Let P, Q be Sep formulas such that $P \vdash Q$, let Z be a recursive predicate and $\vec{E}$ be a vector of heap terms. We define $\mathsf{intro}(Z, \vec{E}, P, Q)$ as follows: if $P \vdash (\exists \emptyset.\ Z(\vec{E}) * [A]) \vdash Q$ has a solution and $A \nvdash Q$, define $\mathsf{intro}(Z, \vec{E}, P, Q) = Z(\vec{E}) * A$. Otherwise, define $\mathsf{intro}(Z, \vec{E}, P, Q) = Q$.

Intuitively, the abduction problem has a solution when P implies $Z(\vec{E})$ and $Z(\vec{E})$ can be *excised* from Q. The condition $A \nvdash Q$ is used to ensure that the

---

[2] But if I' *does* contain recursive predicates, then $\mathsf{itp}(S, \mathsf{c}, I')$ may also.

excision from Q is non-trivial (i.e., the part of the heap that satisfies $Z(\vec{E})$ "consumes" some heaplet of Q).

To define the interpolation rule for assumptions, suppose c is $\mathtt{assume(E \neq F)}$ (the case of equality assumptions is similar). Letting $\{\langle Z_i, \vec{E}_i \rangle\}_{i \leqslant n}$ be an enumeration of the (finitely many) possible choices of Z and $\vec{E}$, we define a formula M to be the result of applying intro to I' over all possible choices of Z and $\vec{E}$:

$$M = \mathsf{intro}(Z_1, \vec{E}_1, S \wedge E \neq F, \mathsf{intro}(Z_2, \vec{E}_2, S \wedge E \neq F, \dots))$$

where the innermost occurrence of intro in this definition is $\mathsf{intro}(Z_n, \vec{E}_n, S \wedge E \neq F, I')$. Since intro preserves entailment (in the sense that if $P \vdash Q$ then $P \vdash \mathsf{intro}(Z, \vec{E}, P, Q)$), we have that $S \wedge E \neq F \vdash M$. From a proof of $S \wedge E \neq F \vdash M$, we can construct a formula M' which is entailed by S and differs from M only in that it renames variables and exposes additional equalities and disequalities implied by S, and take $\mathsf{itp}(S, \mathsf{c}, I')$ to be this M'.

The construction of M' from M is straightforward but tedious. *The procedure is detailed in the extended version [2]; here, we will just give an example to give intuition on why it is necessary.* Suppose that S is $x = w : y \mapsto z$ and I' is $\mathsf{ls}(w, z)$, and c is $\mathtt{assume(x = y)}$. Since there is no opportunity to introduce new recursive predicates in I', M is simply $\mathsf{ls}(w, z)$. However, M is not a valid interpolant since $S \not\models M$, so we must expose the equality $x = w$ and rename w to y in the list segment in $M' \equiv x = w : \mathsf{ls}(y, z)$.

In practice, it is undesirable to enumerate all possible choices of Z and $\vec{E}$ when constructing M (considering that if there are k in-scope data terms, a recursive predicate of arity n requires enumerating k^n choices for $\vec{E}$). A reasonable heuristic is to let Π be the strongest pure formula implied by S, and enumerate only those combinations of Z and $\vec{E}$ such that there is some $\Pi' : \Sigma' \in cases(Z(\vec{R}, \vec{x}))$ such that $\underline{\Pi'}[\vec{E}/\vec{x}] \wedge \Pi \wedge x \neq y$ is unsatisfiable. For example, for $\mathtt{assume(x \neq y)}$, this heuristic means that we enumerate only $\langle x, y \rangle$ and $\langle y, x \rangle$ (i.e, we attempt to introduce a list segment from x to y and from y to x).

We conclude this section with a theorem stating the correctness of our spatial interpolation procedure.

Theorem 1. *Let S and I' be Sep formulas and let c be a command such that $exec(\mathsf{c}, S) \vdash I'$. Then $\mathsf{itp}(S, \mathsf{c}, I')$ is a spatial interpolant for S, c, and I'.*

5   Spatial Interpolation Modulo Theories

We now consider the problem of *refining* (or *strengthening*) a given separation logic proof of memory safety with information about (non-spatial) data. This refinement procedure results in a proof of a conclusion stronger than can be proved by reasoning about the heap alone. In view of our example from Fig. 3, this section addresses how to derive the third sequence (Spatial Interpolants Modulo Theories) from the second (Spatial Interpolants).

The input to our spatial interpolation modulo theories procedure is a path π, a separation logic (Sep) proof ζ of the triple $\{true : \mathsf{emp}\}\ \pi\ \{true : \mathsf{true}\}$

──────── *Entailment rules* ────────

STAR
$$\frac{\mathcal{C}_0 \;\blacktriangleright\; \Pi \wedge \Phi : \Sigma_0 \vdash \Pi' \wedge \Phi' : \Sigma_0' \qquad \mathcal{C}_1 \;\blacktriangleright\; \Pi \wedge \Phi : \Sigma_1 \vdash \Pi' \wedge \Phi' : \Sigma_1'}{\mathcal{C}_0;\mathcal{C}_1 \;\blacktriangleright\; \Pi \wedge \Phi : \Sigma_0 * \Sigma_1 \vdash \Pi' \wedge \Phi' : \Sigma_0' * \Sigma_1'}$$

POINTS-TO
$$\frac{\Pi \models \Pi'}{\Phi' \leftarrow \Phi \;\blacktriangleright\; \Pi \wedge \Phi : E \mapsto [\vec{A}, \vec{F}] \vdash \Pi' \wedge \Phi' : E \mapsto [\vec{A}, \vec{F}]}$$

FOLD
$$\frac{\mathcal{C} \;\blacktriangleright\; \Pi : \Sigma \vdash \Pi' : \Sigma' * P[\vec{\tau}/\vec{R}, \vec{E}/\vec{x}]}{\mathcal{C} \;\blacktriangleright\; \Pi : \Sigma \vdash \Pi' : \Sigma' * Z(\vec{\tau}, \vec{E})} \; P \in cases(Z(\vec{R}, \vec{x}))$$

UNFOLD
$$\frac{\mathcal{C}_1 \;\blacktriangleright\; \Pi : \Sigma * P_1[\vec{\tau}/\vec{R}, \vec{E}/\vec{x}] \vdash \Pi' : \Sigma' \quad \cdots}{\mathcal{C}_n \;\blacktriangleright\; \Pi : \Sigma * P_n[\vec{\tau}/\vec{R}, \vec{E}/\vec{x}] \vdash \Pi' : \Sigma'} \quad \{P_1, \ldots, P_n\} =$$
$$\overline{\mathcal{C}_1; \ldots; \mathcal{C}_n \;\blacktriangleright\; \Pi : \Sigma * Z(\vec{\tau}, \vec{E}) \vdash \Pi' : \Sigma'} \quad cases(Z(\vec{R}, \vec{x}))$$

PREDICATE
$$\frac{\Pi \models \Pi'}{\substack{\Phi' \leftarrow \Phi; \Psi_1' \leftarrow \Psi_1 \wedge \Phi; \ldots; \Psi_{|\vec{\tau}|}' \leftarrow \Psi_{|\vec{\tau}|} \wedge \Phi \;\blacktriangleright\; \\ \Pi \wedge \Phi : Z(\vec{\tau}, \vec{E}) \vdash \Pi' \wedge \Phi' : Z(\vec{\tau'}, \vec{E})}} \quad \substack{\text{Where } \tau_i = (\lambda\vec{a}_i.\Psi_i) \\ \text{and } \tau_i' = (\lambda\vec{a}_i.\Psi_i')}$$

──────── *Execution rules* ────────

DATA-ASSUME
$$\frac{\mathcal{C} \;\blacktriangleright\; P \wedge \varphi \vdash Q}{\mathcal{C} \;\blacktriangleright\; \{P\} \, \mathsf{assume}(\varphi) \, \{Q\}}$$

FREE
$$\frac{\mathcal{C} \;\blacktriangleright\; P \vdash \Pi \wedge \Phi : \Sigma * x \mapsto [\vec{A}, \vec{E}]}{\mathcal{C} \;\blacktriangleright\; \{P\} \, \mathsf{free(x)} \, \{\Pi \wedge \Phi : \Sigma\}}$$

SEQUENCE
$$\frac{\mathcal{C}_0 \;\blacktriangleright\; \{P\} \, \pi_0 \, \{\widehat{O}\} \qquad \mathcal{C}_1 \;\blacktriangleright\; \{\widehat{O}\} \, \pi_1 \, \{Q\}}{\mathcal{C}_0;\mathcal{C}_1 \;\blacktriangleright\; \{P\} \, \pi_0; \pi_1 \, \{Q\}}$$

DATA-LOAD
$$\frac{\mathcal{C}_0 \;\blacktriangleright\; P \vdash (\exists X. \, \Pi \wedge \widehat{\Phi} : \widehat{\Sigma} * x \mapsto [\vec{A}, \vec{E}])}{\mathcal{C}_1 \;\blacktriangleright\; (\exists X, a'. \, \Pi[a'/a] \wedge \widehat{\Phi}[a'/a] \wedge a = A_i[a'/a] : (\widehat{\Sigma} * x \mapsto [\vec{A}, \vec{E}])[a'/a]) \vdash Q}{\mathcal{C}_0;\mathcal{C}_1 \;\blacktriangleright\; \{P\} \, \mathsf{a := x\text{-}>D}_i \, \{Q\}}$$

DATA-ASSIGN
$$\frac{\mathcal{C} \;\blacktriangleright\; (\exists a'. \, \Pi \wedge \Phi[a'/a] \wedge a = A[a'/a] : \Sigma[a'/a] \vdash Q)}{\mathcal{C} \;\blacktriangleright\; \{\Pi \wedge \Phi : \Sigma\} \, \mathsf{a := A} \, \{Q\}}$$

DATA-STORE
$$\frac{\mathcal{C}_0 \;\blacktriangleright\; P \vdash (\exists X. \, \Pi \wedge \widehat{\Phi} : \widehat{\Sigma} * x \mapsto [\vec{A}, \vec{E}])}{\mathcal{C}_1 \;\blacktriangleright\; (\exists X, a'. \, \Pi \wedge \widehat{\Phi} \wedge a' = A : \widehat{\Sigma} * x \mapsto [\vec{A}[a'/A_i], \vec{E}]) \vdash Q}{\mathcal{C}_0;\mathcal{C}_1 \;\blacktriangleright\; \{P\} \, \mathsf{x\text{-}>D}_i \, \mathsf{:= A} \, \{Q\}}$$

ALLOC
$$\frac{\mathcal{C} \;\blacktriangleright\; (\exists x', \vec{a}, \vec{x}. \, \Pi[x'/x] \wedge \Phi : \Sigma[x'/x] * x \mapsto [\vec{a}, \vec{x}]) \vdash Q}{\mathcal{C} \;\blacktriangleright\; \{\Pi \wedge \Phi : \Sigma\} \, \mathsf{x := new}(n, m) \, \{Q\}}$$

Fig. 6. Constraint generation

Refined memory safety proof ζ'	Constraint system $\mathcal{C}$	Solution σ
$\{R_0(i) : true\}$	$R_0(i') \leftarrow true$	$R_0(i) : true$
i = nondet(); x = null	$R_1(i') \leftarrow R_0(i)$	$R_1(i) : true$
$\{R_1(i) : \mathsf{ls}((\lambda a.R_{\mathsf{ls}1}(\nu, i)), x, \mathsf{null}) * true\}$	$R_2(i') \leftarrow R_1(i) \wedge i \neq 0 \wedge i' = i+1$	$R_2(i) : true$
assume(i != 0); ...; i—;	$R_3(i) \leftarrow R_2(i) \wedge i = 0$	$R_3(i) : true$
$\{R_2(i) : \mathsf{ls}((\lambda a.R_{\mathsf{ls}2}(\nu, i)), x, \mathsf{null}) * true\}$	$R_4(i, d') \leftarrow R_3(i) \wedge R_{\mathsf{ls}3}(d', i)$	$R_4(i, d') : d' \geqslant 0$
assume(i == 0)	$R_{\mathsf{ls}2}(\nu, i') \leftarrow R_1(i) \wedge R_{\mathsf{ls}1}(\nu, i) \wedge i \neq 0 \wedge i' = i+1$	$R_{\mathsf{ls}1}(\nu, i) : \nu \geqslant i$
$\{R_3(i) : \mathsf{ls}((\lambda a.R_{\mathsf{ls}3}(\nu, i)), x, \mathsf{null}) * true\}$	$R_{\mathsf{ls}2}(\nu, i') \leftarrow R_1(i) \wedge \nu = i \wedge i \neq 0 \wedge i' = i+1$	$R_{\mathsf{ls}2}(\nu, i) : \nu \geqslant i$
assume(x != null)	$R_{\mathsf{ls}3}(\nu, i) \leftarrow R_2(i) \wedge R_{\mathsf{ls}2}(\nu, i) \wedge i = 0$	$R_{\mathsf{ls}3}(\nu, i) : \nu \geqslant 0$
$\{(\exists d', y. R_4(i, d') : x \mapsto [d', y] * true)\}$	$d' \geqslant 0 \leftarrow R_4(i, d')$	

Fig. 7. Example constraints.

(i.e., a memory safety proof for π), and a postcondition φ. The goal is to transform ζ into an RSep proof of the triple $\{true : \mathsf{emp}\} \pi \{\varphi : \mathsf{true}\}$. The high-level operation of our procedure is as follows. First, we traverse the memory safety proof ζ and build (1) a corresponding *refined* proof ζ' where refinements may contain second-order variables, and (2) a constraint system $\mathcal{C}$ which encodes logical dependencies between the second-order variables. We then attempt to find a solution to $\mathcal{C}$, which is an assignment of data formulas to the second-order variables such that all constraints are satisfied. If we are successful, we use the solution to instantiate the second-order variables in ζ', which yields a valid RSep proof of the triple $\{true : \mathsf{emp}\} \pi \{\varphi : \mathsf{true}\}$.

Horn Clauses. The constraint system produced by our procedure is a recursion-free set of Horn clauses, which can be solved efficiently using existing first-order interpolation techniques (see [34] for a detailed survey). Following [18], we define a *query* to be an application $Q(\vec{a})$ of a second-order variable Q to a vector of (data) variables, and define an *atom* to be either a data formula $\varphi \in \mathsf{DFormula}$ or a query $Q(\vec{a})$. A *Horn clause* is of the form $h \leftarrow b_1 \wedge \cdots \wedge b_N$ where each of $h, b_1, \ldots, b_N$ is an atom. In our constraint generation rules, it will be convenient to use a more general form which can be translated to Horn clauses: we will allow constraints of the form $h_1 \wedge \cdots \wedge h_M \leftarrow b_1 \wedge \cdots \wedge b_N$ (shorthand for the set of Horn clauses $\{h_i \leftarrow b_1 \wedge \cdots \wedge b_N\}_{1 \leqslant i \leqslant M}$) and we will allow queries to be of the form $Q(\vec{A})$ (i.e., take arbitrary data terms as arguments rather than variables). If $\mathcal{C}$ and $\mathcal{C}'$ are sets of constraints, we will use $\mathcal{C}; \mathcal{C}'$ to denote their union.

A *solution* to a system of Horn clauses $\mathcal{C}$ is a map σ that assigns each second-order variable Q of arity k a $\mathsf{DFormula}$ Q^σ with free variables drawn from $\vec{\nu} = \langle \nu_1, \ldots, \nu_k \rangle$ such that for each clause $h \leftarrow b_1 \wedge \cdots \wedge b_N$ in $\mathcal{C}$ the implication $\forall A.(h^\sigma \Leftarrow (\exists B. b_1^\sigma \wedge \cdots \wedge b_N^\sigma))$ holds, where A is the set of free variables in h and B the set of variables free in some b_i but not in h. In the above, for any data formula φ, φ^σ is defined to be φ, and for any query $Q(\vec{a})$, $Q(\vec{a})^\sigma$ is defined to be $Q^\sigma[a_1/\nu_1, \ldots, a_k/\nu_k]$ (where k is the arity of Q).

Constraint Generation Calculus. We will present our algorithm for spatial interpolation modulo theories as a calculus whose inference rules mirror the ones of separation logic. The calculus makes use of the same syntax used in recursive predicate definitions in Sec. 3. We use τ to denote a *refinement term* and Φ to denote a *refined formula*. The calculus has two types of judgements.

An *entailment judgement* is of the form

$$\mathcal{C} \blacktriangleright (\exists X.\ \Pi \wedge \Phi : \Sigma) \vdash (\exists X'.\ \Pi' \wedge \Phi' : \Sigma')$$

where Π, Π' are equational pure assertions over heap terms, Σ, Σ' are refined spatial assertions, Φ, Φ' are refined formulas, and $\mathcal{C}$ is a recursion-free set of Horn clauses. Such an entailment judgement should be read as "for any solution σ to the set of constraints $\mathcal{C}$, $(\exists X.\ \Pi \wedge \Phi^\sigma : \Sigma^\sigma)$ entails $(\exists X'.\ \Pi' \wedge \Phi'^\sigma : \Sigma'^\sigma)$," where Φ^σ is Φ with all second order variables replaced by their data formula assignments in σ (and similarly for Σ^σ).

Similarly, an *execution judgement* is of the form

$$\mathcal{C} \blacktriangleright \{(\exists X.\ \Pi \wedge \Phi : \Sigma)\}\ \pi\ \{(\exists X'.\ \Pi' \wedge \Phi' : \Sigma')\}$$

where π is a path and $X, X', \Pi, \Pi', \Phi, \Phi', \Sigma, \Sigma'$, and $\mathcal{C}$ are as above. Such an execution judgement should be read as "for any solution σ to the set of constraints $\mathcal{C}$,

$$\{(\exists X.\ \Pi \wedge \Phi^\sigma : \Sigma^\sigma)\}\ \pi\ \{(\exists X'.\ \Pi' \wedge \Phi'^\sigma : \Sigma'^\sigma)\}$$

is a valid triple."

Let π be a path, let ζ be a separation logic proof of the triple $\{true : \mathsf{emp}\}\ \pi\ \{true : \mathsf{true}\}$ (i.e., a memory safety proof for π), and let $\varphi \in \mathsf{DFormula}$ be a postcondition. Given these inputs, our algorithm operates as follows. We use $\vec{v}$ to denote a vector of all data-typed program variables. The triple is *rewritten with refinements* by letting R and R' be fresh second-order variables of arity $|\vec{v}|$ and conjoining $R(\vec{v})$ and $R'(\vec{v})$ to the pre and post. By recursing on ζ, at each step applying the appropriate rule from our calculus in Fig. 6, we derive a judgement

$$\frac{\zeta'}{\mathcal{C} \blacktriangleright \{true \wedge R(\vec{v}) : \mathsf{true}\}\ \pi\ \{true \wedge R'(\vec{v}) : \mathsf{true}\}}$$

and then compute a solution σ to the constraint system

$$\mathcal{C};\quad R(\vec{v}) \leftarrow true;\quad \varphi \leftarrow R'(\vec{v})$$

(if one exists). The algorithm then returns ζ'^σ, the proof obtained by applying the substitution σ to ζ'.

Intuitively, our algorithm operates by recursing on a separation logic proof, introducing refinements into formulas on the way down, and building a system of constraints on the way up. Each inference rule in the calculus encodes both the downwards and upwards step of this algorithm. For example, consider the FOLD rule of our calculus: we will illustrate the intended reading of this rule with a concrete example. Suppose that the input to the algorithm is a derivation of the following form:

$$\frac{\begin{array}{c}\zeta_0 \\ \hline x \mapsto [a, \mathsf{null}] \vdash (\exists b, y.\ x \mapsto [b, y] * \mathsf{ls}(y, \mathsf{null}))\end{array}}{Q(i) : x \mapsto [a, \mathsf{null}] \vdash R(i) : \mathsf{ls}((\lambda a.S(x, a)), x, \mathsf{null})}\ \text{FOLD}$$

(i.e., a derivation where the last inference rule is an application of FOLD, and the conclusion has already been rewritten with refinements). We introduce refinements in the premise and recurse on the following derivation:

$$\frac{\zeta_0}{Q(i) : x \mapsto [a, \mathsf{null}] \vdash \quad (\exists b, y.\ R(i) \land S(i,b) : x \mapsto [b,y] * \mathsf{ls}((\lambda a.S(x,a)), y, \mathsf{null}))}$$

The result of this recursive call is a refined derivation ζ_0' as well as a constraint system $\mathcal{C}$. We then return both (1) the refined derivation obtained by catenating the conclusion of the FOLD rule onto ζ_0' and (2) the constraint system $\mathcal{C}$.

A crucial point of our algorithm is hidden inside the hat notation in Fig. 6 (e.g., $\hat{O}$ in SEQUENCE): this notation is used to denote the introduction of fresh second-order variables. For many of the inference rules (such as FOLD), the refinements which appear in the premises follow fairly directly from the refinements which appear in the conclusion. However, in some rules entirely new formulas appear in the premises which do not appear in the conclusion (e.g., in the SEQUENCE rule in Fig. 6, the intermediate assertion $\hat{O}$ is an arbitrary formula which has no obvious relationship to the precondition P or the postcondition Q). We refine such formula O by introducing a fresh second-order variable for the pure assertion and for each refinement term that appears in O. The following offers a concrete example.

Example 3. Consider the trace π in Fig. 3. Suppose that we are given a memory safety proof for π which ends in an application of the SEQUENCE rule:

$$\frac{\{true : \mathsf{emp}\}\ \pi_0\ \{true : \mathsf{ls}(x, \mathsf{null})\} \quad \{true : \mathsf{ls}(x, \mathsf{null})\}\ \pi_1\ \{(\exists b, y.\ true : x \mapsto [b,y])\}}{\{Q(i) : \mathsf{emp}\}\ \pi_0; \pi_1\ \{(\exists b, y.\ R(i, b) : x \mapsto [b,y])\}} \ \text{SEQUENCE}$$

where π is decomposed as $\pi_0; \pi_1$, π_0 is the path from 1 to 3, and π_1 is the path from 3 to 4. Let $O = true : \mathsf{ls}(x, \mathsf{null})$ denote the intermediate assertion which appears in this proof. To derive $\hat{O}$, we introduce two fresh second order variables, S (with arity 1) and T (with arity 2), and define $\hat{O} = S(i) : \mathsf{ls}((\lambda a.T(i,a)), x, \mathsf{null})$. The resulting inference is as follows:

$$\frac{\{Q(i) : \mathsf{emp}\}\ \pi_0\ \{S(i) : \mathsf{ls}((\lambda a.T(i,a)), x, \mathsf{null})\} \quad \{S(i) : \mathsf{ls}((\lambda a.T(i,a)), x, \mathsf{null})\}\ \pi_1\ \{(\exists b, y.\ R(i, b) : x \mapsto [b,y])\}}{\{Q(i) : \mathsf{emp}\}\ \pi_0; \pi_1\ \{(\exists b, y.\ R(i, b) : x \mapsto [b,y])\}} \quad \lrcorner$$

The following example provides a simple demonstration of our constraint generation procedure:

Example 4. Recall the example in Fig. 3 of Sec. 2. The row of spatial interpolants in Fig. 3 is a memory safety proof ζ of the program path. Fig. 7 shows the refined proof ζ', which is the proof ζ with second-order variables that act as placeholders for data formulas. ***For the sake of illustration, we have simplified the constraints by skipping a number of intermediate annotations in the Hoare-style proof.***

EMPTY
$$\frac{}{\Pi : [\mathsf{emp}]^c \vdash \Pi' : \langle [\mathsf{emp}]^c \trianglelefteq \mathsf{emp} \rangle} \quad \Pi \models \Pi'$$

STAR
$$\frac{\Pi : \Sigma_0 \vdash \Pi' : \Sigma_0' \quad \Pi : \Sigma_1 \vdash \Pi' : \Sigma_1'}{\Pi : \Sigma_0 * \Sigma_1 \vdash \Pi' : \Sigma_0' * \Sigma_1'}$$

POINTS-TO
$$\frac{\Pi \models \Pi'}{\Pi : [E \mapsto [a, F]]^c \vdash \Pi' : \langle [E \mapsto [a, F]]^c \trianglelefteq E \mapsto [a, F] \rangle}$$

TRUE
$$\frac{\Pi \models \Pi'}{\Pi : \Sigma \vdash \Pi' : \langle [\mathsf{true}]^c \trianglelefteq \mathsf{true} \rangle}$$

SUBSTITUTION
$$\frac{\Pi[E/x] : \Sigma[E/x] \vdash \Pi'[E/x] : \Sigma'[E/x] \quad \Pi \models x = E}{\Pi : \Sigma \vdash \Pi' : \Sigma'}$$

∃-RIGHT
$$\frac{P \vdash Q[\mathcal{E}/x]}{P \vdash (\exists x. \, Q)}$$

Fig. 8. Coloured strengthening. All primed variables are chosen fresh.

The constraint system $\mathcal{C}$ specifies the logical dependencies between the introduced second-order variables in ζ'. For instance, the relation between R_2 and R_3 is specified by the Horn clause $R_3(i) \leftarrow R_2(i) \wedge i = 0$, which takes into account the constraint imposed by `assume (i == 0)` in the path. The Horn clause $d' \geqslant 0 \leftarrow R_4(i, d')$ specifies the postcondition defined by the assertion `assert(x->D >= 0)`, which states that the value of the data field of the node x should be $\geqslant 0$.

Replacing second-order variables in ζ' with their respective solutions in σ produces a proof that the assertion at the end of the path holds (last row of Fig. 3). ⌐

Soundness and Completeness. The key result regarding the constraint systems produced by these judgements is that any solution to the constraints yields a valid refined proof. The formalization of the result is the following theorem.

Theorem 2 (Soundness). *Suppose that π is a path, ζ is a derivation of the judgement $\mathcal{C} \; \blacktriangleright \; \{P\} \, \pi \, \{Q\}$, and that σ is a solution to $\mathcal{C}$. Then ζ^σ, the proof obtained by applying the substitution σ to ζ, is a (refined) separation logic proof of $\{P^\sigma\} \, \pi \, \{Q^\sigma\}$.*

Another crucial result for our counterexample generation strategy is a kind of completeness theorem, which effectively states that the strongest memory safety proof always admits a refinement.

Theorem 3 (Completeness). *Suppose that π is a memory-feasible path and ζ is a derivation of the judgement $\mathcal{C} \; \blacktriangleright \; \{R_0(\vec{v}) : \mathsf{emp}\} \, \pi \, \{R_1(\vec{v}) : \mathsf{true}\}$ obtained by symbolic execution. If φ is a data formula such that $\{\mathsf{true} : \mathsf{emp}\} \, \pi \, \{\varphi : \mathsf{true}\}$ holds, then there is a solution σ to $\mathcal{C}$ such that $R_1^\sigma(\vec{v}) \Rightarrow \varphi$.*

6   Bounded Abduction

In this section, we discuss our algorithm for bounded abduction. Given a bounded abduction problem

$$L \vdash (\exists X. \, M * [\,]) \vdash R$$

we would like to find a formula A such that $L \vdash (\exists X.\ M * A) \vdash R$. Our algorithm is sound but not complete: it is possible that there exists a solution to the bounded abduction problem, but our procedure cannot find it. In fact, there is in general no complete procedure for bounded abduction, as a consequence of the fact that we do not pre-suppose that our proof system for entailment is complete, or even that entailment is decidable.

High Level Description. Our algorithm proceeds in three steps:

1. Find a *colouring* of L. This is an assignment of a colour, either *red* or *blue*, to each heaplet appearing in L. Intuitively, red heaplets are used to satisfy M, and blue heaplets are left over. This colouring can be computed by recursion on a proof of $L \vdash (\exists X.\ M * \mathsf{true})$.

2. Find a *coloured strengthening* $\Pi : [M']^{\mathrm{r}} * [A]^{\mathrm{b}}$ of R. (We use the notation $[\Sigma]^{\mathrm{r}}$ or $[\Sigma]^{\mathrm{b}}$ to denote a spatial formula Σ of red or blue colour, respectively.) Intuitively, this is a formula that (1) entails R and (2) is coloured in such a way that the red heaplets correspond to the red heaplets of L, and the blue heaplets correspond to the blue heaplets of L. This coloured strengthening can be computed by recursion on a proof of $L \vdash R$ using the colouring of L computed in step 1.

3. Check $\Pi' : M * A \models R$, where Π' is the strongest pure formula implied by L. This step is necessary because M may be weaker than M'. If the entailment check fails, then our algorithm fails to compute a solution to the bounded abduction problem. If the entailment check succeeds, then $\Pi'' : A$ is a solution, where Π'' is the set of all equalities and disequalities in Π' which were actually used in the proof of the entailment $\Pi' : M * A \models R$ (roughly, all those equalities and disequalities which appear in the leaves of the proof tree, plus the equalities that were used in some instance of the SUBSTITUTION rule).

First, we give an example to illustrate these high-level steps:

Example 5. Suppose we want to solve the following bounded abduction problem:

$$L \vdash \mathsf{ls}(x, y) * [\ ] \vdash R$$

where $L = x \mapsto [a, y] * y \mapsto [b, \mathsf{null}]$ and $R = (\exists z.\ x \mapsto [a, z] * \mathsf{ls}(y, \mathsf{null}))$. Our algorithm operates as follows:

1. Colour L: $[x \mapsto [a, y]]^{\mathrm{r}} * [y \mapsto [b, \mathsf{null}]]^{\mathrm{b}}$
2. Colour R: $(\exists z.\ [x \mapsto [a, z]]^{\mathrm{r}} * [\mathsf{ls}(y, \mathsf{null})]^{\mathrm{b}})$
3. Prove the entailment

$$x \neq \mathsf{null} \wedge y \neq \mathsf{null} \wedge x \neq y : \mathsf{ls}(x, y) * \mathsf{ls}(y, \mathsf{null}) \models R$$

This proof succeeds, and uses the pure assertion $x \neq y$.
Our algorithm computes $x \neq y : \mathsf{ls}(y, \mathsf{null})$ as the solution to the bounded abduction problem. ⌐

We now elaborate our bounded abduction algorithm. We assume that L is quantifier free (without loss of generality, since quantified variables can be Skolemized) and *saturated* in the sense that for any pure formula Π', if $L \vdash \Pi'$, where $L = \Pi : \Sigma$, then $\Pi \vdash \Pi'$.

Step 1. The first step of the algorithm is straightforward. If we suppose that there exists a solution, A, to the bounded abduction problem, then by definition we must that have $L \models (\exists X. \ M * A)$. Since $(\exists X. \ M * A) \models (\exists X. \ M * \mathsf{true})$, we must also have $L \models (\exists X. \ M * \mathsf{true})$. We begin step 1 by computing a proof of $L \vdash (\exists X. \ M * \mathsf{true})$. If we fail, then we abort the procedure and report that we cannot find a solution to the abduction problem. If we succeed, then we can colour the heaplets of L as follows: for each heaplet $E \mapsto [\vec{A}, \vec{F}]$ in L, either $E \mapsto [\vec{A}, \vec{F}]$ was used in an application of the POINTS-TO axiom in the proof of $L \vdash (\exists X. \ M * \mathsf{true})$ or not. If yes, we colour $E \mapsto [\vec{A}, \vec{F}]$ red; otherwise, we colour it blue. We denote a heaplet H coloured by a colour c by $[H]^c$.

Step 2. The second step is to find a coloured strengthening of R. Again, supposing that there is some solution A to the bounded abduction problem, we must have $L \models (\exists X. \ M * A) \models R$, and therefore $L \models R$. We begin step 2 by computing a proof of $L \vdash R$. If we fail, then we abort. If we succeed, then we define a coloured strengthening of R by recursion on the proof of $L \vdash R$. Intuitively, this algorithm operates by inducing a colouring on points-to predicates in the leaves of the proof tree from the colouring of L (via the POINTS-TO rule in Fig. 8) and then only folding recursive predicates when all the folded heaplets have the same colour.

More formally, for each formula P appearing as the consequent of some sequent in a proof tree, our algorithm produces a mapping from heaplets in P to coloured spatial formulas. The mapping is represented using the notation $\langle \Sigma \trianglelefteq H \rangle$, which denotes that the heaplet H is mapped to the coloured spatial formula Σ. For each recursive predicate Z and each $(\exists X. \ \Pi : H_1 * \cdots * H_n) \in cases(Z(\vec{R}, \vec{x}))$, we define two versions of the fold rule, corresponding to when $H_1, \ldots, H_n$ are coloured homogeneously (FOLD1) and heterogeneously (FOLD2):

FOLD1
$$\frac{(\Pi : \Sigma \vdash \Pi' : \Sigma' * \langle [H_1]^c \trianglelefteq H_1 \rangle * \cdots * \langle [H_n]^c \trianglelefteq H_n \rangle)[\vec{E}/\vec{x}]}{\Pi : \Sigma \vdash \Pi' : \Sigma' * \langle [Z(\vec{E})]^c \trianglelefteq Z(\vec{E}) \rangle}$$

FOLD2
$$\frac{(\Pi : \Sigma \vdash \Pi' : \Sigma' * \langle \Sigma_1' \trianglelefteq H_1 \rangle * \cdots * \langle \Sigma_n' \trianglelefteq H_n \rangle)[\vec{E}/\vec{x}]}{\Pi : \Sigma \vdash \Pi' : \Sigma' * \langle \Sigma_1' * \cdots * \Sigma_n' \trianglelefteq Z(\vec{E}) \rangle}$$

The remaining rules for our algorithm are presented formally in Fig. 8.[3] To illustrate how this algorithm works, consider the FOLD1 and FOLD2 rules. If a given (sub-)proof finishes with an instance of FOLD that folds $H_1 * \cdots * H_n$ into $Z(\vec{E})$, we begin by colouring the sub-proof of

$$\Pi : \Sigma \vdash \Pi' : \Sigma' * H_1 * \cdots * H_n$$

This colouring process produces a coloured heaplet Σ_i for each H_i. If there is some colour c such that each Σ_i' is $[H_i]^c$, then we apply FOLD1 and $Z(\vec{E})$ gets mapped to $[Z(\vec{E})]^c$. Otherwise (if there is some i such that Σ_i is not H_i or there

---

[3] Note that some of the inference rules are missing. This is because these rules are inapplicable (in the case of UNFOLD and INCONSISTENT) or unnecessary (in the case of NULL-NOT-LVAL and *-PARTIAL), given our assumptions on the antecedent.

is some i, j such that Σ_i and Σ_j have different colours), we apply FOLD2, and map $Z(\vec{E})$ to $\Sigma_1 * \cdots * \Sigma_n$.

After colouring a proof, we define A to be the blue part of R. That is, if the colouring process ends with a judgement of

$$\Pi : [\Sigma_1]^r * [\Sigma_2]^b \vdash \Pi' : \langle [\Sigma'_{11}]^r * [\Sigma_{12}]^b \trianglelefteq H_1 \rangle * \cdots * \langle [\Sigma'_{n1}]^r * [\Sigma_{n2}]^b \trianglelefteq H_n \rangle$$

(where for any coloured spatial formula Σ, its partition into red and blue heaplets is denoted by $[\Sigma_1]^r * [\Sigma_2]^b$), we define A to be $\Pi' : \Sigma_{12} * \cdots * \Sigma_{n2}$. This choice is justified by the following lemma:

Lemma 1. *Suppose that*
$$\Pi : [\Sigma_1]^r * [\Sigma_2]^b \vdash \Pi' : \langle [\Sigma'_{11}]^r * [\Sigma_{12}]^b \trianglelefteq H_1 \rangle * \cdots * \langle [\Sigma'_{n1}]^r * [\Sigma_{n2}]^b \trianglelefteq H_n \rangle$$
is derivable using the rules of Fig. 8, and that the antecedent is saturated. Then the following hold:

- *$\Pi' : \Sigma_{11} * \Sigma_{12} * \cdots * \Sigma_{n2} \models \Pi' : H_1 * \cdots * H_n$;*
- *$\Pi : \Sigma_1 \models \Pi' : \Sigma_{11} * \cdots * \Sigma_{n1}$; and*
- *$\Pi : \Sigma_2 \models \Pi' : \Sigma_{12} * \cdots * \Sigma_{n2}$.*

Step 3. The third step of our algorithm is to check the entailment $\Pi : M * A \models R$. To illustrate why this is necessary, consider the following example:

Example 6. Suppose we want to solve the following bounded abduction problem:

$$x \neq y : x \mapsto [a, y] \vdash \mathsf{ls}(x, y) * [\ ] \vdash x \mapsto [a, y] .$$

In Step 1, we compute the colouring $x \neq y : [x \mapsto [a, y]]^r * [\mathsf{emp}]^b$ of the left hand side. In step 2, we compute the colouring $[x \mapsto [a, y]]^r * [\mathsf{emp}]^b$ of the right hand side. However, emp is not a solution to the bounded abduction problem. In fact, there is no solution to the bounded abduction problem. Intuitively, this is because M is too weak to entail the red part of the right hand side.   ⌙

7   Implementation and Evaluation

Our primary goal is to study the feasibility of our proposed algorithm. To that end, we implemented an instantiation of our generic algorithm with the linked list recursive predicate ls (as defined in Sec. 3) and refinements in the theory of linear arithmetic (QF_LRA). The following describes our implementation and evaluation of SPLINTER in detail.

Implementation. We implemented SPLINTER in the T2 safety and termination verifier [38]. Specifically, we extended T2's front-end to handle heap-manipulating programs, and used its safety checking component (which implements McMillan's IMPACT algorithm) as a basis for our implementation of SPLINTER. To enable reasoning in separation logic, we implemented an entailment checker for RSep along with a bounded abduction procedure.

We implemented a constraint-based solver using the linear rational arithmetic interpolation techniques of Rybalchenko and Stokkermans [35] to solve the non-recursive Horn clauses generated by SPLINTER. Although many off-the-shelf tools for interpolation exist (e.g., [27]) we implemented our own solver for experimentation and evaluation purposes to allow us more flexibility in controlling the forms of interpolants we are looking for. We expect that SPLINTER would perform even better using these highly tuned interpolation engines.

Our main goal is to evaluate the feasibility of our proposed extension of interpolation-based verification to heap and data reasoning, and not necessarily to demonstrate performance improvements against other tools. Nonetheless, we note that there are two tools that target similar programs: (1) THOR [23], which computes a memory safety proof and uses off-the-shelf numerical verifiers to strengthen it, and (2) XISA [13], which combines shape and data abstract domains in an abstract interpretation framework. THOR cannot compute arbitrary refinements of recursive predicates (like the ones demonstrated here and required in our benchmarks) unless they are manually supplied with the required theory predicates. Instantiated with the right abstract data domains, XISA can in principle handle most programs we target in our evaluation. (XISA was unavailable for comparison [12].) Sec. 8 provides a detailed comparison with related work.

Benchmarks. To evaluate SPLINTER, we used a number of linked list benchmarks that require heap and data reasoning. First, we used a number of simple benchmarks: `listdata` is similar to Fig. 2, where a linked list is constructed and its data elements are later checked; `twolists` requires an invariant comparing data elements of two lists (all elements in list A are greater than those in list B); `ptloop` tests our spatial interpolation technique, where the head of the list must not be folded in order to ensure its data element is accessible; and `refCount` is a reference counting program, where our goal is to prove memory safety (no double free). For our second set of benchmarks, we used a cut-down version of BinChunker (`http://he.fi/bchunk/`), a Linux utility for converting between different audio CD formats. BinChunker maintains linked lists and uses their data elements for traversing an array. Our property of interest is thus ensuring that all array accesses are within bounds. To test our approach, we used a number of modifications of BinChunker, `bchunk_a` to `bchunk_f`, where `a` is the simplest benchmark and `f` is the most complex one.

Heuristics. We employed a number of heuristics to improve our implementation. First, given a program path to prove correct, we attempt to find a similar proof to previously proven paths that traverse the same control flow locations. This is similar to the *forced covering* heuristic of [26] to force path interpolants to generalize to inductive invariants. Second, our Horn clause solver uses Farkas' lemma to compute linear arithmetic interpolants. We found that minimizing the number of non-zero *Farkas coefficients* results in more generalizable refinements. A similar heuristic is employed by [1].

Results. Table 1 shows the results of running SPLINTER on our benchmark suite. Each row shows the number of calls to ProvePath (number of paths proved), the total time taken by SPLINTER in seconds, the time taken to generate Horn clauses and compute theory interpolants ($\mathcal{T}$ Time), and the time taken to compute spatial interpolants (Sp. Time). SPLINTER proves all benchmarks correct w.r.t. their respective properties. As expected, on simpler examples, the number

Table 1. Results of running SPLINTER on our benchmark set

Benchmark	#ProvePath	Time (s)	$\mathcal{T}$ Time	Sp. Time
listdata	5	1.37	0.45	0.2
twolists	5	3.12	2.06	0.27
ptloop	3	1.03	0.28	0.15
refCount	14	1.6	0.59	0.14
bchunk_a	6	1.56	0.51	0.25
bchunk_b	18	4.78	1.7	0.2
bchunk_c	69	31.6	14.3	0.26
bchunk_d	23	9.3	4.42	0.27
bchunk_e	52	30.1	12.2	0.25
bchunk_f	57	22.4	12.0	0.25

of paths sampled by SPLINTER is relatively small (3 to 14). In the bchunk_*
examples, SPLINTER examines up to 69 paths (bchunk_c). It is important to
note that, in all benchmarks, almost half of the total time is spent in theory
interpolation. We expect this can be drastically cut with the use of a more
efficient interpolation engine. The time taken by spatial interpolation is very
small in comparison, and becomes negligible in larger examples. The rest of the
time is spent in checking entailment of RSep formulas and other miscellaneous
operations.

Our results highlight the utility of our proposed approach. Using our prototype
implementation of SPLINTER, we were able to verify a set of realistic programs
that require non-trivial combinations of heap and data reasoning. We expect
the performance of our prototype implementation of SPLINTER can greatly improve with the help of state-of-the-art Horn clause solvers, and more efficient
entailment checkers for separation logic.

8   Related Work

Abstraction Refinement for the Heap. To the best of our knowledge, the
work of Botincan et al. [8] is the only separation logic shape analysis that employs a form of abstraction refinement. It starts with a family of separation logic
domains of increasing precision, and uses spurious counterexample traces (reported by forward fixed-point computation) to pick a more precise domain to
restart the analysis and (possibly) eliminate the counterexample. Limitations of
this technique include: (1) The precision of the analysis is contingent on the set
of abstract domains it is started with. (2) The refinement strategy (in contrast
to SPLINTER) does not guarantee progress (it may explore the same path repeatedly), and may report false positives. On the other hand, given a program
path, SPLINTER is guaranteed to find a proof for the path or correctly declare it
an unsafe execution. (3) Finally, it is unclear whether refinement with a powerful
theory like linear arithmetic can be encoded in such a framework, e.g., as a set
of domains with increasingly more arithmetic predicates.

Podelski and Wies [31] propose an abstraction refinement algorithm for a
shape-analysis domain with a logic-based view of three-valued shape analysis
(specifically, first-order logic plus transitive closure). Spurious counterexamples
are used to either refine the set of predicates used in the analysis, or refine

an imprecise abstract transformer. The approach is used to verify specifications given by the user as first-order logic formulas. A limitation of the approach is that refinement is syntactic, and if an important recursive predicate (e.g., there is a list from x to null) is not explicitly supplied in the specification, it cannot be inferred automatically. Furthermore, abstract post computation can be expensive, as the abstract domain uses quantified predicates. Additionally, the analysis assumes a memory safe program to start, whereas, in SPLINTER, we construct a memory safety proof as part of the invariant, enabling us to detect unsafe memory operations that lead to undefined program behavior.

Beyer et al. [6] propose using shape analysis information on demand to augment numerical predicate abstraction. They use shape analysis as a backup analysis when failing to prove a given path safe without tracking the heap, and incrementally refines TVLA's [7] three-valued shape analysis [36] to track more heap information as required. As with [31], [6] makes an *a priori* assumption of memory safety and requires an expensive abstract post operator.

Finally, Manevich et al. [24] give a theoretical treatment of counterexample-driven refinement in power set (e.g., shape) abstract domains.

Combined Shape and Data Analyses. The work of Magill et al. [23] infers shape and numerical invariants, and is the most closely related to ours. First, a separation logic analysis is used to construct a memory safety proof of the whole program. This proof is then *instrumented* by adding additional user-defined integer parameters to the recursive predicates appearing in the proof (with corresponding user-defined interpretations). A numerical program is generated from this instrumented proof and checked using an off-the-shelf verification tool, which need not reason about the heap. Our technique and [23]'s are similar in that we both decorate separation logic proofs with additional information: in [23], the extra information is instrumentation variables; in this paper, the extra information is refinement predicates. Neither of these techniques properly subsumes the other, and we believe that they may be profitably combined. An important difference is that we synthesize data refinements automatically from program paths, whereas [23] uses a fixed (though user-definable) abstraction.

A number of papers have proposed abstract domains for shape and data invariants. Chang and Rival [13] propose a separation logic–based abstract domain that is parameterized by programmer-supplied *invariant checkers* (recursive predicates) and a data domain for reasoning about contents of these structures. McCloskey et al. [25] also proposed a combination of heap and numeric abstract domains, this time using 3-valued structures for the heap. While the approaches to combining shape and data information are significantly different, an advantage of our method is that it does not lose precision due to limitations in the abstract domain, widening, and join.

Bouajjani et al. [9,10] propose an abstract domain for list manipulating programs that is parameterized by a data domain. They show that by varying the data domain, one can infer invariants about list sizes, sum of elements, etc. Quantified data automata (QDA) [17] have been proposed as an abstract domain for representing list invariants where the data in a list is described by a regular language. In [16], invariants over QDA have been synthesized using language learning techniques from concrete program executions. Expressive logics have also been proposed for reasoning about heap and data [32], but have thus far been only used for invariant checking, not invariant synthesis. A number of decision procedures for combinations of the singly-linked-list fragment of separation logic with SMT theories have recently been proposed [30,29].

Path-Based Verification. A number of works proposed path-based algorithms for verification. Our work builds on McMillan's IMPACT technique [26] and extends it to heap/data reasoning. Earlier work [20] used interpolants to compute predicates from spurious paths in a CEGAR loop. Beyer et al. [5] proposed *path invariants*, where infeasible paths induce program slices that are proved correct, and from which predicates are mined for full program verification. Heizmann et al. [19] presented a technique that uses interpolants to compute path proofs and generalize a path into a visibly push-down language of correct paths. In comparison with SPLINTER, all of these techniques are restricted to first-order invariants.

Our work is similar to that of Itzhaky et al. [22], in the sense that we both generalize from bounded unrollings of the program to compute ingredients of a proof. However, they compute proofs in a fragment of first-order logic that can only express linked lists and has not yet been extended to combined heap and data properties.

References

1. Albarghouthi, A., McMillan, K.L.: Beautiful interpolants. In: Sharygina and Veith [37]
2. Albarghouthi, A., Berdine, J., Cook, B., Kincaid, Z.: Spatial interpolants. Tech. Rep. MSR-TR-2015-4 (January 2015), http://research.microsoft.com/apps/pubs/default.aspx?id=238328
3. Ball, T., Jones, R.B. (eds.): CAV 2006. LNCS, vol. 4144. Springer, Heidelberg (2006)
4. Berdine, J., Calcagno, C., O'Hearn, P.W.: Symbolic execution with separation logic. In: Yi, K. (ed.) APLAS 2005. LNCS, vol. 3780, pp. 52–68. Springer, Heidelberg (2005)
5. Beyer, D., Henzinger, T.A., Majumdar, R., Rybalchenko, A.: Path invariants. In: Ferrante, J., McKinley, K.S. (eds.) PLDI, ACM (2007)
6. Beyer, D., Henzinger, T.A., Théoduloz, G.: Lazy shape analysis. In: Ball and Jones [3]
7. Bogudlov, I., Lev-Ami, T., Reps, T., Sagiv, M.: Revamping TVLA: Making parametric shape analysis competitive. In: Damm, W., Hermanns, H. (eds.) CAV 2007. LNCS, vol. 4590, pp. 221–225. Springer, Heidelberg (2007)
8. Botincan, M., Dodds, M., Magill, S.: Abstraction refinement for separation logic program analyses, http://www.cl.cam.ac.uk/~mb741/papers/abs_ref_draft.pdf
9. Bouajjani, A., Drăgoi, C., Enea, C., Rezine, A., Sighireanu, M.: Invariant synthesis for programs manipulating lists with unbounded data. In: Touili, T., Cook, B., Jackson, P. (eds.) CAV 2010. LNCS, vol. 6174, pp. 72–88. Springer, Heidelberg (2010)
10. Bouajjani, A., Drăgoi, C., Enea, C., Sighireanu, M.: Abstract domains for automated reasoning about list-manipulating programs with infinite data. In: Kuncak,

V., Rybalchenko, A. (eds.) VMCAI 2012. LNCS, vol. 7148, pp. 1–22. Springer, Heidelberg (2012)

11. Calcagno, C., Distefano, D., O'Hearn, P.W., Yang, H.: Compositional shape analysis by means of bi-abduction. In: Shao, Z., Pierce, B.C. (eds.) POPL. ACM (2009)

12. Chang, B.Y.E.: Personal communication

13. Chang, B.E., Rival, X.: Relational inductive shape analysis. In: Necula, G.C., Wadler, P. (eds.) POPL. ACM (2008)

14. Cook, B., Haase, C., Ouaknine, J., Parkinson, M., Worrell, J.: Tractable reasoning in a fragment of separation logic. In: Katoen, J.-P., König, B. (eds.) CONCUR 2011. LNCS, vol. 6901, pp. 235–249. Springer, Heidelberg (2011)

15. Distefano, D., O'Hearn, P.W., Yang, H.: A local shape analysis based on separation logic. In: Hermanns, H., Palsberg, J. (eds.) TACAS 2006. LNCS, vol. 3920, pp. 287–302. Springer, Heidelberg (2006)

16. Garg, P., Löding, C., Madhusudan, P., Neider, D.: Learning universally quantified invariants of linear data structures. In: Sharygina and Veith [37]

17. Garg, P., Madhusudan, P., Parlato, G.: Quantified data automata on skinny trees: An abstract domain for lists. In: Logozzo, F., Fähndrich, M. (eds.) Static Analysis. LNCS, vol. 7935, pp. 172–193. Springer, Heidelberg (2013)

18. Gupta, A., Popeea, C., Rybalchenko, A.: Solving recursion-free horn clauses over LI+UIF. In: Yang, H. (ed.) APLAS 2011. LNCS, vol. 7078, pp. 188–203. Springer, Heidelberg (2011)

19. Heizmann, M., Hoenicke, J., Podelski, A.: Nested interpolants. In: Hermenegildo and Palsberg [21]

20. Henzinger, T.A., Jhala, R., Majumdar, R., McMillan, K.L.: Abstractions from proofs. In: Jones, N.D., Leroy, X. (eds.) POPL. ACM (2004)

21. Hermenegildo, M.V., Palsberg, J. (eds.): POPL. ACM (2010)

22. Itzhaky, S., Bjørner, N., Reps, T., Sagiv, M., Thakur, A.: Property-directed shape analysis. In: Biere, A., Bloem, R. (eds.) CAV 2014. LNCS, vol. 8559, pp. 35–51. Springer, Heidelberg (2014)

23. Magill, S., Tsai, M., Lee, P., Tsay, Y.: Automatic numeric abstractions for heap-manipulating programs. In: Hermenegildo and Palsberg [21]

24. Manevich, R., Field, J., Henzinger, T.A., Ramalingam, G., Sagiv, M.: Abstract counterexample-based refinement for powerset domains. In: Reps, T., Sagiv, M., Bauer, J. (eds.) Wilhelm Festschrift. LNCS, vol. 4444, pp. 273–292. Springer, Heidelberg (2007)

25. McCloskey, B., Reps, T., Sagiv, M.: Statically inferring complex heap, array, and numeric invariants. In: Cousot, R., Martel, M. (eds.) SAS 2010. LNCS, vol. 6337, pp. 71–99. Springer, Heidelberg (2010)

26. McMillan, K.L.: Lazy abstraction with interpolants. In: Ball and Jones [3]

27. McMillan, K.L.: Interpolants from Z3 proofs. In: Bjesse, P., Slobodová, A. (eds.) FMCAD. FMCAD Inc. (2011)

28. Pérez, J.A.N., Rybalchenko, A.: Separation logic + superposition calculus = heap theorem prover. In: Hall, M.W., Padua, D.A. (eds.) PLDI. ACM (2011)

29. Navarro Pérez, J.A., Rybalchenko, A.: Separation logic modulo theories. In: Shan, C.-c. (ed.) APLAS 2013. LNCS, vol. 8301, pp. 90–106. Springer, Heidelberg (2013)

30. Piskac, R., Wies, T., Zufferey, D.: Automating separation logic using SMT. In: Sharygina and Veith [37]

31. Podelski, A., Wies, T.: Counterexample-guided focus. In: Hermenegildo and Palsberg [21]

32. Qiu, X., Garg, P., Stefanescu, A., Madhusudan, P.: Natural proofs for structure, data, and separation. In: Boehm, H., Flanagan, C. (eds.) PLDI. ACM (2013)

33. Reynolds, J.C.: Separation logic: A logic for shared mutable data structures. In: LICS. IEEE Computer Society Press (2002)
34. Rümmer, P., Hojjat, H., Kuncak, V.: Classifying and solving horn clauses for verification. In: Cohen, E., Rybalchenko, A. (eds.) VSTTE 2013. LNCS, vol. 8164, pp. 1–21. Springer, Heidelberg (2014)
35. Rybalchenko, A., Sofronie-Stokkermans, V.: Constraint solving for interpolation. In: Cook, B., Podelski, A. (eds.) VMCAI 2007. LNCS, vol. 4349, pp. 346–362. Springer, Heidelberg (2007)
36. Sagiv, S., Reps, T.W., Wilhelm, R.: Parametric shape analysis via 3-valued logic. In: Appel, A.W., Aiken, A. (eds.) POPL. ACM (1999)
37. Sharygina, N., Veith, H. (eds.): CAV 2013. LNCS, vol. 8044. Springer, Heidelberg (2013)
38. T2, http://research.microsoft.com/en-us/projects/t2/

Propositional Reasoning about Safety and Termination of Heap-Manipulating Programs*

Cristina David, Daniel Kroening, and Matt Lewis

University of Oxford

Abstract. This paper shows that it is possible to reason about the safety and termination of programs handling potentially cyclic, singly-linked lists using propositional reasoning even when the safety invariants and termination arguments depend on constraints over the lengths of lists. For this purpose, we propose the theory SLH of singly-linked lists with length, which is able to capture non-trivial interactions between shape and arithmetic. When using the theory of bit-vector arithmetic as background theory, SLH is efficiently decidable via a reduction to SAT. We show the utility of SLH for software verification by using it to express safety invariants and termination arguments for programs manipulating potentially cyclic, singly-linked lists with unrestricted, unspecified sharing. We also provide an implementation of the decision procedure and apply it to check safety and termination proofs for several heap-manipulating programs.

Keywords: Heap, SAT, safety, termination.

1 Introduction

Proving safety of heap-manipulating programs is a notoriously difficult task. One of the main culprits is the complexity of the verification conditions generated for such programs. The constraints comprising these verification conditions can be arithmetic (e.g. the value stored at location pointed by x is equal to 3), structural (e.g. x points to an acyclic singly-linked list), or a combination of the first two when certain structural properties of a data structure are captured as numeric values (e.g. the length of the list pointed by x is 3). Solving these combined constraints requires non-trivial interaction between shape and arithmetic.

For illustration, consider the program in Figure 1b, which iterates simultaneously over the lists x and y. The program is safe, i.e. there is no null pointer dereferencing and the assertion after the loop holds. While the absence of null pointer dereferences is trivial to observe and prove, the fact that the assertion after the loop holds relies on the fact that at the beginning of the program and after each loop iteration the lengths of the lists z and t are equal. Thus, the specification language must be capable of expressing the fact that both z and t reach null in the same

---

* Supported by UK EPSRC EP/J012564/1 and ERC project 280053.

number of steps. Note that the interaction between shape and arithmetic constraints is intricate, and cannot be solved by a mere theory combination.

The problem is even more pronounced when proving termination of heap-manipulating programs. The reason is that, even more frequently than in the case of safety checking, termination arguments depend on the size of the heap data structures. For example, a loop iterating over the nodes of such a data structure terminates after all the reachable nodes have been explored. Thus, the termination argument is directly linked to the number of nodes in the data structure. This situation is illustrated again by the loop in Figure 1b.

There are few logics capable of expressing this type of interdependent shape and arithmetic constraint. One of the reasons is that, given the complexity of the constraints, such logics can easily become undecidable (even the simplest use of transitive closure leads to undecidability [8]), or at best inefficient.

The tricky part is identifying a logic that is expressive enough to capture the corresponding constraints and at the same time is efficiently decidable. One work that inspired us in this endeavour is the recent approach by Itzhaky et al. on reasoning about reachability between dynamically allocated memory locations in linked lists using effectively-propositional (EPR) reasoning [9]. This result is appealing as it can harness advances in SAT solvers. The only downside is that the logic presented in [9] is better suited for safety than termination checking, and is best for situations where safety does not depend on the interaction between shape and arithmetic. Thus, our goal is to define a logic that can be used in such scenarios while still being reducible to SAT.

This paper shows that it is possible to reason about the safety and termination of programs handling potentially cyclic, singly-linked lists using propositional reasoning. For this purpose, we present the logic SLH which can express interdependent shape and arithmetic constraints. We empirically show its utility for the verification of heap-manipulating programs by using it to express safety invariants and termination arguments for intricate programs with potentially cyclic, singly-linked lists with unrestricted, unspecified sharing.

SLH is parametrised by the background arithmetic theory used to express the length of lists (and implicitly every numeric variable). The decision procedure reduces validity of a formula in SLH to satisfiability of a formula in the background theory. Thus, SLH is decidable if the background theory is decidable.

As we are interested in a reduction to SAT, we instantiate SLH with the theory of bit-vector arithmetic, resulting in $SLH[\mathcal{T}_{\mathcal{BV}}]$. This allows us to handle non-linear operations on lists length (e.g. the example in Figure 1c), while still retaining decidability. However, SLH can be combined with other background theories, e.g. Presburger arithmetic.

We provide an implementation of our decision procedure for $SLH[\mathcal{T}_{\mathcal{BV}}]$ and test its efficiency by verifying a suite of programs against safety and termination specifications expressed in SLH. Whenever the verification fails, our decision procedure produces a counterexample.

Contributions:

- We propose the theory SLH of singly-linked lists with length. SLH allows *unrestricted* sharing and cycles.
- We define the strongest post-condition for formulae in SLH.
- We show the utility of SLH for software verification by using it to express safety invariants and termination arguments for programs with potentially cyclic singly-linked lists.
- We present the instantiation SLH[$\mathcal{T}_{\mathcal{BV}}$] of SLH with the theory of bit-vector arithmetic. SLH[$\mathcal{T}_{\mathcal{BV}}$] can express non-linear operations on the lengths of lists, while still retaining decidability.
- We provide a reduction from satisfiability of SLH[$\mathcal{T}_{\mathcal{BV}}$] to propositional SAT.
- We provide an implementation of the decision procedure for SLH[$\mathcal{T}_{\mathcal{BV}}$] and test it by checking safety and termination for several heap-manipulating programs (against provided safety invariants and termination arguments).

2   Motivation

Consider the examples in Figure 1. They all capture situations where the safety (i.e. absence of null pointer dereferencing and no assertion failure) and termination of the program depend on interdependent shape and arithmetic constraints. In this section we only give an intuitive description of these examples, and we revisit and formally specify them in Section 7. We assume the existence of the following two functions: (1) $length(x)$ returns the number of nodes on the path from x to NULL if the list pointed by x is acyclic, and MAXINT otherwise; (2) $circular(x)$ returns true iff the list pointed by x is circular (i.e. x is part of a cycle).

In Figure 1a, we iterate over the potentially cyclic singly-linked list pointed by x a number of times equal with the result of $length(x)$. The program is safe (i.e. y is not NULL at loop entry) and terminating. A safety invariant for the loop needs to capture the length of the path from y to NULL.

The loop in Figure 1b iterates over the lists pointed by x and y, respectively, until one of them becomes NULL. In order to check whether the assertion after the loop holds, the safety invariant must relate the length of the list pointed by x to the length of the list pointed by y. Similarly, a termination argument needs to consider the length of the two lists.

The example in Figure 1c illustrates how non-linear arithmetic can be encoded via singly-linked lists. Thus, the loop in $divides(x, y)$ iterates over the list pointed by x a number of nodes equal to the quotient of the integer division $length(x)/length(y)$ such that, after the loop, the list pointed by z has a length equal with the remainder of the division.

The function in Figure 1d returns true iff the list passed in as a parameter is circular. The functional correctness of this function is captured by the assertion after the loop checking that pointers p and q end up being equal iff the list l is circular.

```
                                     List x, y, z = x, t = y;

                                     assume(length(x) == length(y));
List x, y = x;
int n = length(x), i = 0;            while (z != NULL && t != NULL) {
                                       z = z→next;
while (i < n) {                        t = t→next;
  y = y→next;                        }
  i = i+1;
}                                    assert (z == NULL && t == NULL);
```

(a)   (b)

```
int divides(List x, List y) {        int isCircular(List l) {
  List z = y;                          List p = q = l;
  List w = x;
                                       do {
  assume(length(x) != MAXINT &&          if (p != NULL) p = p→next;
         length(y) != MAXINT &&          if (q != NULL) q = q→next;
         y != NULL);                     if (q != NULL) q = q→next;
                                       }
  while (w != NULL) {                  while (p != NULL &&
    if (z == NULL) z = y;                    q != NULL &&
    z = z→next;                              p != q);
    w = w→next;
  }                                    assert(p == q ⇔ circular(l));
                                       return p == q;
  assert(z == NULL ⇔
         length(x)%length(y) == 0);  }
  return z == NULL;
}                                                             (d)
```

(c)

Fig. 1. Motivational examples

3   Theory of Singly Linked Lists with Length

In this section we introduce the theory SLH for reasoning about potentially cyclic singly linked lists.

3.1   Informal Description of SLH

We imagine that there is a set of pointer variables $x, y, \ldots$ which point to heap cells. The cells in the heap are arranged into singly linked lists, i.e. each cell has a "next" pointer which points somewhere in the heap. The lists can be cyclic and two lists can share a tail, so for example the following heap is allowed in our logic:

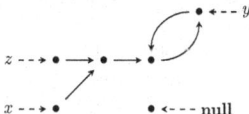

Our logic contains functions for examining the state of the heap, along with the four standard operations for mutating linked lists: *new*, *assign*, *lookup* and *update*. We capture the side-effects of these mutation operators by explicitly naming the current heap – we introduce heap variables h, h' etc. which denote the heap in which each function is to be interpreted. The mutation operators then become pure functions mapping heaps to heaps. The heap functions of the logic are illustrated by example in Figure 3 and have the following meanings:

$alias(h, x, y)$: do x and y point to the same cell in heap h?

$isPath(h, x, y)$: is there a path from x to y in h?

$pathLength(h, x, y)$: the length of the shortest path from x to y in h.

$isNull(h, x)$: is x **null** in h?

$circular(h, x)$: is x part of a cycle, i.e. is there some non-empty path from x back to x in h?

$h' = new(h, x)$: obtain h' from h by allocating a new heap cell and reassigning x so that it points to this cell. The newly allocated cell is not reachable from any other cell and its successor is **null**. This models the program statement $x = new()$. For simplicity, we opt for this allocation policy, but we are not restricted to it.

$h' = assign(h, x, y)$: obtain h' from h by assigning x so that it points to the same cell as y. Models the statement $x = y$.

$h' = lookup(h, x, y)$: obtain h' from h by assigning x to point to y's successor. Models the statement $x = y \rightarrow next$.

$h' = update(h, x, y)$: obtain h' from h by updating x's successor to point to y. Models $x \rightarrow next = y$.

3.2   Syntax of SLH

The theory of singly-linked lists with length, SLH, uses a background arithmetic theory $\mathcal{T}_\mathcal{B}$ for the length of lists (implicitly any numeric variable). Thus, SLH has the following signature:

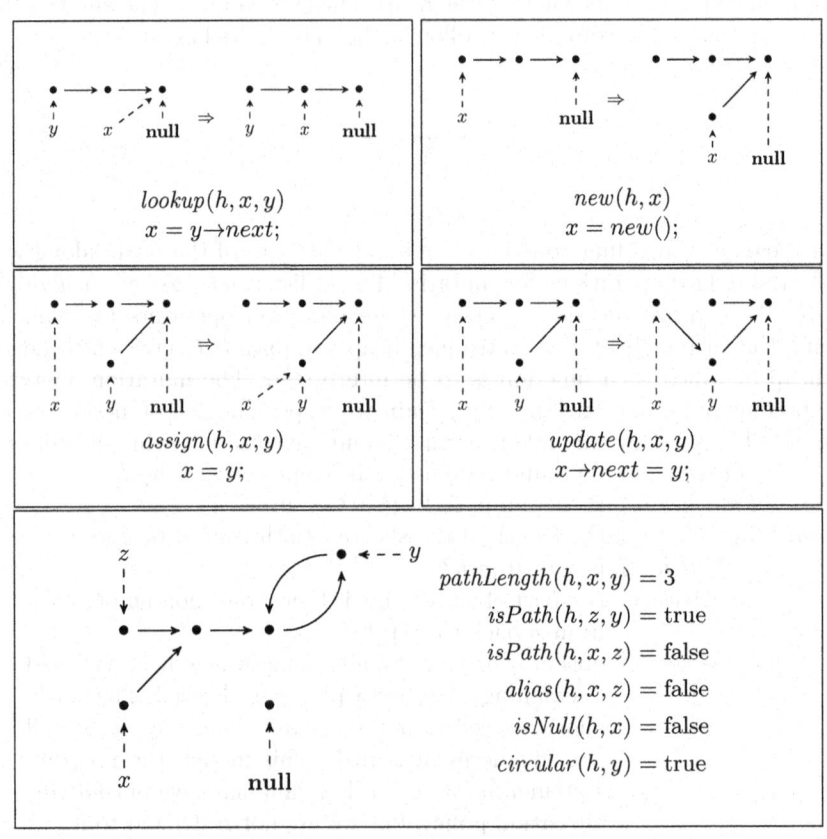

Fig. 3. SLH by example

$$\Sigma_{SLH} = \Sigma_B \cup \{ alias(\cdot,\cdot,\cdot), isPath(\cdot,\cdot,\cdot), isNull(\cdot,\cdot), circular(\cdot,\cdot),$$
$$pathLength(\cdot,\cdot,\cdot), \cdot = new(\cdot,\cdot), \cdot = assign(\cdot,\cdot,\cdot),$$
$$\cdot = lookup(\cdot,\cdot,\cdot), \cdot = update(\cdot,\cdot,\cdot) \}.$$

where the nine new symbols correspond to the heap-specific functions described in the previous section (the first four are actually heap predicates).

Sorts. Heap variables (e.g. h in $alias(h,x,y)$) have sort $\mathcal{S}_{\mathcal{H}}$, pointer variables have sort $\mathcal{S}_{Addr}$ (e.g. x and y in $alias(h,x,y)$), numeric variables have sort $\mathcal{S}_B$ (e.g. n in $n = pathLength(h,x,y)$).

Literal and formula. A literal in SLH is either a heap function (including the negation of the predicates) or a $\mathcal{T}_B$-literal (which may refer to *pathLength*). A formula in SLH is a Boolean combination of SLH-literals.

3.3   Semantics of SLH

We give the semantics of SLH by defining the models in which an SLH formula holds. An interpretation Γ is a function mapping free variables to elements of the appropriate sort. If an SLH formula ϕ holds in some interpretation Γ, we say that Γ *models* ϕ and write $\Gamma \models \phi$.

Interpretations may be constructed using the following substitution rule:

$$\Gamma[h \mapsto H](x) = \begin{cases} H & \text{if } x = h \\ \Gamma(x) & \text{otherwise} \end{cases}$$

Pointer variables are considered to be a set of constant symbols and are thus given a fixed interpretation. The only thing that matters is that their interpretation is pairwise different. We assume that the pointer variables include a special name **null**. The set of pointer variables is denoted by the symbol P.

We will consider the semantics of propositional logic to be standard and the semantics of $\mathcal{T}_B$ given, and thus just define the semantics of heap functions. To do this, we will first define the class of objects that will be used to interpret heap variables.

Definition 1 (Heap). *A heap over pointer variables P is a pair $H = \langle L, G \rangle$. G is a finite graph with vertices $V(G)$ and edges $E(G)$. $L : P \to V(G)$ is a labelling function mapping each pointer variable to a vertex of G. We define the cardinality of a heap to be the cardinality of the vertices of the underlying graph: $|H| = |V(G)|$.*

Definition 2 (Singly Linked Heap). *A heap $H = \langle L, G \rangle$ is a singly linked heap iff each vertex has outdegree 1, except for a single sink vertex that has outdegree 0 and is labelled by **null**:*

$$\forall v \in V(G).(\text{outdegree}(v) = 1 \land L(\textbf{null}) \neq v) \lor$$
$$(\text{outdegree}(v) = 0 \land L(\textbf{null}) = v)$$

Having defined our domain of discourse, we are now in a position to define the semantics of the various heap functions introduced in Section 3.1. We begin with the functions examining the state of the heap and will use a standard structural recursion to give the semantics of the functions with respect to an implicit interpretation Γ, so that $[\![h]\!]\Gamma = \Gamma(h)$. We will use the shorthand $u \xrightarrow{n} v$ to say that if we start at node u, then follow n edges, we arrive at v. We also use $L(H)$ to select the labelling function L from H:

$$u \xrightarrow{n} v \overset{\text{def}}{=} \langle u, v \rangle \in E^n$$

$$u \rightarrow^* v \overset{\text{def}}{=} \exists n \geq 0.u \xrightarrow{n} v$$

$$u \rightarrow^+ v \overset{\text{def}}{=} \exists n > 0.u \xrightarrow{n} v$$

Note that $u \xrightarrow{0} u$. The semantics of the heap functions are then:

$$[\![pathLength(h, x, y)]\!]\Gamma \overset{\text{def}}{=} \min \left(\{n \mid L([\![h]\!]\Gamma)(x) \xrightarrow{n} L([\![h]\!]\Gamma)(y)\} \cup \{\infty\} \right)$$

$$[\![circular(h, x)]\!]\Gamma \overset{\text{def}}{=} \exists v \in V([\![h]\!]\Gamma).L([\![h]\!]\Gamma)(x) \rightarrow^+ v \wedge v \rightarrow^+ L([\![h]\!]\Gamma)(x)$$

$$[\![alias(h, x, y)]\!]\Gamma \overset{\text{def}}{=} [\![pathLength(h, x, y)]\!]\Gamma == 0$$

$$[\![isPath(h, x, y)]\!]\Gamma \overset{\text{def}}{=} [\![pathLength(h, x, y)]\!]\Gamma \neq \infty$$

$$[\![isNull(h, x)]\!]\Gamma \overset{\text{def}}{=} [\![pathLength(h, x, \mathbf{null})]\!]\Gamma == 0$$

Note that since the graph underlying H has outdegree 1, *pathLength* and *circular* can be computed in $O(|H|)$ time, or equivalently they can be encoded with $O(|H|)$ arithmetic constraints.

To define the semantics of the mutation operations, we will consider separately the effect of each mutation on each component of the heap – the labelling function L, the vertex set V and the edge set E. Where a mutation's effect on some heap component is not explicitly stated, the effect is id. For example, *assign* does not modify the vertex set, and so $assign_V =$ id. In the following definitions, we will say that succ(v) is the unique vertex such that $(v, \text{succ}(v)) \in E(H)$.

$$[\![new_V(h, x)]\!]\Gamma \overset{\text{def}}{=} V([\![h]\!]\Gamma) \cup \{q\} \quad \text{where } q \text{ is a fresh vertex}$$

$$[\![new_E(h, x)]\!]\Gamma \overset{\text{def}}{=} E([\![h]\!]\Gamma) \cup \{(q, \mathbf{null})\}$$

$$[\![new_L(h, x)]\!]\Gamma \overset{\text{def}}{=} L([\![h]\!]\Gamma)[x \mapsto q]$$

$$[\![assign_L(h, x, y)]\!]\Gamma \overset{\text{def}}{=} L([\![h]\!]\Gamma)[x \mapsto L([\![h]\!]\Gamma)(y)]$$

$$[\![lookup_L(h, x, y)]\!]\Gamma \overset{\text{def}}{=} L([\![h]\!]\Gamma)[x \mapsto \text{succ}(L([\![h]\!]\Gamma)(y))]$$

$$[\![update_E(h, x, y)]\!]\Gamma \overset{\text{def}}{=} (E([\![h]\!]\Gamma) \setminus \{(L([\![h]\!]\Gamma)(x), \text{succ}(L([\![h]\!]\Gamma)(x)))\}) \cup \{(L([\![h]\!]\Gamma)(x), L([\![h]\!]\Gamma)(y))\}$$

4   Deciding Validity of SLH

We will now turn to the question of deciding the validity of an SLH formula, that is for some formula ϕ we wish to determine whether ϕ is a tautology or if there is some Γ such that $\Gamma \models \neg\phi$. To do this, we will show that SLH enjoys a finite model property and that the existence of a fixed-size model can be encoded directly as an arithmetic constraint.

Our high-level strategy for this proof will be to define progressively coarser equivalence relations on SLH heaps that respect the transformers and observation functions. The idea is that all of the heaps in a particular equivalence class will be equivalent in terms of the SLH formulae they satisfy. We will eventually arrive at an equivalence relation (homeomorphism) that is sound in the above sense and which is also guaranteed to have a small heap in each equivalence class.

From here on we will slightly generalise the definition of a singly linked heap and say that the underlying graph is weighted with weight function $W : E(H) \to \mathbb{N}$. When we omit the weight of an edge (as we have in all heaps until now), it is to be understood that the edge's weight is 1.

4.1   Sound Equivalence Relations

We will say that an equivalence relation $\approx$ is *sound* if the following conditions hold for each pair of pointer variables x, y and transformer τ:

$$\forall H, H' \cdot H \approx H' \Rightarrow pathLength(H, x, y) = pathLength(H', x, y) \;\wedge \quad (1)$$
$$circular(H, x) = circular(H', x) \;\wedge \quad (2)$$
$$\tau(H) \approx \tau(H') \quad (3)$$

The first two conditions say that if two heaps are in the same equivalence class, there is no observation that can distinguish them. The third condition says that the equivalence relation is inductive with respect to the transformers. There is therefore no sequence of transformers and observations that can distinguish two heaps in the same equivalence class.

We begin by defining two sound equivalence relations:

Definition 3 (Reachable Sub-Heap). *The reachable sub-heap $H|_P$ of a heap H is H with vertices restricted to those reachable from the pointer variables P:*

$$V(H|_P) = \{v \mid \exists p \in P . \langle L(p), v \rangle \in E^* \}$$

Then the relation $\{ \langle H, H' \rangle \mid H|_P = H'|_P \}$ is sound.

Definition 4 (heap isomorphism). *Two heaps $H = \langle L, G \rangle, H' = \langle L', G' \rangle$ are isomorphic (written $H \simeq H'$) iff there exists a graph isomorphism $f : G|_P \to G'|_P$ that respects the labelling function, i.e., $\forall p \in P. f(L(p)) = L'(p)$.*

Example 1. H and H' are not isomorphic, even though their underlying graphs are.

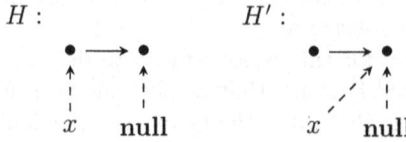

Theorem 1. *Heap isomorphism is a sound equivalence relation.*

4.2   Heap Homeomorphism

The final notion of equivalence we will describe is the weakest. Loosely, we would like to say that two heaps are equivalent if they are "the same shape" and if the shortest distance between pointer variables is the same. To formalise this relationship, we will be using an analogue of topological homeomorphism.

Definition 5 (Edge Subdivision). *A graph G' is a subdivison of G iff G' can be obtained by repeatedly subdividing edges in G, i.e., for some edge $(u, v) \in E(G)$ introducing a fresh vertex q and replacing the edge (u, v) with edges $(u, q), (q, v)$ such that $W'(u, q) + W'(q, v) = W(u, v)$. Subdivision for heaps is defined in terms of their underlying graphs.*

We define a function *subdivide*, which subdivides an edge in a heap. As usual, the function is defined componentwise on the heap:

$$subdivide_V(H, u, v, k) = V \cup \{q\}$$
$$subdivide_E(H, u, v, k) = (E \setminus \{(u, v)\}) \cup \{(u, q), (q, v)\}$$
$$subdivide_W(H, u, v, k) = W(H)[(u, v) \mapsto \infty, (u, q) \mapsto k, (q, v) \mapsto W(H)(u, v) - k]$$

Definition 6 (Edge Smoothing). *The inverse of edge subdivision is called edge smoothing. If G' can be obtained by subdividing edges in G, then we say that G is a smoothing of G'.*

Basically, edge *smoothing* is the dual of edge subdivision – if we have two edges $u \xrightarrow{n} q \xrightarrow{m} v$, where q is unlabelled and has no other incoming edges, we can remove q and add the single edge $u \xrightarrow{n+m} v$.

Example 2. H' is a subdivision of H.

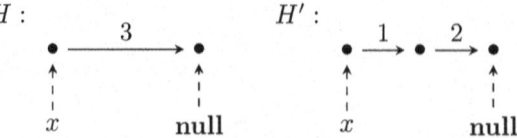

Lemma 1 (Subdividing an Edge Preserves Observations). *If H' is obtained from H by subdividing one edge, then for any x, y we have:*

$$pathLength(H, x, y) = pathLength(H', x, y) \qquad (4)$$
$$circular(H, x) = circular(H', x) \qquad (5)$$

Definition 7 (Heap Homeomorphism). *Two heaps H, H' are homeomorphic (written $H \sim H'$) iff there there is a heap isomorphism from some subdivision of H to some subdivision of H'.*

Intuitively, homeomorphisms preserve the topology of heaps: if two heaps are homeomorphic, then they have the same number of loops and the same number of "joins" (vertices with indegree ≥ 2).

Example 3. H and H' are homeomorphic, since they can each be subdivided to produce S.

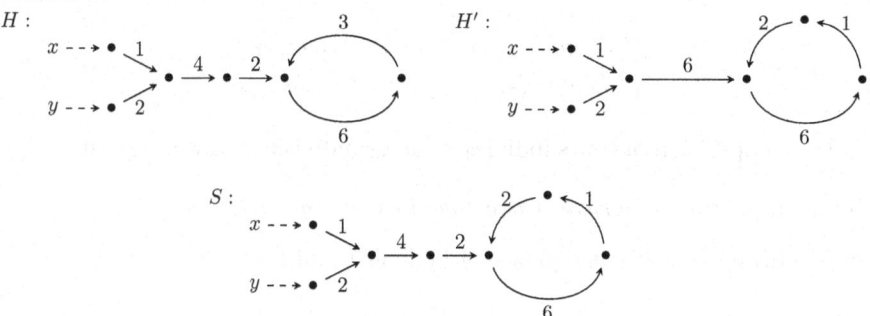

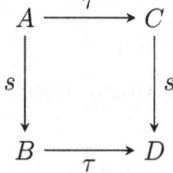

Lemma 2 (Transformers Respect Homeomorphism). *For any heap transformer τ, if $H_1 \sim H_2$ then $\tau(H_1) \sim \tau(H_2)$.*

Proof. It suffices to show that for any transformer τ and single-edge subdivision s, the following diagram commutes:

$$
\begin{array}{ccc}
A & \xrightarrow{\ \tau\ } & C \\
\downarrow{\scriptstyle s} & & \downarrow{\scriptstyle s} \\
B & \xrightarrow[\ \tau\ ]{} & D
\end{array}
$$

We will check that $\tau \circ s = s \circ \tau$ by considering the components of each arrow separately and using the semantics defined in Section 3.3. The only difficult case is for *lookup*, for which we provide the proof in full. This case is illustrative of the style of reasoning used for the proofs of the other transformers.

$\tau = lookup(h, x, y)$: Now that we have weighted heaps, there are two cases for *lookup*: if the edge leaving $L(y)$ does not have weight 1, we need to first subdivide so that it does; otherwise the transformer is exactly as in the unweighted case, which can be seen easily to commute.

In the second (unweighted) case, all of the components commute due to id. Otherwise, *lookup* is a composition of some subdivision s' and then unweighted lookup: $lookup = lookup_U \circ s'$.

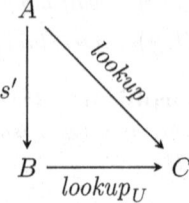

Our commutativity condition is then:

$$(lookup_U \circ s') \circ s = s \circ (lookup_U \circ s')$$

We know that unweighted *lookup* commutes with arbitrary subdivisions, so

$$(lookup_U \circ s') \circ s = s \circ (s' \circ lookup_U)$$
$$lookup_U \circ (s' \circ s) = (s \circ s') \circ lookup_U$$

But the composition of two subdivisions is a subdivision, so we are done.

Theorem 2. *Homeomorphism is a sound equivalence relation.*

Proof. This is a direct consequence of Lemma 1 and Lemma 2.

4.3   Small Model Property

We would now like to show that for each equivalence class induced by $\sim$, there is a unique minimal element. We call that element the *kernel*.

Definition 8 (Kernel). *A kernel is a heap $H = (L, G)$ such that all the vertices in G are either labelled by L, or have at least two incoming edges.*

In other words, a kernel is the maximally smoothed heap.

Theorem 3 (The Kernel is Unique). *Each equivalence class induced by $\sim$ has a unique kernel.*

Proof. We can prove this by contradiction. Let's assume there are two such kernels K_1 and K_2 in an equivalence class. Then $K_1 \sim K_2$, and according to the homeomorphism definition, one is a subdivision of the other. Let's say K_1 is a subdivision of K_2. However, subdividing an edge introduces anonymous vertices with only one incoming edge. Thus K_1 is not a kernel.

As an alternative intuition for this, readers familiar with category theory can consider the category **SLH** of singly linked heaps, with edge subdivisions as arrows. The category **SLH** are singly linked heaps, and there is an arrow from one heap to another if the first can be subdivided into the second. To illustrate, Example 3 is represented in **SLH** by the following diagram:

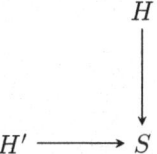

Now for every pair of homeomorphic heaps $H_1 \sim H_2$ we know that there is some X that is a subdivision of both H_1 and H_2. Clearly if we continue subdividing edges, we will eventually arrive at a heap where every edge has weight 1, at which point we will be unable to subdivide any further. Let us call this maximally subdivided heap the *shell*, which we will denote by $\mathrm{Sh}(H_1)$. Then $\mathrm{Sh}(H_1) = \mathrm{Sh}(H_2)$ is the pushout of the previous diagram. Dually, there is some Y that both H_1 and H_2 are subdivisions of, and the previous diagram has a pullback, which we shall call the *kernel*. This is the heap in which all edges have been smoothed. The following diagram commutes, and since a composition of subdivisions and smoothings is a homeomorphism, all of the arrows (and their inverses) in this diagram are homeomorphisms. In fact, the $H_1, H_2, X, Y, \mathrm{Sh}$ and Ke are exactly an equivalence class:

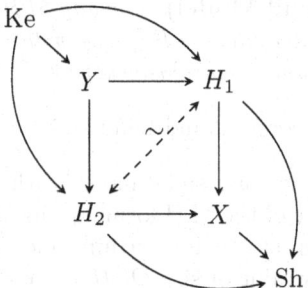

Lemma 3 (Kernels are Small). *For any H, $|\mathrm{Ke}(H)| \leq 2 \times |P|$.*

Proof. Since $\mathrm{Ke}(H)$ is maximally smoothed, every unlabelled vertex has indegree ≥ 2. We will partition the vertices of H into named and unlabelled vertices:

$$N = \{v \in V(H) \mid \exists p \in P.L(p) = v\}$$
$$U = \{u \in V(H) \mid \forall p \in P.L(p) \neq u\}$$
$$V(H) = N \cup U$$

Then let $n = |N|$ and $u = |U|$. Now, the total indegree of the underlying graph must be equal to the total outdegree, so:

$$\sum_{v \in V(H)} \text{out}(v) = \sum_{v \in V(H)} \text{in}(v)$$

$$n + u = \sum_{n \in N} \text{in}(n) + \sum_{u \in U} \text{in}(u)$$

$$= \sum_{n \in N} \text{in}(n) + 2u + k$$

where $k \geq 0$, since $\text{in}(u) \geq 2$ for each u.

$$n = u + \underbrace{\sum_{n \in N} \text{in}(n) + k}_{\geq 0}$$

$$n \geq u$$

So $u \leq n \leq |P|$, hence $|\text{Ke}(H)| = n + u \leq 2 \times |P|$.

Theorem 4 (SLH has Small Model). *For any SLH formula $\forall h.\phi$, if there is a counterexample $\Gamma \models \neg\phi$, then there is $\Gamma' \models \neg\phi$ with every heap-sorted variable in Γ being interpreted by a homeomorphism kernel.*

Proof. This follows from Theorem 2 and Lemma 3.

We can encode the existence of a small model with an arithmetic constraint whose size is linear in the size of the SLH formula, since each of the transformers can be encoded with a constant sized constraint and the observation functions can be encoded with a constraint of size $O(|H|) = O(|P|)$. An example implementation of the constraints used to encode each atom is given in Section 6. We need one constraint for each of the theory atoms, which gives us $O(|P| \times |\phi|)$ constraints in total.

Corollary 1 (Decidability of SLH). *If the background theory $\mathcal{T}_B$ is decidable, then SLH is decidable.*

Proof. The existence of a small model can be encoded with a linear number of arithmetic constraints in $\mathcal{T}_B$.

5   Using SLH for Verification

Our intention is to use SLH for reasoning about the safety and termination of programs with potentially cyclic singly-linked lists:

$$
\begin{array}{ll}
datat := & struct\ C\ \{(typ\ v)^*\} \\
e \quad := & v \mid v{\rightarrow}next \mid \mathrm{new}(C) \mid \mathbf{null} \\
S \quad := & v{=}e \mid v_1{\rightarrow}next{=}v_2 \mid S_1; S_2 \mid \mathit{if}\ (B)\ S_1\ \mathit{else}\ S_2 \mid \\
& \mathit{while}\ (B)\ S \mid \mathrm{assert}(\phi) \mid \mathrm{assume}(\phi)
\end{array}
$$

Fig. 4. Programming Language

- For safety, we annotate loops with safety invariants and generate VCs check-ing that each loop annotation is genuinely a safety invariant, i.e. (1) it is satisfied by each state reachable on entry to the loop, (2) it is inductive with respect to the program's transition relation, and (3) excludes any states where an assertion violation takes place (the assertions include those ensur-ing memory safety). The existence of a safety invariant corresponds to the notion of partial correctness: no assertion fails, but the program may never stop running.
- For termination, we provide ranking functions for each loop and generate VCs to check that the loops do terminate, i.e. the ranking function is mono-tonically decreasing with respect to the loop's body and (2) it is bounded from below. By combining these VCs with those generated for safety, we create a total-correctness specification.

The two additional items we must provide in order to be able to generate these VCs are a programming language and the strongest post-condition for formulae in SLH with respect to statements in the programming language. We do so next.

5.1 Programming Language

We use the sequential programming language in Fig. 4. It allows heap allocation and mutation, with v denoting a variable and *next* a pointer field. To simplify the presentation, we assume each data structure has only one pointer field, *next*, and allow only one-level field access, denoted by $v{\rightarrow}next$. Chained dereferences of the form $v{\rightarrow}next{\rightarrow}next\ldots$ are handled by introducing auxiliary variables. The statement $assert(\phi)$ checks whether ϕ (expressed in the heap theory described in Section 3) holds for the current program state, whereas $assume(\phi)$ constrains the program state.

For convenience when using SLH in the context of safety and termination verification, the SLH functions we expose in the specification language are side-effect free. That is to say, we don't require the explicit heap h to be mentioned in the specifications.

5.2 Strongest Post-condition

To create a verification condition from a specification, we first decompose the specification into Hoare triples and then compute the strongest post-condition to generate a VC in the SLH theory. Since SLH includes primitive operations for heap manipulation, our strongest post-condition is easy to compute:

$$SP(x = y, \phi) \overset{\text{def}}{=} \phi[h'/h] \wedge h = assign(h', x, y)$$

$$SP(x = y{\rightarrow}next, \phi) \overset{\text{def}}{=} \phi[h'/h] \wedge h = lookup(h', x, y)$$

$$SP(x = \text{new}(C), \phi) \overset{\text{def}}{=} \phi[h'/h] \wedge h = new(h', x, y)$$

$$SP(x{\rightarrow}next = y, \phi) \overset{\text{def}}{=} \phi[h'/h] \wedge h = update(h', x, y)$$

In the definitions above, h' is a fresh heap variable. The remaining cases for SP are standard.

5.3   VC Generation Example

```
x = y;

while (x ≠ null) {
   {isPath(y, x)}
   x = x→next;
}

assert (isPath(y, x));
```

Fig. 5. An annotated program

Consider the program in Figure 5, which has been annotated with a loop invariant. In order to verify the partial-correctness condition that the assertion cannot fail, we must check the following Hoare triples:

$$\{\top\}\, x = y \,\{isPath(y, x)\} \tag{6}$$

$$\{isPath(y, x) \wedge \neg isNull(x)\}\, x = x \rightarrow next \,\{isPath(y, x)\} \tag{7}$$

$$\{isPath(y, x) \wedge isNull(x)\}\, \textbf{skip} \,\{isPath(y, x)\} \tag{8}$$

Taking strongest post-condition across each of these triples generates the following SLH VCs:

$$\forall h. h' = assign(h, x, y) \Rightarrow isPath(h', y, x) \tag{9}$$

$$\forall h. isPath(h, y, x) \wedge \neg isNull(x) \wedge h' = lookup(h, x, x) \Rightarrow isPath(h', y, x) \tag{10}$$

$$\forall h. isPath(h, y, x) \wedge isNull(x) \Rightarrow isPath(h, y, x) \tag{11}$$

6   Implementation

For our implementation, we instantiate SLH with the theory of bit-vector arithmetic. Thus, according to Corollary 1, the resulting theory SLH[$\mathcal{T_{BV}}$] is decidable. In this section, we provide details about the implementation of the decision procedure via a reduction to SAT.

To check validity of an SLH[$\mathcal{T_{BV}}$] formula ϕ, we search for a small counterexample heap H. By Theorem 4, if no such small H exists, there is no counterexample and so ϕ is a tautology. We encode the existence of a small counterexample by constructing a SAT formula.

To generate the SAT formula, we instantiate every occurrence of the SLH[$\mathcal{T_{BV}}$] functions with the functions shown in Figure 6. The structure that the functions operate over is the following, where N is the number of vertices in the structure and P is the number of program variables:

```
typedef int node;
typedef int ptr;

struct heap {
  ptr : node[P];
  succ  : (node × int)[N];
  num_nodes: int;
}
```

The heap contains N nodes, of which num_nodes are allocated. Pointer variables are represented as integers in the range $[0, P-1]$ where by convention **null** $= 0$. Each pointer variable is mapped to an allocated node by the ptr array, with the restriction that **null** maps to node 0. The edges in the graph are encoded in the succ array where h.succ[n] $=$ (m, w) iff the edge (n, m) with weight w is in the graph. For a heap with N nodes, this structure requires $3N + 1$ integers to encode.

The implementations of the SLH[$\mathcal{T_{BV}}$] functions described in Section 3.1 are given in Figure 6. Note that only **Alloc** and **Lookup** can allocate new nodes. Therefore if we are searching for a counterexample heap with at most $2P$ nodes, and our formula contains k occurrences of **Alloc** and **Lookup**, the largest heap that can occur in the counterexample will contain no more than $2P + k$ nodes. We can therefore encode all of the heaps using $6P + 3k + 1$ integers each.

When constructing the SAT formula corresponding to the SLH[$\mathcal{T_{BV}}$] formula, each of the functions can be encoded (via symbolic execution) as a formula in the background theory $\mathcal{T_{BV}}$ of constant size, except for **PathLength** which contains a loop. This loop iterates $N = 2P + k$ times and so expands to a formula of size $O(P)$. If the SLH[$\mathcal{T_{BV}}$] formula contains x operations, the final SAT formula in $\mathcal{T_{BV}}$ is therefore of size $x \times P$. We use CBMC [6] to construct and solve the SAT formula.

One important optimisation when constructing the SAT formula involves a symmetry reduction on the counterexamples. Since our encoding assigns names to each of the vertices in the graph, we can have multiple representations for

heaps that are isomorphic. To ensure that the SAT solver only considers a single
counterexample from each homeomorphism class, we choose a canonical representative of each class and add a constraint that the counterexample we are
looking for must be one of these canonical representatives. We define the canonical form of a heap such that the nodes are ordered topologically and so that the
ordering is compatible with the ordering on the program variables. Note that
this canonical form is described in terms of a breadth-first traversal of the graph,
which eliminates cycles.

$$\forall p, p' \in P. p < p' \Rightarrow \forall n, n'. L(p) \rightarrow^* n \wedge L(p') \rightarrow^* n' \Rightarrow n \leq n'$$
$$\forall n, n'. n \rightarrow n' \Rightarrow n \leq n'$$

Where $n \rightarrow^* n'$ means n' is reachable from n.

```
function NEWNODE(heap h)
    n ← h.num_nodes
    h.num_nodes ← h.num_nodes + 1
    h.succ[n] ← (null, 1)
    return n

function SUBDIVIDE(heap h, node a)
    n ← NewNode(h)
    (b, w) ← h.succ[a]
    h.succ[a] ← (n, 1)
    h.succ[n] ← (b, w - 1)
    return n

function UPDATE(heap h, ptr x, ptr y)
    n ← h.ptr[x]
    m ← h.ptr[y]
    h.succ[n] ← (m, 1)

function ASSIGN(heap h, ptr x, ptr y)
    h.ptr[x] ← h.ptr[y]

function LOOKUP(heap h, ptr x, ptr y)
    n ← h.ptr[y]
    (n', w) ← h.succ[n]
    if w ≠ 1 then
        n' ← Subdivide(h, n)
    h.ptr[x] ← n'
```

```
function ALLOC(heap h, ptr x)
    n ← NewNode(h)
    h.ptr[x] ← n

function PATHLENGTH(heap h, ptr x, ptr y)
    n ← h.ptr[x]
    m ← h.ptr[y]
    distance ← 0
    for i ← 0 to h.num_nodes do
        if n = m then
            return distance
        else
            (n, w) ← h.succ[n]
            distance ← distance + w
    return ∞

function CIRCULAR(heap h, ptr x)
    n ← h.ptr[x]
    m ← h.succ[n]
    distance ← 0
    for i ← 0 to h.num_nodes do
        if m = n then
            return True
        else
            if n = null then
                return False
        m ← h.succ[m]
    return False
```

Fig. 6. Implementation of the SLH[$\mathcal{T}_{\mathcal{BV}}$] functions

7   Motivation Revisited

In this section, we get back to the motivational examples in Figure 1 and express their safety invariants and termination arguments in SLH. As mentioned
in Section 5.1, for ease of use, we don't mention the explicit heap h in the
specifications.

In Figure 1a, assuming that the call to the *length* function ensures the state before the loop to be $pathLength(h, x, \mathbf{null}) = n$, then a possible safety invariant is $pathLength(h, y, \mathbf{null}) = n - i$. Note that this invariant covers both the case where the list pointed by x is acyclic and the case where it contains a cycle. In the latter scenario, given that $\infty - i = \infty$, the invariant is equivalent to $pathLength(h, y, \mathbf{null}) = \infty$. A ranking function for this program is $R(i) = -i$.

The program in Figure 1b is safe with a possible safety invariant:

$$pathLength(h, z, \mathbf{null}) == pathLength(h, t, \mathbf{null}).$$

Similar to the previous case, this invariant covers the scenario where the lists pointed by x and y are acyclic, as well as the one where they are cyclic. In the latter situation, the program does not terminate.

For the example in Figure 1c, the *divides* function is safe and a safety invariant is:

$$isPath(x, \mathbf{null}) \wedge isPath(z, \mathbf{null}) \wedge isPath(y, \mathbf{null}) \wedge isPath(y, z) \wedge isPath(x, w) \wedge$$
$$\neg isNull(y) \wedge (pathLength(x, w) + pathLength(z, \mathbf{null})) \% pathLength(y, \mathbf{null}) == 0.$$

Additionally, the function terminates as witnessed by the ranking function $R(w) = pathLength(w, \mathbf{null})$.

Function *isCircular* in Figure 1c is safe and terminating with the safety invariant: $pathLength(l, p) \wedge pathLength(p, q) \wedge isPath(q, p) \neq isPath(l, \mathbf{null})$, and lexicographic ranking function: $R(q, p) = (pathLength(q, \mathbf{null}), pathLength(q, p))$.

8   Experiments

To evaluate the applicability of our theory, we created a tool for verifying that heaps don't lie: SHAKIRA [16]. We ran SHAKIRA on a collection of programs manipulating singly linked lists. This collections includes the standard operations of traversal, reversal, sorting etc. as well as the motivational examples from Section 2. Each of the programs in this collection is annotated with correctness assertions and loop invariants, as well as the standard memory-safety checks. One of the programs (the motivational program from Figure 1b) used a non-linear loop invariant, but this did not require any special treatment by SHAKIRA.

To generate VCs for each program, we generated a Hoare proof and then used CBMC 4.9 [6] to compute the strongest post-conditions for each Hoare triple using symbolic execution. The resulting VCs were solved using Glucose 4.0 [1]. As well as correctness and memory safety, these VCs proved that each loop annotation was genuinely a loop invariant. For four of the programs, we annotated loops with ranking functions and generated VCs to check that the loops terminated, thereby creating a total-correctness specification.

None of the proofs in our collection relied on assumptions about the shape of the heap beyond that it consisted of singly linked lists. In particular, our safety

680   C. David, D. Kroening, and M. Lewis

proofs show that the safe programs are safe even in the presence of arbitrary
cycles and sharing between pointers.

We ran our experiments on a 4-core 3.30 GHz Core i5 with 8 GB of RAM.
The results of these experiments are given in Table 1.

Table 1. Experimental results

	LOC	#VCs	Symex(s)	SAT(s)	C/E
Safe benchmarks (UNSAT VCs)					
SLL (safe)	236	40	18.2	5.9	—
SLL (termination)	113	25	14.7	9.6	—
Counterexamples (SAT VCs)					
CLL (nonterm)	38	14	6.9	1.6	3
Null-deref	165	31	13.6	3.0	3
Assertion Failure	73	11	3.5	0.7	3.5
Inadequate Invariant	37	4	4.9	1.2	6

Legend:
LOC   Total lines of code
#VCs   Number of VCs
Symex(s) Total time spent in symbolic execution to generate VCs
SAT(s)   Total time spent in SAT solver
C/E   Average counterexample size (number of nodes)

The top half of the table gives the aggregate results for the benchmarks
in which the specifications held, i.e., the VCs were unsatisfiable. These "safe"
benchmarks are divided into two categories: partial- and total-correctness proofs.
Note that the total-correctness proofs involve solving more complex VCs – the
partial correctness proofs solved 40 VCs in 5.9 s, while the total correctness
proofs solved only 25 VCs in 9.6 s. This is due to the presence of ranking functions in the total-correctness proofs, which by necessity introduces a higher level
of arithmetic complexity.

The bottom half of the table contains the results for benchmarks in which the
VCs were satisfiable. Since the VCs were generated from a Hoare proof, their
satisfiability only tells us that the purported proof is not in fact a real proof
of the program's correctness. However, SHAKIRA outputs models when the VCs
are satisfiable and these can be examined to diagnose the cause of the proof's
failure. For our benchmarks, the counterexamples fell into four categories:

- Non-termination due to cyclic lists.
- Null dereferences.
- A correctness assertion (not a memory-safety assertion) failing.
- The loop invariant being inadequate, either by being too weak to prove the
  required properties, or failing to be inductive.

A counterexample generated by SHAKIRA is given in Figure 7. This program is a variation on the motivational program from Figure 1c in which the programmer has tried to speed up the loop by unwinding it once. The result is that the program no longer terminates if the list contains a cycle whose size is exactly one, as shown in the counterexample found by SHAKIRA.

```
int has_cycle(list l) {
  list p = l;
  list q = l→n;

  do {
    // Unwind loop to search
    // twice as fast!
    if (p != NULL) p = p→n;
    if (p != NULL) p = p→n;

    if (q != NULL) q = q→n;
    if (q != NULL) q = q→n;
    if (q != NULL) q = q→n;
    if (q != NULL) q = q→n;
  } while (p != q &&
           p != NULL &&
           q != NULL);

  return p == q;
}
```

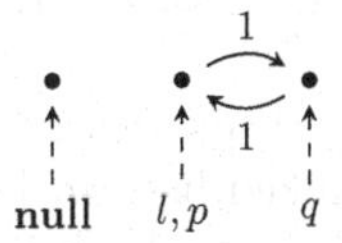

Counterexample heap leading to
non-termination.

Fig. 7. A non-terminating program and the counterexample found by SHAKIRA

These results show that discharging VCs written in SLH is practical with current technology. They further show that SLH is expressive enough to specify safety, termination and correctness properties for difficult programs. When the VCs require arithmetic to be done on list lengths, as is necessary when proving termination, the decision problem becomes noticeably more difficult. Our encoding is efficient enough that even when the VCs contain non-linear arithmetic on path lengths, they can be solved quickly by an off-the-shelf SAT solver.

9 Related Work

Research works on relating the shape of data structures to their numeric properties (e.g. length) follow several directions. For abstract interpretation based analyses, an abstract domain that captures both heap and size was proposed in [3]. The THOR tool [12,13] implements a separation logic [15] based shape analysis and uses an off-the-shelf arithmetic analysis tool to add support for arithmetic reasoning. This approach is conceptually different from ours as it aims to separate the shape reasoning from the numeric reasoning by constructing a numeric program that explicitly tracks changes in data structure sizes. In [4], Boujjani et al. introduce the logic SLAD for reasoning about singly-linked lists and arrays with unbounded data, which allows to combine shape constraints, written in a fragment of separation logic, with data and size constraints. While SLAD is a powerful logic and has a decidable fragment, our main motivation for designing a new logic was its translation to SAT. A second motivation was the unrestricted sharing.

Other recent decidable logics for reasoning about linked lists were developed [9,14,17,11,4]. Piskac et al. provide a reduction of decidable separation logic fragments to a decidable first-order SMT theory [14]. A decision procedure for an alternation-free sub-fragment of first-order logic with transitive closure is described in [9]. Lahiri and Qadeer introduce the Logic of Interpreted Sets and Bounded Quantification (LISBQ) capable to express properties on the shape and data of composite data structures [10]. In [5], Brain et al. propose a decision procedure for reasoning about aliasing and reachability based on Abstract Conflict Driven Clause Learning (ACDCL) [7]. As they don't capture the lengths of lists, these logics are better suited for safety and less for termination proving.

In [2], Berdine et al. present a small model property for a fragment of separation logic with linked lists without explicit lengths. Their small model property says that it suffices to check if lists of lengths zero and two entail the formula (i.e. it suffices to unfold the list predicates 0 and 2 times). However if their fragment allowed imposing minimum lengths for lists, their small model result would be violated. In our case, since SLH allows adding explicit constraints on the lengths of lists (thus, one can impose minimum lengths), their small model property does not hold.

10 Conclusions

We have presented the logic SLH for reasoning about potentially cyclic singly-linked lists. The main characteristics of SLH are the fact that it allows unrestricted sharing in the heap and can relate the structure of lists to their length, i.e. reachability constraints with numeric ones. As SLH is parametrised by the background arithmetic theory used to express the length of lists, we present its instantiation SLH[$\mathcal{T}_{\mathcal{BV}}$] with the theory of bit-vector arithmetic and provide a way of efficiently deciding its validity via a reduction to SAT. We empirically show that SLH is both efficient and expressive enough for reasoning about safety and (especially) termination of list programs.

Limitations. It is not straightforward how to add quantifiers to our approach. Also, extending our technique to other data structures such as trees would break the small model property in its current form, and although we can see ways of adapting it theoretically, it is unclear whether the SAT instances would still be tractable.

References

1. Audemard, G., Simon, L.: Predicting learnt clauses quality in modern SAT solvers. In: Proceedings of the 21st International Joint Conference on Artificial Intelligence, IJCAI 2009, Pasadena, California, USA, July 11-17, pp. 399–404 (2009)

2. Berdine, J., Calcagno, C., W.O'Hearn, P.: A decidable fragment of separation logic. In: Lodaya, K., Mahajan, M. (eds.) FSTTCS 2004. LNCS, vol. 3328, pp. 97–109. Springer, Heidelberg (2004), http://dx.doi.org/10.1007/978-3-540-30538-5_9
3. Bouajjani, A., Drăgoi, C., Enea, C., Sighireanu, M.: Abstract domains for automated reasoning about list-manipulating programs with infinite data. In: Kuncak, V., Rybalchenko, A. (eds.) VMCAI 2012. LNCS, vol. 7148, pp. 1–22. Springer, Heidelberg (2012)
4. Bouajjani, A., Drăgoi, C., Enea, C., Sighireanu, M.: Accurate invariant checking for programs manipulating lists and arrays with infinite data. In: Chakraborty, S., Mukund, M. (eds.) ATVA 2012. LNCS, vol. 7561, pp. 167–182. Springer, Heidelberg (2012)
5. Brain, M., David, C., Kroening, D., Schrammel, P.: Model and proof generation for heap-manipulating programs. In: Shao, Z. (ed.) ESOP 2014 (ETAPS). LNCS, vol. 8410, pp. 432–452. Springer, Heidelberg (2014)
6. Clarke, E., Kroning, D., Lerda, F.: A tool for checking ANSI-C programs. In: Jensen, K., Podelski, A. (eds.) TACAS 2004. LNCS, vol. 2988, pp. 168–176. Springer, Heidelberg (2004)
7. D'Silva, V., Haller, L., Kroening, D.: Abstract conflict driven learning. In: The 40th Annual ACM SIGPLAN-SIGACT Symposium on Principles of Programming Languages, POPL 2013, Rome, Italy, January 23 - 25, pp. 143–154 (2013)
8. Immerman, N., Rabinovich, A., Reps, T., Sagiv, M., Yorsh, G.: The boundary between decidability and undecidability for transitive-closure logics. In: Marcinkowski, J., Tarlecki, A. (eds.) CSL 2004. LNCS, vol. 3210, pp. 160–174. Springer, Heidelberg (2004)
9. Itzhaky, S., Banerjee, A., Immerman, N., Nanevski, A., Sagiv, M.: Effectively-propositional reasoning about reachability in linked data structures. In: Sharygina, N., Veith, H. (eds.) CAV 2013. LNCS, vol. 8044, pp. 756–772. Springer, Heidelberg (2013)
10. Lahiri, S.K., Qadeer, S.: Back to the future: revisiting precise program verification using SMT solvers. In: Proceedings of the 35th ACM SIGPLAN-SIGACT Symposium on Principles of Programming Languages, POPL 2008, San Francisco, California, USA, January 7-12, pp. 171–182 (2008)
11. Madhusudan, P., Parlato, G., Qiu, X.: Decidable logics combining heap structures and data. In: Proceedings of the 38th ACM SIGPLAN-SIGACT Symposium on Principles of Programming Languages, POPL 2011, Austin, TX, USA, January 26-28, pp. 611–622 (2011)
12. Magill, S., Tsai, M.-H., Lee, P., Tsay, Y.-K.: THOR: A tool for reasoning about shape and arithmetic. In: Gupta, A., Malik, S. (eds.) CAV 2008. LNCS, vol. 5123, pp. 428–432. Springer, Heidelberg (2008)
13. Magill, S., Tsai, M.-H., Lee, P., Tsay, Y.-K.: Automatic numeric abstractions for heap-manipulating programs. In: Proceedings of the 37th ACM SIGPLAN-SIGACT Symposium on Principles of Programming Languages, POPL 2010, Madrid, Spain, January 17-23, pp. 211–222 (2010)
14. Piskac, R., Wies, T., Zufferey, D.: Automating separation logic using SMT. In: Sharygina, N., Veith, H. (eds.) CAV 2013. LNCS, vol. 8044, pp. 773–789. Springer, Heidelberg (2013)

15. Reynolds, J.C.: Separation logic: A logic for shared mutable data structures. In: 17th IEEE Symposium on Logic in Computer Science (LICS 2002), Copenhagen, Denmark, July 22-25, pp. 55–74 (2002)
16. Shakira: Hips Don't Lie (2006)
17. Yorsh, G., Rabinovich, A.M., Sagiv, M., Meyer, A., Bouajjani, A.: A logic of reachable patterns in linked data-structures. J.Log.Alg.Prog. 73(1-2) (2007)

Full Reduction in the Face of Absurdity

Gabriel Scherer and Didier Rémy

INRIA
{gabriel.scherer,didier.remy}@inria.fr

Abstract. Core calculi that model the essence of computations use full reduction semantics to be built on solid grounds. Expressive type systems for these calculi may use propositions to refine the notion of types, which allows abstraction over possibly inconsistent hypotheses. To preserve type soundness, reduction must then be delayed until logical hypotheses on which the computation depends have been proved consistent. When logical information is explicit inside terms, proposition variables delay the evaluation by construction. However, logical hypotheses may be left implicit, for the user's convenience in a surface language or because they have been erased prior to computation in an internal language. It then becomes difficult to track the dependencies of computations over possibly inconsistent hypotheses.

We propose an expressive type system with implicit coercions, consistent and inconsistent abstraction over coercions, and *assumption hiding*, which provides a fine-grained control of dependencies between computations and the logical hypotheses they depend on. Assumption hiding opens a continuum between explicit and implicit use of hypotheses, and restores confluence when full and weak reductions are mixed.

Extended version. For reasons of page limits, the proofs have been omitted from this conference version. A full version with additional remarks and all proofs, is available electronically.[1]

1 Introduction

The Curry-Howard isomorphism trained generations of statically-typed-language designers to be able to instantly switch their point of view from *programs* to *proof terms*, and from *types* to *logic statements*. Proof assistants based on type theory let us use our functional programming intuitions to *program* proofs. One example of the merits of such a re-unification is the strikingly simple and natural treatment of *axioms* in the functional languages of those assistants: assuming an axiom P is just abstracting over a variable $(x : P)$ of the corresponding type, and using this assumption is done by applying or pattern matching this bound variable x. These languages generally allow *full reduction*, in particular reducing under unapplied λ-abstractions. For example, a Coq program abstracting over

---

[1] At http://gallium.inria.fr/~remy/coercions/

© Springer-Verlag Berlin Heidelberg 2015
J. Vitek (Ed.): ESOP 2015, LNCS 9032, pp. 685–709, 2015.
DOI: 10.1007/978-3-662-46669-8_28

an axiom P is of the form $\lambda(x : P)\,a$, where a may be computed as usual, but the reduction of subterms depending on x will be blocked.

There is a subtle but important contrast with how logical assumptions have been dealt with in languages designed mostly for programming rather than proving, such as ML or Haskell. Typical examples are the reasoning on type equalities in the ML module system, or in Generalized Algebraic Data Types (GADTs). Consider the following example that implements application up to type equality, given in OCaml-like syntax:

type $(\_,\ \_)$ eq $=$ Refl $: (\alpha,\ \alpha)$ eq

let apply $: \forall \alpha_1 \alpha_2 \beta.\ (\alpha_1 \to \beta) \to \alpha_2 \to (\alpha_1,\ \alpha_2)$ eq $\to \beta$
$=$ **fun** f x Refl $\to$ f x

With GADTs, the equality assumption is present at the term level, marked by a λ-abstraction over the type (α_1, α_2) eq, but the *use* of this equality is *implicit*: equality assumptions introduced by abstraction or pattern-matching can be silently used in the corresponding term clauses. This implicitness can be explained away by translating source terms into an intermediate language, such as System FC [Vytiniotis and Jones, 2011] that marks uses of equality assumptions with explicit coercions – providing a treatment similar to logical assumptions in proof assistants. But it can also be formalized directly, as in the presentation of GADTs extended to arbitrary logic constraints by Simonet and Pottier [2007], or Dependent ML by Xi [2007].

It is well-known however that, with implicit use of potentially-absurd assumptions, it is no longer safe to use full reduction under those assumptions. Assuming (fst $: (\alpha * \beta) \to \alpha$) and (true $:$ bool), the term apply fst true reduces to

fun (Refl $:$ (bool, $(\alpha * \beta)$)) eq) $\to$ fst true

Reducing under this abstraction would mean computing fst true, *i.e.* the application of a destructor to a constructor of an incompatible type, which is called a *runtime error*. Interestingly, this issue does not happen with an *explicit* handling of logical assumptions. In System FC, the above example would reduce to the following normal form (assuming that the assumption γ has been used to convert the type of the argument true rather than the type of the function fst):

fun (Refl $(\gamma :$ bool $\tilde{}\# (\alpha * \beta))) \to$ fst (true $\triangleright \gamma$)

Here, bool $\tilde{}\#$ ('a $*$ 'b) is the type of coercions that prove the equality between bool and ('a $*$ 'b), and (true $\triangleright \gamma$) is the application of the coercion γ to true. This application cannot be reduced until the formal variable γ has been instantiated (that is, never, if we are in an empty context with a sound type system). Meaning, γ remains in between fst and true preventing the application. Although System FC is a weak calculus (abstracting on term or coercion variables blocks reduction), full reduction could be used in a similar, explicit system.

We are convinced that it is important to also study the implicit presentation directly. There is a convergence of designs that indicates that *implicit* use of assumptions is significantly more convenient to the programmer. For a less-obvious example than GADTs in ML or Haskell, the book *Programming in Martin-Löf*

Type Theory [Nordström et al., 1990] uses a type theory with *extensional* equality, which allows implicit use of equality assumptions, especially to simplify programming with quotient types. We want to study λ-calculi that match how users wish to program and define the operational behavior of programs directly at this level.

Besides, we deem unfortunate the absolute reign of *weak reduction* on λ-calculi designed for programming. We argue that while one could have a weak-reduction semantics for reasoning about runtime complexity, a more abstract *full reduction* understanding is better to reason about correctness – as an important step towards equational reasoning on open terms.

In fact, type systems of programming languages are designed for full reduction strategies, and left unchanged when restricting the semantics to weak reduction strategies. For example, the type systems of ML, System F and its derivatives (F_η [Mitchell, 1988], $F_{<:}$ [Cardelli, 1993], MLF [Le Botlan and Rémy, 2003], *etc.*) are all sound for full reduction. While type systems are regularly improved to accept more well-typed programs, they do not try in general to take advantage of weak reduction strategies to accept nonsense under yet unapplied abstractions, *e.g.* $\lambda(x)\,1$ true, on the basis that these errors won't be reachable by a weak-reduction strategy.[2]

We claim that (pure) lambda-calculi for programming languages ought to strive to support *full reduction*; this design pressure should result in a better understanding of programming constructs. For example, soundness of full reduction subsumes soundness for any evaluation strategy such as call-by-name and call-by-value; and full reduction is used in practice in dependently-typed languages such as Coq or Agda, with significant efforts spent to make it practical [Grégoire and Leroy, 2002, Boespflug et al., 2011].

We could summarize the topic of this article with the following question. We know how to design calculi with *explicit* uses of logical assumptions and *full* reduction, or calculi with *implicit* uses of assumptions and *weak* reduction. Can we merge those apparently incompatible feature pairs into a single calculus, close to the non-encumbered terms the programmer wishes to write?

Consistent and Inconsistent Abstraction. Intuitively, an abstraction on a type is *consistent* when we can prove at the point of abstraction that there always exists a possible instantiation for it in the current typing context; otherwise, we say it is *inconsistent*. A typical example of a consistent abstraction is an abstraction over a type variable α that has the kind $\star$ of concrete types, as we know that at least int has kind $\star$ so it is a valid instance for α. Abstraction over a type variable may also be constrained by a proposition that restricts the possible instances of the type variable. An example of an unsatisfiable proposition is the inter-convertibility int $\simeq$ (int $\to \tau$), which is *absurd* for any type τ. Hence, an abstraction over a type variable α such that int $\simeq$ (int $\to \alpha$) is inconsistent.

---

[2] One interesting counterexample is typechecking record concatenation, which delays the resolution of typing constraints based on the evaluation strategy [Pottier, 2000] in order to avoid the heavy cost of early consistency checking.

Previous work by Cretin [2014] and Cretin and Rémy [2014] introduced the calculus F_{cc}, built around *consistent coercion abstraction*, a mechanism that allows implicit abstraction over coercions and uses typing-transforming coercions, provided we prove at their abstraction point that they are instantiable; such coercions are completely erasable – they are not at all present at the term level. This generalizes the traditional ML-style polymorphism in an expressive way, encompassing the type systems of System F, MLF, $F_{<:}$, F_η, and F_ω. To be able to also abstract over hypotheses that may not be consistent, Cretin and Rémy added a distinct mechanism of *inconsistent polymorphism* that is present at the term level and blocks reduction.

If an F_{cc} term a has type τ in the context $\Gamma, \alpha : \kappa$, and we can prove that the kind κ is instantiable by producing some type $\Gamma \vdash \sigma : \kappa$, we will consider that a has type $\forall(\alpha : \kappa)\,\tau$ in context Γ. The Curry-style presentation, with no explicit syntax for type abstraction at the term level – we just write a, not $\Lambda(\alpha : \kappa)\,a$ – highlights that this form of polymorphism is erasable.

If on the contrary we do not know how to prove that κ is instantiable (or do not wish to do so), we may use *inconsistent* abstraction by building the term ∂a at the distinct type[3] $\Pi(\alpha : \kappa)\,\tau$. This form blocks the reduction of a and is thus explicit at the term level.

To the explicit incoherent abstraction corresponds an explicit incoherent application, which unblocks computation: if the type σ has kind κ, then κ is in fact inhabited and $a \,\Diamond$ has type $\tau[\sigma/\alpha]$.

Even in full reduction (when reduction under λ's is allowed), reduction remains forbidden under ∂'s; an inconsistent abstraction is eliminated by the corresponding application, $(\partial a)\,\Diamond$, which reduces to a letting the evaluation of a be resumed. In contrast, consistent abstraction is erasable by construction: it is absent from the term itself, which alone determines reduction.

Issues with F_{cc}. In the absence of inconsistency abstraction, the language F_{cc} has a full reduction semantics and is confluent. However, both properties break when introducing inconsistent abstraction, since inconsistent abstraction blocks the evaluation to maintain soundness. This amounts to introducing a form of weak reduction inside the language. While some reductions under ∂ are unsound and must be blocked, others may be harmless and could be safely reduced—but this is not allowed. This is going against our claim that core calculi ought to support full reduction to the largest possible extent.

Besides, it is well known that mixing weak and strong reductions may break confluence, and this problem affects F_{cc}. If b reduces to b', then $(\lambda(x)\,\partial x)\,b$ can reduce to either $(\lambda(x)\,\partial x)\,b'$ and then $\partial b'$, or to ∂b, which cannot be further reduced and, in particular, does not reduce to $\partial b'$ (or one of its reducts), as confluence would require.

These issues were well-understood by Cretin and Rémy [2014] and left for future work. We present an improved variant of F_{cc} that solves both problems simultaneously. In the course of doing so, we also encountered some more minor

---

[3] The notation $\Pi(\alpha : \kappa)\,\tau$ has nothing to do with dependent types.

issues in the details of F_{cc}, which allowed us to also improve the system as a whole.

Propositional Truths and Hiding. The language F_{cc} uses a blocking construct ∂a to introduce the inconsistent abstraction $\Pi(\alpha : \kappa)\,\tau$. This does not match, however, the way potentially absurd assumptions are handled in dependent type theories, such as Coq, where reduction is blocked at the point of *use* of the assumption, not its point of *introduction*. This distinction is essential, in particular, to allow writing certified programs as Coq program terms: if the axioms (*e.g.* classical logic or proof irrelevance) are only used in logic parts of the formalization (under terms at type Prop), they get removed by extraction; a program whose correctness proof uses axioms can compute – while it would be blocked if we used our ∂ to introduce the axiom.

We therefore split inconsistent abstraction into two more atomic notions. First, an abstraction form *introduces* the assumption, but does not allow its implicit *use* yet; this does not block computation – the assumption is as frozen. Second, an *elimination* construct on frozen assumptions makes them available for implicit use – but blocks reduction.

Since the elimination construct blocks reduction, it needs to be present at the term level; it refers to assumption names introduced by the abstraction construct, which therefore also needs to be in terms – but without blocking reduction. In fact, we just reuse λ-abstraction for that purpose: locked assumptions are term variables at a new type $[P]$ of *propositional truths*, representing the assumption that the proposition P is true.

We write $\diamond$ for the introduction of propositional truths, and $\delta(a, \phi.b)$ for its elimination. Informally,[4] if the proposition P holds in the typing context Γ, then $\diamond$ is a witness of P at type $[P]$. The corresponding elimination rule, $\delta(a, \phi.b)$ computes a at type $[P]$, while blocking the reduction of b, type-checked under the assumption $\phi : P$, until a turns into a concrete witness $\diamond$. Then, $\delta(\diamond, \phi.b)$ can be reduced to the pseudo-substitution $b[\diamond/\phi]$ whose effect is to remove all occurrences of ϕ in b, and, finally, the reduction of b can proceed.

With these new constructions, we may use standard abstraction $\lambda(x : [P])\,a$ to abstract over a proposition P without blocking the evaluation of a, which means that a cannot use P yet. In particular, a may be of the form $a[\delta(x, \phi.b_1), \delta(x, \phi.b_2)]$, allowing the *implicit* use of the proposition P in subterms b_1 and b_2, which cannot be reduced, while full reduction is still allowed in a.

Propositional truth elimination allows the user to express the fact that an assumption P may not actually be used directly at its abstraction site, but only "at some later time". Conversely, there are situations where an elimination on P is needed to type-check parts of a term a and is no longer needed to typecheck some subterm b of a. To enable reduction of b in such a case, we introduce *assumption hiding* hide ϕ in b, which enforces that the proposition variable ϕ will *not* be used implicitly in the subterm b. In exchange for losing this convenience, we regain the full reduction behavior for b.

---

[4] The language is formally defined in §2.

While assumption hiding has been introduced for programming reasons, it is also instrumental in restoring confluence. The loss of confluence happens when a substitution places a reducible term in an irreducible context. We may now restore confluence by inserting appropriate hidings during substitution when traversing proposition eliminators so as to preserve reducibility.

Contributions. The central, novel idea of our work is the interaction of the explicit and implicit modes of use of logical assertions in a programming calculus admitting full-reduction. From a theoretical point of view, implicitness was a somewhat-neglected design choice, and we propose a continuum between implicit and explicit uses thanks to *propositional truths* and *assumption hiding* (Section 2.2). It reveals, for example, that GADTs are fundamentally different from the usual algebraic datatypes. From a practical point of view, this gives the user flexible control over the scope of logical assumptions to prevent them from leaking into unrelated parts of his program—while retaining the convenience of their implicit invocation.

Another, more technical contribution is a new formal full-reduction calculus F_{th}, with inconsistent coercion abstraction that is confluent (Section 3.5). It is notable that the construction that regains confluence (*hiding*) was initially motivated by increasing the programmer's convenience.

Besides, there are several other contributions:

– We improve some details of the existing F_{cc} calculus, updating its mechanized soundness proof accordingly. Although coercion calculi in the spirit of F_{cc} are neither surface nor internal languages, they are good at exploring the design space; hence, even small improvements are valuable in the long term.

– We extend (3.5) the confluence proof technique of Takahashi [1995] so that it scales to larger calculi expressed in the Wright-Felleisen style, using reduction contexts to factor out common patterns and avoid a combinatorial increase in the number of cases.

– When translating between two given calculi, precisely establishing a bisimulation generally requires the use of an administrative variant of the target calculus; in our case, we need an administrative arrow type that is incompatible with the usual arrow type. While this is a common trick in the literature, its soundness proof is not as obvious as one would expect. We provide precise proofs that would be applicable to any calculus with several computational type constructors, *e.g.* arrows and products.

2   A Calculus with Propositional Truths

In this section, we formally present our calculus, F_{th} – with propositional *truths* and *hiding*. As another instance of calculus based on erasable coercions, it is strongly inspired by the previous work on F_{cc} by Cretin and Rémy [2014] and

$$\frac{}{\text{TERMVAR}}$$

TERMVAR
$\Gamma, x : \sigma, \Delta \vdash x : \sigma$

TERMLAM
$$\frac{\Gamma \vdash \tau : \star \quad \Gamma, x : \tau \vdash a : \sigma}{\Gamma \vdash \lambda(x)\, a : \tau \to \sigma}$$

TERMAPP
$$\frac{\Gamma \vdash a : \tau \to \sigma \quad \Gamma \vdash b : \tau}{\Gamma \vdash a\, b : \sigma}$$

TERMPROD
$$\frac{\Gamma \vdash a : \tau_1 \quad \Gamma \vdash b : \tau_2}{\Gamma \vdash (a, b) : \tau_1 * \tau_2}$$

TERMPROJ
$$\frac{\Gamma \vdash a : \tau_1 * \tau_2}{\Gamma \vdash \pi_i\, a : \tau_i}$$

TERMCOERCE
$$\frac{\Gamma, \Sigma \vdash a : \tau \quad \Gamma \vdash (\Sigma \vdash \tau) \triangleright \sigma}{\Gamma \vdash a : \sigma}$$

TERMWIT
$$\frac{\Gamma \vdash Q}{\Gamma \vdash \diamond : [Q]}$$

TERMASSUME
$$\frac{\Gamma \vdash a : [P] \quad \Gamma, \phi : P \vdash b : \sigma}{\Gamma \vdash \delta(a, \phi.b) : \sigma}$$

TERMHIDE
$$\frac{\Gamma \Vdash \Delta \quad \Gamma \vdash \exists \Delta \quad \Gamma, \Delta \vdash a : \tau}{\Gamma, \phi : P, \Delta \vdash \textbf{hide}\, \phi \, \textbf{in}\, a : \tau}$$

Fig. 1. F_{th} term typing judgment $\Gamma \vdash a : \tau$

$$
\begin{aligned}
a, b &::= x, y \ldots \mid \lambda(x)\, a \mid a\, a \mid (a, a) \mid \pi_i\, a &&\text{Terms}\\
&\quad \mid \diamond \mid \delta(a, \phi.a) \mid \textbf{hide}\, \phi \, \textbf{in}\, a\\
\tau, \sigma &::= \alpha, \beta \ldots \mid \tau \to \tau \mid \tau * \tau &&\text{Types}\\
&\quad \mid \forall(\alpha : \kappa)\, \tau \mid (\tau, \sigma) \mid \pi_i\, \tau \mid () \mid [P]\\
\kappa &::= \star \mid 1 \mid \kappa * \kappa \mid \{\alpha : \kappa \mid P\} &&\text{Kinds}\\
P, Q &::= \top \mid P \wedge P \mid \forall(\alpha : \kappa)\, P \mid \exists \kappa \mid (\Sigma \vdash \tau) \triangleright \tau &&\text{Prop.}\\
\Gamma, \Sigma, \Delta &::= \emptyset \mid \Gamma, x : \tau \mid \Gamma, \alpha : \kappa \mid \Gamma, \phi : P &&\text{Contexts}
\end{aligned}
$$

Fig. 2. Syntax of terms, types, kinds and propositions

follows the same global structure of judgments. Yet, we do not assume familiarity with F_{cc}.[5]

We first present the general structure of judgments and the constructs that are common to both F_{cc} and F_{th}, together with their typing rules (§2.1). We then detail the novel features of F_{th}, namely propositional truths and assumption hiding (§2.2). Last, we present the dynamic semantics of F_{th} (§2.3). In §2.4, we introduce a variant of F_{th} that is used to prove the soundness of F_{th} by translation to F_{cc} in several steps (§3).

2.1   Consistent Coercion Calculus

Cretin and Rémy [2014] use a general notion of *erasable coercions* with abstraction over consistent coercions to present different type system features, such as polymorphism, subtyping, and more in a common framework where these features can be easily composed together. The restriction that only *consistent* coercions can be abstracted over is key to *erasability*.

Our calculus has four syntactic categories: terms a, b; types τ, σ; kinds κ; and propositions P, Q. The syntax of each category and that of typing environments Γ, are described in Figure 2.

---

[5] F_{cc} also supports equi-recursive types; we left them out of this presentation as they are orthogonal to reduction under inconsistent assumptions. It is the only feature of F_{cc} as previously described that is absent from F_{th}.

The static semantics is given by four main judgments: a typing judgment $\Gamma \vdash a : \sigma$; a kinding judgment $\Gamma \vdash \sigma : \kappa$; a proposition satisfiability judgment $\Gamma \vdash P$; a coercion judgment $\Gamma \vdash (\Sigma \vdash \tau) \triangleright \sigma$; plus a context consistency judgment $\Gamma \vdash \exists \Delta$ and well-formedness judgments $\Gamma \Vdash t$ where t may be an environment, a kind, a proposition, or a coercion.

Terms. We first describe terms of the consistent subset of F_{th}, which are the terms of the untyped λ-calculus with products, extended with one additional construct for coercions. Other constructs for manipulating inconsistent assumptions, namely, propositional truth and assumption hiding will be presented in §2.2.

The term typing judgment is defined by the rules in Figure 1. The introduction and elimination rules for arrows (TERMVAR, TERMLAM, TERMAPP) and products (TERMPROD, and TERMPROJ) are standard.

A remarkable feature of coercion calculi is that there is exactly one rule that does not change the term (and thus does not influence the dynamic semantics): the coercion rule TERMCOERCE. All runtime-irrelevant typing constructions, such as subtyping conversion and polymorphism introduction and elimination, are factorized into coercions. To express polymorphism, these coercions are *typing* coercions $(\Sigma \vdash \tau) \triangleright \sigma$ rather than *type* coercions $\tau \triangleright \sigma$: they also affect the typing environment Γ in which the coercion is used, by extending Γ with Σ when typechecking the premise of type τ, as described by Rule TERMCOERCE.

This factorization has been explained in previous works of Cretin and Rémy [2014] and is orthogonal to our point of interest in the present paper, namely, the interplay between *program types* and *logical propositions* in a programming system. We thus focus our presentation on propositions in general rather than coercions, and propositional truths would naturally extend to many other program logics, such as arithmetic reasoning or general refinement types. Still, by maintaining a crisp separation between (Curry-style) *program terms* that compute and *derivations* on which we statically reason, consistent coercion calculi are good systems in which to think about implicit versus explicit uses of logic reasoning in program terms.

Coercions. Despite the fact that coercions are included in the syntactic class of propositions, there are still two separate judgments $\Gamma \vdash (\Sigma \vdash \tau) \triangleright \sigma$ and $\Gamma \vdash P$.

The coercion judgment is defined in Figure 3. Besides structural rules of reflexivity (COERREFL) and transitivity (COERTRANS), coercions have rules for polymorphism (type abstraction COERGEN and type application COERINST), and rules COERARROW, COERPROD, and COERWIT for distributivity of coercions under *computational* type constructors (those that describe the shape of terms and appear in the term typing judgment). Formulating rules for both polymorphism and distributivity under computational type constructors as coercions let us easily compose them: the F_η rules for instantiation of polymorphism under constructors naturally fall out as derived rules in consistent coercion calculi. Finally, Rule COERPROP injects any propositional proof of a coercion (seen as a proposition) into the coercion judgment – when the coercion context is consistent. We refer the reader to Cretin and Rémy [2014] for a detailed presentation.

CoerTrans

CoerRefl

$$\frac{\Gamma, \Sigma_1 \vdash (\Sigma_2 \vdash \tau_3) \rhd \tau_2 \qquad \Gamma \vdash (\Sigma_1 \vdash \tau_2) \rhd \tau_1}{\Gamma \vdash (\Sigma_1, \Sigma_2 \vdash \tau_3) \rhd \tau_1}$$

$$\Gamma \vdash \tau \rhd \tau$$

CoerGen

$$\frac{\Gamma \vdash \exists \kappa}{\Gamma \vdash (\alpha : \kappa \vdash \tau) \rhd \forall (\alpha : \kappa) \tau}$$

CoerInst

$$\frac{\Gamma \vdash \sigma : \kappa}{\Gamma \vdash \forall (\alpha : \kappa) \tau \rhd \tau [\sigma / \alpha]}$$

CoerArrow

$$\frac{\Gamma, \Sigma \vdash \tau' \rhd \tau \qquad \Gamma \vdash \tau' : \star \qquad \Gamma \vdash (\Sigma \vdash \sigma) \rhd \sigma'}{\Gamma \vdash (\Sigma \vdash (\tau \to \sigma)) \rhd (\tau' \to \sigma')}$$

CoerProd

$$\frac{\Gamma \vdash (\Sigma \vdash \tau) \rhd \tau' \qquad \Gamma \vdash (\Sigma \vdash \sigma) \rhd \sigma'}{\Gamma \vdash (\Sigma \vdash \tau * \sigma) \rhd \tau' * \sigma'}$$

CoerWit

$$\frac{\Gamma, \phi : P \vdash Q}{\Gamma \vdash [P] \rhd [Q]}$$

CoerProp

$$\frac{\Gamma \vdash (\Sigma \vdash \tau) > \sigma \qquad \Gamma \vdash \exists \Sigma}{\Gamma \vdash (\Sigma \vdash \tau) \rhd \sigma}$$

Fig. 3. Coercion judgment $\Gamma \vdash (\Sigma \vdash \tau) \rhd \sigma$

PropVar

$$\Gamma, \phi : P, \Delta \vdash P$$

PropAnd

$$\frac{\Gamma \vdash P_1 \qquad \Gamma \vdash P_2}{\Gamma \vdash P_1 \wedge P_2}$$

PropProj

$$\frac{\Gamma \vdash P_1 \wedge P_2}{\Gamma \vdash P_i}$$

PropGen

$$\frac{\Gamma \Vdash \kappa \qquad \Gamma, \alpha : \kappa \vdash P}{\Gamma \vdash \forall (\alpha : \kappa) P}$$

PropInst

$$\frac{\Gamma \vdash \forall (\alpha : \kappa) P \qquad \Gamma \vdash \tau : \kappa}{\Gamma \vdash P[\tau / \alpha]}$$

PropTrue

$$\Gamma \vdash \top$$

PropConv

$$\frac{\Gamma \vdash P \qquad P =_\beta P' \qquad \Gamma \Vdash P'}{\Gamma \vdash P'}$$

PropKind

$$\frac{\Gamma \vdash \tau : \{\alpha : \kappa \mid P\}}{\Gamma \vdash P[\tau / \alpha]}$$

PropInh

$$\frac{\Gamma \vdash \sigma : \kappa}{\Gamma \vdash \exists \kappa}$$

PropCoer

$$\frac{\Gamma \vdash (\Sigma \vdash \tau) \rhd \sigma \qquad \Gamma, \Sigma \vdash \tau : \star}{\Gamma \vdash (\Sigma \vdash \tau) > \sigma}$$

Fig. 4. Proposition satisfiability judgment $\Gamma \vdash P$

Notice how the introduction rule for polymorphism CoerGen requires the quantified-over kind κ to be inhabited—the proposition $\exists \kappa$ denoting kind inhabitation. This is the cornerstone of the distinction between *consistent* and *inconsistent* polymorphism: to abstract over a potentially-absurd kind or proposition (you have no inhabitation proof at hand), one must instead use the inconsistent abstraction, which changes the term as it blocks the reduction.

Kinding, Satisfiability, and Consistency. Figures 5, 4, and 6 present those three related judgments.

The proposition satisfiability judgment $\Gamma \vdash P$ is defined in Figure 4. Besides coercions $(\Sigma \vdash \tau) > \sigma$, the propositional features inherited from $\mathsf{F_{cc}}$ are relatively limited: there are the features used to subsume existing System F variants with some form of constrained quantification (F_η, $\mathsf{F}_{<:}$, MLF *etc.*), but more propositions could be added. The trivial true proposition, conjunction of propositions, and type-polymorphic propositions have obvious introduction and elimination rules.

The kind inhabitation proposition $\exists \kappa$ is true whenever kind κ is inhabited by some type σ; we use the judgment $\Gamma \vdash \exists \kappa$ instead of $\Gamma \vdash \sigma : \kappa$ when only

$$\frac{}{\Gamma, \alpha : \kappa, \Delta \vdash \alpha : \kappa} \text{ KindVar}$$

$$\text{KindArrow } \frac{\Gamma \vdash \tau : \star \quad \Gamma \vdash \sigma : \star}{\Gamma \vdash \tau \to \sigma : \star}$$

$$\text{KindProd } \frac{\Gamma \vdash \tau : \star \quad \Gamma \vdash \sigma : \star}{\Gamma \vdash \tau * \sigma : \star}$$

$$\text{KindWit } \frac{\Gamma \Vdash P}{\Gamma \vdash [P] : \star}$$

$$\text{KindAll } \frac{\Gamma \Vdash \kappa \quad \Gamma, \alpha : \kappa \vdash \tau : \star}{\Gamma \vdash \forall(\alpha : \kappa)\, \tau : \star}$$

$$\text{KindUnit } \frac{}{\Gamma \vdash () : 1}$$

$$\text{KindPair } \frac{\Gamma \vdash \tau : \kappa_1 \quad \Gamma \vdash \sigma : \kappa_2}{\Gamma \vdash (\tau, \sigma) : \kappa_1 * \kappa_2}$$

$$\text{KindProj } \frac{\Gamma \vdash \tau : \kappa_1 * \kappa_2}{\Gamma \vdash \pi_i\, \tau : \kappa_i}$$

$$\text{KindRefine } \frac{\Gamma \vdash \tau : \kappa \quad \Gamma, \alpha : \kappa \Vdash P \quad \Gamma \vdash P[\tau/\alpha]}{\Gamma \vdash \tau : \{\alpha : \kappa \mid P\}}$$

$$\text{KindForget } \frac{\Gamma \vdash \tau : \{\alpha : \kappa \mid P\}}{\Gamma \vdash \tau : \kappa}$$

$$\text{KindConv } \frac{\Gamma \vdash \tau : \kappa \quad \kappa =_\beta \kappa' \quad \Gamma \Vdash \kappa'}{\Gamma \vdash \tau : \kappa'}$$

Fig. 5. Kinding judgment $\Gamma \vdash \tau : \kappa$

$$\text{ContEmpty } \frac{}{\Gamma \vdash \exists \emptyset}$$

$$\text{ContTerm } \frac{\Gamma \vdash \exists \Delta \quad \Gamma, \Delta \vdash \tau : \star}{\Gamma \vdash \exists(\Delta, x : \tau)}$$

$$\text{ContType } \frac{\Gamma \vdash \exists \Delta \quad \Gamma, \Delta \vdash \exists \kappa}{\Gamma \vdash \exists(\Delta, \alpha : \kappa)}$$

$$\text{ContProp } \frac{\Gamma \vdash \exists \Delta \quad \Gamma, \Delta \vdash P}{\Gamma \vdash \exists(\Delta, \phi : P)}$$

Fig. 6. Context consistency judgment $\Gamma \vdash \exists \Delta$

consistency matters. It is defined in Figure 4. Inhabitation is lifted to whole contexts in Figure 6, as the judgment $\Gamma \vdash \exists \Delta$.

Kinding rules are defined in Figure 5. Kinding rules for base types are standard. The unit kind 1 is inhabited by the type-level trivial value (). Refinement kinds are the only construction introducing propositions in kinds—and thus in types: a refinement kind $\{\alpha : \kappa \mid P\}$ is inhabited by the types τ of kind κ such that the proposition $P[\tau/\alpha]$ holds. For example, the bounded quantification $\forall(\alpha \le \tau)\, \sigma$ can be expressed as $\forall(\alpha : \{\alpha : \star \mid \alpha \ge \tau\})\, \sigma$.

Product kinds allow quantifying over several kinds at once; in combination with refinement kinds, this gives an expressive and convenient way to use refinement conditions $P(\alpha, \beta)$ that depend on several variables. In particular, $\forall(\gamma : \{\gamma : \kappa_1 * \kappa_2 \mid P(\pi_1\, \gamma, \pi_2\, \gamma)\})\, \tau$ cannot be expressed in the general case as a double abstraction of the form $\forall(\alpha : \kappa_1) \forall(\beta : \{\beta : \kappa_2 \mid P(\alpha, \beta)\})\, \tau$; the consistency proof in the former case requires a witness $\gamma : \kappa_1 * \kappa_2$ that satisfies P, while the consistency proof of the second abstraction in the latter case requires to provide a witness $\beta : \kappa_2$ for *any* fixed (rigid) variable $\alpha : \kappa_1$. Depending on P, the first form may be consistent and the second inconsistent; to split bindings while keeping consistency, one has to constrain the domain of α by writing $\forall(\alpha : \{\alpha : \kappa_1 \mid \exists\{\beta : \kappa_2 \mid P(\alpha, \beta)\}\}) \forall(\beta : \{\beta : \kappa_2 \mid P(\alpha, \beta)\})\, \tau$, which inconveniently duplicates the proposition.

The reader may have recognized in refinement kinds a restricted form of (kind-level) dependent product. Indeed, this would exactly be a dependent product if the propositions were included into the kinds – dependent products would then

$$\Gamma \Vdash \star \qquad \Gamma \Vdash 1 \qquad \frac{\Gamma \Vdash \kappa_1 \quad \Gamma \Vdash \kappa_2}{\Gamma \Vdash \kappa_1 * \kappa_2} \qquad \frac{\Gamma \Vdash \kappa \quad \Gamma, \alpha : \kappa \Vdash P}{\Gamma \Vdash \{\alpha : \kappa \mid P\}} \qquad \Gamma \Vdash \top$$

$$\frac{\Gamma \Vdash \kappa}{\Gamma \Vdash \exists \kappa} \qquad \frac{\Gamma \Vdash P \quad \Gamma \Vdash Q}{\Gamma \Vdash P \wedge Q} \qquad \frac{\Gamma \Vdash \kappa \quad \Gamma, \alpha : \kappa \Vdash P}{\Gamma \Vdash \forall(\alpha : \kappa) P}$$

$$\frac{\Gamma \Vdash \Sigma \quad \Gamma, \Sigma \vdash \tau : \star \quad \Gamma \vdash \sigma : \star}{\Gamma \Vdash (\Sigma \vdash \tau) \triangleright \sigma} \qquad \Gamma \Vdash \emptyset \qquad \frac{\Gamma \Vdash \Delta \quad \Gamma, \Delta \vdash \tau : \star \quad x \notin \Gamma, \Delta}{\Gamma \Vdash \Delta, x : \tau}$$

$$\frac{\Gamma \Vdash \Delta \quad \Gamma, \Delta \Vdash \kappa \quad \alpha \notin \Gamma, \Delta}{\Gamma \Vdash \Delta, \alpha : \kappa} \qquad \frac{\Gamma \Vdash \Delta \quad \Gamma \Vdash P \quad \phi \notin \Gamma, \Delta}{\Gamma \Vdash \Delta, \phi : P}$$

Fig. 7. Well-formedness judgments

unify product kinds, refinement kinds, and conjunction of propositions. However, F_{cc}'s irrelevant handling of proposition proofs gives us very simple, clutter-free elimination rules for the refinement kind, which do not have to appear in the syntax of types. We occasionally benefit from that convenience.

The presence of type-level data structures (in our case product kinds) implies a need for type-level computation and identification of computationally-equal objects, in particular in rules PROPCONV and KINDCONV. The conversion rules for kinds and propositions allow to interchange well-formed objects equal upto β-reduction of projections, $\pi_i (\tau_1, \tau_2) =_\beta \tau_i$, and is closed by congruence and equivalence to all types, propositions, and kinds.

Well-formedness. Figure 7 presents the well-formedness judgments of F_{cc} for contexts, kinds and propositions, which are all standard.

2.2 Propositional Truths and Hiding

The type $[P]$ represents the type of dynamic witnesses that P is satisfied. The type-checking rule for types of the form $[P]$ are listed in Figure 1. This type is introduced by the token $\diamond$, a ground value that inhabits $[P]$ exactly when the proposition P is satisfied in the current typing environment (Rule TERMWIT). It is eliminated by the construction $\delta(a, \phi.b)$, where a must have a propositional truth type $[Q]$, and b is type-checked in an extended context where the assumption $\phi : Q$ is implicitly available (Rule TERMASSUME) – until it is hidden again in some subterm of the form $\mathsf{hide}\,\phi\,\mathsf{in}\,a'$.

As any other computational type, there is a distributivity coercion for propositional truths, Rule COERWIT (Figure 3), which following F_{cc} design principle, can be derived and justified from the context $\delta(\square, \phi.\diamond)$, as an η-expansion of the identity context $\square$. Rule COERWIT tells us that a witness for a proposition P of type $[P]$ can be coerced into a witness for a proposition Q of type $[Q]$ whenever P implies Q as a proposition.

Finally, the typing rule TERMHIDE (Figure 1) for $\mathsf{hide}\,\phi\,\mathsf{in}\,a$ in context $\Gamma, \phi : P, \Delta$ is a form of weakening of ϕ. It is only valid under the condition that

$$(\lambda(x)\, a)\, b \circ\!\!\rightarrow a[b/x]_\emptyset$$
$$\delta(\diamond,\, \phi.b) \circ\!\!\rightarrow b[\diamond/\phi]$$
$$\pi_i\,(a_1, a_2) \circ\!\!\rightarrow a_i$$

CONTEXT
$$\frac{a \circ\!\!\rightarrow b \qquad \mathtt{unguarded}(E)}{E[a] \longrightarrow E[b]}$$

$$c_{th} ::= \lambda(x)\, a \mid (a, b) \mid \diamond$$
$$d_{th} ::= \square\, b \mid \pi_i\, \square \mid \delta(\square,\, \phi.b)$$
$$\mathcal{E}_{th} \triangleq \{E[a] \mid \mathtt{unguarded}(E),\ a = d_{th}[c_{th}],\ a \not\circ\!\!\rightarrow\}$$

Fig. 8. Dynamic semantics of F_{th}

$$\mathtt{unguarded}(E) \triangleq (\mathtt{guard}_\emptyset(E) = \emptyset) \qquad\qquad \mathtt{guard}_S(\delta(E,\, \phi.b)) \triangleq \mathtt{guard}_S(E)$$
$$\mathtt{guard}_S(\lambda(x)\, E) \triangleq \mathtt{guard}_S(E) \qquad\qquad \mathtt{guard}_S(\delta(a,\, \phi.E)) \triangleq \mathtt{guard}_{S\cup\{\phi\}}(E)$$
$$\mathtt{guard}_S(a\, E) \triangleq \mathtt{guard}_S(E) \qquad\qquad \mathtt{guard}_S(\mathtt{hide}\, \phi\, \mathtt{in}\, E) \triangleq \mathtt{guard}_{S\setminus\{\phi\}}(E)$$
$$\mathtt{guard}_S(E\, a) \triangleq \mathtt{guard}_S(E) \qquad\qquad \mathtt{guard}_S(\square) \triangleq S$$

Fig. 9. Guards

$\Gamma \vdash \exists\Delta$ holds. This does not mean that Δ must be consistent (it can depend on variables in Γ that were introduced by blocking elimination), but that it is consistent *relative to* Γ.

Kind-level propositions. Propositional truths are named as such because they are constructed and abstracted over in terms, with an explicit elimination construction; by contrast with the *definitional* judgment $\Gamma \vdash P$ which only lives in typing derivations. Note that it is possible to see propositions as kinds: the kind $\{\alpha : 1 \mid P\}$, which could be abbreviated as $\langle P \rangle$, is inhabited by () exactly when the proposition P is satisfiable.

2.3 Dynamic Semantics

The dynamic semantics of F_{th} is defined in figures 8, 9 and 10. Because of assumption hiding, the notion of elimination contexts is non-standard: irreducible terms may have reducible subterms. In fact, the head β-reduction steps are also non-standard, because of the way hiding constructions are added during substitution of reducible values, so as to preserve confluence.

Reduction and Head Reduction. We define a β-reduction relation ($\longrightarrow$) that is congruent to reduction contexts, and a *head* β-reduction relation ($\circ\!\!\rightarrow$) that only applies to head β-redexes (Figure 8). Distinguishing head reductions is important for the confluence proof (Section 3.5). Those reductions are fairly standard, except for the use of non-standard notions of substitution, and a sidecondition on contexts described below.

Reduction Contexts. Full reduction is meant to allow any reduction path, so in general all one-hole term contexts E are reduction contexts. In F_{th}, subterms

$$x[b/x]_S \stackrel{\triangle}{=} \text{hide } S \text{ in } b \qquad\qquad x[\diamond/\phi] \stackrel{\triangle}{=} x$$

$$y[b/x]_S \stackrel{\triangle}{=} y \qquad (\text{if } y \neq x) \qquad (\lambda(x)\,a)[\diamond/\phi] \stackrel{\triangle}{=} \lambda(x)\,a[\diamond/\phi]$$

$$\diamond[b/x]_S \stackrel{\triangle}{=} \diamond \qquad\qquad (\text{hide } \phi \text{ in } a)[\diamond/\phi] \stackrel{\triangle}{=} a$$

$$(\lambda(y)\,a)[b/x]_S \stackrel{\triangle}{=} \lambda(y)\,a[b/x]_S \quad (\text{if } y \neq x) \qquad (\text{hide } \psi \text{ in } a)[\diamond/\phi] \stackrel{\triangle}{=} \text{hide } \psi \text{ in } a[\diamond/\phi]$$

$$(a\,a')[b/x]_S \stackrel{\triangle}{=} a[b/x]_S\,a'[b/x]_S \qquad\qquad (\text{if } \psi \neq \phi)$$

$$\delta(a,\,\phi.a')[b/x]_S \stackrel{\triangle}{=} \delta(a[b/x]_S,\,\phi.a'[b/x]_{S\cup\{\phi\}})$$

$$(\text{if } \phi \notin S)$$

$$(\text{hide } \phi \text{ in } a)[b/x]_S \stackrel{\triangle}{=} \text{hide } \phi \text{ in } a[b/x]_{S\setminus\{\phi\}}$$

Fig. 10. Hiding and unhiding substitutions

that are in the scope of an implicit assumption, or equivalently of a proposition variable, must still be blocked. We use an auxiliary function $\text{guard}_S(E)$ to compute the set of proposition variables, called the *guards*, under which the hole □ of the single-hole context E is blocked, extended with an initial set S. The predicate $\text{unguarded}(E)$ is then an abbreviation for the emptyness of $\text{guard}_\emptyset(E)$. *Reduction contexts* are the *unguarded* one-hole contexts E. For example, $\delta(a,\,\phi.\square)$ is not a reduction context, whereas $(\lambda(x)\,\square)$ and $\delta(w_1,\,\phi.\delta(w_2,\,\psi.\text{hide } \psi \text{ in hide } \phi \text{ in } \square))$ are. Unguardedness is checked by an additional premise in Rule CONTEXT.

Hiding Substitution $a[b/x]_S$. In order to preserve confluence it is essential that β-reduction preserves reducibility of subterms. A counter-example for confluence in F_{cc}, translated in F_{th}, is the term $(\lambda(x)\,\delta(y,\,\phi.x))\,b$. The problem is that b appears in a reducible position but would become irreducible after one head reduction step, *i.e.* in the term $\delta(y,\,\phi.b)$—with the usual notion of reduction.

Our solution is to define the reduction of λ-redexes using a non-standard notion of substitution, $a[b/x]_\emptyset$ that inserts assumption hidings as necessary for substituted terms to remain reducible. For instance, $\delta(y,\,\phi.x)[b/x]_\emptyset$ is equal to $\delta(y,\,\phi.\text{hide } \phi \text{ in } b)$. In general, this hiding substitution can be indexed by any guard, which is the list of logical assumptions made so far during term traversal.

The hiding substitution is defined on Figure 10 where $\stackrel{\triangle}{=}$ stands for definition equality and $\text{hide } S \text{ in } b$ is syntactic sugar for repeated hiding for all variables in the set S. Some cases that are simple traversals have been omitted.

Un-hiding Substitution $b[\diamond/\phi]$. When the witness a of a propositional elimination $\delta(a,\,\phi.b)$ is blocked over reduces to $\diamond$, we know that the proposition witnessed by a is true, and the reduction of b can proceed. We remove each occurrence of $\text{hide } \phi$ in each subterm of b, as it is not only unnecessary, but could also block now-reducible β-redexes if it remained: $a[\diamond/\phi]$ removes all occurrences of "$\text{hide } \phi$" in the term b while traversing b. It is also defined in Figure 10 – we give only a few representative cases. Note that the typing rule for assumption hiding guarantees that ϕ cannot appear in the subterm of $\text{hide } \phi$ – and this property is preserved by reduction.

Errors. Head reduction occurs when a destructor of some computational type meets a constructor of the same type. An *immediate error* is a term whose head is a destructor applied on a constructor of a different type. Figure 8 defines destructor contexts d_{th} and constructor terms c_{th}; the set $\mathcal{E}_{th}$ of *errors* is then defined as immediate errors occurring in a reduction context. Note that being stuck on a free variable is not an error, that errors may still further reduce (if it contains other reducible positions with valid redexes), and that a non-error term may contain an immediate error blocked under a propositional elimination, such as $\lambda(x)\,\delta(x,\,\phi.\pi_1\,\mathsf{true})$ in our introductory example of abstracting over an equality between `int` and `bool`.

Below, we define errors for variants of our calculus in the same way, generated from a definition of constructor terms, destructor contexts, and the head reduction relation. Given a language of terms with a reduction relation and a set of errors $\mathcal{E}$, we say that a term a is *sound* if no reduction sequence starting from a ends in $\mathcal{E}$.

2.4   Two Variants of F_{th}: F_t and F_{cc}

The soundness of F_{th} is proved by translation into F_{cc}, which has been proved sound [Cretin and Rémy, 2014]. In fact, the translation is in two steps, using an intermediate calculus F_t. Below, we formally define the calculi F_t and F_{cc}.

Removing assumption hiding. The language F_t is obtain from F_{th} by restricting to terms without hiding and by modifying the semantics of β-reduction so that it does not introduce hiding: the λ-reduction rule is $(\lambda(x)\,a)\,b \hookrightarrow a[b/x]$. (As a consequence, F_t is not confluent.)

The rest of the definition is unchanged; in absence of hiding, unguarded contexts $\mathsf{unguarded}(E)$ degenerate to a simpler, context-free definition that includes $\lambda(x)\,\square$ and $\delta(\square,\,\phi.b)$, but not $\delta(a,\,\phi.\square)$; and unhiding substitutions $b[\diamond/\phi]$ leave terms unchanged. Error terms $\mathcal{E}_t$ are the subset of $\mathcal{E}_{th}$ of terms without hiding.

Primitive inconsistent abstraction. F_{cc} uses a different primitive of *inconsistent abstraction* to work with inconsistent propositions, or rather potentially-uninhabited kinds. Its construction ∂a, mentioned in the introduction, blocks reduction immediately and has a type of the form $\Pi(\alpha : \kappa)\,\tau$ stating that it assumes a type $\alpha : \kappa$ while κ may be uninhabited. Conversely, $a \diamond$ unblocks a computation of type $\Pi(\alpha : \kappa)\,\tau$, whenever the kind κ can be shown inhabited. The head reduction rule is $(\partial a)\,\diamond \hookrightarrow a$, and reduction contexts are as before, if we consider that $\mathsf{guard}_S(\partial E)$ is defined as $\mathsf{guard}_{S \cup \phi}(E)$ where ϕ is a fresh propositional variable. The typing rules for those constructs are as follows:

$$
\begin{array}{cc}
\text{INCOHINTRO} & \text{INCOHELIM} \\[4pt]
\dfrac{\Gamma \Vdash \kappa \qquad \Gamma, \alpha : \kappa \vdash a : \tau}{\Gamma \vdash \partial a : \Pi(\alpha : \kappa)\,\tau} & \dfrac{\Gamma \vdash a : \Pi(\alpha : \kappa)\,\tau \qquad \Gamma \vdash \sigma : \kappa}{\Gamma \vdash a \diamond : \tau[\sigma/\alpha]}
\end{array}
$$

KindIncoh

$$\dfrac{\Gamma \Vdash \kappa \qquad \Gamma, \alpha : \kappa \vdash \tau : \star}{\Gamma \vdash \Pi(\alpha : \kappa)\, \tau : \star}$$

CoerIncoh

$$\dfrac{\Gamma, \alpha : \kappa', \Sigma \vdash \sigma : \kappa \qquad \Gamma \vdash \exists \Sigma \qquad \Gamma, \alpha : \kappa' \vdash (\Sigma \vdash \tau[\sigma/\alpha]) \rhd \tau'}{\Gamma \vdash (\Sigma \vdash \Pi(\alpha : \kappa)\, \tau) \rhd \Pi(\alpha : \kappa')\, \tau'}$$

The set $\mathcal{E}_{cc}$ of F_{cc} error terms is generated from its head reduction, its constructor terms (as in F_{th}, but without $\diamond$ and with $(\partial\, a)$) and its destructor contexts (as in F_{th}, but without $\delta(\Box,\ \phi.b)$ and with $(\Box\ \diamondsuit)$).

3   Soundness and Confluence

In this section, we prove our two technical results, confluence and soundness of F_{th}. The proof proceeds by a series of translations, proving that the source language is sound if the target language is sound as well. In §3.1, we recall the F_{cc} soundness result from previous work. In §3.2, we show a translation from the sublanguage F_t to (an administrative variant of) F_{cc}. This establishes soundness of F_t. In §3.5, we prove confluence of F_{th} using parallel reductions. For this we precisely define F_{th} multi-hole contexts, which give convenient tools to reason on its dynamic semantics. Finally, §3.6 proves soundness of F_{th}, using the tools introduced for the confluence proof.

3.1   Soundness of F_{cc}

F_{cc} comes with a (computer-checked) soundness proof for its (non-deterministic) reduction: starting from a well-typed term, no reduction path can lead to an erroneous stuck term. For deep reasons detailed in previous work, subject reduction (preservation of typing by reduction) does not hold for F_{cc}. Therefore, the soundness proof uses more semantic tools, building a model of the type system where types are sets of terms.

Theorem 1 (Previous work, Cretin and Rémy [2014] Soundness of F_{cc}).
Terms that are well-typed in F_{cc} in a consistent environment are sound. That is, if $\emptyset \vdash \exists \Gamma$ and $\Gamma \vdash a : \tau$, then a is sound.

3.2   Translating Propositional Truths to F_{cc}

We now define a translation $[\![\_]\!]$ of terms, types and judgment derivations into F_{cc}. Informally, the idea of the translation is a form of CPS-encoding: we can translate a witness of type $[P]$ into a continuation consuming any inconsistent abstraction $\Pi(\alpha : \langle P \rangle)\, \tau$ to return a τ. Witness construction $\diamond$ would become the elimination continuation $\lambda(x)\,(x\ \diamondsuit)$, while propositional elimination $\delta(a,\ \phi.b)$ uses the translation of a as a continuation: $[\![a]\!]\ (\partial\ [\![b]\!])$.

The actual translation on terms and types is close to the informal description above, with an important difference. The informal translation gives the expected computational behavior to well-typed terms, but has the defect of mapping some terms that are errors in F_t to terms in F_{cc} that may still further reduce: for example, $\delta((\lambda(x)\,x),\ \phi.y)$ is a stuck F_t term, but its translation $(\lambda(x)\,x)\ (\partial\,y)$ can be further reduced.

Because the soundness proof of F_{cc} is done semantically, and subject reduction does not hold for this calculus, it is important that our translation of F_t terms be well-behaved even on ill-typed terms. Indeed, we want to translate whole reduction paths starting from a known F_t term which, even if well-typed, may reduce to ill-typed terms (but, as we prove in this section, not an error). We also want to reason about the translation of those sound, ill-typed reducts.

To get a translation of $\delta((\lambda(x)\,x),\ \phi.y)$ that is stuck, we use a slight variant of F_{cc}, called $\mathsf{F}_{cc}^{\flat}$, for the target language. It is equipped with an "administrative" copy of the arrow type $(\tau \to^{\flat} \sigma)$, of λ-abstraction $(\lambda^{\flat}(x)\,a)$ and application $(a^{\flat}\,b)$. The type system and reduction semantics are exactly those of F_{cc}, with each rule (in the static and dynamic semantics) about λ-abstractions duplicated into an identical "administrative" variant.

The administrative $\lambda^{\flat}$ is entirely separate from the usual λ, and in particular $(\lambda(x)\,a)^{\flat}\,b$ and $(\lambda^{\flat}(x)\,a)\,b$ do not reduce and thus are both errors.

We can now formally define the translation from F_t to $\mathsf{F}_{cc}^{\flat}$, which makes judicious use of administrative constructions to preserve stuck terms. It is defined below on the F_t-specific constructions; it just preserves the structure of other constructions and translate their subterms (we use _ for unused variable bindings):

$$[\![P]\!] \triangleq \forall(\beta : \star)\,(\Pi(\_ : \{\_ : 1 \mid [\![P]\!]\})\,\beta) \to^{\flat} \beta$$
$$[\![\delta(a,\ \phi.b)]\!] \triangleq [\![a]\!]^{\flat}\,(\partial\ [\![b]\!])$$
$$[\![\diamond]\!] \triangleq \lambda^{\flat}(x)\,(x\ \diamond)$$
$$[\![\Gamma,\phi : P]\!] \triangleq [\![\Gamma]\!],\alpha : \{\_ : 1 \mid [\![P]\!]\}$$

For example, the translation of the error $\delta((\lambda(x)\,x),\ \phi.y)$ is now $(\lambda(x)\,x)^{\flat}\,(\partial\,y)$, which is also an error. One cannot build a counter-example of the form $\delta((\lambda^{\flat}(x)\,a),\ \phi.b)$ as the administrative variants are not part of the input language F_t. One can show by induction that the translation preserves errors and typing.

Lemma 1 (Error preservation of F_t). *A term a is an error in F_t if and only if $[\![a]\!]$ is an error in $\mathsf{F}_{cc}^{\flat}$.*

Lemma 2 (Typing preservation of F_t). *If $\Gamma \vdash a : \tau$ in F_t, then $[\![\Gamma]\!] \vdash [\![a]\!] : [\![\tau]\!]$ in $\mathsf{F}_{cc}^{\flat}$.*

Independently of typing preservation, we also prove a bisimulation property between F_t and $\mathsf{F}_{cc}^{\flat}$. It is not quite the case that any single-reduction step in F_t is turned into a single-reduction step of $\mathsf{F}_{cc}^{\flat}$, because the reduction of the translation of $\delta(\diamond,\ \phi.b)$, that is $(\lambda^{\flat}(x)\,(x\ \diamond))^{\flat}\,(\partial\ [\![b]\!])$, does an extra administrative $\lambda^{\flat}$-reduction step before the expected ∂-reduction. We define the relation $(\circ\to_{\flat})$ of *administrative* head β-reductions as the subset, in $\mathsf{F}_{cc}^{\flat}$, of reductions in $(\circ\to)$ of the form $(\lambda^{\flat}(x)\,a)^{\flat}\,b \circ\to a[b/x]$, and $(\longrightarrow_{\flat})$, the administrative β-reductions, its closure $(\circ\to_{\flat})$ under reduction contexts.

For any relation $(\mathcal{R})$, we define the relation $\mathcal{R}^?$ by $a\ \mathcal{R}^?\ b$ if and only if $(a\ \mathcal{R}\ b) \vee (a = b)$.

Lemma 3 (Bisimulation of F_t by $\mathsf{F}_{cc}^\flat$). *For any $a \longrightarrow a'$ in F_t, we have* $[\![a]\!] \longrightarrow_\flat^? b \longrightarrow [\![a']\!]$ *for some b in $\mathsf{F}_{cc}^\flat$.*

Conversely, if $[\![a]\!] \longrightarrow b$ in $\mathsf{F}_{cc}^\flat$, then $a \longrightarrow a'$ for some a' such that either $b = [\![a']\!]$ *or* $[\![a]\!] \longrightarrow_\flat b \longrightarrow [\![a']\!]$.

Corollary 1. *If $[\![a]\!]$ is sound in $\mathsf{F}_{cc}^\flat$, then a is sound in F_t.*

Corollary 2. *If $\mathsf{F}_{cc}^\flat$ is sound, then so is F_t.*

We note that we do not need the bisimulation result to establish soundness (relative to $\mathsf{F}_{cc}^\flat$), but only the forward simulation and the forward translation of errors.

The backward simulation shows that besides having the same soundness property, F_t and $\mathsf{F}_{cc}^\flat$ are also the same in term of number of reductions up to administrative steps: reasoning on program efficiency can therefore also be transposed from one to the other. In this respect, it may be important to remark that the one computation step we allowed to neglect, the administrative $\lambda^\flat$-reduction, never performs arbitrary duplication of its argument: whenever it appears in the translation, the $\lambda^\flat$-variable appears exactly once in the body. We could better enforce this invariant by using a linear type for this administrative construction, but this would require invasive changes to the type system.

3.3   Translating F_{cc} into F_t

Just as we presented a translation from F_t into (an administrative variant of) F_{cc} to prove F_t's soundness, it is possible and enlightening to translate F_{cc} back into (an administrative variant of) F_t – after fixing a minor defect of F_{cc} as previously presented. By lack of space, we have not included this translation in the conference version of this article, but it is available in the full version.

3.4   Soundness of the Administrative Arrow

To conclude, from the two previous sections, that F_{cc}'s soundness implies F_t's soundness and conversely, we need to prove the soundness of the administrative variants relative to their base calculus. While this is a common technique, its soundness proof is actually not as obvious as one would expect. By lack of space, the proof is only available in the full version.

This result proves, in particular, the soundness of $\mathsf{F}_{cc}^\flat$ relative to F_{cc}. Along with Corollary 2, establishing the soundness of F_t relative to $\mathsf{F}_{cc}^\flat$, and the already established soundness of F_{cc} (Theorem 1) this concludes the soundness proof of F_t.

$$\square_i : S \vdash \square_i : S \qquad \emptyset \vdash x : S \qquad \dfrac{\Gamma \vdash E : S\backslash\{\phi\}}{\Gamma \vdash \mathbf{hide}\,\phi\,\mathbf{in}\,E : S} \qquad \dfrac{\Gamma \vdash E_1 : S \qquad \Delta \vdash E_2 : S \cup \{\phi\}}{\Gamma, \Delta \vdash \delta(E_1, \phi.E_2) : S}$$

$$\dfrac{\Gamma \vdash E_1 : S \qquad \Delta \vdash E_2 : S}{\Gamma, \Delta \vdash (E_1\ E_2), (E_1, E_2) : S} \qquad \dfrac{\Gamma \vdash E : S}{\Gamma \vdash (\lambda(x)\,E), (\pi_i\,E) : S}$$

$$\dfrac{a = E[x]^i \qquad x \notin E \qquad (\square_i : S_i) \vdash E : S}{a[b/x]_S \stackrel{\triangle}{=} E[\mathbf{hide}\,S_i\,\mathbf{in}\,b]^i} \qquad \dfrac{a = E[\mathbf{hide}\,\phi\,\mathbf{in}\,b_i]^i \qquad \phi \notin E}{a[\diamond/\phi] \stackrel{\triangle}{=} E[b_i]^i}$$

Fig. 11. Guard analysis of multi-hole contexts

3.5   Confluence of F_{th}

Multi-hole Contexts. Figure 11 introduces a new judgment $(\square_i : S_i)^{i \in I} \vdash E : S$, that is a simple syntactic analysis of the *guards* of a multi-hole context, that is the set of propositional variables that block the reduction of each hole. The judgment can be read as "if the whole term is guarded by S, then the i-th hole $\square_i$ is guarded by S_i". A multi-hole context is just a term whose variables are, by convention, named $\square_i$ for some i in I, and which appear only once in the term; we enforce that latter invariant by using disjoint union for the context union Γ, Δ, which corresponds to a simple linear typing discipline. The notation $E[\_]^{i \in I}$ corresponds to a context with a family of holes indexed by i, and in contexts $\mathrm{II}^{i \in I} \Delta_i$ is the disjoint union of a family of contexts $(\Delta_i)^{i \in I}$. For sake of brevity, we often leave I implicit and just write i instead of $i \in I$.

Notice that $\mathrm{guard}_S(E)$ for a single-hole context is uniquely defined by $\square : \mathrm{guard}_S(E) \vdash E : S$. We also use multi-contexts to re-define the hiding substitution $a[b/x]_S$ defined in §2.3, and the hide-removing substitution $a[\diamond/\phi]$ used in the reduction rule for $\delta(\diamond, \phi.a)$.

Finally, a multi-context E is a *prefix* of E' (or a term, if E' has no holes) if E' can be obtained by substituting sub-contexts into the holes of E.

Parallel Reductions. We prove confluence using the Tait-Martin-Löf technique of parallel reductions, with a simple proof argument inspired by Takahashi [1995]. The idea of Takahashi is that parallel reduction (noted $a \Longrightarrow b$) for the simple λ-calculus with only arrows can be made deterministic by adding a redex-avoidance rule (the $a \neq (\lambda(\_)\,\_)$ hypothesis below meaning that a does not start with an abstraction) to the parallel reduction of application:

$$\dfrac{a \neq (\lambda(\_)\,\_) \qquad a \Longrightarrow a' \qquad b \Longrightarrow b'}{a\,b \Longrightarrow a'\,b'} \qquad \dfrac{a \Longrightarrow a' \qquad b \Longrightarrow b'}{(\lambda(x)\,a)\,b \Longrightarrow a'[b'/x]}$$

Without the redex-avoiding condition $a \neq \lambda(\_)\,\_$ in the application reduction rule, two reduction paths are available to each β-redex, performing the β-reduction or not. This gives a parallel condition that *may* reduce each redex in one step, and can thus subsume the usual single-step reduction relation by choosing to reduce exactly one redex. Takahashi remarks that the condition forces all redexes of the

$$\frac{E \nrightarrow \quad (\square_i : \emptyset)^i \vdash E : \emptyset \quad (a_i \Longrightarrow b_i)^i}{E[a_i]^i \Longrightarrow E[b_i]^i} \qquad \frac{(a_i \Longrightarrow a_i')^i \quad R[a_i']^i \overset{?}{\circ \!\!\rightarrow} b}{R[a_i]^i \circ\!\!\Rightarrow b}$$

$$R ::= (\lambda(x)\,\square_1)\,\square_2 \mid \pi_i\,(\square_1, \square_2) \mid \delta(\diamond,\ \phi.E_\phi[\mathbf{hide}\ \phi\ \mathbf{in}\ \square_i]^i)$$

$$(\text{with } \mathbf{unguarded}(E_\phi) \text{ and } \phi \notin E_\phi)$$

Fig. 12. Parallel reduction

term to be reduced (Gross-Knuth reduction), and that this modified relation trivially forces confluence of the parallel reduction, of which it is a special case.

Adapting Takahashi's idea to a Wright-Felleisen setting of head reduction and elimination contexts suggests a new formulation, which is to decompose a reducible term a into the form $E[b_i]^i$ where the multi-context E *is not reducible* when seen as a term – a generalization of the redex-avoiding condition. For the same reason that Takahashi's reduction was deterministic, the decomposition of a into $E[b_i]^i$ where E is not reducible and the b_i are head redexes is unique, since E is the largest head context that does not contain redexes. This decomposition still let us define parallel reduction, rather than only the Gross-Knuth reduction.

Figure 12 gives the definition of our parallel reduction $a \Longrightarrow b$, mutually defined with the *head* parallel reduction $a \circ\!\!\Rightarrow b$ that reduces head redexes . The notation $E \nrightarrow$ can be understood in term of the single-step reduction relation, when E is seen as a term as any other: $\neg(\exists E',\, E \longrightarrow E')$.

The parallel reduction of $E[a_i]^i$ only happens when the a_i are all redexes, as they must be related to some b_i by the head parallel reduction ($\circ\!\!\Rightarrow$) that only starts from head redexes $R[\square_i]^i$. Not all these redexes need to be reduced, however, as the head beta-reduction step $R[a_i']^i \circ\!\!\rightarrow^? b$ is optional. In particular, taking $R[a_i']^i = b$ for each redex shows that the relation ($\Longrightarrow$) is reflexive.

The restriction that the substituted terms a_i are redexes is crucial to modularly reason about reducibility; for if we substituted the non-redex $\lambda(x)\,a$ into the context ($\square\,b$), we would get a reducible result while neither the term nor the context were. No such situation can happen when the plugged terms are head redexes themselves, as redexes do not overlap.

Lemma 4 (Orthogonality). *Redexes do not overlap: If $(\square_i : \emptyset)^i \vdash E : \emptyset$ is a one-hole irreducible context distinct from $\square$, then for any redex contexts $R[\square_j]^j$ and $(R_i'[\square_k]^k)^i$ and families of terms $(a_j)^j$ and $(b_{i,k})^{i,k}$ we have $R[a_j]^j \neq E[R_i'[b_{i,k}]^k]^i$.*

The other lemma we need, to prove the unicity of the decomposition by irreducible contexts, is about the structure of reducible positions in a term or context.

Lemma 5 (Reducible positions). *For any guard S, any term a has a minimal non-empty prefix F such that $(\square_k : \emptyset)^k \vdash F[\square_k]^k : S$. For any non-empty prefix F' of a with $(\square_{k'})^{k'} \vdash F'[\square_{k'}]^{k'} : S$, F is a prefix of F'. Furthermore, F is irreducible.*

Lemma 6 (Unique decomposition of irreducible contexts). *If two parallel reductions have the same source, then they use the same context-redexes decomposition.*

In the general case of filling a context E with subterms that are not necessarily head redexes, we may still reason on reducibility of the subterms:

Lemma 7 (Composability of parallel reduction). *The following rule, which does not constrain the $(a_i)^i$ to be head redexes or E to be irreducible, is admissible:*

$$\frac{(\Box_i : \emptyset)^i \vdash E : \emptyset \qquad (a_i \Longrightarrow b_i)^i}{E[a_i]^i \Longrightarrow E[b_i]^i}$$

The last technical lemma we need closes a commutative diagram between parallel reduction and one-step head reduction.

Lemma 8 (Commutation $\Longrightarrow$ and $\circ\!\!\rightarrow$). *If $R[a_i]^i \Longrightarrow R[a_i']^i$ and both $R[a_i]^i \circ\!\!\rightarrow b$ and $R[a_i']^i \circ\!\!\rightarrow b'$, then $b \Longrightarrow b'$.*

Note that it is precisely that last lemma that failed with F_t or F_{cc} without a hiding construct. Indeed, with $b \Longrightarrow b'$, reducing $(\lambda(x)\,\delta(y,\ \phi.x))\,b$ to $\delta(y,\ \phi.b)$ does not allow closing the diagram to $\delta(y,\ \phi.b')$, while reducing to $\delta(y,\ \phi.\mathsf{hide}\,\phi\,\mathsf{in}\,b)$ allows closing the diagram by reducing to $\delta(y,\ \phi.\mathsf{hide}\,\phi\,\mathsf{in}\,b')$.

Theorem 2. *The parallel reduction relation $(\Longrightarrow)$ is confluent.*

Corollary 3 (Confluence). *The relation $(\longrightarrow^*)$ is confluent.*

3.6 Soundness of F_{th}

The soundness proof of F_{th} is again a translation from F_{th} to F_t with a forward simulation. Before getting to the translation proper, we need to study two transformations used to define it. *Hide-extrusion* (3.6) removes hiding from a F_{th} term, and its correctness property let us simulate forward reductions of the form $\delta(\diamond,\ \phi.b) \circ\!\!\rightarrow b[\diamond/\phi]$. *Hide-normalization* (3.6) strengthens the structure of hiding in a F_{th} term, in such a way that we can forward-simulate the other F_{th} reductions, despite the mismatch between F_{th}'s hiding substitution $a[b/x]_S$ and F_t's natural substitution. We finally prove F_{th}'s soundness (Theorem 3).

Hide-extrusion. In a language without hiding such as F_t, it is possible for the programmer to emulate the effects of hiding by extruding terms out of a blocking construction. Instead of $\delta(a,\ \phi.E[\mathsf{hide}\,\phi\,\mathsf{in}\,b])$, one can write $\mathtt{let}\ x_b = b\ \mathtt{in}\ \delta(a,\ \phi.E[x_b])$, where b appears in reducible position; we call this transformation *hide-extrusion*. In the general case, E may bind variables or block over other proposition variables, and the translation needs to be refined to preserve b's typing environment; for example, $\delta(a,\ \phi.\lambda(y)\,(f\ (\mathsf{hide}\,\phi\,\mathsf{in}\,b)))$ is hide-extruded into $\mathtt{let}\ x_b = \lambda(y)\,b\ \mathtt{in}\ \delta(a,\ \phi.\lambda(y)\,(f\ (x_b\ y)))$.

$$\delta(b,\ \phi.C[\textrm{hide}\ \phi\ \textrm{in}\ a]) \ \hookrightarrow\ \textrm{let}\ x = \textrm{abs}(C,a)\ \textrm{in}\ \delta(b,\ \phi.C[\textrm{app}(x,C)])$$

$$N ::= \Box\, b \mid a\, \Box \mid (\Box, b) \mid (a, \Box) \mid \pi_i\, \Box \mid \sigma_i\, \Box \mid \delta(\Box,\ \phi.b) \qquad \textrm{Non-binding contexts}$$

$$\textrm{abs}(\Box, a) \triangleq a \qquad\qquad\qquad\qquad \textrm{app}(a, \Box) \triangleq a$$
$$\textrm{abs}(N[C], a) \triangleq \textrm{abs}(C, a) \qquad\qquad\qquad \textrm{app}(a, N[C]) \triangleq \textrm{app}(a, C)$$
$$\textrm{abs}(\lambda(y)\, C, a) \triangleq \lambda(y)\, \textrm{abs}(C, a) \qquad\qquad \textrm{app}(a, \lambda(y)\, C) \triangleq \textrm{app}(a\ y, C)$$
$$\textrm{abs}(\delta(w,\ \psi.C), a) \triangleq \lambda(x_w)\, \delta(x_w,\ \psi.\textrm{abs}(C, a)) \qquad \textrm{app}(a, \delta(x_w,\ \psi.C)) \triangleq \textrm{app}(a \diamond, C)$$

Fig. 13. Hide-extrusion: translating **hide** back into plain F_t

Figure 13 gives a formal definition of the hide-extruding rewrite $a \hookrightarrow a'$ by defining two functions $\textrm{abs}(C, a)$, which abstracts a over all the variables bound in the context C, and $\textrm{app}(b, C)$, which closes such an abstracted b (when applied from under the context C) by applying it to the appropriate variables, accordingly. These definitions are factorized by a grammar N of context frames that do not bind any variable.

Lemma 9 (Typing preservation of hide-extrusion). *If a is well-typed in F_{th} and $a \hookrightarrow b$, then b is well-typed, at the same type.*

Lemma 10 (Hide-extrusion of errors). *If $a \hookrightarrow b$, then a is an error if and only if b is an error.*

Lemma 11 (Extrusion Reduction). $\textrm{app}(\textrm{abs}(C, a), C) \longrightarrow^* a[\diamond/\textrm{guard}_\emptyset(C)]$.

Each hide-extrusion rewrite removes exactly one **hide** ϕ from the source term; in particular, iterating hide-extrusion terminates, and gives a term without any hiding construct. We prove in 3.6 that this gives a forward simulation of F_{th} by F_t. This is easy to see in the simple case of extrusion through a reduction context E without any other **hide** ϕ:

$$\delta(b,\ \phi.E[\textrm{hide}\ \phi\ \textrm{in}\ a]) \ \hookrightarrow\ \textrm{let}\ x = \textrm{abs}(E, a)\ \textrm{in}\ \delta(b,\ \phi.E[\textrm{app}(x, E)])$$

The only reducible subterms of the source term are a and b. The subterms b and a are still reducible in the target term, since in particular $\textrm{guard}_\emptyset(\textrm{abs}(E, \Box))$ is empty. If b eventually reduces to $\diamond$, then the source becomes $E[a]$ which may reduce further. The target can reduce to $\textrm{let}\ x = \textrm{abs}(E, a)\ \textrm{in}\ E[\textrm{app}(x, E)]$, which itself reduces to $E[\textrm{app}(\textrm{abs}(E, a), E)]$, then to $E[a]$ by the previous lemma 11.

Hide-normalization. The remaining issue for a forward simulation of F_{th} by F_t is the difference between the substitutions used in β-reductions. If $(\lambda(x)\, a)\, b$ is related to some $(\lambda(x)\, a')\, b'$ by hide-extrusion, $a[b/x]_\emptyset$ may not be related to $a'[b'/x]$ in the general case, as the substitution in F_{th} may introduce new hiding constructs that have to be extruded again.

The idea of hide-normalization is to rewrite a term so that both substitutions *coincide*, by establishing the invariant that the guard of each bound variable occurrence is equal to the guard of its binder. For example, in $\lambda(x)\,\delta(y,\ \phi.x)$ the guard of x's binding site is $\emptyset$, while its occurrence has guard $\{\phi\}$. β-reducing this λ-abstraction would introduce a $\mathtt{hide}\,\phi$. We can statically rewrite it into $\lambda(x)\,\delta(y,\ \phi.\mathtt{hide}\,\phi\,\mathtt{in}\,x)$, which is equivalent (unblocking free variables doesn't affect reduction), and whose β-reduction doesn't introduce hiding.

In the general case, we define the hide-normalization function $\mathtt{H}(a)$ from F_{th} to F_{th}. It recursively traverses all subterms and is a direct mapping, except :

HideNormLam
$$\frac{(\square_i : S_i)^i \vdash C : \emptyset \qquad x \notin C}{\mathtt{H}(\lambda(x)\,C[x]^i) \triangleq \lambda(x)\,\mathtt{H}(C[\mathtt{hide}\,S_i\,\mathtt{in}\,x]^i)}$$

Lemma 12 (Type preservation of hide-normalization). *If a is well-typed in F_{th}, then $\mathtt{H}(a)$ is also well-typed, at the same type.*

Lemma 13 (Error preservation of hide-normalization). *A F_{th} term a is an error if and only if $\mathtt{H}(a)$ is an error.*

Lemma 14 (Hide-normalization is stable by reduction). *If $\mathtt{H}(a) \longrightarrow b'$, then b' is equal to $\mathtt{H}(b)$ for some b.*

Lemma 15 (Hide-normalization is a forward simulation). *If $a \longrightarrow b$ hen $\mathtt{H}(a) \longrightarrow \mathtt{H}(b)$.*

Soundness. Given a well-typed F_{th} term a, its hide-normalized form $\mathtt{H}(a)$ is still well-typed and has the same reduction behavior – errors included. We can compute the maximal hide-extrusion a' of $\mathtt{H}(a)$; this term is well-typed in both F_t and F_{th}. All that remains, to establish that the original term a is sound, is to forward-simulate any reduction path starting from a' in F_t. This should be done carefully, however, as it is *not* the case that the hide-extrusion of $\mathtt{H}(a)$ is itself hide-normal; it is, except on the subterms created by hide-extrusion. Hide-extrusion introduces linearly-used variables to preserve scoping, and of course does insert the appropriate hiding constructs, as its goal is to remove hiding. Fortunately, we do not need to hide-normalize the terms produced by hide-extrusion: they remain well-separated from other subterms during reduction, and are not affected by β-reduction from other parts of the term.

Theorem 3 (Soundness of F_{th}). *Every well-typed F_{th} term is sound.*

4 Related and Future Work

Related Work

Confluence and Weak Reduction. It appears to be folklore that there are three ways to get confluence in a weak reduction setting. One solution is to allow reduction under weak binders of subterms that do not use the bound variables [Çağman and Hindley, 1998]; we cannot apply this method in F_{th} as *uses* of propositions are not traced in terms. Another solution is to introduce explicit weakening when substituting under a binder, so as to preserve the non-dependency with bound variables. This corresponds to our hiding substitution. Finally, one may use explicit substitutions and forbid them from going through weak bindings, so that the substituted terms remain reducible. Interestingly, this happens to be precisely the computational behavior of terms used in our final soundness proof (from F_{th} to F_t), as a result of hide-normalization followed by hide-extrusion.

Some explicit substitution calculi [Kesner, 2007] also have explicit weakening for the purpose of understanding reduction behavior of substructural systems (*e.g.* linear proof nets) where weakening must be applied *maximally* and this invariant is preserved by reductions and substitutions. This gives a reduction semantics that is different from our more relaxed system.

Another system with explicit weakening is Adbmal by Hendriks and van Oostrom [2003]. Their weakening construct enforces a well-parenthesized order between introduction and weakening by removing not just one variable from scope, but also all variables introduced afterwards. Our hiding construct allows non-bracketed introduction-hiding sequences, which is more convenient for the programmer. Interestingly, we also considered a construct hide $\star$ in a to hide *all* propositional variables in scope and simplify the definition of hiding substitutions, but the local use of hide-normalization in the soundness proof suffices to get a similar effect. The *scope-extrusion* performed before Adbmal's β-reductions, which extrudes the weakening above a bound variable to also weaken its binder is also related to our hide-normalization technique.

System FC. The family of works on System FC [Sulzmann et al., 2007] is related to consistent coercion calculi in general, but also to our specific focus on implicit *v.s.* explicit use of potentially-inconsistent propositions. Sulzmann et al. [2007, §3.8] argue that explicit coercions often simplify understanding of compiler transformations by turning semantically incorrect hard-to-debug optimizations into scope-breaking transformations that are immediately detected. Implicit use of logical hypotheses is for user's convenience, and is not necessary in a compiler intermediate language. Yet, we claim that F_{th} retain some advantages in an explicit setting. The explicit reduction-blocking elimination reifies the semantic boundary into the syntax, which simplifies reasoning for both users and compiler designers. Another relation to our work is the march towards richer kind systems. F_{cc} includes a small set of features to demonstrate its usefulness, but the features studied for System FC, which moves towards a fully dependent type and kind sublanguage [Weirich et al., 2013], would also make sense in our setting. In particular, dependent kinds would make it natural to include propositions directly as kinds and merge product kinds, refinement kinds, and proposition conjunction as a single dependent product constructor. Consistency is known to be a pain point in the metatheory of System FC. It is neither needed nor traced in arbitrary coercion abstractions – they are not quite erasable as coercion abstraction blocks reduction. Yet, it is required for the axioms introduced at the toplevel – *e.g.* to model type families. An F_{cc}-inspired, more explicit treatment of consistency may structure System FC and provide optimization opportunities. We know that the mode of use of coercions corresponding to bounded quantification is consistent and can be erased; but the practical question of how to decide consistency is not answered in our work.

Future Work

Completing Consistent Coercion Calculi. In the process of our work we have encountered small glitches in F_{cc}: rules that we would expect to be derivable, and that were not in the current system. We have fixed them as necessary for F_{th}'s need, but some aspects could still be improved – adding η-expansions in the kind equality, and understanding whether the context consistency requirement of coercions could be removed, and recovered by a semantics argument.

Extraction. Coq's extraction process [Letouzey, 2004] compiles a language with full reduction and explicit uses of hypotheses into OCaml, a language with weak reduction and implicit uses of hypotheses; F_{th} might be a good intermediate language in which to express and study some of the optimizations happening during the translation—which is known to be difficult. More generally, the dependent type community is aware that computation is very different under arbitrary contexts [Brady et al., 2003]. We suspect that context consistency could be a good generalization of the "empty context" assumption. A distinction between *propositional* and *definitional* truths naturally arises in our framework and, interestingly, we have a use for abstracting over definitional truths – while dependent systems don't generally consider abstracting over definitional equalities.

Conclusion

We have introduced F_{th}, a consistent coercion calculus that blocks reductions under implicit inconsistent assumptions in a fine-grained manner. This solves both practical issues (user control over reducibility) and theoretical issues (confluence) with a previous calculus of erasable coercions, F_{cc}, and opens interesting perspectives on the study of full-reduction calculi for programming language design, the interplay between type systems and weak reduction strategies, and an explicit handling of consistency in dependent type systems.

Acknowledgments. Julien Cretin made many helpful remarks on our work; in particular, he suggested to introduce incoherent abstraction on *propositions* instead of *kinds*, which simplifies the presentation. We had fruitful discussions with Luc Maranget and Thibaut Balabonski about weak reduction; Thibaut Balabonski suggested the use of multi-hole contexts to unify guards (Figure 9) and hiding substitution (Figure 10), an idea we used in the proof of confluence.

References

Boespflug, M., Dénès, M., Grégoire, B.: Full reduction at full throttle. In: Certified Programs and Proofs, CPP (2011)

Brady, E., McBride, C., McKinna, J.: Inductive families need not store their indices. In: Berardi, S., Coppo, M., Damiani, F. (eds.) TYPES 2003. LNCS, vol. 3085, pp. 115–129. Springer, Heidelberg (2004)

Çağman, N., Hindley, J.R.: Combinatory weak reduction in lambda calculus. Theoretical Computer Science 198(1), 239–247 (1998)

Cardelli, L.: An implementation of FSub. Research Report 97 (1993)

Cretin, J.: Erasable coercions: a unified approach to type systems. PhD thesis, Université Paris-Diderot, Paris 7 (2014)

Cretin, J., Rémy, D.: System F with Coercion Constraints. In: Logics In Computer Science (LICS), ACM (July 2014)

Grégoire, B., Leroy, X.: A compiled implementation of strong reduction. ACM SIGPLAN Notices 37(9), 235–246 (2002)

Hendriks, D., van Oostrom, V.: adbmal. In: CADE (2003)

Kesner, D.: The theory of calculi with explicit substitutions revisited. In: Duparc, J., Henzinger, T.A. (eds.) CSL 2007. LNCS, vol. 4646, pp. 238–252. Springer, Heidelberg (2007)

Le Botlan, D., Rémy, D.: MLF: Raising ML to the power of System-F. In: ICFP 1980 (August 2003)

Letouzey, P.: A New Extraction for Coq. In: Geuvers, H., Wiedijk, F. (eds.) TYPES 2002. LNCS, vol. 2646, pp. 200–219. Springer, Heidelberg (2003)

Mitchell, J.C.: Polymorphic type inference and containment. Information and Computation 2/3(76) (1988)

Nordström, B., Petersson, K., Smith, J.M.: Programming in Martin-Löfs type theory. Oxford University Press, Oxford (1990)

Pottier, F.: A versatile constraint-based type inference system. Nordic Journal of Computing 7(4), 312–347 (2000)

Simonet, V., Pottier, F.: A constraint-based approach to guarded algebraic data types. TOPLAS 29(1) (2007)

Sulzmann, M., Chakravarty, M.M.T., Jones, S.L.P., Donnelly, K.: System f with type equality coercions. In: TLDI, pp. 53–66 (2007)

Takahashi, M.: Parallel reductions in lambda-calculus. Inf. Comput. 118(1), 120–127 (1995)

Vytiniotis, D., Jones, S.P.: Practical aspects of evidence-based compilation in system FC (2011)

Weirich, S., Hsu, J., Eisenberg, R.A.: System FC with explicit kind equality. In: ICFP, pp. 275–286 (2013)

Xi, H.: Dependent ml an approach to practical programming with dependent types. J. Funct. Program. 17(2), 215–286 (2007)

CoLoSL: Concurrent Local Subjective Logic

Azalea Raad, Jules Villard, and Philippa Gardner

Imperial College London
{azalea,j.villard,pg}@doc.ic.ac.uk

Abstract. A key difficulty in verifying shared-memory concurrent programs is reasoning compositionally about each thread in isolation. Existing verification techniques for fine-grained concurrency typically require reasoning about either the entire shared state or disjoint parts of the shared state, impeding compositionality. This paper introduces the program logic CoLoSL, where each thread is verified with respect to its subjective view of the global shared state. This subjective view describes only that part of the state accessed by the thread. Subjective views may arbitrarily overlap with each other, and expand and contract depending on the resource required by the thread. This flexibility gives rise to small specifications and, hence, more compositional reasoning for concurrent programs. We demonstrate our reasoning on a range of examples, including a concurrent computation of a spanning tree of a graph.

1 Introduction

A key difficulty in verifying properties of fine-grained concurrent programs is being able to reason compositionally about each thread in isolation, even though in reality the correctness of the whole system is the collaborative result of intricately intertwined actions of the threads. Such compositional reasoning is essential for: verifying large concurrent systems, since it allows them to be verified component-wise; verifying library code and incomplete programs, since one does not need to know about the context of execution; and replicating a programmer's intuition about why their implementations are correct, since their informal arguments are typically kept local and do not bring the whole system into the reasoning.

Rely-guarantee [16] and various combinations of rely-guarantee and separation logic reasoning [26,9,10,7,23,3] achieve compositional reasoning for increasingly difficult inter-thread interactions. However, we believe that, despite substantial progress, there are many examples where the specifications and proofs are not as concise as they might be; they are either too coarse or too contrived. We explore a different approach, introducing the program logic CoLoSL, using which we can give small specifications and proofs which correspond to the programmer's intuition of what shared resource is required by the thread.

Small specifications were emphasised in the work of O'Hearn, Reynolds and Yang on separation logic [19]. The original separation logic [21,15] achieves local, compositional reasoning for sequential heap-manipulating programs by splitting the heap into disjoint heaplets for describing the local resources required by a program. Compositionality then rests on two powerful mechanisms: a program can be

© Springer-Verlag Berlin Heidelberg 2015
J. Vitek (Ed.): ESOP 2015, LNCS 9032, pp. 710–735, 2015.
DOI: 10.1007/978-3-662-46669-8_29

specified using only those resources that it actually accesses; and this specification can be simply reused in any context that contains these resources. By making the specification as small as possible, we can ensure that it can be reused in a large set of possible contexts using the *frame rule*. This plays an important role in achieving scalable compositional reasoning, as each block of code can be proved in isolation and its small specification reused in larger contexts.

Concurrent separation logic extends this compositional reasoning to concurrent programs using the *disjoint concurrency rule*, where individual threads use both local resources private to the thread as well as static shared resources, which can be accessed by all threads through critical sections. Since then, there have been many extensions combining rely-guarantee reasoning with ideas from concurrent separation logic to reason about fine-grained concurrency: RGSep [26] introduced local resource and shared disjoint regions which are stable with respect to a static interference relation stating how the current thread and the environment can affect the region; CAP [7] and its extensions [24,23,3] in addition abstract the regions. This work has achieved substantial success; see the related work section for more details. Yet, we believe these approaches are not always able to provide small specifications for concurrent programs, impeding compositionality.

The problem is due to the rigid nature of the shared disjoint regions and their static interference, which limits how we can work with a concurrent data structure. A disjoint region typically either describes the entire data structure such as a linked list, or contains individual components of the data structure such as the nodes of a linked list. However, threads may have shared access to arbitrary parts of the data structure, which cannot be directly expressed in the reasoning. The interference associated with the region is static in that it is defined when the region is created and is fixed throughout its entire lifetime. However, threads having access to parts of a shared region need only know about the interference on those parts. In addition, it is not always possible to predict all future interactions associated with a shared region at the time of creation. Just as the original separation logic uses the frame rule to obtain small specifications for the *local* state, we seek an analogous framing mechanism on both the shared resource and its interference to obtain small specifications of the *shared* state.

For example, consider a linked list consisting of $n+1$ nodes accessed concurrently by n threads where the ith thread requires access to the ith and $(i+1)$st nodes. Current approaches cannot provide a small specification for each thread which captures just these two shared nodes and their interference.[1] Now consider a program whose threads manipulate subgraphs of a graph, such as a recursive concurrent spanning tree algorithm. Current approaches cannot give small specifications capturing just the subgraph manipulated by each thread due to intrinsic, unspecified sharing between subgraphs. Finally, consider a concurrent set implementation. In CoLoSL it is possible to describe the interference associated with a new element as it is added to the set. The interference on new elements need not be the same as before and may

---

[1] In [13], similar problems were encountered in an attempt to provide small specifications for a doubly-linked list implementation of a concurrent tree.

only be known at the time they are added. Existing approaches cannot accommodate such dynamic interference relations.

This paper introduces the program logic CoLoSL, which stands for Concurrent Local Subjective Logic. CoLoSL's semantic model is based on one global shared state, and each thread is verified with respect to its partial *subjective view* of this state. Each subjective (personalised) view comprises an assertion which describes *parts* of the shared global resource used by a thread, and an interference relation which describes how the thread and the environment may affect these parts. Subjective views may arbitrarily overlap with each other, with both their resources and interference expanding and contracting in accordance with what is required by the thread. Interestingly, this sometimes requires rewriting the interference relation so that the interference on the smaller state captures the same information as the interference on the bigger state. This expansion and contraction of subjective views provides small specifications for individual threads and local reasoning about a thread's shared state.

We demonstrate CoLoSL reasoning on a range of examples. The first example in §2 is Dijkstra's token ring mutual exclusion algorithm [5]. Regardless of the size of the ring, we are able to give a small specification to each thread in isolation such that each proof only mentions resources associated with two of its neighbours. This means that the proof can be largely reused when the implementation of the ring is changed to allow dynamic spawning of new participants. In §4, we study two further examples. The first is a concurrent spanning tree algorithm for graphs, where threads are recursively spawned on potentially overlapping subgraphs. We demonstrate that the flexible, overlapping subjective views of CoLoSL are just what we need. The second is a concurrent set module implemented using a hand-over-hand list-locking algorithm. Our CoLoSL reasoning for this set module improves on CAP reasoning [7] in that our small specifications can be dynamically extended to include other behaviour in future whereas the static CAP interference must predict all behaviour from the start.

Most of the technical details have been left out due to space constraints, and are provided in the accompanying technical report [20].

Related work. Jones' rely-guarantee reasoning [16] provided a breakthrough in compositional reasoning about concurrent programs. It described permitted interferences between threads using global rely and guarantee relations on one global shared state. This work is compositional in the sense that the guarantee of one thread is the rely of another. However, O'Hearn [18] demonstrated that, for concurrent reasoning to scale, it is important to work with small specifications based on local state and shared state rather than working with one global state. In particular, he introduced concurrent separation logic based on thread-local state and static invariants for the shared state.

Since then, there has been much recent work on compositional reasoning about fine-grained concurrent algorithms. RGSep [26] combined rely-guarantee reasoning with separation logic, with the state split into thread-local state and disjoint shared regions, and global rely-guarantee relations providing the interference on these regions. Deny-guarantee extends rely-guarantee and RGSep with permissions that can turn pieces of interference on or off during the proof of a program, as would be typically needed in a programming language with fork and join constructs instead of statically-scoped parallel composition. This has influenced the capabilities of CAP. CAP reasoning [7] increased compositionality by, instead of the global rely-guarantee relations, having static interference relations associated with the disjoint shared regions and capabilities in the local state for dynamically controlling the permitted interference, and concurrent abstract predicates for hiding implementation details. Several program logics have adapted this work to incorporate more abstraction [25], higher-order features [23] and more flexible capabilities [3]. We believe that all this work has limited compositional reasoning in the sense that it is only possible to frame off local state and unused shared regions. It is not possible to frame within regions nor frame inside the interference relations. (It is, of course, possible to weaken the region assertion by the logical implication $P * Q \Rightarrow P * true$, but this is not the same as being able to shrink regions *and* their interference to just work with P.)

Meanwhile, a different breakthrough in compositionality came from Feng's local rely-guarantee (LRG) reasoning [10]. The LRG model comprises one global shared state, with assertions describing disjoint but flexible parts of this state and rely-guarantee relations determining how the thread and environment can affect these partial states. With LRG, it is possible to frame off disjoint parts of shared state and their associated disjoint rely-guarantee relations, but, as noted by its author, the strong disjointness restrictions make this approach only applicable for disjoint modules in the code. We took significant inspiration from LRG, combining the flexibility of the LRG approach with our subjective views to reason locally about overlapping shared states.

Finally, Fine-grained Concurrent Separation Logic (FCSL) [17] of Nanevski *et al.* explores a different notion of region called *concurroids*. Like the regions in CAP, concurroids describe disjoint pieces of shared state together with their static interference. It is therefore not possible to frame within concurroids. The term "subjectivity" in FCSL refers to the fact that, unlike CAP, concurroids have three parts: the "joint" part; and the disjoint "self" and "other" parts. Although FCSL did not influence CoLoSL, it does highlight an interesting point regarding resource transfer between regions: in CAP, communication between regions is achieved indirectly via the local state; in FCSL, communication between concurroids is achieved directly through dangling external transitions; and, in CoLoSL, communication between compatible subjective views can be achieved by merging the two views.

2   Informal Development

We sketch a proof of an implementation of Dijkstra's token ring mutual exclusion algorithm, which pioneered *self-stabilising* distributed algorithms [5]. Our proof highlights the main reasoning principles of CoLoSL and results in *small specifications* for each participant in the ring. Besides their concision, we show how small specifications are robust against non-trivial changes to the program.

The algorithm assumes a network of $n+1$ machines arranged in a ring, with a designated *master* machine and n slave machines. Each machine maintains a

$$\boxed{P * Q}_I \Rightarrow \boxed{P}_I \qquad \text{(FORGET)} \qquad\qquad \boxed{P}_{I_1} * \boxed{Q}_{I_2} \Rightarrow \boxed{P \uplus Q}_{I_1 \cup I_2} \quad \text{(MERGE)}$$

$$(P \Rightarrow Q) \Rightarrow \boxed{P}_I \Rightarrow \boxed{Q}_I \quad \text{(WEAKEN)} \qquad I \sqsubseteq^P I' \text{ implies } \boxed{P}_I \Rightarrow \boxed{P}_{I'} \qquad \text{(SHIFT)}$$

$$\boxed{P}_I \Rightarrow \boxed{P}_I * \boxed{P}_I \qquad \text{(COPY)} \qquad P \copyright I \text{ implies } P \Rightarrow \exists A, A'. [A] * \boxed{P * [A']}_I$$
$$\text{(EXTEND)}$$

$$\frac{\{P_1\}\, \mathbb{C}_1\, \{Q_1\} \qquad \{P_2\}\, \mathbb{C}_2\, \{Q_2\}}{\{P_1 * P_2\}\, \mathbb{C}_1 \parallel \mathbb{C}_2\, \{Q_1 * Q_2\}} \; \text{PAR} \qquad \frac{P \Rightarrow P' \quad \vdash \{P'\}\, \mathbb{C}\, \{Q'\} \quad Q' \Rightarrow Q}{\vdash \{P\}\, \mathbb{C}\, \{Q\}} \; \text{CONSQ}$$

Fig. 1. Main reasoning principles and rules of CoLoSL

local counter and has access to the value of its left neighbour's counter; the state of the system consists of all $n+1$ counters. Starting in an arbitrary state, the network eventually stabilises to legitimate states [4], with the following global property: in the ith legitimate state, machines 0 to $i-1$ (machine 0 being the master) have some value $v+1$, and all others have value v. In the ith legitimate state, only the ith machine (indicated by • below) can make progress: it increments its counter by 1 and takes the system to the next legitimate state $(i+1 \bmod n+1)$. For a ring at address x, the ith legitimate state is depicted below (for both $i > 0$ and $i = 0$).

x	$x+1$		$x+i-1$	$x+i$		$x+n$
v+1	v+1	$\cdots$	v+1	v •	$\cdots$	v

x	$x+1$		$x+n$
v •	v	$\cdots$	v

In [20], we outline a proof of the token ring's self-stabilisation phase, which shows that the system always converges to a legitimate state. We also provide a proof of the token ring as a mutual exclusion mechanism, where a machine holding • may gain ownership of a shared resource. For simplicity, here we focus on the case where the token ring is in the 0th legitimate state, with all counters holding value 0. Our proof makes use of the CoLoSL principles laid out in Fig. 1, together with the usual concurrency rule of separation logic [18] and the rule of semantic consequence from the views framework [6]. We introduce them informally as needed, and discuss them in more detail in §3.

CoLoSL introduces a new assertion $\boxed{P}_I$ called a *subjective view*, which comprises a *subjective assertion* P describing a *part* of the global shared state and an *interference assertion* I characterising how this partial shared state may be changed by the thread or the environment. Similar to the interference assertions of CAP [7], I declares actions of the form $[a] : Q \rightsquigarrow R$, where a thread in possession of the $[a]$ capability in its local state may carry out its transition and update parts of the shared state satisfying Q to a state satisfying R. Assertions in Hoare triples must be *stable* with respect to the interference from the environment: that is, robust with respect to the interference assertion I.

Consider the program ring(x) defined in Fig. 2 which represents a token ring with $n+1$ machines. It is written in pseudo-code resembling C with additional constructs for concurrency: atomic sections $\langle \_ \rangle$ which declare that code behaves atomically; and parallel composition $\_ \parallel \_$ which spawns threads then waits until they complete. In our example, the $n+1$ threads run in parallel until all counters

ring (x)

$// \left\{ x \Rightarrow x * [m_x] * [s_{x+1}] * \cdots * [s_{x+n}] * \boxed{n \Rightarrow n * x \mapsto 0 * x{+}1 \mapsto 0 * \cdots * x{+}n \mapsto 0}_I \right\}$

{ master(x) || slave(x+1) || $\cdots$ || slave(x+n);

} $// \left\{ x \Rightarrow x * [m_x] * [s_{x+1}] * \cdots * [s_{x+n}] * \boxed{n \Rightarrow n * x \mapsto 10 * x{+}1 \mapsto 10 * \cdots * x{+}n \mapsto 10}_I \right\}$

master (x)

$// \left\{ \boxed{\begin{array}{l} x \Rightarrow x * [m_x] * \\ \boxed{n \Rightarrow n * x \mapsto 0 * x{+}n \mapsto 0} \end{array}}_{M_x'} \right\}$

{ while (*x != 10)

$// \left\{ \boxed{\begin{array}{l} x \Rightarrow x * [m_x] * \\ \boxed{\begin{array}{l} n \Rightarrow n * \exists v.\, x \mapsto v * \\ (x{+}n \mapsto v \vee x{+}n \mapsto v{-}1) \end{array}} \end{array}}_{M_x'} \right\}$

    { ⟨if (*x == *(x+n))
            *x = *x + 1;⟩ }

} $// \left\{ \boxed{\begin{array}{l} x \Rightarrow x * [m_x] * \\ \boxed{\begin{array}{l} n \Rightarrow n * x \mapsto 10 * \\ (x{+}n \mapsto 10 \vee x{+}n \mapsto 9) \end{array}} \end{array}}_{M_x'} \right\}$

slave (x)

$// \left\{ \boxed{\begin{array}{l} x \Rightarrow x * [s_x] * \\ \boxed{x \mapsto 0 * (x{-}1 \mapsto 0 \vee x{-}1 \mapsto 1)} \end{array}}_{S_x'} \right\}$

{ while (*x != 10)

$// \left\{ \boxed{\begin{array}{l} x \Rightarrow x * [s_x] * \\ \boxed{\begin{array}{l} \exists v.\, x \mapsto v * \\ (x{-}1 \mapsto v \vee x{-}1 \mapsto v{+}1) \end{array}} \end{array}}_{S_x'} \right\}$

    { ⟨if (*x != *(x-1))
            *x = *(x-1);⟩ }

} $// \left\{ \boxed{\begin{array}{l} x \Rightarrow x * [s_x] * \\ \boxed{\begin{array}{l} x \mapsto 10 * \\ (x{-}1 \mapsto 10 \vee x{-}1 \mapsto 11) \end{array}} \end{array}}_{S_x'} \right\}$

$s_x \triangleq [s_x] \colon \exists v.\, x \mapsto v * x{-}1 \mapsto v{+}1 \qquad\qquad \rightsquigarrow x \mapsto v{+}1 * x{-}1 \mapsto v{+}1$

$s_x' \triangleq [s_x] \colon \exists v.\, x{+}1 \mapsto v * x \mapsto v * x{-}1 \mapsto v{+}1 \rightsquigarrow x{+}1 \mapsto v * x \mapsto v{+}1 * x{-}1 \mapsto v{+}1$

$m_x \triangleq [m_x] \colon \exists v, n.\, n \Rightarrow n * x \mapsto v * x{+}n \mapsto v \quad \rightsquigarrow n \Rightarrow n * x \mapsto v{+}1 * x{+}n \mapsto v$

$m_x' \triangleq [m_x] \colon \exists v, n.\, n \Rightarrow n * x{+}1 \mapsto v * x \mapsto v * x{+}n \mapsto v \rightsquigarrow$
$\qquad\qquad\qquad n \Rightarrow n * x{+}1 \mapsto v * x \mapsto v{+}1 * x{+}n \mapsto v$

$l_x' \triangleq [s_x] \colon \exists v, n.\, n \Rightarrow n * x \mapsto v{+}1 * x{+}n \mapsto v * x{+}n{-}1 \mapsto v{+}1 \rightsquigarrow$
$\qquad\qquad\qquad n \Rightarrow n * x \mapsto v{+}1 * x{+}n \mapsto v{+}1 * x{+}n{-}1 \mapsto v{+}1$

$I \triangleq \{m_x, s_{x+1}, \ldots, s_{x+n}\} \qquad M_x' \triangleq \{m_x, l_x'\} \qquad F_x' \triangleq \{s_x, m_{x-1}'\} \qquad S_x' \triangleq \{s_x, s_{x-1}'\}$

Fig. 2. A proof sketch of the token ring in CoLoSL. Assertions in lines starting with //
describe the local state and the subjective shared state at the relevant program points.
The proof of slave(x) applies to all slaves expect the first one (called the foreman) in
the parallel composition, where S_x' is replaced by F_x'.

reach value 10. While the implementation of all the slave threads are identical,
we shall see that the proof of the first slave in the ring (at $x{+}1$) is slightly
different from the others. We henceforth refer to the first slave thread as the
foreman. Let us proceed with the proof of the other slaves.

Proof of the slaves. Let us temporarily forget about the proof outline of Fig. 2
and attempt to prove slave(x) in isolation, in the spirit of local reasoning. Since
slave(x) inspects the value of its counter pointed to by x and compares it against
the counter at x-1 (its left neighbour in the ring), a tempting precondition for
slave(x) would be one describing just these two locations, e.g.

$$ x \Rightarrow x * [s_x] * \boxed{x \mapsto 0 * (x{-}1 \mapsto 0 \vee x{-}1 \mapsto 1)}_{S_x} \qquad S_x = \{s_x, s_{x-1}\} $$

The above assertion comprises: a) a variable assertion stating that the thread
locally owns variable x with value x (using the variables-as-resource model [1]);
b) a *capability* $[s_x]$ that allows it to perform the associated s_x action (see below);
and c) a subjective view of the shared state: x points to 0, and $x{-}1$, its left

neighbour, points to either 0 or 1 since it might already have incremented its own counter. The $*$ connective used between assertions is that of separation logic and means that the assertions describe *disjoint* pieces of state. The interference assertion associated with the subjective view is captured by S_x and consists of two actions: s_x and s_{x-1}, where s_x represents an increment of the contents of x under the condition that its value is one less than the value at address $x-1$ (see Fig. 2). Since the current thread owns $[s_x]$ locally, only it can perform s_x. On the other hand, the capability $[s_{x-1}]$ is not locally owned, thus the environment could potentially perform the associated action whenever its precondition (on the left-hand side of $\rightsquigarrow$) is satisfied. Upon closer inspection, since this subjective view says nothing of the value of the cell at address $x-2$, s_{x-1} could potentially *always* fire. The assertion is thus not *stable*: nothing prevents the counter at $x-1$ from incrementing beyond value 1. A weaker, stable assertion is thus:

$$\mathrm{x} \Rightarrow x * \boxed{[s_x]} * \boxed{\exists v.\, x \mapsto 0 * x-1 \mapsto v}_{S_x}$$

Fortunately, we can do better and obtain a stronger small precondition. Let us first step back and think again at the level of the whole algorithm. As the programmer knows, the situation above cannot happen as $x-2$ can only be at most one ahead of x itself. We can thus replace S_x by S'_x and give a stronger stable precondition that captures just what we want, as in Fig. 2. The proof of slave(x) is now relatively straightforward. By inspection (or using the rules of §3.3), the invariant of the while loop and the postcondition are also stable. The atomic section temporarily "opens the box" to perform action s_x then "closes back the box", and preserves the invariant. The final postcondition of slave(x) follows from the invariant and the boolean expression of the loop.

Proof of the master and foreman. The proof sketches of the master and foreman threads given in Fig. 2 are analogous. As the first slave, the foreman has to account for interference from the master instead of another slave. Moreover, the master and its associated action m_x have access to variable n holding the current size of the ring, since they depend on the value of the counter at address $x + n$.

Proof of the ring. The precondition of ring(x) states that it owns all capabilities, which will be distributed amongst the $n+1$ threads; the global variable n $\Rightarrow n$ is shared, as are all $n+1$ counters, initialised to 0. The interference I associated with the subjective view consists of the actions of the $n+1$ threads. Because ring(x) has a global view of the state of the ring (and moreover all capabilities are held locally), the s_{x+i} actions are enough to guarantee stability.

Let us write $\boxed{P}_I$ for this initial subjective view. This assertion may be freely duplicated using the COPY principle of Fig. 1 and each thread is given a copy together with the appropriate capability using the usual PAR rule of concurrent separation logic. For instance, the thread running slave(x+i), for $i > 1$, gets $[s_{x+i}] * \boxed{P}_I$. This assertion does not match the precondition of slave(x+i) just yet. Using the principles of Fig. 1, we can weaken the assertion as such:

$$\boxed{P}_I \stackrel{(\text{SHIFT})}{\Rightarrow} \boxed{P}_{(I\setminus\{s_{x+i-1}\})\cup\{s'_{x+i-1}\}}$$

$$\stackrel{(\text{FORGET})}{\Rightarrow} \boxed{x+i-1 \mapsto 0 * x+i \mapsto 0}_{(I\setminus\{s_{x+i-1}\})\cup\{s'_{x+i-1}\}}$$

$$\stackrel{(\text{SHIFT})}{\Rightarrow} \boxed{x+i-1 \mapsto 0 * x+i \mapsto 0}_{\{s'_{x+i-1}, s_{x+i}\}}$$

$$\stackrel{(\text{WEAKEN})}{\Rightarrow} \boxed{x+i \mapsto 0 * (x+i-1 \mapsto 0 \vee x+i-1 \mapsto 1)}_{\{s'_{x+i-1}, s_{x+i}\}}$$

We start by exchanging the action s_{x+i-1} of I for the stronger action s'_{x+i-1} using the SHIFT principle. In general, SHIFT allows us to replace I with any interference assertion I' that has the same observable effect as far as the subjective assertion P is concerned (written $I \sqsubseteq^P I'$). In this instance, the actions s_{x+i-1} and s'_{x+i-1} have the same effect according to P, as discussed in the proof of slave(x). As such, rewriting s_{x+i-1} as s'_{x+i-1} merely reflects stronger knowledge about how $x+i$ and $x+i-2$ are related through $x+i-1$. In particular, $I \sqsubseteq^P (I \setminus \{s_{x+i-1}\}) \cup \{s'_{x+i-1}\}$ as required.

Next, because subjective views only describe *parts* of the shared state, we can use the FORGET principle to obtain a weaker view of the shared state, in this case a view that ignores all cells in the ring except for those at addresses $x+i-1$ and $x+i$. With all other cells out of the subjective view, their actions no longer have observable effects on the assertion, since they leave $x+i-1$ and $x+i$ unchanged. We can thus apply the SHIFT principle again to *frame off* those actions and obtain $S'_{x+i} = \{s'_{x+i-1}, s_{x+i}\}$.

Finally, we weaken the resulting subjective view so that it is stable with respect to S'_{x+i}, i.e. preserved by those of its actions that the environment may perform (here, s'_{x+i-1}). This yields the precondition of slave(x+i) as in Fig. 2. The preconditions of the master and foreman threads can be derived analogously.

Once all threads have completed their operations, we join up their postconditions using the MERGE principle, which embodies a crucial feature of CoLoSL: different subjective views *overlap*. The overlapping conjunction ⊎ between two assertions means that the two assertions describe potentially *overlapping* pieces of state. In particular, $A * B \Rightarrow A \uplus B$ and $A \wedge B \Rightarrow A \uplus B$. This connective has been used in the past to reason about sharing in data structures [22,11,14]. Since $\vee$ distributes over ⊎, the subjective view simplifies to $\boxed{x \mapsto 10 * x+1 \mapsto 10 * \cdots * x+n \mapsto 10}_{I'}$, where $I' = M'_x \cup F'_{x+1} \cup S'_{x+2} \cup \cdots \cup S'_{x+n}$. Finally, since $I' \sqsubseteq^{n\rightarrow n * x \mapsto 10 * x+1 \mapsto 10 * \cdots * x+n \mapsto 10} I$, we get the postcondition of ring(x) by the SHIFT principle. This concludes our CoLoSL proof of ring(x).

Small specifications and proof reuse. Our expansion and contraction of subjective views, in particular with the shifting of interference assertions at key places, enables us to confine the specification and verification of each thread to just the resources they need. Such small specifications make proofs robust against changes to each thread's environment, and provide more opportunities for proof reuse.

For instance, let us now add a thread that dynamically grows the ring by spawning extra slaves to the parallel composition of ring(x) (the details can be found in [20]). When the ring has $n+1$ machines, we use the EXTEND principle as follows to add a new slave (at $x+n+1$) to the shared state with the associated interference relation and capability.[2]

$$(x+n+1) \mapsto v \overset{(\text{EXTEND})}{\Rightarrow} \exists \mathsf{s}_{x+n+1}. \, [\mathsf{s}_{x+n+1}] * \boxed{x+n+1 \mapsto v}_{\{s_{x+n+1}\}}$$

Here, the *view shift* [6] (or *repartitioning* [7]) $P \Rrightarrow Q$ means that an instrumented (logical) state satisfying P may be changed to Q as long as the underlying machine state does not change. In particular, $(P \Rrightarrow Q) \Rightarrow (P \Rrightarrow Q)$. The point is that our proof changes only minimally to accommodate the new program: the proof of master(x) accounts for new interference on $\mathsf{n} \Rrightarrow n$ since the environment can grow the ring, hence mutate n; the proofs of other threads can be directly reused.

In contrast, in existing CAP-like approaches [7,3], both $\mathsf{n} \Rrightarrow n$ and the global interference relation are observed by all threads. As such, with the above extension, the global interference relation needs to change (to include the interference on $\mathsf{n} \Rrightarrow n$), and the proofs of *both* master and slave threads need to be adapted.

3   CoLoSL

We now give a more formal overview of how to use CoLoSL for program verification, eschewing some details of the semantics for lack of space, while still providing enough ingredients to carry program proofs. We describe the underlying model of CoLoSL, give the semantics of CoLoSL assertions, then present proof rules to reduce the various obligations typically encountered in proofs (namely, the side-conditions of SHIFT and EXTEND, and stability checks) to classical separation logic entailments that, in particular, do not mention subjective views.

3.1   CoLoSL Worlds

A *world* is a triple $(l, g, \mathcal{I})$ where l and g are *logical states* and $\mathcal{I}$ is an *action model*. The *local logical state* (or local state) l represents the locally owned resources of a thread, in the standard separation logic sense, while the *global logical state* (or shared state) g represents shared resources. The action model $\mathcal{I}$ records all possible interferences on the shared state.

Logical states have two components: one describes machine states (e.g. stacks and heaps); the other represents *capabilities*. The latter are inspired by the capabilities in CAP [7]: a thread in possession of a given capability is allowed to perform the associated actions (as prescribed by the *action model* component of each world, defined below), while any capability *not* owned by a thread means that the environment can perform the action. This can be seen as a unified treatment of the rely and guarantee relations in rely-guarantee reasoning [16]: a capability

---

[2] Unlike CAP [7] and as in iCAP [23], we do not provide an explicit *unsharing* mechanism to claim back shared resources. Instead, this can be simply encoded as an action of the form $\bigstar_{j \in J} [\mathsf{a}_j] \colon P \rightsquigarrow \bigstar_{j \in J} [\mathsf{a}_j]$: a thread holding $\bigstar_{j \in J} [\mathsf{a}_j]$ can move the shared resource P into its local state in exchange for the associated capabilities.

fully owned (resp. fully not owned) during the entire lifetime of a thread represents its guarantee (resp. its rely), while a partially-owned capability means that the corresponding action is both in the rely and the guarantee. Capabilities go beyond the rely-guarantee model [9]; in particular, they may be transferred between a thread and its environment just like any other resource to temporarily block or enable certain actions. See the presentation of CAP [7] and deny-guarantee [9] for further details and motivation.

In general, each component of a logical state is taken from an arbitrary *separation algebra* [2] (i.e. a cancellative, partial commutative monoid) that satisfies the *cross-split* property[3][8] (this is needed for $\uplus$ to be associative [14]). As we demonstrate in the examples of §4, our programs often call for a more complex model of machine states and capabilities than that presented here. For instance, we may need our capabilities to be fractionally owned, where ownership of a *fraction* of a capability grants the right to perform the action to both the thread and the environment, while a fully-owned capability by the thread *denies* the right to the environment to perform the associated action. For ease of presentation, the focus of this paper is on the standard stack and heap model for machine states, and finite sets of *tokens* (which are simple names) for capabilities. We assume a set of program values Val, as well as infinite disjoint sets PVar, Loc, and Token of program variables, memory locations, and tokens, respectively.

Definition 1 (Logical states). *A logical state is a tuple* $((\sigma, h), \kappa) \in$ LState, *also written* (σ, h, κ), *of a finite partial stack* $\sigma \in$ Stack *associating program variables with values, a heap* $h \in$ Heap *associating heap locations with values, and a capability* $\kappa \in \mathbb{K}$:

$$\text{Stack} \triangleq \text{PVar} \rightharpoonup_{fin} \text{Val} \qquad \text{Heap} \triangleq \text{Loc} \rightharpoonup_{fin} \text{Val} \qquad \mathbb{M} \triangleq \text{Stack} \times \text{Heap}$$

$$\mathbb{K} \triangleq \mathcal{P}_{fin}(\text{Token}) \qquad\qquad \text{LState} \triangleq \mathbb{M} \times \mathbb{K}$$

The local and global logical states of a world are always *compatible*: they can be composed with one another. This captures the intuition that locally-owned resources are disjoint from shared resources. The composition of logical states is defined component-wise as disjoint function union $\uplus$ over stacks and heaps, and disjoint set union $\uplus$ on capabilities.

Definition 2 (Logical state composition). *The* composition of logical states $\circ$: LState $\times$ LState $\rightharpoonup$ LState *is defined as:*

$$((\sigma, h), \kappa) \circ ((\sigma', h'), \kappa') \triangleq ((\sigma \uplus \sigma', h \uplus h'), \kappa \uplus \kappa')$$

We write l to range over arbitrary logical states or just local states, and g to range over logical states representing global shared states. The empty logical state $(\emptyset, \emptyset, \emptyset)$ is written $\mathbf{0_L}$. We write $l_1 \leq l_2$ when there exists l such that $l \circ l_1 = l_2$, and

---

[3] A monoid $(\mathbb{A}, +, \mathbf{1})$ satisfies the cross-split property iff, for all $a, b, c, d \in \mathbb{A}$, if $a + b = c + d$, there exists $x, y, z, w \in \mathbb{A}$ s.t. $a = x + y$, $b = z + w$, $c = x + z$ and $d = y + w$.

write $l_2 - l_1$ to denote the unique such l (by cancellativity) when it exists. When $l_1 \circ l_2$ is defined, we say that l_1 and l_2 are *compatible* and write $l_1 \sharp l_2$.

An action is a triple (p, q, c) of logical states where p and q are the action *pre-* and *post-states* describing how the shared state is modified by the action; and c is the action *catalyst*. An action catalyst has to be present for the action to take effect, but is left unchanged by the action. It is maximal in the sense that no further, non-empty catalyst c' can be found, which we write $p \perp q$: $\forall l. l \le p \wedge l \le q \implies l = 0_L$.[4] For instance, as we shall shortly see, s_x in Fig. 2 will be interpreted as the set of actions $S = \{a_v \triangleq ((\emptyset, \{\ell : v\}, \emptyset), (\emptyset, \{\ell : v{+}1\}, \emptyset), (\emptyset, \{\ell{-}1 : v{+}1\}, \emptyset)) \mid v \in \mathbb{N}\}$, where ℓ is the value of x in the current logical environment.

An action model is a partial function from *capabilities* to sets of *actions*. It corresponds to the semantic interpretation of an interference assertion.

Definition 3 (Action models). *The set* Action *of actions, ranged over by* a, *and the set* AMod *of action models, ranged over by* $\mathfrak{I}$, *are defined as:*

$$\text{Action} \triangleq \text{LState} \times \text{LState} \times \text{LState} \qquad \text{AMod} \triangleq \mathbb{K} \rightharpoonup \mathcal{P}\,(\text{Action})$$

Worlds are triples $(l, g, \mathfrak{I}) \in \text{LState} \times \text{LState} \times \text{AMod}$ that satisfy several well-formedness conditions: the local and shared states are compatible; the capabilities owned by l and g are in the domain of $\mathfrak{I}$; and actions in $\mathfrak{I}$ are *confined* to g (written $g \,\copyright\, \mathfrak{I}$). An action $a = (p, q, c)$ is confined to g if and only if, whenever it is enabled ($p \circ c$ agrees with g), then its pre-state p is contained in g ($p \le g$). We motivate the need for the confinement condition in §3.3.

Definition 4 (Well-formedness). *A triple* $(l, g, \mathfrak{I}) \in \text{LState} \times \text{LState} \times \text{AMod}$ *is* well-formed, *written* wf $(l, g, \mathfrak{I})$, *iff* $l \sharp g$, $l_\kappa \cup g_\kappa \subseteq \bigcup dom(\mathfrak{I})$ *and* $g \,\copyright\, \mathfrak{I}$.

Definition 5 (Worlds). *The set* World *of worlds consists of all well-formed triples:*

$$\text{World} \triangleq \{w \in \text{LState} \times \text{LState} \times \text{AMod} \mid \text{wf}\,(w)\}$$

Finally, the composition of two worlds is defined when their local states are compatible, their global shared states and action models are the same, and the resulting tuple is well-formed.

Definition 6 (World composition). *The* composition of worlds, $\bullet$: World $\times$ World $\rightharpoonup$ World, *is defined as:*

$$(l, g, \mathfrak{I}) \bullet (l', g', \mathfrak{I}') \triangleq \begin{cases} (l \circ l', g, \mathfrak{I}) & \text{if } g = g', \text{ and } \mathfrak{I} = \mathfrak{I}', \text{ and } \text{wf}((l \circ l', g, \mathfrak{I})) \\ \text{undefined} & \text{otherwise} \end{cases}$$

---

[4] Alternatively, the catalyst could be computed a posteriori for each action. However, we often need to isolate the part of the state that is modified by an action, hence our technical choice of recording the catalyst in the model.

3.2   CoLoSL Assertions

Our assertion language extends separation logic with *subjective views* and *capability assertions*.

CoLoSL is parametric in the assertions of machine states and capabilities, and can be instantiated with any assertion language over machine states $\mathbb{M}$ and capabilities $\mathbb{K}$. In this paper, we use standard heap and stack assertions for machine state assertions, and single token assertions of the form [a] for capability assertions where a $\in$ Token. We write [A] as a shorthand for $\star_{a \in A}[a]$. We assume an infinite set LVar of *logical variables*, disjoint from PVar.

Definition 7 (Assertion syntax). *Given* $x \in$ LVar, x $\in$ PVar, *and* a $\in$ Token, *the assertions of CoLoSL , Assn, are described by the grammars below:*

$$\text{LAssn} \ni p,q ::= \textit{false} \mid E_1 = E_2 \mid \text{emp} \mid \text{x} \Rightarrow E \mid E_1 \mapsto E_2 \mid [E]$$
$$\mid p \vee q \mid \neg p \mid \exists x.\, p \mid p * q \mid p \uplus q \mid p \mathbin{-\!\circledast} q$$

$$\text{Assn} \ni P,Q ::= p \mid \exists x.\, P \mid P \vee Q \mid P * Q \mid P \uplus Q \mid \boxed{P}_I$$

$$I ::= \emptyset \mid \{[A]\colon \exists \bar{x}.\, P \rightsquigarrow Q\} \cup I \qquad E ::= x \mid a \mid E_1 + E_2 \mid \cdots$$

The syntax and semantics of *local assertions* LAssn are as in standard separation logic with variables-as-resource [1].[5] Local assertions are interpreted over a world's local state. The empty local state $\mathbf{0}_L$ is denoted by emp. The assertion $\text{x} \Rightarrow E$ denotes a singleton stack where x has value E. Similarly, $E_1 \mapsto E_2$ is true of the singleton heap where only address E_1 is allocated and has value E_2. The capability assertion $[E]$ is true of the singleton capability $\{a\}$ if E evaluates to a. The *separating conjunction* $p * q$ is true when the local state can be split into two according to $\circ$ such that one state satisfies p and the other satisfies q. The *overlapping conjunction* $p \uplus q$ is true when the local state can be split three-ways according to $\circ$, such that the $\circ$-composition of the first two states satisfies p and the $\circ$-composition of the last two satisfy q [14,12,22]. *Septraction* (or *existential magic wand*) $p \mathbin{-\!\circledast} q$ is true when there exists a local state satisfying p that can be $\circ$-composed with the current one to yield a state satisfying q. The usual predicates and connectives have their standard classical meaning.

As in RGSep [26], our assertions Assn are defined on top of local assertions. For simplicity, assertions do not include negation nor septraction. The interpretation of assertions is a simple lift from that of local assertions, with the exception of the subjective view $\boxed{P}_I$. First, an *interference assertion* I describes actions enabled by a given capability, in the form of a pre- and postcondition. A subjective view $\boxed{P}_I$ is then true of $(l, g, \mathcal{I})$ when $l = \mathbf{0}_L$ and a *subjective state* s can be found in the global shared state g (i.e. $g = s \circ r$ for some r), such that s satisfies P, and $\mathcal{I}$ and I *agree* given the decomposition s and r, written $\mathcal{I} \downarrow (s, r, [\![I]\!]_\iota)$, in the following sense:

---

[5] Note in particular that expressions E do not allow program variables: they can only appear on the left-hand side of $\text{x} \Rightarrow E$.

1. every action in I is reflected in $\mathcal{J}$;
2. every action in $\mathcal{J}$ that has a visible effect on s is reflected in I;
3. the above holds after any number of actions in $\mathcal{J}$ takes place.

Thus, given a world $(l, g, \mathcal{J})$ and a subjective view $\boxed{P}_I$, P describes a subjective state s that is *a part* of g and I describes *all parts* of $\mathcal{J}$ concerning s, while $\mathcal{J}$ describes the overall interference on g. We refer to the above agreement between the action model and the subjective view as the *action model closure property*. We omit its formal definition for lack of space.

The semantics of CoLoSL assertions is given by a forcing relation $w, \iota \vDash P$ between a world w, a logical environment $\iota \in \mathsf{LEnv}$, and an assertion P. We use two auxiliary forcing relations. The first one $l, \iota \vDash_{\mathsf{SL}} p$ interprets local assertions $p \in \mathsf{LAssn}$ in the usual separation logic sense over a logical state l. The second one $s, \iota \vDash_{g,\mathcal{J}} P$ interprets assertions $P \in \mathsf{Assn}$ over a *subjective state* s that is part of the global shared state g, subject to the action model $\mathcal{J}$. This second relation is needed to deal with the nesting of subjective views.[6] Since logical connectives are interpreted uniformly in all cases, we write $\vDash_{\dagger}$ for any of the three satisfaction relations, u for elements of either World or LState, and $\bullet$ for either $\bullet$ or $\circ$, as appropriate.

Definition 8 (Assertion semantics). *Given a logical environment ι : $\mathsf{LVar} \to \mathsf{Val}$, the semantics of CoLoSL assertions is defined below, using the semantics of interference assertions $[\![.]\!]_{(.)}$: $\mathsf{LEnv} \to \mathsf{AMod}$ also defined below:*

$$
\begin{aligned}
&(l, g, \mathcal{J}), \iota \vDash p && \textit{iff } l, \iota \vDash_{\mathsf{SL}} p \\
&(l, g, \mathcal{J}), \iota \vDash \boxed{P}_I && \textit{iff } l = \mathbf{0_L} \textit{ and } \exists s, r. \, g = s \circ r \textit{ and } s, \iota \vDash_{g,\mathcal{J}} P \textit{ and } \mathcal{J}{\downarrow}\,(s, r, [\![I]\!]_\iota) \\
&\quad s, \iota \vDash_{g,\mathcal{J}} p && \textit{iff } s, \iota \vDash_{\mathsf{SL}} p \\
&\quad s, \iota \vDash_{g,\mathcal{J}} \boxed{P}_I && \textit{iff } (s, g, \mathcal{J}), \iota \vDash \boxed{P}_I \\
&\quad u, \iota \vDash_{\dagger} \exists x. \, P && \textit{iff } \exists v. \, u, [\iota \mid x : v] \vDash_{\dagger} P \\
&\quad u, \iota \vDash_{\dagger} P \vee Q && \textit{iff } u, \iota \vDash_{\dagger} P \textit{ or } u, \iota \vDash_{\dagger} Q \\
&\quad u, \iota \vDash_{\dagger} P_1 * P_2 && \textit{iff } \exists u_1, u_2. \, u = u_1 \bullet u_2 \textit{ and } u_1, \iota \vDash_{\dagger} P_1 \textit{ and } u_2, \iota \vDash_{\dagger} P_2 \\
&\quad u, \iota \vDash_{\dagger} P_1 \mathbin{\talloblong} P_2 && \textit{iff } \exists u', u_1, u_2. \, u = u' \bullet u_1 \bullet u_2 \textit{ and} \\
&\quad && \qquad u' \bullet u_1, \iota \vDash_{\dagger} P_1 \textit{ and } u' \bullet u_2, \iota \vDash_{\dagger} P_2 \\
&\quad l, \iota \vDash_{\mathsf{SL}} \textit{false} && \textit{never} \\
&\quad l, \iota \vDash_{\mathsf{SL}} \mathsf{emp} && \textit{iff } l = \mathbf{0_L} \\
&\quad l, \iota \vDash_{\mathsf{SL}} E_1 = E_2 && \textit{iff } [\![E_1]\!]_\iota = [\![E_2]\!]_\iota \\
&\quad l, \iota \vDash_{\mathsf{SL}} \mathsf{x} \Rightarrow E && \textit{iff } l = (\{\mathsf{x} : [\![E]\!]_\iota\}, \emptyset, \emptyset) \\
&\quad l, \iota \vDash_{\mathsf{SL}} E_1 \mapsto E_2 && \textit{iff } l = (\emptyset, \{[\![E_1]\!]_\iota : [\![E_2]\!]_\iota\}, \emptyset) \\
&\quad l, \iota \vDash_{\mathsf{SL}} [E] && \textit{iff } l = (\emptyset, \emptyset, \{[\![E]\!]_\iota\}) \\
&\quad l, \iota \vDash_{\mathsf{SL}} \neg p && \textit{iff } l, \iota \nvDash_{\mathsf{SL}} p \\
&\quad l, \iota \vDash_{\mathsf{SL}} p \mathbin{-\!\!\circledast} q && \textit{iff } \exists l'. \, l', \iota \vDash_{\mathsf{SL}} p \textit{ and } l \mathbin{\sharp} l' \textit{ and } l \circ l', \iota \vDash_{\mathsf{SL}} q
\end{aligned}
$$

---

[6] This presentation with several forcing relations differs from the usual CAP presentation [7], where assertions are first interpreted over worlds that are not necessarily well-formed, and then cut down to well-formed ones. The advantage of our presentation is that the semantics of assertions is *compositional*, e.g. the semantics of $P * Q$ follows directly from the semantics of P and Q.

$$[\![I]\!]_\iota(A) \triangleq \left\{ (p,q,c) \;\middle|\; \begin{array}{l} [\mathsf{A}]\colon \exists \bar{x}.\, P \rightsquigarrow Q \in I \wedge p \perp q \wedge \exists \bar{v}, \mathfrak{I}, r, r'.\, c = r \circ r' \wedge \\ p \circ r, [\iota \mid \bar{x} : \bar{v}] \vDash_{poc,\mathfrak{I}} P \wedge q \circ r, [\iota \mid \bar{x} : \bar{v}] \vDash_{qoc,\mathfrak{I}} Q \end{array} \right\}$$

Note that, as in the CAP family [7,23,3], CoLoSL cannot ensure that proved programs do not leak memory. This is because of the following property of the semantics with respect to the shared state (sometimes called "intuitionistic semantics" [21]): if $(l, g, \mathfrak{I}), \iota \vDash P$ then $(l, g \circ g', \mathfrak{I}), \iota \vDash P$.

Five of the principles of Fig. 1 are direct consequences of the semantics.

Lemma 1. *The CoLoSL reasoning principles* FORGET, MERGE, SHIFT, WEAKEN, *and* COPY *are valid.*

Proof (sketch). The cases of WEAKEN and COPY are straightforward. For FORGET, MERGE, and SHIFT, we note in [20] that action model closure is preserved by picking a smaller subjective state, taking the union of subjective states and their interference assertions, and shifting the interference assertion, respectively. □

3.3 Reducing CoLoSL Principles to Separation Logic Entailments

We turn to the remaining two principles, EXTEND and SHIFT, and to the stability of assertions. These involve reasoning outside our assertion language, potentially requiring semantic reasoning in the model. Fortunately, it is enough to work with a partial axiomatisation for all three conditions to verify our examples. In this section, we give cut-down versions of these rules for a fragment of the CoLoSL assertion language where the nesting of subjective views is not permitted and interference assertions cannot mention subjective views. This restriction is easily lifted: assertions with nested boxes can always be *flatten* into logically equivalent assertions with no nesting; and interference assertions mentioning other subjective views in their actions may be rewritten into ones that do not. See [20] for the full details.

Confinement. The soundness of CoLoSL hinges on the fact that, given a world $(l, g, \mathfrak{I})$, the action model $\mathfrak{I}$ contains all actions that could possibly affect the shared state g. This was captured by a well-formedness condition in the definition of worlds (Def. 5) as $g \;\circledC\; \mathfrak{I}$, stipulating that the actions in $\mathfrak{I}$ are confined to the shared state g. It is also possible to extend g at any time. Any part l' of the local state can migrate to the shared state under a new set of actions $\mathfrak{I}'$, yielding a new shared state $g \circ l'$ and action model $\mathfrak{I} \cup \mathfrak{I}'$. This migration is only permitted if $l' \;\circledC\; \mathfrak{I}'$. This confinement condition means that the extension does not invalidate the views of the threads.

The technical definition of confinement of an action model on a logical state is given in the technical report [20]. Intuitively, it means that, whenever an action $a = (p, q, c)$ of $\mathfrak{I}'$ is enabled, the pre-state p must be a substate of l'. It is possible for some of the catalyst c to lie outside l', since the fact that it does not change during the course of the action means that it will not have an effect on the views of other threads. For example, recall the interpretation of s_x given by S just before Def. 3. The action $a_0 = ((\emptyset, \{\ell : 0\}, \emptyset), (\emptyset, \{\ell : 1\}, \emptyset), (\emptyset, \{\ell - 1 : 1\}, \emptyset)) \in S$ is confined to

$$\frac{P \vdash f \quad f \blacktriangleright I}{P \, \text{\textcircled{c}} \, I} \qquad \frac{\forall ([A] \colon \exists \bar{x}. \, p \rightsquigarrow q) \in I. \, f \blacktriangleright \{[A] \colon \exists \bar{x}. \, p \rightsquigarrow q\}}{f \blacktriangleright I} \qquad \frac{f \, \text{\reflectbox{$\mathbb{U}$}} \, p \vdash_{\mathsf{SL}} \textit{false}}{f \blacktriangleright \{[A] \colon \exists \bar{x}. \, p \rightsquigarrow q\}}$$

$$\frac{p \vdash_{\mathsf{SL}} p' * r \qquad q \vdash_{\mathsf{SL}} q' * r \qquad \mathsf{exact}(r) \qquad f \blacktriangleright \{[A] \colon \exists \bar{x}. \, p' \rightsquigarrow q'\}}{f \blacktriangleright \{[A] \colon \exists \bar{x}. \, p \rightsquigarrow q\}}$$

$$\frac{(p \mathbin{-\circledast} f) * q \vdash_{\mathsf{SL}} f \qquad f \Leftrightarrow \bigvee_{i \in J} f_i \qquad (\mathsf{precise} \, (f_i) \wedge f_i \, \text{\reflectbox{$\mathbb{U}$}} \, p \vdash_{\mathsf{SL}} f_i \ \text{for } i \in J)}{f \blacktriangleright \{[A] \colon \exists \bar{x}. \, p \rightsquigarrow q\}}$$

Fig. 3. Selected confinement and local fencing rules. We assume that variables in $\bar{x}$ do not appear free in f, and in f_i for any i.

the logical state $l_0 = (\emptyset, \{\ell \colon 0\}, \emptyset)$ because l_0 is the first component of the action, and state $(\emptyset, \{\ell - 1 \colon 0\}, \emptyset)$ because it is incompatible with the action, but not state $(\emptyset, \{\ell - 1 \colon 1\}, \emptyset)$ because the action can potentially affect address ℓ. The definition of confinement also requires that all states resulting from the successive application of actions in $\mathcal{I}'$ themselves confine all actions in $\mathcal{I}'$. For instance, we also require that $l_1 = (\emptyset, \{\ell \colon 1\}, \emptyset)$ resulting from the application of a_0 on l_0, $l_2 = (\emptyset, \{\ell \colon 2\}, \emptyset)$ resulting from the application of a_1 on l_1, and so forth, *all* confine the actions in S. Inspired by the LRG approach [10], we achieve this by first finding a set that is invariant under all actions in $\mathcal{I}'$ (called a *fence*), then checking the confinement condition for each action. We provide the technical details of confinement in [20].

For our examples, it is in fact enough to work with confinement in the logic. We lift the notion of confinement to assertions, and write $P \, \text{\textcircled{c}} \, I$ when the set of states described by P confines the actions of interface assertion I. In the logic, the shared state can be extended by the resources in P under the interference assertion I via the EXTEND principle, which requires that $P \, \text{\textcircled{c}} \, I$ be established.

Fig. 3 presents a set of rules which reduces $P \, \text{\textcircled{c}} \, I$ to a series of entailments in ordinary separation logic. As expressed by the first rule, $P \, \text{\textcircled{c}} \, I$ holds if there is a weaker local assertion f that acts as a *local fence* for I, written $f \blacktriangleright I$. This relation states that f must be invariant under all actions of I and must confine the actions in f. In [20], we show that it is always possible to weaken an arbitrary assertion into a local assertion. This fencing condition is checked for each action in I (see the second rule). For each action $[A] \colon \exists \bar{x}. \, p \rightsquigarrow q$, the three remaining rules of the figure may apply. In the first of these rules, the action cannot possibly fire, because its precondition does not agree with f: no state satisfying f may be extended such that a subpart satisfies p. The second of these rules allows us to trim a *neutral* part r (corresponding to a part of the catalyst in the interpretation of this action) of an action $[A] \colon \exists \bar{x}. \, p \rightsquigarrow q$ appearing both in p and q. This only applies when r is *exact*, i.e. satisfied by at most one logical state:[7] the part of the state denoted by r is then uniquely determined and left unchanged by the action. The last rule finally reduces local fencing to entailment checking, provided the fence f can be expressed as a disjunction of *precise* assertions, i.e. assertions satisfied

---

[7] $\mathsf{exact}(p) \triangleq \forall \iota, l_1, l_2. \, l_1, \iota \vDash_{\mathsf{SL}} p$ and $l_2, \iota \vDash_{\mathsf{SL}} p$ implies $l_1 = l_2$

$$\frac{P \vdash f \quad I \sqsubseteq^f I'}{I \sqsubseteq^P I'} \qquad \overline{I \sqsubseteq^f I} \qquad \frac{f \triangleright I \cup I_1 \quad I_1 \sqsubseteq^f I_2}{I \cup I_1 \sqsubseteq^f I \cup I_2}$$

$$\frac{p \vdash_{\text{SL}} p' * r \quad q \vdash_{\text{SL}} q' * r \quad \text{exact}(r) \quad f \perp p'}{\{[A] : \exists \bar{x}. \, p \rightsquigarrow q\} \sqsubseteq^f \emptyset} \qquad \frac{(p - \circledast (f \uplus p)) * q \vdash_{\text{SL}} false}{\{[A] : \exists \bar{x}. \, p \rightsquigarrow q\} \sqsubseteq^f \emptyset}$$

$$\frac{f \uplus p \vdash_{\text{SL}} \bigvee_{i \in J} f \uplus (p * r_i) \quad \text{exact}(r_i) \text{ for } i \in J}{\{[A] : \exists \bar{x}. \, p \rightsquigarrow q\} \equiv^f \bigcup_{i \in J} \{[A] : \exists \bar{x}. \, p * r_i \rightsquigarrow q * r_i\}}$$

$$\overline{\bigcup_{i \in K, j \in J} \{[A] : \exists \bar{x}. \, p_i \rightsquigarrow q_j\} \equiv^{true} \left\{[A] : \exists \bar{x}. \, \bigvee_{i \in K} p_i \rightsquigarrow \bigvee_{j \in J} q_j\right\}} \qquad \overline{true \triangleright I}$$

$$\frac{\forall [A] : \exists \bar{x}. \, p \rightsquigarrow q \in I. \, f \triangleright \{[A] : \exists \bar{x}. \, p \rightsquigarrow q\}}{f \triangleright I} \qquad \frac{p \perp q \quad (p - \circledast (f \uplus p)) * q \vdash_{\text{SL}} f}{f \triangleright \{[A] : \exists \bar{x}. \, p \rightsquigarrow q\}}$$

$$\frac{f \perp p}{f \triangleright \{[A] : \exists \bar{x}. \, p \rightsquigarrow q\}} \qquad \frac{p \vdash_{\text{SL}} p' * r \quad q \vdash_{\text{SL}} q' * r \quad \text{exact}(r) \quad f \triangleright \{[A] : \exists \bar{x}. \, p' \rightsquigarrow q'\}}{f \triangleright \{[A] : \exists \bar{x}. \, p \rightsquigarrow q\}}$$

Fig. 4. Selected action shifting rules. We write $I \equiv^f I'$ for $I \sqsubseteq^f I' \wedge I' \sqsubseteq^f I$, and $p \perp q$ to denote that states satisfying p and q have empty intersections. We assume that variables in $\bar{x}$ do not appear free in f.

by at most one substate of each logical state.[8] The first premise states that f is invariant under the action $[A]: \exists \bar{x}. \, p \rightsquigarrow q$, similar to the way that RGSep encodes stability checks as separation logic entailments. Informally, it reads: for any state in f, remove a part satisfying p, add a state satisfying q, and the result should still be in f. The third premise checks the confinement condition: given a state in local assertion f_i ($l_1 \circ l_2 \vDash f_i$), and a state in p ($l_2 \circ l_3 \vDash p$ with l_1 and l_3 disjoint), the combined state ($l_1 \circ l_2 \circ l_3$) must also be in f_i. Hence, by precision of f_i, we have $l_1 \circ l_2 \circ l_3 = l_1 \circ l_2$, i.e. $l_2 \circ l_3 \leq l_1 \circ l_2$ as required.

Shifting and fencing. Fig. 4 presents a partial axiomatisation of the shifting condition $I \sqsubseteq^P I'$ required by the SHIFT principle. As with confinement, we omit the direct semantic definition. Intuitively, the relation $I \sqsubseteq^P I'$ means that interference assertion I can be replaced by I', because I and I' describe the same interference with respect to the states described by P. The first rule weakens the shifting condition from assertions to local fence assertions. The third rule reduces the shifting judgement $I \cup I_1 \sqsubseteq^f I \cup I_2$ to the simpler $I_1 \sqsubseteq^f I_2$, provided that f is invariant with respect to $I \cup I_1$, written $f \triangleright I \cup I_1$. This invariant fencing condition is necessary: $I_1 \sqsubseteq^f I_2$ only states that I_1 and I_2 have the same effect with respect to f; $f \triangleright I \cup I_1$ states that f is an *invariant* of the shared state under the combined interferences of I and I_1.

The next two rules describe situations where it is impossible to apply the action to f: when the precondition of the action is entirely outside f; or when the

---

[8] $\text{precise}(p) \triangleq \forall \iota, l, l_1, l_2. \, l_1 \leq l, \, l_2 \leq l, \, l_1, \iota \vDash_{\text{SL}} p$, and $l_2, \iota \vDash_{\text{SL}} p$ implies $l_1 = l_2$

postcondition is incompatible with that part of f not associated with the precondition. The notation $p \perp q$ asserts that states described by p and q have *empty intersections*: whenever $l_1, \iota \vDash_{\text{SL}} p$ and $l_2, \iota \vDash_{\text{SL}} q$, we have $l_1 \perp l_2$ as defined in §3.1. This is expressible in standard separation logic as follows:

$$p \perp q \Leftrightarrow p \vdash_{\text{SL}} \neg (true * (\neg\text{emp} \wedge (true \mathbin{-\!\circledast} q)))$$

The next rule is a shifting equivalence that uses the knowledge embodied by f of all the possible states that the subjective shared state may be in to rewrite an action into an equivalent one. More precisely, if whenever the precondition p of the action agrees with f then one of the r_i's is also true, then adding r_i as a neutral part of the action produces the same behaviour. We can use this rule (with the single $r_0 = x \mapsto v$) to justify the shiftings of §2. The fact that the r_i's are exact guarantees that no piece of state in r_i is mutated by the "larger" action. In general, it may not be the case that a single exact assertion can be added, but it may be the case that a disjunction of exact facts holds. The last shifting equivalence is straightforward.

The last five rules partially axiomatise the *fencing* relation $f \rhd I$. Most are similar to those for local fencing $f \blacktriangleright I$. The first two state that *true* fences any interference assertion, and fencing can be checked per action. The last one states that, as for local fencing, neutral parts of actions may be ignored. The one before that states that, contrarily to local fencing, actions are allowed to have effects outside of the fence. If the action precondition does not intersect with the fence, then its effect is entirely outside the fence and the action may be ignored.

Let us now focus on the third of these rules, which states that whenever p and q do not intersect in an action $[\mathsf{A}]: \exists \bar{x}.\, p \rightsquigarrow q$ (e.g. when their common parts have been removed using the last rule), then the application of the action must preserve the fence f. Contrarily to the case of local fencing, the action is allowed to act partly outside of f, hence the state on which the action is applied is $f \uplus p$. However, the whole of the postcondition q of the action is then added, and the resulting state must still be in f. One might instead have expected that only parts of the resulting state need to be represented in f, to mimic the relationship between f and p (and indeed, this is all that is required for stability, as we shall see next). However, we do need the full q. Recall that shifting asserts that two interference assertions have the same effect even after an arbitrary number of steps. Doing otherwise would be unsound because there would be no guarantee that I accounts for all possible actions on that part of q that would be discarded, since it would not be part of f. Hence, we could end up with a new interference assertion I' that breaks the original action model closure property. Finally, the reason why p and q must not intersect stems from similar considerations. If p and q had a non-empty intersection c, such that c is not part of f, then this would force the fence to account for c which would prevent us from forgetting actions associated with it using shifting. For instance, it would make it impossible to forget actions as we do in §2.

Stability. Fig. 5 partially axiomatises the stability of assertions of the form $\exists \bar{x}.\, p * \mathlarger{\mathlarger{*}}_{i \in J} \boxed{q_i}_{I_i}$. We work with a more restricted form of assertions than our two previous axiomatisations for simplicity; see [20] for the general case. The first two

$$\frac{}{\mathsf{Stable}(p)} \qquad \frac{\mathsf{Stable}(P)}{\mathsf{Stable}(\exists x.\, P)} \qquad \frac{\forall j \in J.\, \mathsf{Stable}(q_j, I_j, p, \biguplus_{i \in J} q_i)}{\mathsf{Stable}(p * \divideontimes_{i \in J} \boxed{q_i}_{I_i})}$$

$$\frac{\forall [A]\colon \exists \bar{x}.\, p_1 \rightsquigarrow p_2 \in I.\, \mathsf{Stable}(q, \{[A]\colon \exists \bar{x}.\, p_1 \rightsquigarrow p_2\}, p, r)}{\mathsf{Stable}(q, I, p, r)} \qquad \frac{[A] * p * r \vdash_{\mathsf{SL}} \mathit{false}}{\mathsf{Stable}(q, \{[A]\colon \exists \bar{x}.\, p_1 \rightsquigarrow p_2\}, p, r)}$$

$$\frac{p * (r \uplus p_1) \vdash_{\mathsf{SL}} \mathit{false}}{\mathsf{Stable}(q, \{[A]\colon \exists \bar{x}.\, p_1 \rightsquigarrow p_2\}, p, r)} \qquad \frac{(p_1 -\circledast (q \uplus p_1)) * p_2 \vdash_{\mathsf{SL}} q * \mathit{true}}{\mathsf{Stable}(q, \{[A]\colon \exists \bar{x}.\, p_1 \rightsquigarrow p_2\}, p, r)}$$

Fig. 5. Selected rules for stability checks. We assume that variables in $\bar{x}$ do not appear free in p, q, and r.

rules state that local assertions are always stable and existentials can be eliminated. The next rule states that checking the stability of $p * \divideontimes_{i \in J} \boxed{q_i}_{I_i}$ boils down to establishing, for each local assertion q_j for $j \in J$, that the four-place predicate $\mathsf{Stable}(q_j, I_j, p, \biguplus_{i \in J} q_i)$ holds: this means that q_i is stable under the interference assertion I_i, a local context p, and a shared context made of the $\uplus$-combination of all the subjective assertions (including q_j). In turn, checking this fact reduces to checking stability for each action of I_j. The last three rules deal with checking stability for a single action, in a similar way to the fencing rules above.

The first of these rules is unfamiliar. Unlike fencing, stability checking for assertions only has to be checked against actions for which the environment may have the capability. If the capability required by the action cannot exist separately from those held by the assertion (that is, those in $p * r$) then the environment cannot possibly own the capability to perform the action. Similarly, an action whose precondition is incompatible with the assertion $p * r$ cannot possibly fire, as stated by the next rule. The last rule checks that q is preserved by the effect of an action. Again, there is a crucial difference with the corresponding check for fencing: in p_2, the action may bring in some newly-shared state, hence the result $q * \mathit{true}$; but the FORGET rule allows us to immediately discard it, if appropriate.

Proof reuse. Note that, once the local fencing judgement $f \blacktriangleright I$ has been established, it automatically establishes the weaker fencing judgement $f \triangleright I$, which in turn implies that f is stable under I. Also, fencing is preserved by action shifting. These observations provide the admissible rules:

$$\frac{f \blacktriangleright I}{f \triangleright I} \qquad \frac{p \triangleright I}{\mathsf{Stable}(\boxed{p}_I)} \qquad \frac{f \triangleright I \qquad I \sqsubseteq^f I'}{f \triangleright I'}$$

The other directions are not valid in general: fencing lacks the confinement condition required by local fencing; stability of p under I may omit part of the state resulting from an action application to re-establish p, which is not allowed in fencing; and a fence f for a smaller interference assertion need not be a fence for a larger interference assertion.

3.4   Soundness

We prove the soundness of CoLoSL, parametrised by the underlying models of machine states ($\mathbb{M}$) and capabilities ($\mathbb{K}$). We appeal to the general soundness result of the views framework, providing parameters such as the reification function in Def. 9 and proving lemmas required to make the result hold.

The main part of the proof establishes the soundness of the following rule for atomic commands, where $\vdash_{\mathsf{SL}}$ represents standard sequential separation logic:

$$\frac{\vdash_{\mathsf{SL}} \{p\}\ \mathbb{C}\ \{q\} \quad P \Rrightarrow^{\{p\}\{q\}} Q}{\vdash \{P\}\ \langle \mathbb{C} \rangle\ \{Q\}} \ \text{ATOM}$$

This rule is present in logics arising from CAP [7,23]. For CoLoSL, the *repartitioning* $P \Rrightarrow^{\{p\}\{q\}} Q$ holds if, from any world $(l, g, \mathfrak{I})$ satisfying P, whenever parts of $l \circ g$ satisfies p then substituting that part for any other satisfying q will yield a state $l' \circ g'$ such that there exists $\mathfrak{I}'$ such that $(l', g', \mathfrak{I}')$ satisfies Q. Moreover, the passage from $(l, g, \mathfrak{I})$ to $(l', g', \mathfrak{I}')$ can be achieved via a succession of valid updates from $\mathfrak{I}$ and valid extension steps. The details can be found in the accompanying technical report [20].

Semantic validity of Hoare triples depends on the definition of an operational semantics $\mathbb{C}, m \to^{*} \mathbb{C}', m'$, where $m, m' \in \mathbb{M}$, and a reification function that relates a CoLoSL world to a concrete machine state.

Definition 9 (Reification). *The* reification, $\lfloor . \rfloor_{\mathsf{W}} : \text{World} \to \mathbb{M}$ *is defined as:*

$$\lfloor ((\sigma_l, h_l, \kappa_l), (\sigma_g, h_g, \kappa_g), \mathfrak{I}) \rfloor_{\mathsf{W}} \triangleq (\sigma_l \uplus \sigma_g, h_l \uplus h_g)$$

Definition 10 (Valid triple). *A triple is* valid, *written* $\vDash \{P\}\ \mathbb{C}\ \{Q\}$, *if and only if, for all* $\iota \in \mathsf{LEnv}$, $w \in \text{World}$ *and* $m, m' \in \mathbb{M}$,

$(w, \iota \vDash P \text{ and } \mathbb{C}, \lfloor w \rfloor_{\mathsf{W}} \to^{*} \text{skip}, m')$ *implies* $\exists w'.\ w', \iota \vDash Q \text{ and } m' = \lfloor w' \rfloor_{\mathsf{W}}$

Theorem 1 (Soundness). *If* $\vdash \{P\}\ \mathbb{C}\ \{Q\}$ *then* $\vDash \{P\}\ \mathbb{C}\ \{Q\}$.

4   Examples

4.1   Concurrent Spanning Tree

Programs manipulating arbitrary graphs present a significant challenge for compositional reasoning, because deep, *unspecified* sharing between different components of a graph results in changes to one subgraph affecting other subgraphs pointing into it. This makes it hard to reason about updates to each subgraph in isolation. In a concurrent setting, this difficulty is compounded by the fact that threads working on different parts of the graph may affect each other in ways that are difficult to reason about locally to each subgraph. Using a concurrent spanning tree algorithm, we demonstrate that CoLoSL reasoning might be just the right approach, as subjective views naturally provide arbitrary overlapping views of the shared state where interferences can be naturally tailored to a given subjective view. With CoLoSL, we have achieved local reasoning about the shared state for this challenging program.

Our example, presented in Fig. 8, operates on a *directed binary graph* (henceforth simply *graph*): that is, a directed graph where each node has at most two successors, called its left and right children. The program concurrently computes an in-place spanning tree of the graph (i.e. a tree that covers all nodes of the graphs from a given root), as follows: each time a new node is encountered, two new threads are created that each prune the edges of its left and right children recursively. A mark bit is associated with each node to keep track of which nodes have already been visited. Each thread returns whether it marked the node it was called on itself or whether somebody else did. In the latter case, the parent thread removes the link from its own root node to the corresponding child. Intuitively, it is allowed to do so because the child has already been reached via some other path in the graph since it was marked by another thread.

We will prove that, given a shared graph as input, the program always returns a tree, i.e. all sharing and cycles have been appropriately removed. Pleasingly, the CoLoSL specification captures just the subgraph manipulated by the thread, instead of the whole graph.

To reason about this program, following [14] we use two representations of graphs. The first is a mathematical representation $\gamma = (V, E)$ where V is a finite set of vertices and $E : V \to (V \uplus \{\text{null}\}) \times (V \uplus \{\text{null}\})$ is a function associating each vertex with at most two successors, where null denotes the absence of a child. We write $n \in \gamma$ for $n \in V$, $\gamma(n)$ for $E(n)$ which also assumes $n \in \gamma$, $n \leadsto_\gamma n'$ for $n' \in \gamma(n)$, and $\leadsto_\gamma^*$ for the reflexive and transitive closure of $\leadsto_\gamma$.

Mathematical graphs are connected to their in-memory representations by a predicate graph (x, γ) shown in Fig. 6, denoting a spatial (in-heap) graph rooted at address x corresponding to the mathematical graph γ. This predicate uses the overlapping conjunction to account for potential sharing between the left and right children and for potential cycles in the graph. Each vertex is represented as three consecutive cells in the heap tracking the mark bit and the addresses of the left and right subgraphs. We write $x \mapsto m, l, r$ for $x \mapsto m * x + 1 \mapsto l * x + 2 \mapsto r$, and $x.l$ and $x.r$ for $x + 1$ and $x + 2$, respectively. When vertex x is in the unmarked state $\mathsf{U}(x, l, r)$, the whole cell $x \mapsto 0, l, r$ resides in the shared state. In the marked state $\mathsf{M}(x)$, only $x \mapsto 1$ is owned. In both cases, the shared state also contains the left and right subgraphs represented by $\mathsf{G}(l, \gamma)$ and $\mathsf{G}(r, \gamma)$.

To understand the difference in ownership between $\mathsf{U}(x, l, r)$ and $\mathsf{M}(x)$, let us look at the interference associated with the graph, which is the union of interferences pertaining each vertex in the graph. For each vertex $n \in \gamma$, the only permitted action is that of marking n. (For simplicity, this action does not require any capability.) When changing the mark field of n from 0 to 1, the current thread also claims ownership of its left and right pointers. Indeed, we observe that other threads need not access the children of n once they see that it has already been marked. The atomic CAS (compare-and-swap) instruction prevents two threads from concurrently marking the same node and claiming ownership of the same resource.

$$\mathsf{graph}\,(x,\gamma) \triangleq \boxed{\mathsf{G}\,(x,\gamma)}_{I_\gamma} \qquad I_\gamma \triangleq \bigcup_{n\in\gamma} I(n) \qquad I(n) \triangleq \{[\emptyset] : \exists l,r.\, \mathsf{U}\,(n,l,r) \rightsquigarrow \mathsf{M}\,(n)\}$$

$$\mathsf{G}\,(x,\gamma) \triangleq (x = \mathsf{null} \wedge \mathsf{emp}) \vee \exists l,r.\, \gamma(x) = (l,r) \wedge \qquad \mathsf{U}\,(x,l,r) \triangleq x \mapsto 0, l, r$$
$$(\mathsf{U}\,(x,l,r) \vee \mathsf{M}\,(x)) \uplus \mathsf{G}\,(l,\gamma) \uplus \mathsf{G}\,(r,\gamma) \qquad\qquad \mathsf{M}\,(x) \triangleq x \mapsto 1$$

Fig. 6. Globally-shared graph predicate

$$\mathsf{g}\,(x,\gamma) \triangleq (x = \mathsf{null} \wedge \mathsf{emp}) \vee \exists l,r.\, \gamma(x) = (l,r) \wedge$$
$$\boxed{\mathsf{U}\,(x,l,r) \vee \mathsf{M}\,(x)}_{I(x)} * \mathsf{g}\,(l,\gamma) * \mathsf{g}\,(r,\gamma)$$
$$\mathsf{t}\,(x,\gamma) \triangleq (x = \mathsf{null} \wedge \mathsf{emp}) \vee (\exists l,r.\, \gamma(x) = (l,r) \wedge \exists l' \in \{l, \mathsf{null}\}, r' \in \{r, \mathsf{null}\}.$$
$$\boxed{\mathsf{M}\,(x)}_{I(x)} * x.l \mapsto l' * \mathsf{t}\,(l',\gamma) * x.r \mapsto r' * \mathsf{t}\,(r',\gamma))$$

Fig. 7. Locally-shared graph predicate

The $\mathsf{graph}\,(x,\gamma)$ predicate defined in Fig. 6 is a *global* subjective view of the graph that contains all vertices and the interference associated with γ. However, our spanning tree algorithm operates *locally* as it is called upon recursively for each node. That is, for each span(n) call (where $n \in \gamma$), the footprint of the call is limited to the subgraph rooted at n. Moreover, in order to reason about the concurrent recursive calls span(x->l) || span(x->r), we need to *split* the state into two using ∗, pass each constituent state to the relevant thread and ∗-combine the resulting states, as required by the PAR rule. We thus conduct our proof with respect to a *local* description of the graph, $\mathsf{g}\,(x,\gamma)$ as defined in Fig. 7. The definition of the $\mathsf{g}\,(x,\gamma)$ predicate is similar to that of $\boxed{\mathsf{G}\,(x,\gamma)}_{I_\gamma}$ except that the single view $\boxed{\mathsf{G}\,(x,\gamma)}_{I_\gamma}$ has been broken down into individual views for each vertex n reachable from x. Moreover, the interference assertion of each local view concerning a vertex $n \in \gamma$ has been shifted from I_γ to $I(n)$ so as to reflect only those actions that affect n.

The $\mathsf{t}\,(x,\gamma)$ predicate, also given in Fig. 7, represents a *tree* rooted at x, as is standard in separation logic [21], and consists, once fully unfolded, of one subjective view for each vertex n reachable from x in γ. The assertion of the subjective view for x reflects the fact that x has been marked, and its left and right pointers have been claimed by the marking thread and moved into its local state. The vertex l' addressed by the left pointer of x corresponds to either the initial value l prior to marking, or to null when l has more than one predecessor and has been marked by another thread.

Our goal is to prove that the following specification holds using the global graph predicate:

$$\{\mathsf{graph}\,(x,\gamma)\}\, \mathsf{b} = \mathsf{span}(x)\, \{(b = 0 \wedge \mathsf{emp}) \vee (b = 1 \wedge \mathsf{t}\,(x,\gamma))\} \qquad (1)$$

This is achieved below by giving a proof of the analogous specification using the local graph predicate in Fig. 8, and demonstrating that the global graph specification implies the local one using the principles of CoLoSL from Fig. 1. The proof sketched in Fig. 8 is mostly straightforward. One subtlety is the consequence step

```
b = span(x)  //{g(x,γ)}
```
$\{ \ // \{(x = \text{null} \land \text{emp}) \lor (\exists l, r.\, \gamma(x) = (l, r) \land \boxed{U(x, l, r) \lor M(x)}_{I(x)} * g(l, \gamma) * g(r, \gamma)) \}$

```
  if (x == null) {
```
$// \{x = \text{null} \land \text{emp}\}$
$// \{t(x, \gamma)\}$
```
    return 1;
```
$\}// \left\{ \exists l, r.\, \gamma(x) = (l, r) \land \boxed{U(x, l, r) \lor M(x)}_{I(x)} * g(l, \gamma) * g(r, \gamma) \right\}$
```
  res = ⟨ CAS(x, 0, 1) ⟩;
```
$// \left\{ \begin{array}{l} \exists l, r.\, \gamma(x) = (l, r) \land \boxed{M(x)}_{I(x)} * g(l, \gamma) * g(r, \gamma) * \\ ((\text{res} = 0 \land \text{emp}) \lor (\text{res} = 1 \land x.l \mapsto l * x.r \mapsto r)) \end{array} \right\}$

$// \left\{ (\text{res} = 0 \land \text{emp}) \lor \left(\begin{array}{l} \text{res} = 1 \land \exists l, r.\, \gamma(x) = (l, r) \land \\ \boxed{M(x)}_{I(x)} * g(l, \gamma) * g(r, \gamma) * x.l \mapsto l * x.r \mapsto r \end{array} \right) \right\}$
```
  if (res) {
```
$// \left\{ \text{res} = 1 \land \exists l, r.\, \gamma(x) = (l, r) \land \boxed{M(x)}_{I(x)} * g(l, \gamma) * g(r, \gamma) * x.l \mapsto l * x.r \mapsto r \right\}$

$// \{g(l, \gamma)\}$ $\| \quad // \{g(r, \gamma)\}$
```
    b1 = span(x->l);              ||    b2 = span(x->r);
```
$// \{(b1 = 0 \land \text{emp}) \lor (b1 = 1 \land t(l, \gamma))\}$ $\| \quad // \{(b2 = 0 \land \text{emp}) \lor (b2 = 1 \land t(r, \gamma))\}$

$// \left\{ \begin{array}{l} \text{res} = 1 \land \exists l, r.\, \gamma(x) = (l, r) \land \boxed{M(x)}_{I(x)} * x.l \mapsto l * x.r \mapsto r * \\ ((b1 = 0 \land \text{emp}) \lor (b1 = 1 \land t(l, \gamma))) * ((b2 = 0 \land \text{emp}) \lor (b2 = 1 \land t(r, \gamma))) \end{array} \right\}$
```
    if (!b1) { x->l = null; }
    if (!b2) { x->r = null; }
```
$// \left\{ \begin{array}{l} \text{res} = 1 \land \exists l, r.\, \gamma(x) = (l, r) \land \boxed{M(x)}_{I(x)} * \\ \exists l' \in \{l, \text{null}\}.\, x.l \mapsto l' * t(l', \gamma) * \exists r' \in \{r, \text{null}\}.\, x.r \mapsto r' * t(r', \gamma) \end{array} \right\}$
$// \{\text{res} = 1 \land t(x, \gamma)\}$
```
  } //{(res = 0 ∧ emp) ∨ (res = 1 ∧ t(x,γ))}
  return res;
} //{(b = 0 ∧ emp) ∨ (b = 1 ∧ t(x,γ))}
```

Fig. 8. Code and proof sketch of the concurrent spanning tree program. We omit the obvious variables as resource assertions.

just before the second **if** statement. There, we first distribute the shared state $\boxed{M(x)}_{I(x)} * g(l, \gamma) * g(r, \gamma)$ over the disjunction, then use the fact that it implies emp to discard it in the left disjunct. This proof demonstrates that our CoLoSL reasoning really is compositional, in the sense that we are doing local reasoning on the shared state (the subgraphs).

Let us show how to derive the specification (1) from the one obtained in Fig. 8. We introduce the *iterative star* operator ⊛; when iterating over the empty set, it denotes emp by convention (needed below, when x is null). We define

$$P(x, \gamma) \triangleq \underset{x \rightsquigarrow^*_\gamma n}{\circledast} \exists l, r.\, \gamma(n) = (l, r) \land (U(n, l, r) \lor M(n))$$

$$Q(x, \gamma) \triangleq \underset{x \rightsquigarrow^*_\gamma n}{\circledast} \exists l, r.\, \gamma(n) = (l, r) \land \boxed{U(n, l, r) \lor M(n)}_{I(n)}$$

From the definitions of $G(x, \gamma)$ and $g(x, \gamma)$, one can show that

$$\text{graph}(x, \gamma) \iff \boxed{P(x, \gamma)}_{I_\gamma} \qquad\qquad g(x, \gamma) \iff Q(x, \gamma)$$

The specification (1) then follows from CONSQ, Fig. 8, and the derivation below:

$$\text{graph}\,(x,\gamma) \overset{(\text{Copy})}{\Rightarrow} \underset{x \leadsto_\gamma^* n}{\circledast} \boxed{P(x,\gamma)}_{I_\gamma}$$

$$\overset{(\text{Forget})}{\Rightarrow} \underset{x \leadsto_\gamma^* n}{\circledast} \left(\exists l,r.\,\gamma(n) = (l,r) \wedge \boxed{\mathsf{U}\,(n,l,r) \vee \mathsf{M}\,(n)}_{I_\gamma} \right)$$

$$\overset{(\text{Shift})}{\Rightarrow} \underset{x \leadsto_\gamma^* n}{\circledast} \left(\exists l,r.\,\gamma(n) = (l,r) \wedge \boxed{\mathsf{U}\,(n,l,r) \vee \mathsf{M}\,(n)}_{I(n)} \right) \overset{\text{def}}{\Longleftrightarrow} \mathsf{g}\,(x,\gamma)$$

4.2   Set Module

Finally, we give a pictorial description of our reasoning about a concurrent set module implemented as a singly-linked list. We compare our CoLoSL reasoning with the original CAP reasoning of [7], demonstrating that our CoLoSL reasoning provides more concise proofs using our local reasoning about shared state.

Consider the following diagram which illustrates the CAP set predicate of [7]:

$$\boxed{\begin{array}{l} \overset{x}{v_1} \to \overset{y}{v_2} \to \overset{z}{v_3} \to \cdots * \underset{(x,y)\notin S}{\circledast} \underset{v}{\circledast} [\mathsf{U}(x,y,v)] * \underset{(x,y)\in S}{\circledast} \exists w.\ \underset{v \neq w}{\circledast} [\mathsf{U}(x,y,v)] \end{array}}^{s}_{I_x \cup I_y \cup I_z \cup \cdots}$$

The set is represented as a sorted singly-linked list with no duplicate elements. The list starts at address x with value v_1, points to the next element at address y with value v_2, and so forth. Hereafter, we write $\mathsf{node}\,(x,v,y)$ to denote a node at address x, with value v and successor y.

All nodes of the list reside in a single shared region labelled s and the interference on the list is the combined interference associated with each constituent node. Each node at a given address x is associated with a set of update capabilities of the form $[\mathsf{U}(x,y,v)]$ for *all* possible addresses y and *all* possible values v. This is to capture all potential successor addresses y and all potential values v that may be stored at address x. In order to modify a node, a thread can acquire the lock associated with the node and subsequently claim the relevant update capability.

Since in CAP the capabilities associated with a region can only be generated upon its creation, the shared region is required to keep track of all possible update capabilities $[\mathsf{U}(x,y,v)]$ associated with all addresses x (including those not currently in the domain of the list), all addresses y and all values v. At any one point, given $\mathsf{node}\,(x,v,y)$, the only update capability that can be claimed by a thread (through locking) is the one that reflects its current status, namely $[\mathsf{U}(x,y,v)]$. As a result, an auxiliary mathematical set S is used to track those nodes of the list that are currently locked and thus infer which $[\mathsf{U}]$ capabilities have been claimed. The distribution of update capabilities is captured by the two assertions written as the *infinite multiplicative star operator* $\circledast$. The first part of the assertion states that given any node at address x with successor y, if it is not locked, i.e. $(x,y) \notin S$, then all of its update capabilities of the form $[\mathsf{U}(x,y,v)]$ lie in the shared region for all values v. Dually, if it is locked, i.e. $(x,y) \in S$, then the update capabilities for all values v but one $(w \neq v)$ are in the shared region.

This CAP set predicate is unnecessarily complicated. It is counter-intuitive to have to account for the capabilities associated with addresses not in the domain of the list. Moreover, each thread observes all nodes in the list and thus needs to account for their associated interference.

TaDA [3] took the first steps towards addressing the above shortcomings of the CAP approach. TaDA regions are parametric in the separation algebra of capabilities (called guards). As such, one can choose a more suitable algebra to axiomatise the desired behaviour of capabilities. While TaDA's approach is much cleaner than that of CAP, it nevertheless requires the foresight of specifying all desired interference associated with the region upon its creation. As such, interference specifications are *static* and cannot be extended with new behaviour even when the existing resources are left untouched. On the other hand, as well as being parametric in its capability separation algebra, the dynamic subjective views of CoLoSL provide local reasoning about the shared resource and its interference.

We proceed with the CoLoSL proof of the set implementation. Recall from §3 that CoLoSL is parametric in the separation algebra of capabilities. We thus instantiate it with a heap-like capability separation algebra that is *stateful* and demonstrate that this allows for a more concise proof.

We specify the set predicate as the *-composition of subjective views associated each node in the singly-linked list as illustrated by:

$$\left[x.\mathsf{N}{\to}y\right]*\overset{x}{\underbrace{\overbrace{\textcircled{v_1}}}}_{I_x}*\left[y.\mathsf{N}{\to}z\right]*\overset{y}{\underbrace{\overbrace{\textcircled{v_2}}}}_{I_y}*\left[z.\mathsf{N}{\to}\dots\right]*\overset{z}{\underbrace{\overbrace{\textcircled{v_3}}}}_{I_z}*\cdots$$

The interference on each subjective view is limited to the node in question. Associated with each node at address x is a "next" capability $[x.\mathsf{N}{\to}y]$ that tracks its successor y. This is analogous to the $[\mathsf{U}(x, y, v)]$ capability of CAP and we shortly demonstrate how it is utilised in our reasoning.

Since CoLoSL allows for dynamic extension of the shared state, we do not need to account for capabilities associated with all addresses. Instead, fresh capabilities are generated dynamically as needed. We demonstrate this by giving a reasoning outline of the $\mathrm{add}(v')$ method that adds value v' to the set by inserting it in the sorted list. Suppose $v_2 < v' < v_3$, and thus a new node w with value v' is to be inserted after node y. The operating thread proceeds by traversing the list by hand-over-hand locking until it reaches node y. It then locks y and claims its next pointer and moves it to its local state, as allowed by I_y. Subsequently, the shared state is extended by the resources associated with the new node and its associated capabilities ($[w.\mathsf{N}{\to}z]$) are generated on the fly as illustrated by:

$$\left[x.\mathsf{N}{\to}y\right]*\overset{x}{\underbrace{\overbrace{\textcircled{v_1}}}}_{I_x}*\left[y.\mathsf{N}{\to}z\right]*\overset{y}{\underbrace{\overbrace{\textcircled{v_2}}}}_{I_y}*\left[w.\mathsf{N}{\to}z\right]*\overset{w}{\underbrace{\overbrace{\textcircled{v'}}}}_{I_w}*\left[z.\mathsf{N}{\to}\dots\right]*\overset{z}{\underbrace{\overbrace{\textcircled{v_3}}}}_{I_z}*\cdots$$

Since the locking thread holds the next pointer of y in its local state, it modifies it to point to the new node w. It then unlocks y and returns its next pointer to the shared state. When inserting a new node between y and z, the associated interference assertion I_y allows y to be unlocked only if it has been directed to a new node whose successor is z. As such, the unlocking thread must demonstrate that the new node w does indeed point to z. In order to establish this, we use the MERGE principle to combine the subjective views of y and w as follows:

$$\boxed{[x.\text{N}{\to}y] * \textcircled{v_1}}^x_{I_x} {}^* \boxed{[y.\text{N}{\to}z] * \textcircled{v_2}}^y {}^* [w.\text{N}{\to}z] * \textcircled{v'}}^w_{I_y \cup I_w} {}^* \boxed{[z.\text{N}{\to}\dots] * \textcircled{v_3}}^z_{I_z} {}^* \cdots$$

Finally, y is unlocked; its next pointer is returned to the shared state and its next capability is modified to reflect its new successor. Using the COPY, FORGET and SHIFT principles in order, we obtain the set predicate with w inserted.

$$\boxed{[x.\text{N}{\to}y] * \textcircled{v_1}}^x_{I_x} {}^* \boxed{[y.\text{N}{\to}w] * \textcircled{v_2}}^y_{I_y} {}^* \boxed{[w.\text{N}{\to}z] * \textcircled{v'}}^w_{I_w} {}^* \boxed{[z.\text{N}{\to}\dots] * \textcircled{v_3}}^z_{I_z} {}^* \cdots$$

We can reason about the remove operation in a similar fashion. The dynamic extension afforded by the EXTEND principle allows us to generate new capabilities only when needed and thus gives way to a concise proof. Moreover, rather than having a distinct capability to modify the element at address x, for each possible successor address y (as with $[U(x, y, v)]$ in CAP), we appeal to a single capability of the form $[x.\text{N}{\to}y]$ that is modified to $[x.\text{N}{\to}y']$ whenever x's successor changes from y to y'. Lastly, using the reasoning principles of MERGE, FORGET, SHIFT and COPY, we can grow and shrink our subjective views as needed. This means that, at any one point, we only view the relevant parts of the shared state. The technical details can be found in [20].

Concluding Remarks. We have introduced CoLoSL, a new program logic for reasoning locally about the shared state. We focus on subjective views, which expand and contract to provide a flexible treatment of both the shared resource and its interference. However, CoLoSL is still young, and lacks many features of its various cousins. There are many interesting ideas present in the literature: e.g. abstract states governed by state transition systems [25]; higher-order reasoning [23]; and abstract atomicity [3]. All these ideas require further investigation. Here, our aim was to simply introduce subjective views as a fundamental new way of underpinning such reasoning.

Acknowledgements. We are grateful to Aquinas Hobor for providing us with the example of §2, and to Matthew Parkinson for suggesting to record catalysts in the model. We would also like to thank Pedro da Rocha Pinto for his continuous feedback on earlier versions of this paper. This research was funded by EPSRC grants K008528/1 and H008373/2.

References

1. Bornat, R., Calcagno, C., Yang, H.: Variables as resource in separation logic. ENTCS 155 (2006)
2. Calcagno, C., O'Hearn, P.W., Yang, H.: Local action and abstract separation logic. In: LICS (2007)
3. da Rocha Pinto, P., Dinsdale-Young, T., Gardner, P.: TaDA: A logic for time and data abstraction. In: Jones, R. (ed.) ECOOP 2014. LNCS, vol. 8586, pp. 207–231. Springer, Heidelberg (2014)
4. Dijkstra, E.: A belated proof of self-stabilization. Distributed Computing 1(1) (1986)

5. Dijkstra, E.W.: Self-stabilizing systems in spite of distributed control. Commun. ACM 17(11) (1974)
6. Dinsdale-Young, T., Birkedal, L., Gardner, P., Parkinson, M.J., Yang, H.: Views: compositional reasoning for oncurrent programs. In: POPL (2013)
7. Dinsdale-Young, T., Dodds, M., Gardner, P., Parkinson, M.J., Vafeiadis, V.: Concurrent abstract predicates. In: D'Hondt, T. (ed.) ECOOP 2010. LNCS, vol. 6183, pp. 504–528. Springer, Heidelberg (2010)
8. Dockins, R., Hobor, A., Appel, A.W.: A fresh look at separation algebras and share accounting. In: Hu, Z. (ed.) APLAS 2009. LNCS, vol. 5904, pp. 161–177. Springer, Heidelberg (2009)
9. Dodds, M., Feng, X., Parkinson, M., Vafeiadis, V.: Deny-guarantee reasoning. In: Castagna, G. (ed.) ESOP 2009. LNCS, vol. 5502, pp. 363–377. Springer, Heidelberg (2009)
10. Feng, X.: Local rely-guarantee reasoning. In: POPL. ACM (2009)
11. Gardner, P., Maffeis, S., Smith, G.D.: Towards a program logic for JavaScript. In: POPL. ACM (2012)
12. Gardner, P., Maffeis, S., Smith, G.D.: Towards a program logic for JavaScript. In: POPL. ACM (2012)
13. Gardner, P., Raad, A., Wheelhouse, M., Wright, A.: Abstract local reasoning for concurrent libraries: mind the gap. In: MFPS (2014)
14. Hobor, A., Villard, J.: The ramifications of sharing in data structures. In: POPL (2013)
15. Ishtiaq, S.S., O'Hearn, P.W.: BI as an assertion language for mutable data structures. In: POPL (2001)
16. Jones, C.B.: Specification and design of (parallel) programs. In: IFIP Cong (1983)
17. Nanevski, A., Ley-Wild, R., Sergey, I., Delbianco, G.A.: Communicating state transition systems for fine-grained concurrent resources. In: Shao, Z. (ed.) ESOP 2014 (ETAPS). LNCS, vol. 8410, pp. 290–310. Springer, Heidelberg (2014)
18. O'Hearn, P.W.: Resources, concurrency, and local reasoning. TCS 375(1-3) (2007)
19. O'Hearn, P.W., Reynolds, J.C., Yang, H.: Local reasoning about programs that alter data structures. In: Fribourg, L. (ed.) CSL 2001 and EACSL 2001. LNCS, vol. 2142, p. 1. Springer, Heidelberg (2001)
20. Raad, A., Villard, J., Gardner, P.: CoLoSL: Concurrent Local Subjective Logic (2014), http://www.doc.ic.ac.uk/~azalea/ESOP2015/CoLoSL-TR.pdf
21. Reynolds, J.C.: Separation logic: A logic for shared mutable data structures. In: LICS. IEEE Computer Society (2002)
22. Reynolds, J.C.: A short course on separation logic (2003), http://www.cs.cmu.edu/afs/cs.cmu.edu/project/fox-19/member/jcr/wwwaac2003/notes7.ps
23. Svendsen, K., Birkedal, L.: Impredicative concurrent abstract predicates. In: Shao, Z. (ed.) ESOP 2014 (ETAPS). LNCS, vol. 8410, pp. 149–168. Springer, Heidelberg (2014)
24. Svendsen, K., Birkedal, L., Parkinson, M.: Modular reasoning about separation of concurrent data structures. In: Felleisen, M., Gardner, P. (eds.) ESOP 2013. LNCS, vol. 7792, pp. 169–188. Springer, Heidelberg (2013)
25. Turon, A., Dreyer, D., Birkedal, L.: Unifying refinement and hoare-style reasoning in a logic for higher-order concurrency. SIGPLAN Not 9 (2013)
26. Vafeiadis, V., Parkinson, M.: A marriage of rely/Guarantee and separation logic. In: Caires, L., Vasconcelos, V.T. (eds.) CONCUR 2007. LNCS, vol. 4703, pp. 256–271. Springer, Heidelberg (2007)

A Separation Logic for Fictional Sequential Consistency

Filip Sieczkowski[1], Kasper Svendsen[1],
Lars Birkedal[1], and Jean Pichon-Pharabod[2]

[1] Aarhus University
{filips,ksvendsen,birkedal}@cs.au.dk
[2] University of Cambridge
Jean.Pichon@cl.cam.ac.uk

Abstract. To improve performance, modern multiprocessors and programming languages typically implement *relaxed* memory models that do not require all processors/threads to observe memory operations in the same order. To relieve programmers from having to reason directly about these relaxed behaviors, languages often provide efficient synchronization primitives and concurrent data structures with stronger high-level guarantees about memory reorderings. For instance, locks usually ensure that when a thread acquires a lock, it can observe all memory operations of the releasing thread, prior to the release. When used correctly, these synchronization primitives and data structures allow clients to recover a *fiction* of a *sequentially consistent* memory model.

In this paper we propose a new proof system, iCAP-TSO, that captures this fiction formally, for a language with a TSO memory model. The logic supports reasoning about libraries that directly exploit the relaxed memory model to achieve maximum efficiency. When these libraries provide sufficient guarantees, the logic hides the underlying complexity and admits standard separation logic rules for reasoning about their more high-level clients.

1 Introduction

Modern multiprocessors and programming languages typically implement *relaxed* memory models that allow the processor and compiler to reorder memory operations. While these reorderings cannot be observed in a sequential setting, they *can* be observed in the presence of concurrency. Relaxed memory models help improve performance by allowing more agressive compiler optimizations and avoiding unnecessary synchronization between processes. However, they also make it significantly more difficult to write correct and efficient concurrent code: programmers now have to explicitly enforce the orderings they rely on, but enforcing too much ordering negates the performance benefits of the relaxed memory model.

To help programmers, several languages [2,1] provide standard libraries that contain efficient synchronization primitives and concurrent data structures.

© Springer-Verlag Berlin Heidelberg 2015
J. Vitek (Ed.): ESOP 2015, LNCS 9032, pp. 736–761, 2015.
DOI: 10.1007/978-3-662-46669-8_30

These constructs restrict the reordering of low-level memory operations in order to express more high-level concepts, such as acquiring or releasing a lock, or pushing and popping an element from a stack. For instance, the collections provided by `java.util.concurrent` enforce that memory operations in a first thread prior to adding an element to a collection cannot be reordered past the subsequent removal by a second thread. Provided the library is used correctly, these high-level guarantees suffice for clients to recover a fiction of a sequentially consistent memory model, without introducing unnecessary synchronization in client code.

The result is a two-level structure: At the low-level we have libraries that directly exploit the relaxed memory model to achieve maximum efficiency, but enforce enough ordering to provide a fiction of sequential consistency; at the high-level we have clients that use these libraries for synchronization. While we have to reason about relaxed behaviors when reasoning about low-level libraries, ideally we should be able to use standard reasoning for the high-level clients. In this paper we propose a new proof system, iCAP-TSO, specifically designed to support this two-level approach, for a language with a TSO memory model.

We focus on TSO for two reasons. Firstly, while the definition of TSO is simple, reasoning about TSO programs is difficult, especially modular reasoning. Reasoning therefore greatly benefits from a program logic, in particular with the fiction of sequential consistency we provide. Moreover, a logic specifically tailored for TSO allows us to reason about idioms that are valid under TSO but not necessarily under weaker memory models, such as double-checked initialization (see examples).

In the TSO memory model, each thread is connected to main memory via a FIFO store buffer, modeled as a sequence of (address, value) pairs, see, e.g., [20]. When a value is written to an address, the write is recorded in the writing thread's store buffer. Threads can commit these buffered writes to main memory at any point in time. When reading from a location, a thread first consults its own store buffer; if it contains buffered writes to that location, then the thread reads the value of its last buffered write to that location; otherwise, it consults main memory. Each thread thus has its own *subjective* view of the current state of memory, which might differ from other threads'.

In contrast, in a sequentially consistent memory model, threads read and write directly to main memory and thus share an *objective* view of the current state of the memory. In separation logics for languages with sequentially consistent memory models we thus use assertions such as $x \mapsto 1$, which express an *objective* property of the value of location x. Since in the TSO setting each thread has a subjective view of the state, in order to preserve the standard proof rules for reading and writing, we need a *subjective* interpretation of pre- and postconditions. The first component of our proof system, the SC logic (for sequentially consistent), provides exactly this kind of subjective interpretation.

In the SC logic we use specifications of the form $\{P\}\ e\ \{r.Q\}$, which express that if e is executed by some thread t from an initial state that satisfies P *from the point of view* of t and e terminates with some value v, then the terminal state satisfies Q[v/r] from the point of view of thread t. Informally, an assertion P holds from the point of view of a thread t if the property holds in a heap updated with t's store buffer. Additionally, to ensure that other threads' store buffers cannot invalidate the property, no store buffer other than t's can contain buffered writes to the parts of the heap described by P. In particular, $x \mapsto v$ holds from the point of view of thread t, if the value of x that t can observe is v. We shall see that this interpretation justifies the standard separation logic read and write rules.

What about transfer of resources? In separation logics for sequentially consistent memory models, assertions about resources are objective and can thus be transferred freely between threads. However, since assertions in the SC logic are interpreted subjectively, they may not hold from the point of view of other threads. To transfer resources between threads, their views of the resources must match. Thus, the SC logic is not expressive enough to reason about implementations of low-level concurrency primitives. To verify such data structures, we use the TSO logic, which allows us to reason about the complete TSO machine state, including store buffers. Importantly, in cases where the data structure provides enough synchronization to transfer resources between two threads, we can verify the implementation against an SC specification. This gives us a *fiction of sequential consistency* and allows us to reason about the *clients* of such data structures using the SC logic.

Example. To illustrate, consider a simple spin-lock library with acquire and release methods. We can specify the lock in the SC logic as follows.

$$\exists \text{isLock}, \text{locked} : \text{Prop}_{sc} \times \text{Val} \to \text{Prop}_{sc}. \ \forall R : \text{Prop}_{sc}. \ \text{stable}(R) \ \Rightarrow$$
$$\{R\} \ \text{Lock}() \ \{r. \ \text{isLock}(R, r)\}$$
$$\wedge \ \{\text{isLock}(R, \textbf{this})\} \ \text{Lock.acquire}() \ \{\text{locked}(R, \textbf{this}) * R\}$$
$$\wedge \ \{\text{locked}(R, \textbf{this}) * R\} \ \text{Lock.release}() \ \{\top\}$$
$$\wedge \ \text{valid}(\forall x : \text{Val}. \ \text{isLock}(R, x) \Leftrightarrow \text{isLock}(R, x) * \text{isLock}(R, x))$$
$$\wedge \ \forall x : \text{Val}. \ \text{stable}(\text{isLock}(R, x)) \wedge \text{stable}(\text{locked}(R, x))$$

Here Prop_{sc} is the type of propositions of the SC logic, and isLock and locked are thus abstract representation predicates. The predicate $\text{isLock}(R, x)$ expresses that x is a lock protecting the resource invariant R, while $\text{locked}(R, x)$ expresses that the lock x is indeed locked. Acquiring the lock grants ownership of R, while releasing the lock requires the client to relinquish ownership of R. Since the resource invariant R is universally quantified, this is a very strong specification; in particular, the client is free to instantiate R with any SC proposition. This specification requires the resource invariant to be stable, $\text{stable}(R)$. The reason is that R could in general refer to shared resources and to reason about shared resources we need to ensure we only use assertions that are closed under interference from the environment. This is what stability expresses.

Note that this specification is expressed in the SC logic and the specification of the acquire method thus grants ownership of the resource R from the caller's point of view. Likewise, the release method only requires that R holds from the caller's point of view. This specification thus provides a fiction of sequential consistency, by allowing transfer of SC resources. Crucially, since the lock specification is an SC specification, we can reason about the clients that use it to transfer resources *entirely* using the standard proof rules of the SC logic. We illustrate this by verifying a shared bag in Section 3.

Using the TSO logic we can verify that an *efficient* spin-lock implementation satisfies this specification. The spin-lock that we verify is inspired by the Linux spin-lock implementation [3], which allows the release to be buffered. To verify the implementation we must prove that between releasing and acquiring the lock, the releasing and acquiring threads' views of the resource R match. Intuitively, this is the case because if R holds from the point of view of the releasing thread, once the buffered release makes it to main memory, R holds objectively. This style of reasoning relies on the ordering of buffered writes. To capture this intuition, we introduce a new operator in the TSO logic for expressing such ordering dependencies. This operator supports modular reasoning about many of the ordering dependencies that arise naturally in TSO-optimized data structures. In Section 5 we illustrate how to use the TSO logic to verify the spin-lock and briefly discuss other case studies we have verified.

iCAP. iCAP-TSO builds on iCAP [22], a recent extension of higher-order separation logic [5] for modular reasoning about concurrent higher-order programs with shared mutable state. While the meta-theory of iCAP is intricate, understanding it is not required to understand this paper. By building on iCAP, we can use higher-order quantification to express abstract specifications that abstract over both internal data representations and client resources (such as the resource invariant R in the lock specification). This is crucial for modular verification, as it allows libraries and clients to be verified *independently* against abstract specifications and thus to scale verification to large programs. In addition, by abstractly specifying possible interference from the environment, iCAP allows us to reason about shared mutable state without having to consider all possible interleavings.

Summary of Contributions. We provide a new proof system, iCAP-TSO, for a TSO memory model, which features:

- a novel logic for reasoning about low-level *racy* code, called the TSO logic; this logic features new connectives for expressing ordering dependencies introduced by store buffers;
- a *standard* separation logic, called the SC logic, that allows *simple* reasoning for clients that transfer resources through libraries that provide sufficient synchronization;
- a notion of *fiction of sequential consistency* which allows us to provide SC specifications for libraries that provide synchronized resource transfer, even if the implementations exhibit relaxed behaviors.

Moreover, we prove soundness of iCAP-TSO. We use the logic to verify *efficient* spin-lock and bounded ticket lock implementations, double-checked initialization

$$\text{Val} \ni v ::= \ x \mid \textbf{null} \mid \textbf{this} \mid o \mid n \mid b \mid ()$$
$$\text{Exp} \ni e ::= \ v \mid \textbf{let } x = e_1 \textbf{ in } e_2 \mid \textbf{if } v \textbf{ then } e_1 \textbf{ else } e_2 \mid \textbf{new } C(\bar{v})$$
$$\mid v.f \mid v_1.f := v_2 \mid v.m(\bar{v}) \mid \textbf{CAS}(v_1.f, v_2, v_3) \mid \textbf{fence} \mid \textbf{fork}(v.m)$$

Fig. 1. Syntax of the programming language. In the definition of values, n ranges over machine integers, b over booleans, and o over object references. In the definition of expressions, f ranges over the field names, and m over the method names.

that uses a spin-lock internally, a circular buffer, and Treiber's stack against SC specifications. Crucially, this means that we can reason about clients of these libraries *entirely* using standard separation logic proof rules!

Outline. In Section 2 we introduce the programming language that we reason about and its operational semantics. Section 3 illustrates how the fiction of sequential consistency allows us to reason about shared resources using standard separation logic. Section 4 introduces the TSO logic and connectives for reasoning about store buffers. In Section 5 we illustrate the use of the TSO logic to verify an efficient spin-lock and briefly discuss the other case-studies we have verified. In Section 6 we discuss the iCAP-TSO soundness theorem. Finally, in Sections 7 and 8 we discuss related work and future work and conclude. Details and proofs can be found in the accompanying technical report [21]. The technical report is available online at http://cs.au.dk/~filips/icap-tso-tr.pdf.

2   Language

We build our logic for a simple, class-based programming language. For simplicity of both semantics and the logic, the language uses let-bindings and expressions, but we keep it relatively low-level by ensuring that all the values are machine-word size. The values include variables, natural numbers, booleans, unit, object references (pointers), the null pointer and the special variable **this**. The expressions include values, let bindings, conditionals, constructor and method calls, field reads and writes, atomic compare-and-swap expressions, a fork call and an explicit fence instruction. The syntax of values and expressions is shown in Figure 1. The class and method definitions are standard and therefore omitted; they can be found in the accompanying technical report.

To simplify the construction of the logic, we follow the Views framework [10] and split the operational semantics into two components. The first is a thread-local small-step semantics labeled with actions that occur during the step, the second — an action semantics that defines the effect of each action on the machine state, which in our case consists of the heap and the store buffer pool. In the thread-local semantics, a thread, which consists of a thread identifier and an expression, takes a single step of evaluation to a finite set of threads that contains besides the original thread also the threads spawned by this step.

$$\frac{\quad\quad\quad\quad\quad\quad\quad\quad\quad\quad\quad}{(t, E[\mathsf{o}.\mathsf{f} := \mathsf{v}]) \xrightarrow{write(t,o,f,v)} \{(t, E[\mathsf{()}])\}} \text{ WRITE} \quad \frac{\quad\quad\quad\quad\quad\quad\quad\quad\quad}{(t, E[\mathsf{o}.\mathsf{f}]) \xrightarrow{read(t,o,f,v)} \{(t, E[\mathsf{v}])\}} \text{ READ}$$

$$\frac{\quad\quad\quad\quad\quad\quad\quad\quad\quad\quad\quad}{(t, E[\mathsf{CAS}(\mathsf{o}.\mathsf{f}, \mathsf{v_o}, \mathsf{v_n})]) \xrightarrow{cas(t,o,f,v_o,v_n,r)} \{(t, E[\mathsf{r}])\}} \text{ CAS} \quad \frac{\quad\quad\quad\quad}{(t, \mathsf{e}) \xrightarrow{flush(t)} \{(t, \mathsf{e})\}} \text{ FLUSH}$$

$$\frac{\mathsf{body}(C, m) = (\mathsf{unit}\ m() = e) \quad t \neq t'}{(t, E[\mathsf{fork}(\mathsf{o}.m)]) \xrightarrow{fork(t,o,C,t')} \{(t, E[\mathsf{()}]), (t', e[o/\mathsf{this}])\}} \text{ FORK}$$

Fig. 2. Selected cases of the thread-local operational semantics

It also emits the action that describes the interaction with the memory. For instance, the WRITE rule in Figure 2 applies when the expression associated with thread t is an assignment (possibly in some evaluation context). It reduces by replacing the assignment with a unit value, and emits a write action that states that thread t wrote the value v to the field f of object o.

The non-fault memory state consists of a heap — a finite map from pairs of an object reference and a field to semantic values (i.e., all the values that are not variables) — and a store buffer pool, which contains a sequence of buffered updates for each of the finitely many thread identifiers. The memory can also be in a fault state (written $\maltese$), which means that an error in the execution of the program has occurred. The action semantics interprets the actions as functions from memory states to sets of memory states: if it is impossible for the action to occur in a given state, the result is an empty set; if, however, the action may occur in the given state but it would be erroneous, the result is the fault state. Consider the write action emitted by reducing an assignment. In Figure 3 we can see the interpretation of the action: there are three distinct cases. The action is successful if there is a store buffer associated with the thread that emitted the action and the object is allocated in the heap, and has the appropriate field. In this case, the write gets added to the end of the thread's buffer. However, the write action can have two additional outcomes: if there is no store buffer associated with the thread in the store buffer pool, the initial state had to be ill-formed, and so the interpretation of the action is an empty set; however, if the thread is defined, but the reference to the field $o.f$ is not found in the heap, the execution will fault.

The state of a complete program consists of the thread pool and a memory state, and is consistent if the memory state is a fault, or the domain of the store buffer pool equals the domain of the thread pool. The complete semantics proceeds by reducing one of the threads using the thread-local semantics, then interpreting the resulting action with the action semantics, and reducing to a memory state in the resultant set:

$$\frac{t \in \mathrm{dom}\,T \quad (t, T(t)) \xrightarrow{a} T' \quad \mu' \in [\![a]\!](\mu)}{(\mu, T) \to (\mu', (T - t) \uplus T')}$$

Note how in some cases, notably read, this might require "guessing" the return value, and checking that the guess was right using the action semantics. Some of the

$\llbracket read(t,o,f,v) \rrbracket(h,U) =$

$$
\begin{cases}
\{(h,U)\} & \text{if } (o,f) \in \text{dom } h \text{ and lookup}(o.f,U(t),h) = v \\
\emptyset & \text{if } t \notin \text{dom } U \text{ or } (o,f) \in \text{dom } h \text{ and lookup}(o.f,U(t),h) \neq v \\
\{\lightning\} & \text{if } (o,f) \notin \text{dom } h
\end{cases}
$$

$\llbracket write(t,o,f,v) \rrbracket(h,U) =$

$$
\begin{cases}
\{(h,U[t \mapsto U(t) \cdot (o,f,v)])\} & \text{if } (o,f) \in \text{dom } h \text{ and } t \in \text{dom } U \\
\{\lightning\} & \text{if } (o,f) \notin \text{dom } h \\
\emptyset & \text{if } t \notin \text{dom } U
\end{cases}
$$

$\llbracket cas(t,o,f,v_o,v_n,r) \rrbracket(h,U) =$

$$
\begin{cases}
\{(\text{flush}(h,U(t) \cdot (o,f,v_n)), U[t \mapsto \varepsilon])\} & \text{if } (o,f) \in \text{dom } h, \; r = \textbf{true} \\
& \text{and lookup}(o.f,U(t),h) = v_o \\
\{(\text{flush}(h,U(t)), U[t \mapsto \varepsilon])\} & \text{if } (o,f) \in \text{dom } h, \; r = \textbf{false} \\
& \text{and lookup}(o.f,U(t),h) \neq v_o \\
\{\lightning\} & \text{if } (o,f) \notin \text{dom } h \\
\emptyset & \text{otherwise}
\end{cases}
$$

$\llbracket flush(t) \rrbracket(h,U) =$

$$
\begin{cases}
\{(h[(o,f) \mapsto v], U[t \mapsto \alpha])\} & \text{if } U(t) = (o,f,v) \cdot \alpha \text{ and } (o,f) \in \text{dom } h \\
\emptyset & \text{if } t \notin \text{dom}(U), \; U(t) = \varepsilon, \text{ or } (o,f) \notin \text{dom } h
\end{cases}
$$

Fig. 3. Selected cases of the action semantics. The lookup function finds the newest value associated with the field, including the store buffer, while the flush function applies all the updates from the store buffer to the heap in order.

cases of the semantics are written out in Figures 2 and 3. In particular, we show the reduction and action semantics that correspond to the (nondeterministic) flushing of a store buffer: a flush action can be emitted by a thread at any time, and the action is interpreted by flushing the oldest buffered write to the memory. Note also the rules for the compare-and-swap expression: similarly to reading, the return value has to be guessed by the thread-local semantics. However, whether the guess matches the state of the memory or not, the *whole* content of the store buffer is written to main memory. Moreover, if the compare-and-swap succeeds, the update resulting from it is also written to main memory. Thus, this expression can serve as a synchronization primitive.

Note that our operational semantics is the usual TSO semantics of the x86 [20] adapted to a high-level language. The only difference is that we have a notion of allocation of objects, which does not exist in the processor-level semantics. Our semantics allocates the new object directly on the heap to avoid different threads trying to allocate the same object.

3   Reasoning in the SC Logic

The SC logic of iCAP-TSO allows us to reason about code that always uses synchronization to transfer resources *using standard separation logic*, without having to reason about store buffers. Naturally, this also includes standard mutable data structures without any sharing. We can thus easily verify a list library in the SC logic against the standard separation logic specification as it enforces a *unique* owner. Crucially, within the SC logic we can also use libraries that provide synchronized resource transfer. For instance, we can use the specification of the spin-lock from the Introduction and the fiction of sequential consistency that it provides. We illustrate this point by verifying a shared bag library, implemented as a list protected by a lock.

The SC Logic. The SC logic is an intuitionistic higher-order separation logic. Recall that the SC logic features Hoare triples of the form $\{P\}$ e $\{r.\ Q\}$, where P and Q are SC assertions. Formally, SC assertions are terms of type $\mathsf{Prop_{sc}}$. SC assertions include the usual connectives and quantifiers of higher-order separation logic and language specific assertions such as points-to, $x.f \mapsto v$, for asserting the value of field f of object x.

Recall that SC triples employ a subjective interpretation of the pre- and postcondition: $\{P\}$ e $\{r.\ Q\}$ expresses that if thread t executes the expression e from an initial state where P holds from the point of view of thread t and e terminates with value v then $Q[v/r]$ holds for the terminal state from the point of view of thread t. An assertion P holds from the point of view of a thread t if P's assertions about the heap hold from the point of view of t's store buffer and main memory *and* no other thread's store buffer contains a buffered write to these parts of the heap. The assertion $x.f \mapsto v$ thus holds from the point of view of thread t if

- the value of the most recently buffered write to x.f in t's store buffer is v
- or t's store buffer does not contain any buffered writes to x.f and the value of x.f in main memory is v

and no other threads store buffer contains a buffered write to x.f. The condition that no other thread's store buffer can contain a buffered write to x.f ensures that flushing of store buffers cannot invalidate $x.f \mapsto v$ from the point of view of a given thread.

If $x.f \mapsto v$ holds from the point of view of thread t and thread t attempts to read x.f it will thus read the value v either from main memory or its own store buffer. Likewise, if $x.f \mapsto v_1$ holds from the point of view of thread t and thread t writes v_2 to x.f, afterwards $x.f \mapsto v_2$ holds from the point of view of thread t. We thus get the standard rules for reading and writing to a field in our SC logic:

$$\frac{}{\{x.f \mapsto v\}\ x.f\ \{r.\ x.f \mapsto v * r = v\}}\ \text{S-Read} \qquad \frac{}{\{x.f \mapsto v_1\}\ x.f := v_2\ \{r.\ x.f \mapsto v_2\}}\ \text{S-Write}$$

Using SC specifications: a shared bag To illustrate how we can use the lock specification from the Introduction, consider a shared bag implemented using a list. Each shared bag maintains a list of elements and a lock to ensure exclusive access to the list of elements. Each bag method acquires the lock and calls the corresponding method of the list library before releasing the lock.

We take the following specification, which allows unrestricted sharing of the bag, to be our specification of a shared bag. This is not the most general specification we can express — we discuss a more general specification of a stack in the technical report — but it suffices to illustrate that verification of the shared bag against such specifications is standard. Since the specification allows unrestricted sharing (the bag predicate is duplicable), no client can know the contents of the bag; instead, the specification allows clients to associate ownership of additional resources (expressed using the predicate P) with each element in the bag.

$$\exists \mathsf{bag} : \mathsf{Val} \times (\mathsf{Val} \to \mathsf{Prop}_{\mathsf{sc}}) \to \mathsf{Prop}_{\mathsf{sc}}. \ \forall \mathsf{P} : \mathsf{Val} \to \mathsf{Prop}_{\mathsf{sc}}.$$
$$(\forall \mathsf{x} : \mathsf{Val}. \ \mathsf{stable}(\mathsf{P}(\mathsf{x}))) \ \Rightarrow$$
$$\{\top\} \ \mathsf{Bag}() \ \{\mathsf{r}. \ \mathsf{bag}(\mathsf{r}, \mathsf{P})\} \ \wedge$$
$$\{\mathsf{bag}(\mathbf{this}, \mathsf{P}) * \mathsf{P}(\mathsf{x})\} \ \mathsf{Bag.push}(\mathsf{x}) \ \{\top\} \ \wedge$$
$$\{\mathsf{bag}(\mathbf{this}, \mathsf{P})\} \ \mathsf{Bag.pop}() \ \{\mathsf{r}. \ (\mathsf{P}(\mathsf{r}) \vee \mathsf{r} = \mathbf{null})\} \ \wedge$$
$$\forall \mathsf{x} : \mathsf{Val}. \ \mathsf{valid}(\mathsf{bag}(\mathsf{x}, \mathsf{P}) \Leftrightarrow \mathsf{bag}(\mathsf{x}, \mathsf{P}) * \mathsf{bag}(\mathsf{x}, \mathsf{P}))$$

Pushing an element x thus requires the client to transfer ownership of P(x) to the bag. Likewise, either **pop** returns **null** or the client receives ownership of the resources associated with the returned element.

To verify the implementation against this specification, we first have to define the abstract bag representation predicate. To define bag we first need to choose the resource invariant of the lock. Intuitively, the lock owns the list of elements and the resources associated with the elements currently in the list. This is expressed by the following resource invariant $R_{\mathsf{bag}}(\mathsf{xs}, \mathsf{P})$, where xs refers to the list of elements.

$$R_{\mathsf{bag}}(\mathsf{xs}, \mathsf{P}) \overset{\mathrm{def}}{=} \exists \mathsf{l} : \mathrm{list} \ \mathsf{Val}. \ \mathsf{lst}(\mathsf{xs}, \mathsf{l}) * \circledast_{\mathsf{y} \in mem(\mathsf{l})} \mathsf{P}(\mathsf{y})$$

The bag predicate asserts *read-only* ownership of the lock and elms fields, and that the lock field refers to a lock with the above resource invariant.

$$\mathsf{bag}(\mathsf{x}, \mathsf{P}) \overset{\mathrm{def}}{=} \exists \mathsf{y}, \mathsf{xs} : \mathsf{Val}. \ \mathsf{x.lock} \mapsto \mathsf{y} * \mathsf{x.elms} \mapsto \mathsf{xs} * \mathsf{isLock}(R_{\mathsf{bag}}(\mathsf{xs}, \mathsf{P}), \mathsf{y})$$

Now, we are ready to verify the bag methods. The most interesting method is pop, as it actually returns the resources associated with the elements it returns. A proof outline of pop is presented below. The crucial thing to note is that since locks introduce sufficient synchronization, they can mediate ownership transfer. Thus, once a thread t acquires the lock, it receives the resource invariant $R_{\mathsf{bag}}(\mathsf{xs}, \mathsf{P})$ *from the point of view of t*. Since it now owns the list, t can call the List.pop method, and finally — again using the fiction of sequential consistency provided by the lock — release the lock.

```
class Bag {
  Lock lock; List elms;
  Object pop() =
     {bag(this, P)}
     let x = this.lock in let xs = this.elms in x.acquire();
     {this.elms ↦ xs * locked(R_bag(xs, P), x) * R_bag(xs, P)}
     let z = xs.pop() in
     {locked(R_bag(xs, P), x) * R_bag(xs, P) * (z = null ∨ P(z))}
     x.release();
     {isLock(R_bag(xs, P), x) * (z = null ∨ P(z))}
     z
     {r. r = null ∨ P(r)}
   ...
}
```

This example illustrates the general pattern that we can use to verify clients of libraries that provide fiction of sequential consistency. As long as these clients only transfer resources using libraries that provide sufficient synchronization, the verification can proceed entirely within the SC logic.

4   TSO Logic and Connectives

In this section we describe the TSO logic and introduce our new TSO connectives that allow us to reason about the kinds of relaxed behaviors that occur in low-level concurrency libraries.

We can express the additional reorderings that the memory model allows by extending the space of states over which the assertions are built. In the case of our TSO model, we include the store buffer pool as an additional component of the memory state. However, reasoning about the buffers directly would be extremely unwieldy and contrary to the spirit of program logics. Hence, we introduce new logical connectives that allow us to specify this interference abstractly, and provide appropriate reasoning rules.

The Triples and Assertions of the TSO Logic. First, however, we need to consider how the TSO logic is built. As mentioned in the Introduction, its propositions extend the propositions of SC logic by adding the store buffer pool component. Just like SC assertions, this space forms a higher-order intuitionistic separation logic, with the usual rules for reasoning about assertion entailment. However, we are still reasoning about the code running in a particular thread and we often need to state properties that hold of its own store buffer. Thus, formally, the typing rule for the TSO logic triples is as follows:

$$\frac{P : \mathsf{Tld} \to \mathsf{Prop}_{\mathsf{TSO}} \qquad Q : \mathsf{Tld} \to \mathsf{Val} \to \mathsf{Prop}_{\mathsf{TSO}}}{[P] \; e \; [Q] : \mathsf{Spec}}$$

where Tld is the type of thread identifiers and Spec is the type of specifications. We usually keep this quantification over thread identifiers implicit, by introducing appropriate syntactic sugar for the TSO-level connectives. The logic also includes another family of Hoare triples, the atomic triples, with the following typing rule:

$$\frac{\text{atomic}(e) \qquad P : \text{TId} \to \text{Prop}_{\text{TSO}} \qquad Q : \text{TId} \to \text{Val} \to \text{Prop}_{\text{TSO}}}{\langle P \rangle\ e\ \langle Q \rangle : \text{Spec}}$$

As the rule states, these triples can only be used to reason about atomic expressions — read, write, fence, and compare-and-swap. This feature is inherited from iCAP, as a means of reasoning about the way the shared state changes at the atomic updates. We give an example of such reasoning in Section 5.

Note that the triples above use a different space of assertions than the SC triples introduced in Section 3. Hence, in order to provide the fiction of sequential consistency and prove SC specifications for implementations whose correctness involves reasoning about buffered updates, we need to use of both of these spaces in the TSO logic. To this end we define two embeddings of Prop_{SC} into Prop_{TSO}.

The Subjective Embedding. The subjective embedding is denoted $\ulcorner - \textbf{ in } - \urcorner$: $\text{Prop}_{\text{SC}} \times \text{TId} \to \text{Prop}_{\text{TSO}}$. Intuitively, $\ulcorner P \textbf{ in } t \urcorner$ means that P holds from the perspective of thread t — including the possible buffered updates in the store buffer of t, but forbidding buffered updates that "touch" P by other threads. Thus, it means that if the buffer of thread t is flushed to the memory, P will hold in the resulting state. Note that this corresponds to the interpretation of the assertions in the SC triples.

For a concrete example of what this embedding means, consider an assertion $x.f \mapsto v : \text{Prop}_{\text{SC}}$. Clearly, we can use our embedding to get $\ulcorner x.f \mapsto v \textbf{ in } t \urcorner$. This assertion requires that the reference x.f is defined, and there are no buffered updates in store buffers of threads other than t. As for t's store buffer, the last update of x.f has to set its value to v or, if there are no buffered updates of x.f, the value associated with it in main memory is v. This means that from the point of view of thread t, $x.f \mapsto v$ holds, but from the point of view of the other threads, the only information is that x.f is defined.

The Objective Embedding. The objective embedding is denoted $\ulcorner - \urcorner$: $\text{Prop}_{\text{SC}} \to \text{Prop}_{\text{TSO}}$. The idea is that $\ulcorner P \urcorner$ holds in a state that does include store buffers if P holds in the state where we ignore the buffers *and* none of the buffers contain buffered updates to the locations mentioned by P. The intuition behind this embedding is that P should hold *in main memory*, and as such from the point of view of *all* threads. This makes it very useful to express resource transfer: an assertion that holds for all threads can be transferred to any of them.

Using the points-to example again, $\ulcorner x.f \mapsto v \urcorner$ means precisely that the reference x.f is defined, its associated value in the heap is v, and there are no buffered updates in any of the store buffers to the field x.f.

Semantics of assertions and embeddings In the following, we provide a simplified presentation of parts of the model for the interested reader, to flesh out the intuitions given above. We concentrate on the interpretation of TSO-specific constructs and elide the parts inherited from iCAP, which are orthogonal.

Following the Views framework [10], TSO assertions (terms of type $\mathsf{Prop}_{\mathsf{TSO}}$) are modeled as predicates over *instrumented states*. In addition to the underlying machine state, instrumented states contain shared regions, protocols and phantom state. The instrumented states form a Kripke model in which worlds consist of allocated regions and their associated protocols. Since iCAP-TSO inherits iCAP's impredicative protocols [22], worlds need to be recursively defined. Hence we use a meta-theory that supports the definition of sets by guarded recursion, namely the so-called internal language of the topos of trees [6]. We refer readers to the accompanying technical report [21] for details and proofs.

Propositions are interpreted as subsets of instrumented states upwards-closed wrt. extension ordering. The states are instrumented with shared regions and protocols which we inherit from iCAP. In the following these are denoted with X, and we elide their definition.

$$[\![\mathsf{Prop}_{\mathsf{TSO}}]\!] \stackrel{\text{def}}{=} \mathcal{P}^{\uparrow}(LState \times SPool \times X) \qquad [\![\mathsf{Prop}_{\mathsf{SC}}]\!] \stackrel{\text{def}}{=} \mathcal{P}^{\uparrow}(LState \times X)$$

In these definitions $LState$ denotes the local state, including the partial physical heap, while $SPool$ is the store-buffer pool that directly corresponds to the operational semantics. Note that the interpretation of $\mathsf{Prop}_{\mathsf{SC}}$ does not consider store buffer pools, only the local state and the instrumentation. This allows us to interpret the connectives at this level in a standard way.

At the level of $\mathsf{Prop}_{\mathsf{TSO}}$, we have several important connectives, namely separating conjunction, and both embeddings we have introduced before. These are defined as follows:

$$\mathrm{lfd}(l, U) \stackrel{\text{def}}{=} \forall t, v.\ \forall (o, f) \in \mathrm{dom}(l).\ (o, f, v) \notin U(t)$$

$$[\![\ulcorner \mathsf{P} \urcorner]\!] \stackrel{\text{def}}{=} \{(l, U, x) \mid \exists l' \leq l.\ (l', x) \in [\![\mathsf{P}]\!] \wedge \mathrm{lfd}(l', U)\}$$

$$[\![\ulcorner \mathsf{P}\ \mathbf{in}\ t \urcorner]\!] \stackrel{\text{def}}{=} \{(l, U, x) \mid (\mathrm{flush}(l, U(t)), U[t \mapsto \varepsilon], x) \in [\![\ulcorner \mathsf{P} \urcorner]\!]\}$$

$$[\![\mathsf{P} * \mathsf{Q}]\!] \stackrel{\text{def}}{=} \{(l, U, x) \mid \exists l_1, l_2.\ l = l_1 \bullet l_2 \wedge (l_1, U, x) \in [\![\mathsf{P}]\!] \wedge (l_2, U, x) \in [\![\mathsf{Q}]\!]\}.$$

The embeddings are defined using the auxiliary "locally flushed" $\mathrm{lfd}(l, U)$ predicate, which ensures that no updates to $\mathrm{dom}(l)$ are present in U. We only require this on a sub-state of the local state to ensure good behavior with respect to the extension ordering. The subjective embedding is then defined in terms of the objective one, with all the updates in the corresponding store-buffer flushed. Finally, the separating conjunction is defined as a composition of local states. Separating conjunction does not split the instrumentation or store-buffer pool and both conjuncts thus have to hold with the same pools of buffered updates.

Reasoning About Buffered Updates. To effectively reason about the store buffers, we need an operator that describes how the state *changes* due to an update. To this end, we define $-\,\mathcal{U}_{-}\,- : \mathsf{Prop}_{\mathsf{TSO}} \times \mathsf{TId} \times \mathsf{Prop}_{\mathsf{TSO}} \to \mathsf{Prop}_{\mathsf{TSO}}$. Because of its role, this connective has a certain temporal feel: in fact, it behaves in a way that is somewhat similar to the classic "until" operator. Intuitively, $\mathsf{P}\,\mathcal{U}_t\,\mathsf{Q}$ means that there exists a buffered update in the store buffer of thread t, such that until this update is flushed the assertion P holds, while after the update gets written to memory, the assertion Q holds. Thus, it can be used to describe the ordering dependencies introduced by the presence of store buffers. This intuition should become clearer by observing the proof rules in Figure 4 (explained in the following).

Again, let us consider a simple example. In the state described by $\ulcorner x.f \mapsto 1 \urcorner \mathcal{U}_t$ $\ulcorner x.f \mapsto 2 \urcorner$, we know that the value of x.f in the heap is 1, and that there exists a buffered update in thread t. Before that update there are no updates to x.f, due to the use of $\ulcorner - \urcorner$, so it has to be the first update to x.f in the store buffer of t. Additionally, after it gets flushed $\ulcorner x.f \mapsto 2 \urcorner$ holds — so the update must set x.f to 2. Since the right-hand side of $\mathcal{U}_t$ also uses $\ulcorner - \urcorner$, we also know that there are no further buffered updates to x.f. This means that the thread t can observe the value of x.f to be 2, while all of the other threads can observe it to be 1. Note that, since $\mathcal{U}_t$ is a binary operator on $\mathsf{Prop}_{\mathsf{TSO}}$, it is possible to use it to express multiple buffered updates.

The semantics of the until operator follow very closely the intuition given above. Note that for some assertions and states, several choices of the update would validate the conditions. However, this rarely occurs in practice due to the use of the objective embedding, which requires no updates in its footprint.

$$[\![P \, \mathcal{U}_t \, Q]\!] \stackrel{\text{def}}{=} \{ (l, U, x) \mid \exists \alpha, \beta, o, f, v. \ U(t) = \alpha \cdot (o, f, v) \cdot \beta \ \wedge$$
$$(l, U[t \mapsto \alpha], x) \in [\![P]\!] \wedge (\text{flush}(l, \alpha \cdot (o, f, v)), U[t \mapsto \beta], x) \in [\![Q]\!] \}$$

Relating the Two Embeddings. The two embeddings we have defined are in fact quite related. Since an assertion under an objective embedding holds from the perspective of any thread, we get $\ulcorner P \urcorner \Rightarrow \ulcorner P \text{ in } t \urcorner$. We also have $P \, \mathcal{U}_t \, \ulcorner Q \urcorner \Rightarrow \ulcorner Q \text{ in } t \urcorner$: since there is a buffered update at which $\ulcorner Q \urcorner$ starts to hold, Q holds from t's perspective.

Since most of the time we are reasoning from the perspective of a particular thread, we also include some syntactic sugar: $\mathcal{U}$ is a shorthand for an update in the current thread, while $\mathcal{U}^\circ$ is a shorthand for an update in some thread *other than the current one*. We also use $\overline{P}$ as a shorthand for $\ulcorner P \text{ in } t \urcorner$, where t is the current thread. To make the syntax simpler, whenever we need to refer to the thread identifier explicitly, we use an assertion $\mathsf{iam}(t)$. This is just syntactic sugar for a function $\lambda t'. \ t = t' : \mathsf{Tld} \to \mathsf{Prop}_{\mathsf{TSO}}$, which allows us to bind the thread identifier of the thread we are reasoning about to a logical variable.

Reading and Writing State. The presence of additional connectives that mention the state makes reading fields of an object and writing to them more involved in the TSO logic than in standard separation logic. We deal with this by introducing additional judgments that specify when we can read a value and what the result of flushing a store buffer will be. Intuitively, $P \vdash_{\mathsf{rd}(t)} x.f \mapsto v$ specifies that thread t can read the value v from the reference x.f — precisely what we need for reading the state. The other new judgment, $P \vdash_{\mathsf{fl}(t)} Q$, means that if we flush thread t's store buffer in a state specified by P, the resulting state will satisfy Q. This action judgment is clearly useful for specifying actions that flush the store buffer:

$$\frac{}{\rhd\ulcorner\mathsf{x.f} \mapsto \mathsf{v}\ \mathbf{in}\ t\urcorner \vdash_{\mathrm{rd}(t)} \mathsf{x.f} \mapsto \mathsf{v}}\ \text{RD-Ax} \qquad \frac{\mathsf{P} \vdash_{\mathrm{rd}(t)} \mathsf{x.f} \mapsto \mathsf{v} \qquad t \neq t'}{\mathsf{P}\ \mathcal{U}_{t'}\ \mathsf{Q} \vdash_{\mathrm{rd}(t)} \mathsf{x.f} \mapsto \mathsf{v}}\ \text{RD-}\mathcal{U}\text{-Neq}$$

$$\frac{}{\rhd\ulcorner\mathsf{P}\ \mathbf{in}\ t\urcorner \vdash_{\mathrm{fl}(t)} \ulcorner\mathsf{P}\urcorner}\ \text{FL-Ax} \qquad \frac{\mathsf{P}(t) \vdash_{\mathrm{fl}(t)} \ulcorner\mathsf{x.f} \mapsto -\urcorner * \mathsf{Q}(t)}{\langle \mathsf{P} * \mathrm{iam}(t)\rangle\ \mathsf{x.f} := \mathsf{v}\ \langle \_.\ \mathsf{P}\ \mathcal{U}\ (\mathsf{Q} * \ulcorner\mathsf{x.f} \mapsto \mathsf{v}\urcorner)\rangle^{C}}\ \text{A-Write}$$

$$\frac{\mathsf{Q} \vdash_{\mathrm{fl}(t)} \mathsf{R}}{\mathsf{P}\ \mathcal{U}_{t}\ \mathsf{Q} \vdash_{\mathrm{fl}(t)} \mathsf{R}}\ \text{FL-}\mathcal{U}\text{-Eq} \qquad \frac{\mathsf{P}(t) \vdash_{\mathrm{rd}(t)} \mathsf{x.f} \mapsto \mathsf{v}}{\langle \mathsf{P} * \mathrm{iam}(t)\rangle\ \mathsf{x.f}\ \langle \mathsf{r.}\ \mathsf{P} * \mathsf{r} = \mathsf{v}\rangle^{C}}\ \text{A-Read}$$

$$\frac{\mathsf{P}(t) \vdash_{\mathrm{fl}(t)} \ulcorner\mathsf{x.f} \mapsto \mathsf{v_o}\urcorner * \mathsf{Q}(t)}{\langle \mathsf{P} * \mathrm{iam}(t)\rangle\ \mathbf{CAS}(\mathsf{x.f}, \mathsf{v_n}, \mathsf{v_o})\ \langle \mathsf{r.}\ \mathsf{r} = \mathbf{true} * \mathsf{Q} * \ulcorner\mathsf{x.f} \mapsto \mathsf{v_n}\urcorner\rangle^{C}}\ \text{A-CAS-True}$$

$$\frac{\mathsf{P}(t) \vdash_{\mathrm{rd}(t)} \mathsf{x.f} \mapsto \mathsf{v} \qquad \mathsf{P}(t) \vdash_{\mathrm{fl}(t)} \mathsf{Q}(t) \qquad \mathsf{v} \neq \mathsf{v_o}}{\langle \mathsf{P} * \mathrm{iam}(t)\rangle\ \mathbf{CAS}(\mathsf{x.f}, \mathsf{v_n}, \mathsf{v_o})\ \langle \mathsf{r.}\ \mathsf{r} = \mathbf{false} * \mathsf{Q}\rangle^{C}}\ \text{A-CAS-False}$$

$$\frac{\langle \mathsf{P}\rangle\ \mathsf{e}\ \langle \mathsf{Q}\rangle^{C} \qquad \mathrm{atomic}(\mathsf{e})}{[\mathsf{P}]\ \mathsf{e}\ [\mathsf{Q}]}\ \text{A-Start} \qquad \frac{[\mathsf{P}]\ \mathsf{e_1}\ [\mathsf{r.}\ \mathsf{Q}(\mathsf{r})] \qquad [\mathsf{Q}(\mathsf{x})]\ \mathsf{e_2}\ [\mathsf{r.}\ \mathsf{R}(\mathsf{r})]}{[\mathsf{P}]\ \mathbf{let}\ \mathsf{x} = \mathsf{e_1}\ \mathbf{in}\ \mathsf{e_2}\ [\mathsf{r.}\ \mathsf{R}(\mathsf{r})]}\ \text{Bind}$$

$$\frac{[\mathsf{P}]\ \mathsf{e}\ [\mathsf{Q}] \qquad \mathrm{stable}(\mathsf{R})}{[\mathsf{P} * \mathsf{R}]\ \mathsf{e}\ [\mathsf{Q} * \mathsf{R}]}\ \text{Frame} \qquad \frac{[\overline{\mathsf{P}}]\ \mathsf{e}\ [\overline{\mathsf{Q}}]}{\{\mathsf{P}\}\ \mathsf{e}\ \{\mathsf{Q}\}}\ \text{S-Shift}$$

Fig. 4. Selected rules of the TSO logic

compare-and-swap and fences. However, it is also used to specify the non-flushing writes. To see this, consider the rule A-WRITE in Figure 4. Since the semantics of assignment will introduce a buffered update, we know that *after* this new update reaches main memory, all the other updates will also have reached it. Thus, at that point in time, Q will also hold, since it is disjoint from the reference x.f. The other interesting rules are related to the CAS expression. In A-CAS-TRUE, we do not need to establish the read judgment, since the form of the flush judgment ensures that the value we can observe is $\mathsf{v_o}$. Aside from that, the rule behaves like a combination of writing and flushing. The rule A-CAS-FALSE, on the other hand, requires a separate read judgment. This is because it does not perform an assignment, and so the current value does not need to appear in the right-hand side of the flush assumption, like in A-WRITE and A-CAS-TRUE rules.

Also of interest are some of the proof rules for the read and flush judgments. Note how in rules RD-AX and FL-AX the *later* operator ($\rhd$) appears. This arises from the fact that the model is defined using guarded recursion to break circularities, and later is used as a guard. However, since the guardedness is tied to operational semantics through step-indexing and atomic expressions always take one evaluation step, *later* can be removed at the atomic steps of the proof, as expressed by the rules. Moreover, the rules also match the intuition we gave about the store buffer related connectives. First, the judgment means that all the updates in t's store buffer are flushed: thus, it is enough to know $\ulcorner\mathsf{P}\ \mathbf{in}\ t\urcorner$ holds to get $\ulcorner\mathsf{P}\urcorner$ in FL-AX, and similarly we only look to the right-hand side of $\mathcal{U}$ in the rule FL-$\mathcal{U}$-EQ. Note also, that we can reason about updates buffered in other threads, as evidenced by the rule RD-$\mathcal{U}$-NEQ, where we "ignore" the buffered update and read from the left-hand side of $\mathcal{U}$.

Stability and Stabilization. There is one potentially worrying issue in the definition of the $\mathcal{U}_t$ operator given in this section: since at any point in the program a *flush* action can occur nondeterministically, how can we know that there still exists a buffered update as asserted by $\mathcal{U}_t$? After all, it might have been flushed to the memory. This is the question of *stability*[1] of the until operator — and the answer is that it is unstable by design. The rationale behind this choice is simple: Suppose we had made it stable by allowing the possibility that the buffered update has already been flushed. Then, if we were to read a field that had a buffered write to it in a different thread, we would not know whether the write was still buffered or had been flushed, and so we would not know what value we read. With the current definition, when we read, we know that the update is still buffered and so the result of the read is known. However, we only allow reasoning with unstable assertions in the *atomic* triples, i.e., when reasoning about a single read, write or compare-and-swap expression. Hence, we need a way to make $\mathcal{U}$ stable. For this reason, we define an explicit *stabilization* operator, $(\![-]\!)$. It is a closure operator, which means we have $P \vdash (\![P]\!)$. Moreover, for stable assertions, the other direction, $(\![P]\!) \vdash P$, also holds. The important part, however, is how stabilization behaves with respect to $\mathcal{U}$: provided P and Q are stable, we have $(\![P \, \mathcal{U}_t \, Q]\!) \dashv\vdash (P \, \mathcal{U}_t \, Q) \vee Q$. This does indeed correspond to our intuition — even for stable assertions P and Q, the interference can flush the buffered update that is asserted in the definition of $\mathcal{U}$, which would transition to a state in which Q holds. However, since P and Q are stable, this is also the *only* problem that the interference could cause.

Explicit stabilization has been explored before in separation logic, most often in connection with rely-guarantee reasoning. In particular, Wickerson studies explicit stabilization in RGSep in his PhD thesis [25, Chapter 3], and Ridge [18] uses it to reason about x86-TSO.

Semantically, stability is defined through the notion of *interference*, which expresses the effect that the environment can have on a state. In iCAP-TSO there are two classes of interference. Firstly, other threads can concurrently change the state of shared regions. This source of interference is inherited from iCAP; we reason about it by considering the states of shared regions, and the transitions the environment is allowed to make. As an example, after releasing a lock we cannot be certain it remains unlocked, since other threads could concurrently acquire it. The protocol for a lock is described in Section 5. A second class of interference is related to the TSO nature of our semantics, and includes the interference that arises in the memory system: we refer to this class as *store-buffer* interference. It is defined through three possible actions of the memory system: allocation of a new store-buffer (which happens when a fork command gets executed), adding a new buffered update to a location outside the assertion's footprint to a store-buffer, and committing the oldest buffered update from one

---

[1] Recall an assertion is stable, if it cannot be invalidated by the environment.

of the buffers. Stability under allocation of new store-buffers and under buffering new updates is never a problem. Most of the connectives we use are also stable under flushing — both embeddings are specifically designed in this way. As we mentioned above, $\mathcal{U}$ is unstable under flushing by design, and we stabilize it explicitly. For the formal definition of the interference relation we refer the reader to the technical appendix [21].

Interpretation of the SC Logic. As we have already mentioned, the intuition that lies behind the SC logic, discussed in the previous section, is precisely expressed by the $\ulcorner -$ **in** $t \urcorner$ embedding. This is more formally captured by the rule S-SHIFT in Figure 4 (recall $\overline{P}$ is syntactic sugar for $\lambda t. \ulcorner P$ **in** $t \urcorner$), which states that the two ways of expressing that a triple holds from the perspective of the current thread are equivalent. In fact, we take this rule as the *definition* of the SC triples, and so we can prove that the SC triples actually form a standard separation logic by proving that the proof rules of the SC logic correspond to admissible rules in the TSO logic. This is expressed by the following theorem:

Theorem 1 (Soundness of SC logic). *The SC logic is sound wrt. its interpretation within the TSO logic, i.e., the proof rules of the SC logic composed with the rule S-SHIFT are admissible rules of the TSO logic.*

For most of the proof rules, the soundness follows directly; the only ones that require additional properties to be proved are the frame, consequence, and standard quantifier rules, which additionally require the following property:

Lemma 1. *The embeddings $\ulcorner - \urcorner$ and $\ulcorner -$ **in** $t \urcorner$ distribute over quantifiers and separating conjunction, and preserve entailment.*

The formal statement of this property, along with the proof, can be found in the accompanying technical report.

5 Reasoning in the TSO Logic

In Section 3 we illustrated that the fiction of sequential consistency provided by the lock specification allows us to reason about shared mutable data structures shared through locks, without explicitly reasoning about the underlying relaxed memory model. Of course, to verify a lock implementation against this lock specification, we *do* have to reason about the relaxed memory model. In this section we illustrate how to achieve this using our TSO logic. We focus on the use of the TSO connectives introduced in Section 4 to describe the machine states of the spin-lock and elide the details related to the use of concurrent abstract predicates.

The spin-lock implementation that we wish to verify is given in Figure 5. It uses a compare-and-swap (CAS) instruction to attempt to acquire the lock, but only a primitive write instruction to release the lock. While CAS flushes the store buffer of the thread that executes the CAS, a plain write does not. To verify this implementation, we thus have to explicitly reason about the possibility of buffered releases in store buffers.

```
Lock {
   bool locked;
   Lock() = this.locked := false; fence; this
   unit acquire() =
      let x = CAS(this.locked, true, false) in
         if x then () else acquire()
   unit release() = this.locked := false
}
```

Fig. 5. Left: spin-lock implementation. Right: lock protocol

Specification. In the Introduction we introduced a lock specification expressed in our SC logic. When verifying the spin-lock implementation, we actually verify the implementation against the following slightly stronger specification, from which we can easily derive the SC specification.

$$\exists \mathsf{isLock}, \mathsf{locked} : \mathsf{Prop}_{sc} \times \mathsf{Val} \to \mathsf{Prop}_{sc}. \ \forall R : \mathsf{Prop}_{sc}. \ \mathsf{stable}(R) \ \Rightarrow$$

$$[\overline{R}] \ \mathsf{Lock}() \ [r. \ \overline{\mathsf{isLock}(R, r)}]$$

$$\wedge \ [\overline{\mathsf{isLock}(R, x)}] \ \mathsf{Lock.acquire}() \ [\overline{\mathsf{locked}(R, x)} * \ulcorner R \urcorner]$$

$$\wedge \ [\overline{\mathsf{locked}(R, x) * R}] \ \mathsf{Lock.release}() \ [\top]$$

$$\wedge \ \mathsf{valid}(\forall x : \mathsf{Val}. \ \mathsf{isLock}(R, x) \Leftrightarrow \mathsf{isLock}(R, x) * \mathsf{isLock}(R, x))$$

$$\wedge \ \forall x : \mathsf{Val}. \ \mathsf{stable}(\mathsf{isLock}(R, x)) \wedge \mathsf{stable}(\mathsf{locked}(R, x))$$

Note that this stronger specification is expressed using TSO triples. This specification of the acquire method is slightly stronger: this specification asserts that upon termination of acquire, the resource invariant R holds in main memory and there are no buffered writes affecting R in *any* store buffer ($\ulcorner R \urcorner$). The weaker SC specification only asserts that the resource invariant R holds from the point of view of the acquiring thread and that there are no buffered writes affecting R in *any of the other threads'* store buffers ($\overline{R}$).

Lock Protocol. To verify the spin-lock implementation against the above specification, we first need to define the abstract representation predicates isLock and locked. Following CAP [11] and iCAP [22], to reason about sharing iCAP-TSO extends separation logic with shared regions, with protocols governing the resources owned by each shared region. In the case of the spin-lock, upon allocation of a new spin-lock the idea is to allocate a new shared region governing the state of the spin-lock and ownership of the resource invariant.

Conceptually, a spin-lock can be in one of two states: locked and unlocked. In iCAP-TSO we express this formally using the transition system in Figure 5. This labeled transition system specifies an abstract model of the lock. To relate it to the concrete implementation, for each abstract state (L and U), we choose an assertion that describes the resources the lock owns in that state.

Since acquiring the lock flushes the store buffer of the acquiring thread, the locked state is fairly simple. In the locked state the spin-lock owns the locked field, which contains the value true in main memory and there are no buffered writes to locked in any store buffer. The spin-lock x with resource invariant R thus owns the resources described by $I_L(x, R, n)$ in the abstract locked state.

$$I_L(x, R, n) = \ulcorner x.\text{locked} \mapsto \text{true} \urcorner$$

Due to the possibility of buffered releases in store buffers, the unlocked state is more complicated. In the unlocked state,

- either locked is false in main memory and there are no buffered writes to locked in any store buffer
- or locked is true in main memory, and there is *exactly one* store buffer with a buffered write to locked, and the value of this buffered write is false

Furthermore, in case there is a buffered write to locked that changes its value from true to false, then, once the buffered write reaches main memory, the resource invariant holds in main memory. Since the resource invariant must hold from the point of view of the releasing thread before the lock is released any buffered writes affecting the resource invariant must reach main memory before the buffered release. We can express this ordering dependency using the until operator:

$$I_U(x, R, n) = \exists t : \text{TId.} \left(\ulcorner x.\text{locked} \mapsto \text{true} \urcorner \, \mathcal{U}_t \, \ulcorner x.\text{locked} \mapsto \text{false} * R * [\text{REL}]_1^n \urcorner \right)$$

Here $[\text{REL}]_1^n$ is a CAP action permission used to ensure that only the current holder of the lock can release it. Since this is orthogonal to the underlying memory model, we refer the interested reader to the technical report [21] for details.

Since both arguments of $\mathcal{U}_t$ are stable, as explained in Section 4, $I_U(x, R, n)$ is equivalent to the following assertion.

$$\exists t : \text{TId.} \ulcorner x.\text{locked} \mapsto \text{false} * R * [\text{REL}]_1^n \urcorner \lor$$
$$\left(\ulcorner x.\text{locked} \mapsto \text{true} \urcorner \, \mathcal{U}_t \, \ulcorner x.\text{locked} \mapsto \text{false} * R * [\text{REL}]_1^n \urcorner \right)$$

The first disjunct corresponds to the case where the release has made its way to main memory and the second disjunct to the case where it is still buffered.

The definition of isLock in terms of I_L and I_U now follows iCAP.[2] The isLock predicate asserts the existence of a shared region governed by the above labeled transition system, where the resources owned by the shared region in the two abstract states are given by I_L and I_U. It further asserts that the abstract state of the shared region is either locked or unlocked and also a non-exclusive right to acquire the lock.

Proof Outline. To verify the spin-lock implementation, it remains to verify each method against the specification instantiated with the concrete isLock and locked predicates. To illustrate the reasoning related to the relaxed memory model, we focus on the verification of the acquire method and the compare-and-swap

---

[2] See the accompanying technical report for a formal definition of isLock.

instruction in particular. The full proof outline is given in the accompanying technical report.

As the name suggests, the resources owned by a shared region are shared between all threads. Atomic instructions are allowed to access and modify resources owned by shared regions, provided they follow the protocol imposed by the region. In the case of the spin-lock, the spin-lock region owns the shared locked field and we thus need to follow the spin-lock protocol to access and modify the locked field. Since the precondition of acquire asserts that the lock is either in the locked or unlocked state, we need to consider two cases.

If the spin-lock region is already locked, then the compare-and-swap fails and we remain in the locked state. This results in the following proof obligation:

$$\langle \triangleright I_L(\textbf{this}, R, n) * \text{iam}(t)\rangle \ \textbf{CAS}(\textbf{this}.\text{locked}, \textbf{true}, \textbf{false}) \ \langle r. \ \triangleright I_L(\textbf{this}, R, n) * \text{iam}(t) * r = \textbf{false}\rangle$$

That is, if locked contains the value true from the point of view of a thread t, then CAS'ing from false to true in thread t will fail. This is easily shown to hold by rule A-CAS-FALSE.

If the spin-lock region is unlocked, then the compare-and-swap may or may not succeed, depending on whether the buffered release has made it to main memory and which thread performed the buffered release. If it succeeds, the acquiring thread transitions the shared region to the locked state and takes ownership of the resource invariant; otherwise, the shared region remains in the unlocked state. This results in the following proof obligation:

$$\langle \triangleright I_U(\textbf{this}, R, n) * \text{iam}(t)\rangle \ \textbf{CAS}(\textbf{this}.\text{locked}, \textbf{true}, \textbf{false}) \ \langle r. \ \exists y. \ \triangleright I_y(\textbf{this}, R, n) * \text{iam}(t) * Q(y, r, n)\rangle$$

where $Q(y, r, n) \stackrel{\text{def}}{=} (y = U * r = \textbf{false}) \vee (y = L * [\text{Rel}]_1^n * \ulcorner R \urcorner * r = \textbf{true})$. Rewriting the explicit stabilization to a disjunction and commuting in $\triangleright$, this reduces to the following proof obligation:

$$\langle \text{iam}(t) * (\exists t' : \text{Tld.} \ \triangleright \ulcorner x.\text{locked} \mapsto \textbf{false} * R * [\text{Rel}]_1^n \urcorner \vee$$
$$(\triangleright \ulcorner x.\text{locked} \mapsto \textbf{true} \urcorner \mathcal{U}_{t'} \ \triangleright \ulcorner x.\text{locked} \mapsto \textbf{false} * R * [\text{Rel}]_1^n \urcorner)))$$
$$\textbf{CAS}(\textbf{this}.\text{locked}, \textbf{true}, \textbf{false})$$
$$\langle r. \ \exists y \in \{U, L\}. \ \triangleright I_y(\textbf{this}, R, n) * \text{iam}(t) * Q(y, r, n)\rangle$$

In case the second disjunct holds and there exist buffered releases in the store buffer of t', the CAS will succeed if executed by thread t' and fail if executed by any other thread. To prove this obligation, we thus do case analysis on whether t' is our thread or not, i.e., whether $t = t'$. This leaves us with three proof obligations (after strengthening the post-condition):

- either the buffered release is in our store buffer

$$\langle (\triangleright \ulcorner \textbf{this}.\text{locked} \mapsto \textbf{true} \urcorner \mathcal{U} \ \triangleright \ulcorner x.\text{locked} \mapsto \textbf{false} * R * [\text{Rel}]_1^n \urcorner) * \text{iam}(t)\rangle$$
$$\textbf{CAS}(\textbf{this}.\text{locked}, \textbf{true}, \textbf{false})$$
$$\langle r. \ \triangleright I_L(\textbf{this}, R, n) * [\text{Rel}]_1^n * \ulcorner R \urcorner * \text{iam}(t) * r = \textbf{true}\rangle$$

- or in some other thread's store buffer

$$\langle (\triangleright \ulcorner \textbf{this}.\text{locked} \mapsto \textbf{true} \urcorner \mathcal{U}^o \ \triangleright \ulcorner x.\text{locked} \mapsto \textbf{false} * R * [\text{Rel}]_1^n \urcorner) * \text{iam}(t)\rangle$$
$$\textbf{CAS}(\textbf{this}.\text{locked}, \textbf{true}, \textbf{false})$$
$$\langle r. \ \triangleright I_U(\textbf{this}, R, n) * \text{iam}(t) * r = \textbf{false}\rangle$$

— or it has already been flushed

$\langle \triangleright^{\ulcorner} \mathsf{x.locked} \mapsto \mathsf{false} * \mathsf{R} * [\textsc{Rel}]_1^{\mathsf{n}\urcorner} * \mathsf{iam}(t)\rangle$
   CAS(this.locked, **true, false)**
$\langle \mathsf{r}.\ \triangleright I_L(\mathbf{this}, \mathsf{R}, \mathsf{n}) * [\textsc{Rel}]_1^{\mathsf{n}} * {}^{\ulcorner}\mathsf{R}^{\urcorner} * \mathsf{iam}(t) * \mathsf{r} = \mathsf{true}\rangle$

These three proof obligations are easily discharged using rules A-CAS-TRUE and A-CAS-FALSE.

Note that our logic makes us consider exactly those four cases that intuitively one has to consider when reasoning operationally in TSO.

Logical Atomicity and Relaxed Memory

Although shared-memory concurrency introduces opportunity for threads to interfere, concurrent data structures are often written to ensure that all operations provided by the library are *observably*, or *logically atomic*. That is, for clients of the concurrent data structure, any concurrent execution of operations provided by the library should behave *as if* it occurred in *some* sequential order. This property immensely simplifies client-side reasoning, since the clients need not reason about any internal states of the library. One way of ensuring logical atomicity is by using coarse-grained synchronization, for instance by wrapping the whole data structure in a lock. However, this is far from efficient, and many real-life concurrent data structures opt to use fine-grained synchronization, such as compare-and-swap, while still being logically atomic. Since the simplification of the client-side reasoning one can obtain by exploiting the logical atomicity can be significant, any truly modular proof system that supports fine-grained concurrency should support logical atomicity. This is a known and well-researched problem in the sequentially consistent setting; here we discuss its interplay with relaxed memory and sketch how our system tackles it.

One of the approaches to express logical atomicity is to develop a program logic that *internalizes* the concept, i.e., in which one can express atomicity as a specification and prove that implementations satisfy such a spec within the logic. Several of the more recent program logics go this route, in particular TaDA and iCAP [9,22]. In this work we follow iCAP, which uses a reasonably simple specification pattern to encode abstract atomicity. The crux of the idea is for the data structure to provide an *abstract* mathematical model of its state, and to model the (possibly non-atomic) updates of the concrete state with an atomic update of the abstract state. Since the abstract state is only a model, it can be updated after *any* atomic step of the program, and thus any update of the abstract state can be considered atomic.

Since iCAP-TSO inherits most of the properties of iCAP, one could imagine that we inherit iCAP's specification pattern for logical atomicity verbatim. This, however, would lead to problems. If we ported the pattern to the relaxed setting directly, we would gain a way to express logical atomicity, but lose all the information about the flushing behavior of the data structure—and in effect we would not be able to derive an SC specification, even if the data structure provided a fiction of sequential consistency. The idea for how one can adjust the specification pattern hinges on using the SC assertions and the embeddings described earlier to describe the state of the store-buffers when the abstract update happens. We refer the interested reader to the accompanying technical report [21] for an explanation of how to extend the iCAP pattern to the TSO setting, as well as the formal proofs and technical details.

Other Case Studies

In addition to the spin-lock, we have verified several other algorithms in the TSO logic against SC specifications. Below we discuss the challenges of each case-study. Full proofs are included in the accompanying technical report.

Treiber's Stack. Treiber's stack is a classic fine-grained concurrent stack implementation. We verify this data structure against a specification that provides logical atomicity, based on the one given in [22]. From this general specification we derive two classic specifications: a single-owner stack and a shared-bag that provides a fiction of sequential consistency. The challenge, as explained in the preceding section, is to provide a specification pattern that provides both logical atomicity and fiction of sequential consistency.

Double-Checked Initialization. Double-checked initialization [19] is a design pattern that reduces the cost of lazy initialization by having clients only use a lock if the wrapped object has not been initialized yet, to their knowledge. We verify this algorithm against a specification that ensures that the wrapped object is initialized only once. The challenge is to capture the fact that holding the lock ensures that there are no buffered updates to the object.

Ticket Lock. A bounded ticket lock [16] is a fair locking algorithm where threads obtain a ticket number and wait for it to be served, and where the ticket number goes back to zero when it reaches its bound. We verify this algorithm against a specification that allows a bounded number of clients to transfer resources. The challenge is to ensure that a thread's ticket will not be skipped and reissued to another thread, despite the fact that in TSO, the increment to the serving number in the release can be buffered, as for the spinlock.

Circular Buffer. A circular buffer [14] is a single-writer single-reader resource ownership transfer mechanism based on an array viewed circularly. This algorithm is interesting in TSO because it does not need any synchronisation: the FIFO behavior of store buffers is enough. Because there are no synchronisation operations, a thread can be ahead of main memory in the array, and the challenge is to ensure that despite that, the writer does not overtake the reader.

6   Soundness

We prove soundness of iCAP-TSO with respect to the TSO model of section 2. Soundness is proven by relating the machine semantics to an instrumented semantics that, for instance, enforces that clients obey the chosen protocols when accessing shared state. This relation is expressed through an erasure function, $\lfloor - \rfloor$, that erases an instrumented state to a set of machine states.

The soundness theorem is stated in terms of the following $eval(\mu, T, q)$ predicate, which asserts that for any terminating execution of the thread pool T from initial state μ, the predicate q must hold for the terminal state and thread pool. The $eval$ predicate is defined as a guarded recursive predicate (the recursive occurrence of $eval$ is guarded by $\triangleright$), to express that each step of evaluation in the machine semantics corresponds to a step in the topos of trees.

$$eval(\mu, T, q) \stackrel{\text{def}}{=} (\text{irr}(\mu, T) \Rightarrow (\mu, T) \in q) \wedge$$
$$(\forall T', \mu'. (\mu, T) \to (\mu', T') \Rightarrow \triangleright eval(\mu', T', q))$$

Here $\text{irr}(\mu, T)$ means that (μ, T) is irreducible. We can now state the soundness of iCAP-TSO.

Theorem 2 (Soundness). *If* $[\mathsf{P}]$ e $[\mathsf{r}. \mathsf{Q}]$ *and* $\mu \in \lfloor [\![\mathsf{P}]\!](t) \rfloor$ *then*

$$eval(\mu, [t \mapsto e], \lambda(\mu', T). \ \mu' \in \lfloor [\![\mathsf{Q}]\!](t)(T(t)) \rfloor)$$

This theorem expresses that if a specification $[\mathsf{P}]$ e $[\mathsf{r}. \mathsf{Q}]$ holds and the execution of the thread pool $[t \mapsto e]$ with a single thread t from an initial state μ in the erasure of P terminates (including threads spawned by t), then the execution has finished in a proper terminal state (i.e., did not fault), which is in the erasure of Q instantiated with the return value $T(t)$ of thread t.

7 Related Work

Our work builds directly on iCAP [22], which is an extension of separation logic for modular reasoning about concurrent higher-order programs with shared mutable state. Our work extends the model of iCAP with store buffers to implement a TSO memory model, extends the iCAP logic with TSO-connectives for reasoning about these store buffers and crucially, it reduces to standard concurrent separation logic for sequentially consistent clients.

Rely/Guarantee Reasoning Over Operational Models. Conceptually, iCAP-TSO is a Rely/Guarantee-based proof system for reasoning about an operational semantics with a relaxed memory model. This approach has also been explored by Ridge [18], Wehrman [24], and Jacobs [15].

Ridge [18] and Wehrman [24] both propose proof systems for low-level reasoning about racy TSO programs based on Rely/Guarantee reasoning. In Ridge's system [18] the Rely/Guarantee is explicit, while in Wehrman's system [24] it is expressed implicitly through a separation logic. To reason in the presence of a relaxed memory model, both systems enforce a rely that includes possible interference from write buffers. Consequently, both systems support reasoning about racy code. However, in the case where a library includes sufficient synchronization, neither system is able to take advantage of the stronger rely provided by this synchronization to simplify client proofs. This is exactly what our fiction of sequential consistency allows.

Jacobs [15] proposes to extend separation logic with "TSO spaces" for reasoning about shared resources in a TSO setting. While the exact goals of his approach remain a bit unclear, it seems that Jacobs is also aiming for a system that reduces to standard separation logic reasoning when possible. However, to ensure soundness Jacobs' proof system lacks the usual structural rules for disjunction and existentials. This results in non-standard reasoning even for non-racy clients.

Recovering Sequential Consistency. There are several other approaches for recovering sequentially consistent reasoning about clients in the presence of a relaxed memory model.

Cohen and Schirmer [8] propose a programming discipline based on ownership, which ensures that all TSO program behaviors can be simulated by a sequentially consistent machine. Unfortunately, the proposed discipline enforces too much synchronization. In particular, an efficient spin-lock implementation with a buffered release, like the one we verify in Section 5, does not obey their programming discipline. Their approach is thus unable to deal with such code without introducing additional synchronization.

Owens [17] defines a trace property on the set of SC behaviors of a program which ensures that all TSO behaviors can be simulated by an SC machine. Owens shows how this property allows clients of synchronization primitives to reason using SC semantics, despite *racy* implementations of these synchronization primitives. However, in contrast to our approach, Owens' approach is non-compositional: while Owens proves similar results for multiple synchronization primitives *in isolation*, these results do not apply to clients that *combine* two or more of these synchronization primitives.

Gotsman et al. [13] propose another approach for providing clients with a fiction of sequential consistency, based on linearizability. By relating racy library implementations on a TSO architecture with abstract specifications on an SC architecture, they can reason about data-race free clients that call racy libraries using an SC memory model. Their approach is only compositional for *non-interacting* libraries (libraries that do not interact through the heap) and further requires libraries and clients to be non-interacting. Their approach can also relate fine-grained implementations with coarse-grained implementations, which provides similar advantages to our logical atomicity.

Our approach does not suffer from the compositionality problems of [17,13] or the need for unnecessary and potentially expensive synchronization required by [8]. In particular, iCAP-TSO allows *racy* libraries that interact through the heap to be verified *independently*.

Reasoning Over Axiomatic Models. Relaxed memory models are often defined using relations over read and write events that enforce certain consistency/visibility constraints.

Alglave et al. [4] proposes the use of such axiomatic models to support efficient model-checking in the context of relaxed memory models. The use of an axiomatic semantics avoids the need to consider all the possible interleavings introduced by operational models with explicit buffers and caches. Alglave et al.'s approach supports fully automatic verification of simple correctness properties of realistic C code. Alglave et al.'s approach is non-modular in the sense that it only supports whole-program verification and thus lacks support for verifying modules independently.

While our logic is based on an operational model with explicit buffers, we use Rely/Guarantee reasoning to avoid the explosion in interleavings observed by Alglave et al. Our fiction of sequential consistency is specifically designed to strengthen the rely (and implicitly, reduce the number of possible interleavings that have to be considered) when the code enforces sufficient synchronization.

More recently, Turon et al. [23] has proposed GPS, a proof system over the axiomatic C11 memory model. GPS extends separation logic with per-location protocols which internalize some of the properties of the underlying visibility properties between read and write events. GPS supports two of the C11 access modes: non-atomics and release/acquire. Reasoning about non-atomics reduces to standard separation logic. However, ownership transfer requires the use of release/acquire and explicit reasoning about visibility of memory events. GPS lacks support for logical atomicity and thus cannot express canonical specifications for concurrent data structures such as the specification of Treiber's stack in the technical report.

8   Conclusion and Future Work

We have presented a new proof system, iCAP-TSO, to support modular and scalable reasoning for a language with a TSO memory model. The proof system consists of two logics. The TSO logic supports reasoning about libraries with low-level racy code. In cases where the libraries provide sufficient synchronization, they can be verified against SC specifications. Clients that only do resource transfer through such libraries can then be verified entirely within the SC logic, which uses standard separation logic rules.

We use the TSO logic to verify an efficient spin-lock implementation against an SC specification. We use this to verify a shared bag library, implemented using a spin-lock, in the SC logic. We also verify a double-checked initialization wrapper, a bounded ticket lock, and a circular buffer against SC specifications. Lastly, we verify Treiber's stack against a specification that showcases how logical atomicity can be extended to TSO.

We think of iCAP-TSO as a first step towards more automated/interactive tools for reasoning about the TSO memory model. In this paper we have focused on the foundational issues of constructing a logic that allows simple reasoning for well-behaved code. As future work it would be interesting to try to extend tools like [7,12] to support mostly automated verification in the SC logic and interactive verification in the TSO logic. We believe that the fiction of sequential consistency could be really beneficial in this area: one could imagine that the lock-free concurrency libraries would be verified by hand, while automated tools could verify properties of client programs. Since the rules of the SC logic are standard, the whole range of techniques developed for automating separation logic could be applicable. The open question here is how one could infer the instantiations of higher-order specifications, for instance the invariants for locks, and this should be investigated.

Acknowledgements. We gratefully acknowledge the comments and suggestions from our anonymous reviewers. We thank Mark Batty, Aleš Bizjak, Susmit Sarkar, and Peter Sewell for helpful discussions on this work. This research was supported in part by the ModuRes Sapere Aude Advanced Grant from The Danish Council for Independent Research for the Natural Sciences (FNU).

References

1. Intel threading building blocks documentation: Fenced data transfer, https:// software.intel.com/en-us/node/506122 (accessed: June 25, 2014)
2. java.util.concurrent API, http://docs.oracle.com/javase/7/docs/api/java/ util/concurrent/package-summary.html (accessed: June 25, 2014)
3. Linux kernel mailing list, spin_unlock optimization(i386) (November 1999)
4. Alglave, J., Kroening, D., Tautschnig, M.: Partial orders for efficient bounded model checking of concurrent software. In: Sharygina, N., Veith, H. (eds.) CAV 2013. LNCS, vol. 8044, pp. 141–157. Springer, Heidelberg (2013)
5. Biering, B., Birkedal, L., Torp-Smith, N.: BI-Hyperdoctrines, Higher-order Separation Logic, and Abstraction. ACM TOPLAS (2007)
6. Birkedal, L., et al.: First Steps in Synthetic Guarded Domain Theory: Step-Indexing in the Topos of Trees. In: Proc. of LICS (2011)
7. Chlipala, A.: Mostly-automated Verification of Low-level Programs in Computational Separation Logic. In: Proc. of PLDI (2011)
8. Cohen, E., Schirmer, B.: From total store order to sequential consistency: A practical reduction theorem. In: Kaufmann, M., Paulson, L.C. (eds.) ITP 2010. LNCS, vol. 6172, pp. 403–418. Springer, Heidelberg (2010)
9. da Rocha Pinto, P., Dinsdale-Young, T., Gardner, P.: TaDA: A logic for time and data abstraction. In: Jones, R. (ed.) ECOOP 2014. LNCS, vol. 8586, pp. 207–231. Springer, Heidelberg (2014)
10. Dinsdale-Young, T., Birkedal, L., Gardner, P., Parkinson, M., Yang, H.: Views: Compositional Reasoning for Concurrent Programs. In: Proc. of POPL (2013)
11. Dinsdale-Young, T., Dodds, M., Gardner, P., Parkinson, M.J., Vafeiadis, V.: Concurrent abstract predicates. In: D'Hondt, T. (ed.) ECOOP 2010. LNCS, vol. 6183, pp. 504–528. Springer, Heidelberg (2010)
12. Gotsman, A., Berdine, J., Cook, B., Sagiv, M.: Thread-modular Shape Analysis. In: Proc. of PLDI (2007)
13. Gotsman, A., Musuvathi, M., Yang, H.: Show No Weakness: Sequentially Consistent Specifications of TSO Libraries. In: Proc. of DISC (2012)
14. Howells, D., McKenney, P.E.: Circular buffers, https://www.kernel.org/doc/Documentation/circular-buffers.txt
15. Jacobs, B.: Verifying TSO Programs. Technical report, Report CW660 (May 2014)

16. Mellor-Crummey, J.M., Scott, M.L.: Algorithms for scalable synchronization on shared-memory multiprocessors. ACM TOCS 9(1), 21–65 (1991)
17. Owens, S.: Reasoning about the implementation of concurrency abstractions on x86-TSO. In: D'Hondt, T. (ed.) ECOOP 2010. LNCS, vol. 6183, pp. 478–503. Springer, Heidelberg (2010)
18. Ridge, T.: A rely-guarantee proof system for x86-TSO. In: Leavens, G.T., O'Hearn, P., Rajamani, S.K. (eds.) VSTTE 2010. LNCS, vol. 6217, pp. 55–70. Springer, Heidelberg (2010)
19. Schmidt, D.C., Harrison, T.: Double-checked locking - an optimization pattern for efficiently initializing and accessing thread-safe objects (1997), http://www.dre.vanderbilt.edu/~schmidt/PDF/DC-Locking.pdf
20. Sewell, P., Sarkar, S., Owens, S., Zappa Nardelli, F., Myreen, M.O.: x86-TSO: A Rigorous and Usable Programmers Model for x86 Multiprocessors. In: Comm. ACM (2010)
21. Sieczkowski, F., Svendsen, K., Birkedal, L., Pichon-Pharabod, J.: A Separation Logic for Fictional Sequential Consistency. Technical report, Aarhus University (2014), http://cs.au.dk/~filips/icap-tso-tr.pdf
22. Svendsen, K., Birkedal, L.: Impredicative concurrent abstract predicates. In: Shao, Z. (ed.) ESOP 2014 (ETAPS). LNCS, vol. 8410, pp. 149–168. Springer, Heidelberg (2014)
23. Turon, A., Vafeiadis, V., Dreyer, D.: GPS: Navigating Weak Memory with Ghosts, Protocols, and Separation. In: Proc. of OOPSLA (2014)
24. Wehrman, I.: Weak-Memory Local Reasoning. PhD thesis, University of Texas, Dissertation draft (2012)
25. Wickerson, J.: Concurrent verification for sequential programs. PhD thesis, University of Cambridge (2012)

Binding Structures as an Abstract Data Type

Wilmer Ricciotti

IRIT – Institut de Recherche en Informatique de Toulouse
Université de Toulouse
Wilmer.Ricciotti@irit.fr

Abstract. A long line of research has been dealing with the representation, in a formal tool such as an interactive theorem prover, of languages with binding structures (e.g. the lambda calculus). Several concrete encodings of binding have been proposed, including de Bruijn dummies, the locally nameless representation, and others. Each of these encodings has its strong and weak points, with no clear winner emerging. One common drawback to such techniques is that reasoning on them discloses too much information about what we could call "implementation details": often, in a formal proof, an unbound index will appear out of nowhere, only to be substituted immediately after; such details are never seen in an informal proof. To hide this unnecessary complexity, we propose to represent binding structures by means of an abstract data type, equipped with high level operations allowing to manipulate terms with binding with a degree of abstraction comparable to that of informal proofs. We also prove that our abstract representation is sound by providing a de Bruijn model.

1 Introduction

The techniques for reasoning on languages with binders are a very popular topic in both programming and logic ([20,5]). Especially in logic, the choice of a representation of binding structures is one of the most significant issues when formalizing the metatheory of a programming language. Over the years, a number of different styles have been proposed to deal with binding, roughly divided in two different categories: first order encodings, also called *concrete* encodings, and higher-order encodings like higher-order abstract syntax (HOAS). In interactive theorem provers based on a strong type theory, like Coq, Matita, or Agda, trivial implementations of HOAS by means of inductive types are rejected because they do not satisfy the positivity checks required by those systems to ensure consistency and, more importantly, adequacy concerns related to the appearance of *exotic terms* arise; thus, concrete encodings are more usually employed (notable exceptions include two-level approaches [8] and weak HOAS [10]).

Concrete encodings include some of the best known styles, like the de Bruijn nameless encoding [16] (which represents variables using indices pointing to the binder that declares them), the locally nameless encoding (a variant of the de Bruijn encoding where only bound variables are represented by indices, whereas free variables still use names) and the canonically named encoding of

© Springer-Verlag Berlin Heidelberg 2015
J. Vitek (Ed.): ESOP 2015, LNCS 9032, pp. 762–786, 2015.
DOI: 10.1007/978-3-662-46669-8_31

Pollack and Sato [23], where a bound variable is represented by means of a name that is programmatically chosen depending on the structure of the term within scope. All of these styles are described as *canonical* because terms that are equal up to α-renaming are identified. We have studied these styles in [17,3] and drawn a comparison in [22].

Our experience tells us that every concrete encoding has its own disadvantages, but more importantly that all of them share one problem: they force the formalizer to deal with the intricacy of the inner representation of binding, something that in an informal proof is never seen. In a formal proof based on a concrete approach, it is only a matter of time before nameless dummies, lifting operations, or name choosing operations come to the surface.

We should ask ourselves whether this inconvenience is inherent to the concrete representation of binding. Our understanding is that very often (if not *always*) the internal representation of binding must be treated explicitly because of the lack of an infrastructure designed to keep it hidden. We have very good access to the *implementation* but, crucially, we lack an *abstract* view on binding.

This paper describes a project which aims at representing binding only by means of abstract operations (similar to the ones employed in a pencil-and-paper proof), keeping the implementation details hidden from the user. More precisely, we will represent the terms of the object language as an abstract data type, which can only be manipulated by means of the operations and logical properties provided by its module. Based on this, we will prove two kinds of results:

- soundness properties, showing the existence of an implementation, or model, of the abstract data type which validates all the stated properties;
- theorems about the object language (e.g.: subject reduction), whose are carried out within the abstract data type, without resorting to any property specific to the model.

While other proposals to treat binding structures axiomatically exist ([11,19]), in this paper we will address some topics that have been neglected, particularly the treatment of inductively defined predicates over binding structures. All of the proofs presented here have been proved valid in the Matita theorem prover.[1]

The paper is structured as follows: Section 2 presents an abstract data type representing the term language of the simply typed lambda calculus; in Section 3 we provide an implementation of the abstract data type in the form of a locally nameless model; after recalling the problem of induction principle strengthening in the context of typing rules (Section 4), we extend our abstract data type to the level of type systems (Section 5) and beta reduction (Section 6), showing that the technique is sufficiently powerful to carry out common proofs like weakening and subject reduction; finally Section 7 concludes.

---

[1] The Matita formalization can be found at
http://www.irit.fr/~Wilmer.Ricciotti/publications.html .

2 An Abstract View of Binding

We present in this section a collection of abstract data types describing a simple language with binding: the simply typed lambda calculus (or, for brevity, $\lambda_\rightarrow$). Similarly to the axiomatization in [11], the operations working on our data type include a set of opaque constants acting as "constructors" for the terms of the language and a principle allowing to define functions by structural recursion on the terms. However, instead of a primitive substitution function, we provide facilities to form contexts (terms with holes) and apply them to variables, which we regard as more basic. An operation to retrieve the list of the free variables in a term or context is also given. In addition, properties asserting the computational behavior of the aforementioned operations are provided.

Signature. Our module defines the abstract data types of $\lambda_\rightarrow$ as follows:

$$
\begin{aligned}
&tp && : \textbf{Type}\\
&\mathsf{Atom} && : tp\\
&\mathsf{Arr} && : tp \Rightarrow tp \Rightarrow tp\\[6pt]
&(\Lambda_i)_{i\in\mathbb{N}} && : \textbf{Type}\\
&\mathsf{Par} && : \mathbb{A} \Rightarrow \Lambda_0\\
&\mathsf{App} && : \Lambda_0 \Rightarrow \Lambda_0 \Rightarrow \Lambda_0\\
&\mathsf{Lam} && : \mathbb{A} \Rightarrow tp \Rightarrow \Lambda_0 \Rightarrow \Lambda_0\\[6pt]
&\nu && : \mathbb{A} \Rightarrow \Lambda_i \Rightarrow \Lambda_{i+1}\\
&-\lceil-\rceil && : \Lambda_{i+1} \Rightarrow \mathbb{A} \Rightarrow \Lambda_i\\[6pt]
&\mathrm{FV} && : (\Lambda_i)_{i\in\mathbb{N}} \Rightarrow \mathit{list}\ \mathbb{A}\\[6pt]
&\mathcal{R}_{\Lambda_0} && : \forall T : \Lambda_0 \Rightarrow \textbf{Type}, C : \mathit{list}\ \mathbb{A}.\\
&&&\quad (\forall x : \mathbb{A}.T\ (\mathsf{Par}\ x)) \Rightarrow\\
&&&\quad (\forall u, v : \Lambda_0.T\ u \Rightarrow T\ v \Rightarrow T\ (\mathsf{App}\ u\ v)) \Rightarrow\\
&&&\quad \left(\begin{array}{l} \forall x : \mathbb{A}, \sigma : tp, v : \Lambda_1.x \notin \mathrm{FV}(v), C \Rightarrow T\ (v\lceil x\rceil)\\ \quad \Rightarrow T\ (\mathsf{Lam}\ x\ \sigma\ (v\lceil x\rceil)) \end{array}\right) \Rightarrow\\
&&&\quad \forall u : \Lambda_0.T\ u
\end{aligned}
$$

We call this presentation of binding *ostensibly named* because at the external level we always manipulate terms as entities containing names, including bound variables: we never see bound variables represented as nameless dummies, or pointers to their binder. The concrete implementation of binding structures may or may not use names, but this is hidden from the user.

The set of types tp of the simply typed lambda calculus is of no particular interest and is here provided for reference only: it is the free algebra obtained from the zeroary constructor of the atomic type Atom and the binary constructor of arrow (function) types Arr. Types of the simply typed lambda calculus will be denoted by $\sigma, \tau, \ldots$

Λ_i will represent the type of terms with i holes, or i-ary contexts. Zeroary contexts are taken as the terms of the calculus. We will denote terms and contexts alike by $u, v, \ldots$. The type of names $\mathbb{A}$ is an arbitrary infinite type with decidable equality. We assume the existence of an operation $\varphi : list\ \mathbb{A} \Rightarrow \mathbb{A}$ allowing us to choose a name which is *fresh* with respect to any given finite list (i.e., $\varphi(C) \notin C$ holds for all finite lists of names C).

The constructors of terms include Par, encapsulating a name to represent a free variable or parameter, applications App, and lambda abstractions Lam. Just as in informal syntax, lambda abstractions bear a type and bind a name inside a subterm. For example, the identity function $\lambda x : \mathsf{Atom}.x$ is expressed as

$$\mathsf{Lam}\ x\ \mathsf{Atom}\ (\mathsf{Par}\ x)$$

provided that x is a name in $\mathbb{A}$.

Crucially, to put our representation to some use we need to be able to talk about *contexts*. Two operations ν and $-\lceil-\rceil$ (respectively *variable closing* or *context formation* and *variable opening* or *context application*) are provided to build and apply contexts: $\nu x.u$ substitutes a hole for all (free) occurrences of Par x in u, thus increasing its arity, whereas $u\lceil x\rceil$ replaces the last created hole in u with Par x, decreasing its arity. It is worth noting that, since it cancels out a free variable, ν acts like a binder (the notation was chosen in analogy to the "new channel" operator of the π-calculus). Closing and opening can be combined (in this order) to rename a variable: we will use the following special notation for *variable renaming*:

$$u\ \langle y/x \rangle \triangleq (\nu x.u)\lceil y\rceil$$

If $p = (x, y)$ is a pair of variable names, we will write $u\ \langle p \rangle$ for $u\ \langle y/x \rangle$. This notation is further extended to vectors: if $\overrightarrow{p} = [p_1, \ldots, p_n]$ is a vector of pairs, we will write $u\ \langle \overrightarrow{p} \rangle$ for $u\ \langle p_1 \rangle \cdots \langle p_n \rangle$.

An abstract operation FV takes as input a term or a context and returns the list of free names used in that term or context. Lastly, $\mathcal{R}_{\Lambda_0}$ is a primitive recursion principle over terms. Recursion principles on inductive types have a well defined shape, which is followed by $\mathcal{R}_{\Lambda_0}$, except for the Lam case, which provides a special treatment for the bound variable. We will make this clear in the following paragraphs.

Properties of terms and contexts. The following properties of terms and context forming operations are assumed:

$$
\begin{aligned}
x\ \langle y/x \rangle &= y \\
z\ \langle y/x \rangle &= z &&\text{if } z \neq x \\
(\mathsf{App}\ u\ v)\ \langle y/x \rangle &= \mathsf{App}\ (u\ \langle y/x \rangle)\ (v\ \langle y/x \rangle) \\
(\mathsf{Lam}\ z\ \sigma\ u)\ \langle y/x \rangle &= \mathsf{Lam}\ z\ \sigma\ (u\ \langle y/x \rangle) &&\text{if } z \neq x, y
\end{aligned}
$$

$$
\begin{aligned}
u\ \langle x/x \rangle &= u &&(*) \\
\nu x.(u\lceil x\rceil) &= u &&\text{if } x \notin \mathrm{FV}(u) \\
\mathsf{Lam}\ x\ \sigma\ (u\lceil x\rceil) &= \mathsf{Lam}\ y\ \sigma\ (u\lceil y\rceil) &&\text{if } x, y \notin \mathrm{FV}(u)
\end{aligned}
$$

The first four lines fall logically into the same group, they describe the computational behaviour of renaming. The last line is also remarkable, since it expresses the fact that Λ_0 is canonical, i.e. α-convertible terms are provably equal. The second-to-last property expresses a sort of an "η-equivalence" on contexts: opening a context and then closing it with respect to the same variable yields the original context, provided that the variable involved is fresh.

The property marked with (*) has a special status since, assuming a suitable induction principle on Λ_0, it could be proved from the other properties whenever u is a term; we will provide such an induction principle on terms, but if we want (*) to be valid not just for terms, but for proper contexts as well, we will still have to assume it as part of the abstract data type.

Recursion. We employ contexts to express a recursion principle $\mathcal{R}_{\Lambda_0}$ for Λ_0, allowing us to define functions over terms by structural recursion.

The lines 2–5 of the type of $\mathcal{R}_{\Lambda_0}$ express the types of the arguments of the principle which will provide its behaviour in the Par, App, and Lam case. To better understand how $\mathcal{R}_{\Lambda_0}$ works, we use it to define the usual operation of substitution of terms for free variables. Informally, substitution is often defined as follows:

$$u\,[v/x] \triangleq \begin{cases} (\mathsf{Par}\ x)\,[v/x] & = v \\ (\mathsf{Par}\ y)\,[v/x] & = \mathsf{Par}\ y & \text{if } x \neq y \\ (\mathsf{App}\ u_1\ u_2)\,[v/x] & = \mathsf{App}\ (u_1\,[v/x])\ (u_2\,[v/x]) \\ (\mathsf{Lam}\ y\ \sigma\ u_1)\,[v/x] & = \mathsf{Lam}\ y\ \sigma\ (u_1\,[v/x]) & \text{if } y \notin \{x\} \cup \mathrm{FV}(v) \end{cases}$$

This is not a regular pattern matching over an inductive type: while the Par and App cases do not look special (and the same can be said about the types of the associated clauses in $\mathcal{R}_{\Lambda_0}$) the Lam case hides an implicit α-conversion in order to make the bound variable different from both x and any free variable occurring in v, to prevent variable capture. More generally, an effective recursion principle over lambda abstractions should allow us to retrieve, for a bound variable, a name that is fresh with respect to an arbitrary list: for this reason, we add a "freshness context" C to the principle $\mathcal{R}_{\Lambda_0}$ (similarly to what is done in Nominal Isabelle [25]).

Thus, we can express the substitution operation as a structurally recursive function over ostensibly named terms as follows.

Definition 1 (substitution). *For all terms u, v and parameter names x, the function subst is defined as follows:*

$$\begin{aligned} subst\ u\ x\ v \triangleq\ &\mathcal{R}_{\Lambda_0}\ (\lambda\_.\Lambda_0)\ (x, \mathrm{FV}(v)) \\ &(\lambda y.\mathit{if}\,(x = y)\ \mathit{then}\ v\ \mathit{else}\ (\mathsf{Par}\ y)) \\ &(\lambda u_1, u_2, r_1, r_2.\mathsf{App}\ r_1\ r_2) \\ &(\lambda y, \sigma, u^*, \_, r^*.\mathsf{Lam}\ y\ \sigma\ r^*)\ u \end{aligned}$$

We will use the notation $u\,[v/x]$ as a short form for subst $u\ x\ v$.

In this definition, variables r_1, r_2, r^* are used to represent the result of recursion on the subterms $u_1, u_2, u^*\lceil y \rceil$ respectively. The abstraction operation is special: the recursion principle unpacks it as $\mathsf{Lam}\ y\ \sigma\ (u^*\lceil y \rceil)$, where u^* is a unary context and y is taken to be fresh with respect to the list $x, \mathrm{FV}(v)$ we provided as an argument and also with respect to $\mathrm{FV}(u^*)$ (a proof that $y \notin x, \mathrm{FV}(v), \mathrm{FV}(u^*)$ is also provided as the underscore "_" argument, that is irrelevant to the definition of the substitution, but may be employed in a proof of correctness).

Properties of FV. For FV, we assume that the following properties hold:

$$\mathrm{FV}(\mathsf{Par}\ x) = [x]$$
$$\mathrm{FV}(\mathsf{App}\ u\ v) = \mathrm{FV}(u) \cup \mathrm{FV}(v)$$
$$\mathrm{FV}(\mathsf{Lam}\ x\ \sigma\ u) = \mathrm{FV}(\nu x.u)$$

$$x \in \mathrm{FV}(\nu y.u) \iff (x \neq y \wedge x \in \mathrm{FV}(u))$$

$$\mathrm{FV}(u\lceil x \rceil) \subseteq \{x\} \cup \mathrm{FV}(u)$$
$$\mathrm{FV}(u) \subseteq \mathrm{FV}(u\lceil x \rceil)$$

Properties of Recursion. Since the recursion principle is part of our ostensibly named interface, it is opaque. This means its algorithmic behaviour must be expressed explicitly. Let Rec be short for $\mathcal{R}_{\Lambda_0}\ T\ C\ f_{\mathsf{Par}}\ f_{\mathsf{App}}\ f_{\mathsf{Lam}}$ (where T, C, f_{Par}, f_{App}, f_{Lam} have a suitable type). We will assume that the following properties hold:

$$Rec\ (\mathsf{Par}\ x) = f_{\mathsf{Par}}\ x$$

$$Rec\ (\mathsf{App}\ u\ v) = f_{\mathsf{App}}\ u\ v\ (Rec\ u)\ (Rec\ v)$$

$$\forall U : (\forall u : \Lambda_0.T\ u \Rightarrow \mathbf{Type}).x \notin \mathrm{FV}(u) \Rightarrow$$
$$(\forall y, H_y.U\ (u\lceil y \rceil)\ (f_{\mathsf{Lam}}\ C\ y\ \sigma\ u\ H_y\ (Rec\ (u\lceil y \rceil)))) \Rightarrow$$
$$U\ (\mathsf{Lam}\ x\ \sigma\ (u\lceil x \rceil))\ (Rec\ (\mathsf{Lam}\ x\ \sigma\ (u\lceil x \rceil)))$$

The first two lines are equations stating that on parameters and applications, $\mathcal{R}_{\Lambda_0}$ behaves as a normal recursion operator on an inductive type: it can be rewritten as an application of the appropriate branch (f_{Par} or f_{App}) to the arguments of the constructor and, in the case of App, to the result of recursion on its subterms. The last line expresses the behaviour in the Lam case, which is more complicated: if we allowed the same scheme as with Par and App we could take $f_{\mathsf{Lam}} = \lambda y, \_, \_, \_, \_.y$ and use $\mathcal{R}_{\Lambda_0}$ to expose the variable bound by Lam as follows:

$$Rec\ (\mathsf{Lam}\ x\ \sigma\ (u\lceil x \rceil))$$
$$= (\lambda y, \_, \_, \_, \_.y)\ x\ \sigma\ u\ H\ (Rec\ (u\lceil x \rceil))$$
$$= x$$

(where H is any proof that $x \notin \mathrm{FV}(u), C$). As it turns out, this equation would make it possible to look into the name bound by Lam: this would in turn enable us to discriminate abstractions in terms of their bound variables, which is clearly inconsistent with the α-equivalence hypothesis.

The fact that we should not be able to extract naming information from binders prevents us from expressing the computational behaviour of the recursion principle explicitly in the Lam case. The property we stated is a "constrained rewriting principle" which does not allow, in general, to compute the result of a structurally recursive function in the Lam $x\ \sigma\ (u\lceil x\rceil)$ case. However, if we employ it in a proof whose goal involves such a function, we will get a new variable y, together with a proof H_y that $y \notin C, \mathrm{FV}(u)$ (both universally quantified in the property) and the goal will be rewritten in such a way that we have the illusion that Lam $x\ \sigma\ (u\lceil x\rceil)$ has been renamed to Lam $y\ \sigma\ (u\lceil y\rceil)$ and a computation step on the recursive definition has occurred. Notice the difference with Nominal Isabelle, which does not allow one to define functions exposing bound variables: in contrast, not only can we write a function that given a term Lam $x\ \sigma\ (u\lceil x\rceil)$ (with $x \notin \mathrm{FV}(u)$) will return the tuple $\langle y, \sigma, u\lceil y\rceil\rangle$ (for some $y \notin \mathrm{FV}(u)$), but we can also prove that putting together those items we obtain the original term, i.e. Lam $x\ \sigma\ (u\lceil x\rceil) = $ Lam $y\ \sigma\ (u\lceil y\rceil)$.

2.1 Some Derived Properties

Since $\mathcal{R}_{\Lambda_0}$ provides case analysis, it can be used to prove many of the properties that we expect from Par, App and Lam as constructors. Among these, an important one concerns injectivity and discrimination:

Lemma 2. *The following properties hold:*

- *if* Par $x = $ Par y, *then* $x = y$;
- *if* App $u_1\ u_2 = $ App $v_1\ v_2$, *then* $u_1 = v_1$ *and* $u_2 = v_2$;
- *if* Lam $x\ \sigma\ u = $ Lam $x\ \tau\ v$, *then* $\sigma = \tau$ *and* $u = v$;
- *different constructors always yield different terms:* Par $x \neq$ App $u\ v$, App $u_1\ u_2 \neq$ Lam $x\ \sigma\ v$, Par $x \neq$ Lam $y\ \sigma\ v$.

Proof (sketch). McBride's generic proof for inductive types ([13]) only requires pattern matching and reasoning by cases: it is thus easy to adapt it to our abstract data type.

One thing to notice is that the injectivity property for Lam requires the bound variable to be the same in both abstractions. If this is not the case, the two should be made equal by α-equivalence before applying injectivity.

Other properties cannot be proved as easily. One reason for this is that the constrained rewriting approach for the recursion principle is not completely satisfying: we are giving up the possibility to compute directly the result of functions defined by means of $\mathcal{R}_{\Lambda_0}$ because some of those functions (like those that try to expose bound variables) are ill-behaved. But other functions, like substitution, are well-behaved: we expect to know that whenever the bound variable x is not in $y, \mathrm{FV}(v)$, then the following equality holds:

$$(\text{Lam } x\ \sigma\ u)\,[v/y] = \text{Lam } x\ \sigma\ (u\,[v/y])$$

As a matter of fact, the equality holds in the context of our abstract data type. To prove it, we need the following induction principle:

Theorem 3. *The following induction principle $\mathcal{E}_{\Lambda_0}$ is provable:*

$$\mathcal{E}_{\Lambda_0} : \forall P : \Lambda_0 \Rightarrow \textbf{Prop.}$$
$$(\forall x.P \; (\textsf{Par} \; x)) \Rightarrow$$
$$(\forall u_1, u_2.P \; u_1 \Rightarrow P \; u_2 \Rightarrow P \; (\textsf{App} \; u_1 \; u_2)) \Rightarrow$$
$$(\forall x, \sigma, v.x \notin \text{FV}(v) \Rightarrow$$
$$(\forall y.y \notin \text{FV}(v) \Rightarrow P \; (v\lceil y\rceil)) \Rightarrow P \; (\textsf{Lam} \; x \; \sigma \; (v\lceil x\rceil))) \Rightarrow$$
$$\forall u.P \; u$$

Proof (sketch). We assume the branches of $\mathcal{E}_{\Lambda_0}$ as hypotheses and we subsequently prove $\forall \overrightarrow{p}.P \; (u \langle \overrightarrow{p}\rangle)$ using the recursion principle $\mathcal{R}_{\Lambda_0}$ on u. This implies $P \; u$ by instantiating $\overrightarrow{p}$ with the empty list $[]$.

That $\mathcal{E}_{\Lambda_0}$ can be proved using $\mathcal{R}_{\Lambda_0}$ should not be surprising, as in type theory recursion and induction are intimately related. Actually, when we ignore the computational content of $\mathcal{R}_{\Lambda_0}$ and only consider its type, we see that its form is very similar to that of an induction principle (where the result of recursion on a subterm corresponds to the induction hypothesis). Besides the fact that $\mathcal{E}_{\Lambda_0}$ returns a proof of a proposition rather than an object of a given type[2], the biggest difference between $\mathcal{R}_{\Lambda_0}$ and $\mathcal{E}_{\Lambda_0}$ is that the latter, in the Lam case, provides a different induction hypothesis for all possible choices of the bound variable, while $\mathcal{R}_{\Lambda_0}$ only considers a single variable (which one is not under our control: at most, we can require that it should be sufficiently fresh): this is usually enough to define a recursive function, but not in proofs by induction like the following ones. We will come back to the topic of universally quantified induction hypotheses in Sections 4 and 5.

Lemma 4. *For all u, x, v, $\text{FV}(u\,[v/x]) \subseteq \text{FV}(u) \cup \text{FV}(v)$.*

Lemma 5. *For all u, x, v, if $\overrightarrow{p}$ is a list of pairs of variable names not in $x, \text{FV}(v)$, then*

$$u\,[v/x] \langle \overrightarrow{p}\rangle = (u \langle \overrightarrow{p}\rangle)\,[v/x]$$

Both proofs are by induction on u, with Lemma 5 using Lemma 4 in the Lam case. These two properties are what we need to prove the commutation property for subst in the Lam case.

Fact 6 *If $x \notin y, \text{FV}(v)$, then $(\textsf{Lam} \; x \; \sigma \; u)\,[v/y] = \textsf{Lam} \; x \; \sigma \; (u\,[v/y])$*

Proof. Notice that $u = u \langle x/x\rangle = (\nu x.u)\lceil x\rceil$. Thus we can apply the constrained rewriting property: this leaves us with the goal

$$\textsf{Lam} \; z \; \sigma \; ((u \langle z/x\rangle)\,[v/y]) = \textsf{Lam} \; x \; \sigma \; (u\,[v/y])$$

---

[2] Induction principles are usually given for **Prop**; however we could as well derive a similar principle for **Type**, at no additional formalization cost.

where z is not free in $\nu x.u$ or $y, \mathrm{FV}(v)$. The goal can be proved by rewriting the left-hand side as follows:

$$
\begin{aligned}
&\mathsf{Lam}\ z\ \sigma\ ((u\ \langle z/x\rangle)\ [v/y]) \\
&= \mathsf{Lam}\ z\ \sigma\ (u\ [v/y]\ \langle z/x\rangle) && \text{(by Lemma 5)} \\
&= \mathsf{Lam}\ z\ \sigma\ ((\nu x.u\ [v/y])\lceil z\rceil) && \text{(by def. of renaming)} \\
&= \mathsf{Lam}\ x\ \sigma\ ((\nu x.u\ [v/y])\lceil x\rceil) && \text{(by } \alpha\text{-equivalence)} \\
&= \mathsf{Lam}\ x\ \sigma\ (u\ [v/y]\ \langle x/x\rangle) && \text{(by def. of renaming)} \\
&= \mathsf{Lam}\ x\ \sigma\ (u\ [v/y]) && \text{(by axiom)}
\end{aligned}
$$

where the α-equivalence holds because $z \notin \nu x.u\ [v/y]$, which is easily proved using Lemma 4.

It is worth noting that the recursion principle in Gordon and Melham's axiomatization ([11]) handles abstractions differently, allowing direct computation of functions employing it. This, however, comes at the price of requiring explicit treatment of variable renaming in the function definition. As a consequence, if we expressed substitution in that style, the following computation property would trivially hold in the Lam case:

$$
(\mathsf{Lam}\ x\ \sigma\ u)\ [v/y] = \mathsf{Lam}\ \varphi(y, \mathrm{FV}(v))\ \sigma\ ((u\ \langle \varphi(y, \mathrm{FV}(v))/x\rangle)\ [v/y])
$$

However, this is a weaker property than Fact 6 (which remains provable, with a similar argument as ours).

3   A Locally Nameless Model

A locally nameless representation [9] of a language with binders is a variant of de Bruijn's nameless representation where names are allowed to represent free parameters, but indices are always used to express bound variables. A locally nameless representation of the simply typed lambda calculus can be given as the following *pretm* inductive type of *pre-terms*:

$$
\begin{aligned}
&\text{inductive } pretm : \mathbf{Type} \triangleq \\
&\quad \mathsf{var} : \mathbb{N} \Rightarrow pretm \\
&\quad \mathsf{par} : \mathbb{A} \Rightarrow pretm \\
&\quad \mathsf{app} : pretm \Rightarrow pretm \Rightarrow pretm \\
&\quad \mathsf{abs} : tp \Rightarrow pretm \Rightarrow pretm.
\end{aligned}
$$

Notice we use lowercase identifiers to distinguish the constructors of *pretm* from the similar operations discussed in the previous section. The constructor var is used to construct indices and par for named parameters; abs is a nameless abstraction that is used as the counterpart of Lam abstractions, binding an index rather than a named variable: by convention, our indices are zero-based, so that index var k is considered to be bound to the $(k+1)$-th outer abstraction.

In such a representation, indices whose values are too high and thus do not point to any binder are said to be *dangling*: a dangling index is neither a bound variable nor a free, named parameter, thus it is often an unwanted situation. Most formalizations employing this style adopt a validity predicate on pre-terms that is verified only for real terms, i.e. those that do not contain dangling indices (also called *locally closed*).

However, in our case the type of pre-terms will have a much more substantial value as the interpretation of both terms and n-ary contexts. We regard dangling indices as holes implicitly bound at the outermost level, waiting for a context application to fill a free variable in them.

The following function checks whether a pre-term can be the interpretation of a k-ary context by verifying that the value of all dangling indices is less than k:

$$\text{check } u \ k \triangleq \begin{cases} \text{check (var } n) \ k = \begin{cases} \text{true if } n < k \\ \text{false else} \end{cases} \\ \text{check (par } x) \ k = \text{true} \\ \text{check (app } u_1 \ u_2) \ k = (\text{check } u_1 \ k) \wedge (\text{check } u_2 \ k) \\ \text{check (abs } \sigma \ u_1) \ k = \text{check } u_1 \ (k+1) \end{cases}$$

We thus define the interpretation of i-ary ostensibly named contexts in the locally nameless model as the dependent pair associating a pre-term u to the proof that check $u \ i = \text{true}$:

$$[\![\Lambda_i]\!] = ctx \ i \triangleq \Sigma u : pretm.\text{check } u \ i = \text{true}$$

We define algorithmically in the model two contextual operations that are a counterpart to the similar operations of the ostensibly named presentation. They employ a parameter k that is used, in recursive calls, to keep track of the number of abstractions crossed.

$$\nu_k x.u \triangleq \begin{cases} \nu_k x.\text{var } n = \begin{cases} \text{var } n & \text{if } n < k \\ \text{var } (n+1) \text{ else} \end{cases} \\ \nu_k x.\text{par } y = \begin{cases} \text{var } k & \text{if } x = y \\ \text{par } y & \text{else} \end{cases} \\ \nu_k x.\text{app } u_1 \ u_2 = \text{app } (\nu_k x.u_1) \ (\nu_k x.u_2) \\ \nu_k x.\text{abs } \sigma \ u_1 = \text{abs } \sigma \ (\nu_{k+1} x.u_1) \end{cases}$$

$$u\lceil x \rceil_k \triangleq \begin{cases} (\text{var } n)\lceil x \rceil_k = \begin{cases} \text{par } x & \text{if } n = k \\ \text{var } n & \text{if } n < k \\ \text{var } (n-1) & \text{if } n > k \end{cases} \\ (\text{par } y)\lceil x \rceil_k = \text{par } y \\ (\text{app } u_1 \ u_2)\lceil x \rceil_k = \text{app } (u_1\lceil x \rceil_k) \ (u_2\lceil x \rceil_k) \\ (\text{abs } \sigma \ u_1)\lceil x \rceil_k = \text{abs } \sigma \ (u_1\lceil x \rceil_{k+1}) \end{cases}$$

The definition of $[\![\Lambda_i]\!]$ as a dependent pair implies that, in the model, every term or context is composed of a structural part – a pre-term – together with a proof object asserting that the pre-term has the expected arity. This is reflected in the interpretation:

$$[\![\text{Par } x]\!] \quad\quad = (\text{par } x, \ldots)$$
$$[\![\text{App } u_1\ u_2]\!] = (\text{app } \pi_1([\![u_1]\!])\ \pi_1([\![u_2]\!]), \ldots)$$
$$[\![\text{Lam } x\ \sigma\ u_1]\!] = (\text{abs } \sigma\ (\nu_0 x.\pi_1([\![u_1]\!])), \ldots)$$

$$[\![\nu x.u]\!] \triangleq (\nu_0 x.\pi_1([\![u]\!]), \ldots)$$
$$[\![u\lceil x\rceil]\!] \triangleq (\pi_1([\![u]\!])\lceil x\rceil_0, \ldots)$$

where π_1 is the left projection of a dependent pair (here used to extract a pre-term from the interpretation of a term) and the ellipses "..." must be filled with appropriate proof objects. When we limit ourselves to the structural part, most of the interpretations are straightforward, but that of Lam is worth looking into: the name-carrying lambda is transformed by interpreting its body u_1 first, then turning all the occurrences of the parameter x into a dangling index that is immediately bound by a nameless abstraction.

We have formalized the existence of proof objects such that the interpretation of terms and contexts satisfies the following lemma, stating its soundness.

Lemma 7.

1. $[\![\text{Par } x]\!] : ctx\ 0$
2. *if* $[\![u]\!] : ctx\ 0$ *and* $[\![v]\!] : ctx\ 0$, *then* $[\![\text{App } u\ v]\!] : ctx\ 0$
3. *if* $[\![u]\!] : ctx\ 0$, *then* $[\![\text{Lam } x\ \sigma\ u]\!] : ctx\ 0$
4. *if* $[\![u]\!] : ctx\ i$, *then* $[\![\nu x.u]\!] : ctx\ (i+1)$
5. *if* $[\![u]\!] : ctx\ (i+1)$, *then* $[\![u\lceil x\rceil]\!] : ctx\ i$

We omit the trivial interpretation of the FV operation and state some of the remaing properties we proved to ensure the validity of the model.

Lemma 8.

1. $[\![x\ \langle y/x\rangle]\!] = [\![y]\!]$
2. *if* $x \neq y$, *then* $[\![x\ \langle z/y\rangle]\!] = [\![x]\!]$
3. $[\![(\text{App } u\ v)\ \langle y/x\rangle]\!] = [\![\text{App } (u\ \langle y/x\rangle)\ (v\ \langle y/x\rangle)]\!]$
4. *if* $z \neq x, y$, *then* $[\![(\text{Lam } z\ \sigma\ u)\ \langle y/x\rangle]\!] = [\![\text{Lam } z\ \sigma\ (u\ \langle y/x\rangle)]\!]$

Lemma 9. *(α-conversion)*
If $x, y \notin \text{FV}(u)$, *then* $[\![\text{Lam } x\ \sigma\ (u\lceil x\rceil)]\!] = [\![\text{Lam } y\ \sigma\ (u\lceil y\rceil)]\!]$

Lemma 10.

1. $[\![(\nu x.u)\lceil x\rceil]\!] = [\![u]\!]$
2. *if* $x \notin \text{FV}(u)$, *then* $[\![\nu x.(u\lceil x\rceil)]\!] = [\![u]\!]$

A final piece is missing to complete the model: an intepretation of the recursion principle $\mathcal{R}_{\Lambda_0}$, and the proof that its equational properties are valid. We provide such an interpretation as a recursive function *pretm_rec* on pre-terms, which is later lifted to proper terms.

The function *pretm_rec* receives similar arguments to the abstract $\mathcal{R}_{\Lambda_0}$, plus an additional f_{var} for dangling indices (which are missing from the ostensibly named presentation) that is not of particular interest here.

When dealing with a term of the form abs $\sigma\, u$, we generate a new fresh name $x = \varphi(C, \mathrm{FV}(u))$ and open u with respect to that name; we then perform the recursive call on the opened $u\lceil x\rceil_0$. The full *pretm_rec* is defined by recursion on the *height* of the syntax tree of a pre-term, rather than structural recursion on the pre-term, because not all the recursive calls are on a pre-term which is structurally smaller than the one received in input (something that is beyond the capabilities of the termination heuristics found in Matita):

let rec *pretm_rec_aux* $(P : \text{pretm} \Rightarrow \textbf{Type})$
  $(C : \text{list } \mathbb{A})\ (f_{\mathsf{par}} : \forall x.P\ (\mathsf{par}\ x))\ (f_{\mathsf{var}} : \forall n.P\ (\mathsf{var}\ n))$
  $(f_{\mathsf{app}} : \forall v1, v2.P\ v1 \Rightarrow P\ v2 \Rightarrow P\ (\mathsf{app}\ v1\ v2))$
  $(f_{\mathsf{abs}} : \forall x, s, v.x \notin \mathrm{FV}\ v \Rightarrow x \notin C \Rightarrow P\ (v\lceil x\rceil) \Rightarrow P\ (\mathsf{abs}\ s\ (\nu x.(v\lceil x\rceil))))$
  $(h : \mathbb{N})\ u \text{ on } h : (\text{height}(u) < h \Rightarrow P\ u) \triangleq \text{match } h \text{ with}$
  $[0 \Rightarrow \ldots\ (\text{* absurd: height is always > 0 *})$
  $|S\ h_0 \Rightarrow \text{let } rcall \triangleq \text{pretm_rec_aux } P\ C\ f_{\mathsf{par}}\ f_{\mathsf{var}}\ f_{\mathsf{app}}\ f_{\mathsf{abs}}\ h_0 \text{ in}$
    match u with
    $[\mathsf{par}\ x \Rightarrow \lambda\_.f_{\mathsf{par}}\ x$
    $|\mathsf{var}\ n \Rightarrow \lambda\_.f_{\mathsf{var}}\ n$
    $|\mathsf{app}\ v_1\ v_2 \Rightarrow \lambda p.f_{\mathsf{app}}\ldots(rcall\ v_1\ldots)\ (rcall\ v_2\ldots)$
    $|\mathsf{abs}\ \sigma\ v \Rightarrow \text{let } x \triangleq \varphi(C, \mathrm{FV}(v)) \text{ in}$
      $f_{\mathsf{abs}}\ldots(rcall\ (v\lceil x\rceil_0)\ldots)\ldots)]]$

pretm_rec $P\ C\ f_{\mathsf{par}}\ f_{\mathsf{var}}\ f_{\mathsf{app}}\ f_{\mathsf{abs}}\ u \triangleq$
  pretm_rec_aux $P\ C\ f_{\mathsf{par}}\ f_{\mathsf{var}}\ f_{\mathsf{app}}\ f_{\mathsf{abs}}\ (S\ \text{height}(u))\ u\ldots$

The ellipses in *pretm_rec* and in the app and abs cases of *pretm_rec_aux* must be filled with proofs that the value provided for h is an upper bound to the height of the term on which we are performing recursion (in our formalization, those proofs were filled in interactively).

Since proper terms are a subset of pre-terms, expressing $\mathcal{R}_{\Lambda_0}$ in terms of *pretm_rec* is conceptually simple, although in practice the related proofs are technical, due to the handling of dependent types. The interested reader can check the details of the proof in the formalization, within the module `model.ma`.

4   Intermezzo: Formalizing Typing Rules

We now turn our attention to the formalization of more complex structures: typing judgments and their derivations by means of inductive rules. We chose the simply typed lambda-calculus as our setting, because even in its simplicity some of the issues of the representation of binding are already quite visible.

Its formalization in the most common representations of binding is well understood. Most locally nameless formalizations employ the following concrete introduction rule for lambda abstractions:

$$\frac{x \notin FV(u) \qquad \langle x, \sigma\rangle, \Gamma \vdash u\ \{\mathsf{var}\ 0 \mapsto \mathsf{par}\ x\} : \tau}{\Gamma \vdash \mathsf{abs}\ \sigma\ u : \sigma \to \tau}\ (\text{LN-T-Abs})$$

where $u \{v \mapsto v'\}$ is a generalized substitution operator, replacing a subterm v in u with v', preserving scopes. Following [23], we call rules in this style "backward", as they are most easily read from the bottom upwards: if the term which we intend to type can be deconstructed as abs σ u, then we should first get a typing derivation for u in an extended typing environment. However, since unboxing an abstraction yields a term where the index var 0 is possibly dangling, we are supposed to substitute a fresh name x for it, which must also be used in the extended context.

An alternative "forward" representation of the abstraction rule has a more familiar look:

$$\frac{\langle x, \sigma \rangle, \Gamma \vdash u : \tau}{\Gamma \vdash \text{Lam } x \ \sigma \ u : \sigma \to \tau} \ (\text{LN-T-Lam})$$

In this case, the substitution is hidden inside the Lam operator: Lam x σ u is syntactic sugar for abs σ $u \{\text{par } x \mapsto \text{var } 0\}$. Although this rule is more pleasant to read, in practice it is seldom used in formalizations because the associated induction principle is more difficult to use, due to the fact that Lam is not a real constructor: on the contrary, the algorithmic interpretation of the backward rule is immediate, as we argued some lines above.

As it turns out, even if we formalize a type system by means of backward rules, we get an induction principle which is weaker than what a formalizer expects. For example, suppose that we write a type checker for the simply typed calculus: we can verify its soundness with respect to the formalized type system (type-checking does not succeed for ill-typed terms) quite easily by induction; however verifying completeness (all well-typed terms typecheck successfully) turns out to be challenging for a naive formalizer.

As originally noted by McKinna and Pollack [15], the reason behind this difficulty lies in the fact that the LN-T-ABS rule is quite liberal: x can be any sufficiently fresh parameter. Given the typing judgment associated to an abstraction, we get a different derivation for every choice of a suitable x. All the derivations are isomorphic, but contain, so to say, a "hardcoded" parameter name: in other words, when we view typing derivations as data structures, they are not canonical.

The problem with typing derivations being not canonical is that, in a proof by induction, the hardcoded fresh parameter x makes its return as part of the induction hypothesis associated with the abstraction case:

$$\forall P.$$
$$...$$
$$\left(\begin{array}{l} \forall \Gamma, x, \sigma, u, \tau. \\ \quad x \notin \text{FV}(u) \Rightarrow \\ \quad \langle x, \sigma \rangle, \Gamma \vdash u \{\text{var } 0 \mapsto \text{par } x\} : \tau \Rightarrow \\ \quad \boxed{P(\langle x, \sigma \rangle, \Gamma, u \{\text{var } 0 \mapsto \text{par } x\}, \tau)} \Rightarrow \\ \quad P(\Gamma, \text{abs } \sigma \ u, \sigma \to \tau) \end{array} \right) \Rightarrow$$
$$..$$
$$\forall \Gamma, u, \sigma. \Gamma \vdash u : \sigma \Rightarrow P(\Gamma, u, \sigma)$$

However on many occasions we will need our induction hypothesis to refer to an *arbitrary* $y \notin \mathrm{dom}(\Gamma)$ (or even *all* such ys).

We can force typing derivations to be canonical (independent of arbitrary choices of parameter names) by means of a universally quantified premise:

$$\frac{\left(\begin{array}{l} \forall x. x \notin \mathrm{dom}(\Gamma), \mathrm{FV}(u) \Rightarrow \\ \langle x, \sigma \rangle, \Gamma \vdash u \{\mathsf{var}\ 0 \mapsto \mathsf{par}\ x\} : \tau) \end{array}\right)}{\Gamma \vdash \mathsf{abs}\ \sigma\ u : \sigma \to \tau} \quad (\text{LN-T-Abs'})$$

This yields a *strong* induction principle, where the induction hypothesis associated to the abstraction case is similarly quantified over all suitable xs. However, the rule LN-T-Abs' itself is actually weaker: to derive a typing judgment for abstractions, one now needs to prove an infinite number of judgments, one for every choice of x! This is not how typecheckers work and is thus usually not considered a good formalization of a typing rule.

Still, it must be noted that all the rules presented in this section are equivalent. In particular, it is possible to prove the "strong" induction principle for a formalization using LN-T-Abs by showing that the typing judgment is *equivariant*, i.e. stable under arbitrary finite permutations of names π:

$$\Gamma \vdash u : \sigma \iff \forall \pi. \pi \cdot \Gamma \vdash \pi \cdot u : \sigma$$

5 Ostensibly Named Representation of Typing

We employ the ostensibly named style presented in Section 2 to express the type system of the simply typed lambda calculus.

$$\frac{\langle x, \sigma \rangle \in \Gamma \qquad \mathrm{dom}(\Gamma)\ \text{is duplicate-free}}{\Gamma \vdash_O \mathsf{Par}\ x : \sigma} \quad (\text{ON-T-Par})$$

$$\frac{\langle x, \sigma \rangle, \Gamma \vdash_O u : \tau}{\Gamma \vdash_O \mathsf{Lam}\ x\ \sigma\ u : \sigma \to \tau} \quad (\text{ON-T-Lam})$$

$$\frac{\Gamma \vdash_O u : \sigma \to \tau \qquad \Gamma \vdash_O v : \sigma}{\Gamma \vdash_O \mathsf{App}\ u\ v : \tau} \quad (\text{ON-T-App})$$

Fig. 1. Typing rules for $\lambda_\to$, ostensibly named style

The typing rules, shown in Figure 1, look quite unremarkable. The rule ON-T-Lam, in particular, looks the same as the rule LN-T-Lam of the previous section, although in this case Lam is opaque and, more importantly, the rules themselves must not be intended as the constructors of the concrete inductive type of typing derivations, but as operations provided by the abstract data type of typing derivations. We postpone the discussion about the internal representation of typing to the next section.

$$\forall P.$$
$$(\forall \Gamma, x, \sigma.\langle x, \sigma \rangle \in \Gamma \Rightarrow P(\Gamma, \mathsf{Par}\, x, \sigma)) \Rightarrow$$
$$\left(\begin{array}{l} \forall \Gamma, x, \sigma, u, \tau.x \notin \mathrm{dom}(\Gamma), \mathrm{FV}(u) \Rightarrow \\ (\forall y.y \notin \mathrm{dom}(\Gamma), \mathrm{FV}(u) \Rightarrow \langle y, \sigma \rangle, \Gamma \vdash_O u\lceil y \rceil : \tau) \Rightarrow \\ (\forall y.y \notin \mathrm{dom}(\Gamma), \mathrm{FV}(u) \Rightarrow P(\langle y, \sigma \rangle, \Gamma, u\lceil y \rceil, \tau)) \Rightarrow \\ P(\Gamma, \mathsf{Lam}\, x\, \sigma\, (u\lceil x \rceil), \sigma \to \tau) \end{array} \right) \Rightarrow$$
$$\left(\begin{array}{l} \forall \Gamma, u, \sigma, \tau. \\ \Gamma \vdash_O u : \sigma \to \tau \Rightarrow \Gamma \vdash_O v : \sigma \Rightarrow \\ P(\Gamma, u, \sigma \to \tau) \Rightarrow P(\Gamma, v, \sigma) \Rightarrow \\ P(\Gamma, \mathsf{App}\, u\, v, \tau) \end{array} \right) \Rightarrow$$
$$\forall \Gamma, u, \sigma.\Gamma \vdash_O u : \sigma \Rightarrow P(\Gamma, u, \sigma)$$

Fig. 2. Rule induction for the $\lambda_\to$ typing derivations, ostensibly named style

The ostensibly named induction principle we associate to these rules (Figure 2) is more interesting. The induction hypothesis of the lambda case (highlighted in the figure) is quantified over all suitable parameter names, as in a strong principle; however, we use the variable opening operation, both in the induction hypothesis and in the conclusion, to avoid exposing the internal structure of the terms. To prevent variable capture, the new names are chosen to be fresh with respect to the context u being opened.

Ostensibly Named Inversion. We can use the ostensibly named induction principle to derive an *inversion* principle in the style of McBride ([14]). Together with Lemma 2, inversion principles provide an effective tool to perform case analysis on the last rule used in a derivation tree. The inversion principle we obtain is *strong* (as in [6]) in the sense that, for instance, given a derivation of $\Gamma \vdash \mathsf{Lam}\, x\, \sigma\, (u\lceil x \rceil) : \sigma \to \tau$ with $x \notin \mathrm{FV}(u)$, we can deduce $\langle y, \sigma \rangle, \Gamma \vdash u\lceil y \rceil : \tau$ for all $y \notin \mathrm{FV}(u), \mathrm{dom}(\Gamma)$.

5.1 Internal Representation of Typing Rules

As we argued in Section 4, the weak or strong induction principle dilemma, in the context of typing, stems from the fact that the natural typing rules mentioning a specific variable in the binder case, yield a plurality of derivations for the same typing judgment; but to have a single derivation and thus a strong induction principle, one has to employ an infinitary typing rule.

In essence, names are the origin of the dilemma: so it is just natural to look at a de Bruijn formalization of the typing rules, shown in Figure 3. In the nameless encoding, typing environments are just lists of types: we denote them as $\gamma, \gamma', \dots$.

$$\frac{\gamma(n) = \sigma}{\gamma \vdash_D \text{var } n : \sigma} \quad (\text{DB-T-VAR})$$

$$\frac{\sigma, \gamma \vdash_D u : \tau}{\gamma \vdash_D \text{abs } \sigma\, u : \sigma \to \tau} \quad (\text{DB-T-ABS})$$

$$\frac{\gamma \vdash_D u : \sigma \to \tau \qquad \gamma \vdash_D v : \sigma}{\gamma \vdash_D \text{app } u\, v : \tau} \quad (\text{DB-T-APP})$$

Fig. 3. Typing rules for $\lambda_\to$, pure de Bruijn style

Since in this presentation no named parameter appears, context references are by position (rule DB-T-VAR, where $\gamma(n)$ returns the $n+1$-th type in γ). In the abstraction rule, unboxing an abstraction yields, in the premise, a new dangling index, whose type is referenced in an extended context.

The nice thing about going nameless is the following: the rule DB-T-ABS is finitary (in fact, unary), but at the same time it is also canonical! For every well-typed abstraction, there is exactly one derivation, because we do not have the freedom of choosing *any* fresh name: in fact, we choose *none*. This desirable situation comes from the fact that, in a nameless setting, not only abstractions, but also the typing environment γ of the judgments and even the rule DB-T-ABS are treated as binders.

These properties make the de Bruijn style rules, together with the associated induction principle (Figure 4), an ideal model for the abstract rules of the previous section.

$$
\begin{aligned}
&\forall P. \\
&(\forall \gamma, n, \sigma. \gamma(n) = \sigma \Rightarrow P(\gamma, \text{var } n, \sigma)) \Rightarrow \\
&\left(\begin{array}{l} \forall \gamma, \sigma, u, \tau. \\ \quad \sigma, \gamma \vdash_D u : \tau \Rightarrow \\ \quad P(\sigma, \gamma, u, \tau) \Rightarrow \\ \quad P(\gamma, \text{abs } \sigma\, u, \sigma \to \tau) \end{array} \right) \Rightarrow \\
&\left(\begin{array}{l} \forall \gamma, u, \sigma, \tau. \\ \quad \gamma \vdash_D u : \sigma \to \tau \Rightarrow \gamma \vdash_D v : \sigma \Rightarrow \\ \quad P(\gamma, u, \sigma \to \tau) \Rightarrow P(\gamma, v, \sigma) \Rightarrow \\ \quad P(\gamma, \text{App } u\, v, \tau) \end{array} \right) \Rightarrow \\
&\forall \gamma, u, \sigma. \gamma \vdash_D u : \sigma \Rightarrow P(\gamma, u, \sigma)
\end{aligned}
$$

Fig. 4. Rule induction for the $\lambda_\to$ typing derivations, de Bruijn style

To model the ostensibly named presentation of $\lambda_\to$, we first need to interpret $\vdash_O$ in terms of $\vdash_D$. For this purpose, we give an interpretation of the ostensibly named representations of types, typing environments, and terms into the corresponding concepts of the de Bruijn representation. As usual, the interpretation of types is the identity. For what concerns typing environments, all we need to do is to throw away the names, keeping the types in the same order: this is best done by projecting the second component of each pair in the list. Finally the interpretation of terms is given by taking the interpretation we used in Section 3 and subsequently closing the obtained locally-nameless term with respect to the names in its typing context: assuming all the names referenced in the term have an entry in the environment, the resulting interpretation is nameless. In symbols:

$$[\![\sigma]\!] \triangleq \sigma$$
$$[\![\Gamma]\!] \triangleq \mathrm{cod}(\Gamma)$$
$$[\![u]\!]_{\vec{x}} \triangleq \nu_0 \vec{x}.[\![u]\!]$$

$$[\![\Gamma \vdash_O u : \sigma]\!] \triangleq df(\mathrm{dom}(\Gamma)) \wedge [\![\Gamma]\!] \vdash_D [\![u]\!]_{\mathrm{dom}(\Gamma)} : [\![\sigma]\!]$$

where:
$$\mathrm{dom}([x_1 : \sigma_1; \ldots; x_n : \sigma_n]) \triangleq [x_1; \ldots; x_n]$$
$$\mathrm{cod}([x_1 : \sigma_1; \ldots; x_n : \sigma_n]) \triangleq [\sigma_1; \ldots; \sigma_n]$$
$$\nu_k[x_1; \ldots; x_n].u \triangleq \nu_k x_1 \ldots \nu_k x_n.u$$

The model of an ostensibly named judgment contains, in addition to its nameless counterpart, a proof that the domain of Γ is duplicate-free (a property which we expect to be able to prove, and which is not implied by the nameless judgment). We have used the predicate df to assert that a certain list of names is duplicate-free. The vector notation $\vec{x}$ is employed as a compact way of referring to lists, in this case to a list of names. Our second task is to model the rules of Figure 1 as instances of their nameless counterparts. This is expressed by the following lemma:

Lemma 11.

1. If $\langle x, \sigma \rangle \in \Gamma$ and $\mathrm{dom}(\Gamma)$ is duplicate-free, then $[\![\Gamma \vdash_O \mathsf{Par}\ x : \sigma]\!]$.
2. If $[\![\langle x, \sigma \rangle, \Gamma \vdash_O u : \tau]\!]$, then $[\![\Gamma \vdash_O \mathsf{Lam}\ x\ \sigma\ u : \sigma \to \tau]\!]$.
3. If $[\![\Gamma \vdash_O u : \sigma \to \tau]\!]$ and $[\![\Gamma \vdash_O v : \sigma]\!]$, then $[\![\Gamma \vdash_O \mathsf{App}\ u\ v : \tau]\!]$.

Finally, we provide an interpretation of the ostensibly named induction principle as follows:

Theorem 12. *Let P be a predicate over named typing environments, terms, and types. Assume the following properties:*

1. *for all Γ, x, σ, $\langle x, \sigma \rangle \in \Gamma$ implies $P(\Gamma, \mathsf{Par}\ x, \sigma)$;*
2. *for all $\Gamma, x, \sigma, u, \tau$ such that*
   - *$x \notin \mathrm{dom}(\Gamma), \mathrm{FV}(u)$*
   - *$\forall y. y \notin \mathrm{dom}(\Gamma), \mathrm{FV}(u) \Rightarrow [\![\langle y, \sigma \rangle, \Gamma \vdash_O u[y] : \tau]\!]$*
   - *$\forall y. y \notin \mathrm{dom}(\Gamma), \mathrm{FV}(u) \Rightarrow P(\langle y, \sigma \rangle, \Gamma, u\lceil y \rceil, \tau)$*
   then $P(\Gamma, \mathsf{Lam}\ x\ \sigma\ (u\lceil x \rceil), \sigma \to \tau)$ holds;

3. for all $\Gamma, u, v, \sigma, \tau$ such that
- $[\![\Gamma \vdash_O u : \sigma \to \tau]\!]$
- $[\![\Gamma \vdash_O v : \sigma]\!]$
- $P(\Gamma, u, \sigma \to \tau)$
- $P(\Gamma, v, \sigma)$

then $P(\Gamma, \mathsf{App}\ u\ v, \tau)$.

Then for all Γ, u, σ such that $[\![\Gamma \vdash_O u : \sigma]\!]$, $P(\Gamma, u, \sigma)$ holds.

Proof (sketch). Assume $[\![\Gamma \vdash_O u : \sigma]\!]$. By definition, we know that the domain of Γ is duplicate-free and that $[\![\Gamma]\!] \vdash_D [\![u]\!]_{\mathrm{dom}(\Gamma)} : \sigma$.

Let $\hat{P}$ be the augmented predicate:

$$\hat{P}(\gamma, u, \sigma) \triangleq \forall \overrightarrow{x_{|\gamma|}}.df(\overrightarrow{x_{|\gamma|}}) \Rightarrow P(\gamma \lceil \overrightarrow{x_{|\gamma|}} \rceil, u \lceil \overrightarrow{x_{|\gamma|}} 0 \rceil, \sigma)$$

where we have extended the definition of $-\lceil - \rceil$ as follows:

$$[\sigma_1; \ldots; \sigma_n] \lceil [x_1; \ldots; x_n] \rceil \triangleq [\langle x_1, \sigma_1 \rangle; \ldots; \langle x_n, \sigma_n \rangle]$$

$$u \lceil \overrightarrow{x} \rceil_k \triangleq \begin{cases} u \lceil [] \rceil_k = u \\ u \lceil y, \overrightarrow{z} \rceil_k = u \lceil y \rceil_k \lceil \overrightarrow{z} \rceil_k \end{cases}$$

The notation $|\gamma|$ indicates the length of the list γ. The extended vector notation in the form $\overrightarrow{x_n}$ is used to express lists of length n. We now proceed by induction on $[\![\Gamma]\!] \vdash_D [\![u]\!]_{\mathrm{dom}(\Gamma)} : \sigma$ to prove $\hat{P}([\![\Gamma]\!], [\![u]\!]_{\mathrm{dom}(\Gamma)}, \sigma)$. By instantiating the augmented predicate over the list $\mathrm{dom}(\Gamma)$, we finally obtain $P(\Gamma, u, \sigma)$ (using lemma 10).

The most difficult part of the induction is the abstraction case: given $\gamma_0, \sigma_0, \tau_0, u_0$ and a duplicate free list of names $\overrightarrow{x_{|\gamma_0|}}$, we need to prove

$$P(\gamma_0 \lceil \overrightarrow{x_{|\gamma_0|}} \rceil, (\mathsf{abs}\ \sigma_0\ u_0) \lceil \overrightarrow{x_{|\gamma_0|}} \rceil, \sigma_0 \to \tau_0)$$

under the hypotheses

$$\sigma_0, \gamma_0 \vdash_D u_0 : \tau_0$$
$$\hat{P}(\sigma_0, \gamma_0, u_0, \tau_0)$$

where the latter is the induction hypothesis. Then we take a fresh name y and rewrite in the thesis

$$(\mathsf{abs}\ \sigma_0\ u_0) \lceil \overrightarrow{x_{|\gamma_0|}} \rceil$$
$$= \mathsf{abs}\ \sigma_0\ (u_0 \lceil \overrightarrow{x_{|\gamma_0|}} 1 \rceil)$$
$$= \mathsf{Lam}\ y\ \sigma_0\ (u_0 \lceil \overrightarrow{x_{|\gamma_0|}} 1 \lceil y \rceil)$$

This allows us to apply the second lemma hypothesis (to fulfill the guards of the hypothesis, we exploit the equality $u_0 \lceil \overrightarrow{x_{|\gamma_0|}} 1 \lceil y \rceil = u_0 \lceil y, \overrightarrow{x_{|\gamma_0|}} \rceil$).

6   Beta Reduction

The ostensibly named technique we used in the previous section to define typing judgments extends to other types of judgments. The trick is to make explicit the environment where all the free parameters appearing in the judgment are defined. Other than that, we axiomatize reduction rules that are close to informal syntax (Fig. 5). These introduction rules are completed by a strong induction principle, providing a universally quantified induction hypothesis for the case where reduction happens under a lambda (Fig. 6).

6.1   De Bruijn Model of Beta Reduction

A de Bruijn-style model of beta reduction is given in Figure 7. The main difference with the ostensibly named rules is that the list of free parameters is replaced by an integer stating the number of dangling indices possibly appearing in the judgment. Thus we define the interpretation of an ostensibly named reduction as:

$$\llbracket \overrightarrow{x_k} \vdash_O u \triangleright v \rrbracket \triangleq k \vdash_D \llbracket u \rrbracket_{\overrightarrow{x_k}} \triangleright \llbracket v \rrbracket_{\overrightarrow{x_k}}$$

All the preterms involved in the de Bruijn judgment must be contexts of a suitable arity containing no parameters. When this property is not implied by

$$\frac{\overrightarrow{x} \text{ is duplicate-free} \qquad \text{FV}(\mathsf{App}\ (\mathsf{Lam}\ y\ \sigma\ u)\ v) \subseteq \overrightarrow{x}}{\overrightarrow{x} \vdash_O \mathsf{App}\ (\mathsf{Lam}\ y\ \sigma\ u)\ v \triangleright u\,[^v/_y]} \quad \text{(ON-B-Red)}$$

$$\frac{\overrightarrow{x} \vdash_O u \triangleright u' \qquad \text{FV}(v) \subseteq \overrightarrow{x}}{\overrightarrow{x} \vdash_O \mathsf{App}\ u\ v \triangleright \mathsf{App}\ u'\ v} \quad \text{(ON-B-App1)}$$

$$\frac{\overrightarrow{x} \vdash_O v \triangleright v' \qquad \text{FV}(u) \subseteq \overrightarrow{x}}{\overrightarrow{x} \vdash_O \mathsf{App}\ u\ v \triangleright \mathsf{App}\ u\ v'} \quad \text{(ON-B-App2)}$$

$$\frac{y, \overrightarrow{x} \vdash_O u \triangleright u'}{\overrightarrow{x} \vdash_O \mathsf{Lam}\ y\ \sigma\ u \triangleright \mathsf{Lam}\ y\ \sigma\ u'} \quad \text{(ON-B-Xi)}$$

Fig. 5. Beta reduction: ostensibly named encoding

$$\forall P. \left\{ \begin{array}{l} \left(\begin{array}{l} \forall \overrightarrow{x}, y, \sigma, u, v.df(\overrightarrow{x}) \Rightarrow \\ \text{FV}(\mathsf{App}\ (\mathsf{Lam}\ y\ \sigma\ u)\ v) \subseteq \overrightarrow{x} \Rightarrow \\ P(\overrightarrow{x}, \mathsf{App}\ (\mathsf{Lam}\ y\ \sigma\ u)\ v, u\,[^v/_y]) \end{array} \right) \Rightarrow \\[6pt] \left(\begin{array}{l} \forall \overrightarrow{x}, u, u', v.\,\text{FV}(v) \subseteq \overrightarrow{x} \Rightarrow \\ \overrightarrow{x} \vdash_O u \triangleright u' \Rightarrow P(\overrightarrow{x}, u, u') \Rightarrow \\ P(\overrightarrow{x}, \mathsf{App}\ u\ v, \mathsf{App}\ u'\ v) \end{array} \right) \Rightarrow \\[6pt] \left(\begin{array}{l} \forall \overrightarrow{x}, u, v, v'.\,\text{FV}(u) \subseteq \overrightarrow{x} \Rightarrow \\ \overrightarrow{x} \vdash_O v \triangleright v' \Rightarrow P(\overrightarrow{x}, v, v') \Rightarrow \\ P(\overrightarrow{x}, \mathsf{App}\ u\ v, \mathsf{App}\ u\ v') \end{array} \right) \Rightarrow \\[6pt] \left(\begin{array}{l} \forall \overrightarrow{x}, y, \sigma, u, u'.y \notin \overrightarrow{x}, \text{FV}(u), \text{FV}(u') \Rightarrow \\ (\forall z.z \notin \overrightarrow{x}, \text{FV}(u), \text{FV}(u') \Rightarrow z, \overrightarrow{x} \vdash_O u\lceil z\rceil \triangleright u'\lceil z\rceil) \Rightarrow \\ (\forall z.z \notin \overrightarrow{x}, \text{FV}(u), \text{FV}(u') \Rightarrow P(z, \overrightarrow{x}, u\lceil z\rceil, u'\lceil z\rceil)) \Rightarrow \\ P(\overrightarrow{x}, \mathsf{Lam}\ y\ \sigma\ (u\lceil y\rceil), \mathsf{Lam}\ y\ \sigma\ (u'\lceil y\rceil)) \end{array} \right) \Rightarrow \\[6pt] \forall \overrightarrow{x}, u, u'.\overrightarrow{x} \vdash_O u \triangleright u' \Rightarrow P(\overrightarrow{x}, u, u') \end{array} \right.$$

Fig. 6. Induction principle for beta reduction: ostensibly named encoding

a recursive premise of a rule, we have to specify it as an extra premise: for this purpose, we use the notation

$$k \vdash_D u \ \mathbf{ok} \triangleq \text{check\_tm } u \ k = \textit{true} \wedge \text{FV}(u) = \emptyset$$

$$\frac{y, \overrightarrow{x_k} \text{ is duplicate-free} \qquad k \vdash_D \text{app (abs } \sigma \ u) \ v \ \mathbf{ok}}{k \vdash_D \text{app (abs } \sigma \ u) \ v \triangleright [\![u \ulcorner y, \overrightarrow{x_k}\urcorner] \ [v\ulcorner\overrightarrow{x_k}\urcorner/y]]\!]_{\overrightarrow{x_k}}} \ \text{(DB-B-Red)}$$

$$\frac{k \vdash_D u \triangleright u' \qquad k \vdash_D v \ \mathbf{ok}}{k \vdash_D \text{app } u \ v \triangleright \text{app } u' \ v} \ \text{(DB-B-App1)}$$

$$\frac{k \vdash_D v \triangleright v' \qquad k \vdash_D u \ \mathbf{ok}}{k \vdash_D \text{app } u \ v \triangleright \text{app } u \ v'} \ \text{(DB-B-App2)}$$

$$\frac{k + 1 \vdash_D u \triangleright u'}{k \vdash_D \text{abs } \sigma \ u \triangleright \text{abs } \sigma \ u'} \ \text{(DB-B-Xi)}$$

Fig. 7. Beta reduction: de Bruijn encoding

While most other adaptations are trivial, rule DB-B-Red is slightly upsetting: de Bruijn terms u and v are opened in an arbitrary environment to become ostensibly named terms; then we use ostensibly named substitution and finally convert the result back to a de Bruijn term. This round-trip is entirely unnecessary if we define a substitution operation on de Bruijn terms; however for our purpose – justifying the ostensibly named rules and induction principle – this is not required: thus we decided not to bother dealing with two notions of substitution and their equivalence.

The following properties show that the de Bruijn rules are a model of the ostensibly named rules. Their proofs are similar to those relative to the typing judgment.

Lemma 13.

1. If $\overrightarrow{x}$ is duplicate-free and $\text{FV}(\text{App (Lam } y \ \sigma \ u) \ v) \subseteq \overrightarrow{x}$ then $[\![\overrightarrow{x} \vdash_O \text{App (Lam } y \ \sigma \ u) \ v \triangleright u \ [v/y]]\!]$.
2. If $[\![\overrightarrow{x} \vdash_O u \triangleright u']\!]$ and $\text{FV}(v) \subseteq \overrightarrow{x}$ then $[\![\overrightarrow{x} \vdash_O \text{App } u \ v \triangleright \text{App } u' \ v]\!]$.
3. If $[\![\overrightarrow{x} \vdash_O v \triangleright v']\!]$ and $\text{FV}(u) \subseteq \overrightarrow{x}$ then $[\![\overrightarrow{x} \vdash_O \text{App } u \ v \triangleright \text{App } u \ v']\!]$.
4. If $[\![y, \overrightarrow{x} \vdash_O u \triangleright u']\!]$ then $[\![\overrightarrow{x} \vdash_O \text{Lam } y \ \sigma \ u \triangleright \text{Lam } y \ \sigma \ u')]\!]$.

Theorem 14. *Let P be a predicate over named lists of variable names and pairs of terms. Assume the following properties:*

1. *for all $\overrightarrow{x}$,y,σ,u,v such that $\overrightarrow{x}$ is duplicate-free,* $\mathrm{FV}(\mathsf{App}\ (\mathsf{Lam}\ y\ \sigma\ u)\ v) \subseteq \overrightarrow{x}$
   implies $P(\overrightarrow{x}, \mathsf{App}\ (\mathsf{Lam}\ y\ \sigma\ u)\ v, u\ [v/y])$;
2. *for all $\overrightarrow{x}$,u,u',v such that*
   - $\mathrm{FV}(v) \subseteq \overrightarrow{x}$
   - $\llbracket \overrightarrow{x} \vdash u \triangleright u' \rrbracket$
   - $P(\overrightarrow{x}, u, u')$
   then $P(\overrightarrow{x}, \mathsf{App}\ u\ v, \mathsf{App}\ u'\ v)$ *holds;*
3. *for all $\overrightarrow{x}$,u,v,v' such that*
   - $\mathrm{FV}(u) \subseteq \overrightarrow{x}$
   - $\llbracket \overrightarrow{x} \vdash v \triangleright v' \rrbracket$
   - $P(\overrightarrow{x}, v, v')$
   then $P(\overrightarrow{x}, \mathsf{App}\ u\ v, \mathsf{App}\ u\ v')$ *holds;*
4. *for all $\overrightarrow{x}$,y,σ,u,u' such that*
   - $y \notin \overrightarrow{x}, \mathrm{FV}(u), \mathrm{FV}(u')$;
   - $\forall z.z \notin \overrightarrow{x}, \mathrm{FV}(u), \mathrm{FV}(u') \Rightarrow \llbracket y, \overrightarrow{x} \vdash_O u\lceil z \rceil \triangleright u'\lceil z \rceil \rrbracket$
   - $\forall z.z \notin \overrightarrow{x}, \mathrm{FV}(u), \mathrm{FV}(u') \Rightarrow P(y, \overrightarrow{x}, u\lceil z \rceil, u'\lceil z \rceil)$
   then $P(\overrightarrow{x}, \mathsf{Lam}\ y\ \sigma\ (u\lceil y \rceil), \mathsf{Lam}\ y\ \sigma, u'\lceil y \rceil)$ *holds;*

Then for all $\overrightarrow{x}$,u,u' such that $\llbracket \overrightarrow{x} \vdash_O u \triangleright u' \rrbracket$, $P(\overrightarrow{x}, u, u')$ holds.

6.2   Some Formalized Results

The machinery we have presented in the previous sections is all we need to prove metatheoretical properties of $\boldsymbol{\lambda}_\rightarrow$ such as weakening of typing judgments and subject reduction.

Theorem 15 (weakening of typing). *If $\Gamma \vdash_O u : \sigma$ and $\Gamma \subseteq \Delta$, then $\Delta \vdash_O u : \sigma$.*

Proof. Routine induction on the derivation of $\Gamma \vdash_O u : \sigma$, closely resembling the corresponding proof in a locally nameless setting. This is remarkable when considering that the underlying implementation of our typing judgments uses a pure nameless approach: normally, a proof of weakening in a nameless setting requires relatively complex arguments about lifting and permutations of indices and typing environments. Such unnecessary technicalities are completely hidden in our proof because the ostensibly named approach allows for a more adequate degree of abstraction.

Lemma 16 (substitutivity of typing). *If $\Delta, \langle x, \sigma \rangle, \Gamma \vdash_O u : \tau$ and $\Gamma \vdash_O v : \sigma$, then $\Delta, \Gamma \vdash_O u\ [v/x] : \tau$.*

Theorem 17 (preservation of typing). *If $\Gamma \vdash_O u : \sigma$ and $\mathrm{dom}(\Gamma) \vdash_O u \triangleright u'$, then $\Gamma \vdash_O u' : \sigma$.*

Proof. We proceed by induction on the derivation of $\Gamma \vdash u : \sigma$, followed, for each case, by an inversion on the reduction judgment. These are the interesting cases:

- ON-T-App and ON-B-Red: we have $u = \mathsf{App}\ (\mathsf{Lam}\ x\ \sigma_0\ u_0)\ v_0$, $u' = u_0\ [v_0/x]$ and $\sigma = \sigma_0 \to \tau_0$, and we also know that $\Gamma \vdash_O \mathsf{Lam}\ x\ \sigma_0\ u_0 : \sigma_0 \to \tau_0$ and $\Gamma \vdash_O v_0 : \sigma_0$; we must prove $\Gamma \vdash_O u_0\ [v_0/x]$. By inversion we obtain z, u_1 such that $z \notin \mathrm{FV}(u_1), \mathrm{dom}(\Gamma)$, $\mathsf{Lam}\ \sigma_0\ x\ u_0 = \mathsf{Lam}\ z\ \sigma_0\ (u_1\lceil z\rceil)$ and $\langle z : \sigma_0\rangle, \Gamma \vdash u_1\lceil z\rceil : \tau_0$ (this implies $u_1\lceil z\rceil = u_0\ \langle z/x\rangle$); thus, by Lemma 16 we get $\Gamma \vdash_O u_1\lceil z\rceil\ [v/z]$; it is then easy to prove $u_1\lceil z\rceil\ [v/z] = u_0\ \langle z/x\rangle\ [v/z] = u_0\ [v/x]$, as needed.

- ON-T-Lam and ON-B-Xi: we have $u = \mathsf{Lam}\ x\ \sigma_0\ (u_0\lceil x\rceil)$, $u' = \mathsf{Lam}\ y\ \sigma_0\ (u_0'\lceil y\rceil)$ and $\sigma = \sigma_0 \to \tau_0$, where $x \notin \mathrm{dom}(\Gamma), \mathrm{FV}(u_0)$ and $y \notin \mathrm{dom}(\Gamma), \mathrm{FV}(u_0), \mathrm{FV}(u_0')$, and we also know that $\mathrm{dom}(\langle y, \sigma_0\rangle, \Gamma) \vdash_O u_0\lceil y\rceil \triangleright u_0'\lceil y\rceil$. By induction hypothesis, for all $z \notin \mathrm{dom}(\Gamma), \mathrm{FV}(u_0)$, for all u'' such that $\mathrm{dom}(\langle z, \sigma\rangle, \Gamma) \vdash u_0\lceil z\rceil \triangleright u''$, we have $\langle z, \sigma_0\rangle, \Gamma \vdash_O u'' : \tau_0$; thus, by taking $z = y$ and $u'' = u_0'\lceil y\rceil$, we have $\langle y, \sigma_0\rangle, \Gamma \vdash_O u_0'\lceil y\rceil : \tau_0$. From this we immediately derive $\Gamma \vdash_O \mathsf{Lam}\ y\ \sigma_0\ (u_0'\lceil y\rceil) : \sigma_0 \to \tau_0$, as needed.

Theorem 18 (progress). *If $\vdash_O u : \sigma$ and u is not a value (i.e. it is not in the form $\mathsf{Lam}\ x\ \tau\ v$), then there exists u' such that $\vdash_O u \triangleright u'$.*

Proof. By induction on u. By hypothesis, u is not a Lam abstraction and, since it is well typed in the empty environment, it is not a Par either. Therefore, we have $u = \mathsf{App}\ u_1\ u_2$. By inversion of typing, we also know that $\vdash_O u_1 : \tau \to \sigma$ and $\vdash_O u_2 : \tau$, for some type τ. Then:

- if u_1 is in the form $\mathsf{Lam}\ y\ \tau\ u_1'$, we have a redex: we can take $u' = u_1'\ [u_2/y]$ and obtain the thesis by rule ON-B-Red;
- otherwise u_1 is not a value and by induction hypothesis there exists a u_1' such that $\vdash_O u_1 \triangleright u_1'$: then we can take $u' = \mathsf{App}\ u_1'\ u_2$ and obtain the thesis by rule ON-B-App1 (the side condition about the free variables of u_2 is easily closed knowing that u_2 is well typed in the empty environment).

7 Conclusions

In this paper we have presented an ostensibly named abstract data type for the formalization of languages with binding, which enables the user of an interactive theorem prover to only deal with familiar concepts like named binders and terms with holes. Our work can be likened to other axiomatic or abstract approaches ([11,19,21] just to list a few). While other authors have focused especially on the representation of terms and recursively defined functions, our technique extends to inductively defined judgments. In the representation of judgments, an important role is played by our ability to express contexts (terms with holes).

To show the soundness of our axiomatization, we provided and fully formalized a constructive model employing de Bruijn indices. The term language of the model is locally nameless, with non-locally closed terms used to represent contexts. Judgments, instead, are represented in a pure de Bruijn fashion. Since the model is formalized, we retain the possibility of extracting code from all the definitions and proofs based on the ostensibly named ADT.

Even though our internal representation of binding structures employs nameless dummies, other models are possible as long as they are canonical, the most obvious alternatives being the canonical locally named representation [17] and nested datatypes [7]. However, users do not need to worry about this, since they only deal with an abstract data type that does not expose such inner details.

In the long run, every representation of binding should be expected to scale up to dependently-typed object languages with generalized binders (i.e. binders declaring multiple variables simultaneously) and possibly other complex operations. The system $\lambda_\rightarrow$ that we formalized does not include dependent types; however multiple binders *are* part of our formalization, if only in the constrained form of typing judgments. Other recent efforts to accomodate generalized binding have been made in the context of Nominal Isabelle [12,26]. Among the biggest challenges in the formalization of binding, we include languages combining generalized binding with dependent types and hereditary substitution ([1]). In practice formalizations of such rich languages are attempted rarely and require non-trivial adaptations; however, it is with complex languages that abstract approaches like ours give the most ample benefits. For this reason, we are working on an extension of ostensibly named syntax to languages with signatures expressed in a generic way, in the style of [2].

In perspective, the ostensibly named approach seems to enjoy very desirable properties that would recommend its adoption as an alternative to more established techniques. These properties, however, come at a cost: without the help of automated tools, the burden of providing two formalization levels (concrete nameless and abstract ostensibly named) together with the associated proofs, will scare away most formalizers. Secondly, abstract recursion principles whose computational behaviour is expressed by an equational theory are not as convenient as the concrete ones available for inductive types. Lastly, defining recursive functions as instances of a recursion principle is quite unusual and can be tricky, although syntactic sugar can be used to make definitions more readable.

Such drawbacks could be greatly mitigated, if not completely eliminated, by means of specialized tool support. For this purpose, we plan to investigate in the future whether it is feasible to produce the overhead to an ostensibly named formalization programmatically from a declarative specification of a language with binding (similarly to what tools like DBgen [18] and LNgen [4] provide for pure de Bruijn and locally nameless formalizations). Taking advantage of recent works on specialized automation tactics for concrete encodings of binding (e.g. Autosubst [24]), we will also study the design of tactics and syntactic constructs to allow interactive theorem provers to present ostensibly named interfaces almost as if they were inductive types, automating computation of recursively defined functions and allowing definitions by pattern matching.

Acknowledgements. We would like to thank Benedikt Ahrens and Ralph Matthes for their valuable remarks. We also thank Randy Pollack for his useful comments about a preliminary version of this work.

References

1. Adams, R.: A Modular Hierarchy of Logical Frameworks. Ph.D. thesis, University of Manchester (2004)
2. Ahrens, B., Zsido, J.: Initial semantics for higher-order typed syntax in Coq. Journal of Formalized Reasoning 4(1) (2011), http://jfr.unibo.it/article/view/2066
3. Asperti, A., et al.: Formal metatheory of programming languages in the Matita interactive theorem prover. Journal of Automated Reasoning: Special Issue on the Poplmark Challenge 49(3), 427–451 (2012)
4. Aydemir, B., Weirich, S.: LNgen: Tool support for locally nameless representations. Tech. Rep. MS-CIS-10-24, University of Pennsylvania, Department of Computer and Information Science (2010)
5. Aydemir, B.E., et al.: Mechanized metatheory for the masses: The POPLMARK challenge. In: Hurd, J., Melham, T. (eds.) TPHOLs 2005. LNCS, vol. 3603, pp. 50–65. Springer, Heidelberg (2005)
6. Berghofer, S., Urban, C.: Nominal inversion principles. In: Mohamed, O.A., Muñoz, C., Tahar, S. (eds.) TPHOLs 2008. LNCS, vol. 5170, pp. 71–85. Springer, Heidelberg (2008), http://dx.doi.org/10.1007/978-3-540-71067-7_10
7. Bird, R.S., Paterson, R.: De Bruijn notation as a nested datatype. Journal of Functional Programming (1999)
8. Capretta, V., Felty, A.: Higher-order abstract syntax in type theory. In: Cooper, S.B., Geuvers, H., Pillay, A., Väänänen, J. (eds.) Logic Colloquium 2006. Lecture Notes in Logic, vol. 32, pp. 65–90. Cambridge University Press (2009)
9. Charguéraud, A.: The locally nameless representation. Journal of Automated Reasoning 49(3), 363–408 (2012), http://dx.doi.org/10.1007/s10817-011-9225-2
10. Ciaffaglione, A., Scagnetto, I.: A weak HOAS approach to the POPLmark challenge. In: Kesner, D., Viana, P. (eds.) Proceedings Seventh Workshop on Logical and Semantic Frameworks, with Applications, Rio de Janeiro, Brazil, September 29-30. Electronic Proceedings in Theoretical Computer Science, vol. 113, pp. 109–124. Open Publishing Association (2013)
11. Gordon, A.D., Melham, T.: Five axioms of alpha-conversion. In: von Wright, J., Harrison, J., Grundy, J. (eds.) TPHOLs 1996. LNCS, vol. 1125, pp. 173–190. Springer, Heidelberg (1996), http://dx.doi.org/10.1007/BFb0105404
12. Huffman, B., Urban, C.: A new foundation for nominal isabelle. In: Kaufmann, M., Paulson, L.C. (eds.) ITP 2010. LNCS, vol. 6172, pp. 35–50. Springer, Heidelberg (2010), http://dx.doi.org/10.1007/978-3-642-14052-5_5
13. McBride, C.: Dependently Typed Functional Programs and their Proofs. Ph.D. thesis, University of Edinburgh (1999)
14. McBride, C.: Elimination with a motive. In: Callaghan, P., et al. (eds.) TYPES 2000. LNCS, vol. 2277, pp. 197–216. Springer, Heidelberg (2002)
15. McKinna, J., Pollack, R.: Some lambda calculus and type theory formalized. Journal of Automated Reasoning 23(3), 373–409 (1999), http://dx.doi.org/10.1023/A
16. de Bruijn, N.G.: Lambda calculus notation with nameless dummies, a tool for automatic formula manipulation, with application to the Church-Rosser theorem. Indagationes Mathematicae 34, 381–392 (1972)
17. Pollack, R., Sato, M., Ricciotti, W.: A canonical locally named representation of binding. Journal of Automated Reasoning 49(2), 185–207 (2012), http://dx.doi.org/10.1007/s10817-011-9229-y

18. Polonowski, E.: Automatically generated infrastructure for de bruijn syntaxes. In: Blazy, S., Paulin-Mohring, C., Pichardie, D. (eds.) ITP 2013. LNCS, vol. 7998, pp. 402–417. Springer, Heidelberg (2013), http://dx.doi.org/10.1007/978-3-642-39634-2_29

19. Popescu, A., Gunter, E.L.: Recursion principles for syntax with bindings and substitution. In: Proceedings of the 16th ACM SIGPLAN International Conference on Functional Programming. pp. 346–358. ICFP 2011 ACM, New York (2011), http://doi.acm.org/10.1145/2034773.2034819

20. Pottier, F.: An overview of Caml. Electronic Notes in Theoretical Computer Science 148(2), 27–52 (2006), Proceedings of the ACM-SIGPLAN Workshop on ML (ML 2005) ACM-SIGPLAN Workshop on ML 2005 (2005)

21. Pouillard, N.: Nameless, painless. In: Proceedings of the 16th ACM SIGPLAN International Conference on Functional Programming, ICFP 2011, pp. 320–332. ACM, New York (2011), http://doi.acm.org/10.1145/2034773.2034817

22. Ricciotti, W.: Theoretical and Implementation Aspects in the Mechanization of the Metatheory of Programming Languages. Ph.D. thesis, Università di Bologna (2011)

23. Sato, M., Pollack, R.: External and internal syntax of the lambda-calculus. J. Symb. Comput. 45(5), 598–616 (2010), http://dx.doi.org/10.1016/j.jsc.2010.01.010

24. Schäfer, S., Tebbi, T.: Autosubst: Automation for de Bruijn substitutions. In: 6th Coq Workshop (July 2014)

25. Urban, C.: Nominal techniques in Isabelle/HOL. Journal of Automated Reasoning 40(4), 327–356 (2008), http://dx.doi.org/10.1007/s10817-008-9097-2

26. Urban, C., Kaliszyk, C.: General bindings and alpha-equivalence in nominal isabelle. In: Barthe, G. (ed.) ESOP 2011. LNCS, vol. 6602, pp. 480–500. Springer, Heidelberg (2011), http://dx.doi.org/10.1007/978-3-642-19718-5_25

Type-Based Allocation Analysis for Co-recursion in Lazy Functional Languages

Pedro Vasconcelos[1], Steffen Jost[2], Mário Florido[1], and Kevin Hammond[3]

[1] LIACC, Universidade do Porto, Porto, Portugal
{pbv,amf}@dcc.fc.up.pt
[2] Ludwig Maximillians Universität, Munich, Germany
jost@tcs.ifi.lmu.de
[3] University of St Andrews, St Andrews, UK
kevin@kevinhammond.net

Abstract. This paper presents a novel type-and-effect analysis for predicting upper-bounds on memory allocation costs for co-recursive definitions in a simple lazily-evaluated functional language. We show the soundness of this system against an instrumented variant of Launchbury's semantics for lazy evaluation which serves as a formal cost model. Our soundness proof requires an intermediate semantics employing indirections. Our proof of correspondence between these semantics that we provide is thus a crucial part of this work.

The analysis has been implemented as an automatic inference system. We demonstrate its effectiveness using several example programs that previously could not be automatically analysed.

1 Introduction

Co-recursion can be treated as a construction principle for infinite data structures: whereas recursion progressively deconstructs (finite) data structures, co-recursion progressively constructs (possibly infinite) data structures through synthesis from some base case [1]. In lazy functional programming, co-recursion allows concise and elegant definitions by separating data generation from control. For example, an infinite sequence of Fibonacci numbers, `fibs`, can be defined in Haskell by zipping a list with its own tail [2]:

```
fibs = 0 : 1 : zipWith (+) fibs (tail fibs)
```

There are two co-recursive base cases (0 and 1). The `zipWith` operation then builds the remainder of `fibs` constructively using both these base cases. Thanks to lazy evaluation, the above definition is efficient: each successive Fibonacci number is produced in constant cost. Furthermore, the flow of demand will ensure that each number is evaluated once only when it is needed. However, reasoning about execution costs requires a detailed understanding of the *operational* properties of lazy evaluation, particularly how intermediate results are shared. Moreover, apparently innocuous changes may have a significant impact on execution costs. As a simple example, consider two definitions of the function `cycle` that produces an infinite list by repeated concatenation:

© Springer-Verlag Berlin Heidelberg 2015
J. Vitek (Ed.): ESOP 2015, LNCS 9032, pp. 787–811, 2015.
DOI: 10.1007/978-3-662-46669-8_32

```
cycle  xs = xs' where xs' = xs++xs'
cycle' xs = xs ++ cycle' xs
```

Although the two definitions are denotationally equivalent, the evaluation of `cycle'` will allocate space that is proportional to the number of elements that are demanded from the result, whereas `cycle` will generate a circular list and thus only use constant space. Difficulties in reasoning about space usage are often mentioned as a hindrance to the practical use of lazily evaluated languages, such as Haskell, especially in domains where predictability is a primary concern.

This paper presents a new static analysis for obtaining *a-priori* bounds on the dynamic costs of *co-recursive definitions* for a foundational subset of Haskell. The analysis is formulated as a proof system for inferring annotated types that express upper bounds on the costs of program fragments. For concreteness, we chose to bound the number of heap allocations performed by a standard operational semantics for lazy evaluation. Note that the semantics and our analysis do not model *deallocation* — hence our cost model does not account for residency of an implementation using e.g. garbage collection. Nonetheless, measuring allocations has the foundational benefit of being directly derivable from a standard semantics. Furthermore, the number of allocations has been shown to be a good predictor in practice for the effects of optimizations in real implementations of lazy languages [3].

The work presented here complements our previous analysis for lazy functional programs [4]. We have previously shown that amortisation allows cost bounds to be determined for recursive definitions over finite data, but also that it does not contribute to the analysis of co-recursion over infinite data. For clarity of presentation, we thefore omit amortisation here. Any automated analysis that is aimed at practical use could obviously combine both methods for improved precision. We do not foresee any problems in doing this (in fact, our implementation includes amortisation), but the technical complexity of the presentation and proofs is likely to increase substantially.

2   Language and Cost Semantics

We consider the λ-calculus extended with local bindings, data constructors and pattern matching:

$$e ::= x \mid \lambda x.e \mid e\ x \mid \text{let } x = e_1 \text{ in } e_2 \mid c(\overline{x}) \mid \text{match } e_0 \text{ with } \{c_i(\overline{x_i}) \text{->} e_i\}_{i=1}^{n}$$

Our semantics is built on Sestoft's revision [5] of Launchbury's natural semantics for lazy evaluation [6], which is one of the earliest and most widely-used operational accounts of lazy evaluation for the λ-calculus. As in Launchbury's semantics, we restrict arguments of applications to simple variables; nested applications must translated into nested let-bindings.[1] *Let*-expressions bind variables

---

[1] This transformation does not increase worst-case costs because, in a call-by-need setting, function arguments must, in general, be heap-allocated in order to allow in-place update and sharing of normal forms.

$$\frac{}{\mathcal{H},\mathcal{S},\mathcal{L} \vdash_{m}^{m} w \Downarrow w, \mathcal{H}} \quad (\mathrm{WHNF}_\Downarrow)$$

$$\frac{\ell \notin \mathcal{L} \qquad \mathcal{H}[\ell \mapsto e], \mathcal{S}, \mathcal{L} \cup \{\ell\} \vdash_{m'}^{m} e \Downarrow w, \mathcal{H}'[\ell \mapsto e]}{\mathcal{H}[\ell \mapsto e], \mathcal{S}, \mathcal{L} \vdash_{m'}^{m} \ell \Downarrow w, \mathcal{H}'[\ell \mapsto w]} \quad (\mathrm{VAR}_\Downarrow)$$

$$\frac{\ell \text{ is fresh} \qquad \mathcal{H}[\ell \mapsto e_1[\ell/x]], \mathcal{S}, \mathcal{L} \vdash_{m'}^{m} e_2[\ell/x] \Downarrow w, \mathcal{H}'}{\mathcal{H}, \mathcal{S}, \mathcal{L} \vdash_{m'}^{m+1} \text{let } x = e_1 \text{ in } e_2 \Downarrow w, \mathcal{H}'} \quad (\mathrm{LET}_\Downarrow)$$

$$\frac{\mathcal{H}, \mathcal{S}, \mathcal{L} \vdash_{m'}^{m} e \Downarrow \lambda x.e', \mathcal{H}' \qquad \mathcal{H}', \mathcal{S}, \mathcal{L} \vdash_{m''}^{m'} e'[\ell/x] \Downarrow w, \mathcal{H}''}{\mathcal{H}, \mathcal{S}, \mathcal{L} \vdash_{m''}^{m} e\, \ell \Downarrow w, \mathcal{H}''} \quad (\mathrm{APP}_\Downarrow)$$

$$\frac{\mathcal{H}, \mathcal{S} \cup \left(\bigcup_{i=1}^{n}\{\overline{x}_i\} \cup \mathrm{BV}(e_i)\right), \mathcal{L} \vdash_{m'}^{m} e_0 \Downarrow c_k(\overline{\ell}), \mathcal{H}' \qquad \mathcal{H}', \mathcal{S}, \mathcal{L} \vdash_{m''}^{m'} e_k[\overline{\ell}/\overline{x}_k] \Downarrow w, \mathcal{H}''}{\mathcal{H}, \mathcal{S}, \mathcal{L} \vdash_{m''}^{m} \text{match } e_0 \text{ with } \{c_i(\overline{x}_i)\text{->}e_i\}_{i=1}^{n} \Downarrow w, \mathcal{H}''}$$
$$(\mathrm{MATCH}_\Downarrow)$$

Fig. 1. Instrumented version of Launchbury's operational semantics

to possibly (co)recursive terms. In line with common practice in non-strict functional languages, we do not have a separate *letrec* form, as in ML. For simplicity, we consider only single-variable let-bindings: multiple let-bindings can be encoded, if needed, using pairs and projections. Unlike [4], we do not require a distinguished let-construct for introducing constructors here.

Figure 1 defines an instrumented version of Launchbury's semantics, using a simple cost counting mechanism, against which we prove the soundness of our cost analysis. Our semantics is given as a relation $\mathcal{H}, \mathcal{S}, \mathcal{L} \vdash_{m'}^{m} e \Downarrow w, \mathcal{H}'$, where e is an expression; the *heap* $\mathcal{H}$ is a finite mapping from variables to possibly-unevaluated expressions (*thunks*):

$$\mathcal{H} ::= \emptyset \mid \mathcal{H}[x \mapsto e]$$

Some notation conventions: we will write $\mathrm{dom}(\mathcal{H})$ for the set of variables occurring in the left-hand side of all mappings in $\mathcal{H}$. We also assume that variables are assigned at most once, i.e. the notation $\mathcal{H}[x \mapsto e]$ requires $x \notin \mathrm{dom}(\mathcal{H})$, and we will use heaps as partial functions, i.e. use $\mathcal{H}(x)$ for the (possibly-undefined) expression associated with x in $\mathcal{H}$. The set $\mathcal{S}$ contains bound variables that are used to ensure the freshness condition in the $\mathrm{LET}_\Downarrow$ rule; and $\mathcal{L}$ is a set of variables used to record thunks that are under evaluation, thereby preventing cyclic evaluation (similar to the well-known "black-hole" technique used in [6]). The result of evaluation is an expression w in weak head normal form *(whnf)* and a final heap $\mathcal{H}'$. Note that we use lowercase letters $x, y, \ldots$ for bound variables in the original expression and $\ell, \ell', \ldots$ for "fresh" variables (designated *locations*) introduced by the evaluation of let-expressions. The parameters m, m' are non-negative integers representing the number of available heap locations before and after evaluation, respectively. The choice of instrumenting the semantics with before-and-after resources, as opposed to net costs, is conceptually simpler because it does rely on an a-priori assumption of cost additivity. Furthermore, it also makes it easier to adapt to cost models that allow e.g. for deallocation in the future.

The purpose of the analysis that will be developed in Section 3 is to obtain static bounds on m and m' that will allow the execution to proceed. For readability, we may omit the resource information from judgements when they are not otherwise mentioned, writing simply $\mathcal{H}, \mathcal{S}, \mathcal{L} \vdash e \Downarrow w, \mathcal{H}'$ instead of $\mathcal{H}, \mathcal{S}, \mathcal{L} \vdash_{m'}^{m} e \Downarrow w, \mathcal{H}'$.

Under a lazy evaluation model, expressions are evaluated *only* when they are *demanded* (that is when their value is needed in order to progress evaluation). In our operational semantics, this happens: i) when we need the value of a variable in VAR$_\Downarrow$ (which is looked up from the environment); ii) when we need the value of a function (a λ-expression) in APP$_\Downarrow$; or iii) when we need the value of the constructor argument in a match-expression. LET$_\Downarrow$ is the only rule that augments the heap with a new expression bound to a "fresh" location. Accordingly, it is the only rule that requires a positive heap cost in the annotation above the turnstile; all other rules simply "thread" costs from sub-expressions to the outermost one. For simplicity, but without loss of generality, we choose to use a uniform cost model where each freshly allocated location is counted as a single cost unit. More complex cost model, e.g. for determining the usage of other resources such as execution time, or stack usage (as in [7]), could be easily substituted, if required.

The WHNF$_\Downarrow$ rule deals with weak-head normal forms (λ-expressions and constructors) that require no further evaluation, and hence it incurs no cost.

The VAR$_\Downarrow$ and APP$_\Downarrow$ rules are identical to their equivalents in Launchbury's semantics, except for passing on m,m',etc. Note that the VAR$_\Downarrow$ rule is restricted to locations that are not marked as being under evaluation, $\ell \notin \mathcal{L}$ (so enforcing "black-holing" that explicitly excludes some non-terminating evaluations).

The MATCH$_\Downarrow$ rule deals with pattern matching against a constructor. The variables bound in the matching pattern are replaced in the corresponding branch expression e_k by the locations within the heap (also just variables, but we use the meta-variable ℓ to range over variables within the domain of the heap), which is then evaluated. Regardless of the actual branch taken, all possibly bound variables are added to $\mathcal{S}$; this is done solely to ensure the freshness condition in subsequent applications of the LET$_\Downarrow$ rule.

For the sake of completeness, we state the auxiliary definition that formalises the notion of variable freshness. This is due to de La Encina and Peña-Marí [8].

Definition 1 (Freshness). *A variable x is* fresh *in judgement* $\mathcal{H}, \mathcal{S}, \mathcal{L} \vdash e \Downarrow w, \mathcal{H}'$ *if x does not occur in either* $dom(\mathcal{H})$, $\mathcal{L}$ *or* $\mathcal{S}$ *nor does it occur bound in either e or the range of the heap $\mathcal{H}$.*

3 Type and Effect Analysis

The syntax of annotated types is as follows:

$$A \quad ::= \quad X \mid A_1 \xrightarrow{q} A_2 \mid \mathsf{T}^p(A) \mid \mu X.\{c_i : \overline{A}_i\}_{i=1}^n$$

```
let one  = 1 in
in letcons ones = Cons(one,ones)
in let map = \f xs -> match xs with   Nil () -> letcons r = Nil() in r
                              | Cons (x,xs') -> let y = f x
                                                in let ys = map f xs'
                                                in letcons r = Cons(y,ys) in r
in let f  = (\x -> let two=2 in two * x)
in map f ones
```

Fig. 2. Map over an infinite cyclic list

We use meta-variables A, B, C for types, X, Y for type variables and p, q for *cost annotations* (i.e. non-negative rational numbers). Function types $A_1 \xrightarrow{q} A_2$ are annotated with a cost q of evaluating the function; thunk types $\mathsf{T}^p(A)$ are annotated with a cost p of evaluating the thunk to *whnf*.

Both recursive and non-recursive algebraic data types are encoded as μ-types $\mu X.\{c_i : \overline{A}_i\}$ where the c_i are constructors, $\overline{A}_i$ is a sequence of argument types and and X is a recursively-bound type variable. For example, the type of lists with elements of type A can be encoded as $\mu X.\{\mathsf{Nil} : () \mid \mathsf{Cons} : (A, X)\}$. Note that we do not distinguish co-recursive data types syntactically, hence finite and infinite lists have the same type.

3.1 Worked Example

Before introducing the formal type rules we start by providing some intuition for the type annotations through a simple example. This is also available through the web version of our analysis at `http://kashmir.dcc.fc.up.pt/cgi/lazy.cgi`.

The example in Fig. 2 multiplies by two every integer in an infinite list, using the canonical `map` function. To this end we augment expressions with primitive integer constants and associated arithmetic operations and a primitive type for integers. Our prototype analysis infers the following annotated types:

$$f : \mathsf{T}^0(\mathsf{Int}) \xrightarrow{1} \mathsf{Int}$$

$$ones : \mu X.\{\mathsf{Nil} : () \mid \mathsf{Cons} : (\mathsf{T}^0(\mathsf{Int}), \mathsf{T}^0(X))\}$$

$$map \ f \ ones : \mu X.\{\mathsf{Nil} : () \mid \mathsf{Cons} : (\mathsf{T}^1(\mathsf{Int}), \mathsf{T}^3(X))\}$$

From these types we observe the following:

1. the type for function f has a cost of 1 on the arrow; this is because each evaluation of f allocates one integer (`let two=2 in...`);
2. the type of *ones* has 0 costs assigned to the head and tail thunks; this implies that forcing elements from this infinite list does not incur more allocations (because *ones* is a cyclic list);
3. the type of the result expression *map f ones* has costs of 1 and 3 units assigned to the head and tail thunks, respectively; this means that evaluating each tail of of the result list costs 3 allocations (for the *map*) while evaluating each head costs 1 allocation (for the argument function f);

4. from the type of *map f ones* we can also read a closed formula of $3n + 1$ for the cost of evaluating the n-th element of the result list (n times the evaluation of the tail thunks plus one evaluation of a head thunk).

Note, however, that if we transform the program by floating the *let*-binding for *two* outwards (i.e. `let two=2 in let f = \x -> two * x`) we infer annotated types that reflect this optimization.

$$f : \mathsf{T}^0(\mathsf{Int}) \xrightarrow{\ \ 0\ \ } \mathsf{Int}$$
$$map\ f\ ones : \mu X.\{\mathsf{Nil} : ()\ |\ \mathsf{Cons} : (\mathsf{T}^0(\mathsf{Int}), \mathsf{T}^3(X))\}$$

After let-floating, f does not incur any allocation (a single allocation is done for all evaluations). Hence, the type for the result list has now only positive costs for the each successive tail. Note that the cost of *3 per successive list node* is accurate because under lazy evaluation, applying *map* to a cyclic list produces an infinite acyclic list.

3.2   Formal Description of Type Rules

Our analysis is presented in Figures 3 and 4 as a proof system that derives annotated typing judgments for expressions. The rules use two auxiliary relations on types (subtyping and lowering thunk costs) defined in Figures 5 and 6.

An annotated type judgment has the form $\Gamma \vdash^{p}_{p'} e : A$ where Γ is a context assigning types to variables,[2] e is an expression, A is an annotated type and p, p' are non-negative numbers approximating the available resources before and after evaluation of e, respectively; these annotations are used for "threading" resources through sub-expressions in rules LET and MATCH. As with the operational semantics, we omit the annotations on the turnstile whenever they are not further referenced.

Because variables reference heap expressions, rules dealing with the introduction and elimination of variables also deal with the introduction and elimination of thunk types: VAR eliminates an assumption of a thunk type, i.e. of the form $x : \mathsf{T}^q(A)$. Dually, LET introduce an assumption of a thunk type. Note how the cost of evaluating a thunk is deferred from LET to VAR. Similarly, the cost of evaluating the body of a λ-abstraction is deferred to application. Rules ABS and APP are otherwise standard.

The type rules CONS and MATCH for constructors and pattern matching are straightforward.[3] The CONS rule just ensures consistency between the arguments to a constructor and its result type. In correspondence with our operational semantics, there is no extra cost for constructors, since allocation is accounted for in rule LET. The MATCH rule deals with pattern-matching over an expression of a (possibly recursive) data type. The rule requires that all branches admit an

---

[2]   We use the standard notation $x : A$ to denote the singleton context mapping variable x to type A, and a comma between two contexts denotes disjoint union.

[3]   Note that these rules are simpler than in our earlier work [4], since data constructors do not carry *potential* as required for the amortisation technique.

$$\overline{\Gamma,\ x{:}\mathsf{T}^q(A) \vdash_0^q x : A}\qquad\text{(VAR)}$$

$$\frac{\Gamma,x{:}\mathsf{T}^0(A') \vdash_0^q e_1 : A \qquad A' \lhd A \qquad \Gamma,x{:}\mathsf{T}^q(A) \vdash_{p'}^p e_2 : C}{\Gamma \vdash_{p'}^{1+p} \text{let } x = e_1 \text{ in } e_2 : C}\qquad\text{(LET)}$$

$$\frac{\Gamma,x{:}A \vdash_0^q e : C}{\Gamma,x{:}A \vdash_0^0 \lambda x.e : A \xrightarrow{q} C}\qquad\text{(ABS)}$$

$$\frac{\Gamma \vdash_{p'}^p e : A \xrightarrow{q} C}{\Gamma,y{:}A \vdash_{p'}^{p+q} e\,y : C}\qquad\text{(APP)}$$

$$\frac{B = \mu X.\{\cdots | c : \overline{A} | \cdots\}}{\Gamma,\ \overline{y}{:}\overline{A}[B/X] \vdash_0^0 c(\overline{y}) : B}\qquad\text{(CONS)}$$

$$\frac{|\overline{A}_i| = |\overline{x}_i| \qquad B = \mu X.\{c_i : \overline{A}_i\} \qquad \Gamma \vdash_{p'}^p e_0 : B \qquad \Gamma,\overline{x}_i{:}\overline{A}_i[B/X] \vdash_{p''}^{p'} e_i : C}{\Gamma \vdash_{p''}^p \text{match } e_0 \text{ with } \{c_i(\overline{x}_i)\text{->}e_i\} : C}$$
$$\text{(MATCH)}$$

Fig. 3. Syntax directed type rules

$$\frac{q \geq p \qquad p - p' \geq q - q' \qquad \Gamma \vdash_{p'}^p e : A}{\Gamma \vdash_{q'}^q e : A}\qquad\text{(RELAX)}$$

$$\frac{\Gamma \vdash_{p'}^p e : A \qquad A <: B}{\Gamma \vdash_{p'}^p e : B}\qquad\text{(SUBTYPE)}$$

$$\frac{\Gamma,x{:}B \vdash_{p'}^p e : C \qquad A <: B}{\Gamma,\ x{:}A \vdash_{p'}^p e : C}\qquad\text{(SUPERTYPE)}$$

$$\frac{\Gamma,\ x{:}\mathsf{T}^{q_0}(A) \vdash_{p'}^p e : C}{\Gamma,\ x{:}\mathsf{T}^{q_0+q_1}(A) \vdash_{p'}^{p+q_1} e : C}\qquad\text{(PREPAY)}$$

Fig. 4. Structural type rules

$$\overline{X <: X}$$

$$\frac{|\overline{A}_i| = |\overline{B}_i| \qquad A_{ij} <: B_{ij}}{\mu X.\{c_i : \overline{A}_i\}_{i=1}^n <: \mu X.\{c_i : \overline{B}_i\}_{i=1}^n}$$

$$\frac{A' <: A \qquad B <: B' \qquad p \le q}{A \xrightarrow{p} B <: A' \xrightarrow{q} B'}$$

$$\frac{A <: B \qquad p \le q}{\mathsf{T}^p(A) <: \mathsf{T}^q(B)}$$

$$\overline{A \lhd A}$$

$$\frac{|\overline{A}_i| = |\overline{B}_i| \qquad A_{ij} \lhd_x B_{ij}}{\mu X.\{c_i : \overline{A}_i\}_{i=1}^n \lhd \mu X.\{c_i : \overline{B}_i\}_{i=1}^n}$$

$$\overline{A \lhd_x A}$$

$$\frac{p \le q}{\mathsf{T}^p(X) \lhd_x \mathsf{T}^q(X)}$$

Fig. 5. Subtyping relation   **Fig. 6.** Lowering thunk costs

identical result type and that estimated resources after execution of any of the branches are equal; fulfilling such a condition may require relaxing type and/or cost information using the structural rules below.

A significant difference from our previous work [4] lies in the typing of let-expressions. Typing let $x = e_1$ in e_2 allows lower costs for the bound variable x within the recursively-defined expression e_1. Specifically, it allows zero costs for both the thunk itself and also for its recursive references; this is justified because any recursive access to the defined value cannot incur evaluation costs, since either the thunk is already in *whnf*, or the access would cause a self-referential loop (which is prevented by the "black-holing" in the operational semantics).

The type rule LET uses an auxiliary ordering relation $\lhd$ on annotated types defined in Figure 6. Note that relation $\lhd$ is different from subtyping ($<:$); the former is used in LET for lowering *only* the cost for the recursively-defined thunks in a μ-type (thereby increasing precision) while the latter is used in a structural rule to lower *any* thunk or arrow costs to allow common types for branches with distinct costs. Note also that it would be enough for $\lhd$ to lower μ-thunks costs to zero (since that will always lead to a more precise analysis), but because the type system is only constraining admissible type annotations, we choose to express this as an ordering relation and let the constraint solver choose suitable annotations in order to minimize costs.

The structural rules of Figure 4 allow the analysis to be relaxed in various ways: RELAX allows the relaxing of cost bounds. SUBTYPE and SUPERTYPE allow subtyping in the conclusion and supertyping in a hypothesis, respectively; these make use of an separate relation defined in Figure 5. Informally, $A <: B$ means that the A and B have identical type structure but A has lower cost annotations in both thunk and function types.

The crucial rule PREPAY allows (part of) the cost of a thunk to be paid in advance, thus reducing the cost of further uses of the same variable. Rule VAR requires the cost of the thunk to be paid for *every* use, as in call-by-name evaluation. However, PREPAY allows the cost of a thunk to be shared, which models the effect of memoization in call-by-need evaluation.

Note that weakening and contraction are implicitly allowed without any restrictions, so type assumptions may be freely duplicated without requiring the application of an explicit type rule.[4]

4   Experimental Results

We have constructed a prototype implementation of our analysis as an inference algorithm for the type system of Section 3. A publicly accessible web version with several editable examples (including the ones presented here) is available at `http://kashmir.dcc.fc.up.pt/cgi/lazy.cgi`. The implementation combines the analysis presented in this paper with our earlier amortised analysis [4]. These techniques complement each other: amortisation deals with recursive definitions over finite data, while our new system deals with co-recursive definitions on infinite data. In this paper, of course, we focus only on examples of co-recursive definitions here.

The analysis is fully automatic, i.e. it does not require type annotations from the programmer and either produces an annotated typing or fails when no cost bounds can be found. Inference for a whole program is currently performed in three steps:

1. We first perform Damas-Milner type inference to obtain an unannotated version of the type derivation. The unannotated types form a free algebra and can be determined using standard first-order unification.
2. We then decorate types with fresh variables and perform a traversal of the type derivation gathering linear constraints among annotations following the type rules.
3. Finally, we feed the linear constraints to a standard linear programming solver[5] with the objective of minimizing the overall expression cost on the turnstile. Any solution gives rise to a valid annotated typing derivation.

As in Standard ML or Haskell, we associate constructors with specific data types (e.g. Cons and Nil with lists). This ensures that the use of the CONS rule is syntax-directed. Also, the implementation includes some minor language extensions, namely, primitive integers and associated arithmetic operations.

It remains to explain how to decide the use of the structural rules from Figure 4. PREPAY is used immediately whenever bound variables are introduced, namely, in the body of a lambda, let-expression or match alternative. This can be done uniformly because the rule allows any part of the cost to be paid (including zero); hence, we defer to the LP solver the choice of how much should each individual thunk be prepaid in order to achieve an overall optimal solution. Finally, we allow the use of RELAX at every node of the derivation and SUBTYPE at the application rule (to enforce compatibility between the function and its argument) and at the MATCH rule (to obtain a compatible result type).

---

[4] This is again quite different to [4], where restrictions on weakening and contraction are needed because of the amortisation technique.

[5] We use the GLPK library: `http://www.gnu.org/software/glpk`.

4.1 Zipping Streams

Our first example is a co-recursive *zipWith* function that combines two infinite lists by applying a function to corresponding elements:

```
let zipWith = \f xs ys -> match xs with
  Cons(x,xs') -> match ys with
    Cons(y,ys') -> let t = f x y
                    in let r = zipWith f xs' ys'
                    in let s = Cons(t,r) in s
```

The analysis infers the following annotated type:

```
zipWith : T(T(a) -> T(b) -> c) ->
   T(Rec{Cons:(T(a),T(#)) | Nil:()}) ->
    T(Rec{Cons:(T(b),T(#)) | Nil:()}) ->@3
      Rec{Cons:(T(c),T@3(#)) | Nil:()}
```

Some remarks on the analysis output: μ-types are written Rec{...} with an implicit bound type variable represented by an '#'-sign; annotations in thunk and arrow types are marked by an '@'-sign; for readability, zero annotations are omitted. Hence, the type above ensures that *zipWith* yields a list where each successive tail costs (at most) 3 allocations (T@3(#)) plus 3 for the application itself (->@3); thus the cost for obtaining n elements is bounded by $3 + 3n$.

Note that the inference algorithm outputs only one of an infinite set of admissible solutions. Because *zipWith* was analysed in isolation, we obtained a type with zero costs for the function argument and, therefore, where all costs of the result are assigned to the list spine. If *zipWith* was used in a context where the argument function requires positive costs, we might instead obtain a type with costs in both head and spine thunks, e.g.:

```
zipWith : T(T(a) -> T(b) ->@1 c) ->
   T(Rec{Cons:(T(a),T(#)) | Nil:()}) ->
    T(Rec{Cons:(T(b),T(#)) | Nil:()}) ->@3
      Rec{Cons:(T@1(c),T@3(#)) | Nil:()}
```

4.2 Fibonacci Numbers

Our next example is the infinite list of Fibonacci numbers from the introduction; this can be defined using the *zipWith* function shown before:

```
let zero = 0 in
let one  = 1 in
let plus = \x y -> x+y in
let fibs = (let t = match fibs with
                      Cons(x,fibs') -> zipWith plus fibs fibs'
              in let r = Cons(one,t)
              in let s = Cons(zero,r)
              in s)
```

Here we extend the language with a type for integers by adding suitable constructors for each constant and primitive arithmetic operators. As in the STG machine [9], operators must be fully applied; higher-order values can be obtained using explicit lambda-expressions (`plus` in the example). We also assume that arithmetic operations have no intrinsic allocation costs, but since arguments of applications are restricted to be variables, compound results have to be let-bound (and thus heap allocated) anyway. The type inferred for *fibs* is as follows:

```
fibs : Rec{Cons:(T(Int),T@3(#)) | Nil:()}
```

From the type above we see that the infinite list evaluating each successive of Fibonacci requires at most 3 allocations. This matches exactly the cost of *zipWith* because *plus* has zero cost in our model. Note that it is essential that `fibs'` is the tail of `fibs`, for otherwise one would have to pay twice for evaluating each argument of `zipWith`. Thanks to our novel LET typing rule, our analysis can recognise this reduction in cost due to aliasing.

4.3   The Hamming Problem

Our final example is the *Hamming problem*: produce an infinite list of numbers in ascending order and without duplicates, starting with 1 and such that, whenever x occurs in the list, so do $2 \times x$, $3 \times x$ and $5 \times x$. One elegant Haskell solution (from Bird and Wadler's textbook [2]) uses a function that merges infinite lists in ascending order:

```
merge (x:xs) (y:ys) | x==y = x : merge    xs      ys
                    | x<y  = x : merge    xs   (y:ys)
                    | x>y  = y : merge (x:xs)    ys
```

The Hamming numbers can then be defined co-recursively using *merge* and the standard list *map*:

```
hamming = 1 : merge (map (2*) hamming)
                (merge (map (3*) hamming) (map (5*) hamming))
```

Using some informal reasoning about the sharing properties of the cyclic list above, Bird and Wadler argue that n elements can be computed with bounded $O(n)$ cost [2]. We will see that our analysis can confirm this with a precise bound.

However, a direct translation of the above definitions into our core language does *not* admit an annotated type in our system: the two uses of *merge* in the definition of *hamming* require different cost annotations. Because of this, the constraints generated by the reconstruction algorithm will not admit a solution.[6]

One work around for this limitation is to simply duplicate the definition of *merge* so that each use can be assigned a precise type[7]:

---

[6] This does not happen for *map* because, in this particular problem, all the uses have identical costs.

[7] A more general solution would be to extend the analysis to include *effect polymorphism* – we leave this as further work (see Section 7).

Table 1. Comparison of analysis with the operational semantics

	evaluation demand	0	1	2	3	4	5	6	7	8	9	10
Fibs	analysis	8	8	11	14	17	20	23	26	29	32	35
	semantics	8	8	8	11	14	17	20	23	26	29	32
Hamming	analysis	17	17	30	43	56	69	82	95	108	121	134
	semantics	17	17	30	35	42	47	54	64	69	76	86

```
hamming = 1 : merge1 (map (2*) hamming)
               (merge2 (map (3*) hamming) (map (5*) hamming))
```

With this translation, the reconstruction algorithm is able to obtain the following annotated types:

```
merge1 : T@3(Rec{Cons:(T(Int),T@3(#)) | Nil:()}) ->
            T@3(Rec{Cons:(T(Int),T@3(#)) | Nil:()}) ->@8
            Rec{Cons:(T(Int),T@8(#)) | Nil:()}
merge2 : T@3(Rec{Cons:(T(Int),T@3(#)) | Nil:()}) ->
            T@8(Rec{Cons:(T(Int),T@8(#)) | Nil:()}) ->@13
            Rec{Cons:(T(Int),T@13(#)) | Nil:()}
hamming : Rec{Cons:(T(Int),T@13(#)) | Nil:()}
```

The type inferred for *hamming* confirms Bird and Wadler's reasoning and provides a precise bound: each successive Fibonacci number requires (at most) 13 allocations.

Finally, we note that the revised LET type rule that is presented in this paper is essential for obtaining annotated types for the *fibs* and *hamming* examples above. In fact, these two examples do not admit annotated types using just the amortised analysis described in [4].

4.4   Comparison with the Instrumented Semantics

Table 1 presents a short assessment of the quality of the upper-bounds obtained from our analysis by comparison with the exact costs obtained from an implementation of the operational semantics of Section 2. Figures are grouped by the evaluation demand from the result infinite lists, where 0 evaluates the list to *whnf* (i.e. just a Cons cell), 1 evaluates the first element to *whnf*, 2 evaluates the second, etc. We first note that the analysis is indeed producing upper-bounds; this is true in general as shown by the soundness theorem proved in Section 5.

The results for the *fibs* are quite accurate: the inferred cost of 3 allocation for each successive elements is exact. There is a fixed overestimation of 3 allocations because a recursive type Rec{Cons:(T(Int),T@3(#))|...} cannot distinguish the lower cost of the first two elements. The results for *hamming* are less accurate; this is because the exact cost for successive elements is not constant, instead varying between 0 and 10 allocations; however, our annotated types assign identical cost for the entire spine (`Rec{Cons:(T(Int),T@13(#)|...}` — i.e. 13 allocations), hence the overestimation.

5 Soundness

In this section we formulate the soundness of our analysis from Section 3. The structure of our proof is as follows:

1. in Section 5.1, we define a variant of the operational semantics which uses *indirections*;
2. in Section 5.5, we establish the soundness of the type rules against the indirection semantics.
3. finally, in Section 5.6, we show the equivalence of the original semantics and the revised indirection semantics (including preservation of resource bounds).

5.1 Indirection Semantics

To facilitate proving the soundness of the type analysis of Section 3 we will consider a variant of the operational semantics. We exploit a new syntactic form for *indirections*. These do not occur in the original program, but are used internally by the evaluation mechanism.

$$e ::= \cdots \mid \mathsf{ind}(x) \qquad\qquad w ::= \lambda x.e \mid c(\overline{x}) \mid \mathsf{ind}(x)$$

Operationally, an indirection $\mathsf{ind}(x)$ will be treated similarly to the variable x (i.e. it references some expression in the heap). However, evaluation of an indirection will *not* force the evaluation of a thunk; instead, it succeeds immediately. This will be crucial for establishing the soundness of the type rule LET.

Figure 7 presents the revised semantics as a relation $\mathcal{H}, \mathcal{S}, \mathcal{L} \vdash_{\overline{m}}^{m} e \Downarrow^{\mathrm{I}} w, \mathcal{H}'$ where the components play identical roles to the semantics of Section 2. Note that, in this revised judgment, both expressions and *whnfs* may be indirections; they may also occur in heaps, $\mathcal{H}$ or $\mathcal{H}'$.

The WHNF$_{\Downarrow\mathrm{I}}$ rule is revised to allow indirections as results. Indirections are used to mark recursive self-references, and thus this revised rule allows justifying lower costs for such cases in the soundness proof. The VAR$_{\Downarrow\mathrm{I}}$ rule is identical to the previous one. The revised rule for let-expressions LET$_{\Downarrow\mathrm{I}}$ substitutes the bound variable in e_1 by an indirection instead of a self-reference; this will allow the costs of (co-)recursive uses to be distinguished in the soundness proof.

The revised rules WHNF$_{\Downarrow\mathrm{I}}$, APP$_{\Downarrow\mathrm{I}}$ and MATCH$_{\Downarrow\mathrm{I}}$ make use of a auxiliary partial function $\mathcal{H}@w$ for de-referencing a result w with respect to a heap $\mathcal{H}$. For constructors and abstractions this function is the identity; and in the case of indirections it dereferences a heap location. It is undefined for other expressions.

$$\mathcal{H}@\lambda x.e \overset{\mathrm{def}}{=} \lambda x.e \qquad \mathcal{H}@c(\overline{x}) \overset{\mathrm{def}}{=} c(\overline{x}) \qquad \mathcal{H}@\mathsf{ind}(\ell) \overset{\mathrm{def}}{=} \mathcal{H}(\ell) \text{ if } \ell \in \mathrm{dom}(\mathcal{H})$$

$$\frac{\mathcal{H}@w \text{ is defined}}{\mathcal{H}, \mathcal{S}, \mathcal{L} \vdash^m_m w \Downarrow^{\mathrm{I}} w, \mathcal{H}} \qquad (\textsc{Whnf}_{\Downarrow \mathrm{I}})$$

$$\frac{\ell \notin \mathcal{L} \qquad \mathcal{H}[\ell \mapsto e], \mathcal{S}, \mathcal{L} \cup \{\ell\} \vdash^m_{m'} e \Downarrow^{\mathrm{I}} w, \mathcal{H}'[\ell \mapsto e]}{\mathcal{H}[\ell \mapsto e], \mathcal{S}, \mathcal{L} \vdash^m_{m'} \ell \Downarrow^{\mathrm{I}} w, \mathcal{H}'[\ell \mapsto w]} \qquad (\textsc{Var}_{\Downarrow \mathrm{I}})$$

$$\frac{\ell, \ell' \text{ are fresh} \qquad \mathcal{H}[\ell \mapsto e_1[\ell'/x], \ell' \mapsto \mathrm{ind}(\ell)], \mathcal{S}, \mathcal{L} \vdash^m_{m'} e_2[\ell/x] \Downarrow^{\mathrm{I}} w, \mathcal{H}'}{\mathcal{H}, \mathcal{S}, \mathcal{L} \vdash^{m+1}_{m'} \text{let } x = e_1 \text{ in } e_2 \Downarrow^{\mathrm{I}} w, \mathcal{H}'} \qquad (\textsc{Let}_{\Downarrow \mathrm{I}})$$

$$\frac{\mathcal{H}, \mathcal{S}, \mathcal{L} \vdash^m_{m'} e \Downarrow^{\mathrm{I}} u, \mathcal{H}' \qquad \mathcal{H}'@u = \lambda x.\, e' \qquad \mathcal{H}', \mathcal{S}, \mathcal{L} \vdash^{m'}_{m''} e'[\ell/x] \Downarrow^{\mathrm{I}} w, \mathcal{H}''}{\mathcal{H}, \mathcal{S}, \mathcal{L} \vdash^m_{m''} e\, \ell \Downarrow^{\mathrm{I}} w, \mathcal{H}''} \qquad (\textsc{App}_{\Downarrow \mathrm{I}})$$

$$\frac{\begin{array}{c} \mathcal{H}, \mathcal{S} \cup \left(\bigcup_{i=1}^n \{\overline{x}_i\} \cup \mathrm{BV}(e_i) \right), \mathcal{L} \vdash^m_{m'} e_0 \Downarrow^{\mathrm{I}} u, \mathcal{H}' \\ \mathcal{H}'@u = c_k(\overline{\ell}) \qquad \mathcal{H}', \mathcal{S}, \mathcal{L} \vdash^{m'}_{m''} e_k[\overline{\ell}/\overline{x}_k] \Downarrow^{\mathrm{I}} w, \mathcal{H}'' \end{array}}{\mathcal{H}, \mathcal{S}, \mathcal{L} \vdash^m_{m''} \text{match } e_0 \text{ with } \{c_i(\overline{x}_i)\text{->}e_i\}_{i=1}^n \Downarrow^{\mathrm{I}} w, \mathcal{H}''} \qquad (\textsc{Match}_{\Downarrow \mathrm{I}})$$

Fig. 7. Indirection semantics

5.2   Typing Rule for Indirections

We introduce the following typing rule IND for indirections:

$$\frac{A' \lhd A}{\Gamma, x{:}\mathsf{T}^q(A) \vdash^0_0 \mathrm{ind}(x) : A'} \qquad (\textsc{Ind})$$

This rule is similar to VAR except that it allows lowering the thunk costs both on the judgment and on the recursive type; we use the relation $\lhd$ of Figure 6 for the latter. The rule will be needed in the soundness proof solely for establishing well-typing of intermediate heap configurations (since indirections may not occur within source programs).

5.3   Global Types and Balance

The *global types* are given by a mapping $\mathcal{M}$ from locations to (annotated) types. The intuition is that when $\mathcal{M}(\ell) = \mathsf{T}^q(A)$ then q is (an upper bound of) the cost of evaluating ℓ and the resulting *whnf* admits type A. Furthermore, we introduce an auxiliary *balance* function $\mathcal{B}$ mapping locations to non-negative numbers. This keeps track of the partial costs that have been paid in advance by uses of the PREPAY rule. We also define the *balance sum* over a heap configuration as the sum of the balance associated with all thunks that are not under evaluation:

$$\sum_{\mathcal{H}, \mathcal{L}} \mathcal{B} \stackrel{\text{def}}{=} \sum \{\mathcal{B}(\ell) : \ell \in \mathrm{dom}(\mathcal{H}) \text{ and } \ell \notin \mathcal{L} \text{ and } \mathcal{H}(\ell) \text{ is not a } \textit{whnf}\}$$

Note that the balance is needed to prove the soundness of the analysis, but is *not* part of the operational semantics — in particular, it does not incur runtime costs.

5.4 Consistency and Compatibility

We can now define the principal soundness invariants of our analysis, namely, a *consistency* relation for typing heap configurations and a *compatibility* relation between global types and contexts. We proceed by first defining typing of a single location and then extend it to typing a heap configuration.

Definition 2 (Typing of locations). *We say that location ℓ admits type $\mathsf{T}^q(A)$ under context Γ, balance $\mathcal{B}$, heap configuration $(\mathcal{H}, \mathcal{L})$, and write $\Gamma, \mathcal{B}; \mathcal{H}, \mathcal{L} \vdash_{\mathrm{Loc}} \ell : \mathsf{T}^q(A)$ if one of the following cases holds:*

(Loc1) $\mathcal{H}(\ell)$ *is in* whnf *and* $\Gamma \vdash^0_0 \mathcal{H}(\ell) : A$;

(Loc2) $\mathcal{H}(\ell)$ *is not in* whnf *and* $\Gamma \vdash^{\frac{q + \mathcal{B}(\ell)}{0}} \mathcal{H}(\ell) : A$;

The two cases above are mutually exclusive: Loc1 applies when the expression in the heap is already in *whnf*; otherwise Loc2 applies. For Loc2, the balance $\mathcal{B}(\ell)$ associated with location ℓ is added to the available resources for typing the thunk $\mathcal{H}(\ell)$, effectively reducing its cost by the prepaid amount. Note that in [4], we additionally distinguished whether a location was under evaluation. Since we do not use the notion of *potential* here, this is no longer needed.

Definition 3 (Typing of heap configurations). *We say that a heap configuration $(\mathcal{H}, \mathcal{L})$ is consistent with context Γ, global types $\mathcal{M}$ and balance $\mathcal{B}$, and write $\Gamma, \mathcal{B} \vdash_{\mathrm{Mem}} (\mathcal{H}, \mathcal{L}) : \mathcal{M}$, if and only if for all $\ell \in dom(\mathcal{H})$ we have $\Gamma, \mathcal{B}; \mathcal{H}, \mathcal{L} \vdash_{\mathrm{Loc}} \ell : \mathcal{M}(\ell)$.*

The compatibility relation enforces that the global types of locations are supertypes (i.e. have lower costs) of the types occurring in a context.

Definition 4 (Compatibility). *We say that a global types $\mathcal{M}$ are compatible with a context Γ, written $\mathcal{M} <: \Gamma$, if and only if $\mathcal{M}(\ell) <: A$ for all $\ell{:}A \in \Gamma$.*

5.5 Soundness of the Proof System

We state the soundness of our analysis as an augmented type preservation result.

Theorem 1 (Soundness). *Let $t \geq 0$ be fixed but arbitrary. If the following statements hold*

$$\Gamma \vdash^p_{p'} e : A \tag{1}$$

$$\Gamma, \mathcal{B} \vdash_{\mathrm{Mem}} (\mathcal{H}, \mathcal{L}) : \mathcal{M} \tag{2}$$

$$\mathcal{M} <: \Gamma \tag{3}$$

$$\mathcal{H}, \mathcal{S}, \mathcal{L} \vdash e \Downarrow^{\mathrm{I}} w, \mathcal{H}' \tag{4}$$

then for all m *such that* $m \geq t + p + \sum_{\mathcal{H},\mathcal{L}} \mathcal{B}$, *there exists* m', Γ', $\mathcal{B}'$ *and* $\mathcal{M}'$ *such that*

$$\Gamma' \vdash^0_0 w : A \tag{5}$$

$$\Gamma', \mathcal{B}' \vdash_{\text{MEM}} (\mathcal{H}', \mathcal{L}) : \mathcal{M}' \tag{6}$$

$$\mathcal{M}' <: \Gamma' \tag{7}$$

$$\mathcal{H}, \mathcal{S}, \mathcal{L} \vdash^m_{m'} e \Downarrow^{\mathrm{I}} w, \mathcal{H}' \tag{8}$$

$$m' \geq t + p' + \sum_{\mathcal{H}',\mathcal{L}} \mathcal{B}' \tag{9}$$

Informally, the soundness theorem reads as follows: if an expression e admits type A (1), the heap can be typed (2) (3), and the evaluation is successful (4), then the result *whnf* also admits type A (5). Furthermore, the final heap can also be typed (6) (7) and the static bounds that are obtained from the typing of e give safe resource estimates for evaluation (8) (9). Because of space limitations, we will only present the cases that differ significantly from our previous work [4], particularly the revised typing rule for *let* and for indirections.

Proof. The proof is by induction on the lengths of the derivations of evaluation (4) and typing (1) ordered lexicographically, with the former taking priority over the later.

We proceed by case analysis of the typing rule used in premise (1), considering just some representative cases.

Case LET. The typing premise (1) instantiates as

$$\Gamma \vdash^{1+p}_{p} \text{let } x = e_1 \text{ in } e_2 : C$$

By inversion of rule LET together with the substituition lemma, we get

$$\Gamma, \ell':\mathrm{T}^0(A') \vdash^q_0 e_1[\ell'/x] : A \tag{10}$$

$$\Gamma, \ell:\mathrm{T}^q(A) \vdash^p_{p'} e_2[\ell/x] : C \tag{11}$$

where $A' \lhd A$. The evaluation premise (4) instantiates as

$$\mathcal{H}, \mathcal{S}, \mathcal{L} \vdash^{1+m}_{m'} \text{let } x = e_1 \text{ in } e_2 \Downarrow^{\mathrm{I}} w, \mathcal{H}'$$

from which we get $\mathcal{H}_0, \mathcal{S}, \mathcal{L} \vdash^m_{m'} e_2[\ell/x] \Downarrow^{\mathrm{I}} w, \mathcal{H}'$ where $\mathcal{H}_0 = \mathcal{H}[\ell \mapsto e_1[\ell'/x], \ell' \mapsto \text{ind}(\ell)]$. Define:

$$\mathcal{B}_0 = \mathcal{B}[\ell \mapsto 0, \ell' \mapsto 0]$$
$$\mathcal{M}_0 = \mathcal{M}[\ell \mapsto \mathrm{T}^q(A), \ell' \mapsto \mathrm{T}^0(A')]$$
$$\Gamma_0 = \Gamma, \ell:\mathrm{T}^q(A), \ell':\mathrm{T}^0(A')$$

To apply induction to the evaluation of $e_2[\ell/x]$ we first need to re-establish type consistency and compatibility. Type consistency for ℓ follows from (10) and (LOC2); and for ℓ' follows directly from the type rule IND and (LOC1).

Compatibility is immediate because the types for ℓ and ℓ' in Γ_0 are exactly $\mathcal{M}_0(\ell)$ and $\mathcal{M}_0(\ell')$. Applying induction to (11) then yields all required conclusions. Note that the lower thunk costs for A' are only allowed for the recursive reference ℓ' introduced in the let-expression; crucially this is sound only because the recursive reference is introduced in the heap as in indirection whose cost is ignored by the typing rule IND. Otherwise compatibility would not hold.

Case IND. This case is immediate: taking $\Gamma' = \Gamma$, $\mathcal{B}' = \mathcal{B}$, $\mathcal{M}' = \mathcal{M}$ and $m' = m$ yields all required conclusions.

Case VAR. The typing premise is $\Gamma, \ell{:}\mathsf{T}^q(A) \vdash^{0}_{0} \ell : A$ and the evaluation premise is $\mathcal{H}, \mathcal{S}, \mathcal{L} \vdash^{m}_{m'} \ell \Downarrow^{\mathrm{I}} w, \mathcal{H}'[\ell \mapsto w]$; by inversion of rule VAR$_{\Downarrow^{\mathrm{I}}}$ we get $\mathcal{H}, \mathcal{S}, \mathcal{L} \cup \{\ell\} \vdash^{m}_{m'} \mathcal{H}(\ell) \Downarrow w, \mathcal{H}'$ and $\ell \notin \mathcal{L}$. By the type compatibility hypothesis we get that

$$\mathcal{M}(\ell) <: \mathsf{T}^q(A)$$

We now distinguish the two applicable cases:

$\mathcal{H}(\ell)$ **is in** *whnf.* The evaluation succeeds immediately by either WHNF$_{\Downarrow^{\mathrm{I}}}$ and we have $w = \mathcal{H}(\ell)$, $\mathcal{H} = \mathcal{H}'$ and $m' = m$, i.e. the update is without effect. Taking $\Gamma' = \Gamma$, $\mathcal{B}' = \mathcal{B}$, $\mathcal{M}' = \mathcal{M}$. By type consistency, we get

$$\Gamma \vdash^{0}_{0} \mathcal{H}(\ell) : A$$

which is equivalent to the required conclusion

$$\Gamma' \vdash^{0}_{0} w : A$$

The remaining conclusions are immediate because $\mathcal{H}' = \mathcal{H}$.

$\mathcal{H}(\ell)$ **is not in** *whnf.* Let $\mathsf{T}^r(\widehat{A}) = \mathcal{M}(\ell)$. In this case, type consistency for ℓ requires (LOC2), which instantiates as

$$\Gamma, \ell{:}\mathsf{T}^q(A) \vdash^{r + \mathcal{B}(\ell)}_{0} \mathcal{H}(\ell) : \widehat{A}$$

Recall that from compatibility for location ℓ we get $\mathsf{T}^r(\widehat{A}) <: \mathsf{T}^q(A)$, which implies $\widehat{A} <: A$. By applying the type rule SUBTYPE we obtain

$$\Gamma, \ell{:}\mathsf{T}^q(A) \vdash^{r + \mathcal{B}(\ell)}_{0} \mathcal{H}(\ell) : A$$

By inversion of the evaluation premise we get

$$\mathcal{H}, \mathcal{S}, \mathcal{L} \cup \{\ell\} \vdash^{m}_{m'} \mathcal{H}(\ell) \Downarrow^{\mathrm{I}} w, \mathcal{H}'$$

and $\ell \notin \mathcal{L}$. We now apply the induction hypothesis to the evaluation of $\mathcal{H}(\ell)$. Note that according to rule VAR$_{\Downarrow^{\mathrm{I}}}$, we apply the induction hypothesis for

$\mathcal{L}' = \mathcal{L} \cup \{\ell\}$. Observe that $m \geq t + p + \mathcal{B}(\ell) + \sum_{\mathcal{H},\mathcal{L} \cup \{\ell\}} \mathcal{B}$ holds as required. We thus obtain m', Γ', $\mathcal{B}'$ and $\mathcal{M}'$ such that

$$\Gamma' \vdash^{0}_{0} w : A \tag{12}$$

$$\Gamma', \mathcal{B}' \vdash_{\text{MEM}} (\mathcal{H}', \mathcal{L} \cup \{\ell\}) : \mathcal{M}' \tag{13}$$

$$\mathcal{M}' <: \Gamma' \tag{14}$$

$$m' \geq t + p' + \sum_{\mathcal{H}',\mathcal{L} \cup \{\ell\}} \mathcal{B} \tag{15}$$

The only remaining proof obligations is to re-establish these statements for the updated heap $\mathcal{H}'' = \mathcal{H}'[\ell \mapsto w]$. In particular, we need

$$\Gamma', \mathcal{B}' \vdash_{\text{MEM}} (\mathcal{H}'', \mathcal{L}) : \mathcal{M}' \tag{16}$$

$$m' \geq t + p' + \sum_{\mathcal{H}',\mathcal{L} \cup \{\ell\}} \mathcal{B} \tag{17}$$

The only location changed from (13) to (16) is ℓ were the applicable case changes from (LOC2) to (LOC1). But the latter is immediate from (12) because $\mathcal{H}''(\ell) = w$. Since the balance sum skips locations mapped to whnf, we have $\sum_{\mathcal{H}',\mathcal{L} \cup \{\ell\}} \mathcal{B} = \sum_{\mathcal{H}'',\mathcal{L}} \mathcal{B}$, thus establishing (17) as required.

Case APP. The typing and evaluation premises in this case are

$$\Gamma, y{:}A \vdash^{p+q}_{p'} e\, y : C \tag{18}$$

$$\mathcal{H}, \mathcal{S}, \mathcal{L} \vdash^{m}_{m''} (e\,\ell) \Downarrow^{\text{I}} u, \mathcal{H}'' \tag{19}$$

By inversion of the type rule (APP) applied to (18) we obtain

$$\Gamma, y{:}A \vdash^{p}_{p'} e : A \xrightarrow{q} C \tag{20}$$

By inversion of the evaluation rule APP$_{\Downarrow\text{I}}$ applied to (19) we get

$$\mathcal{H}, \mathcal{S}, \mathcal{L} \vdash^{m}_{m'} e \Downarrow^{\text{I}} u, \mathcal{H}' \tag{21}$$

$$\mathcal{H}'@u = \lambda x.e' \tag{22}$$

$$\mathcal{H}', \mathcal{S}, \mathcal{L} \vdash^{m'}_{m''} e'[\ell/x] \Downarrow^{\text{I}} w, \mathcal{H}'' \tag{23}$$

Taking $t' = t + q$, we show that we verify the conditions for applying induction to the evaluation of e because

$$m \geq \underbrace{(t + q)}_{t'} + p + \sum_{\mathcal{H},\mathcal{L}} \mathcal{B} \tag{24}$$

By induction we obtain $\Gamma', \mathcal{B}', \mathcal{M}'$ such that

$$\Gamma' \vdash^{0}_{0} u : A \xrightarrow{q} C \tag{25}$$

$$\Gamma', \mathcal{B}' \vdash_{\text{MEM}} (\mathcal{H}', \mathcal{L}) : \mathcal{M}' \tag{26}$$

$$\mathcal{M}' <: \Gamma' \tag{27}$$

$$m' \geq (t + q) + p' + \sum_{\mathcal{H}',\mathcal{L}} \mathcal{B}'$$

$$= \underbrace{(t + p')}_{t''} + q + \sum_{\mathcal{H}',\mathcal{L}} \mathcal{B}' \tag{28}$$

By $\mathcal{H}'@u = \lambda x.\, e'$ we either have $u = \lambda x.\, e'$ or $u = \text{ind}(\kappa)$. In either case the type remains unchanged, due to rule IND and due to $\lhd$ not altering function types, we thus get

$$\Gamma' \vdash^{0}_{0} \lambda x.e' : A \xrightarrow{q} C$$

Using a standard ABS inversion lemma, we get

$$\Gamma', x{:}A \vdash^{a}_{0} e' : C$$

By the substitution lemma we get

$$\Gamma', \ell{:}A \vdash^{a}_{0} e'[\ell/x] : C$$

We can now apply induction again to (23) (evaluation of $e'[\ell/x]$) and obtain $m'', \Gamma'', \mathcal{M}'', \mathcal{B}''$ satisfying all desired conclusions:

$$\Gamma'' \vdash^{0}_{0} w : C \tag{29}$$

$$\Gamma'', \mathcal{B}'' \vdash_{\text{MEM}} (\mathcal{H}'', \mathcal{L}) : \mathcal{M}'' \tag{30}$$

$$\mathcal{M}'' <: \Gamma'' \tag{31}$$

$$\begin{aligned} m'' &\geq (t + p') + 0 + \sum_{\mathcal{H}'', \mathcal{L}} \mathcal{B}'' \\ &= t + p' + \sum_{\mathcal{H}'', \mathcal{L}} \mathcal{B}'' \end{aligned} \tag{32}$$

Corollary 1. *If the evaluation of a closed expression e with an initially empty memory succeeds, and e is well-typed $\emptyset \vdash^{p}_{p'} e : A$, then the total amount of allocations during this evaluation is bounded by p.*

Proof. This is a direct consequence of Theorem 1, by choosing $t = 0$. Observe that preconditions 2 and 3 are trivial in an empty memory configuration. Furthermore, the sum over the balance is an empty sum and thus equal to zero. Thus, for all m such that $m \geq p$, there exists some $m', \mathcal{H}'$ with $\emptyset, \emptyset, \emptyset \vdash_{\frac{m}{m'}} e \Downarrow^{\text{I}} w, \mathcal{H}'$

5.6 Relationship with Launchbury's Semantics

In this section we sketch the correspondence between the indirection semantics and Launchbury's standard semantics, which justifies our cost model. More precisely, we prove for every evaluation in the standard semantics that there is a corresponding one in the indirection semantics and that the conversion preserves cost (Theorem 2). The reverse correspondence also holds, but is not required for the soundness result, so we do not pursue it here. The development of the relationship follows [10]. We start by defining an auxiliary function to remove indirections from a heap.

Definition 5. *Consider a heap $\mathcal{H}$ such that $\mathcal{H}(\ell) = \text{ind}(\ell')$. The indirection erasure of ℓ from $\mathcal{H}$, written $\mathcal{H} \ominus \ell$, is defined as follows:*

$$\emptyset[\ell \mapsto \text{ind}(\ell')] \ominus \ell \overset{def}{=} \emptyset$$

$$\mathcal{H}[\kappa \mapsto e,\, \ell \mapsto \text{ind}(\ell')] \ominus \ell \overset{def}{=} (\mathcal{H}[\ell \mapsto \text{ind}(\ell')] \ominus \ell)[\kappa \mapsto e[\ell'/\ell]]$$

Note that we remove not just the indirection $\ell \mapsto \mathrm{ind}(\ell')$ but also rename all occurrences of ℓ to ℓ' in the remaining heap expressions.

Using indirection erasure, we can now define a relation on heaps $\mathcal{H} \succcurlyeq \mathcal{H}'$ which informally says that we obtain $\mathcal{H}'$ from $\mathcal{H}$ by removing a sequence of indirections.

Definition 6. *We say that $\mathcal{H}$ is indirection-related to $\mathcal{H}'$ and write $\mathcal{H} \succcurlyeq \mathcal{H}'$ iff there exists a (possibly empty) sequence of locations $\overline{\ell}$ such that $\mathcal{H}(\ell_i)$ is an indirection and $\mathcal{H} \ominus \overline{\ell} = \mathcal{H}'$.*

The next two lemmas state some auxiliary results about $\succcurlyeq$; the proofs are similar to the corresponding results from [10].

Lemma 1. $\succcurlyeq$ *is reflexive and transitive (i.e. a pre-order relation on heaps).*

Lemma 2. *If $\mathcal{H} \succcurlyeq \mathcal{H}'$ then $\mathrm{dom}(\mathcal{H}) \supseteq \mathrm{dom}(\mathcal{H}')$.*

Because expressions have free variables which must be interpreted in the context of a heap, it is convenient to extend the indirection relation to pairs $(\mathcal{H}, e)$ of a heap and associated expression; we do so by simply introducing the expression in a fresh location.

Definition 7. *We say that $(\mathcal{H}, e)$ is indirection-related to $(\mathcal{H}', e')$ and write $(\mathcal{H}, e) \succcurlyeq (\mathcal{H}', e')$ iff there exists $\ell \notin \mathrm{dom}(\mathcal{H}) \cup \mathrm{dom}(\mathcal{H}') \cup FV(\mathcal{H}) \cup FV(\mathcal{H}')$ such that $\mathcal{H}[\ell \mapsto e] \succcurlyeq \mathcal{H}'[\ell \mapsto e']$.*

Before presenting the correspondence result we state some auxiliary lemmas; for space restrictions we omit most proofs.

Lemma 3. *If $(\mathcal{H}, e) \succcurlyeq (\mathcal{H}', e')$ then e' is a renaming of e i.e. there exist variables $\overline{x}$ and $\overline{y}$ such that $e[\overline{y}/\overline{x}] = e'$.*

Lemma 4. *If $(\mathcal{H}, e) \succcurlyeq (\mathcal{H}', e')$ and $\ell \in \mathrm{dom}(\mathcal{H}')$ and $\ell \in FV(e')$ then $\ell \in FV(e)$.*

Lemma 5. *If $(\mathcal{H}, u) \succcurlyeq (\mathcal{H}', w)$ and w is in whnf, then $\mathcal{H}@u$ is defined (e.g. u is either a whnf or an indirection from which a whnf can be reached in a finite number of steps).*

Proof (Sketch.). By the definition of $\succcurlyeq$, there is a sequence of locations $\overline{\ell}$ such that $\mathcal{H} \ominus \overline{\ell} = \mathcal{H}'$. The proof is by induction on the length of $\overline{\ell}$.

Lemma 6. *If $(\mathcal{H}, \mathsf{let}\ x = e_1\ \mathsf{in}\ e_2) \succcurlyeq (\mathcal{H}', \mathsf{let}\ x = e_1'\ \mathsf{in}\ e_2')$ and ℓ is a fresh location then*

$$(\mathcal{H}[\ell \mapsto e_1[\ell/x]],\ e_2[\ell/x]) \succcurlyeq (\mathcal{H}'[\ell \mapsto e_1'[\ell/x]],\ e_2'[\ell/x])\ .$$

We can now finally establish the correspondence between $\Downarrow$ and $\Downarrow^{\mathrm{I}}$.

Theorem 2. *If $\mathcal{H}, \mathcal{S}, \mathcal{L} \vdash\!\!\frac{m}{m'}\!\!\!-\ e \Downarrow w, \mathcal{H}'$ then for all $\widehat{\mathcal{H}}$ such that $\widehat{\mathcal{H}} \succcurlyeq \mathcal{H}$ there exists $\widehat{\mathcal{H}}'$ and $\widehat{w}$ such that:*

$$\widehat{\mathcal{H}}, \mathcal{S}, \mathcal{L} \vdash\!\!\frac{m}{m'}\!\!\!-\ e \Downarrow^{\mathrm{I}} \widehat{w}, \widehat{\mathcal{H}}'$$
$$(\widehat{\mathcal{H}}', \widehat{w}) \succcurlyeq (\mathcal{H}', w)$$

In order to prove the above theorem by induction on the evaluation we need to strengthen the statement by allowing the evaluations to start from indirection-related heap-expression pairs.

Proposition 1. *If* $\mathcal{H}, \mathcal{S}, \mathcal{L} \vdash_{m'}^{m} e \Downarrow w, \mathcal{H}'$ *then for all* $\widehat{\mathcal{H}}$ *and* $\widehat{e}$ *with*

$$(\widehat{\mathcal{H}}, \widehat{e}) \succcurlyeq (\mathcal{H}, e) \tag{33}$$

there exists $\widehat{\mathcal{H}}'$ *and* $\widehat{w}$ *such that:*

$$\widehat{\mathcal{H}}, \mathcal{S}, \mathcal{L} \vdash_{m'}^{m} \widehat{e} \Downarrow^{\mathrm{I}} \widehat{w}, \widehat{\mathcal{H}}' \tag{34}$$

$$(\widehat{\mathcal{H}}', \widehat{\mathcal{H}}'@\widehat{w}) \succcurlyeq (\mathcal{H}', w) \tag{35}$$

Proof. By induction on the derivation of the evaluation $\mathcal{H}, \mathcal{S}, \mathcal{L} \vdash_{m'}^{m} e \Downarrow w, \mathcal{H}'$; we proceed by case-analysis of the last rule used. Due to space limitations, we only present selected cases.

Case WHNF$_\Downarrow$. The premises are $\mathcal{H}, \mathcal{S}, \mathcal{L}, \vdash_{m}^{m} w \Downarrow w, \mathcal{H}$ and $(\widehat{\mathcal{H}}, u) \succcurlyeq (\mathcal{H}, w)$ for some $\widehat{\mathcal{H}}$ and expression u. By Lemma 5, we get that $\widehat{\mathcal{H}}@u$ is defined. This satisfies the side condition of rule WHNF$_{\Downarrow^{\mathrm{I}}}$. Hence, we application of this evaluation rule yields the required conclusion (34). Conclusion (35) follows by transitivity of $\succcurlyeq$.

Case VAR$_\Downarrow$. The evaluation premise is $\mathcal{H}[\ell \mapsto e], \mathcal{S}, \mathcal{L} \vdash_{m'}^{m} \ell \Downarrow w, \mathcal{H}'[\ell \mapsto w]$. By inversion of rule VAR$_\Downarrow$ we get $\ell \notin \mathcal{L}$ and

$$\mathcal{H}[\ell \mapsto e], \mathcal{S}, \mathcal{L} \cup \{\ell\} \vdash_{m'}^{m} e \Downarrow w, \mathcal{H}'[\ell \mapsto e] \tag{36}$$

The premise (33) is $(\widehat{\mathcal{H}}_0, \widehat{e}_0) \succcurlyeq (\mathcal{H}[\ell \mapsto e], \ell)$; by Lemma 2 this implies $\widehat{\mathcal{H}}_0 = \widehat{\mathcal{H}}[\ell \mapsto \widehat{e}]$ for some $\widehat{e}$; and by Lemmas 3 and 4 we get $\widehat{e}_0 = \ell$. Thus the premise instantiates in this case as:

$$(\widehat{\mathcal{H}}[\ell \mapsto \widehat{e}], \ell) \succcurlyeq (\mathcal{H}[\ell \mapsto e], \ell) \tag{37}$$

We can now apply induction to (36) and (37) and obtain

$$\widehat{\mathcal{H}}[\ell \mapsto \widehat{e}], \mathcal{S}, \mathcal{L} \vdash_{m'}^{m} \widehat{e} \Downarrow^{\mathrm{I}} \widehat{w}, \widehat{\mathcal{H}}'[\ell \mapsto \widehat{e}] \tag{38}$$

$$(\widehat{\mathcal{H}}'[\ell \mapsto \widehat{e}], \widehat{\mathcal{H}}'[\ell \mapsto \widehat{e}]@\widehat{w}) \succcurlyeq (\mathcal{H}'[\ell \mapsto e], w) \tag{39}$$

Conclusion (39) fulfils proof obligation (35). Applying VAR$_{\Downarrow^{\mathrm{I}}}$ to (38) yields the remaining obligation (34). This concludes the proof of the VAR$_\Downarrow$ case.

Case LET$_\Downarrow$. The evaluation premise is $\mathcal{H}, \mathcal{S}, \mathcal{L} \vdash_{m'}^{1+m} \mathrm{let}\ x = e_1\ \mathrm{in}\ e_2 \Downarrow w, \mathcal{H}'$. By inversion of rule LET$_\Downarrow$ we get for some fresh ℓ:

$$\mathcal{H}[\ell \mapsto e_1[\ell/x]], \mathcal{S}, \mathcal{L} \vdash_{m'}^{m} e_2[\ell/x] \Downarrow w, \mathcal{H}' \tag{40}$$

By Lemma 3, the premise (33) instantiates as

$$(\widehat{\mathcal{H}}, \mathrm{let}\ x = \widehat{e}_1\ \mathrm{in}\ \widehat{e}_2) \succcurlyeq (\mathcal{H}, \mathrm{let}\ x = e_1\ \mathrm{in}\ e_2) \tag{41}$$

By Lemma 6 and (41) we get

$$(\widehat{\mathcal{H}}[\ell \mapsto \widehat{e}_1[\ell/x]], \widehat{e}_2[\ell/x]) \succcurlyeq (\mathcal{H}[\ell \mapsto e_1[\ell/x], e_2[\ell/x]) \tag{42}$$

By the definition of $\succcurlyeq$ (erasing the indirection $\ell' \mapsto \text{ind}(\ell)$) it is immediate that:

$$\widehat{\mathcal{H}}[\ell \mapsto \widehat{e}_1[\ell'/x], \ell' \mapsto \text{ind}(\ell)] \succcurlyeq \widehat{\mathcal{H}}[\ell \mapsto \widehat{e}_1[\ell/x]] \tag{43}$$

By transitivity of $\succcurlyeq$ (Lemma 1) and (43) plus (42) we get

$$\begin{aligned}(\widehat{\mathcal{H}}[\ell \mapsto \widehat{e}_1[\ell'/x], \ell' \mapsto \text{ind}(\ell)], \widehat{e}_2[\ell/x]) \\ \succcurlyeq (\mathcal{H}[\ell \mapsto e_1[\ell/x], e_2[\ell/x])\end{aligned}$$

which is the premise needed for applying induction to the evaluation (40). As a result of induction we get

$$\widehat{\mathcal{H}}[\ell \mapsto \widehat{e}_1[\ell'/x], \ell' \mapsto \text{ind}(\ell)], \mathcal{S}, \mathcal{L} \xmapsto{\frac{m}{m'}} \widehat{e}_2[\ell/x] \Downarrow^I \widehat{w}, \widehat{\mathcal{H}}' \tag{44}$$

$$(\widehat{\mathcal{H}}', \widehat{\mathcal{H}}'@\widehat{w}') \succcurlyeq (\mathcal{H}', w) \tag{45}$$

Applying rule LET$_{\Downarrow^I}$ to (44) together with (45) yields the required conclusions.

6   Related Work

This paper extends our previous work on type-based static analysis of resource bounds for lazy functional programs using amortisation [4]. Unlike that system, here we focus on co-recursive infinite data structures and show that a simpler type-and-effect system without amortisation suffices to obtain static resource bounds. As described in Section 4, this type-and-effect system successfully produces resource bounds for examples that could not previously be analysed.

Our cost model is based on Launchbury's natural semantics for lazy evaluation [6], as subsequently refined by Sestoft [5], de la Encina and Peña-Marì [11,12]. The proof technique used in Section 5.6 for establishing correspondence between the indirections semantics and the standard one is based on work by Sánchez-Gil, Hidalgo-Herrero and Ortega-Mallén [10]. The first work on cost analysis for lazy evaluation of higher-order functional programs was by Sands [13,14]. This used *evaluation contexts* [15] and *projections* [16] to capture the degree of evaluation of data structures. This was intended to aid manual reasoning about program costs but is not directly automatable for use in a compiler or static analysis tool.

Several authors have proposed *symbolic profiling* approaches, where programs are annotated with additional cost parameters. For example, Wadler [17] has used a state monad to count reduction costs through a tick-counting operation. Danielsson extends this work using a cost-annotated monad [18] that allows expressing machine-checkable complexity annotations through dependent types in the Agda programming language. Unlike the work presented here, this system allows checking but not automatic inference of complexity annotations.

Turner's *elementary strong functional programming* [19] explores issues of guaranteed termination in a purely functional programming language. Turner's approach separates inductive data structures from *co-data* structures such as streams. This ensures that functions on both finite and infinite structures are total by construction using only primitive recursive definitions. However, this work does not consider evaluation costs, and does not provide an analysis.

Hughes, Pareto and Sabry [20] describe a *sized type* system for a simple higher-order, non-strict functional language, that guarantees *termination* and *productivity* of recursive and co-recursive definitions. This work was subsequently developed to ensure bounded space usage in the strict functional language Embedded ML [21], which lacks co-recursion. Brady and Hammond [22] have also developed an embedding of sized types in a dependently typed framework. However, all three approaches require the programmer to provide explicit size information, that is checked rather than inferred. Finally, a combination of sized types with memory regions has been suggested by Peña and Segura [23], building on information provided by ancillary analyses on termination and safe destruction [24]. However, this does not deal with co-recursive costs.

7   Conclusions and Further Work

This paper presents a type-and-effect system for predicting upper-bounds on allocation costs for co-recursive definitions in a lazy functional language. The analysis is formally based on a standard operational semantics for lazy evaluation and we present a detailed proof sketch of soundness. We have also implemented this type system as a fully automatic static analysis. Initial experimental results show it can deal with non-trivial examples (the Fibonacci sequence and the Hamming problem). We are not aware of any previous automatic analysis that is capable of dealing with these examples.

A number of future research directions are left open by this work. For simplicity we presented a type system without either type polymorphism or effect polymorphism. This limits the compositionality of the analysis (cf. the duplication of definitions required in Hamming example of Section 4.3). Let-bound polymorphism could, in principle, be added simply by capturing constraints in type schemes as in [25,26,27]. It then remains an open question whether our system admits a notion of principal types schemes [28] (although this concerns only the completeness of the inference algorithm and not soundness).

Again for simplicity we chose a uniform cost model (each *let*-expression costs one unit). It should be straightforward to extend this to a more realistic cost model by allowing variable costs, derived from e.g. the STG abstract machine [9,29]. Another option would be to focus on resources other than heap, e.g. time or stack usage.

We have considered annotated types that express linear bounds for co-recursion, i.e. where the cost for each successive value is bounded by a constant. Hoffmann *et al.* have previously demonstrated successful extensions to multivariate polynomial bounds, in the context of amortised cost analysis for recursion [30]. It would be interesting to explore whether their techniques could also be applied in our work, to allow for non-linear costs with respect to evaluation depth.

Finally, the presented analysis and cost model do not yet consider *deallocation* of resources. In order to reason about memory residency in a type-based system, we would need, for example, to express deallocation using some syntax-directed primitives. It should be possible to extend our language and type-system with a deallocation primitive (e.g. the deallocating match in [31] or a region-based mechanism [32,33]) to accommodate this. This would pave the way for an inter-mediate language for compiling lazily evaluated programs with static residency guarantees.

References

1. Barwise, J., Moss, L.: Vicious Circles. CSLI Publications (1996)
2. Bird, R., Wadler, P.: Introduction to Functional Programming. Prentice-Hall, Englewood Cliffs (1988)
3. Coutts, D.: Stream Fusion: Practical shortcut fusion for coinductive sequence types. PhD thesis, Worcester College, University of Oxford (2010)
4. Simoes, H., Vasconcelos, P., Jost, S., Hammond, K., Florido, M.: Automatic amortised analysis of dynamic memory allocation for lazy functional programs. In: Proc. of ACM Intl. Conf. Func. Programming (ICFP 2012), pp. 165–176. ACM (2012)
5. Sestoft, P.: Deriving a Lazy Abstract Machine. J. Functional Programming 7(3), 231–264 (1997)
6. Launchbury, J.: A Natural Semantics for Lazy Evaluation. In: Proc. POPL 1993: Symp. on Princ. of Prog. Langs., pp. 144–154 (1993)
7. Jost, S., Loidl, H.W., Hammond, K., Scaife, N., Hofmann, M.: "Carbon Credits" for Resource-Bounded Computations Using Amortised Analysis. In: Cavalcanti, A., Dams, D.R. (eds.) FM 2009. LNCS, vol. 5850, pp. 354–369. Springer, Heidelberg (2009)
8. de la Encina, A., Peña, R.: Proving the Correctness of the STG Machine. In: Arts, T., Mohnen, M. (eds.) IFL 2002. LNCS, vol. 2312, pp. 88–104. Springer, Heidelberg (2002)
9. Peyton Jones, S.L.: Implementing Lazy Functional Languages on Stock Hardware – the Spineless Tagless G-machine. J. Functional Programming 2(2), 127–202 (1992)
10. Sánchez-Gil, L., Hidalgo-Herrero, M., Ortega-Mallén, Y.: The role of indirections in lazy natural semantics. Technical Report TR-13-13, Departamento de Sistemas Informticos y Computacin, Universidad Complutense de Madrid (2013)
11. de la Encina, A., Peña-Marí, R.: Formally Deriving an STG Machine. In: Proc. 5th International ACM SIGPLAN Conference on Principles and Practice of Declarative Programming, Uppsala, Sweden, August 27-29, pp. 102–112. ACM (2003)
12. de la Encina, A., Peña-Marí, R.: From Natural Semantics to C: a Formal Derivation of two STG Machines. J. Funct. Program. 19(1), 47–94 (2009)
13. Sands, D.: Calculi for Time Analysis of Functional Programs. PhD thesis, Imperial College, University of London (September 1990)
14. Sands, D.: Complexity Analysis for a Lazy Higher-Order Language. In: Jones, N.D. (ed.) ESOP 1990. LNCS, vol. 432, pp. 361–376. Springer, Heidelberg (1990)

15. Wadler, P.: Strictness Analysis aids Time Analysis. In: Proc. POPL 1988: ACM Symp. on Princ. of Prog. Langs, pp. 119–132 (1988)
16. Wadler, P., Hughes, J.: Projections for Strictness Analysis. In: Kahn, G. (ed.) FPCA 1987. LNCS, vol. 274, pp. 385–407. Springer, Heidelberg (1987)
17. Wadler, P.: The Essence of Functional Programming. In: Proc. POPL 1992: ACM Symp. on Principles of Prog. Langs., pp. 1–14 (January 1992)
18. Danielsson, N.A.: Lightweight Semiformal Time Complexity Analysis for Purely Functional Data Structures. In: Proc. POPL 2008: Symp. on Principles of Prog. Langs., San Francisco, USA, January 7-12, pp. 133–144. ACM (2008)
19. Turner, D.: Elementary Strong Functional Programming. In: Hartel, P.H., Plasmeijer, R. (eds.) FPLE 1995. LNCS, vol. 1022, pp. 1–13. Springer, Heidelberg (1995)
20. Hughes, R., Pareto, L., Sabry, A.: Proving the Correctness of Reactive Systems Using Sized Types. In: ACM Symp. on Principles of Prog. Langs (POPL 1996), St. Petersburg Beach, USA, pp. 410–423. ACM (January 1996)
21. Hughes, R., Pareto, L.: Recursion and Dynamic Data Structures in Bounded Space: Towards Embedded ML Programming. In: Proc. 1999 ACM Intl. Conf. on Functional Programming (ICFP 1999), pp. 70–81 (1999)
22. Brady, E., Hammond, K.: A Dependently Typed Framework for Static Analysis of Program Execution Costs. In: Butterfield, A., Grelck, C., Huch, F. (eds.) IFL 2005. LNCS, vol. 4015, pp. 74–90. Springer, Heidelberg (2006)
23. Pena, R., Segura, C.: A First-Order Functl. Lang. for Reasoning about Heap Consumption. In: Draft Proc. Intl. Workshop on Impl. and Appl. of Functl. Langs. (IFL 2004), pp. 64–80 (2004)
24. Montenegro, M., Pena, R., Segura, C.: An Inference Algorithm for Guaranteeing Safe Destruction. In: Draft Proc. Trends in Functional Programming (TFP 2007), New York, April 2-4 (2007)
25. Talpin, J.P., Jouvelot, P.: Polymorphic type, region and effect inference. J. Funct. Program. 2(3), 245–271 (1992)
26. Nielson, F., Nielson, H.R., Amtoft, T.: Polymorphic subtyping for effect analysis: The algorithm. In: Logical and Operational Methods in the Analysis of Programs and Systems, pp. 207–243 (1996)
27. Nielson, H.R., Nielson, F., Amtoft, T.: Polymorphic subtyping for effect analysis: The static semantics. In: Logical and Operational Methods in the Analysis of Programs and Systems, pp. 141–171 (1996)
28. Damas, L., Milner, R.: Principal type-schemes for functional programs. In: ACM Symp. on Principles of Prog. Langs (POPL 1982), pp. 207–212. ACM, New York (1982)
29. Marlow, S., Jones, S.P.: Making a fast curry: push/enter vs. eval/apply for higher-order languages. In: Proc. of the ACM SIGPLAN 2004 Intl. Conf. on Functional Programming (ICFP 2004), pp. 4–15. ACM Press (January 2004)
30. Hoffmann, J., Aehlig, K., Hofmann, M.: Multivariate Amortized Resource Analysis. ACM Trans. Program. Lang. Syst. 34(3), 14:1–14:62 (2012)
31. Hofmann, M., Jost, S.: Static prediction of heap space usage for first-order functional programs. In: ACM Symp. on Principles of Prog. Langs (POPL 2003), pp. 185–197. ACM (January 2003)
32. Tofte, M., Talpin, J.P.: Region-based memory management. Information and Computation 132(2), 109–176 (1997)
33. Tofte, M., et al.: Programming with regions in the ml kit, IT University of Copenhagen (April 2002), http://www.itu.dk/research/mlkit/

Type Targeted Testing

Eric L. Seidel, Niki Vazou, and Ranjit Jhala

UC San Diego

Abstract. We present a new technique called *type targeted testing*, which translates precise *refinement types* into comprehensive test-suites. The key insight behind our approach is that through the lens of SMT solvers, refinement types can also be viewed as a high-level, declarative, test generation technique, wherein types are converted to SMT queries whose models can be decoded into concrete program inputs. Our approach enables the systematic and exhaustive testing of implementations from high-level declarative specifications, and furthermore, provides a gradual path from testing to full verification. We have implemented our approach as a Haskell testing tool called TARGET, and present an evaluation that shows how TARGET can be used to test a wide variety of properties and how it compares against state-of-the-art testing approaches.

1 Introduction

Should the programmer spend her time writing *better types* or *thorough tests*? Types have long been the most pervasive means of describing the intended behavior of code. However, a type signature is often a very coarse description; the actual inputs and outputs may be a subset of the values described by the types. For example, the set of ordered integer lists is a very sparse subset of the set of all integer lists. Thus, to validate functions that produce or consume such values, the programmer must painstakingly enumerate these values by hand or via ad-hoc generators for unit tests.

We present a new technique called *type targeted testing*, abbreviated to TARGET, that enables the generation of unit tests from precise *refinement types*. Over the last decade, various groups have shown how refinement types – which compose the usual types with logical refinement predicates that characterize the subset of actual type inhabitants – can be used to specify and formally verify a wide variety of correctness properties of programs [29,7,23,27]. Our insight is that through the lens of SMT solvers, refinement types can be viewed as a high-level, declarative, test generation technique.

TARGET tests an implementation function against a refinement type specification using a *query-decode-check* loop. First, TARGET translates the argument types into a logical *query* for which we obtain a satisfying assignment (or model) from the SMT solver. Next, TARGET *decodes* the SMT solver's model to obtain concrete input values for the function. Finally, TARGET executes the function on the inputs to get the corresponding output, which we *check* belongs to the specified result type. If the check fails, the inputs are returned as a counterexample, otherwise TARGET refutes the given model to force the SMT solver to return a different set of inputs. This process is repeated for a given number of iterations, or until *all* inputs up to a certain size have been tested.

TARGET offers several benefits over other testing techniques. Refinement types provide a succinct description of the input and output requirements, eliminating the need

© Springer-Verlag Berlin Heidelberg 2015
J. Vitek (Ed.): ESOP 2015, LNCS 9032, pp. 812–836, 2015.
DOI: 10.1007/978-3-662-46669-8_33

to enumerate individual test cases by hand or to write custom generators. Furthermore, TARGET generates *all* values (up to a given size) that inhabit a type, and thus does not skip any corner cases that a hand-written generator might miss. Finally, while the above advantages can be recovered by a brute-force generate-and-filter approach that discards inputs that do not meet some predicate, we show that our SMT-based method can be significantly more efficient for enumerating valid inputs in a highly-constrained space.

TARGET paves a *gradual path* from testing to verification, that affords several advantages over verification. First, the programmer has an *incentive* to write formal specifications using refinement types. TARGET provides the immediate gratification of an automatically generated, exhaustive suite of unit tests that can expose errors. Thus, the programmer is rewarded without paying, up front, the extra price of annotations, hints, strengthened inductive invariants, or tactics needed for formally verifying the specification. Second, our approach makes it possible to use refinement types to formally verify *some* parts of the program, while using tests to validate other parts that may be too difficult to verify TARGET integrates the two modes by using refinement types as the uniform specification mechanism. Functions in the verified half can be formally checked *assuming* the functions in the tested half adhere to their specifications. We could even use refinements to generate dynamic contracts [9] around the tested half if so desired. Third, even when formally verifying the type specifications, the generated tests can act as valuable *counterexamples* to help *debug* the specification or implementation in the event that the program is rejected by the verifier.

Finally, TARGET offers several concrete advantages over previous property-based testing techniques, which also have the potential for gradual verification. First, instead of specifying properties with arbitrary code [4,21] which complicates the task of subsequent formal verification, with TARGET the properties are specified via refinement types, for which there are already several existing formal verification algorithms [27]. Second, while symbolic execution tools [12,22,28] can generate tests from arbitrary code contracts (*e.g.* assertions) we find that highly constrained inputs trigger path explosion which precludes the use of such tools for gradual verification.

In the rest of this paper, we start with an overview of how TARGET can be used and how its query-decode-check loop is implemented (§ 2). Next, we formalize a general framework for type-targeted testing (§ 3) and show how it can be instantiated to generating tests for lists (§ 4), and then automatically generalized to other types (§ 4.6). All the benefits of TARGET come at a price; we are limited to properties that can be specified with refinement types. We present an empirical evaluation that shows TARGET is efficient and expressive enough to capture a variety of sophisticated properties, demonstrating that type-targeted testing is a sweet spot between automatic testing and verification (§ 5).

2   Overview

We start with a series of examples pertaining to a small grading library called `Scores`. The examples provide a bird's eye view of how a user interacts with TARGET, how TARGET is implemented, and the advantages of type-based testing.

Refinement Types.   A refinement type is one where the basic types are decorated with logical predicates drawn from an efficiently decidable theory. For example,

```
type Nat   = {v:Int | 0 <= v}
type Pos   = {v:Int | 0 <  v}
type Rng N = {v:Int | 0 <= v && v < N}
```

are refinement types describing the set of integers that are non-negative, strictly positive, and in the interval [0, N) respectively. We will also build up function and collection types over base refinement types like the above. In this paper, we will not address the issue of *checking* refinement type signatures [27]. We assume the code is typechecked, *e.g.* by GHC, against the standard type signatures obtained by erasing the refinements. Instead, we focus on using the refinements to synthesize tests to *execute* the function, and to find *counterexamples* that violate the given specification.

2.1 Testing with Types

Base Types. Let us write a function `rescale` that takes a source range [0,r1), a target range [0,r2), and a score n from the source range, and returns the linearly scaled score in the target range. For example, `rescale 5 100 2` should return 40. Here is a first attempt at `rescale`

```
rescale :: r1:Nat -> r2:Nat -> s:Rng r1 -> Rng r2
rescale r1 r2 s = s * (r2 `div` r1)
```

When we run TARGET, it immediately reports

```
Found counter-example: (1, 0, 0)
```

Indeed, `rescale 1 0 0` results in 0 which is not in the target Rng 0, as the latter is empty! We could fix this in various ways, *e.g.* by requiring the ranges are non-empty:

```
rescale :: r1:Pos -> r2:Pos -> s:Rng r1 -> Rng r2
```

Now, TARGET accepts the function and reports

```
OK. Passed all tests.
```

Thus, using the refinement type *specification* for `rescale`, TARGET systematically tests the *implementation* by generating all valid inputs (up to a given size bound) that respect the pre-conditions, running the function, and checking that the output satisfies the post-condition. Testing against random, unconstrained inputs would be of limited value as the function is not designed to work on all Int values. While in this case we could filter invalid inputs, we shall show that TARGET can be more effective.

Containers. Let us suppose we have normalized all scores to be out of 100

```
type Score = Rng 100
```

Next, let us write a function to compute a *weighted* average of a list of scores.

```
average      :: [(Int, Score)] -> Score
average []   = 0
average wxs  = total `div` n
  where
     total  = sum [w * x | (w, x) <- wxs ]
     n      = sum [w     | (w, _) <- wxs ]
```

It can be tricky to *verify* this function as it requires non-linear reasoning about an unbounded collection. However, we can gain a great degree of confidence by systematically testing it using the type specification; indeed, TARGET responds:

```
Found counter-example: [(0,0)]
```

Clearly, an unfortunate choice of weights can trigger a divide-by-zero; we can fix this by requiring the weights be non-zero:

```
average :: [({v:Int | v /= 0}, Score)] -> Score
```

but now TARGET responds with

```
Found counter-example: [(-3,3),(3,0)]
```

which also triggers the divide-by-zero! We will play it safe and require positive weights,

```
average :: [(Pos, Score)] -> Score
```

at which point TARGET reports that all tests pass.

Ordered Containers. The very nature of our business requires that at the end of the day, we order students by their scores. We can represent ordered lists by requiring the elements of the tail t to be greater than the head h:

```
data OrdList a = [] | (:) {h :: a, t :: OrdList {v:a | h <= v}}
```

Note that erasing the refinement predicates gives us plain old Haskell lists. We can now write a function to insert a score into an ordered list:

```
insert :: (Ord a) => a -> OrdList a -> OrdList a
```

TARGET automatically generates all ordered lists (up to a given size) and executes insert to check for any errors. Unlike randomized testers, TARGET is not thwarted by the ordering constraint, and does not require a custom generator from the user.

Structured Containers. Everyone has a few bad days. Let us write a function that takes the best k scores for a particular student. That is, the output must satisfy a *structural* constraint – that its size equals k. We can encode the size of a list with a logical measure function [27]:

```
measure len :: [a] -> Nat
len []     = 0
len (x:xs) = 1 + len xs
```

Now, we can stipulate that the output indeed has k scores:

```
best       :: k:Nat -> [Score] -> {v:[Score] | k = len v}
best k xs = take k $ reverse $ sort xs
```

Now, TARGET quickly finds a counterexample:

```
Found counter-example: (2,[])
```

Of course – we need to have at least k scores to start with!

```
best :: k:Nat -> {v:[Score]|k <= len v} -> {v:[Score]|k = len v}
```

and now, TARGET is assuaged and reports no counterexamples. While randomized testing would suffice for `best`, we will see more sophisticated structural properties such as height balancedness, which stymie random testers, but are easily handled by TARGET.

Higher-order Functions. Perhaps instead of taking the k best grades, we would like to pad each individual grade, and, furthermore, we want to be able to experiment with different padding functions. Let us rewrite `average` to take a functional argument, and stipulate that it can only increase a `Score`.

```
padAverage          :: (s:Score -> {v:Score | s <= v})
                    -> [(Pos, Score)] -> Score
padAverage f []   = f 0
padAverage f wxs  = total `div` n
  where
    total   = sum [w * f x | (w, x) <- wxs ]
    n       = sum [w        | (w, _) <- wxs ]
```

TARGET automatically checks that `padAverage` is a safe generalization of `average`. Randomized testing tools can also generate functions, but those functions are unlikely to satisfy non-trivial constraints, thereby burdening the user with custom generators.

2.2 Synthesizing Tests

Next, let us look under the hood to get an idea of how TARGET synthesizes tests from types. At a high-level, our strategy is to: (1) *query* an SMT solver for satisfying assigments to a set of logical constraints derived from the refinement type, (2) *decode* the model into Haskell values that are suitable inputs, (3) *execute* the function on the decoded values to obtain the output, (4) *check* that the output satisfies the output type, (5) *refute* the model to generate a different test, and repeat the above steps until all tests up to a certain size are executed. We focus here on steps 1, 2, and 4 – query, decode, and check – the others are standard and require little explanation.

Base Types. Recall the initial (buggy) specification

```
rescale :: r1:Nat -> r2:Nat -> s:Rng r1 -> Rng r2
```

TARGET *encodes* input requirements for base types directly from their corresponding refinements. The constraints for multiple, related inputs are just the *conjunction* of the constraints for each input. Hence, the constraint for `rescale` is:

$$C_0 \doteq 0 \leq r1 \wedge 0 \leq r2 \wedge 0 \leq s < r1$$

In practice, C_0 will also contain conjuncts of the form $-N \leq x \leq N$ that restrict `Int`-valued variables x to be within the size bound N supplied by the user, but we will omit these throughout the paper for clarity.

Note how easy it is to capture dependencies between inputs, *e.g.* that the score s be in the range defined by `r1`. On querying the SMT solver with the above, we get a model $[r1 \mapsto 1, r2 \mapsto 1, s \mapsto 0]$. TARGET decodes this model and executes `rescale 1 1 0` to obtain the value `v = 0`. Then, TARGET validates v against the post-condition by checking the validity of the output type's constraint:

$$r2 = 1 \wedge v = 0 \wedge 0 \leq v \wedge v < r2$$

As the above is valid, TARGET moves on to generate another test by conjoining C_0 with a constraint that refutes the previous model:

$$C_1 \doteq C_0 \wedge (r1 \neq 1 \vee r2 \neq 1 \vee s \neq 0)$$

This time, the SMT solver returns a model: $[r1 \mapsto 1, r2 \mapsto 0, s \mapsto 0]$ which, when decoded and executed, yields the result 0 that does *not* inhabit the output type, and so is reported as a counterexample. When we fix the specification to only allow Pos ranges, each test produces a valid output, so TARGET reports that all tests pass.

Containers. Next, we use TARGET to test the implementation of average. To do so, TARGET needs to generate Haskell lists with the appropriate constraints. Since each list is recursively either "nil" or "cons", TARGET generates constraints that symbolically represent *all* possible lists up to a given depth, using propositional *choice variables* to symbolically pick between these two alternatives. Every (satisfying) assignment of choices returned by the SMT solver gives TARGET the concrete data and constructors used at each level, allowing it to decode the assignment into a Haskell value.

For example, TARGET represents valid [(Pos, Score)] inputs (of depth up to 3), required to test average, as the conjunction of C_{list} and C_{data}:

$$
\begin{aligned}
C_{list} \doteq\ & (c_{00} \Rightarrow xs_0 = [\,]) \wedge (c_{01} \Rightarrow xs_0 = x_1 : xs_1) \wedge (c_{00} \oplus c_{01}) \\
\wedge\ & (c_{10} \Rightarrow xs_1 = [\,]) \wedge (c_{11} \Rightarrow xs_1 = x_2 : xs_2) \wedge (c_{01} \Rightarrow c_{10} \oplus c_{11}) \\
\wedge\ & (c_{20} \Rightarrow xs_2 = [\,]) \wedge (c_{21} \Rightarrow xs_2 = x_3 : xs_3) \wedge (c_{11} \Rightarrow c_{20} \oplus c_{21}) \\
\wedge\ & (c_{30} \Rightarrow xs_3 = [\,]) \wedge (c_{21} \Rightarrow c_{30})
\end{aligned}
$$

$$
\begin{aligned}
C_{data} \doteq\ & (c_{01} \Rightarrow x_1 = (w_1, s_1) \ \wedge\ 0 < w_1 \ \wedge\ 0 \leq s_1 < 100) \\
\wedge\ & (c_{11} \Rightarrow x_2 = (w_2, s_2) \ \wedge\ 0 < w_2 \ \wedge\ 0 \leq s_2 < 100) \\
\wedge\ & (c_{21} \Rightarrow x_3 = (w_3, s_3) \ \wedge\ 0 < w_3 \ \wedge\ 0 \leq s_3 < 100)
\end{aligned}
$$

The first set of constraints C_{list} describes all lists up to size 3. At each level i, the *choice* variables c_{i0} and c_{i1} determine whether at that level the constructed list xs_i is a "nil" or a "cons". In the constraints $[\,]$ and $(\,:\,)$ are *uninterpreted* functions that represent "nil" and "cons" respectively. These functions only obey the congruence axiom and hence, can be efficiently analyzed by SMT solvers [19]. The data at each level x_i is constrained to be a pair of a positive weight w_i and a valid score s_i.

The choice variables at each level are used to *guard* the constraints on the next levels. First, if we are generating a "cons" at a given level, then exactly one of the choice variables for the next level must be selected; *e.g.* $c_{11} \Rightarrow c_{20} \oplus c_{21}$. Second, the constraints on the data at a given level only hold if we are generating values for that level; *e.g.* c_{21} is used to guard the constraints on x_3, w_3 and s_3. This is essential to avoid over-constraining the system which would cause TARGET to miss certain tests.

To *decode* a model of the above into a Haskell value of type [(Int, Int)], we traverse constraints and use the valuations of the choice variables to build up the list appropriately. At each level, if $c_{i0} \mapsto true$, then the list at that level is [], otherwise $c_{i1} \mapsto true$ and we decode x_{i+1} and xs_{i+1} and "cons" the results.

We can iteratively generate *multiple* inputs by adding a constraint that refutes each prior model. As an important optimization, we only refute the relevant parts of the model, *i.e.* those needed to construct the list (§ 4.5).

Ordered Containers. Next, let us see how TARGET enables automatic testing with highly constrained inputs, such as the *increasingly ordered* OrdList values required by insert. From the type definition, it is apparent that ordered lists are the same as the usual lists described by C_{list}, except that each unfolded *tail* must only contain values that are greater than the corresponding *head*. That is, as we unfold x1:x2:xs :: OrdList

- At level 0, we have OrdList {v:Score| true}
- At level 1, we have OrdList {v:Score| x1 <= v}
- At level 2, we have OrdList {v:Score| x2 <= v && x1 <= v}

and so on. Thus, we encode OrdList Score (of depth up to 3) by conjoining C_{list} with C_{score} and C_{ord}, which capture the valid score and ordering requirements respectively:

$$C_{ord} \doteq (c_{11} \Rightarrow x_1 \leq x_2) \wedge (c_{21} \Rightarrow x_2 \leq x_3 \wedge x_1 \leq x_3)$$
$$C_{score} \doteq (c_{01} \Rightarrow 0 \leq x_1 < 100) \wedge (c_{11} \Rightarrow 0 \leq x_2 < 100) \wedge (c_{21} \Rightarrow 0 \leq x_3 < 100)$$

Structured Containers. Recall that best k requires inputs whose *structure* is constrained – the size of the list should be no less than k. We specify size using special measure functions [27], which let us relate the size of a list with that of its unfolding, and hence, let us encode the notion of size inside the constraints:

$$C_{size} \doteq (c_{00} \Rightarrow len\ xs_0 = 0) \wedge (c_{01} \Rightarrow len\ xs_0 = 1 + len\ xs_1)$$
$$\wedge\ (c_{10} \Rightarrow len\ xs_1 = 0) \wedge (c_{11} \Rightarrow len\ xs_1 = 1 + len\ xs_2)$$
$$\wedge\ (c_{20} \Rightarrow len\ xs_2 = 0) \wedge (c_{21} \Rightarrow len\ xs_2 = 1 + len\ xs_3)$$
$$\wedge\ (c_{30} \Rightarrow len\ xs_3 = 0)$$

At each unfolding, we instantiate the definition of the measure for each alternative of the datatype. In the constraints, $len\ \cdot$ is an uninterpreted function derived from the measure definition. All of the relevant properties of the function are spelled out by the unfolded constraints in C_{size} and hence, we can use SMT to search for models for the above constraint. Hence, TARGET constrains the input type for best as:

$$0 \leq k \wedge C_{list} \wedge C_{score} \wedge C_{size} \wedge k \leq len\ xs_0$$

where the final conjunct comes from the top-level refinement that stipulates the input have at least k scores. Thus, TARGET only generates lists that are large enough. For example, in any model where $k = 2$, it will *not* generate the empty or singleton list, as in those cases, $len\ xs_0$ would be 0 (resp. 1), violating the final conjunct above.

Higher-order Functions. Finally, TARGET's type-directed testing scales up to higher-order functions using the same insight as in QuickCheck [4], namely, to generate a function it suffices to be able to generate the *output* of the function. When tasked with the generation of a functional argument f, TARGET returns a Haskell function that when executed checks whether its inputs satisfy f's pre-conditions. If they do, then

```
-- Manipulating Refinements
refinement :: RefType -> Refinement
subst      :: RefType -> [(Var, Var)] -> RefType

-- Manipulating Types
unfold     :: Ctor  -> RefType -> [(Var, RefType)]
binder     :: RefType -> Var
proxy      :: RefType -> Proxy a
```

Fig. 1. Refinement Type API

f uses TARGET to dynamically query the SMT solver for an output that satisfies the constraints imposed by the concrete inputs. Otherwise, f's specifications are violated and TARGET reports a counterexample.

This concludes our high-level tour of the benefits and implementation of TARGET. Notice that the property specification mechanism – refinement types – allowed us to get immediate feedback that helped debug not just the code, but also the specification itself. Additionally, the specifications gave us machine-readable documentation about the behavior of functions, and a large unit test suite with which to automatically validate the implementation. Finally, though we do not focus on it here, the specifications are amenable to formal verification should the programmer so desire.

3   A Framework for Type Targeted Testing

Next, we describe a framework for type targeted testing, by formalizing an abstract representation of refinement types (§ 3.1), describing the operations needed to generate tests from types (§ 3.2), and then using the above to implement TARGET via a query-decode-check loop (§ 3.3). Subsequently, we instantiate the framework to obtain tests for refined primitive types, lists, algebraic datatypes and higher-order functions (§ 4).

3.1   Refinement Types

A refinement type is a type, where each component is decorated with a predicate from a refinement logic. For clarity, we describe refinement types and refinements abstractly as RefType and Refinement respectively. We write Var as an alias for Refinement that is typically used to represent logical variables appearing within the refinement.

Notation. In the sequel, we will use double brackets $[\![\,]\!]$ to represent the various entities in the meta-language used to describe TARGET. For example, $[\![k]\!]$, $[\![k \leq len\ v]\!]$, and $[\![\{v : [Score] \mid k \leq len\ v\}]\!]$ are the Var, Refinement, and RefType representing the corresponding entities written in the brackets.

Next, we describe the various operations over them needed to implement TARGET. These operations, summarized in Figure 1, fall into two categories: those which manipulate the *refinements* and those which manipulate the *types*.

Operating on Refinements. To generate constraints and check inhabitation, we use the function `refinement` which returns the (top-level) refinement that decorates the given refinement type. We will generate fresh `Var`s to name values of components, and will use `subst` to replace the free occurrences of variables in a given `RefType`. Suppose that t is the `RefType` represented by $[\{v : [Score] \mid k \leq len\ v\}]$. Then,

- `refinement` t evaluates to $[k \leq len\ v]$ and
- `subst` t `[ (`$[k]$`, `$[x_0]$`) ]` evaluates to $[\{v : [Score] \mid x_0 \leq len\ v\}]$.

Operating on Types. To build up compound values (*e.g.* lists) from components (*e.g.* an integer and a list), `unfold` breaks a `RefType` (*e.g.* a list of integers) into its constituents (*e.g.* an integer and a list of integers) at a given constructor (*e.g.* "cons"). `binder` simply extracts the `Var` representing the value being refined from the `RefType`. To write generic functions over `RefType`s and use Haskell's type class machinery to `query` and `decode` components of types, we associate with each refinement type a *proxy* representing the corresponding Haskell type (in practice this must be passed around as a separate argument). For example, if t is $[\{v : [Score] \mid k \leq len\ v\}]$,

- `unfold` $[:]$ t evaluates to $[($$[x]$`, `$[Score]$`), (`$[xs]$`, `$[[Score]]$`) ]`,
- `binder` t evaluates to $[v]$, and
- `proxy` t evaluates to a value of type `Proxy [Int]`.

3.2 The `Targetable` Type Class

Following QuickCheck, we encapsulate the key operations needed for type-targeted testing in a type class `Targetable` (Figure 2). This class characterizes the set of types

```
class Targetable a where
    query  :: Proxy a -> Int -> RefType -> SMT Var
    decode :: Var -> SMT a
    check  :: a -> RefType -> SMT (Bool, Var)
    toReft :: a -> Refinement
```

Fig. 2. The class of types that can be tested by TARGET

which can be tested by TARGET. All of the operations can interact with an external SMT solver, and so return values in an `SMT` monad.

- `query` takes a *proxy* for the Haskell type for which we are generating values, an integer *depth* bound, and a *refinement type* describing the desired constraints, and generates a set of logical constraints and a `Var` that represents the constrained value.
- `decode` takes a `Var`, generated via a previous `query` and queries the model returned by the SMT solver to construct a Haskell value of type a.
- `check` takes a value of type a, translates it back into logical form, and verifies that it inhabits the output type t.
- `toReft` takes a value of type a and translates it back into logical form (a specialization of `check`).

3.3   The Query-Decode-Check Loop

Figure 3 summarizes the overall implementation of TARGET, which takes as input a function f and its refinement type specification t and proceeds to test the function against the specification via a *query-decode-check* loop: (1) First, we translate the refined inputTypes into a logical *query*. (2) Next, we *decode* the model (*i.e.* satisfying assignment) for the query returned by the SMT solver to obtain concrete inputs. (3) Finally, we execute the function f on the inputs to get the corresponding output, which we check belongs to the specified outputType. If the check fails, we return the inputs as a counterexample. After each test, TARGET, refutes the given test to force the SMT solver to return a different set of inputs, and this process is repeated until a user specified number of iterations. The checkSMT call may fail to find a model meaning that we have exhaustively tested all inputs upto a given testDepth bound. If all iterations succeed, *i.e.* no counterexamples are found, then TARGET returns Ok, indicating that f satisfies t up to the given depth bound.

```
target f t = do
  let txs = inputTypes t
  vars  <- forM txs $ \tx ->
              query (proxy tx) testDepth tx -- Query
  forM [1 .. testNum] $ \_ -> do
    hasModel <- checkSMT
    when hasModel $ do
      inputs <- forM vars decode          -- Decode
      output <- execute f inputs
      let su = zip (map binder txs) (map toReft inputs)
      let to = outputType t 'subst' su
      (ok,_) <- check output to           -- Check
      if ok then
        refuteSMT
      else
        throw (CounterExample inputs)
  return Ok
```

Fig. 3. Implementing TARGET via a *query-decode-check* loop

4   Instantiating the TARGET Framework

Next, we describe a concrete instantiation of TARGET for lists. We start with a constraint generation API (§ 4.1). Then we use the API to implement the key operations query (§ 4.2), decode (§ 4.3), check (§ 4.4), and refuteSMT (§ 4.5), thereby enabling TARGET to automatically test functions over lists. We omit the definition of toReft as it follows directly from the definition of check. Finally, we show how the list instance can be generalized to algebraic datatypes and higher-order functions (§ 4.6).

4.1 SMT Solver Interface

Figure 4 describes the interface to the SMT solvers that TARGET uses for constraint generation and model decoding. The interface has functions to (a) generate logical variables of type `Var`, (b) constrain their values using `Refinement` predicates, and (c) determine the values assigned to the variables in satisfying models.

```
fresh       :: SMT Var
guard       :: Var -> SMT a        -> SMT a
constrain   :: Var -> Refinement -> SMT ()

apply       :: Ctor -> [Var] -> SMT Var
unapply     :: Var   -> SMT (Ctor, [Var])

oneOf       :: Var -> [(Var, Var)] -> SMT ()
whichOf     :: Var -> SMT Var

eval        :: Refinement -> SMT Bool
```

Fig. 4. SMT Solver API

- `fresh` allocates a new logical variable.
- `guard b act` ensures that all the constraints generated by `act` are *guarded by* the choice variable `b`. That is, if `act` generates the constraint p then `guard b act` generates the (implication) constraint $b \Rightarrow p$.
- `constrain x r` generates a constraint that `x` satisfies the refinement predicate `r`.
- `apply c xs` generates a new `Var` for the folded up value obtained by applying the constructor `c` to the fields `xs`, while also generating constraints from the measures. For example, `apply [:] [[x_1], [xs_1]]` returns $[x_1 : xs_1]$ and generates the constraint $len\ (x_1 : xs_1) = 1 + len\ xs_1$.
- `unapply x` returns the `Ctor` and `Var`s from which the input `x` was constructed.
- `oneOf x cxs` generates a constraint that `x` equals exactly one of the elements of `cxs`. For example, `oneOf [xs_0] [([c_{00}], [[]]), ([c_{01}], [x_1 : xs_1])]` yields:

$$(c_{00} \Rightarrow xs_0 = []) \land (c_{01} \Rightarrow xs_0 = x_1 : xs_1) \land (c_{00} \oplus c_{01})$$

- `whichOf x` returns the particular alternative that was assigned to `x` in the current model returned by the SMT solver. Continuing the previous example, if the model sets $[c_{00}]$ (resp. $[c_{01}]$) to *true*, `whichOf [xs_0]` returns $[[]]$ (resp $[x_1 : xs_1]$).
- `eval r` checks the validity of a refinement with no free variables. For example, `eval [len\ (1 : []) > 0]` would return `True`.

```
query p d t = do
  let cs = ctors d
  bs <- forM cs (\_ -> fresh)
  xs <- zipWithM (queryCtor (d-1) t) bs cs
  x  <- fresh
  oneOf x      (zip bs xs)
  constrain x (refinement t)
  return x

queryCtor d t b c = guard b (do
  let fts = unfold c t
  fs'      <- scanM (queryField d) [] fts
  x        <- apply c fs'
  return x)

queryField d su (f, t) = do
  f' <- query (proxy t) d (t 'subst' su)
  return ((f, f') : su, f')
ctors d
  | d > 0      = [ [:], [[]] ]
  | otherwise = [ [[]] ]
```

Fig. 5. Generating a Query

4.2 Query

Figure 5 shows the procedure for constructing a query from a refined list type, *e.g.* the one required as an input to the best or insert functions from § 2.

Lists query returns a Var that represent *all* lists up to depth d that satisfy the logical constraints associated with the refined list type t. To this end, it invokes ctors to obtain all of the suitable constructors for depth d. For lists, when the depth is 0 we should only use the [[]] constructor, otherwise we can use either [:] or [[]]. This ensures that query terminates after encoding all possible lists up to a given depth d. Next, it uses fresh to generate a distinct *choice* variable for each constructor, and calls queryCtor to generate constraints and a corresponding symbolic Var for each constructor. The choice variable for each constructor is supplied to queryCtor to ensure that the constraints are *guarded*, *i.e.* only required to hold *if* the corresponding choice variable is selected in the model and not otherwise. Finally, a fresh x represents the value at depth d and is constrained to be oneOf the alternatives represented by the constructors, and to satisfy the top-level refinement of t.

Constructors queryCtor takes as input the refined list type t, a depth d, a particular constructor c for the list type, and generates a query describing the *unfolding* of t at the constructor c, guarded by the choice variable b that determines whether this alternative is indeed part of the value. These constraints are the conjunction of those describing the values of the individual fields which can be combined via c to obtain a t value. To do

```
decode x = do                    decodeCtor [[]] [] = return []
    x'        <- whichOf x       decodeCtor [:] [x,xs] = do
    (c,fs') <- unapply x'            v  <- decode x
    decodeCtor c fs'                 vs <- decode xs
                                     return (v:vs)
```

Fig. 6. Decoding Models into Haskell Values

so, queryCtor first unfolds the type t at c, obtaining a list of constituent fields and their respective refinement types fts. Next, it uses

```
scanM :: Monad m => (a -> b -> m (a, c)) -> a -> [b] -> m [c]
```

to traverse the fields from left to right, building up representations of values for the fields from their unfolded refinement types. Finally, we invoke apply on c and the fields fs' to return a symbolic representation of the constructed value that is constrained to satisfy the measure properties of c.

Fields queryField generates the actual constraints for a single field f with refinement type t, by invoking query on t. The proxy enables us to resolve the appropriate type-class instance for generating the query for the field's value. Each field is described by a new symbolic name f' which is substituted for the formal name of the field f in the refinements of subsequent fields, thereby tracking dependencies between the fields. For example, these substitutions ensure the values in the tail are greater than the head as needed by OrdList from § 2.

4.3 Decode

Once we have generated the constraints we query the SMT solver for a model, and if one is found we must *decode* it into a concrete Haskell value with which to test the given function. Figure 6 shows how to decode an SMT model for lists.

Lists decode takes as input the top-level symbolic representation x and queries the model to determine which alternative was assigned by the solver to x, *i.e.* a nil or a cons. Once the alternative is determined, we use unapply to destruct it into its constructor c and fields fs', which are recursively decoded by decodeCtor.

Constructors decodeCtor takes the constructor c and a list of symbolic representations for fields, and decodes each field into a value and applies the constructor to obtain the Haskell value. For example, in the case of the [[]] constructor, there are no fields, so we return the empty list. In the case of the [:] constructor, we decode the head and the tail, and cons them to return the decoded value. decodeCtor has the type

```
Targetable a => Ctor -> [Var] -> SMT [a]
```

i.e. if a is a decodable type, then decodeCtor suffices to decode lists of a. Primitives like integers that are directly encoded in the refinement logic are the base case – *i.e.* the value in the model is directly translated into the corresponding Haskell value.

```
check v t = do
  let (c,vs) = splitCtor v
  let fts    = unfold c t
  (bs, vs') <- fmap unzip (scanM checkField [] (zip vs fts))
  v'         <- apply c vs'
  let t'      = t 'subst' [(binder t, v')]
  b'         <- eval (refinement t')
  return (and (b:bs), v')

checkField su (v, (f, t)) = do
  (b, v') <- check v (t 'subst' su)
  return ((f, v') : su, (b, v'))

splitCtor []     = ([[]] , [])
splitCtor (x:xs) = ([:] , [x,xs])
```

Fig. 7. Checking Outputs

4.4 Check

The third step of the query-decode-check loop is to verify that the output produced by the function under test indeed satisfies the output refinement type of the function. We accomplish this by *encoding* the output value as a logical expression, and evaluating the output refinement applied to the logical representation of the output value.

check, shown in Figure 7, takes a Haskell (output) value v and the (output) refinement type t, and recursively verifies each component of the output type. It converts each component into a logical representation, substitutes the logical expression for the symbolic value, and evaluates the resulting Refinement.

4.5 Refuting Models

Finally, TARGET invokes refuteSMT to *refute* a given model in order to force the SMT solver to produce a different model that will yield a different test input. A naïve implementation of refutation is as follows. Let X be the set of all variables appearing in the constraints. Suppose that in the current model, each variable x is assigned the value $\sigma(x)$. Then, to refute the model, we add a *refutation constraint* $\bigvee_{x \in X} x \neq \sigma(x)$. That is, we stipulate that *some* variable be assigned a different value.

The naïve implementation is extremely inefficient. The SMT solver is free to pick a different value for some *irrelevant* variable which was not even used for decoding. As a result, the next model can, after decoding, yield the *same* Haskell value, thereby blowing up the number of iterations needed to generate all tests of a given size.

TARGET solves this problem by forcing the SMT solver to return models that yield *different decoded tests* in each iteration. To this end TARGET restricts the refutation constraint to the set of variables that were actually used to decode the Haskell value. We track this set by instrumenting the SMT monad to log the set of variables and choice-variables that are transitively queried via the recursive calls to decode. That is, each call to decode logs its argument, and each call to whichOf logs the choice variable corresponding to the alternative that was returned. Let R be the resulting set of *decode-relevant* variables. TARGET refutes the model by using a *relevant refutation constraint* $\bigvee_{x \in R} x \neq \sigma(x)$ which ensures that the next model decodes to a different value.

4.6   Generalizing TARGET To Other Types

The implementation in § 4 is for List types, but ctors, decodeCtor, and splitCtor are the only functions that are List-specific. Thus, we can easily generalize the implementation to:

- *primitive datatypes*, *e.g.* integers, by returning an empty list of constructors,
- *algebraic datatypes*, by implementing ctors, decodeCtor, and splitCtor for that type.
- *higher-order functions*, by lifting instances of a to functions returning a.

Algebraic Datatypes   Our List implementation has three pieces of type-specific logic:

- ctors, which returns a list of constructors to unfold;
- decodeCtor, which decodes a specific Ctor; and
- splitCtor, which splits a Haskell value into a pair of its Ctor and fields.

Thus, to instantiate TARGET on a new data type, all we need is to implement these three operations for the type. This implementation essentially follows the concrete template for Lists. In fact, we observe that the recipe is entirely mechanical boilerplate, and can be fully automated for *all* algebraic data types by using a *generics* library.

Any algebraic datatype (ADT) can be represented as a *sum-of-products* of component types. A generics library, such as GHC.Generics [15], provides a *univeral* sum-of-products type and functions to automatically convert any ADT to and from the universal representation. Thus, to obtain Targetable instances for *any* ADT it suffices to define a Targetable instance for the *universal* type.

Once the universal type is Targetable we can automatically get an instance for any new user-defined ADT (that is an instance of Generic) as follows: (1) to generate a *query* we simply create a query for GHC.Generics' universal representation of the refined type, (2) to *decode* the results from the SMT solver, we decode them into the universal representation and then use GHC.Generics to map them back into the user-defined type, (3) to *check* that a given value inhabits a user-defined refinement type, we check that the universal representation of the value inhabits the type's universal counterpart.

The Targetable instance for the universal representation is a generalized version of the List instance from § 4, that relies on various technical details of GHC.Generics.

Higher Order Functions   Our type-directed approach to specification makes it easy to extend TARGET to higher-order functions. Concretely, it suffices to implement a type-class instance:

```
instance (Targetable input, Targetable output)
   => Targetable (input -> output)
```

In essence, this instance uses the `Targetable` instances for `input` and `output` to create an instance for functions from `input -> output`, after which Haskell's type class machinery suffices to generate concrete function values.

To create such instances, we use the insight from QuickCheck, that to generate (constrained) functions, we need only to generate *output* values for the function. Following this route, we generate functions by creating new lambdas that take in the inputs from the calling context, and use their values to create queries for the output, after which we can call the SMT solver and decode the results to get concrete outputs that are returned by the lambda, completing the function definition. Note that we require `input` to also be `Targetable` so that we can encode the Haskell value in the refinement logic, in order to constrain the output values suitably. We additionally memoize the generated function to preserve the illusion of purity. It is also possible to, in the future, extend our implementation to refute functions by asserting that the output value for a given input be distinct from any previous outputs for that input.

5   Evaluation

We have built a prototype implementation of TARGET[1] and next, describe an evaluation on a series of benchmarks ranging from textbook examples of algorithms and data structures to widely used Haskell libraries like CONTAINERS and XMONAD. Our goal in this evaluation is two-fold. First, we describe micro-benchmarks (*i.e.* functions) that *quantitatively compare* TARGET with the existing state-of-the-art, property-based testing tools for Haskell – namely SmallCheck and QuickCheck – to determine whether TARGET is indeed able to generate highly constrained inputs more effectively. Second, we describe macro-benchmarks (*i.e.* modules) that evaluate the amount of *code coverage* that we get from type-targeted testing.

5.1   Comparison with QuickCheck and SmallCheck

We compare TARGET with QuickCheck and SmallCheck by using a set of benchmarks with highly constrained inputs. For each benchmark we compared TARGET with Small-Check and QuickCheck, with the latter two using the generate-and-filter approach, wherein a value is generated and subsequently discarded if it does not meet the desired constraint. While one could possibly write custom "operational" generators for each property, the point of this evaluation is compare the different approaches ability to enable "declarative" specification driven testing. Next, we describe the benchmarks and then summarize the results of the comparison (Figure 8).

Inserting into a sorted List   Our first benchmark is the `insert` function from the homonymous sorting routine. We use the specification that given an element and a sorted list, `insert x xs` should evaluate to a sorted list. We express this with the type

```
type Sorted a = List <{\hd v -> hd < v}> a
insert :: a -> Sorted a -> Sorted a
```

[1] http://hackage.haskell.org/package/target-0.1.1.0

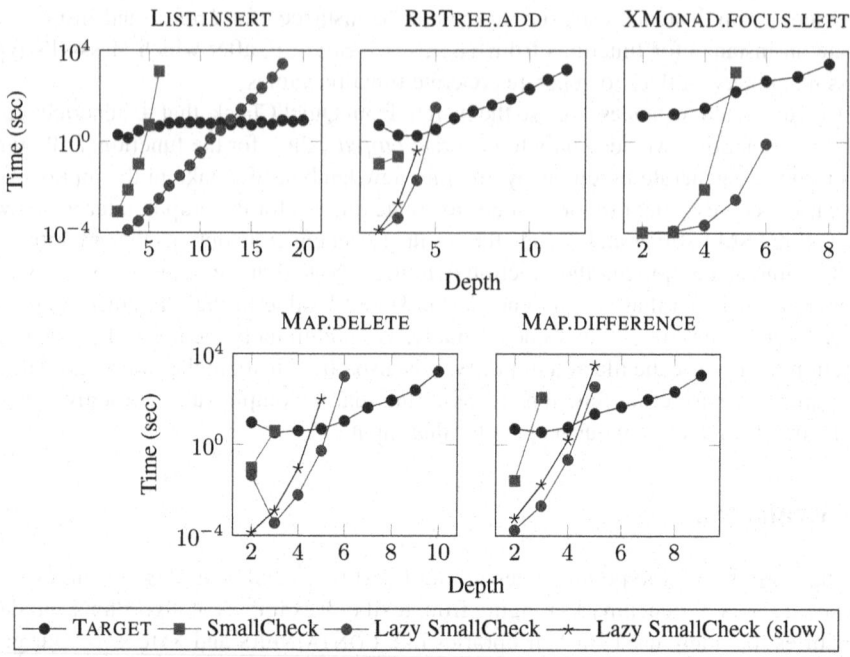

Fig. 8. Results of comparing TARGET with QuickCheck, SmallCheck, and Lazy SmallCheck on a series of functions. TARGET, SmallCheck, and Lazy SmallCheck were both configured to check the first 1000 inputs that satisfied the precondition at increasing depth parameters, with a 60 minute timeout per depth; QuickCheck was run with the default settings, *i.e.* it had to produce 100 test cases. TARGET, SmallCheck, and Lazy SmallCheck were configured to use the same notion of depth, in order to ensure they would generate the same number of valid inputs at each depth level. QuickCheck was unable to successfully complete any run due to the low probability of generating valid inputs at random.

where the ordering constraint is captured by an abstract refinement [25] which states that *each* list head hd is less than every element v in its tail.

Inserting into a Red-Black Tree Next, we consider insertion into a Red-Black tree.

```
data RBT a = Leaf | Node Col a (RBT a) (RBT a)
data Col   = Black | Red
```

Red-black trees must satisfy three invariants: (1) red nodes always have black children, (2) the black height of all paths from the root to a leaf is the same, and (3) the elements in the tree should be ordered. We capture (1) via a measure that recursively checks each Red node has Black children.

```
measure isRB :: RBT a -> Prop
isRB Leaf              = true
isRB (Node c x l r) = isRB l && isRB r &&
                       (c == Red => isBlack l && isBlack r)
```

We specify (2) by defining the `Black` height as:

```
measure bh :: RBT a -> Int
bh Leaf         = 0
bh (Node c x l r) = bh l + (if c == Red then 0 else 1)
```

and then checking that the `Black` height of both subtrees is the same:

```
measure isBH :: RBT a -> Prop
isBH Leaf         = true
isBH (Node c x l r) = isBH l && isBH r && bh l == bh r
```

Finally, we specify the (3), the ordering invariant as:

```
type OrdRBT a = RBT <{\r v -> v < r}, {\r v -> r < v}> a
```

i.e. with two abstract refinements for the left and right subtrees respectively, which state that the root r is greater than (resp. less than) each element v in the subtrees. Finally, a valid Red-Black tree is:

```
type OkRBT a = {v:OrdRBT a | isRB v && isBH v}
```

Note that while the specification for the *internal* invariants for Red-Black trees is tricky, the specification for the public API – *e.g.* the add function – is straightforward:

```
add :: a -> OkRBT a -> OkRBT a
```

Deleting from a Data.Map Our third benchmark is the delete function from the `Data.Map` module in the Haskell standard libraries. The `Map` structure is a balanced binary search tree that implements purely functional key-value dictionaries:

```
data Map k a = Tip | Bin Int k a (Map k a) (Map k a)
```

A valid `Data.Map` must satisfy two properties: (1) the size of the left and right subtrees must be within a factor of three of each other, and (2) the keys must obey a binary search ordering. We specify the balancedness invariant (1) with a measure

```
measure isBal :: Map k a -> Prop
isBal (Tip)         = true
isBal (Bin s k v l r) = isBal l && isBal r &&
                        (sz l + sz r <= 1 ||
                         sz l <= 3 * sz r <= 3 * sz l)
```

and combine it with an ordering invariant (like `OrdRBT`) to specify valid trees.

```
type OkMap k a = {v : OrdMap k a | isBal v}
```

We can check that delete preserves the invariants by checking that its output is an `OkMap k a`. However, we can also go one step further and check the functional correctness property that delete removes the given key, with a type:

```
delete :: Ord k => k:k -> m:OkMap k a
        -> {v:OkMap k a | MinusKey v m k}
```

where the predicate `MinusKey` is defined as:

```
predicate MinusKey M1 M2 K
  = keys M1 = difference (keys M2) (singleton K)
```

830   E.L. Seidel, N. Vazou, and R. Jhala

using the measure `keys` describing the contents of the `Map`:

```
measure keys :: Map k a -> Set k
keys (Tip)         = empty ()
keys (Bin s k v l r) = union (singleton k)
                             (union (keys l) (keys r))
```

Refocusing XMonad StackSets Our last benchmark comes from the tiling window manager XMonad. The key invariant of XMonad's internal `StackSet` data structure is that the elements (windows) must all be *unique, i.e.* contain no duplicates. XMonad comes with a test-suite of over 100 QuickCheck properties; we select one which states that moving the focus between windows in a `StackSet` should not affect the *order* of the windows.

```
prop_focus_left_master n s =
   index (foldr (const focusUp) s [1..n]) == index s
```

With QuickCheck, the user writes a custom generator for valid `StackSet`s and then runs the above function on test inputs created by the generator, to check if in each case, the result of the above is `True`.

With TARGET, it is possible to test such properties *without* requiring custom generators. Instead the user writes a declarative specification:

```
type OkStackSet = {v:StackSet | NoDuplicates v}
```

(We refer the reader to [26] for a full discussion of how to specify `NoDuplicates`). Next, we define a refinement type:

```
type TTrue = {v:Bool | Prop v}
```

that is only inhabited by `True`, and use it to type the QuickCheck property as:

```
prop_focus_left_master :: Nat -> OkStackSet -> TTrue
```

This property is particularly difficult to *verify*; however, TARGET is able to automatically generate valid inputs to *test* that `prop_focus_left_master` always returns `True`.

Results Figure 8 summarizes the results of the comparison. QuickCheck was unable to successfully complete *any* benchmark to the low probability of generating properly constrained values at random.

List Insert TARGET is able to test `insert` all the way to depth 20, whereas Lazy SmallCheck times out at depth 19.

Red-Black Tree Insert TARGET is able to test `add` up to depth 12, while Lazy SmallCheck times out at depth 6.

Map Delete TARGET is able to check `delete` up to depth 10, whereas Lazy SmallCheck times out at depth 7 if it checks ordering first, or depth 6 if it checks balancedness first.

StackSet Refocus TARGET and is able to check this property up to depth 8, while Lazy SmallCheck times out at depth 7.

TARGET sees a performance hit with properties that require reasoning with the theory of Sets *e.g.* the no-duplicates invariant of StackSet. While Lazy SmallCheck times out at a higher depths, when it completes *e.g.* at depth 6, it does so in 0.7s versus TARGET's 9 minutes. We suspect this is because the theory of sets are a relatively recent addition to SMT solvers [18], and with further improvements in SMT technology, these numbers will get significantly better.

Overall, we found that for *small inputs* Lazy SmallCheck is substantially faster as exhaustive enumeration is tractable, and does not incur the overhead of communicating with an external general-purpose solver. Additionally, Lazy SmallCheck benefits from pruning predicates that exploit laziness and only force a small portion of the structure (*e.g.* ordering). However, we found that constraints that force the entire structure (*e.g.* balancedness), or composing predicates in the wrong *order*, can force Lazy SmallCheck to enumerate the entire exponentially growing search space.

TARGET, on the other hand, scales nicely to larger input sizes, allowing systematic and exhaustive testing of larger, more complex inputs. This is because TARGET eschews *explicit* enumeration-and-filtering (which results in searching for fewer needles in larger haystacks as the sizes increas), in favor of *symbolically* searching for valid models via SMT, making TARGET robust to the strictness or ordering of constraints.

5.2 Measuring Code Coverage

The second question we seek to answer is whether TARGET is suitable for testing entire libraries, *i.e.* how much of the program can be automatically exercised using our system? Keeping in mind the well-known issues with treating code coverage as an indication of test-suite quality [16], we consider this experiment a negative filter.

To this end, we ran TARGET against the entire user-facing API of Data.Map, our RBTree library, and XMonad.StackSet – using the constrained refined types (*e.g.* OkMap, OkRBT, OkStackSet) as the specification for the exposed types – and measured the expression and branch coverage, as reported by hpc [11]. We used an increasing timeout ranging from one to thirty minutes per exported function.

Results The results of our experiments are shown in Figure 9. Across all three libraries, TARGET achieved at least 70% expression and 64% alternative coverage at the shortest timeout of one minute per function. Interestingly, the coverage metrics for RBTree and Data.Map remain relatively constant as we increase the timeouts, with a small jump in expression coverage between 10 and 20 minutes. XMonad on the other hand, jumps from 70% expression and 64% alternative coverage with a one minute timeout, to 96% expression and 94% alternative with a ten minute timeout.

There are three things to consider when examining these results. First is that some expressions are not evaluated due to Haskell's laziness (*e.g.* the values contained in a Map). Second is that some expressions *should not* be evaluated and some branches *should not* be taken, as these only happen when an unexpected error condition is triggered (*i.e.* these expressions should be dead code). TARGET considers any inputs that trigger an uncaught exception a valid counterexample; the pre-conditions should rule out these inputs, and so we expect not to cover those expressions with TARGET.

The last remark is not intrinsically related to TARGET, but rather our means of collecting the coverage data. hpc includes **otherwise** guards in the "always-true"

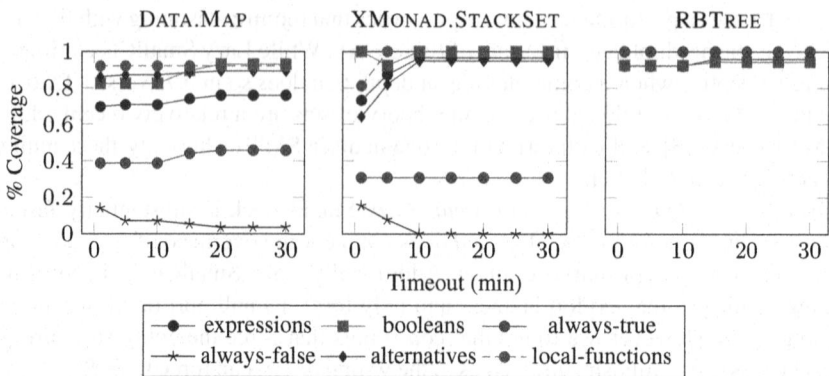

Fig. 9. Coverage-testing of `Data.Map.Base`, RBTree, and `XMonad.StackSet` using TAR-GET. Each exported function was tested with increasing depth limits until a single run hit a time-out ranging from one to thirty minutes. Lower is better for "always-true" and "always-false", higher is better for everything else.

category, even though they cannot evaluate to anything else. `Data.Map` contained 56 guards, of which 24 were marked "always-true". We manually counted 21 **otherwise** guards, the remaining 3 "always-true" guards compared the size of subtrees when re-balancing to determine whether a single or double rotation was needed; we were unable to trigger the double rotation in these cases. XMonad contained 9 guards, of which 4 were "always-true". 3 of these were **otherwise** guards; the remaining "always-true" guard dynamically checked a function's pre-condition. If the pre-condition check had failed an error would have been thrown by the next case, we consider it a success of TARGET that the error branch was not triggered.

5.3 Discussion

To sum up, our experiments demonstrate that TARGET generates valid inputs: (1) where QuickCheck fails outright, due to the low probability of generating random values sat-isfying a property; (2) more efficiently than Lazy SmallCheck, which relies on lazy pruning predicates; and (3) providing high code coverage for real-world libraries with no hand-written test cases.

Of course our approach is not without drawbacks; we highlight five classes of pitfalls the user may encounter.

Laziness in the function or in the output refinement can cause exceptions to go un-thrown if the output value is not fully demanded. For example, TARGET would decide that the result `[1, undefined]` inhabits `[Int]` but not `[Score]`, as the latter would have to evaluate `0 <= undefined < 100`. This limitation is not specific to our sys-tem, rather it is fundamental to any tool that exercises lazy programs. Furthermore, TARGET only generates inductively-defined values, it cannot generate infinite or cyclic structures, nor will the generated values ever contain ⊥.

Polymorphism. Like any other tool that actually runs the function under scrutiny, TAR-GET can only test monomorphic instantiations of polymorphic functions. For example, when testing XMonad we instantiated the "window" parameter to Char and all other type parameters to (), as the properties we were testing only examined the window. This helped drastically reduce the search space, both for TARGET and SmallCheck.

Advanced type-system features such as GADTs and Existential types may prevent GHC from deriving a Generic instance, which would force the programmer to write her own Targetable instance. Though tedious, the single hand-written instance allows TARGET to automatically generate values satisfying disparate constraints, which is still an improvement over the generate-and-filter approach.

Refinement types are less expressive than properties written in the host language. If the pre-conditions are not expressible in TARGET's logic, the user will have to use the generate-and-filter approach, losing the benefits of symbolic enumeration.

Input explosion. TARGET excels when the space of valid inputs is a sparse subset of the space of all inputs. If the input space is not sufficiently constrained, TARGET may spend lose its competitive advantage over other tools due to the overhead of using a general-purpose solver.

6 Related Work

TARGET is closely related to a number of lines of work on connecting formal specifications, execution, and automated constraint-based testing. Next, we describe the closest lines of work on test-generation and situate them with respect to our approach.

6.1 Model-Based Testing

Model-based testing encompasses a broad range of black-box testing tools that facilitate generating concrete test-cases from an abstract model of the system under test. These systems generally (though not necessarily) model the system at a holistic level using state machines to describe the desired behavior [6], and may or may not provide fully automatic test-case generation. In addition to generating test-cases, many model-based testing tools, *e.g.* Spec Explorer [28] will produce extra artifacts like visualizations to help the programmer understand the model. One could view property-based testing, including our system, as a subset of model-based testing focusing on lower-level properties of individual functions (unit-testing), while using the type-structure of the functions under scrutiny to provide fully automatic generation of test-cases.

6.2 Property-Based Testing

Many property-based testing tools have been developed to automatically generate test-suites. QuickCheck [4] randomly generates inputs based on the property under scrutiny, but requires custom generators to consistently generate constrained inputs. [3] extends QuickCheck to randomly generate constrained values from a uniform distribution. In contrast SmallCheck [21] enumerates all possible inputs up to some depth, which allows it to check existential properties in addition to universal properties; however, it too has difficulty generating inputs to properties with complex pre-conditions. Lazy Small-Check [21] addresses the issue of generating constrained inputs by taking advantage of the inherent laziness of the property, generating *partially-defined* values (*i.e.* values containing ⊥) and only filling in the holes if and when they are demanded. Korat [2] instruments a programmer-supplied repOk method, which checks class invariants and method pre-conditions, to monitor which object fields are accessed. The authors observe that unaccessed fields cannot have had an effect on the return value of repOk and are thereby able to exclude from the search space any objects that differ only in the values of the unaccessed fields. While Lazy SmallCheck and Korat's reliance on functions in the source language for specifying properties is convenient for the programmer (specification and implementation in the same language), it makes the method less amenable to formal verification, the properties would need to be re-specified in another language that is restricted enough to facilitate verification.

6.3   Symbolic Execution and Model-Checking

Another popular technique for automatically generating test-cases is to analyze the source code and attempt to construct inputs that will trigger different paths through the program. DART [12], CUTE [22], and Pex [24] all use a combination of symbolic and dynamic execution to explore different paths through a program. While executing the program they collect *path predicates*, conditions that characterize a path through a program, and at the end of a run they negate the path predicates and query a constraint solver for another assignment of values to program variables. This enables such tools to efficiently explore many different paths through a program, but the technique relies on the path predicates being expressible symbolically. When the predicates are not expressible in the logic of the constraint solver, they fall back to the values produced by the concrete execution, at a severe loss of precision. Instead of trying to trigger all paths through a program, one might simply try to trigger erroneous behavior. Check 'n' Crash [5] uses the ESC/Java analyzer [10] to discover potential bugs and constructs concrete test-cases designed to trigger the bugs, if they exist. Similarly, [1] uses the BLAST model-checker to construct test-cases that bring the program to a state satisfying some user-provided predicate.

In contrast to these approaches, TARGET (and more generally, property-based testing) treats the program as a *black-box* and only requires that the pre- and post-conditions be expressible in the solver's logic. Of course, by expressing specifications in the source language, *e.g.* as contracts, as in PEX [24], one can use symbolic execution to generate tests directly from specifications. One concrete advantage of our approach over the symbolic execution based method of PEX is that the latter generates tests by *explicitly enumerating* paths through the contract code, which suffers from a similar combinatorial problem as SmallCheck and QuickCheck. In contrast, TARGET performs the same search *symbolically* within the SMT engine, which performs better for larger input sizes.

6.4   Integrating Constraint-Solving and Execution

TARGET is one of many tools that makes specifications executable via constraint solving. An early example of this approach is TestEra [17] that uses specifications written in the Alloy modeling language [13] to generate all non-isomorphic Java objects that satisfy method pre-conditions and class invariants. As the specifications are written in Alloy, one can use Alloy's SAT-solver based model finding to symbolically enumerate candidate inputs. Check 'n' Crash uses a similar idea, and SMT solvers to generate inputs that satisfy a given JML specification [5]. Recent systems such as SBV [8] and Kaplan [14] offer a monadic API for writing SMT constraints within the program, and use them to synthesize program values at *run-time*. SBV provides a thin DSL over the logics understood by SMT solvers, whereas Kaplan integrates deeply with Scala, allowing the use of user-defined recursive types and functions. Test generation can be viewed as a special case of value-synthesis, and indeed Kaplan has been used to generate test-suites from preconditions in a similar manner to TARGET.

However, in all of the above (and also symbolic execution based methods like PEX or JCrasher), the specifications are *assertions* in the Floyd-Hoare sense. Consequently, the techniques are limited to testing first-order functions over monomorphic data types. In contrast, TARGET shows how to view *types* as executable specifications, which yields several advantages. First, we can use types to compositionally lift specifications about flat values (*e.g.* `Score`) over collections (*e.g.* `[Score]`), without requiring special recursive predicates to describe such collection invariants. Second, the compositional nature of types yields a compositional method for generating tests, allowing us to use type-class machinery to generate tests for richer structures from tests for sub-structures. Third, (refinement) types have proven to be effective for *verifying* correctness properties in modern modern languages that make ubiquitous use of parametric polymorphism and higher order functions [29,7,20,23,26] and thus, we believe TARGET's approach of making refinement types executable is a crucial step towards our goal of enabling *gradual verification* for modern languages.

Acknowledgements. This work was supported by NSF grants CCF-1422471, CNS-0964702, CNS-1223850, CCF-1218344, CCF-1018672, and a generous gift from Microsoft Research. We thank Lee Pike and the reviewers for their excellent feedback on a draft of this paper.

References

1. Beyer, D., Chlipala, A.J., Henzinger, T.A., Jhala, R., Majumdar, R.: Generating tests from counterexamples. In: ICSE 2004: Software Engineering, pp. 326–335 (2004)
2. Boyapati, C., Khurshid, S., Marinov, D.: Korat: Automated testing based on Java predicates. In: ISSTA 2002: Software Testing and Analysis, pp. 123–133. ACM (2002)
3. Claessen, K., Duregård, J., Palka, M.H.: Generating constrained random data with uniform distribution. In: FLOPS 2014, pp. 18–34 (2014)
4. Claessen, K., Hughes, J.: QuickCheck: a lightweight tool for random testing of haskell programs. In: ICFP. ACM (2000)
5. Csallner, C., Smaragdakis, Y.: Check 'n' crash: combining static checking and testing. In: ICSE, pp. 422–431 (2005)
6. Neto, A.C.D., Subramanyan, R., Vieira, M., Travassos, G.H.: A survey on model-based testing approaches: A systematic review. In: WEASELTech 2007. ACM (2007)
7. Dunfield, J.: Refined typechecking with Stardust. In: PLPV (2007)

8. Erkök, L.: SBV: SMT based verification in haskell, http://leventerkok.github.io/sbv/
9. Findler, R.B., Felleisen, M.: Contract soundness for object-oriented languages. In: OOPSLA, pp. 1–15 (2001)
10. Flanagan, C., Leino, K.R.M., Lillibridge, M., Nelson, G., Saxe, J.B., Stata, R.: Extended static checking for Java. In: PLDI (2002)
11. Gill, A., Runciman, C.: Haskell program coverage. In: Haskell 2007. ACM (2007)
12. Godefroid, P., Klarlund, N., Sen, K.: Dart: directed automated random testing. In: PLDI, pp. 213–223 (2005)
13. Jackson, D.: Alloy: a lightweight object modelling notation. ACM Transactions on Software Engineering and Methodology (TOSEM) 11(2), 256–290 (2002)
14. Köksal, A.S., Kuncak, V., Suter, P.: Constraints as control. In: POPL 2012, pp. 151–164. ACM, New York (2012)
15. Magalhães, J.P., Dijkstra, A., Jeuring, J., Löh, A.: A generic deriving mechanism for haskell. In: Haskell Symposium. ACM (2010)
16. Marick, B.: How to misuse code coverage. In: Proceedings of the 16th Interational Conference on Testing Computer Software, pp. 16–18 (1999)
17. Marinov, D., Khurshid, S.: Testera: A novel framework for automated testing of java programs. In: ASE 2001. IEEE Computer Society, Washington, DC (2001)
18. de Moura, L., Bjørner, N.: Generalized, efficient array decision procedures. In: FMCAD (2009)
19. Nelson, G.: Techniques for program verification. Tech. Rep. CSL81-10, Xerox Palo Alto Research Center (1981)
20. Nystrom, N., Saraswat, V.A., Palsberg, J., Grothoff, C.: Constrained types for object-oriented languages. In: OOPSLA, pp. 457–474 (2008)
21. Runciman, C., Naylor, M., Lindblad, F.: Smallcheck and lazy smallcheck: Automatic exhaustive testing for small values. In: Haskell Symposium. ACM (2008)
22. Sen, K., Marinov, D., Agha, G.: Cute: A concolic unit testing engine for c. In: ESEC/FSE. ACM (2005)
23. Swamy, N., Chen, J., Fournet, C., Strub, P.-Y., Bhargavan, K., Yang, J.: Secure distributed programming with value-dependent types. In: ICFP (2011)
24. Tillmann, N., de Halleux, J.: Pex–white box test generation for.NET. In: Beckert, B., Hähnle, R. (eds.) TAP 2008. LNCS, vol. 4966, pp. 134–153. Springer, Heidelberg (2008)
25. Vazou, N., Rondon, P.M., Jhala, R.: Abstract refinement types. In: Felleisen, M., Gardner, P. (eds.) ESOP 2013. LNCS, vol. 7792, pp. 209–228. Springer, Heidelberg (2013)
26. Vazou, N., Seidel, E.L., Jhala, R.: Liquidhaskell: Experience with refinement types in the real world. In: Haskell Symposium (2014)
27. Vazou, N., Seidel, E.L., Jhala, R., Vytiniotis, D., Peyton-Jones, S.: Refinement types for haskell. In: ICFP (2014)
28. Veanes, M., Campbell, C., Grieskamp, W., Schulte, W., Tillmann, N., Nachmanson, L.: Model-based testing of object-oriented reactive systems with spec explorer. In: Hierons, R.M., Bowen, J.P., Harman, M. (eds.) FORTEST. LNCS, vol. 4949, pp. 39–76. Springer, Heidelberg (2008)
29. Xi, H., Pfenning, F.: Eliminating array bound checking through dependent types. In: PLDI (1998)

Author Index

838   Author Index